Probiotics and Prebiotics

Current Research and Future Trends

Edited by

Koen Venema

Beneficial Microbes Consultancy
Wageningen
The Netherlands

and

Ana Paula do Carmo

Instituto Federal do Espírito Santo - IFES
Soteco
Vila Velha
Brazil

Caister Academic Press

Caister Academic Press
Norfolk, UK

www.caister.com

British Library Cataloguing-in-Publication Data
A catalogue record for this book is available from the British Library

ISBN: 978-1-910190-09-8 (hardback)
ISBN: 978-1-910190-10-4 (ebook)

Cover design adapted from Figure 7.6.

Contents

Current Books of Interest

Microalgae: Current Research and Applications	2016
Arboviruses: Molecular Biology, Evolution and Control	2016
Shigella: Molecular and Cellular Biology	2016
Aquatic Biofilms: Ecology, Water Quality and Wastewater Treatment	2016
Alphaviruses: Current Biology	2016
Thermophilic Microorganisms	2015
Flow Cytometry in Microbiology: Technology and Applications	2015
Epigenetics: Current Research and Emerging Trends	2015
Corynebacterium glutamicum: From Systems Biology to Biotechnological Applications	2015
Advanced Vaccine Research Methods for the Decade of Vaccines	2015
Antifungals: From Genomics to Resistance and the Development of Novel Agents	2015
Bacteria-Plant Interactions: Advanced Research and Future Trends	2015
Aeromonas	2015
Antibiotics: Current Innovations and Future Trends	2015
Leishmania: Current Biology and Control	2015
Acanthamoeba: Biology and Pathogenesis (2nd edition)	2015
Microarrays: Current Technology, Innovations and Applications	2014
Metagenomics of the Microbial Nitrogen Cycle: Theory, Methods and Applications	2014
Pathogenic *Neisseria*: Genomics, Molecular Biology and Disease Intervention	2014
Proteomics: Targeted Technology, Innovations and Applications	2014
Biofuels: From Microbes to Molecules	2014
Human Pathogenic Fungi: Molecular Biology and Pathogenic Mechanisms	2014
Applied RNAi: From Fundamental Research to Therapeutic Applications	2014
Halophiles: Genetics and Genomes	2014
Molecular Diagnostics: Current Research and Applications	2014
Phage Therapy: Current Research and Applications	2014
Bioinformatics and Data Analysis in Microbiology	2014
The Cell Biology of Cyanobacteria	2014
Pathogenic *Escherichia coli*: Molecular and Cellular Microbiology	2014
Campylobacter Ecology and Evolution	2014
Burkholderia: From Genomes to Function	2014
Myxobacteria: Genomics, Cellular and Molecular Biology	2014
Next-generation Sequencing: Current Technologies and Applications	2014
Omics in Soil Science	2014
Applications of Molecular Microbiological Methods	2014
Mollicutes: Molecular Biology and Pathogenesis	2014

Full details at www.caister.com

Contributors

Célia Alencar de Moraes
Universidade Federal de Viçosa (UFV)
Viçosa, MG
Brazil

camoraes101@gmail.com

Irene Aloisio
Department of Agricultural Sciences
University of Bologna
Bologna
Italy

irene.aloisio@unibo.it

Juliet Ansell
Food, Nutrition and Health
New Zealand Institute for Plant & Food Research
Palmerston North;
Zespri International Limited
Mt Maunganui; and
Riddet Institute
Massey University
Palmerston North
New Zealand

juliet.ansell@zespri.com

Patricia Aparecida Pimenta Pereira
Department of Food
Federal University of Ouro Preto
Ouro Preto, MG
Brazil

patricia@enut.ufop.br

Tyler E. Askelson
Department of Poultry Science
Texas A&M University
College Station, TX
USA

teaskelson@gmail.com

Vasco Azevedo
Institute of Biological Sciences
Federal University of Minas Gerais (UFMG-ICB)
Belo Horizonte, MG
Brazil

vasco@icb.ufmg.br

Loredana Baccigalupi
Department of Biology
Federico II University of Naples
Naples
Italy

lorbacci@unina.it

Luis G. Bermudez-Humaran
INRA
UMR1319 Micalis; and
AgroParisTech
UMR Micalis
Jouy-en-Josas
France

luis.bermudez@jouy.inra.fr

Paul A. Blatchford
Food, Nutrition and Health
New Zealand Institute for Plant & Food Research
Palmerston North
New Zealand

paul.blatchford@plantandfood.co.nz

Michael Blaut
Department of Gastrointestinal Microbiology
German Institute of Human Nutrition Potsdam-Rehbrücke
Nuthetal
Germany

blaut@dife.de

Marc R. Bomhof
Faculty of Kinesiology
University of Calgary
Calgary, AB
Canada

mrbomhof@ucalgary.ca

Natalia Breyner
INRA
UMR1319 Micalis;
AgroParisTech
UMR1319 Micalis
Jouy-en-Josas
France; and
Institute of Biological Sciences
Federal University of Minas Gerais (UFMG-ICB)
Belo Horizonte, MG
Brazil

nbreyner@jouy.inra.fr

Peter A. Bron
TI Food & Nutrition
Wageningen; and
NIZO food research
Ede
The Netherlands

peter.bron@nizo.com

Robert-Jan Brummer
School of Health and Medical Sciences
Faculty of Medicine and Health
Örebro University
Örebro
Sweden

robert.brummer@oru.se

Jose M. Bruno-Bárcena
Department of Plant and Microbial Biology
North Carolina State University
Raleigh, NC
USA

jose_bruno-barcena@ncsu.edu

Camila Carvalho Menezes
Federal University of Ouro Preto
Department of Food
Ouro Preto, MG
Brazil

camilamenezes@enut.ufop.br

Jean-Marc Chatel
INRA
UMR1319 Micalis; and
AgroParisTech
UMR1319 Micalis
Jouy-en-Josas
France

jean-marc.chatel@jouy.inra.fr

Sy-Bing Choi
Natural Product and Drug Discovery Centre
Malaysian Institute of Pharmaceuticals and Nutraceuticals
National Institutes of Biotechnology Malaysia
Ministry of Science, Technology and Innovation
Block 5-C, Halaman Bukit Gambir
Pulau Pinang; and
School of Industrial Technology
Universiti Sains Malaysia
Penang
Malaysia

sybing@ipharm.gov.my
sybing@gmail.com

Eric Banan-Mwine Daliri
School of Biotechnology
Jiangnan University
Wuxi
China

ericdaliri@yahoo.com
ericdaliri@gmail.com

Vicky De Preter
Translational Research Center for Gastrointestinal Disorders (TARGID); and
Leuven Food Science and Nutrition Research Centre (LFoRCe)
KU Leuven; and
Department of Health and Technology, UCLL
KHLeuven
Leuven
Belgium

vicky.depreter@med.kuleuven.be

Parneet Kaur Deol
University Institute of Pharmaceutical Sciences
Panjab University
Chandigarh
India

parneetdeol5988@gmail.com

Diana Di Gioia
Department of Agricultural Sciences
University of Bologna
Bologna
Italy

diana.digioia@unibo.it

Ana Paula Do Carmo
Instituto Federal do Espírito Santo (IFES)
Espírito Santo
Brazil

carmoap@icloud.com

Flaviano dos Santos Martins
Departamento de Microbiologia
Instituto de Ciências Biológicas
Universidade Federal de Minas Gerais
Belo Horizonte, MG
Brazil

flaviano@icb.ufmg.br

Tri Duong
Department of Poultry Science
Texas A&M University
College Station, TX
USA

t-duong@poultry.tamu.edu

Hanna Edebol
School of Health and Medical Sciences
Faculty of Medicine and Health
Örebro University
Örebro
Sweden

hanna.edebol-carlman@oru.se

Francesca Fava
Nutrition and Nutrigenomics Group
Department of Food Quality and Nutrition
Research and Innovation Centre
Fondazione Edmund Mach
Trento
Italy

francesca.fava@fmach.it

Hanne Frøkiær
Section for Experimental Animal Models
Department of Veterinary Disease Biology
Copenhagen University
Copenhagen
Denmark

hafr@sund.ku.dk

John-Peter Ganda Mall
School of Health and Medical Sciences
Faculty of Medicine and Health
Örebro University
Örebro
Sweden

john-peter.ganda-mall@oru.se

Emilia Ghelardi
Department of Translational Research and New Technologies in Medicine and Surgery
University of Pisa
Pisa
Italy

emilia.ghelardi@med.unipi.it

Uday C. Ghoshal
Department of Gastroenterology
Sanjay Gandhi Post Graduate Institute of Medical Sciences
Lucknow
India

udayghoshal@gmail.com

Ujjala Ghoshal
Department of Microbiology
Sanjay Gandhi Post Graduate Institute of Medical Sciences
Lucknow
India

ujjalaghoshal@yahoo.co.in

Débora L.R. Gomes
INRA
UMR1319 Micalis;
AgroParisTech
UMR1319 Micalis
Jouy-en-Josas
France; and
Faculty of Pharmacy
Realengo Campus
Federal Institute of Rio de Janeiro (IFRJ)
Rio de Janeiro
Brazil

debora.gomes@ifrj.edu.br

Caroline T. Gottlieb
CED-Fermentation
Chr. Hansen A/S
Hørsholm
Denmark

dkctg@chr-hansen.com

Hosni M. Hassan
Prestage Department of Poultry Science
North Carolina State University
Raleigh, NC
USA

hmhassan@ncsu.edu

Gunnard K. Jacobson
Lesaffre Human Care
Milwaukee, WI
USA

gunnard.jacobson@lsaf.com

Eric Johansen
CED-Innovation Management
Chr. Hansen A/S
Hørsholm
Denmark

dkejo@chr-hansen.com

Mette Rose Jørgensen
Department of Odontology
Faculty of Health and Medical Sciences
University of Copenhagen
Copenhagen
Denmark

mrj@sund.ku.dk

Indu Pal Kaur
University Institute of Pharmaceutical Sciences
Panjab University
Chandigarh
India

indupalkaur@yahoo.com

Mette Kirstine Keller
Department of Odontology
Faculty of Health and Medical Sciences
University of Copenhagen
Copenhagen
Denmark

mke@sund.ku.dk

Caroline C. Kim
Food, Nutrition and Health
New Zealand Institute for Plant & Food Research
Palmerston North
New Zealand

caroline.kim@plantandfood.co.nz

Michiel Kleerebezem
TI Food & Nutrition
Wageningen;
NIZO food research
Ede; and
Host-Microbe Interactomics
Wageningen University and Research Centre
Wageningen
The Netherlands

michiel.kleerebezem@wur.nl

GwangPyo Ko
Graduate School of Public Health
Seoul National University
Seoul
South Korea

gko@snu.ac.kr

Julia König
School of Health and Medical Sciences
Faculty of Medicine and Health
Örebro University
Örebro
Sweden

julia.konig@oru.se

Philippe Langella
INRA
UMR1319 Micalis; and
AgroParisTech
UMR Micalis
Jouy-en-Josas
France

philippe.langella@jouy.inra.fr

Amy-Sie-Yik Lau
School of Industrial Technology
Universiti Sains Malaysia
Penang
Malaysia

amylau1028@hotmail.com

Byong-H. Lee
School of Biotechnology
Jiangnan University
Wuxi
China; and
Departments of Food Science and Microbiology and Immunology
McGill University
Montreal, QC
Canada

byonghlee@gmail.com
byong.lee@mcgill.ca

Marta C.T. Leite
Federal Institute of Education, Science and Technology of Espirito Santo
Itapina, ES
Brazil

marta.leite@ifes.edu.br

Thomas D. Leser
Health and Nutrition
Innovation
Chr. Hansen A/S
Hørsholm
Denmark

dktdl@chr-hansen.com

Lee-Ching Lew
School of Industrial Technology
Universiti Sains Malaysia
Penang
Malaysia

gem_ching@hotmail.com

Luciene Lignani Bitencourt
Instituto Federal do Espírito Santo (IFES)
Colatina, ES
Brazil

luciene.lignani@ifes.edu.br

Min-Tze Liong
School of Industrial Technology
Universiti Sains Malaysia
Penang
Malaysia

mintze.liong@usm.my

Rebeca Martin
INRA
UMR1319 Micalis; and
AgroParisTech
UMR Micalis
Jouy-en-Josas
France

rebeca.martinrosique@jouy.inra.fr

Luana Martins Perin
Veterinary Department
Universidade Federal de Viçosa
Viçosa, MG
Brazil

luana.mperin@gmail.com

Giuseppe Mazzola
Department of Agricultural Sciences
University of Bologna
Bologna
Italy

giuseppe.mazzola3@unibo.it

Marjolein Meijerink
Wageningen University
Host-Microbes Interactomics
Wageningen
The Netherlands

marjolein.meijerink@wur.nl

Stine B. Metzdorff
Section for Experimental Animal Models
Department of Veterinary Disease Biology
Copenhagen University
Copenhagen
Denmark

broeng@sund.ku.dk

Sylvie Miquel
INRA
UMR1319 Micalis; and
AgroParisTech
UMR Micalis
Jouy-en-Josas
France

sylmiquel@gmail.com

Anderson Myioshi
Institute of Biological Sciences
Federal University of Minas Gerais (UFMG-ICB)
Belo Horizonte, MG
Brazil

miyoshi@icb.ufmg.br

Luís Augusto Nero
Veterinary Department Universidade Federal de Viçosa
Viçosa, MG
Brazil

nero@ufv.br

Jacques Robert Nicoli
Departamento de Microbiologia
Instituto de Ciências Biológicas
Universidade Federal de Minas Gerais
Belo Horizonte, MG
Brazil

jnicoli@icb.ufmg.br

Maïwenn Olier
Neurogastroenterology & Nutrition
UMR INRA 1331-Toxalim; and
Molecular & Cellular Pathogenesis of Escherichia coli Infections
USC INRA 1360-UMR INSERM 1043
Toulouse
France

maiwenn.olier@toulouse.inra.fr

Shanthi G. Parkar
Food, Nutrition and Health
New Zealand Institute for Plant & Food Research
Palmerston North
New Zealand

shanthi.parkar@plantandfood.co.nz

Seema Patel
San Diego State University
San Diego, CA
USA

seemabiotech83@gmail.com

Adriana Perez Chaia
Centro de Referencias para Lactobacilos (CERELA)
Consejo Nacional de Investigaciones Científicas y Técnicas (CONICET)
San Miguel de Tucumán
Argentina

apchaia@cerela.org.ar

Uelinton M. Pinto
Department of Food and Experimental Nutrition
School of Pharmaceutical Sciences
University of São Paulo
São Paulo
Brazil

uelintonpinto@usp.br

Ignacio Rangel
School of Health and Medical Sciences
Faculty of Medicine and Health
Örebro University
Örebro
Sweden

ignacio.rangel@oru.se

Raylene A. Reimer
Faculty of Kinesiology and Department of Biochemistry and Molecular Biology
University of Calgary
Calgary, AB
Canada

reimer@ucalgary.ca

Ezio Ricca
Department of Biology
Federico II University of Naples
Naples
Italy

ericca@unina.it

Vanessa Rios de Souza
Federal University of Lavras
Department of Food Science
Lavras, MG
Brazil

vanessardsouza@gmail.com

Praveen Rishi
Department of Microbiology
Panjab University
Chandigarh
India

rishipraveen@yahoo.com

Luciana Rodrigues Cunha
Federal University of Ouro Preto
Department of Food
Ouro Preto, MG
Brazil

lrcunha@enut.ufop.br

Douglas I. Rosendale
Food, Nutrition and Health
New Zealand Institute for Plant & Food Research
Palmerston North
New Zealand

doug.rosendale@plantandfood.co.nz

Simarjot Kaur Sandhu
University Institute of Pharmaceutical Sciences
Panjab University
Chandigarh
India

simarjots@gmail.com

Delphine M. Saulnier
Department of Gastrointestinal Microbiology
German Institute of Human Nutrition Potsdam-Rehbrücke
Nuthetal
Germany

delphine.saulnier@dife.de

Nicola Segata
Centre for Integrative Biology (CIBIO)
University of Trento
Trento
Italy

nicola.segata@unitn.it

Ratnakar Shukla
Department of Microbiology
Sanjay Gandhi Post Graduate Institute of Medical Sciences
Lucknow
India

ratnakarshukla2021@gmail.com

Jiyeon Si
Graduate School of Public Health
Seoul National University
Seoul
South Korea

sijy@snu.ac.kr

Juliana Teixeira de Magalhães
Universidade Federal de São João Del Rei (UFSJ)
Divinópolis, MG
Brazil

julimagaufsj@gmail.com

Marta Cristina Teixeira Leite
Instituto Federal do Espírito Santo (IFES)
Itapina, ES
Brazil

mctleite@yahoo.com.br

Muriel Thomas
INRA
UMR1319 Micalis; and
AgroParisTech
UMR Micalis
Jouy-en-Josas
France

muriel.thomas@jouy.inra.fr

Svetoslav Dimitrov Todorov
Veterinary Department
Universidade Federal de Viçosa
Viçosa, MG
Brazil

slavi31057@abv.bg

Bryan Troxell
Prestage Department of Poultry Science
North Carolina State University
Raleigh, NC
USA

bryan_troxell@ncsu.edu

Kieran M. Tuohy
Nutrition and Nutrigenomics Group
Department of Food Quality and Nutrition
Research and Innovation Centre
Fondazione Edmund Mach
Trento
Italy

kieran.tuohy@fmach.it

Svante Twetman
Department of Odontology
Faculty of Health and Medical Sciences
University of Copenhagen
Copenhagen
Denmark

stwe@sund.ku.dk

Hermien van Bokhorst-van de Veen
TI Food & Nutrition
Wageningen;
NIZO food research
Ede; and
Laboratory of Microbiology
Wageningen University and Research Centre
Wageningen
The Netherlands

Present address:
Food and Biobased Research
Wageningen University and Research Centre
Wageningen
The Netherlands

hermien.vandeveen@wur.nl

Koen Venema
Beneficial Microbes Consultancy
Wageningen
The Netherlands

koen.venema@outlook.com

Kristin Verbeke
Translational Research Center for Gastrointestinal Disorders (TARGID); and
Leuven Food Science and Nutrition Research Centre (LFoRCe)
KU Leuven
Leuven
Belgium

kristin.verbeke@med.kuleuven.be

Hyun Ju You
Graduate School of Public Health
Seoul National University
Seoul
South Korea

dhlover1@snu.ac.kr

Gabriela Zárate
Centro de Referencias para Lactobacilos (CERELA)
Consejo Nacional de Investigaciones Científicas y Técnicas (CONICET)
San Miguel de Tucumán
Argentina

gzarate@cerela.org.ar

Preface

Probiotics and prebiotics have received increasing attention over the past few decades as ways to increase the health of the host. The definitions of both probiotics and prebiotics include a 'health benefit for the host', which is why regulatory bodies do not allow the use of the phrases on commercial products. These organizations are not (yet) convinced by the ever growing evidence that shows the benefit of the exogenous beneficial microbes (probiotics) or the compounds (prebiotics) that selectively increase the numbers and/or activity of the endogenous beneficial microbes in our gut. Despite the reservation by regulatory bodies to allow the use of these words, scientists still use the phrases in scientific publications when a benefit to the host has been shown. Significant advances have been made in recent years in the development of products with probiotics and/or prebiotics to prevent numerous diseases and disorders. Here, after a brief introduction to the field, first the different probiotic genera are discussed. Next, topics related to prebiotics are highlighted. Furthermore, effects of probiotics and prebiotics on health are included, focusing on areas in which these dietary ingredients have recently been used. Finally, an outlook to the future is given. The contributions take a critical view on the evidence for health benefits of probiotics and prebiotics. The authors are believers, although we remain critical about certain aspects of the research, such as dose, duration of intake, combinations of probiotics, and combination of probiotics with prebiotics. Nevertheless, we believe there is a future for these health beneficial components of the diet as natural ways to combat the diseases and disorders that perturb our modern society.

Koen Venema and Ana Paula do Carmo

Part I

General Introduction

Probiotics and Prebiotics: Current Status and Future Trends

1

Koen Venema and Ana Paula do Carmo

Abstract

Over the past several decades the research into the health benefits of probiotics and prebiotics has rocketed sky high. There are several new applications and diseases and disorders for these healthy dietary components that were previously unthinkable. However, the efficacy has not been scientifically substantiated for all these applications yet, and care needs to be taken that pro- and prebiotics are not considered as a cure for everything. For starters, probiotic effects are strain dependent, and hence not all strains are beneficial for all disorders. In fact, some strains may be detrimental when given to certain patients, and it may aggravate the problems that these patients have. Similarly, prebiotics are not identical, and will stimulate different microorganisms in different individuals, in some case leading to worsening of the disease. Moreover, dose dependency has rarely been studied and, in the case of probiotics, culture conditions may affect their efficacy as well. In addition, although numerous positive results have been obtained with several well-studied probiotic strains, the mechanism of action usually is still completely unclear, let alone what the molecular molecule is that is responsible for the benefit. So, despite several decades of intense research, there is still much to be discovered.

Definition of pro-, pre-, syn-, post- and other -biotics

The definition of *probiotics* has undergone several revisions over the past two decades. The definition has recently been refined by an expert panel commissioned by the Food and Agriculture Organization of the United Nations (FAO) and the World Health Organization (WHO), and is now commonly accepted to be 'Live microorganisms which when administered in adequate amounts confer a health benefit on the host' (FAO/WHO, 2002). This definition is preferred over historical definition as it includes both intestinal and other forms of probiotics, such as vaginal or topical probiotics. This definition requires that the term 'probiotic' only be applied to live microbes having a substantiated beneficial effect (Reid *et al.*, 2003), although preparations of dead cells and cell components may also exert some health-promoting physiological effects, e.g. (Adams, 2010; Lau, 2014). These have recently been referred to as postbiotics (Tsilingiri and Rescigno, 2013). *Postbiotics* have been defined as non-viable bacterial products or metabolic by-products from probiotic microorganisms that have biological activity in the host (Tsilingiri and Rescigno, 2013). For some physiological benefit it is clear that the cells need to be active. However, for other benefits, e.g. lactose intolerance, it is hypothesized that lysis of the cells within the gastrointestinal tract is required to increase digestion of lactose in lactose intolerant individuals (Venema, 2012).

New trends that may soon be reality are the development of new therapeutic strategies, such as the development of phagebiotics, psychobiotics and (genetically modified) *pharmabiotics* (Eeckhaut *et al.*, 2014; Paton *et al.*, 2012; Shanahan *et al.*, 2012).

Like for probiotics, the original definition for a *prebiotic* was limited to a benefit to the gastrointestinal (GI) tract, and hence it was originally defined as 'a nondigestible food ingredient that beneficially affects the host by selectively stimulating the growth and/or activity of one or a limited number of bacteria in the colon, and thus improves host health' (Gibson and Roberfroid, 1995). This definition was later refined by the same authors to include other bodily areas of the host that may benefit from selective targeting of particular microorganisms (Gibson *et al.*, 2004). A more recently adopted definition by the International Scientific Association of Probiotics and Prebiotics (ISAPP) is: 'a selectively fermented ingredient that results in specific changes in the composition and/or activity of the gastrointestinal microbiota, thus conferring benefit(s) upon host health' (Blatchford *et al.*, 2014), in which the phrase 'gastrointestinal' re-appeared, indicating that the jury is still out on non-GI prebiotics. Even though other areas of the body are being targeted by prebiotics, this is usually done giving prebiotics orally, e.g. (pro- and) prebiotics for skin health (Foolad and Armstrong, 2014; Miyazaki *et al.*, 2014).

To be classified as a genuine prebiotic, a dietary compound historically had to fulfil three criteria: (1) resistance to gastric acidity, hydrolysis by mammalian enzymes and gastrointestinal (GI) absorption (2) fermentation by intestinal microbiota, and (3) selective stimulation of the growth and/or activity of intestinal bacteria associated with health and well-being. Given the fact that other areas of the host besides the GI tract are now also targeted, these criteria are no longer all applicable. In fact, they are not applicable at all any more, unless the phrase 'intestinal' is deleted from criteria 2 and 3. However, most prebiotics developed up to date are geared towards the GI tract.

Most prebiotics that have been developed so far (and that have sufficient scientific proof) are carbohydrates or fibres. However, though most prebiotics (so far) are fibres, not all fibres are prebiotics (Slavin, 2013). Although the concept of prebiotics is relatively new, foods high in prebiotics have been consumed since prehistoric times: Archaeological evidence from dry cave deposits in the northern Chihuahuan Desert, that straddles the US–Mexico border, show intensive utilization of desert plants that were high in what we now consider prebiotic fructans (Slavin, 2013). Dietary intake of fructans was estimated to be about 135 g/day for the typical adult male hunter–forager. The fructans inulin and fructo-oligosaccharides (FOS) occur naturally in foods such as leeks, asparagus, chicory, Jerusalem artichokes, garlic, and onions. Nowadays, consumption in typical US and European diets has been estimated to be only several grams per day (Slavin, 2013). Despite the possibility of ingesting high quantities of inulin (135 g/day for hunter–foragers), in our current time inulin and FOS consumption in large quantities is associated with impaired gastrointestinal tolerance (Grabitske and Slavin, 2009; Lied *et al.*, 2011). Other prebiotic fibres (e.g. wheat dextrin, polydextrose) exhibit higher gastrointestinal tolerability (up to 30–45 g per day) (Pasman *et al.*, 2006). Examples of well-established prebiotics are inulin and FOS, galacto-oligosaccharides (GOS; a.k.a. transgalacto-oligosaccharides or TOS) and lactulose. Less well-studied prebiotics include lactitol, isomalto-oligosaccharides (IMO), gluco-oligosaccharides (Gl-OS), lactosucrose, xylo-oligosaccharides (XOS), arabinoxylo-oligosaccharides (AXOS), resistant starch, raffinose (Gal-Glc-Fru) and stachyose (Gal-Gal-Glc-Fru) from soy and (acidic) pectic oligosaccharides, although there is accumulating evidence for these.

Inulin and FOS are composed of linear chains of β-2,1-linked fructose units, with or without a terminal glucose. GOS are a mixture of oligosaccharides derived from lactose by enzymatic transglycosylation. The mixture generally consists of oligosaccharides from tri- to pentasaccharides with β-1,6, β-1,3 and β-1,4 linkages (Coulier *et al.*, 2009), which make it refractive to digestion. Lactulose is manufactured by the isomerization of lactose to generate the disaccharide galactosyl- β–1,4-fructose. Lactitol is a sugar alcohol and is the hydrogenated form of lactose. IMO are manufactured from starch, which is first hydrolysed by the combined action of α-amylase and pullulanase, and the resultant malto-oligosaccharides are acted upon by α-glucosidase to create IMO. IMO is partly digestible by pancreatic enzymes, but is digested slowly. Depending on the individual enzyme activity and food transit, variable amounts of IMO may reach the colon. Gl-OS are synthesized by the action of the enzyme dextran-sucrase on sucrose in the presence of maltose. Due to the presence of glycosidic bonds uncharacteristic of starch, it is not hydrolysed by pancreatic α-amylase. Lactosucrose is produced from a mixture of lactose and sucrose using the enzyme β-fructofuranosidase. XOS and AXOS are manufactured by the enzymatic hydrolysis of (arabino)xylan from various cereal sources. The above brief expose on the chemical nature of several different prebiotics indicates the wide range of different carbohydrates that are being or have been evaluated for their prebiotic potential. This is partly due to the fact that we have increased our insight in the health beneficial aspects (and at the same time the detrimental activities) of the endogenous gut microbiota. Whereas in the past a prebiotic activity was synonymous with a bifidogenic activity (Gibson and Roberfroid, 1995), currently the microbiota has been implicated in numerous diseases and disorders and hence modifying the microbiota composition beyond a bifidogenic stimulation with different prebiotics is now also considered to be beneficial, although even in recent applications, bifidogenic prebiotics seems to function as well, as has been shown for instance for FOS in obesity (Geurts *et al.*, 2014).

A large part of the adult world population cannot digest lactose (Venema, 2012), as the expression of the lactase enzyme in the gut wall is decreased with increasing age. This phenomenon is called lactase deficiency (no lactase) or hypolactasia (some remaining activity) (Venema, 2012). If these individuals would ingest lactose (Gal-Glc), the disaccharide would make it to the colon. As such it can be considered a prebiotic. However, lactose ingestion is very frequently associated with gastrointestinal complaints. This lactose intolerance is thought to occur due to fermentation of lactose by the gut microbiota (He *et al.*, 2008). There is a lot of anecdotal evidence and some scientific data that slowly increasing the dose of lactose would diminish the intolerance. Adaptation to long-term lactose ingestion in lactose maldigesters has been thought to be related to adaptation of the colonic microbiota and colonic function (He *et al.*, 2008; Hertzler and Savaiano, 1996). The mechanism underlying lactose intolerance is hypothesized to be the very quick fermentation of the disaccharide by the microbiota, likely leading to accumulation of microbial metabolites and gas, which leads to the experience of bloating, rumbling and diarrhoea (He *et al.*, 2008). Adaptation of the microbiota, or adaptation of the colonic epithelium to cope with the increased metabolites, is thought to relieve the symptoms. A similar mechanism may also be in play for oligosaccharides such as inulin and FOS, which also have been shown to cause GI complaints after high doses. Experiments in validated *in vitro* models of the colon have indicated that these oligosaccharides [degree of polymerization (DP) ranging from 3 to 9] are also very quickly fermented (van Nuenen *et al.*, 2003), and might thus also lead to accumulation of microbial metabolites and gas, with consequent GI complaints. A very long-chain inulin (average DP > 55, with a maximum DP of ~75) extracted from globe artichoke had pronounced prebiotic effects in human subjects, but was well tolerated (Costabile *et al.*, 2010). Besides DP also the biochemical structure of the oligosaccharides determines (speed of) fermentability. GOS contains many different molecules with different DP and glycosidic linkages (Coulier *et al.*, 2009), which cannot all be degraded by the same microbial species. Similarly, fermentation of linear α-1,6 dextrans and dextrans with α-1,2 branching by the human microbiota was different (Sarbini *et al.*, 2011). In addition, a branched inulin from agave shows different properties than the linear inulin from, for example, chicory: fermentation of the agave fructan was slower (Koenen, M.E., Netherlands Organisation for Applied Scientific Research (TNO); Cruz Rubio, J.M., Nekutli; and Venema, K., Beneficial Microbes Consultancy, personal communication). This may be (part of) the reason why the hunter-foragers in the Chihuahuan Desert could ingest higher quantities of these fructans, as the agava fructans are highly branched (López and Urías-Silvas, 2007). It has been shown that different bifidobacterial strains fall into one of four clusters with reference to their metabolism of inulin and shorter chain FOS, or the monosaccharide fructose. Some strains only metabolized fructose; other strains preferred fructose but could also metabolize FOS, but with decreasing affinity as molecular weight (or DP) increased. Some strains preferred FOS with little ability to grow on the monosaccharide or inulin, while the last cluster had equal affinity to grow on fructose and FOS and have some ability to grow on inulin (Falony *et al.*, 2009). Similarly, structurally different wheat-derived arabinoxylo-oligosaccharides with different DP showed different fermentation characteristics (Van Craeyveld *et al.*, 2008). This shows that the structure, but also the dose defines the beneficial properties of the prebiotic that is ingested. Whilst we do not have a clear and comprehensive understanding of the molecular mechanisms involved in the bacterial fermentation of complex carbohydrates in the gut, it is clear that these processes will involve the action of glycosylhydrolases and transport systems to take up prebiotics or their breakdown products. This is discussed in more detail in Chapter 15.

Synbiotics

Prebiotics function complementary to, and possibly synergistically with, probiotics. Gibson and Roberfroid first defined synbiotics as 'a mixture of probiotics and prebiotics that beneficially affects the host by improving the survival and implantation of live microbial dietary supplements (read: probiotics) in the gastrointestinal tract, by selectively stimulating the growth and/ or by activating the metabolism of one or a limited number of health-promoting bacteria, and thus improving host welfare' (Gibson and Roberfroid, 1995). Synbiotics are covered in detail in Chapter 20.

The prebiotics present in synbiotic mixtures are generally considered to have one of two possible functions. First and foremost they act as prebiotics, in the sense that they stimulate the numbers and activity of beneficial microbes endogenous to the (gut of the) host, independent of the activity of the probiotic (complementary approach). Secondly, they are considered a 'lunch-box' for the probiotics and should increase their survival while the beneficial microbes travel through the hostile environment of the GI tract (synergistic approach) (Kolida and Gibson, 2011). However, the lunch-box function of prebiotics assumes that the probiotics are able to use the prebiotics as a substrate, and this is certainly not always true, as some products combine probiotic lactobacilli with a bifidogenic prebiotic. In this case the probiotic will not benefit from the ability to use the prebiotic as a substrate for growth. Not all currently commercially available probiotics for instance are capable of fermenting FOS or GOS. Indications on the potential growth of a probiotic on a selected substrate can be obtained through a genomic scan of the glycosidase spectra of the strain (assuming that the genome sequence is available), which may allow for a rational selection of the potential prebiotics that this particular strain can use (Kolida and Gibson, 2011). In those cases where the probiotic can ferment the prebiotic, a second assumption is that the presence of a fermentable substrate indeed does increase the probiotic survival. Although there is evidence that this is the case (Martinez *et al.*, 2011), we also have some yet unpublished data that this depends on the dose of prebiotics added. At relative low doses it appears that survival of the probiotic strain may in fact be lower than that without the presence of the prebiotic. We hypothesize that this is caused by the following: Most probiotics when tested for survival have been cultured until the stationary phase of growth before they are harvested. When grown *in situ* in a product, this harvesting is simply distribution over individual packages and shipping out for retailing. When the strains are to be incorporated into tablets, capsules or other formulations, they are usually freeze- or spray-dried. When entering the stationary growth phase, bacteria express stress-proteins that help them cope this phase of low substrate availability, etc. This has been generally shown in lactic acid bacteria and is also true for probiotics (Mills *et al.*, 2011). These stress-proteins protect the cells against stresses they can encounter in the production pipeline, such as low or high temperature, low pH, high pressure, etc. When cells that have expressed these stress proteins encounter the hostile environment in the GI tract, with low gastric acidity and high bile concentrations in the small intestine, the cells are usually cross-protected to a certain extent by the stress proteins. However, this never is a full protection, and some cells will die when travelling through the GI tract. This has been nicely shown using a dynamic *in vitro* model of the stomach and small intestine (TIM-1), that accurately mimics the dynamic parameters occurring in the upper GI tract, such as gastric acidity, bile and pancreatic enzyme concentrations, and food transit (Marteau *et al.*, 1997). These models are excellent tools to study the mechanisms underlying optimal survival of probiotics in the GI tract. For instance, by changing growth conditions of the probiotics, the effect of protective stress proteins on their survival can be established. Also, some strains are more sensitive to acid, others are more sensitive to bile. When this is known, specific measures can be taken to increase survival of these strains, e.g. by screening for optimal growth conditions, or by developing a protective coating for strains that are sensitive to gastric acid, whether with or without a prebiotic, e.g. (Chaluvadi *et al.*, 2012; Chavarri *et al.*, 2010; Eiberger *et al.*, submitted; Riaz and Masud, 2013).

Now what happens when these probiotic strains which have expressed (increased) stress proteins are allowed to have access to a prebiotic during GI transit? When the probiotics start to feed on the prebiotic they leave the stationary phase and enter the logarithmic growth phase. In this log-phase there is no need to express the stress proteins, and hence cells are less able to cope with the stresses encountered in the GI tract. Hence survival drops. However, one could argue that since the prebiotic is a substrate for growth, one would get multiplication of the cells as well. One would be right. However, there appears to be a delicate balance between the dose of prebiotics added that destroys cross-protection by the stress proteins (and leads to lower survival) and the dose that allows for sufficient growth to occur such that the number of viable cells increases. This increase in viable cells then appears to be increased survival, but in reality is the sum of increased death due to loss of protection by stress proteins and increased numbers of viable cells due to growth. As researchers, we have no easy way of distinguishing this. From this, two things follow: (a) there should be a dose of prebiotics that exactly balances increased death and increased cell numbers and that for the researcher does not seem to affect survival; and (b) apparent increased survival can be obtained by choosing (usually be trial and error) the proper dose of the prebiotic. The TIM-1 system referred to above is a perfect tool to study this under standardized conditions, as we will see in the section on potential mechanisms of synbiotics.

Not all studies that investigated a synbiotic have compared the effects with those of the single ingredients, i.e. the probiotic or prebiotic alone (de Vrese and Schrezenmeir, 2008). Moreover, not all studies show that the combination of probiotics and prebiotics was more effective than the single ingredients alone, at least not on the parameters that were studied. Not all studies evaluated the effect of the synbiotic on microbiota composition and activity, but rather focused on parameters such as immunomodulation (by the probiotic).

Recent applications of pro- and prebiotics, alone or in combination

It is beyond the scope of this chapter to discuss in detail all past applications of pro- and prebiotics. We have chosen to highlight some of the recent hot topics that these dietary ingredients are used for, as well as some promising areas that are under development. Most of these areas are covered in separate chapters elsewhere in this book.

Excellent recent reviews have appeared before on the beneficial effects of probiotics on, for example, diarrhoea (McFarland, 2014), functional constipation (Dimidi *et al.*, 2014), inflammatory bowel disease (IBD) (Ford *et al.*, 2014), allergic disease (Kim *et al.*, 2013), respiratory infections (King *et al.*, 2014), vaginal health (Reid, 2014), animal growth (Chaucheyras-Durand and Durand, 2010), and the effects of prebiotics on changes in the gut microbiota related to health (Ford *et al.*, 2014; Sanders *et al.*, 2014).

Obesity

Recently, there is a lot of focus on the beneficial effects of both pro- and prebiotics in obesity. Since it became clear that the gut microbiota plays a role in obesity, numerous studies have looked into this. However, it now appears that numerous mechanisms may be involved (Delzenne *et al.*, 2011; Geurts *et al.*, 2014). These include gut barrier function and low grade systemic inflammation (due to LPS leakage into the system), modulation of gut hormones such as GLP-1 and PYY, modulation of activity of lipases, energy extraction, adipocyte function, and the endocannabinoid system. Microorganisms that have been implicated to play a role are for instance bifidobacteria (Geurts *et al.*, 2014), and *Akkermansia muciniphila* (Everard *et al.*, 2013), although in energy extraction from dietary components likely multiple members of the microbiota participate. Although there are studies that have shown a difference in composition of the microbiota between lean and obese individuals (Bervoets *et al.*, 2013; Duncan *et al.*, 2008; Ferrer *et al.*, 2013; Santacruz *et al.*, 2009; Turnbaugh *et al.*, 2006), if you take these studies together there does not seem to be a common microorganism that can be implicated. Obesity seems to be correlated with a reduced microbial gene richness (Cotillard *et al.*, 2013; Le Chatelier *et al.*, 2013). Owing to the multiple possible mechanisms involved, pro- and prebiotics may be of benefit for obese individuals (Delzenne *et al.*, 2011). An interesting philosophy is whether the gut microbiota drives our eating behaviour (Alcock *et al.*, 2014).

Diabetes

Related to obesity, the effect of pro- and prebiotics on type 1 and type 2 diabetes is studied as well. Diabetes is a condition of multifactorial origin, linked through several molecular mechanisms to the gut microbiota. For instance, in type 2 diabetes, systemic inflammation, which occurs in obesity, induces the phosphorylation of serine residues in insulin receptor substrate-1, reducing insulin sensitivity. In type 1 diabetes, a greater microbiota-induced immune response may aggravate the autoimmunity. A large body of evidence suggests that certain probiotic strains reduce the host inflammatory response, and/or reduce intestinal permeability. Such effects increase insulin sensitivity and reduce autoimmune responses (Gomes *et al.*, 2014). Similarly, prebiotics may affect the gut microbiota composition and/or activity, by this indirectly influencing diabetes. However, further investigations are needed to investigate mechanisms and to elucidate whether the administration of pro- and prebiotics can be effective for the prevention and management of both types of diabetes (Petschow *et al.*, 2013).

Cardiovascular disease

Also linked to obesity are cardiovascular diseases, such as hypercholesterolaemia and arterial hypertension, and metabolic disturbances, such as hyperhomocysteinaemia and oxidative stress. Several recent reviews have addressed this (DiRienzo, 2014; Ebel *et al.*, 2014), and also this book contains a contribution on the effects of pro- and prebiotics on plasma cholesterol levels. However, there is limited clinical evidence. Most research has been done in animals, with few exceptions (see DiRienzo, 2014, for review).

Gut–brain axis

Another hot topic in the research on pro- and prebiotics is the gut–brain axis. Several aspects of this axis are investigated, including anxiety/stress, behaviour, cognition (Lyte and Cryan, 2014), and autism (MacFabe *et al.*, 2011), with potential benefits in neurological disorders like Parkinson's disease and Alzheimer's disease. Usually irritable bowel syndrome (IBS) is also incorporated in the gut–brain axis. There is ample evidence in several rodent models that the gut microbiota influences anxiety, behaviour and cognition. A nice example is how faecal transfer from the more aggressive (less anxious) NIH Swiss mice into the more anxious BALB/c mice lead the mice to become more aggressive (Bercik *et al.*, 2011). Symptoms of depression and anxiety in anorexia nervosa have been linked with plasma tryptophan and serotonin metabolism (Gauthier *et al.*, 2014). The gut microbiota can synthesize these compounds (Neufeld *et al.*, 2011), and in that manner influence depression and anxiety. It has been shown that germ-free mice (mice without any microbiota) show reduced anxiety-like behaviour coupled to central neurochemical changes (Neufeld *et al.*, 2011). Influencing gut microbiota composition and metabolism may thus have an effect on anxiety and behaviour. Recent studies in germ-free mice have also demonstrated impaired (non-spatial) memory and social cognition, suggesting a link between cognitive processes and the presence of the gut microbiota (Desbonnet *et al.*, 2014; Gareau *et al.*, 2011). Treatment with probiotics in mice enhanced fear memory and reversed memory deficits observed in *Citrobacter rodentium*-infected mice after acute stress exposure (Bravo *et al.*, 2011; Gareau *et al.*, 2011). Autism or rather autism spectrum disorders (ASDs) are neurodevelopmental disorders characterized by presence of stereotypical behaviour and social interaction deficits (MacFabe *et al.*, 2011). Studies in germ-free mice demonstrated robust and reproducible social deficits and increases in repetitive behaviours similar to that observed in ASD (Desbonnet *et al.*, 2014), suggesting that the microbiota is a critical factor in the development of social behaviour and the aetiology of ASDs. In a series of elegant experiments the group of MacFabe in Canada has shown that the microbial metabolite propionate induces this repetitive behaviour in mice, when infused in the brain. Upon metabolism or clearance of the infused propionate, the behaviour returns to normal. This assumes that in man propionate may be one of the triggers for ASD, which seems to correlate with anecdotal evidence that an increase in carbohydrate intake, which would lead to increased propionate production by the gut microbiota in the colon, leads to a worsening of the behaviour of autistic children.

Oral health

The oral cavity (strictly also part of the GI tract) also contains a complex microbiota. The role of this microbiota in oral disease has long been acknowledged. Recently, the effect of pro- and prebiotics to increase oral health (or prevent disease) has been studied (Pham *et al.*, 2009; see also Chapter 28). Some *in vitro* studies regarding inhibition of potential probiotics towards *S. mutans* (Gungor *et al.*, 2013), or use of dietary sugars by potential probiotics (Stamatova *et al.*, 2012) is ongoing as well.

Skin health

Another area of the body that benefits from pro- and prebiotics is the skin (Al-Ghazzewi and Tester, 2014). In a recent special issue of the journal Beneficial Microbes (June 2014), several aspects of skin health are reviewed. Interestingly, these pro- and prebiotics are used orally, but have an effect on different aspects of the skin, such as glow of the hair of mice (Erdman and Poutahidis, 2014), skin sensitivity and transepidermal water loss, related to wrinkling of the skin (Gueniche *et al.*, 2014). Interestingly, in one study the mechanism has been elucidated. Phenols, for example phenol and p-cresol, as metabolites of aromatic amino acids produced by gut microbiota, are regarded as bioactive toxins and serum biomarkers of a disturbed gut environment. Recent studies have demonstrated that these phenols disturb the differentiation of keratinocytes *in vitro*, and disrupt keratinocyte differentiation in hairless mice. *Bifidobacterium breve* strain Yakult fermented milk in combination with GOS led to improved skin health by decreasing production of phenols by the gut microbiota (Miyazaki *et al.*, 2014).

Probiotics do not always influence the composition of the gut microbiota. No significant changes in bacterial species composition or in the proportional representation of genes encoding known enzymes were observed in the faeces of humans consuming a yoghurt containing *B. animalis* subsp. *lactis*. In a mouse model containing a defined 15-strain microbial community, only minimal changes in microbiota configuration were noted following probiotic yoghurt consumption. However, RNA-Seq analysis of faecal samples and follow-up mass spectrometry of urinary metabolites disclosed that feeding the probiotic yoghurt to mice resulted in significant changes in expression of defined community-encoded enzymes involved in numerous metabolic pathways, most prominently those related to carbohydrate metabolism (McNulty *et al.*, 2011).

Pitfalls

Strain dependency of probiotic activity

Although to most researchers it is clear that probiotic activity is strain dependent and that not all probiotics function against all disorders and diseases studied, the strain number is usually not mentioned when people talk about the effect of probiotics. Sentences like '*L. acidophilus* showed a benefit in stool consistency', is meaningless without the mentioning of the strain number, e.g. NCFM. However, this concept of strain dependency has led researchers to believe that the probiotic functionality could be deciphered from genome information of a collection of strains that showed a range of activities. One example using comparative genome hybridization showed that *L. plantarum* WCFS1 carried several genetic loci that modulated the immune response of dendritic cells. Deletions of these genes or loci from the chromosome confirmed that these genes were at least partly involved in immunomodulation (Meijerink *et al.*, 2010). Such quests for probiotic effector molecules should lead to unravelling strain specificity at the molecular level (Lee *et al.*, 2013). However, nature appears to be slightly more complicated than anticipated: the 42 *L. plantarum* strains showed an activity range against DCs of several orders of magnitude (measured as cytokine production), and based on that the different loci were selected. Yet, deleting of individual loci had at most a two-fold effect on immunomodulation, and not an effect of several orders of magnitude. Bacterial background may be one factor at play here (Lee *et al.*, 2013).

Similarly there are differences between different prebiotics that have a different monosaccharide composition or vary in glycosidic linkages. Even molecules with the same saccharides and linkages, but simply differing in degree of polymerization have different effects as shown for speed of fermentation with different inulin fractions with different DP (van Nuenen *et al.*, 2003).

Dose

Dose–response relationships of probiotics are rarely studied, not even in animal trials. This is very likely attributed to the high costs of clinical trials, but failure of choosing the proper dose my lead either to results in the trial that are not significant or to no benefit at all, which would mean that the money spent was wasted too. In one of the few dose–response studies, *B. animalis* subsp. *lactis* BB-12 and *L. paracasei* subsp. *paracasei* CRL-341 in healthy young adults, doses of 10^8, 10^9, 10^{10} and 10^{11} cfu/day were tested on blood lipids, recovery from faeces and bowel habits (Larsen *et al.*, 2006). In the group receiving 10^{11} cfu/day BB-12 was recovered in faecal samples of 13 out of 15 volunteers. CRL-431 was not recovered in any of the faecal samples. Supplementation with probiotics at the level of 10^{10} cfu/day did not change the faecal bacterial composition. Unfortunately, this was not tested for the other doses, but is in agreement with the study of McNulty *et al.* (McNulty *et al.*, 2011) reported above. A significant linear increase in faecal consistency (looser stool) with increasing probiotic dose was observed. No overall dose–response effect was found on the blood lipids. In another study the two strains *L. acidophilus* CL1285 and *L. casei* LBC80R were provided to volunteers in capsules to test the effect on antibiotic-associated diarrhoea (AAD) or *Clostridium difficile*-associated diarrhoea (CDAD) (Gao *et al.*, 2010). Volunteers either took one or two capsules containing 5×10^{10} cfu/capsule. The high-dose group had both lower AAD and lower CDAD incidences, as well as a shorter duration of diarrhoea than the low-dose group, and both groups scored better than the placebo. Mechanistically, dose–response was studied by incubating *L. rhamnosus* Lcr35 with human monocyte-derived immature dendritic cells (DCs), using a wide range of bacterial concentrations (multiplicity of infection, MOI, from 0.01 to 100). Flow cytometry analysis also revealed a dose-dependent maturation of the DCs, while DC-secreted cytokine analysis showed that Lcr35 induced a strong dose-dependent increase of certain proinflammatory cytokine levels, but only a small increase in IL-10.

Similarly, different doses of prebiotic will lead to different effects. For example, at low doses a prebiotic may be specific for, let's say, bifidobacteria, due to the high specificity of bifidobacteria for that particular prebiotic. But when the dose is increased, this would leave some substrate for other microorganisms that have lower specificity for the substrate, but are essentially able to ferment it. The outcome of a high dose would show less specificity than that of the low dose.

Laboratory vs. industrial conditions, stability of probiotics

Practically all research with probiotics starts with simple *in vitro* experiments in the laboratory. These strains are usually cultured in commercial media, such as MRS. Under these culturing conditions, the probiotic cells are incubated under conditions simulating the GI tract, or interacted with, for example, immune cells. Based on these initial results, some promising isolates make it to the market and are then cultured in large quantities. However, since MRS is an expensive media, probiotics are grown on different media in industrial processes. It is often neglected that under these different culturing conditions the gene expression, and hence the probiotic functionality of the probiotics, may be completely different. Although entirely speculative, this may be an additional factor explaining why certain clinical trials show a benefit, whereas others, with the same probiotic, fail (Kalliomaki *et al.*, 2001, 2003; Kopp *et al.*, 2008; Rose *et al.*, 2010).

It proves very difficult to translate *in vitro* and animal work to humans (Meijerink *et al.*, 2013). Consider the following: in *in vitro* experiments, apart from being grown in laboratory media, the cells are perhaps a little stressed (depending on the growth phase), but otherwise healthy. Yet, in the GI tract, cells are exposed to acid, bile and digestive enzymes. The cell surface of these GI-exposed cells, when studied under the electron microscope, are entirely different than those of laboratory-media grown cells (Pavan, S., Netherlands Organisation for Applied Scientific Research (TNO); Venema, K., Beneficial Microbes Consultancy, unpublished results). Since, the consensus is that surface structures of probiotics interact with the host, it is perhaps not surprising that *in vitro* experiments performed with laboratory-media grown cells are not predictive. This should not be forgotten when performing initial screening assays of collections of isolates for potential probiotic activity. Apart from difficulties in translating *in vitro* research to clinical trials, also animal data is difficult to interpret (Meijerink *et al.*, 2013).

The latter is also true for prebiotics, since animals have a different microbiota than humans. In fact, our unpublished results show that even pigs, which are considered to look like humans physiologically, have a different microbiota. A pig microbiota produced esters of methanol or ethanol and short-chain fatty acids (SCFA), rather than SCFA alone, as is the case for a human microbiota. These esters, when interacting with the host, will have a different functionality, and this may not translate to the human situation.

The production of viable functional probiotics presupposes strain stability in the final product, i.e. during storage (in the production plant, supermarket, at home) and transport. It is very difficult to control this. For example, low temperatures during transport from the supermarket to home are usually not available. How this affects probiotic survival is unknown. For over-the-counter applications, probiotics are provided in sachets, tablets or capsules, containing probiotics in freeze-dried form. How freeze-drying affected probiotic functionality was recently tested (du Toit *et al.*, 2013). These authors also investigated the effect of acquiring heat-tolerance, by providing a heat-shock at elevated temperatures; ranging from 70 to 85°C, depending on the strain. Freeze-dried strains showed a significant reduction in probiotic functionality (measured here as inhibition of pathogen adhesion to colonic mucus). Some strains in which heat tolerance was induced prior to freeze-drying showed a much better remaining probiotic functionality. The same group studied the effect of suspending (freeze-dried) *L. rhamnosus* GG cells in oil on survival (Endo *et al.*, 2014). Freeze-dried cells suspended in oils showed limited reduction in viability after 5 min and 20 min at 80°C and oils still contained viable cells after 5 min at 95°C. Long-term storage indicated that freeze-dried cells suspended in oils were still viable after 4 months. Cells that were not freeze-dried did not show viability after suspension in oils. These results clearly indicate that moisture and cell conditions have a great impact on survival of *L. rhamnosus* GG under severe heat stress in processing and during long-term storage. Combination of freeze-dried cells and oils as carrier provided the best benefit to preserve viability of this strain in food processes and during storage.

Combinations of strains, or with prebiotics

Several commercial products consist of multiple strains of probiotics. In fact, it seems to be a recent development to create mixtures of potential probiotic isolates, rather than to rely on single strains. However, it is usually unclear how these strains interact with one another, and whether they act synergistically or antagonistically. Some combinations are based on results of *in vitro* experiments, but even then interactions between strains are usually not studied. Some mixtures consist of 6–8 strains, and it is imperative that the individual activities of the components is established (preferably even clinically), if the mode of action of probiotics, and certainly of mixtures, is to be deciphered.

As discussed earlier in the section on synbiotics, combinations of pro- and prebiotics are also arbitrarily chosen. Sometimes the selected probiotic cannot even ferment the selected prebiotic. It is unusual that the components of synbiotics are tested individually in clinical trials, although this does sometimes occur in preclinical experiments. As mentioned above, it is imperative that the individual components are tested separately to be able to decipher mechanisms.

Mechanism of action

Although numerous efforts have already been undertaken to investigate the mode of action of probiotics, it is largely unclear what molecular effector molecule determines whether one strain is active whereas another is not (Lee *et al.*, 2013; Meijerink *et al.*, 2013). However, it is imperative that efforts to understand how these probiotics function molecularly are continued, despite the complexity of the interaction between probiotics and the host. Without this molecular understanding the field will not advance.

For prebiotics it is easier to determine their mechanism of action. They usually stimulate the amounts and activity of a limited number of members of the microbiota. Nevertheless, how this stimulation leads to the underlying health effects is not yet clear. General actions such as acidification of the colonic lumen, inhibition of pathogens, etc. are likely valid for all prebiotics. They also all increase production of SCFA. Nevertheless, since prebiotics are chemically different, they will no doubt have their own individual activity, and these need to be studied to determine which prebiotic is beneficial in which disorder.

Future trends

Rather than repeat the future perspectives that are given by the various authors in this book, here we refer to these sections in the individual contributions. Moreover, in our last chapter 'Is the sky the limit?', we hypothesize on future benefits that have not been (fully) explored. Here we focus on those aspects that require further research and that have been addressed above in the section on pitfalls.

For both pro- and prebiotics only a handful of studies have been performed into what is the proper dose for efficacy. However, some of these studies have shown that there is an optimal dose, with more not always being better. This being the case for both pro- and prebiotics, future experiments should take this into consideration, despite the fact that this will increase costs of a clinical trial. If a dose is chosen arbitrarily, the outcome may not be as expected. Moreover, researchers should take into account the individuality response of the (patient) population they investigate. As for drugs, there will be non-responders. It would be good if we could stratify the volunteers, to get more consistent and better results. To be able to do this however, the mechanism of action of pro- and prebiotics need to be deciphered. Only then can we screen the population for (still to be identified) biomarkers, and provide them with the proper, personalized, pro- or prebiotic. We advise that research into the mechanism of action gets high priority, to prevent a lot of trial-and-error from occurring.

It should be emphasized that probiotic effects are strain dependent. Not every strain is good for every disease or disorder. For example, a proinflammatory, immunostimulatory strain may be beneficial in increasing the antibody titres of a vaccine, but the same strain is likely to be detrimental in autoimmunity. On top of that, care needs to be taken in interpreting and translating results that are obtained in the lab to the human (or animal) situation. As an example, most *in vitro* experiments in the lab are performed with cultures grown in commercial media. However, in industry these media are much too expensive and alternative culture media are used. It is usually neglected that gene expression in these industrial media is completely different than that on laboratory media, and hence gene-expression within the probiotic cells is different too. This may dramatically affect its probiotic potential, as on one of the media the probiotic molecule may be expressed, while it is not on the other media. More dramatically, this may even hold true for industrial processes at two different sites in the world that culture the same probiotic, with the same processing parameters. However, if the ingredients for the media are purchased from different suppliers (which is likely when the industry plants are located in, for example, The Netherlands and Brazil, the home-countries of the two co-authors of this contribution), then slight differences in ingredients may already lead to differences in gene-expression and hence probiotic functionality. Different batches from different plants therefore need to be compared in a head to head comparison. Moreover, any change in industrial processing conditions within a single production plant may have similar effects, and thus we advocate that after such a change in conditions, the efficacy of the resulting product should be tested. In fact, there is evidence that different batches of the probiotic *E. coli* Nissle 1917 from the same plant result in different immunomodulatory profiles (Schmolz, M., HOT Screen GmbH). We believe that there should be a research focus on this.

Then again, if the probiotic molecule within a probiotic microorganism is known, industrial processes could be optimized to get enhanced production of this molecule. This however, requires the mechanism of action to be known. For immunomodulatory activity the current consensus is that structures on the surface of the probiotic cells interact with receptors on immune cells. Although this may indeed still be the case, the mechanism usually is a little more complex than one molecule interacting with one receptor, as for example indicated by Lee *et al.* (2013). Moreover, when the effector molecule is known, it is probably easier to purify this molecule and add the purified compound as a postbiotic, rather than the original probiotic microorganism.

There are many multi-strain (and even multi-species and multi-genera) probiotics. It is usually unclear what the individual contribution of these multiple strains to the probiotic activity is. It may well be that one of the strains in the mixtures counteracts the effect of another. In some cases where the individual strains have been tested, a single strain dominated the effect and perhaps this one particular strain could do the trick on its own. However, since this has not been tested, this remains speculative. But without doubt these are things that need to be investigated. The results would aid to a better understanding of the mechanism of action of probiotics, either alone or in combination. Similarly, the probiotic strain and prebiotic carbohydrate in synbiotics are frequently arbitrarily chosen on the basis of their individual activities. Rarely are studies performed where the individual components are tested separately and together. And then we are not even discussing optimal dose of either component in the synbiotic.

It is clear from the above that despite several decades of research, there still is a lot to be done, before we really start to grasp the potential and the benefits of pro- and prebiotics. Once we properly understand their molecular mode of action, we can use them in a targeted manner.

References

Adams, C.A. (2010). The probiotic paradox: live and dead cells are biological response modifiers. Nutr. Res. Rev. 23, 37–46.

Al-Ghazzewi, F.H., and Tester, R.F. (2014). Impact of prebiotics and probiotics on skin health. Benef. Microbes 5, 99–107.

Alcock, J., Maley, C.C., and Aktipis, C.A. (2014). Is eating behavior manipulated by the gastrointestinal microbiota? Evolutionary pressures and potential mechanisms. Bioessays 36, 940–949.

Bercik, P., Denou, E., Collins, J., Jackson, W., Lu, J., Jury, J., Deng, Y., Blennerhassett, P., Macri, J., McCoy, K.D., *et al.* (2011). The intestinal microbiota affect central levels of brain-derived neurotropic factor and behavior in mice. Gastroenterology *141*, 599–609, 609 e591–593.

Bervoets, L., Van Hoorenbeeck, K., Kortleven, I., Van Noten, C., Hens, N., Vael, C., Goossens, H., Desager, K.N., and Vankerckhoven, V. (2013). Differences in gut microbiota composition between obese and lean children: a cross-sectional study. Gut Pathog. 5, 10.

Blatchford, P., Ansell, J., de Godoy, M.R.C., Fahey, G., Garcia-Mazcorro, J.F., Gibson, G.R., Goh, Y.J., Hotchkiss, A.T., Hutkins, R., Lacroix, C., *et al.* (2014). Prebiotic mechanisms, functions and applications. Int. J. Probiotics Prebiotics (in press).

Bravo, J.A., Forsythe, P., Chew, M.V., Escaravage, E., Savignac, H.M., Dinan, T.G., Bienenstock, J., and Cryan, J.F. (2011). Ingestion of *Lactobacillus* strain regulates emotional behavior and central GABA receptor expression in a mouse via the vagus nerve. Proc. Natl. Acad. Sci. U.S.A. *108*, 16050–16055.

Chaluvadi, S., Hotchkiss, A.T., Jr., Call, J.E., Luchansky, J.B., Phillips, J.G., Liu, L., and Yam, K.L. (2012). Protection of probiotic bacteria in a synbiotic matrix following aerobic storage at 4 degrees C. Benef. Microbes 3, 175–187.

Chaucheyras-Durand, F., and Durand, H. (2010). Probiotics in animal nutrition and health. Benef. Microbes *1*, 3–9.

Chavarri, M., Maranon, I., Ares, R., Ibanez, F.C., Marzo, F., and Villaran Mdel, C. (2010). Microencapsulation of a probiotic and prebiotic in alginate-chitosan capsules improves survival in simulated gastro-intestinal conditions. Int. J. Food Microbiol. *142*, 185–189.

Costabile, A., Kolida, S., Klinder, A., Gietl, E., Bauerlein, M., Frohberg, C., Landschutze, V., and Gibson, G.R. (2010). A double-blind, placebo-controlled, cross-over study to establish the bifidogenic effect of a very-long-chain inulin extracted from globe artichoke (Cynara scolymus) in healthy human subjects. Br. J. Nutr. *104*, 1007–1017.

Cotillard, A., Kennedy, S.P., Kong, L.C., Prifti, E., Pons, N., Le Chatelier, E., Almeida, M., Quinquis, B., Levenez, F., Galleron, N., *et al.* (2013). Dietary intervention impact on gut microbial gene richness. Nature *500*, 585–588.

Coulier, L., Timmermans, J., Bas, R., Van Den Dool, R., Haaksman, I., Klarenbeek, B., Slaghek, T., and Van Dongen, W. (2009). In-depth characterization of prebiotic galacto-oligosaccharides by a combination of analytical techniques. J. Agric. Food Chem. *57*, 8488–8495.

de Vrese, M., and Schrezenmeir, J. (2008). Probiotics, prebiotics, and synbiotics. Adv. Biochem. Eng. Biotechnol. *111*, 1–66.

Delzenne, N.M., Neyrinck, A.M., Backhed, F., and Cani, P.D. (2011). Targeting gut microbiota in obesity: effects of prebiotics and probiotics. Nat. Rev. Endocrinol. *7*, 639–646.

Desbonnet, L., Clarke, G., Shanahan, F., Dinan, T.G., and Cryan, J.F. (2014). Microbiota is essential for social development in the mouse. Mol. Psychiatry *19*, 146–148.

Dimidi, E., Christodoulides, S., Fragkos, K.C., Scott, S.M., and Whelan, K. (2014). The effect of probiotics on functional constipation in adults: a systematic review and meta-analysis of randomized controlled trials. Am. J. Clin. Nutr. *100*, 1075–1084.

DiRienzo, D.B. (2014). Effect of probiotics on biomarkers of cardiovascular disease: implications for heart-healthy diets. Nutr. Rev. *72*, 18–29.

du Toit, E., Vesterlund, S., Gueimonde, M., and Salminen, S. (2013). Assessment of the effect of stress-tolerance acquisition on some basic characteristics of specific probiotics. Int. J. Food Microbiol. *165*, 51–56.

Duncan, S.H., Lobley, G.E., Holtrop, G., Ince, J., Johnstone, A.M., Louis, P., and Flint, H.J. (2008). Human colonic microbiota associated with diet, obesity and weight loss. Int. J. Obes. (Lond.). *32*, 1720–1724.

Ebel, B., Lemetais, G., Beney, L., Cachon, R., Sokol, H., Langella, P., and Gervais, P. (2014). Impact of probiotics on risk factors for cardiovascular diseases. A review. Crit. Rev. Food Sci. Nutr. *54*, 175–189.

Eeckhaut, V., Ducatelle, R., Sas, B., Vermeire, S., and Van Immerseel, F. (2014). Progress towards butyrate-producing pharmabiotics: Butyricicoccus pullicaecorum capsule and efficacy in TNBS models in comparison with therapeutics. Gut *63*, 367.

Eiberger, I., Fuerst, T., Maathuis, A.J.H., and Venema, K. (2015). Use of a dynamic, computer-controlled *in vitro* model of the stomach and small intestine (TIM-1) as a tool in the development of probiotic products, to study survival of the bacteria and their delivery to the site of action. Benef. Microbes, submitted.

Endo, A., Terasjarvi, J., and Salminen, S. (2014). Food matrices and cell conditions influence survival of *Lactobacillus rhamnosus* GG under heat stresses and during storage. Int. J. Food Microbiol. *174*, 110–112.

Erdman, S.E., and Poutahidis, T. (2014). Probiotic 'glow of health': it's more than skin deep. Benef. Microbes *5*, 109–119.

Everard, A., Belzer, C., Geurts, L., Ouwerkerk, J.P., Druart, C., Bindels, L.B., Guiot, Y., Derrien, M., Muccioli, G.G., Delzenne, N.M., *et al.* (2013). Cross-talk between Akkermansia muciniphila and intestinal epithelium controls diet-induced obesity. Proc. Natl. Acad. Sci. U.S.A. *110*, 9066–9071.

Falony, G., Lazidou, K., Verschaeren, A., Weckx, S., Maes, D., and De Vuyst, L. (2009). *In vitro* kinetic analysis of fermentation of prebiotic inulin-type fructans by Bifidobacterium species reveals four different phenotypes. Appl. Environ. Microbiol. *75*, 454–461.

FAO/WHO (2002). Guidelines for the Evaluation of Probiotics in Food: Joint FAO/WHO Working Group meeting, London Ontario, Canada, 30 April-1 May 2002.

Ferrer, M., Ruiz, A., Lanza, F., Haange, S.B., Oberbach, A., Till, H., Bargiela, R., Campoy, C., Segura, M.T., Richter, M., *et al.* (2013). Microbiota from the distal guts of lean and obese adolescents exhibit partial functional redundancy besides clear differences in community structure. Environ. Microbiol. *15*, 211–226.

Foolad, N., and Armstrong, A.W. (2014). Prebiotics and probiotics: the prevention and reduction in severity of atopic dermatitis in children. Benef. Microbes *5*, 151–160.

Ford, A.C., Quigley, E.M., Lacy, B.E., Lembo, A.J., Saito, Y.A., Schiller, L.R., Soffer, E.E., Spiegel, B.M., and Moayyedi, P. (2014). Efficacy of prebiotics, probiotics, and synbiotics in irritable bowel syndrome and chronic idiopathic constipation: systematic review and meta-analysis. Am. J. Gastroenterol. *109*, 1547–1561.

Gao, X.W., Mubasher, M., Fang, C.Y., Reifer, C., and Miller, L.E. (2010). Dose-response efficacy of a proprietary probiotic formula of *Lactobacillus* acidophilus CL1285 and *Lactobacillus* casei LBC80R for antibiotic-associated diarrhea and *Clostridium difficile*-associated diarrhea prophylaxis in adult patients. Am. J. Gastroenterol. *105*, 1636–1641.

Gareau, M.G., Wine, E., Rodrigues, D.M., Cho, J.H., Whary, M.T., Philpott, D.J., Macqueen, G., and Sherman, P.M. (2011). Bacterial infection causes stress-induced memory dysfunction in mice. Gut *60*, 307–317.

Gauthier, C., Hassler, C., Mattar, L., Launay, J.M., Callebert, J., Steiger, H., Melchior, J.C., Falissard, B., Berthoz, S., Mourier-Soleillant, V., *et al.* (2014). Symptoms of depression and anxiety in anorexia nervosa: links with plasma tryptophan and serotonin metabolism. Psychoneuroendocrinology *39*, 170–178.

Geurts, L., Neyrinck, A.M., Delzenne, N.M., Knauf, C., and Cani, P.D. (2014). Gut microbiota controls adipose tissue expansion, gut barrier and glucose metabolism: novel insights into molecular targets and interventions using prebiotics. Benef. Microbes *5*, 3–17.

Gibson, G.R., Probert, H.M., Loo, J.V., Rastall, R.A., and Roberfroid, M.B. (2004). Dietary modulation of the human colonic microbiota: updating the concept of prebiotics. Nutr. Res. Rev. *17*, 259–275.

Gibson, G.R., and Roberfroid, M.B. (1995). Dietary modulation of the human colonic microbiota: introducing the concept of prebiotics. J. Nutr. *125*, 1401–1412.

Gomes, A.C., Bueno, A.A., de Souza, R.G., and Mota, J.F. (2014). Gut microbiota, probiotics and diabetes. Nutr. J. *13*, 60.

Grabitske, H.A., and Slavin, J.L. (2009). Gastrointestinal effects of low-digestible carbohydrates. Crit. Rev. Food Sci. Nutr. *49*, 327–360.

Gueniche, A., Philippe, D., Bastien, P., Reuteler, G., Blum, S., Castiel-Higounenc, I., Breton, L., and Benyacoub, J. (2014). Randomised double-blind placebo-controlled study of the effect of *Lactobacillus paracasei* NCC 2461 on skin reactivity. Benef. Microbes *5*, 137–145.

Gungor, O.E., Kirzioglu, Z., Dincer, E., and Kivanc, M. (2013). Who will win the race in children's oral cavities? Streptococcus mutans or beneficial lactic acid bacteria? Benef. Microbes *4*, 237–245.

He, T., Venema, K., Priebe, M.G., Welling, G.W., Brummer, R.J., and Vonk, R.J. (2008). The role of colonic metabolism in lactose intolerance. Eur. J. Clin. Invest. *38*, 541–547.

Hertzler, S.R., and Savaiano, D.A. (1996). Colonic adaptation to daily lactose feeding in lactose maldigesters reduces lactose intolerance. Am. J. Clin. Nutr. *64*, 232–236.

Kalliomaki, M., Salminen, S., Arvilommi, H., Kero, P., Koskinen, P., and Isolauri, E. (2001). Probiotics in primary prevention of atopic disease: a randomised placebo-controlled trial. Lancet *357*, 1076–1079.

Kalliomaki, M., Salminen, S., Poussa, T., Arvilommi, H., and Isolauri, E. (2003). Probiotics and prevention of atopic disease: 4-year follow-up of a randomised placebo-controlled trial. Lancet *361*, 1869–1871.

Kim, H.J., Kim, H.Y., Lee, S.Y., Seo, J.H., Lee, E., and Hong, S.J. (2013). Clinical efficacy and mechanism of probiotics in allergic diseases. Korean J. Pediatr. *56*, 369–376.

King, S., Glanville, J., Sanders, M.E., Fitzgerald, A., and Varley, D. (2014). Effectiveness of probiotics on the duration of illness in healthy children and adults who develop common acute respiratory infectious conditions: a systematic review and meta-analysis. Br. J. Nutr. *112*, 41–54.

Kolida, S., and Gibson, G.R. (2011). Synbiotics in health and disease. Annu. Rev. Food. Sci. Technol. 2, 373–393.

Kopp, M.V., Hennemuth, I., Heinzmann, A., and Urbanek, R. (2008). Randomized, double-blind, placebo-controlled trial of probiotics for primary prevention: no clinical effects of *Lactobacillus* GG supplementation. Pediatrics *121*, e850–856.

Larsen, C.N., Nielsen, S., Kaestel, P., Brockmann, E., Bennedsen, M., Christensen, H.R., Eskesen, D.C., Jacobsen, B.L., and Michaelsen, K.F. (2006). Dose-response study of probiotic bacteria Bifidobacterium animalis subsp. lactis BB-12 and *Lactobacillus paracasei* subsp. paracasei CRL-341 in healthy young adults. Eur. J. Clin. Nutr. *60*, 1284–1293.

Lau, S. (2014). Oral application of bacterial lysate in infancy diminishes the prevalence of atopic dermatitis in children at risk for atopy. Benef. Microbes *5*, 147–149.

Le Chatelier, E., Nielsen, T., Qin, J., Prifti, E., Hildebrand, F., Falony, G., Almeida, M., Arumugam, M., Batto, J.M., Kennedy, S., *et al.* (2013). Richness of human gut microbiome correlates with metabolic markers. Nature *500*, 541–546.

Lee, I.C., Tomita, S., Kleerebezem, M., and Bron, P.A. (2013). The quest for probiotic effector molecules – unraveling strain specificity at the molecular level. Pharmacol. Res. *69*, 61–74.

Lied, G.A., Lillestol, K., Lind, R., Valeur, J., Morken, M.H., Vaali, K., Gregersen, K., Florvaag, E., Tangen, T., and Berstad, A. (2011). Perceived food hypersensitivity: a review of 10 years of interdisciplinary research at a reference center. Scand. J. Gastroenterol. *46*, 1169–1178.

López, M.G., and Urías-Silvas, J.E. (2007). Agave fructans as prebiotics. In Recent Advances in Fructooligosaccharides Research, S. Norio, B. Noureddine, and O. Shuichi, eds. (Kerala, India, Research Signpost), pp. 297–310.

Lyte, M., and Cryan, J.F., eds. (2014). Microbial Endocrinology: The Microbiota-Gut-Brain Axis in Health and Disease (New York, Springer).

MacFabe, D.F., Cain, N.E., Boon, F., Ossenkopp, K.P., and Cain, D.P. (2011). Effects of the enteric bacterial metabolic product propionic acid on object-directed behavior, social behavior, cognition, and neuroinflammation in adolescent rats: Relevance to autism spectrum disorder. Behav. Brain Res. *217*, 47–54.

Marteau, P., Minekus, M., Havenaar, R., and Huis in't Veld, J.H. (1997). Survival of lactic acid bacteria in a dynamic model of the stomach and small intestine: validation and the effects of bile. J. Dairy Sci. *80*, 1031–1037.

Martinez, R.C., Aynaou, A.E., Albrecht, S., Schols, H.A., De Martinis, E.C., Zoetendal, E.G., Venema, K., Saad, S.M., and Smidt, H. (2011). *In vitro* evaluation of gastrointestinal survival of *Lactobacillus* amylovorus DSM 16698 alone and combined with galactooligosaccharides, milk and/or Bifidobacterium animalis subsp. lactis Bb-12. Int. J. Food Microbiol. *149*, 152–158.

McFarland, L.V. (2014). Deciphering meta-analytic results: a mini-review of probiotics for the prevention of paediatric antibiotic-associated diarrhoea and *Clostridium difficile* infections. Benef. Microbes, 1–6.

McNulty, N.P., Yatsunenko, T., Hsiao, A., Faith, J.J., Muegge, B.D., Goodman, A.L., Henrissat, B., Oozeer, R., Cools-Portier, S., Gobert, G., *et al.* (2011). The impact of a consortium of fermented milk strains on the gut microbiome of gnotobiotic mice and monozygotic twins. Sci. Transl. Med. 3, 106ra106.

Meijerink, M., Mercenier, A., and Wells, J.M. (2013). Challenges in translational research on probiotic lactobacilli: from *in vitro* assays to clinical trials. Benef. Microbes 4, 83–100.

Meijerink, M., van Hemert, S., Taverne, N., Wels, M., de Vos, P., Bron, P.A., Savelkoul, H.F., van Bilsen, J., Kleerebezem, M., and Wells, J.M. (2010). Identification of genetic loci in *Lactobacillus plantarum* that modulate the immune response of dendritic cells using comparative genome hybridization. PLoS One 5, e10632.

Mills, S., Stanton, C., Fitzgerald, G.F., and Ross, R.P. (2011). Enhancing the stress responses of probiotics for a lifestyle from gut to product and back again. Microb. Cell Fact. *10* (Suppl. 1), S19.

Miyazaki, K., Masuoka, N., Kano, M., and Iizuka, R. (2014). Bifidobacterium fermented milk and galacto-oligosaccharides lead to improved skin health by decreasing phenols production by gut microbiota. Benef. Microbes 5, 121–128.

Neufeld, K.M., Kang, N., Bienenstock, J., and Foster, J.A. (2011). Reduced anxiety-like behavior and central neurochemical change in germ-free mice. Neurogastroenterol. Motil. 23, 255–264, e119.

Pasman, W., Wils, D., Saniez, M.H., and Kardinaal, A. (2006). Long-term gastrointestinal tolerance of NUTRIOSE FB in healthy men. Eur. J. Clin. Nutr. *60*, 1024–1034.

Paton, A.W., Morona, R., and Paton, J.C. (2012). Bioengineered microbes in disease therapy. Trends Mol. Med. *18*, 417–425.

Petschow, B., Dore, J., Hibberd, P., Dinan, T., Reid, G., Blaser, M., Cani, P.D., Degnan, F.H., Foster, J., Gibson, G., *et al.* (2013). Probiotics, prebiotics, and the host microbiome: the science of translation. Ann. N. Y. Acad. Sci. *1306*, 1–17.

Pham, L.C., van Spanning, R.J., Roling, W.F., Prosperi, A.C., Terefework, Z., Ten Cate, J.M., Crielaard, W., and Zaura, E. (2009). Effects of probiotic *Lactobacillus salivarius* W24 on the compositional stability of oral microbial communities. Arch. Oral Biol. *54*, 132–137.

Reid, G. (2014). Modulating the vaginal microbiome: the need for a bridge between science and practice. Semin. Reprod Med. *32*, 28–34.

Reid, G., Sanders, M.E., Gaskins, H.R., Gibson, G.R., Mercenier, A., Rastall, R., Roberfroid, M., Rowland, I., Cherbut, C., and Klaenhammer, T.R. (2003). New scientific paradigms for probiotics and prebiotics. J. Clin. Gastroenterol. *37*, 105–118.

Riaz, Q.U., and Masud, T. (2013). Recent trends and applications of encapsulating materials for probiotic stability. Crit. Rev. Food Sci. Nutr. *53*, 231–244.

Rose, M.A., Stieglitz, F., Koksal, A., Schubert, R., Schulze, J., and Zielen, S. (2010). Efficacy of probiotic *Lactobacillus* GG on allergic sensitization and asthma in infants at risk. Clin. Exp. Allergy *40*, 1398–1405.

Sanders, M.E., Lenoir-Wijnkoop, I., Salminen, S., Merenstein, D.J., Gibson, G.R., Petschow, B.W., Nieuwdorp, M., Tancredi, D.J., Cifelli, C.J., Jacques, P., *et al.* (2014). Probiotics and prebiotics: prospects for public health and nutritional recommendations. Ann. N.Y. Acad. Sci. *1309*, 19–29.

Santacruz, A., Marcos, A., Warnberg, J., Marti, A., Martin-Matillas, M., Campoy, C., Moreno, L.A., Veiga, O., Redondo-Figuero, C., Garagorri, J.M., *et al.* (2009). Interplay between weight loss and gut microbiota composition in overweight adolescents. Obesity (Silver Spring) *17*, 1906–1915.

Sarbini, S.R., Kolida, S., Naeye, T., Einerhand, A., Brison, Y., Remaud-Simeon, M., Monsan, P., Gibson, G.R., and Rastall, R.A. (2011). *In vitro* fermentation of linear and alpha-1,2-branched dextrans by the human fecal microbiota. Appl. Environ. Microbiol. *77*, 5307–5315.

Shanahan, F., Dinan, T.G., Ross, P., and Hill, C. (2012). Probiotics in transition. Clin. Gastroenterol. Hepatol. *10*, 1220–1224.

Slavin, J. (2013). Fiber and prebiotics: mechanisms and health benefits. Nutrients *5*, 1417–1435.

Stamatova, I., Kari, K., Hervonen, L., and Meurman, J.H. (2012). Dietary sugar utilisation by putative oral probiotics. Benef. Microbes *3*, 221–227.

Tsilingiri, K., and Rescigno, M. (2013). Postbiotics: what else? Benef. Microbes *4*, 101–107.

Turnbaugh, P.J., Ley, R.E., Mahowald, M.A., Magrini, V., Mardis, E.R., and Gordon, J.I. (2006). An obesity-associated gut microbiome with increased capacity for energy harvest. Nature *444*, 1027–1031.

Van Craeyveld, V., Swennen, K., Dornez, E., Van de Wiele, T., Marzorati, M., Verstraete, W., Delaedt, Y., Onagbesan, O., Decuypere, E., Buyse, J., *et al.* (2008). Structurally different wheat-derived arabinoxylooligosaccharides have different prebiotic and fermentation properties in rats. J. Nutr. *138*, 2348–2355.

van Nuenen, M.H.M.C., Meyer, P.D., and Venema, K. (2003). The effect of various inulins and *Clostridium difficile* on the metabolic activity of the human colonic microbiota *in vitro*. Microb. Ecol. Health Dis. *15*, 137–144.

Venema, K. (2012). Intestinal fermentation of lactose and prebiotic lactose derivatives, including human milk oligosaccharides. Int. Dairy J. *22*, 123–140.

Functional Aspects of Prebiotics and the Impact on Human Health

2

Vicky De Preter and Kristin Verbeke

Abstract

Owing to the recent developments in analytical techniques to analyse the composition of complex microbial ecosystems, our understanding of the intestinal microbiota has tremendously increased. Several disorders have been associated with an altered composition of the gut bacteria. As a consequence, the microbiota is increasingly recognized as a therapeutic target to improve health. Besides having a trophic and protective function, the microbiota is a metabolically very active ecosystem. Amongst the wide variety of metabolites produced, short-chain fatty acids (SCFA) constitute the most relevant compounds in relation to health. As far as we know to date, administration of prebiotics selectively modifies the composition of the intestinal microbiota through several mechanisms and favours the saccharolytic fermentation resulting in increased SCFA production. These SCFA play a pivotal role in the health benefits associated with prebiotic intake as they acidify the colonic lumen, which influences metabolic pathways and inhibits pathogens, and act as signalling molecules on specific receptors. In the future, more detailed information on the exact role of each individual SCFA and on the proportion of the SCFA produced from different prebiotic substrates will be essential to further exploit the benefits of prebiotic use.

Introduction

The recent development of high-throughput methodologies such as metagenome sequencing has allowed more detailed investigations of the gut microbiota composition and its functional capacities (Van Baarlen *et al.*, 2013). More than 100 trillion bacteria and archaea, consisting of approximately 500–1000 different species, make up the gut microbiota (Huttenhower *et al.*, 2012). During the last decades, the impact of the interplay between these intestinal bacteria, the host metabolism, the diet and the immune system on the human health status has been increasingly recognized. An altered composition of the intestinal microbiota or dysbiosis has been associated with pathological conditions including inflammatory bowel diseases (Manichanh *et al.*, 2006; Frank *et al.*, 2007; Sartor, 2008; Machiels *et al.*, 2013), colorectal cancer (Sobhani *et al.*, 2011; Marchesi *et al.*, 2011), liver diseases such as non-alcoholic fatty liver disease (Henao-Mejia, *et al.*, 2012) and alcoholic steatosis (Bull-Otterson *et al.*, 2013; Hartmann *et al.*, 2012), obesity (Turnbaugh *et al.*, 2009; Tremaroli and Bäckhed, 2012) and rheumatoid arthritis (Yeoh *et al.*, 2013). Simultaneously, this awareness raised the potential for the microbiota to be exploited as a therapeutic target to improve health. Strategies to deliberately manipulate the composition and metabolic activity of the colonic microbiota include administration of probiotics, prebiotics or their combination, synbiotics. In this chapter, we will describe the impact of prebiotic intake on the metabolic and functional interactions within the microbiota.

The colonic microbiota and its functions

Microbiota composition

The human gastrointestinal tract hosts a complex and dynamic collection of microbial species. Metagenomic analyses show that in adults, the major constituents of the colonic microbiota are represented by the phyla *Bacteroidetes*, *Firmicutes*, *Actinobacteria* and *Proteobacteria*. Nevertheless, there is a substantial variation in the species composition between individuals, indicating that the composition of the microbiota is host specific (Ventura *et al.*, 2009). In 2011, Arumugam and colleagues found that the microbiome forms 'enterotypes', or discrete, stable clusters of similar microbiome configurations. Three distinct clusters of gut composition were found through comparison of the phylogenetic and functional composition of the gut microbiota in three independent international cohorts (Arumugam *et al.*, 2011).

Furthermore, the human microbiome composition appears to be particularly dependent on age and geography, although both are confounded by a variety of dietary, developmental, environmental, and genetic factors (Morgan *et al.*, 2013). The microbiota of infants differs from those of adults with more bifidobacteria and bacteroides species present. In elderly, the diversity of the bacteria is reduced, which may be due to dietary changes and changes in digestive physiology (Woodsmansey *et al.*, 2007). The major functions of the microbiota include a metabolic function, trophic function and protective function.

Metabolic function

The intestinal bacteria live in a symbiotic relationship with the host which implicates that both parties benefit from the relationship. The number of microbial genes outnumbers the host's genetic repertoire by several hundred-fold and provides a diverse range of metabolic activities to complement the host's physiology (Methé *et al.*, 2012; Qin *et al.*, 2010). In this way, the microbiota provide us with metabolic pathways that we do not have to develop ourselves. In return, the human host guarantees a nutrient-rich environment for the colonic microbiota. Dietary components that escaped digestion or cannot be digested by endogenous enzymes in the upper gastrointestinal tract become available for fermentation in the colon. The major substrates at the bacteria's disposal include undigested carbohydrates and proteins or peptides. It is estimated that about 20–60 g dietary carbohydrates [such as resistant starches (RS), dietary fibre, poly- and oligosaccharides and plant cell wall material], and 5–20 g protein pass daily into the colon (Cummings and Macfarlane, 1991; Silvester *et al.*, 1995). The digestion of fat is more complete (around 90–95%), resulting in about 3.5 g fat entering the colon per day (assuming a load of 70 g/day) (Priebe *et al.*, 2002). Besides food components, also pancreatic enzymes, mucus and desquamated intestinal cells enter the colon (Macfarlane *et al.*, 1986). Through the fermentation of this variety of substrates, colonic bacteria produce a wide range of compounds that are in close contact with the intestinal cells and may influence the physiological processes within the host. The human microbiota is characterized by a significant degree of functional redundancy, meaning that different bacteria can perform similar functions and metabolize the same substrate, thereby producing similar metabolites (Mahowald *et al.*, 2009). A number of factors influence the composition and the metabolic activity of the colonic microbiota. Nutrient availability and in particular the ratio of available carbohydrate to nitrogen is believed to be the most important regulator of bacterial metabolism (Macfarlane *et al.*, 1992). In the proximal colon, substrates are abundantly available. The microbiota will preferentially ferment carbohydrates since it is energetically more favourable to produce ATP from carbohydrates than from proteins (Van Loo *et al.*, 2004). As a consequence, saccharolytic activity is predominant in the proximal part of the colon. Along the length of the large intestine, the ratio of available carbohydrate to nitrogen progressively declines and bacterial activity changes towards a more proteolytic, methanogenic and sulphate-reducing type of microbiota (Cummings and Macfarlane, 1991; Macfarlane *et al.*, 1992).

Overall outcomes of this complex host–microbial interaction are recovery of metabolic energy and absorbable substrates for the host, the shaping of the immune system and supply of energy and nutritive products for bacterial growth and proliferation (Hooper *et al.*, 2002; Vieira *et al.*, 2013).

An overview of the end products of bacterial fermentation is presented in Fig. 2.1.

Short-chain fatty acids

Short-chain fatty acids (SCFA) are the major anions within the intestinal lumen and are mainly produced by anaerobic fermentation of undigested carbohydrates and, to a lesser extent, proteins (Topping and Clifton, 2001). The most abundant SCFA (acetate, propionate and butyrate) are usually produced in a molar ratio of 60:20:20, respectively (Cummings, 1981) although proportions and yields of the individual SCFA depend on the type of substrate. Starches yield relatively high proportions of butyrate (Englyst *et al.*, 1987), while fructooligosaccharides result in higher molar ratios of propionate and butyrate than lactulose (Luo *et al.*, 1996; Rycroft *et al.*, 2001). Most of the SCFA formed by the intestinal bacteria are rapidly absorbed and used

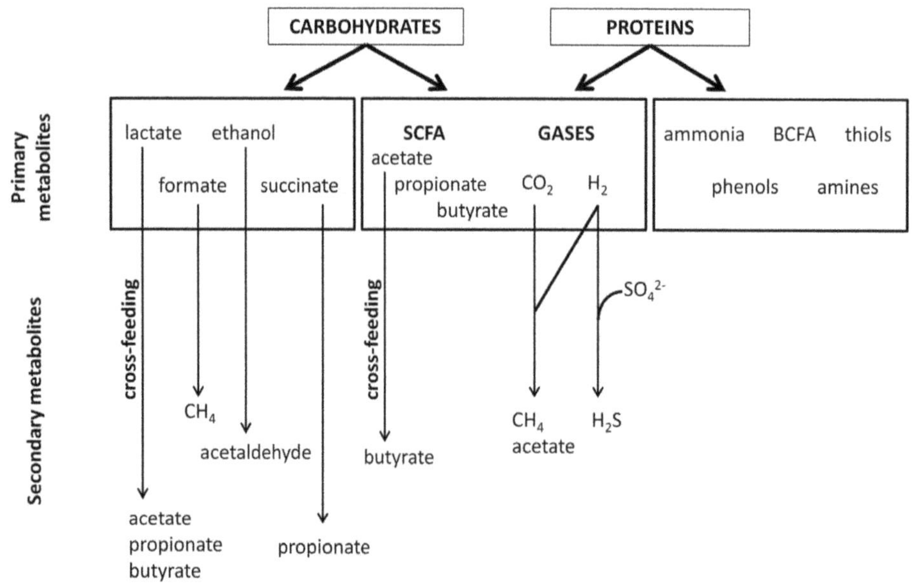

Figure 2.1 Overview of the metabolites of bacterial fermentation of non-digestible carbohydrates and proteins. Primary metabolites are directly produced from the available substrates and might be converted to secondary metabolites by other bacteria. This process is referred to as cross-feeding.

to some degree as energy substrates by mucosal epithelial cells. In this way, SCFA provide in humans about 10% of the daily caloric requirements (Bergman, 1990) with butyrate being the preferred energy source for the colonocytes. If not metabolized in the colonocytes, propionate is transported to the liver for gluconeogenesis whereas acetate is used as a fuel in various tissues. Increased SCFA synthesis also contributes to host homeostasis by acidifying the luminal pH which inhibits the growth of pathogens (Scheppach *et al.*, 2001), reduces the formation of secondary bile acids (Zampa *et al.*, 2004) and impairs the activity of specific enzymes such as proteases (Macfarlane *et al.*, 1988). Furthermore, SCFA have been shown to possess anti-inflammatory capacities, affect satiety hormones and play a role in insulin resistance (Hamer *et al.*, 2008). A role for SCFA in prevention of some human pathological conditions such as ulcerative colitis and colon carcinogenesis has been presumed. Diversion colitis, which occurs in diverted segments of the large bowel excluded from faecal transit, improves after treatment with a local perfusion of SCFA (Kiely *et al.*, 2001).

Gases

During bacterial fermentation of glucose (glycolysis pathway), pyruvate is produced and is consequently converted into acetyl-CoA, thereby producing the gases carbon dioxide and hydrogen (Calloway *et al.*, 1966). Carbon dioxide equilibrates with the body bicarbonate pool and is excreted in breath or is further metabolized into methane or acetate. Although hydrogen is partly excreted in breath and flatus, the majority is metabolized by methanogenic, sulfate-reducing and, to a lesser extent, acetogenic bacteria to methane, hydrogen sulfide and acetate, respectively (Gibson *et al.*, 1990). Excretion of hydrogen in breath is often used as diagnostic biomarker for carbohydrate malabsorption (Pimentel *et al.*, 2013). The most common methanogen in the human gastrointestinal tract is *Methanobrevibacter smithii* (Weaver *et al.*, 1986; Miller and Wolin, 1982) which uses H_2 from neighbouring microbes as a substrate to produce methane. Over the last years, methane has been associated with constipation (Chatterjee *et al.*, 2007) and in fact may be the cause of constipation (Pimentel *et al.*, 2003; Pimentel *et al.*, 2006) although the exact mechanism by which methane slows intestinal transit is unknown. More recently methane has also been associated with obesity (Zhang *et al.*, 2009). Like methanogens, sulfate-reducing bacteria (SRB) also consume H_2 that is converted to hydrogen sulfide. The toxic effects associated with H_2S, when present in millimolar concentrations, are mediated by its binding to cytochrome C oxidase resulting in inhibition of the butyrate oxidation (Rowan *et al.*, 2009). In contrast, at micromolar concentrations, H_2S displays protective effects which may be related to its ability to neutralize reactive species, the modulation of intracellular caspases or kinase pathways and the up-regulation of endogenous antioxidant systems (Szabo, 2007).

Protein fermentation metabolites

Proteins and peptides are the main nitrogen source for bacterial growth and protein synthesis. If not incorporated into biomass, anaerobic metabolism of undigested dietary proteins and proteins from desquamated epithelial cells are fermented to SCFA and branched-chain fatty acids (isobutyrate, methylbutyrate, isovalerate) as well as to a range of substances such as ammonia, amines, phenols, thiols, and indoles (Rastall, 2004). Proteolysis mainly occurs in the distal part of the colon where less carbohydrate is available and the pH is close to neutral (Macfarlane *et al.*, 1992). Some of these metabolites are re-used as nitrogen source for bacterial growth, whereas others are taken up by colonocytes and transported into the blood stream.

Host–microbial co-metabolism

Microbial metabolites formed during colonic fermentation are removed from the colonic lumen either by excretion in the faeces, or by absorption through the colonic mucosa after which they enter the systemic circulation where they can be further modified by human metabolism. For instance, p-cresol is a unique bacterial fermentation product of the amino acid tyrosine that is effectively absorbed by the colonocytes. It is conjugated in the colon mucosa or liver to p-cresol sulfate or p-cresol glucuronide which improves the water solubility and facilitates its urinary excretion (Evenepoel *et al.*, 2009). Also the opposite occurs. A number of metabolites that are derived from host metabolism are returned to the gut via biliary excretion where they can be further metabolized by the microbiota. For instance, bile acids that have escaped absorption in the terminal ileum can be deconjugated and converted to secondary bile acids by microbial metabolism (Bajor *et al.*, 2010).

Trophic function

Another role of SCFA on colonic physiology is their trophic effect on the intestinal epithelium. It had been shown that only in the presence of gut bacteria, readily fermentable dietary fibre stimulates epithelial cell proliferation in the intestine suggesting that end products of fermentation are responsible for this effect. In rats, administration of SCFA resulted in increased crypt cell production rates, which is a measure of epithelial cell proliferation (Sakata, 1987). Whereas acetate and butyrate were active at physiological concentrations, supraphysiological concentrations of propionate were required. However, it is unlikely that SCFA at physiological levels promote epithelial cell proliferation beyond the normal range as the maximal crypt cell production rate was still within the physiological range after stimulation with extremely high doses of butyrate.

Frankel *et al.* (1994) showed that direct colonic infusion of SCFA in rats increased colonic mucosal DNA and crypt depth. In addition, caecal infusion of SCFAs into innervated rats also increased jejunal DNA, villous height, surface area, crypt depth, and gastrin whereas these effects were not observed in rats that underwent caecal denervation. These results suggested that the jejunotrophic effects of SCFA were mediated by the autonomic nervous system and are associated with increased jejunal gastrin.

However, the most specific activities can be attributed to butyrate. Butyrate inhibits cell proliferation and stimulates cell differentiation in epithelial cell lines of neoplastic origin *in vitro*. Moreover, butyrate promotes reversion of cells from neoplastic to non-neoplastic phenotypes (Hamer *et al.*, 2008).

Protective function

The intestinal mucosa is the interface between the interior and the external environment and forms a crucial line of defence to prevent luminal pathogens and harmful substances from entering into the internal milieu. This barrier function is ensured by protection mechanisms at multiple levels (Koboziev *et al.*, 2014). Certain bacteria have the ability to secrete antimicrobial substances, such as bacteriocins and organic acids, that inhibit the growth of other bacteria. The secretion of mucus and immunoglobulin A (IgA) by different epithelial cells minimizes the chances for direct contact of bacteria with the epithelial cells. Commensals have been shown to limit pathogen colonization through competition for nutrients and adhesion sites, a process called colonization resistance (Kamada *et al.*, 2013).

In addition, the immune system of the intestinal tract referred to as the gut-associated lymphoid tissues (GALT) contains the largest pool of immunocompetent cells in the human body (Pabst *et al.*, 2008). The major function of the GALT is to control our relationship with the microbiota. A central strategy is to minimize contact between microorganisms and the epithelial cell surface, thereby limiting tissue inflammation and microbial translocation (Belkaid and Hand, 2014). Commensal microorganisms that have been able to penetrate the epithelial barrier will be rapidly phagocytosed and destroyed by intestinal macrophages. This dialogue between the gut microbiota and the immune system allows the host to tolerate a large amount of antigens present in the gut. Intestinal bacteria at the mucosal surface can create signals called microbial-associated molecular patterns (MAMPs) that stimulate pattern recognition receptors, such as Toll-like receptors, expressed on the surface of epithelial cells and trigger a cascade of immunological defence mechanisms including the production of antimicrobial peptides, pro- and anti-inflammatory cytokines, or triggers for apoptosis (Sharma *et al.*, 2010). In turn, to protect their ecological niche, a dominant action of the healthy microbiota on the immune system is aimed at reinforcing barrier immunity and therefore their own containment.

The prebiotic concept

The prebiotic concept, introduced by Gibson and Roberfroid in 1995 (Gibson and Roberfroid, 1995), is based on the idea that the composition and metabolic activity of the colonic microbiota is largely determined by the energy sources at the disposal of the bacteria. Manipulating the available substrates for the bacteria through dietary changes will result in changes in microbiota composition. The most recent definition was agreed at the 2010 Meeting of the International Scientific Association of Probiotics and Prebiotics (ISAPP) and defines a prebiotic as a selectively fermented ingredient that results in specific changes in the composition and/or activity of the gastrointestinal microbiota, thus conferring benefit(s) upon host health (Guarner *et al.*, 2012). To reach the target site, it is a prerequisite that a prebiotic is not hydrolysed by pancreatic or brush border enzymes (Roberfroid *et al.*, 2010). Within the colon, prebiotic substrates serve as energy source for specific groups of bacteria favouring proliferation of some advantageous indigenous species over others. Indeed, a large number of studies have demonstrated that dietary consumption of certain food products result in statistically significant changes in the composition of the gut microbiota in line with the prebiotic concept (Roberfroid *et al.*, 2010).

Prebiotics are usually undigestible carbohydrates such as non-starch polysaccharides or oligosaccharides. The physical and chemical properties of the substrates, such as monomer composition, monomer linkage, the degree of polymerization (DP) and the degree of substitution (DS) determine the rate and extent of fermentation and in this way also the impact on the microbiota composition and activity (Table 2.1). For instance, plant cell wall polysaccharides and some storage polysaccharides like inulin contain bonds that are not hydrolysed by mammalian hydrolytic enzymes. A key condition that needs to be demonstrated for a substrate to be considered as a prebiotic is the selectivity of fermentation. In addition, substantiation of a prebiotic effect requires the demonstration of a causal link or at least a correlation between the 'selective stimulation of growth and/or activity of one or a limited number of bacteria' and a 'health benefit' (Roberfroid *et al.*, 2010). Inulin and fructo-oligosaccharides are by far the most studied non-digestible carbohydrates and are considered as reference prebiotics (Roberfroid *et al.*, 2010). Inulin-type fructans are naturally available in a wide variety of food products such as chicory root, wheat, barley, onion and garlic. GOS are available as commercially products derived from lactose, but some of them are naturally present in human milk. Lactulose is a synthetic disaccharide that cannot be hydrolysed by human intestinal β-galactosidases. Also the evidence related to xylooligosaccharides (XOS) and arabinoxylanoligosaccharides (AXOS) seems sufficient to consider them as prebiotics (Table 2.2). Other compounds like isomaltooligosaccharides and soybean oligosaccharides have shown promising results but need further investigations before achieving the status of prebiotics (Slavin, 2013). Table 2.2 provides an overview of the available evidence for different undigestible carbohydrates to be classified as a prebiotic substrate.

Prebiotic modulation of the microbiota composition

As mentioned above, the need to be a selective substrate for one or more beneficial bacteria is a key criterion for a non-digestible carbohydrates to be classified as a prebiotic. To confirm this selectivity, careful monitoring of the microbiota composition during prebiotic supplementation during both *in vitro* and *in vivo* intervention studies is required. Selectivity of fermentation can be achieved by choosing a carbohydrate that stimulates specific microorganisms with a high affinity and degradative activity for that specific substrate (Flint, 2012). Originally, the effect of prebiotics was evaluated by monitoring the growth of lactobacilli and/or bifidobacteria. Bifidobacteria and lactobacilli are Gram-positive bacteria that ferment carbohydrates to acids. Both genera have similar metabolic properties and a longstanding association with health. With the advances in metagenome sequencing techniques, it is likely that we will identify additional bacterial species that are stimulated by prebiotic substrates. Emerging genera

Table 2.1 Classification of (potential) prebiotics [Goffin *et al.* (2001), Roberfroid *et al.* (2010), Aachary *et al.* (2011), Panesar and Kumari (2011), Broekaert *et al.* (2011)]

	Origin	Composition	Structure (DP-DS)*
Disaccharides			
Lactulose	Synthetic: enzymatically from lactose	4-O-b-D-galactopyranosyl-D-fructose (β(1–4)-linkage)	
Oligosaccharides			
Fructo-oligosaccharides (FOS)	Enzymatic hydrolysis of inulin	Short chains of fructose units β(2–1)-linkage	DP between 3 and 7
Galactooligosaccharides (GOS)	Naturally in human breast milk Enzymatic conversion of lactose (GOS-LA) or lactulose (GOS-LU)	α-D-Glucose-(1–4)-β-D-galactose-(1–6)$_n$	DP between 3 and 6
Xylooligosaccharides (XOS)	Naturally in bamboo shoots, fruits, vegetables, milk, and honey Enzymatic hydrolysis of xylan-containing lignocellulosic materials	Xylose residues with β(1–4)-linkages	DP between 3 and 9
Isomaltooligosaccharides (IMOS)	Naturally in sake, soy sauce, honey, sugarcane juice and derived products	Glucose residues with α(1,6) linkages	DP between 3 and 9
Arabinoxylanoligosaccharides (AXOS)	Enzymatic hydrolysis of arabinoxylan (part of fibre fraction of cereal grain)	Xylose backbone with β(1–4)-linkage and arabinose branches through α(1–3) or α(1–2) linkages	DP between 3–6 DS around 0.26–0.27
Polysaccharides			
Inulin	Chicory root, wheat, barley, onion and garlic	Long chains of fructose units β(2–1) and one terminal glucose unit	DP between 10 and 60

*DP=average degree of polymerization, DS=average degree of arabinose substitution.

Table 2.2 Available evidence to classify a food compound as a prebiotic (non-exhaustive reference list)

	Non-digestibility		Selective fermentation						Beneficial health effects		
Substrate		Reference	*In vitro*	Reference	Animals	Reference	Humans	Reference	Humans	Reference	Prebiotic
Lactulose	+	Panesar and Kumari (2011)	+	Cardelle-Cobas *et al.* (2012), Rycroft *et al.* (2001)	+/–	Nilsson and Nyman (2005)	+	Schumann (2002)	+	Panesar and Kumari (2011)	Yes
FOS	+	Cummings *et al.* (2001), Roberfroid *et al.* (2010)	+	Roberfroid *et al.* (2010), Rycroft *et al.* (2001)	+	Roberfroid *et al.* (2010), Macfarlane *et al.* (2008)	+	Roberfroid *et al.* (2010), Macfarlane *et al.* (2008)	+	Roberfroid *et al.* (2010), Macfarlane *et al.* (2008)	Yes
GOS	+	Hernández *et al.* (2012)	+	Maathuis *et al.* (2012), Rycroft *et al.* (2001)	+	Macfarlane *et al.* (2008)	+	Macfarlane *et al.* (2008)	+	Macfarlane *et al.* (2008)	Yes
XOS	+	Broekaert *et al.* (2011), Aachary *et al.* (2011)	+	Broekaert *et al.* (2011), Rycroft *et al.* (2001)	+	Broekaert *et al.* (2011)	+	Broekaert *et al.* (2011)	+	Broekaert *et al.* (2011)	Yes
IMOS	–	Goffin *et al.* (2001)	+	Rycroft *et al.* (2001)	+	Goffin *et al.* (2001)	+	Goffin *et al.* (2001)	+/–	Goffin *et al.* (2001), Yen *et al.* (2011)	No
AXOS	+	Broekaert *et al.* (2011)	+	Broekaert *et al.* (2011)	+	Broekaert *et al.* (2011)	+	Broekaert *et al.* (2011)	+	Broekaert *et al.* (2011)	Yes
Inulin	+	Cummings *et al.* (2001), Roberfroid *et al.* (2010)	+	Rycroft *et al.* (2001), Roberfroid *et al.* (2010)	+	Roberfroid *et al.* (2010), Macfarlane *et al.* (2008)	+	Roberfroid *et al.* (2010), Macfarlane *et al.* (2008)	+	Roberfroid *et al.* (2010), Macfarlane *et al.* (2008)	Yes

(+) effect proven in several studies (+/–) limited data available, and (–) unproven effect. The indicated references are mainly review papers that summarize the available evidence.

include *Eubacterium, Faecalibacterium* and *Roseburia* (Roberfroid *et al.*, 2010).

The modulation of the microbiota composition by dietary fructans has been more extensively studied compared to other (potential) prebiotic substrates. Although many studies have shown that inulin and fructo-oligosaccharides promote the growth of different *Bifidobacterium* species and *Lactobacillus* species (Roberfroid *et al.*, 2010), the impact of prebiotics on other gut bacteria is still under-researched. Yet, a few studies have shown that inulin-type fructans can also be degraded by other bacterial species such as *Roseburia* spp., *Bacteroides* spp. and some enterobacteria, which are also abundantly present in the large intestine (Falony *et al.*, 2009; Scott *et al.*, 2014). In *ob/ob* mice, intervention with inulin-type fructans induced changes in more than 100 distinct bacterial taxa, 16 of which displayed a 10-fold change in abundance (Everard *et al.*, 2011). Scott and colleagues showed in a recent *in vitro* study that long-chain inulin was utilized by *Roseburia inulinivorans*, a butyrate-producing low G+C Firmicute bacterium from the *Clostridium* cluster XIVa, isolated from human faeces by the same group (Scott *et al.*, 2011; Duncan *et al.*, 2006). Previously, Van der Meulen *et al.* (2006) showed that *Bacteroides thetaiotaomicron* LMG 11262 and *Bacteroides fragilis* LMG 10263 were able to grow on oligofructose in an *in vitro* laboratory fermentation. By comparing the kinetics of oligofructose consumption by pure cultures of *Bifidobacterium* and *Bacteroides* spp., they revealed a preferential metabolism of the short oligofructose fractions for *Bifidobacterium longum* BB536. Only after depletion of the short fractions, the larger oligofructose fractions were metabolized. Both Bacteroides strains did not display such a preferential metabolism and degraded all oligofructose fractions simultaneously. Whereas bifidobacteria preferentially degraded shorter oligosaccharides (composed of only two or three monomers) intracellularly, probably due to the presence of an appropriate uptake system, *Bacteroides* strains degraded all fructose fractions extracellularly, thereby putting the digestion products at the disposal of opportunistic competitors. Even within the *Bifidobacterium* genus, differences in degradation abilities and metabolic activities exist (Falony *et al.*, 2009; Scott *et al.*, 2014). Falony *et al.* (2009) identified four phenotypes amongst 18 bifidobacteria strains based on the differential capacity of the strains to grow on fructose, oligofructose and inulin.

The ability of bacterial species to compete for substrates clearly depends on the presence and regulation of degradative enzymes and transport systems. In addition, the chain length of the substrate is a determining factor for the fermentability and even the site of degradation (Van de Wiele *et al.*, 2007). The increasing availability of whole bacterial genome sequences will increase our understanding of how the composition of the microbial community is influenced by dietary intervention with compounds such as prebiotics.

A second mechanism that contributes to the selective stimulation of bacterial groups by prebiotics relates to cross-feeding. Cross-feeding implicates that partial degradation products released by primary degraders stimulate the growth of other bacterial groups. For instance, lactate produced by bifidobacteria can be converted by lactate-utilizing bacteria such as *Eubacterium* and *Anaerostipes*, to produce butyrate (Munoz-Tamayo *et al.*, 2011). Other butyrate-producing bacteria mostly belong to the Firmicutes phylum, *Faecalibacterium prausnitzii* and *Roseburia* spp., and convert acetate into butyrate (Duncan *et al.*, 2004; Falony *et al.*, 2006). A stimulation of *F. prausnitzii* was observed after dietary intervention with inulin (10 g/day for 16 days) in healthy subjects (Ramirez-Farias *et al.*, 2009). This mechanism also explains the butyrogenic effect observed after prebiotic administration. Bifidobacteria, stimulated by the prebiotics, do not produce butyrate but provide lactate or acetate for cross-feeding to those other bacterial groups.

Recently, in a study in rats colonized with a human microbiota, inulin and long-chain arabinoxylan supplementation increased the caecal mucin levels three- and sixfold, respectively, and shifted the predominant habitat of the mucin-degrading *Akkermansia muciniphila* from the caecum to more distal regions (Van den Abbeele *et al.*, 2011). Mucin-degrading species have the capacity to convert mucins into acetate and also propionate (Collado *et al.*, 2007). This characteristic of prebiotic supplementation shifting mucin degradation to distal colonic regions, where mucin degraders may produce beneficial metabolites, indicates that prebiotics may potentially improve gut health along the entire length of the intestine. More studies, especially human intervention studies, are required to confirm these results.

Finally, supplementation of prebiotics results in a decrease of the colonic luminal pH due to the production of SCFA which affects the composition of the microbiota due to the differential sensitivity of bacterial species to acidic pH. *Bacteroides* spp. have been shown to be relatively sensitive to mildly acid pH whereas Firmicutes and bifidobacteria are more acid tolerant and therefore less affected by a decrease in pH (Duncan *et al.*, 2009).

Besides selectively stimulating the growth of beneficial bacteria, prebiotics also influence the composition of the microbiota by inhibiting the growth of pathogens. A recent study investigated the role of inulin and FOS on the growth of *Pseudomonas aeruginosa* PAO1, an opportunistic pathogen to humans that is ubiquitously present in the environment. FOS was shown to inhibit the growth, biofilm formation and motility of *P. aeruginosa*, to limit secretion of NF-κB-dependent cytokine production and to decrease the production of exotoxin A by *P. aeruginosa* (Ortega-Gonzalez *et al.*, 2014). Similar but less pronounced effects were observed with inulin indicating the importance of the chain length of the prebiotic substrate.

Prebiotic mechanisms

Many health benefits in and outside the gut associated with prebiotic fermentation have been attributed to the increased production of SCFA by the stimulated beneficial bacteria (Fig. 2.2). Simple acidification of the colonic lumen by the production of those acids, can already explain some of the observed benefits of prebiotics. In addition, SCFA are considered as a class of bacterial products that mediate the interactions between the diet, the intestinal microbiota and the host. Two major SCFA signalling

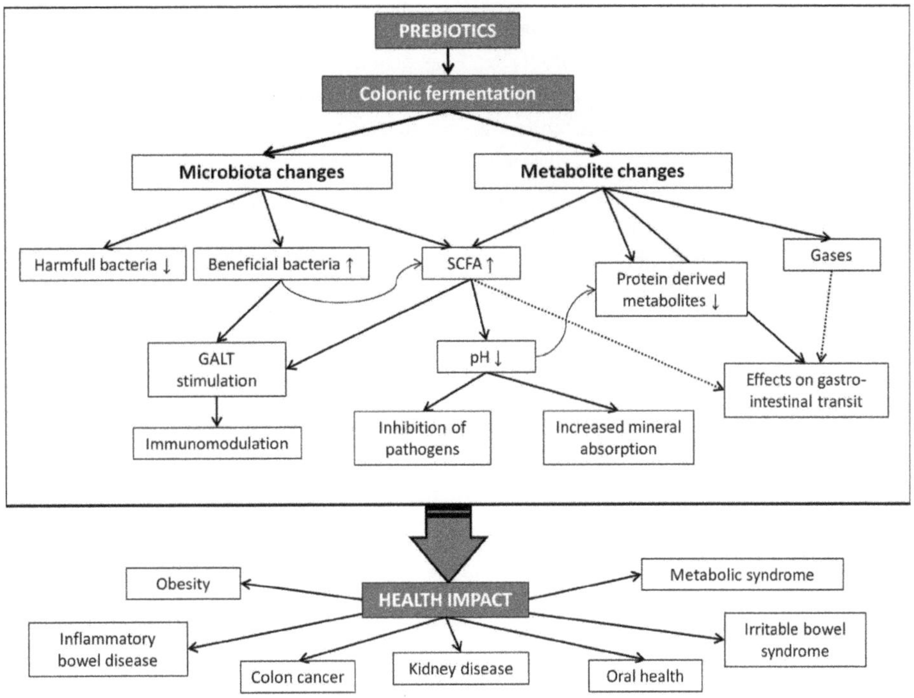

Figure 2.2 Overview of the main mechanisms and health benefits in and outside the gut associated with prebiotic fermentation.

mechanisms have been identified which include the inhibition of histone deacetylase (HDACs) and the activation of G-protein coupled receptors (GPCRs). In addition, SCFA easily enter the cells through passive diffusion or receptor-mediated transport and can internally act at other sites (Layden *et al.*, 2013).

Inhibition of HDACs

Gene expression is modulated by epigenetic modification of the DNA which involves mainly two interdependent mechanisms, i.e. DNA methylation and histone acetylation. Acetylation of the lysine residues in histones results in a less compact chromatin structure that is more easily transcribed. In contrast, removal of the acetyl groups by HDACs results in a more condensed DNA structure and silencing of the gene expression. Mammalian cells contain 18 different HDACs which are grouped in four classes (I to IV) (Santini *et al.*, 2007). The SCFA butyrate and propionate, but not acetate, are inhibitors of the HDACs and induce hyperacetylation of the histones which can in this way affect gene expression. HDAC inhibitors display anti-inflammatory properties and induce growth arrest, maturation and apoptosis of colon cancer cell lines (Schilderink *et al.*, 2013). The transduction pathway of butyrate-induced apoptosis has been shown to involve the activation of the caspase cascade. Butyrate activates p38 mitogen-activated protein kinase (p38 MAPK), which in turn up-regulates expression and receptor activity of the peroxisome proliferator0activated receptor (PPAR) γ. PPARγ activates caspase-8 and -9 leading to increased caspase-3 activity which will eventually result in cell death (Schwab *et al.*, 2006). PPARγ is a ligand-dependent nuclear transcription factor that controls the expression of a very large number of genes. PPARγ plays a fundamentally important role in the immune response through its ability to inhibit the expression of inflammatory cytokines and to direct the differentiation of immune cells towards anti-inflammatory phenotypes (Martin, 2009). Activation of PPARγ has been shown effective in the prophylaxis and to a lesser extent in the treatment of several animal models of acute or chronic colitis (Viladomiu *et al.*, 2013; Annese *et al.*, 2012).

The impact of the SCFA butyrate on gene expression was shown in a human intervention study in healthy subjects. Administration of butyrate enemas for 2 weeks induced a differential expression of 500 genes in biopsies taken from the sigmoid colon (Vanhoutvin *et al.*, 2009). The concentration of butyrate that was administered (100 mM) was physiologically achievable by consumption of a high-fibre diet. Pathway analysis showed that the pathways representing fatty acid oxidation, electron transport chain and oxidative stress were up-regulated. Also several genes associated with epithelial integrity and apoptosis, were differentially expressed after the butyrate intervention.

Activation of G-protein-coupled receptors (GPCRs)

In 2003, SCFA were identified as ligands for the G-protein-coupled receptors GPR43 and GPR41 (Brown *et al.*, 2003; Le Poul *et al.*, 2003; Nilsson *et al.*, 2003). Since that time, those receptors have been renamed as free fatty acid receptor 2 (FFAR2) and FFAR3, respectively (Kaji *et al.*, 2014). Originally, FFAR2 was shown to be highly expressed in human immune cells such as monocytes and neutrophils. Additional sites of expression include adipose tissue, the spleen, pancreas and bone marrow (den Besten *et al.*, 2013a; Regard *et al.*, 2008). Besides, FFAR2 is expressed in human colonic mucosa cells with a particularly strong expression in enteroendocrine L cells that produce glucagon-like peptide-1

(GLP-1) and peptide YY (PYY) (Karaki *et al.*, 2006; Tazoe *et al.*, 2009). Consequently, the functions affected by stimulation of the FFAR2 depend on the cell type. Inflammation is suppressed by SCFA in immune cells whereas GLP-1 secretion is modulated in the distal small intestine and colon. Also FFAR3 shows a widespread expression pattern including expression in human colonic L-cells (Tazoe *et al.*, 2009). Whereas acetate preferentially activates FFAR2, butyrate preferentially activates FFAR3 and propionate has a comparable potency for FFAR2 and FFAR3 (Kaji *et al.*, 2014).

The involvement of the FFAR2 in the inflammatory response was evidenced from the observation that GPR43-deficient (Gpr43–/–) mice showed exacerbated or unresolving inflammation in models of colitis, arthritis and asthma (Maslowski *et al.*, 2009). When colitis was induced in either GPR43 knockout (–/–) and wild-type mice by adding dextran sulfate sodium to the drinking water, treatment with acetate (150 mM) improved indices of disease in the wild-type mice but was ineffective in the GPR43 knockout (–/–) mice (Masui *et al.*, 2013). In a mouse model, activation of FFAR2 by SCFA regulated the size and function of the colonic regulatory T-cell (T_{reg}) pool and protected in this way against colitis (Smith *et al.*, 2013).

In rats treated with oligofructose for 4 weeks, the number of GLP-1-expressing cells was doubled in the proximal colon by a mechanism involving up-regulation of the differentiation factors neurogenin 3 and NeuroD (Cani *et al.*, 2007). In a similar intervention study, it was shown that in parallel with an increase in GLP-1 cells, the densities of FFAR2-positive enteroendocrine cells increased in FOS-treated rat, demonstrating the co-expression of FFAR2 with GLP-1 (Kaji *et al.*, 2011).

More recently, the niacin receptor GPR109a (also known as the hydroxycarboxylic acid receptor 2 or HCA2) was identified as another receptor on colonic cells important for colonic health. GPR109a recognizes butyrate with low affinity requiring millimolar concentrations of butyrate to activate the receptor. In colon cancer in humans, the expression of GPR109a is silenced. However, re-expression of GPR109a in colon cancer cells induces apoptosis, but only in the presence of its ligands butyrate and nicotinate (Thangaraju *et al.*, 2009). Singh *et al.* (2014) showed that GPR109a signalling by butyrate promoted anti-inflammatory properties in colonic macrophages and dendritic cells and enabled them to induce differentiation of T_{reg} cells and IL-10-producing T-cells.

Absorption of minerals

The production of SCFA is also involved in the positive effects on the absorption of minerals particularly calcium and magnesium observed after prebiotic fermentation. Three different mechanisms have been suggested for the increased mineral absorption. Improved absorption of minerals may indirectly be caused by the production of SCFA, resulting in a lower luminal pH which improves the solubility of minerals (Raschka and Daniel, 2005). The addition of fructo-oligosaccharides to the diet may also increase the length of the villi and depth of the crypts, which then results in a larger absorptive surface of the colonic mucosa and improves mineral absorption in this way (Kleessen *et al.*, 2003). The underlying mechanism for this change in mucosal structure is probably the production of butyrate, which fuels the mucosal cells. Another possible mechanism is an increased production of mucosal calcium-binding proteins after fructo-oligosaccharides intake (Ohta *et al.*, 1998).

Effects of SCFA on barrier function

In an *in vitro* Caco-2 cell monolayer model, it was demonstrated that butyrate in physiological concentrations promotes intestinal barrier function as measured by increases in TER and decreases in inulin permeability (Peng *et al.*, 2007). This regulation of barrier function was induced by facilitating the assembly of tight junctions via activation of AMP-activated protein kinase (AMPK), a serine/threonine kinase that is activated by binding of AMP (Peng *et al.*, 2009).

Another potential mechanism in the enhanced barrier function involves glucagon-like-peptide 2 (GLP-2), a 33 amino acid enteroendocrine peptide secreted from the distal ileal L cells in response to nutrients and mucosal injury. Like GLP-1, GLP-2 is formed from the cleavage of proglucagon peptide (Dube *et al.*, 2007). When genetically obese mice (*ob/ob* mice) were fed a prebiotic, intestinal proglucagon mRNA and plasma GLP-2 levels were increased and inversely correlated to markers of gut permeability (Cani *et al.*, 2009). However, concomitant treatment of the mice with a GLP-2 antagonist completely blocked the features of the prebiotic treatment. The mechanism through which GLP-2 improves gut barrier function need to be further elucidated. Data suggest that the downstream molecular mechanism by which GLP-2 receptor activation controls barrier function could be related to activation of the β-catenin signalling pathway (Dube *et al.*, 2009) or to the intestinal secretion of insulin-like growth factor (IGF)-I (Dube *et al.*, 2006). Among the produced SCFA, butyrate appears to be responsible for increasing plasma GLP-2 concentration (Tappenden *et al.*, 2003).

Effects of prebiotics on motility

Several studies in humans indicate that colonic fermentation of indigestible carbohydrates could stimulate the human colonic motor activity. In healthy subjects, ingestion or intracolonic infusion of 15 g lactulose significantly prolonged tonic contractions whereas phasic contractions remained unchanged (Jouet *et al.*, 2006). In another study, colonic fermentation of a physiological malabsorbed amount of starch had no effect on the tonic and phasic colonic motor activities, but produced a significant increase in the number of high amplitude propagated contractions (Jouet *et al.*, 2011). Whether these effects of carbohydrates on motor function in humans are mediated by SCFA production remains to be elucidated. In animal studies the effect of SCFA on colonic motor function is controversial. Recently, it was shown in humans that the stimulation of the colonic motor activity resulting from fermentation of a physiological amount of carbohydrate could not be explained by the acidification of the colonic contents or the resulting production of SCFAs (Jouet *et al.*, 2013). It was suggested that other mechanisms such as the production of gasses or other by-products of carbohydrate fermentation might be responsible for the effects on motor activity.

Reduction of protein fermentation

Concomitantly to an increase of the saccharolytic activity, prebiotics reduce the proteolytic fermentation. Degradation of proteins mainly occurs in the distal part of the colon, where less carbohydrate substrates are available, and starts with hydrolysis of proteins and amino acids by bacterial proteases and peptidases.

Different mechanisms can explain how prebiotics reduce protein fermentation in the colon. Due to the rapid fermentation of prebiotic substrates with production of SCFA, the pH in the colonic lumen declines. As most bacterial proteases have a neutral to slightly alkaline pH optimum, they do not function well at low pH and the breakdown of proteins is reduced (Rastall, 2004). A second mechanisms is a process of so-called catabolite repression which implicates that the transcription of genes involved in the amino acids metabolism is repressed in the presence of carbohydrates resulting in an inhibition of the deamination of amino acids (Vince and Burridge, 1980). Thirdly, the uptake of amino acid and intermediary metabolites by bacteria is enhanced because of their increased needs for biosynthesis (Cummings and Bingham, 1987). The impact of dietary interventions on protein fermentation has mainly been evidenced from excretion of phenolic compounds or ammonia. Several human intervention studies have indicated that protein fermentation, by means of the measurement of the urinary excretion of phenols, is reduced after prebiotic administration (De Preter *et al.*, 2007; Cloetens *et al.*, 2010).

Typical metabolites resulting from protein fermentation include ammonia, amines, phenols, thiols, and indoles. Proteolytic fermentation is often considered harmful to the host. However, this presumption is mainly based on *in vitro* evidence suggesting that some of these metabolites are carcinogenic or co-carcinogenic (Windey *et al.*, 2012a). For example, ammonia has been shown to increase colonic epithelial cell proliferation, to alter the morphology and intermediary metabolism of intestinal cells, affect DNA synthesis and reduce the life span of cells (Ichikawa and Sakata, 1998; Visek, 1978). However, whether a reduction in protein fermentation and in the production of those metabolites effectively confers a health benefit *in vivo* remains to be elucidated. In a number of rat intervention studies, prebiotic administration was associated with a reduction in faecal water genotoxicity (Toden *et al.*, 2005; Klinder *et al.*, 2004), a surrogate marker to assess the risk for colorectal cancer. In contrast, human epidemiological studies do not support an association between protein intake or protein fermentation and CRC (Windey *et al.*, 2012a), at least suggesting that other dietary or lifestyle risk factors may be more important. Similarly, a recent human intervention study investigating the impact of protein intake on DNA damage failed to proof a harmful effect of protein fermentation (Windey *et al.*, 2012b). Although the protein fermentation was effectively modulated by varying the protein intake, the data exposed no link between protein fermentation and faecal water toxicity. These results implicate that although individual protein fermentation metabolites induce detrimental effects *in vitro*, their contribution to the toxicity of faecal water appears to be limited. Further research is necessary to determine the major determinants of faecal water toxicity.

Health implications associated to prebiotics

We have increasingly realized that a well-balanced intestinal microbiota plays a critical role in sustaining health and that a disruption of this balance can lead to disease. Therefore, prebiotics offer potential in the treatment of disorders that have been associated with an altered composition or activity of the gut bacteria. Remarkably, most of the evidence regarding the potential health benefits of prebiotics is derived from experimental animal studies and human trials in small numbers of subjects.

The potential usefulness of prebiotics has been evaluated in gastrointestinal disorders including inflammatory bowel disease (Benjamin *et al.*, 2011; Joossens *et al.*, 2012) and irritable bowel disorders (Olesen and Gudmand-Hoyer, 2000; Silk *et al.*, 2009), but also in haemodialysis patients (Meijers *et al.*, 2010) or obesity (Dewulf *et al.*, 2013; Parnell and Reimer, 2009). In addition, prebiotics have been investigated in infant nutrition to reduce gastrointestinal infections (Aslanoglu *et al.*, 2007; Szajewska *et al.*, 2012) or to prevent atopic dermatitis (Moro *et al.*, 2006) and allergy (van Hoffen *et al.*, 2009). The clinical applications of prebiotics will be discussed in detail in the further chapters of this book.

Future needs to substantiate prebiotic functionality

At present, technological difficulties to measure *in vivo* SCFA production in humans hamper the exhaustive and thorough evaluation of the role of prebiotic fermentation and SCFA in health. Due to the inaccessibility of the proximal colon, it is not possible to sample SCFA at the site of production. In most human studies, the production of SCFA is estimated from quantification of the SCFA concentrations in faeces. However, it is important to realize that faecal measurements of SCFA are not representative for the actual SCFA production as more than 90% of SCFA produced in the colonic lumen are directly absorbed through the colonic wall. In addition, the absorption may be affected by the transit time (Cummings and Macfarlane, 1991). Therefore, faecal levels of SCFA more likely reflect distal colonic fermentation activity, rather than the total SCFA production. Similarly, SCFA levels in blood are not representative for the colonic SCFA production as SCFA are extensively metabolized in the colonic mucosa and/or the liver resulting in very low plasma concentrations.

Alternatively, information on production rates of SCFA is derived from *in vitro* incubation studies using faecal microbiota as inoculum. Experimental set-ups vary from simple batch fermentations to more sophisticated models simulating the gastrointestinal tract. After supplementation of faecal slurries from 10 healthy subjects with oligofructose-enriched inulin (OF-IN), a dose and time dependent increase in SCFA levels could be demonstrated (De Preter *et al.*, 2009). Each mg of OF-IN resulted in the production of 0.3mg of acetate, 0.025mg of propionate and 0.04 mg of butyrate within 24 hours. The Simulator of the Human Intestinal Microbial Ecosystem (SHIME; University of Gent, Belgium) consists of a series of five temperature and pH-controlled vessels that simulate the stomach, small intestine, ascending, transverse and descending colon, respectively (Molly *et al.*, 1993). The SHIME harbours a microbial community resembling that from the human colon both in fermentation activity and composition. This model has been applied to investigate the effects of prebiotics on microbiota composition and fermentation properties (Terpend *et al.*, 2013; Van den Abbeele *et al.*, 2013). The TNO-intestinal model of the colon (TIM-2, TNO, The Netherlands) is a more sophisticated *in vitro* model of fermentation in the proximal large intestine (Minekus *et al.*, 1999). It consists of a series of glass vessels linked with flexible walls, which allows simulation of peristalsis. The vessels are further equipped with a hollow fibre membrane in the lumen to simulate absorption of water and SCFAs (Maathuis *et al.*, 2012).

The application of stable isotope labelled substrates might provide a suitable and non-invasive alternative to quantify the *in vivo* production of SCFA. After consumption of ^{13}C-labelled prebiotics, those substrates will be fermented in the colon into ^{13}C-labelled SCFA that will appear in the systemic circulation. The use of stable isotopes allows to quantify in plasma in a very sensitive and selective way those SCFA that originate from the colon. Using ^{13}C-labelled barley either in the form of kernels (containing high amounts of resistant starch) or porridge (containing low amounts of resistant starch) differences in the kinetics of SCFA appearance in the systemic circulation could be demonstrated (Verbeke *et al.*, 2010). In a recent study in mice, the quantitative role of SCFA as a catabolic or anabolic substrate for the host was evaluated by infusing stable isotope labelled SCFA directly into the caecum and quantification of their incorporation in carbohydrates and lipids. Whereas 62% of infused propionate was used in whole body glucose production, acetate and butyrate were used in the liver for the synthesis of fatty acids and cholesterol (den Besten *et al.*, 2013b).

Conclusion

Owing to intensive research and technological advances over the last decades, the mechanisms by which prebiotics exerts beneficial effects on human health have been gradually elucidated. Production of SCFA seems to play a crucial role as they acidify the colonic lumen, which influences metabolic pathways and inhibits pathogens. More recently, these metabolites have been shown to behave as signalling molecules that link the diet to the microbiota and the host. More detailed information of the contribution of each individual SCFA and the molar proportion of the SCFA produced from different prebiotics, might facilitate the development of tailored prebiotics for specific health applications.

References

Aachary, A.A., and Prapulla, S.G. (2011). Xylooligosaccharides (XOS) as an emerging prebiotic: microbial synthesis, utilization, structural characterization, bioactive properties, and applications. Comp. Rev. Food Sci. Food Safety *10*, 2–16.

Annese, V., Rogai, F., Settesoldi, A., and Bagnoli, S. (2012). PPAR gamma in inflammatory bowel disease. Ppar Res. *2012*, 620839.

Arslanoglu, S., Moro, G.E.,and Boehm, G. (2007). Early supplementation of prebiotic oligosaccharides protects formula-fed infants against infections during the first 6 months of life. J. Nutr. *137*, 2420–2424.

Arumugam, M., Raes, J., Pelletier, E., Le Paslier, D., Yamada, T., Mende, D.R., Fernandes, G.R., Tap, J., Bruls, T., Batto, J.M., *et al.* (2011). Enterotypes of the human gut microbiome. Nature *473*, 174–180.

Bajor, A., Gillberg, P.G., and Abrahamsson, H. (2010). Bile acids: short and long term effects in the intestine. Scand. J. Gastroenterol. *45*, 645–664.

Belkaid, Y., and Hand, T.W. (2014). Role of the microbiota in immunity and inflammation. Cell *157*, 121–141.

Bergman, E.N. (1990). Energy contributions of volatile fatty-acids from the gastrointestinal tract in various species. Physiol. Rev. *70*, 567–590.

Broekaert, W.F., Courtin, C.M., Verbeke, K., Van de Wiele, T., Verstraete, W., and Delcour, J.A. (2011). Prebiotic and other health-related effects of cereal-derived arabinoxylans, arabinoxylan-aligosaccharides, and xylooligosaccharides. Crit. Rev. Food Sci. Nutr. *51*, 178–194.

Brown, A.J., Goldsworthy, S.M., Barnes, A.A., Eilert, M.M., Tcheang, L., Daniels, D., Muir, A.I., Wigglesworth, M.J., Kinghorn, I., Fraser, N.J., *et al.* (2003). The Orphan G protein-coupled receptors GPR41 and GPR43 are activated by propionate and other short chain carboxylic acids. J. Biol. Chem. *278*, 11312–11319.

Bull-Otterson, L., Feng, W., Kirpich, I., Wang, Y., Qin, X., Liu, Y., Gobejishvili, L., Joshi-Barve, S., Ayvaz, T., Petrosino, J., *et al.* (2013). Metagenomic analyses of alcohol induced pathogenic alterations in the intestinal microbiome and the effect of *Lactobacillus rhamnosus* GG treatment. PLoS One *8*, e53028.

Calloway, D.H., Colasito, D.J., and Matthews, R.D. (1966). Gases produced by human intestinal microflora. Nature *212*, 122–127.

Cani, P.D., Hoste, S., Guiot, Y., and Delzenne, N.M. (2007). Dietary non-digestible carbohydrates promote L-cell differentiation in the proximal colon of rats. Br. J. Nutr. *98*, 32–37.

Cani, P.D., Possemiers, S., van de Wiele, T., Guiot, Y., Everard, A., Rottier, O., Geurts, L., Naslain, D., Neyrinck, A., Lambert, D.M., *et al.* (2009). Changes in gut microbiota control inflammation in obese mice through a mechanism involving GLP-2-driven improvement of gut permeability. Gut *58*, 1091–1103.

Cardelle-Cobas, A., Olano, A., Corzo, N., Villamiel, M., Collins, M., Kolida, S., and Rastall, R.A. (2012). *In vitro* fermentation of lactulose-derived oligosaccharides by mixed faecal microbiota. J. Agric. Food Chem. *60*, 2024–2032.

Chatterjee, S., Park, S., Low, K., Kong, Y., and Pimentel, M. (2007). The degree of breath methane production in IBS correlates with the severity of constipation. Am. J. Gastroenterol. *102*, 837–841.

Cloetens, L., Broekaert, W.F., Delaedt, Y., Ollevier, F., Courtin, C.M., Delcour, J.A., Rutgeerts, P., and Verbeke, K. (2010). Tolerance of arabinoxylan-oligosaccharides and their prebiotic activity in healthy subjects: a randomized, placebo-controlled cross-over study. Br. J. Nutr. *103*, 703–713.

Collado, M.C., Derrien, M., Isolauri, E., de Vos, W.M., and Salminen, S. (2007). Intestinal integrity and Akkermansia muciniphila, a mucin-degrading member of the intestinal microbiota present in infants, adults, and the elderly. Appl. Environ. Microbiol. *73*, 7767–7770.

Cummings, J.H. (1981). Short chain fatty acids in the human colon. Gut *22*, 763–779.

Cummings, J.H., and Bingham, S.A. (1987). Dietary fiber, fermentation and large bowel cancer. Cancer Surv. *6*, 601–621.

Cummings, J.H., and Macfarlane, G.T. (1991). The control and consequences of bacterial fermentation in the human colon. J. Appl. Bacteriol. *70*, 443–459.

Cummings, J.H., Macfarlane, G.T., and Englyst, H.N. (2001). Prebiotic digestion and fermentation. Am. J. Clin. Nutr. *73*, 415s-420s.

den Besten, G., van Eunen, K., Groen, A.K., Venema, K., Reijngoud, D.J., and Bakker, B.M. (2013a). The role of short-chain fatty acids in the interplay between diet, gut microbiota, and host energy metabolism. J. Lipid Res. *54*, 2325–2340.

den Besten, G., Lange, K., Havinga, R., van Dijk, T.H., Gerding, A., van Eunen, K., Müller, M., Groen, A.K., Hooiveld, G.J., Bakker, B.M., *et al.* (2013b). Gut-derived short-chain fatty acids are vividly assimilated into host carbohydrates and lipids. Am. J. Physiol. Gastrointest. Liver Physiol. *305*, G900–G910.

De Preter, V., Falony, G., Windey, K., Hamer, H.M., De Vuyst, L., and Verbeke, K. (2010). The prebiotic oligofructose-enriched inulin, modulates the faecal metabolite profile, an *in vitro* analysis. Mol. Nutr. Food Res. *54*, 1791–1801.

De Preter, V., Vanhoutte, T., Huys, G., Swings, J., De Vuyst, L., Rutgeerts, P., and Verbeke, K. (2007). Effects of *Lactobacillus* casei Shirota, *Bifidobacterium breve*, and oligofructose-enriched inulin on colonic nitrogen-protein metabolism in healthy humans. Am. J. Physiol. Gastrointest. Liver Physiol. 292, G358-G368.

Dewulf, E.M., Cani, P.D., Claus, S.P., Fuentes, S., Puylaert, P.G., Neyrinck, A.M., Bindels, L.B., De Vos, W.M., Gibson, G.R., Thissen, J.P., *et al.* (2013). Insight into the prebiotic concept: lessons from an exploratory, double blind intervention study with inulin-type fructans in obese women. Gut *62*, 1112–1121.

Dube, P.E., Forse, C.L., Bahrami, J., and Brubaker, P.L. (2006). The essential role of insulin-like growth factor-1 in the intestinal tropic effects of glucagon-like peptide-2 in mice. Gastroenterology *131*, 589–605.

Dube, P.E., and Brubaker, P.L. (2007). Frontiers in glucagon-like peptide-2: multiple actions, multiple mediators. Am. J. Physiol. Endocrinol. Metab. 93, E460–E465.

Dube, P.E., Rowland, K.J., and Brubaker, P.L. (2008). Glucagon-like peptide-2 activates beta-catenin signaling in the mouse intestinal crypt: Role of insulin-like growth factor-I. Endocrinology *149*, 291–301.

Duncan, S.H., Holtrop, G., Lobley, G.E., Calder, A.G., Stewart, C.S., and Flint, H.J. (2004). Contribution of acetate to butyrate formation by human faecal bacteria. Br. J. Nutr. *91*, 915–923.

Duncan, S.H., Aminov, R.I., Scott, K.P., Louis, P., Stanton, T.B., and Flint, H.J. (2006). Proposal of Roseburia faecis sp. nov., Roseburia hominis sp. nov. and *Roseburia inulinivorans* sp. nov., based on isolates from human faeces. Int. J. Syst. Evol. Microbiol. *56*, 2437–2441.

Duncan, S.H., Louis, P., Thomson, J.M., and Flint, H.J. (2009). The role of pH in determining the species composition of the human colonic microbiota. Environ. Microbiol. *11*, 2112–2122.

Englyst, H.N., Hay, S., and Macfarlane, G.T. (1987). Polysaccharide breakdown by mixed populations of human fecal bacteria. FEMS Microbiol. Ecol. *45*, 163–171.

Evenepoel, P., Meijers, B.K., Bammens, B.R., and Verbeke, K. (2009). Uremic toxins originating from colonic microbial metabolism. Kidney. Int. Suppl. *114*, S12–19.

Everard, A., Lazarevic, V., Derrien, M., Girard, M., Muccioli, G.M., Neyrinck, A.M., Possemiers, S., Van Holle, A., Francois, P., De Vos, W.M., *et al.* (2011). Responses of gut microbiota and glucose and lipid metabolism to prebiotics in genetic obese and diet-induced leptin-resistant mice. Diabetes *60*, 2775–2786.

Falony, G., Lazidou, K., Verschaeren, A., Weckx, S., Maes, D., and De Vuyst, L. (2009). *In vitro* kinetic analysis of fermentation of prebiotic inulin-type fructans by Bifidobacterium species reveals four different phenotypes. Appl. Environm. Microbiol. *75*, 454–461.

Falony, G., Vlachou, A., Verbrugghe, K., and De Vuyst, L. (2006). Cross-feeding between *Bifidobacterium longum* BB536 and acetate-converting, butyrate-producing colon bacteria during growth on oligofructose. Appl. Environ. Microbiol. *72*, 7835–7841.

Flint, H.J. (2012). The impact of nutrition on the human microbiome. Nutr. Rev. *70*, S10–S13.

Frank, D.N., Amand, A.L.S., Feldman, R.A., Boedeker, E.C., Harpaz, N., and Pace, N.R. (2007). Molecular-phylogenetic characterization of microbial community imbalances in human inflammatory bowel diseases. Proc. Natl. Acad. Sci. U.S.A. *104*, 13780–13785.

Frankel, W.L., Zhang, W., Singh, A., Klurfeld, D.M., Don, S., Sakata, T., Modlin, I., and Rombeau, J.L. (1994). Mediation of the trophic effects of short-chain fatty acids on the rat jejunum and colon. Gastroenterology *106*, 375–380.

Gibson, G.R., Cummings, J.H., Macfarlane, G.T., Allison, C., Segal, I., Vorster, H.H., and Walker, A.R. (1990). Alternative pathways for hydrogen exposal during fermentation in the human colon. Gut *31*, 679–683.

Gibson, G.R., and Roberfroid, M.B. (1995). Dietary modulation of the human colonic microbiota – introducing the concept of prebiotics. J. Nutr. *125*, 1401–1412.

Goffin, D., Delzenne, N., Blecker, C., Hanon, E., Deroanne, C., and Paquot, M. (2011). Will isomalto-oligosaccharides, a well-established functional food in Asia, break through the European and American market? The status of knowledge on these prebiotics. Crit. Rev. Food Sci. Nutr. *51*, 394–409.

Guarner, F., Khan, A.G., Garisch, J., Eliakim, R., Gangl, A., Thomson, A., Krabshuis, J., Lemair, T., Kaufmann, P., de Paula, J.A., *et al.* (2012). World Gastroenterology Organization Global Guidelines: Probiotics and prebiotics. J. Clin. Gastroenterol. *46*, 468–481.

Hamer, H.M., Jonkers, D., Venema, K., Vanhoutvin, S., Troost F.J., and Brummer, R.J. (2008). Review article: the role of butyrate on colonic function. Aliment. Pharmacol. Ther. *27*, 104–119.

Hartmann, P., Chen, W.C., and Schnabl, B. (2012). The intestinal microbiome and the leaky gut as therapeutic targets in alcoholic liver disease. Front Physiol. 3, 402. eCollection 2012.

Henao-Mejia, J., Elinav, E., Jin, C., Hao, L., Mehal, W.Z., Strowig, T., Thaiss, C.A., Kau, A.L., Eisenbarth, S.C., Jurczak, M.J., *et al.* (2012). Inflammasome-mediated dysbiosis regulates progression of NAFLD and obesity. Nature *482*, 179–185.

Hernández-Hernández, O., Marín-Manzano, M.C., Rubio, L.A., Moreno, F.J., Sanz, M.L., and Clemente, A. (2012). Monomer and linkage type of galacto-oligosaccharides affect their resistance to ileal digestion and prebiotic properties in rats. J. Nutr. *142*, 1232–1239.

Hooper, L.V., Midtvedt, T., and Gordon, J.I. (2002). How host–microbial interactions shape the nutrient environment of the mammalian intestine. Annu. Rev. Nutr. *22*, 283–307.

Huttenhower, C., Human Microbiome Project Consortium (2012). Structure, function and diversity of the healthy human microbiome. Nature *486*, 207–214.

Ichikawa, H., and Sakata, T. (1998). Stimulation of epithelial cell proliferation of isolated distal colon of rats by continuous colonic infusion of ammonia or short-chain fatty acids is nonadditive. J. Nutr. *128*, 843–847.

Jouet, P., Sabate, J.M., Cuillerier, E., Coffin, B., Lemann, M., Jian, R., and Flourie, B. (2006). Low-dose lactulose produces a tonic contraction in the human colon. Neurogastroenterol. Motil. *18*, 45–52.

Jouet, P., Sabate, J.M., Coffin, B., Lemann, M., Jian, R., and Flourie, B. (2011). Fermentation of starch stimulates propagated contractions in the human colon. Neurogastroenterol. Motil. 23, 450–456.

Jouet, P., Moussata, D., Duboc, H., Boschetti, G., Attar, A., Gorbatchef, C., Sabate, J.M., Coffin, B., and Flourie, B. (2013). Effect of short-chain fatty acids and acidification on the phasic and tonic motor activity of the human colon. Neurogastroenterol. Motil. *25*, 943–949.

Kaji, I., Karaki, S., Tanaka, R., and Kuwahara, A. (2011). Density distribution of free fatty acid receptor 2 (FFA2)-expressing and GLP-1 producing enteroendocrine L cells in human and rat lower intestine, and increased cell numbers after ingestion of fructo-oligosaccharide. J. Mol. Histol. *42*, 27–38.

Kaji, I., Karaki, S., and Kuwahara, A. (2014). Short-chain fatty acid receptor and its contribution to glucagon-like peptide-1 release. Digestion *89*, 31–36.

Kamada, N., Chen, G.Y., Inohara, N., and Nunez, G. (2013). Control of pathogens and pathobionts by the gut microbiota. Nat. Immunol. *14*, 685–690.

Karaki, S., Tazoe, H., Hayashi, H., Kashiwabara, H., Tooyama, K., Suzuki, Y., and Kuwahara, A. (2008). Expression of the short-chain fatty acid receptor, GPR43, in human colon. J. Mol. Histol. *39*, 135–142.

Kiely, E.M., Ajayi, N.A., Wheeler, R.A., and Malone, M. (2001). Diversion procto-colitis: response to treatment with short-chain fatty acids. J. Pediatr. Surg. *36*, 1514–1517.

Kleessen, B., Hartmann, L., and Blaut, M. (2003). Fructans in the diet cause alterations of intestinal mucosal architecture, released mucins and mucosa-associated bifidobacteria in gnotobiotic rats. Br. J. Nutr. *89*, 597–606.

Klinder, A., Förster, A., Caderni, G., Femia, A.P., and Pool-Zobel, B.L. (2004). Fecal water genotoxicity is predictive of tumor preventive activities by inulin-like oligofructoses, probiotics (*Lactobacillus rhamnosus* and *Bifidobacterium lactis*) and their symbiotic combination. Nutr. Cancer *49*, 144 155.

Koboziev, I., Webb, C.R., Furr, K.L., and Grisham, M.B. (2014). Role of the enteric microbiota in intestinal homeostasis and inflammation. Free Radic. Biol. Med. *68*, 122–133.

Layden, B.T., Angueira, A.R., Brodsky, M., Durai, V., and Lowe, W.L. (2013). Short chain fatty acids and their receptors: new metabolic targets. Transl. Res. *161*, 131–140.

Le Poul, E., Loison, C., Struyf, S., Springael, J.Y., Lannoy, V., Decobecq, M.E., Brezillon, S., Dupriez, V., Vassart, G., Van Damme, J., *et al.* (2003). Functional characterization of human receptors for short-chain fatty acids and their role in polymorphonuclear cell activation. J. Biol. Chem. *278*, 25481–25489.

Luo, J., Rizkalla, S.W., Alamowitch, C., Boussairi, A., Blayo, A., Barry, J.L., Laffitte, A., Guyon, F., Bornet, F.R., and Slama, G. (1996). Chronic consumption of short-chain fructooligosaccharides by healthy subjects decreased basal hepatic glucose production but had no effect on insulin-stimulated glucose metabolism. Am. J. Clin. Nutr. *63*, 939–945.

Maathuis, A.J., van den Heuvel, E.G., Schoterman, M.H., and Venema, K. (2012). Galacto-oligosaccharides have prebiotic activity in a dynamic *in vitro* colon model using a ^{13}C-labeling technique. J. Nutr. *142*, 1205–1212.

Macfarlane, G.T., and Allison, C. (1986). Utilization of protein by human gut bacteria. FEMS Microbiol. Ecol. *38*, 19–24.

Macfarlane, G.T., Allison, C., and Gibson, G.R. (1988). Effect of pH on protease activities in the large-intestine. Lett. Appl. Microbiol. *7*, 161–164.

Macfarlane, G.T., Gibson, G.R., and Cummings, J.H. (1992). Comparison of fermentation reactions in different regions of the human colon. J. Appl. Bacteriol. *72*, 57–64.

Macfarlane, G.T., Steed, H., and Macfarlane, S. (2008). Bacterial metabolism and health-related effects of galacto-oligosaccharides and other prebiotics. J. Appl. Microbiol. *104*, 305–344.

Machiels, K., Joossens, M., Sabino, J., De Preter, V., Arijs, I., Eeckhaut, V., Ballet, V., Claes, K., Van Immerseel, F., Verbeke, K., *et al.* (2013). A decrease of the butyrate-producing species *Roseburia hominis* and *Faecalibacterium prausnitzii* defines dysbiosis in patients with ulcerative colitis. Gut. doi: 10.1136/gutjnl-2013-304833. [Epub ahead of print].

Mahowald, M.A., Rey, F.E., Seedorf, H., Turnbaugh, P.J., Fulton, R.S., Wollam, A., Shah, N., Wang, C., Magrini, V., Wilson, R.K., *et al.* (2009). Characterizing a model human gut microbiota composed of members of its two dominant bacterial phyla. Proc. Natl. Acad. Sci. U.S.A. *106*, 5859–5864.

Manichanh, C., Rigottier-Gois, L., Bonnaud, E., Gloux, K., Pelletier, E., Frangeul, L., Nalin, R., Jarrin, C., Chardon, P., Marteau, P., *et al.* (2006). Reduced diversity of faecal microbiota in Crohn's disease revealed by a metagenomic approach. Gut *55*, 205–211.

Marchesi, J.R., Dutilh, B.E., Hall, N., Peters, W.H., Roelofs, R., Boleij, A., and Tjalsma, H. (2011). Towards the human colorectal cancer microbiome. PLoS One *6*, e20447.

Martin, H. (2009). Role of PPAR-gamma in inflammation. Prospects for therapeutic intervention by food components. Mutat. Res. *669* (1–2), 1–7.

Maslowski, K.M., Vieira, A.T., Ng, A., Kranich, J., Sierro, F., Yu, D., Schilter, H.C., Rolph, M.S., Mackay, F., Artis, D., *et al.* (2009). Regulation of inflammatory responses by gut microbiota and chemoattractant receptor GPR43. Nature *461*, 1282-U119.

Masui, R., Sasaki, M., Funaki, Y., Ogasawara, N., Mizuno, M., Iida, A., Izawa, S., Kondo, Y., Ito, Y., Tamura, Y., *et al.* (2013). G Protein-Coupled Receptor 43 moderates gut inflammation through cytokine regulation from mononuclear cells. Inflamm. Bowel Dis. *19*, 2848–2856.

Meijers, B.K.I., De Preter, V., Verbeke, K., Vanrenterghem, Y., and Evenepoel, P. (2010). p-Cresyl sulfate serum concentrations in haemodialysis patients are reduced by the prebiotic oligofructose-enriched inulin. Nephrol. Dial. Transplant. *25*, 219–224.

Methé, B.A., Human Microbiome Project Consortium. (2012). A framework for human microbiome research. Nature *486*, 215–221.

Miller, T.L., and Wolin, M.J. (1982). Enumeration of *Methanobrevibacter smithii* in human feces. Arch. Microbiol. *131*, 14–18.

Minekus, M., Smeets-Peeters, M., Bernalier, A., Marol-Bonnin, S., Havenaar, R., Marteau, P., Alric, M., Fonty, G., and Huis in't Veld, J.H. (1999). A computer-controlled system to simulate conditions of the large intestine with peristaltic mixing, water absorption and absorption of fermentation products. Appl. Microbiol. Biotechnol. *53*, 108–114.

Molly, K., Vande Woestyne, M., and Verstraete, W. (1993). Development of a 5-step multi-chamber reactor as a simulation of the human intestinal microbial ecosystem. Appl. Microbiol. Biotechnol. *39*, 254–258.

Morgan, X.C., Segata, N., and Huttenhower, C. (2013). Biodiversity and functional genomics in the human microbiome. Trends Genet. *29*, 51–58.

Moro, G., Arslanoglu, S., Stahl, B., Jelinek, J., Wahn, U., and Boehm, G. (2006). A mixture of prebiotic oligosaccharides reduces the incidence of atopic dermatitis during the first six months of age. Arch. Dis. Child. *91*, 814–819.

Munoz-Tamayo, R., Laroche, B., Walter, E., Doré, J., Duncan, S.H., Flint, H.J., and Leclerc, M. (2011). Kinetic modelling of lactate utilization and butyrate production by key human colonic bacterial species. FEMS Microbiol. Ecol. *76*, 615–624.

Nilsson, U., and Nyman, M. (2005). Short-chain fatty acid formation in the hindgut of rats fed oligosaccharides varying in monomeric composition, degree of polymerization and solubility. Br. J. Nutr. *94*, 705–713.

Nilsson, N.E., Kotarsky, K., Owman, C., and Olde, B. (2003). Identification of a free fatty acid receptor, FFA2R, expressed on leukocytes and activated by short-chain fatty acids. Biochem. Biophys. Res. Commun. *303*, 1047–1052.

Ohta, A., Motohashi, Y., Ohtsuki, M., Hirayama, M., Adachi, T., and Sakuma, K. (1998). Dietary fructooligosaccharides change the concentration of calbindin-D9k differently in the mucosa of the small and large intestine of rats. J. Nutr. *128*, 934–939.

Olesen, M., and Gudmand-Hoyer, E. (2000). Efficacy, safety, and tolerability of fructo-oligosaccharides in the treatment of irritable bowel syndrome. Am. J. Clin. Nutr. *72*, 1570–1575.

Ortega-Gonzalez, M., de Medina, F.S., Molina-Santiago, C., Lopez-Posadas, R., Pacheco, D., Krell, T., Martinez-Augustin, O., and Abdelali, D. (2014). Fructooligosacharides reduce *Pseudomonas aeruginosa* PAO1 pathogenicity through distinct mechanisms. PLoS One 9, e85772.

Pabst, R., Russell, M.W., and Brandtzaeg, P. (2008). Tissue distribution of lymphocytes and plasma cells and the role of the gut. Trends Immunol. *29*, 206–208.

Panesar, P.S., and Kumari, S. (2011). Lactulose: Production, purification and potential applications. Biotechnol. Advan. *29*, 940–948.

Parnell, J.A., and Reimer, R.A. (2009). Weight loss during oligofructose supplementation is associated with decreased ghrelin and increased peptide YY in overweight and obese adults. Am. J. Clin. Nutr. *89*, 1751–1759.

Peng, L.Y., He, Z.J., Chen, W., Holzman, I.R., and Lin, J. (2007). Effects of butyrate on intestinal barrier function in a Caco-2 cell monolayer model of intestinal barrier. Pediatr. Res. *61*, 37–41.

Peng, L.Y., Li, Z.R., Green, R.S., Holzman, I.R., and Lin, J. (2009). Butyrate enhances the intestinal barrier by facilitating tight junction assembly via activation of AMP-activated protein kinase in Caco-2 cell monolayers. J. Nutr. *139*, 1619–1625.

Pimentel, M., Mayer, A.G., Park, S., Chow, E.J., Hasan, A., and Kong, Y. (2003). Methane production during lactulose breath test is associated with gastrointestinal disease presentation. Dig. Dis. Sci. *48*, 86–92.

Pimentel, M., Lin, H.C., Enayati, P., Van den Burg, B., Lee, H.R., Chen, J.H., Park, S., Kong, Y., and Conklin, J. (2006). Methane, a gas produced by enteric bacteria, slows intestinal transit and augments small intestinal contractile activity. Am. J. Physiol. Gastrointest. Liver Physiol. *290*, G1089-G1095.

Pimentel, M., Mathur, R., and Chang, C. (2013). Gas and the microbiome. Curr. Gastroenterol. Rep. *15*, 356. doi:10.1007/s11894-013-0356-y.

Priebe, M.G., Vonk, R.J., Sun, X., He, T., Harmsen, H.J., and Welling, G.W. (2002). The physiology of colonic metabolism. Possibilities for interventions with pre and probiotics. Eur. J. Nutr. *41* (Suppl. 1), I2–10.

Qin, J., Li, R., Raes, J., Arumugam, M., Burgdorf, K.S., Manichanh, C., Nielsen, T., Pons, N., Levenez, F., Yamada, T., *et al.* (2010). A human gut microbial gene catalogue established by metagenomic sequencing. Nature *464*, 59–65.

Ramirez-Farias, C., Slezak, K., Fuller, Z., Duncan, A., Holtrop, G., and Louis, P. (2009). Effect of inulin on the human gut microbiota: stimulation of *Bifidobacterium adolescentis* and *Faecalibacterium prausnitzii*. Br. J. Nutr. *101*, 541–550.

Raschka, L., and Daniel, H. (2005). Diet composition and age determine the effects of inulin-type fructans on intestinal calcium absorption in rat. Eur. J. Nutr. *44*, 360–364.

Rastall, R.A. (2004). Bacteria in the gut: Friends and foes and how to alter the balance. J. Nutr. *134*, 2022S–2026S.

Regard, J.B., Sato, I.T., and Coughlin, S.R. (2008). Anatomical profiling of G protein-coupled receptor expression. Cell *135*, 561–571.

Roberfroid, M., Gibson, G.R., Hoyles, L., McCartney, A.L., Rastall, R., Rowland, I., Wolvers, D., Watzl, B., Szajewska, H., Stahl, B., *et al.* (2010). Prebiotic effects: metabolic and health benefits. Br. J. Nutr. *104*, S1–S63.

Rowan, F.E., Docherty, N.G., Coffey, J.C., and O'Connell, P.R. (2009). Sulphate-reducing bacteria and hydrogen sulphide in the aetiology of ulcerative colitis. Br. J. Surg. *96*, 151–158.

Rycroft, C.E., Jones, R.M., Gibson, G.R., and Rastall, R.A. (2001). A comparative *in vitro* evaluation of the fermentation properties of prebiotic oligosaccharides. J. Appl. Microbiol. *91*, 878–887.

Sakata, T. (1987). Stimulatory effect of short-chain fatty-acids on epithelial-cell proliferation in the rat intestine – A possible explanation for trophic effects of fermentable fibre, gut microbes and luminal trophic factors. Br. J. Nutr. *58*, 95–103.

Santini, V., Gozzini, A., and Ferrari, G. (2007). Histone deacetylase inhibitors: molecular and biological activity as clinical application. Curr. Drug Metab. *8*, 383–394.

Sartor, R.B. (2008). Microbial influences in inflammatory bowel diseases. Gastroenterology *134*, 577–594.

Scheppach, W., Lueh, H., and Menzel, T. (2001). Beneficial health effects of low-digestible carbohydrate consumption. Br. J. Nutr. *85*, S23-S30.

Schilderink, R., Verseijden, C., and de Jonge, W.J. (2013). Dietary inhibitors of histone deacetylases in intestinal immunity and homeostasis. Front. Immunol. *4*. doi: 10.3389/fimmu.2013.00226).

Schumann, C. (2002). Medical, nutritional and technological properties of lactulose. An update. Eur. J. Nutr. *41*, 17–25.

Schwab, M., Reynders, V., Ulrich, S., Zahn, N., Stein, J., and Schroder, O. (2006). PPAR gamma is a key target of butyrate-induced caspase-3 activation in the colorectal cancer cell line Caco-2. Apoptosis *11*, 1801–1811.

Scott, K.P., Martin, J.C., Chassard, C., Clerget, M., Potrykus, J., Campbell, G., Mayer, C.D., Young, P., Rucklidge, G., Ramsay, A.G., *et al.* (2011). Substrate-driven gene expression in *Roseburia inulinivorans*: importance of inducible enzymes in the utilization of inulin and starch. Proc. Natl. Acad. Sci. U.S.A. *108*, 4672–4679.

Scott, K.P., Martin, J.C., Duncan, S.H., and Flint, H.J. (2014). Prebiotic stimulation of human colonic butyrate-producing bacteria and Bifidobacteria, *in vitro*. FEMS Microbiol. Ecol. *87*, 30–40.

Sharma, R., Young, C., and Neu, J. (2010). Molecular modulation of intestinal epithelial barrier: contribution of microbiota. J. Biomed. Biotechnol. *2010*. doi: 10.1155/2010/305879.

Silk, D.B., Davis, A., Vulevic, J., Tzortzis, G., and Gibson, G.R. (2009). Clinical trial: the effects of a trans-galactooligosaccharide prebiotic on faecal microbiota and symptoms in irritable bowel syndrome. Aliment. Pharmacol. Ther. *29*, 508–518.

Silvester, K.R., Englyst, H.N., and Cummings, J.H. (1995). Ileal recovery of starch from whole diets containing resistant starch measured *in vitro* and fermentation of ileal effluent. Am. J. Clin. Nutr. *62*, 403–411.

Singh, N., Gurav, A., Sivaprakasam, S., Brady, R., Padia, H., Shi, D., Thangaraju, M., Prasad, P.D., Manicassamy, S., Munn, D.H., *et al.* (2014). Activation of Gpr109a, receptor for niacin and the commensal metabolite butyrate, suppresses colonic inflammation and carcinogenesis. Immunity *40*, 128–139.

Slavin, J. (2013). Fiber and probiotics: mechanisms and health benefits. Nutrients 5, 1417–1435.

Smith, P.M., Howitt, M.R., Panikov, N., Michaud, M., Gallini, C.A., Bohlooly, Y., Glickman, J.N., and Garrett, W.S. (2013). The microbial metabolites, short-chain fatty acids, regulate colonic T-reg cell homeostasis. Science *341*, 569–573.

Sobhani, I., Tap, J., Roudot-Thoraval, F., Roperch, J.P., Letulle, S., Langella, P., Corthier, G., Tran Van Nhieu, J., and Furet, J.P. (2011). Microbial dysbiosis in colorectal cancer (CRC) patients. PLoS One *6*, e16393.

Szabo, C. (2007). Hydrogen sulphide and its therapeutic potential. Nat. Rev. Drug Discov. *6*, 917–935.

Szajewska, H., Weizman, Z., Abu-Zekry, M., Kekez, A.J., Braegger, C.P., Kolacek, S., Micetic-Turk, D., Ruszczyński, M., Vukavic, T., and ESPGHAN Working Group on Probiotics and Prebiotics (2012). Inulin and fructo-oligosaccharides for the prevention of antibiotic-associated diarrhea in children: report by the ESPGHAN Working Group on Probiotics and Prebiotics. J. Pediatr. Gastroenterol. Nutr. *54*, 828–829.

Tappenden, K.A., Albin, D.M., Bartholome, A.L., and Mangian, H.F. (2003). Glucagon-like peptide-2 and short-chain fatty acids: A new twist to an old story. J. Nutr. *133*, 3717–3720.

Tazoe, H., Otomo, Y., Karaki, S., Kato, I., Fukami, Y., Terasaki, M., and Kuwahara, A. (2009). Expression of short-chain fatty acid receptor GPR41 in the human colon. Biomed. Res. *30*, 149–156.

Terpend, K., Possemiers, S., Daguet, D., and Marzorati, M. (2013). Arabinogalactan and fructo-oligosaccharides have a different fermentation profile in the Simulator of the Human Intestinal Microbial Ecosystem (SHIME®). Environ. Microbiol. Rep. *5*, 595–603.

Thangaraju, M., Cresci, G.A., Liu, K., Ananth, S., Gnanaprakasam, J.P., Browning, D.D., Mellinger, J.D., Smith, S.B., Digby, G.J., Lambert, N.A., *et al.* (2009). GPR109A is a G-protein-coupled receptor for the bacterial fermentation product butyrate and functions as a tumor suppressor in colon. Cancer Res. *69*, 2826–2832.

Toden, S., Bird, A.R., Topping, D.L., and Conlon, M.A. (2005). Resistant starch attenuates colonic DNA damage induced by higher dietary protein in rats. Nutr. Cancer *51*, 45 51.

Topping, D.L., and Clifton, P.M. (2001). Short-chain fatty acids and human colonic function: roles of resistant starch and nonstarch polysaccharides. Physiol. Rev. *81*, 1031–1064.

Tremaroli, V., and Bäckhed, F. (2012). Functional interactions between the gut microbiota and host metabolism. Nature *489*, 242–249.

Turnbaugh, P.J., Hamady, M., Yatsunenko, T., Cantarel, B.L., Duncan, A., Ley, R.E., Sogin, M.L., Jones, W.J., Roe, B.A., Affourtit, J.P., *et al.* (2009). A core gut microbiome in obese and lean twins. Nature *457*, 480–484.

Van Baarlen, P., Kleerebezem, M., and Wells, J.M. (2013). Omics approaches to study host–microbiota interactions. Curr. Opin. Microbiol. *16*, 270–277.

Van de Wiele, T., Boon, N., Possemiers, S., Jacobs, H., and Verstraete, W. (2007). Inulin-type fructans of longer degree of polymerization exert more pronounced *in vitro* prebiotic effect. J. Appl. Microbiol. *102*, 452–460.

Van den Abbeele, P., Gérard, P., Rabot, S., Bruneau, A., El Aidy, S., Derrien, M., Kleerebezem, M., Zoetendal, E.G., Smidt, H., Verstraete, W., *et al.* (2011). Arabinoxylans and inulin differentially modulate the mucosal and luminal gut microbiota and mucin-degradation in humanized rats. Environ. Microbiol. *13*, 2667–2680.

Van den Abbeele, P., Venema, K., Van de Wiele, T., Verstraete, W., and Possemiers, S. (2013). Different human gut models reveal the distinct fermentation patterns of arabinoxylan versus inulin. J. Agric. Food Chem. *61*, 9819–9827.

Van der Meulen, R., Makras, L., Verbrugghe, K., Adriany, T., and De Vuyst, L. (2006). *In vitro* kinetic analysis of oligofructose consumption by *Bacteroides* and *Bifidobacterium* spp. indicates different degradation mechanisms. Appl. Environ. Microbiol. 72, 1006–1012.

van Hoffen, E., Ruiter, B., Faber, J., M'Rabet, L., Knol, E.F., Stahl, B., Arslanoglu, S., Moro, G., Boehm, G., and Garssen, J.A. (2009). Specific mixture of short-chain galacto-oligosaccharides and long-chain fructo-oligosaccharides induces a beneficial immunoglobulin profile in infants at high risk for allergy. Allergy. *64*, 484–487.

Van Loo, J. (2004). The specificity of the interaction with intestinal bacterial fermentation by prebiotics determines their physiological efficacy. Nutr. Res. Rev. *17*, 89–98.

Vanhoutvin, S.A.L.W., Troost, F.J., Hamer, H.M., Lindsey, P.J., Koek, G.H., Jonkers, D.M.A.E., Kodde, A., Venema, K., and Brummer, R.J.M. (2009). Butyrate-induced transcriptional changes in human colonic mucosa. PLoS One *4*, e6759.

Ventura, M., Turroni, F., Canchaya, C., Vaughan, E.E., O'Toole, P.W., and van Sinderen, D. (2009). Microbial diversity in the human intestine and novel insights from metagenomics. Front. Biosci. *14*, 3214–3221.

Verbeke, K., Ferchaud-Roucher, V., Preston, T., Small, A.C., Henckaerts, L., Krempf, M., Wang, H., Vonk, R.J., and Priebe, M.G. (2010). Influence of the type of indigestible carbohydrate on plasma and urine short-chain fatty acid profiles in healthy human volunteers. Eur. J. Clin. Nutr. *64*, 678–684.

Vieira, A.T., Teixera, M.M., and Martins, F.S. (2013). The role of probiotics and prebiotics in inducing gut immunity. Front. Immunol. *4*, 445. eCollection 2013.

Viladomiu, M., Hontecillas, R., Yuan, L.J., Lu, P.Y., and Bassaganya-Riera, J. (2013). Nutritional protective mechanisms against gut inflammation. J. Nutr. Biochem. *24*, 929–939.

Vince, A.J., and Burridge, S.M. (1980). Ammonia production by intestinal bacteria: the effects of lactose, lactulose and glucose. J. Med. Microbiol. *13*, 177–191.

Visek, W.J. (1978). Diet and cell growth modulation by ammonia. Am. J. Clin. Nutr. *31*, S216–S220.

Weaver, G.A., Krause, J.A., Miller, T.L., and Wolin, M.J. (1986). Incidence of methanogenic bacteria in a sigmoidoscopy population – an association of methanogenic bacteria and diverticulosis. Gut 27, 698–704.

Windey, K., De Preter, V., and Verbeke, K. (2012a). Relevance of protein fermentation to gut health. Mol. Nutr. Food Res. *56*, 184–196.

Windey, K., De Preter, V., Louat, T., Schuit, F., Herman, J., Vansant, G., and Verbeke, K. (2012b). Modulation of protein fermentation does not affect fecal water toxicity: A randomized cross-over study in healthy subjects. PLoS One 7, e52387.

Woodsmansey, E.J. (2007). Intestinal bacteria and ageing. J. Appl. Microbiol. *102*, 1178–1186.

Yen, C.H., Kuo, Y.W., Lee, M.C., and Chen, H.L. (2011). Long-term supplementation of isomalto-oligosaccharides improved colonic microflora profile, bowel function, and blood cholesterol levels in constipated elderly people – A placebo-controlled, diet-controlled trial. Nutrition *24*, 445–450.

Yeoh, N., Burton, J.P., Suppiah, P., Reid, G., and Stebbings, S. (2013). The role of the microbiome in rheumatic diseases. Curr. Rheumatol. Rep. *15*, 314.

Zampa, A., Silvi, S., Fabiani, R., Morozzi, G., Orpianesi, C., and Cresci, A. (2004). Effects of different digestible carbohydrates on bile acid metabolism and SCFA production by human gut micro-flora grown in an *in vitro* semi-continuous culture. Anaerobe *10*, 19–26.

Zhang, H.S., DiBaise, J.K., Zuccolo, A., Kudrna, D., Braidotti, M., Yu, Y.S., Parameswaran, P., Crowell, M.D., Wing, R., Rittmann, B.E., and Krajmalnik-Brown, R. (2009). Human gut microbiota in obesity and after gastric bypass. Proc. Nat. Acad. Sci. U.S.A. *106*, 2365–2370.

Part II

Probiotics

Lactobacilli as Probiotics: Discovering New Functional Aspects and Target Sites

3

Koen Venema and Marjolein Meijerink

Abstract

Probiotic lactobacilli have been in use for several decades now. Still, we hardly know the molecular mechanisms underlying the probiotic effect. Two strains, *L. rhamnosus* GG and *L. plantarum* WCFS1 have been studied in great detail, and mutants of these strains have greatly aided in our understanding of the interaction with the host. However, several surprising results were obtained as well, and leave more questions than answers. The first part of this chapter lists the recent advances in the molecular understanding of interaction of probiotic lactobacilli with the host. Especially surface molecules are thought to play a crucial role in this interaction. In the second half of the chapter we briefly highlight some of the newest applications. Although there have not been a lot of studies with these novel approaches, the initial results are promising and require further research, not only to confirm the results found, but also to deduce the mode of action of these probiotics.

Introduction

Over the last few decades there has been an increasing awareness of the role played by commensal bacteria in modulating mucosal immune responses which has greatly increased interest in the therapeutic potential of probiotics for a range of immune disorders. Probiotics are defined as living organisms that, when administered in sufficient numbers, are beneficial to the host (FAO and WHO, 2001). Current evidence indicates that probiotic strains mediate their effects by a variety of different mechanisms that are dependent on the dose as well as the route and frequency of delivery (Sherman *et al.*, 2009). Probiotics may act directly on the host or indirectly via modulation of the composition and/or activity of the commensal microbiota, and/or their stimulation of the immune system. Probiotics can change the conditions in the gut and improve colonization resistance, make the gut environment more favourable to the growth of genera generally considered as beneficial (e.g. lactobacilli and bifidobacteria) and less favourable for pathogens and other harmful microbes. Some probiotics act in the lumen of the gut by elaborating antibacterial molecules such as bacteriocins; others enhance the mucosal barrier by increasing the production of innate immune molecules, including goblet cell–derived mucins and trefoil factors, or defensins produced by intestinal Paneth cells. Probiotics may also mediate beneficial effects by promoting adaptive immune responses [secretory immune globulin A, regulatory T-cells (Tregs), interleukin (IL) 10].

Many probiotic strains are lactic acid bacteria (LAB) which are bacteria that produce lactic acid as their major fermentation product. This includes species of *Lactobacillus*, *Streptococcus*, *Enterococcus*, *Lactococcus*, *Bifidobacterium* and *Leuconostoc*. They are widely distributed in the intestinal tract. The largest genus in this order is *Lactobacillus* which are traditionally used in different food fermentation products such as pickles, sauerkraut, beer, wine, juices, cheese, yoghurt and sausage (Prescott *et al.*, 2002). Dietary LAB are generally regarded as safe (GRAS status) owing to their long history or association with humans and food (Bernardeau *et al.*, 2008) and are therefore often selected as candidate probiotics.

The genus *Lactobacillus*

The number of species of the genus *Lactobacillus* has increased considerably during the last 10–15 years. There are currently > 200 *Lactobacillus* species. The latest overview, published recently but compiled in March 2013, lists 151 species (Vandamme *et al.*, 2014), but recently *Paralactobacillus selangorensis* has been reclassified as *Lactobacillus selangorensis* (Pot *et al.*, 2014), although this is disputed (Mattarelli *et al.*, 2014). The website http://www.bacterio.net/*lactobacillus*.html at this moment (October 2014) lists 204 species, with 29 subspecies. In order to enable taxonomists to correctly identify and allocate a new strain to a specific taxon, a clear understanding of the key criteria that need to be applied on a genus and species level is extremely important. Minimal standards for description of novel taxa within the genera *Bifidobacterium*, *Lactobacillus* and related genera have recently been described (Mattarelli *et al.*, 2014).

The genus *Lactobacillus* shows a remarkable variation of catabolic activities. This and the fact that lactobacilli are generally considered 'safe' for food applications, has been the basis of their very broad use in food fermentations. The most important end-product of fermentation by lactobacilli is lactic acid (> 50% of the metabolites produced), although different parameters will change the composition of the final end-products. In addition to the great variation in carbohydrate metabolism, other metabolic properties of the lactobacilli, such as their proteolytic and lipolytic activities, has resulted in a vast variation of end-products important for a food fermentation, both for safety and organoleptic reasons.

Besides food fermentation products, or perhaps due to their presence in such products, lactobacilli also can be found in the gastro-intestinal tract. Over the past few decades, several strains have been shown to be probiotic, i.e. they show a health benefit for the host upon administration (usually consumption) of adequate numbers of life bacteria (Hill *et al.*, 2014). Just a few species of the > 200 known to date have been screened for probiotic properties. While identification at species level yields outcome on the suitability of bacterial strains for industrial application, their performance as, for example, starter cultures or probiotics, is strain dependent. Therefore, it is of primary importance to use techniques that are able to discriminate among strains. This is especially important for their probiotic functionality, where it has clearly been demonstrated that the capacity to carry out a probiotic functionality is strain-dependent. Unfortunately, the genes, proteins or other cell components responsible for the probiotic activity have not yet been identified, despite substantial efforts by many researchers. In many cases genes and factors have been identified that may contribute to the probiotic functionality, but are in itself not sufficient to explain the full beneficial activity of the microbes. Recent advances in these factors will be discussed in more detail below.

One way to study which factors within strains are probiotic is by genome comparison, or comparative genomics. While this is best done on similar strains from the same species as will be discussed below, Wassenaar and Lukjancenko recently compared the genomes of all known lactobacilli (and other lactic acid bacteria) (Wassenaar and Lukjancenko, 2014). At the time of writing their review approximately 1800 microbial genomes were completely sequenced and over 5000 microbial genomes were presented as 'in progress'. There were 40 finished and 142 incomplete *Lactobacillus* genomes, nine complete and five incomplete *Lactococcus* genomes, as well as four complete and 109 incomplete *Enterococcus* genomes. The numbers for *Streptococcus* genomes were even higher, though most of these are derived from pathogens. Publication of an incomplete genome does not always mean that the sequence will eventually be completed, as the authors notices that quite a few entries have not been updated for several years (Wassenaar and Lukjancenko, 2014). The authors elegantly describe how comparative genomic analysis is not only observational (which genes are present and what are their potential function?), but can also be used for hypothesis-driven analyses. These hypotheses can then be tested in the laboratory. When the genomes of 27 *Lactobacillus* species were compared, a large spread in genome size was observed, ranging from 3.349 kb for *L. plantarum* WCFS1, isolated from human saliva, to 1.376 kb the smallest belonging to *L. sanfranciscensis* strain TMW1.1304, isolated from sourdough. Even within a single *Lactobacillus* species, genome size can vary considerably, as in *L. plantarum* or *L. reuteri*. Using 10 of the sequenced *Lactobacillus* genomes, the pan-genome was composed of 7248 gene families, far more than the 2000 to 3000 gene families that a single *Lactobacillus* genome typically contains. Their core only contains 465 gene families; by definition, members of these core gene families are represented in each of the 10 genomes that were analysed. The pan-genome increased significantly to 16,663 genes when all 37 sequenced *Lactobacillus* genomes were included in this analysis. As expected the core genome only was reduced slightly to 365. The difference in size of the pan-genome and core genome of a genus can serve as a measure for the variation within a genus. For *Lactobacillus*, this difference is 16,298 gene families when 37 genomes belonging to 17 species were analysed (Wassenaar and Lukjancenko, 2014). A similar comparative genomics approach has been taken by Douillard and de Vos to determine the function of lactobacilli present in food and those used for health benefits (Douillard and de Vos, 2014). The fraction of LAB in the gastrointestinal (GI) tract is low (< 1% in most individuals). The highest fractions of LAB are found in the oral and vaginal cavities. As many of the genomes of human-derived lactobacilli have been determined, functional genomics studies of these strains have revealed functions (and the genes corresponding to these) that are important for and contribute to (but never are the sole factors responsible for) colonization, survival and probiotic functionality. Some of these are described below.

Probiotic effector molecules

It is generally accepted that secreted molecules and cell surface molecules interact with the host and aid to probiotic functionality. Most of the probiotic effector molecules are present in the bacterial cell envelope, which is the first contact side to interact with intestinal host cells (Remus *et al.*, 2011; Wells *et al.*, 2011).

Peptidoglycan

In lactobacilli the cytoplasmic membrane is covered by pentapeptide stem-connected layers of peptidoglycan (PG), a polymer composed of alternating residues of N-acetyl-glucosamine (GlcNAc) and β-1-4-linked N-acetyl muramic acid (MurNAc) (Kleerebezem *et al.*, 2010), which is essential for maintaining the cellular shape and integrity. PG has been shown to modulate immune responses (Asong *et al.*, 2009; Dziarski, 2003). This layer also serves as a platform for the anchoring of cell surface molecules, such as wall teichoic acids (WTA), wall polysaccharides and surface proteins (Delcour *et al.*, 1999; Kleerebezem *et al.*, 2010; Vollmer *et al.*, 2008). In lactobacilli, the disaccharide unit of peptidoglycan can undergo a wide range of modifications, which have important consequences for bacterial physiology. For example, N-deacetylation of GlcNAc and/or MurNAc in *L. fermentum* (Logardt and Neujahr, 1975) and 6-O-acetylation of MurNAc in *L. casei*, *L. acidophilus* and *L. fermentum* (Delcour *et al.*, 1999; Kleerebezem *et al.*, 2010; Vollmer, 2008) have been described. The N-deacetylation and 6-O-acetylation in GlcNAc/MurNAc have shown to reduce the susceptibility for autolysis by lysozyme, which is a bacteriolytic enzyme present in the host innate immune system (Kleerebezem *et al.*, 2010; Vollmer, 2008). O-acetylation of peptidoglycan occurs in the majority of *Lactobacillus* species (Kleerebezem *et al.*, 2010; Vollmer and Tomasz, 2000). There are two paralogues of O-acetyltransferase found in *L. plantarum* WCFS1 and *L. sakei* 23 K (Kleerebezem *et al.*, 2010). In *L. plantarum* WCFS1 these enzymes, OatA and OatB, are responsible for O-acetylation of MurNAc and GlcNAc, and consequently results in resistance against lysozyme and endogenous autolysins, respectively (Bernard *et al.*, 2011). The impact of these two modifications on peptidoglycan degradation and remodelling, as well as immunomodulation is still unknown. Peptidoglycan is recognized by TLR-2, which recognizes the diaminopimelic acid-containing peptidoglycan of Gram-negative bacteria and to a lesser extend lysine-containing peptidoglycan motifs of Gram-positive bacteria.

Lipoteichoic acid

In addition to PG, all lactobacilli produce lipoteichoic acid (LTA), a glycosyl-substituted and/or d-Ala-substituted glycerolphosphate (GroP) polymer which is anchored in the cytoplasmic membrane through a glycolipid anchor (Lebeer *et al.*, 2012c). The LTAs contain di-acylated and/or tri-acylated glycolipids (Claes *et al.*, 2012; Jang *et al.*, 2011) that are thought to signal via the heterodimeric TLR complexes TLR-2/6 and TLR-2/1, respectively (Wells *et al.*, 2011). The degrees of D-alanylation of the GroP backbone is important in this interaction (Claes *et al.*, 2012; Grangette *et al.*, 2005). Deletion mutants of *L. rhamnosus* GG and *L. plantarum* WCFS1 unable to D-alanylate LTA induced more anti-inflammatory cytokine responses, and elicited enhanced relief of symptoms in murine colitis models, than the wild-type strains (Claes *et al.*, 2010; Grangette *et al.*, 2005). In addition, a deletion mutant of *L. acidophilus* NCFM lacking LTA was able to down-regulate IL-12 and TNF-α responses but also significantly enhanced IL-10 in dendritic cells (DCs). The more anti-inflammatory cytokine profiles induced by this LTA mutant also resulted in higher efficacy in the treatment of mice in a chemically induced colitis model compared with the wild-type (Mohamadzadeh *et al.*, 2011), and oral treatment with the LTA-deficient *L. acidophilus* NCFM-mutant normalized innate and adaptive pathogenic immune responses and caused regression of established colonic polyps. Even though the LTA backbone polymer structure appears conserved in lactobacilli, it is highly likely that the LTA molecules produced vary between different species or even strains of the same species. Structural variations are anticipated to include different glycolipid anchors, chain length, and the degree and composition of the polymer substitutions with glycosyl moieties and D-alanyl residues (Bron *et al.*, 2013).

Wall teichoic acid

Most lactobacilli also synthesize wall teichoic acid (WTA), which typically consists of GroP or ribitolphosphate (RboP) backbone polymers that are covalently anchored to the MurNAc residue of PG via a phosphodiester bond (Bron *et al.*, 2012a), with the exception of certain specific *Lactobacillus* species, *L. rhamnosus* and *L. reuteri*, which do not have the genetic capacity to produce WTA (Kleerebezem *et al.*, 2010).

In contrast to LTA, the role of WTA in TLR-2-mediated immune signalling is debated, as WTA is not acylated (Xia *et al.*, 2010) and thereby lacks a molecular characteristic that is considered essential for LTA signalling (Wells *et al.*, 2011). In a recent study by Bron *et al.*, the impact of WTA backbone in *L. plantarum* WCFS1 mutants was determined (Bron *et al.*, 2012a). Derivatives of *L. plantarum* WCFS1 were constructed that synthesize alternative WTA variants and these were used for WTA isolation and biochemical analysis. The mutants were shown to either (i) completely lack WTA in their cell envelope (ii) produce an alternative WTA (1,5-linked poly[RboP] with 28% of α-D-glucosyl substitution at C-2 of the ribitol residues) (Tomita *et al.*, 2013) compared to the wild-type molecule (3,6′-linked poly[1-α-D-glucosyl-sn-GroP]) (Tomita *et al.*, 2012) (iii) produce LTA that lacks D-alanine substitution, or (iv) ribitol-backbone WTA instead of the wild-type glycerol-containing backbone. A transcriptome analysis revealed which genetic determinants were involved in this alternative WTA backbone production, and pinpointed ribose and arabinose as the possible biosynthetic precursors. Co-culture of these mutants with human DCs secreted drastically decreased levels of pro-inflammatory cytokines compared to the wild type (Bron *et al.*, 2012a). Moreover, human embryonic kidney NF-κB reporter cell line experiments indicated that WTA dampens TLR-2 and TLR-1/2 signalling to different degrees, strongly depending on the backbone alditol, and to some extend on D-alanyl substitutions, whereas LTA signalling occurs via a TLR-2/6-dependent mechanism completely dependent on D-alanyl substitutions. This study shows that the WTA structure–function can also contribute to the high variability in the host immune response to different strains of lactobacilli, as WTA may actually shield envelope associated ligands and prevent their interaction with their cognate receptor. Alternatively, WTA mutant strains might differ in surface charge, possibly affecting phagocytosis in these immune assays.

Bacterial DNA

In addition to cell wall associated molecules, bacterial genomic DNA can also interact with the host. TLR-9 recognizes bacterial genomic DNA which unlike eukaryotic DNA contains a high frequency of unmethylated CpG motifs (Hemmi *et al.*, 2000). Different species of lactic acid bacteria might differ in their capacity to elicit TLR-9 signalling due to differences in C+G composition and the frequency of stimulatory motifs in the DNA. The expression of TLR-9 by immune cells is intracellular and endosomal, and in polarized epithelial cells (ECs) it is expressed on both on the apical and basolateral membrane. In polarized ECs TLR-9 has been shown to have tolerogenic effects to chronic TLR challenges depending on the location of the stimulus (Lee *et al.*, 2006). Human epithelial HT-29 monolayers treated *in vitro* with bacterial DNA, including DNA from lactobacilli, delayed NF-κB activation, stabilized levels of IκB and attenuated IL-8 secretion in response to pro-inflammatory stimuli (Jijon *et al.*, 2004). Additionally, a recent study reported similar effects for HT29 and T84 cell monolayers using DNA from *L. rhamnosus* GG (LGG) (Ghadimi *et al.*, 2010). In addition, IL-10-deficient mice orally treated with purified DNA from VSL#3 reduced mucosal secretion of TNF-α and IFN-γ and improved histological colitis disease scores (Jijon *et al.*, 2004). Applying neutralizing antibodies against type I interferon (IFN) abolished the anti-inflammatory effects induced by TLR-9 agonists, whereas the administration of recombinant IFN-β mimicked the anti-inflammatory effects induced by TLR-9 agonists (Katakura *et al.*, 2005). In contrast to anti-inflammatory effects, pro-inflammatory effects have been observed using bacterial DNA and synthetic oligonucleotides containing CpG sequences *in vitro* and *in vivo* in combination with LPS challenge (Crabtree *et al.*, 2001). Interestingly, RAW 264.7 cells pre-exposed for 1–3 h to CpG-oligodeoxynucleotide (ODN), augmented a pro-inflammatory cytokine response to LPS, whereas prolonged exposure of the cells for 6–9 h induced the suppression of inflammatory cytokines via the induction of IL-10 expression (Crabtree *et al.*, 2001). *In vivo* studies with experimental colitis showed conflicting effects of CpG-ODN, showing that CpG-ODN in active disease worsened inflammation (Obermeier *et al.*, 2002) and prophylactic administration diminished colitis (Katakura *et al.*, 2005).

Glycoproteins

Of the many post-translational modifications proteins can undergo, glycosylation is the most prevalent and the most diverse. Today, it is clear that both N-glycosylation and O-glycosylation, which was believed to be only restricted to eukaryotes, also happens in bacteria and that bacteria are capable of forming protein attached O-glycans via Thr and/or Ser residues, as well as N-glycans via Asn residues (Abu-Qarn *et al.*, 2008; Kleerebezem *et al.*, 2010; Nothaft and Szymanski, 2010). It has been demonstrated that the purified surface layer protein A (SlpA) of *L. acidophilus* NCFM was recognized by the fucose-recognizing and mannose-recognizing lectins *Aleuria aurantia* lectin (AAL) and concanavalin A (ConA), respectively (Konstantinov *et al.*, 2008). This is supporting the role of SlpA as ligand for the fucose and mannose recognizing C type lectin DC-specific ICAM3-gabbing non-integrin (DC-SIGN) receptor, which is a receptor present on the surface of both macrophages and dendritic cells and recognizes and binds to mannose type carbohydrates. The analysis of a derivative of *L. acidophilus* that lack the slpA established that the signalling of SlpA via DC-SIGN is important, and thereby elicits the production of increased IL10/IL12p70 ratios by maturing DCs (Konstantinov *et al.*, 2008).

The first evidence that glycosylation of proteinaceous effector molecules occurs in probiotic lactobacilli has recently been shown by studies that directly evidence this post-translational modification. The secreted *L. rhamnosus* GG protein Msp1, which functions as a peptidoglycan (PG) hydrolase but also promotes epithelial homeostasis (Yan *et al.*, 2007), was shown to be O-glycosylated at Ser residues 106 and 107 by mass spectrometry (Lebeer *et al.*, 2012b). LGG might express a DC-SIGN ligand, despite the fact it does not produce a SlpA homologue, as these glycan residues are recognized by ConA. This is in agreement with the observation that several lactobacilli that do not produce SlpA homologues can be recognized by the DC-SIGN receptor (Smits *et al.*, 2005).

Remarkably, although ConA-recognized Msp1 glycosylation appears conserved among *L. rhamnosus* strains, the closely related species *L. casei* produces an Msp1 homologue that lacks the Ser residue signature and seems to be unglycosylated in lectin-blotting experiments, suggesting species-specific Msp1 glycosylation (Lebeer *et al.*, 2012b). Likewise, the major autolysin of *L. plantarum* WCFS1, Acm2, is O-glycosylated as revealed by mass spectrometry (Fredriksen *et al.*, 2012).

Although several studies have revealed that protein O-glycosylation is widespread in lactobacilli and impacts on protein stability and activity (Lebeer *et al.*, 2012b), there is no clear evidence that there is a link towards probiotic functionality. However, protein glycosylation analogous to that identified for SlpA of *L. acidophilus* NCFM, have shown to be implicated in periodontitis in pathogenic bacteria like *Tannerella forsythia*, and to impact on modulation of the host immune response (Settem *et al.*, 2013). Moreover, the fact that human cells, like macrophages and dendritic cells, express a variety of C-type lectin receptors that recognize glycans rather than proteins (Rabinovich *et al.*, 2012), is suggesting that glycan motives could play a pivotal role in the immunomodulating activity of lactobacilli in general. Interestingly, it was recently reported that specific Ser-Thr-rich peptides derived from *L. plantarum* BMCM12 were absent from the intestinal microbiota of inflammatory bowel disease patients, and these potentially O-glycosylated peptides were able to stimulate regulatory responses in human intestinal DCs, which suggests that these peptides may play a role in this disease with O-glycosylation being a prerequisite (Bernardo *et al.*, 2012). Overall, the recently emerging field of probiotic glycoproteins is bound to reveal novel probiotic ligand–host receptor interactions and is likely to provide further molecular insight in strain-specific probiotic functionality.

Recently, in *L. plantarum* WCFS1 two genes encoding glycosyl-transferases have been identified. Both enzymes are required for glycosylation of a number (at least six) of *L. plantarum* surface proteins (Lee *et al.*, 2014), including Acm2, discussed above. Lectin blotting experiments using a panel of eight other *L. plantarum* strains revealed that protein glycosylation is a common feature in this species.

Exopolysaccharides (EPS)

Bacterial EPS are a large group of carbohydrate polymers which could be either covalently associated with the cell surface forming a capsule, or be loosely attached, or totally secreted into the environment of the cell (Ruas-Madiedo *et al.*, 2008). Exopolysaccharides (EPS) produced by probiotics have been shown to have various beneficial physiological effects on humans, such as anti-ulcer action (Fanning *et al.*, 2012; Nagaoka *et al.*, 1994), modulation of the immune response (Kitazawa *et al.*, 1998), and improvement of intestinal adhesion of probiotics (Ouwehand *et al.*, 1999). In addition, EPS has been shown to protect the producing cells from the harsh conditions in the GI tract, by protecting against bile in the small intestine (Suzuki *et al.*, 2014) and possibly low pH in the stomach. Similarly EPS from the probiotic strain *L. rhamnosus* GG (LGG) was identified as a shielding molecule that reduces exposure of other cell wall compounds on the microbial surface and also contributed to the protection of the strain against host defence activity to antimicrobial peptide LL-37 (Lebeer *et al.*, 2011; Lebeer *et al.*, 2009). Similar results have been shown for *L. plantarum* WCFS1 (our unpublished results). Moreover, it has been reported that the EPS produced by probiotics may be important for their persistence in the gut ecosystem (Fanning *et al.*, 2012) and interactions with specific receptors and effectors of the host defence system (Welman and Maddox, 2003). Purified EPS from *Lactobacillus rhamnosus* E/N has also been shown to be bifidogenic and have antioxidant capacity (Polak-Berecka *et al.*, 2013). Other EPS have also been shown to modify the composition of the gut microbiota, as these molecules are not digested during transit through the upper GI tract (Salazar *et al.*, 2009). The EPS from *L. rhamnosus* KL37C, isolated from the faeces of a human newborn, has been shown to ameliorate arthritis (Nowak *et al.*, 2012). The polymer produced by LGG has been shown to protect human intestinal cells against the toxic effect of bacterial toxins *in vitro* (Ruas-Madiedo *et al.*, 2010) and *in vivo* (Chen *et al.*, 2014). EPS produced by lactobacilli show a high diversity in terms of their structure and physicochemical properties (Salazar *et al.*, 2009). This makes it currently difficult to decipher the molecular mode of action of these molecules. Moreover, since the EPS may shield other cell wall compounds on the bacterial surface, or function through modifying the composition and/or activity o the gut microbiota, the effects may in fact be indirect (Lebeer *et al.*, 2011; Ruas-Madiedo *et al.*, 2010).

Pili

An interesting finding has been the discovery of pili on lactobacilli (Kankainen *et al.*, 2009; Lebeer *et al.*, 2012a) and bifidobacteria (Foroni *et al.*, 2011; O'Connell Motherway *et al.*, 2011), which until then had only been discovered in pathogenic Gram$^+$ bacteria. For LGG it has been shown that these pili are essential for probiotic functionality, as they are involved in adhesion of LGG to the epithelial cells of the gut, through the presence of a mucus-binding domain (Lebeer *et al.*, 2012a). LGG cells contain multiple pili, averaging 10 to 50 per cell, and with lengths of up to 1 μm (Kankainen *et al.*, 2009). LGG pili are composed of three pilin subunits: SpaABC. SpaA is the major component building up the pilus shaft, while SpaB and SpaC are the minor fibre components. Mechanistically it was shown that SpaC, the key adhesion protein, is a multifunctional adhesin with broad specificity. SpaC is engaged in both bacterial aggregation as well as binding specifically to mucin and collagen, two major extracellular components of host epithelial layers (Tripathi *et al.*, 2013). Apart from EPS and pili (discussed in this chapter), LGG has also other molecular structures that may interact with the host. These have recently been reviewed by Segers and Lebeer (2014).

An interesting observation by Sybesma *et al.* (2013) showed that commercial dairy product may contain variants of LGG that have lost the capacity to produce pili. Two genetic deletions have been found in these isolates that cover the region coding for *spaABC*. Up to 70% of the isolates missed the genomic island coding for the pili (Sybesma *et al.*, 2013). Given the believe that these pili are essential for probiotic functionality, it is clear that genome stability should be checked in the final probiotic product to ensure that the cells present can still perform their function. It is likely that certain culture, processing or perhaps storage conditions may influence the genetic drift of LGG, because in another study in LGG isolates from two other commercial products no genome reorganization was found (Douillard *et al.*, 2013). Similar genomic deletions have been shown for. The authors proposed to expand the requirements on probiotic strains, namely, to control retention of the key probiotic genes using molecular biological methods (Averina *et al.*, 2012). Apart from genetic deletions, handling of probiotic cultures by researchers is also a factor that may influence probiotic functionality. For example, centrifugation of LGG cells at a high *g*-factor leads to release of pili from cells (Lebeer, pers. comm.). This, together with gene loss, might in fact be two of the reasons why there are conflicting results in the literature for LGG in experiments and clinical trials with an otherwise very similar set-up.

Effects of genetic, processing, or product formulation changes on efficacy and safety of probiotics

The heading of this paragraph is the exact title of a paper by Sanders *et al.* (2014) and shows that both genetics and processing, which were discussed in the previous paragraph, are now also recognized to play an important factor for probiotic efficacy (and safety) by the international scientific community. The authors highlight key issues important to determining the need to re-confirm efficacy or safety after (i) strain improvement (ii) process optimization, or (iii) product formulation changes. To this may be added, the genetic stability, although this may be part of the efficacy assessment. The authors conclude that research pinpointing the mechanisms of action for probiotic function and the development of assays to measure them are greatly needed to better understand if such changes in industrial manufacturing have a substantive impact on probiotic efficacy.

Gene-trait matching

Gene-trait matching is an approach that makes use of genetic variations among a series of bacterial strains from the same species. This approach has been proven successful in identifying specific genes that are involved in eliciting immunomodulatory responses in dendritic cells (DCs) or peripheral blood mononuclear cells (PBMCs), or are required for mannose adhesion that is associated with specific innate immune responses in pig mucosal tissues *in vivo* (Gross *et al.*, 2008; Meijerink *et al.*, 2010; Pretzer *et al.*, 2005; van Hemert *et al.*, 2010). For instance, to identify novel strain-specific genes and factors that modulate innate recognition by the host, the natural diversity in the immune response to different strains of *L. plantarum* was used. Microarray-based comparative genome hybridization of several *L. plantarum* strains against the *L. plantarum* WCFS1 genome was used to provide a one-directional comparison of gene-content profiles per strain. At the same time strains were co-incubated with PBMCs. The secreted cytokine amounts or cytokine ratios for the *L. plantarum* strains were correlated with the presence or absence of specific genes by regression using the Random Forest algorithm. Several genes or gene loci were identified to be involved in the immunomodulation. These loci included genes encoding an *N*-acetyl-glucosamine/galactosamine phosphotransferase system, the LamBDCA quorum sensing system, a predicted transcriptional regulator gene (*lp_2991*) and components of the plantaricin (bacteriocin) biosynthesis and transport pathway. Deletion mutants of the candidate genes were constructed in the WCFS1 strain in order to validate their anticipated effect on cytokine induction (Meijerink *et al.*, 2010).

One of the confirmed gene clusters with immunomodulatory capacities in PBMCs was the *pts19ADCBR* locus coding for a cell membrane-associated N-acetyl-galactosamine/glucosamine phosphotransferase system, that might be important for utilization of different carbohydrates in the intestine. Another confirmed gene which correlated with immunomodulation of PBMCs was the *L. plantarum* WCFS1 lamB, encoding a processing/export protein of the autoinducer peptide quorum sensing two-component system (AIP-based QS-TCS) LamBDCA. The observed PBMC IL-10/IL-12 ratios for *L. plantarum* might either be mediated directly through the LamBDCA system and the cognate secreted peptide, or indirectly through cell products (e.g. polysaccharides) under the control of this regulatory system. A previously unknown transcriptional regulator *lp_2991* that influenced the immune response was also identified. A deletion mutant of *lp_2991* led to a significantly higher secretion of IL-10, IL-12p70 and TNF-α in DCs compared with the wild-type control. This gene lies upstream of GtcA3, a putative teichoic acid glycosylation protein. Transcriptome analysis and qPCR data supported the hypothesis that the protein encoded by *lp_2991* is a repressor of *gtcA3* transcription and points to this enzyme as being a prime candidate for the altered immune response. Interestingly GtcA3 is predicted to glycosylate teichoic acids, including LTA, which is a known TLR-2 agonist capable of modulating immune cell responses. LTA mutants and mutants affecting D-alanine substitution in LTA have already been found to have effects on the immune response (Grangette *et al.*, 2005; Mohamadzadeh *et al.*, 2011). All of the mutations introduced into the locus encoding the bacteriocin plantaracin affected the immune response and were predicted to inhibit plantaracin production. Thus, it is possible that the secreted plantaracins directly affect the immune response of DCs.

These papers showed that *in silico* analysis led to the identification of novel factors that are involved in the immunomodulating capacities of *L. plantarum* WCSF1. For the selection of new candidate probiotics, the presence or absence of these genes can be used. The differences found between the WCFS1 and the deletion mutants of genes/loci identified in the described papers were much smaller than found among the 42 different *L. plantarum* strains that were screened originally, indicating that other factors play an important role in immunomodulation too, and/or that there are synergistic effects between one or more factors. Thus, it might be insightful to sequence the other strains used in the study described above and apply the gene-trait matching approach to identify other genetic loci.

Gene-trait matching strategies have been applied successfully to increase our understanding of probiotic functionality. However, this approach is limited as it does not allows the identification of conserved mechanisms. The fact that approximately 90% of all genes are conserved within the species *L. plantarum* (Siezen *et al.*, 2010) further exemplifies the limited power of gene-trait matching. This is illustrated by a transcriptome-trait matching approach, where 9 out of the 13 survival genes identified were conserved among all 42 *L. plantarum* strains (Bron *et al.*, 2012b; van Bokhorst-van de Veen *et al.*, 2012) and could thus not have been identified with gene-trait matching. As industrial strains are generally selected on the basis of a combination of properties, such as flavour formation, immunomodulation, or robustness, the transcriptome-trait matching approach seems therefore more applicable than gene-trait matching efforts. This approach identifies the possibilities for fermentation-enhanced improvement of a specific property while applying a single strain, whereas industrial implementation of gene-trait matching results might require tedious selection of alternative strains, on the basis of identified robustness markers, that also express other desired properties. Moreover, when applying this transcriptome-trait matching strategy, trait specific biomarkers can be identified rather than universal biological markers.

Novel applications

Probiotic lactobacilli have been used and studied for several decades now. They beneficially affect a long list of diseases and disorders. These have been described before in excellent reviews and we will not repeat that here. It is hard to list a good review for probiotic lactobacilli, as nowadays reviews are split into disease areas. But a quick search in PubMed for the terms '*Lactobacillus*', 'probiotic', 'review' and '2014' (September 2014) yielded ~35 hits with papers describing about as many different disorders and diseases. Below we briefly discuss recent progress in the field over the past year or so, in areas that are not yet extensively studied but show promising results.

Longevity

Many mechanisms are thought to contribute to ageing, such as telomere shortening in replicative cells, cumulative damage to DNA over time leading to genetic instability, oxidative damage to molecules by reactive oxygen species and chronic low-grade inflammation (inflammageing). Especially the prevention of inflammageing seems to be one of the most effective approaches to increase longevity (Matsumoto and Kurihara, 2011). In a mouse model Matsumoto and colleagues have shown that the administration of *Bifidobacterium animalis* subsp. *lactis* LKM512 for 11 months increased luminal concentrations of polyamines in the large intestine of 21-month-old mice (Matsumoto *et al.*, 2011). LKM512 administration also down-regulated the expression of ageing- and inflammation-associated genes. Gene expression levels in the 21-month-old LKM512-treated mice were very similar to those in the 10-month-old untreated mice. The authors hypothesized that the reduction of low-grade inflammation was due to the increase in polyamines, as these possess anti-inflammatory activity and the ability to maintain intestinal mucosal barrier functions.

Although these results were obtained using a *Bifidobacterium* strain, similar results have been observed in another model, the fruit fly *Drosophila melanogaster*, where lactobacilli have been shown to influence longevity (Erkosar and Leulier, 2014; Erkosar *et al.*, 2013). It is interesting to note that lactobacilli also promote growth of the fruit fly in the larval stage (Storelli *et al.*, 2011). During the larval stages, the microbiota promotes systemic growth by modulating several growth hormones. In the adult stage, the *Drosophila* microbiota impacts intestinal stem cell proliferation and through this tissue homeostasis, as well as life span (Erkosar and Leulier, 2014). Even mating preference is affected by the fly's mirobiota. The mechanisms underlying these effects remain unknown, but *Drosophila* may be an excellent model to dissect this, as the fly has a very simple gut microbiota composition.

Vaginal and urogenital application

A recent meta-analysis assessed the safety and efficacy of lactobacilli on recurrent urogenital infections (Grin *et al.*, 2013). Five studies were included with total of 294 volunteers. There was no statistically significant difference in the overall risk for recurrent urinary tract infections in patients receiving *Lactobacillus* versus controls. When excluding studies using ineffective strains and studies testing for safety, data from 127 patients in two studies remained. A statistically significant decrease was found in patients given *Lactobacillus*, denoted by the pooled risk ratio of 0.51 (95% confidence interval 0.26–0.99). The authors concluded that more random controlled trials should be performed. One such recent study that was not included in the meta-analysis studied the effect of the two probiotic strains *L. rhamnosus* GR-1 and *L. reuteri* RC-14 on the vaginal microbiome in addition to host responses in post-menopausal women (Bisanz *et al.*, 2014). The Nugent score, a measure for total bacteria in the samples, was not changed after probiotic use. However, upon probiotic administration, the relative abundance of *Lactobacillus* was increased, which was weakly associated with an increase in lactate levels. A decrease in *Atopobium* was observed. Analysis of host responses showed that the probiotics had an immune-modulatory response while also affecting epithelial barrier function. The authors proposed that multifaceted approaches are required to detect subtle molecular changes induced by instillation of probiotic strains.

Energy intake

In a piglet model, Bjerg *et al.* (2014a) have shown that *L. paracasei* subsp. *paracasei* W8 supplementation for 2 weeks suppresses energy intake. In addition, an isolated pig intestine was perfused with *L. casei* W8. In both cases the GLP-1 response was increased, suggesting a lowering in appetite. Since the effect was already observed in the perfused intestine the authors hypothesized that the effect was acute. When trying to confirm the results in 21 human volunteers, they only found an effect on energy intake, while no effects on subjective appetite sensation, overall glucose and insulin response and on GLP-1 release were observed. The difference between the pig and human trials was the duration of providing the probiotic, which was 14 days for the pigs and only 3 days for the human volunteers. This may have affected the results. Also the dose may not have been optimal. Nevertheless, the results are promising and add to the wealth of studies that look into the effect of the gut microbiota and/or probiotics on weight gain, in an attempt to halt the global obesity problem. In a follow-up study, the same probiotic was given for 4 weeks to healthy volunteers (Bjerg *et al.*, 2014b). Compared to placebo, *L. casei* W8 did not affect appetite, *ad libitum* energy intake, GLP-1, glucose and insulin response nor total, high-density or low-density lipoprotein cholesterol levels after the four weeks intervention. However, triacylglycerol decreased in the probiotic group compared to placebo at week 4. The same team recently also showed that the effect of *L. paracasei* subsp. *paracasei* W8® on blood levels of fatty acids is independent of colonization (Bjerg *et al.*, 2014c).

Drug metabolism

A fascinating area is the effect of the gut microbiota on drug metabolism. For instance, studies have shown that a particular member of the gut microbiota, *Eggerthella lenta*, inactivates the cardiac drug digoxin and affects its pharmacokinetics (Haiser *et al.*, 2013, 2014). Similarly, dietary modulation of the gut microbiota augments the glucose control and improves gastrointestinal tolerability of metformin (Greenway *et al.*, 2014). A mechanism for this comes from studies performed in *Caenorhabditis elegans* co-cultured with *E. coli* (Cabreiro *et al.*, 2013). *E. coli* exerts complex nutritional and pathogenic effects on its nematode predator/host that impact health and ageing. Metformin was shown to increase lifespan, by altering microbial folate and methionine metabolism. Alterations in metformin-induced longevity by mutation of worm genes encoding methionine synthase or S-adenosylmethionine synthase imply metformin-induced methionine restriction in the host. Metformin increases or decreases worm lifespan, depending on *E. coli* strain metformin sensitivity and glucose concentration. In mammals, metformin-induced alteration of microbial metabolism could contribute to therapeutic efficacy of type II diabetes if the proper strain were selected (Cabreiro *et al.*, 2013). So far, to our knowledge, there are no examples with lactobacilli influencing drug, and through this, host metabolism.

Radiation injury

As indicated already several times, certain strains of *Lactobacillus* have an anti-inflammatory activity. This was one of the reasons why Ciorba and colleagues have used *L. rhamnosus* GG and tested its effect on radiation-induced injury of the intestinal epithelium in mice (Ciorba *et al.*, 2012). Previously it was already shown that lipopolysaccharide (LPS) and flagellin, Gram-negative TLR-4 and TLR- 5 ligands respectively, were radioprotective in the small intestine. This new study showed that both orally administered LGG and its conditioned medium supernatant protected the murine small intestinal epithelium from radiation injury. This effect was shown to be dependent on MyD88, TLR-2 and COX-2 and occurs without significantly altering the composition of the microbiota of the small intestine when analysed at the bacterial family level. This indicates that the effect might be a direct effect of LGG and/or its metabolites, rather than an indirect effect through modulation of the gut microbiota. Interestingly, the probiotic was provided during the three days prior to irradiation, indicating a prophylactic effect of LGG.

Coeliac disease

Coeliac disease is an immune-mediated enteropathy occurring in genetically predisposed individuals and is triggered by the intake of several gluten proteins present in cereals such as wheat (gliadin and glutenin), barley (hordein) and rye (secalin). This disease is often manifested in early childhood after the first exposures to gluten with small intestinal villous atrophy giving rise to malabsorption. It is thought that an overload of gluten peptides in the lamina propria may lead to a loss of tolerance to these epitopes in predisposed subjects. Peptide transport through intestinal mucosa, which is regulated by tight junction assembly, may be an important step in the development of coeliac disease (de Sousa Moraes *et al.*, 2014). Bacterial-derived molecules can control protein trafficking across the epithelium (Konishi *et al.*, 2013) and can protect against the inflammatory properties of invasive *Salmonella* on healthy tissue also by reducing its invasiveness (Tsilingiri *et al.*, 2012; Tsilingiri and Rescigno, 2013). Probiotics may be beneficial, because they reduce entrance of toxic and/or immunogenic peptides, and hence can prevent the immunological cascade. This can occur in (at least) two different ways: the probiotics may induce a tightening of the tight junctions, or they may break down the epitopes in the gluten protein, either during food fermentation, or *in situ* in the GI tract. The former has been shown in the past, where *L. plantarum* WCFS1 was shown to regulate epithelial tight junctions protein in humans to provide a protective effect on the gut barrier (Karczewski *et al.*, 2010). Duar *et al.* (2014) recently showed the ability of lactobacilli strains to degrade immunotoxic peptides present in gliadin. Four strains from the species *L. ruminis*, *L. johnsonii*, *L. amylovorus*, and *L. salivarius* showed the highest peptide-degrading activities. Strains displayed different degradation rates and cleavage patterns that resulted in reduction but not complete removal of immunotoxic epitopes. Whether or not this results in reduced severity or prevention of coeliac disease remains to be seen. In a recent study a novel mechanism was shown. In this study, *L. paracasei* CBA L74 was tested for its ability to prevent the entrance of gliadin peptides in epithelial cells (Sarno *et al.*, 2014). Rather than degrading the epitopes, CBA L74 (and its supernatant) were shown to inhibit endocytosis of small peptides suggesting that it acts on complex endocytic mechanisms. The activity was inactivated by high temperature (100°C) and not by proteases. Moreover, the biological effect was still present 30 min after removal of the supernatant but is reduced after 1 h and 3 h, suggesting that the metabolite does not permanently inhibit endocytosis.

Intense exercise

Prolonged bouts of strenuous exercise have a temporary negative impact on immune function. Post-exercise immune function depression is most pronounced when the exercise is continuous, prolonged (> 1.5 h), of moderate to high intensity, and performed without food intake (Gleeson and Williams, 2013). Several recent studies have shown a benefit of taking probiotic lactobacilli on health of sport athletes, although not all studies show a positive outcome. For example, regular ingestion of a *L. salivarius* strain does not appear to be beneficial in reducing the frequency of urinary tract infection in an athletic cohort ($n=27$) and does not affect blood leucocyte counts or levels of salivary antimicrobial proteins during a spring period of training and competition (Gleeson *et al.*, 2012). In contrast, *L. fermentum* PCC showed some benefit in 50 competitive cyclists, although more in men than in women (West *et al.*, 2011). The load, expressed as duration × severity, of lower respiratory illness symptoms was on average three-fold lower in males taking the probiotic than in those taking placebo but increased by a factor of 2.2 in females. Differences in use of cold and flu medication mirrored these symptoms. There were, however, clear reductions in the magnitude of acute exercise-induced changes in some cytokines. Also, in men GI symptoms at high training loads were reduced. It is at present unclear what caused the gender effect. In a recent study by Haywood *et al.* (2014) the effect of 4 weeks' probiotic supplementation on the incidence, duration and severity of infections during a month of competitive rugby union was investigated. Three strains, one of *L. gasseri*, one of *B. bifidum* and one of *B. longum* (strain numbers not provided), were given to 30 healthy young male Rugby Union players. Significantly fewer participants in the probiotic group experienced a single upper respiratory tract illness or gastrointestinal episode. However, there was no significant difference in the severity of the symptoms. The average number of days of illness tended to be lower for probiotic group as well, but this did not reach statistical significance. A recent study looking into the gut microbiota of another team of rugby players showed that athletes had a higher diversity of gut micro-organisms, representing 22 distinct phyla (Clarke *et al.*, 2014). The authors concluded that the results provide evidence for a beneficial impact of exercise on gut microbiota diversity but also indicate that the relationship is complex and is very likely also related to accompanying dietary extremes, such as excessive protein intake.

Effects on autoimmunity

A mixture of five probiotics, named IRT5, has been previously shown to suppress diverse experimental inflammatory disorders (Kwon *et al.*, 2010; So *et al.*, 2008, 2011). In a recent study, the same team evaluated the prophylactic and therapeutic effects of the IRT5 mixture in experimental autoimmune encephalomyelitis (EAE), a T-cell-mediated inflammatory autoimmune disease of the central nervous system. Pretreatment with IRT5 probiotics before disease induction significantly suppressed EAE development. In addition, treatment of ongoing EAE with IRT5 probiotics delayed disease onset. Administration of IRT5 probiotics was shown to inhibit the pro-inflammatory Th1/Th17 polarization, while inducing IL10 producing or/and Foxp3+ regulatory T-cells, both in the peripheral immune system and at the site of inflammation (Kwon *et al.*, 2013). This is another mixture, next to VSL#3, that shows benefits on inflammatory disorders.

Beneficial microbes for the skin – beauty from within

Recent studies show effects of oral supplementation of lactobacilli on skin health. This concept has been named beauty from within and has been reviewed recently (Al-Ghazzewi and Tester, 2014; Erdman and Poutahidis, 2014; Foolad and Armstrong, 2014). Based on the results of *in vitro* studies, where *Lactobacillus paracasei* NCC 2461 was shown to diminish neurogenic inflammation and promoted the recovery of skin barrier function, this strain was fed daily to human female volunteers in a randomized, double-blind, placebo-controlled clinical trial. The results showed that NCC 2461 exerted its effects beyond the gut and conferred benefits at the skin level. It contributed to the reinforcement of skin barrier function, decreased skin sensitivity and modulated the skin immune system. This lead to the preservation of skin homeostasis (Benyacoub *et al.*, 2014; Gueniche *et al.*, 2014). Thus, the data provide evidence that intake of NCC 2461 could improve reactive skin condition. In a more recent study, Ra *et al.* (2014) studied the effect of *Lactobacillus plantarum* HY7714 on skin hydration in human dermal fibroblasts and in hairless mice. In human dermal fibroblasts, *L. plantarum* HY7714 not only increased serine palmitoyltransferase (SPT) mRNA level, but also decreased ceramidase mRNA level, leading to increased hydration of the cells. In order to confirm the hydrating effects of *L. plantarum* HY7714 *in vivo*, *L. plantarum* HY7714 was administered orally to hairless mice for 8 weeks. In hairless mice, *L. plantarum* HY7714 decreased ultraviolet-B (UVB)-induced epidermal thickness. In addition, *L. plantarum* HY7714 suppressed the increase in transepidermal water loss (TEWL) and decrease in skin hydration, which are the result of barrier function fluctuations following UV irradiation. In particular, *L. plantarum* HY7714 administration increased the ceramide level compared to that in the UVB group. In the experiment on SPT and ceramidase mRNA expressions, *L. plantarum* HY7714 administration improved the reduction in SPT mRNA levels and suppressed the increase in ceramidase mRNA levels caused by UVB in the hairless mice skins, similar to the effects observed in the fibroblast. Collectively, these results provide further insight that oral feeding of this strain can be a potential candidate for preserving skin hydration levels against UV irradiation.

Stress reduction

It is well known that stress affects the intestinal barrier. Early life trauma and stress predispose to the development of functional intestinal disorders, such as irritable bowel syndrome (IBS). Interestingly, increased intestinal epithelial permeability has been incriminated in the pathophysiology of IBS (Spiller *et al.*, 2000), leading to low grade inflammation in the gut. This has also been found in animal models of chronic stress. In these IBS models, increased gut permeability has been positively correlated with visceral pain (Zhou *et al.*, 2009) and in rodents the stress-induced hyperpermeability was responsible for visceral hypersensitivity (Ait-Belgnaoui *et al.*, 2005). Despite these well-known intestinal epithelial barrier impairment and visceral hypersensitivity in IBS patients, structural and physical changes in the mucus layer remain poorly understood. In a recent study, Da Silva and colleagues used the water avoidance stress (WAS) model in Wistar rats to evaluate whether WAS modified gut permeability, visceral sensitivity, mucin expression, and biochemical and physical mucus properties. In addition, they tested whether a 14-day oral treatment with the probiotic *Lactobacillus farciminis* CIP 103136 treatment prevented these alterations. WAS was shown to induce gut hyperpermeability and visceral hypersensitivity but did not modify either the number of intestinal goblet cells or Muc2 expression. However, O-glycosylation of mucins was strongly affected, leading to flattening and loss of the cohesive properties of the mucus layer. Strain CIP 103136 bound to intestinal Muc2 and prevented WAS-induced functional alterations and changes in mucin O-glycosylation and mucus physical properties. *L. farciminis* treatment therefore conferred epithelial and mucus barrier strengthening (Da Silva *et al.*, 2014).

Recombinant LAB

Lactic acid bacteria (LAB), including *Lactobacillus*, have been generally recognized as safe (GRAS). They have a long history of use in fermented food fermentations since before Christ. This is one of the reasons why they are frequently used as probiotics. However, because of their intrinsic characteristics, several members of LAB have also been identified as good candidates for mucosal vaccines and/or therapeutic delivery vehicles for chronic diseases and infections (Trombert, 2014). Recombinant LAB emerged from the hypothesis that commensal or non-pathogenic bacteria could express or deliver antigens, cytokines and/or proteins in the gut, such that they may be used as oral vaccines or drugs containing carriers for the treatment of several (inflammatory) conditions. Two of the major advantages of recombinant LAB is their potential to elicit antigen-specific IgA at mucosal surfaces and promote mixed Th1/Th2 responses that depend on both the antigen and the bacterial host itself (Wells and Mercenier, 2008). Most research on recombinant LAB has centred around *Lactococcus lactis*, but recently *Lactobacillus* has also been used. For example, the pneumococcal surface antigen A (PsaA), a membrane-anchored virulence factor obtained from *S. pneumoniae*, was expressed in *L. lactis*, *L. plantarum*, *L. casei* or *L. helveticus*. The recombinant *L. plantarum* and *L. helveticus* triggered higher antibody titres and a significant reduction of *S. pneumoniae* colonization in challenged mice than the recombinant *L. lactis* and *L. casei* (del Rio *et al.*, 2008; Oliveira *et al.*, 2006). Jiang *et al.* (2014) expressed the antigenic site of the spike protein of transmissible gastroenteritis virus together with an adjuvant (muramyl dipeptide) in *L. casei*. Treatment of mice with recombinant *L. casei* induced higher levels of IgA in the presence of the muramyl dipeptide than in its absence. The immunopotentiating properties of the adjuvant suggest that the use of mucosal adjuvants in recombinant LAB may be useful in oral vaccines inducing humoral and T-cell-mediated immune responses. This is an exciting area that no doubt will get more attention in the future. However, containment of the recombinant bacteria needs to be assured.

Deleterious effects

Not all probiotics are equal. That much we know already. However, sometimes strains (or their metabolites) may also have a deleterious effect on the host. As an example, Xu and colleagues showed that purified cell wall fragment from the probiotic *L. gasseri* ATC33323 activates systemic inflammation and, at higher doses, produces lethality in a rat model. Here, the proposed beneficial effect of probiotic bacterial cell-wall fragments binding to intestinal cell pattern-recognition receptors actually over-activates the normally protective innate immunity (Xu *et al.*, 2014). The *L. gasseri* cell wall was equally or more lethal than a *Staphylococcus aureus* cell wall preparation. Such effects deserve further study, especially with regard to the use of probiotics in critically ill patients, who may be predisposed to a certain inflammatory reaction.

Weight gain

Since lactobacilli can influence energy intake as stated above, they thus also regulate weight gain. Whereas above we have focused on studies that showed a beneficial impact on weight gain, there are also some studies that show the opposite effect. There is considerable controversy in the literature about the effects of lactobacilli on weight gain (Million and Raoult, 2013). Without choosing sides, one should realize that both effects may be wanted. In humans we prefer to reduce weight gain, or induce even weight loss. In farm animals that are raised for meat, lactobacilli could (should) have a weight gain effect (preferably in meat, rather than fat tissue), as this increases the economic value of these animals. Nevertheless, due to difficulties in translation of *in vitro* and animal results to human clinical trials (Meijerink *et al.*, 2013), one ought to be careful in the choice of probiotic strain.

Future perspectives

In this chapter we have highlighted recent advances in both molecular understanding of the mode of action of probiotics as well as novel applications, that require more detailed studies (does not science always?). As voiced by the expert panel of ISAPP (International Scientific Association for Probiotics and Prebiotics) this is exactly what is required to further our understanding of probiotics, certainly if industrial manufacturing conditions chance, without us knowing what effect this has on probiotic functionality: 'Research pinpointing the mechanisms of action for probiotic function and the development of assays to measure them are greatly needed to better understand if such changes have a substantive impact on probiotic efficacy (Sanders *et al.*, 2014)'. We feel that the future will bring more mechanistic studies into the effects of (certain) probiotic(s) (strains), which will have to be translated into human relevance, one way or another, no matter the challenges that we currently face there (Meijerink *et al.*, 2013). Only upon detailed understanding of the interaction of probiotics with the host are we able to carefully select strains for specific disorders, and prevent deleterious effects from occurring.

References

Abu-Qarn, M., Eichler, J., and Sharon, N. (2008). Not just for Eukarya any more: protein glycosylation in Bacteria and Archaea. Curr. Opin, Struct Biol. *18*, 544–550.

Ait-Belgnaoui, A., Bradesi, S., Fioramonti, J., Theodorou, V., and Bueno, L. (2005). Acute stress-induced hypersensitivity to colonic distension depends upon increase in paracellular permeability: role of myosin light chain kinase. Pain *113*, 141–147.

Al-Ghazzewi, F.H., and Tester, R.F. (2014). Impact of prebiotics and probiotics on skin health. Benef. Microbes 5, 99–107.

Asong, J., Wolfert, M.A., Maiti, K.K., Miller, D., and Boons, G.J. (2009). Binding and cellular activation studies reveal that Toll-like Receptor 2 can differentially recognize peptidoglycan from Gram-positive and Gram-negative bacteria. J. Biol. Chem. *284*, 8643–8653.

Averina, O.V., Nezametdinova, V.Z., Alekseeva, M.G., and Danilenko, V.N. (2012). [Genetic instability of probiotic characteristics in the *Bifidobacterium longum* subsp. longum B379M strain during cultivation and maintenance]. Genetika *48*, 1287–1296.

Benyacoub, J., Bosco, N., Blanchard, C., Demont, A., Philippe, D., Castiel-Higounenc, I., and Gueniche, A. (2014). Immune modulation property of *Lactobacillus paracasei* NCC2461 (ST11). strain and impact on skin defences. Benef. Microbes 5, 129–136.

Bernard, E., Rolain, T., Courtin, P., Guillot, A., Langella, P., Hols, P., and Chapot-Chartier, M.P. (2011). Characterization of O-acetylation of N-acetylglucosamine: a novel structural variation of bacterial peptidoglycan. J. Biol. Chem. *286*, 23950–23958.

Bernardeau, M., Vernoux, J.P., Henri-Dubernet, S., and Gueguen, M. (2008). Safety assessment of dairy microorganisms: the *Lactobacillus* genus. Int. J. Food Microbiol. *126*, 278–285.

Bernardo, D., Sanchez, B., Al-Hassi, H.O., Mann, E.R., Urdaci, M.C., Knight, S.C., and Margolles, A. (2012). Microbiota/host crosstalk biomarkers: regulatory response of human intestinal dendritic cells exposed to *Lactobacillus* extracellular encrypted peptide. PLoS One 7, e36262.

Bisanz, J.E., Seney, S., McMillan, A., Vongsa, R., Koenig, D., Wong, L., Dvoracek, B., Gloor, G.B., Sumarah, M., Ford, B., *et al.* (2014). A systems biology approach investigating the effect of probiotics on the vaginal microbiome and host responses in a double blind, placebo-controlled clinical trial of post-menopausal women. PLoS One *9*, e104511.

Bjerg, A.T., Kristensen, M., Ritz, C., Holst, J.J., Rasmussen, C., Leser, T.D., Wellejus, A., and Astrup, A. (2014a). *Lactobacillus paracasei* subsp. paracasei L. casei W8 suppresses energy intake acutely. Appetite *82C*, 111–118.

Bjerg, A.T., Kristensen, M., Ritz, C., Stark, K.D., Holst, J.J., Leser, T.D., Wellejus, A., and Astrup, A. (2014b). Four weeks supplementation with *Lactobacillus paracasei* subsp. paracasei L. casei W8(R) shows modest effect on triacylglycerol in young healthy adults. Benef. Microbes, 1–11.

Bjerg, A.T., Sorensen, M.B., Krych, L., Hansen, L.H., Astrup, A., Kristensen, M., and Nielsen, D.S. (2014c). The effect of *Lactobacillus paracasei* subsp. paracasei L. casei W8(R) on blood levels of triacylglycerol is independent of colonization. Benef. Microbes, 1–7.

Bron, P.A., Tomita, S., Mercenier, A., and Kleerebezem, M. (2013). Cell surface-associated compounds of probiotic lactobacilli sustain the strain-specificity dogma. Curr. Opin. Microbiol. *16*, 262–269.

Bron, P.A., Tomita, S., van Swam, I.I., Remus, D.M., Meijerink, M., Wels, M., Okada, S., Wells, J.M., and Kleerebezem, M. (2012a). *Lactobacillus plantarum* possesses the capability for wall teichoic acid backbone alditol switching. Microb. Cell. Fact. *11*, 123.

Bron, P.A., Wels, M., Bongers, R.S., van Bokhorst-van de Veen, H., Wiersma, A., Overmars, L., Marco, M.L., and Kleerebezem, M. (2012b). Transcriptomes reveal genetic signatures underlying physiological variations imposed by different fermentation conditions in *Lactobacillus plantarum*. PLoS One 7, e38720.

Cabreiro, F., Au, C., Leung, K.Y., Vergara-Irigaray, N., Cocheme, H.M., Noori, T., Weinkove, D., Schuster, E., Greene, N.D., and Gems, D. (2013). Metformin retards aging in C. elegans by altering microbial folate and methionine metabolism. Cell *153*, 228–239.

Chen, X.Y., Woodward, A., Zijlstra, R.T., and Ganzle, M.G. (2014). Exopolysaccharides synthesized by *Lactobacillus reuteri* protect against enterotoxigenic *Escherichia coli* in piglets. Appl. Environ. Microbiol. *80*, 5752–5760.

Ciorba, M.A., Riehl, T.E., Rao, M.S., Moon, C., Ee, X., Nava, G.M., Walker, M.R., Marinshaw, J.M., Stappenbeck, T.S., and Stenson, W.F. (2012). *Lactobacillus* probiotic protects intestinal epithelium from radiation injury in a TLR-2/cyclo-oxygenase-2-dependent manner. Gut *61*, 829–838.

Claes, I.J., Lebeer, S., Shen, C., Verhoeven, T.L., Dilissen, E., De Hertogh, G., Bullens, D.M., Ceuppens, J.L., Van Assche, G., Vermeire, S., *et al.* (2010). Impact of lipoteichoic acid modification on the performance of the probiotic *Lactobacillus rhamnosus* GG in experimental colitis. Clin. Exp. Immunol. *162*, 306–314.

Claes, I.J., Segers, M.E., Verhoeven, T.L., Dusselier, M., Sels, B.F., De Keersmaecker, S.C., Vanderleyden, J., and Lebeer, S. (2012).

Lipoteichoic acid is an important microbe-associated molecular pattern of *Lactobacillus rhamnosus* GG. Microb. Cell. Fact. *11*, 161.

Clarke, S.F., Murphy, E.F., O'Sullivan, O., Lucey, A.J., Humphreys, M., Hogan, A., Hayes, P., O'Reilly, M., Jeffery, I.B., Wood-Martin, R., *et al.* (2014). Exercise and associated dietary extremes impact on gut microbial diversity. Gut *63*, 1913–1920.

Crabtree, T.D., Jin, L., Raymond, D.P., Pelletier, S.J., Houlgrave, C.W., Gleason, T.G., Pruett, T.L., and Sawyer, R.G. (2001). Pre-exposure of murine macrophages to CpG oligonucleotide results in a biphasic tumor necrosis factor alpha response to subsequent lipopolysaccharide challenge. Infect. Immun. *69*, 2123–2129.

Da Silva, S., Robbe-Masselot, C., Ait-Belgnaoui, A., Mancuso, A., Mercade-Loubiere, M., Salvador-Cartier, C., Gillet, M., Ferrier, L., Loubiere, P., Dague, E., *et al.* (2014). Stress disrupts intestinal mucus barrier in rats via mucin O-glycosylation shift: prevention by a probiotic treatment. Am. J. Physiol. Gastrointest. Liver Physiol. *307*, G420–429.

de Sousa Moraes, L.F., Grzeskowiak, L.M., de Sales Teixeira, T.F., and Gouveia Peluzio Mdo, C. (2014). Intestinal microbiota and probiotics in celiac disease. Clin. Microbiol. Rev. *27*, 482–489.

del Rio, B., Dattwyler, R.J., Aroso, M., Neves, V., Meirelles, L., Seegers, J.F., and Gomes-Solecki, M. (2008). Oral immunization with recombinant *lactobacillus plantarum* induces a protective immune response in mice with Lyme disease. Clin. Vaccine Immunol. *15*, 1429–1435.

Delcour, J., Ferain, T., Deghorain, M., Palumbo, E., and Hols, P. (1999). The biosynthesis and functionality of the cell-wall of lactic acid bacteria. Antonie van Leeuwenhoek *76*, 159–184.

Douillard, F.P., and de Vos, W.M. (2014). Functional genomics of lactic acid bacteria: from food to health. Microb. Cell Fact. *13 Suppl. 1*, S8.

Douillard, F.P., Ribbera, A., Jarvinen, H.M., Kant, R., Pietila, T.E., Randazzo, C., Paulin, L., Laine, P.K., Caggia, C., von Ossowski, I., *et al.* (2013). Comparative genomic and functional analysis of *Lactobacillus* casei and *Lactobacillus rhamnosus* strains marketed as probiotics. Appl. Environ. Microbiol. *79*, 1923–1933.

Duar, R.M., Clark, K., Patil, P.B., Hernández, C., Brüning, S., Burkey, T.E., Madayiputhiya, N., Taylor, S.L., and Walter, J. (2014). Identification and characterization of intestinal lactobacilli strains capable of degrading immunotoxic peptides present in gluten. J. Appl. Microbiol. (in press).

Dziarski, R. (2003). Recognition of bacterial peptidoglycan by the innate immune system. Cellular and molecular life sciences: CMLS *60*, 1793–1804.

Erdman, S.E., and Poutahidis, T. (2014). Probiotic 'glow of health': it's more than skin deep. Benef. Microbes *5*, 109–119.

Erkosar, B., and Leulier, F. (2014). Transient adult microbiota, gut homeostasis and longevity: Novel insights from the Drosophila model. FEBS Lett. *588*, 4250–4257.

Erkosar, B., Storelli, G., Defaye, A., and Leulier, F. (2013). Host-intestinal microbiota mutualism: 'learning on the fly'. Cell Host Microbe *13*, 8–14.

Fanning, S., Hall, L.J., Cronin, M., Zomer, A., MacSharry, J., Goulding, D., Motherway, M.O., Shanahan, F., Nally, K., Dougan, G., *et al.* (2012). Bifidobacterial surface-exopolysaccharide facilitates commensal–host interaction through immune modulation and pathogen protection. Proc. Natl. Acad. Sci. U.S.A. *109*, 2108–2113.

FAO, and WHO (2001). Report of a Joint FAO/WHO Expert Consultation on Evaluation of Health and Nutritional Properties of Probiotics in Food Including Powder Milk with Live Lactic Acid Bacteria.

Foolad, N., and Armstrong, A.W. (2014). Prebiotics and probiotics: the prevention and reduction in severity of atopic dermatitis in children. Benef. Microbes *5*, 151–160.

Foroni, E., Serafini, F., Amidani, D., Turroni, F., He, F., Bottacini, F., O'Connell Motherway, M., Viappiani, A., Zhang, Z., Rivetti, C., *et al.* (2011). Genetic analysis and morphological identification of pilus-like structures in members of the genus *Bifidobacterium*. Microb. Cell Fact. *10 Suppl. 1*, S16.

Fredriksen, L., Mathiesen, G., Moen, A., Bron, P.A., Kleerebezem, M., Eijsink, V.G., and Egge-Jacobsen, W. (2012). The major autolysin Acm2 from *Lactobacillus plantarum* undergoes cytoplasmic O-glycosylation. J. Bacteriol. *194*, 325–333.

Ghadimi, D., Vrese, M., Heller, K.J., and Schrezenmeir, J. (2010). Effect of natural commensal-origin DNA on toll-like receptor 9 (TLR9). signaling cascade, chemokine IL-8 expression, and barrier integritiy of polarized intestinal epithelial cells. Inflamm. Bowel Dis. *16*, 410–427.

Gleeson, M., Bishop, N.C., Oliveira, M., McCauley, T., Tauler, P., and Lawrence, C. (2012). Effects of a *Lactobacillus salivarius* probiotic intervention on infection, cold symptom duration and severity, and mucosal immunity in endurance athletes. Int. J. Sport Nutr. Exerc. Metab. *22*, 235–242.

Gleeson, M., and Williams, C. (2013). Intense exercise training and immune function. Nestle Nutr. Inst Workshop Ser *76*, 39–50.

Grangette, C., Nutten, S., Palumbo, E., Morath, S., Hermann, C., Dewulf, J., Pot, B., Hartung, T., Hols, P., and Mercenier, A. (2005). Enhanced antiinflammatory capacity of a *Lactobacillus plantarum* mutant synthesizing modified teichoic acids. Proc. Natl. Acad. Sci. U.S.A. *102*, 10321–10326.

Greenway, F., Wang, S., and Heiman, M. (2014). A novel cobiotic containing a prebiotic and an antioxidant augments the glucose control and gastrointestinal tolerability of metformin: a case report. Benef. Microbes *5*, 29–32.

Grin, P.M., Kowalewska, P.M., Alhazzan, W., and Fox-Robichaud, A.E. (2013). *Lactobacillus* for preventing recurrent urinary tract infections in women: meta-analysis. Can. J. Urol. *20*, 6607–6614.

Gross, G., van der Meulen, J., Snel, J., van der Meer, R., Kleerebezem, M., Niewold, T.A., Hulst, M.M., and Smits, M.A. (2008). Mannose-specific interaction of *Lactobacillus plantarum* with porcine jejunal epithelium. FEMS immunology and medical microbiology *54*, 215–223.

Gueniche, A., Philippe, D., Bastien, P., Reuteler, G., Blum, S., Castiel-Higounenc, I., Breton, L., and Benyacoub, J. (2014). Randomized double-blind placebo-controlled study of the effect of *Lactobacillus paracasei* NCC 2461 on skin reactivity. Benef. Microbes *5*, 137–145.

Haiser, H.J., Gootenberg, D.B., Chatman, K., Sirasani, G., Balskus, E.P., and Turnbaugh, P.J. (2013). Predicting and manipulating cardiac drug inactivation by the human gut bacterium *Eggerthella lenta*. Science *341*, 295–298.

Haiser, H.J., Seim, K.L., Balskus, E.P., and Turnbaugh, P.J. (2014). Mechanistic insight into digoxin inactivation by Eggerthella lenta augments our understanding of its pharmacokinetics. Gut Microbes *5*, 233–238.

Haywood, B.A., Black, K.E., Baker, D., McGarvey, J., Healey, P., and Brown, R.C. (2014). Probiotic supplementation reduces the duration and incidence of infections but not severity in elite rugby union players. J. Sci. Med. Sport *17*, 356–360.

Hemmi, H., Takeuchi, O., Kawai, T., Kaisho, T., Sato, S., Sanjo, H., Matsumoto, M., Hoshino, K., Wagner, H., Takeda, K., *et al.* (2000). A Toll-like receptor recognizes bacterial DNA. Nature *408*, 740–745.

Hill, C., Guarner, F., Reid, G., Gibson, G.R., Merenstein, D.J., Pot, B., Morelli, L., Canani, R.B., Flint, H.J., Salminen, S., *et al.* (2014). Expert consensus document. The International Scientific Association for Probiotics and Prebiotics consensus statement on the scope and appropriate use of the term probiotic. Nat. Rev. Gastroenterol. Hepatol. *11*, 506–514.

Jang, K.S., Baik, J.E., Han, S.H., Chung, D.K., and Kim, B.G. (2011). Multi-spectrometric analyses of lipoteichoic acids isolated from *Lactobacillus plantarum*. Biochemical and biophysical research communications *407*, 823–830.

Jiang, X., Yu, M., Qiao, X., Liu, M., Tang, L., Jiang, Y., Cui, W., and Li, Y. (2014). Up-regulation of MDP and tuftsin gene expression in Th1 and Th17 cells as an adjuvant for an oral *Lactobacillus casei* vaccine against anti-transmissible gastroenteritis virus. Appl. Microbiol. Biotechnol. *98*, 8301–8312.

Jijon, H., Backer, J., Diaz, H., Yeung, H., Thiel, D., McKaigney, C., De Simone, C., and Madsen, K. (2004). DNA from probiotic bacteria modulates murine and human epithelial and immune function. Gastroenterology *126*, 1358–1373.

Kankainen, M., Paulin, L., Tynkkynen, S., von Ossowski, I., Reunanen, J., Partanen, P., Satokari, R., Vesterlund, S., Hendrickx, A.P., Lebeer, S., *et al.* (2009). Comparative genomic analysis of *Lactobacillus rhamnosus* GG reveals pili containing a human- mucus binding protein. Proc. Natl. Acad. Sci. U.S.A. *106*, 17193–17198.

Karczewski, J., Troost, F.J., Konings, I., Dekker, J., Kleerebezem, M., Brummer, R.J., and Wells, J.M. (2010). Regulation of human epithelial tight junction proteins by *Lactobacillus plantarum in vivo* and protective effects on the epithelial barrier. Am. J. Physiol. Gastrointest Liver Physiol. *298*, G851–859.

Katakura, K., Lee, J., Rachmilewitz, D., Li, G., Eckmann, L., and Raz, E. (2005). Toll-like receptor 9-induced type I IFN protects mice from experimental colitis. J. Clin. Invest. *115*, 695–702.

Kitazawa, H., Harata, T., Uemura, J., Saito, T., Kaneko, T., and Itoh, T. (1998). Phosphate group requirement for mitogenic activation of lymphocytes by an extracellular phosphopolysaccharide from *Lactobacillus delbrueckii* subsp. *bulgaricus*. Int. J. Food Microbiol. *40*, 169–175.

Kleerebezem, M., Hols, P., Bernard, E., Rolain, T., Zhou, M., Siezen, R.J., and Bron, P.A. (2010). The extracellular biology of the lactobacilli. FEMS microbiology reviews *34*, 199–230.

Konishi, H., Fujiya, M., and Kohgo, Y. (2013). Traffic control of bacteria-derived molecules: a new system of host–bacterial crosstalk. Int. J. Cell Biol. *2013*, 757148.

Konstantinov, S.R., Smidt, H., de Vos, W.M., Bruijns, S.C., Singh, S.K., Valence, F., Molle, D., Lortal, S., Altermann, E., Klaenhammer, T.R., *et al.* (2008). S layer protein A of *Lactobacillus acidophilus* NCFM regulates immature dendritic cell and T-cell functions. Proc. Natl. Acad. Sci. U.S.A. *105*, 19474–19479.

Kwon, H.K., Kim, G.C., Kim, Y., Hwang, W., Jash, A., Sahoo, A., Kim, J.E., Nam, J.H., and Im, S.H. (2013). Amelioration of experimental autoimmune encephalomyelitis by probiotic mixture is mediated by a shift in T helper cell immune response. Clin. Immunol. *146*, 217–227.

Kwon, H.K., Lee, C.G., So, J.S., Chae, C.S., Hwang, J.S., Sahoo, A., Nam, J.H., Rhee, J.H., Hwang, K.C., and Im, S.H. (2010). Generation of regulatory dendritic cells and CD4+Foxp3+ T-cells by probiotics administration suppresses immune disorders. Proc. Natl. Acad. Sci. U.S.A. *107*, 2159–2164.

Lebeer, S., Claes, I., Tytgat, H.L., Verhoeven, T.L., Marien, E., von Ossowski, I., Reunanen, J., Palva, A., Vos, W.M., Keersmaecker, S.C., *et al.* (2012a). Functional analysis of *Lactobacillus rhamnosus* GG pili in relation to adhesion and immunomodulatory interactions with intestinal epithelial cells. Appl. Environ. Microbiol. *78*, 185–193.

Lebeer, S., Claes, I.J., Balog, C.I., Schoofs, G., Verhoeven, T.L., Nys, K., von Ossowski, I., de Vos, W.M., Tytgat, H.L., Agostinis, P., *et al.* (2012b). The major secreted protein Msp1/p75 is O-glycosylated in *Lactobacillus rhamnosus* GG. Microb. Cell Fact. *11*, 15.

Lebeer, S., Claes, I.J., and Vanderleyden, J. (2012c). Anti-inflammatory potential of probiotics: lipoteichoic acid makes a difference. Trends Microbiol. *20*, 5–10.

Lebeer, S., Claes, I.J., Verhoeven, T.L., Vanderleyden, J., and De Keersmaecker, S.C. (2011). Exopolysaccharides of *Lactobacillus rhamnosus* GG form a protective shield against innate immune factors in the intestine. Microb. Biotechnol. 4, 368–374.

Lebeer, S., Verhoeven, T.L., Francius, G., Schoofs, G., Lambrichts, I., Dufrene, Y., Vanderleyden, J., and De Keersmaecker, S.C. (2009). Identification of a gene cluster for the biosynthesis of a long, galactose-rich exopolysaccharide in *Lactobacillus rhamnosus* GG and functional analysis of the priming glycosyltransferase. Appl. Environ. Microbiol. *75*, 3554–3563.

Lee, I.C., van Swam, I.I., Tomita, S., Morsomme, P., Rolain, T., Hols, P., Kleerebezem, M., and Bron, P.A. (2014). GtfA and GtfB are both required for protein O-glycosylation in *Lactobacillus plantarum*. J. Bacterol. *196*, 1671–1682.

Lee, J., Mo, J.H., Katakura, K., Alkalay, I., Rucker, A.N., Liu, Y.T., Lee, H.K., Shen, C., Cojocaru, G., Shenouda, S., *et al.* (2006). Maintenance of colonic homeostasis by distinctive apical TLR9 signalling in intestinal epithelial cells. Nat. Cell Biol. *8*, 1327–1336.

Logardt, I.M., and Neujahr, H.Y. (1975). Lysis of modified walls from *Lactobacillus fermentum*. J. Bacteriol. *124*, 73–77.

Matsumoto, M., and Kurihara, S. (2011). Probiotics-induced increase of large intestinal luminal polyamine concentration may promote longevity. Med. Hypotheses 77, 469–472.

Matsumoto, M., Kurihara, S., Kibe, R., Ashida, H., and Benno, Y. (2011). Longevity in mice is promoted by probiotic-induced suppression of colonic senescence dependent on upregulation of gut bacterial polyamine production. PLoS One 6, e23652.

Mattarelli, P., Holzapfel, W., Franz, C.M., Endo, A., Felis, G.E., Hammes, W., Pot, B., Dicks, L., and Dellaglio, F. (2014). Recommended minimal standards for description of new taxa of the genera Bifidobacterium, *Lactobacillus* and related genera. Int. J. Syst Evol. Microbiol. *64*, 1434–1451.

Meijerink, M., Mercenier, A., and Wells, J.M. (2013). Challenges in translational research on probiotic lactobacilli: from *in vitro* assays to clinical trials. Benef. Microbes 4, 83–100.

Meijerink, M., van Hemert, S., Taverne, N., Wels, M., de Vos, P., Bron, P.A., Savelkoul, H.F., van Bilsen, J., Kleerebezem, M., and Wells, J.M. (2010). Identification of genetic loci in *Lactobacillus plantarum* that modulate the immune response of dendritic cells using comparative genome hybridization. PLoS One 5, e10632.

Million, M., and Raoult, D. (2013). Species and strain specificity of *Lactobacillus* probiotics effect on weight regulation. Microb. Pathog. *55*, 52–54.

Mohamadzadeh, M., Pfeiler, E.A., Brown, J.B., Zadeh, M., Gramarossa, M., Managlia, E., Bere, P., Sarraj, B., Khan, M.W., Pakanati, K.C., *et al.* (2011). Regulation of induced colonic inflammation by *Lactobacillus acidophilus* deficient in lipoteichoic acid. Proc. Natl. Acad. Sci. U.S.A. *108* (Suppl. 1), 4623–4630.

Nagaoka, M., Hashimoto, S., Watanabe, T., Yokokura, T., and Mori, Y. (1994). Anti-ulcer effects of lactic acid bacteria and their cell wall polysaccharides. Biol. Pharm. Bull. *17*, 1012–1017.

Nothaft, H., and Szymanski, C.M. (2010). Protein glycosylation in bacteria: sweeter than ever. Nat. Rev. Microbiol. *8*, 765–778.

Nowak, B., Ciszek-Lenda, M., Srottek, M., Gamian, A., Kontny, E., Gorska-Fraczek, S., and Marcinkiewicz, J. (2012). *Lactobacillus rhamnosus* exopolysaccharide ameliorates arthritis induced by the systemic injection of collagen and lipopolysaccharide in DBA/1 mice. Arch. Immunol. Ther. Exp. (Warsz) *60*, 211–220.

O'Connell Motherway, M., Zomer, A., Leahy, S.C., Reunanen, J., Bottacini, F., Claesson, M.J., O'Brien, F., Flynn, K., Casey, P.G., Munoz, J.A., *et al.* (2011). Functional genome analysis of *Bifidobacterium breve* UCC2003 reveals type IVb tight adherence (Tad) pili as an essential and conserved host-colonization factor. Proc. Natl. Acad. Sci. U.S.A. *108*, 11217–11222.

Obermeier, F., Dunger, N., Deml, L., Herfarth, H., Scholmerich, J., and Falk, W. (2002). CpG motifs of bacterial DNA exacerbate colitis of dextran sulfate sodium-treated mice. Eur. J. Immunol. *32*, 2084–2092.

Oliveira, M.L., Areas, A.P., Campos, I.B., Monedero, V., Perez-Martinez, G., Miyaji, E.N., Leite, L.C., Aires, K.A., and Lee Ho, P. (2006). Induction of systemic and mucosal immune response and decrease in Streptococcus pneumoniae colonization by nasal inoculation of mice with recombinant lactic acid bacteria expressing pneumococcal surface antigen A. Microbes Infect. *8*, 1016–1024.

Ouwehand, A.C., Kirjavainen, V., Shortt, C., and Salminen, S. (1999). Probiotics: mechanisms and established effects. Int. Dairy J. *9*, 43–52.

Polak-Berecka, M., Wasko, A., Szwajgier, D., and Chomaz, A. (2013). Bifidogenic and antioxidant activity of exopolysaccharides produced by *Lactobacillus rhamnosus* E/N cultivated on different carbon sources. Pol. J. Microbiol. *62*, 181–188.

Pot, B., Felis, G.E., De Bruyne, K., Tsakalidou, E., Papadimitriou, K., Leisner, J., and Vandamme, P. (2014). The genus *Lactobacillus*. In Lactic Acid Bacteria: Biodiversity and Taxonomy, W.H. Holzapfel, and B.J.B. Wood, eds. (Wiley), pp. 249–353.

Prescott, L.M., Harley, J.P., and Kelin, D.A. (2002). Microbiology, Bacteria: The Low G+ C Gram positives. (5th edn). Boston: McGraw Hill, 529–530.

Pretzer, G., Snel, J., Molenaar, D., Wiersma, A., Bron, P.A., Lambert, J., de Vos, W.M., van der Meer, R., Smits, M.A., and Kleerebezem, M. (2005). Biodiversity-based identification and functional characterization of the mannose-specific adhesin of *Lactobacillus plantarum*. J. Bacteriol. *187*, 6128–6136.

Ra, J., Lee, D.E., Kim, S.H., Jeong, J.W., Ku, H.K., Kim, T.Y., Choi, I.D., Jeung, W., Sim, J.H., and Ahn, Y.T. (2014). Effect of oral administration of *Lactobacillus plantarum* HY7714 on epidermal hydration in ultraviolet B-Irradiated hairless mice. J. Microbiol. Biotechnol. *24*, 1736–1743.

Rabinovich, G.A., van Kooyk, Y., and Cobb, B.A. (2012). Glycobiology of immune responses. Ann. N.Y. Acad. Sci. *1253*, 1–15.

Remus, D.M., Kleerebezem, M., and Bron, P.A. (2011). An intimate tete-a-tete – how probiotic lactobacilli communicate with the host. Eur. J. Pharmacol. *668* (Suppl. 1), S33–42.

Ruas-Madiedo, P., Abraham, A., Mozzi, F., and de los Reyes-Gavilán, C.G. (2008). Functionality of exopolysaccharides produced by lactic acid bacteria. In Molecular Aspects of Lactic Acid Bacteria for Traditional and New Applications, B. Mayo, P. López, and G. Pérez-Martínez, eds. (Research Signpost, Kerala India), pp. 137–166.

Ruas-Madiedo, P., Medrano, M., Salazar, N., De Los Reyes-Gavilan, C.G., Perez, P.F., and Abraham, A.G. (2010). Exopolysaccharides produced by *Lactobacillus* and *Bifidobacterium* strains abrogate *in vitro* the cytotoxic effect of bacterial toxins on eukaryotic cells. J. Appl. Microbiol. *109*, 2079–2086.

Salazar, N., Ruas-Madiedo, P., Kolida, S., Collins, M., Rastall, R., Gibson, G., and de Los Reyes-Gavilan, C.G. (2009). Exopolysaccharides produced by *Bifidobacterium longum* IPLA E44 and *Bifidobacterium animalis* subsp. *lactis* IPLA R1 modify the composition and metabolic activity of human faecal microbiota in pH-controlled batch cultures. Int. J. Food Microbiol. *135*, 260–267.

Sanders, M.E., Klaenhammer, T.R., Ouwehand, A.C., Pot, B., Johansen, E., Heimbach, J.T., Marco, M.L., Tennila, J., Ross, R.P., Franz, C., *et al.* (2014). Effects of genetic, processing, or product formulation changes on efficacy and safety of probiotics. Ann. N. Y. Acad. Sci. *1309*, 1–18.

Sarno, M., Lania, G., Cuomo, M., Nigro, F., Passannanti, F., Budelli, A., Fasano, F., Troncone, R., Auricchio, S., Barone, M.V., *et al.* (2014). *Lactobacillus paracasei* CBA L74 interferes with gliadin peptides entrance in Caco-2 cells. Int. J. Food Sci. Nutr., 1–7.

Segers, M.E., and Lebeer, S. (2014). Towards a better understanding of *Lactobacillus rhamnosus* GG–host interactions. Microb. Cell Fact. *13 Suppl. 1*, S7.

Settem, R.P., Honma, K., Nakajima, T., Phansopa, C., Roy, S., Stafford, G.P., and Sharma, A. (2013). A bacterial glycan core linked to surface (S)-layer proteins modulates host immunity through Th17 suppression. Mucosal Immunol. *6*, 415–426.

Sherman, P.M., Ossa, J.C., and Johnson-Henry, K. (2009). Unraveling mechanisms of action of probiotics. Nutr. Clin. Pract *24*, 10–14.

Siezen, R.J., Tzeneva, V.A., Castioni, A., Wels, M., Phan, H.T., Rademaker, J.L., Starrenburg, M.J., Kleerebezem, M., Molenaar, D., and van Hylckama Vlieg, J.E. (2010). Phenotypic and genomic diversity of *Lactobacillus plantarum* strains isolated from various environmental niches. Environ. Microbiol. *12*, 758–773.

Smits, H.H., Engering, A., van der Kleij, D., de Jong, E.C., Schipper, K., van Capel, T.M., Zaat, B.A., Yazdanbakhsh, M., Wierenga, E.A., van Kooyk, Y., *et al.* (2005). Selective probiotic bacteria induce IL-10-producing regulatory T-cells *in vitro* by modulating dendritic cell function through dendritic cell-specific intercellular adhesion molecule 3-grabbing nonintegrin. J. Allergy Clin. Immunol. *115*, 1260–1267.

So, J.S., Lee, C.G., Kwon, H.K., Yi, H.J., Chae, C.S., Park, J.A., Hwang, K.C., and Im, S.H. (2008). *Lactobacillus* casei potentiates induction of oral tolerance in experimental arthritis. Mol. Immunol. *46*, 172–180.

So, J.S., Song, M.K., Kwon, H.K., Lee, C.G., Chae, C.S., Sahoo, A., Jash, A., Lee, S.H., Park, Z.Y., and Im, S.H. (2011). *Lactobacillus* casei enhances type II collagen/glucosamine-mediated suppression of inflammatory responses in experimental osteoarthritis. Life Sci. *88*, 358–366.

Spiller, R.C., Jenkins, D., Thornley, J.P., Hebden, J.M., Wright, T., Skinner, M., and Neal, K.R. (2000). Increased rectal mucosal enteroendocrine cells, T lymphocytes, and increased gut permeability following acute *Campylobacter* enteritis and in post-dysenteric irritable bowel syndrome. Gut *47*, 804–811.

Storelli, G., Defaye, A., Erkosar, B., Hols, P., Royet, J., and Leulier, F. (2011). *Lactobacillus plantarum* promotes Drosophila systemic growth by modulating hormonal signals through TOR-dependent nutrient sensing. Cell Metab. *14*, 403–414.

Suzuki, S., Kimoto-Nira, H., Suganuma, H., Suzuki, C., Saito, T., and Yajima, N. (2014). Cellular fatty acid composition and exopolysaccharide contribute to bile tolerance in *Lactobacillus brevis* strains isolated from fermented Japanese pickles. Can. J. Microbiol. *60*, 183–191.

Sybesma, W., Molenaar, D., van, I.W., Venema, K., and Kort, R. (2013). Genome instability in *Lactobacillus rhamnosus* GG. Appl. Environ. Microbiol. *79*, 2233–2239.

Tomita, S., de Waard, P., Bakx, E.J., Schols, H.A., Kleerebezem, M., and Bron, P.A. (2013). The structure of an alternative wall teichoic acid produced by a *Lactobacillus plantarum* WCFS1 mutant contains a 1,5-linked poly(ribitol phosphate) backbone with 2-alpha-d-glucosyl substitutions. Carbohydrate Res. *370*, 67–71.

Tomita, S., Furihata, K., Tanaka, N., Satoh, E., Nukada, T., and Okada, S. (2012). Determination of strain-specific wall teichoic acid structures in *Lactobacillus plantarum* reveals diverse alpha-d-glucosyl substitutions and high structural uniformity of the repeating units. Microbiology *158*, 2712–2723.

Tripathi, P., Beaussart, A., Alsteens, D., Dupres, V., Claes, I., von Ossowski, I., de Vos, W.M., Palva, A., Lebeer, S., Vanderleyden, J., *et al.* (2013). Adhesion and nanomechanics of pili from the probiotic *Lactobacillus rhamnosus* GG. ACS Nano *7*, 3685–3697.

Trombert, A. (2014). Recombinant lactic acid bacteria as delivery vectors of heterologous antigens: the future of vaccination? Benef. Microbes, 1–12.

Tsilingiri, K., Barbosa, T., Penna, G., Caprioli, F., Sonzogni, A., Viale, G., and Rescigno, M. (2012). Probiotic and postbiotic activity in health and disease: comparison on a novel polarized ex-vivo organ culture model. Gut *61*, 1007–1015.

Tsilingiri, K., and Rescigno, M. (2013). Postbiotics: what else? Benef. Microbes *4*, 101–107.

van Bokhorst-van de Veen, H., Lee, I.C., Marco, M.L., Wels, M., Bron, P.A., and Kleerebezem, M. (2012). Modulation of *Lactobacillus plantarum* gastrointestinal robustness by fermentation conditions enables identification of bacterial robustness markers. PLoS One *7*, e39053.

van Hemert, S., Meijerink, M., Molenaar, D., Bron, P.A., de Vos, P., Kleerebezem, M., Wells, J.M., and Marco, M.L. (2010). Identification of *Lactobacillus plantarum* genes modulating the cytokine response of human peripheral blood mononuclear cells. BMC Microbiol. *10*, 293.

Vandamme, P., De Bruyne, K., and Pot, B. (2014). Phylogenetics and systematics. In Lactic Acid Bacteria: Biodiversity and Taxonomy, W.H. Holzapfel, and B.J.B. Wood, eds. (Wiley), pp. 31–44.

Vollmer, W. (2008). Structural variation in the glycan strands of bacterial peptidoglycan. FEMS microbiol. Rev. *32*, 287–306.

Vollmer, W., Blanot, D., and de Pedro, M.A. (2008). Peptidoglycan structure and architecture. FEMS microbiology reviews *32*, 149–167.

Vollmer, W., and Tomasz, A. (2000). The pgdA gene encodes for a peptidoglycan N-acetylglucosamine deacetylase in Streptococcus pneumoniae. J. Biol. Chem. *275*, 20496–20501.

Wassenaar, T.M., and Lukjancenko, O. (2014). Comparative genomics of *Lactobacillus* and other LAB. In Lactic Acid Bacteria: Biodiversity and Taxonomy, W.H. Holzapfel, and B.J.B. Wood, eds. (Wiley), pp. 55–69.

Wells, J.M., and Mercenier, A. (2008). Mucosal delivery of therapeutic and prophylactic molecules using lactic acid bacteria. Nat. Rev. Microbiol. *6*, 349–362.

Wells, J.M., Rossi, O., Meijerink, M., and van Baarlen, P. (2011). Epithelial crosstalk at the microbiota–mucosal interface. Proc. Natl. Acad. Sci. U.S.A. *108* (Suppl. 1), 4607–4614.

Welman, A.D., and Maddox, I.S. (2003). Exopolysaccharides from lactic acid bacteria: perspectives and challenges. Trends Biotechnol. *21*, 269–274.

West, N.P., Pyne, D.B., Cripps, A.W., Hopkins, W.G., Eskesen, D.C., Jairath, A., Christophersen, C.T., Conlon, M.A., and Fricker, P.A. (2011). *Lactobacillus fermentum* (PCC(R)). supplementation and gastrointestinal and respiratory-tract illness symptoms: a randomized control trial in athletes. Nutr. J. *10*, 30.

Xia, G., Kohler, T., and Peschel, A. (2010). The wall teichoic acid and lipoteichoic acid polymers of *Staphylococcus aureus*. Int. J. Med. Microbiol. *300*, 148–154.

Xu, X., Hicks, C., Li, Y., Su, J., Shiloach, J., Kaufman, J.B., Fitz, Y., Eichacker, P.Q., and Cui, X. (2014). Purified cell wall from the probiotic bacterium *Lactobacillus* gasseri activates systemic inflammation and, at higher doses, produces lethality in a rat model. Crit. Care *18*, R140.

Yan, F., Cao, H., Cover, T.L., Whitehead, R., Washington, M.K., and Polk, D.B. (2007). Soluble proteins produced by probiotic bacteria regulate intestinal epithelial cell survival and growth. Gastroenterology *132*, 562–575.

Zhou, Q., Zhang, B., and Verne, G.N. (2009). Intestinal membrane permeability and hypersensitivity in the irritable bowel syndrome. Pain *146*, 41–46.

Bifidobacteria – Regulators of Intestinal Homeostasis

4

Thomas D. Leser, Caroline T. Gottlieb and Eric Johansen

Abstract

Bifidobacteria are natural inhabitants of the gastrointestinal tract possessing genetic adaptations that enable colonization of this harsh and complex habitat. Owing to their recognized benefits to human health bifidobacteria are used as probiotics; however industrial-scale production of bifidobacteria is a challenge. Bifidobacteria interact with key elements of intestinal functioning and contribute to maintaining homeostasis. Recent scientific progress has demonstrated that bifidobacteria, through strain-dependent interactions with the host may reduce mucosal antigen load, improve the intestinal barrier, and induce regulation of local and systemic immune responses. Continued research on *Bifidobacterium*–host interactions is expected to bring knowledge on the mechanisms involved in these health effects, and to support the identification of even more efficacious strains that will increase the variety of commercially available products.

Introduction

Historically, bifidobacteria have been associated with human health since their discovery and they are often included in commercial probiotic products. Bifidobacteria were first isolated by Henry Tissier, a French paediatrician, at the beginning of the 20th century, at the time when Metchnikoff proposed that it would be possible to replace harmful gut microbes with useful microbes to improve human health and longevity. Tissier observed that healthy children had an abundant type of bacteria in their stool, while children suffering from diarrhoea had low numbers of these characteristically Y-shaped bacteria. Tissier isolated these Gram-positive, non-spore-forming, pleomorphic rods, which he named *Bacillus bifidus communis*, and recommended their administration to treat diarrhoea in infants. Later multiple biotypes of bifidobacteria were isolated and their taxonomy resolved, and in the 1960s *Bifidobacterium* was accepted as an independent genus that today encompasses more than 30 species.

Most *Bifidobacterium* species have been isolated from the gastrointestinal (GI) tract of humans and animals, illustrating that bifidobacteria are evolutionarily adapted to this environment. Bifidobacteria are saccharolytic and can ferment various complex carbon sources such as inulin, oligofructose, arabinoxylan-oligosaccharides, pectin, milk oligosaccharides, and gastric mucin, but the ability to metabolize carbohydrates is species and strain specific. Thus bifidobacteria can utilize plant-derived oligo- and polysaccharides that are not degraded by the host's digestive enzymes in the upper part of the GI tract. Bifidobacteria are furnished with a great proportion and variety of carbohydrate-modifying enzymes as well as ABC transporters to facilitate uptake and metabolism of carbohydrates. Polymeric carbohydrates are degraded to oligosaccharides which are subsequently degraded to monosaccharides. In bifidobacteria glucose and fructose are degraded through the fructose-6-phosphate phosphoketolase pathway to acetate and lactate, in a theoretical ratio of 3:2. The actual ratio, however, depends on the available energy source and the consumption rate. Cross-feeding between bifidobacteria and butyrate-producing bacteria has been observed where bifidobacteria provide oligosaccharides and monosaccharides, as well as acetate and lactate that serves as substrate for butyrate-producing secondary degraders (De Vuyst *et al.*, 2013). By making substrates available to other physiological groups of colonic bacteria, bifidobacteria influence structure and function of the microbiota.

Bifidobacteria are Gram positive with a thick peptidoglycan cell wall containing polysaccharides, proteins and teichoic acids. Optimum temperature for growth is 37–41°C. Bifidobacteria are obligate anaerobes, but sensitivity to oxygen, as well as acid and bile tolerance varies between species and between strains of the same species (Andriantsoanirina *et al.*, 2013). Strains of *B. longum* and *B. breve* isolated from humans were generally most tolerant to these stresses, while strains of *B. adolescentis* were very sensitive. The low oxygen tolerance of bifidobacteria hampers their use as probiotics because stabilization of *Bifidobacterium* products to meet requirements for shelf-life is difficult compared with, for example, lactobacilli.

About a dozen *Bifidobacterium* strains with clinically documented effects are commercially available. Half of these are *Bifidobacterium animalis* subsp. *lactis* strains and the remaining are *Bifidobacterium longum* subsp. *longum* or *infantis* and *Bifidobacterium breve* strains. *B. animalis* subsp. *lactis* strains are mostly tolerant to oxygen, a fact that probably explains why *B. animalis* subsp. *lactis* is the most widely used *Bifidobacterium* species in probiotic products. *B. animalis* subsp. *lactis* is also relatively tolerant to gastric and intestinal conditions resulting in better survival of *B. animalis* subsp. *lactis* strains after exposure to gastric and small intestinal solutions compared to other bifidobacteria (Masco *et al.*, 2007). Evidently this species has protective mechanisms that maintain viability when cells are exposed to various types of stress.

Bifidobacteria are often mixed with lactobacilli in commercial probiotic products. The rationale behind multistrain products is that combining the beneficial effects from several strains increases the overall effect of the product. This can certainly be true if strains are selected based on a scientific understanding of their individual effects and interactions with humans. Health benefits of bifidobacteria and other types of probiotic microorganisms are strongly strain-dependent and generic functions of particular *Bifidobacterium* species rarely exists. The practical implication is that documentation of clinical effects has to be at the strain level and those beneficial effects of a strain belonging to a certain species do not imply that another strain of that species will have similar effects.

In recent years *in vitro*, animal, and clinical studies of *Bifidobacterium*–human interactions have provided a much better understanding of how these bacteria provide health benefits and has led to targeted screenings of strain collections for new strains with probiotic potential. Although it is not certain that a strain selected for a specific function in *in vitro* settings will actually have the predicted effect in animal models or in humans, this is the only approach available. Improved understanding of host–microbe interactions and the experience gained from various model uses will eventually help to improve the models and develop new models with higher predictive value. An excellent example of this route for successfully selecting a new probiotic strain is *B. longum* subsp. *infantis* 35624. This strain was isolated from a healthy human and selected for its capacity to adhere to intestinal epithelial cells (IECs) without inducing a cellular response. In dendritic cells (DCs) the strain elicits a regulatory response that promotes generation of regulatory T-cells. In subsequent animal models *B. longum* subsp. *infantis* 35624 was shown to modulate inflammatory activity locally and systemically. Finally, in humans the strain was shown to reduce symptom severity in patients with irritable bowel syndrome owing to its immunoregulatory function (Konieczna *et al.*, 2012a).

Maintaining intestinal homeostasis is essential to human health. Thus, gastrointestinal conditions must remain stable and relatively constant in spite of fluctuating inputs from the surrounding environment. The intestinal epithelium must maintain optimal absorption of nutritional compounds, yet at the same time be ready to exclude, neutralize, or detoxify harmful components in a controlled manner. Gut mucosal architecture and function has evolved to create a barrier that strives to retain undesired compounds in the gut lumen as well as defence mechanisms that eliminate antagonists breaching the barrier. Through these mechanisms, humans and animals are able to live in close contact with the immense amount of microorganisms inhabiting the gut and to tolerate exposure to a plethora of ingested harmful compounds. Research has substantiated that bifidobacteria influence regulation of these structures and processes in a strain-dependent manner, and thus, contribute to gut homeostasis. When reviewing literature for this chapter we have with a few exceptions only included studies where the strain(s) investigated was clearly stated due to this strain- dependency. Furthermore, we have excluded studies involving multi-strain products where effects cannot be clearly attributed to the *Bifidobacterium* content.

Genomics of *Bifidobacterium*

Genomics, the determination and analysis of the sequence of all the genetic material in an organism, has become a primary tool for the characterization and comparison of microorganisms including probiotics and industrial starter cultures (Garrigues *et al.*, 2013). The first genome sequence of a *Bifidobacterium* strain was published in 2002 (Schell *et al.*, 2002) and since then genome sequences have been published at an increasing pace. Today, the complete genome sequence of 30 *Bifidobacterium* strains representing 10 species/subspecies are publicly available (Table 4.1). This includes a number of strains used in commercial probiotic

Table 4.1 Published *Bifidobacterium* genome sequences

Species	Number of sequenced genomes	Average genome size (Mb)
B. adolescentis	1	2.09
B. animalis subsp. *animalis*	1	1.93
B. animalis subsp. *lactis*	11	1.94
B. asteroides	1	2.17
B. bifidum	3	2.21
B. breve	2	2.38
B. dentium	1	2.64
B. longum subsp. *infantis*	2	2.61
B. longum subsp. *longum*	7	2.36
B. thermophilus	1	2.29

Data obtained from http://www.ebi.ac.uk/genomes/ January 2014 products. This genus, as represented by the sequenced strains has relatively small genomes (1.93–2.64 Mb in size) encoding between 1600 and 2100 proteins. Plasmids are not commonly encountered in *Bifidobacterium* and the ones that have been detected are small (2.1–10.2 kb) with a limited coding capacity (Lee and O'Sullivan, 2010).

Initially, genome sequence data were used to enable a number of molecular techniques for further strain characterization. These include transcriptomics, proteomics and comparative genome hybridization (Garrigues *et al.*, 2005). Subsequent uses of genomic information provide a better understanding of the properties of *Bifidobacterium* including those which enable survival in the GI tract and a better understanding of the interaction between this genus and the host itself.

Changes in the level of expression of all of the genes in an organism can be conveniently measured by transcriptomics using microarrays or direct sequencing of mRNA. *B. animalis* subsp. *lactis* BB-12, a widely consumed probiotic strain, was grown in the presence of fructo-oligosaccharides or bile and the pattern of gene expression was analysed using microarrays developed based on the complete genome sequence of BB-12 (Garrigues *et al.*, 2005). In the presence of fructo-oligosaccharides, only two genes showed significant changes in expression levels; both were up-regulated. These genes are involved in the transport and degradation of fructo-oligosaccharides. Gene expression in the presence of bile was considerably more complicated with several genes involved in stress tolerance being up-regulated and more than 200 genes showing significant changes in expression. Interestingly, the gene encoding bile salt hydrolase was not significantly changed in expression; it is highly expressed, even in the absence of bile.

One of the challenges in proteomics research is identification of the specific spots on a protein gel. This can be done by extraction of the protein followed by peptide mass fingerprinting using matrix-assisted laser desorption ionization–time-of-flight mass spectrometry (MALDI-TOF MS) analysis. Identification of the specific protein is facilitated by possession of the complete genome sequence as this allows a prediction of all the peptides potentially encoded in the genome as well as the proteins they could be derived from. Sánchez *et al.* (2007) used a database of hypothetical peptides derived from the complete genome sequence of BB-12 to identify specific proteins in wild-type and mutant *B. animalis* subsp. *lactis* strains grown in the presence of bile. This work also showed that gene expression in the presence of bile is very complicated and that bile causes a considerable change in the overall metabolism of *Bifidobacterium*. The bile salt hydrolase protein was present in cells grown in the presence or absence of bile as predicted from the transcriptome analysis.

Comparative genome hybridization can be used to explore the genetic content of strains closely related to the sequenced strain. Today, with the reduced cost of genome sequencing, it is easier, less expensive and more accurate to simply determine the genome sequence of all strains of interest. For example, Garrigues *et al.* (2005) used pulsed-field gel electrophoresis and comparative genome hybridization to compare a number of strains of *B. animalis* subsp. *lactis*. From these results, it became clear that one strain, ATCC27673 was considerably different from other members of this subspecies. This was subsequently confirmed by determination of the complete genome sequence of ATCC27673 by Loquasto *et al.* (2013). The overall genome architecture was the same as in all members of this subspecies; the main differences were in genomic islands which were either unique to ATCC27673 or absent in ATCC27673 but present in the other members of this subspecies. This suggests the main differences are due to gene acquisition or gene loss. Strains also differ in a number of single nucleotide polymorphisms (SNPs). However, careful analysis is required to determine which SNPs cause changes to genes or gene expression as well as to eliminate those that are actually the result of sequencing errors.

Genome sequence data can also be used as part of the safety assessment of strains to be used as probiotics or intentionally added to the food chain. One of the main safety concerns is the presence of undesirable genetic content such as virulence factors and transmissible antibiotic resistance genes as these may pose a public health risk. Bennedsen *et al.* (2011) created a publicly available database of more than 2500 gene sequences representing virulence factors and antibiotic resistance genes and used these to screen 28 strains of lactic acid bacteria and bifidobacteria for undesirable gene content. Very few genes of concern were detected in these strains. One gene, present in all known members of *B. animalis* subsp. *lactis* is the gene *tet*(W), encoding low-level tetracycline resistance (Gueimonde *et al.*, 2010). Since this subspecies has never been implicated in an infection (Gueimonde *et al.*, 2013), the fact it is somewhat resistant to tetracycline is not of any concern. Many attempts have been made to demonstrate transfer of the *tet*(W) gene from *B. animalis* subsp. *lactis* to other species. To date, all attempts have been negative i.e. this gene has never been shown to be transmissible when present in *B. animalis* subsp. *lactis* (Gueimonde *et al.*, 2013). These facts combined with the ubiquitous presence of *tet*(W) in the gut microbiota led Wassenaar and Alter (2012) among others to conclude that this gene does not increase any risk associated with this subspecies. The *tet*(W) gene is present in the majority of *Bifidobacterium* strains for which complete genome sequences have been published (Table 4.1). *B. longum* strain F8 has a *tet*(W) gene but it is in a genomic context that is completely different from that in *B. animalis* subsp. *lactis* (Kazimierczak *et al.*, 2006). The *tet*(W) gene is located on a structure resembling a transposon in *B. longum* strain F8 and in this strain, *tet*(W) has indeed been shown to be transmissible, albeit at a very low rate (Kazimierczak *et al.*, 2006). Use of this strain as a probiotic would therefore not be advised.

Pangenomics is the study of the entire complement of genes present within a species or genus (Garrigues *et al.*, 2013). Genes can be divided into two groups; the core genome which includes the genes present in all members of a species or genus and the variable genome which includes the genes present in some but not all members. The pan-genome of *Bifidobacterium* was estimated to consist of more than 5000 genes while the core genome contains less than 1000 genes based on the analysis of 9 complete genome sequences (Bottacini *et al.*, 2010). As the number of genomes sequences analysed increases, the number of genes in the pangenome will increase while the number of genes in the core genome will decrease. Lukjancenko *et al.* (2012) analysed 19 complete genome sequences and found the pangenome to consist of nearly 7000 genes and the core genome to comprise 724 genes. Genes in the core genome include a large number of genes involved in 'housekeeping' functions such as translation, transcription and DNA replication but also include genes involved in the transport and metabolism of specific carbohydrates and signal transduction reflecting the overall adaptation of the genus to a certain environments. Specific adaptations to particular environments are reflected in the variable genome. As mentioned previously, a typical *Bifidobacterium* strain contains around 2000 genes; the fact the *Bifidobacterium* pan-genome has nearly 7000 genes indicates that this is a diverse genus with strains adapted to a variety of specific environments.

In order to exert a probiotic effect, it is expected that a probiotic strain should be metabolically active in the GI tract. This requires the ability to metabolize the carbon sources that might be present. The lower GI tract of humans contains a large variety of carbohydrates that resist digestion in the upper GI tract. These include dietary components such as resistant starch, human milk oligosaccharides and cellulose, host-derived components like mucin as well as carbohydrates produced by other members of the GI tract microbial community (Pokusaeva *et al.*, 2011). The genomes of *Bifidobacterium* contain a high proportion of genes involved in metabolism of diverse carbon sources. Pokusaeva *et al.*, estimate than nearly 9% of the genes in *Bifidobacterium* have this role, compared to less than 5% in bacteria in general. Roughly half these genes are involved in the uptake of the carbohydrate into the cell via specific ABC transporters, permeases, proton symporters and occasionally phosphotransferase systems and half are involved in specific degradation of the carbohydrate into compounds that can readily enter the central metabolism of the cell. Activities involved in carbohydrate degradation include hydrolysis, phosphorylation, deacetylation, and trans-glycosylation.

Of course, the metabolic capabilities of a cell in the lower GI tract are only relevant if the organism is able to arrive at that site in a viable form. *Bifidobacterium* is particularly adept at surviving the harsh conditions present in the upper GI tract; specifically the acidic conditions of the stomach and the antibacterial effects of bile in the small intestine. Jin *et al.* (2012) used transcriptomic analysis by RNA sequencing to study the development of acid tolerance in *B. longum* subsp. *longum*. A number of important mechanisms were identified. These include: strengthening the integrity of the cell wall, reducing the permeability of the membrane to protons, increasing the export of protons by the F_1F_0 -ATPase, neutralization of acid by increased production of ammonia and minimizing and repairing damage to internal cellular components caused by acidic conditions. As mentioned previously, bile salt hydrolase, a key enzyme in detoxifying bile is constitutively expressed to a high level in *B. animalis* subsp. *lactis* so cells are able to detoxify bile as soon as it is encountered. In addition, a number of other mechanisms are activated when *Bifidobacterium* is exposed to bile. For example, Sánchez *et al.* (2006) reported that the F_1F_0-ATPase has a role in bile tolerance. Thus, exposure of cells to the acidic conditions of the stomach actually prepares them for their upcoming encounter with bile in the small intestine, increasing their chance of survival.

Persistence in the lower GI tract should further enhance the ability of a probiotic strain to exert its beneficial effect. Attachment to host cells will reduce the risk of rapid elimination. One common mechanism of attachment involves the formation of pili which specifically bind to structures on the host cell surface. Recently, Foroni *et al.* (2012) identified structures on the surface of several different *Bifidobacterium* isolates which resemble pili. Analysis of the genome sequences showed that all had genes expected to be involved in pilus production and that these genes indeed were expressed under relevant growth conditions. Gilad *et al.* (2011) identified a number of extracellular proteins produced by *B. animalis* subsp. *lactis* BB-12 potentially involved in adhesion to host cells. These included proteins predicted to be involved in pilus formation confirming that these proteins indeed are produced and exported from the cell.

Our knowledge of the genus *Bifidobacterium* has been greatly enhanced by the availability of complete genome sequences. Important contributions have been made in understanding the basic biology of these bacteria as well and giving knowledge on the mechanisms which allow survival in the GI tract facilitating the interaction between *Bifidobacterium* and the host and the host's microbiota thereby conferring the health benefits described elsewhere in this chapter.

Ecology of bifidobacteria

Bifidobacteria are natural inhabitants of the GI tract. Being strictly anaerobic the largest bifidobacterial densities are found in the distal gut where reducing conditions prevail. However, the abundance of bifidobacteria in the human colon varies substantially with extremes from < 1% to up to 80% of the total bacterial microbiota, dependent on age and geographical location of individuals sampled (Lin *et al.*, 2013; Turroni *et al.*, 2012). Analyses of the faecal microbiota using culturing as well as molecular methods has shown that bifidobacteria are abundant and dominant in infants (Turroni *et al.*, 2012; Yatsunenko *et al.*, 2012; Tannock *et al.*, 2013), and remain present in high numbers even after weaning (Fallani *et al.*, 2011). Following weaning the proportion of bifidobacteria in the human large intestine starts to decline and reaches low levels in adults where just a few per cent of the total bacterial population are bifidobacteria (Ringel-Kulka *et al.*, 2013). This decline continues, and in the elderly population bifidobacteria constitute less than 1% of the microbiota (Biagi *et al.*, 2010). Obviously, these numbers are inferred from human faecal samples and may not accurately reflect the abundance and diversity of bifidobacteria in the different parts of the large intestine. Thus, proximal/distal and luminal/mucosal associated *Bifidobacterium* populations are not homogeneous and these populations may not be equally well preserved and represented in faecal samples.

Most bifidobacterial species have been isolated from the GI tract of mammals, insects, or birds (Turroni *et al.*, 2009). More than 30 *Bifidobacterium* species have been identified, but only a few of these species are found in the human GI tract. In human infants *B. longum* subsp. *infantis*, *B. bifidum*, and *B. breve* dominate the *Bifidobacterium* population (Sanchez *et al.*, 2013; Turroni *et al.*, 2012). The majority of bifidobacteria in adults are affiliated with the *B. longum* group (*B. longum* subsp. *longum*, *B. breve*), the *B. adolescentis* group (*B. adolescentis*, *B. catenulatum*, *B. pseudocatenulatum*), and *B. pseudolongum* (Turroni *et al.*, 2009). In addition *B. bifidum* was abundant in an elderly population (Ouwehand *et al.*, 2008). Other bifidobacterial species frequently found in human faecal samples include *B. animalis* subsp. *lactis* and *B. dentium* (Turroni *et al.*, 2009).

Bifidobacteria are adapted to cope with the challenges imposed by the gastrointestinal environment. Before reaching the large intestine, bacteria entering via the oral route are exposed to several stresses in the stomach and small intestine of which the low gastric pH and intestinal bile are the most important. Digestive enzymes also affect bifidobacterial survival although only to a minor extent (Masco *et al.*, 2007). The host innate immune system may also limit survival of bifidobacteria by secretion of immunoglobulins and antimicrobial compounds that are important elements of the mucosal first-line of defence. Antimicrobial peptides and secreted immunoglobulins accumulate in the mucus covering the epithelium (Vaishnava *et al.*, 2011) and limit direct contact between bacteria approaching the epithelium and the epithelial cells, while their effects on bacteria inhabiting the intestinal lumen is limited. Once in the large intestine, bifidobacteria have to compete for nutritional resources with the massive bacterial community dominated by *Firmicutes* and *Bacteroidetes* residing in this part of the intestines ($>10^{11}$ bacteria per gram of intestinal content). Bifidobacteria have evolved mechanisms enabling them to survive these hostile conditions including mechanisms of acid and bile tolerance as well as a variety of glycosidases that hydrolyse indigestible oligosaccharides and glycoconjugates.

Bifidobacteria are early colonizers of the newborn infant GI tract and dominate the gut microbiota just 3 days after birth and continue to dominate till after weaning (Favier *et al.*, 2002). Differences in the bifidobacterial community have been found between breast-fed and formula-fed babies with higher numbers and a greater diversity of *Bifidobacterium* species in breast-fed than formula-fed infants. *B. breve* is particularly common in breast-fed infants (Boesten *et al.*, 2011; Roger *et al.*, 2010).

Breast-feeding favours bifidobacteria in the newborn possibly due to bifidobacterial growth factors present in breast milk (Coppa *et al.*, 2004). Human milk oligosaccharides (HMO) are a nutritional source, at least to some *Bifidobacterium* species promoting their growth (Ward *et al.*, 2007). Thus, the genome of *B. longum* subsp. *infantis* possesses genes encoding transporters and enzymes that specifically allow this species to grow on HMO as the sole carbon source (Garrido *et al.*, 2012). Strains of *B. breve* are also able to utilize HMO to various extent and some strains are adapted to consumption of fucosylated and sialylated HMO (Ruiz-Moyano *et al.*, 2013).

In addition to being a nutrient source, HMO may increase bifidobacterial adhesion to IECs by inducing a transcriptional response resulting in a more adhering phenotype thereby promoting gut colonization (Kavanaugh *et al.*, 2013).

Adhesion to the mucosa is considered important for bacteria to colonize the GI tract and adhesion has been defined as a basic probiotic feature, which is often used to screen bacterial collections for potential new probiotic strains. The entire gastrointestinal epithelium is covered by two mucus layers that separate luminal bacteria from the epithelium. The inner mucus layer is devoid of bacteria while the outer layer, facing the gut lumen, is densely populated by bacteria (Atuma *et al.*, 2001; Johansson *et al.*, 2008). Thus, in healthy individuals direct contact between bacteria and host epithelial cells is the exemption rather than the rule. Whether mucosal adhesion is actually necessary for bifidobacteria to persist in the GI tract is uncertain. Intuitively it makes sense that adhering bacteria would be able to remain in the gut for longer time; however, the constant renewal of mucus and epithelial cells makes the mucosa an unsteady foundation to adhere to. While the rate of flow in the small intestine is high and adhesion to the mucosa potentially would be advantageous to bacteria attempting to colonize this part of the GI tract, the large intestine transit time is slow and the advantage of mucosal adhesion is limited. The main ecological adaptation needed to colonize the large intestine is probably not mucosal adhesion, but rather the ability to utilize specific nutritional niches in a densely populated and highly competitive environment.

Several mechanisms have been identified that allow bifidobacteria to adhere to mucus or epithelial cells, *in vitro*. Adhesion occurs through non-specific mechanisms such as electrostatic forces and hydrophobic interactions, as well as through specific binding of adhesins to mucosal receptors. Secreted or surface-attached extracellular proteinaceous compounds are important adhesion factors. BopA is a lipoprotein involved in adhesion of *B. bifidum* and *B. longum* to Caco-2 monolayers (Gleinser *et al.*, 2008, 2012), although these findings have recently been questioned (Kainulainen *et al.*, 2013). Pili-like proteins adhering to carbohydrate moieties in glycoprotein or glycolipid receptors have been discovered in *B. breve* strains (O'Connell Motherway *et al.*, 2011). The pili-like proteins are up-regulated in the GI tract and they are considered an essential host-colonization factor. Exopolysaccharides (EPS) present on the surface of bifidobacteria are also important for persistence of the bacteria in the gut. Thus significantly lower numbers of EPS-negative strains of *B. breve* compared to the wild-type strain were detected in faecal samples of mice after oral administration of the strains (Fanning *et al.*, 2012).

HMO and intestinal glycoconjugates share structural similarity and the transport mechanisms and catalytic enzymes evolved in some bifidobacteria to utilize HMO also support the ability to metabolize host derived mucins. In particular strains of *B. bifidum* and *B. longum* are able to metabolize mucin through extracellular, membrane bound, and intracellular mucinases (Killer and Marounek, 2011). Transcriptional and proteomic regulation of oligosaccharide transporters and glycosylases by the presence of mucins have been described for both species (Garrido *et al.*, 2011; Ruiz *et al.*, 2011). ABC transporters associated with the import of oligosaccharides in *B. longum* subsp. *infantis* also have high affinity to intestinal mucin oligosaccharides and blood group structures and may be involved in bacterial adhesion to IECs and mucus (Garrido *et al.*, 2011). The evolutionary adaptation of bifidobacteria to import and metabolize complex host-derived oligosaccharides, such as HMO and intestinal mucins, highlight that this group of bacteria is well-adapted to the gastrointestinal habitat.

In addition to ensuring better colonization, adhesion factors are also considered important to pathogen displacement. By competing with pathogens for adhesion sites, non-pathogenic bacteria may displace or reduce binding of the pathogens (Wittmann *et al.*, 2013). On the other hand EPS produced by bifidobacteria have been shown to enhance adhesion of pathogens to mucus. This effect may be caused by EPS binding to the pathogen surface and to mucus (Ruas-Madiero *et al.*, 2006).

Health benefits of bifidobacteria

Since their first discovery, bifidobacteria have been considered beneficial to humans and their health promoting effects have been frequently documented. Examples are shown in Table 4.2.

Low abundance and diversity of bifidobacteria in the gut has been associated with intestinal discomfort such as abdominal pain and bloating in otherwise healthy individuals (Jalanka-Tuovinen *et al.*, 2011). In this study, subjects who experienced pain had five-fold less total bifidobacteria compared to those without pain, and most of the *Bifidobacterium* phylotypes identified were negatively correlated with abdominal pain. In infants with colic, reduced numbers of bifidobacteria (and lactobacilli) were found concomitant with significantly increased abundance of *Proteobacteria* in the first weeks of life (de Weerth *et al.*, 2013). Distress, defined as crying and fussing, was lower in infants with abundant *B. longum*, whereas low numbers of *B. breve* was associated with higher levels of distress (Pärtty *et al.*, 2012).

The gut microbiota has recently been identified as a contributing factor to the development of obesity and metabolic disorders. Bifidobacteria may also play a role in metabolic diseases; prebiotic stimulation of bifidobacteria improves high-fat diet-induced glucose intolerance, insulin secretion and low-grade inflammation (Cani *et al.*, 2007). Type 2 diabetic patients had approximately 10-fold lower faecal numbers of bifidobacteria and higher numbers of lactobacilli than healthy controls. Decreased numbers of *B. adolescentis* explained most of the *Bifidobacterium* reduction in diabetic patients (Le *et al.*, 2013).

Gastrointestinal disturbance is common in children with autism spectrum disorders. The gastrointestinal symptoms have been associated with microbial dysbiosis including lower relative abundance of bifidobacteria (Finegold *et al.*, 2010; Wang *et al.*, 2011).

From such findings of lower numbers of bifidobacteria associated with specific disorders it is tempting to speculate that individuals may recover from discomfort or disease by replenishing the *Bifidobacterium* populations. However, reduced bifidobacteria is often associated with other microbiota abnormalities and discomfort may be caused by the general dysbiosis. Further, it is not clear if dysbiosis is the cause of discomfort and disease, or a consequence thereof. Other circumstances may indeed trigger the disease or modify the intestinal environment and cause dysbiosis.

GI disorders

The majority of beneficial effects of bifidobacteria investigated are related to gut health. The interface between bacteria and their host is strongest in the gut and the influence of bacteria on host physiology is likely strongest close to this interface. However, there are examples that bifidobacteria may also induce systemic effects.

Irritable bowel syndrome

Symptoms of irritable bowel syndrome (IBS) include abdominal cramping and pain that is often concurrent with abnormal bowel habits with diarrhoea, constipation, or alternating episodes of both. The aetiology and pathophysiology of IBS are unknown; however, excessive stimulation of the digestive tract by the central nervous system during stress, the disturbance of sensory and cognitive mechanisms by digestive disorders, intestinal infection or inflammation, and changes in the intestinal microbiota have been implicated. IBS is not a life-threatening disease, but quality of life is impaired.

A recent meta-analysis of the effect of probiotic species on IBS concluded that probiotics are an effective therapeutic option for IBS patients, and the effects on each IBS symptom are likely species-specific (Ortiz-Lucas *et al.*, 2013). The analysis included a range of probiotic products and consequently different *Bifidobacterium* species and strains. In addition most products were multistrain products containing up to eight individual strains and it is impossible to ascribe effects to individual strains. Nevertheless, the authors found that all products containing bifidobacteria improved flatulence, *B. breve* and *B. longum* improved pain scores, and *B. breve* improved distension scores. No clear positive effect of probiotics on quality of life was discerned.

Two clinical trials have documented the efficacy of *B. infantis* 35624 to relieve symptoms of IBS. When administered in a dose of 1×10^{10} live bacterial cells in a malted milk drink, *B. infantis* 35624 significantly reduced IBS symptom scores of abdominal pain/discomfort, bloating/distension, and bowel movement difficulty (O'Mahony *et al.*, 2005). When applied as an encapsulated, freeze-dried product (10^8 cfu/dose), this strain also reduced IBS symptom scores at the end of the 4-week study. Interestingly, the efficacy of a higher and a lower dose were not significantly different from placebo (Whorwell *et al.*, 2006). *B. bifidum* MIMBb75 also alleviate IBS symptoms including pain/discomfort, distension/bloating, urgency and digestive disorder. In addition treated individuals reported a significant gain in quality of life scores (Guglielmetti *et al.*, 2011).

Visceral hypersensitivity

Abdominal pain in IBS is caused by visceral hypersensitivity. The reasons for increased visceral hypersensitivity in IBS patients are unknown but the inverse correlation between bifidobacterial abundance and pain suggests that the gut microbiota is involved. Anxiety is thought to play a role in IBS with symptoms often exacerbated during periods of anxiety.

In rats, acute stress induces visceral hypersensitivity and stress models have been used to show that *B. animalis* subsp. *lactis*

Table 4.2 Examples of health beneficial effects of bifidobacteria in humans and animal models

	Bifidobacterium strain	Model	Health effect	Reference
IBS	*B. infantis* 35624	Human	Improve global IBS scores	Whorwell *et al.* (2006), Brenner *et al.* (2009)
	B. bifidum MIMBb75	Human	Reduce IBS symptoms	Guglielmetti *et al.* (2011)
	B. animalis subsp. *lactis* DN-173 010	Human	Accelerate colonic transit and reduce abdominal distension in IBS constipation; improve quality of life in IBS	Agrawal *et al.* (2009), Guyonnet *et al.* (2007)
IBD	Bifidobacteria	Human	Bifidobacteria lower in IBD	Fyderek *et al.* (2009)
	Bifidobacteria	Human	Bifidobacteria lower in active IBD	Sokol *et al.* (2009)
	B. infantis 35624	Human	Reduce plasma CRP and IL-6 in UC	Groeger *et al.* (2013)
CD	*B. infantis* NSL	Human	Alleviate symptoms in CD	Smecuol *et al.* (2013)
	B. longum CECT 7347	Rat	Ameliorate inflammation caused by gliadin	Olivares *et al.* (2012), Laparra *et al.* (2012)
NEC	*B. bifidum* OLB6378	Rat	Reduce NEC incidence	Khailova *et al.* (2009)
	B. infantis BB-02	Mouse	Reduce NEC incidence. Preserve TJs	Bergmann *et al.* (2013)
Peristalsis	*B. animalis* subsp. *lactis* BB-12, *B. pseudocatenulatum* LKM10070	Mouse	Promote peristalsis	Matsumoto *et al.* (2012)
	B. animalis subsp. *lactis* HN019, *B. animalis* subsp. *lactis* DN-173 010	Meta-analysis	Decrease transit time	Miller *et al.* (2013)
	B. animalis subsp. *lactis* DN-173 010	Human	Decrease transit time	Marteau *et al.* (2002), Agrawal *et al.* (2009)
Visceral sensitivity	*B. animalis* subsp. *lactis* DN-173 010	Rat	Inhibit stress-induced visceral hypersensitivity	Agostini *et al.* (2012)
	B. infantis 35624	Rat	Inhibit stress-induced visceral hypersensitivity	McKernan *et al.* (2010)
	B. infantis 35624	Rat	Inhibit TNBS induced visceral hypersensitivity	Johnson *et al.* (2011)
Colic/infant crying	Bifidobacteria	Human	Bifidobacteria lower in colic infants and crying infants	De Weerth *et al.* (2013), Pärtty *et al.* (2012)
Pathogen exclusion	*B. longum* AH1205	E12, mucus	Reduce *C. jejuni* translocation	Alemka *et al.* (2010)
	Collection	Mucus	Inhibit pathogen adhesion	Collado *et al.* (2005)
	Collection	Mucus	Displace pathogens	Collado *et al.* (2006)
	B. bifidum M6	Mucus	Displace pathogens	Gueimonde *et al.* (2007)
	B. longum subsp. *longum* JCM 1217^T	Mouse	Prevent *E. coli* O157:H7 induced mouse death	Fukuda *et al.* (2011)
	B. animalis subsp. *lactis* 420	Mouse	Decrease Enterobacteriaceae translocation	Amar *et al.* (2011a)
	B. infantis 157F *B. longum* NCC2705	Mouse	Reduce Shiga toxin production	Yoshimura *et al.* (2010)
	B. breve BR03, *B. breve* B631	n.a.	Inhibit growth of *E. coli*	Mogna *et al.* (2012)
	B. bifidum CECT 7366	Mouse	Inhibit growth of *H. pylori*	Chenoll *et al.* (2011)
	B. animalis subsp. *lactis* A and B	Mouse	Inhibit growth of *S.* Typhimurium	Martins *et al.* (2010)
	B. longum VK1, VK2	Mouse	Inhibit growth of *S. aureus*	Lazarenko *et al.* (2012)
	B. animalis subsp. *lactis* BB-12	Human	Inhibit salivary mutans streptococci	Caglar *et al.* (2008)
	B. adolescentis ATCC 15705	Mouse	Prevent dissemination of *Y. enterocolitica* to spleen	Wittmann *et al.* (2013)
Antiviral	*B. infantis* CECT 7210	MA-104, HT-29	Inhibit rotavirus infection	Munoz *et al.* (2011)
Antioxidant	*B. animalis* subsp. *lactis* DSMZ 23032	Rat	Reduce oxidative stress	Amaretti *et al.* (2013)
Allergy	*B. animalis* subsp. *lactis* BB-12	Mouse	Prevent allergen-induced sensitization	Feleszko *et al.* (2006)
	B. animalis subsp. *lactis* NCC 2818	Human	Reduce allergic symptoms	Singh *et al.* (2013)
	B. longum subsp. *longum* CCM 7952	Mouse	Reduce allergic sensitization	Schwarzer *et al.* (2013)
Metabolism/obesity	Bifidobacteria, *B. adolescentis*	Human T2D	Bifidobacteria low in T2D	Le *et al.* (2013)
	B. animalis subsp. *lactis* 420	Mouse	Normalize characteristics of T2D	Amar *et al.* (2011a)
	B. animalis subsp. *lactis* BB-12	Obese rats	Improved glucose tolerance	Bomhof *et al.* (2013)
	B. pseudocatenulatum CECT 7765	Mouse	Control metabolic disorders	Cano *et al.* (2013)
	B. longum BIF CGMCC NO.2107	Mouse	Improve high fat-induced MS	Chen *et al.* (2011)
Cholesterol lowering	*B. bifidum* MB107 *B. breve* MB113 *B. animalis* subsp. *lactis* MB 2409	Rat	Lower total and LDL cholesterol	Bordoni *et al.* (2013)

Table 4.2 Continued

	Bifidobacterium strain	Model	Health effect	Reference
	B. pseudocatenulatum CECT 7765	Mouse	Modify cholesterol metabolism	Moya-Perez *et al.* (2013)
Ageing	*B. bifidum* ATCC 86321	Mouse	Delay senescence	Fu *et al.* (2010)
	B. animalis subsp. *lactis* BB-12	Mouse	Increase longevity	Matsumoto *et al.* (2011)
	B. animalis subsp. *lactis* BB-12	Mouse	Reduce age-related metabolic dysfunction	Brasili *et al.* (2013)
Systemic inflammation	*B. infantis* 35624	Human	Systemic immunomodulation	Groeger *et al.* (2013)
Infant growth	*B. bifidum* OLB6378	Infants	Promote infant growth	Yamasaki *et al.* (2012)
Skin health	*B. breve* strain Yakult	Human	Maintain a healthy skin	Miyazaki *et al.* (2013)
Autism	Bifidobacteria	Human	Bifidobacteria low in autism	Wang *et al.* (2011)
	Bifidobacteria	Human	Divergent bifidobacteria in autism	Finegold *et al.* (2010)
Stress/anxiety	*B. infantis* 35624	Rat, stress	Normalize behavioural effects	Desbonnet *et al.* (2010)
	B. longum NCC3001	Mouse, DSS	Normalize anxiety-like behaviour	Bercik *et al.* (2011)

IBS, irritable bowel syndrome; IBD, inflammatory bowel disease; CD, coeliac disease; NEC, necrotizing enterocolitis; TNBS, trinitrobenzenesulfonic acid; UC, ulcerative colitis; T2D, type 2 diabetes; TJs, tight junctions; MS, metabolic syndrome; LDL, low-density lipoprotein.

CNCM I-2494 and *B. infantis* 35624 reduce stress-induced hypersensitivity. *B. animalis* subsp. *lactis* CNCM I-2494 reduced visceral sensitivity by normalizing intestinal epithelial barrier function and decreasing endotoxin translocation (Agostini *et al.*, 2012). As described above, in human IBS patients *B. infantis* 35624 reduced scores of abdominal pain/discomfort significantly when ingested at a dose of 10^8 cfu/day over a 4-week period (Whorwell *et al.*, 2006). The clinical response was associated with normalization of the ratio of an anti-inflammatory to a pro-inflammatory cytokine, suggesting that the effect occurred through immune modulation (O'Mahony *et al.*, 2005).

Inflammatory bowel disease

Inflammatory bowel disease (IBD) includes ulcerative colitis (UC) and Crohn's disease. IBD is characterized by chronic inflammation of the intestinal tissue. The pathogenesis of IBD remains unclear; however, there is increasing support for the idea that the gut microbiota is involved in the onset and persistence of inflammation. Dysbiosis of the mucosal microbiota has been found in inflamed compared to non-inflamed mucosa of IBD patients and compared to healthy controls. Significantly less bifidobacteria were found in the inflamed tissue of both IBD types (Fyderek *et al.*, 2009).

Few clinical trials have been performed with probiotics to treat or maintain remission in IBD. A strain of *B. longum* (no strain designation provided), isolated from a healthy rectal mucosa significantly reduced interleukin (IL) 1a produced by HT-29 cells *in vitro*. This was found to improve clinical scores in patients with UC when the strain was provided as a synbiotic product also containing a fructo-oligosaccharide and inulin. Four weeks consumption of the synbiotic significantly reduced mucosal inflammatory markers in active UC concurrent with a reduction in colitis. Pro-inflammatory cytokines were significantly lower in rectal biopsies after synbiotic treatment (Furrie *et al.*, 2005). In this study the mucosal numbers of the administered *B. longum* strain was not determined, but a 42-fold increase in the abundance of total mucosal bifidobacteria was found after synbiotic treatment whereas the increase was only 4.6-fold in the placebo group. Therefore, the effects on the inflamed epithelium could not be decisively attributed to the *B. longum* strain, but could result from the general stimulation of bifidobacteria in the mucosa.

Reduced circulating levels of pro-inflammatory markers have been found in UC patients after treatment with *B. infantis* 35624. C-reactive protein (CRP), tumour necrosis factor (TNF-α), and IL-6 are elevated in UC patients but *B. infantis* 35624 feeding for 6 weeks reduced systemic levels of these inflammatory markers (Groeger *et al.*, 2013). *B. infantis* 35624 also improved systemic inflammatory markers in patients suffering from chronic fatigue syndrome and psoriasis. These non-gastrointestinal disorders are also associated with increased systemic inflammation and the results show that immunomodulatory effects of the microbiota occur not only in the gut mucosa but extend to the systemic immune system.

Peristalsis and transit time

Colonic transit disturbances such as constipation are frequent symptoms of gastrointestinal disorders in the human population. In addition, decreasing intestinal transit time by food supplements in the non-diseased general population is considered a beneficial physiological effect by the European Food Safety Authority. Some strains of bifidobacteria reduce colonic transit time. In a recent meta-analysis, medium to large treatment effects on intestinal transit time were identified for *B. animalis* subsp. *lactis* HN019 and *B. animalis* subsp. *lactis* DN-173 010, while smaller effects were found for other strains (Miller and Ouwehand, 2013). Both oro-caecal and colonic transit times were significantly reduced by *B. animalis* subsp. *lactis* DN-173 010 treatment in IBS patients with constipation and a concomitant reduction in abdominal distension and pain was documented (Agrawal *et al.*, 2009). When supplemented in a fermented oat drink *B. animalis* subsp. *lactis* BB-12 or *B. longum* 46 and 2C increased bowel movement in elderly nursing home residents significantly without increasing diarrhoea. The effect was stronger than with the fermented drink alone (Pitkala *et al.*, 2007).

The mechanisms of bifidobacterial stimulation of peristalsis and intestinal transit time are unknown, with one exception. Sennosides, the laxative constituents of *Daio*, an ancient Japanese medicine, are converted by the intestinal microbiota to the active component rheinanthrone, promoting peristalsis. Owing to variances in gut microbiota composition between individuals considerable differences in the efficacy of this drug are found. *B. animalis* subsp. *lactis* BB-12 was found to efficiently hydrolyse sennosides to rheinanthrone *in vitro*, and to reduce faecal sennoside contents after oral administration of BB-12 to mice followed by sennoside administration. Intestinal peristalsis was significantly stronger and transit time lower in the treated mice (Matsumoto *et al.*, 2012).

Diarrhoea

Cochrane reviews have concluded that probiotics have preventative effects on acute infectious, paediatric antibiotic-associated, and *Clostridium difficile*-associated diarrhoea (Allen *et al.*, 2010; Goldenberg *et al.*, 2013; Johnston *et al.*, 2011). Nearly all studies included in these reviews were conducted with mixtures of probiotic bacteria including lactobacilli and the specific effects of bifidobacteria on diarrhoea cannot be revealed.

Acute infectious diarrhoea may be of bacterial or viral origin. In industrialized countries the majority of diarrheal incidences are caused by rotavirus and norovirus infection. Bifidobacteria may limit viral infections. *B. longum* subsp. *infantis* CECT 7210 is a strain capable of inhibiting rotavirus infections *in vitro*. The strain provided some protection against rotavirus infection when applied in a mouse model, where viral shedding in stools occurred later and decreased faster than in control mice possibly due to a reduction in viral replication. In addition, CECT 7210 temporarily increased faecal concentration of SIgA (Munoz *et al.*, 2011). In children diagnosed with rotavirus gastroenteritis, oral rehydration therapy and rapid refeeding with a normal diet with *B. animalis* subsp. *lactis* B94 significantly shortened duration of diarrhoea (Erdogan *et al.*, 2012). Acidified milk formula supplemented with *B. animalis* subsp. *lactis* BB-12 provided to children in residential care settings lowered the mean number of days with diarrhoea and significantly reduced the daily probability of diarrhoea (Chouraqui *et al.*, 2004).

Bifidobacteria may reduce diarrhoea caused by bacterial infections through antagonistic effects limiting pathogen growth, adhesion, toxin production, and translocation (discussed later).

Coeliac disease

Coeliac disease (CD) is an autoimmune intestinal disorder triggered by gluten proteins in genetically predisposed individuals. Gliadins are the soluble and toxic fraction of gluten. Small intestinal mucosal damage results from the uncontrolled innate and adaptive immunological responses to the protein. Reduced numbers of bifidobacteria were found in the duodenum (Nadal *et al.*, 2007) and faeces (Di Cagno *et al.*, 2012) of children with CD compared to healthy controls. Aberrant *Bifidobacterium* abundance in CD was part of a general dysbiosis with greater amounts of *Proteobacteria* that could contribute to the pro-inflammatory cytokine profile as well as to the maturation of monocytes and increased T-cell activation.

Intestinal barrier disruption is one of the toxic effects of gliadins. Increased intestinal permeability and actin cytoskeleton reorganization in cell monolayers after exposure to gliadins have been observed. When monolayers were incubated with a strain of *B. animalis* subsp. *lactis* (no strain designation provided) the effects of gliadins were partially inhibited. Whether this was due to enzymatic breakdown of gliadins by *B. animalis* subsp. *lactis* peptidases, or via interaction between *B. animalis* subsp. *lactis* and cells was not determined (Lindfors *et al.*, 2008). *In vitro* digestion of gliadins inoculated with *Bifidobacterium* strains (*B. bifidum* IATA-ES2, *B. longum* IATA-ES1, and *B. animalis* subsp. *animalis* IATA-A2) isolated from healthy infants reduced toxicity to Caco-2 cells. Bifidobacterial peptidases degraded gliadins and the derived peptides were less cytotoxic to epithelial cells and reduced pro-inflammatory activation (Laparra and Sanz, 2010).

Gliadin reduced the numbers of goblet cells but increased mucin secretion and caused damage to tight junctions in germ-free rats. The effect was more pronounced when gliadin was combined with pathogenic enterobacteria, which increased gliadin translocation into the mucosa. These effects were reversed in the presence of *B. bifidum* IATA-ES2, which also up-regulated chemotactic factors and inhibitors of metalloproteinases contributing to mucosal protection (Cinova *et al.*, 2011).

In a small human clinical study involving CD patients who consumed at least 12 g of gluten/day, 3 weeks of treatment with *B. infantis* NLS significantly improved perception of indigestion, constipation, and gastroesophageal reflux, while no effect was recorded on diarrhoea and abdominal pain. Intestinal permeability was not affected by this treatment, but immunological changes with reduced serum antibodies were found (Smecuol *et al.*, 2013). In weanling rats, gliadin feeding induced changes in the jejunal proteome, indicative of an inflammatory response and alteration of levels of proteins involved in ionic homeostasis, lipid turnover, cell motility, and redox regulation. Administering *B. longum* CECT 7347 partially counteracted the gliadin-induced changes (Olivares *et al.*, 2012). The strain reduced inflammatory cytokine production and increased regulatory cytokines in the jejunal mucosa and reduced activation of T helper type 1 (T_H1) cells (Laparra *et al.*, 2012).

Necrotizing enterocolitis

Necrotizing enterocolitis (NEC) is a common gastrointestinal disease among premature infants. NEC is a multifactorial disease, but it was recently shown in a neonatal NEC mouse model that increased intestinal permeability precedes NEC due to tight junction disruption (Bergmann *et al.*, 2013). Treatment of neonatal mice with *B. infantis* BB-02 preserved tight junctions, prevented the increase in intestinal permeability and decreased incidence and severity of NEC. Protection against NEC was also observed with *B. bifidum* OLB6378 in a rat model. This strain normalized the expression and localization of cellular junction proteins and reduced the inflammatory reaction in the ileum (Khailova *et al.*, 2009).

Although probiotic treatment of premature infants is now routine in many hospitals and the outcome of such treatments is generally very positive, strains with even better functions could possibly be identified through *in vitro* screening of strain collections for intestinal barrier improvement, and subsequent confirmation in animal studies.

Metabolic effects

The total number of faecal bifidobacteria in early childhood is lower in children who later become overweight than in children who maintain a normal weight. Low *Bifidobacterium* abundance was associated with higher numbers of *Staphylococcus aureus* and it was speculated that abundant *S. aureus* in early life could trigger a low-grade inflammation contributing to the development of obesity (Kalliomäki *et al.*, 2008).

Feeding rats a high-fat diet modifies the intestinal microbiota, increases circulating endotoxin concentrations, induces low-grade inflammation (metabolic endotoxaemia), and leads to insulin resistance and type 2 diabetes. Translocation of Gram-negative bacteria from the gut lumen into the mucosa, adipose tissue, and blood precedes the development of insulin resistance in mice fed a fat-enriched diet (Amar *et al.*, 2011a), and in the human general population blood concentrations of bacterial DNA, which is more than 90% derived from *Proteobacteria* is a marker of risk of diabetes and abdominal adiposity (Amar *et al.*, 2011b). Furthermore, these studies demonstrated that six weeks of treatment with *B. animalis* subsp. *lactis* 420 reduced bacterial translocation and improved the inflammatory and metabolic status of mice fed the fat-enriched diet (Amar *et al.*, 2011a).

While *B. animalis* subsp. *lactis* BB-12 had no effect on energy intake, weight gain and fat mass in diet-induced obese rats, the strain improved insulin sensitivity and glycaemia. This effect was possibly associated with increased intestinal barrier function mediated by increased levels of GLP-2 (Bomhof *et al.*, 2013). Improved gut barrier function induced by *B. longum* BIF CGMCC NO.2107 treatment decreased high-fat diet-induced plasma endotoxins and intestinal inflammation, and reversed metabolic effects of the high-fat diet. *Bifidobacterium* supplementation increased intestinal expression of regenerating islet-derived 1 (REG I), a protein involved in regulation of cell growth (Chen *et al.*, 2011). A common denominator of the effects of bifidobacteria in these studies is the improvement of the intestinal barrier. Barrier improvement leading to reduced endotoxin and bacterial translocation could be a primary mode of action of bifidobacterial probiotics, although different signalling mechanisms and downstream pathways may be involved. Another possible beneficial effect of bifidobacteria could be reduction of intestinal *Enterobacteriaceae* abundance, reducing the inflammatory signals coming from the gut as demonstrated in high-fat diet-fed mice supplemented with *B. pseudocatenulatum* CECT 7765 (Cano *et al.*, 2013).

Cholesterol-lowering is an established metabolic effect of probiotics including bifidobacteria. The lowering of cholesterol may occur through bile salt hydrolase (BSH) activity which deconjugates bile acids. Deconjugated bile acids are less efficiently reabsorbed in the intestines and more bile salts are excreted in faeces. Consequently, the liver increases *de novo* synthesis of bile salts leading to a reduction in serum cholesterol. Assimilation of cholesterol in the membrane phospholipid layer is another mechanism of cholesterol removal by bacteria. To identify bifidobacteria with a high cholesterol-lowering potential Bordoni *et al.* (2013) screened a collection of strains for BSH activity and cholesterol absorption. BSH activity and cholesterol assimilation was strain specific and could not be linked to the *Bifidobacterium* species tested. Three strains were selected from this screening and combined in a probiotic mixture that was administered to hypercholesterolaemic rats. A significant reduction of total cholesterol and low-density cholesterol was found whereas high-density cholesterol was unaffected. Another mechanism by which bifidobacteria may affect cholesterol metabolism is by modifying expression of liver proteins. Administering *B. pseudocatenulatum* CECT 7765 to mice fed a high-fat diet significantly counteracted the effects caused by the diet on expression of liver proteins involved in regulation of metabolism and transport of lipids. Accordingly, *B. pseudocatenulatum* CECT 7765 treatment decreased levels of serum cholesterol (Moya-Perez *et al.*, 2013).

Allergy

Allergic infants have been found to harbour an adult-like *Bifidobacterium* microbiota dominated by *B. adolescentis*, whereas *B. bifidum* dominates in healthy infants, and it was suggested that the *B. adolescentis* dominated microbiota is unable to provide adequate maturation signals for developing humoral immune defence mechanisms against allergy. Further, it was hypothesized that supplementing a microbiota typical of healthy infants could prevent allergy (Ouwehand *et al.*, 2001). Prevention of allergy by bifidobacteria has been confirmed in animal models, however, the strains applied were not selected based on the criteria suggested above.

Administering *B. animalis* subsp. *lactis* BB-12 to newborn mice during systemic sensitization and airway challenge with ovalbumin suppressed all aspects of the asthmatic phenotype. Total IgE antibodies as well as allergen-specific IgE and IgG2a were significantly inhibited in BB-12 treated mice compared with untreated controls. Increased regulatory T-cell (T_{REG}) activity was observed causing allergen-specific immunosuppressive effects and tolerance. Interestingly, BB-12 applied to adult mice failed to prevent allergy (Feleszko *et al.*, 2006). Neonatal colonization of germ-free mice with *B. longum* subsp. *longum* CCM 7952 prevented the development of allergic sensitization to the major birch pollen allergen by inducing a regulatory response. Thus, total IgE and allergen-specific IgG2a in serum was reduced compared to sensitized germ-free controls, and the birch pollen induced production of both T_H1- and T_H2-associated cytokines in spleen cell cultures was significantly reduced in treated mice (Schwarzer *et al.*, 2013).

Two important conclusions can be made from these studies. They clearly demonstrate that orally ingested bifidobacteria have effects not only in the gastrointestinal region, but may also affect other organs such as the respiratory tract. Since no preventative effect of probiotic administration on asthma was observed when the bacteria where applied to adult mice, it can be concluded that early-life modulation of the immune system, possibly during the first 'hit' with allergen, is required in order to successfully induce tolerance.

Nonetheless, administering *B. animalis* subsp. *lactis* NCC2818 for 8 weeks to adult humans suffering from seasonal allergic rhinitis to grass pollen improved allergic symptoms and immune parameters during the pollen season. Total nasal symptom scores were significantly lower after 2 months of treatment compared with placebo. At this time-point effects on immunological parameters, mainly T_H2 cytokine levels and basophil activation was found, and it was concluded that a continuous 8-week oral administration was required to reach a beneficial effect on allergy (Singh *et al.*, 2013).

Senescence

Metchnikoff proposed that the ageing process results from the activity of proteolytic bacteria producing toxic substances in the large intestine and that this 'intestinal auto-intoxication' caused the physical changes associated with old age, and further, that replacing the proteolytic with saccharolytic bacteria could delay the ageing process. Interestingly, a few publications report a senescence suppressive potential of bifidobacteria.

Supplementing 10-month-old mice with *B. animalis* subsp. *lactis* BB-12 for 45 weeks significantly increased longevity of the treated mice. After 45 weeks, 75% of the treated mice were still alive, while less than 25% were alive in the control group. Skin ulcers and tumours were more common among the control mice (Matsumoto *et al.*, 2011). BB-12 supplementation resulted in a host of physiological changes, but the overall effect was to suppress the senescence-associated modifications of the colonic environment. Thus, gene expression patterns of the colonic mucosa as revealed by microarray analysis, resembled that of much younger mice and were different from untreated aged mice. The transcriptional changes in untreated ageing mice were characterized by up-regulation of genes associated with several functional categories, and the effect of BB-12 was to suppress this gene regulation and maintain the levels seen in young mice. Regulating mucosal gene expression inhibited morphological disruption of the intestinal barrier and suppressed systemic and colonic inflammation during the ageing process. The proposed mechanism of life extension was through increased microbial production of polyamines that promote intestinal health and inhibit systemic low-grade inflammation. Significantly higher faecal concentrations of polyamines, particularly spermine were found in BB-12 treated mice. However, the BB-12 genome sequence indicates that this strain does not possess the pathways for spermine synthesis and other bacteria must have been responsible for the increased polyamine production. BB-12 supplementation modified the colonic microbiota and suppressed age-related changes of the microbiota. *Prevotella* spp. and the *Bacteroides fragilis* group were significantly more abundant in aged mice treated with BB-12, while *Clostridium* cluster XIVa, enterobacteria and *Enterococcus* were present in lower numbers resembling the microbiota of younger mice (Matsumoto *et al.*, 2011). Thus, indirect effects of BB-12 supplementation on mouse longevity via microbiota modifications are conceivable.

B. bifidum ATCC 86321 potentially delays senescence in mice. Although the actual effect on longevity was not determined for this strain, it appears to counter age-related physiological changes. Treatment of aged mice with *B. bifidum* ATCC 86321 increased body weight as well as thymus and spleen indexes, improved antioxidation activity, and reduced lipid peroxidation in thymus and spleen. Furthermore, this strain ameliorated age-induced immunosenescence by increasing circulating levels of IL-2 and IFN-γ and reducing pro-inflammatory cytokines (Fu *et al.*, 2010). So far, the antioxidant activity of bifidobacteria has been sparsely investigated. By screening a collection of bifidobacteria and lactobacilli for antioxidant activity, this capacity was found to be strain dependent, but a mixture of strains including *B. animalis* subsp. *lactis* DSNZ 23032 effectively reduced doxorubicin-induced oxidative stress in a rat model (Amaretti *et al.*, 2013).

In a study of the potential effect of *B. animalis* subsp. *lactis* BB-12 in combination with *Lactobacillus acidophilus* LA-5 on senescence-associated metabolic disorders it was found that this probiotic product modulated homocysteine and NAD metabolism pathways in aged mice as revealed by ^{1}H-NMR spectroscopy of urine samples. These metabolic pathways regulate fundamental cellular processes and are altered in senescence and it was suggested that probiotic supplementation may reduce age-related metabolic dysfunctions (Brasili *et al.*, 2013).

These studies identify an emerging research field that is anticipated to provide basic information on *Bifidobacterium*–host interactions and potentially also probiotic products with beneficial effects on senescence. The goal of such products may not be to increase longevity, but rather to improve quality of life in the elderly population.

Other beneficial effects of bifidobacteria

Less studied effects of *Bifidobacterium* supplementation include acceleration of growth of premature infants (Yamasaki *et al.*, 2012), prevention of skin dryness and disruption of keratinization during winter (Miyazaki *et al.*, 2013), and reversal of behavioural deficits and restoration of brainstem noradrenaline levels in a rat stress model (Desbonnet *et al.*, 2010).

Interactions between bifidobacteria and the host

The healthy gut

A healthy gut is characterized by a homeostatic symbiosis between the host and the gastrointestinal microbiota where the host provides the microbiota with nutrients and a stable environment and the microbiota helps shaping the host's gut mucosa and provides nutritional contributions. Maintaining the homeostasis involves interplay between at least three elements: the epithelial barrier, the antigen load, and the mucosal immune barrier (Fig. 4.1).

The epithelial barrier acts as a physical and chemical blockade against antigen penetration into the mucosa. Elements of the epithelial barrier include the mucus layers covering the epithelium and the constant renewal of mucus and epithelial cells, cellular junctions connecting epithelial cells, and secretory IgA and

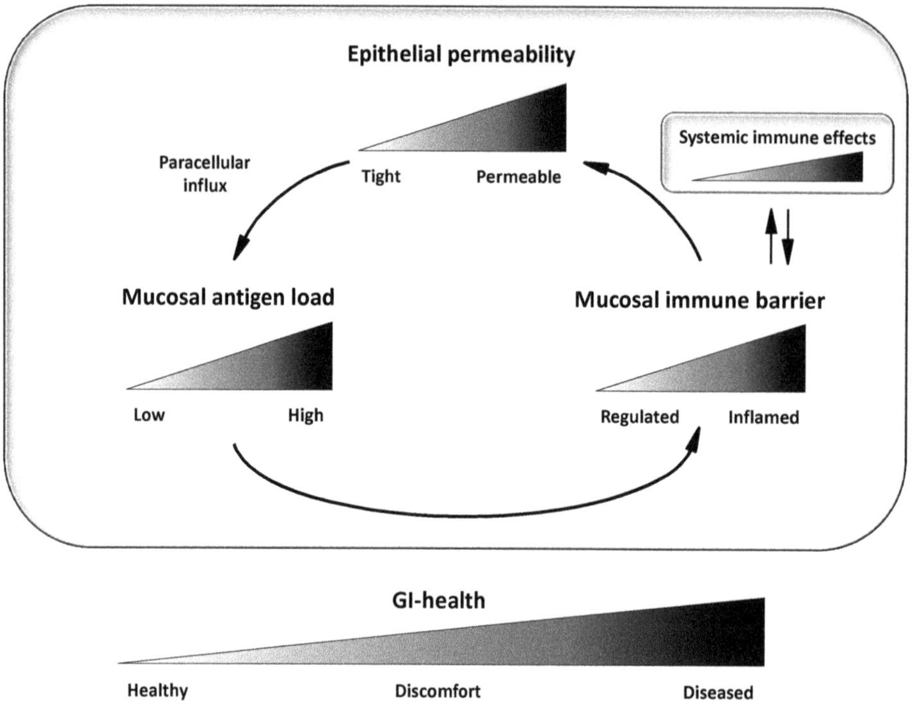

Figure 4.1 Key elements of the gastrointestinal epithelium essential to maintaining homeostasis and gut health. Epithelial permeability, mucosal antigen load, and immune barrier function are mutually dependent and control the immunological state of the epithelium. In the healthy epithelium permeability is low and limits paracellular antigen influx to the mucosa even in the presence of luminal antigens. Accordingly, the mucosal immune system is in a regulated, non-inflamed state. If for any reason one of these elements is perturbed it affects the other elements and homeostasis starts to disrupt leading to gastrointestinal discomfort and eventually disease. Sustained intake of a high-fat diet, alcohol, or non-steroidal anti-inflammatory drugs increase epithelial paracellular permeability causing more luminal antigens to penetrate into the lamina propria. Increased mucosal antigen load forces a local immune response to eliminate the antigens. Unfortunately, pro-inflammatory cytokines released locally after immune activation further increase permeability and a vicious circle has started. If the inflamed state is maintained systemic immune effects may result. Microorganisms with a pathogenic, invasive lifestyle also trigger this chain of events by breaching the mucosal barrier, which activates an inflammatory response leading to barrier break-down and more antigens entering the mucosa. When antigens are eliminated the immune activation must terminate to avoid further damage to the epithelium and chronic inflammation. Bifidobacteria may prevent homeostasis breakdown by preserving epithelial low permeability, reducing the mucosal antigen load, and/or by promoting a more regulated immune response.

antimicrobial compounds secreted by the epithelium. The antigen load originating from food intake may be of microbial origin such as intact bacteria or endotoxins, or foreign proteins derived from the food. When maintained in the gut-lumen the antigens represent no direct threat to intestinal homeostasis. However, when antigens penetrate into the mucosa due to a pathogenic lifestyle or a compromised intestinal barrier, the mucosal immune barrier is forced to respond. In a healthy mucosal-defence, protective immune responses against invasive pathogenic organisms are allowed to proceed to efficiently eliminate pathogens without exaggerating the inflammatory response that would otherwise cause inflammatory damage to the tissue. While pathogens are eliminated, the indigenous microbiota is tolerated. Disturbance of this host–microbiota homeostasis is associated with chronic intestinal inflammatory diseases.

Bifidobacteria may interfere with these elements and contribute to maintaining gut homeostasis. Thus, bifidobacteria can reinforce the epithelial barrier by strengthening tight junctions, stimulating secretion of IgA and antimicrobial peptides, and stimulating mucus production by goblet cells. Bifidobacteria may also reduce the mucosal antigen load through competitive exclusion of pathogens. By secretion of bactericidal or bacteriostatic compounds such as acetate or bacteriocins, or by competing for nutrients or mucosal attachment sites, bifidobacteria can reduce numbers of pathogenic bacteria, pathogen attachment to the mucosa, or pathogen penetration into the mucosa. Finally, bifidobacteria may direct the intestinal immune system into a more regulated and tolerant state.

It is important to realize that effects of bifidobacteria on gut homeostasis are highly strain dependent and usually cannot be generalized among strains even of the same species.

Barrier function

The intestinal epithelium is specialized to ensure optimal absorption of nutritional compounds, yet at the same time to exclude and neutralize or detoxify harmful components of the intestinal contents including microorganisms. In a healthy gut these functions are optimized and a healthy epithelium is essential to maintain a healthy gut. The epithelial lining consists of a single layer of epithelial cells covered by layers of mucus produced by specialized goblet cells. Beneath the epithelial cells is the lamina propria containing a variety of immune cells (gut-associated lymphoid tissue; GALT). Epithelial cells are joined together by cell junctions of which tight junctions (TJs) play a major role in preventing molecules from entering the epithelium between cells. TJs are responsible for restricting paracellular (between cells) diffusion of proteins, lipids and small solutes. Thus, in a healthy epithelium transport of molecules is regulated by cellular uptake mechanisms. TJs consist of proteins spanning the space between adjacent IECs. It is now apparent that TJs are dynamic structures that are involved in developmental, physiological and pathological processes. Various stressors may cause weakening of tight junctions and increase un-regulated paracellular transport of macromolecules into the mucosa. The uncontrolled diffusion of intraluminal toxins, antigens and enteric microbiota to the underlying tissue results in local and systemic inflammation.

The potential of bifidobacteria to strengthen the epithelial barrier can be evaluated *in vitro* by determining the trans-epithelial electrical resistance (TER) in epithelial cell monolayers such as Caco-2 or HT29 epithelial cells. TER is dependent on the paracellular flux of ions which is regulated primarily by TJs. A relaxation of TJs results in a higher flux of ions between the apical and basolateral side of the monolayers and decreased electrical resistance. Changes in TER correlate well with the permeability of monolayers to macromolecules. In a comparison of *B. breve* IPLA 20004 and *B. bifidum* LMG13195, it was found that IPLA 20004 increased TER faster and to a greater extent when applied on fully differentiated HT29 monolayers. This effect of IPLA 20004 was already significant after 3 h of co-culturing and maximal effect was reached after 8 h co-culture (Lopez *et al.*, 2012). Stimulation of TJs may be caused by short-chain fatty acids produced by bifidobacteria. Fermentation products of bifidobacteria of various types of prebiotic carbohydrates increased TER in Caco-2 monolayers, and the effect was strain and prebiotic dependent. Fermentation of Raftilose by *B. animalis* subsp. *lactis* BB-12 produced high acetate concentrations and this combination induced the strongest TER increase (Commane *et al.*, 2005). In primary keratinocyte cell cultures, lysates of *B. longum* ATCC 51870 stimulated TER as well as expression of TJs proteins claudin 1, claudin 4, ZO-1, and occludin in a dose-dependent manner. Peptidoglycan, which is a cell-wall component of Gram-positive bacteria, had a similar effect. Stimulation of TJs by *B. longum* ATCC 51870 and peptidoglycan was mediated through Toll-like receptor 2 (TLR-2) activation (Sultana *et al.*, 2013).

Improvement of barrier function by bifidobacteria and the consequences for disease have been demonstrated in animal models. Administration of *B. infantis* BB-02 in a neonatal mouse NEC model attenuated the effects induced by the NEC protocol and reduced NEC incidence. Intestinal permeability in the treated group was significantly lower than in untreated controls and localization of proteins involved in formation of TJs, claudin 4 and occludin at TJs was preserved. The beneficial effect of *B. infantis* BB-02 was, at least in part, due to its TJ and barrier-preserving properties (Bergmann *et al.*, 2013).

Acute stress in rats induces visceral hypersensitivity (pain) as a consequence of increased intestinal permeability. Increased paracellular permeability allows microorganisms and other antigens to penetrate into the mucosa where they activate the local immune system and cause abdominal pain by sensitizing afferent nerve fibres in the mucosa. Feeding a fermented milk product containing *B. animalis* subsp. *lactis* CNCM I-2494 to rats prevented the stress-induced intestinal permeability and restored gene expression of occludin and JAM-A to control levels. Circulating endotoxin levels remained normal in treated rats and stress-induced visceral hypersensitivity decreased in a dose-dependent manner. The dose required to obtain significant effects of *B. animalis* subsp. *lactis* CNCM I-2494 when provided in saline (10^{10} cfu/ml) by gavage was 100-fold higher than when provided in fermented milk (10^{8} cfu/ml) potentially because this matrix protected viability of the strain (Agostini *et al.*, 2012).

These results demonstrate that strains of bifidobacteria can increase barrier function and decrease epithelial permeability through preservation of TJ protein localization and up-regulating gene expression.

Mucus production

Minimizing bacterial contact with the intestinal mucosa is an important part of maintaining gut homeostasis and several mechanisms have evolved to constrain bacterial access to mucosal cells. This separation allows the co-existence of an enormous amount and variety of antigens in the gut without eliciting excessive and damaging mucosal immune responses. The mucus lining of the epithelium, secreted immunoglobulins and antimicrobial compounds are part of the barrier.

The gut epithelium is covered by mucus layers. The mucus layers function as a mechanical barrier separating luminal bacteria from the epithelium. Shear forces and the constant shedding of epithelial cells and mucus reduce the number of mucus-associated bacterial cells and in healthy individuals there is little contact between luminal microorganisms and epithelial cells. Chronic intestinal inflammation is often associated with an excessive mucosal antigen load and strengthening of the intestinal mucus layers to reduce antigen contact with epithelial cells could potentially relieve such symptoms.

Two mucus layers have been identified in the GI tract of rats: an inner layer firmly attached to the epithelium and a loosely adherent layer facing the intestinal lumen. Bacteria colonize the outer mucus layer, while the inner layer is devoid of bacteria (Swidsinski *et al.*, 2008; van der Waaij *et al.*, 2005). Mucus is produced by specialized epithelial goblet cells. At least nine human mucin genes have been identified, however; MUC2 is the major gel-forming mucin and is the main structural component of the mucus gel (Deplancke and Gaskins, 2001).

Literature on the effects of bifidobacteria on mucus structure and/or secretion is limited, but a few studies have shown stimulation of mucin gene expression and mucus secretion in rats. *B. bifidum* R0071 treatment of rats increased Muc3 protein and mRNA expression in jejunum and ileum. However, stimulation was not sustained when probiotics were administered daily and declined in just a few days (Dykstra *et al.*, 2011). In contrast, the probiotic mixture VSL#3, containing four *Lactobacillus* strains and three *Bifidobacterium* strains, stimulated mucin secretion and mucin gene expression after 7 days of probiotic treatment. MUC2 gene expression in colonic loops was highly up-regulated after VSL#3 treatment, while only a modest stimulation of MUC1 and MUC3 was seen. When the stimulatory efficacy of individual

VSL#3 strains was analysed it became evident that the three bifidobacteria had little effect and that mucin stimulation could mainly be ascribed to the lactobacilli (Caballero-Franco *et al.*, 2007). Two other studies showed no stimulation of MUC2 gene expression in rat intestines after treatment with bifidobacteria. In an NEC neonatal rat model, ileal expression of MUC2 was significantly decreased in rats with NEC while MUC3 expression was increased. Oral administration of *B. bifidum* OLB6378 reduced NEC incidence and severity, however MUC2 gene expression was unaffected and MUC3 mRNA levels decreased significantly in the treated group (Khailova *et al.*, 2009). *In vitro*, co-culturing of LS174T-cells with a collection of bifidobacteria had no effect on MUC2 expression, whereas MUC1 expression was elevated (Becker *et al.*, 2013).

These studies suggest that bifidobacteria have little stimulatory effects on mucin gene expression and mucus secretion and in particular that effects on MUC2, the major gel-forming mucin, are limited or absent. It is possible that a temporary discharge of mucins from goblet cell vesicles can take place without influencing mucin gene expression (Caballero-Franco *et al.*, 2007)

Secretory immunoglobulin A

Administering bifidobacteria to animals or humans increases faecal contents of secretory IgA (SIgA) (Kabeerdoss *et al.*, 2011; Plaza-Diaz *et al.*, 2013). After cessation of probiotic intake SIgA returns to normal levels. SIgA is an important part of the first line of defence that provides protection by binding and excluding microbial and dietary antigens from the mucosal cell surfaces. SIgA is produced by mucosal B-cells in the lamina propria and translocates across epithelial cells to the luminal side. In the gut lumen SIgA accumulates in the epithelial mucus lining where it rapidly covers antigens situated in the mucus. Immune exclusion of antigens by SIgA covering prevents microbial attachment to epithelial cells and subsequent invasion of the mucosa.

Bifidobacteria can stimulate general and specific SIgA production after pathogen challenge. Thus, faecal levels of anti-*Escherichia coli* O157:H7-specific SIgA was higher in mice that had been pretreated with *B. thermacidophilum* RBL 71 for 1 week prior to *E. coli* O157:H7 challenge than in mice that were not treated. Faecal *E. coli* O157:H7 counts, mucosal invasion and local inflammation were lower in the treated group, potentially due to the improved SIgA response (Gagnon *et al.*, 2006). In humans, consumption of *B. animalis* subsp. *lactis* BB-12 daily for 6 weeks before seasonal influenza vaccination significantly increased vaccine-specific SIgA in saliva as well as total antibody concentrations (Rizzardini *et al.*, 2011). This strain also increased faecal concentrations of anti-polio-specific SIgA in infants fed a supplemented infant formula compared with non-supplemented formula following immunization, and tended to increase anti-rotavirus-specific SIgA in caesarean-delivered infants (Holscher *et al.*, 2012). The mechanism by which bifidobacteria (and other probiotics) boost SIgA production remains unclear.

Antimicrobial peptides

Antimicrobial peptides (AMPs) that are produced by epithelial cells and secreted into the intestinal lumen also function to separate luminal bacteria from the epithelium. AMPs are small peptides that affect viability of bacteria by disrupting the bacterial cell wall or inhibiting cell wall synthesis. Biosynthesis and secretion of AMPs is induced by pathogenic microorganisms as well as commensal bacteria, including bifidobacteria.

In vitro, *B. breve* Ha6/14c and *B. adolescentis* TSD, but not *B. longum* DSM 20219^{T} stimulated human β-defensin 2 mRNA expression in the colon adenocarcinoma cell line LS174T (Becker *et al.*, 2013). Human β-defensin 2 is a bacteriostatic peptide which is active against Gram-negative bacteria. Gene activation of epithelial cells occurred via NF-κB and AP-1 signalling.

Chemokines share structural properties with defensins and many chemokines exhibit defensin-like antimicrobial activity *in vitro*. An example is 'chemokine (C-C motif) ligand 20' (CCL20), which is produced by epithelial cells and secreted basolaterally where it attracts lymphocytes and DCs. CCL20 is also secreted apically and accumulates in the inner mucus layer where it exerts strong antimicrobial effects on certain Gram-positive and Gram-negative bacteria. CCL20 gene expression and peptide secretion is up-regulated by many *Bifidobacterium* strains (unpublished results). Contradictory to these findings *B. infantis* 35624 has been found not to induce CCL20 secretion in HT29 cells and in fact attenuates pathogen or flagellin induced CCL20 secretion in this cell line (Sibartie *et al.*, 2009). The importance of this needs to be evaluated *in vivo* to determine the net outcome of two functions associated with CCL20 secretion: chemoattraction of immune cells on the mucosal side, and anti-microbial effects on the luminal side.

Regenerating islet-derived III (REGIII) proteins, which belong to the family of C-type lectins, are antimicrobials expressed in the intestine. Bifidobacteria can increase ileal and colonic REGIII production in mice via the MyD88-Ticam1 pathway and potentially enhance barrier integrity and protect against inflammation (Chen *et al.*, 2011; Natividad *et al.*, 2013).

Bifidobacteria may improve the epithelial first line of defence by stimulating epithelial synthesis and secretion of a range of antimicrobial peptides that maintain mucosal barrier integrity and protect against inflammation by keeping microbes from direct contact with the epithelium. The significance of this activity *in vivo*, however, needs further investigation.

Pathogen exclusion

Stimulation of resistance to enteric pathogen infections is a classic probiotic mode of action and several mechanisms have been proposed to explain this feature. Probiotic pathogen exclusion may be conferred through competition for nutrients, competitive adhesion to mucosal attachment sites, the production of antimicrobial compounds, or stimulation of the mucosal immune system.

Inhibition of enteric pathogen adhesion to intestinal mucus of various bifidobacteria has been demonstrated *in vitro* (Collado *et al.*, 2005; Collado *et al.*, 2006). The proposed mechanism for pathogen exclusion by bifidobacteria is competition for attachment sites with pathogens and/or coaggregation with the pathogens. Diverse molecules seem to be involved including proteins, carbohydrates and lipids. Adaptation of bifidobacteria to acid or bile stress may change their ability to prevent pathogen adhesion to mucus or to displace pre-attached pathogens, presumably due to modified surface properties of the adapted strains (Collado *et al.*, 2006; Gueimonde *et al.*, 2007).

Several studies have investigated the effects of bifidobacteria on enterohaemorrhagic *E. coli* O157:H17 infection in mouse models. Shiga toxins are involved in the high mortality rate in mice after *E. coli* O157:H17 infection and certain *Bifidobacterium* strains (*B. breve* Yakult, *B. pseudocatenulatum* DSM 20439, *B. infantis* 157F, and *B. longum* NS) can completely abolish mortality in infected mice, while other strains have no effect. The preventative strains reduce toxin production without affecting viability of *E. coli* O157:H17 (Asahara *et al.*, 2004; Yoshimura *et al.*, 2010). Inhibition of Shiga toxin production by bifidobacteria may be due to production of acetate and pH lowering in the intestines (Asahara *et al.*, 2004). Acetate produced by bifidobacteria may also inhibit translocation of Shiga toxin from the gut lumen into the blood stream (Fukuda *et al.*, 2011). *In vitro* acetate prevented epithelial barrier breakdown caused by *E. coli* O157:H17 in Caco-2 monolayers and prevented translocation of Shiga toxin from the apical to the basolateral side of the cell layers. *In vivo*, acetate production in the distal colon was restricted to those *Bifidobacterium* strains possessing specific carbohydrate transporters that allow uptake and catabolism of fructose when other, readily absorbable, sugars are fully consumed. Such strains can improve the epithelial barrier and inhibit Shiga toxin and *E. coli* O157:H17 penetration into the mucosa through their acetate production in the lower large intestine.

Reduced mucosal translocation of *Enterobacteriaceae* after treatment of mice with *B. animalis* subsp. *lactis* 420 was elegantly demonstrated by Amar and co-workers using fluorescently labelled *E. coli*. After gavage, rapid adhesion of GFP-labelled *E. coli* to the intestinal mucosa was found in mice fed a high-fat-diet, but not in mice fed a normal chow diet. Translocation of fluorescent bacteria followed immediately after and GFP-labelled *E. coli* could be detected in the lamina propria of intestinal segments already 2h after gavage. Accumulation of *Enterobacteriaceae* in mucosal tissues, mesenteric adipose tissue, and in blood was confirmed by specific 16S rDNA PCR. Substantial numbers of DCs in lamina propria and lymph nodes co-localized with the fluorescent *E. coli* and generated a moderate, but significant inflammatory state. Six-weeks pretreatment with *B. animalis* subsp. *lactis* 420 effectively prevented adherence and translocation of *Enterobacteriaceae* and abolished the low-grade inflammation otherwise seen in mice fed the high-fat diet (Amar *et al.*, 2011a). It was not investigated whether reduced adherence and translocation of *Enterobacteriaceae* was caused by acetate production of *B. animalis* subsp. *lactis* 420.

In vitro and *in vivo* antagonistic effects of bifidobacteria against growth of vaginal *Staphylococcus aureus* (Lazarenko *et al.*, 2012), intestinal *E. coli* (Mogna *et al.*, 2012), *Salmonella* Typhimurium and *Clostridium difficile* (Tejero-Sariñena *et al.*, 2013), and *Helicobacter pylori* (Chenoll *et al.*, 2011), internalization and translocation of *Campylobacter jejuni* (Alemka *et al.*, 2010), and numbers of oral salivary mutans streptococci (Caglar *et al.*, 2008) have been found.

Thus, bifidobacteria may have antagonistic effects on diverse pathogenic bacteria and can potentially limit pathogen growth, adhesion to the intestinal mucosa, toxin production, and translocation to the lamina propria.

Immunomodulation

The gut-associated lymphoid tissues (GALT) is the largest collection of lymphoid tissues in the body. The GALT is comprised of Peyer's patches (PPs), interspersed lymphocytes, plasma cells and lymphocytes present in the lamina propria, and mesenteric lymph nodes. The key function of the GALT is to manage the immune response to antigens in the gut. IECs are closely linked to the regulation of innate and adaptive defence mechanisms through their contact with luminal antigens and with immune cells in lamina propria. Thus, activation of Toll-like receptors (TLRs) on the apical surface of IECs, typical of commensal, non-invasive microbes, induces partial and protective $NF\text{-}_{K}B$ activation, while basolateral activation of TLRs by invasive pathogens leads to a robust inflammatory response.

The main interaction between orally administered bifidobacteria and the host immune system occurs in PPs in the small intestine. Luminal bifidobacteria, or cellular components, are taken up by specialized M-cells in PPs where they can be found surrounded by $CD11c^{+}$ cells (DC). Bifidobacteria induce DC maturation and secretion of cytokines such as IL-10 and IL-12 (Hiramatsu *et al.*, 2011).

The microbiota has profound effects on the immune system and drives inflammatory responses as well as induces immune tolerance. Some bacterial strains will induce protective signals while others stimulate strong and potentially destructive immune responses.

Expressed microbe-associated molecular patterns (MAMPs), i.e. bacterial surface exposed molecules, which are recognized by host cell pattern recognition receptors (PRRs) activates the immune system. MAMPs vary between bifidobacterial strains causing different host immune responses. Membrane bound PRRs responsive to bifidobacteria on antigen presenting cells (DCs and macrophages) located in the intestinal mucosa include Toll-like receptors (TLR) and C-type lectin receptors (CLR). Upon specific receptor binding, PRRs activate intracellular signalling pathways resulting in immune cell maturation and cytokine secretion. Cytokines secreted by antigen presenting cells subsequently direct the maturation of naive T-cells into the different subsets of T-helper cells (T_H1, T_H2, T_H17, and T_{REG}).

Some bifidobacteria are anti-inflammatory/immunoregulatory

Although bifidobacterial MAMPs are strain specific, and strains elicit specific immune responses (Lopez *et al.*, 2010), a general trend to promote immunotolerance has been observed for bifidobacteria. The mechanism involves activation and maturation of local DCs that produce high levels of anti-inflammatory IL-10 (Dong *et al.*, 2012; Herfel *et al.*, 2013; Hiramatsu *et al.*, 2011; O'Hara *et al.*, 2006; Plaza-Diaz *et al.*, 2013; You and Yaqoob, 2012), which direct maturation of naïve T-cells towards the T_{REG} subtype (Donkor *et al.*, 2012; Konieczna *et al.*, 2012b; Lopez *et al.*, 2011; Lopez *et al.*, 2012; O'Mahony *et al.*, 2008; Wittmann *et al.*, 2013). The concomitant low polarization of T-cells towards T_H1 and T_H2 types by bifidobacteria results in the overall regulated and tolerogenic intestinal immune system needed to maintain intestinal homeostasis. Bifidobacteria are more likely to induce high IL–10/IL-12 ratios in peripheral blood mononuclear cells (PBMCs) and bone marrow derived DCs (BMDCs) than lactobacilli. Thus, in BMDCs bifidobacteria induced only little IL-12 and TNF-α production contrasting most of the lactobacilli investigated and the bifidobacteria were found to inhibit IL-12 production induced by *L. acidophilus* (Weiss *et al.*, 2011). Consequently bifidobacteria tended to attenuate the inflammatory response. Similar results have been found for several *Bifidobacterium* strains *in vitro*, and *in vivo*. In PBMCs *B. bifidum* MF 20/5, *B. longum* SP 07/3 and *B. longum* subsp. *infantis* CCUG 52486 all induced a high IL–10/IL-12 ratio while none of the *Lactobacillus* strains tested did so (Dong *et al.*, 2012; You and Yaqoob, 2012). This effect was more pronounced in PBMCs derived from older subjects than from young subjects (You and Yaqoob, 2012). In DCs isolated from PBMCs, *B. infantis* 35624, or spent culture medium of *B. breve* C50 induced full maturation of the DCs and induced high IL-10 secretion (O'Hara *et al.*, 2006). Activation by *B. breve* C50 spent medium was TLR-2 dependent (Hoarau *et al.*, 2006).

In a neonatal piglet model supplementation with *B. longum* AH1206 increased ileal mucosal expression of IL-10 and TNF-α (Herfel *et al.*, 2013). The increased IL-10 expression found in this study suggests that *B. longum* AH1206 promotes immune tolerance, however, the concomitant TNF-α stimulation suggests a pro-inflammatory response, but the net result in terms of inflammatory tone was not investigated. Administration of *B. breve* CNCM I-4035 to human volunteers resulted in increased serum levels of IL-10 and IL-4, while serum IL-12 decreased resulting in a regulatory immune response in human volunteers (Plaza-Diaz *et al.*, 2013).

Unrestrained T_H1, T_H2, or T_H17 effector responses will cause tissue damage but is controlled by T_{REG} cells. There is increasing evidence that strains of bifidobacteria induce maturation of T_{REG} from naïve precursors. T_{REG} inducing probiotics could benefit patients suffering from autoimmune and allergic diseases (Kwon *et al.*, 2010; Lavasani *et al.*, 2010). Strains of *B. bifidum* are particularly effective in T_{REG} polarization through IL-10 and IL-17 induction in monocyte-derived DCs (Lopez *et al.*, 2011). IECs are also involved in generation of T_{REG} cells (Lopez *et al.*, 2011). Co-culturing HT29 epithelial cells with *B. bifidum* LMG13195 increased HT29 expression of chemokines including CXCL10, CCL20, CXCL11, and CCL22. When PBMCs were exposed to supernatants from HT29 cells co-cultured with *B. bifidum* LMG13195 the proportion of matured T_{REG} increased significantly and chemokine receptors were highly expressed on T_{REG} (Lopez *et al.*, 2012). These findings suggest that IECs exposed to antigens can generate soluble factors that polarize and recruit regulatory T-cells to the intestinal mucosa.

Transforming growth factor beta (TGF-β) is another cytokine important for T_{REG} differentiation and TGF-β is involved in polarization of T_{REG} or T_H17 effector cells together with IL-6 and IL-21. *B. longum* BL536 and *B. animalis* subsp. *lactis* B94 were found to induce high levels of TGF-β and IL-10, and to induce T_{REG} polarization (Donkor *et al.*, 2012). In human volunteers IL-10 and T_{REG} was significantly enhanced in peripheral blood after 8 weeks' feeding with *B. infantis* 35624 (Konieczna *et al.*, 2012b). The same strain protected mice from the effects of *Salmonella* Typhimurium infection. Administering *B. infantis* 35624 to mice prior to *Salmonella* infection increased numbers of T_{REG} which attenuated NF-κB activation, reduced pro-inflammatory cytokine secretion and controlled the effects of infection (O'Mahony *et al.*, 2008). Protection against *Yersinia* infection by *B. adolescentis* Reuter 1963 in mice has also been demonstrated. Protection was associated with increased frequencies of intestinal plasmacytoid DCs and T_{REG} (Wittmann *et al.*, 2013).

The cytokine profiles induced by bifidobacteria in antigen presenting cells is far from uniform. The immune response of human-monocyte derived DCs to a collection of 12 *Bifidobacterium* strains belonging to 4 species (*B. longum, B. breve, B. bifidum, and B. animalis* subsp. *lactis*) was investigated by Lopez and co-workers. All *Bifidobacterium* strains induced DC maturation, but different cytokine profiles were found. Strains of *B. animalis* and *B. longum* generated high TNF-α/IL-10 and low IL-10/IL-12 ratios, thus promoting T_H1 polarization, while *B. bifidum* strains induced a high IL–10/IL-12 ratio and significant amounts of IL-17 suggesting promotion of T_H2 and potentially T_H17 polarization. Very low induction of DC maturation was found when cell-free culture supernatants were tested indicating low secretion of the bacterial immune activating molecules and the need for direct contact between bacteria and DCs (Lopez *et al.*, 2010).

Dendritic cell activation by bifidobacteria is TLR-2 dependent (Hoarau *et al.*, 2006). TLR-2 is involved in the recognition of Gram-positive and Gram-negative bacteria and known agonists include peptidoglycan, lipoteichoic acid and lipoproteins. Surface expressed, or secreted lipoproteins from bifidobacteria appear to be ligands for dendritic cell activation, possibly in TLR-2/TLR-1, or TLR–2/TLR-6 complexes (Zeuthen *et al.*, 2008). TLR-9 recognizing bacterial CpG DNA is also involved in recognition of bifidobacteria and stimulates pro-inflammatory cytokine secretion, but the simultaneous stimulation of TLR-2 induces an inhibitory signal to downstream pathways resulting in a lower inflammatory profile than from lactobacilli (Plantinga *et al.*, 2011).

Sortase-dependent pili of *B. bifidum* PRL2010 is another ligand displaying immunomodulatory activity. PRL2010 pili expressed in *Lactococcus lactis* stimulated TNF-α expression and lowered the IL-10 response in the mouse intestine compared to non-piliated *L. lactis* (Turroni *et al.*, 2013). Soluble compounds secreted by bifidobacteria may also interact with mucosal DCs as demonstrated by the effects of a *B. breve* C50 supernatant. This supernatant, without viable bacteria, induced DC maturation and prolonged DC survival through TLR2 and stimulated high IL-10 production (Hoarau *et al.*, 2006).

Immunomodulatory effects of bifidobacteria have been demonstrated frequently, *in vitro*, but only few studies have shown that such regulatory strains have beneficial effects on intestinal inflammatory conditions in animals or humans. *B. longum* subsp.

infantis 35624 is a bacterial strain isolated from a healthy human. The immunomodulatory effects of 35624 have been thoroughly investigated *in vitro*, as well as in animal models and humans. In DCs, *B. infantis* 35624 induces secretion of IL-10, but not IL-12 or IFN-α through activation of TLR-2/6, TLR-9, and DC-SIGN (Konieczna *et al.*, 2012b). *B. infantis* 35624 stimulated DCs up-regulate expression of Foxp3 (a marker of T_{REG}) in naïve lymphocytes thus promoting T_{REG} and a regulatory immune response (Konieczna *et al.*, 2012b). When applied in different colitis models *B. infantis* 35624 reduced the severity of symptoms (McCarthy *et al.*, 2003; O'Mahony *et al.*, 2008). In a *Salmonella* Typhimurium infection model which is characterized by a potent inflammatory response, *B. infantis* 35624 attenuated this response in a T_{REG}-dependent manner and protected mice from the effects of the infection (O'Mahony *et al.*, 2008). In humans *B. longum* subsp. *infantis* 35624 is able to induce IL-10 secretion and mount a regulatory immune response. Volunteers fed *B. infantis* 35624 had a higher IL-10 secretion and enhanced Foxp3 expression in peripheral blood. DCs secreted IL-10, but not IL-12 in response to the strain. Activation of TLR-2/6, DC-SIGN, and TLR-9 receptors on human DCs was partially responsible for inducing high levels of IL-10 secretion and lymphocyte Foxp3 expression. (Konieczna *et al.*, 2012b). In IBS patients *B. infantis* 35624 provided a significant reduction in overall symptoms including abdominal pain/discomfort (Brenner and Chey, 2009). In a recent study *B. infantis* 35624 was shown to modulate not only the local mucosal immune response but also the systemic immune system. In patients with UC, chronic fatigue syndrome, or psoriasis, 6–8 weeks oral administering of *B. infantis* 35624 significantly reduced plasma levels of C-reactive protein and proinflammatory TNF-α in chronic fatigue syndrome and psoriasis, and tended to decrease IL-6 in UC and chronic fatigue syndrome. (Groeger *et al.*, 2013). All 3 immune markers were elevated in the patient groups compared with healthy individuals. These findings show that orally ingested bifidobacteria can produce systemic immune regulation and that this potentially occurs through T_{REG} polarization.

Serum inflammatory markers such as IL-6 and MCP-1 are known to increase in obesity and related metabolic disorders. In a mouse model of high-fat diet-induced obesity, supplementation with *B. pseudocatenulatum* CECT 7765 reduced serum concentrations of IL-6, MCP-1, and IL-4, which were increased in the high-fat diet group compared with mice receiving the standard diet. Changes in inflammatory markers correlated with improvements in glucose tolerance and insulin sensitivity (Cano *et al.*, 2013). This study also showed that the high-fat diet impaired DCs antigen-presenting ability and T-cell proliferation stimulation, and that *B. pseudocatenulatum* CECT 7765 could restore DC functioning.

Industrial-scale production of bifidobacteria

Apart from survival through the GI tract, the success of a probiotic product is also dependent on the ability of the specific *Bifidobacterium* strain to survive the stresses encountered during industrial production and storage. In brief, industrial scale production of bifidobacteria begins with the production of cell biomass by fermentation. When cell growth has reached an optimum, the biomass is concentrated by centrifugation or membrane filtration and the concentrated cell mass is typically preserved by freezing and/or drying. Before freezing and drying, a protective matrix may be added that serves to increase survival during downstream processing and storage. The production process imposes various stresses to the cells, including oscillations in pH, temperature, and nutrient availability due to inhomogeneous mixing during batch fermentation, as well as cold and osmotic stress during freezing, heat and osmotic stress during drying and oxidative stress during storage (Saarela *et al.*, 2005; Corcoran *et al.*, 2008; Forssten *et al.*, 2011; Makinen *et al.*, 2012; Wallenius *et al.*, 2012).

Bifidobacteria are less resistant than lactobacilli to the stressful environments encountered in the manufacturing process (Dunne *et al.*, 1999, Ruiz *et al.*, 2011). This may explain the predominance of lactobacilli in commercial probiotic products. Historically, dairy based growth media and food matrices have been preferred for production and delivery of probiotic bacteria. However, the low proteolytic activity of bifidobacteria in milk results in poor growth compared to other lactic acid bacteria and consequently impedes the production and use of bifidobacteria as probiotics (Collins and Hall, 1984; Samona and Robinson, 1994; Gomes and Malcata, 1999). In addition, the low pH resulting from starter culture acidification in fermented milk and the accumulation of acetate and lactate in pH-controlled fermentations limit bifidobacterial biomass production (Desjardins *et al.*, 1990). The low tolerance of bifidobacteria to acidic pH may partly be explained by the membrane fatty acid composition. Cyclopropane fatty acids (CFA) are present in relatively low amounts in bifidobacteria and do not increase in stationary growth phase as observed in other lactic acid bacteria (Veerkamp 1971; Poupard *et al.*, 1973). An increased amount of membrane CFA has been associated with increased survival during acid stress as well as during freezing and freeze drying (Beal *et al.*, 2001; Broadbent *et al.*, 2010; Jin *et al.*, 2012; Li *et al.*, 2009; Muños-Rojas *et al.*, 2006; Smittle *et al.*, 1974). Consequently, the low content of CFA in bifidobacteria may explain their poor survival during freezing, freeze drying and gastric transit.

Different intrinsic tolerances to stress have been found among *Bifidobacterium* species. The predominance of *B. animalis* subsp. *lactis* strains in probiotic products is probably explained by their high tolerance to oxygen, acid and heat compared to other bifidobacteria (Masco *et al.*, 2005; Matsumoto *et al.*, 2004; Mattö *et al.*, 2006; Maus and Ingham 2003; Simpson *et al.*, 2005). However, intrinsic stress resistance is also strain-specific, for instance a strain of *B. longum* subsp. *infantis* was shown to be the most resistant to oxidative stress in a study that also included three *B. animalis* subsp. *lactis* strains (Oberg *et al.*, 2011).

With the shortcomings of bifidobacteria described above, how is it even possible to produce bifidobacterial probiotic products? Bifidobacteria are not as fastidious as other lactic acid bacteria and simple, defined media containing only a few amino acids or even ammonium as the sole nitrogen source have been described (Hassinen *et al.*, 1951; Poch and Bezkorovainy, 1988).

In addition, their anaerobic nature can be largely accommodated by addition of cysteine (or another reducing agent) to the growth medium to lower the redox potential (Bolduc *et al.*, 2006; Collins and Hall, 1984). The optimal growth medium is, of course, strain dependent. A typical *Bifidobacterium* industrial medium contains a complex nitrogen source such as yeast extract or peptone (3–4% w/w), an easily accessible carbon source such as glucose (2–4% w/w), an optimized combination of salts (e.g. $MgSO_4$, $MnSO_4$, phosphates), Tween 80 (0.1% w/w), and cysteine (0.05% w/w) (Kiviharju *et al.*, 2005; Saarela *et al.*, 2005).

Viability of probiotic cultures during processing and storage can be improved by stress adaptation and cross-protection. Tolerance to acidic pH is a critical factor determining bacterial survival during storage in foods and during gastric transit. A comparison of survival of lactobacilli and bifidobacteria in response to acid and heat stress after adaptation to sub-lethal stresses, showed most pronounced improvements for lactobacilli compared with bifidobacteria (Saarela *et al.*, 2004). Nevertheless, several studies describe successful adaptation of bifidobacteria to acid, bile, cold, and heat stresses (Collado and Sanz, 2007; du Toit *et al.*, 2013; Maus and Ingham, 2003; Schmidt and Zink, 2000). For instance, exposure of *B. longum* subsp. *longum* BBMN68 to pH 4.5 for 2 hours elicited an acid tolerance response that increased survival at lethal pH 3.5 for 2 hours by 70-fold (Jin *et al.*, 2012). However, apart from improving stability (i.e. the viability) of bifidobacteria, stress adaptation may also give rise to phenotypic changes that affect their probiotic functionality. In addition, production processes applied by different manufacturers may impose stress adaptations, and it is of crucial importance that the properties of *Bifidobacterium* strains after stress adaptation, or production process modifications are evaluated (Collado and Sanz, 2007; du Toit, 2013; Grześkowiak *et al.*, 2011).

Technological approaches have been applied to improve the processing stability of probiotic bifidobacteria. One approach is to optimize the environment during storage, e.g. by decreasing the temperature and relative humidity, or replacing oxygen with nitrogen (Celik and Sullivan, 2013; Champagne *et al.*, 1996). However, when storage conditions cannot be controlled, improving the protective matrix that serves to enhance survival during freezing, drying and storage as well as the drying process itself can increase bacterial viability (Kiviharju *et al.*, 2005; Mattila-Sandholm *et al.*, 2002). For instance, optimizing the temperature profile during freeze drying increased the bacterial survival with 160% and reduced the overall process time by 50% (Kiviharju *et al.*, 2005). Stability can also be increased by microencapsulation, a technique that entraps the live bacteria within a protective film that prevents entry of damaging substances such as oxygen or gastric juice. Microencapsulation which can be obtained by spray drying a mixture of probiotic cells with carrier media such as milk, starch, gelatin or gum Arabic has been shown to enhance the survival both during storage and gastric transit. However, the effect is widely dependent on the carrier components used for encapsulation (Collado and Sanz, 2007; Ding and Shah, 2007; Lian *et al.*, 2003; O'Riordan *et al.*, 2001; Thantsha *et al.*, 2014).

Many of the approaches to improve stress tolerance and, thus, the industrial compatibility of bifidobacteria proposed in the literature are interesting from a scientific point of view, but may not be applicable to industrial production. As an example, lactose and skimmed milk powder are widely used as both nutrients and protective agents; however, these components are also allergens and cannot be applied in hypoallergenic probiotic formulations. Moreover, the food industry is subjected to strict safety regulations and may choose to comply with specific requirements of Kosher, or Halal food products limiting the number and nature of compounds that can be used for optimization of growth and stability of probiotics (Pedersen *et al.*, 2005). Implementing of laboratory-scale production solutions may also be limited by the much longer processing times inherent to industrial-scale production. Thus, lowering the temperature from 37°C during fermentation to 5–10°C at the end of fermentation can easily take an hour in a cubic metre scale fermenter. A better small-scale resemblance of industrial production may be obtained by approaches such as those by Maus and Ingham (2003) where an 'industrial standard' was applied for sublethal stress adaptation. In these studies, cooling was not initiated until 30 minutes after entry into stationary growth phase and a cooling gradient was applied that reduced the temperature slowly over a period of 30 minutes, mimicking large-scale processing times and providing better indication of the robustness of the phenotype in industrial production.

In conclusion, maintaining viability of bifidobacteria during product manufacture and subsequent storage in food products or in dietary supplements is fundamental for their use as probiotics. The somewhat fragile nature of bifidobacteria poses particular challenges to industrial production and technological advancements for enhancing fermentation and downstream processes are highly needed to fully exploit the potential of this group of health-beneficial bacteria.

Conclusions and future perspectives

Although health benefits of bifidobacteria have long been recognized and bifidobacteria are readily available in commercial probiotic products, we have not yet realized the full potential of this group of beneficial bacteria. Animal and human trials have demonstrated that bifidobacteria are not only key players in maintaining gastrointestinal homeostasis and health, but also have the potential to induce systemic effects that affects organs distant from the GI tract. Genomic, transcriptomic and proteomic studies have revealed ecological adaptations of bifidobacteria to gastrointestinal conditions and are just beginning to provide insights into the nature of *Bifidobacterium*–host interactions at the molecular level.

Metagenomic investigations of the intestinal microbiota have indicated that reduced numbers of bifidobacteria are part of a general dysbiosis associated with various disorders. A concomitant increase in numbers of *Proteobacteria* is often found. *Proteobacteria* are Gram-negative bacteria with an outer membrane containing lipopolysaccharides, which triggers strong pro-inflammatory immune responses. It appears that exaggerated translocation of lipopolysaccharides and/or *Proteobacteria* into the mucosa, e.g. caused by long-term intake of a high-fat diet is a major driver of sustained inflammation in the gut and systemically. There are many unknowns in this simplified view. What is causing dysbiosis, and is dysbiosis the trigger of disease, and further, can the apparent inverse proportional relationship between bifidobacteria and *Proteobacteria* be exploited to reverse dysbiosis by providing bifidobacteria to the perturbed microbiota?

The studies of Amar *et al.* (2011a,b) show that increased intestinal adhesion and translocation of *Proteobacteria* into the mucosa and further into the adipose tissue and blood is caused by the intake of a high-fat containing diet, at least in mice, and that translocation induces chronic low-grade inflammation that precedes the onset of insulin resistance and hyperglycaemia. This is an important observation because it suggests that limiting translocation of lipopolysaccharides and *Proteobacteria* may prevent inflammation. Indeed translocation and the subsequent chain of events could be abolished by treatment of mice with a *Bifidobacterium* strain.

In the GI tract *Bifidobacterium* strains alleviate symptoms of diseases associated with inflammation, such as irritable bowel syndrome, ulcerative colitis, necrotizing enterocolitis in preterm infants, and coeliac disease. Various strain-dependent mechanisms are involved, but common themes of symptom relief are found.

Bifidobacteria may attenuate mucosal inflammation by limiting antigen penetration into the mucosa. This is accomplished by inhibition of pathogen growth, adhesion, or toxin production where bifidobacteria act directly on pathogenic microorganisms, e.g. via bacteriocin production, short-chain fatty acid production, competition for mucosal attachment sites, and coaggregation with pathogens. In coeliac disease bifidobacteria may degrade gliadins to less cytotoxic peptides.

Another mechanism by which bifidobacteria limits mucosal antigen load is through stimulation of the mucosal barrier. Bifidobacteria can induce epithelial production of antimicrobial peptides and secretory IgA that accumulate in the inner mucus layer covering the epithelium. Antimicrobial peptides limit pathogen colonization close to the epithelium, whereas secretory IgA restricts bacterial access to the underlying epithelial cells. Moreover, bifidobacteria can preserve cellular junctions between epithelial cells limiting paracellular transport of macromolecules and microorganisms. Although the mechanisms are not fully understood, bifidobacteria appear to increase expression of tight junction proteins and to preserve their localization. The ability to limit mucosal antigen load and improve intestinal barrier function are important health promoting effects of bifidobacteria that should be further investigated. *In vitro* models are available for such studies and should be used to increase our understanding of the molecular mechanisms involved, and to select for strains with a high potential for maintaining intestinal homeostasis and health.

Some *Bifidobacterium* strains promote regulation of the mucosal immune system and reduce the inflammatory state. Several studies have demonstrated activation and maturation by bifidobacteria of dendritic cells producing high levels of anti-inflammatory cytokines that direct naive T-cells towards the T_{REG} subtype. At the same time these bifidobacterial strains induce low polarization towards T_H1 and T_H2 cell types resulting essentially in tolerance and control of inflammation. Regulation of the immune system, locally and systemically is attractive because it could potentially benefit conditions and disorders associated with chronic inflammation. Many such conditions are linked to a modern lifestyle of a high-fat containing diet and physical inactivity, which promotes chronic low-grade inflammation. Again, *in vitro* models are available for mechanistic studies and for screening of *Bifidobacterium* isolates to identify strains with strong immunoregulatory effects.

Probiotic multi-strain products combining *Bifidobacterium* strains capable of reducing antigen-load, improving the intestinal barrier, and eliciting a regulated immune response could potentially have stronger overall effects on gastrointestinal homeostasis and health than single-strain products. Through directed screening and combining strains that each influence these key elements the most, a collective effect may be obtained realizing the full health promoting potential of bifidobacteria. Studies have concluded that low numbers of bifidobacteria are associated with diseases, and even low abundance of specific *Bifidobacterium* species has been observed in type 2 diabetic patients, children with a high degree of distress, and allergic children. Whether low abundance of particular *Bifidobacterium* species is really part of disease remains uncertain, however, these observations could additionally be included when strains are selected for probiotic multi-strain products targeting specific conditions.

More research is required to improve our knowledge on the molecular mechanisms of *Bifidobacterium*–host interactions. Our current understanding of signalling mechanisms is very limited. Some interactions require the direct contact with epithelial or immune cells, while others are induced through bifidobacterial secretion of bioactive molecules. Ligands have rarely been identified, but the few known examples display a molecular diversity extending from surface embedded pili to acetic acid produced in fermentation. This line of research is far more advanced for *Lactobacillus* where gene- and transcriptome-trait matching, and targeted mutagenesis has identified effector molecules involved in host interactions located in the cell envelope (Lee *et al.*, 2013). Thus, comparative genomics and transcriptomics of active versus non-active *Bifidobacterium* strains could help to identify molecules interacting with host receptors. Mutagenesis could subsequently be applied to verify the importance of those molecules for specific interactions. However, such approaches are hampered by the lack of efficient methods for targeted gene mutagenesis in bifidobacteria, although recent progress has been made. Identification of effector molecules is an important step to improve our knowledge on *Bifidobacterium* mechanisms of action and to understand conditions that promote their expression. Moreover, it will help to appreciate the molecular basis for strain-specific functionality and improve the selection of new strains with better efficacy as well as provide molecular measures for quality control in production. In addition, identification of effector molecules and knowledge of bifidobacterial mode of action might help to separate potential responders to a specific treatment from non-responders in the human population, and support stratification of populations in clinical investigations (Bron *et al.*, 2013).

References

Agostini, S., Goubern, M., Tondereau, V., Salvador-Cartier, C., Bezirard, V., Leveque, M., Keranen, H., Theodorou, V., Bourdu-Naturel, S., Goupil-Feuillerat, N., et al. (2012). A marketed fermented dairy product containing Bifidobacterium lactis CNCM I-2494 suppresses gut hypersensitivity and colonic barrier disruption induced by acute stress in rats. Neurogastroenterol. Motil. *24*, 376-e172.

Agrawal, A., Houghton, L.A., Morris, J., Reilly, B., Guyonnet, D., Goupil, F.N., Schlumberger, A., Jakob, S., and Whorwell, P.J. (2009). Clinical trial: the effects of a fermented milk product containing Bifidobacterium lactis DN-173 010 on abdominal distension and gastrointestinal transit in irritable bowel syndrome with constipation. Aliment. Pharmacol. Ther. *29*, 104–114.

Alemka, A., Clyne, M., Shanahan, F., Tompkins, T., Corcionivoschi, N., and Bourke, B. (2010). Probiotic colonization of the adherent mucus layer of HT29MTXE12 cells attenuates *Campylobacter jejuni* virulence properties. Infect. Immun. *78*, 2812–2822.

Allen, S.J., Martinez, E.G., Gregorio, G.V., and Dans, L.F. (2010). Probiotics for treating acute infectious diarrhoea. Cochrane Database Syst. Rev. CD003048.

Amar, J., Chabo, C., Waget, A., Klopp, P., Vachoux, C., Bermudez-Humaran, L.G., Smirnova, N., Berge, M., Sulpice, T., Lahtinen, S., *et al.* (2011a). Intestinal mucosal adherence and translocation of commensal bacteria at the early onset of type 2 diabetes: molecular mechanisms and probiotic treatment. EMBO. Mol. Med. *3*, 559–572.

Amar, J., Serino, M., Lange, C., Chabo, C., Iacovoni, J., Mondot, S., Lepage, P., Klopp, C., Mariette, J., Bouchez, O., *et al.* (2011b). Involvement of tissue bacteria in the onset of diabetes in humans: evidence for a concept. Diabetologia *54*, 3055–3061.

Amaretti, A., di Nunzio, M., Pompei, A., Raimondi, S., Rossi, M., and Bordoni, A. (2013). Antioxidant properties of potentially probiotic bacteria: *in vitro* and *in vivo* activities. Appl. Microbiol. Biotechnol. *97*, 809–817.

Andriantsoanirina, V., Allano, S., Butel, M.J., and Aires, J. (2013). Tolerance of *Bifidobacterium* human isolates to bile, acid and oxygen. Anaerobe *21*, 39–42.

Asahara, T., Shimizu, K., Nomoto, K., Hamabata, T., Ozawa, A., and Takeda, Y. (2004). Probiotic bifidobacteria protect mice from lethal infection with Shiga toxin-producing *Escherichia coli* O157:H7. Infect. Immun. *72*, 2240–2247.

Atuma, C., Strugala, V., Allen, A., and Holm, L. (2001). The adherent gastrointestinal mucus gel layer: thickness and physical state *in vivo*. Am. J. Physiol. Gastrointest. Liver Physiol. *280*, G922-G929.

Beal, C., Fonseca, F., and Corrieu, G. (2001). Resistance to freezing and frozen storage of *Streptococcus thermophilus* is related to membrane fatty acid composition. J. Dairy Sci. *84*, 2347–2356.

Becker, S., Oelschlaeger, T.A., Wullaert, A., Vlantis, K., Pasparakis, M., Wehkamp, J., Stange, E.F., and Gersemann, M. (2013). Bacteria regulate intestinal epithelial cell differentiation factors both *in vitro* and *in vivo*. PLoS One *8*, e55620.

Bennedsen, M., Stuer-Lauridsen, B., Danielsen, M., and Johansen, E. (2011). Screening for antimicrobial resistance genes and virulence factors via genome sequencing. Appl. Environ. Microbiol. *77*, 2785–2787.

Bercik, P., Park, A.J., Sinclair, D., Khosdel, A., Lu, J., Huang, X., Deng, Y., Blennerhassett, P.A., Fahnestock, M., Moine, D., *et al.* (2011). The anxiolytic effect of *Bifidobacterium longum* NCC3001 involves vagal pathways for gut–brain communication. Neurogastroenterol. Motil. *23*, 1132–1139.

Bergmann, K.R., Liu, S.X., Tian, R., Kushnir, A., Turner, J.R., Li, H.L., Chou, P.M., Weber, C.R., and De Plaen, I.G. (2013). Bifidobacteria stabilize claudins at tight junctions and prevent intestinal barrier dysfunction in mouse necrotizing enterocolitis. Am. J. Pathol. *182*, 1595–1606.

Biagi, E., Nylund, L., Candela, M., Ostan, R., Bucci, L., Pini, E., Nikkila, J., Monti, D., Satokari, R., Franceschi, C., *et al.* (2010). Through ageing, and beyond: gut microbiota and inflammatory status in seniors and centenarians. PLoS One *5*, e10667.

Boesten, R., Schuren, F., Wind, R.D., Knol, J., and De Vos, W.M. (2011). Analysis of infant isolates of *Bifidobacterium breve* by comparative genome hybridization indicates the existence of new subspecies with marked infant specificity. Res. Microbiol. *162*, 664–670.

Bolduc, M.-P., Raymond, Y., Fustier, P., Champagne, C.P., and Vuillemard, J.-C. (2006). Sensitivity of bifidobacteria to oxygen and redox potential in non-fermented pasteurized milk. Int. Dairy J. *16*, 1038–1048.

Bomhof, M.R., Saha, D.C., Reid, D.T., Paul, H.A., and Reimer, R.A. (2013). Combined effects of oligofructose and *Bifidobacterium animalis* on gut microbiota and glycemia in obese rats. Obesity (Silver Spring) *22*, 763–771.

Bordoni, A., Amaretti, A., Leonardi, A., Boschetti, E., Danesi, F., Matteuzzi, D., Roncaglia, L., Raimondi, S., and Rossi, M. (2013). Cholesterollowering probiotics: *in vitro* selection and *in vivo* testing of bifidobacteria. Appl. Microbiol. Biotechnol. *97*, 8273–8281.

Bottacini, F., Medini, D., Pavesi, A., Turroni, F., Foroni, E., Riley, D., Giubellini, V., Tettelin, H., Van Sinderen, D., and Ventura, M. (2010). Comparative genomics of the genus *Bifidobacterium*. Microbiology *156*, 3243–3254.

Brasili, E., Mengheri, E., Tomassini, A., Capuani, G., Roselli, M., Finamore, A., Sciubba, F., Marini, F., and Miccheli, A. (2013). *Lactobacillus acidophilus* La5 and *Bifidobacterium lactis* Bb12 induce different agerelated metabolic profiles revealed by 1H-NMR spectroscopy in urine and feces of mice. J. Nutr. *143*, 1549–1557.

Brenner, D.M., and Chey, W.D. (2009). *Bifidobacterium infantis* 35624: a novel probiotic for the treatment of irritable bowel syndrome. Rev. Gastroenterol. Disord. *9*, 7–15.

Broadbent, J.R., Larsen, R.L., Deibel, V., and Steele, J.L. (2010). Physiological and transcriptional response of *Lactobacillus casei* ATCC 334 to acid stress. J. Bacteriol. *192*, 2445–2458.

Bron, P.A., Tomita, S., Mercenier, A., and Kleerebezem, M. (2013). Cell surface-associated compounds of probiotic lactobacilli sustain the strai specificity dogma. Curr. Opin. Microbiol. *16*, 262–269.

Caballero-Franco, C., Keller, K., De Simone, C., and Chadee, K. (2007). The VSL#3 probiotic formula induces mucin gene expression and secretion in colonic epithelial cells. Am. J. Physiol. Gastrointest. Liver Physiol. *292*, G315-G322.

Caglar, E., Kuscu, O.O., Selvi, K.S., Kavaloglu, C.S., Sandalli, N., and Twetman, S. (2008). Short-term effect of ice-cream containing *Bifidobacterium lactis* Bb-12 on the number of salivary mutans streptococci and lactobacilli. Acta Odontol. Scand. *66*, 154–158.

Cani, P.D., Amar, J., Iglesias, M.A., Poggi, M., Knauf, C., Bastelica, D., Neyrinck, A.M., Fava, F., Tuohy, K.M., Chabo, C., *et al.* (2007). Metabolic endotoxemia initiates obesity and insulin resistance. Diabetes *56*, 1761–1772.

Cano, P.G., Santacruz, A., Trejo, F.M., and Sanz, Y. (2013). *Bifidobacterium* CECT 7765 improves metabolic and immunological alterations associated with obesity in high-fat diet-fed mice. Obesity (Silver Spring) *21*, 2310–2321.

Celik, O.F., and O'Sullivan, D.J. (2013). Factors influencing the stability of freeze-dried stress-resilient and stress-sensitive strains of bifidobacteria. J. Dairy Sci. *96*, 3506–3516.

Champagne, C.P., Mondou, F., Raymond, Y., and Roy, D. (1996). Effect of polymers and storage temperature on the stability of freeze-dried lactic acid bacteria. Food Res. Int. *29*, 555–562.

Chen, J.J., Wang, R., Li, X.F., and Wang, R.L. (2011). *Bifidobacterium longum* supplementation improved high-fat-fed-induced metabolic syndrome and promoted intestinal Reg I gene expression. Exp. Biol. Med. *236*, 823–831.

Chenoll, E., Casinos, B., Bataller, E., Astals, P., Echevarria, J., Iglesias, J.R., Balbarie, P., Ramon, D., and Genoves, S. (2011). Novel probiotic *Bifidobacterium bifidum* CECT 7366 strain active against the pathogenic bacterium *Helicobacter pylori*. Appl. Environ. Microbiol. *77*, 1335–1343.

Chouraqui, J.P., Van Egroo, L.D., and Fichot, M.C. (2004). Acidified milk formula supplemented with *Bifidobacterium lactis*: impact on infant diarrhoea in residential care settings. J. Pediatr. Gastroenterol. Nutr. *38*, 288–292.

Cinova, J., De Palma, G., Stepankova, R., Kofronova, O., Kverka, M., Sanz, Y., and Tuckova, L. (2011). Role of intestinal bacteria in gliadin-induced changes in intestinal mucosa: study in germ-free rats. PLoS One *6*, e16169.

Collado, M.C., Gueimonde, M., Hernandez, M., Sanz, Y., and Salminen, S. (2005). Adhesion of selected *Bifidobacterium* strains to human intestinal mucus and the role of adhesion in enteropathogen exclusion. J. Food Prot. *68*, 2672–2678.

Collado, M.C., Gueimonde, M., Sanz, Y., and Salminen, S. (2006). Adhesion properties and competitive pathogen exclusion ability of bifidobacteria with acquired acid resistance. J. Food Prot. *69*, 1675–1679.

Collado, M.C., and Sanz, Y. (2007). Induction of acid resistance in Bifidobacterium: a mechanism for improving desirable traits of potentially probiotic strains. J. Appl. Microbiol. *103*, 1147–1157.

Collins, E.B., and Hall, B.J. (1983). Growth of bifidobacteria in milk and preparation of *Bifidobacterium infantis* for a dietary adjunct. J. Dairy Sci. *67*, 1376–1380.

Commane, D.M., Shortt, C.T., Silvi, S., Cresci, A., Hughes, R.M., and Rowland, I.R. (2005). Effects of fermentation products of pro- and prebiotics on trans-epithelial electrical resistance in an *in vitro* model of the colon. Nutr. Cancer *51*, 102–109.

Coppa, G.V., Bruni, S., Morelli, L., Soldi, S., and Gabrielli, O. (2004). The first prebiotics in humans: human milk oligosaccharides. J. Clin. Gastroenterol. *38*, S80-S83.

Corcoran, B.M., Stanton, C., Fitzgerald, G., and Ross, R.P. (2008). Life under stress: the probiotic stress response and how it may be manipulated. Curr. Pharm. Des. *14*, 1382–1399.

De Vuyst, L., Moens, F., Selak, M., Riviere, A., and Leroy, F. (2013). Summer Meeting 2013: growth and physiology of bifidobacteria. J. Appl. Microbiol. *116*, 477-491.

De Weerth, C., Fuentes, S., and De Vos, W.M. (2013). Crying in infants: on the possible role of intestinal microbiota in the development of colic. Gut Microbes *4*, 416–421.

Deplancke, B., and Gaskins, H.R. (2001). Microbial modulation of innate defense: goblet cells and the intestinal mucus layer. Am. J. Clin. Nutr. *73*, 1131S-1141S.

Desjardins, M.-L., Roy, D., Toupin, C., and Goulet, J. (1990). Uncoupling of growth and acids production in *Bifidobacterium* subsp. J. Dairy Sci. *73*, 1478–1484.

Desbonnet, L., Garrett, L., Clarke, G., Kiely, B., Cryan, J.F., and Dinan, T.G. (2010). Effects of the probiotic *Bifidobacterium infantis* in the maternal separation model of depression. Neuroscience *170*, 1179–1188.

Di Cagno, R., De Angelis, M., De Pasquale, I., Ndagijimana, M., Vernocchi, P., Ricciuti, P., Gagliardi, F., Laghi, L., Crecchio, C., Guerzoni, M.E., *et al.* (2011). Duodenal and faecal microbiota of celiac children: molecular, phenotype and metabolome characterization. BMC Microbiol. *11*, 219.

Ding, W.K., and Shah, N.P. (2007). Acid, bile, and heat tolerance of free and microencapsulated probiotic bacteria. J. Food Sci. *72*, M446–50.

Dong, H., Rowland, I., and Yaqoob, P. (2012). Comparative effects of six probiotic strains on immune function *in vitro*. Br. J. Nutr. *108*, 459–470.

Donkor, O.N., Ravikumar, M., Proudfoot, O., Day, S.L., Apostolopoulos, V., Paukovics, G., Vasiljevic, T., Nutt, S.L., and Gill, H. (2012). Cytokine profile and induction of T helper type 17 and regulatory T-cells by human peripheral mononuclear cells after microbial exposure. Clin. Exp. Immunol. *167*, 282–295.

Dunne, C., Murphy, L., Flynn, S., O'Mahony, L., O'Halloran, S., Feeney, M., Morrissey, D., Thornton, G., Fitzgerald, G., Daly, C., *et al.* (1999). Probiotics: from myth to reality. Demonstration of functionality in animal models of disease and in human clinical trials. Antonie Van Leeuwenhoek *76*, 279–292.

Dykstra, N.S., Hyde, L., Adawi, D., Kulik, D., Ahrne, S., Molin, G., Jeppsson, B., Mackenzie, A., and Mack, D.R. (2011). Pulse probiotic administration induces repeated small intestinal Muc3 expression in rats. Pediatr. Res. *69*, 206–211.

du Toit, E., Vesterlund, S., Gueimonde, M., and Salminen, S. (2013). Assessment of the effect of stress-tolerance acquisition on some basic characteristics of specific probiotics. Int. J. Food Microbiol. *165*, 51–56.

Erdogan, O., Tanyeri, B., Torun, E., Gonullu, E., Arslan, H., Erenberk, U., and Oktem, F. (2012). The comparition of the efficacy of two different probiotics in rotavirus gastroenteritis in children. J. Trop. Med. *2012*, 787240.

Fallani, M., Amarri, S., Uusijarvi, A., Adam, R., Khanna, S., Aguilera, M., Gil, A., Vieites, J.M., Norin, E., Young, D., *et al.* (2011). Determinants of the human infant intestinal microbiota after the introduction of first complementary foods in infant samples from five European centres. Microbiology *157*, 1385–1392.

Fanning, S., Hall, L.J., Cronin, M., Zomer, A., Macsharry, J., Goulding, D., Motherway, M.O., Shanahan, F., Nally, K., Dougan, G., and Van Sinderen, D. (2012). Bifidobacterial surface-exopolysaccharide facilitates commensal–host interaction through immune modulation and pathogen protection. Proc. Natl. Acad. Sci. U.S.A. *109*, 2108–2113.

Favier, C.F., Vaughan, E.E., De Vos, W.M., and Akkermans, A.D. (2002). Molecular monitoring of succession of bacterial communities in human neonates. Appl. Environ. Microbiol. *68*, 219–226.

Feleszko, W., Jaworska, J., Rha, R.D., Steinhausen, S., Avagyan, A., Jaudszus, A., Ahrens, B., Groneberg, D.A., Wahn, U., and Hamelmann, E. (2007). Probiotic-induced suppression of allergic sensitization and airway inflammation is associated with an increase of T regulatory-dependent mechanisms in a murine model of asthma. Clin. Exp. Allergy *37*, 498–505.

Finegold, S.M., Dowd, S.E., Gontcharova, V., Liu, C., Henley, K.E., Wolcott, R.D., Youn, E., Summanen, P.H., Granpeesheh, D., Dixon, D., *et al.* (2010). Pyrosequencing study of faecal microflora of autistic and control children. Anaerobe *16*, 444–453.

Foroni, E., Serafini, F., Amidani, D., Turroni, F., He, F., Bottacini, F., O'Connell, M.M., Viappiani, A., Zhang, Z., Rivetti, C., *et al.* (2011). Genetic analysis and morphological identification of pilus-like structures in members of the genus *Bifidobacterium*. Microb. Cell. Fact. *10 Suppl. 1*, S16.

Forssten, S.D., Sindelar, C.W., and Ouwehand, A.C. (2011). Probiotics from an industrial perspective. Anaerobe *17*, 410–413.

Fu, Y.R., Yi, Z.J., Pei, J.L., and Guan, S. (2010). Effects of *Bifidobacterium bifidum* on adaptive immune senescence in ageing mice. Microbiol. Immunol. *54*, 578–583.

Fukuda, S., Toh, H., Hase, K., Oshima, K., Nakanishi, Y., Yoshimura, K., Tobe, T., Clarke, J.M., Topping, D.L., Suzuki, T., *et al.* (2011). Bifidobacteria can protect from enteropathogenic infection through production of acetate. Nature *469*, 543–547.

Furrie, E., Macfarlane, S., Kennedy, A., Cummings, J.H., Walsh, S.V., O'neil, D.A., and Macfarlane, G.T. (2005). Synbiotic therapy (*Bifidobacterium longum*/Synergy 1). initiates resolution of inflammation in patients with active ulcerative colitis: a randomized controlled pilot trial. Gut *54*, 242–249.

Fyderek, K., Strus, M., Kowalska-Duplaga, K., Gosiewski, T., Wedrychowicz, A., Jedynak-Wasowicz, U., Sladek, M., Pieczarkowski, S., Adamski, P., Kochan, P., *et al.* (2009). Mucosal bacterial microflora and mucus layer thickness in adolescents with inflammatory bowel disease. World J. Gastroenterol. *15*, 5287–5294.

Gagnon, M., Kheadr, E.E., Dabour, N., Richard, D., and Fliss, I. (2006). Effect of *Bifidobacterium thermacidophilum* probiotic feeding on enterohemorrhagic *Escherichia coli* O157:H7 infection in BALB/c mice. Int. J. Food Microbiol. *111*, 26–33.

Garrido, D., Barile, D., and Mills, D.A. (2012). A molecular basis for bifidobacterial enrichment in the infant gastrointestinal tract. Adv. Nutr. *3*, 415S–421S.

Garrido, D., Kim, J.H., German, J.B., Raybould, H.E., and Mills, D.A. (2011). Oligosaccharide binding proteins from *Bifidobacterium longum* subsp. *infantis* reveal a preference for host glycans. PLoS One *6*, e17315.

Garrigues, C., Johansen, E., and Pedersen, M.B. (2005). Complete genome sequence of *Bifidobacterium animalis* subsp. *lactis* BB-12, a widely consumed probiotic strain. J. Bacteriol. *192*, 2467–2468.

Garrigues, C., Johansen, E., and Crittenden, R. (2013). Pangenomics – an avenue to improved industrial starter cultures and probiotics. Curr. Opin. Biotechnol. *24*, 187–191.

Gilad, O., Svensson, B., Viborg, A.H., Stuer-Lauridsen, B., and Jacobsen, S. (2011). The extracellular proteome of *Bifidobacterium animalis* subsp. *lactis* BB-12 reveals proteins with putative roles in probiotic effects. Proteomics *11*, 2503–2514.

Gleinser, M., Grimm, V., Zhurina, D., Yuan, J., and Riedel, C.U. (2012). Improved adhesive properties of recombinant bifidobacteria expressing the *Bifidobacterium bifidum*-specific lipoprotein BopA. Microb. Cell. Fact. *11*, 80.

Goldenberg, J.Z., Ma, S.S., Saxton, J.D., Martzen, M.R., Vandvik, P.O., Thorlund, K., Guyatt, G.H., and Johnston, B.C. (2013). Probiotics for the prevention of *Clostridium difficile*-associated diarrhea in adults and children. Cochrane Database Syst. Rev. *5*, CD006095.

Gomes, A.M.P., and Malcata, F.X. (1999). *Bifidobacterium* spp. and *Lactobacillus acidophilus*: Biological, biochemical, technological and therapeutical properties relevant for use as probiotics. Trends Food Sci. Tech. *10*, 139–157.

Groeger, D., O'Mahony, L., Murphy, E.F., Bourke, J.F., Dinan, T.G., Kiely, B., Shanahan, F., and Quigley, E.M. (2013). *Bifidobacterium infantis* 35624 modulates host inflammatory processes beyond the gut. Gut Microbes *4*, 325–339.

Grześkowiak, Ł., Isolauri, E., Salminen, S., and Gueimonde, M. (2011). Manufacturing process influences properties of probiotic bacteria. Br. J. Nutr. *105*, 887–894.

Gueimonde, M., Margolles, A., de los Reyes-Gavilan C.G., and Salminen, S. (2007). Competitive exclusion of enteropathogens from human intestinal mucus by *Bifidobacterium* strains with acquired resistance to bile – a preliminary study. Int. J. Food Microbiol. *113*, 228–232.

Gueimonde, M., Florez, A.B., van Hoek, A.H., Stuer-Lauridsen, B., Stroman, P., de los Reyes-Gavilan C.G., and Margolles, A. (2010). Genetic basis of tetracycline resistance in *Bifidobacterium animalis* subsp. *lactis*. Appl. Environ. Microbiol. *76*, 3364–3369.

Gueimonde, M., Sanchez, B., de los Reyes-Gavilan C.G., and Margolles, A. (2013). Antibiotic resistance in probiotic bacteria. Frontiers in Microbiol. *4*, 1–6.

Guglielmetti, S., Mora, D., Gschwender, M., and Popp, K. (2011). Randomized clinical trial: *Bifidobacterium bifidum* MIMBb75 significantly alleviates irritable bowel syndrome and improves quality of life – a double-blind, placebo-controlled study. Aliment. Pharmacol. Ther. *33*, 1123–1132.

Guyonnet, D., Chassany, O., Ducrotte, P., Picard, C., Mouret, M., Mercier, C.-H., and Matuchansky, C. (2007). Effect of a fermented milk containing *Bifidobacterium animalis* DN-173 010 on the health-related quality of life and symptoms in irritable bowel syndrome in adults in primary care: a multicentre, randomized, double-blind, controlled trial. Aliment. Pharmacol. Ther. *26*, 475–486.

Hassinen, J.B., Durbin, G.T., Tomarelli, R.M., and Bernhart, F.W. (1951). The minimal nutritional requirements of *Lactobacillus bifidus*. J. Bacteriol. *62*, 771–777.

Herfel, T.M., Jacobi, S.K., Lin, X., Jouni, Z.E., Chichlowski, M., Stahl, C.H., and Odle, J. (2013). Dietary supplementation of *Bifidobacterium longum* strain AH1206 increases its cecal abundance and elevates intestinal interleukin-10 expression in the neonatal piglet. Food Chem. Toxicol. *60*, 116–122.

Hiramatsu, Y., Hosono, A., Konno, T., Nakanishi, Y., Muto, M., Suyama, A., Hachimura, S., Sato, R., Takahashi, K., and Kaminogawa, S. (2011). Orally administered *Bifidobacterium* triggers immune responses following capture by CD11c(+). cells in Peyer's patches and caecal patches. Cytotechnology *63*, 307–317.

Hoarau, C., Lagaraine, C., Martin, L., Velge-Roussel, F., and Lebranchu, Y. (2006). Supernatant of *Bifidobacterium breve* induces dendritic cell maturation, activation, and survival through a Toll-like receptor 2 pathway. J. Allergy Clin. Immunol. *117*, 696–702.

Holscher, H.D., Czerkies, L.A., Cekola, P., Litov, R., Benbow, M., Santema, S., Alexander, D.D., Perez, V., Sun, S., Saavedra, J.M., *et al.* (2012). *Bifidobacterium lactis* Bb12 enhances intestinal antibody response in formula-fed infants: a randomized, double-blind, controlled trial. JPEN. J. Parenter. Enteral. Nutr. *36*, 106S-117S.

Jalanka-Tuovinen, J., Salonen, A., Nikkila, J., Immonen, O., Kekkonen, R., Lahti, L., Palva, A., and De Vos, W.M. (2011). Intestinal microbiota in healthy adults: temporal analysis reveals individual and common core and relation to intestinal symptoms. PLoS One *6*, e23035.

Jin, J., Zhang, B., Guo, H., Cui, J., Jiang, L., Song, S., Sun, M., and Ren, F. (2012). Mechanism analysis of acid tolerance response of *Bifidobacterium longum* subsp. *longum* BBMN 68 by gene expression profile using RNAsequencing. PLoS One *7*, e50777.

Johansson, M.E., Phillipson, M., Petersson, J., Velcich, A., Holm, L., and Hansson, G.C. (2008). The inner of the two Muc2 mucin-dependent mucus layers in colon is devoid of bacteria. Proc. Natl. Acad. Sci. U.S.A. *105*, 15064–15069.

Johnson, A.C., Beverley Greenwood-Van Meerveld, B., and McRorie, J. (2011). Effects of *Bifidobacterium infantis* 35624 on post-inflammatory visceral hypersensitivity in the rat. Dig. Dis. Sci. *56*, 3179–3186.

Johnston, B.C., Goldenberg, J.Z., Vandvik, P.O., Sun, X., and Guyatt, G.H. (2011). Probiotics for the prevention of pediatric antibiotic-associated diarrhea. Cochrane Database Syst. Rev. CD004827.

Kabeerdoss, J., Shobana, D.R., Regina, M.R., Prabhavathi, D., Vidya, R., Mechenro, J., Mahendri, N.V., Pugazhendhi, S., and Ramakrishna, B.S. (2011). Effect of yoghurt containing *Bifidobacterium lactis* Bb12 – on faecal excretion of secretory immunoglobulin A and human beta-defensin-2 in healthy adult volunteers. Nutr. J. *10*, 138.

Kainulainen, V., Reunanen, J., Hiippala, K., Guglielmetti, S., Vesterlund, S., Palva, A., and Satokari, R. (2013). BopA does not have a major role in the adhesion of *Bifidobacterium bifidum* to intestinal epithelial cells, extracellular matrix proteins, and mucus. Appl. Environ. Microbiol. *79*, 6989–6997.

Kalliomaki, M., Collado, M.C., Salminen, S., and Isolauri, E. (2008). Early differences in fecal microbiota composition in children may predict overweight. Am. J. Clin. Nutr. *87*, 534–538.

Kavanaugh, D.W., O'Callaghan, J., Butto, L.F., Slattery, H., Lane, J., Clyne, M., Kane, M., Joshi, L., and Hickey, R.M. (2013). Exposure of *Bifidobacterium longum* subsp. *infantis* to milk oligosaccharides increases adhesion to epithelial cells and induces a substantial transcriptional response. PLoS One *8*, e67224.

Kazimierczak, K.A., Flint, H.J., and Scott, K.P. (2006). Comparative analysis of sequences flanking tet(W) resistance genes in multiple species of gut bacteria. Antimicrob. Agents Chemother. *50*, 2632–2639.

Khailova, L., Dvorak, K., Arganbright, K.M., Halpern, M.D., Kinouchi, T., Yajima, M., and Dvorak, B. (2009). *Bifidobacterium bifidum* improves intestinal integrity in a rat model of necrotizing enterocolitis. Am. J. Physiol. Gastrointest. Liver Physiol. *297*, G940-G949.

Killer, J., and Marounek, M. (2011). Fermentation of mucin by bifidobacteria from rectal samples of humans and rectal and intestinal samples of animals. Folia Microbiol. (Praha) *56*, 85–89.

Kiviharju, K., Leisola, M., and Eerikäinen, T. (2005). Optimization of a *Bifidobacterium longum* production process. J. Biotechnol. *117*, 299–308.

Konieczna, P., Akdis, C.A., Quigley, E.M., Shanahan, F., and O'Mahony, L. (2012a). Portrait of an immunoregulatory *Bifidobacterium*. Gut Microbes *3*, 261–266.

Konieczna, P., Groeger, D., Ziegler, M., Frei, R., Ferstl, R., Shanahan, F., Quigley, E.M., Kiely, B., Akdis, C.A., and O'Mahony, L. (2012b). *Bifidobacterium infantis* 35624 administration induces Foxp3 T regulatory cells in human peripheral blood: potential role for myeloid and plasmacytoid dendritic cells. Gut *61*, 354–366.

Kwon, H.K., Lee, C.G., So, J.S., Chae, C.S., Hwang, J.S., Sahoo, A., Nam, J.H., Rhee, J.H., Hwang, K.C., and Im, S.H. (2010). Generation of regulatory dendritic cells and CD4+Foxp3+ T-cells by probiotics administration suppresses immune disorders. Proc. Natl. Acad. Sci. U.S.A. *107*, 2159–2164.

Laparra, J.M., and Sanz, Y. (2010). Bifidobacteria inhibit the inflammatory response induced by gliadins in intestinal epithelial cells via modifications of toxic peptide generation during digestion. J. Cell. Biochem. *109*, 801–807.

Laparra, J.M., Olivares, M., Gallina, O., and Sanz, Y. (2012). *Bifidobacterium longum* CECT 7347 modulates immune responses in a gliadin-induced enteropathy animal model. PLoS One *7*, e30744.

Lavasani, S., Dzhambazov, B., Nouri, M., Fak, F., Buske, S., Molin, G., Thorlacius, H., Alenfall, J., Jeppsson, B., and Westrom, B. (2010). A novel probiotic mixture exerts a therapeutic effect on experimental autoimmune encephalomyelitis mediated by IL-10 producing regulatory T-cells. PLoS One *5*, e9009.

Lazarenko, L., Babenko, L., Sichel, L.S., Pidgorskyi, V., Mokrozub, V., Voronkova, O., and Spivak, M. (2012). Antagonistic action of lactobacilli and bifidobacteria in relation to *Staphylococcus aureus* and their influence on the immune response in cases of intravaginal Staphylococcosis in mice. Probiotics Antimicrob. Proteins *4*, 78–89.

Lê, K.A., Li, Y., Xu, X., Yang, W., Liu, T., Zhao, X., Tang, Y.G., Cai, D., Go, V.L., Pandol, S., *et al.* (2012). Alterations in fecal *Lactobacillus* and *Bifidobacterium* species in type 2 diabetic patients in Southern China population. Front. Physiol. *3*, 496.

Lee, I.C., Tomita, S., Kleerebezem, M., and Bron, P.A. (2013). The quest for probiotic effector molecules – unraveling strain specificity at the molecular level. Pharmacol. Res. *69*, 61–74.

Lee, J.H., and O'Sullivan, D.J. (2010). Genomic insights into bifidobacteria. Microbiol. Mol. Biol. Rev. *74*, 378–416.

Li, C., Zhao, J.-L., Wang, Y.-T., Han, X., and Liu, N. (2009). Synthesis of cyclopropane fatty acid and its effect on freeze-drying survival of *Lactobacillus bulgaricus* L2 at different growth conditions. World J. Microbiol. Biotechnol. *25*, 1659–1565.

Lian, W.C., Hsiao, H.C., and Chou, C.C. (2003). Viability of microencapsulated bifidobacteria in simulated gastric juice and bile solution. Int. J. Food Microbiol. *86*, 293–301.

Lin, A., Bik, E.M., Costello, E.K., Dethlefsen, L., Haque, R., Relman, D.A., and Singh, U. (2013). Distinct distal gut microbiome diversity and composition in healthy children from Bangladesh and the United States. PLoS One *8*, e53838.

Lindfors, K., Blomqvist, T., Juuti-Uusitalo, K., Stenman, S., Venalainen, J., Maki, M., and Kaukinen, K. (2008). Live probiotic *Bifidobacterium lactis* bacteria inhibit the toxic effects induced by wheat gliadin in epithelial cell culture. Clin. Exp. Immunol. *152*, 552–558.

Lopez, P., Gueimonde, M., Margolles, A., and Suarez, A. (2010). Distinct *Bifidobacterium* strains drive different immune responses *in vitro*. Int. J. Food Microbiol. *138*, 157–165.

Lopez, P., Gonzalez-Rodriguez, I., Gueimonde, M., Margolles, A., and Suarez, A. (2011). Immune response to *Bifidobacterium bifidum* strains support Treg/Th17 plasticity. PLoS One *6*, e24776.

Lopez, P., Gonzalez-Rodriguez, I., Sanchez, B., Ruas-Madiedo, P., Suarez, A., Margolles, A., and Gueimonde, M. (2012). Interaction of *Bifidobacterium bifidum* LMG13195 with HT29 cells influences regulatory-T-cellassociated chemokine receptor expression. Appl. Environ. Microbiol. *78*, 2850–2857.

Loquasto, J.R., Barrangou, R., Dudley, E.G., Stahl, B., Chen, C., and Roberts, R.F. (2013). *Bifidobacterium animalis* subsp. *lactis* ATCC 27673 is a genomically unique strain within its conserved subspecies. Appl. Environ. Microbiol. *79*, 6903–6910.

Lukjancenko, O., Ussery, D.W., and Wassenaar, T.M. (2012). Comparative genomics of *Bifidobacterium, Lactobacillus* and related probiotic genera. Microb. Ecol. *63*, 651–673.

Makinen, K., Berger, B., Bel-Rhlid, R., and Ananta, E. (2012). Science and technology for the mastership of probiotic applications in food products. J. Biotechnol. *162*, 356–365.

Marteau, P., Cuillerier, E., Meance, S., Gerhardt, M.F., Myara, A., Bouviers, M., Bouley, C., Tondu, F., Bommelaer, G. and Grimauds, J.C. (2002). *Bifidobacterium animalis* strain DN-173 010 shortens the colonic transit time in healthy women: a double-blind, randomized, controlled study. Aliment. Pharmacol. Ther. *16*, 587-593.

Martins, A.K., Martins, F.S., Gomes, D.A., Elian, S.D., Vieira, A.T., Teixeira, M.M., Cara, D.C., Nardi, R.M., and Nicoli, J.R. (2010). Evaluation of in vitro antagonism and of in vivo immune modulation and protection against pathogenic experimental challenge of two probiotic strains of *Bifidobacterium animalis* var. *lactis*. Arch. Microbiol. *192*, 995–1003.

Masco, L., Huys, G., De Brandt, E., Temmerman, R., and Swings, J. (2005). Int. J. Food Microbiol. *102*, 221–230.

Masco, L., Crockaert, C., Van Hoorde, K., Swings, J., and Huys, G. (2007). *In vitro* assessment of the gastrointestinal transit tolerance of taxonomic reference strains from human origin and probiotic product isolates of *Bifidobacterium*. J. Dairy Sci. *90*, 3572–3578.

Matsumoto, M., Kurihara, S., Kibe, R., Ashida, H., and Benno, Y. (2011). Longevity in mice is promoted by probiotic-induced suppression of colonic senescence dependent on upregulation of gut bacterial polyamine production. PLoS One *6*, e23652.

Matsumoto, M., Ishige, A., Yazawa, Y., Kondo, M., Muramatsu, K., and Watanabe, K. (2012). Promotion of intestinal peristalsis by *Bifidobacterium* spp. capable of hydrolysing sennosides in mice. PLoS One 7, e31700.

Matsumoto, M., Ohishi, H., and Benno, Y. (2004). H+-ATPase activity in *Bifidobacterium* with special reference to acid tolerance. Int. J. Food Microbiol. *93*, 109–113.

Mattila-Sandholm, T., Myllärinen, P., Crittenden, R., Mogensen, G., Fondén, R., and Saarela, M. (2002). Technological challenges for future probiotic foods. Int. Dairy J. *12*, 173–182.

Mattö, J., Alakomi, H.L., Vaari, A., Virkajarvi, I., and Saarela, M. (2006). Influence of processing conditions on *Bifidobacterium animalis* subsp. *lactis* functionality with a special focus on acid tolerance and factors affecting it. Int. Dairy J. *16*, 1029–1037.

Maus, J.E., and Ingham, S.C. (2003). Employment of stressful conditions during vulture production to enhance subsequent cold- and acidtolerance of bifidobacteria. J. Appl. Microbiol. *95*, 146–154.

McCarthy, J., O'Mahony, L., O'Callaghan, L., Sheil, B., Vaughan, E.E., Fitzsimons, N., Fitzgibbon, J., O'Sullivan, G.C., Kiely, B., Collins, J.K., *et al.* (2003). Double blind, placebo controlled trial of two probiotic strains in interleukin 10 knockout mice and mechanistic link with cytokine balance. Gut. *52*, 975–980.

McKernan, D.P., Fitzgerald, P., Dinan, T.G., and Cryan, J.F. (2010). The probiotic *Bifidobacterium infantis* 35624 displays visceral antinociceptive effects in the rat. Neurogastroenterol. Motil. *22*, 1029–1035.

Miller, L.E., and Ouwehand, A.C. (2013). Probiotic supplementation decreases intestinal transit time: meta-analysis of randomized controlled trials. World J. Gastroenterol. *19*, 4718–4725.

Miyazaki, K., Masuoka, N., Kano, M., and Iizuka, R. (2013). *Bifidobacterium* fermented milk and galacto-oligosaccharides lead to improved skin health by decreasing phenols production by gut microbiota. Benef. Microbes. *5*, 121–128.

Mogna, L., Del Piano, M., Deidda, F., Nicola, S., Soattini, L., Debiaggi, R., Sforza, F., Strozzi, G., and Mogna, G. (2012). Assessment of the *in vitro* inhibitory activity of specific probiotic bacteria against different *Escherichia coli* strains. J. Clin. Gastroenterol. *46 Suppl*, S29-S32.

Moya-Perez, A., Romo-Vaquero, M., Tomas-Barberan, F., Sanz, Y., and Garcia-Conesa, M.T. (2013). Hepatic molecular responses to *Bifidobacterium pseudocatenulatum* CECT 7765 in a mouse model of dietinduced obesity. Nutr. Metab. Cardiovasc. Dis. *24*, 57-64.

Munoz, J.A., Chenoll, E., Casinos, B., Bataller, E., Ramon, D., Genoves, S., Montava, R., Ribes, J.M., Buesa, J., Fabrega, J., *et al.* (2011). Novel probiotic *Bifidobacterium longum* subsp. *infantis* CECT 7210 strain active against rotavirus infections. Appl. Environ. Microbiol. *77*, 8775–8783.

Munoz-Rojas, J., Bernal, P., Duque, E., Godoy, P., Segura, A., and Ramos, J.L. (2006). Involvement of cyclopropane fatty acids in the response of *Pseudomonas putida* KT2440 to freeze-drying. Appl. Environ. Microbiol. *72*, 472–477.

Nadal, I., Donat, E., Ribes-Koninckx, C., Calabuig, M., and Sanz, Y. (2007). Imbalance in the composition of the duodenal microbiota of children with coeliac disease. J. Med. Microbiol. *56*, 1669–1674.

Natividad, J.M., Hayes, C.L., Motta, J.P., Jury, J., Galipeau, H.J., Philip, V., Garcia-Rodenas, C.L., Kiyama, H., Bercik, P., and Verdu, E.F. (2013). Differential Induction of antimicrobial REGIII by the intestinal microbiota and *Bifidobacterium breve* NCC2950. Appl. Environ. Microbiol. *79*, 7745–7754.

Oberg, T.S., Steele, J.L., Ingham, S.C., Smeianov, V.V., Briczinski, E.P., Abdalla, A., and Broadbent, J.R. (2011). Intrinsic and inducible resistance to hydrogen peroxide in *Bifidobacterium* species. J. Ind. Microbiol. Biotechnol. *38*, 1947–1953.

O'Connell, M.M., Zomer, A., Leahy, S.C., Reunanen, J., Bottacini, F., Claesson, M.J., O'Brien, F., Flynn, K., Casey, P.G., Munoz, J.A., *et al.* (2011). Functional genome analysis of *Bifidobacterium breve* UCC2003 reveals type IVb tight adherence (Tad) pili as an essential and conserved host-colonization factor. Proc. Natl. Acad. Sci. U.S.A. *108*, 11217–11222.

O'Hara, A.M., O'Regan, P., Fanning, A., O'Mahony, C., Macsharry, J., Lyons, A., Bienenstock, J., O'Mahony, L., and Shanahan, F. (2006). Functional modulation of human intestinal epithelial cell responses by *Bifidobacterium infantis* and *Lactobacillus salivarius*. Immunology *118*, 202–215.

O'Mahony, L., McCarthy, J., Kelly, P., Hurley, G., Luo, F., Chen, K., O'Sullivan, G.C., Kiely, B., Collins, J.K., Shanahan, F., *et al.* (2005). *Lactobacillus* and *Bifidobacterium* in irritable bowel syndrome: symptom responses and relationship to cytokine profiles. Gastroenterology. *128*, 541–551.

O'Mahony, C., Scully, P., O'Mahony, D., Murphy, S., O'Brien, F., Lyons, A., Sherlock, G., Macsharry, J., Kiely, B., Shanahan, F., *et al.* (2008). Commensal-induced regulatory T-cells mediate protection against pathogen-stimulated NF-kappaB activation. PLoS Pathog. *4*, e1000112.

O'Riordan, K., Andrews, D., Buckle, K., and Conway, P. (2001). Evaluation of microencapsulation of a *Bifidobacterium* strain with starch as an approach to prolonging viability during storage. J. Appl. Microbiol. *91*, 1059–1066.

Olivares, M., Laparra, M., and Sanz, Y. (2012). Oral administration of *Bifidobacterium longum* CECT 7347 modulates jejunal proteome in an *in vivo* gliadin-induced enteropathy animal model. J. Proteomics *77*, 310–320.

Ortiz-Lucas, M., Tobias, A., Saz, P., and Sebastian, J.J. (2013). Effect of probiotic species on irritable bowel syndrome symptoms: A bring up to date meta-analysis. Rev. Esp. Enferm. Dig. *105*, 19–36.

Ouwehand, A.C., Isolauri, E., He, F., Hashimoto, H., Benno, Y., and Salminen, S. (2001). Differences in *Bifidobacterium* flora composition in allergic and healthy infants. J. Allergy Clin. Immunol. *108*, 144–145.

Ouwehand, A.C., Bergsma, N., Parhiala, R., Lahtinen, S., Gueimonde, M., Finne-Soveri, H., Strandberg, T., Pitkala, K., and Salminen, S. (2008). *Bifidobacterium* microbiota and parameters of immune function in elderly subjects. FEMS Immunol. Med. Microbiol. *53*, 18–25.

Pärtty, A., Kalliomaki, M., Endo, A., Salminen, S., and Isolauri, E. (2012). Compositional development of *Bifidobacterium* and *Lactobacillus* microbiota is linked with crying and fussing in early infancy. PLoS One *7*, e32495.

Pedersen, M.B., Iversen, S.L., Sorensen, K.I., and Johansen, E. (2005). The long and winding road from the research laboratory to industrial applications of lactic acid bacteria. FEMS Microbiol. Rev. *29*, 611–624.

Pitkala, K.H., Strandberg, T.E., Finne Soveri, U.H., Ouwehand, A.C., Poussa, T., and Salminen, S. (2007). Fermented cereal with specific bifidobacteria normalizes bowel movements in elderly nursing home residents. A randomized, controlled trial. J. Nutr. Health Aging *11*, 305–311.

Plantinga, T.S., van Maren, W.W., van Bergenhenegouwen, J., Hameetman, M., Nierkens, S., Jacobs, C., de Jong, D.J., Joosten, L.A., van't Land, B., Garssen, J., Adema, G.J., *et al.* (2011). Differential Toll-like receptor recognition and induction of cytokine profile by *Bifidobacterium breve* and *Lactobacillus* strains of probiotics. Clin. Vaccine Immunol. *18*, 621–628.

Plaza-Diaz, J., Gomez-Llorente, C., Campana-Martin, L., Matencio, E., Ortuno, I., Martinez-Silla, R., Gomez-Gallego, C., Periago, M.J., Ros, G., Chenoll, E., *et al.* (2013). Safety and immunomodulatory effects of three probiotic strains isolated from the feces of breast-fed infants in healthy adults: SETOPROB Study. PLoS One *8*, e78111.

Poch, M., and Bezkorovainy, A. (1988). Growth-enhancing supplements for various species of the genus Bifidobacterium. J. Dairy Sci. *71*, 3214–3221.

Pokusaeva, K., Fitzgerald, G.F., and Van Sinderen, D. (2011). Carbohydrate metabolism in bifidobacteria. Genes Nutr. *6*, 285–306.

Poupard, J.A., Husain, I., and Norris, R.F. (1973). Biology of the bifidobacteria. Bacteriol. Rev. *37*, 136–165.

Ringel-Kulka, T., Cheng, J., Ringel, Y., Salojarvi, J., Carroll, I., Palva, A., De Vos, W.M., and Satokari, R. (2013). Intestinal microbiota in healthy U.S. young children and adults – a high throughput microarray analysis. PLoS One *8*, e64315.

Rizzardini, G., Eskesen, D., Calder, P.C., Capetti, A., Jespersen, L., and Clerici, M. (2012). Evaluation of the immune benefits of two probiotic strains *Bifidobacterium animalis* subsp. *lactis*, BB-12® and *Lactobacillus paracasei* subsp. *paracasei*, L. casei 431® in an influenza vaccination model: a randomized, double-blind, placebo-controlled study. Br. J. Nutr. *107*, 876–884.

Roger, L.C., Costabile, A., Holland, D.T., Hoyles, L., and McCartney, A.L. (2010). Examination of faecal *Bifidobacterium* populations in breast- and formula-fed infants during the first 18 months of life. Microbiology *156*, 3329–3341.

Ruas-Madiedo, P., Gueimonde, M., Margolles, A., de los Reyes-Gavilan C.G., and Salminen, S. (2006). Exopolysaccharides produced by probiotic strains modify the adhesion of probiotics and enteropathogens to human intestinal mucus. J. Food Prot. *69*, 2011–2015.

Ruiz, L., Ruas-Madiedo, P., Gueimonde, M., de los Reyes-Gavilan C.G., Margolles, A., and Sanchez, B. (2011). How do bifidobacteria counteract environmental challenges? Mechanisms involved and physiological consequences. Genes Nutr. *6*, 307–318.

Ruiz-Moyano, S., Totten, S.M., Garrido, D.A., Smilowitz, J.T., German, J.B., Lebrilla, C.B., and Mills, D.A. (2013). Variation in consumption of human milk oligosaccharides by infant gut-associated strains of *Bifidobacterium breve*. Appl. Environ. Microbiol. *79*, 6040–6049.

Saarela, M., Rantala, M., Hallamaa, K., Nohynek, L., Virkajärvi, I., and Mättö, J. (2004). Stationary-phase acid and heat treatments for improvement of the viability of probiotic lactobacilli and bifidobacteria. J. Appl. Microbiol. *96*, 1205–1214.

Saarela, M., Virkajärvi, I., Alakomi, H.L., Mattila-Sandholm, T., Vaari, A., Suomalainen, T., and Mättö, J. (2005). Influence of fermentation time, cryoprotectant and neutralization of cell concentrate on freeze-drying survival, storage stability, and acid and bile exposure of *Bifidobacterium animalis* subsp. *lactis* cells produced without milk-based ingredients. J. Appl. Microbiol. *99*, 1330–1339.

Samona, A., and Robinson, R.K. (1994). Effect of yoghurt cultures on the survival of bifidobacteria in fermented milks. J. Soc. Dairy Tech. *47*, 58–60.

Sanchez, B., de los Reyes-Gavilan C.G., and Margolles, A. (2006). The F1F0-ATPase of *Bifidobacterium animalis* is involved in bile tolerance. Environ. Microbiol. *8*, 1825–1833.

Sanchez, B., Champomier-Verges, M.C., Stuer-Lauridsen, B., Ruas-Madiedo, P., Anglade, P., Baraige, F., de los Reyes-Gavilan C.G., Johansen, E., Zagorec, M., and Margolles, A. (2007). Adaptation and response of *Bifidobacterium animalis* subsp. *lactis* to bile: a proteomic and physiological approach. Appl. Environ. Microbiol. *73*, 6757–6767.

Sanchez, B., Ruiz, L., Gueimonde, M., Ruas-Madiedo, P., and Margolles, A. (2013). Adaptation of bifidobacteria to the gastrointestinal tract and functional consequences. Pharmacol. Res. *69*, 127–136.

Schell, M.A., Karmirantzou, M., Snel, B., Vilanova, D., Berger, B., Pessi, G., Zwahlen, M.C., Desiere, F., Bork, P., Delley, M., *et al.* (2002). The genome sequence of *Bifidobacterium longum* reflects its adaptation to the human gastrointestinal tract. Proc. Natl. Acad. Sci. U.S.A. *99*, 14422–14427.

Schmidt, G., and Zink, R. (2000). Basic features of the stress response in three species of bifidobacteria: *B. longum*, *B. adolescentis*, and *B. breve*. Int. J. Food Microbiol. *55*, 41–45.

Simpson, P.J., Stanton, C., Fitzgerald, G.F., and Ross, R.P. (2005). Intrinsic tolerance of *Bifidobacterium* species to heat and oxygen and survival following spray drying and storage. J. Appl. Microbiol. *99*, 493–501.

Schwarzer, M., Srutkova, D., Schabussova, I., Hudcovic, T., Akgun, J., Wiedermann, U., and Kozakova, H. (2013). Neonatal colonization of germ-free mice with *Bifidobacterium longum* prevents allergic sensitization to major biArch. pollen allergen Bet v 1. Vaccine *31*, 5405–5412.

Sibartie, S., O'Hara, A.M., Ryan, J., Fanning, A., O'Mahony, J., O'Neill, S., Sheil, B., O'Mahony, L., and Shanahan, F. (2009). Modulation of pathogen-induced CCL20 secretion from HT-29 human intestinal epithelial cells by commensal bacteria. BMC Immunol. *10*, 54.

Singh, A., Hacini-Rachinel, F., Gosoniu, M.L., Bourdeau, T., Holvoet, S., Doucet-Ladeveze, R., Beaumont, M., Mercenier, A., and Nutten, S. (2013). Immune-modulatory effect of probiotic *Bifidobacterium lactis* NCC2818 in individuals suffering from seasonal allergic rhinitis to grass pollen: an exploratory, randomized, placebo-controlled clinical trial. Eur. J. Clin. Nutr. *67*, 161–167.

Smecuol, E., Hwang, H.J., Sugai, E., Corso, L., Chernavsky, A.C., Bellavite, F.P., Gonzalez, A., Vodanovich, F., Moreno, M.L., Vazquez, H., *et al.* (2013). Exploratory, randomized, double-blind, placebo-controlled study on the effects of *Bifidobacterium infantis* natren life start strain super strain in active coeliac disease. J. Clin. Gastroenterol. *47*, 139–147.

Smittle, R.B., Gilliland, S.E., Speck, M.L., and Walter, W.M. Jr. (1974). Relationship of cellular fatty acid composition to survival of *Lactobacillus bulgaricus* in liquid nitrogen. Appl. Microbiol. *27*, 738–743.

Sokol, H., Seksik, P., Furet, J.P., Firmesse, O., Nion-Larmurier, I., Beaugerie, L., Cosnes, J., Corthier, G., Marteau, P., and Doré J. (2009). Low counts of *Faecalibacterium prausnitzii* in colitis microbiota. Inflamm. Bowel. Dis. *15*, 1183-1189.

Sultana, R., McBain, A.J., and O'Neill, C.A. (2013). Strain-dependent augmentation of tight-junction barrier function in human primary epidermal keratinocytes by *Lactobacillus* and *Bifidobacterium* lysates. Appl. Environ. Microbiol. *79*, 4887–4894.

Swidsinski, A., Loening-Baucke, V., Verstraelen, H., Osowska, S., and Doerffel, Y. (2008). Biostructure of faecal microbiota in healthy subjects and patients with chronic idiopathic diarrhea. Gastroenterology *135*, 568–579.

Tannock, G.W., Lawley, B., Munro, K., Gowri, P.S., Zhou, S.J., Makrides, M., Gibson, R.A., Sullivan, T., Prosser, C.G., Lowry, D., *et al.* (2013). Comparison of the compositions of the stool microbiotas of infants fed goat milk formula, cow milk-based formula, or breast milk. Appl. Environ. Microbiol. *79*, 3040–3048.

Tejero-Sarinena, S., Barlow, J., Costabile, A., Gibson, G.R., and Rowland, I. (2013). Antipathogenic activity of probiotics against *Salmonella* Typhimurium and *Clostridium difficile* in anaerobic batch culture systems: Is it due to synergies in probiotic mixtures or the specificity of single strains? Anaerobe *24*, 60–65.

Thantsha, M.S., Labuschagne, P.W., and Mamvura, C.I. (2014). Supercritical CO2 interpolymer complex encapsulation improves heat stability of probiotic bifidobacteria. World J. Microbiol. Biotechnol. *30*, 479–486.

Turroni, F., Foroni, E., Pizzetti, P., Giubellini, V., Ribbera, A., Merusi, P., Cagnasso, P., Bizzarri, B., de'Angelis, G.L., Shanahan, F., *et al.* (2009). Exploring the diversity of the bifidobacterial population in the human intestinal tract. Appl. Environ. Microbiol. *75*, 1534–1545.

Turroni, F., Peano, C., Pass, D.A., Foroni, E., Severgnini, M., Claesson, M.J., Kerr, C., Hourihane, J., Murray, D., Fuligni, F., *et al.* (2012). Diversity of bifidobacteria within the infant gut microbiota. PLoS One 7, e36957.

Turroni, F., Serafini, F., Foroni, E., Duranti, S., O'Connell, M.M., Taverniti, V., Mangifesta, M., Milani, C., Viappiani, A., Roversi, T., *et al.* (2013). Role of sortase-dependent pili of *Bifidobacterium bifidum* PRL2010 in modulating bacterium–host interactions. Proc. Natl. Acad. Sci. U.S.A. *110*, 11151–11156.

Vaishnava, S., Yamamoto, M., Severson, K.M., Ruhn, K.A., Yu, X., Koren, O., Ley, R., Wakeland, E.K., and Hooper, L.V. (2011). The antibacterial lectin RegIIIgamma promotes the spatial segregation of microbiota and host in the intestine. Science *334*, 255–258.

van der Waaij, L.A., Harmsen, H.J., Madjipour, M., Kroese, F.G., Zwiers, M., van Dullemen, H.M., de Boer, N.K., Welling, G.W., and Jansen, P.L. (2005). Bacterial population analysis of human colon and terminal ileum biopsies with 16S rRNA-based fluorescent probes: commensal bacteria live in suspension and have no direct contact with epithelial cells. Inflamm. Bowel. Dis. *11*, 865–871.

Veerkamp, J.H. (1971). Fatty acid composition of *Bifidobacterium* and *Lactobacillus* strains. J. Bacteriol. *108*, 861–867.

Wallenius, J., Barth, D., and Eerikäinen, T. (2012). The effects of pH oscillation on *Lactobacillus rhamnosus* batch cultivation. Appl. Microbiol. Biotechnol. *95*, 1265–1273.

Wang, L., Christophersen, C.T., Sorich, M.J., Gerber, J.P., Angley, M.T., and Conlon, M.A. (2011). Low relative abundances of the mucolytic bacterium *Akkermansia muciniphila* and *Bifidobacterium* spp. in faeces of children with autism. Appl. Environ. Microbiol. *77*, 6718–6721.

Ward, R.E., Ninonuevo, M., Mills, D.A., Lebrilla, C.B., and German, J.B. (2007). *In vitro* fermentability of human milk oligosaccharides by several strains of bifidobacteria. Mol. Nutr. Food Res. *51*, 1398–1405.

Wassenaar, T.M., and Alter, T. (2012). Virulence genes in risk assessment of beneficial microorganisms: what do genome sequences tell us? In Beneficial microorganisms in agriculture, food and the environment, I. Sundh, A. Wilcks, and M. Goettel, eds. (CAB International, Oxfordshire, UK), pp. 180–196.

Weiss, G., Christensen, H.R., Zeuthen, L.H., Vogensen, F.K., Jakobsen, M., and Frokiaer, H. (2011). Lactobacilli and bifidobacteria induce differential interferon-beta profiles in dendritic cells. Cytokine. *56*, 520–530.

Whorwell, P.J., Altringer, L., Morel, J., Bond, Y., Charbonneau, D., O'Mahony, L., Kiely, B., Shanahan, F., and Quigley, E.M. (2006). Efficacy of an encapsulated probiotic *Bifidobacterium infantis* 35624 in women with irritable bowel syndrome. Am. J. Gastroenterol. *101*, 1581–1590.

Wittmann, A., Autenrieth, I.B., and Frick, J.S. (2013). Plasmacytoid dendritic cells are crucial in *Bifidobacterium adolescentis*-mediated inhibition of Yersinia enterocolitica infection. PLoS One *8*, e71338.

Yamasaki, C., Totsu, S., Uchiyama, A., Nakanishi, H., Masumoto, K., Washio, Y., Shuri, K., Ishida, S., Imai, K., and Kusuda, S. (2012). Effect of *Bifidobacterium* administration on very-low-birthweight infants. Pediatr. Int. *54*, 651–656.

Yatsunenko, T., Rey, F.E., Manary, M.J., Trehan, I., Dominguez-Bello, M.G., Contreras, M., Magris, M., Hidalgo, G., Baldassano, R.N., Anokhin, A.P., *et al.* (2012). Human gut microbiome viewed across age and geography. Nature *486*, 222–227.

Yoshimura, K., Matsui, T., and Itoh, K. (2010). Prevention of *Escherichia coli* O157:H7 infection in gnotobiotic mice associated with *Bifidobacterium* strains. Antonie Van Leeuwenhoek 97, 107–117.

You, J., and Yaqoob, P. (2012). Evidence of immunomodulatory effects of a novel probiotic, *Bifidobacterium longum* bv. *infantis* CCUG 52486. FEMS Immunol. Med. Microbiol. *66*, 353–362.

Zeuthen, L.H., Fink, L.N., and Frokiaer, H. (2008). Toll-like receptor 2 and nucleotide-binding oligomerization domain-2 play divergent roles in the recognition of gut-derived lactobacilli and bifidobacteria in dendritic cells. Immunology *124*, 489–502.

Propionibacteria also have Probiotic Potential

5

Gabriela Zárate and Adriana Perez Chaia

Abstract

Propionibacteria were first described by the end of the nineteenth century and named some years later by Orla-Jensen (1909) who proposed the genus *Propionibacterium* for referring to bacteria that produce propionic acid as their main fermentation end-product. Based on habitat of origin, they are conventionally divided into 'classical or dairy' and 'cutaneous' microorganisms which mainly inhabit dairy/silage environments and the skin/intestine of human and animals, respectively.

Historically, the economic relevance of *Propionibacterium* has been related to the industrial application of classical species as dairy starters for cheeses manufacture and as biological producers of propionic acid. However, propionibacteria also display probiotic potential. Over the last two decades, the ability of these microorganisms to improve the health of humans and animals by being used as dietary microbial adjuncts has been extensively demonstrated. Both *in vitro* and *in vivo* studies revealed that propionibacteria are able to modulate in a favourable way gut physiology, microbiota composition and immunity. Much of these health benefits could be related to the ability of propionibacteria to remain in high numbers in the gastrointestinal tract by surviving the adverse environmental conditions and adhering to the intestinal mucosa. In addition, other promising properties like the production of nutraceuticals and relevant biomolecules such as vitamins B and K, conjugated linoleic acid (CLA), exopolysaccharides (EPS), trehalose, bifidogenic factors, bacteriocins, etc. have been reported. In recent years, the availability of genome sequences of different propionibacteria species have allowed to deep insight into the metabolism and physiology of these microorganisms and became a useful tool for selecting appropriate strains for technological, functional or probiotic applications.

In the present chapter, we review exhaustively the evidences that support the potential of propionibacteria to be used as probiotic supplements for human and animal nutrition. Besides the positive results on health obtained by us and others, the hardiness and adaptability of propionibacteria to both technological and physiological stresses encourage their usage for designing new functional foods.

Introduction: description of propionibacteria

Phenotypic characteristics of *Propionibacterium*: ecology and metabolism

The members of the genus *Propionibacterium* are Gram-positive, catalase-positive, non-spore-forming and non-motile pleomorphic bacteria with high G+C content (53–68%) (Cummins and Johnson, 1986; Stackebrandt *et al.*, 2006). They are anaerobic to slightly aerotolerant and morphologically heterogeneous microorganisms including rod-shaped and filamentous branched cells that may occur singly, in pairs, forming a V or a Y shape, or arranged as 'Chinese characters'.

Propionibacteria are also mesophilic microorganisms, with optimal growth conditions at 30°C and pH 6.8. However, they can grow at a temperature range between 15 and 40°C and tolerate pH variations between 5.1 and 8.5 (Cummins and Johnson, 1986).

Besides these characteristics, the key feature of propionibacteria is their peculiar heterofermentative metabolism that leads to the formation of propionic and acetic acids and CO_2 as main end-products from different substrates like carbohydrates, alcohols and organic acids, through a pathway commonly known as the Wood–Werkman cycle or propionic fermentation.

Although this microbial fermentation was demonstrated by Pasteur as early 1861, the first studies about the morphology and physiology of propionic acid bacteria were carried out by Albert Fitz (1879) being the microorganisms responsible for the 'eyes' formation in Emmental cheeses. The microorganism was finally isolated in 1906 by E. Von Freudenreich and Sigurd Orla-Jensen. A few years later Orla Jensen (1909) proposed the genus *Propionibacterium* for referring to bacteria that produced large amounts of propionic acid.

Since then, several strains were isolated and different species were described based on phenotypic and genotypic characteristics and included in the successive editions of *Bergey's Manuals of Determinative or Systematic Bacteriology* (for the historical evolution of the taxonomy of propionibacteria see Zárate *et al.*, 2011).

At present, the taxonomic position of propionibacteria is the following (Stackebrandt *et al.*, 2006): phylum *Actinobacteria*; class *Actinobacteria*; subclass *Actinobacteridae*; order *Actinomycetales*; suborder *Propionibacterineae*; family *Propionibacteriaceae*; genus *Propionibacterium*.

To date, 13 species have been classified under this genus: *Propionibacterium acidipropionici*, *P. freudenreichii* (subsp. *freudenreichii* and *shermanii*), *P. jensenii* and *P. thoenii*, were isolated from dairy environments between 1906 and 1928; *P. acnes*, *P. avidum*, *P. granulosum* and *P. propionicum* (formerly *Arachnia propionica*) came from human skin and were isolated between 1900 and 1969. In the last two decades five new species were isolated: *P. cyclohexanicum* was obtained from spoiled orange juice (Kusano *et al.*, 1997); *P. microaerophilum* was isolated from olive mill wastewater (Koussemon *et al.*, 2001); *P. australiense* came from granulomatous bovine lesions (Bernard *et al.*, 2002) and *P. acidifaciens* from human dental caries (Downes and Wade, 2009). A new species, *P. humerusii*, isolated from human humerus, has been proposed recently (Butler-Wu *et al.*, 2011).

Based on their natural habitats, propionibacteria were conventionally classified in two distinct groups: 'dairy or classical propionibacteria', which inhabit dairy environments and silages, and 'cutaneous propionibacteria', which inhabit the skin and the intestine of humans and animals (Johnson and Cummins, 1972; Cummis and Johnson, 1986). Although new species do not fit exactly to these biotopes, homology studies have allowed allocating them in either group (Table 5.1).

Among the main habitats of classical propionibacteria it could be mentioned: raw milk and cheese (Cummins and Johnson, 1986; Stackebrandt *et al.*, 2006) silages and vegetables for human consumption (Mantere-Alhonen and Ryhänen, 1994), ruminal content and faeces of cows and calves (Rinta-Koski *et al.*, 2001) and the intestine of pigs and laying hens (Argañaraz Martinez *et al.*, 2013). Bacteria of this group are recognized as safe microorganisms and have a long history of application on industrial processes like cheeses manufacture and biological production of propionic acid among other relevant metabolites.

On the other side, cutaneous species are found mainly in the human skin, but have been isolated also from the intestine of humans, chicken and pigs (Cummins and Johnson, 1986; Macfarlane *et al.*, 1988; Stackebrandt *et al.*, 2006; Argañaraz Martinez *et al.*, 2013), and are generally considered opportunistic pathogens in compromised hosts.

With respect to metabolism and physiology, propionibacteria have low nutritional requirements and almost the same for all the species. Regarding carbon metabolism, they are heterotrophic microorganisms that ferment carbohydrates like glucose, galactose, lactose, fructose and other sugars; polyols such as glycerol; erythritol and others; and organic acids such as lactic and gluconic acids producing propionic acid, acetic acid and CO_2 as the main fermentation end-products (Hettinga and Reinbold, 1972; Cummins and Johnson, 1986; Piveteau, 1999; Falentin *et al.*, 2010).

The carbon metabolic pathway in propionibacteria begins with transformation of substrates to pyruvate via glycolysis, pentose phosphate or the Entner–Doudoroff pathways. Pyruvate is then oxidized to acetate and CO_2 or reduced to propionate via the Wood–Werkman cycle or transcarboxylase cycle which represents the key component of the carbon metabolism in propionibacteria (Hettinga and Reinbold, 1972).

The most important reaction of this cycle is transcarboxylation catalysed by methylmalonyl-CoA carboxytransferase (Hervé *et al.*, 2007; Falentin *et al.*, 2010) that transfers a carboxyl group from methylmalonyl-CoA to pyruvate to form oxaloacetate and propionyl-CoA, without ATP consumption. Oxaloacetate is reduced to succinate (via malate and fumarate) which is then converted to succinyl-CoA, concomitantly with the conversion

Table 5.1 Current species of the genus *Propionibacterium*

Species	Origin	Type strain
Dairy or classical propionibacteria		
P. acidicpropionici	Dairy environments	ATCC 25562
P. cyclohexanicum	Spoiled off-flavour orange juice	Strain TA-12^T (= IAM 14535 = NRIC 0247)
P. freudenreichii	Dairy environments: raw milk, Swiss cheese	ATCC 6207
P. jensenii	Dairy products and silage	ATCC 4868
P. microaerophilum	Olive mill wastewater	M5^T (=CNCM I-2360^T=DSM 13435^T)
P. thoenii	Dairy environments. Red spots in Emmental cheese	ATCC 4874
Cutaneous propionibacteria		
P. acidifaciens	Human mouth, carious lesions	Strain C3M_31^T (=DSM 21887^T =CCUG 57100^T).
P. acnes	Normal skin, comedos of acne vulgaris, intestinal contents, wounds, tissue abscesses	ATCC 6919
P. australiense	Granulomatous bovine lesions	Strain 98A072 (=ATCC BAA-264^T, CCUG 46075^T)
P. avidum	Moister areas of the skin such as the axilla and perineum	ATCC 25577
P. granulosum	Sebum rich oily areas of skin	ATCC 25564
P. humerusii	Human humeral membrane	Strain P08
P. propionicum	Lacrimal canaliculitis	ATCC 14157 (= DSM 43307).

of propionyl-CoA to propionate, by a CoA-transferase. Finally, succinyl-CoA is isomerized to methylmalonyl-CoA by a methylmalonyl-CoA mutase (EC 5.4.99.2) which is a unique coenzyme B_{12}-dependent enzyme found in propionibacteria. It must be emphasized that the Wood Werkman cycle used by propionibacteria to produce propionate is coupled to oxidative phosphorylation and yields more ATP than in the other bacteria that produce propionic acid (Piveteau, 1999; Thierry *et al.*, 2011).

Although propionate, acetate and CO_2 are the common metabolic end-products of the cycle their proportions could vary depending on the strains, the substrate used, and the environmental conditions, just to maintain the redox balance (Thierry *et al.*, 2011). As examples, it has been observed that the oxidation of glucose and lactic acid leads to a molar ratio of propionate to acetate of 2:1 whereas the oxidation of glycerol leads to the formation of propionate only. If lactate is co-metabolized with aspartate/asparagine more pyruvate is oxidized to acetate + CO_2 than converted to propionate (Crow, 1986).

With respect to nitrogen metabolism, it has been reported that propionibacteria are able to synthesize all amino acids (Falentin *et al.*, 2010). If proteins are the substrates, cutaneous propionibacteria are able to hydrolyse some of them, like gelatin and fibronectin, promoting damages and inflammation of the host tissues. On the contrary, dairy propionibacteria grow scarcely in milk as they do not possess proteases capable of hydrolysing milk caseins (Dupuis *et al.*, 1995). However, some proteinases have been described for these microorganisms, one cell wall associated and one intracellular or membrane bound with low activity. Different peptidases such as aminopeptidases, proline iminopeptidase, proline imidopeptidase, X-prolyl-dipeptidyl-amino-peptidase, endopeptidases and carboxypeptidase, have been reported and characterized in *Propionibacterium*. Some amino acids, such as aspartic acid, alanine, serine and glycine, are degraded but with variations among species and strains (Langsrud *et al.*, 1995).

Finally, with respect to vitamins requirements, all propionibacteria need pantothenate and biotin, whereas some strains require thiamine and p-aminobenzoic acid (Hettinga and Reindbold, 1972; Falentin *et al.*, 2010).

Two features that deserve particular attention are the sturdiness and adaptability of propionibacteria that enable them to survive to different stresses and to remain viable for long periods of time in adverse environments (Thierry *et al.*, 2011). In the case of classical propionibacteria, technological processing by dairy industry and consumption may represent different and successive stresses to withstand. In this sense, lyophilization for preparation of starters and the manufacture of a Swiss type cheese represent for microbial cells successive stresses that begin with desiccation, followed by acidification of the curd, heating during cooking, osmotic stress due to brining, and low temperature (4 to 12°C) during cheese ripening. The transit through the digestive tract after consumption adds further stresses such as the exposure of bacteria to gastrointestinal fluids. Dairy propionibacteria (*P. freudenreichii*, *P. acidipropionici* and *P. jensenii*) have shown to withstand these harsh conditions (Jan *et al.*, 2002a; Huang *et al.*, 2003; Pérez Chaia and Zárate, 2005; Cousin *et al.*, 2012a).

The stress adaptation and long-term survival has been investigated in dairy propionibacteria at the genomic and expression levels (Falentin *et al.*, 2010; Parizzi *et al.*, 2012) finding out that both stress adaptation proteins and the accumulation of various energy carbon storage compounds are involved in the ability to survive and remain active in unfavourable environments (Leverrier *et al.*, 2003, 2004; Anastasiou *et al.*, 2006).

In dairy propionibacteria such as *P. freudenrechii* the cell machinery involved in general stress adaptation is encoded by multicopy stress-induced genes (Falentin *et al.*, 2010) with the specific proteins differentially expressed depending on the strain and the stress (Falentin *et al.*, 2010; Leverrier *et al.*, 2003, 2004; Anastasiou *et al.*, 2006).

For example, gastrointestinal stresses such as acid and bile induce the synthesis of pyruvate–flavodoxin oxidoreductase and succinate dehydrogenase which are involved in electron transport and ATP synthesis, as well as glutamate decarboxylase and aspartate ammonia-lyase, which are involved in intracellular pH homeostasis. In addition, bile induces oxidative stress so that survival of *P. freudenreichii* within the intestine is due to the remediation of oxidative damages by the expression of genes for disulfide-reduction and elimination of reactive oxygen species. In response to bile, P. freudenreichii overexpresses the iron/manganese superoxide dismutase, glutathione-S-transferase, two cysteine synthases and S-adenosylmethionine synthetase (Falentin *et al.*, 2010). Other inducible proteins involved in protection and repair of DNA damages include Ssb protein which is involved in DNA recombination and repair, as well as Dps which protects DNA against oxidative stress (Leverrier *et al.*, 2004).

As mentioned above, technological processing comprises both thermal and osmotic stresses. With respect to thermotolerance, survival seems to be related to the overexpression of constitutive stress-related molecular chaperones and ATP-dependent proteases as well as the induction of the dihydroxyacetone kinase locus (dhaKL, PFREUD_07980 and PFREUD_07990) (Falentin *et al.*, 2010; Anastasiou *et al.*, 2006). In the case of osmotic stress propionibacteria are able to withstand it by accumulation of compatible solutes such as glycine betaine and trehalose (Boyaval *et al.*, 1999).

All dairy propionibacteria are able, in a strain-dependent manner, to synthesize and accumulate trehalose from glucose and pyruvate (Deborde *et al.*, 1996). Pathways for trehalose metabolism have been identified in P. freudenreichii NIZO B365 and P. freudenreichii CIRM-BIA1^T. Trehalose-6-phosphate synthase/phosphatase (OtsA–OtsB) pathway is involved in trehalose synthesis whereas Trehalose synthase (TreS) pathway is proposed to operate in trehalose degradation (Cardoso *et al.*, 2007; Falentin *et al.*, 2010). Synthesis and storage of this sugar are also enhanced at stationary phase and under oxidative and acid stress conditions (Cardoso *et al.*, 2004).

Long term survival of propionibacteria on extreme environments could be due to the accumulation of storage compounds, compatible solutes, and the induction of a multi-tolerance response under carbon starvation (Falentin *et al.*, 2010). It has been demonstrated that propionibacteria accumulate inorganic polyphosphate (polyP) as energy reserve, in contrast to other bacteria that use ATP. Polyphosphates (polyPs) are linear polymers composed of orthophosphate residues. The synthesis of PolyP is catalysed by polyphosphate kinase (PPK) that transfers reversibly the terminal phosphate of ATP to polyP. The genes related to polyphosphate metabolism were found in the genomes of *P. freudenreichii, P acidipropionici* and *P. acnes* (Falentin *et al.*, 2010; Parizzi *et al.*, 2012). In addition to the one copy of PPK per genome, propionibacteria possess another copy of PPK2 that differs from PPK by its use of either GTP or ATP, its preference for Mn^{2+} over Mg^{2+} and for being stimulated by polyP. It has been demonstrated that PolyPs enable microorganisms to tolerate adverse conditions since ppk mutants are unable to survive during stationary phase (Clark *et al.*, 1986; Parizzi *et al.*, 2012).

Propionibacteria are also able to synthesize glycogen and the genes related to glycogen metabolism were identified in the genome of *P. freudenreichii* and *P. acnes*. Since both species are unable to ferment extracellular glycogen all the enzymes encoded by these genes seem to be involved in intracellular accumulation and hydrolysis of glycogen.

Finally, it has been observed that dairy propionibacteria survive for many months at room temperature under conditions of carbon starvation and the majority of the strains are non-lytic (Stackebrandt *et al.*, 2006). This prolonged survival in stationary phase or dormant phase could be the consequence of a multitolerance response that involves the synthesis and accumulation of polyP, glycogen, trehalose and the overexpression of molecular protein chaperones. A gene encoding an Rpf (resuscitation promoting factor) protein which is essential for the growth of dormant cells from actinobacteria has been described in the genome of *P. freudenreichii* and is probably involved also in the long-term survival of propionibacteria (Falentin *et al.*, 2010).

Genotypic characteristics of *Propionibacterium*

Propionibacteria, like the majority of bacteria, possess a circular-shaped chromosome that varies in size between 2.1 and 3.6 Mb depending on the species, with a G+C DNA content of 53–68 mol% (Gautier *et al.*, 1992; Parizzi *et al.*, 2012). In general, they do not possess plasmids but their existence has been reported in strains of *P. acidipropionici, P. freudenreichii, P. jensenii* and *P. granulosum* (Rehberger and Glatz, 1990; Miescher *et al.*, 2000; van Luijk *et al.*, 2002, Farrar *et al.*, 2007). In a recent study a plasmid of the species *P. acnes* was described for the first time (Kasimatis *et al.*, 2013. The presence of two types of bacteriophages has also been described for propionibacteria: the bacteriophage B22 and the bacteriophage B5, which would be the first infectious filamentous virus described in a Gram-positive bacterium (Gautier *et al.*, 1999).

In last years the biology of propionibacteria was investigated at the molecular level mainly through genome sequencing and its comparative analysis with other close related and fully sequenced bacterial species. Since 2004, the complete genomes of the following species belonging to both taxonomic groups of propionibacteria were sequenced: *Propionibacterim acnes* (represented in Table 5.2 by the sequence of strain KPA171202) (Brüggemann *et al.*, 2004) and *P. avidum* (strain 44067) (Ordogh *et al.*, 2013) representing cutaneous propionibacteria and *P. freudenreichii* (subsp. *shermanii*, strain CIRM-BIA1) (Falentin *et al.*, 2010) and *P. acidipropionici* (strain ATCC4875) (Parizzi *et al.*, 2012) coming from the dairy group.

The general features of these genomes are presented in Table 5.2.

Genome sequencing has led to a better understanding of the basic biology of propionibacteria and allowed to gain insight into their physiology and adaptation to different environments. Different metabolic pathways have been elucidated as well as different features that may explain the plasticity and sturdiness of these species.

Genome analysis has revealed in all species the enzymatic machinery involved in the Embden–Meyerhof pathway, the pentose phosphate pathway and the Wood–Werkman cycle, genes encoding for vitamin B_{12} biosynthesis, transport proteins (mainly ABC transporters), two-component regulatory systems, oxidoreductases and transcriptional regulators. Although propionibacteria are usually described as anaerobes, all the genes related to aerobic respiration and oxidative stress are present, such as NADH dehydrogenase, succinate dehydrogenase, cytochrome oxidases, ATPase and catalase (Bruggeman *et al.*, 2004; Falentin *et al.*, 2010, Parizzi *et al.*, 2012).

As a remarkable feature, *P. freudenreichii, P. acnes* and *P. avidum* harbour approximately 1000 genes fewer than *P. acidipropionici*, a fact that could reflect the specialization due to the continuous use of *P. freudenreichii* as a cheese starter and that of cutaneous propionibacteria as opportunistic pathogens of human skin. On the contrary, the higher number of genes of *P. acidipropionici* reveals less specialization and a more flexible metabolism that could confer competitive advantages and allow the bacterium to adapt to a wide variety of environments (Parizzi *et al.*, 2012).

In the case of dairy propionibacteria, the genome annotations

Table 5.2 Comparison of main characteristics of genomes of dairy and cutaneous propionibacteria (selected genomes of each species are shown)

Microorganism	Genome Size (Mb)	G+C content	rRNA operons	No. of genes	No. of predicted proteins	No. of tRNA	Plasmids
Propionibacterium freudenreichii subsp. *shermanii* CIRM-BIA1^T	2.6	67.3	2	2435	2326	45	0
Propionibacterium acidipropionici ATCC4875	3.6	68.8	4	3433	3336	53	1
Propionibacterium acnes KPA171202	2.6	60.0	3	2368	2297	45	0
Propionibacterium avidum 44067	2.5	63.5	9	2297	2242	46	0

showed that both species sequenced possess all the enzymes needed for de novo biosynthesis of amino acids and vitamins (except pantothenate and biotin), genes for the defence of genome integrity and immunity against phages by the CRISPR/Cas system, chaperones for stress resistance, and accumulation of inorganic polyphosphate, glycogen and compatible solutes like trehalose that confer to these bacteria a long survival in stationary phase (Falentin *et al.*, 2010; Parizzi *et al.*, 2012).

The *P. acidipropionici* genome possess a notable set of genes encoding for transporters and enzymes involved in uptake and degradation of a higher number of substrates (including carbohydrates, polar and hydrophobic amino acids) than *P. freudenreichii*. Among these, one invertase coding gene (PACID_33010) should be mentioned that enables *P. acidipropionici* to use sucrose as carbon source opening the possibility of developing industrial fermentations at low cost from renewable feedstocks such as sugarcane (Parizzi *et al.*, 2012).

Besides the uptake and metabolism of carbohydrates, *P. acidipropionici* displays some other unique traits related to acetic and propionic acids production. By difference from literature, the enzyme phosphoenolpyruvate carboxytransphosphorylase (PEPC) that would lead to the formation of oxalacetate from CO_2 and PEP is not present in sequenced propionibacteria. Instead, a pyruvate carboxylase could be responsible for heterotrophic CO_2 fixation in *P. acidipropionici* since this gene was identified in the genome and the presence of the corresponding enzyme was confirmed in proteome analysis. Similarly, the phosphate acetyltransferase (PTA) and acetate kinase (ACK) genes, which are involved in acetate assimilation or dissimilation, were not present in the genome of *P. acidipropionici*. This function seems to be performed by an ADP forming acetyl-CoA synthetase gene and its corresponding enzyme which was confirmed in the proteome (ADP-ACS, EC 6.2.1.13 PACID_02150). Unlike PTA and ACK, ADP-ACS converts acetyl-CoA, inorganic phosphate and ADP into acetate, ATP and CoA in one step. The acetate assimilation is probably performed by AMP-forming ACS enzymes (EC 6.2.1.1; PACID_13940 and PACID_13950) (Parizzi *et al.*, 2012).

With respect to technological application in dairy industries, *P. freudenreichii* subsp. *shermanii* CIRM-BIA1^{T} is able to metabolize lactose, but this trait seems to be strain-dependent. CIRM-BIA1^{T} possesses a lactose utilization locus encoding a β-galactosidase, a galactose transporter and an UDP-glucose isomerase. This locus is surrounded by transposable elements, and is believed to have been acquired through a horizontal transfer event mediated by phage infection that facilitated the adaptation of the microorganism to the dairy environment.

Regarding the formation of cheese flavour compounds, various pathways were identified in the genome of this strain such as the enzymes involved in the production of volatile branched chain fatty acids from branched amino acid degradation, and free fatty acids and esters from lipid catabolism (Falentin *et al.*, 2010). Two esterases, one secreted (PF#279) and other cell wall-anchored (PF#774), seem to be involved in lipolysis of milk glycerides, the first key action of Swiss cheese lipolysis (Dherbécourt *et al.*, 2010; Abeijon Mukdsi *et al.*, 2014). In addition, 10 other intracellular esterases were found in the genome of *P. freudenreichii* that could be involved in the synthesis of the volatile esters associated with the fruity flavour of cheese (Dupuis *et al.*, 1993; Dherbécourt *et al.*, 2008, 2010).

In relation to probiotic functionality, the complete biosynthesis pathway for a bifidogenic compound (DHNA) has been identified, as well as sequences corresponding to a high number of surface proteins involved in the interactions with the host (like adhesion and immunomodulation). None of the reported probiotic related genes (gluconate kinase, S-layer proteins, cell-wall peptidase, microcin resistance) were found in the *P. acidipropionici* genome (Parizzi *et al.*, 2012). By comparative genomics with *P. acnes*, no pathogenicity factors were identified in *P. freudenreichii*, which is consistent with the generally recognized as safe (GRAS) and qualified presumption of safety (QPS) status of this species.

As previously mentioned, cutaneous propionibacteria are commensal components of normal human skin microbiota that could be occasionally associated with inflammatory diseases (acne vulgaris; corneal ulceration, endocarditis, sarcoidosis, osteomyelitis, post surgical infections, among others). At present, the genomes of at least 10 strains of *P. acnes* coming from healthy or acne-affected skin were completely sequenced, showing genetic variations among strains that may explain their differences in phenotypic and virulent properties (Brüggemann *et al.*, 2004; Brzuszkiewicz *et al.*, 2011; Hunyadkürti *et al.*, 2011; Horvath *et al.*, 2012; Voros *et al.*, 2012; Kasimatis *et al.*, 2013, Minegishi *et al.*, 2013).

Some protein families that were identified only in *P. acnes* genomes are probably related to pathogenicity and host interaction. Examples of these host interacting factors include hyaluronidase, neuraminidases, endoglycoceramidases, haemolysins, sialidases, and lysophospholipase probably involved in in host tissue degradation. Surface-associated and other immunogenic factors (dermatan-sulphate adhesins, cAMP factors) might be responsible of triggering acne inflammation and the other associated diseases.

From comparative genomic and transcriptomic analysis it has been stated that *P. acnes* strains may contribute differently to skin health and disease and that the pathogenic potential of the different strains is determined not only by the phylotype-specific genome content but also by variable gene expression (Brzuszkiewicz *et al.*, 2011). In this respect it has been determined that genes encoding components of the energy-conserving respiratory chain as well as virulence secreted and cell-associated factors were transcribed during the exponential phase, while genes involved in stress responses and amino acid metabolism were up-regulated at the stationary growth phase. These results evidence the important role that growth status may play in the inflammation-inducing activity of *P. acnes* (Brzuszkiewicz *et al.*, 2011).

As it is not clear if virulence factors are shared by all cutaneous propionibacteria a recent study compared the genomes of *P. granulosum* and *P avidum* with *P. acnes* in order to ascertain inter- species differences (Mak *et al.*, 2013). Results revealed that the three organisms have evolved distinct structures to interact with the host: *P. avidum* and *P. granulosum* produce an EPS-like surface structure and pili-like appendices, respectively, whereas *P. acnes* possesses a number of unique surface-exposed proteins with host-interacting properties. The different surface features of the three cutaneous propionibacteria suggest that each species interacts differently with the human host, which likely results in distinct pathogenic potentials.

It is expected in the future, that the genome and transcriptomic information combined with metabolic approaches will increase our knowledge on physiology of propionibacteria and may help to design strategies for increasing and optimizing their biotechnological and clinical management.

Propionibacterium as human and animal probiotics

Selection of propionibacteria for probiotic purposes

Probiotics are live microorganisms that confer health benefits to the host when administered in adequate amounts. (FAO/WHO, 2002). Over the last few decades many researchers have focused on the selection of appropriate strains for developing probiotic-containing products in order to satisfy the increasing demands of consumers for healthy foods and alternatives to traditional drug therapy. In this respect, the majority of evidence concerning the beneficial effects of microorganisms both in human and animal health come from studies dealing with lactic acid bacteria and bifidobacteria since they are common components of the gastrointestinal microbiota. However, in recent years other less conventional genera such as *Bacillus, Enterococcus, Leuconostoc, Pediococcus, Saccharomyces, Streptococcus* and *Propionibacterium* have began to been considered for this purposes. As result, an increasing number of reports on potential probiotic properties of propionibacteria have been published (Vorobjeva *et al.*, 2008; Zárate *et al.*, 2011, Cousin *et al.*, 2011; Poonam *et al.*, 2012; Zárate, 2012). In the following sections, the selection criteria and all the reported probiotic effects, evaluated through *in vitro* and *in vivo* studies, respectively, that support the potential of *Propionibacterium* to be used as supplements for humans and animals are reviewed.

In order to be incorporated as probiotics in either a food or pharmaceutical product, strains must fulfil some important requirements that have been taken as selection criteria (FAO/WHO, 2002). The most important aspects to assess include: origin and safety of the strains, survival and activity in the environmental conditions of the target organ, technological tolerance to product processing and storage; and at least one health benefit.

As mentioned, the first step in the long way towards the development of a probiotic-containing product is assessment of safety of the strains selected on the basis of their potential beneficial effects. First studies emphasized the host-specificity of strains as a mandatory criterion for safety guarantee but more recent evidences suggest that it is the specificity of the action and not the source of the microorganism that is really important. In this respect, dairy propionibacteria could be considered a safe component of human and animal diets. Classical propionibacteria have been isolated from silage, vegetables, raw milk, secondary microbiota of cheese and other naturally fermented foods. In addition, *P. freudenreichii* is worldwide used as a Swiss-type cheese starter, reaching at the end of maturation concentrations close to 10^9 bacteria/g. Therefore, it could be considered that propionibacteria would arrive to the gut of man and animals at least once in their lives. Moreover, recently propionibacteria have been shown to be part of the endogenous microbiota (Suomalainen *et al.*, 2008; Argañaraz Martinez *et al.*, 2013).

Different studies have assessed and established safety of dairy propionibacteria (Meile *et al.*, 2008) in animal models (Pérez Chaia *et al.*, 1995; Huang *et al.*, 2003; Lan *et al.*, 2007a; Uchida *et al.*, 2011a) as well as in humans (Kukkonen *et al.*, 2008; Uchida *et al.*, 2011b; Cousin *et al.*, 2011). It has been observed that propionibacteria did not translocate to blood, liver or spleen and no adverse effects on body weight gain and general health status was observed after short (Perez Chaia *et al.*, 1995; Bougle *et al.*, 1999) and long term (Huang *et al.*, 2003; Hattaka *et al.*, 2007; Kajander *et al.*, 2008) administration of strains of *P. freundenreichii, P. acidipropionici* and *P. jensenii*. No toxicity or side-effects associated with the ingestion of propionibacteria have been reported either. The majority of trials have been performed with the common dairy starter *P. freudenreichii* and *P. acidipropionici* being the first species granted the generally recognized as safe (GRAS) status from the US Food and Drug Administration and both with the qualified presumption of safety (QPS) of the European Food Safety Authority (Meile *et al.*, 2008; EFSA, 2009).

With respect to cutaneous propionibacteria, their application as probiotics is not completely discouraged but needs more careful studies in reference to safety and virulence factors. Although these microorganisms are considered normal commensal microbiota of skin and intestine of human and animals they could behave as opportunistic pathogens in immunocompromised hosts.

As cited above, other aspects closely related with safety of the microorganisms are the expression of virulence factors and the antibiotic resistance due to the risk of spreading resistance to other members of the native microbiota. It has been stated that dairy propionibacteria have intrinsic resistance to some antibiotics like oxacillin, aminoglycosides, 1st and 2nd generation quinolones, colistin, metronidazole and fosfomycin and this resistance does not appear to be encoded by plasmids or other mobile genetic elements (Rehberger *et al.*, 1990; Meile *et al.*, 2008; Suomalainen *et al.*, 2008; EFSA, 2009). In a recent study, Darilmaz and Beyatli (2012) isolated 29 strains of dairy propionibacteria (*P. freundenreichii* subsp. *freudenreichii, P. jensenii* and *P. thoenii*) from Turkish traditional home-made cheeses and screened *in vitro* some selection criteria for probiotics. All propionibacterial strains were sensitive to the majority of antibiotics used in clinical investigations except nalidixic acid. In the same manner, no atypical antibiotic resistances were detected for *P. freudenreichii* strain JS or the strain *P. acidipropionici* LET103 isolated from the intestine of laying hens (Babot, 2014).

For cutaneous propionibacteria a chromosomal resistance to aminoglycosides and metronidazole has been reported (Funke *et al.*, 1997). Besides, a worldwide increase in the prevalence of antibiotic resistance in *P. acnes* strains has been observed due to the prolonged use of antibiotics in acne therapy (Leyden, 2004).

By comparative genomics, no virulence factors found in *P. acnes* or in other pathogenic species were identified in *P. freudenreichii* or *P acidipropionici* genomes, although some strains of *P. thoenii* and *P. jensenii* possess β-haemolytic activity (Meile *et al.*, 2008; Falentin *et al.*, 2010; Parizzi *et al.*, 2012).

The next criterion to consider when assessing strains for probiotic application is their ability to remain viable and active in their site of action. In this sense, probiotic strains intended to be delivered through food systems should survive the passage through the upper gastrointestinal tract and arrive alive and active to the intestine where they should persist long enough to provide their beneficial effects to the host.

Many studies have demonstrated by *in vitro* assays the ability of dairy propionibacteria to survive and tolerate the hostile conditions of the gastrointestinal tract represented by the low pH of the stomach and aggressive intestinal fluids like bile and pancreatic enzymes (Zárate *et al.*, 2000a,b; Jan *et al.*, 2002a; Huang and Adams., 2004; Leverrier *et al.*, 2005; Lan *et al.*, 2007a; Suomalainen *et al.*, 2008; Darilmaz and Beyatli, 2012). This tolerance could be improved by a pre-adaptation of the microorganisms to the adverse conditions of the gut by a brief exposure to the stressful conditions at a non-lethal level (Jan *et al.*, 2001; Leverrier *et al.*, 2003; Darilmaz and Beyatli, 2012).

Both acid and bile tolerance have shown to be strain-dependent properties. In previous studies (Zárate *et al.*, 2000a,b) we observed that dairy propionibacteria grown in a medium containing bile (0–0.5%) could be classified as 'bile-tolerant' and 'non bile-tolerant' strains and that there were differences among *P. freudenreichii* and *P. acidipropionici* strains in their tolerance to pancreatic enzymes when subjected to sequential digestion with artificial gastric and intestinal fluids.

It must be emphasized that the vehicle used for delivery of probiotics is important for improving digestive stress tolerance since cells included in food matrices like milk or cheese had better tolerance to acid challenge than free cultures (Zárate *et al.*, 2000b). Similar results to ours were obtained by Huang and Adams (2004), who protected propionibacteria from acid and bile stresses with a soymilk and a cereal beverage, Leverrier *et al.* (2005), which used a yoghurt-type fermented milk and Ranadheera *et al.* (2012) with a goat's milk ice cream. Immobilization and storage in alginate beads was also effective for protecting cells of *P. freudenreichii* against digestive stresses simulated *in vitro* (Leverrier *et al.*, 2005).

In a recent study Cousin *et al.* (2012), developed a dairy product exclusively fermented by dairy propionibacteria by supplementing milk ultrafiltrate with lactate and casein hydrolysate in which the probiotic strain *P. freudenreichii* BIA138 remained viable and resisted gastrointestinal stress *in vitro*. The milk fermented by propionibacteria was an efficient vector that delivered active cells to the colon of piglets since viable propionibacteria and higher SCFA concentrations were evidenced.

Some properties of strains, such as production of EPS, could also help to protect cells during digestive transit and may be important in the selection of probiotic propionibacterial strains able to resist the upper gastrointestinal system conditions (Darilmaz *et al.*, 2013).

Survival of propionibacteria in the digestive tract has also been demonstrated by *in vivo* studies in rats (Huang *et al.*, 2003; Lan *et al.*, 2007a), mice (Pérez Chaia *et al.*, 1995; Pérez Chaia and Zárate, 2005; Zárate and Pérez Chaia, 2012), pigs (Cousin *et al.*, 2012a) and humans (Bougle *et al.*, 1999; Suomalainen *et al.*, 2008; Jan *et al.*, 2002a; Hervé *et al.*, 2007). In most studies, a high level of propionibacteria was detected in the intestinal contents and faeces during the feeding period but this concentration gradually declined and returned to the initial levels a few weeks after consumption ceased.

Some studies (Hervé *et al.*, 2007; Lan *et al.*, 2007a; Saraoui *et al.*, 2013) have revealed by molecular methods that propionibacteria not only survive but remain metabolically active in the gut. Hervé *et al.* (2007) developed efficient non-invasive methods (real time PCR and real-time reverse transcription (RT) PCR) which specifically monitor *P. freudenreichii* concentration and transcriptional activity within the intestine by detecting the propionibacteria-specific transcarboxylase mRNA in faeces of humans and in caecal contents of human microbiota-associated rats (Lan *et al.*, 2007a).

In a more recent study, Saraoi *et al.* (2013) demonstrated in an *in vivo* experimental approach, the activity and metabolic adaptation of probiotic *P. freudenreichii* CIRM BIA1^{T} to the colon environment of pigs. A transcriptomic analysis of gene expression revealed drastic changes in the catabolism of sugars and amino acids. Glycolysis, the Wood–Werkman cycle and the oxidative phosphorylation pathways were down-regulated but induction of specific carbohydrate catabolisms and alternative pathways were induced to produce NADH, NADPH, ATP and precursors (using propanediol, gluconate, lactate, purine and pyrimidine and amino acids as substrates). Genes involved in stress response were down-regulated and genes specifically expressed during cell division were induced, suggesting that *P. freudenreichii* adapts its metabolism to the conditions found in the colon.

Bacterial adhesion to intestinal cells and mucus is generally recognized as the initial step in the colonization of the gut and has been related to many of the health effects of probiotics, as it prolongs the time that beneficial bacteria can influence the gastrointestinal microbiota and immune system. Since propionibacteria grow slowly in natural environments and culture media, adhesion ability becomes an important feature in order to avoid wash-out by gut peristalsis and could be considered a selection criterion of propionibacteria for probiotic purposes.

Both *in vitro* and *in vivo* assays have demonstrated the adhesion capabilities of propionibacteria. Dairy species have shown to adhere to immobilized mucus (Ouwehand *et al.*, 2000; Thiel *et al.*, 2004); to isolated intestinal cells of mouse and chicken (Zárate *et al.*, 2002a,b; Argañaraz Martinez *et al.*, 2013), poultry intestinal tissue explants (Argañaraz Martínez *et al.*, 2013) and to human intestinal cell lines (Huang and Adams, 2003; Lehto and Salminen, 1997; Zárate *et al.*, 2010). Different strains such as *P. freudenreichii* subsp. *shermanii* JS, *P. jensenii* 702 and *P. acidipropionici* Q4 were able to adhere to Caco-2, C2BBe1 and HT29 cell lines respectively, that resemble the colonic epithelium.

Adhesion *in vivo* to the intestinal mucosa was evidenced in murine models with the probiotic strain *P. acidipropionici* CRL

1198 and *P. jensenii* 702. At intervals during treatments the adhering propionibacteria were effectively counted on the intestinal walls of mice and faeces by a plate count method with a selective culture medium (Pérez Chaia *et al.*, 1995; Huang *et al.*, 2003; Pérez Chaia and Zárate, 2005; Zárate and Pérez Chaia, 2012).

In other studies, we have correlated the *in vitro* and *in vivo* abilities of dairy *Propionibacterium* strains to adhere to intestinal epithelial cells and observed by scanning electron microscopy, that *P. acidipropionici* CRL 1198 adheres well to intestinal epithelial cells (IEC) or the mucus layer covering them (Zárate *et al.*, 2002a,b). Microscopic examination revealed two adhesion patterns in propionibacteria: autoaggregating strains adhere in clusters, with adhesion being mediated by only a few bacteria, whereas non autoaggregating propionibacteria adhere individually making contact with each epithelial cell with the entire bacterial surface (Zárate *et al.*, 2002a).

In a recent study, the *in vitro* adhesion ability of probiotic propionibacteria (*P. jensenii* 702) was found to be influenced by the carrier food matrix, with fruit yoghurt providing the most favourable outcomes relative to goat's milk ice cream or plain yoghurt (Ranadheera *et al.*, 2012). It seems evident that food components and their physicochemical properties such as fat content or pH may influence viability, adhesion and functional properties of probiotics and should be considered when choosing an appropriate carrier food matrix. Interactions with the host gut mucosa are also suggested by the analysis of the genome of *P. freudenreichii* since genes encoding for surface proteins involved in adhesion (also present in other probiotic bacteria) have been reported (Falentin *et al.*, 2010a). At least 161 surface exposed proteins have been suggested by SurfG+ software (Barinov *et al.*, 2009) including seven distinct S-layer proteins (Slp), one of which, SlpA, was identified by surface proteomics (Falentin *et al.*, 2010a). Slp proteins are involved in lactobacilli in adhesion and immunomodulation (Hynönen and Palva, 2013). In addition, two parts of a disrupted gene which are similar to an internalin A-like gene, involved in adhesion by other probiotic organisms, exist in the *P. freudenreichii* genome and the protein was also identified by proteomics (Falentin *et al.*, 2010a). However, the involvement of *P. freudenreichii* surface proteins in adhesion and immunomodulatory properties should be investigated more exhaustively.

A final criterion to take into consideration is the tolerance of the selected microorganisms to the industrial processing and storage of the product designed as probiotic-delivering vehicle.

Differences in food manufacturing or pharmaceutical processing can affect probiotic viability as well as their functional properties. Modifications of the physicochemical and organoleptic characteristics of the carrier by the addition of probiotics should also be assessed in order to address favourable changes and avoid undesired technological alterations. On the other side, the performance of the probiotic features should be adequately tested in the designated product. In this respect, potential probiotic strains of propionibacteria have shown to be resistant to Swiss-type cheese and fermented milk manufacturing processes by remaining viable in high numbers and without changes in their probiotic properties (Jan *et al.*, 2002; Zárate and Pérez Chaia, 2005; Cousin *et al.*, 2012a).

Equally important is to maintain viability and probiotic efficacy during the shelf-life of the product that could vary between a few days to several months at refrigeration or ambient temperature. The stability of dairy propionibacteria population during cold storage has previously been evidenced in dairy products. Maity *et al.* (2008) showed that fermented whey containing 2.9×10^8 cfu/ml of dairy propionibacteria at the end of the fermentation retained 9.3×10^7 cfu/ml after 15 days at 5°C. In addition, Kalfirtova and Sovjak (2005) showed that *P. freudenreichii* population stayed at the same level in different dairy products during 14 days of storage at 4°C. Farhadi *et al.* (2013) developed a dairy beverage fermented by *P. freudenreichii* subsp. *shermanii* DSM20270 and *L. acidophilus* LA5 that retained high counts (ca. 10^7 cfu/ml) after 28 days of cold storage suggesting that this dairy drink would provide a stable environment for these microorganisms. Further studies are needed to assess any relevance of this product on human health. Cousin *et al.* (2012a) assessed the survival of dairy propionibacteria at low temperature in the new fermented milk they designed finding out that all strains tested remained viable during at least 15 days at 4°C, which is the usual shelf-life of fermented dairy products. After 6-month of storage at 4°C, the propionibacteria concentrations were still over 10^9 cfu/ml in the fermented milk. We also observed that strains of *P. acidipropionici* and *P. avidum* were able to retain high populations (more than 10^8 cfu/ml) over 12 months suspended in skim milk at −20°C (the storage temperature of frozen dairy products such as ice cream) (Zárate *et al.*, unpublished results) whereas in Emmental cheese conditions, dairy propionibacteria remained alive and metabolically active after 7 weeks of ripening (Perez Chaia and Zárate, 2005) and after 9 days of cold storage (4°C),contributing to the aroma compound production in cheese (Dalmasso *et al.*, 2012).

Mode of action and type of studies

To date, a multitude of health promoting properties have been reported for the genus *Propionibacterium* (Mantere-Alhonen, 1995; Pérez Chaia *et al.*, 1999; Vorobjeva *et al.*, 2008; Cousin *et al.*, 2011; Poonam *et al.*, 2012; Zárate, 2012). Numerous studies have revealed that propionibacteria administered alone or combined with other microorganisms, can exert beneficial effects in the host.

Although the list of health benefits is extensive all of them could be ascribed to one or more of the following mechanisms: (i) immunomodulation; (ii) influence on native microbial composition and exclusion of pathogens; and (iii) modulation of the microbial/host physiology.

Different experimental strategies were employed to assess probiotic potential of propionibacteria. Positive evidence was obtained from: *in vitro* and *ex vivo* assays, *in vivo* trials in animal models with probiotics intended for humans, clinical trials in humans, and assays in animals considered as the probiotic target. The majority of studies have focused on dairy propionibacteria due to their recognized safety. However cutaneous propionibacteria were also used in some studies to improve health of humans and animals. The most documented probiotic effects for both groups of propionibacteria are summarized below.

Overview of reported probiotic effects of dairy propionibacteria

Immunomodulation

Many studies have reported the ability of propionibacteria to modulate in a beneficial manner the innate and adaptive responses of the human immune system (Cousin *et al.*, 2012b). Much research has been done *in vitro* and *in vivo* with the strain *P. freudenreichii* subsp. *shermanii* JS. This microorganism was able to stimulate the proliferative activity of B and T lymphocytes depending on doses and treatment duration in mice (Kirjavainen *et al.*, 1999), to induce TNF-α and IL-10 production in human PBMCs (Kekkonen *et al.*, 2008) and to inhibit the *H. pylori*-induced IL-8 and PGE2 release in human intestinal epithelial cells (Myllyluoma *et al.*, 2008).

Further beneficial results with *P. freudenreichii* JS were obtained with different randomized, placebo-controlled, double-blind trials in humans such as: reduction in the serum level of C-reactive protein (an inflammation marker) (Kekkonen *et al.*, 2008); induction of IL-4 and IFN-g production in PBMCs of infants with cow's milk allergy (Pohjavuori *et al.*, 2004); prevention of IgE-associated allergy in caesarean-delivered children (Kukkonen *et al.*, 2007) and increase in the resistance to respiratory infections during the first two years of life (Kukkonen *et al.*, 2008).In a more recent study Oksaharju *et al.* (2013) showed that consumption of probiotic *Lactobacillus rhamnosus* GG and *P. freudenreichii* subsp. *shermanii* JS by high–fat fed susceptible mice led to reduced levels of inflammation markers together with reduced gonadal adipose tissue.

Other dairy *P. freudenreichii* strains also showed promising immunomodulatory properties by strongly inducing the synthesis of anti-inflammatory IL-10 by human PBMCs with very low induction of IL-12, TNF-a, and IFNγ (Foligné *et al.*, 2010). Two strains of *P. freudenreichii* selected by this assay of cytokine release showed a protective effect against experimental colitis in mice. Some surface compounds should be involved in these immunomodulatory effects of propionibacteria since removal of surface layer proteins decreased the *in vitro* induction of anti-inflammatory cytokines (Foligné *et al.*, 2010).

In a recent study, Cousin *et al.* (2012a) screened *in vitro* 10 strains of propionibacteria grown in a newly developed fermented milk ultrafiltrate, for immunomodulatory properties. The most anti-inflammatory strain, *P. freudenreichii* BIA129, was further tested on piglets which showed that colonic mucosa explants of treated pigs secreted less IL- 8 and TNFα either in basal conditions or after a lipopolysaccharide challenge. Results suggest that dairy propionibacteria-fermented products could be helpful in the prevention and treatment of inflammatory bowel diseases. However, production of surface EPS should be taken into account when selecting strains for immunomodulation since surface b-glucan appears to partly hide the anti-inflammatory properties of *P. freudenreichii* (Deutsch *et al.*, 2012).

With respect to other dairy species, an increase in the phagocytic activity of peritoneal macrophages and the phagocytic function of the reticuloendothelial system was observed in mice fed with *P. acidipropionici* CRL 1198 (Pérez Chaia *et al.*, 1995). In addition, administration of this strain prior to infection of mice with *Salmonella* Typhimurium led to an increase of the anti-*Salmonella* IgA level and the number of IgA producing cells (Alvarez *et al.*, 1996).

Dairy propionibacteria may also act as safe adjuvant for development of oral vaccines. Adams *et al.* (2005) found that *P. jensenii* 702 co-administered orally with soluble *Mycobacterium tuberculosis* antigens to mice stimulate T-cell proliferation of splenic lymphocytes in a significant manner so that this strain could act as a potential living vaccine vector to be used against mucosal transmitted diseases.

Influence on native microbial composition and exclusion of pathogens

A major probiotic effect attributed to propionibacteria is their ability to modulate intestinal microbiota by enhancing the growth of bifidobacteria, a major gut component associated with human health. This property seems to be widespread among propionibacteria (Warminska-Radyko *et al.*, 2002) and has been demonstrated by both *in vitro* (Mori *et al.*, 1997, Kaneko, 1999; Isawa *et al.*, 2002; Warminska-Radyko *et al.*, 2002), and *in vivo* (Roland *et al.*, 1998; Bougle *et al.*, 1999; Satomi *et al.*, 1999; Hojo *et al.*, 2002) studies. The bifidogenic growth stimulators (BGS) present in supernatants of *P. freudenreichii*, *P. jensenii* and *P. acidipropionici* strains were purified and identified as 2-amino-3-carboxy-1,4-naphtoquinone (ACNQ) (Mori *et al.*, 1997; Kaneko *et al.*, 1999) and 1,4-dihydroxy-2-naphthoic acid (DHNA) a precursor of menaquinone (vitamin K2) biosynthesis (Isawa *et al.*, 2002). It has been proposed that these compounds may serve as electron transfer mediators for NADP regeneration in bifidobacteria (Yamazaki *et al.*, 1999), thus favouring their growth.

Besides their prebiotic effect on bifidobacteria it has been reported that DHNA exerts anti-inflammatory effects in the gastrointestinal tract so that oral BGS therapy may represent a non-toxic way to treat ulcerative colitis. Okada *et al.* (2006) demonstrated in a mouse model of colitis induced by dextran sodium sulphate (DSS), that oral administration of DHNA attenuated colonic mucosal inflammation. In addition, the decreased number of *Lactobacillus* and *Enterobacteriaceae* induced by DSS was recovered by DHNA. The mechanism proposed for the protective effect observed, was restoring of microbiota intestinal balance and suppressing of inflammatory lymphocyte infiltration. In a human trial, patients with active ulcerative colitis that consumed BGS daily for 4 weeks showed improvement in their clinical condition (Suzuki *et al.*, 2006).

In a recent study, Nagata *et al.* (2010) demonstrated that DHNA and the culture filtrates of propionibacterial strains specifically inhibited *in vitro* the growth of *Helicobacter pylori* including clarithromycin-resistant strains but did not affect other Gram-negative or anaerobic bacteria of the human intestinal microbiota. The bactericidal activity of DHNA was due to inhibition of cellular respiration and ATP generation of the pathogen. *In vivo* assays showed that oral administration of DHNA could eradicate *H. pylori* in infected germ-free mice avoiding its colonization activity.

Human trials performed with healthy volunteers also evidenced the bifidogenic effect of selected strains of *P. freudenreichii* (Roland *et al.*, 1998; Bouglé *et al.*, 1999; Hojo *et al.*, 2002) or purified BGS (Satomi *et al.*, 1999; Seki *et al.*, 2004) since they showed increased faecal bifidobacterial populations even after 14 days after stopping the consumption of BGS or propionibacteria. As additional effects, a reduced colonic transit time and a reduction in the numbers of clostridia were evidenced in some studies (Bouglé *et al.*, 1999; Seki *et al.*, 2004).

Several studies have reported that dairy propionibacteria are able to inhibit exogenous and opportunistic pathogens by competitive exclusion and production of antimicrobials. *In vitro* studies showed that *P. freudenreichii* strain JS, alone or combined with other probiotics inhibited the adhesion of different pathogens including *H. pylori* to intestinal mucus and Caco2 cell line also improving the epithelial barrier function (Collado *et al.*, 2007; Myllyluoma *et al.*, 2008). Other dairy species like *P. acidipropionici* strain Q4 was able to prevent the adhesion of *Salmonella enteritidis* and *Escherichia coli* to HT29 cells (Zárate *et al.*, 2010) whereas *P. acidipropionici* CRL 1198 regulates *in vitro* the growth of *Bacteroides* and *Clostridium* in caecal homogenates of mice supplemented with propionibacteria and/or inulin (Lorenzo Pisarello *et al.*, 2010). Mice consuming this strain delivered in water, milk or cheese showed a decrease in the number of anaerobes and coliforms in the caecal content one week after feeding (Perez Chaia *et al.*, 1995; Perez Chaia and Zárate, 2005; Zarate and Perez Chaia, 2012). *P. acidipropionici* CRL 1198 also prevented tissue colonization by *Salmonella* Typhimurium in mice (Alvarez *et al.*, 1996).

In humans, probiotic mixtures containing propionibacteria were used in the treatment of intestinal disorders and regulation of gut microbiota and motility. Infants and children fed with propiono-acido-bifido (PAB) milk (manufactured with *L. acidophilus* R, *B. bifidum* NDRI and *P. freundenreichii* subsp. *shermanii* NDRI) (Sarkar and Misra, 1998) or milk containing *P. freudenreichii* subsp. *shermanii* T-73 and *L. acidophilus* (strain not known) (Sidorchuk *et al.*, 1984), showed a reduction in coliforms with an increase in lactobacilli and bifidobacteria populations whereas consumption of *P. freudenreichii* JS reduced oral *Candida* in elderly (Hatakka *et al.*, 2007). Kajander *et al.* (2008) reported that the multispecies probiotic mixture containing *Propionibacterium freudenreichii* JS normalizes intestinal microbiota leading to alleviation of irritable bowel syndrome symptoms.

Propionibacteria produce a wide variety of antimicrobials such as organic acids and bacteriocins that inhibit *in vitro* different food-spoiling bacteria and pathogens. However, both microorganisms and/or the inhibitory substances are mainly used as biopreservatives since their synthesis and action *in situ* in the gut has not been demonstrated yet.

Modulation of the microbial and host physiology

Propionibacteria have the ability to influence the host metabolism mainly through their metabolites and many positive effects on health have been reported in this regard.

It is well known that probiotics may improve lactose malabsorption by supplying their own β-galactosidase for the intraintestinal hydrolysis of lactose (Montalto *et al.*, 2006). Several studies have suggested that *P. acidipropionici* strains may have potential on this subject: they posses high β-galactosidase activity that remain unaltered in the conditions of the human intestine and their cells are permeabilized by bile, which in turn may favour the hydrolysis of lactose within the intestine (Zárate *et al.*, 2000a, Zarate *et al.*, 2000b). The main properties of the β-galactosidase of *P. acidipropionici* CRL 1198 and Q4 (Zárate *et al.*, 2003) as well as its regulation in the presence of lactose and lactate (Zárate and Pérez Chaia, 2012) were determined in order to decide the most appropriate vehicle to deliver propionibacteria to the host. Based on these results an experimental Swiss-type cheese containing *P. acidipropionici* CRL 1198 was elaborated showing no decrease in the synthesis and activity of the β-galactosidase of this strain (Pérez Chaia and Zárate, 2005). When mice were fed with *P. acidipropionici* CRL 1198 included in milk or cheese, both β-galactosidase levels in the small bowels and propionic acid concentrations in the caecum were significantly increased. It is known that short-chain fatty acids (SCFA) absorption favours the colonic recovery of water and electrolytes counteracting the osmotic diarrhoea induced by lactose and/or other unabsorbed carbohydrates (Holtug *et al.*, 1992). However clinical reports on this property are still missing for dairy propionibacteria.

The use of propionibacteria for cancer prevention has gained much attention in the last years due to positive results coming from *in vitro* and *in vivo* studies (mainly in animal models). Several mechanisms have been proposed including the production *in situ* of SCFA, inhibition of enzyme activities that convert procarcinogens into carcinogens, binding of toxic compounds and antimutagenic and antigenotoxic activities.

Vorobjeva (2001, 2008) demonstrated the antimutagenic activity (AM) of *P. freudenreichii* subsp. *shermanii* VKM-103 against the mutations induced by 4-nitro-quinoline and N-nitro-N-nitrosoguanidine (transition mutations), and by 9-aminoacridine and 2-nitrofluorene (frame-shift mutations). This AM activity was exerted by live and dead cells and by the spent culture medium. The active compound responsible for this activity was identified as a cysteine synthase which is induced by some stress factors.

Regarding control of harmful enzymes activities in the gut, it was shown that *P. acidipropionici* CRL1198 fed to mice prevented the induction of azoreductase, nitroreductase and β-glucuronidase activities caused by a cooked red-meat supplemented diet. Furthermore, feeding with propionibacteria resulted in a remarkable reduction of β-glucuronidase activity and slight reductions of azo- and nitroreductase activities when a commercial non-meat containing diet was used (Perez Chaia *et al.*, 1999). In human studies, consumption of *P. freudenreichii* subsp. *shermanii* JS decreased to different extents faecal azoreductase activity in elderly subjects, β-glucosidase and urease in healthy young men and β-glucuronidase activity of irritable bowel syndrome patients (Ouwehand *et al.*, 2002, Hatakka *et al.*, 2008).

Other researchers have focused on the effectiveness of propionibacterial metabolites like SCFA for killing cancer cells. *In vitro* assays showed that dairy propionibacteria kill human colorectal adenocarcinoma cells through apoptosis triggered by propionate and acetate (Jan *et al.*, 2002b, Lan *et al.*, 2007b). In the same way apropionibacteria-fermented milk recently designed with *P. freudenreichii* subsp. *shermani* strain ITG P9 (also known as CIRM-BIA138 and TL133), induced apoptosis of HGT-1 human gastric cancer cells (Cousin *et al.*, 2012c). Similar results were obtained *in vivo* in animals, since consumption of *P. freudenreichii* TL133 by human microbiota associated rats increased the number of apoptotic cells in the colon of 1,2-dimethylhydrazine treated rats but had no effect on healthy colonic mucosa (Lan *et al.*, 2008). It was proposed that, after a genotoxic insult, propionibacteria may help to eliminate damaged cells within the colon by apoptosis. Clinical studies assessing the protective role of propionibacteria against colon cancer are still missing.

Other beneficial effects have also been reported for propionibacteria. Somkuti and Johnson (1990) reported that strains of *P. freudenreichii* remove by cell surface adsorption up to 70% of the cholesterol from the medium, whereas Perez Chaia *et al.* (1995) demonstrated, in an animal model, that *P. acidipropionici* CRL 1198 reversed the hyperlipaemic effect of a diet with high lipid content. Although the mechanisms underlying this effect were not determined, uptake and/or inhibition of endogenous cholesterol synthesis by propionic acid production could be involved (Hara *et al.*, 1999).

Propionibacteria have been related to alleviation of inflammatory bowel diseases (Michel *et al.*, 2005; Uchida *et al.*, 2005; Okada *et al.*, 2006; Suzuki *et al.*, 2006; Mitsuyama *et al.*, 2007; Kajander *et al.*, 2008). Among the mechanisms responsible of the effect both immunomodulation and regulation of intestinal microbiota have been cited. However other studies have ascribed the positive results observed with *P. freudenreichii* and *P. acidipropionici* to propionate (Uchida *et al.*, 2005) and nitric oxide (NO) synthesized from nitrate reduction (Michel *et al.*, 2005), respectively.

Numerous studies have reported the ability of dairy propionibacteria to bind and remove toxic compounds from different environments such as the gut and food. It has been proposed that probiotics may reduce by binding, the availability of free toxic compounds within the intestinal tract which in turn decrease their negative effects in the host. The capacity of *P. freudenreichii* strain JS used alone and combined with lactobacilli (*L. rhamnosus* GG or LC705) to remove mycotoxins has been studied by *in vitro* (El-Nezami *et al.*, 2002; Gratz *et al.*, 2005; Niderkorn *et al.*, 2006), *ex vivo* (Gratz *et al.*, 2005) and *in vivo* assays (El-Nezami *et al.*, 2000, 2006). It was determined that both viable and heat-killed forms of propionibacteria were able to remove efficiently aflatoxin B1 (AFB1), fumonisins and trichotecenes from liquid media. In clinical trials it has been observed that the consumption of a probiotic mixture containing both *P. freudenreichii* JS and *L. rhamnosus* LC-705 reduced in a significant manner the levels AFB_1 in faecal samples (El-Nezami *et al.*, 2000) and the concentration of urinary AFB-N7-guanine (El-Nezami *et al.*, 2006) of healthy volunteers during treatment and even several days after probiotic consumption ceased. Results suggest that the probiotic bacteria could block the intestinal absorption of AFB1.

Other studies have reported that *P. freudenreichii* subsp. *shermanii* JS alone and combined with other probiotics have the ability to remove cyanotoxins and also cadmium and lead from aqueous solution (Ibrahim *et al.*, 2006; Halttunen *et al.*, 2008) being promising microorganisms for decontamination in food and gut.

In recent years we have been investigating the potential of dairy propionibacteria to protect the intestinal mucosa from the toxic and antinutritional effects of plant lectins from the *Leguminosae* family. By *in vitro* and *in vivo* studies we determined that certain strains of *P. acidipropionici* and *P. freudenreichii* were able to bind and remove different dietary lectins such as concanavalin A (Con A), peanut agglutinin (PNA) and jacalin (AIL) from media, preventing their cytotoxic effects on intestinal epithelial cells (Zárate and Pérez Chaia, 2009). Daily ingestion of *P. acidipropionici* CRL 1198, at the same time as ConA prevented the deleterious effects caused by this lectin on some morphological and physiological parameters related to intestinal functionality in mice (Zárate and Pérez Chaia, 2012). Propionibacteria reduced the incidence of colonic lesions, the enlargement of organs, the disruption of brush border membranes and the decrease of their disaccharidase activities. These results suggest that consumption of food containing these bacteria would be a beneficial tool to protect intestinal epithelia.

Dairy propionibacteria for improving animal health

In a similar manner as probiotics intended for humans, the use of probiotics to enhance welfare and performance of animals has increased in recent years. Different microorganisms have been proposed to replace antibiotics as growth promoters for farm animals, being applied mainly to ruminants (cows, calves, steer), chicken, and pigs. More recently their benefits have been extended to horses, pets and aquaculture.

As a general result, dairy propionibacteria administered alone or in a mixture of probiotic microorganisms increased the weight gain, feed efficiency and health of different animals such as lactating cows, calves (Cerna *et al.*, 1991; Adams *et al.*, 2007), pigs (Mantere Alhonen, 1995) and chicken (Babot, 2014). One desired effect for ruminant probiotics is an improvement in propionate production as it is considered the major precursor for hepatic gluconeogenesis that provides substrate for lactose synthesis in lactating dairy cows. Various strains of *Propionibacterium* have shown to increase the molar proportion of ruminal propionate when fed to ruminants and may hold potential as a natural feed alternative to hormones and antibiotics to enhance lactational performance (Kim *et al.*, 2000; Stein *et al.*, 2006). In this respect, *Propionibacterium acidipropionici* P169 could be an effective direct-fed microorganism (DFM) to increase milk production in dairy cows. It has been reported that this strain administered to beef cattle, was able to increase hepatic glucose production via enhanced ruminal propionate production and absorption. Lehloenya *et al.* (2007) reported a 9% increase in milk yield when a mixture of yeast and strain P169 was fed to dairy cows from 2 weeks pre-partum to 30 weeks post partum. When directly fed to early lactating dairy cows, propionibacteria tended to improve milk protein content and energy efficiency during early lactation, without affecting the reproductive function (Francisco *et al.*, 2002; Stein *et al.*, 2006; Weiss *et al.*, 2008).

In addition, it has been proposed that a DFM-product containing *L. acidophilus* NP51 plus *P. freudenreichii* NP24 could be used to enhance the performance of cows subject to heat stress during hot weather (Boyd *et al.*, 2011).

Recently, Lettat *et al.* (2012) evidenced that probiotic mixtures containing *Propionibacterium* strainP63 may be useful to minimize the occurrence of butyric and propionic subacute ruminal acidosis (SARA) in sheep.

Regarding calves, a blended culture of lactic acid bacteria and one strain of *P. freudenreichii* (known as Proma), and a DFM product containing *P. jensenii* 702 were effective to improve weight gain during pre-weaning and weaning periods (Cerna *et al.*, 1991; Adams *et al.*, 2007).

With respect to monogastric animals, Mantere Alhonen (1995) was the first to demonstrate the probiotic and growth promoting effects of pure propionibacteria in young piglets. The most effective species was *P. freudenreichii* subsp. *shermanii.* When fed to the test animals in a daily dose of 2×10^9 cfu/g, the weekly weight gain of piglets was 9.2–14.5% higher and their fodder demand 7.2–46.1% lower than in the control group. In addition the animals treated with propionibacteria had a lower incidence of diarrhoea. In older pigs the effects were even more pronounced. With respect to chickens, both undefined and defined 'Nurmi Cultures' have been used to promote growth, to stimulate immune system and to prevent colonization by avian pathogens, like *Salmonella,* in young animals (Khan *et al.*, 2007; Higgins *et al.*, 2008) Some researchers have reported the presence of propionibacteria in the ileum and caecum of chickens (Salanitro *et al.*, 1978) and in caecal Nurmi cultures (Waters *et al.*, 2006). However, propionibacteria have not been widely studied in this ecological niche.

In a recent study (Argañaraz Martinez *et al.*, 2013), the presence of *Propionibacterium* strains in hens of conventional production systems and backyard hens was investigated. Bacteria from this genus were evidenced in the large intestine of 26% of the animals sampled, with all the isolates identified as *P. acidipropionici* or *P. avidum.* Some *in vitro* studies on the probiotic potential of the strains, showed that *P. acidipropionici* LET 105 was resistant to gastrointestinal digestion, and inhibited *Salmonella enterica* serovar Enteritidis by SCFA production and competence for adhesion sites in the intestinal mucosa. Results provide evidence of the potential application of dairy propionibacteria as probiotic to prevent intestinal infections in poultry.

Overview of reported probiotic effects of cutaneous propionibacteria

Cutaneous propionibacteria have been used as probiotics much less than dairy species due to their potential behaviour as opportunistic pathogens. In contrast to the recognized safe status of dairy propionibacteria, the genomes of cutaneous propionibacteria have revealed some virulence factors that discourage their probiotic application or require at least much more care.

It is well documented that cutaneous propionibacteria are potent immunomodulators, since they have been tested in several assays both in humans and rodents used as animal models (Roszkowski *et al.*, 1990). Administration of cutaneous propionibacteria (*P. avidum, P. granulosum, P. acnes*) have shown to be beneficial in the treatment of neoplastic and infectious diseases (Mara *et al.*, 1994; Isenberg *et al.*, 1994; 1995). Besides, dead *P. acnes* or a polysaccharide extracted from its cell wall have proven to be effective in the induction of macrophages with an antitumour effect (Ananias *et al.*, 2007) and in modulating an experimental immunization against *Trypanosoma cruzi* (Mussalem *et al.*, 2006).

Cutaenous propionibacteria have also been used to improve the health of pigs. *P. avidum* KP-40 showed to be a potent immunomodulator that stimulated granulopoiesis as well as a faster body weight gain in pregnant pigs and their offspring (Markowska-Daniel *et al.*, 1992). The usefulness of the prophylactic application of this strain, against porcine microbial infections was tested in pigs showing that *P. avidum* application caused positive immunoregulation of porcine innate immune system effectors, non-specific activation of lymphocytes and antibody production that resulted in milder clinical symptoms, faster recovery and a greater body weight gain (Markowska-Daniel *et al.*, 1993). In a more recent trial, Lo *et al.* (2009) evaluated the immunomodulating activities of inactivated *P. granulosum* cell walls plus *E. coli* lipopolysaccharide (PG/LPS) (Inmodulen®) on porcine immunity. Results showed that the cytotoxicity of natural killer cells and proliferation of lymphocytes in response to mitogen stimulation were significantly enhanced in pigs that were intramuscularly administered PG/LPS (1 ml/10 kg of body weight) once or twice compared with the controls injected with saline solution. PG/LPS also displayed adjuvant effects with classical swine fever virus (CSFV) and *Mycoplasma hyopneumoniae* vaccines. Moreover, pigs receiving PG/LPS showed a 20.8% growth enhancement compared with untreated pigs. Thus, PG/LPS caused positive immunoregulation of porcine innate immune system effectors, non-specific activation of lymphocytes and antibody production.

Bacterial interference, or bacteriotherapy, in which commensal bacteria are used to prevent colonization of the host by pathogens, has been shown to be a promising way for prevention and treatment of infections. In an interesting approach, Shu *et al.* (2013) demonstrated that fermentation of glycerol with *P. acnes* ATCC6919, considered a skin commensal bacterium, functioned as a skin probiotic for *in vitro* and *in vivo* growth suppression of USA300, the most prevalent community-acquired methicillin-resistant *Staphylococcus aureus* (CA-MRSA). In this sense, the concept of fermentation of skin commensals against pathogens creates potentially a new skin industry and also generates a new area of investigation into the biological function of skin microbiome for promoting human health.

In Table 5.3 we summarize the beneficial effects reported for both dairy and cutaneous propionibacteria through *in vivo* studies in animals and humans.

Other relevant properties with impact on health: potential of propionibacteria as nutraceutical factories

Propionibacteria produce a wide variety of metabolites that have shown positive effects on health so they could be considered as

Table 5.3 Probiotic effects of propionibacteria demonstrated *in vivo* in animals or humans trials

Mode of action	Species	Reported effect	Reference
Dairy propionibacteria			
Immunomodulation	*P. freudenreichii* JS	Reduction in the serum level of CRP (an inflammation marker)	Kekkonen *et al.* (2008)
	P. freudenreichii JS	induction of IL-4 and IFN-g production in PBMCs of infants with cow's milk allergy	Pohjavuori *et al.* (2004)
	P. freudenreichii JS	prevention of IgE-associated allergy in caesarean-delivered children and increase in the resistance to respiratory infections during the first two years of life	Kukkonen *et al.* (2007, (2008)
	P. freudenreichii JS	Reduction of inflammation markers together with reduced gonadal adipose tissue in mice	Oksaharju *et al.* (2013)
	P. freudenreichii BIA118 and SI48	Protective effect against experimental colitis in mice	Foligné *et al.* (2010)
	P. acidipropionici CRL1198	In the phagocytic function of the reticuloendothelial system, anti-*Salmonella* IgA level and the number of IgA producing cells in mice	P. Chaia *et al.* (1995), Alvarez *et al.* (1996)
	P. jensenii 702	Potential living vaccine vector: stimulation of lymphocyte proliferation in mice	Adams *et al.* (2005)
Microbiota modulation	*P. freudenreichii* SI26, SI41,ET-3	Bifidogenic effect: Increased faecal bifidobacterial populations	Roland *et al.* (1998), Bougle *et al.* (1999), Satomi *et al.* (1999), Hojo *et al.* (2002
	P. freudenreichii JS	ø oral *Candida* in elderly	Hatakka *et al.* (2007)
	P. freudenreichii T-73 and NDRI	ø coliforms and - lactobacilli in bifidobacteria in humans	Sidorchuk *et al.* (1984), Sarkar and Misra (1998)
	P. acidipropionici CRL1198	ø anaerobes and coliforms in the caecum of mice and prevent colonization by *Salmonella* Typhimurium	Pérez Chaia *et al.* (1995), Alvarez *et al.* (1996
Metabolism modulation			
Cancer prevention	*P. acidipropionici* CRL1198 *P. freudenreichii* JS	Control of carcinogenic enzyme activities in the gut of mice and humans	P. Chaia *et al.* (1999), Ouwehand *et al.* (2002), Hatakka *et al.* (2008)
	P. freudenreichii TL133	Apoptosis in the colon of 1,2-dimethylhydrazine treated rats	Lan *et al.* (2008)
Hypocholesterolaemic effect	*P. acidipropionici* CRL1198	ø hyperlipaemic effect of a diet with high lipid content	Pérez Chaia *et al.* (1995)
Alleviation of IBS	*P. freudenreichii* ET-3, JS *P. acidipropionici* TL15 and TL233	Alleviation of symptoms, stabilization of microbiota	Michel *et al.* (2005), Uchida *et al.* (2005), Kajander *et al.* (2008)
Binding of toxic compounds	*P. freudenreichii* JS, *P. acidipropionici* CRL1198	Removal in the gut of mycotoxins and vegetable lectins in the gut	El-Nezami *et al.* (2000), 2006), Zárate and Pérez Chaia (2012)
Enhancement of animal's welfare	*P. acidipropionici* DH42, *Propionibacterium* strain P169	Increase of ruminal propionate. Improvement in milk protein content and energy efficiency during early lactation	Kim et al (2000), Francisco *et al.* (2002), Stein *et al.* (2006), Weiss *et al.* (2008)
	Propionibacterium P63	ø occurrence of butyric and propionic subacute ruminal acidosis (SARA) in sheep	Lettat *et al.* (2012)
	P. freudenreichii	Promoted growth of piglets	Mantere-Alhonen *et al.* (1995)
Cutaneous propionibacteria			
Immunomodulation	*P. acnes* (strain from Inst A. Lutz, São Paulo, Brazil)	Adjuvant on immunization against *Trypanosoma cruzi*	Mussalem *et al.* (2006)
	P. granulosum (Inmodulen®)	Positive immunoregulation of porcine innate immune system effectors, non-specific activation of lymphocytes and antibody production	Lo *et al.* (2009)
	P. granulosum KP-45	Immunostimulation in neoplastic and infectious diseases	Isenberg *et al.* (1994, 1995)
	P. acnes (strain from Inst A. Lutz, São Paulo, Brazil)	Immunomodulation on peritoneal cells exudate of mice: major NKT-cell recruitment and increased cytotoxicity	Ananias *et al.* (2007)
	P. avidum KP-40	Positive immunoregulation of porcine innate immune system effectors, non-specific activation of lymphocytes and antibody production against microbial infections. Greater body weight gain	Markowska-Daniel *et al.* (1992, 1993)
Microbiota modulation	*P. acnes* ATCC 6919	Growth suppression of USA300, the most prevalent community-acquired methicillin-resistant *Staphylococcus aureus* (CA-MRSA)	Shu *et al.* (2013)

'nutraceuticals factories' for food enrichment. Among the most relevant biological compounds synthesized by propionibacteria are vitamins B and K, conjugated linoleic acid (CLA), exopolysaccharides (EPS) and trehalose.

Vitamin B_{12}, also known as cobalamin, is essential for the normal functioning of the body. Animals, plants and fungi are incapable of synthesizing this vitamin and it is exclusively produced by microorganisms. Then, humans have to obtain cobalamin from diet. Among B_{12} producers, propionibacteria are the only safe microorganisms (GRAS status) being in consequence the preferred for vitamin B_{12} food enrichment. With this aim, they have been employed as adjunct cultures for elevating the B_{12} levels of different fermented foods (Babuchowski *et al.*, 1999; Van Wyk *et al.*, 2011) and for correcting systemic vitamin B_{12} levels in deficient animals (Adams *et al.*, 2002)

Propionibacteria synthesize cobalamin as a cofactor for propionic acid fermentation and the biosynthesis pathway has been completely elucidated in *P. freudenreichii* (Murooka *et al.*, 2005; Falentin *et al.*, 2010). In order to improve industrial use of propionibacteria as vitamin producers, many efforts have been made to optimize the production process and to increase the yields of cobalamin through genetic engineering (Murooka *et al.*, 2005) and other biotechnological strategies like fermentation manipulations (Piao *et al.*, 2004; Murooka *et al.*, 2005).

Several B_{12} pathway genes were cloned and expressed in *P. freudenreichii* revealing that the recombinant strain displayed a significant improvement in its vitamin synthesis ability (Piao *et al.*, 2004). Other approaches such as random mutagenesis by UV radiation (Ali and Mohd, 2011) and nitrosoguanidine (NTG) (Khan and Mir, 2011) have also been used to improve vitamin B_{12} production by *P. freudenreichii*.

Regarding culture manipulations, a semi-continuous process for large scale production was described that involves, first, an anaerobic growth phase in which most of the biomass and precursors, like precorrins and corbyrinic acid, are produced followed by supplementation of culture media with cobalt and 5,6-dimethylbenzimidazole and aerobic incubation in the latter phase of fermentation (Ye *et al.*, 1996, Hunik, 2000). Addition of vitamin B_{12} analogues in the fermentation medium, proposed to overcome the feedback inhibition, also resulted in a considerable increase in vitamin B_{12} production by *P freudenreichii* subsp. *shermanii* (Thirupathaiah *et al.*, 2012). In other recent approaches, spent media from LAB fermentation (Gardner and Champagne, 2005), glucose (Chiliveri *et al.*, 2010) and crude glycerol (Kosmider *et al.*, 2012) have shown to support high biomass and vitamin B_{12} production by propionibacteria under optimized conditions.

Vitamin B_2, or riboflavin, is the central component of the cofactors FAD and FMN, and is required by all flavoproteins. Moreover, it is needed for numerous cellular reactions and is involved in vital metabolic processes in the body. Although present in many foods (milk, meat, some vegetables), ariboflavinosis could be present in developing countries. Hence, riboflavin-producing starter strains can be used for supplementing this vitamin in human diet. Selection of spontaneous roseoflavin-resistant mutants has been proposed as a reliable method to obtain natural riboflavin-overproducing strains (Burgess *et al.*, 2006). In fact mutant strains of *P. freudenreichii* have shown to produce higher amounts of riboflavin than the wild type strains (Burgess *et al.*, 2006). It has been reported that a riboflavin enriched yoghurt containing *P. freudenreichii* NIZO2336 (a roseoflavin resistant mutant) as adjunct culture, alleviate the symptoms of experimentally induced ariboflavinosis in rats (LeBlanc *et al.*, 2006).

Propionibacteria also produce vitamin B_7 (biotin) and vitamin B_9 (folic acid), with some strains producing as high folate as *S. thermophilus* strains (known folate producers) (Hugenholtz *et al.*, 2002).

Overall, results suggest that propionibacteria-containing products could be expected to be good sources of B-group vitamins. Some functional food products using folate and vitamin B_{12} producing *Propionibacterium* strains have been already developed to counteract vitamin deficiencies (Van Wyk *et al.*, 2011).

Vitamin K (a group of 2-methyl-1,4-naphthoquinone derivatives) is an essential cofactor for the formation of γ-carboxyglutamic acid-containing proteins that bind calcium ions and are involved in blood coagulation and tissue calcification. Vitamin K_1 or phylloquinone is present in plants, and vitamin K_2, also called menaquinone, is produced in animals and bacteria that live within the intestine.

It has been reported that *P. freudenreichii* produces large amounts of tetrahydromenaquinone-9 (MK-9 (4H)) and the precursor 1,4-dyhidroxy-2-naphthoic acid (DHNA) which is a known bifidogenic factor (Mori *et al.*, 1997; Furuichi *et al.*, 2006, 2007; Isawa *et al.*, 2002).

Hojo *et al.* (2007) assessed the concentration of MK-9 (4H) in commercial propionibacteria-fermented cheeses showing a positive correlation between the increase in propionibacteria and the generation of MK-9 (4H) in cheese. Owing to their high MK-9 (4H) concentrations (200 to 650 ng/g), some cheeses, such as Emmental and Jarlsberg, should be important sources of vitamin K for humans.

In addition, attempts on enhancement of DHNA production by propionibacteria have been made by modifying culture conditions such as the supply of oxygen and carbon source. As result, an improvement in DHNA production has been obtained by a cultivation method that combines anaerobic fed-batch and aerobic batch cultures (Furuichi *et al.*, 2006; 2007).

CLA is a family of linoleic acid isomers (cis-9, cis-12-octadecadienoic acid) commonly found in food products like milk and meat of ruminants. Many health benefits have been ascribed to CLA consumption such as anticarcinogenic, antiatherogenic, antidiabetogenic, antiadipogenic, antioxidative properties and immune system modulation (Benjamin and Spener, 2009). Some microorganisms like *Bifidobacterium*, *Lactobacillus*, *Enterococcus* and *Propionibacterium* display linoleic acid isomerase activity and form CLA as a detoxification mechanism (Sieber *et al.*, 2004). In consequence, they have been proposed as starter or adjunct cultures for increasing the CLA content and nutritional value of some fermented products.

Regarding propionibacteria, both growing and resting cells of dairy (*P. freudenreichii*) (Rainio *et al.*, 2002; Wang *et al.*, 2007) and cutaneous propionibacteria (*P. acnes*) (McIntosh *et al.*, 2009) produce cis-9, trans-11 and trans-10, cis-12, the major isomers with biological activity, on different growth media: culture broths (Wang *et al.*, 2007), lipid containing plant materials (Vahvaselka *et al.*, 2010), milk and ripening cheese (Xu *et al.*, 2005). The optimal conditions that favour the accumulation of CLA by *P. freudenreichii* were determined by varying the source of LA for conjugation and the fermentation conditions leading to a high efficiency (50–90%) of conversion of free LA to mainly extracellular CLA (Xu *et al.*, 2004; Wang *et al.*, 2007).

In a recent study, propionibacteria were assessed for their ability to conjugate a range of other unsaturated fatty acids during fermentation (Hennessy *et al.*, 2012). Results demonstrated that propionibacteria were able to bio-convert linoleic, a-linolenic, γ-linolenic and stearidonic acids to their conjugated isomers (CLA, CLNA, CGLA and CSA respectively) through the activity of its linoleic acid isomerase. These conjugated fatty acids may offer the combined health promoting properties of conjugated fatty acids such as CLA and CLNA, along with those of the unsaturated fatty acids from which they are formed. The encouraging results obtained support the potential of propionibacteria for producing CLA enriched products.

EPS have relevance both in health and food industry as they have been related to immunomodulation and the texture-viscosity of dairy products. Hence EPS-producing bacteria have been proposed either as biological thickeners and probiotics (Badel *et al.*, 2011).

Dairy propionibacteria are able to produce both homopolysaccharide (Nordmark *et al.*, 2005; Deutsch *et al.*, 2010) and heteropolymers (Dobruchowskaa *et al.*, 2007) in a strain-dependent manner that is influenced by the medium composition and the fermentation conditions (Racine *et al.*, 1991; Gorret *et al.*, 2001). In addition, the single gene *gtf* encoding for a β-D-glucan synthase that is responsible for the synthesis of surface polysaccharide has been identified in *P. freudenreichii* subsp. *shermanii* (Falentin *et al.*, 2010; Deutsch *et al.*, 2010).

Trehalose is an important nutraceutical of wide commercial interest in the food and pharmaceutical industries. Applications include its use as healthy sugar substitute in foods, cryopreservative, stabilizer in clinical reagents and component in cosmetics.

Among the known microbial trehalose producers (*Escherichia, Mycobacterium, Corynebacterium, Propionibacterium*), dairy propionibacteria are the only food grade bacteria capable of trehalose biosynthesis (Deborde *et al.*, 1996). Propionibacteria synthesize trehalose as a reserve compound and as a stress-response metabolite (Deborde *et al.*, 1996; Boyaval *et al.*, 1999; Cardoso *et al.*, 2004, 2007). Different studies have manipulated the environmental conditions in order to enhance the trehalose production by dairy propionibacteria (Deborde *et al.*, 1996; Cardoso *et al.*, 2004). Trehalose accumulation increased by decreasing the pH from 7.0 to 4.7, by increasing the concentration of NaCl to 2% (w/v), or during growth under aerobic conditions. Regarding carbon sources, lactose was the best substrate for trehalose production, whereas lactate, was a poor precursor. Recently, crude glycerol was reported as a new and effective medium for achieving high yields of trehalose and propionic acid from *P. freudenreichii* subsp. *shermanii*. Presence of KCl in the crude glycerol medium seemed to be responsible of the higher level of intracellular trehalose accumulation (Ruhal *et al.*, 2011). Further improvements in trehalose production were obtained with a chemically derived osmotically sensitive mutant due to its higher ADP-glucose pyrophosphorylase activity (Ruhal and Choudhury, 2012a) and by employing a novel aerobic growth strategy (Ruhal and Choudhury, 2012b).

With respect to the production of this sugar *in situ* in food products, it has been observed that *P. freudenreichii* subsp. *shermanii* NIZO B365 produces high levels of trehalose in skim milk (Cardoso *et al.*, 2004). Recently, otsB gene from this strain was cloned and expressed in *L. lactis*. The recombinant strain produced and released trehalose in the medium and was remarkably tolerant to acid, heat and cold shock treatments (Carvalho *et al.*, 2011). These properties could in turn, favour survival of strains against technological and digestive stresses.

Development of a trehalose enriched functional products using *Propionibacterium* strains may represent an attractive approach to deliver this important molecule into human system.

Industrial application of propionibacteria: dairy starters and biopreservatives

In the present chapter, we focused on the evidence that relate propionibacteria with human and animal health. However propionibacteria possess other important technological properties that make them economically relevant. We briefly summarize them in the following paragraphs.

The most traditional industrial application of the genus *Propionibacterium* is the usage of 'classical propionibacteria' as dairy starters for the manufacture and ripening of Swiss type cheeses. In these products, they play an important role in the development of taste and opening, through the production of flavour compounds and the release of CO_2 responsible for the characteristic 'eyes' formation (Langsrud and Reinbold, 1973a).

The species most commonly used as starter is *P. freudenreichii* which grows during ripening and reaches at the end of maturation stable levels over 10^9 cfu/g of cheese. It has been demonstrated that propionibacteria play a major role in the formation of cheese flavour (Thierry *et al.*, 2011). Recent molecular studies have shown that *P. freudenreichii* maintain viability and metabolic activity up to the end of ripening, producing flavour compounds during growth in cheeses at 24°C and further cold storage (about 4_C) (Dalmasso *et al.*, 2012). Propionibacteria produce flavour compounds from three main pathways: lactate and aspartate fermentation (that release propionic and acetic acids and CO_2), amino acid catabolism (that results mainly in 2-methylbutanoic acid and 3-methylbutanoic acid from isoleucine and leucine degradation, respectively) and fat hydrolysis (which generates free fatty acids from milk glycerides). Although *P. freudenreichii* possesses diverse intracellular peptidases, it does not lyse in cheese. Hence, it plays a limited role in secondary proteolysis, compared to starter and non-starter lactic acid bacteria (NSLAB) (Gagnaire *et al.*, 2001).

Other dairy products such as yoghurt and fermented milks are less used as carriers for propionibacteria due to the presence of inhibitory substances, the low pH attained by lactic fermentation and their weak proteolytic activities that do not allow them to grow. Hence, other possibilities have been considered. In this respect, the combinational use of *Bifidobacterium* and *Propionibacterium* has been proposed for the development of fermented milk or soy milk products (Wu *et al.*, 2012). As mentioned, a novel milk-based product exclusively fermented with *P. freudenreichii* that ensures survival and metabolic activity of propionibacteria during storage and intestinal transit has been developed recently (Cousin *et al.*, 2012a).

Another recognized potential application for propionibacteria is the biological production of propionic acid. This compound displays a great variety of industrial applications such as manufacturing of cellulose-based plastics, herbicides, perfumes and pharmaceuticals. Propionic acid and its salts as well as *Propionibacterium* strains are also used in food industry to prolong the shelf-life of many products by suppressing the growth of spoilage microorganisms (Danilova *et al.*, 2012). In addition, many health benefits have been ascribed to propionic acid in recent years

Propionic acid is mainly produced through petrochemical route but ecological demands for natural preservatives have raised interest in their biological production using low cost raw materials (Feng *et al*, 2011). Since biosynthesis by bacterial fermentation is limited by low productivity, low conversion efficiency, by-product formation and end-product inhibition different attempts have been made to improve propionibacterial production of propionic acid for industrial applications (Liu *et al.*, 2012).

Most studies have focused on the production of propionic acid by *P. acidipropionici* and *P. freudenreichii.* Enhanced production has been obtained by manipulating fermentation conditions and by genetic engineering (Zhang and Yang 2009, Liu *et al.*, 2011; Zhu *et al.*, 2012; Liu *et al.*, 2012). In order to reduce the cost of production process different substrates (glucose, lactose, lactate, whey, glycerol) and industrial and agricultural wastes (sugar cane molasses and bagasse among others) have been assayed as raw material for fermentation (Coral *et al.*, 2008; Zhang and Yang, 2009; Liu *et al.*, 2011; Feng *et al.*, 2011; Zhu *et al.*, 2012;). Recently, glycerol has become the substrate of choice due to its low cost and the ecological relevance of using an important waste from biodiesel production (Zhu *et al.*, 2012; Liu *et al.*, 2012). Besides, it has been shown that fermentation of glycerol leads to a higher propionic acid yield and much lower acetic acid formation compared to glucose (Coral *et al.*, 2008; Zhang and Yang, 2009; Zhu *et al.*, 2010; Ruhal and Choudhury, 2012).Then studies have focused now on the optimization mainly by metabolic engineering of this particular fermentation by propionibacteria (Liu *et al.*, 2012). In a recent approach, Ammar *et al.* (2013) reported the synthesis of n-propanol with a simultaneous improvement of propionic acid production through metabolically engineered propionibacteria offering a new alternative for the biosynthesis of this promising biofuel.

Finally, propionibacteria also possess potential as biopreservatives through the production of bacteriocins. Both starters and naturally occurring bacteria on food have the ability to produce these antimicrobial peptides. Hence, they may have potentially important applications as food preservatives or bacteriocin-producer probiotics to inhibit intestinal pathogens (Gillor *et al.*, 2008).

Different bacteriocins produced by both dairy and cutaneous propionibacteria have been reported and characterized, such as: Propionicin PLG-1 and GBZ-1 produced by *P. thoenii* 127 (Lyon *et al.*, 1993); Jenseniin G isolated from *P. thoenii* P126 (Grinstead *et al.*, 1992); Propionicins SM1 and SM2 produced by *P. jensenii* DF1 (Miescher *et al.*, 2000); Propionicin T1 synthesized by *P. thoenii* 419 and LMG2792 (Faye *et al.*, 2000); Thoenicin 447 isolated from *P. thoenii* 447 (Van der Merwe *et al.*, 2004); Acnecin produced by a strain of *P. acnes* (Paul *et al.*, 1988; and several other propionicins (Ratnam *et al.*, 1999; Brede *et al.*, 2004; Gwiazdowska *et al.*, 2006).

These bacteriocins are active against other propionibacteria, lactic acid bacteria (*Lactobacillus, Lactococcus* and *Streptococcus*), other Gram-positive bacteria (*Clostridium botulinum* types A, B and E), Gram-negative bacteria (*C. jejuni, E. coli, Ps. fluorescens, Ps. aeruginosa, Vibrio parahaemolyticus, Salmonella typhimurium, Yersinia enterocolitica*); yeasts (*Saccharomyces, Candida* and *Scopularopsis* sp.) and moulds (*Aspergillus ventii, Apiotrichum curvatum, Fusarium tricinctum, Phialophora gregata*).

Propionibacteria also produce other peptides and organic acids (2-pyrrolidone-5-carboxylic acid, 3-phenyllactic acid, hydroxyphenyl lactic acid 3-phenyllactic acid) with antiviral, antiyeast and antifungal activities (Ramanathan *et al.*, 1968; Faye *et al.*, 2002; Lind *et al.*, 2007).

Further studies dealing with the production *in situ* in food and gut and optimization of the production process are needed for the application of propionibacteria as biopreservatives based on these antimicrobials.

Future trends and concluding remarks

The research on the genus *Propionibacterium* has expanded in many different fields in the last decades, but considerable attention has focused on the health promoting activities of propionibacteria and the different mechanisms involved.

The first reports in this field deal with the use of mixed preparations that contained dairy propionibacteria and species of lactobacilli and bifidobacteria. Different combinations were assayed for curing intestinal disorders in calves, piglets, hens and also in humans with good success. Later, beneficial effects on weight gain and general welfare of piglets could be attributed to strains of *P. freudenreichii* used in single preparations. At present, different dairy propionibacteria are mainly included in food products alone or in combinations with lactic acid bacteria and bifidobacteria to provide a wider spectrum of activity in the host.

In the last twenty years, researchers have developed different strategies for determining the viability of propionibacteria intended as probiotics, including the formulation of new selective culture media and, more recently, the use of molecular tools to evidence biologically active propionibacteria in the complex microbial consortium that human and animal intestine harbour. Molecular characterization of new species and strains with probiotic potential *in vitro* and the elucidation of mechanisms responsible for the beneficial effects observed *in vivo* in animals trials provided robust information about the genus *Propionibacterium* to support clinical studies in humans, using safe, well-identified and effective strains.

Summarizing the present knowledge, we could briefly characterize propionibacteria as microorganisms with the ability to overcome several stressful environmental conditions, which include the exposure to acids, bile salts and enzymes in the gastrointestinal tract of humans and animals and the cooking, pressing and salting of the cheese processing. Their versatile metabolism, allow them a long survival at low temperatures in fermented milks and frozen dairy products and at room temperature under carbon starvation. Their ability to ferment sugars and organic acids like lactic and pyruvic acids, ensure their adaptation to different environments where they may develop at expenses of metabolic products of other bacterial populations like lactobacilli. In this overview, propionibacteria reveal very particular potential to adapt to the intestinal ecosystem. Several probiotic properties of propionibacteria are mediated by the propionic acid production, an intestinal metabolite with a wide range of physiological effects both in humans and animals, which is naturally produced by the intestinal anaerobic microbiota and normally absorbed by the intestinal mucosa. Other probiotic properties are mainly related to the production of vitamins, some secondary metabolites and the cell wall structure and properties.

The recent genome sequencing of *P. freudenreichii* and *P. acidipropionici* represent a great progress towards elucidation of the survival strategies of probiotic propionibacteria within the intestine and in different stressful environmental conditions. Moreover, the comparison of genomic sequences of propionibacteria, other probiotic bacteria and commensal intestinal bacteria, would provide in the future valuable information on the functional properties and behaviour of probiotics in the gastrointestinal tract.

Bearing in mind their very remarkable resistance to different environmental conditions and versatile metabolism, it could be expected that new food and pharmaceutical products will be developed in the future to deliver propionibacteria to their target site in the host.

Acknowledgements

This review was supported by grants of CONICET (PIP 0043 and PIP 0319) and ANPCyT (PICT 2012-2871).

References

Abeijón Mukdsi, M.C., Falentin, H., Maillard, M.B., Chuat, V., Medina, R.B., Parayre, S., and Thierry, A. (2014). The secreted esterase of *Propionibacterium freudenreichii* has a major role in cheese lipolysis. Appl. Environ. Microbiol. *80(2)*, 751–756.

Adams, M., Luo, J., Rayward, D., King, S., Gibson, R., and Moghaddam, G.H. (2007). Selection of a novel direct-fed microbial to enhance weight gain in intensively reared calves. Anim. Feed Sci. Technol. *145*, 41–52.

Adams, M.C., Huang, Y., Kotula, L., Blake, R.J., and Garg, M.L. (2002). The efficacy of a potential new probiotic, *Propionibacterium jensenii* 702, to correct vitamin B_{12} levels in an *in vivo* deficient animal model. Asia Pac. J. Clin. Nutr. *11*, S261.

Adams, M.C., Lean, M.L., Hitchick, N.C., and Beagley, K.W. (2005). The efficacy of *Propionibacterium jensenii* 702 to stimulate a cell-mediated response to orally administered soluble *Mycobacterium tuberculosis* antigens using a mouse model. Lait *85*, 75–84.

Ali, M.N., and Mohd, M.K. (2011). Enhancement in vitamin B_{12} production by mutant strains of *Propionibacterium freudenreichii*. Int. J. Eng. Sci. *3*, 4921–4925.

Alvarez, S., Medici, M., Vintini, E., Oliver, G., de Ruiz Holgado, A.P., and Perdigon, G. (1996). Effect of the oral administration of *Propionibacterium acidipropionici* on IgA levels and on the prevention of enteric infection in mice. Microbiol. Alim. Nutr. *14*, 237–243.

Ananias, R.Z., Rodrigues, E.G., Braga, E.G., Squaiella, C.C., Mussalem, J.S., Longhini, A.L., Travassos, L.R., and Longo-Maugéri, I.M. (2007). Modulatory effect of killed *Propionibacterium acnes* and its purified soluble polysaccharide on peritoneal exudate cells from C57Bl/6 mice: major NKT-cell recruitment and increased cytotoxicity. Scand. J. Immunol. *65*, 538–548.

Anastasiou, R., Leverrier, P., Krestas, I., Rouault, A., Kalantzopoulos, G., Boyaval, P., Tsakalidou, E., and Jan, G. (2006). Changes in protein synthesis during thermal adaptation of *Propionibacterium freudenreichii* subsp. *shermanii*. Int. J. Food Microbiol. *108*, 301–314.

Argañaraz-Martínez, E., Babot, J.D., Apella, M.C., and Perez Chaia, A. (2013). Physiological and functional characteristics of *Propionibacterium* strains of the poultry microbiota and relevance for the development of probiotic products. Anaerobe *23*, 27–37.

Babot, J.D. (2014). Bacterias lácticas y microorganismos relacionados como protectores de la función digestiva intestinal de aves de criadero PhD Thesis. Universidad Nacional de Tucumán.

Babuchowski, A., Laniewska-Moroz, L., and Warminska-Radyko, I. (1999). Propionibacteria in fermented vegetables. Lait *79*, 113–124.

Badel, S., Bernardi, T., and Michaud, P. (2011). New perspectives for Lactobacilli exopolysaccharides. Biotechnol. Adv. *29*, 54–66.

Benjamin, S., and Spener, F. (2009). Conjugated linoleic acids as functional food: an insight into their health benefits. Nutr. Metab. *6*, 36.

Bernard, K.S.L., Munroa, C., Forbes-Faulknerb, J.C., Pittb, D., Nortonb, J.H., and Thomas, A.D. (2002). *Propionibacterium australiense* sp. nov. Derived from Granulomatous Bovine Lesions. Taxonomy/Systematics *8*, 41–47.

Bougle, D., Roland, N., Lebeurrier, F., and Arhan, P. (1999). Effect of propionibacteria supplementation on fecal bifidobacteria and segmental colonic transit time in healthy human subjects. Scand. J. Gastroenterol. *34*, 144–148.

Boyaval P., and Corre, C. (1995). Production of propionic acid. Lait *75*, 453–461.

Boyaval, P., Deborde, C., Corre, C., Blanco, C., and Begue, E. (1999). Stress and osmoprotection in propionibacteria. Lait *79*, 59–69.

Boyd, J., West, J.W., and Bernard, J.K. (2011). Effects of the addition of direct-fed microbials and glycerol to the diet of lactating dairy cows on milk yield and apparent efficiency of yield. J. Dairy Sci. *94*, 4616–4622.

Brede, D.A., Faye, T., Johnsborg, O., Odegard, I., Nes, I.F., and Holo, H. (2004). Molecular and genetic characterization of propionicin F, a bacteriocin from *Propionibacterium freudenreichii*. Appl. Environ. Microbiol. *70*, 7303–7310.

Brüggemann, H., Henne, A., Hoster, F., Liesegang, H., Wiezer, A., Strittmatter, A., Hujer, S., Dürre, P., and Gottschalk, G. (2004). The complete genome sequence of *Propionibacterium acnes*, a commensal of human skin. Science *305*, 671–673.

Brzuszkiewicz, E., Weiner, J., Wollherr, A., Thürmer, A., Hüpeden, J., Lomholt, H.B., Kilian, M., Gottschalk, G., Daniel, R., Mollenkopf, H.J., Meyer, T.F., and Brüggemann, H. (2011). Comparative genomics and transcriptomics of *Propionibacterium acnes*. PLoS One *6*, e21581.

Burgess, C.M., Smid, E.J., Rutten, G., and Van Sinderen, D. (2006). A general method for selection of riboflavin-overproducing food grade microorganisms. Microb. Cell Fact. *5*, 24.

Butler-Wu, S.M., Sengupta, D.J., Kittichotirat, W., Matsen, F.A. III, and Bumgarner, R.E. (2011). Genome sequence of a novel species, *Propionibacterium humerusii*. J. Bacteriol. *193*, 3678.

Cardoso, F.S., Gaspar, P., Hugenholtz, J., Ramos, A., and Santos, H. (2004). Enhancement of trehalose production in dairy propionibacteria through manipulation of environmental conditions. Int. J. Food Microbiol. *91*, 195–204.

Cardoso, F.S., Castro, R.F., Borges, N., and Santos, H. (2007). Biochemical and genetic characterization of the pathways for trehalose metabolism in *Propionibacterium freudenreichii*, and their role in stress response. Microbiology *153*, 270–280.

Carvalho, A.L., Cardoso, F.S., Bohn, A., Neves, A.R., and Santos, H. (2011). Engineering trehalose synthesis in *Lactococcus lactis* for improved stress tolerance. Appl. Environ. Microb. *77*, 4189–4199.

Cerna, B., Cerny, M., Betkova, H., Patricny, P., Soch, M., and Opatrna, I. (1991). Effect of the Proma probiotics on calves. From Dairy Sci. Abstr. *55*, 1735.

Chiliveri, S.R., Yeruva, T., Panda, S.H., and Linga, V.R. (2010). Optimization of fermentation parameters for vitamin B_{12} production using *Propionibacterium freudenreichii* subsp. *shermanii* OLP-5 by Taguchi method. J. Pure Appl. Microbiol. *4*, 647–658.

Clark, J.E., Beegen, H., and Wood, H.G. (1986). Isolation of intact chains of polyphosphate from *Propionibacterium shermanii* grown on glucose or lactate. J. Bacteriol. *168*, 1212–1219.

Collado, M.C., Meriluoto, J., and Salminen, S. (2007). *In vitro* analysis of probiotic strain combinations to inhibit pathogen adhesion to human intestinal mucus. Food Res. Int. *40*, 629–636.

Coral, J., Karp, S.G., Porto de Souza Vandenberghe, L., Parada, J.L., Pandey, A., and Soccol, C.R. (2008). Batch fermentation model of propionic acid production by *Propionibacterium acidipropionici* in different carbon sources. Appl. Biochem. Biotechnol. *151*, 333–341.

Cousin, F., Mater, D.D.G., Foligné, B., and Jan, G. (2011). Dairy propionibacteria as human probiotics: a review of recent evidence. Dairy Sci. Technol. *91*, 1–26.

Cousin, F.J., Louesdon, S., Maillard, M.-B., Parayre, S., Falentin H., Deutsch, S.-M., Boudry, G., and Jan, G. (2012a). The first dairy product exclusively fermented by *Propionibacterium freudenreichii*: A new vector to study probiotic potentialities *in vivo*. Food Microbiol. 32, 135–146.

Cousin, F.J., Deutsch, S.M., Perez Chaia, A., Foligné, B., and Jan, G. (2012b). Interactions between probiotic dairy propionibacteria and the intestinal epithelium. Curr. Immunol. Rev. *8*, 216–226.

Cousin, F.J., Jouan-Lanhouet, S., Dimanche-Boitrel, M.-T., Corcos, L., and Jan, G. (2012c). Milk Fermented by *Propionibacterium freudenreichii* Induces Apoptosis of HGT-1 Human Gastric Cancer Cells. PLoS One *7(3)*, e31892. doi:10.1371/journal.pone.0031892.

Crow, V. (1986). Utilization of lactate isomers by *Propionibacterium freudenreichii* subsp. *shermanii*: regulatory role for intracellular pyruvate. Appl. Environ. Microbiol. *52*, 352–358.

Cummins, C.S., and Johnson, J.L. (1986). Propionibacterium. In Bergey's Manual of Systematic Bacteriology, J.G. Holt, ed. (Baltimore, MD, USA: The Williams & Wilkins Co.), pp 1346–1353.

Dalmasso, M., Aubert, J., Briard-Bion, V., Chuat, V., Deutsch, S.M., Even, S., Falentin, H., Jardin, J., Maillard, M.B., Parayre, S. *et al.* (2012). A temporal-omic study of *Propionibacterium freudenreichii* CIRM-BIA1 adaptation strategies in conditions mimicking cheese ripening in the cold. PLoS One 7, p.e29083.

Danilova, I.V, Lee, H., Tourova, T.P., Ryzhkova, E.P., and Netrusov, A.I. (2012). *Propionibacterium freudenreichii* strains as antibacterial agents at neutral pH and their production on food-grade media fermented by some lactobacilli. J. Food Safety 32, 48–58.

Darilmaz, D.O., and Beyatli, Y. (2012). Acid-bile, antibiotic resistance and inhibitory properties of propionibacteria isolated from Turkish traditional home-made cheeses. Anaerobe. *18(1)*,122–127.

Darilmaz, D.O. (2013). Relationship between gastrointestinal tolerance and exopolysaccharide production of propionibacteria strains under different pH and bile conditions. Int. J. Dairy Technol. *66*, 194–201.

Deborde, C., Corre, C., Rolin, D.B., Nadal, L., De Certaines, J.D., and Boyaval, P. (1996). Trehalose biosynthesis in dairy *Propionibacterium.* J. Magn. Reson. Analysis 2, 297–304.

Deutsch, S.M., Le Bivic, P., Herve, C., Madec, M.N., LaPointe, G., Jan, G., Le Loir, Y., and Falentin, H. (2010). Correlation of the capsular phenotype in *Propionibacterium freudenreichii* with the level of expression of gtf, a unique polysaccharide synthase encoding gene. Appl. Environ. Microbiol. *76*, 2740–2746.

Deutsch, S.M., Parayre, S., Bouchoux, A., Guyomarc'h, F., Dewulf, J., Dols-Lafargue, M., Baglinière, F., Cousin, F.J., Falentin, H., Jan, G., and Foligné, B. (2012). Contribution of surface β-glucan polysaccharide to physicochemical and immunomodulatory properties of Propionibacterium freudenreichii. Appl. Environ. Microbiol. *78*,1765–1775.

Dherbécourt, J., Falentin, H., Canaan, S., and Thierry, A. (2008). A genomic search approach to identify esterases in *Propionibacterium freudenreichii* involved in the formation of flavour in Emmental cheese. Microb. Cell Fact. *7*, 16.

Dherbécourt, J., Falentin, H., Jardin, J., Maillard, M.B., Baglinière, F., Barloy-Hubler, F., and Thierry, A. (2010). Identification of a secreted lipolytic esterase in *Propionibacterium freudenreichii*, a ripening process bacterium involved in Emmental cheese lipolysis. Appl. Environ. Microbiol. *76*, 1181–1188.

Dobruchowskaa, J.M., Gerwig, G.J., Babuchowski, A., and Kamerling, J.P. (2007). Structural studies on exopolysaccharides produced by three different propionibacteria strains. Carbohydr. Res. *343*, 726–745.

Downes, J., and Wade, W.G. (2009). *Propionibacterium acidifaciens* sp. nov., isolated from the human mouth. Int. J. Syst Evol. Microbiol. *59*, 2778–2781.

Dupuis, C., Corre, C., and Boyaval, P. (1993). Lipase and esterase of *Propionibacterium freudenreichii* subsp. *freudenreichii.* Appl. Environ. Microbiol. *59*, 4004–4009.

Dupuis, C., Corre, C., and Boyaval, P. (1995). Proteinase activity of dairy *Propionibacterium.* Appl. Microbiol. Biotechnol. *40*, 750–755.

EFSA (2009). Scientific opinion of the panel on biological hazards on the maintenance of the list of QPS microorganisms intentionally added to food or feed. The EFSA Journal 7, 1–93.

El-Nezami, H., Mykkanen, H., Kankaanpaa, P., Suomalainen, T., Salminen, S., and Ahokas, J. (2000). Ability of a mixture of *Lactobacillus* and *Propionibacterium* to influence the faecal aflatoxin content in healthy Egyptian volunteers:a pilot clinical study. Biosci. Microflora *19*, 41–45.

El-Nezami, H.S., Chrevatidis, A., Auriola, S., Salminen, S., and Mykkanen, H. (2002). Removal of common *Fusarium* toxins *in vitro* by strains of *Lactobacillus* and *Propionibacterium.* Food Addit. Contam. *19*, 680–686.

El-Nezami, H.S., Polychronaki, N.N., Ma, J., Zhu, H., Ling, W., Salminen, E.K., Juvonen, R.O., Salminen, S.J. Poussa, T., and Mykkänen, H.M. (2006). Probiotic supplementation reduces a biomarker for increased risk of liver cancer in young men from Southern China. Am. J. Clin. Nutr. *83*, 1199–1203.

Falentin H., Deutsch, S.M., Jan, G., Loux, V., Thierry, A., Parayre, S., Maillard, M.B., Dherbécourt, J., Cousin, F.J., Jardin, J., *et al.* (2010). The complete genome of *Propionibacterium freudenreichii* CIRM-BIA1, a hardy actinobacterium with food and probiotic applications. PLoS One *5*, e11748.

FAO/WHO (2002). Guidelines for the Evaluation of Probiotics in Food. Food and Agriculture Organization of the United Nations – World Health Organization. London, Ontario.

Farhadi, S., Khosravi-Darani, K., Mashayekh, M., Mortazavian, A.M., Mohammadi, A and Shahraz, F. (2013). Production of propionic acid in a fermented dairy beverage. Int. J. Dairy Technol. *66*, 127–134.

Farrar, M.D., Howson, K.M., Emmott, J.E., Bojar, R.A., and Holland, K.T. (2007). Characterization of cryptc plasmid pPG01 from Propionibacterium granulosum, the first plasmid to be isolated from a member of the cutaneous propionibacteria. Plasmid *58*, 68–75.

Faye, T., Langsrud, T., Nes, I.F., and Holo H. (2000). Biochemical and genetic characterization of propionicin T1, a new bacteriocin from *Propionibacterium thoenii.* Appl. Environ. Microbiol. *66*, 4230–4236.

Faye, T., Brede, D.A., Langsrud, T., Nes, I.F., and Holo, H. (2002). An antimicrobial peptide is produced by extracellular processing of a protein from *Propionibacterium jensenii.* J. Bacteriol. *184*, 3649–3656.

Feng, X., Chen, F., Xu, H., Wu, B., Li, H., Li, S., and Ouyang, P. (2011). Green and economical production of propionic acid by *Propionibacterium freudenreichii* CCTCC M207015 in plant fibrous-bed bioreactor. Bioresour. Technol. *102*, 6141–6146.

Fitz, A. (1879). Ueber Spaltpilzgährungen. Berichte Dtsch. Chem. Ges. *12*, 474–481.

Foligné, B., Deutsch, S.M., Breton, J., Cousin, F., Dewulf, J., Samson, M., Pot, B., and Jan, G. (2010). Promising immunomodulatory effects of selected strains of dairy propionibacteria evidenced *in vitro* and *in vivo*. Appl. Environ. Microbiol. *76(24)*, 8259–8264.

Francisco, C.C., Chamberlain, C.S., Waldner, D.N., Wettemann, R.P., and Spicer, L.J. (2002). Propionibacteria fed to dairy cows: effects on energy balance, plasma metabolites and hormones, and reproduction. J. Dairy Sci. *85*, 1738–1751.

Funke, G., von Graevenitz, A., Clarridge, J.E.3rd. and Bernard, K.A. (1997). Clinical microbiology of coryneform bacteria. Clin. Microbiol. Rev. *10*, 125–159.

Furuichi, K., Hojo, K., Katakura, Y., Ninomiya, K., and Shioya, S. (2006). Aerobic culture of *Propionibacterium freudenreichii* ET-3 can increase production ratio of 1,4-dihydroxy-2-naphthoic acid to menaquinone. J. Biosci. Bioeng. *101*, 464–470.

Furuichi, K., Katakura, Y., Ninomiya, K., and Shioya, S. (2007). Enhancement of 1,4-dihydroxy-2-naphthoic acid production by *Propionibacterium freudenreichii* ET-3 fed-batch culture. Appl. Environ. Microbiol. 73, 3137–3143.

Gagnaire, V., Thierry, A., and Léonil, J. (2001). Propionibacteria and facultatively heterofermentative lactobacilli weakly contribute to secondary proteolysis of Emmental cheese. Lait *81*, 339–353.

Gardner, N., and Champagne, C.P. (2005). Production of *Propionibacterium shermanii* biomass and vitamin B_{12} on spent media. J. Appl. Microbiol. *99*, 1236–1245.

Gautier, M., Mouchel, N., Rouault, A., and Sanseau, P. (1992). Determination of genome size of four *Propionibacterium* species by pulsed-field gel electrophoresis. Lait *72*, 421–446.

Gautier, M., Rouault, A., Hervé, C., Sommer, P., Leret, V., Jan, G., Fraslin, J.M., Prévot, F., and Coste, A. (1999). Bacteriophage of dairy propionibacteria. Lait *79*, 93–104.

Gillor, O., Etzion, A., and Riley, M.A. (2008). The dual role of bacteriocins as anti- and probiotics. Appl. Microbiol. Biotechnol. *81*, 591–606.

Gorret, N., Maubois, J.L., Engasser, J.M., and Ghoul, M. (2001). Study of the effects of temperature, pH and yeast extract on growth and exopolysaccharides production by *Propionibacterium acidipropionici* on milk microfiltrate using a response surface methodology. J. Appl. Microbiol. *90*, 788–796.

Gratz, S., Mykkanen, H., and El-Nezami, H. (2005). Aflatoxin B1 binding by a mixture of *Lactobacillus* and *Propionibacterium*: *in vitro* versus *ex vivo*. J. Food Prot. *68*, 2470–2474.

Grinstead, D.A., Barefoot, S.F., and Jenseniin, G. (1992). A heat-stable bacteriocin produced by *Propionibacterium jensenii* P126. Appl. Environ. Microbiol. *58*, 215–220.

Gwiazdowska, D., and Trojanowska, K. (2006). Antimicrobial activity and stability of partially purified bacteriocins produced by *Propionibacterium freudenreichii* subsp. *freudenreichii* and subsp. *shermanii*. Lait *86*, 141–154.

Halttunen, T., Collado, M.C., El-Nezami, H., Meriluoto, J., and Salminen, S. (2008). Combining strains of lactic acid bacteria may reduce their toxin and heavy metal removal efficiency from aqueous solution. Lett. Appl. Microbiol. *46*, 160–165.

Hara, H., Haga, S., Aoyama, Y., and Kiriyama, S. (1999). Short-chain fatty acids suppress cholesterol synthesis in rat liver and intestine. J. Nutr. *129*, 942–948.

Hatakka, K., Ahola, A.J., Yli-Knuuttila, H., Richardson, M., Poussa, T., Meurman, J.H., and Korpela, R. (2007). Probiotics reduce the prevalence of oral candida in the elderly – a randomized controlled trial. J. Dent. Res. *86*, 125–130.

Hatakka, K., Holma, R., El-Nezami H., Suomalainen, T., Kuisma, M., Saxelin, M., Polussa, T., Mykkänen, H., and Korpela, R. (2008). The influence of *Lactobacillus rhamnosus* LC705 together with *Propionibacterium freudenreichii* subsp. *shermanii* JS on potentially carcinogenic bacterial activity in human colon. Int. J. Food Microbiol. *128*, 406–410.

Hennessy, A.A., Barrett, E., Paul Ross, R., Fitzgerald, G.F., Devery, R., and Stanton, C. (2012). The production of conjugated a-linolenic, g-linolenic and stearidonic acids by strains of bifidobacteria and propionibacteria. Lipids *47*, 313–327.

Hervé, C., Fondrevez, M., Cheron, A., Barloy-Hubler, F., and Jan, G. (2007). Transcarboxylase mRNA: a marker which evidences *P. freudenreichii* survival and metabolic activity during its transit in the human gut. Int. J. Food Microbiol. *113*, 303–314.

Hettinga, D.H., and Reinbold, G.W. (1972). The propionic acid bacteria – a review. II. Metabolism. J. Milk Food Technol. 35, 358–372.

Higgins, S.E., Higgins, J.P., Wolfenden, A.D., Henderson, S.N., Torres-Rodriguez, A., Tellez, G., and Hargis,B. (2008). Evaluation of a *Lactobacillus*-based probiotic culture for the reduction of *Salmonella enteritidis* in neonatal broiler chicks. Poult. Sci. *87*, 27–31.

Hynönen, U., and Palva, A. (2013). *Lactobacillus* surface layer proteins: structure, function and applications. Appl. Microbiol. Biotechnol. 97, 5225–5243. Hojo, K., Yoda, N., Tsuchita, H., Ohtsu, T., Seki, K., *et al.* (2002). Effect of ingested culture of *Propionibacterium freudenreichii* ET-3 on fecal microflora and stool frequency in healthy females. Biosci. Microflora *21*, 115–120.

Hojo, K., Watanabe, R., Mori, T., and Taketomo, N. (2007). Quantitative measurement of tetrahydromenaquinone-9 in cheese fermented by propionibacteria. J. Dairy Sci. *90*, 4078–4083.

Holtug, K., Clausen, M.R., Hove, N., Christiansen, J., and Mortensen, P.B. (1992). The colon in carbohydrate malabsorption: Short Chain Fatty Acids, pH, and osmotic diarrhea. Scand. J. Gastroenterol. *27*, 545–552.

Horváth, B., Hunyadkürti, J., Vörös, A., Fekete, C., Urbán, E., Kemény, L., and Nagy, I. (2012). Genome Sequence of *Propionibacterium acnes* Type II Strain ATCC 11828. J. Bacteriol. *194*, 202–203.

Huang, Y., Kotula, L., and Adams, M.C. (2003a). The *in vivo* assessment of safety and gastrointestinal survival of an orally administered novel probiotic, *Propionibacterium jensenii* 702, in a male Wistar rat model. Food Chem. Toxicol. *41*, 1781–1787.

Huang, Y., and Adams, M.C. (2003b). An *in vitro* model for investigating intestinal adhesion of potential dairy propionibacteria probiotic strains using cell line C2BBe1. Lett. Appl. Microbiol. *36*, 213–216.

Huang, Y., and Adams, M.C. (2004). *In vitro* assessment of the upper gastrointestinal tolerance of potential probiotic dairy propionibacteria. Int. J. Food Microbiol. *91*, 253–260.

Hugenholtz, J., Hunik, J., Santos, H., and Smid, E. (2002). Nutraceutical production by propionibacteria. Lait *82*, 103–112.

Hunik, J.H. (2000). Improved process for the production of vitamin B_{12}. Patent WO00/37669.

Hunyadkürti, J., Feltóti, Z., Horvát, B., Nagymihály, M., Vörös, A., McDowell, A., Patrick, S., Urbán, E., and Nagy, I. (2011). Complete genome sequence of *Propionibacterium acnes* type IB strain 6609. J. Bacteriol. *193*, 4561–4562.

Ibrahim, F., Halttunen, T., Tahvonen, R., and Salminen, S. (2006). Probiotic bacteria as potential detoxification tools: assessing their heavy metal binding isotherms. Can. J. Microbiol. *52*, 877–885.

Isawa, K., Hojo, K., Yoda, N., Kamiyama, T., Makino, S., *et al.* (2002). Isolation and identification of a new bifidogenic growth stimulator produced by *Propionibacterium freudenreichii* ET-3. Biosci. Biotechnol. Biochem. *66*, 679–681.

Isenberg, J., Ko, H., Pulverer, G., Grundmann, R., Stutzer, H., and Pichlmaier, H. (1994). Preoperative immunostimulation by *Propionibacterium granulosum* KP-45 in colorectal cancer. Anticancer Res. *14*, 1399–1404.

Isenberg, J., Stoffel, B., Wolters, U., Beuth, J., Stutzer, H., Ko, H.L., and Pichlmaier, H. (1995). Immunostimulation by propionibacteria – effects on immune status and antineoplastic treatment. Anticancer Res. *15*, 2363–2368.

Jan, G., Leverrier, P., Pichereau, V., and Boyaval, P. (2001). Changes in protein synthesis and morphology during acid adaptation of *Propionibacterium freudenreichii*. Appl. Environ. Microbiol. *67*, 2029–2036.

Jan, G., Leverrier, P., and Proudy, I. (2002a). Survival and beneficial effects of propionibacteria in the human gut: *in vivo* and *in vitro* investigations. Lait *82*, 131–144.

Jan, G., Belzacq, A.S., Haouzi, D., Rouault, A., Metivier, D., Kroemer, G., and Brenner, C. (2002b). Propionibacteria induce apoptosis of colorectal carcinoma cells via short-chain fatty acids acting on mitochondria. Cell Death Differ. *9*, 179–188.

Kajander, K., Myllyluoma, E., Rajilic-Stojanovic, M., Kyronpalo, S., Rasmussen, M., Järvenpää, S., Zoetendal, E.G., de Vos, W.M., Vapaatalo, H., and Korpela, R. (2008). Clinical trial: multispecies probiotic supplementation alleviates the symptoms of irritable bowel syndrome and stabilizes intestinal microbiota. Aliment. Pharmacol. 27, 48–57.

Kalfirtova, P., and Sovjak, R. (2005). Improve the nutritional value of fermented dairy products by nutraceutical-producing food-grade microorganisms. Agric. Trop. Subtrop. 38, 22–27.

Kaneko, T. (1999). A novel bifidogenic growth stimulator produced by *Propionibacterium freudenreichii*. Biosci. Microflora *18*, 73–80.

Kasimatis, G., Fitz-Gibbon, S., Tomida, S., Wong, M.' and Li, H. (2013). Analysis of complete genomes of *Propionibacterium acnes* reveals a novel plasmid and increased pseudogenes in an acne associated strain. BioMed. Res. Int., vol 2013, Article ID 918320, 11 pages. doi:10.1155/2013/918320

Kekkonen, R.A., Lummela, N., Karjalainen, H., Latvala, S., Tynkkynen, S., Jarvenpaa, S., Kautiainen, H., Julkunen, I., Vapaatalo, H., and Korpela,

R. (2008). Probiotic intervention has strain-specific anti-inflammatory effects in healthy adults. World J. Gastroenterol. *14*, 2029–2036.

Khan, M., Raoult, D., Richet, H., Lepidi, H., and La Scola, B. (2007). Growth-promoting effects of single-dose intragastrically administered probiotics in chickens. Br. Poult. Sci. *48*, 732–735.

Khan, M.M., and Mir, N.A. (2011). Production of vitamin B_{12} by improved strains of *Propionibacterium freudenreichii*. Biotechnol. Bioinf. Bioeng. *1*, 19–24.

Kim, S.W., Standorf, D.G., Roman-Rosario, H., Yokoyama, M.T., and Rust, S.R. (2000). Potential use of *Propionibacterium acidipropionici*, strain DH42 as a direct-fed microbial for cattle. J. Dairy S. *83(1)*, 292 (Abstr.).

Kirjavainen, P.V., El Nezami, H.S., Salminen, S.J., Ahokas, J.T., and Wright, P.F. (1999). Effects of orally administered viable *Lactobacillus rhamnosus* GG and *Propionibacterium freudenreichii* subsp. shermanii JS on mouse lymphocyte proliferation. Clin. Diagn Lab Immunol. *6*, 799–802.

Kośmider, A., Białas, W., Kubiak, P., Drozdzyńska, A., and Czaczyk, K. (2012). Vitamin B_{12} production from crude glycerol by *Propionibacterium freudenreichii* subsp. *shermanii:* Optimization of medium composition through statistical experimental designs. Bioresource Technol. *105*, 128–133.

Koussemon, M., Combet-Blanc, Y., Patel, B.K., Cayol, J.L., Thomas, P., García, J.L., and Ollivier, B. (2001). *Propionibacterium microaerophilum* sp. nov., a microaerophilic bacterium isolated from olive mill wastewater. Int. J. Syst. Evol. Microbiol. *51*, 1373–1382.

Kukkonen, K., Savilahti, E., Haahtela, T., Juntunen-Backman, K., Korpela, R., Poussa, T., Tuure, T., and Kuitunen, M. (2007). Probiotics and prebiotic galacto-oligosaccharides in the prevention of allergic diseases: a randomized, double-blind, placebo-controlled trial, J. Allergy Clin. Immunol. *119*, 192–198.

Kukkonen, K., Savilahti, E., Haahtela, T., Juntunen-Backman, K., Korpela, R., Poussa, T., Tuure, T., and Kuitunen, M. (2008). Long-term safety and impact on infection rates of postnatal probiotic and prebiotic (synbiotic) treatment: randomized, double-blind, placebo-controlled trial. Pediatrics *122*, 8–12.

Kusano, K., Yamada, H., Niwa, M., and Yamasato, K. (1997). *Propionibacterium cyclohexanicum* sp. nov., a new acid-tolerant omega-cyclohexyl fatty acid-containing propionibacterium isolated from spoiled orange juice. Int. J. Syst. Bacteriol. *47*, 825–831.

Lan, A., Bruneau, A., Philippe, C., Rochet, V., Rouault, A., Hervé, C., Roland, N., Rabot, S., and Jan, G. (2007a). Survival and metabolic activity of selected strains of *Propionibacterium freudenreichii* in the gastrointestinal tract of human microbiota-associated rats. Br. J. Nutr. *97*, 714–724.

Lan, A., Lagadic-Gossmann, D., Lemaire, C., Brenner, C., and Jan, G. (2007b). Acidic extracellular pH shifts colorectal cancer cell death from apoptosis to necrosis upon exposure to propionate and acetate, major end-products of the human probiotic propionibacteria. Apoptosis *12*, 573–591.

Lan, A., Bruneau, A., Bensaada, M., Philippe, C., Bellaud, P., Rabot, S., and Jan, G. (2008). Increased induction of apoptosis by *Propionibacterium freudenreichii* TL133 in colonic mucosal crypts of human microbiota-associated rats treated with 1,2-dimethylhydrazine. Br. J. Nutr. *100*, 1251–1259.

Langsrud, T., and Reinbold, G.W. (1973a). Flavour development and microbiology of Swiss cheese. A review. II Starters, manufacturing process and procedure. J. Milk Food Technol. *36*, 531–542.

Langsrud, T., and Reinbold, G.W. (1973b). Flavor development and microbiology of Swiss cheese. A review. III Ripening and flavor production. J. Milk Food Technol. *36*, 593–609.

Langsrud, T., Serhauq, T., and Vegarud, G.E. (1995). Protein degradation and amine acid metabolism by propionibacteria. Lait 75, 325–330.

LeBlanc, J.G., Rutten, G., Bruinenberg, P., Sesma, F., de Giori, G.S., and Smid, E.J. (2006). A novel dairy product fermented with *Propionibacterium freudenreichii* improves the riboflavin status of deficient rats. Nutrition *22*, 645–651.

Lehto, E.M., and Salminen, S. (1997). Adhesion of two *Lactobacillus* strains, one *Lactococcus* and one *Propionibacterium* strain to cultured human intestinal Caco-2 cell line. Biosci. Microflora *16*, 13–17.

Lettat, A., Nozière, P., Silberberg, M., Morgavi, D.P., Berger, C., and Martin, C. (2012). Rumen microbial and fermentation characteristics are affected differently by bacterial probiotic supplementation during induced lactic and subacute acidosis in sheep. BMC Microbiol. *12*, 142.

Leverrier, P., Dimova, D., Pichereau, V., Auffray, Y., Boyaval, P., and Jan, G.L. (2003). Susceptibility and adaptive response to bile salts in *Propionibacterium freudenreichii*: physiological and proteomic analysis, Appl. Environ. Microbiol. *69*, 3809–3818.

Leverrier, P., Vissers, J.P.C., Rouault, A., Boyaval, P., and Jan, G. (2004). Mass spectrometry proteomic analysis of stress adaptation reveals both common and distinct response pathways in *Propionibacterium freudenreichii*. Arch. Microbiol. *181*, 215–230.

Leverrier, P., Fremont, Y., Rouault, A., Boyaval, P., and Jan, G. (2005). *In vitro* tolerance to digestive stresses of propionibacteria: influence of food matrices. Food Microbiol. *22*, 11–18.

Leyden, J.J. (2004). Antibiotic resistance in the topical treatment of acne vulgaris. Cutis 73, 6–10.

Lind, H., Sjogren, J., Gohil, S., Kenne, L., Schnurer, J., and Broberg, A. (2007). Antifungal compounds from cultures of dairy propionibacteria type strains. FEMS Microbiol. Lett. *271*, 310–315.

Liu, L., Zhu, Y., Li, J., Wang, M., Lee, P., Du, G., and Chen, J. (2012). Microbial production of propionic acid from propionibacteria: Current state, challenges and perspectives. Crit. Rev. Biotechnol. *32(4)*, 374–381.

Liu, Y., Zhang, Y.G., Zhang, R.B., Zhang, F., and Zhu, J. (2011). Glycerol/glucose co-fermentation: one more proficient process to produce propionic acid by *Propionibacterium acidipropionici*. Curr. Microbiol. *62*, 152–158.

Lo, D.Y., Hung, C.N., Lee, W.C., Liao, J.W., Blacklaws, B.A., Chen, T.H., Chien, M.S., and Hsuan, S.L. (2009). Effect of immunostimulation by detoxified E. coli lipopolysaccharide combined with inactivated *Propionibacterium granulosum* cells on porcine immunity. J. Vet. Med. Sci. *71*, 897–903.

Lorenzo-Pisarello, M.J., Gultemirian, M.L., Nieto-Peñalver, C., and Perez Chaia, A. (2010). *Propionibacterium acidipropionici* CRL 1198 influences the production of acids and the growth of bacterial genera stimulated by inulin in a murine model of cecal slurries. Anaerobe *16*, 345–354.

Lyon, W.J., Sethi, J.K., and Glatz, B.A. (1993). Inhibition of psychrotrophic organisms by propionicin PLG-1, a bacteriocin produced by *Propionibacterium thoenii*. J. Dairy Sci. *76*, 1506–1513.

Macfarlane, G.T., Allison, C., Gibson, S.A., and Cummings, J.H. (1988). Contribution of the microflora to proteolysis in the human large intestine. J. Appl. Bacteriol. *64*, 37–46.

Maity, T.K., Rakesh Kumar and Misra, A.K. (2008). Development of healthy whey drink with *Lactobacillus rhamnosus, Bifidobacterium bifidum* and *Propionibacterium freudenreichii* subsp. *shermanii*. Mljekarstvo *58*, 315–325.

Mak, T.N., Schmid, M., Brzuszkiewicz, E., Zeng, G., Meyer, R., Sfanos, K.S., Brinkmann, V., Meyer, T.F., and Brüggemann, H. (2013). Comparative genomics reveals distinct host-interacting traits of three major human-associated propionibacteria. BMC Genomics *14*, 640. http://www.biomedcentral.com/1471-2164/14/640

Mantere-Alhonen, S., and Ryhänen, E.L. (1994). Lactobacilli and Propionibacteria in living food. Microbiologie Aliments Nutrition *12*, 399–405.

Mantere Alhonen, S. (1995). Propionibacteria used as probiotics. A review. Lait 75, 447–452.

Mara, M., Julak, J., Bednar, M., Ocenaskova, J., Mikova, Z., and Dohnalová, A. (1994). The influence of *Propionibacterium acnes* (Corynebacterium parvum) fractions on immune response *in vivo*. Zentralbl. Bakteriol. *281*, 549–555.

Markowska-Daniel, I., Pejsak, Z., Szmigielski, S., Jeljaszewicz, J., and Pulverer, G. (1992). Stimulation of granulopoiesis in pregnant swine and their offspring by *Propionibacterium avidum* KP-40. Br. Vet. J. *148*, 133–145.

Markowska-Daniel, I., Pejsak, Z., Szmigielski, S., Jeljaszewicz, J., and Pulverer, G. (1993). Prophylactic application of *Propionibacterium avidum* KP-40 in swine with acute experimental infections. II. Bacterial infections: pleuropneumonia and swine erysipelas. Dtsch. Tierarztl. Wochenschr. *100*, 185–188.

McIntosh, F.M., Shingfield, K.J., Devillard, E., Russell, W.R., and Wallace, R.J. (2009). Mechanism of conjugated linoleic acid and vaccenic acid formation in human faecal suspensions and pure cultures of intestinal bacteria. Microbiology *155*, 285–294.

Meile, L., Le Blay, G., and Thierry, A. (2008). Safety assessment of dairy microorganisms: *Propionibacterium* and *Bifidobacterium*. Int. J. Food Microbiol. *126*, 316–320.

Michel, C., Roland, N., Lecannu, G., Herve, C., Avice, J.C., Rival, M., and Cherbut, C. (2005). Colonic infusion with Propionibacterium acidipropionici reduces severity of chemically induced colitis in rats. Lait 85, 99–111.

Miescher, S., Stierli, M.P., Teuber, M., and Meile, L. (2000). Propionicin SM1, a bacteriocin from *Propionibacterium jensenii* DF1: isolation and characterization of the protein and its gene. Syst. Appl. Microbiol. 23, 174–184.

Minegishi, K., Aikawa, C., Furukawa, A., Watanabe, T., Nakano, T., Ogura, Y., Ohtsubo, Y., Kurokawa, K., Hayashi, T., Maruyama, F., Nakagawa, I., and Eishi, Y. (2013). Complete genome sequence of a *Propionibacterium acnes* Isolate from a sarcoidosis patient. Genome Announc. *1(1)*, e00016-12.

Mitsuyama, K., Masuda, J., Yamasaki, H., Kuwaki, K., Kitazaki, S., Koga, H., Uchida, M., and Sata, M. (2007). Treatment of ulcerative colitis with milk whey culture with *Propionibacterium freudenreichii*. J. Intestinal Microbiol. *21*, 143–147.

Montalto, M., Curigliano, V., Santoro, L., Vastola, M., Cammarota, G., Manna, R., Gasbarrini, A., and Gasbarrini, G. (2006). Management and treatment of lactose malabsorption. World J. Gastroenterol. *14*, 187–191.

Mori, H., Sato, Y., Taketomo, N., Kamiyama, T., Yoshiyama, Y., Meguro, S., Sato, H., and Kaneko, T. (1997). Isolation and structural identification of bifidogenic growth stimulator produced by *Propionibacterium freudenreichii*. J. Dairy Sci. *80*, 1959–1964.

Murooka, Y., Piao, Y., Kiatpapan, P., and Yamashita, M. (2005). Production of tetrapyrrole compounds and vitamin B_{12} using genetically engineering of *Propionibacterium freudenreichii* An overview. Lait *85*, 9–22.

Mussalem, J.S., Vasconcelos, J.R., Squaiella, C.C., Ananias, R.Z., Braga, E.G., Rodrigues, M.M., and Longo-Maugéri, I.M. (2006). Adjuvant effect of the *Propionibacterium acnes* and its purified soluble polysaccharide on the immunization with plasmidial DNA containing a Trypanosoma cruzi gene. Microbiol. Immunol. *50*, 253–263.

Myllyluoma, E., Ahonen, A.M., Korpela, R., Vapaatalo, H., and Kankuri, E. (2008). Effects of multispecies probiotic combination on Helicobacter pylori infection *in vitro*. Clin. Vaccine Immunol. *15*, 1472–1482.

Nagata, K., Inatsu, S., Tanaka, M., Sato, H., Kouya, T., Taniguchi, M., and Fukuda, Y. (2010). The bifidogenic growth stimulator inhibits the growth and respiration of Helicobacter pylori. Helicobacter *15*, 422–429.

Niderkorn V., Boudra, H., and Morgavi, D.P. (2006). Binding of *Fusarium* mycotoxins by fermentative bacteria *in vitro*. J. Appl. Microbiol. *101*, 849–856.

Nordmark, E.L., Yang, Z., Huttunen, E., and Widmalm, G. (2005). Structural studies of the exopolysaccharide produced by *Propionibacterium freudenreichii* subsp. *shermanii* JS. Biomacromolecules *6*, 521–523.

Okada, Y., Tsuzuki, Y., Miyazaki, J., Matsuzaki, K., Hokari, R., Komoto, S., Kato, S., Kawaguchi, A., Nagao, S., Itoh, K., *et al.* (2006). *Propionibacterium freudenreichii* component 1.4-dihydroxy-2-naphthoic acid (DHNA) attenuates dextran sodium sulphate induced colitis by modulation of bacterial flora and lymphocyte homing. Gut *55*, 681–688.

Oksaharju, A., Kooistra, T., Kleemann, R., van Duyvenvoorde, W., Miettinen, M., Lappalainen, J., Lindstedt, K.A., Kovanen, P.T., Korpela, R., and Kekkonen. R.A. (2013). Effects of probiotic *Lactobacillus rhamnosus* GG and *Propionibacterium freudenreichii* subsp. *shermanii* JS supplementation on intestinal and systemic markers of inflammation in ApoE*3Leiden mice consuming a high-fat diet. Br. J. Nutr. *110*, 77–85.

Ördögh, L., Hunyadkürti, J., Vörös, A., Horváth, B., Szűcs, A., Urbán, E., Kereszt, A., Kondorosi, E., and Nagy, I. (2013). Complete Genome Sequence of *Propionibacterium avidum* Strain 44067, Isolated from a Human Skin Abscess. Genome Announc. *1*, e00337-13.

Orla-Jensen, S. (1909). Die hauptlien des natürlichen bakteriensystems. Centrlbl. Bakteriol. Hyg. II Abt. *22*, 305–346.

Ouwehand, A.C., Tölkkö, S., Kulmala, J., Salminen, S., and Salminen, E. (2000). Adhesion of inactivated probiotic strains to intestinal mucus. Lett. Appl. Microbiol. *31*, 82–86.

Ouwehand, A.C., Lagstrom, H., Suomalainen, T., and Salminen, S. (2002). Effect of probiotics on constipation, fecal azoreductase activity and fecal mucin content in the elderly. Ann. Nutr. Metab. *46*, 159–162.

Parizzi, L.P., Grassi, M.C.B., Llerena, L.A., Carazzolle, M.F., Queiroz, V.L., Lunardi, I., Zeidler, A.F., Teixeira, P.J.P.L., Mieczkowski, P., Rincones, J., and Pereira, G.A.G. (2012). The genome sequence of *Propionibacterium acidipropionici* provides insights into its biotechnological and industrial potential. BMC Genomics *13*, 562.

Paul, G.E., and Booth, S.J. (1988). Properties and characteristics of a bacteriocin-like substance produced by *Propionibacterium acnes* isolated from dental plaque. Can. J. Microbiol. *34*, 1344–1347.

Perez Chaia, A., Nader de Macias, M.E., and Oliver, G. (1995). Propionibacteria in the gut: effect on some metabolic activities of the host. Lait 75, 435–445.

Perez Chaia, A., Zárate, G., and Oliver, G. (1999). The probiotic properties of propionibacteria. Lait *79*, 175–185.

Perez Chaia, A., and Zárate, G. (2005). Dairy propionibacteria from milk or cheese diets remain viable and enhance propionic acid production in the mouse cecum. Lait *85*, 85–98.

Piao, Y., Yamashita, M., Kawaraichi, N., Asegawa, R., Ono, H., and Murooka, Y. (2004). Production of vitamin B_{12} in genetically engineered *Propionibacterium freudenreichii*. J. Biosci. Bioeng. *98*, 167–173.

Piveteau, P. (1999). Metabolism of lactate and sugars by dairy propionibacteria: a review. Lait *79*, 23–41.

Pohjavuori, E., Viljanen, M., Korpela, R., Kuitunen, M., Tiittanen, M., Vaarala, O., and Savilathi, E. (2004). *Lactobacillus* GG effect in increasing IFN-gamma production in infants with cow's milk allergy. J. Allergy Clin. Immunol. *114*, 131–136.

Poonam, Pophaly, S.D., Tomar, S.K., De, S., Singh, R. (2012). Multifaceted attributes of dairy propionibacteria: a review. World J. Microbiol. Biotechnol. *28*, 3081–3095.

Racine, M., Dumont, J., Champagne, C.P., and Morin, A. (1991). Production and characterization of the exopolysaccharide from *Propionibacterium acidipropionici* on whey-based media. J. Appl. Bacteriol. *71*, 233–238.

Rainio, A., Vahvaselkä, M., Suomalainen, T., and Laakso, S. (2002). Production of conjugated linoleic acid by *Propionibacterium*. Lait *82*, 91–101.

Ramanathan, S., Wolynec, C., and Cutting, W. (1968). Antiviral principles of propionibacteria – isolation and activity of propionins B and C. Proc. Soc. Exp. Biol. Med. *129*, 73–77.

Ranadheera, C., Evans, C.A., Adams, M.C., and Baines, S.K. (2012). *In vitro* analysis of gastrointestinal tolerance and intestinal cell adhesion of probiotics in goat's milk ice cream and yoghurt. Food Res. Int. *49*, 619–625.

Ratnam, P., Barefoot, S.F., Prince, I.U., Bodine, A.B., and McCaskill, L.H. (1999). Partial purification of the bacteriocin produced by *Propionibacterium jensenii* B_{12}64. Lait *79*, 125–136.

Rehberger, T.G., and Glatz, B.A. (1990). Characterization of *Propionibacterium* plasmids. Appl. Environ. Microbiol. *56*, 864–871.

Rinta-Koski, M., Montoten, L., and Mantere-Alhonen, S. (2001). Propionibacteria isolated from rumen used as possible probiotics together with bifidobacteria. Milchwissenschaft *56(1)*, 11–13.

Roland, N., Bougle, D., Lebeurrier, F., Arhan, P., and Maubois, J. (1998). *Propionibacterium freudenreichii* stimulates the growth of *Bifidobacterium bifidum in vitro* and increases fecal bifidobacteria in healthy human volunteers. Int. Dairy J. *8*, 587–588.

Roszkowski, W., Roszkowski, K., Ko, H.L., Beuth, J., and Jeljaszewicz, J. (1990). Immunomodulation by propionibacteria. Zentralbl. Bakteriol. *274*, 289–298.

Ruhal, R., Aggarwal, S., and Choudhury, B. (2011). Suitability of crude glycerol obtained from biodiesel waste for the production of trehalose and propionic acid. Green Chem. *13*, 3492–3498.

Ruhal, R., and Choudhury, B. (2012a). Improved trehalose production from biodiesel waste using parent and osmotically sensitive mutant of *Propionibacterium freudenreichii* subsp. *shermanii* under aerobic conditions. J. Ind. Microbiol. Biotechnol. *39*, 1153–1160.

Ruhal, R., and Choudhury, B. (2012b). Use of an osmotically sensitive mutant of *Propionibacterium freudenreichii* subspp. *shermanii* for the simultaneous productions of organic acids and trehalose from biodiesel waste based crude glycerol. Bioresour. Technol. *109*, 131–139.

Salanitro, J.P., Blake, I.G., Muirehead, P.A., Maglio, M., and Goodman, J.R. (1978). Bacteria isolated from the duodenum, ileum, and cecum of young chicks. Appl. Environ. Microbiol. *35*, 782–790.

Saraoui, T., Parayre, S., Guernec, G., Loux, V., Montfort, J., Cam, A.L., Boudry, G., Jan, G., and Falentin, H. (2013). A unique *in vivo* experimental approach reveals metabolic adaptation of the probiotic *Propionibacterium freudenreichii* to the colon environment. BMC Genomics *14*, 911.

Sarkar, S., and Misra, A.K. (1998). Effect of feeding Propiono-Acido-Bifido (PAB) milk on the nutritional status and excretory pattern in rats and children. Milchwissenschaft 53, 666–668.

Satomi, K., Kurihara, H., Isawa, K., Mori, H., and Kaneco, T. (1999). Effects of culture-powder of *Propionibacterium freudenreichii* ET-3 on fecal microflora of normal adults. Biosci. Microflora. *18*, 27–30.

Seki, K., Nakao, H., Umino, H., Isshiki, H., Yoda, N., Tachihara, R., Ohuchi, T., Saruta, H., Suzuki, K., and Mitsouka, T. (2004). Effects of fermented milk whey containing novel bifidogenic growth stimulator produced by Propionibacterium on fecal bacteria, putrefactive metabolite, defecation frequency and fecal properties in senile volunteers needed serious nursing-care taking enteral nutrition by tube feeding, properties in senile volunteers needed serious nursing-care taking enteral nutrition by tube feeding. J. Intestinal Microbiol. *18*, 107–115.

Shu, M., Wang, Y., Yu, J., Kuo, S., Coda, A., Jiang, Y., Gallo, R.L., and Huang, C.M. (2013). Fermentation of *Propionibacterium acnes*, a commensal bacterium in the human skin microbiome, as skin probiotics against methicillin-resistant *Staphylococcus aureus*. PLoS One *8*, e55380.

Sidorchuk, I., and Bondarenko, V.M. (1984). Selection of a biologically active mutant of *Propionibacterium shermanii* and the possibility of its use in complex therapy of enteral dysbacteriosis. J. Hyg. Epidemiol. Microbiol. Immunol. *28*, 331–338.

Sieber, R., Collomb, M., Aeschlimann, A., Jelen, P., and Eyer, H. (2004). Impact of microbial cultures on conjugated linoleic acid in dairy products. Int. Dairy J. *14*, 1–15.

Somkuti, G.A., and Johnson, T.L. (1990). Cholesterol uptake by *Propionibacterium freudenreichii*. Curr. Microbiol. *20*, 305–309.

Stackebrandt, E., Cummins, C.S., and Johnson, J.L. (2006). Family Propionibacteriacea: The genus *Propionibacterium*. In The Prokaryotes: A handbook on the Biology of Bacteria, M. Dworkin, S. Falkow, E. Rosemberg, K.H. Schleifer and E. Stackebrandt, eds (Singapur: Springer), pp 400–418.

Stein, D.R., Allen, D.T., Perry, E.B., Bruner, J.C., Gates, K.W., Rehberger, T.G., Mertz, K., Jones, D., and Spicer, L.J. (2006). Effects of feeding propionibacteria to dairy cows on milk yield, milk components, and reproduction. J. Dairy Sci. *89*, 111–125.

Suomalainen, T., Sigvart-Mattila, P., Matto, J., and Tynkkynen, S. (2008). *In vitro* and *in vivo* gastrointestinal survival, antibiotic susceptibility and genetic identification of *Propionibacterium freudenreichii* subsp. *shermanii* JS. Int. Dairy J. *18*, 271–278.

Suzuki, A., Mitsuyama, K., Koga, H., Tomiyasu, N., Masuda, J., Takaki, K., Tsuruta, O., Toyonaga, A., and Sata, M. (2006). Bifidogenic growth stimulator for the treatment of active ulcerative colitis: a pilot study. Nutrition *22*, 76–81.

Thiel, A., Eikmanns, B., Salminen, S., and Ouwehand, A.C. (2004). *In vitro* adhesion of propionibacteria to human intestinal mucus. Ital. J. Food Sci. *16*, 245–254.

Thierry, A., Deutsch, S.-M., Falentin, H., Dalmasso, M., Cousin, F.J., and Jan, G. (2011). New insights into physiology and metabolism of *Propionibacterium freudenreichii*. Int. J. Food Microbiol. *149*, 19–27.

Thirupathaiah, Y., Rani, C.S., Reddy, M.S., and Rao, L.V. (2012). Effect of chemical and microbial vitamin B_{12} analogues on production of vitamin B_{12}. World J. Microbiol. Biotechnol. *28*, 2267–2271.

Uchida, M., and Mogami, O. (2005). Milk whey culture with *Propionibacterium freudenreichii* ET-3 is effective on the colitis induced by 2,4,6- trinitrobenzene sulfonic acid in rats. J. Pharmacol. Sci. *99*, 329–334.

Uchida, M., Yoda, N., Terahara, M., Seki, K., Choi, S.S., and Roberts, A. (2011a). Safety evaluation of *Propionibacterium freudenreichii* ET-3 culture. Regul. Toxicol. Pharmacol. *60*, 249–261.

Uchida, M., Tsuboi, H., Takahashi Arita, M., Nemoto, A., Seki, K., Tsunoo, H., Martyres, S., and Roberts, A. (2011b). Safety of high doses of *Propionibacterium freudenreichii* ET-3 culture in healthy adult subjects. Regul. Toxicol. Pharmacol. *60*, 262–267.

Vahvaselka, M., and Laakso, S. (2010). Production of cis-9,trans-11-conjugated linoleic acid in camelina meal and okara by an oat-assisted microbial process. J. Agric. Food Chem. *58*, 2479–2482.

Van der Merwe, I.R., Bauer, R., Britz, T.J., and Dicks, L.M. (2004). Characterization of thoeniicin 447, a bacteriocin isolated from *Propionibacterium thoenii* strain 447. Int. J. Food Microbiol. *92*, 153–160.

van Luijk, N., Stierli, M.P., Miescher Schwenninger, S., Hervé, C., Dasen, G., Jore, J.P.M., Pouwels, P.H., Van Der Werf, M.J., Teuber, M., and Meile, L. (2002). Review: Genetics and molecular biology of propionibacteria. Lait *82*, 45–57.

Van Wyk, J., Witthuhn, R.C., and Britz, T.J. (2011). Optimisation of vitamin B_{12} and folate production by *Propionibacterium freudenreichii* strains in kefir. Int. D. Journal *21*, 69–74.

Von freudenreich, E., and Orla-Jensen, S. (1906). Über die im Emmentalerkase staffindene Propionsauregarung. Zbl. Bakteriol. *2*, 529–546.

Vorobjeva, L.I., Iljasova, O.V., Khodjaev, E., Ponomareva, G.M., and Varioukhina, S.Y. (2001). Inhibition of Induced Mutagenesis in *Salmonella typhimurium* by the Protein of *Propionibacterium freudenreichii* subsp. *shermanii*. Anaerobe *7(1)*, 37–44.

Vorobjeva, L.I., Khodjaev, E.Y., and Vorobjeva, N.V. (2008). Propionic acid bacteria as probiotics. Microb. Ecol. Health Dis. *20*, 109–112.

Vörös, A., Horváth, B., Hunyadkürti, J., McDowell, A., Barnard, E., Patrick, S., and Nagy, I. (2012). Complete genome sequences of three *Propionibacterium acnes* isolates from the Type IA_2 cluster. J. Bacteriol. *194*, 1621–1622.

Wang, L.M., Lv, J.P., Chu, Z.Q., Cui, Y.Y., and Ren, X.H. (2007). Production of conjugated linoleic acid by *Propionibacterium freudenreichii*. Food Chem. *103*, 313–318.

Warminska-Radyko, I., Laniewska-Moroz, L., and Babuchowski, A. (2002). Possibilities for stimulation of *Bifidobacterium* growth by propionibacteria. Lait *82*, 113–121.

Waters, S.M., Murphy, R.A., and Power, R.F. (2006). Characterisation of prototype Nurmi cultures using culture-based microbiological techniques and PCR-DGGE. Int. J. Food Microbiol. *110*, 268–277.

Weiss, W.P., Wyatt, D.J., and McKelvey T.R. (2008). Effect of feeding propionibacteria on milk production by early lactation dairy cows. J. Dairy Sci. *91*, 646–652.

Wu, Q.Q., You, H.J., Ahn, H.J., Kwon, B., and Ji, G.E. (2012). Changes in growth and survival of *Bifidobacterium* by coculture in soy milk, cow's milk, and modified MRS medium. Int. J. Food Microbiol. *157*, 65–72.

Xu S., Boylston, T.D., and Glatz, B.A. (2004). Effect of lipid source on probiotic bacteria and conjugated linoleic acid formation in milk model systems. J. Am. Oil Chem. *81*, 589–595.

Xu, S., Boylston, T.D., and Glatz, B.A. (2005). Conjugated linoleic acid content and organoleptic attributes of fermented milk products produced with probiotic bacteria. J. Agric. Food. Chem. *53*, 9064–9072.

Yamazaki, S., Kano, K., Ikeda, T., Isawa, K., and Kaneko, T. (1999). Role of 2-amino-3-carboxy-1,4-naphthoquinone, a strong growth stimulator for bifidobacteria, as an electron transfer mediator for $NAD(P)^+$ regeneration in *Bifidobacterium longum*. Biochim. Biophys. Acta-General Subjects *1428*, 241–250.

Ye, K., Shijo, M., Jin, S., and Shimizu, K. (1996). Efficient production of vitamin B_{12} from propionic acid bacteria under periodic variation of dissolved oxygen concentration J. Ferment. Bioeng. *82*, 484–491.

Zárate, G., González, S., Pérez-Chaia, A., and Oliver, G. (2000a). Effect of bile on the β-galactosidase activity of dairy propionibacteria. Lait *80*, 267–276.

Zárate, G., Chaia, A.P., Gonzalez, S., and Oliver, G. (2000b). Viability and beta-galactosidase activity of dairy propionibacteria subjected to digestion by artificial gastric and intestinal fluids. J. Food Prot. *63*, 1214–1221.

Zarate, G., Morata de Ambrosini, V.I., Pérez Chaia, A., and Gonzalez, S.N. (2002a). Adhesion of dairy propionibacteria to intestinal epithelial tissue *in vitro* and *in vivo*. J. Food. Prot. *65*, 534–539.

Zarate, G., Morata De Ambrosini, V., Perez Chaia, A., and Gonzalez, S. (2002b). Some factors affecting the adherence of probiotic *Propionibacterium acidipropionici* CRL 1198 to intestinal epithelial cells. Can. J. Microbiol. *48*, 449–457.

Zarate G., Perez Chaia A., and Oliver G. (2003). Some Characteristics of Practical Relevance of the b-Galactosidase from Potential Probiotic Strains of *Propionibacterium acidipropionici*. Anaerobe *8*, 259–267.

Zárate, G., and Perez-Chaia, A. (2009). Dairy bacteria remove *in vitro* dietary lectins with toxic effects on colonic cells. J. Appl. Microbiol. *106(3)*, 1050–1057.

Zárate, G., and Perez Chaia, A. (2012). Feeding with dairy *Propionibacterium acidipropionici* CRL 1198 reduces the incidence of Concanavalin-A induced alterations in mouse small intestinal epithelium. Food Res. Int. *47*, 3–22.

Zárate, G., Villena, J., Zúñiga-Hansen, M.E., and Pérez-Chaia, A. (2010). Inhibition of adhesion of enteropathogens to human enterocyte-like HT29 cell line by a dairy strain of Propionibacterium acidipropionici. 3rd International Symposium on Propionibacteria and Bifidobacteria: Dairy and Probiotic Applications (Oviedo, España.)

Zárate G., Lorenzo-Pisarello M.J., Babot, J., Argañaraz-Martinez E., Pérez Chaia A. (2011). Dairy Propionibacteria: Technological importance and Probiotic Potential for Application on Human and Animal Nutrition. In: Multidisciplinary Approaches on Food Science and Nutrition for the XXI Century. R. Filip, ed. (Research Signpost, India), pp 175–213.

Zárate, G., and Pérez Chaia A. (2012). Influence of lactose and lactate on growth and b-galactosidase activity of potential probiotic *Propionibacterium acidipropionici*. Anaerobe *18*, 25–30.

Zárate, G. (2012). Dairy propionibacteria: less conventional probiotics to improve the human and animal health. In: Probiotic in Animals, E. Rigobelo, ed. (Intech Open Access), pp 153–202.

Zhang, A., and Yang, S. (2009). Propionic acid production from glycerol by metabolically engineered *Propionibacterium acidipropionici*. Proc. Biochem. *44*, 1346–1351.

Zhu, Y., Li, J., Tan, M., Liu, L., Jiang, L., Sun, J., Lee, P., Du, G., and Chen, J. (2010). Optimization and scale-up of propionic acid production by propionic acid-tolerant Propionibacterium acidipropionici with glycerol as the carbon source. Bioresour. Technol. *101*, 8902–8906.

Zhu, L., Wei, P., Cai, J., Zhu, X., Wang, Z., Huang, L., and Xu, Z. (2012). Improving the productivity of propionic acid with FBB-immobilized cells of an adapted acid-tolerant Propionibacterium acidipropionici. Bioresour. Technol. *112*, 248–253.

Non-LAB Probiotics: Spore Formers

6

Loredana Baccigalupi, Ezio Ricca and Emilia Ghelardi

Abstract

A growing body of evidence suggests that probiotics can be efficiently used to treat/prevent some illnesses, from gastrointestinal or urogenital disorders to allergies, cardiovascular and autoimmune diseases and even to prevent the onset of certain cancers. Although lactic acid bacteria (LAB) and bifidobacteria are the most common microbes used in probiotic preparations, yeasts and other bacteria are also widely used. This chapter focus on the use of bacterial spore formers as probiotics. Spore formers are a group of bacteria able to form an endospore (spore), an exceptionally resistant cell that contains all of the necessary genetic information needed to regenerate a new vegetative cell. Bacterial spores have been commercialized as probiotics for more than 50 years and are now extensively used in humans for the treatment of intestinal disorders and as dietary supplements, in animals as growth promoters and competitive exclusion agents and in aquaculture for enhancing the growth and disease-resistance of cultured fish and shrimps. This chapter will first describe the group of spore-forming bacteria, the sporulation process, the structure of the spore and its interactions with human intestinal and immune cells and then summarize the use of some spore former species as probiotics for human and animal use.

Spore formers

Spore formers are organisms able to form an endospore (spore), a quiescent cell with a peculiar structure that allow survival at conditions that are lethal for standard cells. Spore production (sporulation) is induced by a variety of environmental conditions that limit cell growth and/or block DNA replication. Therefore, producing a spore is a long-term survival strategy evolved by spore formers to preserve their species. More than 1000 bacterial species are able to form spores. These species are grouped within the phylum *Firmicutes* (low GC Gram-positive bacteria) and commonly fall into two genera: the strict anaerobic *Clostridia* and the aerobic/facultative anaerobic *Bacillus* (Fritze, 2004). However, several spore formers do not belong to either one of these two groups but belong to other, less-known, genera. An example comes from members of the very unusual genus *Sporomusa*, that form typical round or oval, heat-resistant endospores, but stain Gram-negative. A recent detailed analysis of the genomes of members of the *Sporomusa* genus has confirmed that they belong to the *Firmicutes* phylum, forming a well-defined separate group within the clostridial lineage (Yutin and Galperin, 2013).

In most species of the *Bacillus* and *Clostridium* genera the sporulation process occurs through a series of generally similar morphological changes. However, substantial differences in the regulation of the onset of sporulation, the engulfment process, and the structure of the protective layers of the mature spore have been observed between various species (Stragier, 2002; Onyenwoke *et al.*, 2004; Paredes *et al.*, 2005; de Hoon *et al.*, 2010; Galperin *et al.*, 2012). Genomic analysis have shown that a common set of 56 key sporulation genes are present in the genomes of all spore formers while an approximally additional 60 genes are found in all bacilli but are absent in all *Clostridia* (Galperin *et al.*, 2012).

Spore formation

As long as water and nutrients are available and environmental conditions allow cell growth, spore formers grow and divide by binary fission (vegetative cycle) (Fig. 6.1). When growth is no longer allowed *B. subtilis*, the model organism for spore formers, responds with a variety of alternative strategies, including the activation of flagellar motility to move to a more favourable environment, production of antimicrobial molecules to kill competing bacteria, secretion of enzymes to scavenge unusual nutrient sources, and induction of competence for uptake of exogenous DNA (Lopez *et al.*, 2009). If these responses do not relieve the growth-blocking condition, *B. subtilis* cells can follow two alternative developmental pathways with some cells forming long chains, producing a polymeric matrix rich in sugars and proteins and assembling into multicellular biofilms, and others entering the irreversible program of spore formation (Vlamakis *et al.*, 2008; Chai *et al.*, 2010; Chai *et al.*, 2011). In dispersed cell populations matrix and spore production are mutually exclusive fates (Vlamakis *et al.*, 2008; Lopez *et al.*, 2009a) and are both bimodal processes in which cells follow either one or the other pathway (Chai *et al.*, 2008; Veening *et al.*, 2008). Both developmental

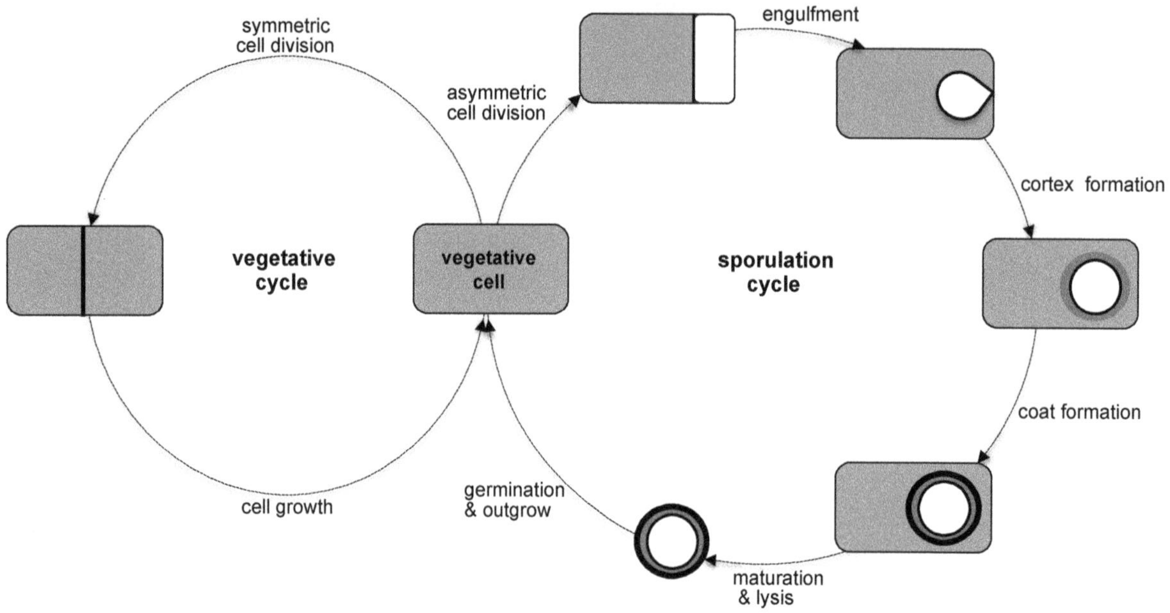

Figure 6.1 Vegetative and sporulation cycles of spore formers. The vegetative cycle occurs by binary fission: a cell divides symmetrically originating in two identical daughter cells that grow and divide again. The sporulation cycle starts with an asymmetrical cell division that originates in a large mother cell and a small forespore. The following morphological modifications (engulfment, cortex and coat formation, maturation and mother cell lysis) are indicated and lead to the release of the free, mature spore in the environment. The spore can germinate originating in a new vegetative cell able to grow and, eventually sporulate again.

cell fates are governed by the regulatory protein Spo0A-P that directly activates genes of the sporulation pathway (Fujita and Losick, 2005) and indirectly acts on matrix synthesis, relieving the repression of genes for matrix production (Branda *et al.*, 2006; Chu *et al.*, 2006; Romero *et al.*, 2010). Induction of one or the other process then depends on the intracellular concentration of Spo0A-P, with low concentrations inducing matrix formation and high concentrations favouring sporulation. Like *B. subtilis*, also other *Bacillus* species can respond to stress conditions by a variety of strategies and follow alternative developmental pathways. An example is a marine isolate of *B. pumilus* that is able to produce a water-soluble pigment. This pigment is involved in the cell response to oxygen-reactive forms and its production only occurs in cell not devoted to sporulation (Manzo *et al.*, 2013).

When the intracellular levels of Spo0A-P are high the sporulation pathway is activated. The first morphological evidence of the induction of sporulation is the formation of an asymmetrically localized cell division septum. As a consequence, a large mother cell and a small forespore are formed (Fig. 6.1). The mother cell contributes to the spore development but undergoes autolysis at the end of sporulation to release the spore into the environment (Fig. 6.1). Soon after the asymmetric cell division, the mother cell initiates a membrane migration process that engulfs the forespore, surrounding it within a double membrane derived from the septum (Fig. 6.1). Synthesis of the protective layers (cortex, coat and, in some species, exosporium) starts early during spore formation and depends on the mother cell, with structural components synthesized in the mother cell and then assembled around the forming spore. The mature spore is characterized by a dehydrated cytoplasm containing a condensed and inactive chromosome and surrounded by various protective layers. Because of such a peculiar structure the spore can survive the absence of water and nutrients and the presence of unfavourable conditions (extremes of heat and pH, UV radiations, solvents, hydrogen peroxide and lytic enzymes) for very long periods. In the presence of water, nutrients and favourable environmental conditions the mature spore can germinate generating a cell able to duplicate vegetatively and, eventually, to re-sporulate (Fig. 6.1). In order to convert the quiescent spore into a metabolically active cell the spore cytoplasm must be hydrated. The associated increase of the cytoplasmic volume can occur only after that the rigid exoskeleton formed by the protective layers is disrupted. The processes of sporulation and germination have been recently reviewed (Higgins and Dworkin, 2012; Dworkin and Shah, 2010).

Spore structure

The peculiar structure of the spore is characterized by several protective layers surrounding a dehydrated cytoplasm (Fig. 6.2) and is responsible for the extreme resistance of the spore to external physical and chemical insults. The **core**, the innermost part of the spore, contains the cytoplasm with proteins, stable RNAs and DNA. The core cytoplasm is dehydrated and has a water content of only 30–50%, as opposed to the 70–90% typical of vegetative cell cytoplasm. This dehydrated physical state of the core plays an important role in spore dormancy and heat resistance (Henriques and Moran, 2007). The core contains large quantities of small acid-soluble proteins (SASPs) which form complexes with the spore DNA protecting it against damages caused by heat and UV irradiation (Setlow, 2005). Within the core the spore-specific compound pyridine-2,6-dicarboxylic acid (dipicolinic acid or DPA) is also highly abundant and accounts for about 15% of the

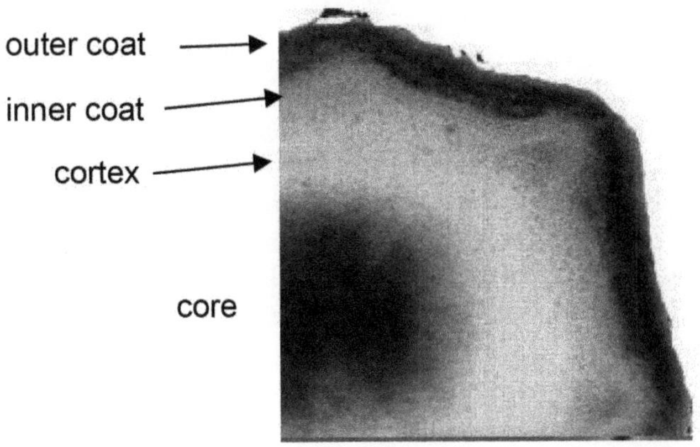

Figure 6.2 A B. subtilis spores as it appears by transmission electron microscopy analysis. The dehydrated core, cortex, inner and outer layers of the coat are indicated.

total spore weight and forms complexes with divalent cations, mainly Ca^{2+} (Paidhungat *et al.*, 2000). DPA plays an important role in spore resistance to wet heat (Paidhungat *et al.*, 2000) and UV irradiation (Slieman and Nicholson, 2001). DPA and calcium are excreted from the core during the first step of spore germination (de Vries *et al.*, 2004). The dehydrated core is surrounded by a membrane (inner membrane) which upon germination will become the plasma membrane of the new vegetative cell and that is the site of localization of several germinant receptors (Paidhungat *et al.*, 2001).

The *cortex*, a thick peptidoglycan-like layer (Fig. 6.2), surrounds the inner membrane, is essential for maintenance of the dehydrated state of the core and contributes to thermal resistance and dormancy (Henriques and Moran, 2007). In the cortex of *B. subtilis* the peptidoglycan has half of the N-acetylmuramic acid residues modified in muramic delta-lactams (Gilmore *et al.*, 2004). In *B. subtilis* the muramoyl-L-alanine amidase CwlD removes the peptide side chain of the muramic acid and the deacetylase PadA carries out both deacetylation and lactam ring formation (Gilmore *et al.*, 2004). Part of the remaining N-acetylmuramic acid residues have an additional modification and have only an L-alanine side chain (Popham *et al.*, 1996). During germination, the cortex peptidoglycan is rapidly degraded by spore cortex lytic enzymes, present but inactive in the dormant spore. The inner part of the cortex, the germ cell wall, lacks the two cortex-specific modifications at N-acetylmuramic acid residues. It is not degraded during germination but instead forms the initial cell wall of the freshly germinated spore, suggesting that the modified peptidoglycan is preferentially recognized by the cortex lytic enzymes activated during germination (Popham, 2002).

Outside of the cortex is the coat, a proteinaceous structure organized in a lamellar inner fraction and an electron-dense outer layer (Fig. 6.2). More than 70 different proteins, about 10% of the spore dry weight, form the coat, whose function is to protect the spore from toxic chemicals and lytic enzymes (Henriques and Moran, 2007). In addition, the coat also provides 'eat resistance' during phagocytic predation by protozoans (Klobutcher *et al.*, 2006). Some coat proteins are enzymes, examples are a superoxide dismutase (Henriques *et al.*, 1997), laccases (Martins *et al.*, 2002) and transglutaminases (Zilhao *et al.*, 2005), all associated to the coat layers. These enzymes are active even when the spore metabolism is totally quiescent and are believed to contribute to the extensive cross-linking observed between various coat components (Henriques and Moran, 2007). Other coat-associated enzymes, present in the inner coat, are involved in spore germination by contributing to cortex degradation (Bagyan and Setlow, 2002; Ragkousi *et al.*, 2003).

In some species, external to the coat is a further structure, the *exosporium*, which takes the form of an additional protective layer of irregular shape, loosely attached to the coat and mainly formed of glycoproteins (Henriques and Moran, 2007). In *B. subtilis*, that does not have an exosporium, an additional coat layer has been recently observed and named crust. The structure of the coat in *B. subtilis* and related species has been recently reviewed (Henriques and Moran, 2007; McKenney *et al.*, 2013).

Aerobic spore formers are intestinal bacteria

While anaerobic spore-formers belonging to the *Clostridium* genus are well-recognized and abundant components of the human and animal gut microbiota (Eckburg *et al.*, 2005; Mahowald *et al.*, 2009), aerobic/facultative anaerobic *Bacilli* are generally considered soil bacteria. Spores of *Bacilli* can be isolated from a variety of different environments including rocks, dust, aquatic environments, but also the gut of various insects and animals (Nicholson, 2002; Nicholson, 2004; Hong *et al.*, 2009). The exceptional longevity and resistance of the spore are the main reasons for the ubiquitous distribution of these organisms and their presence in the gut of insects and animal has long been attributed to the ingestion of contaminated food, water and air. However, in recent years it has been suggested that those organisms are gut inhabitants (Hong *et al.*, 2009; Cutting *et al.*, 2009). The first indication that *Bacillus* spores are intestinal bacteria comes from the observation that large numbers of spores are usually isolated from animal faeces and other intestinal samples. An average of $1.0x10^4$ colony forming units (cfu) of aerobic spore formers have been isolated from human faeces collected in different countries, from people with different dietary habits (Hong *et al*, 2009). Such a large number of spores could not be explained even by a daily intake of heavily contaminated foods. In addition, *in vivo* studies with a murine model have shown that when different doses of spores of a laboratory strain of *B. subtilis* were orally administered to groups of inbred (Balb/C) or outbred mice and faecal samples analysed, the total number of excreted spores was 6-fold higher than the number of spores used to feed the animals, suggesting that spores had germinated, proliferated then re-sporulated in the gut (Hoa *et al.*, 2001). Spore germination, growth and sporulation in the mouse gastrointestinal tract have been unambiguously demonstrated by molecular approaches (Casula and Cutting, 2002; Tam *et al.*, 2006). Taken together these results indicate that a domesticated strain of *B. subtilis* is able to germinate and perform its entire life cycle within the animal gut through five main steps: (i) oral entry in the spore form; (ii) safe transit across the stomach

(Spinosa *et al.*, 2000; Fakhry *et al.*, 2008); (iii) spore germination and vegetative growth in the upper part of the small intestine (Casula and Cutting, 2002); (iv) sporulation in the lower part of the intestine (Tam *et al.*, 2006); and finally (v) excretion in the spore form. The ability to germinate, grow and sporulate in the gut is not restricted to members of the *B. subtilis* species and other *Bacilli* have been shown to germinate and proliferate in the gut of insects (Margulis *et al.*, 1998; Feinberg *et al.*, 1999), poultry and pigs (Jadamus *et al.*, 2001, 2002).

Aerobic spore formers interact with intestinal cells

The interaction between spore formers and intestinal cells is complex and not totally understood. Two independent *in vitro* studies showed that *B. subtilis* spores are internalized by murine (Duc *et al.*, 2004) and human macrophages (Ceragioli *et al.*, 2009). In both systems spore uptake was rather inefficient (2.5% of phagocytosis) and internalized spores were induced to germinate and killed before any round of cell duplication (Duc *et al.*, 2004; Ceragioli *et al.*, 2009). It is worthwhile to report that, in both *in vitro* systems, the analysis of infected macrophages did not reveal any cytotoxic effect exerted by *B. subtilis* spores, supporting the safety of *B. subtilis* spores (Duc *et al.*, 2004; Ceragioli *et al.*, 2009).

In vitro studies have also been used to analyse the interaction of vegetative cells of *B. subtilis* with model intestinal cells. Fujita *et al.* (2007) has shown that the culture supernatant of a domesticated strain of *B. subtilis* added to Caco-2 cells induced the synthesis of the heat-shock (HS) proteins in the intestinal cells. The quorum-sensing pentapeptide ERGMT of *B. subtilis*, known as CSF (competence and sporulation factor), was found responsible for HS induction, which in turn is believed to prevent oxidant-induced intestinal epithelial cell injuries and loss of barrier function by the epithelial cells (Fujita *et al.*, 2007). The ability of intestinal cells to recognize, internalize and respond to a *B. subtilis* molecule is an important finding that highlights the existence of a cross-talk between this bacterial species and the host. Since CSF is produced during the late exponential phase of growth and since *B. subtilis* cells are sensitive to gastric conditions (Fakhry *et al.*, 2008), the interaction between CSF and epithelial cells most likely occurs after ingested spores have germinated in the intestine.

In combination with *Bacteroides fragilis*, *B. subtilis* has also been shown to interact with intestinal immune cells contributing to the development of the gut-associated lymphoid tissue (GALT) and the pre-immune antibody repertoire in rabbits (Rhee *et al.*, 2004). The ability to contribute to GALT development was observed with a wild type strain of *B. subtilis* and with isogenic mutants impaired in general stress responses, flagellar movement or biofilm formation. Only mutants unable to sporulate failed to induce GALT development indicating that sporulation or possibly molecules produced and secreted during sporulation, were essential for interaction with the GALT of the host (Rhee *et al.*, 2004). In the same study other spore formers were also tested but neither *Clostridium subterminale* nor *B. licheniformis* nor *B. pumilus* was able to affect GALT development (Rhee *et al.*, 2004).

Other examples of interaction between spore formers and intestinal immune cells come from the production of spore-specific faecal sIgA and serum IgG in mice orally immunized with *B. subtilis* spores (Duc *et al.*, 2003) and the proliferation of intestinal cells in chickens due to the ingestion of spores of the *B. subtilis* strain Natto, used to prepare a traditional Japanese fermented food (Hosoi *et al.*, 2003). A more recent study has shown that spores of three different *Bacillus* species, *B. subtilis*, *B. licheniformis* and *B. flexus*, orally dosed to mice, promoted active lymphocyte proliferation within the Peyer's patches and the production of cytokines in mesenteric lymph nodes (MLN) (IL-1a, IL-5, IL-6, IFN-g and TNF-a) and in the spleen (IFN-g and TNF-a) (Huang *et al.*, 2008). The same study also showed that vegetative cells of *B. subtilis* up-regulated the expression of the Toll-like receptors TLR2 and TLR4 in cultured macrophages (Huang *et al.*, 2008).

Probiotic spore formers

A variety of commercial probiotics for human or animal use contains spores of several *Bacillus* species (Table 6.1). Although some of those products have been on the market for over 50 years, their use has so far been mainly empirical. In addition, the use of fermented foods containing spore formers is common and traditional in several Asian countries (Cutting, 2011). The long presence on the market of commercial products containing spores and the traditional use of spore formers in fermented foods then represent an exceptional safety record for the strains involved (Cutting, 2011).

An essential advantage of probiotics containing spores over those containing bacterial cells is that spores are extremely resistant and can experience mild heat-treatments used to sterilize food and long-term storage without any lose of viability. As a consequence, spores can be added to foods and retain both viability and probiotic properties (for examples see: http://www.isabellashealthybakery.com/functional.html; http://www.sporegen.com; http://www.sabinsa.com/newsroom/press-releases/pr20121002.html).

A growing body of evidence is now accumulating on the health beneficial effects of ingested spores and on the mechanisms by which such effects are exerted. An example of health beneficial effect of spores comes from an *in vivo* study performed with a mouse model of infection in which *B. subtilis* spores were shown to reduce the susceptibility to enteric pathogens (D'Arienzo *et al.*, 2006). *Citrobacter rodentium* is a mouse pathogen that, following oral infection, colonizes the distal colon, causing crypt hyperplasia, mucosal thickening and uneven epithelial surface, with T-cell infiltration, a highly polarized Th1 immune response and epithelial cell proliferation (Higgins *et al.*, 1999). Epithelial lesions caused in mice by *C. rodentium* are indistinguishable from those caused in humans by enteropathogenic (EPEC) and enterohaemorrhagic (EHEC) strains of *E. coli* (Schauer and Falkow, 1993). Oral administration of *B. subtilis* spores one day before infection with *C. rodentium* was effective in drastically decreasing colon colonization and preventing the enteropathy. A reduced crypt length, no epithelial damage, normal vascularization and a drastic reduction in the mortality rate were observed

Table 6.1 Examples of commercial products containing spore formers for human or animal use[1]

Product	Species (spores/dose)*	Manufacturer	Host
AlCare	*B. licheniformis* NCTC 13123 (10^9–10^{10})	Alpharma (Australia)	Swine
Bactisubtil	*B. cereus* ATCC14893 (1×10^9)	Marion-Merrel (France)	Human
Bio-Acimin	*B. cereus*[2], and other bacteria	Viet-Duc Pharm. (Vietnam)	Human
Bibactyl	*B. subtilis* var. *natto* (10^7–10^8)	UPHACE (Vietnam)	Human
Bidisubtil	*B. cereus*[2] (1×10^6)	Bidiphar (Vietnam)	Human
Bio-Kult	*B. subtilis*[2] and other bacteria[3]	Probiotics International Ltd (UK)	Human
Biobaby	*B. subtilis*[2] (3×10^6), *C. butyricum* (1×10^7), *B. coagulans*[2] (5×10^7)	Ildong Pharma (Korea)	Human
BioGrow	*B. subtilis*[2] (1.6×10^9), *B. licheniformis*[2] (1.6×10^9)	Provita Eurotech (UK)	Poultry, calves, swine
BioPlus2B	*B. subtilis*[2] (1.6×10^9), *B. licheniformis*[2] (1.6×10^9)	Christian Hansen Hoersholm (Denmark)	Piglets, poultry
Biosporin	*B. subtilis* 2335 and *B. licheniformis* 2336	Biofarm (Ukraine)	Human
Biostart	*B. megaterium*[2], *B. licheniformis*[2], *B. subtilis*[2], *Paenibacillus polymyxa*[2]	Microbial Solutions (South Africa)	Aquaculture
Biosubtyl	*B. cereus*[2] (10^6–10^7)	Biophar Company (VietNam)	Human
Biosubtyl DL	*B. subtilis* (10^7–10^8) and other bacteria	IVAC (VietNam)	Human
Biovicerin	*B. cereus* GM (1×10^6)	Geyer Med. (Brazil)	Human
BioZyme-Aqua	*B. subtilis*[2] (1×10^8)	Sino-Aqua Corp. (Taiwan)	Aquaculture
Bispan	*B. polyfermenticus* SCDc (1.7×10^7)	Binex Co. (Korea)	Human
Enterogermina	*B. clausii* NR, SIN, T, OC (2×10^9)	Sanofi Winthorp (Italy)	Human
Esporafeed Plus	*B. cereus*[2] (1×10^9)	Norel (Spain)	Swine
Flora-Balance	*Brevibacillus laterosporus* BOD ($>10^6$)	Flora-Balance (USA)	Human
Ildong Biovita	*B. subtilis*[2] (3×10^6), *C. butyricum*[2] (1×10^7) and other bacteria[3]	Ildong Pharma (Korea)	Human
Lactipan Plus	*B. subtilis*[2] (2×10^9)	Ist. Bioch. Italiano (Italy)	Human
Lactopure	*B. coagulans*[2]	Pharmed Medicare (india)	Poultry, calves, swine
Lactospore	*B. coagulans*[2] (1×10^9)	Sabinsa Corp. (USA)	Human
Latero-Flora	*B. laterosporus* BOD ($>10^6$)	GHC (USA)	Human
Liqualife	*Bacillus* spp.[2]	Cargill (USA)	Aquaculture
Medilac-Vita	*B. subtilis* RO179 (1×10^8) and other bacteria	Hanmi Pharmac. (China)	Human
Neoferm BS	*B. clausii*	Sanofi Sante Nutrition Animale (France)	Poultry, calves, swine
Neolactoflorene	*B. coagulans*[2] and other bacteria	Montefarmaco (Italy)	Human
Pastylbio	*B. subtilis*[2] (1x10^8)	Pasteur Inst. (VietNam)	Human
Primal defence	*B. subtilis*[2] and other bacteria	Garden of Life (USA)	Human
Promarine	*B. subtilis*[2]	Sino-Aqua (Taiwan)	Aquaculture
Sanocare Sanolife	*Bacillus* spp.[2]	INVE Technologies	Aquaculture
Sanoguard	*Bacillus* spp[2]	(Belgium)	Aquaculture
Subtyl	*B. cereus*[2] (10^6–10^7)	Mekophar (VietNam)	Human
Sustenex	*B. coagulans* GBI-30	Ganeden Biotech (USA)	Human
Toyocerin	*B. cereus*[2] (1×10^{10})	Asahi Vet (Japan)	Calves, poultry, rabbit, swine

[1]Data from Cutting, 2011.
[2]No information available on strain number.
*Spores/dose indicated where information available.

in spore-dosed mice (D'Arienzo *et al.*, 2006). Similar results were observed also in chickens infected with *Escherichia coli* O78:K80, *Salmonella enterica* or *Clostridium perfringens* (La Ragione *et al.*, 2001; La Ragione and Woodward, 2003). The reduced susceptibility to enteric pathogens caused by the ingested spores could be explained by a 'competitive exclusion' effect, with spores physically blocking the interaction between the pathogen and the intestinal cells. An alternative possibility is that the ability of spores and germination-derived cells to interact with the intestinal epithelial and immune cells and to stimulate the GALT, as summarized in the previous paragraph, could strengthen the immune system and prime it for an adaptive immune response.

Probiotics for human use: examples and reported effects

Spore-based probiotics have long been introduced to the international market and used for the bacteriotherapy and bacterioprophylaxis of gastrointestinal disorders in humans. In addition, foods fermented with *Bacillus* spp. have a long history in Korea, Japan, and also some African countries. A substantial number of products containing different *Bacillus* spp. alone or in combination with other bacteria are currently available (Table 6.1). They are being used extensively as dietary supplements, but some of them are licensed as over-the counter (OTC) medicines.

The *Bacillus* strains commonly used in Europe today belong to species granted the *Qualified Presumption of Safety* (QPS) status by the European Food Safety Authority (EFSA 2012). These include *B. subtilis, B. licheniformis, B. pumilus, B. clausii* and *B. coagulans*. Some other products for human use contain *B. cereus* strains. *B. cereus* is a food spoilage contaminant and ingestion of spores can produce a diarrhoeal or emetic-type syndrome. These clinical manifestations are due to the production of up to six enterotoxins by *B. cereus* (Senesi and Ghelardi, 2010). However, not all strains of *B. cereus* carry enterotoxin genes and, therefore, any toxigenic potential must be excluded to use *B. cereus* as probiotic organism (EFSA 2011a) Spore-based products that have been formally approved in the USA are few. In this Country, the use of *Bacillus* species as probiotics is regulated by the FDA, and strains may be granted the GRAS (*Generally Regarded As Safe*) status on an individual case-by-case basis. In this paragraph, examples and reported effects of the major *Bacillus* species used in commercial products for human use is summarized.

Many of the species belonging to the *B. subtilis* group, such as *B. subtilis, B. licheniformis, B. amyloliquefaciens, B. atrophaeus, B. moyavensis* and *B. vallismortis*, have been included by the EFSA in the list of QPS recommended biological agents that can be intentionally added to food and feed (EFSA 2012). However, most spore-based probiotic products contain *B. subtilis* alone or in combination with other species such as *B. licheniformis* (Table 6.1). The reasons that can explain the wide use of *B. subtilis* spores in probiotic preparations are the historical importance of this species in the *Bacillus* genus and the intense research interest that *B. subtilis* received in the last 100 years. In addition, *B. subtilis* var. *natto* has probably been the first *Bacillus* strain used for its probiotic properties. This bacterium is required for the production of natto, a traditional food obtained from fermented boiled soybeans, which is consumed mainly in Japan and possesses a history of more than 1000 years. *B. subtilis* var. *natto*, isolated for the first time in 1906, is the sole microorganism used for natto fermentation. Natto carries as many as 10^8 viable spores per gram of product and for decades health benefits have been associated with its consumption, including stimulation of the immune system (Hosoi and Kiuchi, 2004). *B. subtilis* var. *natto* and, although at lower levels, other *B. subtilis* strains produce nattokinase, a strong fibrinolytic enzyme (Sumi *et al.*, 1987) that activates tissue pro-urokinase and plasminogen activator (Yatagai *et al.*, 2008). Nattokinase supplementation was shown to reduce systolic and diastolic blood pressure, suggesting that increased intake of this enzyme may play an important role in preventing and treating hypertension (Kim *et al.*, 2008).

Although several studies *in vitro* and in animal models have demonstrated the potential of *B. subtilis* spores used as probiotics for humans, data on the effectiveness of preparations based on *B. subtilis* alone or in combination with other bacteria in humans are limited. The product Biosporin, containing spores of *B. subtilis* 2335 and *B. licheniformis* 2336, was shown to be effective in the treatment of patients with acute enteric infections (Gracheva *et al.*, 1996). Interestingly, the *B. subtilis* strain contained in this preparation produces the antibiotic amicoumacin with *in vitro* activity against *Helicobacter pylori* (Pinchuk *et al.*, 2001). In a controlled randomized clinical trial in children, Lactipan (*B. subtilis* spores) caused complete remission of aspecific chronic diarrhoea after administration for 15 days (Roggero *et al.*, 1990). In a review summarizing the clinical data of several trials, the product Medilac (*B. subtilis* RO179 and *Enterococcus faecium*) was reported to be able to provide some relief for patients with chronic diarrhoea and irritable bowel syndrome, as well as to reduce antibiotic-associated diarrhoea (Tompkins *et al.*, 2010).

The Italian product known as Enterogermina, which contains *B. clausii* spores, was registered as a pharmaceutical preparation in 1958 in Italy and has an OTC medicinal status since 1999. Nowadays, two different formulations, i.e. lyophilized capsules and liquid vials, are commercialized in 55 countries around the world for the treatment of diarrhoea and the prevention of infectious gastrointestinal diseases in humans. The product is constituted of four antibiotic resistant *B. clausii* strains (named OC, NR, SIN, T) and it is recommended for restoring intestinal microbial balance particularly during combined antibiotic treatment. The four *B. clausii* strains contained in Enterogermina are supposed to be derived from ATCC 9799, a penicillin-resistant strain originally designated as *B. subtilis*, following multiple step selection (Green *et al.*, 1999; Senesi *et al.*, 2001). In fact, in addition to their common resistance to cephalosporins, macrolides, and chloramphenicol, strain NR is resistant to rifampicin, strain T to tetracyclines, and strain SIN to kanamycin, tobramycin and amikacin (Courvalin, 2006). Several studies have demonstrated that the resistance genes within these *B. clausii* strains are stable and unable to be transferred (Bozdogan *et al.*, 2003, 2004; Courvalin, 2006; Galopin *et al*, 2009).

Clinical studies have shown that the bacteriotherapy with Enterogermina reduces the intestinal disorders associated with alterations in the normal human intestinal microbiota (Mazza, 1994) and represent a promising treatment option for small intestinal bacterial overgrowth decontamination (Gabrielli *et al.*, 2009). This therapy stimulates the synthesis of membrane bound and secreted IgA in elderly subjects (Fiorini *et al.*, 1985), affects the expression of many genes involved in immune response and inflammation (Di Caro *et al.*, 2005), modulates the cytokine pattern in allergic subjects (Ciprandi *et al.*, 2005a) and reduces nasal allergic inflammation in children (Ciprandi *et al.*, 2005b). Oral administration of Enterogermina also helps in reducing the frequency of infection episodes in patients with stones-associated recurrent urinary tract infections and the common side-effects related to anti-*H. pylori* antibiotic therapy (Nista *et al.*, 2004). Recently, administration of another *B. clausii* strain (UBBC-07) was shown to be effective in alleviating the symptoms of diarrhoea in patients with acute diarrheal disease (Sudha *et al.*, 2013).

The species *B. coagulans* was originally named *Lactobacillus sporogenes* and, for this reason, products containing this bacterium can still carry this incorrect nomenclature or being labelled as 'formerly *Lactobacillus sporogenes*'. In a 2007 systematic review of the literature, *B. coagulans* appeared to be among the most promising probiotics for preventing antibiotic-associated diarrhoea (Johnston *et al.*, 2007). However, a more recent study indicated that insignificant results are obtained with the oral bacteriotherapy with *B. coagulans* in the treatment of diarrhoea in children (Dutta *et al*, 2011). The differences in the results obtained in these studies have suggested that strain specificity is important for defining the probiotic potential of *B. coagulans*. In this regard, *B. coagulans* strain GBI-30, 6086, named GanedenBC^{30}, has been shown to reduce bloating and abdominal pain in adults with the irritable bowel syndrome (Hun, 2009; Dolin, 2009) or with post prandial intestinal gas-related symptoms (Kalman *et al.*, 2009). Administration of GanedenBC^{30} has been proven effective in enhancing the immunological response to certain viral respiratory tract infections (Baron, 2009; Kimmel *et al.*, 2010) and for relieving symptoms of rheumatoid arthritis in adults (Mandel *et al.*, 2010). GanedenBC^{30} received the GRAS status by FDA in 2012 and is sold as a dietary ingredient for use in functional foods and dietary supplements (e.g. Sustenex) (www.ganedenBC30.com)

Probiotics for animal use: examples and reported effects

The multitude of diseases among farmed animals is an important problem not only for the loss of productivity due to animal mortality but also for the possible contamination of animal products for human use. This problem has been addressed by dosing animals with large amounts of antibiotics. In addition, antibiotics are widely used in animal feed preparations because of their growth promoter activity. More recently, the distribution of antibiotic-resistant microbial populations has led some European countries to limit the use of antibiotics to the treatment of animal diseases, banning their use as growth promoters. This has prompted the search for alternatives to improve animal health. In this context a lot of studies have been performed to evaluate the effects in terms of growth performance, immune response and disease resistance, of probiotic preparations used as feed additives. Also, for animal nutrition, the use of biological agents (including microorganisms) as sources of feed additives needs to be authorized by the European Food Safety Authority (EFSA). Strains belonging the species *B. subtilis*, *B. amyloliquefaciens* and *B. licheniformis* have been proclaimed as non toxigenic (EFSA 2010), have been found to promote growth and improve intestinal health and have been registered as probiotics for animal feed (Gaggia *et al.*, 2010; Novak *et al.*, 2012; Sun *et al.*, 2010). In Table 6.1 a number of commercially available products used for animal feeding are listed. BioPlus 2B®, is a mixture of *B. licheniformis* and *B. subtilis*, that has been approved for slaughter pigs and chickens (EFSA 2011b). Pigs treated with BioPlus 2B® over a period of 23 weeks showed a lower incidence of diarrhoea and mortality than a control group (Alexopoulos *et al.*, 2004). In another feeding experiment with chickens, diet supplementation with BioPlus 2B®, resulted in an increase of weight gain and improvement in feed conversion (Šabatková *et al.*, 2008).

Another commercial product, Toyocerin®, contains *B. cereus* var. *toyoi*, a non-toxigenic and non-pathogenic strain of *B. cereus*. It has been reported to reduce diarrhoea and morbidity in piglets challenged with *E. coli* and to improve weight gain after 6 months of treatment (Williams *et al.*, 2009). The same *B. cereus* strain was found to affect the intestinal immune system of piglets with beneficial effect on animals health (Scharek *et al.*, 2007). Little information has been reported about the probiotic effects of *B. amyloliquefaciens* and the only product approved for animal use containing this species is Ecobiol® that contains spores from strain CECT5940. It has been reported to significantly improve feed conversion rate in chickens for fattening (EFSA, 2008) and to reduce pathogenic *Clostridia* in dogs (Gonzales-Ortis *et al.*, 2013) during the supplementation period. There were no differences between experimental groups in other parameters such as faecal consistency and nutrient digestibility. The same strain has been used in a feeding trial in poultry artificially infected with *Clostridium perfringens* and no beneficial effects were observed on the course of the disease (Jerzsele *et al.*, 2012).

Bacillus probiotics are largely used also in aquaculture to improve health status and disease resistance of bred fish and to increase farming productivity (Mohapatra *et al.*, 2012). Spores can be used as feed supplement or directly added in the rearing water. The addition of a specific strain during egg hatching and the first stages of larvae development is thought to influence the microbiota composition and make it possible to artificially manipulate the gut microbiota. In a recent study (Liu *et al.*, 2012) the effect of *B. subtilis* E20 (a strain isolated from the human health food, natto) on growth performance, immune response and disease resistance of groupers, *Epinephelus coioides*, has been assessed. Groupers fed a diet containing the probiotic significantly increased the per cent weight gain (PWG) and feeding efficiency (FE) compared to control in a dose-dependent manner. During the feeding period the bacterium was able to colonize the host and protect it against bacterial (*Streptococcus* sp.) and viral (*Iridovirus*) infections. The same strain was also able to improve the survival rate and development of white shrimp, *Litopenaeus vannamei*, larvae (Liu *et al.*, 2010). The probiotic was directly added to the rearing water at different concentrations (10^8 and 10^9 cfu/l) every 3 days during the 14 days of breeding. Shrimp larval development was significantly accelerated during the treatment and also the survival rate after metamorphosis to postlarvae, was significantly higher compared to control. The induction of immune-related gene expression suggested a possible role of this probiotic in immune system development and in protection from infections.

The strain FPTB16 belonging to the species *B. amyloliquefaciens*, has been used during a challenge experiment of Catla

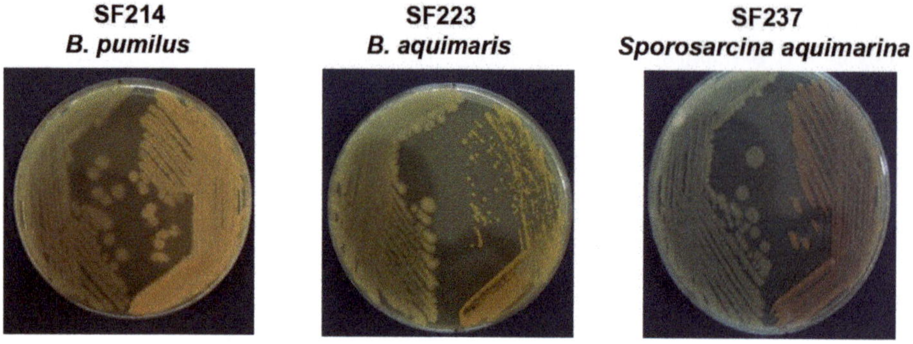

Figure 6.3 Examples of pigmented Bacilli. Those strains are all aerobic Bacilli and have been described by Khaneja *et al.* (2010). Each plate in the figure has a non pigmented B. subtilis strain on the left and the pigmented strain on the right.

aquaculture: a diet containing 10^8 and 10^9 cfu/g of spores significantly enhanced the resistance of fish against *Edwardsiella tarda* infections and improved various systemic and mucosal immunological parameters (Das *et al.*, 2013). On the basis of the reported beneficial effects derived from the use of probiotic spore formers, strong research efforts are now focusing on the identification of new strains, characterization of their activity and on the understanding of the scientific bases of their beneficial effects (Larsen *et al.*, 2013; Jung *et al*, 2012; Thirabunyanon and Thongwittaya, 2012).

Future trends

Although the use of bacterial spores in commercial probiotic products is already common (Table 6.1), it is now further expanding with increasing numbers of studies addressing the health beneficial effects of spores and of germination-derived cells. A challenging goal on which researchers are now focusing is to understand in details the mechanisms by which spores and germination-derived cells exert their beneficial effects on human and animal health.

A promising future development of spore probiotics is to use spores carrying molecules that have additional beneficial properties. Examples are spores of strains that produce pH-stable carotenoids (Duc *et al.*, 2006; Khaneja *et al.*, 2010) (Fig. 6.3). Because of their antioxidant activity carotenoids are of nutritional value and can be used as dietary supplements providing at the same time a source of antioxidants together with the probiotic properties of the live spore formers (Cutting, 2011).

References

Alexopoulos, C., Georgoulakis, I.E., Tzivara, A., Kyriakis, C.S., Govaris, A., and Kyriakis, S.C. (2004). Field evaluation of the effect of a probiotic containing *Bacillus licheniformis* and *Bacillus subtilis* spores on the health status, performance, and carcass quality of grower and finisher pigs. J. Vet. Med. A Physiol. Pathol. Clin. Med. *51*, 306–312.

Bagyan, I., and Setlow, P. (2002). Localization of the cortex lytic enzyme CwlJ. in spores of *Bacillus subtilis*. J. Bacteriol. *184*, 1219–1224.

Baron, M. (2009). A patented strain of *Bacillus coagulans* increased immune response to viral challenge. Postgrad. Med. *121*, 114–118.

Branda, S.S., Chu, F., Kearns, D.B., Losick, R., and Kolter, R. (2006). A major protein component of the *Bacillus subtilis* biofilm matrix. Mol. Microbiol. *59*, 1229–1238.

Bozdogan, B., Galopin, S., Gerbaud, G., Courvalin, P., and Leclercq, R. (2003). Chromosomal *aadD2* encodes an aminoglycoside nucleotidyltransferase in *Bacillus clausii*. Antimicrob. Agents Chemother. *47*, 1343–1346.

Bozdogan, B., Galopin, S., and Leclercq, R. (2004). Characterization of a new *erm*-related macrolide resistance gene present in probiotic strains of *Bacillus clausii*. Appl. Environ. Microbiol. *70*, 280–4.

Casula, G., and Cutting, S.M. (2002). Bacillus probiotics: Spore germination in the gastrointestinal tract. Appl. Environ. Microbiol. *68*, 2344–2352.

Ceragioli, M., Cangiano, G., Esin, S., Ghelardi, E., Ricca, E., and Senesi, S. (2009). Phagocytosis, germination and killing of *Bacillus subtilis* spores presenting heterologous antigens in human macrophages. Microbiology *155*, 338–346.

Chai, Y., Chu, F., Kolter, R., and Losick, R. (2008). Bistability and biofilm formation in *Bacillus subtilis*. Mol. Microbiol. *67*, 254–263.

Chai, Y., Norman, T., Kolter, R., and Losick, R. (2010). An epigenetic switch governing daughter cell separation in *Bacillus subtilis*. Genes Dev. *24*, 754–765.

Chai, Y., Norman, T., Kolter, R., and Losick, R. (2011). Evidence that metabolism and chromosome copy number control mutually exclusive cell fates in *Bacillus subtilis*. EMBO J. *30*, 1402–1413.

Chu, F., Kearns, D.B., Branda, S.S., Kolter, R., and Losick, R. (2006). Targets of the master regulator of biofilm formation in *Bacillus subtilis*. Mol. Microbiol. *59*, 1216–1228.

Ciprandi, G., Vizzaccaro, A., Cirillo, I., and Tosca, M.A. (2005a). *Bacillus clausii* exerts immuno-modulatory activity in allergic subjects: a pilot study. Eur. Ann. Allergy Clin. Immunol. 37,129–134.

Ciprandi, G., Vizzaccaro, A., Cirillo, I., and Tosca, M.A. (2005b). *Bacillus clausii* effects in children with allergic rhinitis. Allergy *60*, 702–703. Coppi, F., Ruoppolo, M., Mandressi, A., Bellorofonte, C., Gonnella, G., and Trinchieri, A. (1985). Results of treatment with *Bacillus subtilis* spores (Enterogermina) after antibiotic therapy in 95 patients with infection calculosis. Chemioterapia. 4, 467–470.

Courvalin, P. (2006). Antibiotic resistance: the pros and cons of probiotics. Dig. Liver Dis. 38 (Suppl. 2), S261–S265.

Cutting, S.M. (2011). *Bacillus* probiotics. Food Microbiol. *28*, 214–220.

Cutting, S.M., Hong, H.A., Baccigalupi, L., and Ricca, E. (2009). Oral vaccine delivery by recombinant spore probiotics. Intern. Rev. Immunol. *28*, 487–505.

D'Arienzo, R., Maurano, F., Mazzarella, G., Luongo, D., Stefanile, R., Ricca, E., and Rossi, M. (2006). *Bacillus subtilis* spores reduce susceptibility to *Citrobacter rodentium*-mediated enteropathy in a mouse model. Res. Microbiol. *157*, 891–897.

Das, A., Nakhro, K., Chowdhury, S., Kamilya, and Fish, D. (2013). Effect of potential probiotic *Bacillus amyloliquefaciens* FPTB16 on systemic and cutaneous mucosal immune responses and desease resistance of catla (*Catla catla*). Shellfish Immunol. *35*, 1547–1553.

de Hoon, M.J., Eichenberger, P., and Vitkup, D. (2010). Hierarchical evolution of the bacterial sporulation network. Curr. Biol. *20*, 735–745.

de Vries, Y.P., de Voort, M.v., van Schaik, W., Hornstra, L.M., de Vos, W.M., and Abee, T. (2004). Progress in food-related research focussing on *Bacillus cereus*. Microb. Environ. *19*, 265–269.

Di Caro, S., Tao, H., Grillo, A., Franceschi, F., Elia, C., Zocco, M.A., Gasbarrini, G., Sepulveda, A.R., and Gasbarrini, A. (2005). *Bacillus clausii* effect on gene expression pattern in small bowel mucosa using DNA microarray analysis. Eur. J. Gastroenterol. Hepatol. *17*, 951–960.

Dolin, B.J. (2009). Effects of a proprietary *Bacillus coagulans* preparation on symptoms of diarrhea-predominant irritable bowel syndrome. Methods Find. Exp. Clin. Pharmacol. *31*, 655–659.

Duc, L.H., Fraser, P., and Cutting, S.M. (2006). Carotenoids present in halotolerant *Bacillus* spore formers. FEMS Microbiol. Lett. *255*, 215–224.

Duc, L.H., Hong, H.A., Fairweather, N., Ricca, E., and Cutting, S.M. (2003). Bacterial spores as vaccine vehicles. Infect. Immun. *71*, 2810–2818.

Duc, L.H., Hong, A.H., Nguyen, Q.U., and Cutting, S.M. (2004). Intracellular fate and immunogenicity of *B. subtilis* spores. Vaccine *22*, 1873–1885.

Dutta, P., Mitra, U., Dutta, S., Rajendran, K., Saha, T.K., and Chatterjee, M.K. (2011). Randomized controlled clinical trial of *Lactobacillus sporogenes* (*Bacillus coagulans*), used as probiotic in clinical practice, on acute watery diarrhoea in children. Trop. Med. Int. Health. *16*, 555–561.

Dworkin, J., and Shah, I.M. (2010). Exit from dormancy in microbial organisms. Nat. Rev. Microbiol. *8*, 890–896.

Eckburg, P.B., Bik, E.M., Bernstein, C.N., Purdom, E., Dethlefsen, Sargent, M., L., Gill, S.R., Nelson, K.E., and Relman, D.A. (2005). Diversity of the human intestinal microbial flora. Science *308*, 1635–1638.

European Food Safety Authority (EFSA). (2008). Scientific Opinion of the Panel on Additives and Products or Substances used in Animal Feed (FEEDAP) on a request from the European Commission on the safety and efficacy of Ecobiol® (*Bacillus amyloliquefaciens*) as feed additive for chickens for fattening. EFSA Journal, 773, 1–13 Doi: 10.2903/j.efsa.2008.773 available on line: www.efsa.europa.eu/it/efsajournal/pub/773.htm

European Food Safety Authority (EFSA). (2010). EFSA Panel on Biological Hazards (BIOHAZ). Scientific Opinion on the maintenance of the list of QPS biological agents intentionally added to food and feed (2010 update). EFSA Journal 2010;8(12):1944 [56 pp.]. doi:10.2903/j.efsa.2010.1944; Available online: www.efsa.europa.eu/efsajournal

European Food Safety Authority (EFSA). (2011a). EFSA Panel on Additives and Products or Substances used in Animal Feed (FEEDAP). Technical Guidance on the assessment of the toxigenic potential of *Bacillus* species used in animal nutrition. EFSA Journal 2011;9(11):2445 [13 pp.]. doi:10.2903/j.efsa.2011.2445; Available online: www.efsa.europa.eu/efsajournal

European Food Safety Authority (EFSA). (2011b). EFSA Panel on Additives and Products or Substances used in Animal Feed (FEEDAP). Scientific Opinion on the safety and efficacy of BioPlus 2B (*Bacillus licheniformis* DSM 5749 and *Bacillus subtilis* DSM 5750). as a feed additive for sows. EFSA Journal 2011;9(9):2356 [10 pp.]. doi:10.2903/j.efsa.2011.2356; Available online: www.efsa.europa.eu/efsajournal

European Food Safety Authority (EFSA). (2012). EFSA Panel on Biological Hazards (BIOHAZ). Scientific Opinion on the maintenance of the list of QPS biological agents intentionally added to food and feed (2012 update). EFSA Journal 2012;10(12):3020 [84 pp.]. doi:10.2903/j.efsa.2012.3020; Available online: www.efsa.europa.eu/efsajournal

Fakhry, S., Sorrentini, I., Ricca, E., De Felice, M., and Baccigalupi, L. (2008). Characterization of spore forming bacilli isolated from the human gastrointestinal tract. J. Appl. Microbiol. *105*, 2178–2186.

Feinberg, L., Jorgensen, J., Haselton, A., Pitt, A., Rudner, R., and Margulis, L. (1999). Arthromitus (*Bacillus cereus*) symbionts in the cockroach *Blaberus giganteus*: Dietary influences on bacterial development and population density. Symbiosis 2, 109–123.

Fiorini, G., Cimminiello, C., Chianese, R., Visconti, G.P., Cova, D., Uberti, T., and Gibelli, A. (1985). *Bacillus subtilis* selectively stimulates the synthesis of membrane bound and secreted IgA. Chemioterapia. *4*, 310–312.

Fritze, D. (2004). Taxonomy and systematics of the aerobic endospore forming bacteria: *Bacillus* and related genera. In Bacterial Spore Formers, E. Ricca, A.O. Henriques, and S.M. Cutting, eds. (Norfolk, UK:Horizon Biosience), pp. 17–34.

Fujita, M., and Losick, R. (2005). Evidence that entry into sporulation in *Bacillus subtilis* is governed by a gradual increase in the level and activity of the master regulator Spo0A. Genes Dev. *19*, 2236–2244.

Fujita, M., Musch, M.W., Nakagawa, Y., Hu, S., Alverdy, J., Kohgo, Y., Schneewind, O., Jabri, B., and Chang, E.B. (2007). The *Bacillus subtilis* quorum-sensing molecule CSF contribute to intestinal homoeostasis via OCTN2, a host cell membrane transporter. Cell Host&Microb. *1*, 299–308.

Gabrielli, M., Lauritano, E.C., Scarpellini, E., Lupascu, A., Ojetti, V., Gasbarrini, G., Silveri, N.G., and Gasbarrini, A. (2009). *Bacillus clausii* as a treatment of small intestinal bacterial overgrowth. Am. J. Gastroenterol. *104*, 1327–1328.

Gaggia, F., Mattarelli, P., and Biavati, B. (2010). Probiotics and prebiotics in animal feeding for safe food production. Int. J. Food Microbiol. 141, S15–S28.

Galperin, M.Y., Mekhedov, S.L., Puigbo, P., Smirnov, S., Wolf, Y.I., and Rigden, D.J. (2012). Genomic determinants of sporulation in Bacilli and Clostridia: towards the minimal set of sporulation-specific genes. Environ. Microbiol. *14*, 2870–2890.

Galopin, S., Cattoir, V., and Leclercq, R. (2009). A chromosomal chloramphenicol acetyltransferase determinant from a probiotic strain of *Bacillus clausii*. FEMS Microbiol. Lett. *296*, 185–189.

Gonzales-Ortis, G., Castillejos, L., Mallo, J.J., Angels Calvo-Torras, M., and Dolores Baucells, M. (2013). Effects of dietary supplementation of *Bacillus amyloliquefaciens* CECT 5940 and *Enterococcus faecium* CECT 4515 in adult healthy dogs. Arch. Anim. Nutr. *67*, 406–415.

Gracheva, N.M., Gavrilov, A.F., Solov'eva, A.I., Smirnov, V.V., Sorokulova, I.B., Reznik, S.R., and Chudnovskaia, N.V. (1996). The efficacy of the new bacterial preparation biosporin in treating acute intestinal infections. Zh. Mikrobiol. Epidemiol. Immunobiol. *1*, 75–77.

Gilmore, M.E., Bandyopadhyay, D., Dean, A.M., Linnstaedt, S.D., and Popham, D.L. (2004). Production of muramic delta-lactam in *Bacillus subtilis* spore peptidoglycan. J. Bacteriol. *186*, 80–89.

Green, D., Wakeley, P., Page, A., Barnes, A., Baccigalupi, L., Ricca, E., and Cutting, S. (1999). Characterization of two *Bacillus* probiotic species. Appl. Environ. Microbiol. *6*, 4288–4291.

Henriques, A.O., Beall, B.W., and Moran, C.P. Jr. (1997). CotM of *Bacillus subtilis*, a member of the alpha-crystallin family of stress proteins, is induced during development and participates in spore outer coat formation. J. Bacteriol. *179*, 1887–1897.

Henriques, A.O., and Moran, C.P. Jr. (2007). Structure, assembly, and function of the spore surface layers. Ann. Rev. Microbiol. *61*, 555–588.

Higgins, L.M., Frankel, G., Douce, G., Dougan, G., and MacDonald, T.T. (1999). *Citrobacter rodentium* infection in mice elicits a mucosal Th1 cytokine response and lesions similar to those in murine inflammatory bowel disease. Infect. Immun. *67*, 3031–3039.

Hong, H.A., To, E., Fakhry, S., Baccigalupi, L., Ricca, E., and Cutting, S.M. (2009). Defining the natural habitat of *Bacillus* spore-formers. Res. Microbiol. *160*, 375–379.

Hoa, T.-T., Duc, L.H., Isticato, R., Baccigalupi, L., Ricca, E., Van, P.H., and Cutting, S.M. (2001). Fate and dissemination of *Bacillus subtilis* spores in a murine model. Appl. Environ. Microbiol. *67*, 3819–3823.

Hosoi, T., Hirose, R., Saegusa, S., Ametani, A., Kiuchi, K., and Kaminogawa, S. (2003). Cytokine responses of human intestinal epithelial-like Caco-2 cells to the nonpathogenic bacterium *Bacillus subtilis* (natto). Int. J. Food Microbiol. *82*, 255–264.

Hosoi, T., and Kiuchi, K. (2004). Production and probiotic effects of Natto. In: Ricca, E., Henriques, A.O., and Cutting, S.M. (Eds.), Horizon Bioscience, pp. 143–154.

Huang, J.-M., La Ragione, R., Nunez, A., and Cutting, S.M. (2008). Immunostimulatory activity of *Bacillus* spores. FEMS Immunol. Med. Microbiol. 53, 195–203.

Hun, L. (2009). *Bacillus coagulans* significantly improved abdominal pain and bloating in patients with IBS. Postgrad. Med. *121*, 119–124.

Jadamus, A., Vahjen, W., and Simon, O. (2001). Growth behaviour of a spore forming probiotic strain in the gastrointestinal tract of broiler chicken and piglets. AArch. Tierernahr. *54*, 1–17.

Jadamus, A., Vahjen, W., Schafer, K., and Simon, O. (2002). Influence of the probiotic strain *Bacillus cereus* var. toyoi on the development of enterobacterial growth and on selected parameters of bacterial metabolism in digested samples of piglets. J. Anim. Physiol. Anim. Nutr. (Berl.). *86*, 42–54.

Jerzsele, A., Szeker, K., Csizinszky, R., Gere, E., Jakab, C., Mallo, J.J., and Galfi, P. (2012). Efficacy of protected sodium butyrate, a protected blend of essential oils, their combination, and *Bacillus amyloliquefaciens* spore suspension against artificially induced necrotic enteritis in broilers Poult. Sci. *91*, 837–843.

Johnston, B.C., Supina, A.L., Ospina, M., and Vohra, S. (2007). Probiotics for the prevention of pediatric antibiotic-associated diarrhea. Cochrane Database Syst. Rev. 2:CD004827.

Jung, J.H., Lee, M.Y., and Chang, H.C. (2012). Evaluation of the probiotic potential of *Bacillus polyfermenticus* CJ6 isolated from Meju, a Korean soybean fermentation J. Microbiol. Biotechnol. 22, 1510–1517.

Kalman, D.S., Schwartz, H.I., Alvarez, P., Feldman, S., Pezzullo, J.C., and Krieger, D.R. (2009). A prospective, randomized, double-blind, placebo-controlled parallel-group dual site trial to evaluate the effects of a *Bacillus coagulans*-based product on functional intestinal gas symptoms. BMC Gastroenterol. 9, 85.

Khaneja, R., Perez-Fons, L., Fakhry, S., Baccigalupi, L., Steiger, S., To, E., Sandmann, G., Dong, T.C., Ricca, E., Fraser, P.D., and Cutting, S.M. (2010). Carotenoids Found in Bacillus. J. Appl. Microbiol. *108*, 1889–1902.

Kim, J.Y., Gum, S.N., Paik, J.K., Lim, H.H., Kim, K.C., Ogasawara, K., Inoue, K., Park, S., Jang, Y., and Lee, J.H. (2008). Effects of nattokinase on blood pressure: a randomized, controlled trial. Hypertens. Res. *31*, 1583–1588.

Kimmel, M., Keller, D., Farmer, S., and Warrino, D.E. (2010). A controlled clinical trial to evaluate the effect of GanedenBC[(30).] on immunological markers. Methods Find. Exp. Clin. Pharmacol. *32*, 129–132.

Klobutcher, L.A., Ragkousi, K., and Setlow, P. (2006). The *Bacillus subtilis* spore coat provides 'eat resistance' during phagocytic predation by the protozoan *Tetrahymena thermophila*. Proc. Natl. Acad. Sci. U.S.A. *103*, 165–170.

La Ragione, R.M., Casula, G., Cutting, S.M., and Woodward, M.J. (2001). *Bacillus subtilis* spores competitively exclude *Escherichia coli* O78:K80 in poultry. Vet. Microbiol. 79, 113–142.

La Ragione, R.M., and Woodward, M.J. (2003). Competitive exclusion by *Bacillus subtilis* spores of *Salmonella enterica* serotype Enteritidis and *Clostridium perfringens* in young chickens. Vet. Microbiol. 94, 245–256.

Larsen, N., Thorsen, L., Kpikpi, E.N., Stuer-Lauridsen, B., Mette Dines Cantor M.D., Bea Nielsen B., Brockmann, E., Derkx, P.M.F., and Jespersen, L. (2013). Characterization of *Bacillus* spp. strains for use as probiotic additives in pig feed. Appl. Microbiol. Biotechnol. *94*, 245–256.

Liu, C., Chiu, C., Wang, S., and Cheng, W. (2012). Dietary administration of the probiotic, *Bacillus subtilis* E20, enhances the growth, innate immune responses, and disease resistance of the grouper, Epinephelus coioides. Fish Shellfish Immunol. *33*, 699–706.

Liu, K.F., Chiu, C.S., Shiu, Y.L., Cheng, W., and Liu, C.H. (2010). Effects of the probiotic, Bacillus subtilis E20, on the survival, development, stress tolerance, and immune status of white shrimp, *Litopenaeus vannamei* larvae. Fish Shellfish Immunol. *28*, 837–844.

Lopez, D., Vlamakis, H., and Kolter, R. (2009). Generation of multiple cell types in *Bacillus subtilis*. FEMS Microbiol. Rev. 33, 152–163.

Lopez, D., Vlamakis, H., Losick, R., and Kolter, R. (2009a). Paracrine signaling in a bacterium. Genes Dev. 23, 1631–1638.

Mandel, D.R., Eichas, K., and Holmes, J. (2010). *Bacillus coagulans*: a viable adjunct therapy for relieving symptoms of rheumatoid arthritis according to a randomized, controlled trial. BMC Complement. Altern. Med. *10*, 1.

Manzo, N., Di Luccia, B., Isticato, R., D'Apuzzo, E., De Felice, M., and Ricca, E. (2013). Pigmentation and sporulation are alternative cell fates in *Bacillus pumilus* SF214. PLoS One *8(4)*, e62093.

Margulis, L., Jorgensen, J.Z., Dolan, S., Kolchinsky, R., Rainey, F.A., and Lo, S.C. (1998). The Arthromitus stage of *Bacillus cereus*: Intestinal symbionts of animals. Proc. Natl. Acad. Sci. U.S.A. 95, 1236–1241.

Martins, L.O., Soares, C.M., Pereira, M.M., Teixeira, M., Costa, T., Jones, G.H., and Henriques, A.O. (2002). Molecular and biochemical characterization of a highly stable bacterial laccase that occurs as a structural component of the *Bacillus subtilis* endospore coat. J. Biol. Chem. 277, 18849–18859.

McKenney, P.T., Driks, A., and Eichenberger, P. (2013). The *Bacillus subtilis* endospore: assembly and functions of the multilayered coat. Nat. Rev. Microbiol. *11*, 33–44.

Mahowald, M.A., Rey, F.E., Seedorf, H., Turnbaugh, P.J., Fulton, R.S., *et al.* (2009). Characterizing a model human gut microbiota composed of members of its two dominant bacterial phyla. Proc. Natl. Acad. Sci. U.S.A. *106*, 5859–5864.

Mazza, P. (1994). The use of *Bacillus subtilis* as an antidiarrhoeal microorganism. Boll. Chim. Farm. *133*, 3–18.

Mohapatra, S., Chakraborty, T., Kumar, V., Deboeck, G., and Mohanta, K.N. (2012). Aquaculture and stress management: a review of probiotic intervention. J. Anim. Physiol. Anim. Nutr., 1–26.

Nicholson, W.L. (2002). Roles of *Bacillus* endospores in the environment. Cell. Mol. Life Sci. *59*, 410–416.

Nicholson, W.L. (2004). Ubiquity, longevity, and ecological roles of *Bacillus* spores. In Bacterial Spore Formers, E. Ricca, A.O. Henriques, and S.M. Cutting, eds. (Norfolk, UK:Horizon Biosience), pp. 1–15.

Nista, E.C., Candelli, M., Cremonini, F., Cazzato, I.A., Zocco, M.A., Franceschi, F., Cammarota, G., Gasbarrini, G., and Gasbarrini, A. (2004). *Bacillus clausii* therapy to reduce side-effects of anti-*Helicobacter pylori* treatment: randomized, double-blind, placebo controlled trial. Aliment. Pharmacol. Ther. *20*, 1181–1188.

Novak, K.N., Davis, E., Wehnes, C.A., Shields, D.R., Coalson, J.A., Smith, A.H., and Rehberger, T.G. (2012). Effect of supplementation with an electrolyte containing a Bacillus-based direct-fed microbial on immune development in dairy calves. Res. Vet. Sci. *92*, 427–434.

Onyenwoke, R.U., Brill, J.A., Farahi, K., and Wiegel, J. (2004). Sporulation genes in members of the low G+C Gram-typepositive phylogenetic branch (Firmicutes). Arch. Microbiol. *182*, 182–192.

Paidhungat, M., Setlow, B., Driks, A., and Setlow, P. (2000). Characterization of spores of *Bacillus subtilis* which lack dipicolinic acid. J. Bacteriol. *182*, 5505–5512.

Paidhungat, M., Ragkousi, K., and Setlow, P. (2001). Genetic requirements for induction of germination of spores of *Bacillus subtilis* by Ca(2+)-dipicolinate. J. Bacteriol. *183*, 4886–4893.

Paredes, C.J., Alsaker, K.V., and Papoutsakis, E.T. (2005). A comparative genomic view of clostridial sporulation and physiology. Nat. Rev. Microbiol. *3*, 969–978.

Pinchuk, I.V., Bressollier, P., Verneuil, B., Fenet, B., Sorokulova, I.B., Mégraud, F., and Urdaci, M.C. (2001). *In vitro* anti-*Helicobacter pylori* activity of the probiotic strain *Bacillus subtilis* 3 is due to secretion of antibiotics. Antimicrob. Agents Chemother. *45*, 3156–3161.

Popham, D.L. (2002). Specialized peptidoglycan of the bacterial endospore: the inner wall of the lockbox. Cell. Mol. Life Sci. *59(3)*, 426–33.

Popham, D.L., Helin, J., Costello, C.E., and Setlow, P. (1996). Muramic lactam in peptidoglycan of *Bacillus subtilis* spores is required for spore outgrowth but not for spore dehydration or heat resistance. Proc. Natl. Acad. Sci. U.S.A. *93*, 15405–15410.

Ragkousi, K., Eichenberger, P., van Ooij, C., and Setlow, P. (2003). Identification of a new gene essential for germination of *Bacillus subtilis* spores with Ca^{2+}-dipicolinate. J. Bacteriol. *185*, 2315–2329.

Rhee, K.-J., Sethupathi, P., Driks, A., Lanning, D.K., and Knight, K.L. (2004). Role of commensal bacteria in development of gut-associated lymphoid tissue and preimmune antibody repertoire. J. Immunol. *172*, 1118–1124.

Roggero, P., Volpe, C., Ceccatelli, M.P., Lambri, A., Giuliani, M.G., Donattini, T., Garavaglia, M.C., De Vincentiis, A. (1990). Crystalline lactulose and oral preparations of micro-organisms for the treatment of chronic aspecific diarrhoea in children. A controlled clinical study. Minerva Pediatr. *42*, 147–150.

Romero, D., Aguilar, C., Losick, R., and Kolter, R. (2010). Amyloid fibres provide structural integrity to *Bacillus subtilis* biofilms. Proc. Natl. Acad. Sci. U.S.A. *107*, 2230–2234.

Šabatková, j., Kumprecht, I., Zoba, P., Suchý, P., and Čermák, B. (2008). The probiotic BioPlus 2B as an alternative to antibiotics in diets for broiler chickens. Acta Vet. Brno 77, 569–574.

Setlow, P. (2005). Mechanisms for the prevention of damage to DNA in spores of *Bacillus* species. Ann. Rev. Microbiol. *49*, 29–54.

Scharek, L., Altherr, B.J., Tölke, C., and Schmidt, M.F.G. (2007). Influence of the probiotic Bacillus cereus var. toyoi on the intestinal immunity of piglets. Vet. Immunol. Immunopathol. *120*, 136–147.

Schauer, D.B., and Falkow, S. (1993). The *eae* gene of *Citrobacter freundii* biotype 4280 is necessary for colonization in transmissible murine colonic hyperplasia. Infect. Immun. *61*, 4654–4661.

Senesi, S., Celandroni, F., Tavanti, A., and Ghelardi, E. (2001). Molecular characterization and identification of *Bacillus clausii* strains marketed for use in oral bacteriotherapy. Appl. Environ. Microbiol. *67*, 834–839.

Senesi, S., and Ghelardi, E. (2010). Production, secretion and biological activity of *Bacillus cereus* enterotoxins. Toxins *2*, 1690–1703.

Slieman, T.A., and Nicholson, W.L. (2001). Role of dipicolinic acid in survival of *Bacillus subtilis* spores exposed to artificial and solar UV radiation. Appl. Environ. Microbiol. *67*,1274–1279.

Spinosa, M.R., Braccini, T., Ricca, E., De Felice, M., Morelli, L., Pozzi, G., and Oggioni, M.R. (2000). On the fate of ingested *Bacillus* spores. Res. Microbiol. *151*, 361–368.

Sudha, M.R., Bhonagiri, S., and Kumar, M.A. (2013). Efficacy of *Bacillus clausii* strain UBBC-07 in the treatment of patients suffering from acute diarrhoea. Benef. Microbes. *4*, 211–216.

Sumi, H., Hamada, H., Tsushima, H., Mihara, H., and Muraki, H. (1987). A novel fibrinolytic enzyme (nattokinase) in the vegetable cheese Natto; a typical and popular soybean food in the Japanese diet. Experientia *43*, 1110–1111.

Sun, P., Wang, J.Q., and Zhang, H.T. (2010). Effects of *Bacillus subtilis* natto on performance and immune function of preweaning calves. J. Dairy Sci. *93*, 5851–5855.

Stragier, P. (2002). A gene odyssey: exploring the genomes of endospore-forming bacteria. In *Bacillus subtilis* and Its Closest Relatives: From Genes to Cells. Sonenshein, A.L., Hoch, J.A., and R. Losick, eds. (Washington, DC, USA: ASM Press), pp. 519–526.

Tam, N.K., Uyen, N.Q., Hong, H.A., Duc, L.H., Hoa, T.T., Serra, C.R., Henriques, A.O., and Cutting, S.M. (2006). The intestinal life cycle of *Bacillus subtilis* and close relatives. J. Bacterol. *188*, 2692–2700.

Thirabunyanon, M., and Thongwittaya, N. (2012). Protection activity of a novel probiotic strain of Bacillus subtilis against Salmonella Enteriditid infection. Res. Vet. Sci. *93*, 74–81.

Tompkins, T.A., Xu, X., and Ahmarani, J. (2010). A comprehensive review of post-market clinical studies performed in adults with an Asian probiotic formulation. Benef. Microbes. *1*, 93–106.

Veening, J.W., Smits, W.K., and Kuipers, O.P. (2008). Bistability, epigenetics, and bethedging in bacteria. Ann. Rev. Microbiol. *62*, 193–210.

Vlamakis, H., Aguilar, C., Losick, R., and Kolter, R. (2008). Control of cell fate by the formation of an architecturally complex bacterial community. Genes Dev. *22*, 945–953.

Williams, L.D., Burdock, G.A., Jimenez, G., and Castillo, M. (2009). Literature review on the safety of Toyocerin, a non-toxigenic and nonpathogenic *Bacillus cereus* var. toyoi preparation. Toxicol. Pharmacol. *55*, 236–246.

Yatagai, C., Maruyama, M., Kawahara, T., and Sumi, H. (2008). Nattokinase-promoted tissue plasminogen activator release from human cells. Pathophysiol. Haemost. Thromb. *36*, 227–232.

Yutin, N., and Galperin, M.Y. (2013). A genomic update on clostridial phylogeny: Gram-negative spore formers and other misplaced clostridia. Environ. Microbiol. *15*, 2631–2641.

Zilhão, R., Isticato, R., Martins, L.O., Steil, L., Völker, U., Ricca, E., Moran Jr., C.P., and Henriques, A.O. (2005). Assembly and function of a transglutaminase at the surface of the developing spore in *Bacillus subtilis.* J. Bacteriol. *187*, 7753–7764.

Mechanisms of Action of Probiotic Yeasts

7

Flaviano dos Santos Martins and Jacques Robert Nicoli

Abstract

Some yeasts, such as *Saccharomyces boulardii* 17 and *Saccharomyces cerevisiae* UFMG 905, can be used as probiotics to prevent or treat various infectious and inflammatory diseases. Similar to bacterial probiotics, beneficial effects of these yeasts are the results of simultaneous action of various mechanisms such as modulation of some aspects of local and systemic immune responses, trapping of bacterial toxin or pathogenic bacterial cells on yeast surface, and maintenance of intestinal epithelium integrity. Acting together, these mechanisms seem to be responsible for a reduction of inflammatory process, intestinal permeability and bacterial translocation observed during infectious and inflammatory diseases.

Beneficial effects of probiotic yeasts as demonstrated by biological and clinical trials

Probiotics are defined as live microorganisms, which when administered in adequate amounts confer a health benefit to the host (FAO/WHO, 2002). These microorganisms are widely used in pharmaceutical preparations or fermented dairy products for treatment or prevention of gastrointestinal, vaginal, neurological or dermatological diseases. Lactobacilli and bifidobacteria are typically found in numerous probiotic products for humans, whereas only few types of yeast, such as *Saccharomyces boulardii*, are used.

The yeast *Saccharomyces cerevisiae* has a long history in preparations of breads and alcoholic beverages, but until a few decades ago, had not been investigated for knowledge of its probiotic properties (McFarland, 2010). In the 1920s, Henri Boulard, a French microbiologist, was in Southeast Asia looking for a yeast strain capable of withstanding high temperatures in order to produce a good wine. At that time there was an epidemic of cholera in Indochina peninsula and Boulard observed that local population ingested an infusion from the bark of a tropical fruit (lychee) to alleviate and even cure diarrhoea. Subsequently, it was found that the fruit was coated by a yeast, and effectiveness against diarrhoea was attributed to the presence of this microorganism, which was then isolated and named *Saccharomyces boulardii*. *Saccharomyces boulardii* is a non-pathogenic yeast and grows at the unusually high temperature of 37°C (MacFarland, 2010). It is often marketed as a probiotic in a lyophilized form and is therefore often referred to as *S. boulardii* lyo 17. *Saccharomyces boulardii* has been used for treatment of different types of diarrheal diseases such as antibiotic-associated diarrhoea (Surawicz, 2003), HIV/AIDS-associated diarrhoea and travellers' diarrhoea (MacFarland, 2010), acute infectious diarrhoea (Corrêa *et al.*, 2011) and *Clostridium difficile*-associated intestinal disease (Surawicz, 2003). More recently, its use has been evaluated for the treatment of inflammatory bowel diseases (IBD) (Thomas *et al.*, 2011).

Saccharomyces boulardii is the only yeast worldwide commercialized as probiotic for humans so far. However, some authors have suggested the use of other yeast species or genera based essentially on *in vitro* assays and some clinical trials (Kumura *et al.*, 2004; Van der Aa Kühle *et al.*, 2005; Tiago *et al.*, 2009; Pedersen *et al.*, 2012). *Saccharomyces boulardii* is genetically nearly identical to *S. cerevisiae*, but very different metabolically, and for this reason sometimes classified as *S. cerevisiae* var. *boulardii* (Fietto *et al.*, 2004). Previous results in our laboratory showed that *S. cerevisiae* strain 905, isolated during sugar cane spirit ('cachaça') production, was able to colonize and survive in the gastrointestinal tract of germ-free and conventional mice, and to protect these animals against oral challenge with *Salmonella enterica* serovar Typhimurium or *Clostridium difficile* (Martins *et al.*, 2005).

Many mechanisms of action have been proposed to explain how bacterial and yeast probiotics act, such as production of antagonistic compounds (organic acids, bacteriocins, H_2S), competition for nutrients, inhibition of pathogen adhesion to epithelium (competition for receptor and spatial exclusion, trapping by adhesion on probiotic surface), inhibition of toxin action (trapping on probiotic surface, proteolytic degradation of toxin or its receptor) and modulation of the immune system (stimulation of IgA production or blood clearance, anti-inflammatory capacity during infection or inflammatory pathologies), interference on bacterial-induced signalling pathways, and actions on bacterial virulence factors (Vieira *et al.*, 2013). Following are some results that demonstrate which of these mechanisms may act in yeasts to explain their beneficial effects observed in biological and clinical assays, and to justify their potential use as probiotics.

Mechanisms of action responsible for beneficial effects of probiotic yeasts

Antagonism

Initially, *S. boulardii* has been shown to have antagonistic activity *in vitro* against various bacterial pathogens (Brugier and Patte, 1975) and *in vivo* against *Candida* species (Ducluzeau and Bensaada, 1982). *Saccharomyces boulardii* 17 not only diminishes *Candida* colonization (Jawhara and Poulain, 2007), but also antagonize *Candida albicans* filamentous growth, adhesion and biofilm formation (Krasowska *et al.*, 2009) through the secretion of capric acid (Murzyn *et al.*, 2010).

However, the *in vitro* results for bacterial pathogens have not been confirmed when the capacity of this yeast to antagonize *S.* Typhimurium and *Shigella flexneri* was evaluated in the intestinal tract of conventional or gnotobiotic mice. Conventional animals were given daily 10 mg doses of *S. boulardii* 17, whereas germ-free animals were given a single 10 mg dose. Both groups were challenged orally 5 days later with the pathogenic bacteria (10^8 or 10^2 viable cells, respectively). Control groups were treated with saline instead of *S. boulardii* 17. Mortality, weight gain and histopathological data showed a protective effect against the pathogenic bacteria in yeast-treated mice (Fig. 7.1). *Saccharomyces boulardii* 17 colonized the digestive tract of gnotobiotic mice and the number of viable cells ranged around 10^{10}/g of faeces. In experimental and control gnotobiotic animals, *S.* Typhimurium and *S. flexneri* became rapidly established at a level of about 10^{10} viable cells/g of faeces and remained at high levels until the animals died or were sacrificed. These results showed that the protection against *S.* Typhimurium and *S. flexneri* obtained in conventional and gnotobiotic mice previously associated with *S. boulardii* 17 is not due to the reduction of the bacterial populations in the intestines through the action of some antagonistic compound (Rodrigues *et al.*, 1996; Martins *et al.*, 2013). These results confirm that production of inhibitory compounds against pathogenic bacteria is typical and more frequently found in lactobacilli and bifidobacteria, and rarely found in yeast (Servin, 2004).

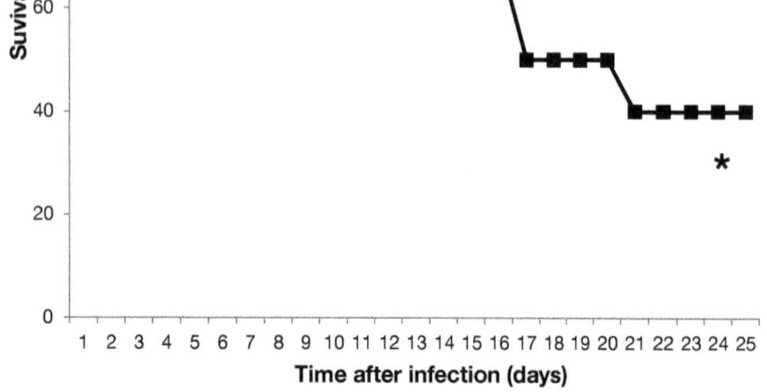

Figure 7.1 Survival of mice treated or not with *Saccharomyces boulardii* 17 before *Salmonella* Typhimurium challenge. Survival was evaluated in control (non-treated and non-challenged) mice (♦), *S.* Typhimurium-challenged mice (■) and mice previously treated with *S. boulardii* 17 and then challenged (▲). *n* = 10 in each group. *Indicates statistically significant difference between experimental groups ($P < 0.05$).

Bacterial motility

Motility is an important component of enteropathogenic bacteria pathogenesis allowing the microorganism to move into appropriate niches, across the mucus layer and invade the intestinal epithelium. A recent study evaluated the possible influence of *S. boulardii* 17 on *S.* Typhimurium motility. Experiments were performed on human colonic T84 cells infected by *S.* Typhimurium alone or in the presence of the yeast, and the *Salmonella* motility was recorded by time-lapse video microscopy and bacteria dynamics analysed by manual tracking. The median curvilinear velocity (CLV) of *Salmonella* incubated alone with T84 decreased from 43.3 μm/s to 31.2 μm/s in the presence of *S. boulardii* 17. Measurement of track linearity showed that *S. boulardii* 17 decreased by 15% the number of bacteria with linear tract and increased by 22% the number of bacteria with rotator tract. Correlation between *S.* Typhimurium motility and invasion was further established by studying a non-motile flagella-deficient *S.* Typhimurium strain, which moved with a significant lower CLV of 0.5 μm/sec and a decrease in invasion properties. This protective mechanism of *S. boulardii* 17 modified the motility of *S.* Typhimurium and significantly decreased invasion of T84 cells by this strain (Pontier-Bres *et al.*, 2012).

Local and systemic immunomodulation (IgA, anti-inflammatory cytokine, clearance)

The first line of defence against the entry of pathogens is represented by the gut epithelial barrier and probiotics may prevent pathogen-induced membrane damage by inhibiting pathogen adhesion and maintaining the correct organization of the tight junction and cytoskeleton proteins. IgA is an antibody playing a critical role in mucosal immunity, acting as a barrier against the adhesion of pathogens to the intestinal mucosa. This immunoglobulin can bind either to the surface of bacteria (trapping them in the mucous layer of the intestinal epithelia) or to bacterial surface molecules that mediate adhesion to epithelial cells (preventing translocation). The ability to stimulate the production of sIgA is not the same for different micro-organisms, and in gnotobiotic mice Gram-negative bacteria and yeast generally induce a higher stimulation than Gram-positive bacteria (Rodrigues *et al.*, 2000; Martins *et al.*, 2007). In a recent study (Martins *et al.*, 2009), the effect of *S. boulardii* 17 on the immune system has been compared with some bacterial probiotics (*Lactobacillus casei*, *Escherichia coli*

EM0, *Bifidobacterium animalis* BB_{12}) using germ-free Swiss/NIH mice monoassociated with each probiotic (Fig. 7.2). An increase in total sIgA production was observed in the intestinal contents of monoassociated mice with all the probiotics when compared with germ-free controls, excepted for *L. casei*. The higher increase in this immunoglobulin was noted for mice associated with the yeast probiotic followed by the *E. coli*-monoassociated mice. Similarly, higher levels ($P < 0.05$) of secretory IgA in intestinal content were observed in *S. cerevisiae* UFMG 905 mono-associated mice when compared to germ-free group (Martins *et al.*, 2007).

Recent studies have shown that probiotics may ameliorate experimental colitis in mice by inducing IL-10 and IL-10-dependent T regulatory cells (Mengheri, 2008). IL-10 is an anti-inflammatory cytokine that is produced primarily by monocytes and, to a lesser extent, by lymphocytes. It has pleiotropic effects in immune-regulation and inflammation. It down-regulates the expression of Th1 cytokines, MHC class II antigens, and co-stimulatory molecules on macrophages. It also enhances B-cell survival, proliferation and antibody production. Studies using knockout mice suggested the function of this cytokine as an essential immune-regulator in the intestinal tract (Grimbaldeston *et al.*, 2007). It is capable of inhibiting synthesis of pro-inflammatory cytokines like IFN-g, IL-2, IL-3, TNF and GM-CSF made by cells such as macrophages and the Th1 cells and also displays potent abilities to suppress the antigen presentation capacity of antigen presenting cells. IL-10 levels were higher in an experimental intestinal obstructed (IO) group of mice when compared to control (CTL) group (50.75 and 29.58 pg/ml serum, respectively) ($P < 0.05$). Treatment with both viable and heat-killed *S. cerevisiae* UFMG 905 increased significantly the levels of IL-10 (102 and 104 pg/ml serum, respectively) when compared with IO group (50.75 pg/ml serum) ($P < 0.05$) (Generoso *et al.*, 2010). In the comparative study cited above, IL-10 levels in intestinal tissue were higher ($P < 0.05$) only for *S. boulardii*-associated mice (38.4 vs. 137.3 pg/ml in germ-free and *S. boulardii*-associated mice, respectively) (Martins *et al.*, 2009) (Fig. 7.3).

In the case of successful translocation, the reticuloendothelial system (RES) plays a key role in removing micro-organisms from the circulatory system, and hepatic phagocytosis by Küpffer cells accounts for more than 80% of the RES function. Glucan and other *Saccharomyces* extracts are important RES stimulators. Proliferation of pluripotent haematopoietic stem cells, granulocyte–macrophage colony-forming cells and erythroid colony-forming cells in murine bone marrow has been demonstrated after glucan administration. Furthermore, after glucan administration, there is an increased production of IL-1 either by a direct effect on macrophages or indirectly by an increase on colony stimulating factor(s) production from glucan stimulated T-cells. It is well known that Küpffer cells represent one of the largest populations of tissue macrophages and, together with neutrophils, have been implicated as responsible for clearance of bacteria from the bloodstream (Gregory and Wing, 2002). The number of Küpffer cells per 100 hepatocytes in liver was

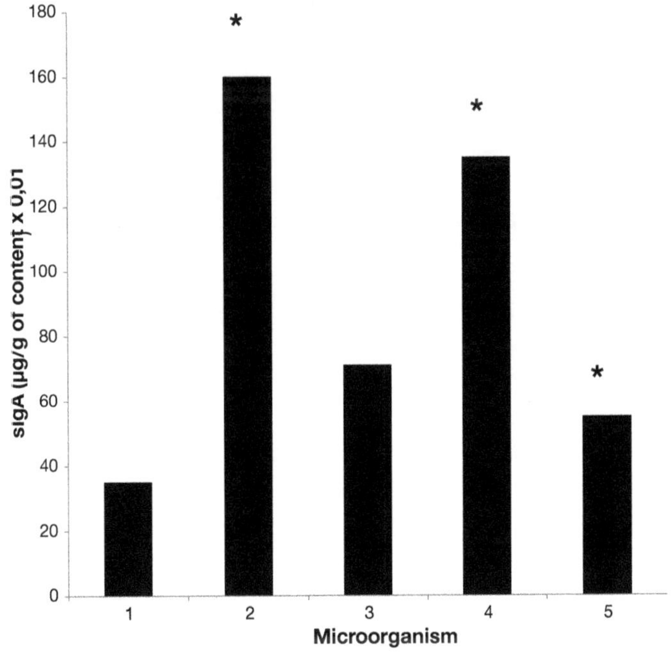

Figure 7.2 Levels of intestinal sIgA of control (1 – germ-free mice), *Saccharomyces boulardii* 17 (2), *Lactobacillus casei* (3), *Escherichia coli* EM0 (4) and *Bifidobacterium animalis* BB_{12} (5) monoassociated mice. The results are expressed as means of immunoglobulin concentration (µg/g of intestinal contents). n=5 in each group. *Indicates statistically significant difference between control and experimental groups ($P < 0.05$).

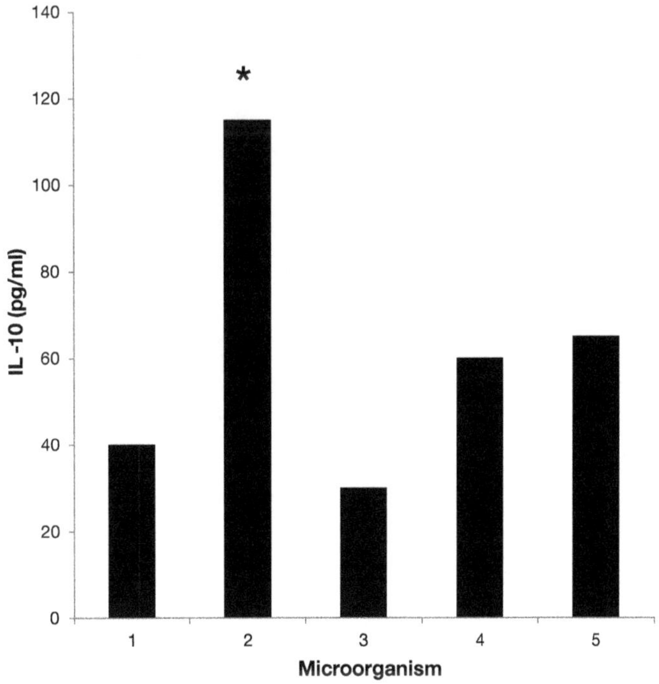

Figure 7.3 Levels of cytokine IL-10 of control (1 – germ-free mice), *Saccharomyces boulardii* 17 (2), *Lactobacillus casei* (3), *Escherichia coli* EM0 (4) and *Bifidobacterium animalis* BB_{12} (5) monoassociated gnotobiotic mice. The results are expressed as means of IL-10 concentration (pg/ml). $n=5$ in each group. *Indicates statistically significant difference between control and experimental groups ($P < 0.05$).

significantly higher in yeast mono-associated mice (42.4 ± 4.4 and 52.9 ± 15.7 with *S. boulardii* 17 and *S. cerevisiae* UFMG 905, respectively) than in germ-free controls (range from 18.2 ± 4.1 to 38.1 ± 9.0). Probably as a consequence, clearance of *E. coli* B_{41} from the bloodstream was more efficient in both *S. boulardii* or *S. cerevisiae* UFMG 905 mono-associated animals when compared to germ-free mice (Rodrigues *et al.*, 2000; Martins *et al.*, 2007) (Fig. 7.4).

Inflammation and permeability of intestinal epithelium

Various studies show that *S. boulardii* 17 and *S. cerevisiae* UFMG 905 modulate the inflammatory transduction pathway induced by enteropathogenic and enterohaemorrhagic *Escherichia coli*

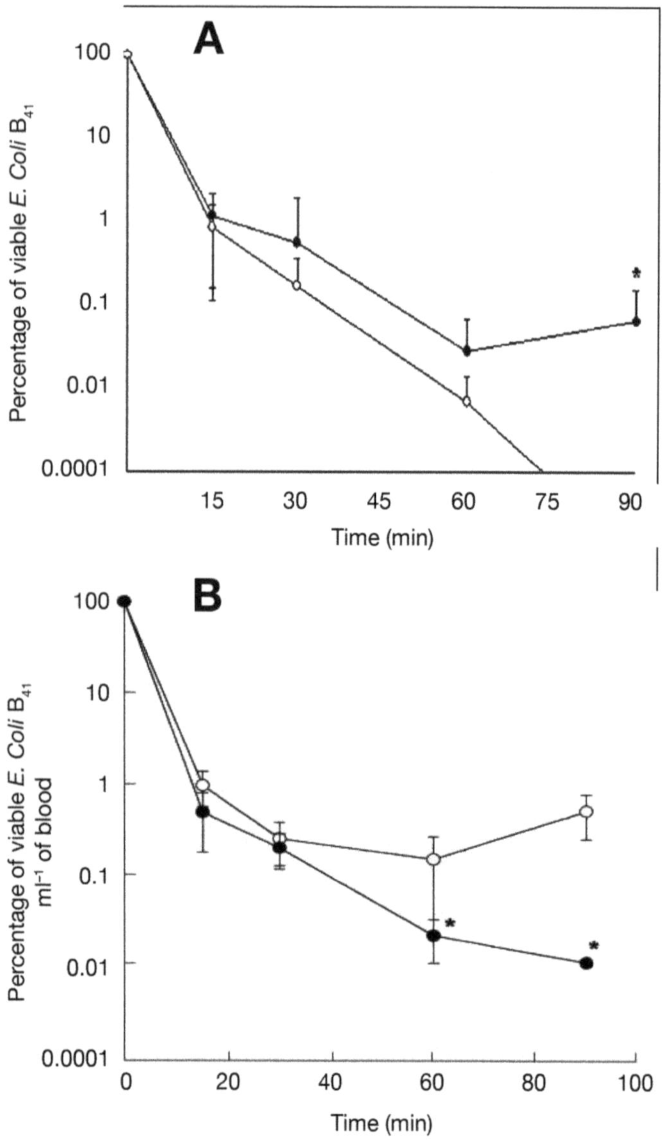

Figure 7.4 Blood clearance of *Escherichia coli* B41 in germ-free mice and mice monoassociated with probiotic yeast. (A) *Saccharomyces cerevisiae* UFMG 905 (○) and germ-free mice (•). (B) *Saccharomyces boulardii* 17 (•) and germ-free mice (○). *Indicates statistically significant difference between control and experimental groups ($P < 0.05$).

(Czerucka *et al.*, 2000; Dahan *et al.*, 2003), *S.* Typhimurium (Martins *et al.*, 2010; 2011; 2013), *C. difficile* (Chen *et al.*, 2006), and *Shigella* (Mumy *et al.*, 2008). Beyond this protective effect against inflammation induced by pathogenic bacteria, *S. boulardii* also presents this anti-inflammatory capacity in inflammatory bowel diseases (Pothoulakis, 2009).

At a molecular level, using T84 human colorectal cancer cells, it was demonstrated that incubation with *S. boulardii* before infection totally abolished *Salmonella* invasion. This correlates with a decrease of activation of Rac1. *Saccharomyces boulardii* 17 preserved T84 barrier function and decreased *S.* Typhimurium-induced IL-8 synthesis. This anti-inflammatory effect was correlated with an inhibitory effect of *S. boulardii* 17 on *S.* Typhimurium-induced activation of the MAPKs ERK1/2, p38 and JNK as well as on activation of NF-κB. Electron and confocal microscopy experiments showed an adhesion of bacteria to yeast cells, which could represent one of the mechanisms by which *S. boulardii* 17 exerts its protective effects. Another explanation could be the production by the yeast of a compound interfering on the bacterial-stimulated cellular signalling pathways. Supernatant of *S. boulardii* 17 growth also exhibits the anti-inflammatory effect suggesting that soluble factor(s) produced by the yeast is (are) implicated. A supernatant fraction containing a small (1 kDa) heat-stable and water soluble anti-inflammatory molecule inhibited NF-κB activation by LPS, IL-1β and TNF-α (Sougioultzis *et al.*, 2006). In the study of Martins *et al.* (2010), the inhibitory effect disappeared after heat treatment of *S. boulardii* 17 but was present in the supernatant containing yeast cell wall. This modulation of inflammation and signal transduction pathway in T84 cells infected by *S.* Typhimurium was confirmed by an *in vivo* protective effect against *S.* Typhimurium infection (Martins *et al.*, 2010). In the case of *C. difficile*, *S. boulardii* 17 inhibits *C. difficile* toxin A-associated enteritis by blocking the activation of Erk1/2 MAP kinases (Chen *et al.*, 2006).

By working with *S. flexneri*, Mumy *et al.* (2008) demonstrated that *S. boulardii* 17 positively impacted the tight junction protein zonula occluden-2 and significantly increased the barrier integrity, also decreasing ERK, JNK, and NF-κB activation in response to *S. flexneri*, events responsible for the observed reduction in IL-8 secretion. Furthermore, a cell-free *S. boulardii* 17 culture supernatant was also capable of reducing IL-8 secretion, suggesting again the secretion of anti-inflammatory compounds by the yeast.

The effects of oral treatment with viable or heat-killed cells of *S. boulardii* 17 and *S. cerevisiae* UFMG 905 on intestinal permeability and histological aspect of the ileum has been evaluated in a murine intestinal obstruction (IO) model described above (Generoso *et al.*, 2010; 2011). Intestinal permeability is considered increased when permeation of molecules > 150 Daltons is observed. For this reason intestinal permeability can be evaluated by measuring blood radioactivity after the intake of ^{99m}Tc-diethylenetriaminopentaacetate (^{99m}Tc-DTPA). This molecule, used as a disodium complex, can be considered adequate for intestinal permeability determination due to its following characteristics: molecular weight of 549 Daltons, water soluble, non-charged, not destroyed in the gut, non-toxic, not metabolized or sequestered, quantitatively cleared by the kidneys, and easily detectable. Results showed that experimental IO in mice induced an increase in intestinal permeability as observed in Fig. 7.5. As demonstrated in the same figure, oral treatment with viable *S. boulardii* 17 or *S. cerevisiae* UFMG 905 maintained the intestinal permeability at physiological levels similar to those observed in the Sham group throughout all of the follow-up periods. Histological analysis showed preservation of the ileum mucosa in mice that received both forms of the yeast when compared to the lesions observed in the IO group (Generoso *et al.*, 2010; 2011). Increase in intestinal permeability was also observed in *in vitro* infectious models, as demonstrated by transepithelial electrical resistance (TER) measure in T84 cell culture infected with enteropathogenic *E. coli* (Czerucka *et al.*, 2000), enterohaemorrhagic *E. coli* (Dahan *et al.*, 2003), *S. flexneri* (Mumy *et al.*, 2008), or with *S.* Typhimurium (Martins *et al.*, 2010), and in all these studies treatment with *S. boulardii* 17 prevents the increase. In a recent clinical trial, Vilela *et al.* (2008) observed important alterations in the intestinal mucosa integrity in patients with Crohn's disease in remission and demonstrated that an oral treatment with *S. boulardii* 17 improved intestinal permeability. The mechanisms that could explain the maintenance of epithelial integrity by the yeasts are not well known. However, the ability of *S. boulardii* 17 and *S. cerevisiae* UFMG 905 to modulate the transduction pathways implicated in the control of tight-junction structure described above could be an explanation (Dahan *et al.*, 2003).

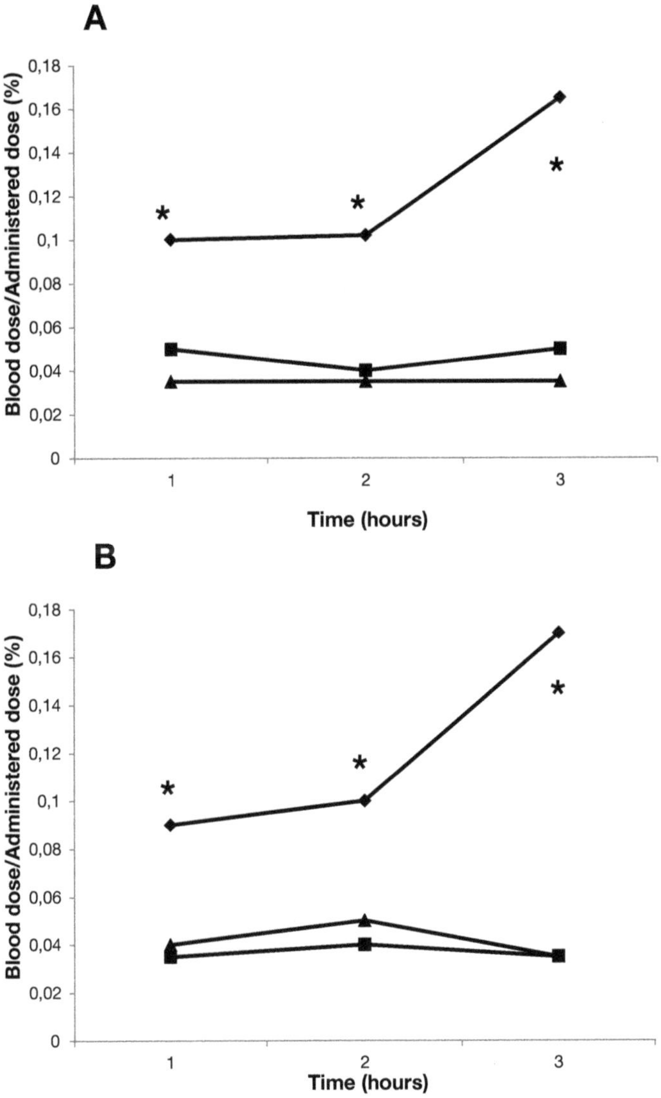

Figure 7.5 Evolution of intestinal permeability as determined by calculating the percentage of radioactivity measured in blood in relation to the oral dose administrated of ^{99m}Tc-DTPA (%) in mice treated or not with *Saccharomyces cerevisiae* UFMG 905 (A) or *Saccharomyces boulardii* 17 (B) and submitted to an intestinal obstruction (IO). Sham (▲), IO (♦), IO + yeast treatment (■) groups. Time: 4 (1), 8 (2) and 18 (3) hours after IO. *Indicates statistically significant difference between control and experimental groups ($P < 0.05$). $n = 5$ in each group.

Inhibition of toxin action

Saccharomyces boulardii produces protease which degrades *C. difficile* toxins A and B and their respective receptors on colonic mucosa (Corthier *et al.*, 1986; Pothoulakis *et al.*, 1993; Castagliuolo *et al.*, 1999; Qamar *et al.*, 2001), and has an 'anti-toxin' action expressed by inhibition of the intestinal water/sodium hypersecretion and of mucosal lesions induced by inoculation of cholera (Czerucka *et al.*, 1989; Czerucka *et al.*, 1994; Dias *et al.*, 1995) and *Escherichia coli* toxins (Massot *et al.*, 1982).

Another protective mechanism against the cholera toxin (CT) could be its trapping by specific receptors onto yeast surface (Brandão *et al.*, 1998). The addition of glucose or dinitrophenol to cells of *S. boulardii* grown on a non-fermentable carbon source activated trehalase in a manner similar to that observed for *S. cerevisiae*. The addition of CT to the same cells also resulted in trehalase activation. Experiments performed separately on the A and B subunits of CT showed that both are necessary for activation. Similarly, the addition of CT but not of its separate subunits led to a cyclic AMP (cAMP) signal in *S. boulardii*. These data suggest that trehalase stimulation by CT probably occurred through the cAMP-mediated protein phosphorylation cascade. The requirement of CT subunit B for both the cAMP signal and trehalase activation indicates the presence of a specific receptor on the yeasts able to bind to the toxin, a situation similar to that observed for mammalian cells. This hypothesis was reinforced by experiments with ^{125}I-labelled CT showing specific binding of the toxin to yeast cells. The adhesion of CT to a receptor on the yeast surface through the B subunit and internalization of the A subunit (necessary for the cAMP signal and trehalase activation) could be one more mechanism explaining protection against the toxin observed for animals treated with yeasts (Brandão *et al.*, 1998).

Trapping onto yeast cell surface

Some studies suggest that adhesion of pathogenic bacteria onto the surface of probiotic yeasts instead of onto intestinal receptors could explain part of a probiotic effect (Gedek, 1999; Pérez-Sotelo *et al.*, 2005; Tiago *et al.*, 2012). The last study evaluated the adhesion of pathogenic bacteria onto the cell wall of *S. boulardii* 17 and *S. cerevisiae* strains UFMG 905, W303 and BY4741. To understand the mechanism of adhesion of pathogens to yeast, cell-wall mutants of the parental strain of *S. cerevisiae* BY4741 were also used because of the difficulty of mutating polyploid yeast, as is the case for *S. cerevisiae* UFMG 905 and *S. boulardii* 17. The tests of adhesion showed that, among 11 enteropathogenic bacteria tested, only *E. coli*, *S.* Typhimurium and *Salmonella* Typhi adhered to the surface of *S. boulardii* 17, *S. cerevisiae* UFMG 905 and *S. cerevisiae* BY4741. The presence of mannose, and to some extent bile salts, inhibited this adhesion, which was not dependent on yeast viability. Among 44 cell-wall mutants of *S. cerevisiae* BY4741, five lost the ability to fix the bacteria. Electron microscopy showed that the phenomenon of yeast–bacteria adhesion occurred both *in vitro and in vivo* (in the digestive tract of dixenic mice) (Fig. 7.6) (Tiago *et al.*, 2012). The localization and mechanisms of binding to the epithelial surface have been studied previously in enteropathogenic bacteria such as *E. coli* and *Salmonella*, and type I fimbriae are frequently involved in this adhesion, generally using mannose as a receptor (Kline *et al.*, 2009). The cell walls of yeasts are known to be rich in mannose, and we have hypothesized that probiotic yeasts may exert their protective effect by the trapping of pathogenic bacteria on their surface (Tiago *et al.*, 2012).

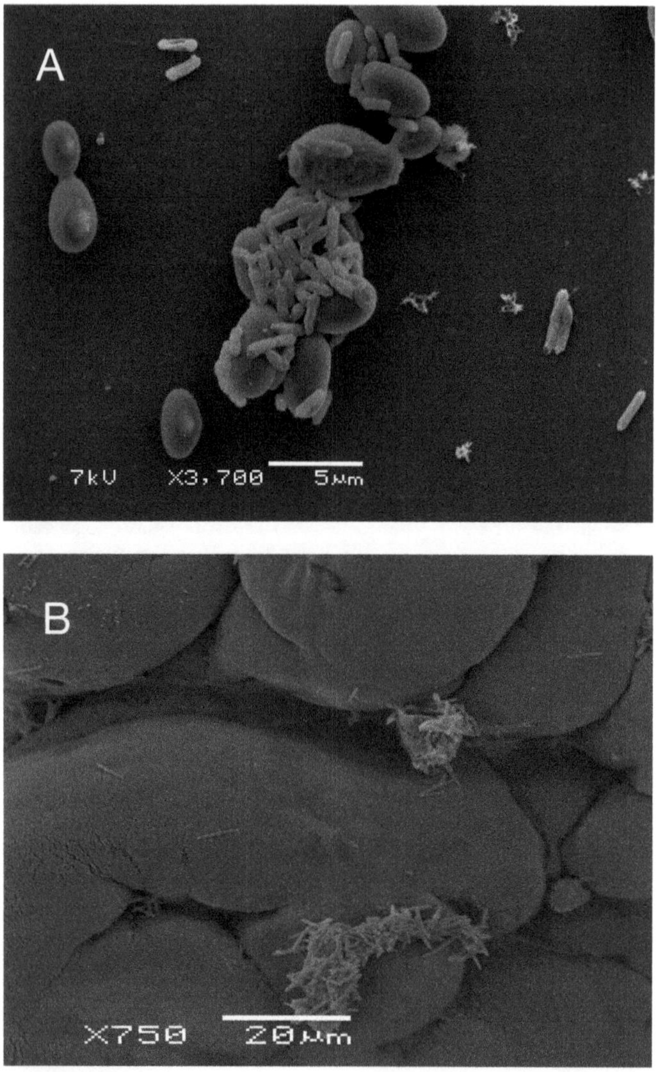

Figure 7.6 Scanning electron micrograph of *in vitro* (A) and *in vivo* (B – intestinal epithelium of di-associated gnotobiotic mice) adhesion between *Saccharomyces boulardii* 17 and *Salmonella* Typhimurium. The magnification is indicated in each panel.

Translocation

Various studies reported that both *S. boulardii* 17 and *S. cerevisiae* UFMG 905 prevented pathogenic bacterial translocation to the liver, mesenteric lymph nodes and spleen, as evaluated in conventional mice. This beneficial effect was probably the consequence of the diverse mechanisms described above acting together. The treatment with the yeast reduced significantly the translocation of *S.* Typhimurium to all the organs tested in conventional mice (Fig. 7.7) (Martins *et al.*, 2007; 2010). The reduced translocation of *S.* Typhimurium observed in mice treated with *S. cerevisiae* UFMG 905 could explain the lower number of hepatic inflammatory foci in these animals 28 days after the challenge with the enteropathogenic bacteria as observed in previous study (Martins *et al.*, 2005). As described by Berg (1995), translocation to liver could occur via intracellular (lymphatic route, via lymph nodes) or extracellular passage (vascular route, due to physical damage to the intestinal epithelium). Reduced levels of translocation have been also observed in various animal models (burn injury, liver injury, experimental immunosuppression, abdominal infection) using probiotic products based on bacteria (Chiva *et al.*, 2002; Gun *et al.*, 2005) and yeast (Peret *et al.*, 1998; Herek *et al.*, 2004). It seems that all the components of the host immune system – including mucosal, cell-mediated and humoral immunities are involved in controlling bacterial translocation. sIgA is thought to inhibit the close association of pathogenic bacteria with the mucosal epithelium, and so to reduce bacterial penetration. Cell-mediated immunity in the lamina propria and in lymphoid organs provides a second line of host defence against bacterial translocation. Serum immunoglobulins probably act as opsonins

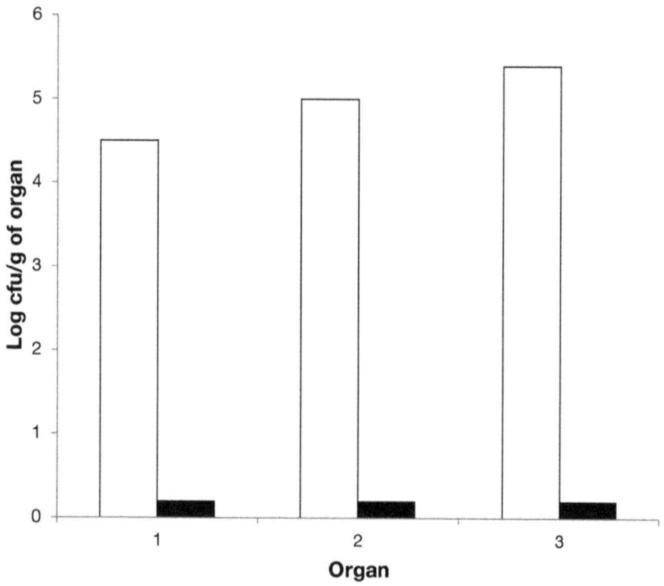

Figure 7.7 Translocation (log_{10} cfu/g of organ) of *Salmonella* Typhimurium in conventional mice previously treated (■) or not (□) with *Saccharomyces cerevisiae* UFMG 905 to liver (1), spleen (2) and mesenteric lymph nodes (3) after 10 days of challenge.

to increase the effectiveness of phagocytosis by macrophages and polymorphonuclear leucocytes, and subsequent killing of translocating bacteria.

Action on bacterial virulence factors

One study reported that *S. boulardii* 17 ameliorates bacteria-induced colitis through actions on bacterial virulence factors (Wu *et al.*, 2008). In mice infected by *Citrobacter rodentium*, *S. boulardii* 17 treatment significantly attenuated weight loss, ameliorated crypt hyperplasia and histological damage score, reduced myeloperoxidase activity, and attenuated increased mannitol flux. The ameliorating effects of *S. boulardii* were associated with significantly reduced numbers of mucosal adherent *C. rodentium*, a marked reduction in Tir (the receptor for bacterial intimin) protein secretion and translocation into mouse colonocytes, and a striking reduction in EspB (a translocation apparatus protein) expression and secretion. Based on the results, the authors conclude that *S. boulardii* 17 attenuated bacterial adherence to host epithelial cells through putative actions on the type III secretion system.

Conclusion

Together, these results show that some strains of yeast such as *S. boulardii* 17 and *S. cerevisiae* UFMG 905 can be used as probiotics to prevent or treat various infectious and inflammatory diseases. As bacterial probiotics, beneficial effects of these yeasts are the results of simultaneous action of various mechanisms such as modulation of some aspects of local and systemic host immune responses, trapping of bacterial toxin or pathogenic bacterial cells onto yeast surface, and improvement of intestinal epithelium integrity. Acting together, these mechanisms seem to be responsible for a reduction of inflammatory process, reduced intestinal permeability and bacterial translocation observed during infectious and inflammatory diseases.

Future perspectives

Yeasts of the species *S. cerevisiae* have several advantages for use as a probiotic in the prevention and treatment of gastrointestinal disturbances (infections and inflammation). Similar to lactic acid bacteria (*Lactobacillus*, *Streptococcus*) and bifidobacteria, they are rarely involved in pathological processes as evidenced by their use in the preparation of fermented foods for millennia. Furthermore, they can be administered simultaneously with a treatment with antibiotics to which they are insensitive (prevention of antibiotic associated diarrhoea). Like all other probiotics, they have more than one mechanism of action that may explain their beneficial effects on the host, which hampers the possible development of resistance to its use by pathogenic microorganisms. Some of these mechanisms appear to be specific to yeast, such as the ability to capture onto its surface toxins and/or cells of pathogenic bacteria by specific receptors. An interesting perspective would be to identify, characterize and clone the genes responsible for the synthesis of these receptors and subsequently try to super-express these products in genetically modified strains of yeast to obtain a probiotic of second generation with more efficient capture capacity. A similar process could be made with the anti-inflammatory compounds synthesized by yeast.

References

Berg, R.D. (1995). Bacterial translocation from the gastrointestinal tract. Trends Microbiol. 3, 149–154.

Brandão, R.L., Castro, I.M., Bambirra, E.A., Amaral, S.C., Fietto, L.O., Tropia, M.J.M., Neves, M.J., Santos, R.G., Gomes, N.C.M., and Nicoli, J.R. (1998). Intracellular signal triggered by cholera toxin in *Saccharomyces boulardii* and *Saccharomyces cerevisiae*. Appl. Environm. Microbiol. *64*, 564–568.

Brugier, S., and Patte, F. (1975). Antagonisme *in vitro* entre l'ultralevure et différent germes bactériens. Médecin Paris 45, 61–66.

Castagliuolo, I., Riegler, M.F., Valenick, L., Lamont, J.T., and Pothoulakis, C. (1999). *Saccharomyces boulardii* protease inhibits the effects of *Clostridium difficile* toxins A and B in human colonic mucosa. Infect. Immun. *67*, 302–307.

Chen, X., Kokkotou, E.G., Mustafa, N., Bhaskar, K.R., Sougioultzis, S., O'Brien, M., Pothoulakis, C., and Kelly, C.P. (2006). *Saccharomyces boulardii* inhibits ERK1/2 mitogen-activated protein kinase activation both *in vitro* and *in vivo* and protects against *Clostridium difficile* toxin A-induced enteritis. J. Biol. Chem. *281*, 24449–24459.

Chiva, M., Soriano, G., Rochat, I., Peralta, C., Rochat, F., Llovet, T., Mirelis, B., Schiffrin, E.J., Guarner, C., and Balanzó, J. (2002). Effect of *Lactobacillus johnsonii* La1 and antioxidants on intestinal flora and bacterial translocation in rats with experimental cirrhosis. J. Hepatol. *37*, 456–462.

Corrêa, N.B.O., Penna, F.J., Lima, F.M.L.S., Nicoli, J.R., and Péret Filho, L.A. (2011). Treatment of acute diarrhoea with *Saccharomyces boulardii* in infants: a double-blind, randomized, placebo controlled trial. J. Pediat. Gastroenterol. Nutr. *53*, 497–501.

Corthier, G., Dubos, P., and Ducluzeau, R. (1986). Prevention of *Clostridium difficile* induced mortality in gnotobiotic mice by *Saccharomyces boulardii*. Can. J. Microbiol. *32*, 894–896.

Czerucka, D., Nano, J.L., Bernasconi, P., and Rampal, P. (1989). Effect of *Saccharomyces boulardii* on cholera toxin-induced CAMP levels in rat epithelial cell lines. Gastroenterol. Clin. Biol. *3*, 383–387.

Czerucka, D., Roux, I., and Rampal, P. (1994). *Saccharomyces boulardii* inhibits secretagogue-mediated adenosine 39,59-cyclic monophosphate induction in intestinal cells. Gastroenterology *106*, 65–72.

Czerucka, D., Dahan, S., Mograbi, B., Rossi, B., and Rampal, P. (2000). *Saccharomyces boulardii* preserves the barrier function and modulates the transduction pathway induced in enteropathogenic *Escherichia coli*-infected T84 cells. Infect. Immun. *68*, 5998–6004.

Dahan, S., Dalmasso, G., Imbert, V., Peyron, J.-F., Rampal, P., and Czerucka, D. (2003). *Saccharomyces boulardii* interferes with enterohaemorrhagic *Escherichia coli*-induced signalling pathways in T84 cells. Infect. Immun. *71*, 766–773.

Dias, R.S., Bambirra, E.A., Silva, M.E., and Nicoli, J.R. (1995). Protective effect of *Saccharomyces boulardii* against the cholera toxin in rats. Braz. J. Biol. Med. Res. *28*, 323–325.

Ducluzeau, R., and Bensaada, M. (1982). Effet comparé de l'administration unique ou en continu de *Saccharomyces boulardii* sur l'établissement de diverses souches de *Candida* dans le tractus digestif de souris gnotoxéniques. Ann. Microbiol. *133B*, 491–501.

FAO/WHO (2002). Working Group, Guidelines for the Evaluation of Probiotics in Food. London, Ontario, Canada.

Fietto, J.L.R., Araújo, R.S., Valadão, F.N., Fietto, L.G., Brandão, R.L., Neves, M.J., Gomes, C.O., Nicoli, J.R., and Castro, I.M. (2004). Molecular and physiological comparisons between *Saccharomyces cerevisiae* and *Saccharomyces boulardii*. Can. J. Microbiol. *50*, 615–621.

Gedek, B.R. (1999). Adherence of *Escherichia coli* serogroup O157 and the *Salmonella typhimurium* mutant DT 104 to the surface of *Saccharomyces boulardii*, Mycoses 42, 261–264.

Generoso, S.V., Viana, M.V., Santos, R., Martins, F.S., Machado, J.A.N., Arantes, R.M.E., Nicoli, J.R., Correia, M.I.T.D., and Cardoso, V.N. (2010). *Saccharomyces cerevisiae* strain UFMG 905 protects against bacterial translocation, preserves gut barrier integrity and stimulates the immune system in a murine intestinal obstruction model. Arch. Microbiol. *192*, 477–484.

Generoso, S.V., Viana, M.V., Santos, R., Arantes, R.M.E., Martins, F.S., Nicoli, J.R., Machado, J.A.N., Correia, M.I.T.D., and Cardoso, V.N. (2011). Protection against increased intestinal permeability and bacterial translocation induced by intestinal obstruction in mice treated with viable and heat-killed *Saccharomyces boulardii*. Eur. J. Nutr. *50*, 261–269.

Gregory, S.H., and Wing, E.J. (2002). Neutrophil-Küpffer cell interaction: a critical component of host defences to systemic bacterial infections. J. Leukoc. Biol. *72*, 239–248.

Grimbaldeston, M., Nakae, S., Kalesnikov, J., Tsai, M., and Galli, S.J. (2007). Mast cell-derived interleukin 10 limits skin pathology in contact dermatitis and chronic irradiation with ultraviolet B. Nat. Immunol. *8*, 1095–1104.

Gun, F., Salman, T., Gurler, N., and Olgac, V. (2005). Effect of probiotic supplementation on bacterial translocation in thermal injury. Surg. Today *35*, 760–764.

Herek, O., Kara, I.G., and Kaleli, I. (2004). Effects of antibiotics and *Saccharomyces boulardii* on bacterial translocation in burn injury. Surg. Today *34*, 256–260.

Jawhara, S., and Poulain, D. (2007). *Saccharomyces boulardii* decreases inflammation and intestinal colonization by *Candida albicans* in a mouse model of chemically-induced colitis. Med. Mycol. *45*, 691–700.

Kline, K.A., Falker, S., Dahlberg, S., Normark, S., and Henriques-Normark, B. (2009). Bacterial adhesins in host–microbe interactions. Cell Host Microb. *5*, 580–592.

Krasowska, A., Murzyn, A., Dyjankiewicz, A., Łukaszewicz, M., and Dziadkowiec, D. (2009). The antagonistic effect of *Saccharomyces boulardii* on *Candida albicans* filamentation, adhesion and biofilm formation. FEMS Yeast Res. *9*, 1312–1321.

Kumura, H., Tanoue, Y., Tsukahara, M., Tanaka, T., and Shimazaki, K. (2004). Screening of dairy yeast strains for probiotic applications. J. Dairy Sci. *87*, 4050–4056.

MacFarland, L.V. (2010). Systematic review and meta-analysis of *Saccharomyces boulardii* in adult patients. World J. Gastroenterol. *16*, 2202–2222.

Martins, F.S., Nardi, R.M.D., Arantes, R.M.E., Rosa, C.A., Neves, M.J., and Nicoli, J.R. (2005). Screening of yeasts as probiotic based on capacities to colonize the gastrointestinal tract and to protect against enteropathogen challenge in mice. J. Gen. Appl. Microbiol. *51*, 83–92.

Martins, F.S., Rodrigues, A.C.P., Tiago, F.C.P., Penna, F.J., Rosa, C.A., Arantes, R.M.E., Nardi, R.M.D., Neves, M.J., and Nicoli, J.R. (2007). *Saccharomyces cerevisiae* strain 905 reduces the translocation of *Salmonella enterica* serotype Typhimurium and stimulates the immune system in gnotobiotic and conventional mice. J. Med. Microbiol. *56*, 352–359.

Martins, F.S., Silva, A., Vieira, A., Barbosa, F.H.F., Arantes, R.M.E., Teixeira, M.M., and Nicoli, J.R. (2009). Comparative study of *Bifidobacterium animalis, Escherichia coli, Lactobacillus casei* and *Saccharomyces boulardii* probiotic properties. Arch. Microbiol. *191*, 623–630.

Martins, F.S., Dalmasso, G., Arantes, M.R.E., Doye, A., Lemichez, E., Lagadec, P., Imbert, V., Peyron, J.-F., Rampal, P., Nicoli, J.R., and Czerucka, D. (2010). Interaction of *Saccharomyces boulardii* with *Salmonella enterica* serovar Typhimurium protects mice and modifies T84 cell response to the infection. PLoS One *5*, e8925.

Martins, F.S., Andrade, S.D.E., Vieira, A.T., Tiago, F.C.P., Martins, A.K.S., Silva, F.C.P., Souza, E.L.S., Sousa, L.P., Araujo, H.R.C., Pimenta, P.F., Bonjardim, C.A., Arantes, R.M.E., Teixeira, M.M., and Nicoli, J.R. (2011). Oral treatment with *Saccharomyces cerevisiae* strain UFMG 905 modulates immune responses and interferes with signal pathways involved in the activation of inflammation in a murine model of typhoid fever. Int. J. Med. Microbiol. *301*, 359–364.

Martins, F.S., Vieira, A.T., Elian, S.D.A., Arantes, R.M.E., Tiago, F.C.P., Sousa, L.P., Araújo, H.R.C., Pimenta, P.F., Bonjardim, C.A., Nicoli, J.R., and Teixeira, M.M. (2013). Inhibition of tissue inflammation and bacterial translocation as ones of the protective mechanisms of *Saccharomyces boulardii* against *Salmonella* infection in mice. Microb. Infect. *15*, 270–279.

Massot, J., Desconclois, M., and Astoin, J. (1982). Protection par *Saccharomyces boulardii* de la diarrhée à *Escherichia coli* du souriceau. Ann. Pharm. Fr. *40*, 445–449.

Mengheri, E. (2008). Health, probiotics, and inflammation. J. Clin. Gastroenterol. *42*, 177–178.

Mumy, K.L., Chen, X., Kelly, C.P., and McCormick, B.A. (2008). *Saccharomyces boulardii* interferes with *Shigella* pathogenesis by post-invasion signalling events. Am. J. Physiol. Gastrointest. Liver Physiol. *294*, G599–G609.

Murzyn, A., Krasowska, A., Stefanowicz, P., Dziadkowiec, D., and Łukaszewicz, M. (2010). Capric acid secreted by *S. boulardii* inhibits *C. albicans* filamentous growth, adhesion and biofilm formation. PLoS One *10*, 5: e12050.

Pedersen, L.L., Owusu-Kwarteng, J., Thorsen, L., and Jespersen, L. (2012). Biodiversity and probiotic potential of yeasts isolated from Fura, a West African spontaneously fermented cereal. Int. J. Food Microbiol. *159*, 144–151.

Peret, L.A., Penna, F.J., Bambirra, E.A., and Nicoli, J.R. (1998). Dose effect of oral *Saccharomyces boulardii* treatments on morbidity and mortality in immunosuppressed mice. J. Med. Microbiol. *47*, 111–116.

Pérez-Sotelo, L.S., Talavera-Rojas, M., Monroy-Salazar, H.G., Lagunas-Bernabé, S., Cuarón-Ibargüengoytia, J.A., Jimenez, R.M., and Vázquez-Chagoyán, J.C. (2005). *In vitro* evaluation of the binding capacity of *Saccharomyces cerevisiae* Sc47 to adhere to the wall of *Salmonella* spp. Rev. Latinoam. Microbiol. *47*, 70–75.

Pontier-Bres, R., Prodon, F., Munro, P., Rampal, P., Lemichez, E., Peyron, J.-F., and Czerucka, D. (2012). Modification of *Salmonella* Typhimurium motility by the probiotic yeast strain *Saccharomyces boulardii*. PLoS One *7*, e33796.

Pothoulakis, C., Kelly, C.P., Joshi, M.A., Gao, N., O'Keane, C.J., Castagliuolo, I., and Lamont, J.T. (1993). *Saccharomyces boulardii* inhibits *Clostridium difficile* toxin A binding and enterotoxicity in rat ileum. Gastroenterology *104*, 1108–1115.

Pothoulakis, C. (2009). Review article: anti-inflammatory mechanisms of action of *Saccharomyces boulardii*. Aliment. Pharmacol. Ther. *30*, 826–833.

Qamar, A., Aboudola, S., Warny, M., Michetti, P., Pothoulakis, C., LaMont, J.T., and Kelly, C.P. (2001). *Saccharomyces boulardii* stimulates intestinal immunoglobulin A immune response to *Clostridium difficile* toxin A in mice. Infect. Immun. *69*, 2762–2765.

Rodrigues, A.C.P., Nardi, R.M., Bambirra, E.A., Vieira, E.C., and Nicoli, J.R. (1996). Protective effect of *Saccharomyces boulardii* against experimental oral infection with *Salmonella typhimurium* and *Shigella flexneri* in gnotoxenic mice. J. Appl. Bacteriol. *81*, 251–256.

Rodrigues, A.C.P., Cara, D.C., Fretez, S.H.G.G., Cunha, F.Q., Vieira, E.C., Nicoli, J.R., and Vieira, L.Q. (2000). *Saccharomyces boulardii* stimulates sIgA production and the phagocytic system of gnotobiotic mice. J. Appl. Microbiol. *89*, 404–414.

Servin, A.L. (2004). Antagonistic activities of lactobacilli and bifidobacteria against microbial pathogens. FEMS Microbiol. Rev. *28*, 405–440.

Sougioultzis, S., Simeonidis, S., Bhaskar, K.R., Chen, X., Anton, P.M., Keates, S., Pothoulakis, C., and Kelly, C.P. (2006). *Saccharomyces boulardii* produces a soluble anti-inflammatory factor that inhibits NF-κB-mediated IL-8 gene expression. Biochem. Biophys. Res. Commun. *343*, 69–76.

Surawicz, C.M. (2003). Probiotics, antibiotic-associated diarrhoea and *Clostridium difficile* diarrhoea in humans. Best Pract. Res. Clin. Gastroenterol. *17*, 775–783.

Thomas, S., Metzke, D., Schmitz, J., Dörffel, Y., and Baumgart, D.C. (2011). Anti-inflammatory effects of *Saccharomyces boulardii* mediated by myeloid dendritic cells from patients with Crohn's disease and ulcerative colitis. Am. J. Physiol. Gastrointest. Liver Physiol. *301*, G1083–1092.

Tiago, F.C.P., Martins, F.S., Rosa, C.A., Nardi, R.M.D., Cara, D.C., and Nicoli, J.R. (2009). Physiological characterization of non-*Saccharomyces* yeasts from agro-industrial and environmental origins with possible probiotic function. World J. Microbiol. Biotechnol. *25*, 657–666.

Tiago, F.C.P., Martins, F.S., Souza, E.L.S., Pimenta, P.F.P., Araujo, H.R.C., Castro, I.M., Brandão, R.L., and Nicoli, J.R. (2012). Adhesion to the yeast cell surface as mechanism for trapping pathogenic bacteria by *Saccharomyces* probiotics. J. Med. Microbiol. *61*, 1194–1207.

Van der Aa Kühle, A., Skovgaard, K., and Jespersen, L. (2005). *In vitro* screening of probiotic properties of *Saccharomyces cerevisiae* var. *boulardii* and food-borne *Saccharomyces cerevisiae* strains. Int. J. Food Microbiol. *101*, 29–39.

Vieira, A.T., Teixeira, M.M., and Martins, F.S. (2013). The role of probiotics and prebiotics in inducing gut immunity. Front. Immunol. *12*, 4: 445.

Vilela, E.G., Abreu Ferrari, M.L., Gama Torres, H.O., Martins, F.P., Goulart, E.M., Lima, A.S., and Cunha, A.S. (2008). Influence of *Saccharomyces boulardii* on the intestinal permeability of patients with Crohn's disease in remission. Scand. J. Gastroenterol. *43*, 842–848.

Wu, X., Vallance, B.A., Boyer, L., Bergstrom, K.S., Walker, J., Madsen, K., O'Kusky, J.R., Buchan, A.M., and Jacobson, K. (2008). *Saccharomyces boulardii* ameliorates *Citrobacter rodentium*-induced colitis through actions on bacterial virulence factors. Am. J. Physiol. Gastrointest. Liver Physiol. *294*, G295-G306.

Yeasts as Probiotics – Established in Animals, but what about Man?

8

Gunnard K. Jacobson

Abstract

Yeasts are single-celled fungi that have been associated with human activity for thousands of years. Yeast strains of *Saccharomyces cerevisiae* have always been a good, and traditional, source of animal and human nutrition because of their B vitamin content. The use of inactive yeasts, live yeasts, and live yeast products such as 'yeast culture' is widespread in animal husbandry. The traditional use of yeast by man has been primarily in food modification (leavening and alcoholic fermentation), but also medicinally, and as a nutritional supplement. Only within the last century have the probiotic properties of some yeast strains been recognized – increased milk yields in dairy cattle and ameliorating diarrheal diseases in man, for example. Some properties may be generic, e.g. oxygen scavenging in the rumen; but other properties are not shared by all strains of *S. cerevisiae* and the explanations for this fact are under intense scientific scrutiny. The probiotic attributes of the *boulardii* subgroup of strains within *S. cerevisiae* have been attributed to effects on enteric pathogens, intestinal barrier function integrity, anti-inflammatory effects, immune stimulation, and trophic effects on the intestinal mucosa. Current research is identifying new yeast probiotics outside of the *S. cerevisiae* subsp. *boulardii* cluster.

What are probiotic microorganisms?

Lilly and Stillwell (1965) originally coined the term 'probiotic' to describe unknown compounds produced by one ciliate that promoted the growth of another. Its usage was later modified to refer to 'A live microbial feed supplement which beneficially affects the host animal by improving its intestinal microbial balance' (Fuller, 1989). Salminen *et al.* (1999) modified this definition to: 'Probiotics are microbial cell preparations or components of microbial cells that have a beneficial effect on the health and well-being of the host.' By this definition the organism does not necessarily need to be viable; and, additionally, extracts or components of the microbe may be considered probiotic if they benefit the host. In 2001, FAO/WHO adapted the following definition of probiotics: 'Live microorganisms which when administered in adequate amounts confer a health benefit on the host.' This latter definition is, perhaps, the most widely accepted.

Microorganisms exist in all three domains of life: Archaea, Bacteria and Eukarya. The first two domains were once all inclusive and was termed Prokarya; virtually all of the members of these two domains are single celled bacteria. However, members of these two bacterial domains are fundamentally different with respect to their 16S RNA structure and membrane composition. Unlike the Eukarya, neither domain contains an organized membrane-bound nucleus or other intracellular organelles. Bacterial probiotic microorganisms are all members of the domain Bacteria, and for the most part, are lactic acid bacteria (LAB).

Eukarya is composed of four kingdoms: Animalia, Plantae, Fungi and Protista. The members of Animalia and Plantae are well known to the layperson. Protista is composed mainly of single-celled organisms (some colonial organisms exist) such as the amoebae and the protozoans. Some are photosynthetic, such as *Euglena*. Eukarya contain a nucleus containing DNA organized into chromosomes; intracellular organelles such as mitochondria which are involved in respiration; and, in the case of photosynthetic organisms, chloroplasts.

Fungi can be defined as eukaryotic organisms typically containing chitin in their cell walls and lacking chloroplasts associated with green plants and algae. Within the kingdom Fungi, the two most familiar divisions are Basidiomycota and Ascomycota. The Basidiomycota bear their sexual spores (sexual in that they are products of meiosis) on the surface of a club-like structure called a basidium. In Ascomycota, the sexual spores are enclosed within a sac-like structure (ascus). The commercial white mushroom, *Agaricus bisporus*, is a familiar basidiomycete; on the other hand, familiar to many, morels, e.g. *Morchella esculenta*, are ascomycetes.

The word yeast is a generic term referring to a fungus that can exist in the single-celled state. In some cases, such as in the jelly fungi (*Tremella* sp.), germinated sexual spores (gametes) exist as yeasts until a yeast of opposite mating type is encountered; the cells then conjugate and complete the sexual cycle by forming a new jelly fungus.

Fungi existing *only* as yeasts are widespread in nature; over 1500 species have been described. In some cases, they can be classified as basidiomycete or ascomycete based on sexual spore formation. For example, *Rhodosporidium* is basidiomycetous yeast; while *Saccharomyces cerevisiae* is ascomycetous. Thus, bakers' yeast is more closely related to the morel, than to the common champignon. Yeasts used for the leavening of bread and the production of ales and distillates are strains of *S. cerevisiae.* Lager beers are fermented with strains of *S. pastorianus* (formerly *S. carlsbergensis*). Consequently, the term brewers' yeast can encompass both species, but in commerce, brewers' yeast generally refers to yeast derived from lager beer fermentations. For the most part, probiotic strains of yeast are confined to the genus and species *S. cerevisiae,* or subspecies within *cerevisiae* such as *boulardii.* Some probiotic yeasts in other genera, especially in the aquaculture arena, are known.

Animal applications: from feed to probiotic

The large scale usage of yeast for animal husbandry was made possible by the industrial revolution and the emergence of large regional breweries. There is a 6- to 8-fold increase in yeast biomass during a typical beer fermentation, and brewers harvest a portion of this yeast to pitch the next fermentation. This can be repeated 4–10 times before a decline in vitality necessitates a propagation of fresh yeast (Reed and Nagodawithana, 1991). For large brewers, this led to the question: 'What are we going to do with all of this yeast?' In the mid-19th century, German brewers, using strains of *Saccharomyces pastorianus* (*carlsbergensis*), had difficulties selling surplus yeast to bakers as lager yeasts are very poor leaveners. Consequently, references to the use of brewers yeast and brewery waste streams in animal feeds appear in the German literature as early as 1889 (Pott, 1889). English ale brewers (using strains of *S. cerevisiae*) could sell yeast to Scottish distilleries, to bakers, or as feed to cattle. The success of a given brewers' yeast in a bakery was dependent on the strain the brewer was using, and a comment of the time was that 'generally that where one found a good baking yeast there was a poor beer' (Baker, 1903). By-products from breweries, spent grains, spent hops, and yeast became important sources of cattle nutrition. While the spent grains portion, because of their bulk, were the most important, the nutritional aspect of dried brewers' yeast was known to increase 'the growth and stamina wherever fed; cows give more milk, and poultry lay more eggs under its influence'. For cattle, palatability issues arose if the yeast was mixed with another waste stream, spent hops (Fuller, 1927). The commodity prices associated with livestock feeds, motivated the search for value added products from spent yeast. At a 1903 meeting of the Midland Counties Institute of Brewing in Birmingham, UK, J.M. Jones lamented that it was 'a blot on our scientific enlightenment to find surplus yeast – a substance so rich in nutritive properties – could be put to no better use than to be turned unceremoniously down the drain.' This meeting was dedicated to the utilization of surplus yeast, and its keynote speaker, J.L. Baker, reviewed over 20 patents for the manufacture of yeast extracts, therapeutic yeast preparations, yeast based fertilizers, bakers' yeast, and an extract of yeast as a decolorant in leather tanning (Baker, 1903).

Ruminants

For decades, yeast was fed to livestock largely as a source of protein and energy, but by the 1920s it was known that yeast was a good source of B vitamins and academic studies began to appear on the use of yeast as dietary supplements (Eckles and Williams, 1925; Eckels *et al.*, 1924). In the United States, regulatory authorities developed standards of identities for the yeast streams entering the animal feed system. The Association of American Feed Control Officials has defined standards of identity for several yeast products:

- 96.1. Primary Dried Yeast (inactive, at least 40% crude protein)
- 96.2. Active Dry Yeast (bulk of fermentation power retained; at least 15 billion live cells/g)
- 96.3. Irradiated Dried Yeast, inactive yeast irradiated with UV light to produce vitamin D_2
- 96.4. Brewers Dried Yeast, non-fermenting yeast derived from ale and beer fermentations
- 96.7. Torula Dried Yeast.

With the exception of Torula Dried Yeast (botanical classification *Candida utilis*) all other yeasts are classified as belonging to the genus *Saccharomyces* (AAFCO, 2009).

Yeast efficacy, as a dietary supplement, with respect to enhanced growth, milk fat content, milk yields, *etc.* was complicated by the introduction of 'yeast culture' products. 'Yeast culture' also has a standard of identity (96.8, Yeast Culture) as '... the dried product composed of yeast (*Saccharomyces cerevisiae* and/or *Kluyveromyces marxianus*) and the media on which it was grown, dried in such a manner as to preserve the fermenting activity of the yeast.' (AAFCO, 2009). Proprietary formulations containing yeast and fermented molasses, dried on grains, are highly palatable to livestock. Poppy, *et al.* (2012) have carried out an meta-analysis of 36 different 'yeast culture' studies and concluded that 'yeast culture' also provides benefits to lactating dairy cattle. In some feeding studies, it is not precisely clear as to which AAFCO defined yeast product is being used. In this chapter, 'yeast culture' will designate an AAFCO 96.8 product.

In dairy cattle, live yeast (*S. cerevisiae*, Biosaf Sc 47) increased fibre digestibility, lowered lactate levels, and reduced ruminal Eh (redox potential) (Marden *et al.*, 2008). de Ondarza *et al.* (2010) used CPM Dairy 3.0.8 to analyse data from 14 research trials, conducted internationally, involving 1615 cows. CPM (Cornell-Penn-Miner) Dairy is a computer model used to evaluate various aspects of ruminant nutrition on dairy cattle performance (Tedeschi *et al.*, 2008). This analysis found that live yeast increased milk fat and milk true protein yields. Live yeast significantly increased 3.5% FCM (fat-corrected milk) yields, especially in cows less than 100 DIM (days in milk).

High-starch diets promote milk yield in dairy cows; but, such diets are associated with subacute ruminal acidosis (SARA). Ferraretto *et al.* (2012) studied the effects of 2 g/day and 4 g/day added live yeast to high starch diets (30% starch in dry matter). The 4 g yeast per cow per day diet tended to increase milk fat content and increased total-tract fibre digestibility. Ruminal pHs were unaffected in this study. Borghese *et al.* (2007) fed dairy buffalo 6 and 60 g/day live yeast (Sc47) for eight months. Yeast supplementation increased milk protein, and consequently, mozzarella cheese yields over control animals. Milewski *et al.* (2012) reported that dried brewers' yeast could influence the ratio of β- to κ-caseins in ewe's milk. This enhanced milk quality.

Live yeast supplementation was also reported to increase nutrient digestibility in growing Kankrej calves (Hossain *et al.*, 2012). Calves were fed 5 or 10 g/day live yeast (Provisacc, Vetcare India, Ltd.) as part of their rations for 100 days. Dry matter intake was slightly higher (but not statistically) in yeast-fed calves. Yeast supplementation reduced ruminal nitrogen levels and increased volatile fatty acid levels. The digestibility coefficients for dry matter, organic matter, crude protein, and fibre were increased at both levels of yeast supplementation due to improved rumen fermentation.

Feed lot conditions are stressful to beef cattle; stresses which result in animals being more susceptible to bovine respiratory disease (BRD). Underweight animals are often given antibiotics prophylactically to minimize the risk of BRD, and this often results in reduced dry matter intake (DMI). A small study with animals receiving an injection of florfenicol suggested that DMI intake returned to normal faster in animals receiving a supplement containing *S. cerevisiae* subsp. *boulardii* CNCM I-1079 (Keyser *et al.*, 2007). Lightweight heifers treated with florfenicol and 0.5 g/day yeast were less likely to need treatment for BRD.

In sheep, *S. cerevisiae* CNCM I-1077 has been shown to enhance colonization of the rumen in newborn lambs by cellulolytic bacteria (Chaucheyras-Durand and Fonty, 2002). Real-time PCR has been used to demonstrate that live yeast (Levucell SC20) fed to sheep resulted in a 2- to 4-fold increase in ruminococci (Mosoni *et al.*, 2007). *In vitro* studies, with clarified ruminal fluid, showed that yeast could stimulate acetogenic bacteria in the presence of methanogens (increased acetate production with reduced methane production). Autoclaved yeast produced similar results, suggesting that, in at least this case, the yeast may be functioning as a nutrient source (Chaucheyras *et al.*, 1995). In lambs, the probiotic yeast, SC I-1077, was shown to be unable to colonize the rumen; the digestive tract was free of yeast about 100 h after ingestion. 17–34% of the yeast was viable by the time they were passed in the faeces. This suggests that the benefits of probiotic yeasts could extend to the digestive tract distal to the rumen (Durand-Chaucheyras *et al.*, 1998).

At the levels administered in feed, the impact of live yeast on rumen microbiota fit the definition of 'Probiotic' as defined by Fuller (1989). Today, live yeast is just one probiotic microorganism considered to be a Direct-Fed Microbial (DFM) for use in ruminants (Denev *et al.*, 2007; Chaucheyras-Durand *et al.*, 2012).

Over the years, hundreds of studies have reported on the benefits or lack of benefits on yeast as a feed additive. However, there has been little consistency in the feed formulations being compared: yeast composition (live vs. dead vs. 'yeast culture'), forage composition, grain composition, feed additives, *etc.*, were highly variable, and animal performance from study to study was mixed (Steckley *et al.*, 1978; Dawson *et al.*, 1990; Williams *et al.*, 1991; Newbold *et al.*, 1995). A meta-analysis of 110 research papers discussing 157 experiments and 376 treatments concluded that yeast increased dry matter intake, milk yield, rumen pH and volatile fatty acid levels, and organic material digestibility in dairy cattle (Desnoyers *et al.*, 2009). It was noted, however, that this analysis included 13 studies thought to be 'live yeast' trials but were actually 'yeast culture' trials (Poppy *et al.*, 2012).

Jouany (2001) summarized the suspected mode(s) of action of live yeast on ruminant (cattle, sheep, goats) performance:

- A Metabolizes glucose released by amylolytic bacteria making less available for fermentative bacteria.
- B Stimulates the metabolism of lactate by *Megasphera elsdenii* and *Selenomonas ruminantium*: rumen pH stabilization and prevention of acidosis.
- C Reduction of rumen oxygen levels thereby benefiting resident cellulolytic bacteria and anaerobic fungi such as *Neocallimastix frontalis*.
- D Accumulation and release of malic acid which stimulates the growth of *S. ruminantium* in the presence of lactic acid.
- E Increases ruminal nitrogen pools.
- F Increases the total number of bacteria and protozoa in the rumen.

New molecular methods have been applied to speciate the microbiota of the rumen as influenced by live yeast. Pinloche, *et al.* (2013) compared two methods of 16S rDNA analysis to determine the bacterial diversity in cattle rumens as a function of added yeast. The authors compared 454 pyrosequencing with SARST (serial analysis of ribosomal sequence tags) of the V1 region of 16S rDNA. The SARST method generates smaller sequences (average 56 nt) than the 454 pyrosequencing (average 248 nt). The SARST method identified 121 operational taxonomic units (OTUs) while the 454 method resulted in 185 OTUs. Cannulated, lactating cows were fed 0.5 or 5.0 g/day of the commercial probiotic yeast, Sc 47. Addition of yeast to the diet was associated with a higher rumen pH, and lower lactate levels and lower E_h. Higher volatile fatty acid (VFA) levels were associated only with the high yeast dosage. Rumen improvements were associated with population changes in fibrolytic bacterial (*Fibrobacter* and *Ruminococcus*) and lactate-utilizing bacteria (*Megasphera* and *Selenomonas*) as suspected more than a decade earlier.

One current interpretation as to the mode of action of live yeast in the modification of the ruminal microbiota concerns its influence on the ruminal redox potential. Recent advances in measuring ruminal redox potential (Marden *et al.*, 2005) have confirmed that live yeast increases ruminal pH and lowers redox potential (Křížová *et al.*, 2011). Live yeast consumes oxygen entering the rumen with feed and mastication. The resulting lower partial pressure of oxygen lowers ruminal fluid redox potential and enhances the conversion of lactic acid to propionic acid by anaerobic bacteria (Moncoulon and Auclair, 2005; Marden *et al.*, 2013).

As noted by Jouany (2001), and others, microbial associated ruminant performance data often have not shown statistical differences from one study to the next (Newbold *et al.*, 1996, 1998; Kung, 2001). In many cases these differences may be due to differences in feed formulations; but in some cases, these differences may be yeast strain related. Newbold, *et al.* (1995) evaluated several strains of *S. cerevisiae*, NCYC 240, NCYC 1088, NCYC 694, and NCYC 1026 in an artificial rumen (Rusitec). They found that NCYC 240 and NCYC 1026 stimulated total cellulolytic bacterial numbers, but NCYC 694 and NCYC 1088 did not. In ruminally cannulated sheep, total and cellulolytic bacterial numbers were stimulated, but the data were only statistically different for strains NCYC 240 and NCYC 1026. Agarwal *et al.* (2000), screened nine strains of *S. cerevisiae* for resistance to low pH, organic acids (acetic, propionic, and butyric in a ratio of 70:20:10), and bile salts. Two strains selected for similar characteristics, NCDC 42 and NCDC 49, were evaluated in an *in vitro* dry matter degradability assay containing ruminal fluid. NCDC 49 enhanced the degradability of wheat straw and berseem (*Trifolium alexandrinum*), but NCDC 42 did not. Chung *et al.* (2011) reported the effects of two yeast strains on ruminal acidosis and methane production in cannulated, non-lactating dairy cows. Strain 1 (Levucell SC) was a commercial probiotic; strain 2 was an experimental strain of *S. cerevisiae* selected for its ability to increase fibre degradation (*in vitro* assay). It was found that rumen pH in cows receiving strain 2 remained below 5.8 longer (acidosis) than cows receiving either strain 1 or no-yeast controls (7.5 h/day vs. 1 h/day for strain 1, and 3.3 h/day for control). Methane emission intensity, measured as grams CH_4/kg dry matter intake, was 10% lower for strain 2 compared with strain 1; but, there was no difference compared with no-yeast controls.

Ruminants are reservoirs for pathogenic strains *of Escherichia coli*; Shiga toxin-producing strains (STEC) are of special concern. While acid resistant strains exist, coliforms are acid sensitive. Chaucheyras-Durand, *et al.* (2010) studied the survival of STECs in a cannulated sheep study. They found that lethality occurred primarily in the rumen and abomasum; acid sensitive STEC strains suffered high lethality in the abomasum. Once in the jejunum and caecum, however, both sensitive and resistant strains were able to reproduce. *Ex vivo* experiments with contents removed from various digestive compartments showed that the addition of *S. cerevisiae* CNCM-1077 or *Lactobacillus acidophilus* BT-1386 to ruminal contents led to the death of STECs. These effects were not seen with contents from other digestive compartments. The authors suggested that these probiotics could limit the number of STECs reaching the hindgut and reduce STEC shedding in faeces.

Monogastrics

Yeast-based and 'yeast culture' products are also used as feed additives for monogastric animals (horses, swine, poultry, *etc.*) where a metabolically active yeast would not necessarily be expected to be important to animal performance. In these animals, yeast may affect the activity of disaccharidases in the intestinal epithelium, have anti-adhesive/antagonistic effects against intestinal pathogens, non-specific immune stimulation, and toxin binding effects (Auclair, 2001; Chaucheyras-Durand and Durand, 2010).

Horses

Like ruminants, horses have a specialized digestive system which relies on the microbial degradation of hemicellulosic and cellulosic feeds. This occurs in the hind gut, not in a rumen as in ruminants. But, like ruminants, microbial digestion is influenced by feeding practices such as the inclusion of grains (starch) in the diet (Jouany *et al.*, 2009). 'Yeast culture' has been shown to modify the microbial populations and fermentation patterns of the equine hindgut (Jouany *et al.*, 2008; Jouany *et al.*, 2009). High starch diets result in a decrease in cellulolytic bacteria and an increase in lactate utilizing bacteria in the caecum. The concentrations of lactobacilli and streptococci increased in both the caecum and right ventral colon. This was associated with a decrease in the (acetate + butyrate)/propionate ratio accompanied by an increase in lactate and a lower pH. 'Yeast culture', YEA SACC^{1026}, ameliorated these adverse effects, but the effect was greater in the caecum (where viable yeast numbers were higher) than in the colon (Medina *et al.*, 2002).

Swine

Swine are monogastrics with a digestive system similar to humans and the conditions under which swine are commercially raised are often stressful. Post-weaning piglets are subject to stress related anorexia and are predisposed to diarrheal diseases. Bontempo *et al.* (2006) conducted a study on 352 post-weanling piglets. *S. cerevisiae* subsp. *boulardii*, CNCM I-1079, at 2 g/kg feed, was trialled in half the animals. Piglets were monitored for 4 weeks. Yeast-supplemented animals had higher live weights and average daily weight gain than control animals. At slaughter, yeast-fed animals had a thinner intestinal adherent mucous layer than controls; proliferating epithelial cell counts were higher in yeast fed animals; and mucosal macrophages were more numerous in yeast fed piglets. Live yeast promoted a 'healthy' intestine in stressed post weaning piglets. Gut-trophic feed additives promoting protection of the anatomically and functionally immature intestinal epithelium in piglets has been reviewed (Domeneghini *et al.*, 2006). Lessard *et al.* (2009) compared the probiotic effects of *Pediococcus acidilactici*, MA 18/5M, and *S. cerevisiae* subsp. *boulardii*, CMCN I-1079, on *E. coli* challenge in piglets. They found that both probiotics reduced bacterial migration to mesenteric lymph nodes similar to the antibiotic tiamulin. *Pediococcus* had a greater effect on modulating intestinal IgA concentration and CD8$^+$ cells than *Saccharomyces*.

In a study on gestating and lactating sows, *S. cerevisiae* Sc 47 was fed as part of the lactating diet or as part of both the gestational diet and lactating diet. There was no effect on sow body weight or reproductive performance (Jang *et al.*, 2013). The IgG content of sow colostrum was slightly higher in sows fed yeast in the lactation diet; this resulted in higher IgG plasma levels in 24-h post-partum piglets. The weaning to oestrus interval decreased in sows fed yeast. An extension of this study looked at three different strains of yeast and their effects on colostrum immunoglobulins (Zanello *et al.*, 2013). Two *S. cerevisiae* strains, CNCM I-4407 (Sc01) and LC 10-05 (Sc02), and one strain of *S. boulardii*, CNCM I-3799 (Sb03), were studied. Colostrum IgG levels were elevated in sows fed Sc01 and Sc03. In milk, IgG levels were elevated in sows fed Sc02 and Sc03. Milk IgA levels were elevated by all three strains, but only Sc01 resulted in elevated colostrum IgG levels, elevated milk IgA levels, with a concomitant decrease in the incidence of piglet diarrhoea.

Studies have shown that live yeast can influence the microbiota of the colon in finishing pigs and faecal coliform counts in nursery pigs (Pinloche *et al.*, 2012; Perez-Sotelo *et al.*, 2011; Monroy-Salazar *et al.*, 2012). Pinloche *et al.* (2012) collected digesta from the caecum and proximal colon of finishing pigs and evaluated the effects of two levels of live yeast, *S. cerevisiae* Sc 47, on gut fermentation and microbiota, *ex vivo*, in an artificial colon (Cositec). Live yeast titres tested were 5×10^{10} (L1) and 5×10^{11} colony-forming units (cfu)/kg (L2) feed. L2, but not L1, increased dry matter digestibility and NH_3 levels over the no yeast control. VFA levels were also higher in L2, but in L1 only valeric acid levels were elevated. Liquid associated (LAB) and solids associated (SAB) bacteria were analysed by terminal RFLP (restriction fragment length polymorphism) analysis. L2 had a significant impact on both populations, L1 only on SAB. Increased dry matter digestibility in L2 was associated with a shift in the SAB population. Perez-Sotelo, *et al.* (2011) fed antibiotic-free pigs Sc47 (*S. cerevisiae*) at 3×10^9 cfu/kg feed for 4 weeks. Yeast fed pigs demonstrated better growth performance and had a 10 fold lower faecal coliform count than control animals. Peripheral blood lymphocyte and monocyte counts were not affected. Monroy-Salazar, *et al.* (2012) fed nursery pigs a diet containing 0.3% Sc47 for 42 days. They also found a reduction in faecal coliform counts but also noted an increase in the peripheral T lymphocytes $CD4^+$, $CD4^+CD8^+$, and $CD8^+$. $CD4^+$ cells are associated with MHC II (major histocompatibility complex class II) immune responses while $CD8^+$ is associated with MHC I responses, suggesting live yeast can modulate both systems. Another study demonstrated that in piglets challenged with *E. coli* K88, Sc47 reduced the levels of the stress related C-reactive protein (CRP) and reduced the expression of Toll-like receptors in the ileum (TLR2 and TLR4). This suggested reduced binding of *E. coli* (components) to the intestinal epithelium (Badia *et al.*, 2012).

An *in vitro* study demonstrated that live *S. cerevisiae* CNCM I-3856 inhibits enterotoxigenic *E. coli* (ETEC) mediated pro-inflammatory response in IPEC-1 porcine intestinal epithelial cells (Zanello *et al.*, 2011a). CNCM I-3856 decreased ETEC induced mRNA expression of IL-6, IL-8, CCL20, CXCL2, and CXCL10. IL-6, CCL20, and CCL12 inhibition was associated with yeast inhibition of the ERK1/2 and p38 MAPK signalling pathways. Yeast viability was essential for this activity. Yeast also up-regulated mRNA levels of the nuclear receptor PPAR-γ, the pro-Th1 cytokine IL-12p35, and CCL25. Yeast also induced agglutination of ETEC. *S. cerevisiae* CNCM I-3856 was compared to *S. boulardii* CNCM I-3799 in porcine IPI-2I cells infected with ETEC expressing $F4^+$ fimbriae. $F4^+$-expressing ETEC are the most prevalent cause of infection in piglets. It was shown that viable, but not non-viable, I-3856 inhibited ETEC-induced TNF-α while *S. boulardii* did not (Zanello *et al.*, 2011b). CNCM I-3856 culture supernatants decreased TNF-a, IL-1a, IL-6, IL-8, CXCL2, and CCL20 ETEC-induced mRNA synthesis. Fractionated supernatants, <10kDa, had the same effects except for TNT-a.

Poultry

The growth promoting properties of 'yeast culture' in poultry rearing was reported as early as the 1940s (Grau and Kamei, 1949). Goa, *et al.* (2008) found that growth performance was optimized at 2.5g 'yeast culture'/kg feed. 'Yeast culture', Diamond V XP, was found to modulate villus height to crypt depth ratios in the duodenum and jejunum; increase calcium and phosphate digestibility; increase IgM and secretory IgA concentrations in the duodenum; and increase antibody titres to Newcastle Virus following vaccination. Some of these effects were most apparent at higher supplementation levels, 5.0g/kg and 7.5g/kg. The authors concluded that the higher 'yeast culture' supplementation levels would be justified in chicks reared under conditions of disease or heat stress. Priya and Babu (2013) compared performance in poultry fed a control diet with those supplemented with 0.5%, 1.0%, and 1.5% of an unspecified dry form of *S. cerevisiae*. Performance was better on all supplemented diets with the 1.5% diet providing the most benefits. Villus height was greater on yeast supplemented diets, confirming yeast's positive effects on intestinal morphology. The live yeast titre of the yeast supplement in this study was 1×10^9 cfu/g. Since the AAFCO definition 96.2 of an active dry yeast feed additive requires 15×10^9 live cells/g; it is possible that the yeast used in this study was a 'yeast culture' product. *S. boulardii* (SB 20) supplementation was shown to reduce caecal *Salmonella* colonization in broiler chicks (Line *et al.*, 1998). Birds were fed feeds containing 1 g yeast/kg feed and 100g yeast/kg feed. After 4 days they were challenged with *S. typhimurium* and *Campylobacter jejuni*; after three weeks the birds were sacrificed and examined for colonization. 70% of the control birds were colonized by *Salmonella* whereas only 20% of the 1g/kg and 5% of the 100g/kg birds were colonized. *S. boulardii* had no effect on *Campylobacter* colonization.

Aquaculture

Probiotic yeasts have found application in aquaculture. Yeasts outside of the genus *Saccharomyces*, such as the autochthonous yeasts *Debaryomyces hansenii* and *Rhodosporidium paludigenum* have been studied. Yeasts have been formulated into feeds either as a traditional feed additive or loaded into live feed organisms such as the nauplii stage of the brine shrimp *Artemia* or rotifers such as *Brachionus* (Fazeli et al., 2006; Zacarias-Soto and Lazo, 2011). These live feed organisms are essentially fed with probiotic yeast prior to being fed to fish. Table 8.1 provides examples of the fish, yeasts, and endpoints studied.

As in other animal systems, strain differences in probiotic effects have been reported in aquaculture. Waché *et al.* (2006) compared a strain of *S. cerevisiae* (Sc 47) with a strain of *S. boulardii* (CMCN I-1079) with respect to their influence on the development of intestinal microbiota and digestive enzymes in trout. They found that the *S. boulardii* fed fish had higher activities of alkaline phosphatase, γ-glutamyl-transpeptidase, and leucine-amino peptidase N than fish fed *S. cerevisiae*. Water temperature was also shown to be important as these fish were reared in spring water at 11–11.5°C and this positive effect on digestive maturation was not seen in fish reared in river water at lower temperatures, 7–8°C. Strain differences were also reported

Table 8.1 Probiotic Yeasts in Aquaculture

Organism	Probiotic yeast	Delivery	End point studied	Reference
Juvenile sea bass	*Saccharomyces cerevisiae**	Feed	Liver antioxidant	Santacroce *et al.* (2012)
Nile tilapia fry	*Saccharomyces cerevisiae* BioSaf™	Feed	Growth	Lara-Flores *et al.* (2003)
Carp fingerlings	*Saccharomyces cerevisiae* (local commercial bakers')	Feed	Growth	Tewary and Patra (2011)
Juvenile shrimp	*Saccharomyces cerevisiae* Procreatin 7™	Feed	Survival, vibriosis resistance	Scholz *et al.* (1999)
Instar II Artemia nauplii	*Saccharomyces boulardii* Biocodex, France	Feed	Vibriosis resistance	Patra and Mohamed (2003)
Sea bass larvae	*Debaryomyces hansenii* CBS 8339	Feed	Antioxidant enzymes	Tovar-Ramírez *et al.* (2010)
Sea bass larvae	*Debaryomyces hansenii* CBS 8339	Feed	Survival, larval development, growth	Tovar-Ramírez *et al.* (2004)
Sea bass larvae	*Debaryomyces hansenii* HF1	Feed	Digestive enzyme activity	Tovar *et al.* (2002)
Seabream	*Debaryomyces hansenii* L2	Feed	Intestinal microbiota, immune function	Tapia-Paniagua *et al.* (2011)
Juvenile rainbow trout	*Phaffia rhodozyma* (KI Chemical)	Feed	Muscle lipid peroxide generation	Nakano *et al.* (1999)
Juvenile shrimp	*Phaffia rhodozyma* (Merck)	Feed	Survival, vibriosis resistance	Scholz *et al.* (1999)
Shrimp	*Rhodosporidium paludigenum* (autochthonous isolate)	Feed	Hepatopancreas antioxidant enzymes	Yang *et al.* (2013)
Pollack larvae	*Saccharomyces cerevisiae* CNCM I-1079	Yeast loaded in Artemia	Growth	Gatesoupe (2002)
Sturgeon larvae	*Saccharomyces cerevisiae* (Thepax®)	Yeast loaded in Artemia	Growth	Iranshani *et al.* (2011)
Halibut larvae	*Saccharomyces cerevisiae** and *S. boulardii* (Levucell SB®)	Yeast loaded in Artemia or Brachionus	Digestive enzyme activity	Zacarias-Soto *et al.* (2011)
Tilapia larvae	Recombinant *Pichia pastoris* SMD 1168H expressing tilapia vitelloginin	Feed	Growth	Lim *et al.* (2005)

*Strain not identified.

between *D. hansenii* (HF1) and *S. cerevisiae* (X2180) on digestive enzyme activity in sea bass larvae (Tovar *et al.*, 2002). The positive attributes of *D. hansenii* were attributed to its production of higher levels of the polyamines spermine and spermidine. It should be noted, however, that X2180 is not an industrial yeast strain. Rather, it is a wild type, haploid, laboratory strain used in academic yeast genetics research with the genotype SUC2, mal, mel, gal2, CUP1 (Mortimer and Contopoulou, 1991). It is not surprising that a lab strain would not produce polyamine levels produced by probiotic yeasts. The astaxanthin-producing yeast, *Phaffia rhodozyma* (KI Chemical, Shizuoka, Japan) was reported to reduce lipid peroxides in trout fed diets containing oxidized oils (Nakano *et al.*, 1999). This specific effect may be due to the anti-oxidative properties of the astaxanthin itself, rather than the probiotic nature of the yeast. Similar effects were reported in salmon (*Salmo salar*) with 60 mg synthetic astaxanthin/kg feed (Christiansen *et al.*, 1995).

Probiotics in aquaculture, yeasts and bacteria, have been the subject of recent reviews (Dimitroglou *et al.*, 2011; Cruz *et al.*, 2012).

Rodents as model monogastrics for the study of probiotic mechanisms

The intestinal effects of yeast have been studied in mouse and rat models. The effects of *S. boulardii* on intestinal villi morphology were studied in 7-day-old BALB/s mice (Cartwright-Shamoon *et al.*, 1996). Mice were intubated with *S. boulardii* (Biocodex, France), rotavirus, or rotavirus + *S. boulardii*; control mice were intubated with saline. Administration of yeast did not prevent rotavirus associated diarrhoea in these animals; nor did it prevent viral related villus changes (stromal separation and increase epithelial vacuolation). However, yeast cells were seen within the intestinal epithelium and lamina propria of treated animals. This suggests that probiotic yeasts can be taken up and translocated by the villous mucosa; the most likely route being between adjacent epithelial cells.

In rodents, *Citrobacter rodentium* causes symptoms similar to enteropathogenic *E. coli* (EPEC) and enterohaemorrhagic *E. coli* (EHEC) infection, and inflammatory colitis disease in man. C57BL/6 mice where challenged with *C. rodentium*; after two days, they were gavaged twice a day with *S. boulardii* (MFI Pharma). Treated animals had lower weight loss; less crypt hyperplasia; lower histological damage scores; reduced mannitol flux; and reduced myeloperoxidase activities relative to untreated mice. These effects correlated with a reduced number of bacteria adhering to the intestinal mucosa. The authors concluded *S. boulardii* helped maintain intestinal barrier function and attenuated inflammatory responses by reducing the numbers of bacteria adhering to the mucosa (Wu *et al.*, 2008). Martins *et al.* (2009a) investigated the protective effects of five commercial probiotic yeast preparations (four *S. boulardii*, one *S. cerevisiae*) on mice challenged with *S. enterica* subsp. *enterica* serovar Typhimurium. The manufacturers of these products was not disclosed. In an experiment determining intestinal wash out, they found that live yeast persisted in the faeces for 8 days following administration of two *S. boulardii* products; the other two *S. boulardii* products were cleared at 7 and 4 days, respectively, while the *S. cerevisiae* yeast was cleared in just two days. Highest 28-day survival rates following challenge correlated with higher faecal yeast counts, with survival being lowest in the *S. cerevisiae* product. Histopathological examination of the liver and ileum at 28 days demonstrated differences between the five yeast preparations with one preparation of *S. boulardii* providing more inflammatory protection than the others. The differences in the probiotic effects of these five products could have been due to yeast strain differences, commercial production methodology, or both. A follow up study showed *S. boulardii* (Biocodex) protected mice from *S. enterica* serovar Typhimurium and prevented bacterial translocation to the liver (Martins *et al.*, 2010). This study also showed that *S. boulardii* preserved barrier function and reduced IL-8 synthesis in T84 cells, a human adenocarcinoma cell line, infected with *S. enterica*. sIgA is secretory immunoglobulin secreted primarily by mucosal membranes, including the intestinal tract. *S. boulardii* (Merck S.A.) elevated levels of sIgA in gnotobiotic (bacteria-free) mice (Rodrigues *et al.*, 2000). Gnotobiotic animals can be used to study the effects of monoassociated probiotics, i.e. the effect of a single microorganism in the gut on animal health. *S. boulardii* monoassociated mice also had higher levels of Küpffer cells (liver specific macrophages); cleared *E. coli* B_{14} from the bloodstream more readily; and, produced infection related cytokines, TNF-α, IL-12, and IFN-γ, faster than non-monoassociated controls when challenged with *E. coli* B_{41}. Martins *et al.* (2009b) compared the probiotic effects of *S. boulardii* (Floratil®, Merck) to those of *Bifidobacterium animalis*, *Lactobacillus casei*, and *Escherichia coli* in gnotobiotic mice challenged with intestinal pathogens. With respect to immunomodulation in the presence of pathogens, the presence of *S. boulardii* induced higher levels of sIgA than *E. coli* EMO and *B. animalis*; and only *S. boulardii* induced significant levels of IL-10.

Transduction pathways consist of a series of proteins which interact sequentially to transmit environmental information from the outside of the cell to the nucleus for appropriate cellular response. Many of these proteins are kinases which phosphorylate specific sites, tyrosine, serine, threonine moieties, for example, on other proteins in the signal cascade. Among these are MAP kinases (Mitogen Activated Protein kinases) which respond to cellular stimuli leading to gene expression, cellular differentiation and/or proliferation, and, in some cases apoptosis (programmed cell death). Toxin A, produced by *Clostridium difficile*, activates ERK and p38 MAP kinases which result in expression of the inflammatory cytokine, IL-8 and cell necrosis in the intestine (Warny *et al.*, 2000). *S. boulardii* secretes large molecular weight proteins during growth. Supernatants from *S. boulardii* (Biocodex) cultures were shown to down-regulate the ERK1/2 MAP kinase pathway both *in vivo* (mouse ileum) and *in vitro* in human NCM460 colonic epithelial cells (Chen *et al.*, 2006). In a follow up study, it was shown that the upstream mediator for ERK1/2 MAP kinase activation is the epidermal growth factor receptor (EGFR). This receptor is deactivated in the presence of *S. boulardii* supernatants and results in the downstream inactivation of both the EGRF-ERK and EGRF-AKT pathways. In Apc^{min} mice, which develop intestinal adenomas spontaneously, orally administered *S. boulardii* (Biocodex) reduced tumour growth and dysplasia. In the human cancer colonic cell lines, HT29 and HCT-116, *S. boulardii* (Biocodex) supernatants inhibited the MAP kinases HER-2 and HER-3 in a similar manner (Chen *et al.*, 2009).

Rats fed either viable or non-viable *S. boulardii* (Biocodex) demonstrated increases in the activities of the disaccharidases lactase, sucrase, and maltase associated with brush border membranes (BBM) (Buts *et al.*, 1986). Similar increases were seen in humans, but no alteration of the intestinal mucosa was seen in this study. A follow-up study showed that live *S. boulardii* stimulated secretory IgA and the secretory component of immunoglobulins in the rat small intestine (Buts *et al.*, 1990). These trophic effects of *S. boulardii* are due, in part, to the release of polyamines (spermine and spermidine) from the yeast (Buts *et al.*, 1994). Other factors shown to be released include invertase, α,α-trehalase, an amino peptidase and a protein phosphatase active against some *E. coli* endotoxins (Buts *et al.*, 2002, 2008; Buts, 2009). Buts and De Keyser (2010a) investigated the transduction pathway activated by *S. boulardii* in rat intestinal epithelium. The pathways induced were the RAS-GAP, RAF-1, and ERK1/2 pathways. The authors conclude that the interaction of *S. boulardii* with the intestinal BBM is key to its probiotic effects (Buts and De Keyser, 2010b). *S. boulardii* (Bioflor®, Kuhnil Pharm.) was shown to up-regulate PPAR-γ (peroxisome proliferator-activated receptor) and down-regulate IL-8 (Interleukin-8) in a trinitrobenzene sulfonic acid (TNBS) induced colitis model in rats (Lee *et al.*, 2009). Its action was via inhibition of TNF-α activation by NF-κB. Expression of pro-inflammatory cytokine genes was suppressed. These authors found similar results, *in vitro*, using human HT-29 colon cells. Other work with human HT-29 colonocytes has shown that *S. boulardii* produces a small (<1 kDa) water-soluble, heat-stable, anti-inflammatory factor (SAIF) active against NF-κB-mediated IL-8 gene expression (Sougioultzis *et al.*, 2006). Soyturk, *et al.* (2012) also studied the effect of *S. boulardii* on TNBS induced colitis in rats. They found using their strain of *S. boulardii* (Biocodex), in their rat model, that there was no improvement in the Disease Activity Index or in colonic inflammation. However, the yeast did reduce tissue Nitric Oxide (NO) levels. Pothoulakis (2009) has reviewed the anti-inflammatory mechanisms of *S. boulardii*.

Golden Syrian hamsters have been used as a model system to study antibiotic induced *C. difficile* infections and pseudomembranous colitis. Hamsters administered clindamycin, or other antibiotics, develop caecitis due to *C. difficile* overgrowth in the intestine. Toothaker and Elmer (1984) conducted 10 studies over the course of 4 months involving 143 hamsters. *S. boulardii* (Biocodex) administered in drinking water 3 days prior, and continuing 10 days after 0.6 mg/kg clindamycin administration, reduced mortality from 81% to 51%. At a clindamycin dose of 0.32 mg/kg, *S. boulardii* was completely protective whereas 15% mortality was seen in the control group. *C. difficile* produces two virulence factors known as toxins A and B which contribute to intestinal inflammation (Sunenshine and McDonald, 2006). A 54 kDa serine protease was isolated from *S. boulardii* (Biocodex) supernatants. Experiments with rat ileal loops and partially purified protease showed that *S. boulardii* protection was via reduction of toxin A binding to brush border membranes and by toxin proteolysis (Castagliuolo *et al.*, 1996).

Fractionated probiotic yeast products and animal health

By the Salminen *et al.* (1999) definition of probiotics, cell preparations or cell components can be considered, although the WHO/FAO definition would not support this. The composition of primary grown dry yeast is roughly 50% protein, 10% nucleic acid, 6% ash, 5% fat, and 4% moisture. The remaining 25% is the yeast cell wall, 80–90% of which is composed of polysaccharides. β-1,3-glucan interspersed with small amounts of chitin forms the inner cell wall; β-1,6-glucan links inner and outer cell wall components. The outer wall is mannans, mostly in the form of mannoproteins. These mannoproteins can be either covalently bound to the cell wall or associated by ionic interactions; the former can be released enzymatically, the latter by detergents and/or heat treatment (Lipke and Ovalle, 1998). The mannooligosaccharides (MOS) associated with the yeast cell wall mannoproteins fall into two classes based on the protein linkage: those linked to a nitrogen atom (N-linked) and those linked to an oxygen atom (O-linked). O-linked mannooligosaccharides can be easily released from protein by treatment with dilute alkali; release of N-linked oligosaccharides requires enzymatic treatment.

Cell walls derived from yeast extract manufacture can be processed into a β-glucan fraction and a MOS fraction. Some classes of β-1,3-glucans are known to stimulate animal immune systems; both soluble and particulate fractions of yeast β-glucans have been evaluated for their immune-potentiating capabilities. Both fractions are able to bind to the Dectin-1 receptor on macrophages and neutrophils, but only the particulate fraction binds and activates this pattern-recognition receptor (Goodridge *et al.*, 2011). Consequently, a properly prepared β-glucan fraction from yeast can be a potent immune stimulant. In fact, yeast derived β-glucans are under evaluation in cancer immunotherapeutic studies (Liu *et al.*, 2009).

Fractionated yeast cell wall material, in various, often unspecified, forms of purity have been used for their probiotic and other properties in animal feeds. They will only be briefly discussed here.

Livestock

Newborn calves have immature immune systems and rely on antibodies in their dam's colostrum for passive immunity. MOS fed to cows (10 g/day) for at least 3 weeks prior to calving demonstrated higher immunoglobulin levels in their colostrum and increased efficiency of vaccine antibody transfer to their calves (van Rensburg, 2009). In feed lot heifers, yeast cell wall has been shown to modify the stress response to administered lipopolysaccharide (LPS) from *E. coli*. The researchers found that in post LPS challenged, cell wall fed heifers, vaginal temperature increase was reduced; stress related serum cortisol levels were lower; serum INF-γ and TNF-α were not affected, but serum levels of IL-6 increased more in control heifers than the cell wall group (Burdick Sanchez *et al.*, 2013). In a study comparing 'yeast culture' (Vi-Cor) with 'yeast culture' plus EHY (enzymatically hydrolysed yeast) in early lactating dairy cattle it was shown that cows fed 'yeast culture' and 'yeast culture' + EHY both demonstrated improved milk production performance. But, those cows fed 'yeast culture' + EHY had less than half the cases of clinical mastitis than 'yeast culture' fed cows (Nocek *et al.*, 2011).

Swine

White *et al.* (2002) investigated the use of dried brewers' yeast as a source of MOS in weanling pigs. After 29 days on diets containing a control diet, diet with antibiotic (carbadox), or a diet with 3% yeast, the pigs were challenged with *E. coli* K88. Only minimal effects on growth, microbial populations, or intestinal health were seen between the antibiotic and yeast diet. The yeast diet reduced total coliforms in the intestines, but did not have a consistent effect on K88 colonization. Yeast fed pigs tended to have higher levels of intestinal lactobacilli and serum IgG levels. IgA and IgM were not affected. Zhao, *et al.* (2012) used a commercial, process enriched MOS to study growth performance in weanling pigs. Control, antibiotic (apramycin), and MOS diets were compared over the course of 28 days. A fructan and a fructan + MOS diet were also compared. During days 0–14, average daily gain was highest in the antibiotic, MOS, and fructan + MOS diets. The average daily feed intake was also higher in these diets. However, there were no differences between diets during days 15–28. Dry matter and nitrogen apparent total tract digestibilities were higher in these diets. While no differences in immunological markers, e.g. IgG, were seen, pigs fed the MOS diet had a lower diarrhoea score.

Poultry

Cell walls from brewers' yeast, *S. cerevisiae* var. *carlsbergensis*, were field tested in 44,000 broiler chicks at a 0.2% dietary inclusion rate (Santin *et al.*, 2001). Cell wall improved weight gain and this was attributed to the trophic effects on the intestinal mucosa, i.e. an increase in villus height during the first 7 days of life. Dietary yeast extracts, whole yeast, and yeast cell walls were compared in a study with broiler chicks (Zhang *et al.*, 2005). From 0 to 3 weeks, the feed to weight gain ratio was lower in cell wall fed chicks; but, from 0 to 5 weeks, cell wall and whole yeast fed chicks had more weight gain. Yeast extract and cell wall fed birds had tissue (thigh and breast meat) oxidation reducing effects. Ileal villus development was slightly better in cell wall fed chicks. Martínez *et al.* (2010) studied cell wall versus antibiotic (zinc bacitracin) fed laying hens housed in two environments: pens and floor with straw litter. Floor egg production systems are dirtier than pen systems, more stressful, and produce eggs of lesser quality. Bacitracin and cell wall fed layers in the floor system had better egg production, feed conversion ratios, egg masses, and fewer dirty eggs than hens on a control diet. It was concluded that yeast cell wall could replace the antibiotics in feeds used for floor system layers.

Challenge studies have also been performed with poultry.

One-day-old chicks were put on diets containing 0 (control) or 500 mg yeast cell wall/kg feed. Chicks were challenged at day 4, intraperitoneally, or day 4, subcutaneously with *E. coli* LPS. Chicks were monitored at days 21 and 28. LPS reduced the relative weight of the bursa of Fabricius except in chicks receiving cell wall. Cell wall also exacerbated cellular immune response to a cutaneous basophilic hypersensitivity test to phytohaemagglutinin-P (Morales-Lopez and Brufau, 2013). The authors concluded yeast cell wall increases the chicks' tolerance to microbial challenges during rearing.

Mycotoxin binding

Mycotoxins are fungal secondary metabolites that exhibit carcinogenic, nephrotoxic, mutagenic, immunosuppressive, teratogenic, and estrogenic properties. They are associated with the genera *Fusarium*, *Penicillium*, *Claviceps*, *Aspergillus*, and *Alternaria*. Toxins include tricothecenes, aflatoxins, ergot alkaloids, ochratoxins, zearalenone, and fumosins. These organisms can infect cereal grains either pathogenically or saprophytically during grain storage; toxin production dramatically affects grain quality. Mycotoxin contamination in animal feed can result in feed refusal, poor feed conversion ratios, decreased disease resistance and reproductive issues. Yeast cell walls and β-glucans fractions derived from yeast cell walls have been shown to bind some mycotoxins associated with animal feeds making them unavailable to intestinal absorption (Huwig *et al.*, 2001; Yiannikouris *et al.*, 2004; Joannis-Cassan *et al.*, 2011).

Yeast probiotics in man

Yeasts, of the genus *Saccharomyces*, are the world's oldest industrial microorganisms. These yeasts, associated with alcoholic fermentations, are thought to have been 'domesticated' over their many millennia of association with human food fermentations (Fay and Benavides, 2005; Pérez-Ortín *et al.*, 2002). Molecular analysis of pottery jar residues from Neolithic China suggests yeast-based wine fermentations date to at least 7000 BC (McGovern *et al.*, 2004). Chemical evidence from jars dating to 5400–5000 BC exists for the production of resinated wines in what is now Iran (McGovern *et al.*, 1996). Yeast cells have been detected in Egyptian brewing and baking residues dating to 2000 to 1200 BC (Samuel, 1996); and by 100 BC, ancient Rome had 250 bakeries (Rose, 1982). Man has encouraged the evolution of these microorganisms by unconsciously selecting for enhanced performance in their designated fermentation niches: the leavening of bread or the alcoholic fermentations leading to wine, ale, lager beer, sake, or distillates. Yeast has been a deliberate part of the human diet for at least 9000 years. It has only been within the last 100 years that the science of microbiology has permitted the conscious selection of improved strains.

Yeast as medicine and food

The medicinal use of yeast, primarily for constipation, derived as a by-product of wine and beer fermentations, has been described in the Ebers Papyrus (Egypt, *ca.* 1550 BC); by Greek and Roman writers; monks of the Middle Ages; and, 19th century physicians (Pierce, 1932; Frey, 1930). In 1944, fresh yeast consumption in 1 drachm doses, 3 times per day, was still advocated for the treatment of boils and furunculosis (Watson Smith, 1944). During the war time meat shortages of World War I, yeast was consumed by the German population for food (Funk *et al.*, 1916; Harrison, 1993). Academic work on yeast nutritional profiles soon followed and the value of yeast as source of water-soluble B-complex vitamins was established by the middle of the last century (Heller, 1923; Parsons *et al.*, 1945). While severe niacin deficiency was not yet recognized as the cause of pellagra, yeast was known to contain 'pellagra preventative factor' and was used to treat the disease in the 1920s (Bollet, 1992). Major yeast-producing companies capitalized on these empirical observations and produced brochures touting the healthy benefits of daily yeast consumption (Anonymous, 1920). While yeast, *per se*, has not become a primary component of the human diet, the use of dried brewers' yeast, primary grown bakers' yeast, and yeasts such as *Candida utilis* (*Torula*) as nutritional supplements has been established for at least a century (Peppler, 1970). Bakers' yeast, grown in proprietary fermentations, has been sold specifically for nutritional supplementation since the 1940s (Keipper, 1950).

Saccharomyces boulardii

Henri Boulard was a French microbiologist/mycologist who was primarily interested in industrial fermentations leading to beer, ethanol, and vinegar. He, according to most accounts, was in French Indochina (now Vietnam) in the early 1920s looking for yeasts that had higher temperature optima for beer. One of his patents, derived from a previous visit (Boulard, 1918), described a simultaneous saccharification and fermentation of grain to ethanol using fungi from the genus *Mucor* as the saccharification agent; a technique he found being used in Annam (now central Vietnam) and elsewhere in southeast Asia. This simultaneous saccharification with fermentation became known as the Boulard Process (Henrici, 1930). It has been reported that he isolated the self-named yeast, *Saccharomyces boulardii*, in 1923 from lychee and mangosteen fruits. The local population either chewed the skins or made tea from the skins of these fruits to treat diarrhoea. He noted during an outbreak of cholera that people drinking tea containing this yeast did not succumb to the disease. In 1947, he sold the rights to the organism to a French commercial enterprise (McFarland, 2010); and beginning in the 1950s, the organism was being used to treat diarrheal diseases in France (McFarland and Bernasconi, 1993). *S. boulardii* has been used successfully for the management of antibiotic-associated diarrhoea (AAD), travellers' diarrhoea (TD), and *C. difficile*-associated diarrheas for several decades (McFarland and Bernasconi, 1993; Buts, 2009; McFarland, 2010).

Taxonomy

Primarily based on carbohydrate fermentation profile (galactose negative), lack of ascospore formation, low pH tolerance, and growth temperature capability (37°C), *S. boulardii* was considered a separate species for decades. Based on recent genomic evidence at the molecular level, *S. boulardii* is now considered conspecific with *S. cerevisiae*. These molecular data include microsatellite typing (Hennequin *et al.*, 2001); restriction fragment length polymorphism (RFLP) analysis of rDNA spacer regions (McCullough *et al.*, 1998; Mitterdorfer *et al.*, 2002); D1/D2 sequences of the 26S rDNA domain and mitochondrial COX2 gene sequences (van der Aa Kühle and Jespersen, 2003); open reading frame (ORF) comparative genome hybridization (Edwards-Ingram *et al.*, 2004); Ty917 retrotransposon hybridization (Posteraro *et al.*, 2005); and chromosome plot number distribution (Edwards-Ingram *et al.*, 2007). Transposable elements (Ty) sequences can constitute up to 3% of yeast genomic DNA and their distribution within the genome varies with strain. Edwards-Ingram *et al.*, 2004; 2007) concluded that strains of *S. boulardii* are stains of *S. cerevisiae* which have lost their Ty1/2 elements by recombination; have become trisomic at chromosome IX; and have altered copy numbers of several genes. Collectively, these data demonstrate conspecificity with *S. cerevisiae*, but they also show that strains of *S. boulardii* form a separate cluster of related strains within the broader definition of the *S. cerevisiae* species. With respect to probiotic yeasts in this chapter, *S. cerevisiae* will be used for non-*boulardii*-like strains.

AAD, TD, and other diarrhoeas

Depending on the patient and antibiotic, diarrhoea ranging in severity from a slight increase in stool frequency to debilitating watery diarrhoea occurs in roughly 20% of patients taking antibiotics (Surawicz *et al.*, 1989). McFarland and Bernasconi (1993) reviewed three clinical trials using *S. boulardii* from Biocodex Laboratories. In the first trial, *S. boulardii* reduced the incidence of AAD from 17.5% to 4.5%; in the second, from 21.8% to 9.5%; and in the third, from 15.2% to 6.7%. Szajewska and Mrukowicz (2005) reviewed 16 clinical trials, five of which were random-controlled, and subjected them to a meta-analysis. They found that *S. boulardii* reduced AAD from 17.2% to 6.7% compared to placebo. Travellers' diarrhoea affects about 12,000,000 people per year. Rates range from 5% to 50% depending upon destination. Agents responsible include *E. coli*, *C. jejuni*, *Salmonella* sp., viruses such as Norwalk-like (norovirus) and rotavirus, and parasites such as *Entamoeba histolytica* and *Giardia lambica*. McFarland (2010) reviewed two randomly controlled trials involving enrolled Austrian tourists travelling to hotter climates. In the first, tourists took either 250 mg/day or 500 mg/day *S. boulardii* starting five days prior to departure and continuing throughout their travels. TD occurred in 43% of the people on placebo; 34% in the low dose *S. boulardii*; and 32% in the high dose *S. boulardii* consuming tourists. In a second trial, tourists were given either 250 mg/day or 1000 mg/day in a similar manner. TD occurred in 39% of those on placebo; 34% in the lose dose group; and, 29% in the high-dose group. A meta-analysis on TD trials showed several probiotics, including *S. boulardii*, to be effective in reducing the incidence of TD (McFarland, 2007). In a study with patients with acute amoebiasis, *S. boulardii* (Biocodex) (250 mg three times a day) was administered with antibiotics (metronidazole and iodoquinol); control subjects received just antibiotics. *S. boulardii* reduced diarrhoea duration from 48 h to 12 h; fever from 24 h to 12 h; and, abdominal pain from 24 h to 12 h. Headache duration was the same in both groups. After 4 weeks, faecal amoebic cysts were still detected in 18.5% of the control subjects, but none in the *S. boulardii* group (Mansour-Ghanaei *et al.*, 2003).

Diarrhoea is a common side-effect in patients receiving enteral nutrition. Intragastric enteral feeding results in abnormal water excretion into the ascending colon and intestinal dysbiosis can occur in those patients developing diarrhoea (Whelan *et al.*, 2009). Complications associated with enteral feeding diarrhoea include electrolyte imbalances and faecal incontinence. Numerous reviews and meta-analyses support the efficacy of *S. boulardii* in ameliorating or preventing these diarrhoeas (Czerucka *et al.*, 2007; Kelesidis and Pothoulakis, 2012; McFarland, 2009, 2010; Zanello *et al.*, 2009).

Acute diarrhoea in children

Infectious diarrheal diseases kill over 2,000,000 children per year. The problem is particularly acute in developing countries where access to proper sanitary facilities, proper nutrition, and treatment are limited. Billoo *et al.* (2006) reported that the administration of 250 mg (twice per day) *S. boulardii* (Hilton Pharma (Pvt) Ltd.) reduced stool frequency and duration of diarrhoea in a cohort of 50 children (median age, 18 months). A similar study with 100 children under the age of two (50 placebo, 50 *S. boulardii*) also showed that an undisclosed commercial preparation of *S. boulardii* reduced stool frequency and diarrhoea duration. Increased efficacy was noted if *S. boulardii* was administered within the first 48 h of the onset of symptoms (Villarruel *et al.*, 2007). In a study on paediatric AAD (ages 1–15 years), it was shown that the incidence of diarrhoea was 18.9% in patients taking antibiotics without *S. boulardii*, but only 5.7% in patients taking an undisclosed source of the probiotic. Interestingly, *S. boulardii* had no effect on diarrhoea rates in children taking azitromycin, but had a significant reduction in the incidence of diarrhoea in patients receiving sulbactam-ampicillin (from 81.1% to 18.9%). Diarrhoea rates were highest in children under 6 years old (Erdeve *et al.*, 2004). A meta-analysis (Szajewska *et al.*, 2007) on five randomly controlled trials concluded that *S. boulardii* provides 'moderate clinical benefits' in acute childhood gastroenteritis, mainly in reducing the duration of the disease. Efficacy is due to toxin inhibition, anti-inflammatory effects, and stimulation of intestinal mucosal brush border enzymes and immune responses (Vandenplas *et al.*, 2009).

Clostridium difficile-associated diarrhoea

C. difficile was first described as a stool isolate from a healthy neonate in 1935. The organism is common in the environment and is carried by 1–4% of healthy adults. It was not recognized as an opportunistic pathogen until the 1970s when its role in antibiotic induced diarrhoea and colitis were noted (Kelley, 2009). *C. difficile* infection (CDI) symptoms can range from mild diarrhoea to pseudomembranous colitis with sepsis, toxic megacolon, organ failure and death. Predisposition to CDI include age > 65, antibiotic therapy and underlying overall health (Martinez *et al.*, 2012). It is the most common cause of nosocomial infectious diarrhoea in North America and Europe. In the US, CDI was associated with 3000 deaths/year in 1999 and 2000; by 2007, 14,000 per year. Nosocomial CDI health care costs were estimated to be between

$897 million to $1.3 billion. 94% of US cases are associated with receiving health care; 75% of which occurred in non-hospitalized patients such as recently released patients, outpatients and nursing home residents (McDonald *et al.*, 2012).

C. difficile produces two virulence factors, toxins A and B. These are responsible for tissue inflammation and damage, and fluid and mucous secretion leading to diarrhoea or colitis. A hypervirulent strain, NAP 1, has recently been responsible for outbreaks of severe diarrhoea. This strain produces elevated levels of toxins A and B, a third toxin, binary toxin, and is resistant to gatifloxacin and moxifloxacin (Sunenshine and McDonald, 2006). Toxin A seems to mediate intestinal damage in animal models whereas toxin B was shown to be more damaging to human colonic mucosa in Ussing chamber assays (Riegler *et al.*, 1995). The protective effects of the *S. boulardii* (Biocodex) secreted 54 kDa serine protease seen in rat ileum (Castagliuolo *et al.*, 1996) has been extended to the human intestinal mucosa (Castagliuolo *et al.*, 1999). In human colonic HT-29 cells, exposure to toxins A and B result in an inhibition of protein synthesis; purified *S. boulardii* protease reversed this inhibition. Toxin A and B exposure results in a reduction of epithelial barrier function in surface colonocytes. The presence of the *S. boulardii* serine protease reversed toxin mediated drops in transepithelial electrical resistance (TER) in excised human colonic mucosal sheets. The protease has activity against both toxin A and B, and antibody raised against the protease negated its effects.

Treatment for CDI includes antibiotic therapy, but once cleared, patients are susceptible to recurring infections. In a clinical study of patients with recurring CDI, *S. boulardii* (Biocodex), with antibiotic treatment, was shown to significantly reduce reoccurrence of CDI (McFarland *et al.*, 1994). A small study (four patients) by a group of private physicians reported commercial bakers' yeast was also effective against relapsing CDI (Chia *et al.*, 1995). A meta-analysis of six randomly controlled trials (RTC) using *S. boulardii* or lactic acid bacteria (LABs: *L. rhamnosus* GG, *L. plantarum* 299v, *L. acidophilus* + *Bifidobacterium bifidum*) as probiotics in the treatment of recurrent CDI showed only *S. boulardii* was effective against recurrent CDI. This meta-analysis also looked at 25 RTCs using LABs or *S. boulardii* in the development of AAD; both LABs and *S. boulardii* were effective in treating AAD (McFarland, 2006).

Antidiarrhoeal effects in other organisms

Vibrio cholerae

V. cholerae produces an 84 kDa heterohexameric (subunit A + 5 smaller identical subunits B, AB_5) toxin (CT). Subunit A is internalized following CT binding to receptors on enterocyte microvilli; this causes activation of adenylate cyclase resulting in a rise in cyclic-AMP (cAMP) levels. This triggers secretion of chloride and bicarbonate ions and a cessation of water absorption. Copious diarrhoea results (Spangler, 1992; Brandão *et al.*, 1998). *S. boulardii* (Floratil®, Merck) is protective in rats against cholera toxin (Brandão *et al.*, 1998). Intact CT toxin, but not subunits A or B, separately, activated a cAMP signal in *S. boulardii* suggesting a receptor on the yeast cell wall surface. The authors reported similar results with *S. cerevisiae* W303, a common academic lab strain, suggesting the receptor may be a general property of the yeast cell wall. A 120 kDa heat-labile protein has been isolated from *S. boulardii* growth medium which reduced CT-induced cAMP levels in IEC-6 intestinal epithelial cells. Activity was not attributable to proteolysis of the toxin (Czerucka *et al.*, 1994; Czerucka and Rampal, 1999).

Shigella subsp.

Like *V. cholerae*, the shigellae, such as *S. flexneri* and *S. dysenteriae*, produce an AB_5 toxin (Spangler, 1992). Infection elicits inflammatory responses via activation of NF-κB and MAPKs such as ERK and JNK. This activation leads to the production of cytokines such as IL-8. *In vitro*, the presence of *S. boulardii* (Biocodex) did not reduce the number of *S. flexneri* cells binding to, or invading T-84 human colonic cell monolayers; nor was the yeast able to prevent perturbations in TER. Presence of the yeast did, however, modestly improve barrier function to permeation by the 40 kDa marker protein, horseradish peroxidase. *S. boulardii* reduced IL-8 production and migration of polymorphonuclear leucocytes (PMNs) into the monolayer. *S. boulardii* growth media, passed through a 10 kDa filter, produced similar results. This dampened inflammatory response was confirmed, *in vivo*, with fetal colonic tissue transplants in *scid* mice (Mumy *et al.*, 2008). The authors concluded that *S. boulardii* would not prevent shigellosis, but may alleviate host inflammatory responses.

Escherichia coli

Enterotoxigenic strains of *E. coli* (ETEC) produce a heat-labile AB_5 toxin (LT) similar to CT. CT antitoxin neutralizes LT toxin (Spangler, 1992). Czerucka *et al.* (1994) demonstrated that *S. boulardii* (Biocodex) supernatants reduce LT-induced cAMP production in human colonic adenocarcinoma derived HT29-D4 cells; but, it was not effective against the AB_5 pertussis toxin from *Bordetella pertussis*.

Enteropathogenic *E. coli* (EPEC) strains initiate disease by (a) binding to intestinal epithelial cells via type IV pili; (b) inducing signalling pathways, e.g. ERK1/2, in host cells; (c) the formation of lesions caused by degeneration of intestinal microvilli and the assembly of cytoskeletal structures beneath the attached bacteria; and (d) internalization of bacteria into the epithelium. In EPEC infected T-84 cells, *S. boulardii* (Biocodex) improved TER; preserved ZO-1 protein distribution (a protein associated with tight cell to cell junctions); and reduced the permeability of infected cells. While *S. boulardii* did not reduce the numbers of adhering EPEC, it reduced the numbers of EPEC internalized by 50%. EPEC-induced activation of the ERK1/2 MAP kinase pathway was diminished in the presence of *S. boulardii* (Czerucka *et al.*, 2000).

Enterohaemorrhagic *E. coli* (EHEC) produce Shiga-like toxins (verotoxins, VT) of the AB_5 type (Spangler, 1992; Dahan *et al.*, 2002). VT may not play a role in diarrheal disease but is nephrotoxic and can lead to potentially fatal haemolytic–uraemic syndrome. EHEC infections of intestinal cells results in inflammation and loss of barrier function. In T-84 cells, infection with EHEC induces a proinflammatory release of IL-8 mediated by the activation of the MAPK AP-1, and NF-κB pathways (Dahan *et al.*, 2002). Coincubation of T-84 cell with *S. boulardii* (Biocodex) and EHEC is sufficient to prevent a decline in TER. But, to significantly reduce the levels of IL-8 production, T-84 cells require preincubation with *S. boulardii*. In EHEC infected cells, TER decline was associated with EHEC induced myosin light chain (MLC) phosphorylation. *S. boulardii* did not affect EHEC adhesion to T-84 cells, but was able to abolish the EHEC induced phosphorylation of MLC. Preincubation with *S. boulardii* reduced IL-8 by inhibiting the NF-κB and MAPK signalling pathways (Dahan *et al.*, 2003). Binding of bacteria to intestinal mucosa is important for pathogenesis. Competitive binding of bacteria by probiotic yeasts may be one mechanism for their efficacy in the treatment of diarrheal disease. Gedek (1999) screened serovars of EHEC for their binding capacity to an undesignated strain of *S. boulardii*. Serogroup O157 strongly bound to the yeast cell surface. Other serogroups O73, O128, and O103, for example, did not bind at all; while others such as O22 demonstrated intermediate binding.

Dendritic cells (DC) are thought to be involved in immunomodulation and contribute to chronic inflammatory disorders in the gut. Lipopolysaccharide from *E. coli* O55:B5 was used to study the effects of a small molecular weight (<3kDa) molecule from *S. boulardii* (Biocodex) supernatants on myeloid DC cells *in vitro*. This small molecule reduced the expression of co-stimulatory molecules CD40 and CD80 known to be involved in T-cell activation. The secretion of proinflammatory cytokines TNF-a and IL-6 were reduced while the secretion of the anti-inflammatory IL-10 increased (Thomas *et al.*, 2009).

Fungaemia

The positive impact of the use of probiotic yeasts in the prevention or amelioration of human diarrheal diseases is overwhelmingly supported by numerous clinical trials. However, there have been several reports of fungaemia associated with *S. boulardii*. Cassone *et al.* (2003) reported on *S. boulardii* fungaemia in three intensive care unit (ICU) patients who shared a unit with patients being treated with this probiotic. The infection probably arose through central venous catheter contamination and was resolved by fluconazole therapy. A biochemical panel performed on the yeasts isolated from two of the patients showed these strains to be *S. cerevisiae*; and, karyotype analysis, by pulsed field gel electrophoresis, demonstrated that these isolates had karyotypes identical to the probiotic. Similar incidences of fungaemia in ICU patients receiving *S. boulardii* by nasogastric tube have been reported (Muñoz *et al.*, 2005; Niault *et al.*, 1999). Fungemia has also been reported in patients on immunosuppressant drugs (Bassetti *et al.*, 1998). Bacteraemia and sepsis has also been reported associated with the use of bacterial probiotics. Major risk factors associated with probiotic sepsis are a debilitated state, immunosuppression and in premature infants. Minor factors include indwelling central venous catheters, impaired intestinal barrier function, administration of the probiotic via jejunostomy, and the use of broad-spectrum antibiotics (Boyle *et al.*, 2006). Consequently, the use of any probiotic in critically ill or immune-compromised patients should be carefully considered.

Emerging yeast probiotics

Strains of *S. cerevisiae* outside of the *S. boulardii* subspecies cluster have also demonstrated probiotic effects. Martins *et al.* (2005) screened 12 strains of *S. cerevisiae* for resistance to simulated gastrointestinal conditions. Strain UFMG 905, a yeast used in cachaça production, was shown to have protective effects in gnotobiotic mice challenged with either *S. enterica* serotype Typhimurium or *C. difficile*. *S. cerevisiae* UFMG 905, live and heat killed, was further studied in an obstructed bowel–*E. coli* challenged mouse model. Bacterial migration from obstructed bowel to lymph nodes, lung, spleen and liver was reduced by both live and dead yeast. Both yeast preparations also limited the uptake of ^{99m}Tc-DTPA (technetium-labelled diethylenetriaminepentaacetic acid) into the blood stream indicating improved intestinal permeability. Live and dead yeast increased levels of IL-10, but only live yeasts resulted in elevated levels of sIgA (Generoso *et al.*, 2010). In an extension of these studies with UFMG 905, it was shown that yeast inhibited the translocation of *S. enterica* serotype Typhimurium to the liver in monoassociated gnotobiotic mice. In conventional mice, UFGM 905 inhibited translocation to liver, spleen, and mesenteric lymph nodes (Martins *et al.*, 2007). UFMG 905 monoassociated mice had higher levels of intestinal IgA and higher serum levels of IgA and IgM. When challenged with *E. coli* B_{41} by tail vein injection, UFMG 905 monoassociated mice were able to clear the *E. coli* from the bloodstream faster than control animals.

van der Aa Kühle *et al.* (2005) screened 18 strains of *S. cerevisiae*, isolated from food sources, and eight isolates of *S. boulardii* from commercial sources. All strains were resistant to pH 2.5 and 0.3% Oxgall, but intestinal adhesion to the porcine jejunal epithelial cell line, IPEC-J2, was variable, even amongst the *S. boulardii* isolates. *E. coli* O139 produces the Shiga-like toxin, 2e, which induces the production of the proinflammatory cytokine IL-1a in IPEC-J2 cells. *S. cerevisiae* A18, isolated from sorghum beer, was as effective as *S. boulardii* in reducing expression of IL-1a in IPEC-J2 cells.

Foligné, *et al.* (2010) selected six strains of *Saccharomyces* sp. yeasts based on their anti-inflammatory potential (induction of cytokines IL-10, IL-12p70, interferon γ, and tumour necrosis factor) in *in vitro* assays using human, immunocompetent, PBMCs (peripheral blood mononuclear cells). The selected strains were then evaluated in a TNBS-induced colitis mouse model. They found that the results of the *in vitro* tests were not good predictors of *in vivo* efficacy. Inflammatory intestinal damage was decreased 27% by strain LV01 (*S. boulardii*) and 18% by LV02 (*S. cerevisiae* CNCM I-3856), while others had no effect at all. By contrast, the administration of prednisone decreased inflammatory damage by 66%.

A similar strain comparison study investigated *Candida albicans* intestinal colonization in a dextran sulfate -sodium (DSS)-induced colitis model (Jawhara *et al.*, 2012). Low doses of DSS were administered for two weeks to induce mild colitis; the mice were then challenged with *C. albicans*. Three days after challenge, the mice were treated with one of five different yeast strains (10^7 cfu/day). *S. boulardii* (CNCM I-3799) and Sc1-1 (*S. cerevisiae* CNCM I-3856) reduced intestinal colonization by *C.*

albicans and significantly reduced inflammation. mRNA levels of the pro-inflammatory cytokine TNF-α were reduced and mRNA levels of protective IL-10 were elevated by the probiotic yeasts. One yeast strain, Scl–4, actually promoted *C. albicans* colonization and mortality. *S. boulardii* SC5314, and its extracts, were shown to inhibit hyphae formation, adhesion, and biofilm formation by *C. albicans*. Capric acid secretion by the probiotic was shown to be responsible for this effect (Murzyn *et al.*, 2010).

The influence of *S. cerevisiae* CNCM I-3856 on *E. coli* O157:H7 has been studied in *in vitro* gastrointestinal models. TIM and ARCOL systems are used to investigate the influence of galenic form, food intake, dose regimen, etc. on viability and resident times of probiotics in the digestive tract (Blanquet-Diot *et al.*, 2012). The TIM model system mimics chyme transit through the stomach, duodenum, jejunum, and ileum. The system allows samples to be taken from each compartment. High bacterial mortality was observed in the stomach and duodenum, but bacteria proliferated in the jejunum and ileum, probably due to the higher pH and lower concentration of bile salts. The co-inoculation of the TIM with CNCM I-3856 did not lead to a reduction in *E. coli* in the stomach and duodenum compartments, but there was a significant reduction of bacteria in the jejunum and ileum. Yeast antagonism was speculated to be due to (a) competition for nutrients (b) alteration of the environment (lower redox potential), or (c) the production of inhibitors such as proteases and ethanol (Etienne-Mesmin *et al.*, 2011). These studies were extended in the ARCOL system which mimics the human colon. This system is a one stage fermentation system using human faeces as an inoculum and can be used to study the impact of treatments on faecal microbiota. With time, O157:H7 was progressively eliminated from the reactor and the presence of either *S. boulardii* (Biocodex) or CNCM I-3856 had no effect on elimination kinetics. *S. boulardii* survived the conditions of the reactor better than CNCM I-3856. Real time quantitative PCR of total genomic DNA extracted from the reactor medium showed that the intestinal microbiota was not significantly altered by the presence of O157:H7, *S. boulardii*, or *S. cerevisiae* CNCM I-3856. The production of short-chain fatty acids (SCFA) in the reactor was measured by gas chromatography. Co-administration of CNCM I-3856, but not *S. boulardii*, led to an increase in acetate production and a decrease in butyrate production. O157:H7 viability was not affected by the change in SCFA profile, but these SCFAs (high acetate, low butyrate) are thought to be modulators of virulence in EHECs (Thévenot *et al.*, 2013). The lack of impact of probiotic yeasts on the profiles of the intestinal microbiota has been demonstrated by other researchers (De Preter *et al.*, 2006; Vanhoutte *et al.*, 2006).

CNCM I-3856 may also have application in ameliorating pain in patients with irritable bowel syndrome. Work with HT-29 cell cultures suggests that I-3856 activates PPARα. The PPARα nuclear receptor has been implicated in pain modulation (Fehernbacher *et al.*, 2009). In a rat model of butyrate enema-induced colonic hypersensitivity, the administration of CNCM I-3856 decreased visceral pain perception similar to morphine (Rousseaux *et al.*, 2010). The translation of this to humans is questionable as in man butyrate has been shown to reduce pain-perception/hypersensitivity rather than induce it (Vanhoutvin *et al.*, 2009)

Hatoum *et al.* (2012) have reviewed the antimicrobial and probiotic properties of *Saccharomyces* and non-*Saccharomyces* yeasts. Some genera such as *Debaryomyces* and *Kluyveromyces* may have antimicrobial properties. Tiago *et al.* (2009) screened 103 strains of non-*Saccharomyces* yeasts using *in vitro* and *in vivo* criteria including growth at 37°C, production of antagonistic compounds against enteric pathogens, and the ability to colonize the intestinal tract of mice. 25 isolates representing the genera *Debaryomyces, Dekkera, Galactomyces, Geotrichum, Hanseniaspora, Issatchenkia, Kluyveromyces, Metschnikowia, Pichia, Rhodotorula, Saccharomycopsis, Starmerella* and *Zygosaccharomyces* were selected for growth at 37°C. These 25 isolates were screened for the production of antagonistic compounds using a panel of 14 bacterial pathogens and one yeast pathogen (*C. albicans*). *Pseudomonas aeruginosa* was susceptible to 23 of the strains; *Enterococcus faecalis* to 14; and *Clostridium perfringens* to 16. However, of these 25 yeast strains, only six were able to colonize gnotobiotic mice at high titres after 10 days of monoassociation: *Debaryomyces hansenii* 403, *Metschnikowia reukaufii* 202, *Pichia guilliermondii* DC-121-1, *Pichia kluyveri* 898, *Pichia membranifaciens* DC65–3, and *Zygosaccharomyces fermentati* 955. Of these, only three could effectively colonize conventional (non-germ free) mice, i.e. detectable in faeces more than 1 day post administration (*P. guilliermondii* was eliminated in 24 h). These three, *M. reukaufii, P. kluyveri* and *Z. fermentati* showed low to medium *in vitro* antagonistic activity to *E. faecalis, P. aeruginosa*, and/or *C. perfringens*. The protective effect of these three yeasts were tested in conventional mice challenged with *C. difficile*. Survival after 28 days post challenge were: control (no yeast), 10%; *P. kluyveri*, 60%, *M. reukaufii*, 50%, *Z. fermentati*, 30%. The yeasts in this study were isolated from insect, agricultural, and food sources suggesting that the potential exists for the development of many additional probiotic yeasts.

Other researchers have attempted to generate novel probiotic yeasts by genetic means. The meiotic spores from sporulating strains of *Saccharomyces* can be used to make inter-strain hybrids. *S. bouldarii*, however, does not sporulate, and other means must be used to affect crosses. Abosereh *et al.* (2007) used protoplast fusion between γ-ray generated, complementing, auxotrophic mutants of *S. boulardii* to produce hybrid yeasts. Fusants were screened with a panel of four bacteria for the production of antimicrobial compounds. The 'genomic shuffling' resulting from protoplast fusion generated a fusant that was more antagonistic to *Bacillus cereus*, more resistant to bile salts, and better able to grow at pH 3 than the parental strains.

Given the diversity of yeasts in the natural environment, in human foodstuffs, and the capabilities of yeast geneticists, the future looks bright for the introduction of new probiotic yeasts outside of the *S. boulardii* cluster of strains. *S. cerevisiae* CNCM I-3856 has recently been commercialized. Some of the probiotic properties of yeasts such as competitive binding and activation of innate immune responses are related to cell wall architecture. An exciting area for future research would be the modification/improvement of cell wall architecture related to probiotic functionality. Cell wall composition is strongly influenced by environmental conditions (Klis *et al.*, 2006; Aguilar-Uscanga and François, 2003). Preliminary studies have reported on changes in cell wall composition of a commercial strain of *S. boulardii* grown on glycerol (Bzducha-Wróbel *et al.*, 2013). The implication is that probiotic yeast producers may be able to influence functionality by manipulating growth conditions during production; perhaps even to the effect that a single strain, depending upon production conditions, may be tailored to multiple clinical applications. Regulatory agencies, such as EFSA and the FDA, have rigid requirements regarding claims of efficacy. For commercialization, promising new yeast probiotics will have to prove efficacy in carefully designed clinical trials.

References

Abosereh, N.A., Mohamed, H.A.-L.A., and El-Chalk, A.B.A. (2007). Genetic construction of potentially probiotic *Saccharomyces boulardii* yeast strains using intraspecific protoplast fusion. J. Appl. Sci. Res. *3*, 209–217.

Agarwal, N., Kamra, D.N., Chaudhary, L.C., Sahoo, A., and Pathak, N.N. (2000). Selection of *Saccharomyces cerevisiae* strains for use as a microbial feed additive. Lett. Appl. Microbiol. *31*, 270–273.

Aguilar-Uscanga, B., and François, J.M. (2003). A study of the yeast cell wall composition and structure in response to growth conditions and mode of cultivation. Lett. Appl. Microbiol. *37*, 268–274.

American Association of Feed Control Officials. Official Publication. (2009). Section 96. Yeast. Mika Alewynse, Section Ed., pp. 430–431.

Anonymous. (1920). Yeast for Health. The Fleischmann Company, New York.

Auclair, E. (2001). Yeast as an example of the mode of action of probiotics in monogastric and ruminant species. In Feed Manufacturing in the Mediterranean Region. Improving Safety: From Feed to Food. J. Brufau, ed. (Zaragoza, ES: CIHEAM), pp. 45–53.

Badia, R., Lizardo, R., Martinez, P., Badiola, I., and Brufau, J. (2012). The influence of dietary locust bean gum and live yeast on some digestive immunological parameters of piglets experimentally challenged with *Escherichia coli*. J. Anim. Sci. *90*, 260–262.

Baker, J.L.J. (1903). The utilization of waste yeast in breweries. Fed. Inst. Brew. *9*, 468–497.

Bassetti, S., Frei, R., and Zimmerli, W. (1998). Fungemia with *Saccharomyces cerevisiae* after treatment with *Saccharomyces boulardii*. Am. J. Med. *105*, 71–72.

Brandão, R.L., Castro, I.M., Bambirra, E.A., Amaral, S.C., Fietto, L.G., Tropia, J.M., Neves, M.J., Dos Santos, R., Gomes, N.C., and Nicoli, J.R. (1998). Intracellular signal triggered by cholera toxin in *Saccharomyces boulardii* and *Saccharomyces cerevisiae*. Appl. Environ. Microbiol. *64*, 564–568.

Billoo, A.B., Memon, M.A., Khaskheli, S.A., Murtaza, G., Iqbal, K., Saeed Shekhani, M., and Siddiqi, A.Q. (2006). Role of a probiotic (*Saccharomyces boulardii*) in management and prevention of diarrhoea. W. J. Gastroenterol. *12*, 4557–4560.

Blanquet-Diot, S., Denis, S., Chalancon, S., Chaira, F., Cardot, J.-M., and Alric, M. (2012). Use of artificial digestive systems to investigate the biopharmaceutical factors influencing the survival of probiotic yeast during gastrointestinal transit in humans. Pharm. Res. *29*, 1444–1453.

Bollet, A.J. (1992). Politics and pellagra: The epidemic of pellagra in the U.S. in the early Twentieth Century. Yale J. Biol. Med. *65*, 211–221.

Bontempo, V., Di Giancomillo, A., Savoini, G., Dell'Orto, V., and Domeneghini, C. (2006). Live yeast dietary supplementation acts upon intestinal morpho-functional aspects and growth in weanling piglets. Anim. Feed Sci. Technol. *129*, 224–236.

Borghese, A., Mazzi, M., Veloccia, C., and Larbier, M. (2007). Effect of Biosaf in dairy buffalo cows. Ital. J. Anim. Sci. *6*, 499–501.

Boulard, H. (1918). Process for the manufacture of alcohol from grain in open vats with the employment of saccharifying mucors. US 1,266,657.

Boyle, R.J., Robins-Browne, R.M., and Tang, M.L.K. (2006). Probiotic use in clinical practice: what are the risks? Am. J. Clin. Nutr. *83*, 1256–1264.

Burdick Sanchez, N.C., Young, T.R., Carroll, J.A., Corley, J.R., Rathmann, R.J., and Johnson, B.J. (2013). Yeast cell wall supplementation alters the metabolic responses of crossbred heifers in an endotoxin challenge. Innate Immun. *19*, 411–419.

Buts, J.-P. (2009). Twenty-five years of research on *Saccharomyces boulardii* trophic effects: updates and perspectives. Dig. Dis. Sci. *54*, 15–18.

Buts, J.-P., Bernasconi, P., Vaerman, J.-P., and Dive, C. (1990). Stimulation of secretory IgA and secretory component of immunoglobulins in small intestine of rats treated with *Saccharomyces boulardii*. Dig. Dis. Sci. *35*, 251–256.

Buts, J.-P., Bernasconi, P., Van Craynest, M.-P., Maldague, P., and De Meyer, R. (1986). Response of human and rat small intestinal mucosa to oral administration of *Saccharomyces boulardii*. Pediatr. Res. *20*, 192–196.

Buts, J.-P., and De Keyser, N. (2010a). Transduction pathways regulating the trophic effects of *Saccharomyces boulardii* in rat intestinal mucosa. Scand. J. Gastroenterol. *45*, 175–185.

Buts, J.-P., and De Keyser, N. (2010b). Interaction of *Saccharomyces boulardii* with intestinal Brush Border Membranes: Key to probiotic effects? J. Pediatr. Gastroenterol. Nutr. *51*, 532–533.

Buts, J.-P., De Keyser, N., and De Raedemaeker, L. (1994). *Saccharomyces boulardii* enhances rat intestinal enzyme expression by endoluminal release of polyamines. Pediatr. Res. *36*, 522–527.

Buts, J.-P., De Keyser, N., Stilmant, C., Sokal, E., and Marandi, S. (2002). *Saccharomyces boulardii* enhances N-terminal peptide hydrolysis in sucking rat small intestine by endoluminal release of a zinc-binding metalloprotease. Pediatr. Res. *51*, 528–534.

Buts, J.-P., Stilmant, C., Bernasconi, P., Neirinck, C., and De Keyser, N. (2008). Characterization of α,α-trehalase released in the intestinal lumen by the probiotic *Saccharomyces boulardii*. Scand. J. Gastroenterol. *43*, 1489–1496.

Bzducha-Wróbel, A., Kieliszek, M., and Błażejak, S. (2013). Chemical composition of the cell wall of probiotic and brewers' yeast in response to cultivation medium with glycerol as a carbon source. Eur. Food Res. Technol. *237*, 489–499.

Cartwright-Shamoon, J., Dickson, G.R., Dodge, J., and Carr, K.E. (1996). Uptake of yeast (*Saccharomyces boulardii*) in normal and rotavirus treated intestine. Gut 39, 204–209.

Cassone, M., Serra, P., Mondello, F., Girolamo, A., Scafetti, S., Pistella, E., and Venditti, M. (2003). Outbreak of *Saccharomyces cerevisiae* subtype *boulardii* fungemia in patients neighbouring those treated with a probiotic preparation of the organism. J. Clin. Microbiol. *41*, 5340–5343.

Castagliuolo, I., LaMont, J.T., Nikulasson, S.T., and Pothoulakis, C. (1996). *Saccharomyces boulardii* protease inhibits *Clostridium difficile* Toxin A effects in the rat ileum. Infect. Immun. *64*, 5225–5232.

Castagliuolo, I., Riegler, M.F., Valenick, L., LaMont, J.T., and Pothoulakis, C. (1999). *Saccharomyces boulardii* protease inhibits the effects of *Clostridium difficile* toxins A and B in human colonic mucosa. Infect. Immun. *67*, 302–307.

Chaucheyras, F., Fonty, G., Bertin, G., and Gouet, P. (1995). *In vitro* H_2 utilization by a ruminal acetogenic bacterium cultivated alone or in association with an Archaea methanogen is stimulated by a probiotic strain of *Saccharomyces cerevisiae*. Appl. Environ. Microbiol. *61*, 3466–3467.

Chaucheyras-Durand, F., and Durand, H. (2010). Probiotics in animal nutrition and health. Benef. Microb. *1*, 3–9.

Chacheyras-Durand, F., Faqir, F., Ameilbonne, A., Rozand, C., and Martin, C. (2010). Fates of acid-resistant and non-acid-resistant Shiga toxin-producing *Escherichia coli* strains in ruminant digestive contents in the absence and presence of probiotics. Appl. Environ. Microbiol. *76*, 640–647.

Chaucheyras-Durand, F., and Fonty, G. (2002). Influence of a probiotic yeast (*Saccharomyces cerevisiae* CNCM I-1077) on microbial colonization and fermentations in the rumen of newborn lambs. Microbiol. Ecol. Health Dis. *14*, 30–36.

Chaucheyras-Durand, F., Piron, A., Chevaux, E., Hall, A., Aguilar, A., Gill, R., and Dussert, L. (2012). How live yeasts used as DFMs can improve feed efficiency in ruminants: The rumen microbiota as a crucial target. In Cornell Nutrition Conference for Feed Manufacturers (Syracuse, NY), pp. 118–129.

Chen, X., Fruehauf, J., Goldsmith, J.D., Xu, H., Katchar, K.K., Koon, H.-W., Zhao, D., Kokkotou, E.G., Pothoulakis, C., and Kelly, C.P. (2009). *Saccharomyces boulardii* inhibits EGF receptor signaling and intestinal tumor growth in Apc^{min} mice. Gastroenterol. *137*, 914–923.

Chen, X., Kokkotou, E.G., Mustafa, N., Bhaskar, K.R., Sougioultzis, S., O'Brien, M., Pothoulkis, C., and Kelly, C. (2006). *Saccharomyces boulardii* inhibits ERK1/2 mitogen-activated protein kinase activation both *in vitro* and *in vivo* and protects against *Clostridium difficile* toxin A-induced enteritis. J. Biol. Chem. *281*, 24449–24454.

Chia, J.K.S., Chan, S.M., and Goldstein, H. (1995). Bakers' yeast as adjunctive therapy for relapses of *Clostridium difficile* diarrhea. Clin. Infect. Dis. *20*, 1581.

Christiansen, R., Glette, J., Lie, Ø., Torrissen, O.J., and Waagbø., R. (1995). Antioxidant status and immunity in Atlantic salmon, *Salmo salar* L., fed semi-purified diets with and without astaxanthin supplementation. J. Fish Dis. *18*, 317–328.

Chung, Y.-H., Walker, N.D., McGinn, S.M., and Beauchemin, K.A. (2011). Differing effects of two active dried yeast (*Saccharomyces cerevisiae*) strains on ruminal acidosis and methane production in nonlactating dairy cows. J. Dairy Sci. *94*, 2431–2439.

Cruz, P.M., Ibáñez, A.L., Monroy Hermosillo, O.A., and Ramírez Saad, H.C. (2012). Use of probiotics in aquaculture. ISRN Microbiol. doi: 10.5402/2012/916845.

Czerucka, D., Dahan, S., Mograbi, B., Rossi, B., and Rampal, P. (2000). *Saccharomyces boulardii* preserves the barrier function and modulates the signal transduction pathway induced in enteropathogenic *Escherichia coli*-infected T84 cells. Infect. Immun. *68*, 5998–6004.

Czerucka, D., Piche, T., and Rampal, P. (2007). Review article: yeast as probiotics – *Saccharomyces boulardii*. Aliment. Pharmacol. Ther. *26*, 767–778.

Czerucka, D., and Rampal, R. (1999). Effect of *Saccharomyces boulardii* on cAMP- and Ca^{2+}-dependent Cl^- secretion in T84 cells. Dig. Dis. Sci. *44*, 2359–2368.

Czerucka, D., Roux, I., and Rampal, P. (1994). *Saccharomyces boulardii* inhibits secretagogue-mediated adenosine 3′,5′-cyclic monophosphate induction in intestinal cells. Gastroenterol. *106*, 65–72.

Dahan, S., Busuttil, V., Imbert, V., Peyron, J.-P., Rampal, P., and Czerucka, D. (2002). Enterohemorrhagic *Escherichia coli* infection induces Interleukin-8 production via activation of mitogen-activated protein kinases and the transcription factors NF-κB, AP-1 in T84 cells. Infect. Immun. *70*, 2304–2310.

Dahan, S., Dalmasso, G., Imbert, V., Peyron, J.-P., Rampal, P., and Czerucka, D. (2003). *Saccharomyces boulardii* interferes with Enterohemorrhagic *Escherichia coli*-induced signaling pathways in T84 cells. Infect. Immun. *71*, 766–773.

Dawson, K.A., Newman, K.E., and Boling, J.A. (1990). Effects of microbial supplements containing yeast and lactobacilli on roughage-fed ruminal microbial activities. J. Anim. Sci. *68*, 3392–3398.

Denev, S.A., Peeva, Tz., Radulova, P., Stancheva, N., Staykova, G., Beev, G., Todorova, P., and Tchobanova, S. (2007). Yeast cultures in ruminant nutrition. Bulg. J. Agricul. Sci. *13*, 357–374.

de Ondarza, M.B., Sniffen, C.J., Dussert, L., Chevaux, E., Sullivan, J., and Walker, N. (2010). Case study: Multiple-study analysis of the effect of live yeast on milk yield, milk component content and yield, and feed efficiency. Prof. Anim. Sci. *26*, 661–666.

De Preter, V., Coopmans, T., Rutgeerts, P., and Verbeke, K. (2006). Influence of long-term administration of lactulose and *Saccharomyces boulardii* on the colonic generation of phenolic compounds in healthy human subjects. J. Am. Coll. Nutr. *25*, 541–549.

Desnoyers, M., Giger-Reverdin, S., Bertin, G., Duvaux-Ponter, C., and Sauvant, D. (2009). Meta-analysis of the influence of *Saccharomyces cerevisiae* supplementation on ruminal parameters and milk production of ruminants. J. Dairy Sci. *92*, 1620–1632.

Dimitroglou, A., Merrifield, D.L., Carnevali, O., Picchietti, S., Avella, M., Daniels, C., Güroy, D., and Davies, S.J. (2011). Microbial manipulations to improve fish health and production – A Mediterranean perspective. Fish Shellfish Immunol. *30*, 1–16.

Domeneghini, C., Di Giancamillo, A., Arrighi, S., and Bosi, G. (2006). Gut-trophic feed additives and their effects upon the gut structure and intestinal metabolism. State of the art in the pig, and perspectives towards humans. Histol. Histopathol. *21*, 273–283.

Durand-Chaucheyras, F., Fonty, G., Bertin, G., Thévenoit, M., and Gouet, P. (1998). Fate of Levucell® SC-1077 yeast additive during digestive transit in lambs. Reprod. Nutr. Dev. *38*, 275–280.

Eckles, C.H., Williams, V.M., Wilbur, J.W., Palmer, L.S., and Harshaw, H.M. (1924). Yeast as a supplementary feed for calves. J. Dairy Sci. *7*, 421–439.

Eckles, C.H., and Williams, V.M. (1925). Yeast as a supplementary feed for lactating cows. J. Dairy Sci. *8*, 89–93.

Edwards-Ingram, L.C., Gent, M.E., Hoyle, D.C., Hayes, A., Stateva, L.I., and Oliver, S.G. (2004). Comparative genomic hybridization provides new insights into the molecular taxonomy of the *Saccharomyces sensu stricto* complex. Genome Res. *14*, 1043–1051.

Edwards-Ingram, L.C., Gitsham, P., Burton, N., Warhurst, G., Clarke, I., Hoyle, D., Oliver, S.G., and Stateva, L. (2007). Genotypic and physiological characterization of *Saccharomyces boulardii*, the probiotic strain of *Saccharomyces cerevisiae*. App. Environ. Microbiol. 73, 5458–5467.

Erdeve, O., Tiras, U., and Dallar, Y. (2004). The probiotic effect of *Saccharomyces boulardii* in a paediatric age group. J. Trop. Pediatr. *50*, 234–238.

Etienne-Mesmin, L., Livrelli, V., Privat, M., Denis, S., Cardot, J.-M., Alric, M., and Blanquet-Diot, S. (2011). Effect of a new probiotic *Saccharomyces cerevisiae* strain on survival of *Escherichia coli* O157:H7 in a dynamic gastrointestinal model. Appl. Environ. Microbiol. *77*, 1127–1131.

FAO/WHO (2001). Health and nutritional properties of probiotics in food including including powder milk with live lactic acid bacteria. Report of a joint FAO/WHO expert consultation on evaluation of health and nutritional properties of probiotics including powder milk with live lactic acid bacteria. p. 2.

Fay, J.C and Benavides, J.A. (2005). Evidence for domesticated and wild populations of *Saccharomyces cerevisiae*. PLoS Genet. doi: 10.1371/journal.pgen.0010005.

Fazeli, Z., and Azari-Takami, G. (2006). The best time and concentration for yeast probiotic enrichment in *Artemia urmiana* nauplii. Pakist. J. Biol. Sci. *9*, 2159–2161.

Fehrenbacher, J.C., LoVerme, J., Clarke, W., Hargreaves, K.M., Piomelli, D., and Taylor, B.K. (2009). Rapid pain modulation with nuclear receptor ligands. Brain Res. Rev. *60*, 114–124.

Ferraretto, L.F., Shaver, R.D., and Bertics, S.J. (2012). Effect of dietary supplementation with live-cell yeast at two dosages on lactation performance, ruminal fermentation, and total-tract nutrient digestibility in dairy cows. J. Dairy Sci. *95*, 4017–4028.

Foligné, B., Dewulf, J., Vandederckove, P., Pignède, G., and Pot, B. (2010). Probiotic yeasts: Anti-inflammatory potential of various non-pathogenic strains and experimental colitis in mice. World J. Gastroenterol. *16*, 2134–2145.

Frey, C.N. (1930). History and development of the modern yeast industry. Indust. Engin. Chem. *22*, 1154–1162.

Fuller, E.A. (1927). Brewing by-products. J. Inst. Brew. 33, 416–429.

Fuller, R. (1989). A Review. Probiotics in man and animals. J. Appl. Bacteriol. *66*, 365–378.

Funk, C., Lyle, W.G., and McCaskey, D. (1916). The nutritive value of yeast, polished rice, and white bread, as determined by experiments on man. J. Biol. Chem. 27, 173–191.

Gao, J., Zhang, H.J., Yu, S.H., Wu, S.G., Yoon, I., Quigley, J., Gao, Y.P., and Qi, G.H. (2008). Effects of yeast culture in broiler diets on performance and immunomodulatory functions. Poult. Sci. *87*, 1377–1384.

Gatesoupe, F.-J. (2002). Probiotic and formaldehyde treatments of *Artemia* nauplii as food for larval Pollack, *Pollachius pollachius*. Aquacult. *212*, 347–360.

Gedek, B.R. (1999). Adherence of *Escherichia coli* serogroup O157 and the *Salmonella typhimurium* mutant DT 104 to the surface of *Saccharomyces boulardii*. Mycoses 42, 261–264.

Generoso, S.V., Viana, M., Santos, R., Martins, F.S., Machado, J.A.N., Arantes, R.M.E., Nicoli, J.R., Correia, M.I.T.D., and Cardoso, V.N. (2010). *Saccharomyces cerevisiae* strain UFMG 905 protects against bacterial translocation, preserves gut barrier integrity and stimulates the immune system in a murine intestinal obstruction model. Arch. Microbiol. *192*, 477–484.

Goodridge, H.S., Reyes, C.N., Becker, C.A., Katsumoto, T.R., Ma, J., Wolf, A.J., Bose, N., Chan, A.S.H., Magee, A.S., Danielson, M.E., Weiss, A., and Vasilakos, J.P. (2011). Activation of the innate immune receptor Dectin-1 upon formation of a 'phagocytic synapse'. Nature 472, 471–476.

Grau, C.R., and Kamei, M. (1949). Yeast culture investigations. Live yeast culture tested for production efficiency as feedstuff for chicks. Calif. Agricult. July. pp. 13.

Harrison, J.S. (1993). Food and fodder yeasts. In The Yeasts, 2nd edn. Vol. 5. A.H. Rose and J. Stuart Harrison, eds. (London, UK: Academic Press), pp. 399–433.

Hatoum, R., Labrie, S., and Fliss, I. (2012). Antimicrobial and probiotic properties of yeasts: from fundamental to novel applications. Front. Microbio. *3*, 421. doi:10.3389/fmicb.2012.00421.

Heller, V.G. (1923). Studies on yeast. V. The vitamin B content of yeast. J. Biol. Chem. *55*, 385–398.

Hennequin, C., Thierry, A., Richard, G.F., Lecointre, G., Nguyen, H.V., Gaillardin, C., and Dujon, B. (2001). Microsatellite typing as a tool for identification of *Saccharomyces cerevisiae* strains. J. Clin. Microbiol. *39*, 551–559.

Henrici, A.T. (1930). Biological activities of yeasts. In Henicri's Molds, Yeasts, and Actinomycetes, 2nd edn (1947). C.E. Skinner, C.W. Emmons, and H.M. Tsuchiya, eds. (New York, USA: John Wiley & Sons), pp 338.

Hossain, S.A., Parnerkar, S., Haque, N., Gupta, R.S., Kumar, D., and Tyagi, A.K. (2012). Influence of dietary supplementation of live yeast (*Saccharomyces cerevisiae*) on nutrient utilization, ruminal and biochemical profiles of Kankrej calves. Int. J. Appl. Anim. Sci. *1*, 30–38.

Huwig, A., Freimund, S., Käppeli, O., and Dutler, H. (2001). Mycotoxin detoxification of animal feed by different absorbents. Toxicol. Lett. *122*, 179–188.

Iranshani, F., Faramarzi, M., Kiaalvandi, S., and Lashkar Boloki, M. (2011). The enhancement of growth and feeding performance of Persian Sturgeon (*Acipenser periscus*) larvae by *Artemia urmiana* nauplii bioencapsulated via bakers' yeast (*Saccharomyces cerevisiae*). J. Anim. Vet. Adv. *10*, 2730–2735.

Jang, Y.D., Kang, K.W., Piao, L.G., Jeong, T.S., Auclair, E., Jonvel, S., D'Inca, R., and Kim, Y.Y. (2013). Effects of live yeast supplementation to gestation and lactation diets on reproductive performance, immunological parameters and milk composition in sows. Livestock Sci. *152*, 167–173.

Jawhara, S., Habib, K., Maggiotto, F., Pignede, G., Vandekerckove, P., Maes, E., Dubuquoy, L., Fontaine, T., Guerardel, Y., and Poulain, D. (2012). Modulation of intestinal inflammation by yeasts and cell wall extracts: Strain dependence and unexpected anti-inflammatory role of glucan fractions. PLoS One *7(7)*, e40648. doi: 10.1371/journal.pone.0040648.

Joannis-Cassan, C., Tozlovanu, M., Hadjeba-Medjdoub, K., Ballet, N., and Pfohl-Leszkowicz, A. (2011). Binding of zearalenone, aflatoxin B, and orchratoxin A by yeast-based products: A method for quantification of adsorption performance. J. Food Protect. *74*, 1175–1185.

Jouany, J.-P. (2001). A new look at yeast cultures as probiotics for ruminants. Feed Mix 9, 17–19.

Jouany, J.-P., Gobert, J., Medina, B., Bertin, G., and Julliand, V. (2008). Effect of live yeast culture supplementation on apparent digestibility and rate of passage in horses fed a high-fiber or high-starch diet. J. Anim. Sci. *86*, 339–347.

Jouany, J.-P., Medina, B., Bertin, G., and Julliand, V. (2009). Effect of live yeast culture supplementation on hindgut microbial communities and their polysaccharidase and glycoside hydrolases activities in horses fed a high-fiber or high-starch. diet. J. Anim. Sci. *87*, 2844–2852.

Keipper, C.H. (1950). Five keys to growing nutritional yeast. Food Indust. 22, 1375–1380.

Kelesidis, T., and Pothoulakis, C. (2012). Efficacy and safety of the probiotic *Saccharomyces boulardii* for the prevention and therapy of gastrointestinal disorders. Ther. Adv. Gastroenterol. *5*, 111–125.

Kelly, C.P. (2009). A 76-year-old man with recurrent *Clostridium difficile*-associated diarrhoea. Review of *C. difficile* infection. JAMA *301*, 954–962.

Keyser, S.A., McMeniman, J.P., Smith, D.R., MacDonald, J.C., and Galyean, M.L. (2007). Effects of *Saccharomyces cerevisiae* subspecies *boulardii* CNCM I-1079 on feed intake by healthy beef cattle treated with florfenicol and on the health and performance of newly received beef heifers. J. Anim. Sci. *85*, 1264–1273.

Klis, F.M., Boorsma, A., and De Groot, P.W. (2006). Cell wall construction in *Saccharomyces cerevisiae*. Yeast 23, 185–202.

Křížová, L., Richter, M., Třináctý, J., Říha, J., and Kumprechtová, D. (2011). The effect of feeding live yeast cultures on ruminal pH and redox potential in dry dairy cows as continuously measured by a new wireless device. Czech. J. Anim. Sci. *56*, 37–45.

Kung Jr., L. (2001). Developing rumen fermentation with direct fed microbials. Feed Mix *9(3)*, 21–24.

Lara-Flores, M., Olvera-Novoa, M.A., Guzmán-Méndez, B.E., and López-Madrid, W. (2003). Use of the bacteria *Streptococcus faecium* and *Lactobacillus acidophilus*, and the yeast *Saccharomyces cerevisiae* as growth promoters in Nile Tilapia (*Oreochromis niloticus*). Aquacult. *216*, 193–201.

Lee, S.K., Kim, Y.W., Chi, S.-G., Joo, Y.-S., and Kim, H.J. (2009). The effect of *Saccharomyces boulardii* on human colon cells and inflammation in rats with trinitrobenzene sulfonic acid-induced colitis. Dig. Dis. Sci. *54*, 255–263.

Lessard, M., Dupuis, M., Gagnon, N., Nadeau, É., Matte, J.J., Goulet, J., and Fairbrother, J.M. (2009). Administration of *Pediococcus acidilactici* or *Saccharomyces cerevisiae boulardii* modulates development of porcine mucosal immunity and reduces intestinal bacterial translocation of *Escherichia coli* challenge. J. Anim. Sci. *87*, 922–934.

Lilly, D.M., and Stillwell, R.H. (1965). Probiotics: Growth-promoting factors produced by microorganisms. Science *147*, 747–748.

Lim, E.M., Lam, T.J., and Ding, J.L. (2005). Single-cell protein diet of a novel recombinant vitelloginin yeast enhances growth and survival of first-feeding Tilapia (*Oreochromis mossambicus*) larvae. J. Nutr. *135*, 513–518.

Line, J.E., Bailey, J.S., Cox, N.A., Stern, N.J., and Tompkins, T. (1998). Effect of yeast-supplementation on *Salmonella* and *Campylobacter* populations in broilers. Poult. Sci. *77*, 405–410.

Lipke, P.L., and Ovalle, R. (1998). Cell wall architecture in yeast: New structure and new challenges. J. Bacteriol. *180*, 3735–3740.

Liu, J., Gunn, L., Hansen, R., and Yan, J. (2009). Combined yeast-derived β-glucan with anti-tumor monoclonal antibody for cancer immunotherapy. Exptl. Molec. Pathol., *86*, 208–214.

Mansour-Ghanaei, F., Dehbashi, N., Yazdanparast, K., and Shafaghi, A. (2003). Efficacy of *Saccharomyces boulardii* with antibiotics in acute amoebiasis. World J. Gastroenterol. *9*, 1832–1833.

Marden, J.P., Bayourthe, C., Auclair, E., and Moncoulon, R. (2013). A bioenergetic-redox approach to the effect of live yeast on ruminal pH during acidosis in dairy cow. Amer. J. Anal. Chem. *4*, 60–68.

Marden, J.P., Bayourthe, C., Enjalbert, F., and Moncoulon, R. (2005). A new device for measuring kinetics of ruminal pH and redox potential in dairy cattle. J. Dairy Sci. *88*, 277–281.

Marden, J.P., Julien, C., Monteils, V., Auclair, E., Moncoulon, R., and Bayourthe, C. (2008). How does live yeast differ from sodium bicarbonate to stabilize ruminal pH in high-yielding dairy cows? J. Dairy Sci. *91*, 3528–3535.

Martínez, B.F., Arias Contreras, and Ávila González, E. (2010). Use of *Saccharomyces cerevisiae* cell wall in diets for two genetic strains of laying hens reared in floor and cages. Int. J. Poult. Sci. *9*, 105–108.

Martinez, F.J., Leffler, D.A., and Kelly, C.P. (2012). *Clostridium difficile* outbreaks: prevention and treatment strategies. Risk Manag. Healthcare Pol. *5*, 55–64.

Martins, F.S., Dalmasso, G., Arantes, R.M.E., Doye, A., Lemichez, E., Lagadec, P., Imbert, V., Peyron, J.-F., Rampal, P., Nicoli, J.R., and Czerucka, D. (2010). Interaction of *Saccharomyces boulardii* with *Salmonella enterica* serovar Typhimurium protects mice and modifies T84 cell response to the infection. PLoS One *5*, e8925. doi:10.1371/journal.pone.0008925.

Martins, F.S., Nardi, R.M.D., Arantes, R.M.E., Rosa, C.A., Neves, M.J., and Nicoli, J.R. (2005). Screening yeasts as probiotic based on capacities to colonize the gastrointestinal tract and to protect against enteropathogenic challenge in mice. J. Gen. Appl. Microbiol. *51*, 83–92.

Martins, F.S., Rodrigues, A.C.P., Tiago, F.C.P., Penna, F.J., Rosa, C.A., Arantes, R.M.E., Nardi, R.M.D., Neves, M.J., and Nicoli, J.R. (2007). *Saccharomyces cerevisiae* strain 905 reduces the translocation of *Salmonella enterica* serotype Typhimurium and stimulates the immune system in gnotobiotic and conventional mice. J. Med. Microbiol. *56*, 352–359.

Martins, F.S., Veloso, L.C., Arantes, R.M.E., and Nicoli, J.R. (2009a). Effects of yeast probiotic formulation on viability, revival and protection against infection with *Salmonella enterica* subsp. *enterica* serovar Typhimurium in mice. Lett. Appl. Microbiol. *49*, 738–744.

Martins, F.S., Silva, A.A., Vieira, A.T., Barbosa, F.H.F., Arantes, R.M.E., Teixeira, M.M., and Nicoli, J.R. (2009b). Comparative study of *Bifidobacterium animalis, Escherichia coli, Lactobacillus casei,* and *Saccharomyces boulardii* probiotic properties. Arch. Microbiol. *191*, 623–630.

McCullough, M.J., Clemons, K.V., McCusker, J.H., and Stevens, D.A. (1998). Species identification and virulence attributes of *Saccharomyces boulardii* (*nom. inval.*). J. Clin. Microbiol. *36*, 2613–2617.

McDonald, L.C., Lessa, F., Sievert, D., Wise, M., Herrera, R., Gould, C., Malpiedi, P., Dudeck, M., Srinivasan, A., Fridkin, S., and Cardo, D. (2012). Vital signs: Preventing *Clostridium difficile* infections. Morb. Mortal. Weekly Rep. *61*, 157–162.

McFarland, L.V. (2006). Meta-analysis of probiotics for the prevention of antibiotic associated diarrhea and the treatment of *Clostridium difficile disease*. Am. J. Gastroenterol. *101*, 812–822.

McFarland, L.V. (2007). Meta-analysis of probiotics for the prevention of traveler's diarrhea. Trav. Med. Infect. Dis. *5*, 97–105.

McFarland, L.V. (2009). Evidence-based review of probiotics for antibiotic-associated diarrhea and *Clostridium difficile* infections. Anaerobe *15*, 274–280.

McFarland, L.V. (2010). Systemic review and meta-analysis of *Saccharomyces boulardii* in adult patients. World J. Gastroenterol. *16*, 2202–2222.

McFarland, L., and Bernasconi, P. (1993). *Saccharomyces boulardii*: A review of an innovative biotherapeutic agent. Microb. Ecol. Health Dis. *6*, 157–171.

McFarland, L., Surawicz, C.M., Greenberg, R.N., Fekety, R., Elmer, G.W., Moyer, K.A., Melcher, S.A., Bowen, K.E., Cox, J.L., Noorani, Z., Harrington, G., Rubin, M., and Greenwald, D. (1994). A randomized placebo-controlled trial of *Saccharomyces boulardii* in combination with standard antibiotics for *Clostridium difficile* disease. JAMA *271*, 1913–1918.

McGovern, P.E., Glusker, D.L., Exner, L.J., and Voigt, M.M. (1996). Neolithic resinated wine. Nature *381*, 480–481.

McGovern, P.E., Zhang, J., Tang, J., Zhang, Z., Hall, G.R., Moreau, R.A., Nuñez, A., Butrym, E.D., Richards, M.P., Wang, C., Cheng, G., Zhao, A., and Wang, C. (2004). Fermented beverages of pre- and proto-historic China. Proc. Natl. Acad. Sci. U.S.A. *101*, 17593–17598.

Medina, B., Girard, I.D., Jacotot, E., and Julliand, V. (2002). Effect of a preparation of *Saccharomyces cerevisiae* on microbial profiles and fermentation patterns in the large intestine of horses fed a high fibre or high starch diet. J. Anim. Sci. *80*, 2600–2609.

Milewski, S., Sobiech, P., Zabek, K., Zarczyńska, K., Antoszkiewicz, Z., and Wielgosz-Groth, Z. (2012). Effect of *Saccharomyces cerevisiae* on milk protein content and composition and serum mineral concentrations in sheep. J. Elementol. *17*, 79–86.

Mitterdorfer, G., Mayer, H.K., Kneifel, W., and Viernstein, H. (2002). Clustering of *Saccharomyces boulardii* strains within the species *S. cerevisiae* using molecular typing techniques. J. Appl. Microbio. *93*, 521–530.

Moncoulon, R., and Auclair, E. (2005). A new concept to explain yeast's effects in ruminants. Feed Mix *13*, 14–16.

Monroy-Salazar, H.G., Pérez-Sotelo, L., González-Hernández, Y., Vaughn, G., Lagunas-Bernabé, S., Cuarón-Ibargüengoytia, J., Montaño-Hirose, J.A., Alonso-Fresán, M.U., Pradal-Roa, P., and Vázquez-Chagoyán, J.C. (2012). Effects of a live yeast supplement on faecal coliform counts and on peripheral blood $CD4^+$, and $CD8^+$ lymphocyte subpopulations in nursery pigs. J. Swine Health Prod. *20*, 276–282.

Morales-Lopez, R., and Brufau, J. (2013). Immune-modulatory effects of dietary *Saccharomyces cerevisiae* cell wall in broiler chickens inoculated with *Escherichia coli* lipopolysaccharide. Brit. Poult. Sci. *54*, 247–251.

Mortimer, R.K., and Contopoulou, R. (1991). Yeast Genetic Stock Center Catalog, 7th edn. University of California, Berkeley. pp. 5.

Mosoni, P., Chaucheyras-Durand, F., Béra-Maillet, C., and Forano, E. (2007). Quantification by real-time PCR of cellulolytic bacteria in the rumen of sheep after supplementation of a forage diet with readily fermentable carbohydrates: effect of a yeast additive. J. Appl. Microbiol. *103*, 2676–2685.

Mumy, K.L., Chen, X., Kelly, C.P., and McCormick, B.A. (2008). *Saccharomyces boulardii* interferes with *Shigella* pathogenesis by postinvasion signalling events. Am. J. Physiol. Gastrointest. Liver Physiol. *294*, G599–G609.

Muñoz, P., Bouza, E., Cuenca-Estrella, M., Eiros, J.M., Pérez, M.J., Sánchez-Somolinos, M., Rincón, C., Hortal, J., and Paláez, T. (2005). *Saccharomyces cerevisiae* fungemia: An emerging infectious disease. Clin. Infect. Dis. *40*, 1625–1634.

Murzyn, A., Krasowska, A., Stefanowicz, P., Dziadkowiec, D., and Łukaszewicz, M. (2010). Capric acid secreted by *Saccharomyces boulardii* inhibits *C. albicans* filamentous growth, adhesion and biofilm formation. PLoS One *5(8)*, e12050. doi: 10.1371/journal.pone.0012050.

Nakano, T., Kanmuri, T., Sato, M., and Takeuchi, M. (1999). Effect of astaxanthin rich red yeast (*Phaffia rhodozyma*) on oxidative stress in Rainbow Trout. Biochim. Biophys. Acta *1426*, 119–125.

Newbold, C.J., McIntosh, F.M., and Wallace, R.J. (1998). Changes in the microbial population of a rumen-stimulating fermenter in response to yeast culture. Can. J. Anim. Sci. *78*, 241–244.

Newbold, C.J., Wallace, R.J., Chen, X.B., and McIntosh, F.M. (1995). Different strains of *Saccharomyces cerevisiae* differ in their effects on ruminal bacterial numbers *in vitro* and in sheep. J. Anim. Sci. *73*, 1811–1818.

Newbold, C.J., Wallace, R.J., and McIntosh, F.M. (1996). The mode of action of *Saccharomyces cerevisiae* as a feed additive for ruminants. Br. J. Nutr. *76*, 249–261.

Niault, M., Thomas, F., Prost, J., Ansari, J.H., and Kalfon, P. (1999). Fungemia due to *Saccharomyces* species in a patient treated with enteral *Saccharomyces boulardii*. Clin. Infect. Dis. *28*, 930.

Nocek, J.E., Holt, M.G., and Oppy, J. (2011). Effects of supplementation with yeast culture and enzymatically hydrolysed yeast on performance of early lactation dairy cows. J. Dairy Sci. *94*, 4046–4056.

Parsons, H.T., Williamson, A., and Johnson, M.L. (1945). The availability of vitamins from yeast. I. The absorption of thiamine by human subjects from various types of bakers' yeast. J. Nutr. *29*, 373–381.

Patra, S.K., and Mohamed, K.S. (2003). Enrichment of *Artemia* nauplii with the probiotic yeast *Saccharomyces boulardii* and its resistance against a pathogenic *Vibrio*. Aquacult. Int. *11*, 505–514.

Peppler, H.J. (1970). Food Yeasts. In The Yeasts. Vol. 3. A.H. Rose and J.S. Harrison, eds. (London: Academic Press), pp. 421–462.

Pérez-Ortín, J.E., Querol, A., Puig, S., and Barrio, E. (2002). Molecular characterization of a chromosomal rearrangement involved in the adaptive evolution of yeast strains. Genome Res. *12*, 1533–1539.

Perez-Soltelo, L., Vaughan, G., Fajardo, R., Gonzalez, Y., Monroy, H., and Vazquez, J.C. (2011). Modulator effects of dietary supplementation with *Saccharomyces cerevisiae* on coliform counts, adaptive general immunologic response and growth performance in pigs. Indian. J. Anim. Nutr. *28*, 191–197.

Pierce, H.B. (1932). The effects of yeast ingestion on the composition of the urine and feces. J. Biol. Chem. *98*, 509–535.

Pinloche, E., McEwan, N., Marden, J.-P., Bayourthe, C., Auclair, E., and Newbold, C.J. (2013). The effects of a probiotic yeast on the bacterial diversity and population structure in the rumen of cattle. PLoS One *8(7)*, e67824. doi: 10.1371/journal.pone.0067824.

Pinloche, E., Williams, M., D'Inca, R., Auclair, E., and Newbold, C.J. (2012). Use of a colon simulation technique to assess the effect of live yeast on fermentation parameters and microbiota of the colon of pig. J. Anim. Sci. *90*, 353–355.

Poppy, G.D., Rabiee, A.R., Lean, I.J., Sanchez, W.K., Dorton, K.L., and Morley, P.S. (2012). A meta-analysis of the effects of feeding yeast culture produced by anaerobic fermentation of *Saccharomyces cerevisiae* on milk production of lactating dairy cows. J. Dairy. Sci. *95*, 6027–6041.

Posteraro, B., Sanguinetti, M., Romano, L., Torelli, R., Novarese, L., and Fadda, G. (2005). Molecular tools for differentiating probiotic and clinical strains of *Saccharomyces cerevisiae*. Int. J. Food Microbiol. *103*, 295–304.

Pothoulakis, C. (2009). Review article: Anti-inflammatory mechanisms of action of *Saccharomyces boulardii*. Aliment. Pharmacol. Ther. *30*, 826–833.

Pott, E. (1889). Nebenprodukte und Abfälle der Bierbrauerei. In Die Landwirtschaftlichen Futtermittel. Handbuch für Tierzüchter und Tierhalter. (Berlin: Verlag Paul von Parey), pp. 565–585.

Priya, B.S., and Babu, S.S. (2013). Effect of different levels of supplemental probiotics (*Saccharomyces cerevisiae*) on performance, haematology, biochemistry, microbiology, histopathology, storage stability and carcass yield of broiler chicken. Int. J. Pharm. Biol. Arch. *4*, 201–207.

Reed, G., and Nagodawithana, T. (1991). Brewers Yeast. In Yeast Technology, 2nd edn (New York: An AVI Book. Van Nostrand Reinhold), pp. 89–149.

Riegler, M., Sedivy, R., Pothoulakis, C., Hamilton, G., Zacherl, J., Bischof, G., Cosentini, E., Fell, W., Schlessel, R., LaMont, J.T., and Wenzel, E. (1995).

Clostridium difficile toxin B is more potent than toxin A in damaging human colonic epithelium *in vitro*. J. Clin. Invest. *95*, 2004–2011.

Rodrigues, A.C.P., Cara, D.C., Fretez, S.H.G.G., Cunha, F.Q., Vieira, E.C., Nicoli, J.R., and Vieira, L.Q. (2000). *Saccharomyces boulardii* stimulates sIgA production and the phagocytic system of gnotobiotic mice. J. Appl. Microbiol. *89*, 404–414.

Rose, A.H. (1982). History and scientific basis of microbial activity in fermented foods. In Economic Microbiology. Vol. 7. A.H. Rose, ed. (London, UK: Academic Press), pp. 1–13.

Rousseaux, C., Bouguen, G., Dubuquoy, C., Dubuquoy, L., Vandekerckove, P., and Desreumaux, P. (2010). *Saccharomyces cerevisiae* CNCM I-3856 decreases intestinal pain through PPAR alpha activation in the gut. Digestive Disease Week. Abstract 274. New Orleans, LA. May 1–5.

Salminen, S., Ouwehand, A., Benno, Y., and Lee, Y.K. (1999). Probiotics: how should they be defined? Trends Food Sci. Technol. *10*, 107–110.

Samuel, D. (1996). Investigation of ancient Egyptian baking and brewing methods by correlative microscopy. Science *273*, 488–490.

Santacroce, J.P., Merra, E., Centoducati, G., Zacchino, V., and Casalino, E. (2012). Effects of dietary yeast *Saccharomyces cerevisiae* on the antioxidant system in the liver of juvenile Sea Bass *Dicentrarchus labrax*. Fish Physiol. Biochem. doi: 10.1007/s10695-012-9640-2.

Santin, E., Maiorka, A., Macari, M., Grecco, M., Sanchez, J.C., Okada, T.M., and Myasaka, A.M. (2001). Performance and intestinal mucosa development of broiler chickens fed diets containing *Saccharomyces cerevisiae* cell wall. J. Appl. Poult. Res. *10*, 236–244.

Scholz, U., Garcia Diaz, G., Ricque, D., Cruz Suarez, L.E., Vargas Albores, F., and Latchford, J. (1999). Enhancement of vibriosis resistance in juvenile *Penaeus vannamei* by supplementation of diets with different yeast products. Aquacult. *176*, 271–283.

Sougioultzis, S., Simeonidis, S., Ramakrishnan Bhaskar, K., Chen, X., Anton, P.M., Keates, S., Pothoulakis, C., and Kelly, C.P. (2006). *Saccharomyces boulardii* produces a soluble anti-inflammatory factor that inhibits NF-κB-mediated IL-8 gene expression. Biochem. Biophys. Res. Comm. *343*, 69–76.

Soyturk, M., Saygili, S.M., Baskin, H., Sagol, O., Yilmaz, O., Saygili, F., and Akpinar, H. (2012). Effectiveness of *Saccharomyces boulardii* in a rat model of colitis. World. J. Gastroenterol. *18*, 6452–6460.

Spangler, B.D. (1992). Structure and function of cholera toxin and the related *Escherichia coli* heat-labile enterotoxin. Microbiol. Rev. *56*, 622–647.

Steckley, J.D., Grieve, D.G., Macleod, G.K., and Moran, E.T. Jr. (1979). Brewer's yeast slurry. II. A source of supplementary protein. J. Dairy Sci. *62*, 947–953.

Sunenshine, R.H., and McDonald, L.C. (2006). *Clostridium difficile* – associated disease: New challenges from an established pathogen. Clev. Clin. J. Med. 73, 187–197.

Surawicz, C.M., Elmer, G.W., Speelman, P., McFarland, L.V., Chinn, J., and van Belle, G. (1989). Prevention of antibiotic-associated diarrhoea by *Saccharomyces boulardii*: A prospective study. Gastroenterol. *96*, 981–988.

Szajewska, H., and Mrukowicz, J. (2005). Meta-analysis: non-pathogenic yeast *Saccharomyces boulardii* in the prevention of anti-biotic associated diarrhoea. Aliment. Pharmacol. Ther. *22*, 365–372.

Szajewska, H., Skórka, A., and Dylag, M. (2007). Meta-analysis: *Saccharomyces boulardii* for treating acute diarrhoea in children. Aliment. Pharmacol. Ther. *25*, 257–264.

Tapia-Paniagua, S.T., Reyes-Becerril, M., Ascencio-Valle, F., Esteban, M.Á., Clavijo, E., Carmen Balebona, M., and Moriñigo, M.A. (2011). Modulation of the intestinal microbiota and immune system of farmed *Sparus aurata* by the administration of the yeast *Debaryomyces hansenii* L2 in conjunction with inulin. J. Aquacult. Res. Devel. doi: 10.4172/2155-9546.S1-S012.

Tedeschi, L.O., Chalupa, W., Janczewski, E., Fox, D.G., Sniffen, C., Munson, R., Kononoff, P.J., and Boston, R. (2008). Evaluation and application of the CPM Dairy nutrition model. J. Agricul. Sci. *146*, 171–182.

Tewary, A., and Patra, B.C. (2011). Oral administration of bakers' yeast (*Saccharomyces cerevisiae*) acts as a growth promoter and immunomodulatory in *Labeo rohita* (Ham.). J. Aquacult. Res. Devel. doi: 10.4172/2155-9546.1000109.

Thévenot, J., Etienne-Mesmin, L., Denis, S., Chalancon, S., Alric, M., Livrelli, V., and Blanquet-Diot, S. (2013). Enterohaemorrhagic *Escherichia coli* O157:H7 survival in a *in vitro* model of the human large intestine and interactions with probiotic yeasts and resident microbiota. Appl. Environ. Microbiol. *79*, 1058–1064.

Thomas, S., Przesdzing, I., Metzke, D., Schmitz, J., Radbruch, A., and Baumgart, D.C. (2009). *Saccharomyces boulardii* inhibits lipopolysaccharide-induced activation of human dendritic cells and T-cell proliferation. Clin. Exptl. Immunol. *156*, 78–87.

Tiago, F.C.P., Martins, F.S., Rosa, C.A., Nardi, R.M.D., Cara, D.C., and Nicoli, J.R. (2009). Physiological characterization of non-*Saccharomyces* yeasts from agro-industrial and environmental origins with possible probiotic function. World J. Microbiol. Biotechnol. *25*, 657–666.

Toothaker, R.D., and Elmer, G.W. (1984). Prevention of clindamycin-induced mortality in hamsters by *Saccharomyces boulardii*. Antimicrob. Agents Chemother. *26*, 552–556.

Tovar, D., Zambonino, J., Cuha, C., Gatesoupe, J.-P., Vázquez-Juárez, R., and Lésel, R. (2002). Effect of live yeast incorporation in compound diet on digestive enzyme activity in Sea Bass (*Dicentrarchus labrax*) larvae. Aquacult. *204*, 113–123.

Tovar-Ramírez, D., Mazurais, D., Gotesoupe, J.-P., Quazuguel, P., Cahu, C.L., and Zambonino-Infante, J.L. (2010). Dietary probiotic live yeast modulates antioxidant enzyme activities and gene expression of Sea Bass (*Dicentrarchus labrax*) larvae. Aquacult. *300*, 142–147.

Tovar-Ramírez, D., Zambonino Infante, J., Cahu, C., Gatesoupe, F.-P., and Vázquez-Juárez, R. (2004). Influence of dietary live yeast on European Sea Bass (*Dicentrarchus labrax*) larval development. Aquacult. *234*, 415–427.

Vandenplas, Y., Brunser, O., and Szajewska, H. (2009). *Saccharomyces boulardii* in childhood. Eur. J. Pediatr. *168*, 253–265.

van der Aa Kühle, A., and Jespersen, L. (2003). The taxonomic position of *Saccharomyces boulardii* as evaluated by sequence analysis of the D1/D2 domain of 26S rDNA, the ITS1-5.8S rDNA-ITS2 region and the mitochondrial cytochrome-c oxidase II gene. Syst. Appl. Microbiol. *26*, 564–571.

van der Aa Kühle, A., Skovgaard, K., and Jespersen, L. (2005). *In vitro* screening of probiotic properties of *Saccharomyces cerevisiae* var. *boulardii* and food-borne *Saccharomyces cerevisiae* strains. Int. J. Food Microbiol. *101*, 29–39.

Vanhoutte, T., De Preter, V., De Brandt, E., Verbeke, K., Swings, J., and Huys, G. (2006). Molecular monitoring of the fecal microbiota of healthy human subjects during administration of lactulose and *Saccharomyces boulardii*. Appl. Environ. Microbiol. 72, 5990–5997.

Vanhoutvin, S.A., Troost, F.J., Kilkens, T.O., Lindsey, P.J., Hamer, H.M., Jonkers, D.M., Venema, K., and Brummer, R.-J. (2009). The effects of butyrate enemas on visceral perception in healthy volunteers. Neurogastroenterol. Motil. *21*, 952–e76.

van Rensburg, L. (2009). Early calf health has big impact on lifetime performance. Feed Management. March/April pp. 22–23.

Villarruel, G., Rubio, D.M., Lopez, F., Cintioni, J., Gurevech, R., Romero, G., and Vandenplas, Y. (2007). *Saccharomyces boulardii* in acute childhood diarrhoea: a randomized, placebo-controlled study. Acta Pediatr. *96*, 538–541.

Waché, Y., Auffray, F., Gatesoupe, F.-J., Zambonino, J., Gayet, V., Labbé, L., and Quentel, C. (2006). Cross effects of the strain of dietary yeast *Saccharomyces cerevisiae* and rearing conditions on the onset of intestinal microbiota and digestive enzymes in Rainbow Trout, *Oncorhynchus mykiss*, fry. Aquacult. *258*, 470–478.

Warny, M., Keates, A.C., Keates, S., Castagliuolo, I., Zacks, J.K., Aboudola, S., Qamar, A., Pothoulakis, C., LaMont, J.T., and Kelly, C.P. (2000). p38 MAP kinase activation by *Clostridium difficile* toxin A mediates monocyte necrosis, IL-8 production, and enteritis. J. Clin. Invest. *105*, 1147–1156.

Watson Smith, S. (1944). Yeast in furunculosis. Br. Med. J. *2*, 256.

Whelan, K., Judd, P.A., Tuohy, K.M., Gibson, G.R., Preedy, V.R., and Taylor, M.A. (2009). Faecal microbiota in patients receiving enteral feeding are highly variable and may be altered in those who develop diarrhea. Am. J. Clin. Nutr. *89*, 240–247.

White, L.A., Newman, M.C., Cromwell, G.L., and Lindemann, M.D. (2002). Brewers dried yeast as a source of mannan oligosaccharides for weanling pigs. J. Anim. Sci. *80*, 2619–2628.

Williams, P.E.V., Tait, C.A.G., Innes, G.M., and Newbold, C.J. (1991). Effects of the inclusion of yeast culture (*Saccharomyces cerevisiae* plus growth medium) in the diet of dairy cows on milk yield and forage degradation and fermentation patterns in the rumen of steers. J. Anim. Sci. *69*, 3016–3026.

Wu. X., Vallance, B.A., Boyer, L., Bergstrom, K.S.B., Walker, J., Madsen, K., O'Kusky, J.R., Buchan, A.M., and Jacobson, K. (2008). *Saccharomyces boulardii* ameliorates *Citrobacter rodentium*-induced colitis through actions on bacterial virulence factors. Am. J. Gastrointest. Liver Physiol. *294*, G295–G306.

Yang, S.-P., Wu, Z.-H., and Jian, J.-C. (2013). Effect of marine red yeast *Rhodosporidium paludigenum* on antioxidant-related gene expression in *Litopenaeus vannemei*. Israeli J. Aquacult. *65*, 817–822.

Yiannikouris, A., François, J., Poughon, L., Dussap, C.-G., Bertin, G., Jeminet, G., and Jouany, J.-P. (2004). Alkalai extraction of β-∂-glucans from *Saccharomyces cerevisiae* cell wall and study of their adsorptive properties toward zearalenone. J. Agric. Food Chem. *52*, 3666–3673.

Zacarias-Soto, M., and Lazo, J.P. (2011). Effect of three probiotics administered through live feed on digestive enzyme activity in California halibut, *Paralichthys californicus*, larvae. J. World Aquacult. Soc. *42*, 321–331.

Zanello, G., Berri, M., Dupont, J., Sizaret, P.-Y., D'Inca, R., Salmon, H., and Meurens, F. (2011a). *Saccharomyces cerevisiae* modulates immune gene expressions and inhibits ETEC-mediated ERK1/2 and p38 signaling pathways in intestinal epithelial cells. PLoS One *6(4)*, e18573. doi: 10.1371/journal.pone.0018573.

Zanello, G., Meurens, F., Berri, M., Chevaleyre, C., Melo, S., Auclair, E., and Salmon, H. (2011b). *Saccharomyces cerevisiae* decreases inflammatory responses induced by $F4^+$ enterotoxigenic *Escherichia coli* in porcine intestinal epithelial cells. Vet. Immunol. Immumopath. *141*, 133–138.

Zanello, G., Meurens, F., Berri, M., and Salmon, H. (2009). *Saccharomyces boulardii* effects on gastrointestinal disease. Curr. Issues Mol. Biol. *11*, 47–58.

Zanello, G., Meurens, F., Serreau, D., Chevaleyre, C., Melo, S., Berri, M., D'Inca, R., Auclair, E., and Salmon, H. (2013). Effects of dietary yeast strains on immunoglobulin in colostrum and milk of sows. Vet. Immunol. Immunopathol. *152*, 20–27.

Zhang, A.W., Lee, B.D., Lee, S.K., Lee, K.W., An, G.H., Song, K.B., and Lee, C.H. (2005). Effects of yeast (*Saccharomyces cerevisiae*) cell components on growth performance, meat quality, and ileal mucosa development in broiler chicks. Poultry Sci. *84*, 1015–1021.

Zhao, P.Y., Jung, J.H., and Kim, I.H. (2012). Effect of mannan oligosaccharides and fructan on growth performance, nutrient digestibility, blood profile, and diarrhea score in weanling pigs. J. Anim. Sci. *90*, 833–839.

Escherichia coli – More than a Pathogen?

9

Maïwenn Olier

Abstract

Escherichia coli is well recognized by the wider public for its ability to cause from self-limiting to life-threatening intestinal and extra-intestinal illnesses. *E. coli* infections remain actually, and primarily, a global public health concern both in developed and developing countries, explaining, at first glance and understandably, the negative connotation associated with this bacterium. Consequently, *E. coli* employment for therapeutic purpose may be considered as an aberrant concept.

We ought however to keep in mind that *E. coli* is not only a pathogen. Originally isolated from neonatal stools, characterized and named *Bacterium coli commune* by Theodor Escherich (1884, reprinted in Escherich, 1988), *E. coli* represents the predominant facultative anaerobic resident of our gut microbiota, where it behaves, at first sight, commensally. Nevertheless, clinical and preclinical beneficial effects of certain *E. coli* strains on the host, i.e. the probiotic *E. coli* strains, have also been demonstrated and they are reviewed herein.

This chapter summarizes current advances in understanding geno- and phenotypic features that allow to discern the pathogen from the commensal or the probiotic *E. coli*. Findings in this field highlighted how complex it sometimes is to establish a clear border between probiotic and pathogenic features within *E. coli* populations. Consequently, ready-made guidelines for the use of probiotic *E. coli* strains by practising physicians are still lacking.

Pathogenic *E. coli* – one pathogen for a broad variety of pathotypes and diseases

Mainly due to genes gain and loss by horizontal transfer associated with recombination events that constantly reshape its genome content, the commensal *E. coli* can become a highly versatile and adapted pathogen (Hacker *et al.*, 2003; Juhas, 2013; Touchon *et al.*, 2009). *E. coli* virulence-associated genes are frequently located on such transmissible genetic elements, e.g. genomic islands, bacteriophages, plasmids and transposons and thus can be easily exchanged between different strains. According to the environment forces and pressures encountered by *E. coli* strains over time, myriads of combinations of virulence genes have been mobilized in their genome, leading to the emergence of several different pathotypes (Kaper *et al.*, 2004). Those different pathotypes each affect a wide range of cellular processes, resulting in a broad spectrum of diseases concerning the gastrointestinal tract and extraintestinal sites such as the urinary tract, bloodstream or central nervous system.

Initially, those pathogenic *E. coli* pathotypes were named according to the clinical syndrome they caused, i.e. diarrhoeagenic *E. coli* (DEC), uropathogenic *E. coli* (UPEC), septicaemic *E. coli* (SePEC), neonatal meningitis-associated *E. coli* (NMEC), or sometimes according to the target host: avian pathogenic *E. coli* (APEC).

Lately cellular microbiology approach has fostered advances in understanding *E. coli* genetics and pathogenesis, leading to the identification of virulence factors and in additional definitions within pathotypes (Kaper *et al.*, 2004):

The enteric pathogenic *E. coli* strains

Among the enteric pathogenic *E. coli* strains, six major categories with distinct features in their interaction with eukaryotic cells have been initially described and reviewed by Kaper *et al.* (2004): enteropathogenic *E. coli* (EPEC), Shiga toxin-producing *E. coli* (STEC including enterohaemorrhagic *E. coli* (EHEC)), enterotoxigenic *E. coli* (ETEC), enteroaggregative *E. coli* (EAEC), diffusely adherent *E. coli* (DAEC) and enteroinvasive *E. coli* (EIEC, with *Shigella*). Croxen *et al.* (2013) have updated that initial description in 2013 and added the description of a seventh pathotype, the adherent invasive *E. coli* (AIEC).

Enteropathogenic *E. coli*

As other diarrhoeagenic *E. coli* strains, EPEC is transmitted primarily from host to host via faecal–oral route. Initially, EPEC infections were an important cause of infantile diarrhoea in the developed world, but it became over the years much more prevalent in developing countries. Fever, vomiting, and dehydration often co-occurs with diarrhoea in children. EPEC is distinguished from the other diarrhoeagenic *E. coli* strains by the ability to form attaching and effacing (A/E) lesions in the ileum (encoded by the locus of enterocyte effacement, LEE genomic island), a phenotype characterized by large-scale cytoskeletal changes in the epithelial cells, resulting in the formation of a pedestal on which the bacteria sits (Nataro and Kaper, 1998).

Shiga toxin-producing *E. coli*

Strains that produce Shiga toxin (also known as verotoxin) are referred to as STEC (or VTEC) whereas the term EHEC is used to denote only the subset of Stx-positive strains that also contain the LEE. STEC causes mild to bloody diarrhoea, which can be associated to fever, abdominal cramps or vomiting. Haemolytic–uraemic syndrome (HUS), which involves thrombocytopenia, haemolytic anaemia and renal damage, can develop several days after the initial onset of diarrhoea. As EPEC, STEC transmission is essentially faecal–oral. Contaminated food and water are responsible of sporadic as well as large outbreak-related illnesses. Unlike EPEC, STEC infectious dose is thought to be very low and viable-but-non-culturable cells present on food would be able to produce Shiga toxin (Dinu and Bach, 2011). Shiga toxin is encoded by stx_1 or stx_2, two genes acquired by a lambdoid bacteriophage, which represents the key virulence factor in STEC causing HUS. The presence of variant combinations of *stx1* and/or *stx2* in conjunction with LEE and other additional virulence genes (*cdtABC*, *hlyA*, etc.) would be indicative of disease severity (reviewed in Croxen *et al.*, 2013).

Enterotoxigenic *E. coli*

As STEC, ETEC-induced watery diarrhoea is mediated by toxin production as the result of contaminated food and water consumption. ETEC is a major cause of infantile and travellers' diarrhoea in developing world regions, and is associated with heat-labile (LT) and heat-stable (ST) enterotoxins production (reviewed in Qadri *et al.*, 2005). In addition to these enterotoxins that are structurally related to cholera enterotoxin produced by *Vibrio cholerae*, most of the human ETEC strains carries specific colonization factors named CS (coli surface antigens) identified by a number (23 characterized at present). 'Animal adhesins' such as K99 (*fanA*) or K88 (*faeG*) are classically associated with animal-specific ETEC strains, which are also an important cause of morbidity and mortality in neonatal and post-weaning piglets (Turner *et al.*, 2006). Unlike other diarrhoeagenic *E. coli* strains, toxins and virulence factors responsible for ETEC pathogenesis are mainly borne by plasmids.

Enteroaggregative *E. coli*

Despite heterogeneity of EAEC (or EAggEC), the pathotype is defined by bacteria ability to display the characteristic 'stacked-brick' pattern when EAEC strains are incubated with Hep-2-epithelial cells in culture (Nataro *et al.*, 1987). EAEC aggregative adherence is associated with both fimbrial (AAF) and afimbrial adhesins but numerous additional toxins and proteins are associated with virulence (reviewed in Estrada-Garcia and Navarro-Garcia, 2012). Multiple additional factors contributing to EAEC-induced inflammation have also been described. A number of plasmid genes coding for virulence factors and at least two genomic islands in the EAEC chromosome are controlled by the AggR regulator. However, none of these factors have been found in every EAEC isolates, and no single one has ever been implicated in EAEC virulence. Since its initial identification in 1987, EAEC has been found to be most commonly associated with paediatric diarrhoea particularly in developing regions but it is also linked to persistent diarrhoea in adults including HIV-positive patients and travellers (Estrada-Garcia and Navarro-Garcia, 2012).

Diffusely adherent *E. coli*

Unlike EAEC that displays aggregative adherence pattern to cultured epithelial Hep-2 cell monolayers, DAEC is dispersed over the surfaces of epithelial cells, resulting in diffuse adherence pattern with little aggregation. This cytopathic effect is considered distinctive from other pathotypes but the lack of a reliable identification method makes DAEC detection difficult in clinical settings. At present, DAEC epidemiology remains unclear. In some studies, DAEC was detected in healthy age-matched controls who did not present diarrhoea at similar rates to those measured in patients (Germani *et al.*, 1996). Consequently, a clear identification of DAEC strains genotype specifically associated with watery diarrhoea, is still lacking. Production of adhesins encoded by a family of *afa/dra/daa*-related operons (Afa/Dr family) has been implicated in the diffusely adherent pattern. Relative risk to develop diarrhoea associated with Afa/Dr-positive DAEC isolates is higher particularly in children aged from 2 to 6 (Le Bouguenec and Servin, 2006). However, such operons are genes that arise and are expressed in a variety of genetic backgrounds (Escobar-Paramo *et al.*, 2004a).

Enteroinvasive *E. coli*

While EIEC and *Shigella spp.* are taxonomically indistinguishable at species level and share a similar invasive phenotype, clinical manifestations are starkly different, explaining the conservation of a distinct nomenclature: an infectious dose of as little of 10 to 100 cfu of *Shigella* has been described whereas higher doses have been reported in the case of EIEC. *Shigella* causes bacillary dysentery with several degrees of severity according mainly to host status or complications, whereas self-limiting diarrhoea and milder symptoms in comparison to those of other diarrhoeagenic *E. coli* strains are reported for EIEC. EIEC/*Shigella* invasion and spreading in intestinal epithelium are restricted to the mucosa layer but trigger an inflammatory response provoking general tissue disruption. Both intracellular lifestyle and inflammation, are linked to multiple virulence factors recently reviewed by Carayol and Tran Van Nhieu (2013). Some virulence determinants are encoded by chromosomal genes (toxins belonging to the serine protease autotransporters family), anyway key virulence factors are plasmid-borne and include a type III secretion system, an apparatus which allows the direct injection into host cells of about 30 type III effectors. Consequently, differences observed in the severity of EIEC or Shigella induced-disease could be explained by the reduced expression and/or the effectiveness of these factors (Moreno *et al.*, 2009). EIEC and *Shigella* identification from stool samples requires mainly the quantification of invasive potential by performing plaque assay in epithelial cultured cells. Additional biochemical typing, serological typing and PCR-based detection of genetic markers, such as the invasion plasmid antigen H gene (*ipaH*), can be used to discriminate this pathotype. Discrimination between EIEC and *Shigella* requires further genotypic analyses such as MLST (multilocus sequence typing) of seven housekeeping genes.

Adherent invasive *E. coli*

AIEC may be associated with Crohn's disease, an inflammatory bowel disease. Microbiota plays an essential role in the perpetuation of such chronic inflammation especially over genetically predisposed subjects. A transient infection by AIEC in mice lacking the flagellin receptor TLR5 has reported to be sufficient to induce chronic spontaneous colitis by definitively altering gut microbiota composition even after AIEC clearance (Chassaing *et al.*, 2013). These recent results strongly support that AIEC could be finally one of the causative agent for Crohn's disease. AIEC strains are also abnormally predominant in human ileal lesions where they are able to adhere, invade epithelial cells and replicate within epithelial cells and macrophages (Darfeuille-Michaud *et al.*, 2004). Genetic determinants accounting for AIEC pathotype remain poorly defined but result probably from the genomic islands acquired by horizontal transfer that have been detected by whole-genome sequencing. AIEC strains do not harbour common virulence factors found in other pathogenic *E. coli* and are genetically more closely related to extraintestinal pathogenic *E. coli* (ExPEC). Potential virulence factors include a type IV secretion system, adhesins (type I pili FimH, long polar fimbriae), siderophores (aerobactin) and transcriptional regulators genes (*phoP-phoQ*, *envZ-ompR*) (Nash *et al.*, 2010). Interestingly, adherence of AIEC is mediated by FimH, which recognizes the CAECAM6 receptor abnormally expressed on ileal epithelial cells of Crohn's disease patients (Barnich *et al.*, 2007). Concomitantly, AIEC-associated mutations in FimH have been reported to confer to AIEC bacteria a significantly higher ability to adhere to CEACAM-expressing intestinal epithelial cells, highlighting a typical signature of photoadaptative evolution of bacterial pathogens (Dreux *et al.*, 2013). Production of OMVs (outer membrane vesicles), which facilitates effectors delivery into host cells, has been described as a mechanism used by AIEC for invasion process (Rolhion *et al.*, 2005). Ability of AIEC to replicate intracellularly has been reported to be mediated by the turnover modulation of the ubiquitin proteasome system within host cells (Cleynen *et al.*, 2013). Alteration of autophagy by AIEC has also been shown to enhance intracellular replication through microRNAs modulation, leading to the increase of inflammatory response and pathology perpetuation (Nguyen *et al.*, 2014).

The extraintestinal pathogenic *E. coli* strains

The *E. coli* pathotypes involved in extraintestinal infections have been classified in human within the group called Extraintestinal pathogenic *E. coli* (ExPEC) (Russo and Johnson, 2000), e.g. uropathogenic *E. coli* (UPEC), neonatal meningitis-associated *E. coli* (NMEC) and sepsis-associated *E. coli* (SePEC). Additional animal specific variants named avian pathogenic *E. coli* (APEC), mammary pathogenic *E. coli* (MPEC) and endometrial pathogenic *E. coli* are incorporated in this pathotype whereas animal strains share common genetic background (Clermont *et al.*, 2011). Johnson and Russo (2002) ironically defined ExPEC as 'the other bad *E. coli*' based on their unique ability to cause disease outside the host intestinal tract. They are also facultative pathogens found in the gut microbiota where they behave as commensals, but once they get access out of the gut, they are able to colonize normally sterile niches where they elicit diseases such as urinary tract infection (UTI), septicaemia or meningitis. Unlike other enteric pathogenic *E. coli*, ExPEC strains are unable to cause gastrointestinal disease in humans. Similarly to enteric pathogenic *E. coli*, many of the ExPEC-associated determinants are encoded on genetic mobile elements. Several studies suggested that some ExPEC-associated features were primarily linked to gut colonization and that virulence is an incidental by-product of commensalism (Diard *et al.*, 2010; Le Gall *et al.*, 2007).

Uropathogenic *E. coli*

UTI are among the most common bacterial infections in human, principally affecting women. About 40–50% of women over the age of 18 years would experience at least one UTI episode in their lifetime (Foxman, 2010). More than 80% of UTI are associated with UPEC strains. Intestinal microbiota is the most common immediate source of UPEC. UPEC takes advantage of host behaviour and host susceptibility by employing virulence factors that facilitates bacterial growth and persistence in urinary tract. These fitness factors reviewed recently in (Ulett *et al.*, 2013) include mainly adhesins (Type I, P, F1C fimbriae), siderophores (salmochelin, yersiniabactin, aerobactin, etc.), and toxins (vacuolating toxin, serine protease, alpha-haemolysin). Serum resistance and encapsulation (production of capsule) are also epidemiologically associated with cystitis and pyelonephritis. Although carriage of corresponding genes is positively associated with UTI, no core set of virulence factors can, however, unambiguously identify UPEC subgroups (Ewers *et al.*, 2007), or even distinguish UPEC strains from faecal commensal strains (Moreno *et al.*, 2008).

Neonatal meningitis-associated *E. coli*

NMEC is one of the primary causes of neonatal meningitis, a disease often associated with important medical complications or even death. Similarly to UPEC, the ability to cause neonatal meningitis is enhanced by a set of virulence/fitness factors implicated in adhesion, invasion and interference with host defence/signalling mechanisms. In addition, the *ibeA* gene implicated in invasion process, has been reported to be significantly more prevalent among NMEC than in UPEC strains (and APEC) as a result of loss of the associated genomic island in UPEC strains (Homeier *et al.*, 2010). Predominance of K1-encapsulated strains, which elicits serum resistance and has antiphagocytic properties, has also been associated particularly with neonatal meningitis, bacteraemia and septicaemia (Sarff *et al.*, 1975). The more representative NMEC strains are plasmid PAI-containing strains (Logue *et al.*, 2012).

Sepsis-associated *E. Coli*

Few reports describe SePEC strains although sepsis and bacteraemia are increasing globally and the emergence of multidrug resistance is becoming problematic in many individuals. Known strains mainly share the archetypal traits found in other ExPEC strains. Severity of clinical outcome is strongly associated with host immune condition or with the primary source of bacteraemia more than with proper bacterial features (Lefort *et al.*, 2011).

How to predict pathogenic potential of *E. Coli* isolates

Considering the earlier description of pathogenic characteristics in *E. coli* populations, it is understandable that the identification of *E. coli* pathogenic potential solely based on virulence profile remains difficult for most of the isolates. Rarely the presence of a single pathogenic factor makes an organism virulent; only a combination of factors will probably determine whether or not a bacterium will cause infection. Because of the high degree of genome flexibility and genetic diversity within this species, the designation of certain isolates into a pathotype is complicated. This genomic dynamism also contributes to *E. coli* evolution (Dobrindt *et al.*, 2010) and consequently to the emergence of new pathogenic hybrids as recently observed in the strain associated with the 2011 German outbreak (Frank *et al.*, 2011a). This strain sickened more than 4300 healthy individuals from 16 countries, where over 900 patients developed HUS (haemolytic–uraemic syndrome). There were more than 50 deaths. Isolates characterization suggested that the identified *E. coli* O104:H4 strain was indeed a hybrid of both EAEC and EHEC, as the result of the carriage of *aggA*, *aggR*, *set1*, *pic*, and *aap* genes, and *stx2* gene respectively (Brzuszkiewicz *et al.*, 2011).

Classical bacterial typing that consists of analysing somatic, capsular and flagellar serogroups (O:K:H), is generally used to further characterize *E. coli* isolates and it is of particular interest for epidemiological investigations. 174 different O and 53 H antigens have been identified but a considerable number of isolates remains untypable. Serogroup typing is informative since a small number of O/H combinations are really associated with disease, O157:H7 and O104:H4 being the most well known. Concerning UTI for example, 75% is caused by *E. coli* strains belonging to 12 of the 174 serogroups O antigens. Nonetheless, serotyping alone remains hugely inadequate to substantiate pathogenic potential of *E. coli*.

Alternatively, phylogenetic typing such as multilocus sequence typing (MLST) is informative and it is currently facilitated by whole-genome sequencing development. While the gene content between pairs of *E. coli* may diverge by more than 30% due to a very high gene flow (Welch *et al.*, 2002), *E. coli* genome remains organized (Touchon *et al.*, 2009) and *E. coli* population remains mostly clonal, allowing identification of at least six phylogenetic groups (A, B1, B2, D, E and F). Genetic background has also been shown to play a role in the acquisition, retention and expression of foreign DNA (Escobar-Paramo *et al.*, 2004a) but no sufficiently close relationship between phylogenetic groups and virulence has been identified, making pathogenic potential of strains with known phylogenetic group impossible to predict. Firstly, not all *E. coli* pathotypes group together. ETEC strains for example, were found to group with phylogroups A and B1 (Sahl *et al.*, 2011), EAEC strains group in phylogroups A, B1, B2 and D (Okeke *et al.*, 2010), whereas ExPEC strains group in phylogroups B2 and D (Picard *et al.*, 1999). Inversely, commensal isolates mostly group in phylogroup A or B1 but some of them belong to the B2 phylogroup (Tenaillon *et al.*, 2010).

Consequently, a typing system that enables by itself precise and unambiguously identification of pathogenic subtypes, as well as anticipation of their pathogenic potential, remains to be developed. The tremendous genome diversity and dynamic within *E. coli* populations are the main hurdles to this typing development and at the same time to vaccine design.

Commensal *E. coli*: the peacefully inhabitant of vertebrate gut – friends or foe?

E. coli is a common and widespread inhabitant of human gastrointestinal tract, that generally appears soon after birth and persists for decades (Favier *et al.*, 2002; Jimenez *et al.*, 2008; Palmer *et al.*, 2007). Unlike the numerous studies performed on pathogenic clones, up to now, only few studies have been focused on the genetics of *E. coli* commensal population (referred to below as 'colibiota'). Commensal strains, and particularly the laboratory K12 strain MG1655, are frequently used in experimental studies but often regarded only as 'negative control' as if K12 strain had necessarily to remain inert to host cells.

Nevertheless, two interesting questions arise from this achievement: whether gut colibiota (i) could be reservoir of virulent and resistant strains, and (ii) could contribute to intestinal homeostasis. Indeed numerous studies evidenced gut microbiota involvement in basic host functions such as trophic effects, nutrients supply, energy balance, metabolic signalling, barrier functions, resistance to pathogen colonization and its key role in promoting postnatal maturation of intestinal mucosal barrier (Backhed *et al.*, 2004; Chung *et al.*, 2012; Hooper *et al.*, 2012; Ivanov *et al.*, 2009; Mazmanian *et al.*, 2005; Stappenbeck *et al.*, 2002; Turnbaugh *et al.*, 2006). Moreover increasing evidence suggests that early host–microbe contacts occurring at birth during gut microbial colonization, partially influence adult phenotype and the subsequent risk of diseases later in life (Rautava *et al.*, 2012; Sommer and Backhed, 2013).

Implication of colibiota, especially neonatal colibiota, on basic host physiological, immune and metabolic phenotypes later in life remains largely unexplored and difficult to apprehend. Firstly, such a study requires following human cohorts from birth to adulthood. Secondly, most human studies exploring gut colibiota have actually evaluated faecal microbiota due to its easy access and collection. It has been postulated that faecal microbiota represents a mixture of shed mucosal bacteria and a separate non-adherent luminal microbiota (Eckburg *et al.*, 2005). Consequently, drawing conclusions from solely faecal analyses may be inappropriate since the luminal and the surface-adherent bacterial populations may be distinct and have different properties.

Here are reported current advances on gut neonatal and adult colibiota resulting from human and/or laboratory animal studies.

Gut colibiota in early life: impact on health

Mother-to-infant transmission of *E. coli* has been described and maternal gut microbiota seems to be an important source of *E. coli* strains for newborns regardless of the delivery mode (Bettelheim and Lennox-King, 1976; de Muinck *et al.*, 2011; Mackie *et al.*, 1999; Vaishampayan *et al.*, 2010). Within a few days after birth, *E. coli* is predominant in immature gut microbiota composition (higher than 10^9 cfu/g faeces (Penders *et al.*, 2006; Tenaillon *et al.*, 2010)) and triggers an anaerobic environment allowing the setting up of *Bacteroides*, *Bifidobacterium* and *Clostridium* (Scholtens *et al.*, 2012).

Although different among individuals and populations, infant faecal microbiota composition achieves a state of relative equilibrium by the end of the first year of life (Palmer *et al.*, 2007) and adopts an adult-like configuration within the 3-year period following birth (Yatsunenko *et al.*, 2012). During childhood, *E. coli* has been detected in half of the children and a concentration of around 10^6 cfu/g faeces has been reported (Collado *et al.*, 2009). In sick children, the prevalence and concentration of *E. coli* have been shown to be increased (Collado *et al.*, 2009; Schwiertz *et al.*, 2010).

In neonates from developed countries the pattern of *E. coli* colonization is characterized by persistence and low strain turnover (Adlerberth *et al.*, 1998; Nowrouzian *et al.*, 2003). Nevertheless, this colonization pattern depends on the phylogroup of the predominant clone, the delivery mode or the human populations studied (Das *et al.*, 2013; Nowrouzian *et al.*, 2009; Nowrouzian *et al.*, 2005):

The majority of strains from neonates born in developing countries belong to phylogroups A and B1, whereas isolates of phylogroup B2 are present in low numbers (8–12%) and particularly associated with neonates delivered vaginally. Predominance of phylogroup B2, in which not only commensal *E. coli* strains but also ExPEC strains such as neonatal meningitis-associated *E. coli* are found, is, however, clearly observed in strains isolated from neonates born in developed countries (46–58% in Sweden, Norway or France) (de Muinck *et al.*, 2011; Nowrouzian *et al.*, 2005; Payros *et al.*, 2014). Comparison of B2 commensal strains from Pakistani infants with B2 commensal strains from Swedish infants showed that 'western' B2 strains displayed higher persistence ability (Nowrouzian *et al.*, 2009). Genomic analyses allowed showing that accumulation of carriage of genomic islands, termed pathogenicity islands (PAI II_{CFT073}, IV_{536}, II_{96}) in 'western' B2 strains correlated with strain persistence time in infant's microbiota (Ostblom *et al.*, 2011), suggesting that PAI, known initially to contribute to extraintestinal infections caused by ExPEC, may be in fact regarded as fitness islands in the gut. Also the *pks* genes contained in the genomic island GI-*asnW* have been reported to enhance long-term persistence in infant gut (Nowrouzian and Oswald, 2012). This genomic island harboured by both commensal and ExPEC strains of phylogroup B2, encodes non-ribosomal peptide synthetases and polyketide synthases required for the production of a genotoxin called colibactin (Nougayrede *et al.*, 2006). Carried by about 15% of French neonates (Payros *et al.*, 2014), human commensal *pks*$^+$ *E. coli* strains maternally acquired at birth, have been shown experimentally to trigger DNA-damage in the gut epithelium of newborn rats only during the first week of life (Payros *et al.*, 2014). DNA damage, consisting in DNA double strand breaks, was no longer observed at adulthood but signs of genotoxic stress were specifically observed at this stage. Indeed anaphase bridges, high rate of crypt fission and accelerated renewal of mature epithelium along with alterations in differentiated epithelial cells and epithelial barrier functions were observed (Payros *et al.*, 2014). However, rats colonized at birth with human commensal *E. coli* strains producing colibactin or not appeared as healthy during experiments. Whether such intestinal epithelial alterations would increase the risk to develop different pathologies such as colorectal cancer or gut dysfunctions, including inflammatory bowel diseases (IBD) or irritable bowel syndrome (IBS) is currently not known. Nevertheless, these results clearly demonstrated that early microbial signatures in the neonatal gut could have long-term impact on intestinal homeostasis at adulthood and illustrate how critical is the genotype of the *E. coli* strain acquired at birth for cellular programming of the host.

Short-term impact of such microbial signatures at birth has not been investigated. Whether maternally acquired *pks*$^+$ *E. coli* may increase in infancy the risk of systemic infections such as sepsis, neonatal meningitis or necrotizing enterocolitis, needs to be further explored, particularly in the context of preterm infants. Relative abundance of *E. coli* within faecal microbiota of premature neonates has been measured (Arboleya *et al.*, 2012a,b) and colibiota implication in infections has been recognized (Gaschignard *et al.*, 2011; Stoll *et al.*, 2011).

Despite predominance of B2 strains among the *E. coli* strains isolated from neonates born in developed countries ('western isolates'), some isolates were also categorized in the A, B1 or D phylogroups (de Muinck *et al.*, 2011; Nowrouzian *et al.*, 2005). Among them, non-B2 isolates from Norwegian neonates showed virulence features and phylogenetic sequence type similar to atypical EPEC strains isolated from Norwegian infants with diarrhoea (Afset *et al.*, 2006). Nonetheless, no episode of diarrhoea was associated with this commensal isolate (de Muinck *et al.*, 2011).

Whatever the phylogroup of *E. coli* strain, studies on gut colibiota suggest that bacterial virulence finally may rather be a matter of context in host and opportunism. As described above, prematurity appears to be one of the factors contributing to disease occurrence in neonates. Maternal deprivation during the first days of life is also an environmental factor that could contribute to bacterial virulence and associated diseases. A single episode of maternal separation (4 hours) in rodent is a stressful event that, particularly in the context of prematurity, has been shown to be sufficient to increase colonic permeability and promote abnormal gut bacterial translocation to systemic organs (Moussaoui *et al.*, 2014). Preliminary experiments on pups whose microbiota was manipulated (spontaneous colonization at birth with a maternally transmitted human commensal reference *E. coli* strain of phylogroup B2), conceptually proved, that maternal separation induced colibiota translocation from the gut to the liver and spleen during neonatal period (N. Moussaoui, M. Olier and E. Houdeau, unpublished). Nevertheless, further investigations are necessary to evaluate short- and long-term impact of such an early contact of bacteria within systemic organs and, consequently, to improve the disease risk evaluation depending on the strain acquired at birth.

Conversely, early acquisition of *E. coli* and bifidobacteria in the first week of life has been reported to promote B-cell maturation in humans (Lundell *et al.*, 2012), suggesting a beneficial effect of early-acquired colibiota on the development of the immune system later in childhood. Geno- and phenotyping of the *E. coli* strains collected in this study in an attempt to identify factors that drive B-cell maturation, would be of particular interest.

The gut colibiota at adulthood: impact on health

E. coli among other *Enterobacteriaceae* represents the predominant facultative anaerobic bacteria within the mature gut microbiota. More than 90% of human beings are colonized by *E. coli* (Smati *et al.*, 2013; Tenaillon *et al.*, 2010). In healthy adult subjects, *E. coli* concentration has been evaluated between 10^4 and 10^8 cfu/g faeces (Bailey *et al.*, 2010b; Pilarczyk-Zurek *et al.*, 2013; Smati *et al.*, 2013). An adult can be colonized by more than one strain at any given time (Smati *et al.*, 2013). Studies performed in rodents showed that the majority of human colibiota sets in the large intestine (caecum and colon), especially in the mucus layer. Growth, colonization and persistence of colibiota within the mammalian gut depend on *E. coli* strain ability to penetrate through mucus layer and to use mucus as nutrient (Conway *et al.*, 2004).

The few studies performed over the previous 50 years ago, showed that gut colibiota colonization and persistence were strain dependent (Sears and Brownlee, 1952; Sears *et al.*, 1950, 1956). Taking into account the increasing knowledge about the population structure of *E. coli*, the evolution of the species due to horizontal gene transfer, and evolution of our lifestyle, *E. coli* strain persistence within human adult populations should be re-examined. Currently knowledge about distribution of human commensal *E. coli* phylogroups originates mainly from a unique sampling and is presented thereafter.

The combined data from 1889 faecal strains isolated from healthy humans over 15 countries around the world reveal that phylogroups A and B2 are more common (respectively 32% and 29.4%) than B1 and D (respectively 17.9% and 20.7%) (Bailey *et al.*, 2010a,b; Tenaillon *et al.*, 2010). However, a large variability in phylogroup prevalence has been found in humans (Tenaillon *et al.*, 2010): age, sex, diet but also year of sampling and country of origin are factors that have been identified. Analyses of these data indicated a shift from the phylogroup A to the phylogroup B2 in the faecal colibiota of urban populations during the past 20 years (Clermont *et al.*, 2013; Duriez *et al.*, 2001; Escobar-Paramo *et al.*, 2004b; Tenaillon *et al.*, 2010; Zhang *et al.*, 2002). Lifestyle, dietary habits or increased level of hygiene are presumably the main factors accounting for this major modification in the phylogroup distribution of the faecal colibiota.

Moreover, the presence of phylogroup B2 strains in faecal samples has been associated with reduction of colibiota diversity, suggesting that strains of phylogroup B2 outcompete strains of other phylogroups (Smati *et al.*, 2013), possibly as a result of higher fitness of B2 strains during intestinal colonization (Diard *et al.*, 2010). Whether prevalence of *pks*$^+$ *E. coli* or other B2 fitness factors increased in urban populations during the past 20 years has not yet examined. In the same way, it would be interesting to study the possible relationship between the occurrence of dominance of the phylogroup B2 and occurrence of the carriage of *pks* genomic island, substantiating an hypothetical antibacterial activity conferred by the production of colibactin.

Skurnik and colleagues (Skurnik *et al.*, 2008) have demonstrated that faecal colibiota was not static and changed dynamically in response to environmental conditions. They compared colibiota of carefully matched subjects that differed in their living environments, i.e. French residents, French expatriates in Guyana and local Wayampi Amerindians in Guyana. The colibiota of the expatriates was intermediate between those of the two other populations, with the same prevalence of phylogroup A as the French residents and the same prevalence of phylogroup B2 as the local Wayampi Amerindians.

Finally, in animal models an intra-individual fluctuation in strain composition has been reported in response to diet. For example, diets differing in their crude fibre content provided *ad libitum* to rats has been reported to alter transit time that presumably in turn, altered colibiota in terms of colonization rates and phylogroup distribution (O'Brien and Gordon, 2011). Phylogroup B2 strains in this study appeared to be unable to maintain colonization rates similar to those of phylogroup B1 strains in rats fed with high-fibre diet.

All environmental factors that shape the population structure of commensal *E. coli* may indirectly play a key role in the antibiotic resistance occurrence (Bailey *et al.*, 2010b) since dominance of the phylogroup A, for example, has been reported to be associated with prevalence of antibiotic resistance (Mammeri *et al.*, 2008; Smati *et al.*, 2013).

Moreover gut colibiota structure and dynamics, can also be affected by pathophysiological conditions. There is compelling evidence that different pathologies including immunodeficiency, metabolic diseases (diabetes, obesity, etc.), inflammatory and autoimmune diseases (IBD, allergy, etc.) or even cancer and psychiatric diseases, are associated with an altered composition of the commensal gut microbiota called dysbiosis (reviewed in Holmes *et al.*, 2012; Peterson *et al.*, 2008; Pflughoeft and Versalovic, 2012). Gut microbiota is considered as an environmental factor that itself is controlled in part by host genetic factors and potentially by interaction between host and microbial genomes. Consequently, susceptible individuals are predisposed to an aberrant pattern of microbial colonization, which contributes to diseases progression (Frank *et al.*, 2011b). A summary concerning the implication of colibiota in this processes is reported below.

Overgrowth of commensal *E. coli* resulting from trauma-associated sepsis has been reported and was related with the release of stress-related neurochemicals. Presence of catecholamine has been shown effectively to directly increase *in vitro* the growth of a gut commensal *E. coli* strain (Freestone *et al.*, 2002).

Mice displaying intestinal overgrowth with commensal colibiota were reported to be more susceptible to *Salmonella enterica* induced gut inflammation (Stecher and Hardt, 2011) as well as to *Campylobacter jejuni* (Haag *et al.*, 2012). Colonization resistance against pathogens, that is normally conferred by microbiota, was supposed to be abrogated by abnormally high colibiota loads resulting either from dysbiosis encountered in genetically susceptible mice (IL-10-deficient mice) or from oral treatment with a live murine commensal *E. coli* strain (Haag *et al.*, 2012).

Altered IgA repertoire or IgA deficiency has also been reported to shift the normal gut microbiota composition by reducing *Bifidobacterium* and *Bacteroides* and increasing colibiota amount (Friman *et al.*, 2002; Kawamoto *et al.*, 2012; Nowrouzian *et al.*, 2007). IgA is a key immunoglobulin in the gut with a specific role in mucosal immunity, tolerance and protection against infections. Individuals with IgA deficiency showed an increased risk of gastrointestinal disorders including coeliac disease and IBD. In IgA-deficient human subjects, phylogroup A strains were more prevalent than in healthy controls and conversely phylogroup B2 strains were lower. Colibiota from IgA-deficient patients has been reported to display low mannose-specific adherence to epithelial cells (Nowrouzian *et al.*, 2013).

Mucosal dysbiosis has been observed in different mouse models of diabetes. In particular, in mice fed with high-fat diet before the onset of diabetes, the accumulation of *E. coli* strains near the mucosa of different gut segments has been reported (Amar *et al.*, 2011). This study established that adherent bacteria translocated from the intestine during the early onset of high-fat diet-induced diabetes into the mesenteric adipose tissue and blood, triggering low-grade metabolic bacteraemia and inflammation.

Similarly, there is also accumulating evidence that mucosal microbiota drives intestinal inflammation in IL-10-deficient mice. Whereas IL-10-deficient mice do not develop colitis under germ-free conditions, Kim and colleagues observed that *E. coli* monoassociated IL-10-deficient mice developed an inflammation of milder intensity, but similar with regards to kinetics and regional localization, to colitis observed in IL-10-deficient mice under conventional conditions (Kim *et al.*, 2005). This murine *E. coli* strain was, with *Enterococcus faecalis*, the only non-pathogenic commensal species reported to be able to induce chronic colitis in monoassociated IL-10-deficient mice. Consistent with these observations, increase of gut colibiota has been described in IL-10-deficient mice (Maharshak *et al.*, 2013; Wohlgemuth *et al.*, 2009). Analysis of colibiota revealed that one single clone of phylogroup B2 was present. Although this strain of phylogroup B2 harboured multiple virulence/fitness traits, it was supposed to be a commensal since it was not able to induce colitis in WT mice. By generating reactive nitrogen species and reactive oxygen species during inflammation, gut has recently reported to confer a marked growth and selective advantage on commensal *E. coli* since, unlike obligate anaerobic members of gut microbiota, colibiota can use nitrate by performing anaerobic respiration (Winter *et al.*, 2013).

As described in genetically susceptible rodents models, IBD patients, including patients suffering from Crohn's disease and ulcerative colitis (particularly extensive UC and pancolitis), have shown to display overgrowth of mucosally associated *Enterobacteriaceae* including *E. coli* (Lopez-Siles *et al.*, 2014; Sokol *et al.*, 2006; Swidsinski *et al.*, 2002). Similarly, isolation of faecal strains of phylogroup B2 was linked to inflammation and strongly correlated with IBD patients suffering from left-sided colitis (Kotlowski *et al.*, 2007; Petersen *et al.*, 2009). Although faecal colibiota of phylogroup B2 appears heterogeneous, genomic profiles of isolated strains display a remarkable resemblance to ExPEC isolates. The majority of isolates collected from biopsies, however, have been shown to adhere to and invade intestinal epithelial cells in culture, suggesting that they are typically pathogenic *E. coli* strains assigned to the AIEC and DAEC pathotypes described earlier (Darfeuille-Michaud *et al.*, 2004; Prorok-Hamon *et al.*, 2014).

Increased mucosa-associated and mucosa-internalized *E. coli* strains have also been reported in colorectal cancer (Martin *et al.*, 2004; Swidsinski *et al.*, 1998). *E. coli* strains were detected in the normal mucosa of patients with colorectal cancer, but also higher numbers of *E. coli* were isolated from tumour tissue (Bonnet *et al.*, 2014). Analysis of these strains revealed that they belonged to the phylogroup B2 and shared the associated-multiple virulence factors encountered in ExPEC, including colibactin (Bonnet *et al.*, 2014; Prorok-Hamon *et al.*, 2014). Use of the Min mouse model (Apc-deficient mice with multiple neoplasia) infected with a representative colon cancer-associated *E. coli* isolate (colibactin-producing *E. coli* of phylogroup B2) allowed to demonstrate that this isolate was able to promote tumorigenesis with an increase in the size and number of adenocarcinoma tumours (Bonnet *et al.*, 2014). Arthur and colleagues, using a mouse model of colitis-associated colorectal cancer, observed similar results, suggesting, in this case, that inflammation created an environment that potentiated both expansion and genotoxicity of the colibactin-producing *E. coli* strain (Arthur *et al.*, 2012).

Summarizing the few studies focusing on the neonatal period, the contribution of gut colibiota in the regular development of host physiology can be considered as major. The neonatal colibiota could also exhibit some positive sides/facets, as exemplified by the contribution of *E. coli* at birth in B-cell maturation (Lundell *et al.*, 2012). At adulthood, only negative impact of colibiota overgrowth on gut homeostasis is highlighted in the literature unlike other health-promoting members of microbiota such as *Bacteroides thetaiotaomicron*, *Bacteroides fragilis* or *Faecalibacterium prausnitzii*. Commensal gut colibiota (at physiological levels) is considered harmless and categorized as inert, although it houses potentially pathogenic and versatile clones that can take advantage from host defects to proliferate and potentiate their intrinsic properties in pathophysiological conditions. However, what would happen in the long-term with human health if gut commensal colibiota was specifically eradicated from microbiota at adulthood?

Probiotic *E. coli* – wolf in sheep's clothing?

One of the most investigated and documented probiotic bacteria is an *E. coli* strain originating from the commensal gut microbiota of a soldier, and sold as a drug under the name 'Mutaflor®'. This strain, named *Escherichia coli* Nissle 1917 (EcN; DSM 6601-German Collection for Microorganisms), was isolated by Alfred Nissle during the First World War from a German officer who did not succumb to any of the intestinal disorders highly prevalent in south-eastern Europe at that time. Historically (and empirically), EcN has been considered as safe, since it appeared to be well tolerated by humans, and effective for the treatment of gastrointestinal infectious disorders or urinary tract infections. However some of the experts who contributed in writing the section 'Experts' Opinion Poll' in the journal *Infection* in 1995, have evaluated EcN use of questionable safety especially in infants (1995). Since no clinical trials concerning the use of living *E. coli* strains in patients were previously published, the experts concluded that further research was urgently needed and they also hoped molecular biology progress would be informative to definitively state about therapy with live *E. coli* bacteria. Now, after around 20 years of post-genomic era, clinical use, safety considerations, and biological mechanisms probably underlying the probiotic nature of *E. coli* strain belonging to B2 phylogroup, are described below:

E. coli Nissle 1917 use in clinical practice

E. coli Nissle 1917 has been used as a probiotic in the preantibiotic era since the early 1920s to treat infectious diseases of the gut on a large scale. Management of infectious gastroenteritis with application of EcN was more recently re-evaluated in infants and toddlers (Henker *et al.*, 2007, 2008a). In these studies, EcN has been found to be safe, well tolerated, and to stop both bacterial and viral acute diarrhoea more quickly than placebo.

Potential applications of EcN for the prevention and/or treatment of a large number of gastrointestinal diseases have been evaluated and considered as suitable especially for the management of IBD (Schultz, 2008). Clinical trials, started in 1997, although with some limitations, showed that daily intake of EcN strain was effective in extending remission period of ulcerative colitis (UC) and this effect was equivalent to standard medication (Kruis *et al.*, 1997, 2004; Rembacken *et al.*, 1999). A dose-dependent efficacy of rectally applied EcN compared with placebo was also described in patients suffering from active mild-to-moderate distal UC (Matthes *et al.*, 2010). Moreover, in children and adolescents EcN strain was reported to be effective in preventing the relapse of inactive UC with no main side effects (Henker *et al.*, 2008b). Although the use of EcN seems promising as an alternative strategy to reduce glucocorticoids need in Crohn's disease (CD) patients, no real benefits in maintaining disease remission were observed by a pilot study (Malchow, 1997).

E. coli Nissle 1917 medical use has been clinically studied in a variety of gastrointestinal disorders, including pouchitis (Kuzela *et al.*, 2001), IBS (Kruis *et al.*, 2012; Plassmann and Schulte-Witte, 2007), collagenous colitis (Tromm *et al.*, 2004), antibiotic-associated diarrhoea/pseudomembranous colitis (Goerg and Schlorer, 1998; Goerg *et al.*, 2008) and chronic constipation (Chmielewska and Szajewska, 2010). For all of these indications, EcN normalized stool frequency and consistency in diarrhoea and constipation, and reduced abdominal pain and cramping. Using an animal model of post-inflammatory visceral hyperalgesia, EcN has been reported to cause a transient decrease of visceral hyperalgesia (Liebregts *et al.*, 2005). Moreover, an improvement of liver functions and decrease of the endotoxaemia in cirrhotic patients treated with EcN have also been observed (Lata *et al.*, 2007).

In experimental animal studies EcN reduced allergic responses, consequently its use as adjuvant therapy has been proposed to induce allergen-specific Th1 responses (Adam *et al.*, 2010; Bickert *et al.*, 2009). Oral administration of this probiotic has also been proposed as an adjuvant in the therapy of allergic inflammatory skin diseases (Weise *et al.*, 2011). Nonetheless, immunomodulation by EcN in grass pollen-allergic patients has failed so far to achieve clinical efficacy (Dolle *et al.*, 2014).

Mutaflor® is widely accepted as a safe probiotic, and it is an over-the-counter drug commercialized in several countries (Canada, Germany, Australia, actually discontinued in USA due to changes in FDA's classification of the product). Consequently, EcN has also been proposed for additional indications such as eczema and urinary infections. However insufficient retrospective data are available to evaluate the outcomes.

Safety considerations based on *E. coli* Nissle 1917 genotype and phenotype

Basically, EcN is considered a non-pathogenic strain of serotype O6:K5:H1 compared with other strains of the O6 serogroup (Blum *et al.*, 1995). The O6 antigen is widely distributed among pathogenic and non-pathogenic faecal *E. coli* strains. *E. coli* O6 isolates frequently are ExPEC isolates such as UPEC strains and to a lesser extent MNEC strains. Unlike other strains of serogroup O6, EcN is serum sensitive because of the expression of a semirough type of O6 lipopolysaccharide (LPS) that results from a point mutation in the O-antigen polymerase *wzy* gene (Grozdanov *et al.*, 2002). Consequently, it was supposed that LPS of this probiotic strain was not able to protect bacteria from the bactericidal effect of serum complement.

Genomic analyses revealed that the overall genome structure and genomic islands organization of EcN, resemble more closely to those of UPEC strains such as the CFT073 strain (urosepsis isolate) than those of the non-pathogenic commensal *E. coli* K-12 strain MG1655 (Grozdanov *et al.*, 2004) or an asymptomatic bacteriuria strain (Vejborg *et al.*, 2010). Moreover, the virulence-associated *iha*, *vat* and *sat* genes as well as the *pks* genomic island were detected within the genome of EcN (Vejborg *et al.*, 2010). Nonetheless, complementary phenotypic tests and analyses of the transcriptomic profile of EcN suggest that few but major differences would be responsible for the non-pathogenic nature of EcN. Indeed important ExPEC-associated virulence features such as P-fimbriae and S-fimbriae, haemolysin or cytotoxic-necrotizing factor (CNF) are absent in EcN (Blum *et al.*, 1995; Grozdanov *et al.*, 2004). Inversely, the expression of fimbrial adhesins such as F1C-fimbriae, and the production of at least six siderophores (aerobactin, enterobactin, yersiniabactin, salmochelin, ferric dicitrate transport system and the *chu* haem transport locus) are considered as key factors for EcN gut colonization (Blum *et al.*, 1995; Grosse *et al.*, 2006; Grozdanov *et al.*, 2004; Hancock *et al.*, 2010; Valdebenito *et al.*, 2006). As result of an insertion within the *fimB* gene, EcN has reported to be unable to produce type 1 fimbriae under selective pressure. The *fimB* insertional mutation might contribute to explain this strain attenuated virulence (Stentebjerg-Olesen *et al.*, 1999).

Given the high genetic and transcriptomic similarities observed between the bona fide pathogenic CFT073 strain and the probiotic strain *E. coli* Nissle 1917, it is tempting to speculate that *E. coli* Nissle 1917 originated initially from a pathogenic clone, which lost its virulence nature as a result of evolution (Vejborg *et al.*, 2010). Indeed, minor genetic lesions, such as point mutations and phenotypic variations, are supposed to trigger a 'degenerated' pathogen as result of some virulence factors expression inhibition or to produce truncated non-functional virulence determinants (Vejborg *et al.*, 2010), as exemplified by mutations detected in the *wzy* gene and *fim*, *pap* or *hly* operons, making EcN unable to produce P or type 1 fimbriae, a complete LPS or haemolysin.

Safety considerations based on the ability of the *E. coli* Nissle 1917 to colonize the gut or to invade the host

E. coli Nissle 1917 survives gastrointestinal passage and successfully colonizes the gut of healthy volunteers, as well as gut of animals (rodents, pigs), without any serious adverse effect on host health (Joeres-Nguyen-Xuan *et al.*, 2010). Additionally, oral administration of EcN in preterm infants has shown to increase specific anti-EcN antibodies of IgA isotype and non specific polyclonal IgM in blood, suggesting a stimulation of systemic humoral response as well as nonspecific natural immunity (Cukrowska *et al.*, 2002). Moreover, neither histological changes nor colitis were observed in the gut of EcN-monoassociated piglet or interleukin-2-deficient mice that normally develop inflammatory bowel disease when colonized with a specific pathogen-free microbiota or with the *E. coli* strain mpk (Bleich *et al.*, 2008; Splichal *et al.*, 2005; Waidmann *et al.*, 2003). The lower virulence of the probiotic strain compared to ExPEC strains of serogroup O6, observed in a rat model of pyelonephritis, also supported the safety of EcN (Blum *et al.*, 1995).

Nevertheless, even though EcN has been advocated for its safety in human, it should be used with caution in high-risk populations in regards to bacterial translocation. Patients with therapy-related or disease-related immunosuppression and intestinal barrier dysfunction or neonates with underdevelopment of mucosal barriers may display a risk of probiotic sepsis. Bacterial sepsis related to the administration of EcN in a preterm infant has also been reported (Guenther *et al.*, 2010), suggesting that manipulation of the gut microbiota during the neonatal period may have potentially severe adverse effects. Potential virulence of the probiotic EcN has also been supported by murine experiments (Bleich *et al.*, 2008; Gronbach *et al.*, 2010). Bleich and colleagues (2008) reported that immunocompetent germ-free mice, regardless of their genetic background, tolerated inoculation of both EcN and the commensal laboratory K-12 strain MG1655 without apparent pathological lesions. Specifically in EcN-monoassociated mice, however, bacterial translocation through intestinal epithelium occurred into mesenteric lymph node, spleen, liver, lung or pancreas. Translocation of EcN led to morbidity and even mortality specifically in C3H/HeJZtm germ-free mice, which carry a defective Toll-like receptor 4 (TLR4) allele (Bleich *et al.*, 2008). Similarly, severe adverse outcomes resulting from exacerbated bacterial translocation were induced in immunocompromised mice fed with EcN only if both microbiota and adaptative immunity were defective (T-cell deficiency) (Gronbach *et al.*, 2010). As previously observed with other probiotic genera, these murine experiments suggest that some factors may increase facultative pathogenicity of the probiotic strain and become virulent factors in immuno-compromised hosts. Contribution of some virulence-associated factors of EcN in triggering sepsis has been investigated by generating EcN isogenic mutants. Complementation of the *wzy* gene to restore the production of a smooth LPS in EcN as well as deletion of *fliC* gene to generate deficiency in its flagella production, did not impact EcN translocation, dissemination or mortality in immunocompromised mice (Gronbach *et al.*, 2010). In a mouse subcutaneous sepsis model, Johnson and colleagues showed that among the seven ExPEC-associated traits linked to virulence (Johnson *et al.*, 2012), 5 are present within EcN genome, i.e. *papA-H*, *papG*, *vat*, *kpsM* and *pks*. Although not investigated in the EcN background, deletion within the *pks* genomic island harboured by an ExPEC strain has been shown to lead to lymphocyte death by apoptosis *in vitro*, and to exacerbated lymphopenia in a mouse footpad sepsis model (Marcq *et al.*, 2014). Altogether, these murine studies suggest that colibactin production might increase the risk of sepsis caused by *E. coli* and could impair chances to survive for susceptible hosts receiving, for example, antibiotic therapy.

Probiotic mechanisms of action

The genome of *E. coli* Nissle 1917 is completely sequenced (Sun *et al.*, 2005). As judged by the larger genome size of EcN, compared to that of the laboratory commensal K-12 strain, the presence of several horizontally acquired DNA regions may contribute to its phenotypic traits (Grozdanov *et al.*, 2004). Nonetheless, comparative analysis of genomes (or transcriptomes) appeared not sufficient to explain mechanisms underlying its probiotic effects, especially because of the multiplicity of putative ORFs coding for hypothetical proteins of unknown functions. Several *in vitro* and *in vivo* studies have demonstrated that EcN mainly exerts colonization resistance by preventing adhesion and/or invasion of pathogens, by uptake of nutrients (iron, sugar), but also by stimulating host innate immune defence (i.e. defensins). Additionally, the strengthening of intestinal barrier as well as anti-inflammatory properties contribute to the probiotic features of EcN.

Proposed probiotic mechanisms of EcN include effects on a wide range of pathogens through inhibition of gut colonization, adhesion and/or invasion of epithelial cells as reported by Boudeau and colleagues using an AIEC strain (Boudeau *et al.*, 2003). The direct inhibition of pathogen colonization is linked to several mechanisms. For example, EcN secretes the microcins H47 and M (for Mutaflor) that inhibit *in vitro* the growth of *Enterobacteriaceae* including pathogenic *E. coli* strains and *Salmonella spp.* (Patzer *et al.*, 2003). Using *in vitro* approaches of cultured cells infected by pathogens, additional inhibitory components produced by EcN were evidenced since an isogenic microcin-negative mutant of EcN displayed similar inhibitory effects on *Salmonella* invasion (Altenhoefer *et al.*, 2004). Unknown secreted components found in EcN supernatant were effectively supposed to affect the secretion of virulence-associated proteins of EPEC such as EspA, EspB or Tir, without regulating their expression (Kleta *et al.*, 2014). Moreover, F1C fimbriae and H1 flagella produced by EcN have been shown to contribute to the inhibitory effects of EcN on EPEC infection, suggesting the use of similar host cell receptors for adhesion (Kleta *et al.*, 2014).

One of the most important factor facilitating EcN colonization and pathogen inhibition is represented by the competition for nutrients. Comparison of *in vivo* sugar-defined niches (i.e. mouse intestine) of representative *E. coli* strains, allowed to demonstrate that the human commensal strain HS and EcN display a unique nutritional program allowing the settlement of distinct nutrient-defined niches in the intestine. Together these two strains prevent colonization of the STEC strain EDL933 as a result of the reduced availability of nutrients required for the pathogen to grow (Maltby *et al.*, 2013). In addition to sugar, iron is an essential element for which EcN in the gut outcompetes pathogens such as *Salmonella* by producing different siderophores (Deriu *et al.*, 2013). In the inflamed gut, Deriu *et al.* showed that EcN can limit *Salmonella* colonization because of its resistance to lipocalin 2, and consequently they evidenced the key role of iron in regulating infection. Induction of lipocalin 2 expression by epithelial cells in inflamed gut inhibits some bacterial siderophores such as enterobactin produced by *Salmonella* and EcN; in any case, EcN harbours additional siderophores that are not targeted by lipocalin 2 (Deriu *et al.*, 2013).

In addition to lipocalin 2, human β-defensin 2 (hBD-2) is a protective host antimicrobial peptide induced by pathogens that may contribute to enhance mucosal barrier. Although EcN is not resistant to hBD-2, this probiotic has been shown to stimulate hBD-2 expression via nuclear factor-κB and to activate protein-1-dependent pathways, in contrast to most commensal *E. coli* strains (Wehkamp *et al.*, 2004). H1 flagellin has been identified as the main contributor for hBD-2 induction but the biological mechanism remains unclear (Schlee *et al.*, 2007). Defects in β-defensin inducibility and/or function are observed in patients suffering from CD but not from UC or pouchitis, therefore this might explain why EcN provides benefit in UC and pouchitis and not in CD.

In addition to the protection against intestinal pathogens through the induction of gut mucosal antimicrobials, EcN showed the ability to preserve and reinforce additional intestinal barrier functions. Bergström and colleagues (Bergstrom *et al.*, 2012) have reported that EcN up-regulates mucin-encoding genes in the host intestinal epithelium in newborn mice, suggesting that colonization at birth with EcN may promote the protective mucus barrier covering the intestinal epithelium. Furthermore EcN has been shown to maintain goblet cell activity in murine models of colitis (Hudcovic *et al.*, 2007). Other investigators have reported that EcN modulates the expression of factors implicated in intestinal epithelial cell differentiation, suggesting changes in the proportion of differentiated epithelial cells in the gut (goblet cells, enteroendocrine cells, Paneth cells or absorptive cells) that could impact intestinal barrier function (Becker *et al.*, 2013). The direct impact of EcN on intestinal barrier function has been investigated by Zyrek *et al.* (2007): barrier disruption induced by infection of cultured epithelial cells with an EPEC strain was restored when co-cultured with EcN. Epithelial barrier integrity restoration was reported to be mediated by ZO-2 protein redistribution, which is known to stabilize tight junction structures *in vitro* (Zyrek *et al.*, 2007). Further investigations are required to confirm the same mechanisms *in vivo*. Contribution of ZO-1 has been reported using healthy EcN-monoassociated mice (Ukena *et al.*, 2007). Moreover, EcN has been shown to normalize the increased intestinal permeability that resulted from DSS-induced colitis (Ukena *et al.*, 2007). Supernatants from EcN were also reported to prevent 5-fluorouracil-induced intestinal cell damage by repairing cell barrier function (Wang *et al.*, 2014). In this *in vitro* model, permeability restoration was associated to an increase of cells viability and a decrease of late-apoptotic cells. *In vivo*, impact of this chemotherapy (5-fluorouracil) on barrier function has not been studied, but diminution of the associated side effects (intestinal mucositis severity) in Dark Agouti rats fed with EcN was also shown (Prisciandaro *et al.*, 2011).

A range of probiotic immune effects have been described, but, nowadays, a direct evidence of the immune mechanisms by which EcN achieves its beneficial effect is not available. The immunomodulatory effect of EcN has been clearly shown in several animal models, in which spontaneous or chemically induced colitis were analysed.

Arribas and colleagues (Arribas *et al.*, 2009) demonstrated in mice that orally administrations of EcN, not only prevent acute chemically induced colitis, as previously observed by other investigators (Hudcovic *et al.*, 2007; Jacobi *et al.*, 2012; Kamada *et al.*, 2005; Kokesova *et al.*, 2006; Olier *et al.*, 2012; Schultz *et al.*, 2004), but also exert systemic anti-inflammatory effects in the LPS-induced organ damage. As observed in other studies, these two animal studies reported that EcN administration decreases proinflammatory cytokines expression such as TNFα (colon and plasma/lung respectively). An inhibition of the production of T-cell proinflammatory cytokines (IL-2, IL-5), as well as a down-regulation of IgG release, were observed after *ex-vivo* activation of splenocytes from LPS-treated mice.

In addition to TNFα, the modulatory effect of EcN on other pro-inflammatory cytokines (INF-γ, IL-2, IL-5, IL-6) as well as regulatory cytokine (IL-10) or chemokines (MIP-2, MCP-1, IL-8, RANTES) has been shown to contribute to the amelioration of experimental acute and chronic colitis (Hafez *et al.*, 2009; Jacobi *et al.*, 2012; Kamada *et al.*, 2005; Olier *et al.*, 2012; Schultz *et al.*, 2004; Sturm *et al.*, 2005). Moreover the decrease of neutrophil recruitment in the inflamed colonic mucosa has been observed in rodents fed with EcN (Arribas *et al.*, 2009; Grabig *et al.*, 2006; Olier *et al.*, 2012). As compared to the laboratory commensal K-12 strain, incubation of cultured cells (Caco-2 cells) with EcN has been reported to reduce NLRP3 inflammasome activation and apoptosis induction (Becker *et al.*, 2014). One mechanism by which EcN limits intestinal inflammation while maintaining intestinal immunological homeostasis, may be the induction of naïve and memory peripheral blood $CD4^+$-T-cell clonal expansion without activation of mucosal and lamina propria T-cells (Sturm *et al.*, 2005). EcN has also been reported to increase activation and cycle progression of γδ T-cells, and, at the same time, to stimulate their apoptosis. This suggests that EcN may limit the ability of γδ T-cells to recognize microbial antigens and to perpetuate inflammatory processes (Guzy *et al.*, 2008). These effects, also observed with inactivated EcN, are mediated by Toll-like receptor-2 (TLR-2) signalling (Guzy *et al.*, 2008; Sturm *et al.*, 2005). TLR-2 and TLR-4 receptors are activated by bacterial lipoteichoic acid (LTA) and LPS respectively and by a study performed on TLR-2 and TLR-4 knockout mice, it has been confirmed that EcN ameliorates DSS-induced colitis through these pathways (Grabig *et al.*, 2006).

Among bacterial factors that have been supposed to drive EcN anti-inflammatory properties, the K5 capsule has been found to be effective in mediating chemokine induction in intestinal epithelial cultured cells (Hafez *et al.*, 2009). Further investigations are required to test the chemokine induction potential of other K5 *E. coli* strains *in vitro* and *in vivo*. The autoinducer AI-2 is produced by EcN and allows it to communicate with other bacterial species, *i.e.* to perform a phenomenon called 'quorum sensing'. Using *luxS* gene silencing method, Jacobi and colleagues showed that AI-2 is able to regulate cytokine expression in DSS-treated mice, suggesting that amelioration of the colitis by EcN could be mediated indirectly by modulating the functional activity of the endogenous microbiota (Jacobi *et al.*, 2012).

In the attempt to improve EcN biosafety, a non-genotoxic EcN mutant has been constructed by deleting one of the genes of the *pks* island responsible for colibactin production (Olier *et al.*, 2012). Unexpectedly, this non-genotoxic mutant has been revealed to no longer be able to prevent both chronic and acute experimental colitis in rodent models. The loss of colibactin production not only abolished probiotic beneficial properties of EcN but also led to the exacerbation of colitis severity. Given that chromosomally complementation of this mutant restored both its genotoxicity and its probiotic properties, it was hypothesized that the genotoxin colibactin also was an immunomodulin by which the probiotic EcN regulates intestinal inflammation. Whether the *pks* island encodes distinct molecules with distinct properties or whether one molecule, colibactin, displays both DNA-damage and immunomodulation, is not known. Interestingly, analyses of microbiota from biopsies of IBD patients revealed higher prevalence of colibiota of phylogroup B2 and D (Kotlowski *et al.*, 2007), and particularly, higher prevalence of B2 *pks*$^+$ *E. coli* strains (Arthur *et al.*, 2012). To highlight whether these *pks*$^+$ strains display anti-inflammatory properties, further investigations are needed. Indeed, these *pks*$^+$ *E. coli* are unlikely to be potential probiotics since IBD patients are characterized by colibiota overgrowth. In case *pks*$^+$ strains from IBD patients do not display anti-inflammatory properties, we may anticipate that one additional factor produced by EcN potentiate anti-inflammatory properties mediated by colibactin in this strain.

Not all *E. coli* bacteria are probiotics, ... but some are

Symbioflor 2® contains both living as well as non-living *E. coli* (DSM17252) that helps with IBS symptoms and common other intestinal disorders (Martens *et al.*, 2010). A volunteer high-exposure study reported no severe side effect (Wassenaar *et al.*, 2014). A prolonged gut colonization was observed in this study after a single high dose intake of Symbioflor®2 (10 to 30 weeks) and was mainly associated with the persistence of the combined genotype G1+ (G1-2, G6-7 and G8) (Wassenaar *et al.*, 2014). One of the six strains found in the probiotic (G3-10) has been shown to produce a novel class of microcin (microcin S), encoded on a large plasmid, that is able to inhibit adherence of an EPEC strain to intestinal epithelial cells (Zschuttig *et al.*, 2012). One other strain of the mixture (genotype G2) has been shown to increase hBD-2 production *in vitro* (as seen with EcN) and in colonized human volunteers faeces (Mondel *et al.*, 2009).

COLINFANT, a preparation of *E. coli* O83, has been shown to colonize the intestine of full-term and preterm infants, to stimulate antibody production in saliva and serum, to decrease the presence of pathogenic strains and nosocomial infections, and to reduce the incidence of allergies (Lodinova-Zadnikova *et al.*, 2010). *In vivo* experiments using rodent models of colitis or arthritis, support its probiotic allegation, but few data concerning the strain are available.

E. coli M-17 strain (EC-M17, BU-239, ATCC 202226) is the probiotic strain of serotype O2:H41 found in ProBactrix. It was originally isolated in Russia more than 80 years ago and widely used for a variety of gastrointestinal infections and disorders, including diarrhoea, IBD and IBS. A controlled clinical trial has been conducted in 2005 to determine strain effectiveness in treating IBS but no results were posted on clinicalTrials.gov. Immunomodulation and attenuation of murine colitis treated with EC-M17 has been reported (Fitzpatrick *et al.*, 2008). To date, no direct analysis has been conducted to compare the genomic sequence of EC-M17 with that of Nissle 1917.

Asymptomatic bacteriuria (ABU) isolate 83972 has been also used for prophylactic purposes in several studies where it was demonstrated to have a beneficial effect in patients with recurrent UTI who are refractory to conventional therapy (Hull *et al.*, 2000). It was originally isolated from a young girl who had carried it for at least 3 years without symptoms. As EcN, strain 83972 is considered a descendant of uropathogens since it closely resembles to uropathogen CFT073 strain. As EcN, strain 83972 belongs to the phylogroup B2 and carries fitness genes that appear as intact. Strain 83972 is also defined as a deconstructed pathogen with high adaptation capabilities under a variety of environments (Hancock *et al.*, 2008).

Interestingly, similarly to *E. coli* Nissle 1917, all the sequenced probiotic *E. coli* (EC-M17 and strain 83972) carry the *pks* genomic island. Whether these strains are genotoxic, and whether colibactin production contributes to their potential immuno-modulatory properties have not yet been elucidated. In the meantime, the contribution of colibactin to the probiotic potential among all colibactin-producing *E. coli* is not known and requires further studies.

Conclusion and future trends

The post-genomic era together with progresses made in molecular biology has enabled researchers to supply a more complete picture of probiotic strain *E. coli* Nissle 1917. EcN is a commensal *E. coli* of phylogroup B2 as many other B2 commensal *E. coli* strains, but, in particular, it is able to outcompete pathogens, to efficiently colonize the gut and to control the exacerbation of mucosal inflammation by expressing a myriad of tools specifically adapted to these aims. Genetically, EcN cannot be distinguished from other ExPEC of phylogroup B2 and it is strongly related to the highly virulent uropathogenic strain CFT073. However, EcN has been hypothesized to have lost functionality of the majority of virulent determinants as the result of genomic evolution, explaining its non-pathogenic character. The matter is nonetheless more complex than it appears at first sight, because living EcN may turn out to be an opportunistic pathogen, even if it can be regarded as a deconstructed pathogen. Indeed, following EcN administration, not only sepsis occurrence in preterm infant and in susceptible mice has been reported, but also the conservation, in the EcN genome, of at least one pathogenic functional factor *i.e.* a genotoxin called colibactin has been observed. Why colibactin production has been conserved may be explained by the necessity for the bacteria to preserve the carriage of the related *pks* genomic island that is probably required for colonization persistence or for an essential unknown role, more than for its probiotic properties, which would be presumably coincidentally conserved.

Concerning colibactin, it remains to deepen the knowledge on the extent of genotoxicity and immunomodulatory properties and to evaluate if they are somehow related and how they operate. Currently this intriguing association highlights, once again, the doubt about the safety of EcN for clinical employment, particularly in preterm infants. Despite benefits linked to EcN administration are partly due to colibactin production, colibactin produced by the human commensal strain M1/5 has been shown to trigger DNA damage in the gut epithelium of rats, specifically during the neonatal period (Payros *et al.*, 2014). This latter Janus-faced role of colibactin may be due to the immaturity of both microbiota and mucosal barrier occurring in neonates, suggesting that perinatal or neonatal manipulation of gut microbiota by applying EcN to prevent diseases such as IBD, should not be recommended. Indeed, colibactin-associated DNA-damage exposure early in life could play a critical role in the initial stages of diseases such as colorectal cancer. Consequently retrospective studies with the aim to re-evaluate the use of EcN early in life, its long-term persistence, and the consequences on health at adult age, would be highly desirable to definitively conclude about EcN safety.

Concerning EcN risk assessment, a particular attention must be paid to genetic manipulations. EcN has been regarded to date as sufficiently safe to be used as a potential carrier able to deliver a variety of drugs. Indeed, EcN has been engineered to secrete an anti-HIV microbicide (Rao *et al.*, 2005). The microbicide secreted was a hybrid peptide including the HIV-gp41 and the haemolysin A (HlyA), which, when naturally carried by *E. coli* strains, induces concentration-dependent cell lyses. Consequently, the engineered strain is equipped with a combination of at least two virulence genes (*hlyA* and *pks*) whose impact and side effects *in vivo* are not known. Given that this genetically modified organism (GMO) has to persist in gastrointestinal and cervicovaginal mucosa in the attempt to block HIV transmission, probably such strategy would have to be reconsidered. Similarly, EcN has been engineered to deliver antitumoral drugs for cancer therapy because it has been reported that intravenous administrated EcN can target and proliferate in solid tumours. This strategy has been reported to be effective in immunocompetent mice but further studies are needed to investigate the mechanism by which EcN targets tumour cells, and to evaluate side effects and efficiency, especially in animal models allowing to mimic more closely the immunological state of patients affected by cancer (Zhang *et al.*, 2012).

Concerning side effects, the clinical use of EcN in order to improve the pharmacokinetic profile of therapeutic drugs, has received, to date, low attention and therefore needs to be studied in detail. A recent publication has also shown how EcN medication may increase in rats the bioavailibility of the anti-arrhythmic drug amidarone (Matuskova *et al.*, 2014). These results reinforce the view that probiotics would have to be referred to as any other medication and not only as 'natural health products'.

Even if some gaps have been filled in our current knowledge about the pathogenic potential of *E. coli* Nissle 1917 when used for therapeutic purpose, additional issues require further research. Up to now, few studies have investigated its effect on gut microbiota composition focusing on the potential indirect effect of EcN on other bacterial species in health benefits (maybe through quorum sensing). In addition to metagenomic data that will provide extensive information about microbiota diversity, gene content and their potential functions, there is no indication on whether DNA comes from viable or dead microorganisms, whether or not the predicted genes are expressed and under which conditions. It would be interesting to my opinion, to explore and to compare the active fraction of both EcN and gut microbiome in different conditions (physiological and pathological). For example, metatranscriptomic approaches have allowed to evaluate the importance of non-coding regulatory small RNA in the regulation of pathogenicity of *Vibrio cholerae* (Mandlik *et al.*, 2011). Generally, the use of the same strategies employed in the understanding of pathogens virulence mechanism would certainly generate promising results if applied for probiotic research. For example, functional pathogenomics-based research that combines high-throughput genomics with context-based functional observations, has been successfully applied to ExPEC strains with the aim to identified new fitness determinants and presumably will allow to find new probiotic determinants in EcN as well as new probiotic strains within commensal colibiota (Wiles *et al.*, 2013).

Disclosure of potential conflicts of interest

The authors have no potential conflicts of interest to disclose.

Acknowledgements

I am greatly indebted to Simona Agostini for her fruitful suggestions in the improvement of this chapter. I am grateful to collaborators who have assisted me with my research over the last several years. I would like to thank Koen Venema and Ana Paula do Carmo for helping with editing this chapter. The research done in the author's laboratory was supported by grants from the non-profit Association F. Aupetit (AFA) and the French National Research Agency (ANR-07-EMPB-009-01, ANR 09-MIEN-005-01).

References

Beutin, L., Seeliger, H., Winberg, J., van der Waaij, D., Nord, C.E., and Marget, W. (1995). Oral administration of a certain strain of live *Escherichia coli* for intestinal disorders? Artificial contamination and manipulation of the digestive tract – our poor GIT! Infection 23, 51–54.

Adam, E., Delbrassine, L., Bouillot, C., Reynders, V., Mailleux, A.C., Muraille, E., and Jacquet, A. (2010). Probiotic *Escherichia coli* Nissle 1917 activates DC and prevents house dust mite allergy through a TLR4-dependent pathway. Eur. J. Immunol. *40*, 1995–2005.

Adlerberth, I., Jalil, F., Carlsson, B., Mellander, L., Hanson, L.A., Larsson, P., Khalil, K., and Wold, A.E. (1998). High turnover rate of *Escherichia coli* strains in the intestinal flora of infants in Pakistan. Epidemiol Infect. *121*, 587–598.

Afset, J.E., Bruant, G., Brousseau, R., Harel, J., Anderssen, E., Bevanger, L., and Bergh, K. (2006). Identification of virulence genes linked with diarrhea due to atypical enteropathogenic *Escherichia coli* by DNA microarray analysis and PCR. J. Clin. Microbiol. *44*, 3703–3711.

Altenhoefer, A., Oswald, S., Sonnenborn, U., Enders, C., Schulze, J., Hacker, J., and Oelschlaeger, T.A. (2004). The probiotic *Escherichia coli* strain Nissle 1917 interferes with invasion of human intestinal epithelial cells by different enteroinvasive bacterial pathogens. FEMS Immunol. Med. Microbiol. *40*, 223–229.

Amar, J., Chabo, C., Waget, A., Klopp, P., Vachoux, C., Bermudez-Humaran, L.G., Smirnova, N., Berge, M., Sulpice, T., Lahtinen, S., *et al.* (2011). Intestinal mucosal adherence and translocation of commensal bacteria at the early onset of type 2 diabetes: molecular mechanisms and probiotic treatment. EMBO Mol. Med. 3, 559–572.

Arboleya, S., Ang, L., Margolles, A., Yiyuan, L., Dongya, Z., Liang, X., Solis, G., Fernandez, N., de Los Reyes-Gavilan, C.G., and Gueimonde, M. (2012a). Deep 16S rRNA metagenomics and quantitative PCR analyses of the premature infant fecal microbiota. Anaerobe *18*, 378–380.

Arboleya, S., Binetti, A., Salazar, N., Fernandez, N., Solis, G., Hernandez-Barranco, A., Margolles, A., de Los Reyes-Gavilan, C.G., and Gueimonde, M. (2012b). Establishment and development of intestinal microbiota in preterm neonates. FEMS Microbiol. Ecol. *79*, 763–772.

Arribas, B., Rodriguez-Cabezas, M.E., Camuesco, D., Comalada, M., Bailon, E., Utrilla, P., Nieto, A., Concha, A., Zarzuelo, A., and Galvez, J. (2009). A probiotic strain of *Escherichia coli*, Nissle 1917, given orally exerts local and systemic anti-inflammatory effects in lipopolysaccharide-induced sepsis in mice. Br. J. Pharmacol. *157*, 1024–1033.

Arthur, J.C., Perez-Chanona, E., Muhlbauer, M., Tomkovich, S., Uronis, J.M., Fan, T.J., Campbell, B.J., Abujamel, T., Dogan, B., Rogers, A.B., *et al.* (2012). Intestinal inflammation targets cancer-inducing activity of the microbiota. Science *338*, 120–123.

Backhed, F., Ding, H., Wang, T., Hooper, L.V., Koh, G.Y., Nagy, A., Semenkovich, C.F., and Gordon, J.I. (2004). The gut microbiota as an environmental factor that regulates fat storage. Proc. Natl. Acad. Sci. U.S.A. *101*, 15718–15723.

Bailey, J.K., J.L., P., S., A., and R.M., H. (2010a). Letter to the Editor-Distribution of human commensal *Escherichia coli* phylogenetic groups. Journal of Clinical Microbiology *48*, 3455–3456.

Bailey, J.K., Pinyon, J.L., Anantham, S., and Hall, R.M. (2010b). Commensal *Escherichia coli* of healthy humans: a reservoir for antibiotic-resistance determinants. J. Med. Microbiol. *59*, 1331–1339.

Barnich, N., Carvalho, F.A., Glasser, A.L., Darcha, C., Jantscheff, P., Allez, M., Peeters, H., Bommelaer, G., Desreumaux, P., Colombel, J.F., *et al.* (2007). CEACAM6 acts as a receptor for adherent-invasive E. coli, supporting ileal mucosa colonization in Crohn disease. J. Clin. Invest. *117*, 1566–1574.

Becker, H.M., Apladas, A., Scharl, M., Fried, M., and Rogler, G. (2014). Probiotic *Escherichia coli* Nissle 1917 and commensal E. coli K12 differentially affect the inflammasome in intestinal epithelial cells. Digestion *89*, 110–118.

Becker, S., Oelschlaeger, T.A., Wullaert, A., Vlantis, K., Pasparakis, M., Wehkamp, J., Stange, E.F., and Gersemann, M. (2013). Bacteria regulate intestinal epithelial cell differentiation factors both *in vitro* and *in vivo*. PLoS One *8*, e55620.

Bergstrom, A., Kristensen, M.B., Bahl, M.I., Metzdorff, S.B., Fink, L.N., Frokiaer, H., and Licht, T.R. (2012). Nature of bacterial colonization influences transcription of mucin genes in mice during the first week of life. BMC Res. Notes *5*, 402.

Bettelheim, K.A., and Lennox-King, S.M. (1976). The acquisition of *Escherichia coli* by new-born babies. Infection *4*, 174–179.

Bickert, T., Trujillo-Vargas, C.M., Duechs, M., Wohlleben, G., Polte, T., Hansen, G., Oelschlaeger, T.A., and Erb, K.J. (2009). Probiotic *Escherichia coli* Nissle 1917 suppresses allergen-induced Th2 responses in the airways. Int. Arch Allergy Immunol. *149*, 219–230.

Bleich, A., Sundberg, J.P., Smoczek, A., von Wasielewski, R., de Buhr, M.F., Janus, L.M., Julga, G., Ukena, S.N., Hedrich, H.J., and Gunzer, F. (2008). Sensitivity to *Escherichia coli* Nissle 1917 in mice is dependent on environment and genetic background. Int. J. Exp. Pathol *89*, 45–54.

Blum, G., Marre, R., and Hacker, J. (1995). Properties of *Escherichia coli* strains of serotype O6. Infection *23*, 234–236.

Bonnet, M., Buc, E., Sauvanet, P., Darcha, C., Dubois, D., Pereira, B., Dechelotte, P., Bonnet, R., Pezet, D., and Darfeuille-Michaud, A. (2014). Colonization of the human gut by E. coli and colorectal cancer risk. Clin. Cancer Res. *20*, 859–867.

Boudeau, J., Glasser, A.L., Julien, S., Colombel, J.F., and Darfeuille-Michaud, A. (2003). Inhibitory effect of probiotic *Escherichia coli* strain Nissle 1917 on adhesion to and invasion of intestinal epithelial cells by adherent-invasive E. coli strains isolated from patients with Crohn's disease. Aliment. Pharmacol. Ther. *18*, 45–56.

Brzuszkiewicz, E., Thurmer, A., Schuldes, J., Leimbach, A., Liesegang, H., Meyer, F.D., Boelter, J., Petersen, H., Gottschalk, G., and Daniel, R. (2011). Genome sequence analyses of two isolates from the recent *Escherichia coli* outbreak in Germany reveal the emergence of a new pathotype: Entero-Aggregative-Haemorrhagic *Escherichia coli* (EAHEC). Arch. Microbiol. *193*, 883–891.

Carayol, N., and Tran Van Nhieu, G. (2013). Tips and tricks about Shigella invasion of epithelial cells. Curr. Opin. Microbiol. *16*, 32–37.

Chassaing, B., Koren, O., Carvalho, F.A., Ley, R.E., and Gewirtz, A.T. (2013). AIEC pathobiont instigates chronic colitis in susceptible hosts by altering microbiota composition. Gut.

Chmielewska, A., and Szajewska, H. (2010). Systematic review of randomized controlled trials: probiotics for functional constipation. World J. Gastroenterol. *16*, 69–75.

Chung, H., Pamp, S.J., Hill, J.A., Surana, N.K., Edelman, S.M., Troy, E.B., Reading, N.C., Villablanca, E.J., Wang, S., Mora, J.R., *et al.* (2012). Gut immune maturation depends on colonization with a host-specific microbiota. Cell *149*, 1578–1593.

Clermont, O., Christenson, J.K., Denamur, E., and Gordon, D.M. (2013). The Clermont *Escherichia coli* phylo-typing method revisited: improvement of specificity and detection of new phylo-groups. Environ. Microbiol. Rep *5*, 58–65.

Clermont, O., Olier, M., Hoede, C., Diancourt, L., Brisse, S., Keroudean, M., Glodt, J., Picard, B., Oswald, E., and Denamur, E. (2011). Animal and human pathogenic *Escherichia coli* strains share common genetic backgrounds. Infect. Genet. Evol. *11*, 654–662.

Cleynen, I., Vazeille, E., Artieda, M., Verspaget, H.W., Szczypiorska, M., Bringer, M.A., Lakatos, P.L., Seibold, F., Parnell, K., Weersma, R.K., *et al.* (2013). Genetic and microbial factors modulating the ubiquitin proteasome system in inflammatory bowel disease. Gut.

Collado, M.C., Donat, E., Ribes-Koninckx, C., Calabuig, M., and Sanz, Y. (2009). Specific duodenal and faecal bacterial groups associated with paediatric coeliac disease. J. Clin. Pathol *62*, 264–269.

Conway, T., Krogfelt, K.A., and Cohen, P.S. (2004). The life of commensal *Escherichia coli* in the mammalian intestine, J. Nataro, ed. (ASP Press, Washington).

Croxen, M.A., Law, R.J., Scholz, R., Keeney, K.M., Wlodarska, M., and Finlay, B.B. (2013). Recent advances in understanding enteric pathogenic *Escherichia coli*. Clin. Microbiol. Rev. *26*, 822–880.

Cukrowska, B., LodInova-ZadnIkova, R., Enders, C., Sonnenborn, U., Schulze, J., and Tlaskalova-Hogenova, H. (2002). Specific proliferative and antibody responses of premature infants to intestinal colonization with nonpathogenic probiotic E. coli strain Nissle 1917. Scand J. Immunol. *55*, 204–209.

Darfeuille-Michaud, A., Boudeau, J., Bulois, P., Neut, C., Glasser, A.L., Barnich, N., Bringer, M.A., Swidsinski, A., Beaugerie, L., and Colombel, J.F. (2004). High prevalence of adherent-invasive *Escherichia coli* associated with ileal mucosa in Crohn's disease. Gastroenterology *127*, 412–421.

Das, P., Singh, A.K., Mukherjee, S., Rajendran, K., Saha, D.R., Koley, H., and Basu, S. (2013). Composition of *Escherichia coli* population in the neonatal gut: phylogroups and virulence determinants. J. Med. Microbiol. *62*, 1680–1687.

de Muinck, E.J., Oien, T., Storro, O., Johnsen, R., Stenseth, N.C., Ronningen, K.S., and Rudi, K. (2011). Diversity, transmission and persistence of *Escherichia coli* in a cohort of mothers and their infants. Environ. Microbiol. Rep 3, 352–359.

Deriu, E., Liu, J.Z., Pezeshki, M., Edwards, R.A., Ochoa, R.J., Contreras, H., Libby, S.J., Fang, F.C., and Raffatellu, M. (2013). Probiotic bacteria reduce salmonella typhimurium intestinal colonization by competing for iron. Cell Host Microbe *14*, 26–37.

Diard, M., Garry, L., Selva, M., Mosser, T., Denamur, E., and Matic, I. (2010). Pathogenicity-associated islands in extraintestinal pathogenic *Escherichia coli* are fitness elements involved in intestinal colonization. J. Bacterol. *192*, 4885–4893.

Dinu, L.D., and Bach, S. (2011). Induction of viable but nonculturable *Escherichia coli* O157:H7 in the phyllosphere of lettuce: a food safety risk factor. Appl. Environ. Microbiol. 77, 8295–8302.

Dobrindt, U., Chowdary, M.G., Krumbholz, G., and Hacker, J. (2010). Genome dynamics and its impact on evolution of *Escherichia coli*. Med. Microbiol. Immunol. *199*, 145–154.

Dolle, S., Berg, J., Rasche, C., and Worm, M. (2014). Tolerability and clinical outcome of coseasonal treatment with *Escherichia coli* strain Nissle 1917 in grass pollen-allergic subjects. Int. Arch Allergy Immunol. *163*, 29–35.

Dreux, N., Denizot, J., Martinez-Medina, M., Mellmann, A., Billig, M., Kisiela, D., Chattopadhyay, S., Sokurenko, E., Neut, C., Gower-Rousseau, C., *et al.* (2013). Point mutations in FimH adhesin of Crohn's disease-associated adherent-invasive *Escherichia coli* enhance intestinal inflammatory response. PLoS Pathog. *9*, e1003141.

Duriez, P., Clermont, O., Bonacorsi, S., Bingen, E., Chaventre, A., Elion, J., Picard, B., and Denamur, E. (2001). Commensal *Escherichia coli* isolates are phylogenetically distributed among geographically distinct human populations. Microbiology *147*, 1671–1676.

Eckburg, P.B., Bik, E.M., Bernstein, C.N., Purdom, E., Dethlefsen, L., Sargent, M., Gill, S.R., Nelson, K.E., and Relman, D.A. (2005). Diversity of the human intestinal microbial flora. Science *308*, 1635–1638.

Escherich, T. (1988). The intestinal bacteria of the neonate and breast-fed infant. 1884. Rev. Infect. Dis. *10*, 1220–1225.

Escobar-Paramo, P., Clermont, O., Blanc-Potard, A.B., Bui, H., Le Bouguenec, C., and Denamur, E. (2004a). A specific genetic background is required for acquisition and expression of virulence factors in *Escherichia coli*. Mol. Biol. Evol. *21*, 1085–1094.

Escobar-Paramo, P., Grenet, K., Le Menac'h, A., Rode, L., Salgado, E., Amorin, C., Gouriou, S., Picard, B., Rahimy, M.C., Andremont, A., *et al.* (2004b). Large-scale population structure of human commensal *Escherichia coli* isolates. Appl. Environ. Microbiol. *70*, 5698–5700.

Estrada-Garcia, T., and Navarro-Garcia, F. (2012). Enteroaggregative *Escherichia coli* pathotype: a genetically heterogeneous emerging foodborne enteropathogen. FEMS Immunol. Med. Microbiol. *66*, 281–298.

Ewers, C., Li, G., Wilking, H., Kiessling, S., Alt, K., Antao, E.M., Laturnus, C., Diehl, I., Glodde, S., Homeier, T., *et al.* (2007). Avian pathogenic, uropathogenic, and newborn meningitis-causing *Escherichia coli*: how closely related are they? Int. J. Med. Microbiol. 297, 163–176.

Favier, C.F., Vaughan, E.E., De Vos, W.M., and Akkermans, A.D. (2002). Molecular monitoring of succession of bacterial communities in human neonates. Appl. Environ. Microbiol. *68*, 219–226.

Fitzpatrick, L.R., Small, J., Hoerr, R.A., Bostwick, E.F., Maines, L., and Koltun, W.A. (2008). *In vitro* and *in vivo* effects of the probiotic *Escherichia coli* strain M-17: immunomodulation and attenuation of murine colitis. Br. J. Nutr. *100*, 530–541.

Foxman, B. (2010). The epidemiology of urinary tract infection. Nat. Rev. Urol. 7, 653–660.

Frank, C., Werber, D., Cramer, J.P., Askar, M., Faber, M., an der Heiden, M., Bernard, H., Fruth, A., Prager, R., Spode, A., *et al.* (2011a). Epidemic profile of Shiga-toxin-producing *Escherichia coli* O104:H4 outbreak in Germany. N. Engl. J. Med. *365*, 1771–1780.

Frank, D.N., Zhu, W., Sartor, R.B., and Li, E. (2011b). Investigating the biological and clinical significance of human dysbioses. Trends Microbiol. *19*, 427–434.

Freestone, P.P., Williams, P.H., Haigh, R.D., Maggs, A.F., Neal, C.P., and Lyte, M. (2002). Growth stimulation of intestinal commensal *Escherichia coli* by catecholamines: a possible contributory factor in trauma-induced sepsis. Shock *18*, 465–470.

Friman, V., Nowrouzian, F., Adlerberth, I., and Wold, A.E. (2002). Increased frequency of intestinal *Escherichia coli* carrying genes for S fimbriae and haemolysin in IgA-deficient individuals. Microb. Pathog. *32*, 35–42.

Gaschignard, J., Levy, C., Romain, O., Cohen, R., Bingen, E., Aujard, Y., and Boileau, P. (2011). Neonatal Bacterial Meningitis: 444 Cases in 7 Years. Pediatr. Infect. Dis. J. *30*, 212–217.

Germani, Y., Begaud, E., Duval, P., and Le Bouguenec, C. (1996). Prevalence of enteropathogenic, enteroaggregative, and diffusely adherent *Escherichia coli* among isolates from children with diarrhoea in new Caledonia. J. Infect. Dis. *174*, 1124–1126.

Goerg, K.J., and Schlorer, E. (1998). [Probiotic therapy of pseudomembranous colitis. Combination of intestinal lavage and oral administration of *Escherichia coli*]. Dtsch Med. Wochenschr *123*, 1274–1278.

Goerg, K.J., Wybierala, M., Rauen-Vossloh, J., and Hader, C. (2008). A new approach in pseudomembranous colitis: probiotic *Escherichia coli* Nissle 1917 after intestinal lavage. Eur. J. Gastroenterol. Hepatol. *20*, 155–156.

Grabig, A., Paclik, D., Guzy, C., Dankof, A., Baumgart, D.C., Erckenbrecht, J., Raupach, B., Sonnenborn, U., Eckert, J., Schumann, R.R., *et al.* (2006). *Escherichia coli* strain Nissle 1917 ameliorates experimental colitis via toll-like receptor 2- and toll-like receptor 4-dependent pathways. Infect. Immun. *74*, 4075–4082.

Gronbach, K., Eberle, U., Muller, M., Olschlager, T.A., Dobrindt, U., Leithauser, F., Niess, J.H., Doring, G., Reimann, J., Autenrieth, I.B., *et al.* (2010). Safety of probiotic *Escherichia coli* strain Nissle 1917 depends on intestinal microbiota and adaptive immunity of the host. Infect. Immun. *78*, 3036–3046.

Grosse, C., Scherer, J., Koch, D., Otto, M., Taudte, N., and Grass, G. (2006). A new ferrous iron-uptake transporter, EfeU (YcdN), from *Escherichia coli*. Mol. Microbiol. *62*, 120–131.

Grozdanov, L., Raasch, C., Schulze, J., Sonnenborn, U., Gottschalk, G., Hacker, J., and Dobrindt, U. (2004). Analysis of the genome structure of the nonpathogenic probiotic *Escherichia coli* strain Nissle 1917. J. Bacterol. *186*, 5432–5441.

Grozdanov, L., Zahringer, U., Blum-Oehler, G., Brade, L., Henne, A., Knirel, Y.A., Schombel, U., Schulze, J., Sonnenborn, U., Gottschalk, G., *et al.* (2002). A single nucleotide exchange in the wzy gene is responsible for the semirough O6 lipopolysaccharide phenotype and serum sensitivity of *Escherichia coli* strain Nissle 1917. J. Bacterol. *184*, 5912–5925.

Guenther, K., Straube, E., Pfister, W., Guenther, A., and Huebler, A. (2010). Sever sepsis after probiotic treatment with *Escherichia coli* NISSLE 1917. Pediatr. Infect. Dis. J. 29, 188–189.

Guzy, C., Paclik, D., Schirbel, A., Sonnenborn, U., Wiedenmann, B., and Sturm, A. (2008). The probiotic *Escherichia coli* strain Nissle 1917 induces gammadelta T-cell apoptosis via caspase- and FasL-dependent pathways. Int. Immunol. *20*, 829–840.

Haag, L.M., Fischer, A., Otto, B., Plickert, R., Kuhl, A.A., Gobel, U.B., Bereswill, S., and Heimesaat, M.M. (2012). Intestinal microbiota shifts towards elevated commensal *Escherichia coli* loads abrogate colonization resistance against *Campylobacter jejuni* in mice. PLoS One 7, e35988.

Hacker, J., Hentschel, U., and Dobrindt, U. (2003). Prokaryotic chromosomes and disease. Science *301*, 790–793.

Hafez, M., Hayes, K., Goldrick, M., Warhurst, G., Grencis, R., and Roberts, I.S. (2009). The K5 capsule of *Escherichia coli* strain Nissle 1917 is important in mediating interactions with intestinal epithelial cells and chemokine induction. Infect. Immun. *77*, 2995–3003.

Hancock, V., Seshasayee, A.S., Ussery, D.W., Luscombe, N.M., and Klemm, P. (2008). Transcriptomics and adaptive genomics of the asymptomatic bacteriuria *Escherichia coli* strain 83972. Mol. Genet. Genomics *279*, 523–534.

Hancock, V., Vejborg, R.M., and Klemm, P. (2010). Functional genomics of probiotic *Escherichia coli* Nissle 1917 and 83972, and UPEC strain CFT073: comparison of transcriptomes, growth and biofilm formation. Mol. Genet. Genomics *284*, 437–454.

Henker, J., Laass, M., Blokhin, B.M., Bolbot, Y.K., Maydannik, V.G., Elze, M., Wolff, C., and Schulze, J. (2007). The probiotic *Escherichia coli* strain Nissle 1917 (EcN) stops acute diarrhoea in infants and toddlers. Eur. J. Pediatr. *166*, 311–318.

Henker, J., Laass, M.W., Blokhin, B.M., Maydannik, V.G., Bolbot, Y.K., Elze, M., Wolff, C., Schreiner, A., and Schulze, J. (2008a). Probiotic *Escherichia coli* Nissle 1917 versus placebo for treating diarrhea of greater than 4 days duration in infants and toddlers. Pediatr. Infect. Dis. J. *27*, 494–499.

Henker, J., Muller, S., Laass, M.W., Schreiner, A., and Schulze, J. (2008b). Probiotic *Escherichia coli* Nissle 1917 (EcN) for successful remission maintenance of ulcerative colitis in children and adolescents: an open-label pilot study. Z Gastroenterol. *46*, 874–875.

Holmes, E., Li, J.V., Marchesi, J.R., and Nicholson, J.K. (2012). Gut microbiota composition and activity in relation to host metabolic phenotype and disease risk. Cell Metab. *16*, 559–564.

Homeier, T., Semmler, T., Wieler, L.H., and Ewers, C. (2010). The GimA locus of extraintestinal pathogenic *E. coli*: does reductive evolution correlate with habitat and pathotype? PLoS One *5*, e10877.

Hooper, L.V., Littman, D.R., and Macpherson, A.J. (2012). Interactions between the microbiota and the immune system. Science *336*, 1268–1273.

Hudcovic, T., Stepankova, R., Kozakova, H., Hrncir, T., and Tlaskalova-Hogenova, H. (2007). Effects of monocolonization with *Escherichia coli* strains O6K13 and Nissle 1917 on the development of experimentally induced acute and chronic intestinal inflammation in germ-free immunocompetent and immunodeficient mice. Folia Microbiol. (Praha) *52*, 618–626.

Hull, R., Rudy, D., Donovan, W., Svanborg, C., Wieser, I., Stewart, C., and Darouiche, R. (2000). Urinary tract infection prophylaxis using *Escherichia coli* 83972 in spinal cord injured patients. J. Urol. *163*, 872–877.

Ivanov, I.I., Atarashi, K., Manel, N., Brodie, E.L., Shima, T., Karaoz, U., Wei, D., Goldfarb, K.C., Santee, C.A., Lynch, S.V., *et al.* (2009). Induction of intestinal Th17 cells by segmented filamentous bacteria. Cell *139*, 485–498.

Jacobi, C.A., Grundler, S., Hsieh, C.J., Frick, J.S., Adam, P., Lamprecht, G., Autenrieth, I.B., Gregor, M., and Malfertheiner, P. (2012). Quorum sensing in the probiotic bacterium *Escherichia coli* Nissle 1917 (Mutaflor) – evidence that furanosyl borate diester (AI-2). is influencing the cytokine expression in the DSS colitis mouse model. Gut Pathog. *4*, 8.

Jimenez, E., Marin, M.L., Martin, R., Odriozola, J.M., Olivares, M., Xaus, J., Fernandez, L., and Rodriguez, J.M. (2008). Is meconium from healthy newborns actually sterile? Res. Microbiol. *159*, 187–193.

Joeres-Nguyen-Xuan, T.H., Boehm, S.K., Joeres, L., Schulze, J., and Kruis, W. (2010). Survival of the probiotic *Escherichia coli* Nissle 1917 (EcN) in the gastrointestinal tract given in combination with oral mesalamine to healthy volunteers. Inflamm. Bowel Dis. *16*, 256–262.

Johnson, J.R., Porter, S.B., Zhanel, G., Kuskowski, M.A., and Denamur, E. (2012). Virulence of *Escherichia coli* clinical isolates in a murine sepsis model in relation to sequence type ST131 status, fluoroquinolone resistance, and virulence genotype. Infect. Immun. *80*, 1554–1562.

Johnson, J.R., and Russo, T.A. (2002). Extraintestinal pathogenic *Escherichia coli*: 'the other bad E coli'. J. Lab Clin. Med. *139*, 155–162.

Juhas, M. (2013). Horizontal gene transfer in human pathogens. Crit. Rev. Microbiol. *41*, 101–108.

Kamada, N., Inoue, N., Hisamatsu, T., Okamoto, S., Matsuoka, K., Sato, T., Chinen, H., Hong, K.S., Yamada, T., Suzuki, Y., *et al.* (2005). Nonpathogenic *Escherichia coli* strain Nissle1917 prevents murine acute and chronic colitis. Inflamm. Bowel Dis. *11*, 455–463.

Kaper, J.B., Nataro, J.P., and Mobley, H.L. (2004). Pathogenic *Escherichia coli*. Nat. Rev. Microbiol. *2*, 123–140.

Kawamoto, S., Tran, T.H., Maruya, M., Suzuki, K., Doi, Y., Tsutsui, Y., Kato, L.M., and Fagarasan, S. (2012). The inhibitory receptor PD-1 regulates IgA selection and bacterial composition in the gut. Science *336*, 485–489.

Kim, S.C., Tonkonogy, S.L., Albright, C.A., Tsang, J., Balish, E.J., Braun, J., Huycke, M.M., and Sartor, R.B. (2005). Variable phenotypes of enterocolitis in interleukin 10-deficient mice monoassociated with two different commensal bacteria. Gastroenterology *128*, 891–906.

Kleta, S., Nordhoff, M., Tedin, K., Wieler, L.H., Kolenda, R., Oswald, S., Oelschlaeger, T.A., Bleiss, W., and Schierack, P. (2014). Role of F1C fimbriae, flagella, and secreted bacterial components in the inhibitory effect of probiotic *Escherichia coli* Nissle 1917 on atypical enteropathogenic E. coli infection. Infect. Immun. *82*, 1801–1812.

Kokesova, A., Frolova, L., Kverka, M., Sokol, D., Rossmann, P., Bartova, J., and Tlaskalova-Hogenova, H. (2006). Oral administration of probiotic bacteria (*E. coli* Nissle, E. coli O83, *Lactobacillus casei*) influences the severity of dextran sodium sulfate-induced colitis in BALB/c mice. Folia Microbiol. (Praha) *51*, 478–484.

Kotlowski, R., Bernstein, C.N., Sepehri, S., and Krause, D.O. (2007). High prevalence of *Escherichia coli* belonging to the B2+D phylogenetic group in inflammatory bowel disease. Gut *56*, 669–675.

Kruis, W., Chrubasik, S., Boehm, S., Stange, C., and Schulze, J. (2012). A double-blind placebo-controlled trial to study therapeutic effects of probiotic *Escherichia coli* Nissle 1917 in subgroups of patients with irritable bowel syndrome. Int. J. Colorectal Dis. *27*, 467–474.

Kruis, W., Fric, P., Pokrotnieks, J., Lukas, M., Fixa, B., Kascak, M., Kamm, M.A., Weismueller, J., Beglinger, C., Stolte, M., *et al.* (2004). Maintaining remission of ulcerative colitis with the probiotic *Escherichia coli* Nissle 1917 is as effective as with standard mesalazine. Gut *53*, 1617–1623.

Kruis, W., Schutz, E., Fric, P., Fixa, B., Judmaier, G., and Stolte, M. (1997). Double-blind comparison of an oral *Escherichia coli* preparation and mesalazine in maintaining remission of ulcerative colitis. Aliment. Pharmacol. Ther. *11*, 853–858.

Kuzela, L., Kascak, M., and Vavrecka, A. (2001). Induction and maintenance of remission with nonpathogenic *Escherichia coli* in patients with pouchitis. Am. J. Gastroenterol. *96*, 3218–3219.

Lata, J., Novotny, I., Pribramska, V., Jurankova, J., Fric, P., Kroupa, R., and Stiburek, O. (2007). The effect of probiotics on gut flora, level of endotoxin and Child–Pugh score in cirrhotic patients: results of a double-blind randomized study. Eur. J. Gastroenterol. Hepatol. *19*, 1111–1113.

Le Bouguenec, C., and Servin, A.L. (2006). Diffusely adherent *Escherichia coli* strains expressing Afa/Dr adhesins (Afa/Dr DAEC): hitherto unrecognized pathogens. FEMS Microbiol. Lett. *256*, 185–194.

Le Gall, T., Clermont, O., Gouriou, S., Picard, B., Nassif, X., Denamur, E., and Tenaillon, O. (2007). Extraintestinal virulence is a coincidental by-product of commensalism in B2 phylogenetic group *Escherichia coli* strains. Mol. Biol. Evol. *24*, 2373–2384.

Lefort, A., Panhard, X., Clermont, O., Woerther, P.L., Branger, C., Mentre, F., Fantin, B., Wolff, M., and Denamur, E. (2011). Host factors and portal of entry outweigh bacterial determinants to predict the severity of *Escherichia coli* bacteremia. J. Clin. Microbiol. *49*, 777–783.

Liebregts, T., Adam, B., Bertel, A., Jones, S., Schulze, J., Enders, C., Sonnenborn, U., Lackner, K., and Holtmann, G. (2005). Effect of E. coli Nissle 1917 on post-inflammatory visceral sensory function in a rat model. Neurogastroenterol. Motil. *17*, 410–414.

Lodinova-Zadnikova, R., Prokesova, L., Kocourkova, I., Hrdy, J., and Zizka, J. (2010). Prevention of allergy in infants of allergic mothers by probiotic *Escherichia coli*. Int. Arch. Allergy Immunol. *153*, 201–206.

Logue, C.M., Doetkott, C., Mangiamele, P., Wannemuehler, Y.M., Johnson, T.J., Tivendale, K.A., Li, G., Sherwood, J.S., and Nolan, L.K. (2012). Genotypic and phenotypic traits that distinguish neonatal meningitis-associated *Escherichia coli* from faecal *E. coli* isolates of healthy human hosts. Appl. Environ. Microbiol. *78*, 5824–5830.

Lopez-Siles, M., Martinez-Medina, M., Busquets, D., Sabat-Mir, M., Duncan, S.H., Flint, H.J., Aldeguer, X., and Garcia-Gil, L.J. (2014). Mucosa-associated *Faecalibacterium prausnitzii* and *Escherichia coli* co-abundance can distinguish irritable bowel syndrome and Inflammatory Bowel Disease phenotypes. Int. J. Med. Microbiol. *304*, 464–475.

Lundell, A.C., Bjornsson, V., Ljung, A., Ceder, M., Johansen, S., Lindhagen, G., Tornhage, C.J., Adlerberth, I., Wold, A.E., and Rudin, A. (2012). Infant B-cell memory differentiation and early gut bacterial colonization. J. Immunol. *188*, 4315–4322.

Mackie, R.I., Sghir, A., and Gaskins, H.R. (1999). Developmental microbial ecology of the neonatal gastrointestinal tract. Am. J. Clin. *69* Nutr. 1035S–1045S.

Maharshak, N., Packey, C.D., Ellermann, M., Manick, S., Siddle, J.P., Huh, E.Y., Plevy, S., Sartor, R.B., and Carroll, I.M. (2013). Altered enteric microbiota ecology in interleukin 10-deficient mice during development and progression of intestinal inflammation. Gut Microbes *4*, 316–324.

Malchow, H.A. (1997). Crohn's disease and *Escherichia coli*. A new approach in therapy to maintain remission of colonic Crohn's disease? J. Clin. Gastroenterol. *25*, 653–658.

Maltby, R., Leatham-Jensen, M.P., Gibson, T., Cohen, P.S., and Conway, T. (2013). Nutritional basis for colonization resistance by human commensal *Escherichia coli* strains HS and Nissle 1917 against E. coli O157:H7 in the mouse intestine. PLoS One *8*, e53957.

Mammeri, H., Eb, F., Berkani, A., and Nordmann, P. (2008). Molecular characterization of AmpC-producing *Escherichia coli* clinical isolates recovered in a French hospital. J. Antimicrob Chemother. *61*, 498–503.

Mandlik, A., Livny, J., Robins, W.P., Ritchie, J.M., Mekalanos, J.J., and Waldor, M.K. (2011). RNA-Seq-based monitoring of infection-linked changes in Vibrio choleraee gene expression. Cell Host Microbe *10*, 165–174.

Marcq, I., Martin, P., Payros, D., Cuevas-Ramos, G., Boury, M., Watrin, C., Nougayrede, J.P., Olier, M., and Oswald, E. (2014). The Genotoxin Colibactin Exacerbates Lymphopenia and Decreases Survival Rate in Mice Infected With Septicaemic *Escherichia coli*. J. Infect. Dis.

Martens, U., Enck, P., and Zieseniss, E. (2010). Probiotic treatment of irritable bowel syndrome in children. Ger Med. Sci. *8*, Doc07.

Martin, H.M., Campbell, B.J., Hart, C.A., Mpofu, C., Nayar, M., Singh, R., Englyst, H., Williams, H.F., and Rhodes, J.M. (2004). Enhanced *Escherichia coli* adherence and invasion in Crohn's disease and colon cancer. Gastroenterology *127*, 80–93.

Matthes, H., Krummenerl, T., Giensch, M., Wolff, C., and Schulze, J. (2010). Clinical trial: probiotic treatment of acute distal ulcerative colitis with rectally administered *Escherichia coli* Nissle 1917 (EcN). BMC Complement. Altern. Med. *10*, 13.

Matuskova, Z., Anzenbacherova, E., Vecera, R., Tlaskalova-Hogenova, H., Kolar, M., and Anzenbacher, P. (2014). Administration of a probiotic can change drug pharmacokinetics: effect of E. coli Nissle 1917 on amidarone absorption in rats. PLoS One *9*, e87150.

Mazmanian, S.K., Liu, C.H., Tzianabos, A.O., and Kasper, D.L. (2005). An immunomodulatory molecule of symbiotic bacteria directs maturation of the host immune system. Cell *122*, 107–118.

Mondel, M., Schroeder, B.O., Zimmermann, K., Huber, H., Nuding, S., Beisner, J., Fellermann, K., Stange, E.F., and Wehkamp, J. (2009). Probiotic E. coli treatment mediates antimicrobial human beta-defensin synthesis and faecal excretion in humans. Mucosal Immunol. *2*, 166–172.

Moreno, A.C., Ferreira, L.G., and Martinez, M.B. (2009). Enteroinvasive *Escherichia coli* vs. Shigella flexneri: how different patterns of gene expression affect virulence. FEMS Microbiol. Lett. *301*, 156–163.

Moreno, E., Andreu, A., Pigrau, C., Kuskowski, M.A., Johnson, J.R., and Prats, G. (2008). Relationship between *Escherichia coli* strains causing acute cystitis in women and the faecal E. coli population of the host. J. Clin. Microbiol. *46*, 2529–2534.

Moussaoui, N., Braniste, V., Ait-Belgnaoui, A., Gabanou, M., Sekkal, S., Olier, M., Theodorou, V., Martin, P.G., and Houdeau, E. (2014). Changes in intestinal glucocorticoid sensitivity in early life shape the risk of epithelial barrier defect in maternal-deprived rats. PLoS One *9*, e88382.

Nash, J.H., Villegas, A., Kropinski, A.M., Aguilar-Valenzuela, R., Konczy, P., Mascarenhas, M., Ziebell, K., Torres, A.G., Karmali, M.A., and Coombes, B.K. (2010). Genome sequence of adherent-invasive *Escherichia coli* and comparative genomic analysis with other E. coli pathotypes. BMC Genomics *11*, 667.

Nataro, J.P., and Kaper, J.B. (1998). Diarrheagenic *Escherichia coli*. Clin. Microbiol. Rev. *11*, 142–201.

Nataro, J.P., Kaper, J.B., Robins-Browne, R., Prado, V., Vial, P., and Levine, M.M. (1987). Patterns of adherence of diarrheagenic *Escherichia coli* to HEp-2 cells. Pediatr. Infect. Dis. J. *6*, 829–831.

Nguyen, H.T., Dalmasso, G., Muller, S., Carriere, J., Seibold, F., and Darfeuille-Michaud, A. (2014). Crohn's disease-associated adherent invasive *Escherichia coli* modulate levels of microRNAs in intestinal epithelial cells to reduce autophagy. Gastroenterology *146*, 508–519.

Nougayrede, J.P., Homburg, S., Taieb, F., Boury, M., Brzuszkiewicz, E., Gottschalk, G., Buchrieser, C., Hacker, J., Dobrindt, U., and Oswald, E. (2006). *Escherichia coli* induces DNA double-strand breaks in eukaryotic cells. Science *313*, 848–851.

Nowrouzian, F., Hesselmar, B., Saalman, R., Strannegard, I.L., Aberg, N., Wold, A.E., and Adlerberth, I. (2003). *Escherichia coli* in infants' intestinal microflora: colonization rate, strain turnover, and virulence gene carriage. Pediatr. Res. *54*, 8–14.

Nowrouzian, F.L., Friman, V., Adlerberth, I., and Wold, A.E. (2007). Reduced phase switch capacity and functional adhesin expression of type 1-fimbriated *Escherichia coli* from immunoglobulin A-deficient individuals. Infect. Immun. *75*, 932–940.

Nowrouzian, F.L., Friman, V., Adlerberth, I., and Wold, A.E. (2013). Different phylogenetic profile and reduced mannose-sensitive adherence capacity characterize commensal *Escherichia coli* in IgA deficient individuals. Microb. Pathog. *61–62*, 62–65.

Nowrouzian, F.L., Ostblom, A.E., Wold, A.E., and Adlerberth, I. (2009). Phylogenetic group B2 *Escherichia coli* strains from the bowel microbiota of Pakistani infants carry few virulence genes and lack the capacity for long-term persistence. Clin. Microbiol. Infect. *15*, 466–472.

Nowrouzian, F.L., and Oswald, E. (2012). *Escherichia coli* strains with the capacity for long-term persistence in the bowel microbiota carry the potentially genotoxic pks island. Microb. Pathog. *53*, 180–182.

Nowrouzian, F.L., Wold, A.E., and Adlerberth, I. (2005). *Escherichia coli* strains belonging to phylogenetic group B2 have superior capacity to persist in the intestinal microflora of infants. J. Infect. Dis. *191*, 1078–1083.

O'Brien, C.L., and Gordon, D.M. (2011). Effect of diet and gut dynamics on the establishment and persistence of *Escherichia coli*. Microbiology *157*, 1375–1384.

Okeke, I.N., Wallace-Gadsden, F., Simons, H.R., Matthews, N., Labar, A.S., Hwang, J., and Wain, J. (2010). Multi-locus sequence typing of enteroaggregative *Escherichia coli* isolates from Nigerian children uncovers multiple lineages. PLoS One *5*, e14093.

Olier, M., Marcq, I., Salvador-Cartier, C., Secher, T., Dobrindt, U., Boury, M., Bacquie, V., Penary, M., Gaultier, E., Nougayrede, J.P., *et al.* (2012). Genotoxicity of *Escherichia coli* Nissle 1917 strain cannot be dissociated from its probiotic activity. Gut Microbes 3, 501–509.

Ostblom, A., Adlerberth, I., Wold, A.E., and Nowrouzian, F.L. (2011). Pathogenicity island markers, virulence determinants malX and usp, and the capacity of *Escherichia coli* to persist in infants' commensal microbiotas. Appl. Environ. Microbiol. *77*, 2303–2308.

Palmer, C., Bik, E.M., DiGiulio, D.B., Relman, D.A., and Brown, P.O. (2007). Development of the human infant intestinal microbiota. PLoS Biol. *5*, e177.

Patzer, S.I., Baquero, M.R., Bravo, D., Moreno, F., and Hantke, K. (2003). The colicin G, H and X determinants encode microcins M and H47, which might utilize the catecholate siderophore receptors FepA, Cir, Fiu and IroN. Microbiology *149*, 2557–2570.

Payros, D., Secher, T., Boury, M., Brehin, C., Ménard, S., Salvador-Cartier, C., Cuevas-Ramos, G., Watrin, C., Marcq, I., Nougayrede, J.P., *et al.* (2014). Maternally acquired genotoxic *Escherichia coli* alters offspring's intestinal homeostasis. Gut Microbes Epub ahead of print.

Penders, J., Thijs, C., Vink, C., Stelma, F.F., Snijders, B., Kummeling, I., van den Brandt, P.A., and Stobberingh, E.E. (2006). Factors influencing the composition of the intestinal microbiota in early infancy. Pediatrics *118*, 511–521.

Petersen, A.M., Nielsen, E.M., Litrup, E., Brynskov, J., Mirsepasi, H., and Krogfelt, K.A. (2009). A phylogenetic group of *Escherichia coli* associated with active left-sided inflammatory bowel disease. BMC Microbiol. *9*, 171.

Peterson, D.A., Frank, D.N., Pace, N.R., and Gordon, J.I. (2008). Metagenomic approaches for defining the pathogenesis of inflammatory bowel diseases. Cell Host Microbe 3, 417–427.

Pflughoeft, K.J., and Versalovic, J. (2012). Human microbiome in health and disease. Annu. Rev. Pathol *7*, 99–122.

Picard, B., Garcia, J.S., Gouriou, S., Duriez, P., Brahimi, N., Bingen, E., Elion, J., and Denamur, E. (1999). The link between phylogeny and virulence in *Escherichia coli* extraintestinal infection. Infect. Immun. *67*, 546–553.

Pilarczyk-Zurek, M., Chmielarczyk, A., Gosiewski, T., Tomusiak, A., Adamski, P., Zwolinska-Wcislo, M., Mach, T., Heczko, P.B., and Strus, M. (2013). Possible role of *Escherichia coli* in propagation and perpetuation of chronic inflammation in ulcerative colitis. BMC Gastroenterol. *13*, 61.

Plassmann, D., and Schulte-Witte, H. (2007). [Treatment of irritable bowel syndrome with *Escherichia coli* strain Nissle 1917 (EcN): a retrospective survey]. Med. Klin. (Munich) *102*, 888–892.

Prisciandaro, L.D., Geier, M.S., Butler, R.N., Cummins, A.G., and Howarth, G.S. (2011). Probiotic factors partially improve parameters of 5-fluorouracil-induced intestinal mucositis in rats. Cancer Biol. Ther. *11*, 671–677.

Prorok-Hamon, M., Friswell, M.K., Alswied, A., Roberts, C.L., Song, F., Flanagan, P.K., Knight, P., Codling, C., Marchesi, J.R., Winstanley, C., *et al.* (2014). Colonic mucosa-associated diffusely adherent afaC+ *Escherichia coli* expressing lpfA and pks are increased in inflammatory bowel disease and colon cancer. Gut *63*, 761–770.

Qadri, F., Svennerholm, A.M., Faruque, A.S., and Sack, R.B. (2005). Enterotoxigenic *Escherichia coli* in developing countries: epidemiology, microbiology, clinical features, treatment, and prevention. Clin. Microbiol. Rev. *18*, 465–483.

Rao, S., Hu, S., McHugh, L., Lueders, K., Henry, K., Zhao, Q., Fekete, R.A., Kar, S., Adhya, S., and Hamer, D.H. (2005). Towards a live microbial microbicide for HIV: commensal bacteria secreting an HIV fusion inhibitor peptide. Proc. Natl. Acad. Sci. U.S.A. *102*, 11993–11998.

Rautava, S., Luoto, R., Salminen, S., and Isolauri, E. (2012). Microbial contact during pregnancy, intestinal colonization and human disease. Nat. Rev. Gastroenterol. Hepatol. *9*, 565–576.

Rembacken, B.J., Snelling, A.M., Hawkey, P.M., Chalmers, D.M., and Axon, A.T. (1999). Non-pathogenic *Escherichia coli* versus mesalazine for the treatment of ulcerative colitis: a randomized trial. Lancet *354*, 635–639.

Rolhion, N., Barnich, N., Claret, L., and Darfeuille-Michaud, A. (2005). Strong decrease in invasive ability and outer membrane vesicle release in Crohn's disease-associated adherent-invasive *Escherichia coli* strain LF82 with the yfgL gene deleted. J. Bacterol. *187*, 2286–2296.

Russo, T.A., and Johnson, J.R. (2000). Proposal for a new inclusive designation for extraintestinal pathogenic isolates of *Escherichia coli*: ExPEC. J. Infect. Dis. *181*, 1753–1754.

Sahl, J.W., Steinsland, H., Redman, J.C., Angiuoli, S.V., Nataro, J.P., Sommerfelt, H., and Rasko, D.A. (2011). A comparative genomic analysis of diverse clonal types of enterotoxigenic *Escherichia coli* reveals pathovar-specific conservation. Infect. Immun. *79*, 950–960.

Sarff, L.D., McCracken, G.H., Schiffer, M.S., Glode, M.P., Robbins, J.B., Orskov, I., and Orskov, F. (1975). Epidemiology of *Escherichia coli* K1 in healthy and diseased newborns. Lancet *1*, 1099–1104.

Schlee, M., Wehkamp, J., Altenhoefer, A., Oelschlaeger, T.A., Stange, E.F., and Fellermann, K. (2007). Induction of human beta-defensin 2 by the probiotic *Escherichia coli* Nissle 1917 is mediated through flagellin. Infect. Immun. *75*, 2399–2407.

Scholtens, P.A., Oozeer, R., Martin, R., Amor, K.B., and Knol, J. (2012). The early settlers: intestinal microbiology in early life. Annu. Rev. Food Sci. Technol. *3*, 425–447.

Schultz, M. (2008). Clinical use of E. coli Nissle 1917 in inflammatory bowel disease. Inflamm. Bowel Dis. *14*, 1012–1018.

Schultz, M., Strauch, U.G., Linde, H.J., Watzl, S., Obermeier, F., Gottl, C., Dunger, N., Grunwald, N., Scholmerich, J., and Rath, H.C. (2004). Preventive effects of *Escherichia coli* strain Nissle 1917 on acute and chronic intestinal inflammation in two different murine models of colitis. Clin. Diagn Lab Immunol. *11*, 372–378.

Schwiertz, A., Jacobi, M., Frick, J.S., Richter, M., Rusch, K., and Kohler, H. (2010). Microbiota in pediatric inflammatory bowel disease. J. Pediatr. *157*, 240–244 e241.

Sears, H.J., and Brownlee, I. (1952). Further observations on the persistence of individual strains of *Escherichia coli* in the intestinal tract of man. J. Bacterol. *63*, 47–57.

Sears, H.J., Brownlee, I., and Uchiyama, J.K. (1950). Persistence of individual strains of *Escherichia coli* in the intestinal tract of man. J. Bacterol. *59*, 293–301.

Sears, H.J., Janes, H., Saloum, R., Brownlee, I., and Lamoreaux, L.F. (1956). Persistence of individual strains of *Escherichia coli* in man and dog under varying conditions. J. Bacterol. *71*, 370–372.

Skurnik, D., Bonnet, D., Bernede-Bauduin, C., Michel, R., Guette, C., Becker, J.M., Balaire, C., Chau, F., Mohler, J., Jarlier, V., *et al.* (2008). Characteristics of human intestinal *Escherichia coli* with changing environments. Environ. Microbiol. *10*, 2132–2137.

Smati, M., Clermont, O., Le Gal, F., Schichmanoff, O., Jaureguy, F., Eddi, A., Denamur, E., and Picard, B. (2013). Real-time PCR for quantitative analysis of human commensal *Escherichia coli* populations reveals a high frequency of subdominant phylogroups. Appl. Environ. Microbiol. *79*, 5005–5012.

Sokol, H., Lepage, P., Seksik, P., Dore, J., and Marteau, P. (2006). Temperature gradient gel electrophoresis of faecal 16S rRNA reveals active *Escherichia coli* in the microbiota of patients with ulcerative colitis. J. Clin. Microbiol. *44*, 3172–3177.

Sommer, F., and Backhed, F. (2013). The gut microbiota – masters of host development and physiology. Nat. Rev. Microbiol. *11*, 227–238.

Splichal, I., Fagerhol, M.K., Trebichavsky, I., Splichalova, A., and Schulze, J. (2005). The effect of intestinal colonization of germ-free pigs with *Escherichia coli* on calprotectin levels in plasma, intestinal and bronchoalveolar lavages. Immunobiology *209*, 681–687.

Stappenbeck, T.S., Hooper, L.V., and Gordon, J.I. (2002). Developmental regulation of intestinal angiogenesis by indigenous microbes via Paneth cells. Proc. Natl. Acad. Sci. U.S.A. *99*, 15451–15455.

Stecher, B., and Hardt, W.D. (2011). Mechanisms controlling pathogen colonization of the gut. Curr. Opin. Microbiol. *14*, 82–91.

Stentebjerg-Olesen, B., Chakraborty, T., and Klemm, P. (1999). Type 1 fimbriation and phase switching in a natural *Escherichia coli* fimB null strain, Nissle 1917. J. Bacterol. *181*, 7470–7478.

Stoll, B.J., Hansen, N.I., Sanchez, P.J., Faix, R.G., Poindexter, B.B., Van Meurs, K.P., Bizzarro, M.J., Goldberg, R.N., Frantz, I.D., 3rd, Hale, E.C., *et al.* (2011). Early onset neonatal sepsis: the burden of group B Streptococcal and E. coli disease continues. Pediatrics *127*, 817–826.

Sturm, A., Rilling, K., Baumgart, D.C., Gargas, K., Abou-Ghazale, T., Raupach, B., Eckert, J., Schumann, R.R., Enders, C., Sonnenborn, U., *et al.* (2005). *Escherichia coli* Nissle 1917 distinctively modulates T-cell cycling and expansion via toll-like receptor 2 signaling. Infect. Immun. 73, 1452–1465.

Sun, J., Gunzer, F., Westendorf, A.M., Buer, J., Scharfe, M., Jarek, M., Gossling, F., Blocker, H., and Zeng, A.P. (2005). Genomic peculiarity of coding sequences and metabolic potential of probiotic *Escherichia coli* strain Nissle 1917 inferred from raw genome data. J. Biotechnol. *117*, 147–161.

Swidsinski, A., Khilkin, M., Kerjaschki, D., Schreiber, S., Ortner, M., Weber, J., and Lochs, H. (1998). Association between intraepithelial *Escherichia coli* and colorectal cancer. Gastroenterology *115*, 281–286.

Swidsinski, A., Ladhoff, A., Pernthaler, A., Swidsinski, S., Loening-Baucke, V., Ortner, M., Weber, J., Hoffmann, U., Schreiber, S., Dietel, M., *et al.* (2002). Mucosal flora in inflammatory bowel disease. Gastroenterology *122*, 44–54.

Tenaillon, O., Skurnik, D., Picard, B., and Denamur, E. (2010). The population genetics of commensal *Escherichia coli*. Nat. Rev. Microbiol. *8*, 207–217.

Touchon, M., Hoede, C., Tenaillon, O., Barbe, V., Baeriswyl, S., Bidet, P., Bingen, E., Bonacorsi, S., Bouchier, C., Bouvet, O., *et al.* (2009). Organized genome dynamics in the *Escherichia coli* species results in highly diverse adaptive paths. PLoS Genet. *5*, e1000344.

Tromm, A., Niewerth, U., Khoury, M., Baestlein, E., Wilhelms, G., Schulze, J., and Stolte, M. (2004). The probiotic E. coli strain Nissle 1917 for the treatment of collagenous colitis: first results of an open-label trial. Z. Gastroenterol. 42, 365–369.

Turnbaugh, P.J., Ley, R.E., Mahowald, M.A., Magrini, V., Mardis, E.R., and Gordon, J.I. (2006). An obesity-associated gut microbiome with increased capacity for energy harvest. Nature *444*, 1027–1031.

Turner, S.M., Scott-Tucker, A., Cooper, L.M., and Henderson, I.R. (2006). Weapons of mass destruction: virulence factors of the global killer enterotoxigenic *Escherichia coli*. FEMS Microbiol. Lett. *263*, 10–20.

Ukena, S.N., Singh, A., Dringenberg, U., Engelhardt, R., Seidler, U., Hansen, W., Bleich, A., Bruder, D., Franzke, A., Rogler, G., *et al.* (2007). Probiotic *Escherichia coli* Nissle 1917 inhibits leaky gut by enhancing mucosal integrity. PLoS One 2, e1308.

Ulett, G.C., Totsika, M., Schaale, K., Carey, A.J., Sweet, M.J., and Schembri, M.A. (2013). Uropathogenic *Escherichia coli* virulence and innate immune responses during urinary tract infection. Curr. Opin. Microbiol. *16*, 100–107.

Vaishampayan, P.A., Kuehl, J.V., Froula, J.L., Morgan, J.L., Ochman, H., and Francino, M.P. (2010). Comparative metagenomics and population dynamics of the gut microbiota in mother and infant. Genome Biol. Evol. *2*, 53–66.

Valdebenito, M., Crumbliss, A.L., Winkelmann, G., and Hantke, K. (2006). Environmental factors influence the production of enterobactin, salmochelin, aerobactin, and yersiniabactin in *Escherichia coli* strain Nissle 1917. Int. J. Med. Microbiol. *296*, 513–520.

Vejborg, R.M., Friis, C., Hancock, V., Schembri, M.A., and Klemm, P. (2010). A virulent parent with probiotic progeny: comparative genomics of *Escherichia coli* strains CFT073, Nissle 1917 and ABU 83972. Mol. Genet. Genomics *283*, 469–484.

Waidmann, M., Bechtold, O., Frick, J.S., Lehr, H.A., Schubert, S., Dobrindt, U., Loeffler, J., Bohn, E., and Autenrieth, I.B. (2003). Bacteroides vulgatus protects against *Escherichia coli*-induced colitis in gnotobiotic interleukin-2-deficient mice. Gastroenterology *125*, 162–177.

Wang, H., Bastian, S.E., Cheah, K.Y., Lawrence, A., and Howarth, G.S. (2014). *Escherichia coli* Nissle 1917-derived factors reduce cell death and late apoptosis and increase transepithelial electrical resistance in a model of 5-fluorouracil-induced intestinal epithelial cell damage. Cancer Biol. Ther. *15*, 560–569.

Wassenaar, T.M., Beimfohr, C., Geske, T., and Zimmermann, K. (2014). Voluntarily exposure to a single, high dose of probiotic *Escherichia coli* results in prolonged colonization. Benef. Microbes, 1–9. In press.

Wehkamp, J., Harder, J., Wehkamp, K., Wehkamp-vonMeissner, B., Schlee, M., Enders, C., Sonnenborn, U., Nuding, S., Bengmark, S., Fellermann, K., *et al.* (2004). NF-kappaB- and AP-1-mediated induction of human beta defensin-2 in intestinal epithelial cells by *Escherichia coli* Nissle 1917: a novel effect of a probiotic bacterium. Infect. Immun. 72, 5750–5758.

Weise, C., Zhu, Y., Ernst, D., Kuhl, A.A., and Worm, M. (2011). Oral administration of *Escherichia coli* Nissle 1917 prevents allergen-induced dermatitis in mice. Exp. Dermatol. *20*, 805–809.

Welch, R.A., Burland, V., Plunkett, G., 3rd, Redford, P., Roesch, P., Rasko, D., Buckles, E.L., Liou, S.R., Boutin, A., Hackett, J., *et al.* (2002). Extensive mosaic structure revealed by the complete genome sequence of uropathogenic *Escherichia coli*. Proc. Natl. Acad. Sci. U.S.A. *99*, 17020–17024.

Wiles, T.J., Norton, J.P., Russell, C.W., Dalley, B.K., Fischer, K.F., and Mulvey, M.A. (2013). Combining quantitative genetic footprinting and trait enrichment analysis to identify fitness determinants of a bacterial pathogen. PLoS Genet. *9*, e1003716.

Winter, S.E., Winter, M.G., Xavier, M.N., Thiennimitr, P., Poon, V., Keestra, A.M., Laughlin, R.C., Gomez, G., Wu, J., Lawhon, S.D., *et al.* (2013). Host-derived nitrate boosts growth of E. coli in the inflamed gut. Science *339*, 708–711.

Wohlgemuth, S., Haller, D., Blaut, M., and Loh, G. (2009). Reduced microbial diversity and high numbers of one single *Escherichia coli* strain in the intestine of colitic mice. Environ. Microbiol. *11*, 1562–1571.

Yatsunenko, T., Rey, F.E., Manary, M.J., Trehan, I., Dominguez-Bello, M.G., Contreras, M., Magris, M., Hidalgo, G., Baldassano, R.N., Anokhin, A.P., *et al.* (2012). Human gut microbiome viewed across age and geography. Nature *486*, 222–227.

Zhang, L., Foxman, B., and Marrs, C. (2002). Both urinary and rectal *Escherichia coli* isolates are dominated by strains of phylogenetic group B2. J. Clin. Microbiol. *40*, 3951–3955.

Zhang, Y., Xia, L., Zhang, X., Ding, X., Yan, F., and Wu, F. (2012). *Escherichia coli* Nissle 1917 targets and restrains mouse B16 melanoma and 4T1 breast tumours through expression of azurin protein. Appl. Environ. Microbiol. *78*, 7603–7610.

Zschuttig, A., Zimmermann, K., Blom, J., Goesmann, A., Pohlmann, C., and Gunzer, F. (2012). Identification and characterization of microcin S, a new antibacterial peptide produced by probiotic *Escherichia coli* G3/10. PLoS One *7*, e33351.

Zyrek, A.A., Cichon, C., Helms, S., Enders, C., Sonnenborn, U., and Schmidt, M.A. (2007). Molecular mechanisms underlying the probiotic effects of *Escherichia coli* Nissle 1917 involve ZO-2 and PKCzeta redistribution resulting in tight junction and epithelial barrier repair. Cell Microbiol. *9*, 804–816.

The Paradoxical Role of *Enterococcus* Species in Foods

10

Luís Augusto Nero, Svetoslav Dimitrov Todorov and Luana Martins Perin

Abstract

Enterococcus species present a diversity of genetic and phenotypic features, allowing them to present a high capacity to survive and grow at different conditions, explaining its constant presence in food production and processing environments and in end products. These microorganisms are able to express a wide variety of virulence factors, highlighting their relevance as safety indicators in foods; however, they are also able to produce bacteriocins, called enterocins, and to promote specific modifications in food during fermentation. Enterococci are present in a variety of ripened cheeses, especially from the Mediterranean region, being responsible for specific aromas and flavours that determine the characteristics of these artisanal foods. Also, many *Enterococcus* species were characterized due to their probiotic potential, being included in commercial products to be consumed by humans and animals, aiming the promotion of health and well-being. Despite being known for possessing virulence genes, many studies demonstrated the absence of expression of such genes, especially isolates obtained from food systems, leading to studies that investigate their real relevance as pathogenic microorganisms. So, this paradoxical role of enterococci in foods must be properly discussed by food microbiologists, which is the focus of this chapter.

Just a general overview about *Enterococcus*

The genus *Enterococcus* includes Gram-positive and catalase-negative cocci, that are facultative anaerobes, and capable of growth at adverse conditions, such as temperatures ranging from 10 to 45°C (their optimal growth temperature is 35°C), NaCl at 6.5%, bile at 40%, and at pH 9.6 (Franz *et al.*, 2003). Also, they are able to survive after heating at 60°C for 30 min, and to hydrolyse esculin in the presence of bile salts (Foulquié Moreno *et al.*, 2006b). Depending on the species, some of these general characteristics can present some minor variations, which can be used as preliminary basis for their identification by biochemical tests (Table 10.1). Along with being considered for identification purposes, the capability of surviving at distinct conditions demonstrates the versatility of the species of this genus, allowing them to survive in different environment conditions. *Enterococcus* species were wrongly included in the *Streptococcus* genus due to their biochemical characteristics, and before the utilization of 16S rRNA sequencing for proper identification of microorganisms (Giraffa, 2002). Since then, a number of methodologies based on the characterization of genome sequences are being employed for differentiation and characterization of isolates, allowing the identification of numerous species belonging to the genus *Enterococcus* (Ogier and Serror, 2008).

Identification of *Enterococcus* isolates based only on biochemical characteristics has always being considered as inadequate, due to their phenotypic diversity and also their particular growth conditions (Table 10.1) (Franz *et al.*, 2011; Franz *et al.*, 2003). Also, many species share identical biochemical characteristics, hindering the proper identification based only on these tests. As detailed, sequencing regions of 16S rRNA allows the proper characterization of this group and other lactic acid bacteria, providing a better taxonomic characterization. However, the conventional protocols based on the sequencing of 16S rRNA allow only the identification of isolates as belonging to the genus *Enterococcus*, jeopardizing the proper identification at the species level (Ogier and Serror, 2008). Despite not allowing the identification of species, the sequencing of 16S rRNA regions allows the grouping of isolates in seven major *Enterococcus* groups (Table 10.1), leading to complementary molecular or phenotypical based protocols to identify their species. With this purpose, alternative methodologies can be employed to provide a definitive identification, such as PCR aiming to amplify specific regions of rRNA intergenic spacers and specific genes (*ddl, van, ace, sod*A), amplification of the putative transcriptional regulator gene Ef027, sequencing of specific genes (*ddl, cpn60, atp*A), and even rep-PCR using GTG_5 primer (Ogier and Serror, 2008). In summary, preliminary identification of novel isolates as *Enterococcus* usually start with characterization of some biochemical characteristics and sequencing of region of 16S rRNA, followed by additional molecular tests to check the presence of specific genes, or even the sequencing of specific amplifications products from typical *Enterococcus* genes. The variety of genetic markers allows the identification of more

Table 10.1 Relevant *Enterococcus* species and their classification in groups based on 16S rRNA gene similarity, and their phenotypical characteristics (adapted from Franz *et al.*, 2003, 2011)

		Growth at		Growth in the presence of					
Group	Species	10°C	45°C	pH 9.6	6.5% NaCl	40% bile	0.04% Sodium azide	Esculin hydrolysis	Other species of the group
Avium	*E. avium*	V	+	+	V	V/+	nd	+	*E. devriesi*, *E. gilvus*
	E. malodoratus	+	–	+	+	+	nd	+	
	E. pseudoavium	+	+	+	+/–	V/+	nd	+	
	E. raffinosus	(+)	+	+	+	V/+	nd	+	
Cecorum	*E. cecorum*	–	+	(+)	–	(+)	–	+	–
	E. columbae	–	nd	nd	–	(+)	–	+	
Dispar	*E. asini*	(+)	(+)	nd	–	+	nd	+	*E. canintestini*, *E. hermanniensis*, *E. pallens*
	E. dispar	+	–	nd	+/–	+	–	+	
Faecalis	*E. faecalis*	+	+	+	+	+	+	+	*E. caccae*, *E. silesiacus*, *E. termitis*, *E. ureasiticus*, *E. quebecensis*
	E. haemoperoxidus	+	–	nd	+	+	+	+	
	E. moraviensis	+	–	nd	+	+	+	+	
Faecium	*E. durans*	+	+	+	+	+	+	+	*E. canis*, *E. phoeniculicola*, *E. thailandicus*
	E. faecium	+	+	+	+	+	+	+	
	E. hirae	+	+	+	+	+	+	+	
	E. mundtii	+	+	+	+	+	+	+	
	E. ratti	+	+	nd	(+)	nd	nd	+	
	E. villorum	nd	nd	nd	+	+	+	+	
Gallinarum	*E. casseliflavus*	+	+	+	V/+	+	+	+	–
	E. gallinarum	+	+	+	+	+	+	+	
Saccharolyticus	*E. saccharolyticus*	+	+	nd	(+)	+	nd	+	*E. acquimarinus*, *E. camelliae*, *E. italicus*
	E. sulfureus	+	–	nd	+	+	+	+	

nd: not determined; (+): weak positive; V: variable; +/–: variable reports; +: positive; –: negative.

than 35 *Enterococcus* species up to date, but this number can change constantly due to the adoption of novel characterization methodologies, determining reclassification as new species or as a previously known one (Franz *et al.*, 2011; Giraffa, 2002; Ogier and Serror, 2008).

Considering the variable phenotypic and genotypic characteristics of *Enterococcus* species, these microorganisms are able to survive in different environmental conditions. A wide variety of species is commonly found in soil, surface and ocean waters, plants, and in the gastrointestinal tract of human and animals (Franz *et al.*, 2011). This ubiquity determines their presence as common contaminants in dairy and meat products, as result of the contamination from the processing environment and also from poor hygienic practices. In addition, *Enterococcus* species are known by their potential association to diseases in humans, due to the production of pathogenic factors and transfer of virulence genes to other bacteria (Franz *et al.*, 2003; Ogier and Serror, 2008). However, in specific situations the presence of *Enterococcus* in foods is desirable, once they are capable of producing specific aromas and flavours, being particularly relevant for the production of fermented foods (Foulquié Moreno *et al.*, 2006b; Giraffa, 2002, 2003). As *Enterococcus* species present tolerance to high concentrations of NaCl and pH values, they are able to survive at typical fermentation conditions of cured meats and ripened cheeses, contributing to the development of organoleptic characteristics of these products (Foulquié Moreno *et al.*, 2006b; Giraffa, 2002). Along with their technological properties, *Enterococcus* species can also be considered beneficial due to the production of natural antimicrobial substances, such as bacteriocins, that can interfere with the development of spoilage and pathogenic bacteria in foods, and also due to their probiotic potential (Franz *et al.*, 2011).

This paradoxical role of *Enterococcus* in foods is a challenging theme to be discussed among food microbiologists. Many studies have demonstrated the autochthonous presence of *Enterococcus* species in many foods, however without direct association to diseases in consumers. Also, many studies have demonstrated the benefits of *Enterococcus* strains obtained from different food sources, due to their production of bacteriocins or even probiotic potential, despite they are presenting different virulence genes and/or expressing pathogenic factors. Here, we intent to present an update of the information related to these characteristics, in order to contribute in the discussion of this paradoxical role of *Enterococcus* species in foods.

The good side of enterococci...

Despite *Enterococcus* could be considered an opportunistic pathogen to humans, the microorganisms from this genus presents also some beneficial aspects. Numerous strains are capable to produce bacteriocins called enterocins, which are very important antimicrobial peptides with activity against some spoilage and pathogenic microorganisms. Enterocins have been exhaustedly studied to be applied in food matrices as biopreservatives. Since *Enterococcus* could predominate the autochthonous microbiota of many food matrices, mainly dairy products, it is supposed that they could interfere positively in the sensory characteristics of the products, due to the production of various substances derived from its metabolism and also being capable to produce several enzymes that interact with the food components, generating diverse beneficial biochemical transformations in the final product (Giraffa, 2003). Enterococci strains are capable to survive in some adverse conditions and may have some health properties that are being explored to be used as probiotics for humans and animals.

Enterococci bacteriocins

Numerous strains from the genera *Enterococcus* are capable to produce one or more enterocins. Enterocins are bacteriocins that are produced exclusively by enterococci. Bacteriocins can be defined as small ribosomally synthesized peptides, which are heat resistant and biologically active ranging in its spectrum and mode of action, molecular weight, genetic, and biochemical properties (Abee *et al.*, 1995). Obviously the producer strain has mechanisms for immunity against their own bacteriocin (De Vuyst and Leroy, 2007; Klaenhammer, 1988). Some enterocins are already very well characterized and could be grouped in four different classes as suggested by Franz *et al.* (2007).

Class I or lantibiotics

Lantibiotics are small peptides that have residues of lanthionine or β-methyllantionine in its structure. Cytolysin is a substance with ambiguous activity: it is virulent due to its haemolytic potential, but it is also considered as a antibacterial/bacteriocin which contains lanthionine residues in its molecular structure, which determines the classification of cytolysin as a class I bacteriocin (Haas *et al.*, 2002). The transfer of the operon encoding cytolysin between strains can occur since that this operon is located on transmissible plasmids and a mobile pathogenicity island (Van Tyne *et al.*, 2013). *E. faecalis* and *E. faecium* isolated from foods were already described as capable of producing cytolysin, despite this characteristic being more common in isolates obtained from clinical samples (Eaton and Gasson, 2001; Semedo *et al.*, 2003). Also, the nisin gene was already identified in some strains of *Enterococcus* (Perin and Nero, 2014). This occurrence can be explained by the capability of this genus to acquire new genetic elements (Riley and Wertz, 2002). However, the presence of the *nis* gene is not necessarily related to the production of nisin by these *Enterococcus* isolates.

Class II.1 or pediocin-like bacteriocins

These peptides contain a highly conserved hydrophilic region (YGNGVXC pediocin-box) and two cysteine residues linked by a sulfur bridge (S-S) which functions to stabilize the structure (Javed *et al.*, 2011). Enterocin A and enterocin P are bacteriocins with typical high inhibitory activity against *Listeria* spp. (Javed *et al.*, 2011). Borrero *et al.* (2011) described that the amount, antimicrobial activity, and specific antimicrobial activity of the enterocin A produced by some recombinant hosts varied depending on the signal peptide, the expression vector, and the host strain. Modifications of the consensus region YGNGVXC can reduce the inhibitory potential of bacteriocin against the target microorganism (Franz *et al.*, 2007). They include Bacteriocin 31 (Tomita *et al.*, 1996) from *E. faecalis*, Enterocin A (Aymerich *et al.*, 1996), Enterocin CRL35 (Farías *et al.*, 1996) and Enterocin P (Cintas *et al.*, 1997) from *E. faecium* and Mundticin from *E. mundtii* (Bennik *et al.*, 1998). Todorov *et al.* (2010) described a bacteriocinogenic strain of *E. mundtii* with antiviral properties.

Class II.2

These are peptides which require two components for their activity. This class includes bacteriocins without the consensus YGNGVXC and are synthesized without leader peptide, requiring a complex system for transport. Enterocin RJ-11 (Yamamoto *et al.*, 2003), Enterocin EJ97 (Gálvez *et al.*, 1998) from *E. faecalis* and Enterocin Q, Enterocin L50A and Enterocin L50B (Cintas *et al.*, 1998) from *E. faecium* (Cintas *et al.*, 2000) belong to this group.

Class II.3

These peptides are linear, non-pediocin-type enterocins that are synthesized with a leader peptide and do not fit into classes II.1 and II.2. An example of enterocin from this class is Enterocin 1071 that has the peculiarity that its antimicrobial activity may be reinforced by the presence of a second peptide (Franz *et al.*, 2007). Enterocin B was first described by Casaus *et al.* (1997), this authors also reported that enterocin B exhibited synergistic activity with enterocin A. Enterocin 1071A and Enterocin 1071B was first describes as produced by *E. faecalis* (Balla *et al.*, 2000).

Class III or cyclic antimicrobial peptides

Enterocin AS-48 was first described as produced by *E. faecalis* in 1986 (Balla *et al.*, 2000), these authors also found that this bacteriocin has a broad spectrum of activity being active against many Gram-positive and some Gram-negative bacteria. It is stable to variations in pH and to heat. This bacteriocin acts by forming pores in the cytoplasmic membrane of susceptible cells allowing the dissipation of the proton motive force and consequent cell death (Cotter *et al.*, 2005).

Class IV

These peptides are large molecular weight and heat labile proteins. Enterolysin A is produced by *E. faecalis* LMG 2333 and DPC5280 (Hickey *et al.*, 2003; Nilsen *et al.*, 2003). Enterolysin A showed activity towards *Lactobacillus, Lactococcus, Pediococcus, Enterococcus, Bacillus, Listeria* and *Staphylococcus* spp. (Franz *et al.*, 2007). Several studies have already characterized enterococci strains that are capable to produce specific bacteriocins. Some examples

Table 10.2 Enterocins and studies that characterize enterococcal bacteriocins-producers.

Bacteriocin	Class	Strain	Source	Reference
Enterocin W	I	*E. faecalis* NKR-4–1	Thai fermented fish	Sawa *et al.* (2012)
Cytolysin	I	*E. faecalis*	Clinical isolates	Booth *et al.* (1996)
Enterocin P	II.1	*E. faecium* P13	Dry fermented sausage	Cintas *et al.* (1997)
Enterocins A and B	II.1 and II.3 respectively	*E. faecium* CTC492	Dry fermented sausage	Aymerich *et al.* (2000)
Enterocins A and B	II.1 and II.3 respectively	*E. faecium* WHE 81	Munster cheese	Izquierdo *et al.* (2009)
Enterocin C	II.2	*E. faecalis* C901	Human colostrum	Maldonado-Barragán *et al.* (2009)
Enterocins L50A and B	II.2	*E. faecium* F58	Goat's milk and Jben	Achemchem *et al.* (2006)
Enterocin L50	II.2	*E. faecium* L50	Dry fermented sausage	Cintas *et al.* (1998)
Enterocin X	II.2	*E. faecium* KU-B5	Sugar apples	Hu *et al.* (2010)
Enterocin 1071	II.3	*E. faecalis* FAIR-E 309	Argentinian cheese	Franz *et al.* (2002)
Enterocin 1071	II.3	*E. faecalis* BFE 1071	Faeces of minipigs	Balla *et al.* (2000)
Enterocin AS-48	III	*E. faecalis* UGRA10	Sheep cheese	Cebrián *et al.* (2012)
Enterocin AS-48	III	*E. faecalis* A-48-32	Milk and cheese	Muñoz *et al.* (2007)
Enterocin M	Probably II.1	*E. faecium* AL41	Environment	Mareková *et al.* (2007)
Enterocin 416K1	Probably II.1	*E. casseliflavus* IM 416K1	Italian sausage	Sabia *et al.* (2002)

are reported in Table 10.2. In general the inhibitory activity of enterocins is due to its interference with the membrane permeability and also on the DNA replication and translocation of the target microorganism (Cleveland *et al.*, 2001). Bacteriocin production is a competitive advantage to the enterococci and as the genetic determinants for some enterocins, such as cytolysin and enterocin AS-48, are encoded on conjugative and pheromone-responsive plasmids this is transferred at high frequency to a narrow range of hosts (Clewell, 1990; Franz *et al.*, 2007). This fact can explain the variety of enterocins and the capability of many strains in production of a variety of bacteriocins. Some studies have already described the production of more than one enterocin by one strain (Dal Bello *et al.*, 2010; Moraes *et al.*, 2012; Perin *et al.*, 2014).

The successful use of biopreservatives stimulates the emergence of studies aiming at the characterization and discovery of new bacteriocins to be applied in various foods. Some studies have already investigated the application and effectiveness of *Enterococcus* or their enterocins in food matrices. Pingitore *et al.* (2012) found that a strain of *E. munditii* (CRL35) has exercised a bacteriostatic effect against *L. monocytogenes* in fresh cheese.

Bacteriocins can be incorporated into foods in their purified or semi-purified form, or bacteriocin-producing LAB can be incorporated directly into the fermented food (replacing the starter culture) (Deegan *et al.*, 2006). However, the application of the *Enterococcus* strains directly in foods should be thoroughly evaluated and safety aspects should be well characterized, such as the presence of virulence factors and antibiotic resistance. They are opportunistic pathogens and even if they are not expressing virulence factors, it could be vertically transferred to a potential pathogen and cause problems related to human health. Also, the interactions with the food matrix can have a large influence on the production of bacteriocins. Therefore, the use of purified enterocins may be the best option.

Technological properties

Lactic acid bacteria have a long history of application as 'starters' in the production of fermented foods and beverages. Starter culture can be defined as microbial preparations containing large numbers of cells of at least one micro-organism that are added to the raw material to accelerate the fermentation of foods such as cheeses, yoghurts, beverages, salami, sausage, among others. Enterococci are the main constituent of the autochthonous microbiota of dairy products (Dal Bello *et al.*, 2010; Ortolani *et al.*, 2010; Perin and Nero, 2014) and become an important part of the microbiota of fermented foods.

Based on sugar fermentation patterns, enterococci are classified as homofermentatives, having lactic acid as a final product of glucose fermentation without gas production. Some strains are capable to generate rapid acidification of the raw material by the production of lactic acid (Franciosi *et al.*, 2009; Leroy and De Vuyst, 2004; Nieto-Arribas *et al.*, 2010; Pingitore *et al.*, 2012; Settanni *et al.*, 2011). The acid production, besides conferring the characteristic acid taste of fermented products, may also inhibit the growth of certain pathogens and spoilage micro-organisms. The majority of enterococcal isolates are slow in producing acids in the first hours of fermentation, but in general acid production by *Enterococcus* tends to be medium (Dal Bello *et al.*, 2012; Franciosi *et al.*, 2009; Morandi *et al.*, 2011). Nieto-Arribas *et al.* (2010) reported that *E. faecalis* were more rapid acidifiers than *E. faecium* after 24h. Callewaert *et al.* (2000) used *E. faecium* CCM 4231 and *E. faecium* RZS C13 as starters in sausage fermentation and they were characterized as strong acidifier and strongly inhibited the growth of *Listeria* spp., being suitable for addition to meat as co-cultures to improve food safety.

Also, some strains are able to produce interesting substances for the technological use, such as acetic acid, ethanol, aromatic compounds and exopolysaccharide, substances that contribute to generating desirable sensory characteristics on the final product (Leroy and De Vuyst, 2004). Franciosi *et al.* (2009), found that some strains of *E. faecalis* and *E. durans* are capable of producing diacetyl at a medium to low level. Diacetyl is a flavour compound generated by the citrate metabolism, which contributes to the characteristic flavour of butter or 'buttermilk' (Leroy and De Vuyst, 2004; Smit *et al.*, 2005). This compound is commonly produced by some LAB strains, mainly by enterococci (Badis *et al.*, 2004; Franciosi *et al.*, 2009) and seems to be strain dependent.

Enterococcus has an important role during cheese maturation, probably due its proteolytic and lipolytic activity and citrate degradation that generate a specific taste and flavour to the final product (Foulquié Moreno *et al.*, 2006; Suzzi *et al.*, 2000), but excessive proteolysis may cause large-scale production of bitter peptides and other undesirable compounds, or even excessive hydrolysis of casein can generate a final product that is too soft (Nieto-Arribas *et al.*, 2010).

The main sources of proteolytic enzymes in cheese are renin and other coagulants, but also some LAB strains which are natural constituents of the microbiota or are purposely added during production. As *Enterococcus* may dominate the non-starter LAB of many artisanal cheeses, it is supposed that they are the main responsible for the proteolyses. The proteolytic activity consists of the action of envelope-associated proteinase and several intracellular peptidases and proteinases, which hydrolyse casein and peptides into small peptides and free essential amino-acids in cheese ripening (Ávila *et al.*, 2005; Piraino *et al.*, 2008). After analyses of the proteolytic activity of enterococci from Manchego cheese, Nieto-Arribas *et al.* (2010) showed that 76% of the isolates had proteolytic activity values ranging between 0.10 and 6.88 mmol glycine/l. High levels of proteolysis were also reported by González *et al.* (2010). However Belgacem *et al.* (2010) analysed *E. faecium* isolates from artisanal Tunisian fermented meat and none were capable to induce proteolysis.

Enterococcus is also capable of producing esterases and lipases that contribute to flavour and texture of ripened cheeses. Hydrolysis of milk triglycerides by enterococci was reported for *E. durans* and *E. faecium*, which seem to be the most lipolytic species, with *E. faecalis* showing lower activity (Durlu-Ozkaya *et al.*, 2001; Giraffa, 2003; Sarantinopoulos *et al.*, 2001). González *et al.* (2010) showed that *E. faecalis* strains had a considerable enzymatic activity towards the substrates tested. Enterococci show higher esterase and lipase activity than strains of most other genera of LAB.

The autolysis of *Enterococcus* will facilitate access of its intracellular peptidases to milk proteins and peptides, and will accelerate cheese ripening (Piraino *et al.*, 2008). This may be caused by autolysins (bacteriolytic enzymes that hydrolyse the peptidoglycan cell wall produced by the bacteria and causes rapid lysis) or endolysins of prophages (peptidoglycan hydrolases produced by phages incorporated in the genome). Although the generated peptides and amino acids contribute to specific flavours in cheese, such as sweet, bitter or malted, it is believed that they give only the basic cheese flavour (Smit *et al.*, 2005; van Kranenburg *et al.*, 2002).

The salt and pH tolerance by *Enterococcus* results in survival and persistence to a high level in fermented plant products. In olive fermentations it contributes to the breakdown of the bitter oleuropein compound (Franz *et al.*, 2011). After smelling tests done by Nieto-Arribas *et al.* (2010) the authors found that *E. faecium* strains gave off a weak yoghurt odour during growth in milk and a strong cheese odour was detected for isolates belonging to *E. faecalis, E. avium* and *E. hirae*. Some studies showed that the acid production, autolysis and proteolysis by enterococci strains from natural sources are strain dependent (Franciosi *et al.*, 2009; Giraffa, 2003). Owing to the beneficial characteristics, such as the development of cheese and fermented meat aroma and oleuropein breakdown in olives, enterococcal strains are very interesting to be used as starter cultures or co-starter cultures in these foods (Franz *et al.*, 2011). Some enterococci strains have been included as starter culture for a variety of European cheeses, such as Mozzarella, Fontina, Feta, Caprino and Manchego (Giraffa, 2003).

Probiotic potential

The current definition of probiotics is 'live microorganisms which when administered in adequate amounts confer a health benefit on the host' (FAO/WHO, 2002). Vergio (1954) wrote the manuscript 'Anti- und Probiotika' that first described the production of beneficial factors by the gut microbiota population. Lilly and Stillwell (1965) were the first to describe probiotics, describing them as an opposite to an antibiotic, being 'a substance produced by one microorganism stimulating the growth of another microorganism' (Franz *et al.*, 2011). These beneficial properties are conferred by some bacteria, especially LAB and have a positive influence on digestion and the immune system.

According FAO/WHO (2002), for humans probiotics may be beneficial in: treating diarrhoea, especially following treatment with various antibiotics; preventing and treating vaginal yeast infections and urinary tract infections; treating irritable bowel syndrome; reducing the recurrence of bladder cancer; shortening the duration of intestinal infections; preventing and treating inflammation following colon surgery; and preventing eczema in children. It should be emphasized that effects are strain dependent, and therefore not all isolates can be considered probiotics. The main uses for enterococcal probiotics in animals are to treat or prevent diarrhoea; for immune stimulation; and to improve growth. A complete overview on the benefits for humans and animals can be found in Franz *et al.* (2011).

Lactobacilli and bifidobacteria are the main species recognized as probiotics and some of them are isolated from the human gut microbiota (Behnsen *et al.*, 2013). Also enterococci are naturally found in the gastrointestinal tract and some strains are established as probiotics (Franz *et al.*, 2011). In fact, there are already a number of probiotics products containing *Enterococcus* spp. on the market. Examples of these are: Symbioflor 1, Cylactin® and ECOFLOR (Nueno-Palop and Narbad, 2011).

The capability of *Enterococcus* to produce antimicrobial substances as bacteriocins, acetic acid and lactic acid, combined with high levels of gut colonization creates a competitive advantage against pathogens (Cotter *et al.*, 2005). To be potentially a probiotic it is very important that the enterococcal strain is capable to remain viable during the shelf life of the food, presents acid tolerance to survive during the passage through the stomach, and is resistant to bile salts and hydrolytic enzymes present in the small intestine (Belma and Gulcin, 2009; Pieniz *et al.*, 2014). Considering that probiotics are used in animal feed and human foods it is very important to ensure the safety aspects of enterococci strains used.

Pieniz *et al.* (2014), demonstrated that the strain *E. durans* LAB18s is acid tolerant, surviving at pH 3 and in the presence of bile salts, does not present haemolytic activity and presents a broad inhibitory spectrum against some pathogens. Rehaiem *et al.* (2014), showed that the strain *E. faecium* MMRA, have a high tolerance to gastrointestinal tract transit, and contains the genes encoding enterocins A, B, P and X. Nueno-Palop and Narbad (2011) demonstrated that the strain *E. faecalis* CP58 has high capacity to adhere to human intestinal Caco-2 cells but also presented some virulent genes and resistance to some antibiotics. Bhardwaj *et al.* (2010) and Miyazaki *et al.* (2010) evaluated the effects of some strains, including *Enterococcus* on the growth, adhesion activity and biofilm formation of *E. coli* enteroaggregative (EAggEC) and found that the supernatants of *E. faecium* had a strong bactericidal effect on EAggEC, but was not capable to inhibit the adhesion of EAggEC to intestinal epithelial cells. Bhardwaj *et al.* (2010) evaluated the probiotic potential of a bacteriocinogenic *E. faecium* KH 24 under *in vitro* and *in vivo* (mouse model) conditions. The authors found that no virulence factors were detected and showed *in vitro* tolerance to biological barriers and *in situ*, the tested mice showed a better weight gain, a decrease in *Salmonella* Enteritidis and *E. coli* counts in the intestine and an enhanced growth of *Lactobacillus*. This *Enterococcus* strain could be considered a good choice to be used as probiotic. Ditu *et al.* (2011) investigated the modulation of virulence and antibiotic susceptibility of enteropathogenic *E. coli* strains by culture fractions of a *Enterococcus faecium* strain and demonstrated that the plurifactorial anti-infective action is also due to the modulation of virulence factors and antibiotic susceptibility expression in *E. coli* pathogenic strains.

When probiotic properties are investigated it is important to know not only the beneficial properties of the probiotic candidate on the host, but also to know about interactions between the bacterial cells and drugs taken by the patients. In the study of Botes *et al.* (2008), the authors investigated the interaction between *Enterococcus mundtii* ST4SA and various commercial anti-inflammatory drugs. However, the authors just recorded this interaction without recommending what doses of the medicaments are important for this interaction/inhibition and if this can be a real scenario *in vivo*.

Cebrián *et al.* (2012) proposed the application of *E. faecalis* UGRA10, a new AS-48-producer strain as potential starter and probiotic culture. The authors showed the sensitivity of *E. faecalis* UGRA10 to most of the antibiotics tested, being resistant only to three antibiotics (aminoglycosides, lincosamide, and pristinamicins). In addition to this, *E. faecalis* UGRA10 developed an ability to form biofilms and to adhere to Caco-2 and HeLa 229 cells. More importantly, this strain showed a high ability to interfere with the adhesion of *Listeria monocytogenes* to Caco-2 cells.

Saavedra *et al.* (2003) investigated 122 strains of *Enterococcus faecium* isolated from homemade traditional Argentinean cheese, with the aim to develop starter cultures for the manufacture of this traditional cheese. In addition to this, the authors investigated the potential probiotic properties of these *E. faecium* isolates. Eleven of the 122 strains showed limited delays in growth in Oxgall and were selected to study their bile salts hydrolase activity (BSH), cholesterol reduction, antimicrobial activity, and virulence determinants. Nine strains were able to remove cholesterol in *in vitro* assays, a property that was closely related to the bile salt hydrolase activity. No virulence factors were detected in any of the 11 selected strains (potential probiotics) of enterococci.

It should be emphasized that all strains described above are at best candidates for probiotics. They can only be considered real probiotics after their clinical efficacy has been shown. Existing enterococcal probiotics have never been reported as causing diseases. The strains used as probiotics such as *E. faecium* SF68 and *E. faecalis* Symbioflor, have their safety aspect well characterized and are on the market for more than 20 years (Franz *et al.*, 2011).

To be good or bad is not only a question of reputation

Microbiology is a relatively young branch of the biological sciences. Pioneers of the Microbiology like van Leeuwenhoek and Pasteur designed scientific bases for microbiology. In the 19th century scientists working in this area have isolated and classified principal species of microorganisms. At this time, very frequently source of isolation have been served as a reference for naming of the strains and to be considered as safe or not. Identification of the microorganisms, and their physiology and biochemistry has served as criteria for the taxonomy and safety evaluation. Frequently strains isolated from food products or plant material has been characterized as safe, and oppositely, clinical isolates as pathogens. Good examples for this are *Lactobacillus plantarum* and *Enterococcus faecalis*. However, with the development of Microbiology, more and more strains have been isolated from much more ecological niches and it was shown that *E. faecalis*, primary isolated from faecal samples can be isolated from fruit, vegetables, and can be a member of the autochthonous microbiota of raw milk and meat (Aslam *et al.*, 2012; Barbosa *et al.*, 2010; Favaro *et al.*, 2014; Foulquié Moreno *et al.*, 2006b; Franz *et al.*, 2001; Jahan and Holley, 2014; Martin *et al.*, 2005; Pieniz *et al.*, 2014; Ribeiro *et al.*, 2011). In addition, based on the biomolecular approaches in the taxonomy of the microorganisms, we have been reclassifying several species and even generating new genera in the last 30 years.

A well-accepted, and in our opinion a not very correct, concept is the fact that to be considered as a good probiotic, a strain needs to be a human (or other animal) isolate. This and following test of the beneficial properties of the strains is the basis of the screening and developing of the new probiotic strains/products. However, in the last decade several reports showed that isolates from fermented and non-fermented food products or from plant material origin can be good probiotic candidates (Cebrián *et al.*, 2012; Ditu *et al.*, 2011; Franz *et al.*, 2011; Pérez-Martín *et al.*, 2014; Rehaiem *et al.*, 2014; Todorov *et al.*, 2014). At the same time, the old concept that human isolates are normally not carrying pathogenic properties is becoming a philosophical conflict. In fact, we need to consider safety evaluation of every strain in order to propose it to be applied in any fermentation process related to human or animal consumption and to be aware of the potential risks. Even if the species has a long safety record, this does not automatically apply to all strains belonging to this species and especially to new isolates.

In the study of Perin *et al.* (2014), *Enterococcus* spp. isolated from raw milk have been evaluated for their safety in order to select potential starter cultures. The authors have paid attention not only to the presence of virulence, antibiotic resistance and biogenic amine production genes, but also on their expression when strains were cultured at various temperatures. From the 18 tested *Enterococcus* spp. (five *E. durans*, six *E. faecalis*, six *E. faecium* and one *E. hirae*) gelatinase production was found only in three *E. faecalis* strains and only when cultured at 25°C, 37°C or 42°C. However, one of the tested *E. durans*, all six *E. faecalis*, all six *E. faecium* and one *E. hirae* strains generated positive amplicons for presence of *gel* gene in a PCR reaction. The question is why this gene was expressed only in 3 from 18 tested *Enterococcus* spp.?

When Perin *et al.* (2014) screened for the presence of *asa* (aggregation substance), 14 of 16 strains generated positive results, but only one was positive for the presence of *cyl*A (activation and expression of cytolysin). The majority of the isolates were positive for the presence of *esp* (enterococcal surface protein, 13 from 16 tested isolates, including three negative E. durans isolates) and 11 from 16 were positive for the presence of *efa*A (endocarditis antigen). Only half of the strains generated positive PCR results for the presence of *ace* (adhesin of collagen protein) and none were carrying *hen* (for production of hyaluronidase). When examined for the presence of vancomycin resistance genes, no presence of *van*B was detected, and only 4 from 16 tested strains generated positive results for *van*A. Regarding the presence of biogenic amines related genes the presence of *hdc*2 (histidine decarboxylase) was only reported in three isolates (all of them *E. faecalis*) while *odc* (ornithine decarboxylase) was absent. However, 14 of studied strains generated positive results for the presence of the *tdc* gene (tyrosine decarboxylase). Only one *E. durans* and one *E. faecalis* strain were negative for *tdc*. When all 16 strains were tested for production of tyramine, one *E. durans*, one *E. faecalis* and four *E. faecium* did not show evidence for production of this biogenic amine even if they generated positive PCR results for the presence of the gene encoding this activity.

In conclusion, Perin *et al.* (2014) stated that some of the investigated *Enterococcus* spp. can be considered as safe based on the presence/absence of certain virulence genes and the expression of the genes based on the physiological tests. In the same study, Perin *et al.* (2014) also investigated 11 *Lactococcus lactis* strains. It was interesting to observe that the spread of the virulence factors in otherwise GRAS *Lactococcus* spp. was similar to the *Enterococcus* spp.

Maietti *et al.* (2007) explored the safety of 30 *E. italicus* strains isolated from milk and dairy products. All isolates were susceptible to vancomycin, gentamicin, erythromycin, ampicillin, chloramphenicol and bacitracin. However, only 17% presented phenotypic resistance to tetracycline. By PCR-based detection, the genetic basis of the TetR phenotype in these strains was linked to the *tet*S gene whereas detection of *tet*L and *tet*M genes and the integrase element *int* of the Tn916/Tn1545 family of transposons were negative. Likewise, none of the strains appeared to contain any of the tested virulence genes *gel*E (gelatinase), *asa*1, *cpd* (sex pheromone), *agg* (aggregation substance), *cyl*A, *cyl*B, and *cyl*M (activation and expression of cytolysin), *ace* and *hyl*Efm (hyaluronidase).

In addition to the previous study, Fortina *et al.* (2008), based on their data regarding antibiotic resistance and the presence of virulence factors, suggested that the presence of *E. italicus* strains in cheeses results in a low health risk, and even some *E. italicus* strains may have application as a new safe adjunct cultures for the dairy industry. Fortina *et al.* (2008) reported that the *E. italicus* strains studied by them were susceptible or poorly resistant to several clinical relevant antibiotics and were associated with low virulence profiles, as verified by screening for the presence of 33 different genes encoding antibiotic resistance and known virulence factors in the genus *Enterococcus*.

Majhenic *et al.* (2005) examined the presence of enterococci in ripened artisanal Tolminc cheese. The authors detected up to 10^6 cfu/g of *E. faecalis*, making it the predominant species among isolates. No PCR evidence for the presence of *van*A and *van*B genes was detected. In addition, the absence of vancomycin resistance was confirmed with both disc and E-tests, as isolates were susceptible to vancomycin according to the National Committee for Clinical Laboratory Standards (NCCLS). However, the results of disc tests with other antimicrobial agents (ampicillin, vancomycin, kanamycin, penicillin, erythromycin, neomycin, chloramphenicol, clindamycin, rifampin) did not differ much among the tested enterococci: they were all very resistant to clindamycin only. Regarding virulence factors, all isolates were detected positive for sex-pheromone encoding genes (*cpd, cob, ccf*) and aggregation protein production (*agg*), *efaAfs* (cell wall adhesion), 90% were positive for *cyl*M (posttranslational modification of cytolysin) and *efaAfm* (cell wall adhesins), 70% for *gel*E, 50% for *cyl*B, *cyl*A and *esp* (Majhenic *et al.*, 2005). If we evaluate these strains based on their antibiotic resistance we can consider them as safe, but at the same time the high presence of *agg, afs, cob* and *ccf* is a problem. However, this cheese is traditionally produced in Central Europe and to the best of our knowledge no clinical incidents have been reported or related to consumption of this kind of cheese.

Favaro *et al.* (2014) studied four non starter *E. faecium* strains isolated from white brine cheese. An interesting finding was that the spread of virulence factors was relatively low. From seven virulence factors, four biogenic amine production and *van*A and *van*B genes, *E. faecium* ST209GB harboured only determinants for *asa*1, *tdc* and *odc*, *E. faecium* ST278GB for *asa*1 and *tdc*, *E. faecium* ST315GB and *E. faecium* ST711GB for *van*B, *tdc* and *odc*. However, even though *E. faecium* ST315GB and *E. faecium* ST711GB harboured the gene for vancomycin resistance, based on the physiological test both were shown to be sensitive to vancomycin. Most probably the gene for vancomycin is present, but under the specific culturing conditions the gene is not expressed.

Ahmadova *et al.* (2013) reported that *E. faecium* AQ71 isolated from Motal cheese, was negative for the tested virulence genes (*asa*1, *ace, esp, efaAfm,* and *cyl*A/B) and did not present multi-resistance to antibiotics.

In contrast to most other tested *E. faecalis* strains, those reported by Cebrián *et al.* (2012), isolated from Spanish sheep's cheese, were sensitive to most of the antibiotics tested, being resistant only to aminoglycosides, lincosamide, and pristinamicins. In addition, PCR amplification revealed the existence of genes encoding for GelE, Asa1, Esp, EfaA, and Ace antigens and for tyrosine decarboxylase (Cebrián *et al.*, 2012).

In a study of Martin-Platero *et al.* (2009) 95 enterococci were isolated from three different goats' milk cheeses and were shown to belonging to the species *E. devriesei, E. faecalis* and *E. malodoratus,* among others. The genes *gel*E, *esp, asa*1, *efa*A and *ace,* all associated to virulence factors in enterococci, were detected at different levels in *E. faecalis* whilst they were not detected in the remaining species studied. In addition, no resistance to vancomycin was detected by PCR screening. The gene for tyrosine decarboxylase (*tdc*) was detected in all the *E. faecalis* and *E. hirae* isolates but in none of the other studied *Enterococcus* spp.

Numerous studies on microbiology of Manchego cheese (Ballesteros *et al.*, 2006; Cabezas *et al.*, 2007; Ramos *et al.*, 1981) have shown that enterococci are the predominant members of the microbiota, with counts sometimes even higher than those corresponding to lactobacilli and lactococci at the last stages of ripening. Nieto-Arribas *et al.* (2011) isolated 132 enterococci from Manchego cheeses. *E. faecalis* was the predominant species, accounting for 81.8% of the total isolates, while *E. faecium, E. hirae* and *E. avium* were present in low proportions. A low antibiotic resistance was found and almost all the strains were susceptible to clinically important antibiotics. On the contrary, only four isolates (two *E. faecalis* and two *E. faecium*) did not harbour some of the *gel*E, *hyl, asa*1, *esp, cyl*A, *ace, efa*A, *van*A and *van*B virulence genes assayed.

The main objective of Chaves-Lopez *et al.* (2011) was to study the enterococcal population in kumis samples traditionally manufactured in some regions of Colombia. The biomolecular identification of 72 isolates evidenced that *E. faecalis* and *E. faecium* were the most dominant species. In 42 isolates, based on PCR analysis, *gel*E, *esp, asa*1, *cyl* and *hyl* genes, all associated with virulence factors in enterococci, were not detected. Based on physiological test, nine *E. faecalis* and two *E. faecium* strains produced fermented milk with ACE-inhibitory activity. Moreover, the strains showed a very low tyrosine decarboxylase activity and did not produce detectable levels of histamine during 48 h fermentation in milk. Based on the results the authors concluded that some *Enterococcus* spp. may be considered as a non-virulent enterococci and potential probiotic due to their ability able to produce fermented milks with ACE-inhibitory activity.

In a study by Klibi *et al.* (2013) *Enterococcus* spp. were isolated from 80% of the meat samples. In total 119 *Enterococcus* spp. (41% *E. faecalis*, 25% *E. faecium*, 10% *E. hirae*, 10% *E. gallinarum*, 5% *E. casseliflavus*, 5% *E. mundtii*, 2% *E. sulfurens* and 2% *E. durans*) isolates were studied for the presence of antibiotic resistance (including presence of related genes and physiological test by antibiotic discs) and some virulence factors related to enterococci. The authors reported that 35% of the isolates showed resistance to erythromycin, 41% to tetracycline, but only six isolates (five *E. faecium* and one *E. gallinarum*) to pristinamycin. Moreover, 14% (11 *E. faecalis*, five *E. faecium* and one *E. gallinarum*) were resistant to streptomycin.

In addition to antibiotic resistance, Klibi *et al.* (2013) have screened all 119 *Enterococcus* spp. for the presence of some virulence genes (*gel*E, *esp* and *hyl*). It is interesting that all isolates have generated positive PCR results for the presence of *gel*E. However, physiological test showed gelatinase activity in only 48 of isolates (40 *E. faecalis*, six *E. faecium* and two *E. gallinarum*). It could be that the *gel*E gene in other isolates is damaged or the experimental conditions did not allow expression. This is something that needs additional attention. However, when the authors explored the presence of *esp* and *hyl*, only two *E. faecalis* and one *E. faecium* showed positive PCR results for presence of *esp* and only one *E. faecium* for *hyl*. Taking in consideration that Klibi *et al.* (2013) explored 119 isolates, this is a relatively low presence.

An example for the presence of *Enterococcus* spp. in traditional meat products has been reported by Barbosa *et al.* (2010). In this study 182 *Enterococcus* spp. have been isolated. Sixty-one isolates have been identified only to the genus level, the other being *E. faecalis* (76), *E. faecium* (44) and *E. casselinoflavus* (1). These results confirm previous reports (Klibi *et al.*, 2013) for the predominance of *E. faecalis* and *E. faecium* in meat products. Interestingly, the authors observed that none of the isolates presented production of DNase, and only 1 was positive for presence of β-haemolytic activity, but 26 isolates were gelatinase positive. When the presence of virulence genes was explored, from 76 *E. faecalis* all generated positive results for *efaAfs*, but this gene was detected only in 1 of 44 *E. faecium* and 5 of 61 *Enterococcus* spp. Opposite results were found for the presence of the *efa*Afm gene. In this case only in 2 from 76 *E. faecalis*, 43 from 44 *E. faecium* and 4 from 62 *Enterococcus* spp. were positive for the *efa*Afm gene. It is interesting to underline that the presence of *esp, agg, cyl*M, *cyl*B, *cyl*A, *cyl*L_L was detected only in 3 from 76 *E. faecalis* and not at all in the other tested enterococci. *Gel*E was found in 51 from 76 *E. faecalis*, in 1 from 44 *E. faecium* and 2 from 62 *Enterococcus* spp.

In an ecological study conducted by Belgacem *et al.* (2010), a collection of Gram-positive bacteria were isolated from Gueddid, an artisanal Tunisian fermented meat. Twenty-four strains from this isolates were identified as *E. faecium* using bio-molecular methods. Investigation of virulence factors by PCR amplification revealed the presence of genes encoding for gelatinase (*gel*E), enterococcal antigen (*efaAfm*), sex pheromone (*cpd* and *ccf*) and expression of cytolysin (the haemolytic component *cyl*B) whereas, other presumed virulence genes encoding for (*agg, esp, cyl*M, *cyl*A, and *cob*) were not detected. In addition, the authors showed that isolates were mostly resistant to erythromycin, rifampicin, ciprofloxacin, clavofloxacin, and nitrofurantoin. However, all of them showed sensitivity to ampicillin, penicillin, vancomycin, chloramphenicol, teicoplanin, gentamicin, streptomycin, and quinupristin/dalfopristin. Based on physiological test tyrosine, lysine, ornithine and histidine were not decarboxylated by any of the tested 24 *E. faecium* strains. Nine of the *E. faecium* strains did not show the presence of any virulence traits or produced biogenic amines, and may be considered as strains with potential technological properties.

Sanchez Valenzuela *et al.* (2008) investigated enterococci isolated from meat, dairy and vegetable foods from Morocco (23 *E. faecalis* and 15 *E. faecium*). Based on the phenotypical test, all isolates were sensitive to ampicillin, penicillin, and gentamicin. 87% of *E. faecalis* isolates were resistant to tetracycline, 78% to rifampicin, 61% to ciprofloxacin, 57% to quinupristin/dalfopristin, 43% to nitrofurantoin, 39% to levofloxacin, 22% erythromycin, 17% to streptomycin, 9% to chloramphenicol, 9% vancomycin and 4% to teicoplanin. In contrast *E. faecium* isolates showed a different antibiotic resistance profile: a high percentage were resistant to nitrofurantoin (73%), followed by erythromycin (67%), ciprofloxacin (67%), levofloxacin (60%), and rifampicin (27%), and only a very low percentage was resistant to tetracycline (7%). One isolate was resistant to both vancomycin and teicoplanin. In the same study the authors evaluated the presence of virulence factors and observed a much higher prevalence among the *E. faecalis* isolates, especially for genes encoding for sex pheromones, collagen adhesin, enterococcal endocarditis antigen, and enterococcal surface protein. Isolates with multiple factors (both antibiotic resistance and virulence traits) were also more frequently detected in *E. faecalis* isolates. One isolate cumulated up to 15 traits (Sanchez Valenzuela *et al.*, 2008). By contrast, several isolates of *E. faecium* investigated in this study had only very few unwanted traits compared with only two isolates in *E. faecalis*.

In a follow up study, Sanchez Valenzuela *et al.* (2009) reported on another 25 enterococcal isolates (nine *E. faecalis* and 16 *E. faecium*) from meat and dairy origin. Investigation of virulence factors based on PCR approach revealed incomplete sets of cytolysin genes both in *E. faecalis* and *E. faecium* isolates. However, in *E. faecalis*, PCR amplification revealed a high incidence of genes encoding for *esp* (seven from nine isolates), *efa*Afs (six from nine), *agg* (two from nine) *ccf*, *cob*, *cpd* (which were detected in nine, five and three out of nine isolates, respectively). In contrast, only *esp* (7 from 16 isolates) and *efaAfm* (10 from 16) were detected in *E. faecium*. Antibiotic resistance detected at high frequencies included rifampicin, nitrofurantoin, ciprofloxacin and levofloxacin. Vancomycin resistance was also detected among these isolates. *E. faecalis* isolates showed decarboxylating activity mostly for tyrosine (five from nine isolates), while *E. faecium* isolates showed a broader decarboxylating capacity, involving tyrosine (11 from 16 isolates), ornithine (6 from 16), lysine (4 from 16) and histidine (3 from 16).

E. faecium (nine isolates), *E. faecalis* (4), *E. avium* (3) and *E. casseliflavus/flavescens* (1) isolates obtained from fermentations of capers (the fruits of *Capparis* sp.) were investigated for the incidence of known virulence determinants, antibiotic resistance and production of biogenic amines (Perez-Pulido *et al.*, 2006). Physiological test on α-haemolytic activity detected its presence only in two *E. avium* and one *E. faecalis* isolates, and β-haemolytic activity was detected in *E. casseliflavus/flavescens*. In addition, the haemolytic component *cyl*B was detected by PCR amplification in three non-haemolytic isolates and in *E. casseliflavus/flavescens*. The collagen adhesin *ace* gene and the endocarditis associated antigen gene *efa*Afm were detected in 2 from 17 tested strains. Genes encoding sex pheromone precursors (*cpd*, *cob*, *ccf*) were detected in *E. faecalis* and *E. casseliflavus/flavescens*. However, other possible virulence genes, such as *agg*, *gel*E, *cyl*M, *cyl*A and *efa*Afs were not detected. All isolates were resistant to rifampicin, erythromycin and ciprofloxacin, and some were also resistant to quinupristin/dalfopristin, tetracycline, levofloxacin, gentamicin and streptomycin. More importantly, vancomycin resistance was not detected in all tested *Enterococcus* spp. Tyrosine decarboxylation was detected in all *E. faecium* isolates. But, on the other hand, of nine tested *E. faecium* isolates, seven showed the absence of virulence genes and two strains carried only two virulence genes. Within the four *E. faecalis* isolates, one strain was free of virulence genes, 1 was carrying only one gene, one gave a positive PCR signal for three genes and one for six genes. None of the three *E. avium* strains generated positive PCR results for the presence of investigated virulence genes. However, *E. casseliflavus/flavescens* generated positive PCR results for the presence of four genes.

The objective of the study of Todorov *et al.* (2014) was to determine some safety aspects of two beneficial strains of *Lactobacillus plantarum*, one strain of *E. faecium*, and three strains of *Lactobacillus sakei* isolated from fermented meat products. According to the results of the diffusion method (physiological test), all six strains demonstrated susceptibility to penicillin G, ampicillin, amoxicillin, amoxicillin/clavulinic acid, imipenem, linezolid and tetracycline. The tested six strains showed a low virulence gene profile as was determined by PCR. It was very interesting to observe that the *E. faecium* isolate did not generate positive results for any of the tested 13 virulence genes. The authors concluded that the *E. faecium* strain can be consider as safe.

Concluding remarks and future perspectives

Enterococcus spp. are well documented to be resistant to various antibiotics. In the last years an emerging problem is the resistance to vancomycin and tobramycin. Various genes can be involved in the antibiotic resistance and sometimes for one antibiotic there can be more than one gene involved in the resistance. It is well documented that vancomycin resistance can be related to at least five different genes and they can be located on the bacterial chromosome or on plasmids. However, the presence of antibiotic resistance genes on plasmids is even more problematic, since this facilitate horizontal gene transfer and spread of these genes in other species in same ecological niche and may spread antibiotic resistance among non-pathogenic organisms. Apart from the notoriety of members of this genus for antibiotic resistance, enterococci may also possess a number of virulence factors which are associated with the severity and duration of infections caused by them. These include genes encoding gelatinase (*gel*), enterococcal surface protein (*esp*), aggregation substance (*asa*), cytolysin (*cyl*), hyaluronidase (*hyl*), endocarditis antigen (*efa*A), adhesion of collagen (*ace*), and vancomycin resistance (*van*A and *van*B). In addition genes related to the production of biogenic amines are part of the genetic heritage of *Enterococcus* spp. Production of histidine decarboxylase (*hdc1* and *hdc2*), *tdc* (tyrosine decarboxylase) and *ord* (ornithine decarboxylase) have been well documented. While the occurrence of virulence traits has been extensively studied among clinical enterococci, studies are only recently beginning to emerge on the elucidation of genes for resistance to antibiotics, virulence factors or production of biogenic amines among enterococci related to food products. Over the years several virulence factors have been identified in food enterococci which include aggregation substances, cytolysin, pheromone, gelatinase, enterococcal surface protein and biofilm formation.

In the last few decades *Enterococcus* spp. have been considered as opportunistic pathogens. Moreover, they are not considered GRAS, and have always been treated as an indicator of faecal contamination and potential health hazard. However, in the last decade also several studies have proposed application of specific enterococcal strains as probiotic strains for human and animal application. In addition, production of numerous antimicrobial peptides (bacteriocins) by various strains belonging to *Enterococcus* is well documented. Traditional fermented products from the Mediterranean part of Europe are using *Enterococcus* spp. as a natural or added starter cultures and they are essential for the organoleptic properties of this products. This includes an enormous variety of salami and other meat products, cheeses and a variety of dairy products, olives fermentation processes, fermented sour bread and other bakery products. *Enterococcus* spp. can be present even during malolactic fermentation of wine. Studies that demonstrate the spread of virulence genes among enterococci isolated from clinical to food samples are very common, highlighting the importance of identifying this potential transfer among isolates in food systems. It likely are environmental factors that regulate expression of the virulence genes. Alternatively, based on the evolutionary process, food related *Enterococcus* isolates have lost some of their virulence potential and can perhaps even be considering safe.

On the other side, what is the chance of one *Enterococcus* spp., free of virulence factors, to become a recipient for the virulence genes if it is going to be in contact with other virulence positive *Enterococcus* spp.? Whether food related and human (probiotic) enterococcal strains are strictly equivalent and share common virulence and antibiotic resistance properties is thus a relevant question. Clewell and Dunny (2002) pointed to the relatively easy exchange of genetic material between different *Enterococcus* spp. But what is the chance for this process to occur in fermented food products or in the gastrointestinal tract of humans and other animals? Maybe we need to draw a line between genetic material exchange in the laboratory conditions and in the food industry. It is important to underline that modern food industry pays special attention on food hygiene and microbial contaminations. On other side, starter cultures have been added to products at high levels. The same is true for levels of application of probiotics. To be considered a probiotic, the strain needs to be applied at 10^8–10^9 cfu/g product (Lahtinen *et al.*, 2012), and this is much higher compared to acceptable potential contaminations.

Another important issue is the relation between the presence (at the DNA level) and the expression of the virulence factors. We can investigate the following scenario. If we investigate a food product fermented by *Enterococcus* spp. and this food product will be consumed only after thermal processing, then it is important that if any virulence factors are present, they are not expressed at the fermentation temperatures of the preparation of the product. In this way, we are not going to have any expression of the virulence genes and after subsequent thermal processing of the food product, the number of cells of the starter culture will be reduced down to nearly zero. Not only thermal processing needs to be considered. Other examples include, changes in water activity (in case of the preparation of various cheeses, salami, olives) which can be a barrier for survival of the starter cultures. Very frequently fermentation processes are conducted by several microorganisms, following different steps in the process and replacing each-other, which is for example the case for olive fermentation (Lucena-Padrós *et al.*, 2014). In all these cases, we need to be aware not only on the presence of the virulence factors, but much more if they can be expressed under these specific conditions in each step of the fermentation process or not.

We postulate that based only on traditional GRAS status, one cannot decide about the safety aspects of new isolated strains. Safety aspects need to be taken into account for each particular strain and to be part of compulsory steps in the evaluation of the new isolates, particularly if these are intended for consumption by humans and animals.

Acknowledgements

CNPq, CAPES and FAPEMIG, due to financial support for academic and scientific activities.

References

Abee, T., Krockel, L., and Hill, C. (1995). Bacteriocins: modes of action and potentials in food preservation and control of food poisoning. Int. J. Food Microbiol. *28*, 169.

Achemchem, F., Abrini, J., Martinez-Bueno, M., Valdivia, E., and Maqueda, M. (2006). Control of *Listeria monocytogenes* in goat's milk and goat's Jben by the bacteriocinogenic *Enterococcus faecium* F58 strain. J. Food Prot. *69*, 2370–2376.

Ahmadova, A., Todorov, S.D., Choiset, Y., Rabesona, H., Zadi, T.M., Kuliyev, A., Franco, B.D.G.M., Chobert, J.-M., and Haertle, T. (2013). Evaluation of antimicrobial activity, probiotic properties and safety of wild strain *Enterococcus faecium* AQ71 isolated from Azerbaijani Motal cheese. Food Control *30*, 631–641.

Aslam, M., Diarra, M.S., Checkley, S., Bohaychuk, V., and Masson, L. (2012). Characterization of antimicrobial resistance and virulence genes in *Enterococcus* spp. isolated from retail meats in Alberta, Canada. Int. J. Food Microbiol. *156*, 222–230.

Ávila, M., Garde, S., Gaya, P., Medina, M., and Nuñez, M. (2005). Influence of a bacteriocin-producing lactic culture on proteolysis and texture of Hispánico cheese. Int. Dairy J. *15*, 145–153.

Aymerich, T., Artigas, M., Garriga, M., Monfort, J., and Hugas, M. (2000). Effect of sausage ingredients and additives on the production of enterocin A and B by *Enterococcus faecium* CTC492. Optimization of *in vitro* production and anti-listerial effect in dry fermented sausages. J. Appl. Microbiol. *88*, 686–694.

Aymerich, T., Holo, H., Håvarstein, L.S., Hugas, M., Garriga, M., and Nes, I.F. (1996). Biochemical and genetic characterization of enterocin A from *Enterococcus faecium*, a new antilisterial bacteriocin in the pediocin family of bacteriocins. Appl. Environ. Microbiol. *62*, 1676–1682.

Badis, A., Guetarni, D., Moussa Boudjema, B., Henni, D., and Kihal, M. (2004). Identification and technological properties of lactic acid bacteria isolated from raw goat milk of four Algerian races. Food Microbiol. *21*, 579–588.

Balla, E., Dicks, L., Du Toit, M., Van Der Merwe, M., and Holzapfel, W. (2000). Characterization and cloning of the genes encoding enterocin 1071A and enterocin 1071B, two antimicrobial peptides produced by *Enterococcus faecalis* BFE 1071. Appl. Environ. Microbiol. *66*, 1298–1304.

Ballesteros, C., Poveda, J.M., Gonzalez-Vinas, M.A., and Cabezas, L. (2006). Microbiological, biochemical and sensory characteristics of artisanal and industrial Manchego cheeses. Food Control *17*, 249–255.

Barbosa, J., Gibbs, P.A., and Teixeira, P. (2010). Virulence factors among enterococci isolated from traditional fermented meat products produced in the North of Portugal. Food Control *21*, 651–656.

Behnsen, J., Deriu, E., Sassone-Corsi, M., and Raffatellu, M. (2013). Probiotics: properties, examples, and specific applications. Cold Spring Harbor Perspectives in Medicine 3, a010074.

Belgacem, Z.B., Abriouel, H., Omar, N.B., Lucas, R., Martínez-Canamero, M., Gálvez, A., and Manai, M. (2010). Antimicrobial activity, safety aspects, and some technological properties of bacteriocinogenic *Enterococcus faecium* from artisanal Tunisian fermented meat. Food Control *21*, 462–470.

Belma, A.S.L.I.M., and Gulcin, A.L.P. (2009). The effect of immobilization on some probiotic properties of *Streptococcus thermophilus* strains. Ann. Microbiol. *59*, 127–132.

Bennik, M.H., Vanloo, B., Brasseur, R., Gorris, L.G., and Smid, E.J. (1998). A novel bacteriocin with a YGNGV motif from vegetable-associated *Enterococcus mundtii*: full characterization and interaction with target organisms. Biochim. Biophys. Acta *1373*, 47–58.

Bhardwaj, A., Gupta, H., Kapila, S., Kaur, G., Vij, S., and Malik, R.K. (2010). Safety assessment and evaluation of probiotic potential of bacteriocinogenic *Enterococcus faecium* KH 24 strain under *in vitro* and *in vivo* conditions. Int. J. Food Microbiol. *141*, 156–164.

Booth, M.C., Bogie, C.P., Sahl, H.G., Siezen, R.J., Hatter, K.L., and Gilmore, M.S. (1996). Structural analysis and proteolytic activation of *Enterococcus faecalis* cytolysin, a novel lantibiotic. Mol. Microbiol. *21*, 1175–1184.

Borrero, J., Jiménez, J.J., Gútiez, L., Herranz, C., Cintas, L.M., and Hernández, P.E. (2011). Protein expression vector and secretion signal peptide optimization to drive the production, secretion, and functional expression of the bacteriocin enterocin A in lactic acid bacteria. J. Biotechnol. *156*, 76–86.

Botes, M., Loos, B., van Reenen, C.A., and Dicks, L.M. (2008). Adhesion of the probiotic strains *Enterococcus mundtii* ST4SA and *Lactobacillus plantarum* 423 to Caco-2 cells under conditions simulating the intestinal tract, and in the presence of antibiotics and anti-inflammatory medicaments. Arch. Microbiol. *190*, 573–584.

Cabezas, L., Sanchez, I., Poveda, J.M., Sesena, S., and Palop, M.L.L. (2007). Comparison of microflora, chemical and sensory characteristics of artisanal Manchego cheeses from two dairies. Food Control *18*, 11–17.

Callewaert, R., Hugas, M., and Vuyst, L.D. (2000). Competitiveness and bacteriocin production of *Enterococci* in the production of Spanish-style dry fermented sausages. Int. J. Food Microbiol. *57*, 33–42.

Casaus, P., Nilsen, T., Cintas, L.M., Nes, I.F., Hernández, P.E., and Holo, H. (1997). Enterocin B, a new bacteriocin from *Enterococcus faecium* T136 which can act synergistically with enterocin A. Microbiol. *143*, 2287–2294.

Cebrián, R., Baños, A., Valdivia, E., Pérez-Pulido, R., Martínez-Bueno, M., and Maqueda, M. (2012). Characterization of functional, safety, and probiotic properties of *Enterococcus faecalis* UGRA10, a new AS-48-producer strain. Food Microbiol. *30*, 59–67.

Chaves-Lopez, C., Serio, A., Martuscelli, M., Paparella, A., Osorio-Cadavid, E., and Suzzi, G. (2011). Microbiological characteristics of kumis, a traditional fermented Colombian milk, with particular emphasis on enterococci population. Food Microbiol. *28*, 1041–1047.

Cintas, L.M., Casaus, P., Håvarstein, L.S., Hernandez, P.E., and Nes, I.F. (1997). Biochemical and genetic characterization of enterocin P, a novel sec-dependent bacteriocin from *Enterococcus faecium* P13 with a broad antimicrobial spectrum. Appl. Environ. Microbiol. *63*, 4321–4330.

Cintas, L.M., Casaus, P., Herranz, C., Håvarstein, L.S., Holo, H., Hernández, P.E., and Nes, I.F. (2000). Biochemical and genetic evidence that *Enterococcus faecium* L50 produces Enterocins L50A and L50B, the sec-dependent Enterocin P, and a novel bacteriocin secreted without an N-Terminal extension termed Enterocin Q. J. Bacterol. *182*, 6806–6814.

Cintas, L.M., Casaus, P., Holo, H., Hernandez, P.E., Nes, I.F., and Håvarstein, L.S. (1998). Enterocins L50A and L50B, two novel bacteriocins from *Enterococcus faecium* L50, are related to staphylococcal hemolysins. J. Bacterol. *180*, 1988–1994.

Cleveland, J., Montville, T.J., Nes, I.F., and Chikindas, M.L. (2001). Bacteriocins: safe, natural antimicrobials for food preservation. Int. J. Food Microbiol. *71*, 1–20.

Clewell, D.B. (1990). Movable genetic elements and antibiotic resistance in enterococci. Eur. J. Clin. Microbiol. Infec Dis. *9*, 90–102.

Clewell, D.B., and Dunny, G.M. (2002). Conjugation and genetic exchange in enterococci. In The Enterococci: Pathogenesis, Molecular Biology, and Antibiotic Resistance, M.S. Gilmore, ed. (Washington, DC: American Society for Microbiology Press), pp. 265–300.

Cotter, P.D., Hill, C., and Ross, R.P. (2005). Bacteriocins: developing innate immunity for food. Nature Rev. Microbiol. 3, 777–788.

Dal Bello, B., Cocolin, L., Zeppa, G., Field, D., Cotter, P.D., and Hill, C. (2012). Technological characterization of bacteriocin producing *Lactococcus lactis* strains employed to control *Listeria monocytogenes* in Cottage cheese. Int. J. Food Microbiol. *153*, 58–65.

Dal Bello, B., Rantsiou, K., Bellio, A., Zeppa, G., Ambrosoli, R., Civera, T., and Cocolin, L. (2010). Microbial ecology of artisanal products from North West of Italy and antimicrobial activity of the autochthonous populations. LWT Food Sci. Technol. *43*, 1151–1159.

De Vuyst, L., and Leroy, F. (2007). Bacteriocins from lactic acid bacteria: production, purification, and food applications. J. Mol. Microbiol. Biotechnol. *13*, 194–199.

Deegan, L.H., Cotter, P.D., Hill, C., and Ross, P. (2006). Bacteriocins: Biological tools for bio-preservation and shelf-life extension. Int. Dairy J. *16*, 1058–1071.

Ditu, L.-M., Chifiriuc, M.C., Bezirtzoglou, E., Voltsi, C., Bleotu, C., Pelinescu, D., Mihaescu, G., and Lazar, V. (2011). Modulation of virulence and antibiotic susceptibility of enteropathogenic *Escherichia coli* strains by *Enterococcus faecium* probiotic strain culture fractions. Anaerobe *17*, 448–451.

Durlu-Ozkaya, F., Xanthopoulos, V., Tunail, N., and Litopoulou-Tzanetaki, E. (2001). Technologically important properties of lactic acid bacteria isolates from Beyaz cheese made from raw ewes' milk. J. Appl. Microbiol. *91*, 861–870.

Eaton, T.J., and Gasson, M.J. (2001). Molecular screening of *Enterococcus* virulence determinants and potential for genetic exchange between food and medical isolates. Appl. Environ. Microbiol. *67*, 1628–1635.

FAO/WHO (2002). Guidelines for the Evaluation of Probiotics in Food. Report of a Joint FAO/WHO Working Group on Drafting Guidelines for the Evaluation of Probiotics in Food. (London Ontario, Canada), pp. 1–11.

Farías, M., Farías, R., Holgado, A., and Sesma, F. (1996). Purification and N-terminal amino acid sequence of Enterocin CRL 35, a 'pediocin-like' bacteriocin produced by *Enterococcus faecium* CRL 35. Lett. Appl. Microbiol. *22*, 417–419.

Favaro, L., Basaglia, M., Casella, S., Hue, I., Dousset, X., Franco, B.D.G.M., and Todorov, S.D. (2014). Bacteriocinogenic potential and safety evaluation of non-starter *Enterococcus faecium* strains isolated from home made white brine cheese. Food Microbiol. *38*, 228–239.

Fortina, M.G., Ricci, G., Borgo, F., Manachini, P.L., Arends, K., Schiwon, K., Abajy, M.Y., and Grohmann, E. (2008). A survey on biotechnological potential and safety of the novel *Enterococcus* species of dairy origin, *E. italicus*. Int. J. Food Microbiol. *123*, 204–211.

Foulquié Moreno, M.R., Sarantinopoulos, P., Tsakalidou, E., and De Vuyst, L. (2006). The role and application of enterococci in food and health. Int. J. Food Microbiol. *106*, 1–24.

Franciosi, E., Settanni, L., Cavazza, A., and Poznanski, E. (2009). Biodiversity and technological potential of wild lactic acid bacteria from raw cows' milk. Int. Dairy J. *19*, 3–11.

Franz, C., Muscholl-Silberhorn, A.B., Yousif, N.M.K., Vancanneyt, M., Swings, J., and Holzapfel, W.H. (2001). Incidence of virulence factors and antibiotic resistance among enterococci isolated from food. Appl. Environ. Microbiol. *67*, 4385–4389.

Franz, C.M., Grube, A., Herrmann, A., Abriouel, H., Stärke, J., Lombardi, A., Tauscher, B., and Holzapfel, W.H. (2002). Biochemical and genetic characterization of the two-peptide bacteriocin enterocin 1071 produced by *Enterococcus faecalis* FAIR-E 309. Appl. Environ. Microbiol. *68*, 2550–2554.

Franz, C.M., Van Belkum, M.J., Holzapfel, W.H., Abriouel, H., and Gálvez, A. (2007). Diversity of enterococcal bacteriocins and their grouping in a new classification scheme. FEMS Microbiol. Rev. *31*, 293–310.

Franz, C.M.A.P., Huch, M., Abriouel, H., Holzapfel, W., and Gálvez, A. (2011). Enterococci as probiotics and their implications in food safety. Int. J. Food Microbiol. *151*, 125–140.

Franz, C.M.A.P., Stiles, M.E., Schleifer, K.H., and Holzapfel, W.H. (2003). Enterococci in foods – a conundrum for food safety. Int. J. Food Microbiol. *88*, 105–122.

Gálvez, A., Valdivia, E., Abriouel, H., Camafeita, E., Mendez, E., Martínez-Bueno, M., and Maqueda, M. (1998). Isolation and characterization of enterocin EJ97, a bacteriocin produced by *Enterococcus faecalis* EJ97. Arch. Microbiol. *171*, 59–65.

Giraffa, G. (2002). Enterococci from foods. FEMS Microbiol. Rev. *26*, 163–171.

Giraffa, G. (2003). Functionality of enterococci in dairy products. Int. J. Food Microbiol. *88*, 215–222.

González, L., Sacristán, N., Arenas, R., Fresno, J.M., and Eugenia Tornadijo, M. (2010). Enzymatic activity of lactic acid bacteria (with antimicrobial properties) isolated from a traditional Spanish cheese. Food Microbiol. *27*, 592–597.

Haas, W., Shepard, B.D., and Gilmore, M.S. (2002). Two-component regulator of *Enterococcus faecalis* cytolysin responds to quorum-sensing autoinduction. Nature *415*, 84–87.

Hickey, R.M., Twomey, D.P., Ross, R.P., and Hill, C. (2003). Production of enterolysin A by a raw milk enterococcal isolate exhibiting multiple virulence factors. Microbiol. *149*, 655–664.

Hu, C.-B., Malaphan, W., Zendo, T., Nakayama, J., and Sonomoto, K. (2010). Enterocin X, a novel two-peptide bacteriocin from *Enterococcus faecium* KU-B5, has an antibacterial spectrum entirely different from those of its component peptides. Appl. Environ. Microbiol. *76*, 4542–4545.

Izquierdo, E., Marchioni, E., Aoude-Werner, D., Hasselmann, C., and Ennahar, S. (2009). Smearing of soft cheese with *Enterococcus faecium* WHE 81, a multi-bacteriocin producer, against *Listeria monocytogenes*. Food Microbiol. *26*, 16–20.

Jahan, M., and Holley, R.A. (2014). Incidence of virulence factors in enterococci from raw and fermented meat and biofilm forming capacity at 25 degrees C and 37 degrees C. Int. J. Food Microbiol. *170*, 65–69.

Javed, A., Masud, T., ul Ain, Q., Imran, M., and Maqsood, S. (2011). Enterocins of *Enterococcus faecium*, emerging natural food preservatives. Ann. Microbiol. *61*, 699–708.

Klaenhammer, T.R. (1988). Bacteriocins of lactic acid bacteria. Biochimie *70*, 337–349.

Klibi, N., Ben Said, L., Jouini, A., Ben Slama, K., Lopez, M., Ben Sallem, R., Boudabous, A., and Torres, C. (2013). Species distribution, antibiotic resistance and virulence traits in enterococci from meat in Tunisia. Meat Sci. 93, 675–680.

Lahtinen, S., Ouwehand, A.C., Salminen, S., and von Wright, A. (2012). Lactic Acid Bacteria: Microbiological and Functional Aspects, 4th edn (New York: CRC Press).

Leroy, F., and De Vuyst, L. (2004). Lactic acid bacteria as functional starter cultures for the food fermentation industry. Trends Food Sci. Technol. *15*, 67–78.

Lilly, D.M., and Stillwell, R.H. (1965). Probiotics: growth-promoting factors produced by microorganisms. Science 147, 747–748.

Lucena-Padrós, H., Caballero-Guerrero, B., Maldonado-Barragán, A., and Ruiz-Barba, J.L. (2014). Microbial diversity and dynamics of Spanish-style green table-olive fermentations in large manufacturing companies through culture-dependent techniques. Food Microbiol. *42*, 154–165.

Maietti, L., Bonvini, B., Huys, G., and Giraffa, G. (2007). Incidence of antibiotic resistance and virulence determinants among *Enterococcus italicus* isolates from dairy products. Syst Appl. Microbiol. *30*, 509–517.

Majhenic, A.C., Rogelj, I., and Perko, B. (2005). Enterococci from Tolminc cheese: Population structure, antibiotic susceptibility and incidence of virulence determinants. Int. J. Food Microbiol. *102*, 239–244.

Maldonado-Barragán, A., Caballero-Guerrero, B., Jiménez, E., Jiménez-Díaz, R., Ruiz-Barba, J.L., and Rodríguez, J.M. (2009). Enterocin C, a class IIb bacteriocin produced by *E. faecalis* C901, a strain isolated from human colostrum. Int. J. Food Microbiol. *133*, 105–112.

Mareková, M., Lauková, A., Skaugen, M., and Nes, I. (2007). Isolation and characterization of a new bacteriocin, termed enterocin M, produced by environmental isolate *Enterococcus faecium* AL41. J. Ind Microbiol. Biotechnol. *34*, 533–537.

Martin-Platero, A.M., Valdivia, E., Maqueda, M., and Martinez-Bueno, M. (2009). Characterization and safety evaluation of enterococci isolated from Spanish goats' milk cheeses. Int. J. Food Microbiol. *132*, 24–32.

Martin, B., Garriga, M., Hugas, M., and Aymerich, T. (2005). Genetic diversity and safety aspects of enterococci from slightly fermented sausages. J. Appl. Microbiol. *98*, 1177–1190.

Miyazaki, Y., Kamiya, S., Hanawa, T., Fukuda, M., Kawakami, H., Takahashi, H., and Yokota, H. (2010). Effect of probiotic bacterial strains of *Lactobacillus*, *Bifidobacterium*, and *Enterococcus* on enteroaggregative *Escherichia coli*. J. Infect. Chemother. *16*, 10–18.

Moraes, P.M., Perin, L.M., Todorov, S.D., Silva, A., Franco, B., and Nero, L.A. (2012). Bacteriocinogenic and virulence potential of *Enterococcus* isolates obtained from raw milk and cheese. J. Appl. Microbiol.

Morandi, S., Brasca, M., and Lodi, R. (2011). Technological, phenotypic and genotypic characterisation of wild lactic acid bacteria involved in the production of Bitto PDO Italian cheese. Dairy Sci. Technol. *91*, 341–359.

Muñoz, A., Ananou, S., Gálvez, A., Martínez-Bueno, M., Rodríguez, A., Maqueda, M., and Valdivia, E. (2007). Inhibition of *Staphylococcus aureus* in dairy products by enterocin AS-48 produced *in situ* and ex situ: Bactericidal synergism with heat. Int. Dairy J. *17*, 760–769.

Nieto-Arribas, P., Sesena, S., Poveda, J.M., Chicon, R., Cabezas, L., and Palop, L. (2011). *Enterococcus* populations in artisanal Manchego cheese: Biodiversity, technological and safety aspects. Food Microbiol. *28*, 891–899.

Nieto-Arribas, P., Seseña, S., Poveda, J.M., Palop, L., and Cabezas, L. (2010). Genotypic and technological characterization of *Leuconostoc* isolates to be used as adjunct starters in Manchego cheese manufacture. Food Microbiol. 27, 85–93.

Nilsen, T., Nes, I.F., and Holo, H. (2003). Enterolysin A, a cell wall-degrading bacteriocin from *Enterococcus faecalis* LMG 2333. Appl. Environ. Microbiol. *69*, 2975–2984.

Nueno-Palop, C., and Narbad, A. (2011). Probiotic assessment of *Enterococcus faecalis* CP58 isolated from human gut. Int. J. Food Microbiol. *145*, 390–394.

Ogier, J.-C., and Serror, P. (2008). Safety assessment of dairy microorganisms: The *Enterococcus* genus. Int. J. Food Microbiol. *126*, 291–301.

Ortolani, M.B.T., Yamazi, A.K., Moraes, P.M., Viçosa, G.N., and Nero, L.A. (2010). Microbiological quality and safety of raw milk and soft cheese and detection of autochthonous lactic acid bacteria with antagonistic activity against *Listeria monocytogenes*, *Salmonella* spp. and *Staphylococcus aureus*. Foodborne Pathog. Dis. 7, 175–180.

Pérez-Martín, F., Seseña, S., Izquierdo, P.M., and Palop, M.L. (2014). Are Enterococcus populations present during malolactic fermentation of red wine safe? Food Microbiol. *42*, 95–101.

Perez-Pulido, R., Abriouel, H., Ben Omar, N., Lucas, R., Martinez-Canamero, M., and Galvez, A. (2006). Safety and potential risks of enterococci isolated from traditional fermented capers. Food Chem. Toxicol. *44*, 2070–2077.

Perin, L.M., Miranda, R.O., Todorov, S.D., Franco, B.D.G.M., and Nero, L.A. (2014). Virulence, antibiotic resistance and biogenic amines of bacteriocinogenic lactococci and enterococci isolated from goat milk. Int. J. Food Microbiol. *185*, 121–126.

Perin, L.M., and Nero, L.A. (2014). Antagonistic lactic acid bacteria isolated from goat milk and identification of a novel nisin variant *Lactococcus lactis*. BMC Microbiol. *14*, 36.

Pieniz, S., Andreazza, R., Anghinoni, T., Camargo, F., and Brandelli, A. (2014). Probiotic potential, antimicrobial and antioxidant activities of *Enterococcus durans* strain LAB18s. Food Control 37, 251–256.

Pingitore, E.V., Todorov, S.D., Sesma, F., and Franco, B.D.G.M. (2012). Application of bacteriocinogenic *Enterococcus mundti* CRL35 and *Enterococcus faecium* ST88Ch in the control of *Listeria monocytogenes* in fresh minas cheese. Food Microbiol. *32*, 38–47.

Piraino, P., Zotta, T., Ricciardi, A., McSweeney, P.L., and Parente, E. (2008). Acid production, proteolysis, autolytic and inhibitory properties of lactic acid bacteria isolated from pasta filata cheeses: A multivariate screening study. Int. Dairy J. *18*, 81–92.

Ramos, M., Barneto, R., and Ordonez, J.A. (1981). Evaluation of a specific starter for Manchego Cheese production. Milk Sci. Int. *36*, 528-&.

Rehaiem, A., Belgacem, Z.B., Edalatian, M.R., Martínez, B., Rodríguez, A., Manai, M., and Guerra, N.P. (2014). Assessment of potential probiotic properties and multiple bacteriocin encoding-genes of the technological performing strain *Enterococcus faecium* MMRA. Food Control *37*, 343–350.

Ribeiro, T., Oliveira, M., Fraqueza, M.J., Laukova, A., Elias, M., Tenreiro, R., Barreto, A.S., and Semedo-Lemsaddek, T. (2011). Antibiotic Resistance and Virulence Factors among Enterococci Isolated from Chourico, a Traditional Portuguese Dry Fermented Sausage. J. Food Prot. *74*, 465–469.

Riley, M.A., and Wertz, J.E. (2002). Bacteriocins: evolution, ecology, and application. Annu. Rev. Microbiol. *56*, 117–137.

Saavedra, L., Taranto, M.P., Sesma, F., and de Valdez, G.F. (2003). Homemade traditional cheeses for the isolation of probiotic *Enterococcus faecium* strains. Int. J. Food Microbiol. *88*, 241–245.

Sabia, C., Manicardi, G., Messi, P., De Niederhäusern, S., and Bondi, M. (2002). Enterocin 416K1, an antilisterial bacteriocin produced by *Enterococcus casseliflavus* IM 416K1 isolated from Italian sausages. Int. J. Food Microbiol. *75*, 163–170.

Sanchez Valenzuela, A., Ben Omar, N., Abriouel, H., Lucas Lopez, R., Ortega, E., Martinez Canamero, M., and Galvez, A. (2008). Risk factors in enterococci isolated from foods in Morocco: Determination of antimicrobial resistance and incidence of virulence traits. Food Chem. Toxicol. *46*, 2648–2652.

Sanchez Valenzuela, A., ben Omar, N., Abriouel, H., Lucas Lopez, R., Veljovic, K., Martinez Canamero, M., Ljubisa Topisirovic, M.K., and Galvez, A. (2009). Virulence factors, antibiotic resistance, and bacteriocins in enterococci from artisan foods of animal origin. Food Control *20*, 381–385.

Sarantinopoulos, P., Andrighetto, C., Georgalaki, M.D., Rea, M.C., Lombardi, A., Cogan, T.M., Kalantzopoulos, G., and Tsakalidou, E. (2001). Biochemical properties of enterococci relevant to their technological performance. Int. Dairy J. *11*, 621–647.

Sawa, N., Wilaipun, P., Kinoshita, S., Zendo, T., Leelawatcharamas, V., Nakayama, J., and Sonomoto, K. (2012). Isolation and characterization of Enterocin W, a Novel Two-Peptide Lantibiotic Produced by *Enterococcus faecalis* NKR-4-1. Appl. Environ. Microbiol. *78*, 900–903.

Semedo, T., Santos, M.A., Lopes, M.F.S., Marques, J.J.F., Crespo, M.T.B., and Tenreiro, R. (2003). Virulence factors in food, clinical and reference enterococci: a common trait in the genus? Syst Appl. Microbiol. *26*, 13–22.

Settanni, L., Franciosi, E., Cavazza, A., Cocconcelli, P.S., and Poznanski, E. (2011). Extension of Tosèla cheese shelf-life using non-starter lactic acid bacteria. Food Microbiol. *28*, 883–890.

Smit, G., Smit, B.A., and Engels, W.J. (2005). Flavour formation by lactic acid bacteria and biochemical flavour profiling of cheese products. FEMS Microbiol. Rev. *29*, 591–610.

Suzzi, G., Caruso, M., Gardini, F., Lombardi, A., Vannini, L., Guerzoni, M., Andrighetto, C., and Lanorte, M. (2000). A survey of the enterococci isolated from an artisanal Italian goat's cheese (semicotto caprino). J. Appl. Microbiol. *89*, 267–274.

Todorov, S.D., Franco, B.D.G.M., and Wiid, I.J. (2014). *In vitro* study of beneficial properties and safety of lactic acid bacteria isolated from Portuguese fermented meat products. Benefic. Microb. *5*, 351–366.

Todorov, S.D., Wachsman, M., Tomé, E., Dousset, X., Destro, M.T., Dicks, L.M.T., de Melo Franco, B.D.G., Vaz-Velho, M., and Drider, D. (2010). Characterisation of an antiviral pediocin-like bacteriocin produced by *Enterococcus faecium*. Food Microbiol. *27*, 869–879.

Tomita, H., Fujimoto, S., Tanimoto, K., and Ike, Y. (1996). Cloning and genetic organization of the bacteriocin 31 determinant encoded on the *Enterococcus faecalis* pheromone-responsive conjugative plasmid pYI17. J. Bacterol. *178*, 3585–3593.

van Kranenburg, R., Kleerebezem, M., van Hylckama Vlieg, J., Ursing, B.M., Boekhorst, J., Smit, B.A., Ayad, E.H., Smit, G., and Siezen, R.J. (2002). Flavour formation from amino acids by lactic acid bacteria: predictions from genome sequence analysis. Int. Dairy J. *12*, 111–121.

Van Tyne, D., Martin, M.J., and Gilmore, M.S. (2013). Structure, function, and biology of the *Enterococcus faecalis* cytolysin. Toxins *5*, 895–911.

Vergio, F. (1954). Anti- und Probiotika. Hippokrates *4*, 116–119.

Yamamoto, Y., Togawa, Y., Shimosaka, M., and Okazaki, M. (2003). Purification and characterization of a novel bacteriocin produced by *Enterococcus faecalis* strain RJ-11. Appl. Environ. Microbiol. *69*, 5746–5753.

Use of rec LABs: Good Bugs to Deliver Molecules of Health Interest: From Mouse to Man

11

Jean-Marc Chatel, Natalia Martins Breyner, Débora Leandro Rama Gomes, Vasco Azevedo, Anderson Miyoshi and Philippe Langella

Abstract

Lactic acid bacteria (LAB) constitute a heterogeneous group of Gram-positive bacteria widely used in the food industry because of their generally regarded as safe (GRAS) status. In this regard, LAB have been widely investigated and used as live vehicles for the production and delivery of heterologous proteins or cDNA of vaccinal, medical or technological interest because of its ease for protein secretion and purification.

Here, we review the expression of heterologous protein and the various delivery systems developed to target heterologous proteins to specific cell locations (cytoplasm, extracellular medium or cell wall), as well as the more recent research on LAB as DNA delivery vehicles and finale with the challenges and future trends to improve the existing strategies and develop new ones.

Choosing the bacteria

In the last decades, different expression systems have been created in order to produce recombinant proteins for diverse biotechnological applications. Although *Escherichia coli* has been the first prokaryotic host model, this microorganism does not present the ideal profile for protein production since it presents reduced secretion (regularly expression is intracellular), proteolytic digestion, complex downstream processes, antibiotic-based maintenance and absence of post-translational modifications required for correct protein folding. In addition, endotoxic lipopolysaccharides (LPS) have to be removed from recombinant proteins before administration to humans, thus restricting pharmaceutical application (Ferrer-Miralles and Villaverde, 2013; Morello *et al.*, 2008).

Consequently, other bacterial microorganisms have been studied in order to expand the options for recombinant protein production as well for their use as live vectors for the delivery of heterologous proteins. In this context, lactic acid bacteria (LAB) represent an alternative to attenuate pathogenic bacteria. They are greatly adapted to extreme diverse environments, such as the acid pH of the stomach, allowing them to survive in the gastrointestinal tract. LAB form a phylogenetically heterogeneous group composed of Gram-positive cocci and bacilli. Moreover, these microorganisms are considered as 'generically recognized as safe' (GRAS) by the United States Food and Drug Administration (FDA) and have a well established safety profile due to their long and extensive use in the food industry. Besides, it is also important to mention that some LAB species are considered probiotics and their use has increased in the last decades, especially for the *Lactobacillus* and *Bifidobacterium* genera (Bahey-El-Din and Gahan, 2010; Wells, 2011).

Lactococcus lactis is a non-invasive facultative Gram-positive bacterium broadly used in the dairy industry (mainly in cheese-making) and, thus, does not possess lipopolysaccharides (LPS) as *Escherichia coli*. LPS, also named endotoxin, has a strong inflammatory effect and has been involved in various diseases as inflammatory bowel disease (IBD) or obesity. The same would be true for, for example, lipoteichoic acid (LTA) of Gram-positive bacteria. However, micromolar concentrations are required for LTA activity, while LPS is effective with nanomolar quantities (Lebeer *et al.*, 2012). Moreover, variations in immunostimulation have already been observed among LTAs from different Gram-positive microorganisms. For instance, LTA from *Lactobacillus plantarum* induced minimal NO production in macrophages and an inferior activation of TLR2/CD14 when compared to LTA from *Staphylococcus aureus* (Ryu *et al.*, 2009). *L. lactis* is the best characterized member of the LAB and plenty of information about its microbiology, physiology, ecology and genetics is publicly available. This microorganism presents a relatively simple and well-known metabolism. In addition, some laboratory strains are plasmid free and secrete only one major protein (Usp45), simplifying downstream purification processes. The genomes of approximately 10 *L. lactis* strains have already been sequenced and as a result, intensive genetic and molecular research has been developed to optimize their potential biomedical applications, such as production and secretion of heterologous proteins, vaccine delivery, gene delivery and therapeutic drug delivery (Bermudez-Humaran, 2009; Bolotin *et al.*, 2001; Mierau and Kleerebezem, 2005; Nouaille *et al.*, 2003; Wegmann *et al.*, 2007).

One of the first reports regarding *L. lactis* as a vehicle to produce immune responses was performed by Iwaki *et al.* (Iwaki *et al.*, 1990) who used killed recombinant cells expressing a surface protective antigen (Pac) from *Streptococcus mutans*. Mice orally immunized with this genetically modified microorganism (GMM) presented Pac-specific serum IgG and mucosal IgA antibodies in salivary secretions and in serum. Nevertheless, the use of *L. lactis* as a live protein delivery vector was first published by Wells *et al.* (1993), who immunized mice via subcutaneous injection with a recombinant *L. lactis* strain producing the tetanus fragment C (TTFC) against a lethal challenge with tetanus toxin (Wells *et al.*, 1993). The effect of either oral (Robinson *et al.*, 1997) or nasal administration of this strain in mice was evaluated (Norton *et al.*, 1997) and although protection efficacy showed to be the same, oral immunization produced lower serum IgG and mucosal IgA antibodies than nasal vaccination.

Since then, *L. lactis* has been genetically engineered to develop live mucosal vaccines expressing antigens derived from viruses and bacteria, as well as allergens, interleukins and enzymes. For example, the bovine β-lactoglobulin (BLG) (Chatel *et al.*, 2001), the rotavirus non-structural protein 4 (Enouf *et al.*, 2001), murine IL-12 (Bermudez-Humaran *et al.*, 2003), the human papillomavirus (HPV) type-16 E7 antigen (Bermudez-Humaran *et al.*, 2002), *Staphylococcus aureus* fibronectin-binding protein A and *Listeria monocytogenes* internalin A (Innocentin *et al.*, 2009), as well as spike-protein subunit VP8 from rotavirus (Marelli *et al.*, 2011), A2 antigen from *Leishmania donovani* (Yam *et al.*, 2011) and one of the best-studied *Leishmania major* antigens, the *Leishmania* homologue of activated C kinase (LACK) (Hugentobler *et al.*, 2012). We have illustrated in Fig. 11.1 the evolution of heterologous protein production in LAB from the first antigens to future molecules.

In this context, lactobacilli, in particular *Lactobacillus casei*, *L. plantarum*, *L. rhamnosus* and *L. acidophilus*, have also been investigated for their use as live delivery vectors. For instance, Hazebrouck *et al.* (2007) described the construction of a recombinant *L. casei* strain that could constitutively deliver BLG to the digestive tract of gnotobiotic mice by applying the LEISS secretion system initially developed for *L. lactis* (Nouaille *et al.*, 2005), leading to a 10-fold increase in LEISS-BLG production (Hazebrouck *et al.*, 2007). Moreover, mice orally immunized with a recombinant *L. casei* strain expressing HPV-16 E7 antigen were able to generate mucosal-specific E7 cytotoxic cellular immune responses (Adachi *et al.*, 2010). Additionally, Kajikawa *et al.* (2012) observed that recombinant *L. acidophilus* displaying HIV-1 Gag at its cell surface induced specific immune responses and that co-expression with the *Salmonella* flagellin presented an adjuvant effect on local IgA production (Kajikawa *et al.*, 2012).

Lactobacilli exhibit interesting features like persisting for

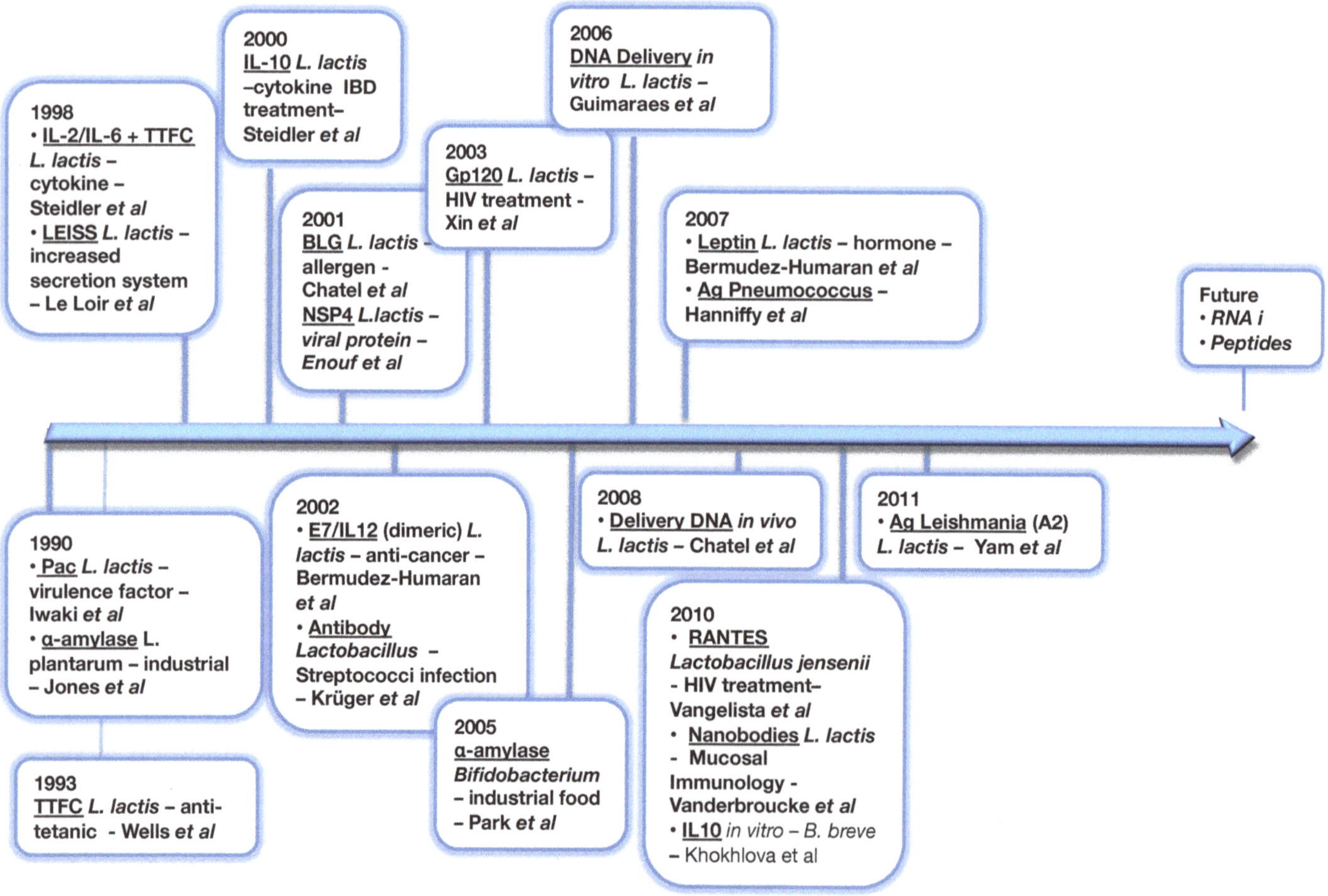

Figure 11.1 Timeline: the landmarks using LAB as vaccine vectors.

longer periods inside the digestive tract as well as intrinsic probiotic attributes, such as enhancing innate and adaptive immune responses and preventing both allergic reactions and gastrointestinal infections (Tsai *et al.*, 2012), although these probiotic attributes are strain dependent. As an example, Ivory and coworkers (2008) demonstrated that oral administration of *L. casei* Shirota reduced the production of IL-5, IL-6 and IFN-gamma, alleviating the severity of allergic rhinitis symptoms (Ivory *et al.*, 2008). In addition, *L. paracasei* subsp. *paracasei* NTU 101 exhibited immunomodulating activity, as it was capable of up-regulating dendritic cells, increase antibody production and suppress the effects of enterohaemorrhagic *E. coli* O157:H7 in mice (Tsai *et al.*, 2010). Conversely, these aspects are not observed in *L. lactis* as this microorganism is rapidly lysed once in the digestive tract and does not present robust adjuvant properties (Hazebrouck *et al.*, 2007). However, regarding the expression of recombinant anti-inflammatory cytokines, *L. lactis* may still be the best choice. It is important to mention that genetic manipulation of lactobacilli presents certain difficulties mainly due to vast molecular and biochemical variation among species, as a result of large differences in the DNA G+C content (LeBlanc *et al.*, 2013; Pouwels *et al.*, 1998).

As for lactobacilli, bifidobacteria are also known for their probiotic properties and commercial formulations containing live cells became very popular in recent years. Strictly speaking, bifidobacteria are not LAB, although they are usually mentioned together. Phylogenetically however they belong to a different phylum, the Actinobacteria. This Gram-positive G+C%-rich microorganism is abundantly present in human intestinal microbiota (especially in breast-fed children) and could be used as mucosal delivery vector of heterologous proteins. Nevertheless, little research has focused on recombinant bifidobacteria as this microorganism is both anaerobic or microaerophilic and has a thick and complex cell wall conferring great recalcitrance for receiving exogenous DNA. Moreover, lack of effective genetic tools, particularly transformation protocols, makes technical procedures difficult (Serafini *et al.*, 2012). Nevertheless, Khokhlova and coworkers (2010) reported the construction of a *Bifidobacterium breve* recombinant strain capable of secreting biologically active human interleukin-10 in buffered tissue culture medium, though in low quantities (1.9 ng/ml) (Khokhlova *et al.*, 2010). Moreover, Ma *et al.* (2012) were capable of expressing two proteins of enterotoxigenic *E. coli* (CfaB, main fimbrial subunit protein, and LTB, the B subunit of heat-labile enterotoxin) in *B. infantis* (Ma *et al.*, 2012).

The quantity and diversity of publications presented above demonstrate that recombinant lactococci, lactobacilli and bifidobacteria are excellent candidates for the production of beneficial compounds and can be used as new therapies for different diseases, although research is still required for improvement.

Choosing the expression system

The expression system is an important aspect that must be considered when developing live bacterial vectors. Various constitutive and inducible expression systems have already been described mainly for *L. lactis*. For example, five constitutive promoters (P21, P23, P32, P44 and P59) were randomly obtained from a genomic library as fusions to the promoter-less *cat-86* reporter gene, which confers chloramphenicol resistance (van der Vossen *et al.*, 1987). Then, these promoters have been used for the production of several heterologous proteins, such as *Streptococcus pyogenes* M6 protein (Piard *et al.*, 1997), *Staphylococcus aureus* clumping factor A (Que *et al.*, 2000) or *Enterococcus faecium* enterocin A (Martinez *et al.*, 2000).

However, the constitutive protein production by these expression systems may lead to intracellular accumulation, increasing toxicity and reducing the system's efficiency. In order to avoid this, inducible expression systems have been developed. The main advantage of these inducible promoters is that they can be regulated by an inducer, a repressor or an environmental factor. Thus, a protein will be expressed only at the right time and in the desired conditions. For example, the constitutive expression of interleukins and their accumulation inside the bacterial cell could cause damaging effects, whereas an induced expression could improve protein delivery, increasing efficiency (Benbouziane *et al.*, 2013; Morello *et al.*, 2008).

The most widely used and best-characterized inducible expression system is the nisin-controlled gene expression (NICE). Nisin is a post-translationally modified broad-spectrum antimicrobial peptide produced by some *L. lactis* strains and is extensively used as preservative in food industry. This molecule induces the *nis* operon (*nisABTCIPRKFEG*) which contains the nisA (the nisin-inducible promoter), nisF and nisRK (the two-component regulator system) promoters. When the gene of interest is located behind the inducible promoter (PnisA) on a plasmid or on the chromosome, expression can be induced by the addition of sub-inhibitory concentrations of nisin (0.1 to 5 ng/ml) to the culture medium. According to the presence or absence of the specific targeting signals, the protein may remain in the cytoplasm, be addressed to the cell wall/cell membrane (CW) or secreted into the medium. Moreover, the NICE system efficiently stimulates expression, leading to high protein yields, as shown for *Staphylococcus simulans* lysostaphin (Mierau *et al.*, 2005). Because of this, it has been broadly used in *L. lactis* to overproduce numerous heterologous proteins. Besides, this system has been employed to study the genetics of pathogenic microorganisms such as *Enterococcus faecalis* (Francia and Clewell, 2002; Waters *et al.*, 2004).

Another tool for controlled protein overexpression in *L. lactis* was described by Miyoshi *et al.* (2004) and called the xylose-inducible expression system (XIES) (Miyoshi *et al.*, 2004). The authors tested the xylose-inducible lactococcal promoter (PxylT) from *L. lactis* NCDO2118 on the *S. aureus* nuclease gene (*nuc*) fused or not to the lactococcal Usp45 signal peptide. The results showed similar high-level and long-term induced Nuc production either to the intracellular or to the extracellular medium. One of the most important advantages of a sugar-dependent system is that most genes involved in carbohydrate catabolism are structured into strongly expressed and controlled operons. Furthermore, considering that nisin is quite expensive, XIES is a viable system, as it requires only two simple sugars (xylose and glucose), enabling its use in large-scale production. Moreover, it can also be used to accomplish *in situ* expression in the GI tract, as co-occurrence of xylose and the genetically modified microorganism (GMM) in the gastrointestinal tract are more likely to occur than the co-occurrence of GMM and nisin. Recently, De Azevedo *et al.* (2012) described a recombinant *L. lactis* strain producing the *Mycobacterium leprae* 65-kDa HSP (Hsp65) antigen intracellularly or secreted via XIES system (de Azevedo *et al.*, 2012b); approximately 7 mg/l of recombinant Hsp65 were detected in the medium and no protein degradation was observed.

Acidification being one of the main properties of LAB, it is important to mention the pH-inducible and growth phase-dependent P170 promoter described for *L. lactis* and developed by Madsen and coworkers (1999). This promoter is up-regulated between pH 6.5 and pH 6.0 during the transition to stationary phase. For protein production, P170 offers a major advantage when compared to the NICE system of being self-inducibility via lactic acid accumulation in the medium during bacterial growth. A concentration of 300 mg/l of the staphylococcal nuclease (Nuc) was reached using the P170 promoter (Madsen *et al.*, 1999).

An alternative method for lactobacilli is the pSIP system: its promoters and regulatory genes are involved in the production of the class II bacteriocins sakacin A and P in *Lactobacillus sakei.* Toxin production in this system is controlled by a three-component system, composed of a secreted peptide pheromone (IP), a membrane-embedded histidine protein kinase (HPK) and a response regulator (RR), all encoded by the same operon. RR is activated by HPK, leading to activation of all promoters of the bacteriocin operon. The pSIP system has been employed to express numerous enzymes, such as β-glucuronidase in several *Lactobacillus* species (Sorvig *et al.*, 2003; Halbmayr *et al.*, 2008; Nguyen *et al.*, 2011).

Finally, a new expression system has recently been described by Benbouziane *et al.* (2013) and called SICE (Stress-Induced Controlled Expression). It operates without the presence of regulatory genes or additives like antibiotics and nisin (Benbouziane *et al.*, 2013). Under the control of the pGroESL promoter, SICE is activated under stressful conditions (temperature, acid pH and bile salts), such as the environment of the digestive tract. This system has already been certified *in vivo* by oral and intranasal administration. Moreover, two different murine models of human pathologies have already been investigated: (i) inflammatory bowel diseases (IBD) and (ii) human papillomavirus type-16 (HPV-16). Considering *in situ* expression in the gut this so far would be the expression system of choice.

Choosing the plasmid

The DNA fragments obtained by cleavage with restriction enzymes should be inserted in another DNA molecule capable of amplifying this genetic information in hundreds of copies. This amplification procedure is obtained through the use of DNA molecules called cloning vectors. Bacterial plasmids are the most common cloning vectors, since they are small circular extrachromosomal self-replicating DNA molecules that can be present in low or multicopy number inside the cells (Lodish, 2012).

The first gene cloning experiment described for *L. lactis* was performed with the pGB301 vector. Nevertheless, this plasmid was quite large (9.8 kb) and contained only a few unique restriction sites for gene cloning (Kondo and McKay, 1984). Since then, much progress has been made in the lactococcal plasmid biology area and a great diversity of plasmids, which vary in size, functions and host range have been described. Moreover, a better comprehension of plasmid replication and stability has improved the construction of efficient vector systems for *L. lactis* (Mills *et al.*, 2006).

Plasmid vectors usually consist of a replicon, a selectable marker and several restriction sites adjacent to the inserted fragment. The minimal replicon of a plasmid is composed of a replication origin (*ori* region) and a gene coding for a replication initiator protein (Rep). The *ori* region contains AT-rich regions generally followed by iterated initiator binding sites (iterons) and two small inverted repeats which overlap the ribosome binding site of the *rep* gene and the *rep* promoter. Iterons are essential not only for initiation of replication but also for regulating the plasmid replication (Mills *et al.*, 2006; Paulsson and Chattoraj, 2006).

Two classes of replicons have already been reported in lactococci: rolling circle and theta. The rolling circle replicon normally works in different bacterial hosts and contains replication regions for both Gram-positive and Gram-negative bacteria. However, replication is always unidirectional and it is segregationally unstable, leading to vector loss — especially when cultivated in non-selective media. On the other hand, theta replicons have limited host range and are generally kept at lower copy number inside the cell. Moreover, DNA synthesis may be uni- or bidirectional and may also begin from different origins. Besides, this replicon presents greater stability, making it the best choice for food-grade vectors (Sridhar *et al.*, 2006). Also from a safety point of view, the limited host range reduces the chance for gene/plasmid transfer to the endogenous microbiota.

In relation to selectable markers, antibiotic resistance genes have been the most widely used in laboratory research vectors, predominantly ampicillin, erythromycin and chloramphenicol. Nevertheless, these genes may be transferable and should not be used in the food or biomedical industry. Thus, alternative selection markers have been developed and currently two types of food-grade vectors exist: those with dominant and those with complementation selection. Dominant markers do not depend on the expression of host genes and can be used to promote plasmid maintenance by means of selection pressure (Emond *et al.*, 2001). Nisin and resistance to heavy metals (principally cadmium and copper) are good examples of dominant markers that have already been widely used in *L. lactis* plasmids (Liu *et al.*, 1997; Liu *et al.*, 1996; Zhou *et al.*, 2006), although the application of heavy metals for selection *in vivo* is undesirable. On the other hand, for complementation markers mutations are produced in the host chromosome in order to be complemented by plasmid-expressed markers. For instance, Bron and coworkers (2002) developed a system for the alanine racemase gene (*alr*) which catalyses the interconversion of D-alanine and L-alanine and hence complements *alr* mutants (Bron *et al.*, 2002). A very stringent selection has been observed with this system and plasmids could be stably cultivated for many generations.

Another important feature is plasmid stability, which is closely related to plasmid replication. Segregational instability occurs when a plasmid is completely lost, while structural instability is defined by the loss of specific DNA sequences. Rolling circle plasmids may present both kinds of instability, probably due to the formation of linear high molecular weight plasmid multimers (Gruss and Ehrlich, 1988). Similarly, theta replication may also produce plasmid multimers by homologous recombination, contributing to segregational instability (Summers and Sherratt, 1984).

Furthermore, plasmids replicate in harmony with their hosts and have negative feedback mechanisms to control their copy number, which may vary from a few to hundred copies per cell. In general, initiator limitation seems to be the central mode of replication control, focusing predominantly on limiting the synthesis or the action of the initiator protein (RepA) by transcriptional autorepression, initiator dimerization, initiator titration or 'handcuffing'. This protein contains DNA binding domains and leucine zipper-like motifs for dimer stabilization. Dimers may bind to iterons and inhibit the interaction between the RepA monomer and the inverted repeats, impeding initiation of replication (Giraldo *et al.*, 2003; Nordstrom and Dasgupta, 2006; Paulsson and Chattoraj, 2006).

Choosing the location of expressed proteins

Scientists have developed LAB capable of expressing heterologous proteins at different locations and observed that the immune response, the amount and/or the quality (solubility) of the protein produced could depend on the targeted protein location (Le Loir *et al.*, 1998; Norton *et al.*, 1995; Robinson *et al.*, 1997; Chatel *et al.*, 2001, 2003; Robinson *et al.*, 2004; Bermudez-Humaran *et al.*, 2005; Karlskas *et al.*, 2014).

Proteins produced in LAB can be secreted, remained inside the cell (cytoplasmic) or anchored to the cell surface, depending on the presence and location (N-terminus or C-terminus) of signal sequences in the construct (de Vos *et al.*, 1995; Le Loir *et al.*, 1998; Norton *et al.*, 1995; Robinson *et al.*, 1997; Chatel *et al.*, 2003; Robinson *et al.*, 2004; Bermudez-Humaran *et al.*, 2005).

Without any signal sequence present, the protein will remain in the cell's cytoplasm. In order to obtain a secreted protein the gene's N-terminus must be fused with the signal peptide of a secreted protein, like Usp45. In this case the protein is handled by the secretory machinery. To be anchored to the cell wall, two different sequence are required: (i) a signal peptide for secretion on the N-terminus and; (ii) a cell wall anchor (CWA) in the C-terminus, like the LPXTG motif (Dieye *et al.*, 2001; Medina *et al.*, 2010; Nouaille *et al.*, 2003).

It is believed that secretion is the best strategy to achieve high production yields of eukaryotic, viral and prokaryotic proteins in *L. lactis*, regardless of their native locations. However, the location, as well as stability, and the amount of recombinant proteins could be crucial to display an important immune response when this recombinant protein is used in medical and animal health areas.

L. lactis has a unique extracellular housekeeping protease (HTR A) involved in propeptide processing, maturation of native proteins and degradation of recombinant ones. In order to improve the production and secretion of heterologous proteins produced in *L. lactis*, a *L. lactis htrA* mutant was developed by inactivating this gene to stabilize the heterologous proteins. Authors confirmed that HTR A was responsible for protein degradation after translocation through the bacterial membrane and the inactivation of *htrA* enabled to increase both yields and stability of different recombinant proteins (Miyoshi *et al.*, 2002). In the same article they found that it was possible to increase the stability and the amount of recombinant proteins by fusion to Nuc.

Enouf and colleagues constructed a *L. lactis* strain able to secrete and to produce an intra-cytoplasmic NSP4 protein by using nisin-inducible expression system. NSP4, from rotavirus, is the major aetiological agent of severe diarrhoea in infants and young children around the world. This protein is supposed to act as an enterotoxin which triggers chloride secretion by calcium-dependent signal transduction pathway. It was the first time that an entire viral antigen was produced in *L. lactis* and they demonstrated that the rNSP4 protein produced in *L. lactis* had the same immunogenic properties as the viral protein. rNSP4 was secreted by using SP_{usp45} and lead to 10-fold-enhanced rNSP4 production compared with the intracellular production. The hypothesis explaining this difference was that the precursor (preUSP:rNSP4) could be more efficiently translated than the intracellular form of rNSP4 produced in *L. lactis* (Enouf *et al.*, 2001).

Le Loir and colleagues have shown that some secreted bacterial proteins are translocated as preproproteins. The LEISSTCDA propeptide, which is processed after translocation, may improve translocation efficiency (Le Loir *et al.*, 1998). To test this, Freitas and coworkers produced secreting Afp-1 (antifungal protein 1; 9.8 kDa) from *Streptomyces tandae* in *L. lactis*, which was the smallest recombinant protein secreted by *L. lactis* at that time. The maturation and folding of this protein was also a challenge as it contains disulfide bonds required for its anti-fungal activity. *L. lactis* was able to secret Afp-1 fused to LEISSTCDA but unfortunately LEISSTCDA:Afp-1 did not show antifungal activity, maybe due to the presence of an additional cysteine added by LEISSTCDA or introduction of some amino acids to the N-terminal moiety of mature Afp-1 (Freitas *et al.*, 2005).

A pioneer study regarding the influence of protein location on production and immune response was performed by Chatel and colleagues. They produced the bovine BLG, the major cow's milk allergen, to study the immune response to food allergy in mice. BLG was produced as cytoplasmic, secreted or fused form either to the LEISSTCDA propeptide or to Nuc. A correlation between the amount of protein secreted and the total amount of protein produced (secreted plus cytoplasmic) was described and the highest production was obtained with the BLG fused to Nuc. This form was able to induce specific IgA production detected in the faeces 3 weeks after oral or intranasal administration to mice (Chatel *et al.*, 2001). Moreover the BLG fused with Nuc was the only form able to protect the mice against any further sensitization (Adel-Patient *et al.*, 2005). Following, Chatel and colleagues constructed a *L. lactis* strain secreting an epitope of BLG fused to Nuc. Oral administration with *L. lactis* BLG and *L. lactis* BLG epitope were tested and both strains were able to induce faecal IgA specific-BLG production (Chatel *et al.*, 2003).

Another well-known study evaluated three recombinant *L. lactis* strains expressing the E7-HPV related protein (Bermudez-Humaran *et al.*, 2002). E7 is considered as a major candidate antigen for vaccination against HPV-related cancer, the second cause of cancer death in woman. This group produced also an E7-HPV protein located in the cytoplasm, secreted or anchored to the cell-wall. The first *in vivo* results of this model demonstrated that E7-specific cellular responses were higher in mice receiving the *L. lactis* strain producing an inducible cell-wall-anchored E7-HPV protein form. They observed that mice immunized with *L. lactis* expressing E7-anchored protein and *L. lactis* secreting IL-12 were able to reduce the progression of a HPV-related tumour elicited by intradermal injection of TC-1, a cancer cell line expressing E7 on the cell surface (Bermudez-Humaran *et al.*, 2002, 2005; Cortes-Perez *et al.*, 2005).

A recent study used *L. lactis* as a live vaccine against leishmaniasis, a parasitic disease affecting more than 12 million individuals worldwide. A modified version of the A2 antigen from *Leishmania donovani* was expressed in *L. lactis* as cytoplasmic, secreted, or cell wall-anchored form and the strains were tested for their capacity to generate A2-specific immune responses, as well as live vaccine against a *L. donovani* infection in mice. Subcutaneous immunization with *L. lactis* expressing the cell wall-anchored form of A2 induced the highest level of antigen-specific serum antibodies, while mice immunized with *L. lactis* producing the cytoplasmic form of A2 demonstrated the highest reduction in liver parasitaemia after visceral *L. donovani* challenge (Yam *et al.*, 2011).

Taken altogether, different strategies can be used to present antigens to the mucosa and it is important to know that the location of the produced protein will influence the amount, the conformation, and the maturation of the protein and could also trigger different types of immune responses. Generally, addition of a secretion signal will increase the amount of protein and guarantee proper folding of the secreted protein. Usually, the most immunogenic form is the anchored protein but regarding the desired immune response to be elicited, humoral or cellular, cytoplasmic localization has to be considered (Fig. 11.2).

Choosing the containment

Live, genetically modified bacteria might be a suitable way for topical delivery of recombinant proteins in the human gastrointestinal tract, but as such strategy starts to be applied, important questions regarding clinical and environmental safety, as well as technological feasibility, are emerging (Braat *et al.*, 2006). In the majority of live bacterial vaccines, the coding antigen is either located on plasmids or integrated into the chromosome. It is important to note that antibiotic resistance markers from plasmids encoding the genes of interest can be transferred to the environment (Lee, 2010). However, even if the gene of interest is integrated in the chromosome, spread and possible release of genetically modified microorganisms into the environment can also occur. Moreover, a substantial disadvantage of the integration in the chromosome is that generally one copy of the antigen gene will be expressed per bacterial cell, thus eliciting low heterologous production. To improve the safe use of LAB, some strategies of biological containment have been developed. Ideal genetically modified microorganism should contain the minimum amount of foreign DNA as possible and should not include any antibiotic resistant markers. In order to use them in the 'real world', some strategies need to be considered. In the meanwhile, no research

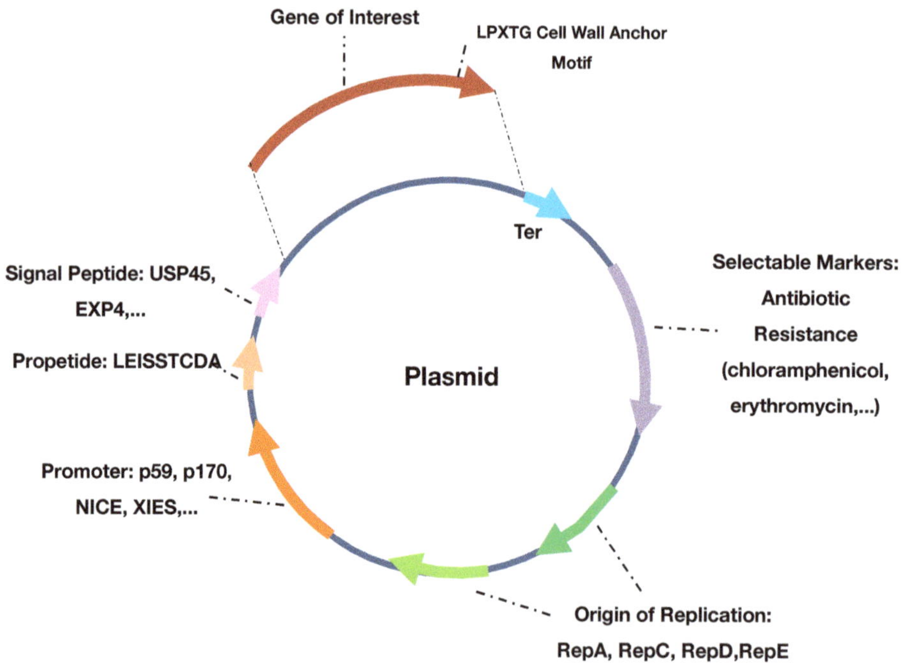

Figure 11.2 General plasmid map.

on human bacterial vaccines that have caused problems among animals has been found in the literature; however, it is possible that some cases exist (Lee, 2010; Bahey-El-Din and Gahan, 2010; Carleton, 2010).

According to the literature, biological containment systems can be generally classified as active or passive. Active systems consist of a conditional genetic control, by an environmental responsive element, or either activation of a killing gene/compound or repression of an essential gene. An inconvenience related with such systems is the large amount of foreign DNA required, which limits their use in humans. However, placing the antigen encoding genes in the bacterial chromosome may limit the spread of these genes. Another consideration is that the route of administration of the vaccine may also be important when evaluating hazards. As live bacterial vaccines are fit for mucosal administration, one must remember that ingestion of foreign DNA occurs. Finally, another drawback is related to mutations or inactivation of a killing gene/compound or a repressor gene that could result in constitutive expression of the essential gene (Lee, 2010; Detmer and Glenting, 2006).

Passive systems are simpler to allow controlling these limitations. These systems are based mainly on complementation of an auxotrophy or other gene defect by supplementation with either the intact gene or the essential metabolite. In this case, the absence of an antibiotic resistance marker in the final product would avoid its potential spread to the environment. The drawback is that these systems are often bacteriostatic rather than bactericidal. For instance, if a system is based on the replacement of a chromosomal thymidylate synthase (*thyA*) by the transgene it can be able to generate a thymine auxotrophy. This choice combines the advantages of passive and active containment systems. The *thyA* gene is essential for the synthesis of thymine and thymidine and in the absence of these compounds bacteria undergo thymine-less death and lysis. Thymidylate synthase is the enzyme that puts a methyl group on uracil or uridine to make thymine or thymidine. There is no bypass around this part of DNA biosynthesis. Additionally, fragmentation of the genomic DNA occurs before lysis, reducing the risk of genetic transfer even further. Even in the unlikely event in which *L. lactis* would acquire another lactococcal *thyA* gene by horizontal DNA transfer, recombination would lead to loss of the transgene (Lee, 2010). Bahey-El-Din *et al.* (2010) developed another biological containment system for a *L. lactis* vaccine strain using a similar pyrimidine synthesis knockout approach but involving the essential cytidine triphosphate (CTP) synthase gene. The *pyrG* gene of *L. lactis* was replaced by the gene encoding the listeriolysin antigen from *L. monocytogenes* through a double cross-over. The resulting auxotroph was effective as a vaccine inducing protective immunity against *L. monocytogenes* in mice (Bahey-El-Din *et al.*, 2010). As mentioned, having the antigen chromosomally integrated avoids the use of antibiotic selection markers and effectively eliminates the need for plasmid-based systems adding more stability (Lee, 2010; Detmer and Glenting, 2006). Steidler and colleagues used auxotrophic transgene containing method (ATCM) in *L. lactis* replacing *thyA* with a synthetic human *il-10* transgene (LL-Thy 12) approved by Dutch authorities as an experimental therapy for humans with IBD (Braat *et al.*, 2006; Steidler *et al.*, 2009).

Clinical trials

As previously mentioned, the pilot trial with LL-Thy 12 in patients with Crohn's disease showed that the containment strategy was highly effective. This kind of vaccine avoids the risk of environmental spread of the GMM (Steidler *et al.*, 2003). A phase I clinical trial was conducted using the LL-Thy12 strain that is thymidine or thymine dependent and unable to survive in the environment without continuous supplementation of either one of these compounds. This bacterium was administered in patients with Crohn's disease and showed highly promising results regarding efficacy, tolerability, and environmental containment (Braat *et al.*, 2006). Consequently, a large-scale randomized, double-blind, and placebo-controlled phase II clinical trial was performed to test this strain in patients with ulcerative colitis and showed that the containment strategy was efficient. However, unfortunately, this assay did not reveal a statistically significant difference in mucosal healing compared with the placebo group (ClinicalTrials.gov – NCT00729872). These *L. lactis* strains capable of delivering biologically active therapeutic molecules are commercially called Actobiotics^{TM}.

Another clinical trial also conducted by Actogenix assessed the safety and tolerability of genetically modified *L. lactis* bacteria engineered to secrete human trefoil factor 1 (hTFF1), and explored the ability of this recombinant bacterium to attenuate the course and severity of oral mucositis, which can develop in subjects receiving chemotherapy for the treatment of head and neck cancer. Trefoil factors have been shown to be important in protecting and healing mucosal tissues, such as the mouth tissues. Results showed that *L. lactis* secreting hTFF1 was safe and well tolerated and 30% of treated patients showed a decrease of oral mucositis symptoms compared with patients receiving the placebo (ClinicalTrials.gov – NCT00938080; Caluwaerts *et al.*, 2010; Limaye *et al.*, 2013).

Another factor that can be used to treat IBD is an anti-TNFα antibody. TNFα has been identified as playing a pivotal role in the inflammatory cascade that causes chronic intestinal inflammation in IBD. Synthetic anti-TNFα antibodies have shown to mitigate this inflammatory process. Curiously, IBD can display extra-intestinal damage, such as neurological disorders associated with IBD and TNFα inhibitors already approved to treat IBD could help to avoid these extra-intestinal disorders (Moris, 2014). Coppieters and coworkers developed formatted anti-TNFα single domain antibody fragments (nanobodies), derived from heavy-chain camelid antibodies as therapeutic agent in murine rheumatoid arthritis (Coppieters *et al.*, 2006). These molecules can be cloned and produced as recombinant protein in bacteria and yeast. *L. lactis* anti-murine TFNα nanobodies treatment efficiently ameliorated both DSS-induced colitis and established enterocolitis in IL10$^{-/-}$ mice. Interestingly, the small quantity of nanobodies delivered by *L. lactis* was highly efficient in reducing signs of enterocolitis. These results encouraged Actogenics to develop an oral capsule formulation controlling the release of metabolically active *L. lactis* bacteria in such a way that it can target inflammation at the lower part of the small intestine and in the colon, making it suitable for treatment of Crohn's disease and ulcerative colitis (Vandenbroucke *et al.*, 2010). These nanobodies could also be used in clinical trials to target other inflammatory diseases.

As presented, LAB can be safely used to treat or prevent diseases, as they are capable to express stable proteins and modulate both mucosal and systemic immune systems. To obtain the best immune response, it is important to choose the correct LAB, the type of plasmid and/or expression system to produce and express the target protein avoiding its spread into the environment and the best cellular location for protein targeting.

Lactic acid bacteria as DNA delivery vehicles

Food and commensal LAB are now also being used as DNA delivery vehicles to develop DNA vaccination strategies. DNA vaccination, also called genetic vaccination, consists in the *in situ* administration of a plasmid containing an expression cassette composed of an eukaryotic promoter driving the expression of a protein of interest. This strategy was first described 20 years ago (Tang *et al.*, 1992; Ulmer *et al.*, 1993). The advantage of DNA vaccines relies in their ability to induce both cellular and humoral Th1 immune responses (Raz *et al.*, 1996). This ability of DNA vaccination to elicit a Th1 profile raised interest of scientists working with allergies, a typically Th2-oriented immune response disease. Bacteria-mediated delivery of DNA vaccines lead to the expression of post-translationally modified antigens by host cells resulting in presentation of conformationally restricted epitopes to the immune system, which could be a problem for bacteria-mediated delivery of proteins. As for protein delivery, the use of food-grade lactococci and lactobacilli as DNA delivery vehicles is a promising alternative to attenuated pathogens.

Genetic immunization using LAB as vectors requires a plasmid which: (i) replicates in the LAB vehicle, and (ii) contains an eukaryotic expression cassette composed of the gene of interest under the control of an eukaryotic promoter region and a transcriptional terminator (poly A) required to establish the primary mRNA transcript. Different functional plasmids have been constructed for DNA delivery using *L. lactis* as a delivery vector (Chatel *et al.*, 2008; Guimaraes *et al.*, 2009; Guimaraes *et al.*, 2006; Tao *et al.*, 2011).

One of the first vectors for DNA delivery by LAB was an *E. coli–L. lactis* shuttle vector called pLIG:BLG1 containing a eukaryotic expression cassette with cDNA of BLG, a major milk allergen (Chatel *et al.*, 2008). The construction resulted from the fusion between the pcDNA3-BLG (Adel-Patient *et al.*, 2001) and pIL253 (Simon and Chopin, 1988) vectors. The BLG coding ORF was under the control of the cytomegalovirus eukaryotic promoter (pCMV). pLIG:BLG1 was transformed into *L. lactis* resulting in LL(pLIG:BLG1). After co-incubation of LL(pLIG:BLG1) with Caco-2 cells, a human epithelial cell line, both BLG cDNA and BLG protein were detected in Caco-2 cells or their supernatant. This was the first demonstration that native *L. lactis* could deliver a DNA plasmid *in vitro* (Guimaraes *et al.*, 2006). The effectiveness of this DNA delivery vehicle was also demonstrated *in vivo* where both BLG cDNA and BLG protein under native conformation were detected in the small intestine in half the mice of the experiment after oral immunization with LL(pLIG:BLG1). BLG expression was transitory as it disappeared 4 days after the last administration, suggesting a plasmid transfer into enterocytes. *In situ* BLG expression elicited a low and transitory BLG-specific Th1-type primary immune response characterized by BLG-specific IgG2a in sera. Moreover, mice that previously received LL(pLIG:BLG1) were protected against further sensitization (Chatel *et al.*, 2008). However, owing to the large size of the pLIG vector (10 kb), cloning and transformation procedures in lactococci proved to be complicated.

To overcome this problem, a new and smaller shuttle vector (3742 bp) called pValac was developed for its use in *L. lactis* (Guimaraes *et al.*, 2009). This plasmid was assembled by the fusion of (i) a eukaryotic region containing the pCMV promoter, to permit expression of the ORF of interest by eukaryotic cells, and the polyadenylation signal of bovine growth hormone (BGH poly A) to establish the mRNA transcript, and (ii) a prokaryotic region containing the replication origin and the resistance marker gene for bacteria selection. The *gfp* ORF was cloned into the pValac to evaluate its capacity as DNA delivery vector and porcine epithelial cells (PK15) were transfected with the pValac:*gfp* and were able to produce GFP protein, demonstrating the functionality of this vector *in vitro* (Guimaraes *et al.*, 2009).

At this step it was hypothesized that plasmid transfer could occur by two mechanisms: (i) lysis of the bacteria, release of the plasmids and capture of the plasmid by eukaryotic cells of the epithelial membrane, or (ii) capture of the bacteria by eukaryotic cells of the epithelial membrane and release of the plasmid in their inside.

In order to test the second hypothesis and improve the efficiency of LAB as DNA delivery vehicles, our research group evaluated the effects of lactococci expressing invasins at their surface. Invasins are proteins from pathogens that confer them invasive abilities like Internalin A (InlA) from *Listeria monocytogenes*, or fibronectin-binding protein A (FnBPA) from *Staphylococcus aureus*. Recombinant lactococci efficiently displayed the cell-wall-anchored form of InlA. Expression of InlA led to internalization of *L. lactis* InlA+ in Caco-2 cells. This strain was capable of invading and delivering a functional eukaryotic *gfp* gene to Caco-2 cells and invade guinea pig intestinal cells after oral inoculation (Bermudez-Humaran *et al.*, 2005). But the use of InlA is restricted *in vivo* to guinea pigs as InlA has low affinity for its receptor, E-cadherin, in mice. In this regard, a *L. lactis* strain was engineered to express FnBPA, which mediates bacterial adhesion to fibrinogen, elastin and fibronectin (Menzies, 2003). Expression of FnBPA conferred invasiveness to lactococci, both *in vitro* and *in vivo* (Innocentin *et al.*, 2009; Que *et al.*, 2005). *In vitro*, the expression of FnBPA increased by 1000 times the invasiveness of the strain compared to the wild type (wt) strain (Innocentin *et al.*, 2009) and also showed to be more efficient of transferring a eukaryotic GFP expression plasmid to Caco-2 cells (Innocentin *et al.*, 2009). More recently it was also shown that LL-(FnBPA+) carrying the pValac:BLG plasmid was able to transfer 30 times more plasmid than the wt lactococci *in vitro* (Pontes *et al.*, 2012). *In vivo*, the use of *L. lactis* expressing FnBPA increased the number of mice producing BLG, but not the level of BLG produced (Pontes *et al.*, 2012). Using GFP as reporter gene and fluorescence microscopy it was demonstrated that plasmid transfer occurs in enterocytes using either strain (FNBPA+ or wt), demonstrating that the production of FnBPA at the surface of lactococci does not modify the localization of plasmid transfer but increases the number of cells that receive it.

A mutated form of InlA (mInlA) developed to bind with a better efficiency than InlA to murine E-cadherin (Wollert *et al.*, 2007) was also tested. mInlA was expressed at the surface of recombinant lactococci strains and its production increased the invasiveness of the recombinant strain at the same rate as the other recombinant invasive strain (LL-FnBPA+ and LL-InlA+). This enhanced invasiveness was correlated to an increased plasmid transfer *in vitro*. Unfortunately, the plasmid transfer did not increase *in vivo* when compared with the wt and LL– mInlA+ strains (de Azevedo *et al.*, 2012a).

Thus, the production of invasins at the lactococcal surface does not drastically change the plasmid transfer, although the LL– FnBPA+ strain increases the number of mice that receive the plasmid. This could mean that direct bacteria–host interactions are not the main way for plasmid transfer. Among various hypothesis it could indicate that plasmid transfer is probably more a result of plasmid release and capture by cells after bacterial lysis than direct plasmid release inside eukaryotic cells after bacterial uptake.

Our research group also tested if the production of FnBPA at the surface of the recombinant strain would modify the level or the immune response type elicited after oral administration. It was shown that oral or intranasal administration of non-invasive lactococci carrying pValac:BLG elicited a Th1 immune response whereas the invasive strain expressing FnBPA elicited a Th2 immune response (Pontes *et al.*, 2014), showing that the expression of FnBPA at the surface of lactococci deeply impacts the immune response. Thus, care should be taken in choosing such an approach.

Plasmid transfer was also used to deliver interleukin 10 (IL-10), a crucial player in gut immune homeostasis. Interleukin-10 has already been produced with success in *L. lactis* (see previous section) interleukin-10 cDNA under the control of the CMV promoter was successfully transferred to epithelial cells by LL-FnBPA+ and evaluated in a TNBS-induced mouse model of experimental colitis. Mice treated with LL-FnBPA+ strain carrying pValac:IL-10 were protected against inflammation regarding macroscopic and histological score or pro-inflammatory/anti-inflammatory cytokine ratios (Del Carmen *et al.*, 2013). This work demonstrated for the first time that plasmid transfer can also be used to deliver active molecules.

A novel *E. coli*-LAB shuttle plasmid vector called pLKV1 was also developed for DNA-delivery into eukaryotic cells (Tao *et al.*, 2011). This plasmid has some advantages over many other existing plasmids for DNA vaccine delivery such as: (i) expression in both *E. coli* and several LAB species, including *Lactococcus, Lactobacillus, Streptococcus* and *Weisseilla*; (ii) very small size (4.4 kb), so it can be used to clone genes of relatively larger sizes and be easily transformed into various bacterial species; (iii) high copy number (about 65 per cell), and (iv) kanamycin resistance marker that expresses in both Gram-negative and Gram-positive bacteria. This feature is desirable as kanamycin is recommended by the FDA for use in DNA vaccine constructions when resistance markers are required due to presenting a better safety record compared to other antibiotic resistance markers. This new plasmid vector has been successfully tested *in vitro* by *L. lactis* delivering a red fluorescent reporter gene into cultured Caco-2 cells (Tao *et al.*, 2011), but as far as we know no *in vivo* results have been published yet.

Lactobacillus acidophilus was used to deliver a plasmid containing the cDNA of VP1, one of the four structural proteins of the foot and mouth disease virus (FMDV) (Li *et al.*, 2007). After oral or intranasal administration the immune response, characterized by serum antibodies, was very low compared to intraperitoneal or intramuscular administration. Surprisingly, authors described that VP1 cDNA could be detected in various tissues including kidney, spleen and heart. This is probably linked to the type of administration which was intramuscular. However, our research group has never been able to detect BLG cDNA in other tissues than the gut when orally administrating the invasive or non-invasive strains.

Future trends

New molecules

RNAi

Recombinant LAB are already able to deliver various sized proteins and plasmids. One of the new leads include RNA interference (RNAi). RNAi was first described in the *Caenorhabditis elegans* worm (Fire *et al.*, 1998) and since then it has rapidly become one of the most powerful and widely used tools for the study of gene function. RNAi can be defined as a mechanism of specific gene-silencing. RNAi is mediated by small RNAs such as microRNA (miRNA), short hairpin RNA (shRNA) or small interfering RNA (siRNA) (Deng *et al.*, 2014). These small RNAs can be synthesized and delivered *in vivo* by viral or non viral systems but also using bacteria. Bacteria can be used to deliver plasmids to mammalian cells with the coding sequence of interfering RNA or to produce the RNA itself, as it has already been done with *E. coli* (Xiang *et al.*, 2006). LAB are ideal vectors for this type of molecules. Plasmid delivery is already available with LAB but production of small RNAs inside the bacteria still requires a lot of work on the selection of the right promoter.

Antibodies

Antibodies are tetrameric molecules composed of two heavy and two light chains and are therefore very difficult to produce heterologously. Over 20 antibodies are now approved by the FDA as therapeutics against cancer, allergy or Crohn's disease and many more are currently being tested in clinical trials. Single chain Fragment variable (scFv) consists of variable regions of heavy (VH) and light (VL) chains, which are joined together by a flexible peptide linker (Ahmad *et al.*, 2012). ScFv have already been successfully produced in lactobacilli (Andersen *et al.*, 2011; Chancey *et al.*, 2006; Kruger *et al.*, 2002; Marcotte *et al.*, 2006; Martin *et al.*, 2011) and have been mainly used to trap pathogens or toxins in a strategy of passive immunity. Moreover, these ScFv could also be used to target recombinant LAB towards specific cellular subtypes of the epithelial membrane such as M cells or DCs. M cells are specialized cells for antigens or bacteria sampling in the small intestine while DCs are antigen-presenting cells (APCs); thus, the targeting of these cells could boost the immunological response

Peptide

Peptides are very powerful molecules involved in all essential life processes including immunology, neurobiology, microbiology, development, metabolism and so on. Peptides are pretty difficult to produce in bacteria, but the fields of application are so huge that it is worth the try. Moreover, peptides could be produced in fusion with a carrier protein or as a multimer linked by a protease sequence.

New bacteria

The idea would be to have a synergy between the produced molecule and the bacterial vector. In this regard, 'traditional' probiotics, such as some species of LAB, could be used, but a new category of probiotics, the gut commensals, are emerging. Commensals from the gut are usually very difficult to cultivate because they are strictly anaerobic or extremely sensitive to oxygen. Recent findings using molecular biology and metagenomic approaches have increased the interest of microbiologists and immunologists for their role in intestinal homeostasis and their potential probiotic effects. Some of them, such as *Faecalibacterium prausnitzii*, *Bacteroides fragilis* or *Akkermansia muciniphila*, have shown strong effects on the immune system in various mice model that range from acute inflammation to obesity (Everard *et al.*, 2013; Mazmanian *et al.*, 2005; Sokol *et al.*, 2008).

New plasmids

The plasmid copy number (PCN) is essential especially for plasmid transfer strategies. Studies have to be performed to determine the best origin of replication, the best host and the best culture conditions to get the highest PCN.

Conclusion

As reviewed, much research has focused on the development of different strategies to improve heterologous protein production and advances have been achieved. Nevertheless, for both areas and depending on the desired output, it is always required to evaluate the chosen bacteria, plasmid, protein location and containment strategy.

Although much progress has been done, many challenges still have to be faced, improved and resolved and many groups are currently focusing on testing new molecules, new bacteria and new plasmids. Despite a lot of effort among many groups just one strain, the IL-10-producing *Lactococcus*, has been used in a clinical trial, unfortunately without success. Gratitude will come if we succeed to perform a clinical trial with positive outcome. To do so, the right molecule has to be produced in the right bacterium. Each step is important and needs to be optimized. That's our challenge for the future if we want to move from mouse to man.

References

Adachi, K., Kawana, K., Yokoyama, T., Fujii, T., Tomio, A., Miura, S., Tomio, K., Kojima, S., Oda, K., Sewaki, T., *et al.* (2010). Oral immunization with a *Lactobacillus* casei vaccine expressing human papillomavirus (HPV) type 16 E7 is an effective strategy to induce mucosal cytotoxic lymphocytes against HPV16 E7. Vaccine 28, 2810–2817.

Adel-Patient, K., Ah-Leung, S., Creminon, C., Nouaille, S., Chatel, J.M., Langella, P., and Wal, J.M. (2005). Oral administration of recombinant *Lactococcus lactis* expressing bovine beta-lactoglobulin partially prevents mice from sensitization. Clin. Exp. Allergy 35, 539–546.

Adel-Patient, K., Creminon, C., Boquet, D., Wal, J.M., and Chatel, J.M. (2001). Genetic immunisation with bovine beta-lactoglobulin cDNA induces a preventive and persistent inhibition of specific anti-BLG IgE response in mice. IntArchAllergy Immunol. *126*, 59–67.

Ahmad, Z.A., Yeap, S.K., Ali, A.M., Ho, W.Y., Alitheen, N.B., and Hamid, M. (2012). scFv antibody: principles and clinical application. Clin. Dev Immunol. *2012*, 980250.

Andersen, K.K., Marcotte, H., Alvarez, B., Boyaka, P.N., and Hammarstrom, L. (2011). In situ gastrointestinal protection against anthrax edema toxin by single-chain antibody fragment producing lactobacilli. BMC Biotechnol. *11*, 126.

Bahey-El-Din, M., Casey, P.G., Griffin, B.T., and Gahan, C.G. (2010). Efficacy of a *Lactococcus lactis* DeltapyrG vaccine delivery platform expressing chromosomally integrated hly from Listeria monocytogenes. Bioeng. Bugs *1*, 66–74.

Bahey-El-Din, M., and Gahan, C.G. (2010). *Lactococcus lactis*: from the dairy industry to antigen and therapeutic protein delivery. Discov Med. *9*, 455–461.

Benbouziane, B., Ribelles, P., Aubry, C., Martin, R., Kharrat, P., Riazi, A., Langella, P., and Bermudez-Humaran, L.G. (2013). Development of a Stress-Inducible Controlled Expression (SICE) system in *Lactococcus lactis* for the production and delivery of therapeutic molecules at mucosal surfaces. J. Biotechnol. *168*, 120–129.

Bermudez-Humaran, L.G. (2009). *Lactococcus lactis* as a live vector for mucosal delivery of therapeutic proteins. Hum. Vaccin. *5*, 264–267.

Bermudez-Humaran, L.G., Cortes-Perez, N.G., Lefevre, F., Guimaraes, V., Rabot, S., Alcocer-Gonzalez, J.M., Gratadoux, J.J., Rodriguez-Padilla, C., Tamez-Guerra, R.S., Corthier, G., *et al.* (2005). A novel mucosal vaccine based on live Lactococci expressing E7 antigen and IL-12 induces systemic and mucosal immune responses and protects mice against human papillomavirus type 16-induced tumors. J. Immunol. *175*, 7297–7302.

Bermudez-Humaran, L.G., Langella, P., Cortes-Perez, N.G., Gruss, A., Tamez-Guerra, R.S., Oliveira, S.C., Saucedo-Cardenas, O., Montes de Oca-Luna, R., and Le Loir, Y. (2003). Intranasal immunization with recombinant *Lactococcus lactis* secreting murine interleukin-12 enhances antigen-specific Th1 cytokine production. Infect. Immun. *71*, 1887–1896.

Bermudez-Humaran, L.G., Langella, P., Miyoshi, A., Gruss, A., Guerra, R.T., Montes de Oca-Luna, R., and Le Loir, Y. (2002). Production of human papillomavirus type 16 E7 protein in *Lactococcus lactis*. Appl. Environ. Microbiol. *68*, 917–922.

Bolotin, A., Wincker, P., Mauger, S., Jaillon, O., Malarme, K., Weissenbach, J., Ehrlich, S.D., and Sorokin, A. (2001). The complete genome sequence of the lactic acid bacterium *Lactococcus lactis* subsp. lactis IL1403. Genome Res. *11*, 731–753.

Braat, H., Rottiers, P., Hommes, D.W., Huyghebaert, N., Remaut, E., Remon, J.P., van Deventer, S.J., Neirynck, S., Peppelenbosch, M.P., and Steidler, L. (2006). A phase I trial with transgenic bacteria expressing interleukin-10 in Crohn's disease. Clin. Gastroenterol. Hepatol. *4*, 754–759.

Bron, P.A., Benchimol, M.G., Lambert, J., Palumbo, E., Deghorain, M., Delcour, J., De Vos, W.M., Kleerebezem, M., and Hols, P. (2002). Use of the alr gene as a food-grade selection marker in lactic acid bacteria. Appl. Environ. Microbiol. *68*, 5663–5670.

Caluwaerts, S., Vandenbroucke, K., Steidler, L., Neirynck, S., Vanhoenacker, P., Corveleyn, S., Watkins, B., Sonis, S., Coulie, B., and Rottiers, P. (2010). AG013, a mouth rinse formulation of *Lactococcus lactis* secreting human Trefoil Factor 1, provides a safe and efficacious therapeutic tool for treating oral mucositis. Oral Oncol. *46*, 564–570.

Carleton, H.A. (2010). Pathogenic bacteria as vaccine vectors: teaching old bugs new tricks. Yale J. Biol. Med. *83*, 217–222.

Chancey, C.J., Khanna, K.V., Seegers, J.F., Zhang, G.W., Hildreth, J., Langan, A., and Markham, R.B. (2006). Lactobacilli-expressed single-chain variable fragment (scFv) specific for intercellular adhesion molecule 1 (ICAM-1). blocks cell-associated HIV-1 transmission across a cervical epithelial monolayer. J. Immunol. *176*, 5627–5636.

Chatel, J.M., Langella, P., Adel-Patient, K., Commissaire, J., Wal, J.M., and Corthier, G. (2001). Induction of mucosal immune response after intranasal or oral inoculation of mice with *Lactococcus lactis* producing bovine beta-lactoglobulin. Clin. DiagnLab Immunol. *8*, 545–551.

Chatel, J.M., Nouaille, S., Adel-Patient, K., Le Loir, Y., Boe, H., Gruss, A., Wal, J.M., and Langella, P. (2003). Characterization of a *Lactococcus lactis* strain that secretes a major epitope of bovine beta-lactoglobulin and evaluation of its immunogenicity in mice. Appl. Environ. Microbiol. *69*, 6620–6627.

Chatel, J.M., Pothelune, L., Ah-Leung, S., Corthier, G., Wal, J.M., and Langella, P. (2008). *In vivo* transfer of plasmid from food-grade transiting lactococci to murine epithelial cells. Gene Ther. *15*, 1184–1190.

Coppieters, K., Dreier, T., Silence, K., de Haard, H., Lauwereys, M., Casteels, P., Beirnaert, E., Jonckheere, H., Van de Wiele, C., Staelens, L., *et al.* (2006). Formatted anti-tumor necrosis factor alpha VHH proteins derived from camelids show superior potency and targeting to inflamed joints in a murine model of collagen-induced arthritis. Arthritis Rheum. *54*, 1856–1866.

Cortes-Perez, N.G., Azevedo, V., Alcocer-Gonzalez, J.M., Rodriguez-Padilla, C., Tamez-Guerra, R.S., Corthier, G., Gruss, A., Langella, P., and Bermudez-Humaran, L.G. (2005). Cell-surface display of E7 antigen from human papillomavirus type-16 in *Lactococcus lactis* and in *Lactobacillus plantarum* using a new cell-wall anchor from lactobacilli. J. Drug Target. *13*, 89–98.

de Azevedo, M., Karczewski, J., Lefevre, F., Azevedo, V., Miyoshi, A., Wells, J.M., Langella, P., and Chatel, J.M. (2012a). *In vitro* and *in vivo* characterization of DNA delivery using recombinant *Lactococcus lactis* expressing a mutated form of L. monocytogenes Internalin A. BMC Microbiol. *12*, 299.

de Azevedo, M.S., Rocha, C.S., Electo, N., Pontes, D.S., Molfetta, J.B., Goncalves, E.D., Azevedo, V., Silva, C.L., and Miyoshi, A. (2012b). Cytoplasmic and extracellular expression of pharmaceutical-grade mycobacterial 65-kDa heat shock protein in *Lactococcus lactis*. Genet. Mol. Res. *11*, 1146–1157.

de Vos, W.M., Kuipers, O.P., van der Meer, J.R., and Siezen, R.J. (1995). Maturation pathway of nisin and other lantibiotics: post-translationally modified antimicrobial peptides exported by Gram-positive bacteria. Mol. Microbiol. *17*, 427–437.

Del Carmen, S., Zurita-Turk, M., Lima, F.A., Coelho Dos Santos, J.S., Leclercq, S.Y., Chatel, J.M., Azevedo, V., De Moreno De LeBlanc, A., Miyoshi, A., and LeBlanc, J.G. (2013). A novel Interleukin-10 DNA mucosal delivery system attenuates intestinal inflammation in a mouse model. Eur. J. Inflamm. *11*, 641–654.

Deng, Y., Wang, C.C., Choy, K.W., Du, Q., Chen, J., Wang, Q., Li, L., Chung, T.K., and Tang, T. (2014). Therapeutic potentials of gene silencing by RNA interference: Principles, challenges, and new strategies. Gene *538*, 217–227.

Detmer, A., and Glenting, J. (2006). Live bacterial vaccines – a review and identification of potential hazards. Microb. Cell Fact. *5*, 23.

Dieye, Y., Usai, S., Clier, F., Gruss, A., and Piard, J.C. (2001). Design of a protein-targeting system for lactic acid bacteria. J. Bacterol. *183*, 4157–4166.

Emond, E., Lavallee, R., Drolet, G., Moineau, S., and LaPointe, G. (2001). Molecular characterization of a theta replication plasmid and its use for development of a two-component food-grade cloning system for *Lactococcus lactis*. Appl. Environ. Microbiol. *67*, 1700–1709.

Enouf, V., Langella, P., Commissaire, J., Cohen, J., and Corthier, G. (2001). Bovine rotavirus nonstructural protein 4 produced by *Lactococcus lactis* is antigenic and immunogenic. Appl. Environ. Microbiol. *67*, 1423–1428.

Everard, A., Belzer, C., Geurts, L., Ouwerkerk, J.P., Druart, C., Bindels, L.B., Guiot, Y., Derrien, M., Muccioli, G.G., Delzenne, N.M., *et al.* (2013). Cross-talk between *Akkermansia muciniphila* and intestinal epithelium controls diet-induced obesity. Proc. Natl. Acad. Sci. U.S.A. *110*, 9066–9071.

Ferrer-Miralles, N., and Villaverde, A. (2013). Bacterial cell factories for recombinant protein production; expanding the catalogue. Microb. Cell Fact. *12*, 113.

Fire, A., Xu, S., Montgomery, M.K., Kostas, S.A., Driver, S.E., and Mello, C.C. (1998). Potent and specific genetic interference by double-stranded RNA in Caenorhabditis elegans. Nature *391*, 806–811.

Francia, M.V., and Clewell, D.B. (2002). Transfer origins in the conjugative Enterococcus faecalis plasmids pAD1 and pAM373: identification of the pAD1 nic site, a specific relaxase and a possible TraG-like protein. Mol. Microbiol. *45*, 375–395.

Freitas, D.A., Leclerc, S., Miyoshi, A., Oliveira, S.C., Sommer, P.S., Rodrigues, L., Correa Junior, A., Gautier, M., Langella, P., Azevedo, V.A., *et al.* (2005). Secretion of Streptomyces tendae antifungal protein 1 by *Lactococcus lactis*. Braz. J. Med. Biol. Res. *38*, 1585–1592.

Giraldo, R., Fernandez-Tornero, C., Evans, P.R., Diaz-Orejas, R., and Romero, A. (2003). A conformational switch between transcriptional repression and replication initiation in the RepA dimerization domain. Nat. Struct. Biol. *10*, 565–571.

Gruss, A., and Ehrlich, S.D. (1988). Insertion of foreign DNA into plasmids from Gram-positive bacteria induces formation of high-molecular-weight plasmid multimers. J. Bacteriol. *170*, 1183–1190.

Guimaraes, V., Innocentin, S., Chatel, J.M., Lefevre, F., Langella, P., Azevedo, V., and Miyoshi, A. (2009). A new plasmid vector for DNA delivery using lactococci. Genet. Vaccines Ther. *7*, 4.

Guimaraes, V.D., Innocentin, S., Lefevre, F., Azevedo, V., Wal, J.M., Langella, P., and Chatel, J.M. (2006). Use of native lactococci as vehicles for delivery of DNA into mammalian epithelial cells. Appl. Environ. Microbiol. *72*, 7091–7097.

Halbmayr, E., Mathiesen, G., Nguyen, T.H., Maischberger, T., Peterbauer, C.K., Eijsink, V.G., and Haltrich, D. (2008). High-level expression of recombinant beta-galactosidases in *Lactobacillus plantarum* and *Lactobacillus* sakei using a Sakacin P-based expression system. J. Agric. Food Chem. *56*, 4710–4719.

Hazebrouck, S., Pothelune, L., Azevedo, V., Corthier, G., Wal, J.M., and Langella, P. (2007). Efficient production and secretion of bovine beta-lactoglobulin by *Lactobacillus* casei. Microb. Cell Fact. *6*, 12.

Hugentobler, F., Di Roberto, R.B., Gillard, J., and Cousineau, B. (2012). Oral immunization using live *Lactococcus lactis* co-expressing LACK and IL-12 protects BALB/c mice against *Leishmania* major infection. Vaccine *30*, 5726–5732.

Innocentin, S., Guimaraes, V., Miyoshi, A., Azevedo, V., Langella, P., Chatel, J.M., and Lefevre, F. (2009). *Lactococcus lactis* expressing either *Staphylococcus aureus* fibronectin-binding protein A or *Listeria monocytogenes* internalin A can efficiently internalize and deliver DNA in human epithelial cells. Appl. Environ. Microbiol. *75*, 4870–4878.

Ivory, K., Chambers, S.J., Pin, C., Prieto, E., Arques, J.L., and Nicoletti, C. (2008). Oral delivery of *Lactobacillus* casei Shirota modifies allergen-induced immune responses in allergic rhinitis. Clin. Exp. Allergy *38*, 1282–1289.

Iwaki, M., Okahashi, N., Takahashi, I., Kanamoto, T., Sugita-Konishi, Y., Aibara, K., and Koga, T. (1990). Oral immunization with recombinant *Streptococcus lactis* carrying the Streptococcus mutans surface protein antigen gene. Infect. Immun. *58*, 2929–2934.

Kajikawa, A., Zhang, L., Long, J., Nordone, S., Stoeker, L., LaVoy, A., Bumgardner, S., Klaenhammer, T., and Dean, G. (2012). Construction and immunological evaluation of dual cell surface display of HIV-1 gag and *Salmonella enterica* serovar Typhimurium FliC in *Lactobacillus* acidophilus for vaccine delivery. Clin. Vaccine Immunol. *19*, 1374–1381.

Karlskas, I.L., Maudal, K., Axelsson, L., Rud, I., Eijsink, V.G., and Mathiesen, G. (2014). Heterologous Protein Secretion in Lactobacilli with Modified pSIP Vectors. PLoS One *9*, e91125.

Khokhlova, E.V., Efimov, B.A., Kafarskaia, L.I., and Shkoporov, A.N. (2010). Heterologous expression of secreted biologically active human interleukin-10 in *Bifidobacterium breve*. Arch. Microbiol. *192*, 769–774.

Kondo, J.K., and McKay, L.L. (1984). Plasmid transformation of *Streptococcus lactis* protoplasts: optimization and use in molecular cloning. Appl. Environ. Microbiol. *48*, 252–259.

Kruger, C., Hu, Y., Pan, Q., Marcotte, H., Hultberg, A., Delwar, D., van Dalen, P.J., Pouwels, P.H., Leer, R.J., Kelly, C.G., *et al.* (2002). In situ delivery of passive immunity by lactobacilli producing single-chain antibodies. Nat. Biotechnol. *20*, 702–706.

Le Loir, Y., Gruss, A., Ehrlich, S.D., and Langella, P. (1998). A nine-residue synthetic propeptide enhances secretion efficiency of heterologous proteins in *Lactococcus lactis*. J. Bacteriol. *180*, 1895–1903.

Lebeer, S., Claes, I.J., and Vanderleyden, J. (2012). Anti-inflammatory potential of probiotics: lipoteichoic acid makes a difference. Trends Microbiol. *20*, 5–10.

LeBlanc, J.G., Aubry, C., Cortes-Perez, N.G., de Moreno de LeBlanc, A., Vergnolle, N., Langella, P., Azevedo, V., Chatel, J.M., Miyoshi, A., and Bermudez-Humaran, L.G. (2013). Mucosal targeting of therapeutic molecules using genetically modified lactic acid bacteria: an update. FEMS Microbiol. Lett. *344*, 1–9.

Lee, P. (2010). Biocontainment strategies for live lactic acid bacteria vaccine vectors. Bioeng. Bugs *1*, 75–77.

Li, Y.G., Tian, F.L., Gao, F.S., Tang, X.S., and Xia, C. (2007). Immune responses generated by *Lactobacillus* as a carrier in DNA immunization against foot-and-mouth disease virus. Vaccine 25, 902–911.

Limaye, S.A., Haddad, R.I., Cilli, F., Sonis, S.T., Colevas, A.D., Brennan, M.T., Hu, K.S., and Murphy, B.A. (2013). Phase 1b, multicenter, single blinded, placebo-controlled, sequential dose escalation study to assess the safety and tolerability of topically applied AG013 in subjects with locally advanced head and neck cancer receiving induction chemotherapy. Cancer *119*, 4268–4276.

Liu, C.Q., Khunajakr, N., Chia, L.G., Deng, Y.M., Charoenchai, P., and Dunn, N.W. (1997). Genetic analysis of regions involved in replication and cadmium resistance of the plasmid pND302 from *Lactococcus lactis*. Plasmid 38, 79–90.

Liu, C.Q., Leelawatcharamas, V., Harvey, M.L., and Dunn, N.W. (1996). Cloning vectors for lactococci based on a plasmid encoding resistance to cadmium. Curr. Microbiol. *33*, 35–39.

Lodish, H.F. (2012). Translational control of protein synthesis: the early years. J. Biol. Chem. *287*, 36528–36535.

Ma, Y., Luo, Y., Huang, X., Song, F., and Liu, G. (2012). Construction of Bifidobacterium infantis as a live oral vaccine that expresses antigens of the major fimbrial subunit (CfaB) and the B subunit of heat-labile enterotoxin (LTB) from enterotoxigenic *Escherichia coli*. Microbiology *158*, 498–504.

Madsen, S.M., Arnau, J., Vrang, A., Givskov, M., and Israelsen, H. (1999). Molecular characterization of the pH-inducible and growth phase-dependent promoter P170 of *Lactococcus lactis*. Mol. Microbiol. *32*, 75–87.

Marcotte, H., Koll-Klais, P., Hultberg, A., Zhao, Y., Gmur, R., Mandar, R., Mikelsaar, M., and Hammarstrom, L. (2006). Expression of single-chain antibody against RgpA protease of *Porphyromonas gingivalis* in *Lactobacillus*. J. Appl. Microbiol. *100*, 256–263.

Marelli, B., Perez, A.R., Banchio, C., de Mendoza, D., and Magni, C. (2011). Oral immunization with live *Lactococcus lactis* expressing rotavirus VP8 subunit induces specific immune response in mice. J. Virol. Methods *175*, 28–37.

Martin, M.C., Pant, N., Ladero, V., Gunaydin, G., Andersen, K.K., Alvarez, B., Martinez, N., Alvarez, M.A., Hammarstrom, L., and Marcotte, H. (2011). Integrative expression system for delivery of antibody fragments by lactobacilli. Appl. Environ. Microbiol. *77*, 2174–2179.

Martinez, J.M., Kok, J., Sanders, J.W., and Hernandez, P.E. (2000). Heterologous coproduction of enterocin A and pediocin PA-1 by *Lactococcus lactis*: detection by specific peptide-directed antibodies. Appl. Environ. Microbiol. *66*, 3543–3549.

Mazmanian, S.K., Liu, C.H., Tzianabos, A.O., and Kasper, D.L. (2005). An immunomodulatory molecule of symbiotic bacteria directs maturation of the host immune system. Cell *122*, 107–118.

Medina, M., Vintini, E., Villena, J., Raya, R., and Alvarez, S. (2010). *Lactococcus lactis* as an adjuvant and delivery vehicle of antigens against pneumococcal respiratory infections. Bioeng. Bugs *1*, 313–325.

Menzies, B.E. (2003). The role of fibronectin binding proteins in the pathogenesis of *Staphylococcus aureus* infections. Curr. Opin. Infect. Dis. *16*, 225–229.

Mierau, I., and Kleerebezem, M. (2005). 10 years of the nisin-controlled gene expression system (NICE) in *Lactococcus lactis*. Appl. Microbiol. Biotechnol. *68*, 705–717.

Mierau, I., Leij, P., van Swam, I., Blommestein, B., Floris, E., Mond, J., and Smid, E.J. (2005). Industrial-scale production and purification of a heterologous protein in *Lactococcus lactis* using the nisin-controlled gene expression system NICE: the case of lysostaphin. Microb. Cell Fact. *4*, 15.

Mills, S., McAuliffe, O.E., Coffey, A., Fitzgerald, G.F., and Ross, R.P. (2006). Plasmids of lactococci – genetic accessories or genetic necessities? FEMS Microbiol. Rev. *30*, 243–273.

Miyoshi, A., Jamet, E., Commissaire, J., Renault, P., Langella, P., and Azevedo, V. (2004). A xylose-inducible expression system for *Lactococcus lactis*. FEMS Microbiol. Lett. *239*, 205–212.

Miyoshi, A., Poquet, I., Azevedo, V., Commissaire, J., Bermudez-Humaran, L., Domakova, E., Le Loir, Y., Oliveira, S.C., Gruss, A., and Langella, P. (2002). Controlled production of stable heterologous proteins in *Lactococcus lactis*. Appl. Environ. Microbiol. *68*, 3141–3146.

Morello, E., Bermudez-Humaran, L.G., Llull, D., Sole, V., Miraglio, N., Langella, P., and Poquet, I. (2008). *Lactococcus lactis*, an efficient cell factory for recombinant protein production and secretion. J. Mol. Microbiol. Biotechnol. *14*, 48–58.

Moris, G. (2014). Inflammatory bowel disease: An increased risk factor for neurologic complications. World J. Gastroenterol. *20*, 1228–1237.

Nguyen, T.T., Nguyen, T.H., Maischberger, T., Schmelzer, P., Mathiesen, G., Eijsink, V.G., Haltrich, D., and Peterbauer, C.K. (2011). Quantitative transcript analysis of the inducible expression system pSIP: comparison of the overexpression of *Lactobacillus* spp. beta-galactosidases in *Lactobacillus plantarum*. Microb. Cell Fact. *10*, 46.

Nordstrom, K., and Dasgupta, S. (2006). Copy-number control of the *Escherichia coli* chromosome: a plasmidologist's view. EMBO Rep. *7*, 484–489.

Norton, P.M., Le Page, R.W.F., and Wells, J.M. (1995). Progress in the development of *Lactococcus lactis* as a recombinant mucosal vaccine delivery system. Folia Microbiol. (Praha) *40*, 225–230.

Norton, P.M., Wells, J.M., Brown, H.W., Macpherson, A.M., and Le Page, R.W. (1997). Protection against tetanus toxin in mice nasally immunized with recombinant *Lactococcus lactis* expressing tetanus toxin fragment C. Vaccine *15*, 616–619.

Nouaille, S., Bermudez-Humaran, L.G., Adel-Patient, K., Commissaire, J., Gruss, A., Wal, J.M., Azevedo, V., Langella, P., and Chatel, J.M. (2005). Improvement of bovine beta-lactoglobulin production and secretion by *Lactococcus lactis*. Braz. J. Med. Biol. Res. *38*, 353–359.

Nouaille, S., Ribeiro, L.A., Miyoshi, A., Pontes, D., Le Loir, Y., Oliveira, S.C., Langella, P., and Azevedo, V. (2003). Heterologous protein production and delivery systems for *Lactococcus lactis*. Genet. Mol. Res. *2*, 102–111.

Paulsson, J., and Chattoraj, D.K. (2006). Origin inactivation in bacterial DNA replication control. Mol. Microbiol. *61*, 9–15.

Piard, J.C., Jimenez-Diaz, R., Fischetti, V.A., Ehrlich, S.D., and Gruss, A. (1997). The M6 protein of *Streptococcus pyogenes* and its potential as a tool to anchor biologically active molecules at the surface of lactic acid bacteria. Adv. Exp. Med. Biol. *418*, 545–550.

Pontes, D., Azevedo, M., Innocentin, S., Blugeon, S., Lefevre, F., Azevedo, V., Miyoshi, A., Courtin, P., Chapot-Chartier, M.P., Langella, P., *et al.* (2014). Immune response elicited by DNA vaccination using *Lactococcus lactis* is modified by the production of surface exposed pathogenic protein. PLoS One 9, e84509.

Pontes, D., Innocentin, S., Del Carmen, S., Almeida, J.F., Leblanc, J.G., de Moreno de Leblanc, A., Blugeon, S., Cherbuy, C., Lefevre, F., Azevedo, V., *et al.* (2012). Production of Fibronectin Binding Protein A at the surface of *Lactococcus lactis* increases plasmid transfer *in vitro* and *in vivo*. PLoS One 7, e44892.

Pouwels, P.H., Leer, R.J., Shaw, M., Heijne den Bak-Glashouwer, M.J., Tielen, F.D., Smit, E., Martinez, B., Jore, J., and Conway, P.L. (1998). Lactic acid bacteria as antigen delivery vehicles for oral immunization purposes. Int. J. Food Microbiol. *41*, 155–167.

Que, Y.A., Haefliger, J.A., Francioli, P., and Moreillon, P. (2000). Expression of *Staphylococcus aureus* clumping factor A in *Lactococcus lactis* subsp. cremoris using a new shuttle vector. Infect. Immun. *68*, 3516–3522.

Que, Y.A., Haefliger, J.A., Piroth, L., Francois, P., Widmer, E., Entenza, J.M., Sinha, B., Herrmann, M., Francioli, P., Vaudaux, P., *et al.* (2005). Fibrinogen and fibronectin binding cooperate for valve infection and invasion in *Staphylococcus aureus* experimental endocarditis. J. Exp. Med. *201*, 1627–1635.

Raz, E., Tighe, H., Sato, Y., Corr, M., Dudler, J.A., Roman, M., Swain, S.L., Spiegelberg, H.L., and Carson, D.A. (1996). Preferential induction of a Th1 immune response and inhibition of specific IgE antibody formation by plasmid DNA immunization. Proc. Natl. Acad. Sci. U.S.A. 93, 5141–5145.

Robinson, K., Chamberlain, L.M., Lopez, M.C., Rush, C.M., Marcotte, H., Le Page, R.W., and Wells, J.M. (2004). Mucosal and cellular immune responses elicited by recombinant *Lactococcus lactis* strains expressing tetanus toxin fragment C. Infect. Immun. 72, 2753–2761.

Robinson, K., Chamberlain, L.M., Schofield, K.M., Wells, J.M., and Le Page, R.W. (1997). Oral vaccination of mice against tetanus with recombinant *Lactococcus lactis*. Nat. Biotechnol. *15*, 653–657.

Ryu, Y.H., Baik, J.E., Yang, J.S., Kang, S.S., Im, J., Yun, C.H., Kim, D.W., Lee, K., Chung, D.K., Ju, H.R., *et al.* (2009). Differential immunostimulatory effects of Gram-positive bacteria due to their lipoteichoic acids. Int. Immunopharmacol. *9*, 127–133.

Serafini, F., Turroni, F., Guglielmetti, S., Gioiosa, L., Foroni, E., Sanghez, V., Bartolomucci, A., Motherway, M.O., Palanza, P., van Sinderen, D., *et al.* (2012). An efficient and reproducible method for transformation of genetically recalcitrant bifidobacteria. FEMS Microbiol. Lett. *333*, 146–152.

Simon, D., and Chopin, A. (1988). Construction of a vector plasmid family and its use for molecular cloning in *Streptococcus lactis*. Biochimie *70*, 559–566.

Sokol, H., Pigneur, B., Watterlot, L., Lakhdari, O., Bermudez-Humaran, L.G., Gratadoux, J.J., Blugeon, S., Bridonneau, C., Furet, J.P., Corthier, G., *et al.* (2008). *Faecalibacterium prausnitzii* is an anti-inflammatory commensal bacterium identified by gut microbiota analysis of Crohn disease patients. Proc. Natl. Acad. Sci. U.S.A. *105*, 16731–16736.

Sorvig, E., Gronqvist, S., Naterstad, K., Mathiesen, G., Eijsink, V.G., and Axelsson, L. (2003). Construction of vectors for inducible gene expression in *Lactobacillus sakei* and *L. plantarum*. FEMS Microbiol. Lett. *229*, 119–126.

Sridhar, V.R., Smeianov, V.V., and Steele, J.L. (2006). Construction and evaluation of food-grade vectors for *Lactococcus lactis* using aspartate aminotransferase and alpha-galactosidase as selectable markers. J. Appl. Microbiol. *101*, 161–171.

Steidler, L., Neirynck, S., Huyghebaert, N., Snoeck, V., Vermeire, A., Goddeeris, B., Cox, E., Remon, J.P., and Remaut, E. (2003). Biological containment of genetically modified *Lactococcus lactis* for intestinal delivery of human interleukin 10. Nat. Biotechnol. *21*, 785–789.

Steidler, L., Rottiers, P., and Coulie, B. (2009). Actobiotics as a novel method for cytokine delivery. Ann. N.Y. Acad. Sci. *1182*, 135–145.

Summers, D.K., and Sherratt, D.J. (1984). Multimerization of high copy number plasmids causes instability: ColE1 encodes a determinant essential for plasmid monomerization and stability. Cell *36*, 1097–1103.

Tang, D.C., DeVit, M., and Johnston, S.A. (1992). Genetic immunization is a simple method for eliciting an immune response. Nature *356*, 152–154.

Tao, L., Pavlova, S.I., Ji, X., Jin, L., and Spear, G. (2011). A novel plasmid for delivering genes into mammalian cells with noninvasive food and commensal lactic acid bacteria. Plasmid *65*, 8–14.

Tsai, Y.T., Cheng, P.C., and Pan, T.M. (2010). Immunomodulating activity of *Lactobacillus paracasei* subsp. *paracasei* NTU 101 in enterohemorrhagic *Escherichia coli* O157H7-infected mice. J. Agric. Food Chem. *58*, 11265–11272.

Tsai, Y.T., Cheng, P.C., and Pan, T.M. (2012). The immunomodulatory effects of lactic acid bacteria for improving immune functions and benefits. Appl. Microbiol. Biotechnol. *96*, 853–862.

Ulmer, J.B., Donnelly, J.J., Parker, S.E., Rhodes, G.H., Felgner, P.L., Dwarki, V.J., Gromkowski, S.H., Deck, R.R., DeWitt, C.M., Friedman, A., *et al.* (1993). Heterologous protection against influenza by injection of DNA encoding a viral protein. Science *259*, 1745–1749.

van der Vossen, J.M., van der, L.D., and Venema, G. (1987). Isolation and characterization of *Streptococcus cremoris* Wg2-specific promoters. ApplEnvironMicrobiol. *53*, 2452–2457.

Vandenbroucke, K., de Haard, H., Beirnaert, E., Dreier, T., Lauwereys, M., Huyck, L., Van Huysse, J., Demetter, P., Steidler, L., Remaut, E., *et al.* (2010). Orally administered L. lactis secreting an anti-TNF Nanobody demonstrate efficacy in chronic colitis. Mucosal Immunol. 3, 49–56.

Waters, C.M., Hirt, H., McCormick, J.K., Schlievert, P.M., Wells, C.L., and Dunny, G.M. (2004). An amino-terminal domain of Enterococcus faecalis aggregation substance is required for aggregation, bacterial internalization by epithelial cells and binding to lipoteichoic acid. Mol. Microbiol. *52*, 1159–1171.

Wegmann, U., O'Connell-Motherway, M., Zomer, A., Buist, G., Shearman, C., Canchaya, C., Ventura, M., Goesmann, A., Gasson, M.J., Kuipers, O.P., *et al.* (2007). Complete genome sequence of the prototype lactic acid bacterium *Lactococcus lactis* subsp. *cremoris* MG1363. J. Bacteriol. *189*, 3256–3270.

Wells, J.M. (2011). Immunomodulatory mechanisms of lactobacilli. Microb. Cell Fact. *10* (Suppl. 1), S17.

Wells, J.M., Wilson, P.W., Norton, P.M., Gasson, M.J., and Le Page, R.W. (1993). *Lactococcus lactis*: high-level expression of tetanus toxin fragment C and protection against lethal challenge. Mol. Microbiol. *8*, 1155–1162.

Wollert, T., Pasche, B., Rochon, M., Deppenmeier, S., van den Heuvel, J., Gruber, A.D., Heinz, D.W., Lengeling, A., and Schubert, W.D. (2007). Extending the host range of Listeria monocytogenes by rational protein design. Cell *129*, 891–902.

Xiang, S., Fruehauf, J., and Li, C.J. (2006). Short hairpin RNA-expressing bacteria elicit RNA interference in mammals. Nat. Biotechnol. *24*, 697–702.

Yam, K.K., Hugentobler, F., Pouliot, P., Stern, A.M., Lalande, J.D., Matlashewski, G., Olivier, M., and Cousineau, B. (2011). Generation and evaluation of A2-expressing *Lactococcus lactis* live vaccines against *Leishmania donovani* in BALB/c mice. J. Med. Microbiol. *60*, 1248–1260.

Zhou, X.X., Li, W.F., Ma, G.X., and Pan, Y.J. (2006). The nisin-controlled gene expression system: construction, application and improvements. Biotechnol. Adv. *24*, 285–295.

The Indigenous Microbiota and its Potential to Exhibit Probiotic Properties

12

Sylvie Miquel, Rebeca Martin, Muriel Thomas, Luis G. Bermudez-Humaran and Philippe Langella

Abstract

Humans harbour a different microbiota depending on the tissue considered. Most of the microorganisms are contained in the gastrointestinal tract (GIT) and this gut microbiota represents approximately 10^{14} cells that correspond to the highest bacterial density for any ecosystem. Our microbiota represents a huge diversity in term of species and functions. A healthy gut microbiota is composed of a well-balanced community of three permanent residents termed symbionts (with beneficial effects), commensals (no effect), and pathobionts (potentially induce pathologies under certa*in* situations), but no pathogens. The term dysbiosis (microbial imbalance) has been related to many different kinds of pathologies although it is not clear whether the imbalance of such a microbiota is a cause or a consequence of the illness. Nowadays, the challenge of linking microbiota to human health and disease is being tackled by different research teams around the world with the aim to investigate the implication of potential beneficial bacteria that could be decreased in the studied microbiota of patients. From this perspective, it could be interesting to use them as potential probiotics to try to resolve dysbioses.

Our indigenous microbiota: definition and composition

Few years ago, microorganisms of interest were essentially human pathogenic bacteria, but today, more and more studies aimed to characterize the interactions between commensal microbial and human physiology (Weinstock, 2012). The term microbiota refers to the microorganisms population present in a specific environment (soil, ocean, bowel, skin, vagina, etc.) including bacteria, viruses, archaea, protozoans and fungi. Humans could be considered as 'meta-organisms' constituted of 10-fold more microorganisms than human cells (Neish, 2009). Humans harbour different microbiota depending on tissues considered; we can thus distinguish the cutaneous, the oral, the intestinal, the vaginal, and the airway microbiota. Nowadays, the area of metagenomics has given a new dimension to the analysis of microbiota not based on classic microbiological techniques of the bacterial culture. In fact, a complete characterization of the microbiota was difficult before the use of high-throughput sequencing and the emergence of 'omics' approaches (metagenomics, metatranscriptomics and metabolomics) (Neish, 2009). This approach has also allowed to define the term 'microbiome' referring to the genetic material of the catalogue of microbial taxa associated with humans (Ursell *et al.*, 2012). Being frequently confused with the term microbiota, the microbiome was first defined by Joshua Lederberg in 2001 (J. Lederberg and McCray, 2001) and is sometimes referred as our 'second genome' consisting of 150-fold more genes than the human genome itself (Bruls and Weissenbach, 2011; Qin *et al.*, 2010). Today, the availability of the reference gene catalogues obtained from MetaHIT project (Qin *et al.*, 2010) and the American Human Microbiome Project (Consortium, 2012) has allowed to evaluate richness of the human gut microbiome. The human microbiota (microbiome by extension), can be considered as a additional human organ from a physiological standpoint with essential roles as an organ (Baquero and Nombela, 2012).

Depending on the considered tissues, the bacterial communities of a same individual have a very different composition (Weinstock, 2012). Most of the microorganisms are contained in the gastrointestinal tract (GIT) and represents approximately 10^{14} cells that correspond to the most important bacterial density for an ecosystem (Backhed *et al.*, 2005; Gill *et al.*, 2006). These microorganisms are for the greater part, around 70%, extremely oxygen sensitive (EOS) bacteria located in the colon (Ley *et al.*, 2006). Firmicutes (the *Clostridum leptum* and *Clostridum coccoides* groups) and Bacteroidetes are the two dominant phyla, representing ~ 90% of the microbiota, and the third most dominant group is Actinobacteria with only ~ 3% (Tap *et al.*, 2009). In the vagina, lactobacilli are the dominant bacteria (70% of the population) even if it could include some microorganisms also present in the GIT (Doderlein, 1982; Martin *et al.*, 2008b). There is a huge diversity in the bacterial species which constitute the intestinal microbiota: approximately 160 species of different bacteria are present within every individual (Qin *et al.*, 2010). Besides, there is variation in bacterial populations within a specific microbiota depending of the micro-localization. For instance, along the GIT, bacterial populations differ from the upper to the lower part of the tract but also between intestinal mucosa and the luminal content (Eckburg *et al.*, 2005; Swidsinski *et al.*, 2008b). Furthermore, even if at the level of phyla, the composition is relatively similar from one individual to another, an important interpersonal variability was observed at the level of species (Eckburg *et al.*, 2005). In spite of this relative heterogeneity, through faecal metagenomic analysis, it is possible to distinguish three main robust clusters named 'Enterotypes' in the gut microbiome, which are driven by species composition. Each of these three enterotypes are identifiable by the variation in the levels of one of three genera *Bacteroides, Prevotella* and *Ruminococcus* (Arumugam *et al.*, 2011). Their abundance and proportions vary between individuals and is associated with long-term dietary habits (Wu *et al.*, 2011).

Our microbiota represents a huge diversity in term of species and functions. It will be interesting to understand how we could study their composition to better understand how they interact with the host. It fact, we will see that this 'organ' participate to the human health. Medicine has developed organ-based specialties such as nephrology, hepatology, cardiology or pneumology. Perhaps, we can envisage 'microbiomology' as a future specialty of or a branch of clinical microbiology, devoted to the study of the physiology, pathology, diagnostics, therapy and prevention of alterations of the community structure of the microbiome (Baquero and Nombela, 2012).

How to study the microbiota

Nowadays, the most common microbiological approaches to study the human microbiota are divided depending on the need or not of the culture of the bacterium object of the study. The possibility to study bacteria without culturing them is a great success in the Microbiology as it has been one of the most important inflexion points in this science.

Classical microbiological approaches

The first analyses of the human microbiota have been developed thanks to culture dependent methodologies (Moore and Holdeman, 1974). For instance, Finegold and co-workers in the early 1970s have already studied the effect of diet on the human gut microbiota using this culture method (Finegold *et al.*, 1974). The protocols of identification were based on microscopic observations and pure cultures (Lagier *et al.*, 2012) joint to phenotypic identification methods. The Gram staining (Biswas *et al.*, 1970) allowed describing a wide number of human ecosystems, such as the faeces population and the vaginal microbiota, and the use of biochemical systems, such as the Analytical Profile Index (API) galleries, is still a method of reference nowadays. These first data were very useful, but as it is known that only 1% of the bacteria can be easily grown *in vitro* (Vartoukian *et al.*, 2010), new approaches have been developed to avoid methodological bias due to culture restrictions.

The main problem of culture dependent methods is that they are focused on aerobic bacteria whereas a wide range of bacteria belonging to the human microbiota prefer micro-aerophilic conditions (Lagier *et al.*, 2012). Furthermore, many microorganisms need special growth conditions, such as the EOS bacteria, that render their culture difficult and their detection even complicated (Miquel *et al.*, 2013). Others have never been grown in culture and may require specific, yet unknown, growth conditions impairing their identification by culture dependent methods (Martin *et al.*, 2014b). For example, the mucolytic bacteria *Akkermansia muciniphila* was isolated by dilution of faeces in anaerobic medium containing gastric mucin as the sole carbon and nitrogen source (Derrien *et al.*, 2004). Consequently, the major populations isolated were those bacteria that grow quickly in aerobic conditions in classical rich nutrient media (Hugenholtz, 2002). In addition, the use of phenotypic identification methods is time consuming and expensive (Seng *et al.*, 2009).

Can we consider that culture is an old method with a limited future? Our feeling is that the use of culture dependent methodologies is still required to achieve complete knowledge about bacterial physiology. We are sure that innovative methods have to be developed to optimize the growth of fastidious bacteria.

'Omics' approaches

Until the 1990s, knowledge of microbiota was limited because bacteriological culture was the only technique available to describe its composition. For example, only a small portion (estimated at < 30%) of the gut microbiota has been cultured to date (Fraher *et al.*, 2012). This is because culture-based approaches, even if being extremely helpful to understand the physiological potential of isolated microorganisms, do not essentially provide complete information on the composition of microbial communities (Orphan *et al.*, 2000). In the 1990s, concomitantly with the expansion of molecular tools, new culture independent methods significantly improved the knowledge about microbiota (Zoetendal *et al.*, 2006). These techniques were mainly based on divergences of the sequence of the small subunit ribosomal RNA (16S rRNA) giving quantitative and qualitative information about the microbiota through time and space (Fraher *et al.*, 2012). Examples of these techniques are denaturing gradient gel electrophoresis (DGGE), terminal restriction fragment length polymorphism (T-RFLP), fluorescence *in situ* hybridization (FISH), DNA microarrays, and next-generation sequencing of the 16S rRNA gene or its amplicons (Fraher *et al.*, 2012).

To study the human microbiota, it is imperative to understand the role of microbes in the host ecosystem and, for this, it is crucial to probe their genetic potential, expression, and ecological status (Lamendella *et al.*, 2012). To achieve this new concept of microbiota as a collection of dynamic ecological communities, new techniques have been developed. These meta-omic or 'systems biology' approaches offer a complementary support to the classical microbiology and are based on high-throughput methods (Del Chierico *et al.*, 2012). The more relevant are those that study the genome, metabolome, transcriptome and proteome of the human microbiota (metagenomics, metabolomics, transcriptomics and proteomics, respectively), although new fields of research show up frequently as the applications of these high-throughput methods are huge. The term metagenomic was originally coined for the shotgun characterization of total DNA and now is also used to name the study of marker genes such as the 16S rRNA gene (Ursell *et al.*, 2012). Its objective is to determine the microbiome which is the catalogue of microbial taxa associated with humans and their genetic material (Ursell *et al.*, 2012). Metabolomics is the study of the set of metabolites synthesized by a biological system (Fiehn, 2002). Transcriptomic refers originally to the identification of the critical messenger RNAs (Kiechle and Holland-Staley, 2003), nowadays the transcriptome is considered the set of all RNA molecules, including mRNA, rRNA, tRNA, and other non-coding RNA molecule produced in one or a population of cells, although most of the interest is in mRNA (Pinto *et al.*, 2011). Proteomics is defined as a large-scale study of proteins, in particular their functions and structures (Ghafourian *et al.*, 2013) being the entire set of proteins expressed by the genome.

All these powerful tools provide information on community diversity and structure (Gans *et al.*, 2005) even if some bias linked to sample extraction and preparation, data acquisition and data mining, may limit some interpretation. So for us, our next challenge is to combine omics methods with culture approaches in order to better understand the functions and the physiological relevance of bacterial communities. The main purpose is to benefit of an integrated view of our physiology involving bacterial and eukaryotic cell communities.

The microbiota: our best companion

The microbial diversity of the GIT results from a co-evolution between the microbial communities and their hosts which represents its ecological niche. For a long time considered as commensal, from the Latin: cum ('with') and mesa ('table'), which means the sharing of the meal, the term mutualism is preferred today to speak about the host-microbiota relation. Indeed, commensalism is a relation between individuals of two species in which one species obtains food or other benefits from the other without either harming or benefiting the latter. Commensalism is then a non-parasitic utilization of a live species by another. However, in the case of the host-intestinal microbiota relation, we speak about a symbiotic mutualism relation because both partners developed an intimate and long-term association with mutual profit (Fig. 12.1). In fact, there is a co-evolution between the human and its microbiota with an acquisition by the microbiota of functionalities absent in human genome. In fact, the intestinal microbiota, by its enzymatic activities, is a complementary metabolic repertoire of the human digestive system. The intestinal bacteria present beneficial properties for the host in terms of synthesis of molecules, degradation of substrates and detoxification. For example, we could quote the acquisition of genes for degradation of the plant polysaccharides and other complex sugars, non-digestible by human enzymes (Ley *et al.*, 2008). Their degradation by bacterial fermentation in short-chain fatty acids (SCFA) constitutes an important source of energy for colonocytes (Macfarlane and Macfarlane, 2011). Moreover, the intestinal bacteria synthesize vitamins, amino acids and are involved in the metabolism of bile salts (Vyas and Ranganathan, 2012). The intestinal microbiota also participates in absorptive capacities of nutriments by the GIT by a direct action on its physiology. For example, the SCFA are not only involved in the energy metabolism but also present properties of regulation of the physiology of the host (differentiation, cellular growth, etc.) (Neish, 2009). A recent publication using gnotobiotic models showed that the colon evolved in parallel with the microbiota composition inducing a homeostasis of the two maturated partners (Tomas *et al.*, 2013). In the same way, during the post natal period, bacteria associated with the mucosa could regulate angiogenesis increasing the absorptive capacity of the intestine (Stappenbeck *et al.*, 2002). Thus, the microbiota supports human nutrition by two main mechanisms: increase of the availability in energy substrates for the host and regulating intestinal absorption capacity.

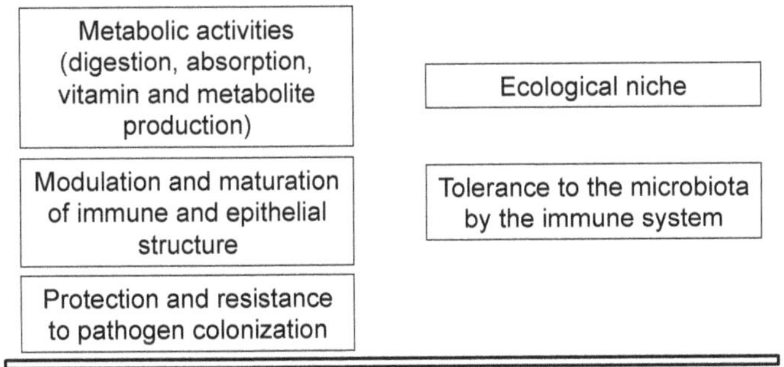

Figure 12.1 Symbiotic mutualism relation forged during the evolution between the host and its microbiota. (Adapted from Claire Cherbuy; http://mediatheque.inra.fr/media/detail/246161/private).

Besides being a nutritive supply, the microbiota also participates in the protection of the host. Indeed, the organism defences at the level of the GIT have the delicate task to distinguish commensal microbiota from pathogenic agents. The mechanisms of the intestinal immune system allowing its tolerance to the microbiota remains poorly known (Littman and Pamer, 2011). To get a glimpse of an answer, it is necessary to better understand the recognition systems of microorganisms developed by the intestinal epithelial cells. Particularly, specific receptors to Microbial-associated molecular patterns (MAMPS) exist, the pattern recognition receptors (PRRs) (Wells *et al.*, 2011). Briefly, the activation of these receptors by commensal bacteria leads to the activation of the innate immunity system and is essential to maintain intestinal homeostasis and the protection of the epithelium against pathogenic bacteria (Lundin *et al.*, 2008; Rakoff-Nahoum and Medzhitov, 2008; Rakoff-Nahoum *et al.*, 2004). So, the microbiota limits the inflammation of the intestinal epithelium and plays a major role in the development of the immune system (Macdonald and Monteleone, 2005). To illustrate the hypothesis, the best example is the use of germ free mice that present an immature immune system at the GIT level and the conventionalisation of these mice (reconstitution of a normal microbiota) restored the immunity of this mucosa (Umesaki *et al.*, 1995). Furthermore, at birth, the immune system is still immature and it was shown that contact with bifidobacteria influences its postnatal development via the production of IgA and an insufficient exposure to these microorganisms could lead to inappropriate immune responses later in the childhood and could induce asthma or allergies (Sjogren *et al.*, 2009).

Thus, the microbiota participates to the protection of the host by supporting the development of the immune system, but commensal bacteria act also directly as a 'microbiological barrier' against pathogens. Indeed, it was shown that germ free mice are more sensitive to intestinal pathogens (Vollaard and Clasener, 1994). This protective capacity of the microbiota could be explained by different mechanisms: a competition with pathogenic bacteria for the access to nutriments and/or for a contact with the epithelium, the production of bactericidal or bacteriostatic molecules and, a modification of the physiology of the epithelium through production of mucus or modulation of epithelial cell tight junctions (Martin *et al.*, 2014b; Yu *et al.*, 2012). For instance, two major commensal bacteria *B. thetaiotaomicron* and *F. prausnitzii* could directly modulate the intestinal mucus barrier by their complementary metabolism by modifying goblet cells and mucin glycosylation (Wrzosek *et al.*, 2013). Moreover, commensal bacteria were able to induce maturation of intestinal barrier function through (i) modulation of the expression of tight junctions that regulate paracellular permeability to maintain separation between tissue compartments by sealing the intercellular space, and (ii) intracellular permeability regulated by toll-like receptor signalling (Jakaitis and Denning, 2014).

It is now clear that the normal intestinal microbiota could influence numerous physiological aspects in the healthy host through the establishment of cross-species homeostatic regulation well detailed in the review of Sommer and Bäckhed (2013). Thus, the microbiota composition and its relation with the gut have resulted from the dynamics of selection and competition (Angelakis *et al.*, 2012). For instance, the bacterial microbiome of the inhabitants of Burkina-Faso was more adapted to the degradation of cellulose suggesting an adaptation to Burkina-Faso's high-fibre diet (De Filippo *et al.*, 2010); One could suggest that our microbiota shelters unknown bacteria adapted to specific functions and represents a real richness for future research concerning human health. Our microbiota takes an important part in human health and an imbalance of its composition could participate to the development of many diseases.

Diseases of microbiota: a break in the homeostasis

Since Robert Koch and Ilya Mechnikov were awarded two Nobel Prizes in physiology and medicine in the 1900s for their discoveries linking microbes and human health, several determinants of host–microbe interactions, including spatial (e.g. skin, vagina, and gut), temporal (e.g. birth and senescence), and environmental (e.g. diet, antibiotics treatments), have been partially unravelled (Cani and Delzenne, 2011). For instance, a mature vaginal microbiota is established only in early adolescence after the hormonal changes typical of this period (Yamamoto *et al.*, 2009). Concerning intestinal microbiota, it varies all along the life, with the particularity to increase its population number and diversity during the first year and decline upon ageing (Claesson *et al.*, 2012; Sekirov *et al.*, 2010). In adults the importance of the microbiome has been highlighted by the microbial 'abnormalities' found in pathological conditions and particularly a disruption of profile or biodiversity. In fact, many diseases are characterized by a dysbiosis, in other words, a microbial imbalance (Round and Mazmanian, 2009). A healthy gut microbiota is composed of a well-balanced community of three permanent residents termed symbionts (with beneficial effects), commensals (no effect), and pathobionts (potentially induce pathologies under certain situations) but no pathogens (Koboziev *et al.*, 2013). The term dysbiosis has been related to many different kinds of pathologies although it is not clear whether the imbalance of microbiota is a cause or a consequence of the illness (Martin *et al.*, 2013). In this way, 'The hygiene hypothesis ', based on the observation of a positive correlation between chronic diseases, hygiene and socio-economic status (Strachan, 1989), could be explained by the creation of dysbioses limiting the regulation of the immunity by the microbiota (von Mutius, 2007).This hypothesis proposes that a lack of early microbial stimulation in industrialized countries results in an aberrant immune response to innocuous antigens later in life (Wills-Karp *et al.*, 2001). In parallel, the 'microbiota hypothesis' claims that the gut microbiota dysbiosis, due to antibiotic use and dietary changes, can disrupt the mechanisms of mucosal immunological tolerance (Noverr and Huffnagle, 2005). Recent epidemiological and clinical data (Droste *et al.*, 2000; Hoskin-Parr *et al.*, 2013), as well as experimental data obtained in mouse models (Noverr *et al.*, 2004; Olszak *et al.*, 2012), support this latter theory. Nowadays, the challenge of linking microbiome to human health and disease is being tackled by different research teams around the world. They are currently studying different disease states to identify potential correlations and ecological models of community structure and function in order to understand the dynamics of all ecosystems that comprise the human microbiome (Martin *et al.*, 2014b).

Impact of intestinal dysbiosis

The GIT is the major site of interaction between environmental microorganisms and host tissues. A high biodiversity of the gut microbiota is associated with a healthy status, while low biodiversity is linked to pathological conditions (Manichanh *et al.*, 2006). More than 30 diseases are associated with intestinal microbiota (dysbiosis) and the list is continuously increased. The pathogenesis of several autoimmune, atherosclerosis or rheumatoid arthritis, and chronic inflammatory diseases could be mentioned (Koboziev *et al.*, 2013). Among these diseases we could distinguish three important classes: metabolic syndromes, intestinal diseases and more prudently psychological syndromes (Fig. 12.2). The common point between all of these patients is that they are suffering of intestinal disorders with different severities. The link between psychological syndrome and gut dysbiosis is a very recent hypothesis present in the literature. It is clear that gut microbiome plays a crucial role in the bidirectional gut–brain axis that integrates the gut and central nervous system activities (Wang and Kasper, 2013). Indeed, it has been shown that the niche-specific microbiome, prominently the gut microbiome, has the capacity to affect both local and distal sites within the host. In fact, intestinal dysbiosis is involved in metabolic syndromes such as diabetes, obesity, liver disease…etc. Interestingly, colonization of germ-free mice with microbiota obtained from obese but not

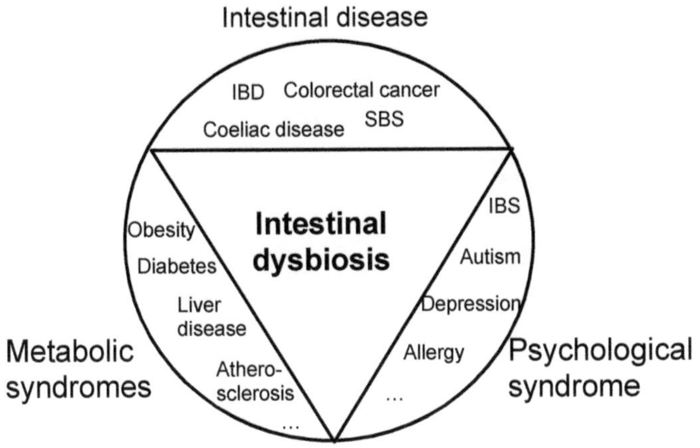

Figure 12.2 Disease associated with intestinal dysbiosis. IBD, inflammatory bowel disease; SBS, short bowel syndrome; IBS, irritable bowel syndrome.

lean mice results in the generation of mice with a significantly greater amount of total body fat (Turnbaugh *et al.*, 2006). Similar results were obtained in non-alcoholic fatty liver disease, the hepatic manifestation of metabolic syndrome (Le Roy *et al.*, 2012), suggesting that these kinds of diseases are transmissible via microbiota. Moreover, a disorder of the crosstalk between innate immune system and intestinal microbiota may promote the development of the metabolic syndrome (Cani and Delzenne, 2011). For instance, a recent study shows that mice genetically deficient in Toll-like receptor 5 (TLR5), a component of the innate immune system that is expressed in the gut mucosa and recognizes pathogen-associated molecular patterns (PAMPs) that are expressed on infectious microbes, develop hallmark features of metabolic syndrome (hyperlipidaemia, hypertension, insulin resistance, and increased adiposity) (Vijay-Kumar *et al.*, 2010).

The best examples of the implication of microbiota in diseases and the importance of the homeostasis between these two partners are the development of intestinal diseases and particularly inflammatory bowel disease (IBD). IBD, including Crohn's disease (CD) and ulcerative colitis (UC), are characterized by a chronic activation of the immune system, of genetically predisposed patients in response to environmental factors and dysbiosis (De Cruz *et al.*, 2012; Lepage *et al.*, 2011). For instance, the proportion of Firmicutes, in particular the symbiotic *Faecalibacterium prausnitzii*, was found to be low in intestinal microbiota of CD patients (Sokol *et al.*, 2008) whereas there was an increased in opportunistic pathogen bacteria, in particular Adherent invasive *Escherichia coli* (AIEC) (Darfeuille-Michaud, 2002; Miquel *et al.*, 2010). A study showed that colitis (IBD like) and colonic hypersensitivity (IBS-like syndrome) can be provoked in healthy rodents by the transfer of microbiota coming from mouse suffering from colitis (Crouzet *et al.*, 2013; Garrett *et al.*, 2007). The excessive inflammation of the GIT would be particularly induced by an inappropriate stimulation of the immune system in the presence of an unbalanced commensal microbiota and its metabolites (Sartor, 2008). Dysbiosis has also been reported for numerous other intestinal patho-physiological contexts such as IBS (Rajilic-Stojanovic *et al.*, 2011), colorectal cancer (Sobhani *et al.*, 2011), short bowel syndrome (SBS) (Joly *et al.*, 2010) and coeliac disease (De Palma *et al.*, 2010).

Vaginal dysbiosis

Although dysbiosis has been considered at length in the GIT, it can also occur on any exposed surface or mucous membrane such as skin (Grice *et al.*, 2008; Grice and Segre, 2011), oral cavity, airway (Charlson *et al.*, 2011; Gollwitzer and Marsland, 2014), and the vagina.

The vaginal microbiota that is mainly formed by lactobacilli, has an important role in human development, physiology, and immunity (Martin *et al.*, 2014b). Abnormal vaginal microbiota can occur because of sexually transmitted pathogens or overgrowth of resident organisms (Lamont *et al.*, 2011). The most common pathologies are bacterial vaginosis (BV), the proliferation of *Candida* sp. (mainly *C. albicans*) (candidiasis) and *Trichomonas vaginalis* (trichomoniasis) (Martin *et al.*, 2008a). The community of mutualistic vaginal bacteria, constitutes the first line of protection for the host by excluding non-indigenous microbes that may cause illness (Martin *et al.*, 2008a).

The lactobacilli exerts their beneficial role by two main mechanisms: (i) exclusion and (ii) inhibition of growth (Boris and Barbes, 2000) (Fig. 12.3). The exclusion is driven by the competition for epithelial cell receptors with urogenital pathogens, of which the most important are: group B *Streptococcus* species, *Staphylococcus aureus*, *Gardnerella vaginalis*, *Neisseria gonorrhoeae*, *Pseudomonas aeruginosa*, *Klebsiella pneumoniae*, *Candida albicans* and *Actinomyces neuii* (Boris *et al.*, 1998; Martin *et al.*, 2012; Osset *et al.*, 2001a,b; Vielfort *et al.*, 2008; Zarate and Nader-Macias, 2006). Furthermore, the ability of some lactobacilli to co-aggregate with pathogens like *E. coli*, *C. albicans*, and *G. vaginalis* has been found (Boris *et al.*, 1998; Osset *et al.*, 2001a). Although the molecular mechanism involved are not completely known and may depend on the *Lactobacillus* strain involved and the pathogen inhibited, shaving and seeding treatments with proteases, lipases or periodic acid point out the proteinic, lipidic or polysaccharide nature of the adhesins and cellular receptors involved (Munoz-Provencio *et al.*, 2010; Sanchez *et al.*, 2008; Tjalsma *et al.*, 2008).

Inhibition of growth is due to the generation of antimicrobial compounds. Lactobacilli are able to produce mainly organic acids from the fermentation of sugars that contribute to the low pH of the vagina, the major factor in the inhibition of pathogen growth (Boskey *et al.*, 2001). Besides, some vaginal lactobacilli are able also to produce other antimicrobial compounds such as bacteriocins, biosurfactants and hydrogen peroxide (H_2O_2) (Velraeds *et al.*, 1998, 2000). The production of H_2O_2 seems to be the second major antimicrobial mechanisms after organic acid production. In fact, the prevalence of H_2O_2-producing strains has been correlated with lower incidence of bacterial vaginosis (BV) and vaginal infections (Eschenbach *et al.*, 1989; Hawes *et al.*, 1996). H_2O_2 exerts its bactericidal effect through generation of oxidizing metabolites such as the radical OH^- that introduces breaks in the DNA of the cell (Klebanoff and Belding, 1974). Furthermore, hydrogen peroxide is able to regulate the lactobacilli population, due to the ability of some lactobacilli to degrade and produce this molecule (Martin *et al.*, 2012) and the lysis of

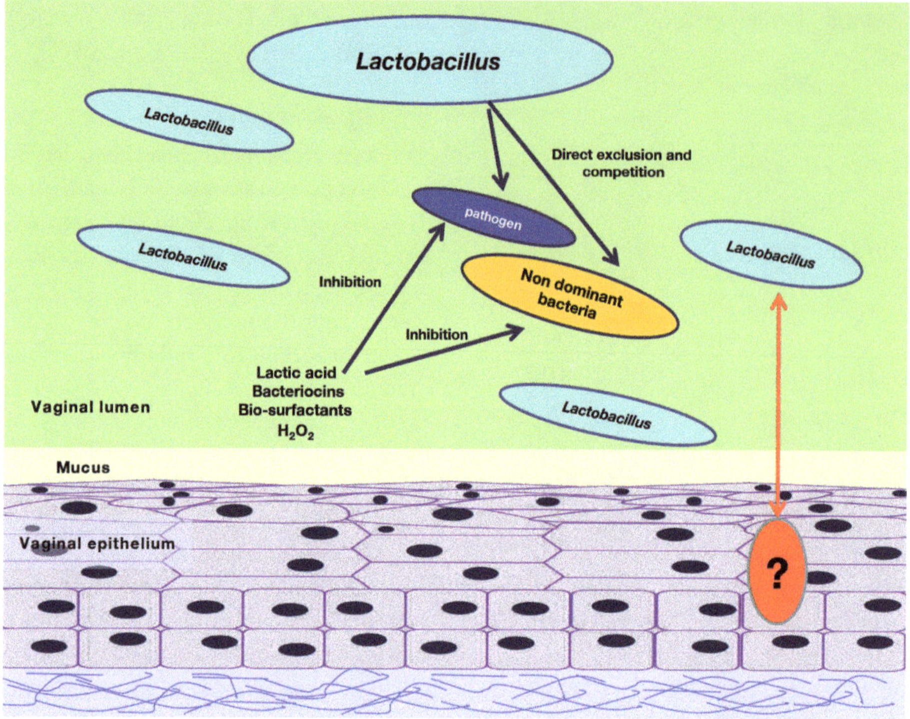

Figure 12.3 Beneficial effects of lactobacilli on the vaginal ecosystem. Lactobacilli protect the host epithelium thanks to two main mechanisms: (i) exclusion, driven by the formation of a biofilm that mask the epithelial cell receptors and (ii) inhibition of growth, due to generation of antimicrobial compounds. The direct interactions between the lactobacilli and the host are not well known. Modified from (Martin *et al.*, 2014b).

lysogenic lactobacilli by a H_2O_2-mediated prophage induction mechanism (Martin *et al.*, 2009; Pavlova *et al.*, 1997). Although, the presence of bacteriocins is thoroughly described in numerous bacterial species (Cotter *et al.*, 2013), only four bacteriocins have been found, to our knowledge, up to day in vaginal *Lactobacillus* strains: Lactocin 160, Salivaricin CRL 1328, L23 and recently a bacteriocin-like substance produced by *Lactobacillus fermentum* CS57 isolated from human vaginal secretions (Aroutcheva *et al.*, 2001; Martin *et al.*, 2012; Ocana *et al.*, 1999; Pascual *et al.*, 2010; Sabia *et al.*, 2014; Simoes *et al.*, 2001).

Despite all the defence mechanisms asserted by lactobacilli, sometimes the proportion of *Lactobacillus* drops under a critical level and this circumstance is used by other microorganisms that can act as opportunistic pathogens (Garcia-Rodriguez and Munoz, 1991). In general these microorganisms can be found in the normal vaginal microbiota, but in a lower abundance due to the antagonist effect of the lactobacilli. Among them, the most common aetiological agents are *Gardnerella vaginalis*, *Mycoplasma hominis*, *Prevotella* and *Peptostreptococcus* that can induce from an asymptomatic infection up to vaginosis (Sobel, 2000; Thorsen *et al.*, 1998). Bacterial vaginosis patients are those who fulfil the Amsel criteria: homogeneous vaginal discharge, amine (fishy) odour when potassium hydroxide solution is added to vaginal secretions (commonly called the 'whiff test'), vaginal pH greater than 4.7 and presence of clue cells (greater than 20%) upon microscopy combined with a significant diminution of Gram-positive lactobacilli replaced by Gram-negative bacteria (Amsel *et al.*, 1983). BV is the most recurrent vaginal imbalance and has been connected with a high diversity microbiota (Pavlova *et al.*, 2002) and the presence of unfamiliar bacteria such as *Mobiluncus* sp., *Atopobium* sp., *Megasphaera* sp. and *Ureaplasma urealyticum* (Doyle *et al.*, 1995; Fredricks *et al.*, 2005; Haggerty *et al.*, 2009; Hyman *et al.*, 2005; Kazor *et al.*, 2003). Thanks to metagenomic approaches some uncultivable microorganisms have also been associated with BV (Fredricks *et al.*, 2005; Lamont *et al.*, 2011; Oakley *et al.*, 2008; Pavlova *et al.*, 2002). Epidemiologically, vaginal dysbiosis such as BV has been associated with preterm birth, development of pelvic inflammatory disease and acquisition of sexually transmitted infections produced by *Neisseria gonorrhoeae*, *Chlamydia* spp. and HIV (Brotman, 2011).

It is interesting to study the characterization of dysbiosis in more detail to investigate the implication of potential beneficial bacteria that could be decreased in the studied microbiota of patients. From this perspective, it could be interesting to use them as potential probiotics to try to resolve dysbioses.

The commensal beneficial bacteria

Commensal microorganisms offer a wide range of benefits to the host. The cross-talk between the host and its microbiota is fundamental for the maintenance of the host homeostasis (Leser and Molbak, 2009). The microbiota is capable to regulate the intestinal epithelium, the presence of pathogenic bacteria and the immune responses as well as to confer nutritional benefits as discussed before (Martin *et al.*, 2013). Both host and indigenous microorganisms, have then adapted to each other to maintain the advantages that this mutualism offers (Sekirov *et al.*, 2010).

Numerous bacterial examples have been analysed in more or less detail. *Bacteroides thetaiotaomicron* and *Faecalibacterium prausnitzii*, both prominent members of the intestinal microbiota of mice and humans, have been widely studied (Miquel *et al.*, 2013; Zocco *et al.*, 2007). *B. thetaiotaomicron* exerts several beneficial effects, such as the modulation of the expression of a large quantity of genes implicated in different aspects of host physiology (Zocco *et al.*, 2007), the modulation of the colon epithelium (Wrzosek *et al.*, 2013) and the defence of the intestinal barrier (Resta-Lenert and Barrett, 2006). Indeed, the acetate produced by *B. thetaiotamicron* seems to be key in the cell differentiation (Wrzosek *et al.*, 2013) and this bacterium prevents cytokine induced increase in permeability (Resta-Lenert and Barrett, 2006). Thanks to the analysis of its genome and proteome, the huge ability to adapt and regulate gene expression in response to a changing ecosystem has been pointed out, highlighting the high level of adaptation of this bacterium to its niche (Comstock and Coyne, 2003).

F. prausnitzii, is a major EOS component of the intestinal microbiota which represents around 5% of the bacterial population of the microbiota and 8% of Firmicutes (Eckburg *et al.*, 2005; Miquel *et al.*, 2014). Its prevalence is low in many health disorders, such as IBS, colorectal cancer, coeliac diseases, obesity, and particularly in IBD patients, suggesting its potential as an indicator of intestinal health (Miquel *et al.*, 2013; Sokol *et al.*, 2008; Swidsinski *et al.*, 2008a). Moreover, *F. prausnitzii* is a butyrate producer and has demonstrated anti-inflammatory effects *in vitro* and *in vivo* using mice colitis models making it a key member of the microbiota that may contribute to intestinal homeostasis (Martin *et al.*, 2014a; Sokol *et al.*, 2008). This bacterium also impacts physical and chemical epithelial barrier functions though modulation of tight junction and mucus layer (Carlsson *et al.*, 2013; Wrzosek *et al.*, 2013). Thus, *F. prausnitzii* is a beneficial commensal bacterium that could be a good candidate of our indigenous microbiota to elaborate new prophylactic or therapeutic applications in human health (Miquel *et al.*, 2013).

Another *Bacteroides* species, *B. fragilis*, has been also widely studied. This species is an important anaerobic gut commensal of humans that prevents and cures intestinal inflammation in mice (Mazmanian *et al.*, 2008). *B. fragilis* NCTC 9343 presents a polysaccharide (PSA) promoting lymphoid organogenesis that increases CD4+ T-cells in germ-free rodents (Mazmanian *et al.*, 2005) and corrects the Th1/Th2 balance as has been shown in cellular and animal models (Mazmanian *et al.*, 2005; Mazmanian *et al.*, 2008). This strain corrects gut permeability, alters microbial composition and ameliorates defects in communicative, stereotypic, anxiety-like and sensorimotor behaviours by being capable to modulate some metabolite levels (Hsiao *et al.*, 2013). *Bacteroides fragilis* ATCC23745 plays a role in maintaining physiological expression of heat-shock protein (Hsp)25 and Hsp72 which are cytoprotective in colonocytes (Kojima *et al.*, 2003).

Escherichia coli Nissle 1917 (ECN) was isolated by Professor Alfred Nissle in 1917 during the First World War. This strain has been used as a probiotic for several years. Its capacities are huge. It is capable to induce the expression of the antimicrobial peptide human beta-defensin-d2 (hBD-2) in the human intestinal epithelial cell line Caco-2 *in vitro*. ECN strengthens the intestinal barrier function by inducing pro-inflammatory pathways, mainly NF-κb and AP-1 as well as MAPKs (Schlee *et al.*, 2007; Wehkamp *et al.*, 2004). This effect has been also demonstrated *in vivo*, where faecal hBD-2 peptide was increased by 78% after 3 weeks of *E. coli* Nissle 1917 administration (Mondel *et al.*, 2009). Furthermore, it is able to inhibit the invasion of various gut epithelial cells lines by adherent-invasive *Escherichia coli* (AIEC), *S. typhimurium*, *Y. enterocolitica*, *S. flexneri*, *Legionella pneumophila* and *L. monocytogenes* (Altenhoefer *et al.*, 2004; Boudeau *et al.*, 2003; Deriu *et al.*, 2013) and to restore the barrier dysfunction induced by *E. coli* (EPEC) and *Salmonella dublin in vitro* (Otte and Podolsky, 2004). Some cellular components of ECN have been proposed in cell signalling. For instance, the host cell internalization of peptidoglycan fragment of ECN leads to interaction with Noll-like receptor (NLR) 1 (Bron *et al.*, 2012) and its flagellins are recognized to interact with the TLR-5 (Hayashi *et al.*, 2001; Ogushi *et al.*, 2004; Schlee *et al.*, 2007). Another mechanism of action could be a competition for iron uptake especially to reduce *Salmonella* typhimurium intestinal colonization (Deriu *et al.*, 2013).

Akkermansia muciniphila is an abundant resident of the human intestinal tract (Derrien *et al.*, 2008) and mucus degrading specialist which correlates with health and disease (Joyce and Gahan, 2014). This bacterium modulates pathways involved in establishing homeostasis for basal metabolism and immune tolerance towards commensal microbiota (Derrien *et al.*, 2011) and has been correlated with inflammation, obesity and diabetic parameters (Collado *et al.*, 2007; Ellekilde *et al.*, 2014; Everard *et al.*, 2013; Kang *et al.*, 2013).

Although the lactic acid bacteria (LAB) are not the major members of the human microbiota, due to their easy isolation and growth they have been thoroughly studied and most of the probiotic products available nowadays are based on bacteria belonging to this group. One of the main strains belonging to this group is *Lactobacillus rhamnosus* GG (LGG). LGG (ATCC 53103) was isolated in 1983 from the intestinal tract of a healthy human being by Sherwood Gorbach and Barry Goldin. LGG has shown positive effects in human patients with pouchitis, ulcerative colitis and Crohn's disease (Hormannsperger and Haller, 2010) and have been found to increase faecal IgA levels (Bakker-Zierikzee *et al.*, 2006; He *et al.*, 2005). A clinical study demonstrated that perinatal administration of this probiotic strain reduced the development of atopic eczema in children (Nermes *et al.*, 2011). This effect may be due to the anti-inflammatory properties of this probiotic bacterium. Consumption of LGG by children with atopic dermatitis has been reported to enhance the production of the anti-inflammatory cytokine IL-10 by the host (Pessi *et al.*, 2000). It has been demonstrated that two soluble factors, proteins p75 and p40, present in the culture supernatant of LGG induce the expression of Hsp in a p38- and JNK/MAPK-dependent way (Tao *et al.*, 2006). Additionally, LGG also prevents cytokine-induced apoptosis activating anti-apoptotic Akt in a phosphatidylinositol-3-kinase dependent manner and inhibiting pro-apoptotic p38/MAPK activation (Yan and Polk, 2002; Yan *et al.*, 2007). These factors are also able to modulate hydrogen peroxide induced damage in Caco-2 cells (Seth *et al.*, 2008). Other molecules produced by lactobacilli have been found to have important characteristics. For example, the analysis of the currently known genomic sequences of *Lactobacillus* strains predicts a broad group of bacteriocins active against Gram-positive bacteria such as *Lactococcus, Streptococcus, Staphylococcus, Listeria* and *Mycobacteria* (Altermann *et al.*, 2005; Chaillou *et al.*, 2005; Makarova *et al.*, 2006; Pridmore *et al.*, 2004). For instance, it is well known that *L. salivarius* UCC118 has the ability to protect mice against infection with *Listeria monocytogenes* thanks to the production of a Class II bacteriocin (Corr *et al.*, 2007). Several other lactobacilli strains have been tested in different *in vivo* and *in vitro* test with positive results. In this regard, *L. salivarius* Ls33 and its peptidoglycan were anti-inflammatory in a murine colitis model (Macho Fernandez *et al.*, 2011a; Macho Fernandez *et al.*, 2011b). *L. farciminis* CIP 103136 prevents stress-induced hypersensitivity, increase in colonic paracellular permeability, and colonocyte myosin light chain phosphorylation (Ait-Belgnaoui *et al.*, 2006). This antinociceptive effect occurs via inhibition of contraction of the cytoskeleton of colonic epithelial cells and the subsequent tight junction opening, and may also involve direct or indirect effects of nitric oxide produced by this strain (Ait-Belgnaoui *et al.*, 2006) or a decrease of the stress-induced activation/sensitization of sensory neurons at the spinal and supraspinal level (Ait-Belgnaoui *et al.*, 2009). Furthermore, *L. farciminis* CIP 103136 also improves TNBS-induced colitis (Lamine *et al.*, 2004a), mainly due to the nitric oxide released, the normalization of colonic microbiota, the prevention of bacterial translocation, the enhancement of barrier integrity and a decrease in the mucosal levels of IL-1β (Lamine *et al.*, 2004a,b).

Contrary to the gut environment, lactobacilli are considered as the dominant organisms of the vaginal cavity (Doderlein, 1982) being more than 70% of all microorganisms isolated (Eschenbach *et al.*, 1989; Redondo-Lopez *et al.*, 1990). In this ecosystem, the success obtained so far with the use of a cocktail containing the spermicide resistant *L. rhamnosus* GR-1 and the H_2O_2 producing *L. reuteri* RC-14 strains by themselves or associated with antimicrobial chemotherapy is well known (Anukam *et al.*, 2006; Reid *et al.*, 2003). These indigenous lactobacilli strains have beneficial effects on the host. GR-1 blocked the *in vitro* attachment of uropathogenic bacteria to human uroepithelial cells and prevented onset of urinary tract infection in murine models (Reid *et al.*, 1985) while RC-14 inhibited *Staphylococcus aureus* infection of surgical implants in rats (Gan *et al.*, 2002). Both strains, employed together, were able to interfere with the opportunistic pathogen *Candida albicans* (Kohler *et al.*, 2012; Martinez *et al.*, 2009) and to prevent or treat BV (Homayouni *et al.*, 2014; Hummelen *et al.*, 2010).

All of presented bacteria are examples of commensal bacteria with beneficial effects in human health and some of them are already used to treat patients. Our microbiota is a real richness for future trend in 'microbiomology' research and could be a source of new probiotics that could impact on the gastrointestinal, nervous, and immune systems.

Future trends: could probiotics be indigenous?

Probiotics are defined as 'live micro-organisms which, when administered in adequate amounts, confers a health benefit on the host' (FAO/WHO, 2001). The use of them began to show clinical evidence of their impact on human health (Hungin *et al.*, 2013). The advent of probiotic treatments appears to be a promising 'pharmaco-nutritional' approach to reverse diseases linked to microbiota dysbiosis. Today, most micro-organisms marketed as probiotics in the food industry are lactic acid bacteria, belonging to the genera *Lactobacillus* and *Bifidobacterium* (Foligne *et al.*, 2013). Surprisingly, most of them, used for intestinal disorders management, are not currently present in our microbiota. However, the literature underscores the need for further understanding of the role of probiotics in health and disease (Klein *et al.*, 2010). This observation, associated with the rapid evolution of the knowledge of this complex ecosystem, suggests a large panel of potential new candidates that could be isolated directly from our indigenous microbiota. A good argument for this hypothesis is the recently gained renewed interest in microbiota transplantation. Proposed as a treatment for *Clostridium difficile* colitis, in a randomized controlled trial, faecal microbiota transplantation was shown to be very efficient in more than 80% of the patients leading to increased bacterial diversity similar to healthy subjects (van Nood *et al.*, 2013). Instead of using stools from healthy donors, a cultured strain mix of sufficiently characterized beneficial commensal bacteria is proposed as an alternative (Petrof *et al.*, 2013).

We speculate that newly discovered intestinal bacteria may be used for development of new micro-organisms containing products that could be novel probiotics with health claim. In fact, host physiology, gut maturation, innate and acquired immune responses and metabolism are largely influenced by the metabolic properties of the (gut) microbiota (Gaboriau-Routhiau *et al.*, 2009; Tomas *et al.*, 2013). Moreover, the activity and the composition of microbiota are modulated by external factors making microbiota a highly 'handleable' tissue in humans (De Filippo *et al.*, 2010). Interestingly, some bacteria, depleted in many intestinal disorders that displayed beneficial effects on the host, could be used to counterbalance the dysbiosis linked to certain diseases. It has been recently proposed that the anti-inflammatory *F. prausnitzii* bacterium, could have prophylactic or therapeutic applications in human health (Miquel *et al.*, 2013; Sokol *et al.*, 2008). Today, we could suggest other species as good candidates: *B. thetaiotaomicron, B. fragilis, L. farciminis*, etc. However, the description of several microbial communities of our environment (particularly thus of our gut microbiota) allowed us to identify new bacterial species that were unknown so far for their health benefit. The recent description of the intestinal metagenome (all genomes of the bacterial populations of an environment), confirmed that the richness concerning bacterial species of the human gut microbiome correlates with metabolic markers (Le Chatelier *et al.*, 2013). Therefore, besides being a reservoir of unexploited bacteria for academic research, our microbiota also present potentially beneficial metabolic capacities for human health that could be exploited industrially. In parallel, it could be speculated that intestinal probiotics prepared from commensal bacteria may be efficient at a lower threshold than an exogenous strain since they could occupy their own ecological niches more easily.

References

Ait-Belgnaoui, A., Eutamene, H., Houdeau, E., Bueno, L., Fioramonti, J., and Theodorou, V. (2009). *Lactobacillus* farciminis treatment attenuates stress-induced overexpression of Fos protein in spinal and supraspinal sites after colorectal distension in rats. Neurogastroenterol. Motil. *21*, 567–573, e518–569.

Ait-Belgnaoui, A., Han, W., Lamine, F., Eutamene, H., Fioramonti, J., Bueno, L., and Theodorou, V. (2006). *Lactobacillus* farciminis treatment suppresses stress induced visceral hypersensitivity: a possible action through interaction with epithelial cell cytoskeleton contraction. Gut 55, 1090–1094.

Altenhoefer, A., Oswald, S., Sonnenborn, U., Enders, C., Schulze, J., Hacker, J., and Oelschlaeger, T.A. (2004). The probiotic *Escherichia coli* strain Nissle 1917 interferes with invasion of human intestinal epithelial cells by different enteroinvasive bacterial pathogens. FEMS immunology and medical microbiology *40*, 223–229.

Altermann, E., Russell, W.M., Azcarate-Peril, M.A., Barrangou, R., Buck, B.L., McAuliffe, O., Souther, N., Dobson, A., Duong, T., Callanan, M., *et al.* (2005). Complete genome sequence of the probiotic lactic acid bacterium *Lactobacillus* acidophilus NCFM. Proc. Natl. Acad. Sci. U.S.A. *102*, 3906–3912.

Amsel, R., Totten, P.A., Spiegel, C.A., Chen, K.C., Eschenbach, D., and Holmes, K.K. (1983). Nonspecific vaginitis. Diagnostic criteria and microbial and epidemiologic associations. Am. J. Med. *74*, 14–22.

Angelakis, E., Armougom, F., Million, M., and Raoult, D. (2012). The relationship between gut microbiota and weight gain in humans. Future Microbiol. *7*, 91–109.

Anukam, K., Osazuwa, E., Ahonkhai, I., Ngwu, M., Osemene, G., Bruce, A.W., and Reid, G. (2006). Augmentation of antimicrobial metronidazole therapy of bacterial vaginosis with oral probiotic *Lactobacillus rhamnosus* GR-1 and *Lactobacillus reuteri* RC-14: randomized, double-blind, placebo controlled trial. Microbes Infect. *8*, 1450–1454.

Aroutcheva, A.A., Simoes, J.A., and Faro, S. (2001). Antimicrobial protein produced by vaginal *Lactobacillus* acidophilus that inhibits Gardnerella vaginalis. Infect. Dis. Obstet. Gynecol. *9*, 33–39.

Arumugam, M., Raes, J., Pelletier, E., Le Paslier, D., Yamada, T., Mende, D.R., Fernandes, G.R., Tap, J., Bruls, T., Batto, J.M., *et al.* (2011). Enterotypes of the human gut microbiome. Nature *473*, 174–180.

Backhed, F., Ley, R.E., Sonnenburg, J.L., Peterson, D.A., and Gordon, J.I. (2005). Host-bacterial mutualism in the human intestine. Science *307*, 1915–1920.

Bakker-Zierikzee, A.M., van Tol, E.A.F., Kroes, H., Alles, M.S., Kok, F.J., and Bindels, J.G. (2006). Faecal SIgA secretion in infants fed on pre- or probiotic infant formula. Pediatr. Allergy Immunol. *17*, 134–140.

Baquero, F., and Nombela, C. (2012). The microbiome as a human organ. Clin. Microbiol. Infect. *18 Suppl. 4*, 2–4.

Biswas, B.B., Basu, P.S., and Pal, M.K. (1970). Gram staining and its molcecular mechanism. International review of cytology *29*, 1–27.

Boris, S., and Barbes, C. (2000). Role played by lactobacilli in controlling the population of vaginal pathogens. Microbes Infect. *2*, 543–546.

Boris, S., Suarez, J.E., Vazquez, F., and Barbes, C. (1998). Adherence of human vaginal lactobacilli to vaginal epithelial cells and interaction with uropathogens. Infect. Immun. *66*, 1985–1989.

Boskey, E.R., Cone, R.A., Whaley, K.J., and Moench, T.R. (2001). Origins of vaginal acidity: high D/L lactate ratio is consistent with bacteria being the primary source. Hum. Reprod. *16*, 1809–1813.

Boudeau, J., Glasser, A.L., Julien, S., Colombel, J.F., and Darfeuille-Michaud, A. (2003). Inhibitory effect of probiotic *Escherichia coli* strain Nissle 1917 on adhesion to and invasion of intestinal epithelial cells by adherent-invasive E-coli strains isolated from patients with Crohn's disease. Alimentary pharmacology and therapeutics *18*, 45–56.

Bron, P.A., van Baarlen, P., and Kleerebezem, M. (2012). Emerging molecular insights into the interaction between probiotics and the host intestinal mucosa. Nat. Rev. Microbiol. *10*, 66-U90.

Brotman, R.M. (2011). Vaginal microbiome and sexually transmitted infections: an epidemiologic perspective. J. Clin. Invest. *121*, 4610–4617.

Bruls, T., and Weissenbach, J. (2011). The human metagenome: our other genome? Hum. Mol. Genet. *20*, R142–148.

Cani, P.D., and Delzenne, N.M. (2011). The gut microbiome as therapeutic target. Pharmacology and therapeutics *130*, 202–212.

Carlsson, A.H., Yakymenko, O., Olivier, I., Hakansson, F., Postma, E., Keita, A.V., and Soderholm, J.D. (2013). *Faecalibacterium prausnitzii* supernatant improves intestinal barrier function in mice DSS colitis. Scand. J. Gastroenterol. *48*, 1136–1144.

Chaillou, S., Champomier-Verges, M.C., Cornet, M., Crutz-Le Coq, A.M., Dudez, A.M., Martin, V., Beaufils, S., Darbon-Rongere, E., Bossy, R., Loux, V., *et al.* (2005). The complete genome sequence of the meat-borne lactic acid bacterium *Lactobacillus sakei* 23K. Nat. Biotechnol. *23*, 1527–1533.

Charlson, E.S., Bittinger, K., Haas, A.R., Fitzgerald, A.S., Frank, I., Yadav, A., Bushman, F.D., and Collman, R.G. (2011). Topographical continuity of bacterial populations in the healthy human respiratory tract. Am. J. Resp. Crit. Care Med. *184*, 957–963.

Claesson, M.J., Jeffery, I.B., Conde, S., Power, S.E., O'Connor, E.M., Cusack, S., Harris, H.M., Coakley, M., Lakshminarayanan, B., O'Sullivan, O., *et al.* (2012). Gut microbiota composition correlates with diet and health in the elderly. Nature *488*, 178–184.

Collado, M.C., Derrien, M., Isolauri, E., de Vos, W.M., and Salminen, S. (2007). Intestinal integrity and *Akkermansia muciniphila*, a mucin-degrading member of the intestinal microbiota present in infants, adults, and the elderly. Appl. Environ. Microbiol. *73*, 7767–7770.

Comstock, L.E., and Coyne, M.J. (2003). Bacteroides thetaiotaomicron: a dynamic, niche-adapted human symbiont. BioEssays25, 926–929.

Consortium, T.H.M.P. (2012). A framework for human microbiome research. Nature *486*, 215–221.

Corr, S.C., Li, Y., Riedel, C.U., O'Toole, P.W., Hill, C., and Gahan, C.G.M. (2007). Bacteriocin production as a mechanism for the antfinfective activity of *Lactobacillus salivarius* UCC118. Proc. Natl. Acad. Sci. U.S.A. *104*, 7617–7621.

Cotter, P.D., Ross, R.P., and Hill, C. (2013). Bacteriocins – a viable alternative to antibiotics? Nat. Rev. *11*, 95–105.

Crouzet, L., Gaultier, E., Del'Homme, C., Cartier, C., Delmas, E., Dapoigny, M., Fioramonti, J., and Bernalier-Donadille, A. (2013). The hypersensitivity to colonic distension of IBS patients can be transferred to rats through their fecal microbiota. Neurogastroenterol. Motil. *25*, e272–282.

Darfeuille-Michaud, A. (2002). Adherent-invasive *Escherichia coli*: a putative new E. coli pathotype associated with Crohn's disease. Int. J. Med. Microbiol. *292*, 185–193.

De Cruz, P., Prideaux, L., Wagner, J., Ng, S.C., McSweeney, C., Kirkwood, C., Morrison, M., and Kamm, M.A. (2012). Characterization of the gastrointestinal microbiota in health and inflammatory bowel disease. Inflammatory bowel diseases *18*, 372–390.

De Filippo, C., Cavalieri, D., Di Paola, M., Ramazzotti, M., Poullet, J.B., Massart, S., Collini, S., Pieraccini, G., and Lionetti, P. (2010). Impact of diet in shaping gut microbiota revealed by a comparative study in children from Europe and rural Africa. Proc. Natl. Acad. Sci. U.S.A. *107*, 14691–14696.

De Palma, G., Nadal, I., Medina, M., Donat, E., Ribes-Koninckx, C., Calabuig, M., and Sanz, Y. (2010). Intestinal dysbiosis and reduced immunoglobulin-coated bacteria associated with coeliac disease in children. BMC Microbiol. *10*, 63.

Del Chierico, F., Vernocchi, P., Bonizzi, L., Carsetti, R., Castellazzi, A.M., Dallapiccola, B., de Vos, W., Guerzoni, M.E., Manco, M., Marseglia, G.L., *et al.* (2012). Early-life gut microbiota under physiological and pathological conditions: the central role of combined meta-omics-based approaches. Journal of proteomics *75*, 4580–4587.

Deriu, E., Liu, J.Z., Pezeshki, M., Edwards, R.A., Ochoa, R.J., Contreras, H., Libby, S.J., Fang, F.C., and Raffatellu, M. (2013). Probiotic bacteria reduce salmonella typhimurium intestinal colonization by competing for iron. Cell Host Microbe *14*, 26–37.

Derrien, M., Collado, M.C., Ben-Amor, K., Salminen, S., and de Vos, W.M. (2008). The mucin degrader *Akkermansia muciniphila* is an abundant resident of the human intestinal tract. Appl. Environ. Microbiol. *74*, 1646–1648.

Derrien, M., Van Baarlen, P., Hooiveld, G., Norin, E., Muller, M., and de Vos, W.M. (2011). Modulation of mucosal immune response, tolerance, and proliferation in mice colonized by the mucin-degrader *Akkermansia muciniphila*. Front. Microbiol. 2, 166.

Derrien, M., Vaughan, E.E., Plugge, C.M., and de Vos, W.M. (2004). Akkermansia muciniphila gen. nov., sp. nov., a human intestinal mucin-degrading bacterium. Int. J. Syst Evol. Microbiol. 54, 1469–1476.

Doderlein, A.S.G. (1982). Das Scheindensekret und seine bedeutung fur das puerperalfieber. Leipzig: O Durr.

Doyle, L.M., McInerney, J.O., Mooney, J., Powell, R., Haikara, A., and Moran, A.P. (1995). Sequence of the gene encoding the 16S rRNA of the beer spoilage organism Megasphaera cerevisiae. J. Ind Microbiol. 15, 67–70.

Droste, J.H., Wieringa, M.H., Weyler, J.J., Nelen, V.J., Vermeire, P.A., and Van Bever, H.P. (2000). Does the use of antibiotics in early childhood increase the risk of asthma and allergic disease? Clin. Exp. Allergy 30, 1547–1553.

Eckburg, P.B., Bik, E.M., Bernstein, C.N., Purdom, E., Dethlefsen, L., Sargent, M., Gill, S.R., Nelson, K.E., and Relman, D.A. (2005). Diversity of the human intestinal microbial flora. Science 308, 1635–1638.

Ellekilde, M., Krych, L., Hansen, C.H., Hufeldt, M.R., Dahl, K., Hansen, L.H., Sorensen, S.J., Vogensen, F.K., Nielsen, D.S., and Hansen, A.K. (2014). Characterization of the gut microbiota in leptin deficient obese mice – Correlation to inflammatory and diabetic parameters. Res. Vet. Sci. 96, 241–250.

Eschenbach, D.A., Davick, P.R., Williams, B.L., Klebanoff, S.J., Young-Smith, K., Critchlow, C.M., and Holmes, K.K. (1989). Prevalence of hydrogen peroxide-producing *Lactobacillus* species in normal women and women with bacterial vaginosis. J. Clin. Microbiol. 27, 251–256.

Everard, A., Belzer, C., Geurts, L., Ouwerkerk, J.P., Druart, C., Bindels, L.B., Guiot, Y., Derrien, M., Muccioli, G.G., Delzenne, N.M., *et al.* (2013). Cross-talk between Akkermansia muciniphila and intestinal epithelium controls diet-induced obesity. Proc. Natl. Acad. Sci. U.S.A. *110*, 9066–9071.

FAO/WHO (2001). Report of a joint FAO/WHO expert consultation on evaluation of health and nutritional properties of probiotics in food including powder milk with live lactic acid bacteria. Cordoba, Argentina, 1–4 October 2001.

Fiehn, O. (2002). Metabolomics – the link between genotypes and phenotypes. Plant Mol. Biol. 48, 155–171.

Finegold, S.M., Attebery, H.R., and Sutter, V.L. (1974). Effect of diet on human faecal flora: comparison of Japanese and American diets. Am. J. Clin. Nutr. 27, 1456–1469.

Foligne, B., Daniel, C., and Pot, B. (2013). Probiotics from research to market: the possibilities, risks and challenges. Curr. Opin. Microbiol. 16, 284–292.

Fraher, M.H., O'Toole, P.W., and Quigley, E.M. (2012). Techniques used to characterize the gut microbiota: a guide for the clinician. Nat. Rev. Gastroenterology and hepatology 9, 312–322.

Fredricks, D.N., Fiedler, T.L., and Marrazzo, J.M. (2005). Molecular identification of bacteria associated with bacterial vaginosis. N. Engl. J. Med. 353, 1899–1911.

Gaboriau-Routhiau, V., Rakotobe, S., Lecuyer, E., Mulder, I., Lan, A., Bridonneau, C., Rochet, V., Pisi, A., De Paepe, M., Brandi, G., *et al.* (2009). The key role of segmented filamentous bacteria in the coordinated maturation of gut helper T-cell responses. Immunity 31, 677–689.

Gan, B.S., Kim, J., Reid, G., Cadieux, P., and Howard, J.C. (2002). *Lactobacillus fermentum* RC-14 inhibits *Staphylococcus aureus* infection of surgical implants in rats. J. Infect. Dis. 185, 1369–1372.

Gans, J., Wolinsky, M., and Dunbar, J. (2005). Computational improvements reveal great bacterial diversity and high metal toxicity in soil. Science *309*, 1387–1390.

Garcia-Rodriguez, J.A., and Munoz, J.L. (1991). [Bacterial vaginosis: infectious disease or ecological change]. Enfermedades infecciosas y microbiologia clinica 9, 265–267.

Garrett, W.S., Lord, G.M., Punit, S., Lugo-Villarino, G., Mazmanian, S.K., Ito, S., Glickman, J.N., and Glimcher, L.H. (2007). Communicable ulcerative colitis induced by T-bet deficiency in the innate immune system. Cell *131*, 33–45.

Ghafourian, S., Sekawi, Z., Raftari, M., and Ali, M.S. (2013). Application of proteomics in lab diagnosis. Clin. Lab. 59, 465–474.

Gill, S.R., Pop, M., Deboy, R.T., Eckburg, P.B., Turnbaugh, P.J., Samuel, B.S., Gordon, J.I., Relman, D.A., Fraser-Liggett, C.M., and Nelson, K.E. (2006). Metagenomic analysis of the human distal gut microbiome. Science 312, 1355–1359.

Gollwitzer, E.S., and Marsland, B.J. (2014). Microbiota abnormalities in inflammatory airway diseases – potential for therapy. Pharmacol. Ther. *141*, 32–39.

Grice, E.A., Kong, H.H., Renaud, G., Young, A.C., Bouffard, G.G., Blakesley, R.W., Wolfsberg, T.G., Turner, M.L., and Segre, J.A. (2008). A diversity profile of the human skin microbiota. Genome Res. 18, 1043–1050.

Grice, E.A., and Segre, J.A. (2011). The skin microbiome. Nat. Rev. 9, 244–253.

Haggerty, C.L., Totten, P.A., Ferris, M., Martin, D.H., Hoferka, S., Astete, S.G., Ondondo, R., Norori, J., and Ness, R.B. (2009). Clinical characteristics of bacterial vaginosis among women testing positive for fastidious bacteria. Sex Transm Infect. 85, 242–248.

Hawes, S.E., Hillier, S.L., Benedetti, J., Stevens, C.E., Koutsky, L.A., Wolner-Hanssen, P., and Holmes, K.K. (1996). Hydrogen peroxide-producing lactobacilli and acquisition of vaginal infections. J. Infect. Dis. *174*, 1058–1063.

Hayashi, F., Smith, K.D., Ozinsky, A., Hawn, T.R., Yi, E.C., Goodlett, D.R., Eng, J.K., Akira, S., Underhill, D.M., and Aderem, A. (2001). The innate immune response to bacterial flagellin is mediated by Toll-like receptor 5. Nature *410*, 1099–1103.

He, F., Morita, H., Kubota, A., Ouwehand, A.C., Hosoda, M., Hiramatsu, M., and Kurisaki, J.I. (2005). Effect of orally administered non-viable *Lactobacillus* cells on murine humoral immune responses. Microbiol. Immunol. 49, 995–999.

Homayouni, A., Bastani, P., Ziyadi, S., Mohammad-Alizadeh-Charandabi, S., Ghalibaf, M., Mortazavian, A.M., and Mehrabany, E.V. (2014). Effects of probiotics on the recurrence of bacterial vaginosis: a review. Journal of lower genital tract disease *18*, 79–86.

Hormannsperger, G., and Haller, D. (2010). Molecular crosstalk of probiotic bacteria with the intestinal immune system: clinical relevance in the context of inflammatory bowel disease. Int. J. Med. Microbiol. *300*, 63–73.

Hoskin-Parr, L., Teyhan, A., Blocker, A., and Henderson, A.J. (2013). Antibiotic exposure in the first two years of life and development of asthma and other allergic diseases by 7.5 yr: a dose-dependent relationship. Pediatr. Allergy Immunol. 24, 762–771.

Hsiao, E.Y., McBride, S.W., Hsien, S., Sharon, G., Hyde, E.R., McCue, T., Codelli, J.A., Chow, J., Reisman, S.E., Petrosino, J.F., *et al.* (2013). Microbiota modulate behavioral and physiological abnormalities associated with neurodevelopmental disorders. Cell 155, 1451–1463.

Hugenholtz, P. (2002). Exploring prokaryotic diversity in the genomic era. Genome biology 3, REVIEWS0003.

Hummelen, R., Changalucha, J., Butamanya, N.L., Cook, A., Habbema, J.D., and Reid, G. (2010). *Lactobacillus rhamnosus* GR-1 and L. reuteri RC-14 to prevent or cure bacterial vaginosis among women with HIV. International journal of gynaecology and obstetrics: the official organ of the International Federation of Gynaecology and Obstetrics *111*, 245–248.

Hungin, A.P., Mulligan, C., Pot, B., Whorwell, P., Agreus, L., Fracasso, P., Lionis, C., Mendive, J., Philippart de Foy, J.M., Rubin, G., *et al.* (2013). Systematic review: probiotics in the management of lower gastrointestinal symptoms in clinical practice – an evidence-based international guide. Alimentary pharmacology and therapeutics 38, 864–886.

Hyman, R.W., Fukushima, M., Diamond, L., Kumm, J., Giudice, L.C., and Davis, R.W. (2005). Microbes on the human vaginal epithelium. Proc. Natl. Acad. Sci. U.S.A. *102*, 7952–7957.

Jakaitis, B.M., and Denning, P.W. (2014). Commensal and probiotic bacteria may prevent NEC by maturing intestinal host defences. Pathophysiology.

Joly, F., Mayeur, C., Bruneau, A., Noordine, M.L., Meylheuc, T., Langella, P., Messing, B., Duee, P.H., Cherbuy, C., and Thomas, M. (2010). Drastic changes in faecal and mucosa-associated microbiota in adult patients with short bowel syndrome. Biochimie 92, 753–761.

Joyce, S.A., and Gahan, C.G. (2014). The gut microbiota and the metabolic health of the host. Curr. Opin. Gastroenterol. 30, 120–127.

Kang, C.S., Ban, M., Choi, E.J., Moon, H.G., Jeon, J.S., Kim, D.K., Park, S.K., Jeon, S.G., Roh, T.Y., Myung, S.J., *et al.* (2013). Extracellular vesicles derived from gut microbiota, especially Akkermansia muciniphila, protect the progression of dextran sulfate sodium-induced colitis. PLoS One *8*, e76520.

Kazor, C.E., Mitchell, P.M., Lee, A.M., Stokes, L.N., Loesche, W.J., Dewhirst, F.E., and Paster, B.J. (2003). Diversity of bacterial populations on the tongue dorsa of patients with halitosis and healthy patients. J. Clin. Microbiol. *41*, 558–563.

Kiechle, F.L., and Holland-Staley, C.A. (2003). Genomics, transcriptomics, proteomics, and numbers. Archives of pathology and laboratory medicine *127*, 1089–1097.

Klebanoff, S.J., and Belding, M.E. (1974). Virucidal activity of H2O2-generating bacteria: requirement for peroxidase and a halide. Journal of Infectious Diseases *129*, 345–348.

Klein, M., Sanders, M.E., Duong, T., and Young, H.A. (2010). Probiotics: from bench to market. Annals of the New York Academy of Sciences *1212* (*Suppl. 1*), E1–14.

Koboziev, I., Reinoso Webb, C., Furr, K.L., and Grisham, M.B. (2013). Role of the enteric microbiota in intestinal homeostasis and inflammation. Free radical biology and medicine *68C*, 122–133.

Kohler, G.A., Assefa, S., and Reid, G. (2012). Probiotic interference of *Lactobacillus rhamnosus* GR-1 and *Lactobacillus reuteri* RC-14 with the opportunistic fungal pathogen *Candida albicans*. Infect. Dis. Obstet. Gynecol. *2012*, 636474.

Kojima, K., Musch, M.W., Ren, H.Y., Boone, D.L., Hendrickson, B.A., Ma, A., and Chang, E.B. (2003). Enteric flora and lymphocyte-derived cytokines determine expression of heat shock proteins in mouse colonic epithelial cells. Gastroenterology *124*, 1395–1407.

Lagier, J.C., Million, M., Hugon, P., Armougom, F., and Raoult, D. (2012). Human gut microbiota: repertoire and variations. Frontiers in cellular and infection microbiology *2*, 136.

Lamendella, R., VerBerkmoes, N., and Jansson, J.K. (2012). 'Omics' of the mammalian gut – new insights into function. Current opinion in biotechnology *23*, 491–500.

Lamine, F., Eutamene, H., Fioramonti, J., Bueno, L., and Theodorou, V. (2004a). Colonic responses to *Lactobacillus farciminis* treatment in trinitrobenzene sulphonic acid-induced colitis in rats. Scand. J. Gastroenterol. *39*, 1250–1258.

Lamine, F., Fioramonti, J., Bueno, L., Nepveu, F., Cauquil, E., Lobysheva, I., Eutamene, H., and Theodorou, V. (2004b). Nitric oxide released by *Lactobacillus* farciminis improves TNBS-induced colitis in rats. Scand. J. Gastroenterol. *39*, 37–45.

Lamont, R.F., Sobel, J.D., Akins, R.A., Hassan, S.S., Chaiworapongsa, T., Kusanovic, J.P., and Romero, R. (2011). The vaginal microbiome: new information about genital tract flora using molecular based techniques. BJOG *118*, 533–549.

Le Chatelier, E., Nielsen, T., Qin, J., Prifti, E., Hildebrand, F., Falony, G., Almeida, M., Arumugam, M., Batto, J.M., Kennedy, S., *et al.* (2013). Richness of human gut microbiome correlates with metabolic markers. Nature *500*, 541–546.

Lederberg, J., and McCray, A. (2001). 'Ome Sweet 'Omics – A Genealogical Treasury of Words. Genealogical Treasury of Words Scientist *15*, 8.

Le Roy, T., Llopis, M., Lepage, P., Bruneau, A., Rabot, S., Bevilacqua, C., Martin, P., Philippe, C., Walker, F., Bado, A., *et al.* (2012). Intestinal microbiota determines development of non-alcoholic fatty liver disease in mice. Gut *62*, 1787–1794.

Lepage, P., Hasler, R., Spehlmann, M.E., Rehman, A., Zvirbliene, A., Begun, A., Ott, S., Kupcinskas, L., Dore, J., Raedler, A., *et al.* (2011). Twin study indicates loss of interaction between microbiota and mucosa of patients with ulcerative colitis. Gastroenterology *141*, 227–236.

Leser, T.D., and Molbak, L. (2009). Better living through microbial action: the benefits of the mammalian gastrointestinal microbiota on the host. Environ. Microbiol. *11*, 2194–2206.

Ley, R.E., Hamady, M., Lozupone, C., Turnbaugh, P.J., Ramey, R.R., Bircher, J.S., Schlegel, M.L., Tucker, T.A., Schrenzel, M.D., Knight, R., *et al.* (2008). Evolution of mammals and their gut microbes. Science *320*, 1647–1651.

Ley, R.E., Peterson, D.A., and Gordon, J.I. (2006). Ecological and evolutionary forces shaping microbial diversity in the human intestine. Cell *124*, 837–848.

Littman, D.R., and Pamer, E.G. (2011). Role of the commensal microbiota in normal and pathogenic host immune responses. Cell Host Microbe *10*, 311–323.

Lundin, A., Bok, C.M., Aronsson, L., Bjorkholm, B., Gustafsson, J.A., Pott, S., Arulampalam, V., Hibberd, M., Rafter, J., and Pettersson, S. (2008). Gut flora, Toll-like receptors and nuclear receptors: a tripartite communication that tunes innate immunity in large intestine. Cell. Microbiol. *10*, 1093–1103.

Macdonald, T.T., and Monteleone, G. (2005). Immunity, inflammation, and allergy in the gut. Science *307*, 1920–1925.

Macfarlane, G.T., and Macfarlane, S. (2011). Fermentation in the human large intestine: its physiologic consequences and the potential contribution of prebiotics. J. Clin. Gastroenterol. *45* (*Suppl.*), S120–127.

Macho Fernandez, E., Pot, B., and Grangette, C. (2011a). Beneficial effect of probiotics in IBD: are peptidogycan and NOD2 the molecular key effectors? Gut microbes *2*, 280–286.

Macho Fernandez, E., Valenti, V., Rockel, C., Hermann, C., Pot, B., Boneca, I.G., and Grangette, C. (2011b). Anti-inflammatory capacity of selected lactobacilli in experimental colitis is driven by NOD2-mediated recognition of a specific peptidoglycan-derived muropeptide. Gut *60*, 1050–1059.

Makarova, K., Slesarev, A., Wolf, Y., Sorokin, A., Mirkin, B., Koonin, E., Pavlov, A., Pavlova, N., Karamychev, V., Polouchine, N., *et al.* (2006). Comparative genomics of the lactic acid bacteria. Proc. Natl. Acad. Sci. U.S.A. *103*, 15611–15616.

Manichanh, C., Rigottier-Gois, L., Bonnaud, E., Gloux, K., Pelletier, E., Frangeul, L., Nalin, R., Jarrin, C., Chardon, P., Marteau, P., *et al.* (2006). Reduced diversity of faecal microbiota in Crohn's disease revealed by a metagenomic approach. Gut *55*, 205–211.

Martin, R., Chain, F., Miquel, S., Lu, J., Gratadoux, J.J., Sokol, H., Verdu, E.F., Bercik, P., Bermudez-Humaran, L.G., and Langella, P. (2014a). The commensal bacterium *Faecalibacterium prausnitzii* is protective in DNBS-induced chronic moderate and severe colitis models. Inflamm. Bowel Dis. *20*, 417–430.

Martin, R., Miquel, S., Langella, P., and Bermudez-Humaran, L.G. (2014b). The role of metagenomics in understanding the human microbiome in health and disease. Virulence *5*.

Martin, R., Miquel, S., Ulmer, J., Kechaou, N., Langella, P., and Bermudez-Humaran, L.G. (2013). Role of commensal and probiotic bacteria in human health: a focus on inflammatory bowel disease. Microb. Cell. Fact. *12*, 71.

Martin, R., Sanchez, B., Suarez, J.E., and Urdaci, M.C. (2012). Characterization of the adherence properties of human Lactobacilli strains to be used as vaginal probiotics. FEMS Microbiol. Lett. *328*, 166–173.

Martin, R., Soberon, N., Escobedo, S., and Suarez, J.E. (2009). Bacteriophage induction versus vaginal homeostasis: role of H(2)O(2). in the selection of *Lactobacillus* defective prophages. Int. Microbiol. *12*, 131–136.

Martin, R., Soberon, N., Vaneechoutte, M., Florez, A.B., Vazquez, F., and Suarez, J.E. (2008a). Characterization of indigenous vaginal lactobacilli from healthy women as probiotic candidates. Int. Microbiol. *11*, 261–266.

Martin, R., Soberon, N., Vazquez, F., and Suarez, J.E. (2008b). [Vaginal microbiota: composition, protective role, associated pathologies, and therapeutic perspectives]. Enfermedades Infecciosas y Microbiologia clinica *26*, 160–167.

Martinez, R.C., Seney, S.L., Summers, K.L., Nomizo, A., De Martinis, E.C., And Reid, G. (2009). Effect of *Lactobacillus rhamnosus* GR-1 and *Lactobacillus reuteri* RC-14 on the ability of *Candida albicans* to infect cells and induce inflammation. Microbiol. Immunol. *53*, 487–495.

Mazmanian, S.K., Liu, C.H., Tzianabos, A.O., and Kasper, D.L. (2005). An immunomodulatory molecule of symbiotic bacteria directs maturation of the host immune system. Cell *122*, 107–118.

Mazmanian, S.K., Round, J.L., and Kasper, D.L. (2008). A microbial symbiosis factor prevents intestinal inflammatory disease. Nature *453*, 620–625.

Miquel, S., Martin, R., Rossi, O., Bermudez-Humaran, L.G., Chatel, J.M., Sokol, H., Thomas, M., Wells, J.M., and Langella, P. (2013). *Faecalibacterium prausnitzii* and human intestinal health. Curr. Opin. Microbiol. *16*, 255–261.

Miquel, S., Martín, R., Bridonneau, C., Robert, V., Sokol, H., Bermúdez-Humarán, L., Thomas, M., and Langella, P. (2014). Ecology and metabolism of the beneficial intestinal commensal bacterium *Faecalibacterium prausnitzii*. Gut microbes.

Miquel, S., Peyretaillade, E., Claret, L., de Vallee, A., Dossat, C., Vacherie, B., Zineb el, H., Segurens, B., Barbe, V., Sauvanet, P., *et al.* (2010). Complete genome sequence of Crohn's disease-associated adherent-invasive E. coli strain LF82. PLoS One 5.

Mondel, M., Schroeder, B.O., Zimmermann, K., Huber, H., Nuding, S., Beisner, J., Fellermann, K., Stange, E.F., and Wehkamp, J. (2009). Probiotic E-coli treatment mediates antimicrobial human beta-defensin synthesis and fecal excretion in humans. Mucosal immunology *2*, 166–172.

Moore, W.E., and Holdeman, L.V. (1974). Human fecal flora: the normal flora of 20 Japanese-Hawaiians. Applied microbiology *27*, 961–979.

Munoz-Provencio, D., Perez-Martinez, G., and Monedero, V. (2010). Characterization of a fibronectin-binding protein from *Lactobacillus* casei BL23. J. Appl. Microbiol. *108*, 1050–1059.

Neish, A.S. (2009). Microbes in gastrointestinal health and disease. Gastroenterology *136*, 65–80.

Nermes, M., Kantele, J.M., Atosuo, T.J., Salminen, S., and Isolauri, E. (2011). Interaction of orally administered *Lactobacillus rhamnosus* GG with skin and gut microbiota and humoral immunity in infants with atopic dermatitis. Clin. Exp. Allergy *41*, 370–377.

Noverr, M.C., and Huffnagle, G.B. (2005). The 'microflora hypothesis' of allergic diseases. Clin. Exp. Allergy 35, 1511–1520.

Noverr, M.C., Noggle, R.M., Toews, G.B., and Huffnagle, G.B. (2004). Role of antibiotics and fungal microbiota in driving pulmonary allergic responses. Infect. Immun. *72*, 4996–5003.

Oakley, B.B., Fiedler, T.L., Marrazzo, J.M., and Fredricks, D.N. (2008). Diversity of human vaginal bacterial communities and associations with clinically defined bacterial vaginosis. Appl. Environ. Microbiol. *74*, 4898–4909.

Ocana, V.S., Pesce De Ruiz Holgado, A.A., and Nader-Macias, M.E. (1999). Characterization of a bacteriocin-like substance produced by a vaginal *Lactobacillus salivarius* strain. Appl. Environ. Microbiol. *65*, 5631–5635.

Ogushi, K., Wada, A., Niidome, T., Okuda, T., Llanes, R., Nakayama, M., Nishi, Y., Kurazono, H., Smith, K.D., Aderem, A., *et al.* (2004). Gangliosides act as co-receptors for Salmonella enteritidis FliC and promote FliC induction of human beta-defensin-2 expression in Caco-2 cells. J. Biol. Chem. *279*, 12213–12219.

Olszak, T., An, D., Zeissig, S., Vera, M.P., Richter, J., Franke, A., Glickman, J.N., Siebert, R., Baron, R.M., Kasper, D.L., *et al.* (2012). Microbial exposure during early life has persistent effects on natural killer T-cell function. Science 336, 489–493.

Orphan, V.J., Taylor, L.T., Hafenbradl, D., and Delong, E.F. (2000). Culture-dependent and culture-independent characterization of microbial assemblages associated with high-temperature petroleum reservoirs. Appl. Environ. Microbiol. *66*, 700–711.

Osset, J., Bartolome, R.M., Garcia, E., and Andreu, A. (2001a). Assessment of the capacity of *Lactobacillus* to inhibit the growth of uropathogens and block their adhesion to vaginal epithelial cells. J. Infect. Dis. *183*, 485–491.

Osset, J., Garcia, E., Bartolome, R.M., and Andreu, A. (2001b). [Role of *Lactobacillus* as protector against vaginal candidiasis]. Med. Clin. (Barc.) *117*, 285–288.

Otte, J.M., and Podolsky, D.K. (2004). Functional modulation of enterocytes by Gram-positive and Gram-negative microorganisms. Am. J. Physiol-Gastr L *286*, G613-G626.

Pascual, L., Ruiz, F., Giordano, W., and Barberis, I.L. (2010). Vaginal colonization and activity of the probiotic bacterium *Lactobacillus fermentum* L23 in a murine model of vaginal tract infection. J. Med. Microbiol. *59*, 360–364.

Pavlova, S.I., Kilic, A.O., Kilic, S.S., So, J.S., Nader-Macias, M.E., Simoes, J.A., and Tao, L. (2002). Genetic diversity of vaginal lactobacilli from women in different countries based on 16S rRNA gene sequences. J. Appl. Microbiol. *92*, 451–459.

Pavlova, S.I., Kilic, A.O., Mou, S.M., and Tao, L. (1997). Phage infection in vaginal lactobacilli: an *in vitro* study. Infect. Dis. Obstet. Gynecol. *5*, 36–44.

Pessi, T., Sutas, Y., Hurme, M., and Isolauri, E. (2000). Interleukin-10 generation in atopic children following oral *Lactobacillus rhamnosus* GG. Clin. Exp. Allergy *30*, 1804–1808.

Petrof, E.O., Gloor, G.B., Vanner, S.J., Weese, S.J., Carter, D., Daigneault, M.C., Brown, E.M., Schroeter, K., and Allen-Vercoe, E. (2013). Stool substitute transplant therapy for the eradication of *Clostridium difficile* infection: 'RePOOPulating' the gut. Microbiome *1*, 3.

Pinto, A.C., Melo-Barbosa, H.P., Miyoshi, A., Silva, A., and Azevedo, V. (2011). Application of RNA-seq to reveal the transcript profile in bacteria. Genetics and molecular research: GMR *10*, 1707–1718.

Pridmore, R.D., Berger, B., Desiere, F., Vilanova, D., Barretto, C., Pittet, A.C., Zwahlen, M.C., Rouvet, M., Altermann, E., Barrangou, R., *et al.* (2004). The genome sequence of the probiotic intestinal bacterium *Lactobacillus* johnsonii NCC 533. Proc. Natl. Acad. Sci. U.S.A. *101*, 2512–2517.

Qin, J., Li, R., Raes, J., Arumugam, M., Burgdorf, K.S., Manichanh, C., Nielsen, T., Pons, N., Levenez, F., Yamada, T., *et al.* (2010). A human gut microbial gene catalogue established by metagenomic sequencing. Nature *464*, 59–65.

Rajilic-Stojanovic, M., Biagi, E., Heilig, H.G., Kajander, K., Kekkonen, R.A., Tims, S., and de Vos, W.M. (2011). Global and deep molecular analysis of microbiota signatures in fecal samples from patients with irritable bowel syndrome. Gastroenterology *141*, 1792–1801.

Rakoff-Nahoum, S., and Medzhitov, R. (2008). Innate immune recognition of the indigenous microbial flora. Mucosal Immunol. *1* (Suppl. 1), S10–14.

Rakoff-Nahoum, S., Paglino, J., Eslami-Varzaneh, F., Edberg, S., and Medzhitov, R. (2004). Recognition of commensal microflora by toll-like receptors is required for intestinal homeostasis. Cell *118*, 229–241.

Redondo-Lopez, V., Cook, R.L., and Sobel, J.D. (1990). Emerging role of lactobacilli in the control and maintenance of the vaginal bacterial microflora. Rev. Infect. Dis. *12*, 856–872.

Reid, G., Chan, R.C., Bruce, A.W., and Costerton, J.W. (1985). Prevention of urinary tract infection in rats with an indigenous *Lactobacillus* casei strain. Infect. Immun. *49*, 320–324.

Reid, G., Charbonneau, D., Erb, J., Kochanowski, B., Beuerman, D., Poehner, R., and Bruce, A.W. (2003). Oral use of *Lactobacillus rhamnosus* GR-1 and L. fermentum RC-14 significantly alters vaginal flora: randomized, placebo-controlled trial in 64 healthy women. FEMS Immunol. Med. Microbiol. *35*, 131–134.

Resta-Lenert, S., and Barrett, K.E. (2006). Probiotics and commensals reverse TNF-alpha- and IFN-gamma-induced dysfunction in human intestinal epithelial cells. Gastroenterology *130*, 731–746.

Round, J.L., and Mazmanian, S.K. (2009). The gut microbiota shapes intestinal immune responses during health and disease. Nat. Rev. Immunol. 9, 313–323.

Sabia, C., Anacarso, I., Bergonzini, A., Gargiulo, R., Sarti, M., Condo, C., Messi, P., de Niederhausern, S., Iseppi, R., and Bondi, M. (2014). Detection and partial characterization of a bacteriocin-like substance produced by *Lactobacillus* fermentum CS57 isolated from human vaginal secretions. Anaerobe.

Sanchez, B., Bressollier, P., and Urdaci, M.C. (2008). Exported proteins in probiotic bacteria: adhesion to intestinal surfaces, host immunomodulation and molecular cross-talking with the host. FEMS Immunol. Med. Microbiol. *54*, 1–17.

Sartor, R.B. (2008). Microbial influences in inflammatory bowel diseases. Gastroenterology *134*, 577–594.

Schlee, M., Wehkamp, J., Altenhoefer, A., Oelschlaeger, T.A., Stange, E.F., and Fellermann, K. (2007). Induction of human beta-defensin 2 by the probiotic *Escherichia coli* nissle 1917 is mediated throuah flagellin. Infect. Immun. *75*, 2399–2407.

Sekirov, I., Russell, S.L., Antunes, L.C., and Finlay, B.B. (2010). Gut microbiota in health and disease. Physiological reviews *90*, 859–904.

Seng, P., Drancourt, M., Gouriet, F., La Scola, B., Fournier, P.E., Rolain, J.M., and Raoult, D. (2009). Ongoing revolution in bacteriology: routine identification of bacteria by matrix-assisted laser desorption ionization time-of-flight mass spectrometry. Clin. Infect. Dis. *49*, 543–551.

Seth, A., Yan, F., Polk, D.B., and Rao, R.K. (2008). Probiotics ameliorate the hydrogen peroxide-induced epithelial barrier disruption by a PKC- and MAP kinase-dependent mechanism. American journal of physiology Gastrointest. Liver Physiol. *294*, G1060–1069.

Simoes, J.A., Aroutcheva, A., Heimler, I., Shott, S., and Faro, S. (2001). Bacteriocin susceptibility of *Gardnerella vaginalis* and its relationship to biotype, genotype, and metronidazole susceptibility. Am. J. Obstet. Gynecol. *185*, 1186–1190.

Sjogren, Y.M., Tomicic, S., Lundberg, A., Bottcher, M.F., Bjorksten, B., Sverremark-Ekstrom, E., and Jenmalm, M.C. (2009). Influence of early gut microbiota on the maturation of childhood mucosal and systemic immune responses. Clin. Exp. Allergy *39*, 1842–1851.

Sobel, J.D. (2000). Bacterial vaginosis. Annual review of medicine *51*, 349–356.

Sobhani, I., Tap, J., Roudot-Thoraval, F., Roperch, J.P., Letulle, S., Langella, P., Corthier, G., Tran Van Nhieu, J., and Furet, J.P. (2011). Microbial dysbiosis in colorectal cancer (CRC) patients. PLoS One *6*, e16393.

Sokol, H., Pigneur, B., Watterlot, L., Lakhdari, O., Bermudez-Humaran, L.G., Gratadoux, J.J., Blugeon, S., Bridonneau, C., Furet, J.P., Corthier, G., *et al.* (2008). *Faecalibacterium prausnitzii* is an anti-inflammatory commensal bacterium identified by gut microbiota analysis of Crohn disease patients. Proc. Natl. Acad. Sci. U.S.A. *105*, 16731–16736.

Sommer, F., and Backhed, F. (2013). The gut microbiota – masters of host development and physiology. Nat. Rev. *11*, 227–238.

Stappenbeck, T.S., Hooper, L.V., and Gordon, J.I. (2002). Developmental regulation of intestinal angiogenesis by indigenous microbes via Paneth cells. Proc. Natl. Acad. Sci. U.S.A. *99*, 15451–15455.

Strachan, D.P. (1989). Hay fever, hygiene, and household size. BMJ *299*, 1259–1260.

Swidsinski, A., Loening-Baucke, V., Vaneechoutte, M., and Doerffel, Y. (2008a). Active Crohn's disease and ulcerative colitis can be specifically diagnosed and monitored based on the biostructure of the fecal flora. Inflammatory bowel diseases *14*, 147–161.

Swidsinski, A., Loening-Baucke, V., Verstraelen, H., Osowska, S., and Doerffel, Y. (2008b). Biostructure of fecal microbiota in healthy subjects and patients with chronic idiopathic diarrhea. Gastroenterology *135*, 568–579.

Tao, Y., Drabik, K.A., Waypa, T.S., Musch, M.W., Alverdy, J.C., Schneewind, O., Chang, E.B., and Petrof, E.O. (2006). Soluble factors from *Lactobacillus* GG activate MAPKs and induce cytoprotective heat shock proteins in intestinal epithelial cells (vol 290, pg C1018, 2006). Am. J. Physiol-Cell Ph *291*, C194-C194.

Tap, J., Mondot, S., Levenez, F., Pelletier, E., Caron, C., Furet, J.P., Ugarte, E., Munoz-Tamayo, R., Paslier, D.L., Nalin, R., *et al.* (2009). Towards the human intestinal microbiota phylogenetic core. Environ. Microbiol. *11*, 2574–2584.

Thorsen, P., Jensen, I.P., Jeune, B., Ebbesen, N., Arpi, M., Bremmelgaard, A., and Moller, B.R. (1998). Few microorganisms associated with bacterial vaginosis may constitute the pathologic core: a population-based microbiologic study among 3596 pregnant women. Am. J. Obstet. Gynecol. *178*, 580–587.

Tjalsma, H., Lambooy, L., Hermans, P.W., and Swinkels, D.W. (2008). Shedding & shaving: disclosure of proteomic expressions on a bacterial face. Proteomics *8*, 1415–1428.

Tomas, J., Wrzosek, L., Bouznad, N., Bouet, S., Mayeur, C., Noordine, M.L., Honvo-Houeto, E., Langella, P., Thomas, M., and Cherbuy, C. (2013). Primocolonization is associated with colonic epithelial maturation during conventionalization. Faseb J. *27*, 645–655.

Turnbaugh, P.J., Ley, R.E., Mahowald, M.A., Magrini, V., Mardis, E.R., and Gordon, J.I. (2006). An obesity-associated gut microbiome with increased capacity for energy harvest. Nature *444*, 1027–1031.

Umesaki, Y., Okada, Y., Matsumoto, S., Imaoka, A., and Setoyama, H. (1995). Segmented filamentous bacteria are indigenous intestinal bacteria that activate intraepithelial lymphocytes and induce MHC class II molecules and fucosyl asialo GM1 glycolipids on the small intestinal epithelial cells in the ex-germ-free mouse. Microbiol. Immunol. *39*, 555–562.

Ursell, L.K., Metcalf, J.L., Parfrey, L.W., and Knight, R. (2012). Defining the human microbiome. Nutr. Rev. *70* (Suppl. 1), S38–44.

van Nood, E., Vrieze, A., Nieuwdorp, M., Fuentes, S., Zoetendal, E.G., de Vos, W.M., Visser, C.E., Kuijper, E.J., Bartelsman, J.F., Tijssen, J.G., *et al.* (2013). Duodenal infusion of donor feces for recurrent *Clostridium difficile*. N. Engl. J. Med. *368*, 407–415.

Vartoukian, S.R., Palmer, R.M., and Wade, W.G. (2010). Strategies for culture of 'unculturable' bacteria. FEMS Microbiol. Lett. *309*, 1–7.

Velraeds, M.M., van de Belt-Gritter, B., Busscher, H.J., Reid, G., and van der Mei, H.C. (2000). Inhibition of uropathogenic biofilm growth on silicone rubber in human urine by lactobacilli – a teleologic approach. World J. Urol. *18*, 422–426.

Velraeds, M.M., van de Belt-Gritter, B., van der Mei, H.C., Reid, G., and Busscher, H.J. (1998). Interference in initial adhesion of uropathogenic bacteria and yeasts to silicone rubber by a *Lactobacillus* acidophilus biosurfactant. J. Med. Microbiol. *47*, 1081–1085.

Vielfort, K., Sjolinder, H., Roos, S., Jonsson, H., and Aro, H. (2008). Adherence of clinically isolated lactobacilli to human cervical cells in competition with Neisseria gonorrhoeae. Microbes Infect. *10*, 1325–1334.

Vijay-Kumar, M., Aitken, J.D., Carvalho, F.A., Cullender, T.C., Mwangi, S., Srinivasan, S., Sitaraman, S.V., Knight, R., Ley, R.E., and Gewirtz, A.T. (2010). Metabolic syndrome and altered gut microbiota in mice lacking Toll-like receptor 5. Science *328*, 228–231.

Vollaard, E.J., and Clasener, H.A. (1994). Colonization resistance. Antimicrob. Agents Chemother. *38*, 409–414.

von Mutius, E. (2007). Allergies, infections and the hygiene hypothesis – the epidemiological evidence. Immunobiology *212*, 433–439.

Vyas, U., and Ranganathan, N. (2012). Probiotics, prebiotics, and synbiotics: gut and beyond. Gastroenterology research and practice *2012*, 872716.

Wang, Y., and Kasper, L.H. (2013). The role of microbiome in central nervous system disorders. Brain Behaviour Immun. *38*, 1–12.

Wehkamp, J., Harder, J., Wehkamp, K., Wehkamp-von Meissner, B., Schlee, M., Enders, C., Sonnenborn, U., Nuding, S., Bengmark, S., Fellermann, K., *et al.* (2004). NF-kappaB- and AP-1-mediated induction of human beta defensin-2 in intestinal epithelial cells by *Escherichia coli* Nissle 1917: a novel effect of a probiotic bacterium. Infect. Immun. *72*, 5750–5758.

Weinstock, G.M. (2012). Genomic approaches to studying the human microbiota. Nature *489*, 250–256.

Wells, J.M., Rossi, O., Meijerink, M., and van Baarlen, P. (2011). Epithelial crosstalk at the microbiota–mucosal interface. Proc. Natl. Acad. Sci. U.S.A. *108* (*Suppl. 1*), 4607–4614.

Wills-Karp, M., Santeliz, J., and Karp, C.L. (2001). The germless theory of allergic disease: revisiting the hygiene hypothesis. Nat. Rev. Immunol. *1*, 69–75.

Wrzosek, L., Miquel, S., Noordine, M.L., Bouet, S., Chevalier-Curt, M.J., Robert, V., Philippe, C., Bridonneau, C., Cherbuy, C., Robbe-Masselot, C., *et al.* (2013). Bacteroides thetaiotaomicron and *Faecalibacterium prausnitzii* influence the production of mucus glycans and the development of goblet cells in the colonic epithelium of a gnotobiotic model rodent. BMC Biol. *11*, 61.

Wu, G.D., Chen, J., Hoffmann, C., Bittinger, K., Chen, Y.Y., Keilbaugh, S.A., Bewtra, M., Knights, D., Walters, W.A., Knight, R., *et al.* (2011). Linking long-term dietary patterns with gut microbial enterotypes. Science *334*, 105–108.

Yamamoto, T., Zhou, X., Williams, C.J., Hochwalt, A., and Forney, L.J. (2009). Bacterial populations in the vaginas of healthy adolescent women. J. Pediatr. Adolesc. Gynecol. *22*, 11–18.

Yan, F., Cao, H., Cover, T.L., Whitehead, R., Washington, M.K., and Polk, D.B. (2007). Soluble proteins produced by probiotic bacteria regulate intestinal epithelial cell survival and growth. Gastroenterology *132*, 562–575.

Yan, F., and Polk, D.B. (2002). Probiotic bacterium prevents cytokine-induced apoptosis in intestinal epithelial cells. J. Biol. Chem. *277*, 50959–50965.

Yu, L.C., Wang, J.T., Wei, S.C., and Ni, Y.H. (2012). Host–microbial interactions and regulation of intestinal epithelial barrier function: From physiology to pathology. World journal of gastrointestinal pathophysiology 3, 27–43.

Zarate, G., and Nader-Macias, M.E. (2006). Influence of probiotic vaginal lactobacilli on *in vitro* adhesion of urogenital pathogens to vaginal epithelial cells. Lett. Appl. Microbiol. *43*, 174–180.

Zocco, M.A., Ainora, M.E., Gasbarrini, G., and Gasbarrini, A. (2007). Bacteroides thetaiotaomicron in the gut: molecular aspects of their interaction. Digestive and liver disease: official journal of the Italian Society of Gastroenterology and the Italian Association for the Study of the Liver 39, 707–712.

Zoetendal, E.G., Vaughan, E.E., and de Vos, W.M. (2006). A microbial world within us. Mol. Microbiol. *59*, 1639–1650.

Improving the Digestive Tract Robustness of Probiotic Lactobacilli

13

Hermien van Bokhorst-van de Veen, Peter A. Bron and Michiel Kleerebezem

Abstract

This chapter describes stress responses of probiotic lactobacilli, in relation to gastrointestinal (GI) tract robustness. An overview is given of some newly developed tools to understand and improve stress responses of the model probiotic *L. plantarum* WCFS1 in relation to its GI persistence. These include a relative simple GI tract assay and the development of transcriptome-trait matching that associates a trait of interest (e.g. GI tract survival) with transcripts levels that enables the identification of genetic robustness markers, that can be modulated by pre-adaptation and/or genetic engineering to enhance robustness. Furthermore, specific additives to the *in situ* delivery matrix may enhance the relative tolerance of specific bacterial strains to detrimental conditions they may encounter in different regions of the GI tract. Moreover, a methodology that allows the molecular quantification of single strains in a mixed bacterial population using engineered sequence-tags or naturally occurring discriminatory intergenic alleles in combination with next-generation sequencing offers another powerful tool for robustness evaluation. Finally, the stress responses of probiotic cultures in relation to improvement of their GI persistence and some future directions for development and GI tract research in the light of probiotic performance are discussed.

Gastrointestinal tract robustness

Probiotics are defined as 'live microorganisms which, when administrated in adequate amounts, confer a health benefit on the host' (FAO/WHO, 2001). To be able to exert their beneficial effects in the intestine, it is a prerequisite for probiotic cultures to counteract the stressful conditions encountered during production, shelf life, and exposure to the harsh conditions of the (upper) digestive tract (Corcoran *et al.*, 2008; Mills *et al.*, 2011). A straightforward strategy to determine GI persistence is to subject probiotic bacteria to *in vitro* GI tract mimicking assays. Several of these assays have already been developed previously, and many of these models characteristically employ relatively simple assays that subject bacteria to a low-pH solution to resemble stomach conditions, and/or neutral pH solutions containing bile salts to resembling duodenal or small intestinal conditions (Bove *et al.*, 2013; Corcoran *et al.*, 2007; Lebeer *et al.*, 2008). Alternatively, also more sophisticated GI tract simulators such as the TNO Intestinal Models and the Simulator of Human Intestinal Microbial Ecosystem can be employed for the same purpose (Kovatcheva-Datchary *et al.*, 2009; Marteau *et al.*, 1997; Molly *et al.*, 1996). Van Bokhorst-van de Veen *et al.* used a relative simple human GI tract simulating assay that consists of initial exposure to stomach mimicking conditions (low pH, pepsin and lipase), followed by neutralization and exposure to pancreatic juice containing pancreatin and bile acids. This assay can monitor survival of bacterial cultures in a standardized and relative high-throughput manner, encompassing exposure to human intestine mimicking conditions and compounds in relevant concentrations [e.g. bile acids (Govers *et al.*, 1994)] (van Bokhorst-van de Veen *et al.*, 2012a). In a generic sense, the GI tract survival assay can readily be employed to compare and contrast the relative survival rates of (i) the same strain grown at different conditions that may adapt it to higher GI tract robustness (ii) wild-type (*L. plantarum* WCFS1) and mutant derivatives of a strain (iii) different strains of a specific species (iv) strains that belong to different species (e.g. *Lactobacillus* or *Bifidobacterium*), or (v) the impact of specific additives or composition characteristics of the matrix in which bacteria are delivered within the GI tract system. The following paragraphs will exemplify each of these possibilities, largely employing *L. plantarum* WCFS1 as an exemplary model microbe for probiotic lactobacilli.

A transcriptome-trait matching approach was applied to associates a trait of interest, such as *in vitro* determined GI tract survival, with transcripts levels. The variation of *in vitro* determined GI tract survival rates of different cultures of *L. plantarum* WCFS1 was correlated to specific fermentation conditions as well as the corresponding *L. plantarum* transcriptomes (van Bokhorst-van de Veen *et al.*, 2012a). The fermentation conditions used encompassed a variety of mild-stress conditions and by combining stress conditions, the number of fermentations could be reduced while still allowing the evaluation of the effects of single or multiple stress conditions in combination (Bron *et al.*, 2012). Using these fermentation conditions, parallel analysis of genome-wide transcriptomics and physiological characteristics (e.g. maximum growth rate, yield, and organic acid profiles) of *L. plantarum* WCFS1, correlations between fermentation conditions and industrially relevant physiological characteristics could be identified (Bron *et al.*, 2012). It appeared that the presence of sodium chloride (NaCl) decreased GI tract survival, while growth at a lower pH positively affected survival. Overall, the work revealed that small changes in growth conditions of *L. plantarum* WCFS1 could elicit large differences in GI survival [up to 7-log (van Bokhorst-van de Veen *et al.*, 2012a)]. Moreover, transcriptome-trait matching enabled the identification of *L. plantarum* WCFS1 transcripts that play a role in survival in the *in vitro* GI tract persistence assay, encoding a sodium–proton pump, a penicillin binding protein, and a transcription factor. The latter gene could subsequently be shown to influence the expression of a capsular polysaccharide gene cluster (van Bokhorst-van de Veen *et al.*, 2012a). Notably, all these three functions appeared to be associated with cell envelope functions but could not *a priori* be predicted to play a role in GI tract survival. Not only transcriptome-trait matching allows the identification of genes involved in complex phenotypes like GI tract persistence, also other 'omics' based technologies (e.g. proteomics) can be employed to unravel stress tolerance related functions (den Besten *et al.*, 2010). These approaches can expand our understanding of the stress-specific and general stress response networks operating in bacteria in relation to their contribution to tolerance to detrimental conditions. The transcriptome-trait matching approaches can also be employed to decipher the involvement of genes in other complex phenotypes.

The GI tract assay also allowed the comparative evaluation of survival rates of wild-type and mutant strains or a panel of different strains of a species. For instance, the *L. plantarum* WCFS1 derivatives lacking the genes encoding CtsR and/or HrcA, canonical stress regulators that play a central role in the adaptations to a plethora of stresses in a variety of organisms, appeared to be unaffected in their GI tract persistence in this assay (Van Bokhorst-van de Veen *et al.*, 2013a). However, as mentioned above, deletion derivatives lacking the *napA3* (encoding a sodium–proton antiporter), *pbp2A* (involved in peptidoglycan biosynthesis), or *lp_1669* (a transcription regulator) genes displayed a significantly increased survival of more than 10-fold (van Bokhorst-van de Veen *et al.*, 2012a).

Comparative analysis of different strains of *L. plantarum* using the GI tract assay, showed that up to 7 log differences in GI persistence exists between the best- and poorest-survivors within a collection of 42 *L. plantarum* strains, and that the model strain *L. plantarum* WCFS1 belongs to the more resistant strains (van Bokhorst-van de Veen *et al.*, 2012b). Another approach that makes use of genetic variations among a series of *L. plantarum* strains is the gene-trait matching method. This approach has been proven successful in identifying specific genes that are involved in eliciting immunomodulatory responses in dendritic or peripheral blood mononuclear cells, or are required for mannose adhesion that is associated with specific innate immune responses in pig mucosal tissues *in vivo* (Gross *et al.*, 2008; Meijerink *et al.*, 2010; Pretzer *et al.*, 2005; van Hemert *et al.*, 2010). However, when employing phenotypic variations between strains of *L. plantarum* that were obtained in the *in vitro* assay for GI tract persistence, the observed strain-specific phenotypic variations could not be credibly linked to a specific gene or a set of genes (van Bokhorst-van de Veen *et al.*, 2012b). This may imply that complex phenotypes (like GI tract survival and persistence) may not be determined by 'simple' gene presence and absence variations, but are related to fine-tuned gene-regulatory responses that involve conserved gene repertoires.

Comparative analyses of different strains of different species supported also this large difference between individual bacteria that belong to the typical 'probiotic' genera *Lactobacillus* and *Bifidobacterium* (Table 13.1). For *L. plantarum* the predominant factor that influenced GI-persistence *in vitro* was the severe acid stress encountered in the stomach, while the exposure to small intestinal conditions that included bile acid exposure hardly affected the strains of this species (van Bokhorst-van de Veen *et al.*, 2012a,b). The *in vivo* persistence curves obtained from a human volunteer study appear to support this notion of predominant stomach-associated killing of *L. plantarum* (van Bokhorst-van de Veen *et al.*, 2012b). However, Table 13.1 shows that specific species or strains tested were more sensitive to small-intestinal conditions, i.e. pancreatic juice and bile acid exposure compared with stomach-like conditions, i.e. gastric juice. This group included most of the *L. rhamnosus* strains that were tested. Notably, the level of sensitivity of these strains for pancreatic juice and bile acid exposure was increased by pre-exposure to the stomach-mimicking acid conditions (Table 13.1). In conclusion, quantitative comparison of relative survival of different bacterial species and strains, revealed substantial differences between these bacteria, which is in agreement with several other studies that reported on species and strain variations in acid and bile tolerance (Berrada *et al.*, 1991; Jacobsen *et al.*, 1999).

The GI tract assay was also employed to illustrate the protective effect that specific delivery matrix constituents may have on the survival of a specific bacterium (Fig. 13.1). Analogously, it has been shown that acid stress survival of *L. casei* strains Zhang and ATCC 334 could be improved by addition of arginine and/or aspartate (Wu *et al.*, 2013a; Zhang *et al.*, 2012), or histidine (Broadbent *et al.*, 2010), respectively. Moreover, addition of glutathione to the media significantly improved the resistance of *L. sanfranciscensis* DSM20451 against cold stress induced by freeze-drying, freeze–thawing, and cold treatment (4°C) (Zhang *et al.*, 2010a), while the same additive also improved the acid stress tolerance of *L. salivarius* strain 200 (Lee *et al.*, 2010) and *Leuconostoc mesenteroides* ATCC 8293 (Kim et al., 2012), and improved the tolerance to osmotic stress of *Lactococcus lactis* SK11 (Zhang *et al.*, 2010b). The observations made with glutathione illustrate the generic impact of oxidative stress tolerance on overall stress resistance and robustness of bacterial cells. Increased tolerance of *L. rhamnosus* GG against (simulated) gastric juice exposure could be achieved by the addition of glucose in the gastric juice (Corcoran *et al.*, 2005) or Tween 80 during culturing (Corcoran *et al.*, 2007). The latter additive induced a 1,000-fold higher survival and the authors suggest that changes in the membrane composition due to incorporation of the fatty acid in the membrane

Table 13.1 Strain-dependent responses of the GI-tract survival assay

Species	Strain	St resistance*	SI tolerance*
L. rhamnosus	GG#	↑	–
	NCIMB 8824#	↑	↓↓↓
	LMG 10772	↓↓↓	↓↓↓
	LMG 6400#	↑	↓↓
L. acidophilus	NCFM	↑↑	↑
	LAFTI-10	–	–
	I233	↓↓↓	↓↓↓
	LMG 9433	↑↑	↑
L. reuteri	DSM20016	↑↑	↑↑
	100-23	↑↑	↑↑
	DSM 17938	↑	↑
	LMG 9213	↑↑	↑↑
L. casei	P2	↓↓	↓↓
	ATCC334	↓↓↓	↓↓↓
	BL23	↓↓	↓↓↓
	LMG 6904	↓↓↓	↓↓↓
L. helveticus	DPC 4571	↓↓	↓↓↓
	CNRZ32	↓↓	↓↓
L. delbrueckii ssp. *bulgaricus*	LMG6901	↓↓↓	↓↓↓
	ATCC BAA365	↓↓↓	↓↓↓
B. animalis ssp. *lactis*	Bb12	↓	↓
	HN019	–	–
B. longum	LMG 13196	↓↓↓	↓↓↓
	LMG 18899	↓↓↓	↓↓

St, Stomach; SI, small intestine; *L.*, *Lactobacillus*; *B.*, *Bifidobacterium*.
* Compared to *L. plantarum* WCFS.
– no change.
These strains were more sensitive to SI juice when they were first challenged with St juice compared to no preceding St juice incubation.

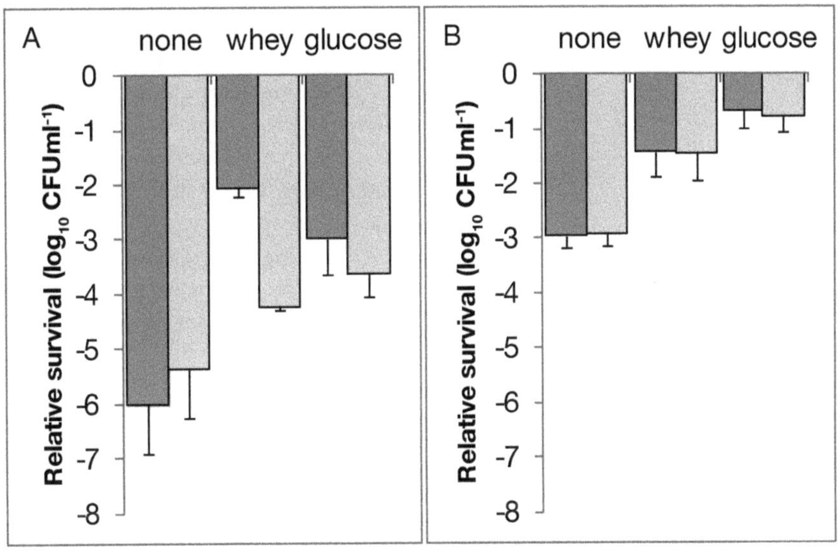

Figure 13.1 Relative survival of *L. plantarum* WCFS1, subjected to an upper GI tract mimicking assay in the absence or presence of additives. L. plantarum cultures were grown in MRS at 37°C. The cultures were harvested at mid-exponential phase (OD_{600}=1.0, A) or stationary phase (25 h after inoculation, B) and subjected to an upper GI tract mimicking assay containing no additive, whey protein isolate (1 mg/ml), or glucose (1.5%). Dark grey bars represent the colony-forming units (CFUs) after 60 min gastric juice incubation and light grey bars represent CFUs after subsequent 60 min small intestine incubation. Input is set at 0 Log_{10} cfu/ml, data presented are averages of technical sextuplicates (– standard deviation).

and the consequences for membrane fluidity are most likely the explanation for the observed enhanced survival. The observation that an energized state is essential for robustness was also made in our studies, as the survival of *L. plantarum* cells during the GI tract assay could be increased over 10-fold by the addition of 1.5% glucose to the bacterial suspension that was subjected to the assay (Fig. 13.1). Moreover, subsequent experiments revealed that the addition of only 0.05% glucose was sufficient to improve GI tract assay survival. Importantly, metabolization of the available carbohydrate is essential for the observed enhanced survival effect, since the stimulation of *L. plantarum* survival in this assay was only achieved by addition of the fermentable D-glucose, and not by the addition of the non-fermentable L-glucose (Fig. 13.2). The fermentable nature of the carbohydrate added implies that ATP generation is most likely the survival enhancing consequence, which is in agreement with the acid tolerance dependency on the F_0F_1-ATPase as suggested by Corcoran and coworkers (Corcoran *et al.*, 2005). Other additives that were evaluated in our studies included whey proteins. This additive also appeared to improve survival rates during the GI tract assay *in vitro* (Fig. 13.1), which was especially apparent for *L. plantarum* WCFS1 when it was taken from the logarithmic phase of growth for which the survival could be enhanced 1,000-fold by the addition of whey proteins. The molecular mechanism by which the whey proteins can improve survival during the GI tract assay remains unknown. To evaluate whether higher survival in the presence of glucose or whey protein is strain specific, several strains of different

Table 13.2 Strain-dependent effect of glucose (1.5%) or whey protein isolate (1 mg/ml) addition during GI-tract survival

Species	Strain	Improved by
L. plantarum	WCFS1	Glucose Whey
L. acidophilus	NCFM	
	LAFTI-10	Glucose
L. rhamnosus	GG	Glucose Whey
	NCIMB8824	Glucose
	LMG10772	Glucose
	LMG6400	Glucose Whey
L. casei	P2	Glucose Whey
	ATCC334	Glucose
	BL23	Glucose Whey
	LMG6904	Glucose Whey
L. helveticus	DPC4571	Glucose Whey
	CNRZ32	Glucose Whey
L. delbrueckii ssp *bulgaricus*	LMG6901	Glucose Whey
	ATCC BAA365	Glucose Whey
B. animalis ssp *lactis*	Bb12	Glucose Whey
	HN019	Glucose Whey
B. longum	LMG18899	Glucose Whey

L., Lactobacillus; B., Bifidobacterium.

lactobacilli and bifidobacteria were subjected to the digestive tract conditions with or without these additives (Table 13.2). All species tested, except *L. acidophilus* NCFM displayed an improved survival rate upon the addition of glucose during the GI tract assay. This indicates that the capacity for energy generation more or less universally protects bacteria against acid-stress conditions in the chosen experimental set-up, which most likely involves the conserved F_0F_1-ATPase (Cotter and Hill, 2003). Analogously, whey proteins improved GI tract survival in more than half of the tested strains and species.

Taken together, the experiments described above showcase that a relatively simple GI tract assay can aid tremendously in revealing the digestive tract survival mechanism of probiotic cultures. However, it should be emphasized that the cultures tested here were grown under laboratory conditions without freeze-drying and other industrial processing. Further research is needed to evaluate whether the assay also is predictive for strains grown under industrial conditions.

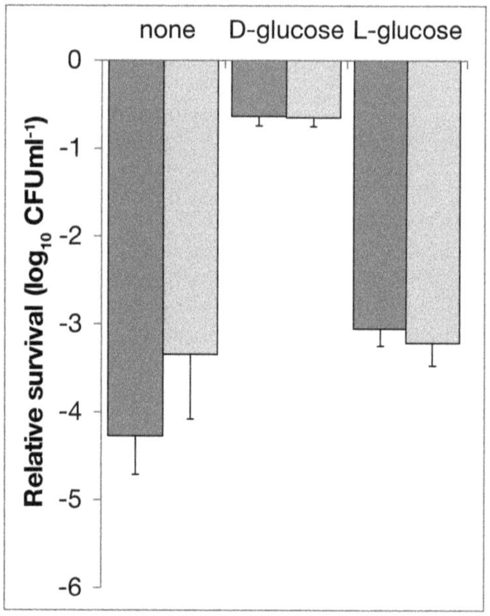

Figure 13.2 Relative survival of *L. plantarum* WCFS1, subjected to an upper GI tract mimicking assay in the absence of L-glucose or D-glucose. *L. plantarum* cultures were grown in 2×CDM at 37°C. The cultures were harvested at stationary phase (25h after inoculation) and subjected to an upper GI tract mimicking assay containing no additive, D-glucose (1.5%) or L-glucose (1.5%). Dark grey bars represent the colony forming units (CFUs) after 60 min gastric juice incubation and light grey bars represent CFUs after subsequent 60 min small intestine incubation. Input is set at 0 Log_{10} cfu/ml, data presented are averages of technical sextuplicates (– standard deviation).

Genetic markers for quantitative, sequence based *in situ* strain-tracking

Employing strain-specific detection methods to differentially quantify survival of individual strains that are mixed is very attractive in animal or human intervention studies, as such an experimental set-up requires fewer animals or volunteers compared with parallel evaluation of individual strains. Several methods are available that discriminate mixed strains and include molecular typing techniques using DNA and restriction enzymes, PCR based fingerprinting techniques, or Real-Time PCR. Less laborious and more high-throughput are techniques that make use of genome-inserted markers such as antibiotic resistance genes, transposons, or sequence-tags in closely related strains. Advantages of these markers are that competition experiments can be performed by following the growth of individual strains in a standardized mixture of strains to evaluate how the strains in the mixture influence and/or compete-with each other. Notably, all genetically modified *L. plantarum* WCFS1 derivatives that are currently constructed in our laboratory contain a unique sequence-tag at the genetic site of mutation, which enables high-throughput, parallel-, sequence-based analysis of population dynamics in mixtures of these mutants. Moreover, similar unique sequence tags can relatively easily be inserted in the genome of *L. plantarum* WCFS1 mutant derivatives that were constructed prior to the implementation of this consistent mutant-tagging procedure. Alternatively, strain specific tags could be inserted into the genomes of different strains of the species *L. plantarum*, especially when the genomic site chosen for insertion is conserved among the strains (allowing one mutagenesis construct to be applied on all strains) and all strains are genetically accessible at a level that is sufficient to allow genetic engineering. As an alternative to engineering strain specific sequence tags into a collection of strains of a species, one could exploit the naturally occurring diversity among strains. This principal of genetic allele variations among strains is already exploited in procedures like multi-locus sequence typing (MLST) that is employed for strain typing. Another next-generation sequencing-based approach that targets a *L. plantarum* variable intergenic region for sequence-based strain discrimination was developed. Notably, this method does not require genetic-engineering and employs wild-type strains. The value of this approach was illustrated in the evaluation of the *in vivo* GI tract persistence of strains of *L. plantarum* when administered in mixtures to healthy human volunteers. Ten volunteers consumed a mixture of 10 *L. plantarum* strains and a remarkable consistency of the strain-specific persistence curves were observed between individual volunteers, which also correlated significantly with the GI tract survival predicted on basis of the *in vitro* assay described above (van Bokhorst-van de Veen *et al.*, 2012b). Up to 10 strains could be combined and this number may be further increased when more sequence varieties of this intergenic region of *L. plantarum* strains are discovered. Alternatively, other hypervariable sequence regions in the genomes of *L. plantarum* strains may be discovered that enable a larger group of strains to be differentially quantified on basis of massive parallel sequence analysis. Notably, the principal approach employed here can be extrapolated and employed to other probiotic species. Especially in the current era where strain specific genome sequencing is becoming feasible through next-generation sequencing approaches and the constantly reducing financial investment required for such efforts.

Stress responses and improving robustness

It is clear that gene-regulatory responses are important for robustness phenotypes. Pre-adaptation occurs when sub-lethal stress conditions elicit the activation of stress response networks that can protect against detrimental conditions, including stress conditions of progressive intensity or stress conditions other than the initial sublethal stress, this latter phenomenon is called cross-protection. Pre-exposure to lower pH growth conditions was shown to improve the survival under more severe acid-stress conditions (such as the GI tract assay) for *L. plantarum* WCFS1 (van Bokhorst-van de Veen *et al.*, 2012a). Pre-adaptation to sub-lethal stress conditions during (initial) production steps can be used to improve the bacterial robustness during subsequent production steps, storage, and/or *in situ* delivery, if applicable (Corcoran *et al.*, 2008). For example, *L. acidophilus* CRL 639 and LA1–1 cells pre-exposed to acid or bile stress were protected against lethal concentrations of the stresses (Kim *et al.*, 2001; Lorca *et al.*, 2002). This pre-adaptation phenomenon was also observed for *L. casei* Zhang (Wu *et al.*, 2013a), *L. sanfranciscensis* CB1 (De Angelis *et al.*, 2001), *Lactococcus lactis* IL1403 (Hartke *et al.*, 1997) and several *Streptococcus* strains (Zotta *et al.*, 2008) during acid challenge. Also sublethal heat stress enhanced stress tolerance of thermophilic streptococci (Zotta *et al.*, 2008). Furthermore, *L. plantarum* strains acclimated in medium containing 6 or 10% ethanol induced an increased survival when challenged with 14% ethanol (Bravo-Ferrada *et al.*, 2013). Parente *et al.*, performed studies on the stress response of 63 strains of the taxonomically related species *L. plantarum*, *L. paraplantarum*, and *L. pentosus* and showed improved heat tolerance when the strains were subjected to mild temperature (Parente *et al.*, 2010). They also found evidence of a generalized stress response in the stationary phase, with increase of starvation stress tolerance, but also improved oxidative and heat tolerance for this group (Parente *et al.*, 2010; Ricciardi *et al.*, 2012). For the adaptive heat stress response also an increased tolerance to acid and detergent was detected (Parente *et al.*, 2010). Latter examples illustrate that pre-adaptation to a certain stress can not only improve survival of bacteria when challenged with the lethal dose of the stress, but can also give protection against other stresses. This cross-protection could allow improvement of robustness without the application of undesired conditions or compounds (e.g. bile acids), during industrial production. *L. plantarum* WCFS1 grown under anaerobic conditions and exposed to low pH showed improved survival compared to cultures grown under respiration conditions (Watanabe *et al.*, 2012). The authors associated this improved survival with the reduced formation of radicals and higher membrane integrity of the cells. Nevertheless, the same *L. plantarum* strain aerobically grown had improved tolerance against heat, oxidative and cold/starvation stress (Zotta *et al.*, 2012). Another example of cross-protection is the lactose starved *L. acidophilus* RD758 cultures that displayed a significantly better resistance to freezing and frozen storage at −20°C compared to the non-starved cells (Wang *et al.*, 2011). Survival of *L. rhamnosus* HN001 during storage at 30°C appeared slightly enhanced when cells were pretreated with osmotic stress conditions (Prasad *et al.*, 2003). In conclusion, these approaches can lead to significant improvement of generic stress robustness of several important starter and probiotic strains. Van Bokhorst-van de Veen *et al.*, unravelled the gene repertoires elicited by solvent stress (ethanol) in *L. plantarum*. Ethanol exposure led to induced expression of genes involved in citrate metabolism and cell envelope architecture, as well as canonical stress response pathways controlled by the central stress regulators HrcA and CtsR. Moreover, a cross protective effect of sublethal ethanol exposure to subsequent 'unrelated' stresses was observed (van Bokhorst-van de Veen *et al.*, 2011). To this end, *L. plantarum* was grown in media that contained sub-lethal ethanol concentrations and were subsequently exposed to other stresses like heat, oxidative and GI-track mimicking stress conditions (van Bokhorst-van de Veen *et al.*, 2011). Sublethal ethanol stress exposure responses in *L. plantarum* were able to induce cross-protection against heat stress, but not against stresses encountered in the GI tract assay. Solvents are predicted to predominantly interact with lipid bilayers (Weber and de Bont, 1996), and can be anticipated to destabilize bacterial membranes. Small heat-shock proteins function as chaperones that assist the protein-folding process by stabilizing unfolded or partially folded proteins (Han *et al.*, 2008) and have been reported to interact and stabilize the phospholipid bilayer (Narberhaus, 2002). Notably, small heat shock protein Lo18 from *Oenococcus oeni* IOB 84.13 is membrane-associated and its expression is induced by addition of benzyl alcohol (Delmas *et al.*, 2001). In addition, Hsp2, another small heat shock protein, can affect membrane fluidity in *L. plantarum* WCFS1 (Capozzi *et al.*, 2011) and was induced upon extended ethanol exposure in this strain (van Bokhorst-van de Veen *et al.*, 2011). In *B. subtilis* 168, the class I stress chaperones DnaK and GroEL were associated with the cell membrane upon short-term ethanol exposure, and displayed enhanced kinase activity under these conditions, suggesting their contribution to membrane-function maintenance upon solvent exposure (Seydlova *et al.*, 2012). In *L. plantarum* WCFS1, GroEL and DnaK are predicted to be regulated by the HrcA repressor (Wels *et al.*, 2006) and thereby belong to the Class I heatshock regulon, which also encompasses the other proteins encoded within the *groE* and *dnaK* operons, GroES, GrpE, and DnaJ (Wels *et al.*, 2006). Next to the Class I regulon, *L. plantarum* also encodes members of the typical Class III stress regulon that is controlled by the dedicated repressor CtsR. The members of the CtsR regulon in *L. plantarum* were experimentally determined and include the predicted members of this class III stress regulon, ClpP, ClpB, ClpE, ClpC, Hsp1, and FtsH (Fiocco *et al.*, 2009; Spano and Massa, 2006; Wels *et al.*, 2011). Although strain specific differences in the regulons associated with stress responses in LAB have been described (Spano and Massa, 2006), the core stress-responses in LAB appear to be conserved, including the predicted regulations of the CtsR and HrcA regulators (Corcoran *et al.*, 2008; De Angelis and Gobbetti, 2004; van de Guchte *et al.*, 2002; Varmanen and Savijoki, 2011). Many reports describe the induction of expression of CtsR and/or HrcA regulon-members upon bacterial exposure to different stresses. Examples include increased expression of GroE and ClpP in *Lactococcus lactis* MG1363 after exposure to heat, acid, or UV-irradiation (Frees and Ingmer, 1999; Hartke *et al.*, 1997), induction of *L. mesenteroides* NRRL B-5 12F DnaK and GroEL upon cold-shock (Salotra *et al.*, 1995), and GroES, GroEL and DnaK induction upon osmotic upshift in *Lactococcus lactis* MG1363 (Kilstrup *et al.*, 1997). To unravel the role of the Class I and Class III stress regulators (HrcA and CtsR, respectively) in *L. plantarum* WCFS1, mutant-derivative strains of *ctsR*, *hrcA* or both, were constructed. Growth of the *ctsR* mutant at elevated temperature (42°C) was impaired in *L. plantarum* (Van Bokhorst-van de Veen *et al.*, 2013a). It appeared that the general impact of CtsR and HrcA transcription factor deficiencies were temperature-dependent and encompassed an impressive network of genes, encompassing many functional categories (Van Bokhorst-van de Veen *et al.*, 2013a). The single mutations of either *hrcA* or *ctsR* led to altered transcription patterns of genes encoding functions involved in transport and binding of sugars and other compounds, primary metabolism, transcription regulation, capsular polysaccharide biosynthesis, as well as fatty acid metabolism. Moreover, mutation of both regulators, elicited expression changes of a large variety of additional genes in a temperature-dependent manner, including many genes that encode functions associated with cell-envelope architecture, suggesting substantial cell envelope remodelling in this mutant. The mutant studies highlighted the interaction of class I and III stress regulons and illustrated the complex gene-regulatory networks involved in adaptive responses to stress conditions in *L. plantarum*.

Nowadays, a vast genetic toolbox for the manipulation of LAB gene expression levels is available, allowing complementary approaches to induce or repress expression levels of genes encoding stress regulon members, regulators or other genes, prior to stress exposure. Examples of such an approach are the heterologous expression of the DNA repair protein RecO in *L. casei* Zhang, which improved its acid, salt and oxygen stress tolerance (Wu *et al.*, 2013b). Also co-expression of superoxide dismutase and catalase in *L. rhamnosus* showed a 400-fold increased survival after incubation with hydrogen peroxide (An *et al.*, 2011). Moreover, viability remained log 6 per ml during 7 days' storage under aerobic conditions, while no living cells of the control strain were detected. Inactivation of the gene encoding a pleiotropic regulator of carbon and nitrogen metabolism (*ccpA*) in *L. plantarum* increased its oxidative, heat, and cold/starvation stress tolerance when grown aerobically (Zotta *et al.*, 2012). An acid-resistant mutant of *L. casei* Zhang was obtained by adaptive evolution and displayed a higher membrane fluidity, higher proportions of unsaturated fatty acids, and higher mean chain length, as well as a more intact cellular structure and lower membrane permeability after environmental acidification (Wu *et al.*, 2012). However, the opposite, which is that mutant strains showed decreased stress tolerance, is also reported. For instance, two H^+-ATPase defective variants of *L. delbrueckii* subsp. *bulgaricus* were more sensitive to acidic and NaCl conditions, indicating that H^+-ATPase might be involved in Na^+ transportation besides H^+. Nevertheless, the post-acidification during chilled storage of yoghurt fermented with the mutant strains was inhibited and this increased the shelf-life of the product. High salt, low pH, or oxidative stress tolerance might be gained by an overproducing polyphosphate strain, since polyphosphate kinase gene (*ppk*)-disrupted lactobacilli showed diminished growth under high salt or low pH and an increased sensitivity to oxidative stress (Alcantara *et al.*, 2014); however, this needs to be validated. The authors suggest that polyphosphate accumulation may play important roles in the physiology of these microorganisms. Heterologous expression of the *shsp* gene, encoding a small heat shock protein, of *Streptococcus thermophilus* in *Lactococcus lactis* exhibited higher tolerance to acid, heat, ethanol, bile salt, and hydrogen peroxide stresses (Tian *et al.*, 2012), while overexpression of *dnaK* induced tolerance to acid, NaCl, and heat stresses (Abdullah Al *et al.*, 2010). Overall, this illustrates the possibilities to improve bacterial robustness by genetic approaches.

In summary, several strategies can be exploited to achieve more robust bacterial strains, ranging from pre-genomic approaches like modifying the growth medium and exploiting cross-protection strategies or genetic engineering to elevate or repress the expression of stress factors or their regulators, to post-genomic strategies that could exploit approaches analogous to the fermentation genomics platform described here for the prediction and improvement of robustness effector molecules.

Future directions

Although the concept of probiotics dates back more than a century (Metchnikoff, 1907) they have only been extensively researched during the past few decades. To move forward in the probiotic field, several things require improvement. The European Food Safety Authority (EFSA) is an organization of the European Union that provides independent scientific advice and communication on existing and emerging risks associated with food and feed. The majority of the currently marketed probiotics targets prophylactic health benefits and thus claims to provide a reduction of disease-risk to healthy subjects. However, some products also claim to treat health-compromises or diseases and thus provide a therapeutic benefit to specific sub-populations. Until now, the EFSA rejected all health claims associated with probiotic products (Guarner *et al.*, 2011), indicating the lack of (evidence for) well-functioning probiotics.

To date it remains largely unknown what molecular mechanism of interaction in the host gastrointestinal tract underlies the proposed health benefit elicited by probiotic strains. Moreover, to measure health beneficial effects in healthy human subjects has proven to be very difficult, since there are no reliable markers for health improvement in healthy populations (Rauch and Lynch, 2012; Remus *et al.*, 2011). Probiotic modes of action have been proposed to include (i) strengthening of epithelial integrity (ii) reduction of infection risk through competitive exclusion of pathogens or via the production of antimicrobial activity (iii) modulation of host immune responses, or (iv) contribution to the *in situ* metabolic conversions (Bron *et al.*, 2011; Lebeer *et al.*, 2010; Remus *et al.*, 2011). Recent studies, have also illustrated that probiotics may influence the functioning of the gut–brain axis, a bidirectional neurohumoral communication system that can be affected by the gut microbiota (Collins *et al.*, 2012). However, the latter domain of probiotic applications is still in its infancy, it is largely based on studies in (germfree) animal models and requires translational studies to evidence effects in humans as well as knowledge of the underlying mechanisms. Nevertheless, this can be a novel area of probiotic application and research and may be fruitful in the treatment of specific behavioural diseases.

Mechanistic insight in probiotic functioning and identification of the bacterial effector molecules involved in the observed health-stimulatory effects is required to improve the predictability of probiotic effects, and to further substantiate the strain specific dogma that is proposed for probiotic effects (Bron *et al.*, 2013). The discovery of probiotic effector molecules and approaches to enhance their expression *in situ* at the site of probiotic action, could strengthen the position of these health-promoting cultures in the functional food market. In addition, methods to accurately monitor the improved health state of the consumer can be deduced from the molecular host responses underlying the proposed health promotion, which could strengthen the efficacy read-out of probiotic clinical trials. A clear prerequisite of probiotic products that can be deduced from the probiotic definition, is their viable state, which implies that the bacteria should survive the stresses that they encounter during for instance product preparation and storage (Corcoran *et al.*, 2008). However, expression of probiotic effector molecules may not correlate with maximum survival of the bacteria, but may depend on specific environmental conditions that are encountered *in situ* in the GI tract. Nevertheless, such *in situ* responses will also depend on the bacterial viability and robustness.

As mentioned above, clinical trials are essential to corroborate probiotic effects and are a prerequisite in probiotic research. Mechanistic insight and analysis of *in situ* expression of probiotic effector molecule is bound to require live bacteria, although depending on the mechanism of action also specific cultivation conditions that maximize effector molecule expression during production may contribute to health benefit effects of the corresponding products. Live delivery to the intestinal tract is a prerequisite when *in situ* expression of effector molecules is the basis of the probiotic effect, and thus may benefit from GI tract adapted strains with improved survival and persistence characteristics. These features can be enhanced in existing cultures through adaptive evolution employing relatively simple regimes of multiple, subsequent passages of the intestinal tract and consistently isolating the more persistent derivatives (van Bokhorst-van de Veen *et al.*, 2013b). To facilitate selective culturing of the administered lactobacilli from faecal material, one could employ intrinsic antibiotic resistances of these microbes (e.g. tetracycline, streptomycin for *L. plantarum*), or introduce specific resistances that do not require genetic modification such as rifampin resistance. The latter resistance marker was employed to selectively recover a panel of lactic acid bacteria, including *L. plantarum*, from saliva to determine their relative persistence in the human oral cavity (Snel *et al.*, 2011) or the human GI tract (van Bokhorst-van de Veen *et al.*, 2012b).

In an ideal world, after measuring several parameters, consumption of the right probiotic strain that is fermented, stored, and delivered under the optimal conditions, will confer the desired beneficial effect on consumer health.

References

Abdullah Al, M., Sugimoto, S., Higashi, C., Matsumoto, S., and Sonomoto, K. (2010). Improvement of multiple-stress tolerance and lactic acid production in *Lactococcus lactis* NZ9000 under conditions of thermal stress by heterologous expression of *Escherichia coli* DnaK. Appl. Environ. Microbiol. *76*, 4277–4285.

Alcantara, C., Blasco, A., Zuniga, M., and Monedero, V. (2014). Accumulation of polyphosphate in *Lactobacillus* spp. and its involvement in stress resistance. Appl. Environ. Microbiol. *80*, 1650–1659.

An, H., Zhai, Z., Yin, S., Luo, Y., Han, B., and Hao, Y. (2011). Coexpression of the superoxide dismutase and the catalase provides remarkable oxidative stress resistance in *Lactobacillus rhamnosus*. J. Agric. Food Chem. *59*, 3851–3856.

Berrada, N., Lemeland, J.F., Laroche, G., Thouvenot, P., and Piaia, M. (1991). *Bifidobacterium* from fermented milks: survival during gastric transit. J. Dairy Sci. *74*, 409–413.

Bove, P., Russo, P., Capozzi, V., Gallone, A., Spano, G., and Fiocco, D. (2013). *Lactobacillus plantarum* passage through an oro-gastro-intestinal tract simulator: Carrier matrix effect and transcriptional analysis of genes associated to stress and probiosis. Microbiol. Res. *168*, 351–359.

Bravo-Ferrada, B.M., Tymczyszyn, E.E., Gomez-Zavaglia, A., and Semorile, L. (2013). Effect of acclimation medium on cell viability, membrane integrity and ability to consume malic acid in synthetic wine by oenological *Lactobacillus plantarum* strains. J. Appl. Microbiol. doi: 10.1111/jam.12372. [Epub ahead of print]

Broadbent, J.R., Larsen, R.L., Deibel, V., and Steele, J.L. (2010). Physiological and transcriptional response of *Lactobacillus casei* ATCC 334 to acid stress. J. Bacterol. *192*, 2445–2458.

Bron, P.A., Tomita, S., Mercenier, A., and Kleerebezem, M. (2013). Cell surface-associated compounds of probiotic lactobacilli sustain the strain-specificity dogma. Curr. Opin. Microbiol. *16*, 262–269.

Bron, P.A., Van Baarlen, P., and Kleerebezem, M. (2011). Emerging molecular insights into the interaction between probiotics and the host intestinal mucosa. Nat. Rev. Microbiol. *10*, 66–78.

Bron, P.A., Wels, M., Bongers, R.S., van Bokhorst-van de Veen, H., Wiersma, A., Overmars, L., Marco, M.L., and Kleerebezem, M. (2012). Transcriptomes reveal genetic signatures underlying physiological variations imposed by different fermentation conditions in *Lactobacillus plantarum*. PLoS One *7*, e38720.

Capozzi, V., Weidmann, S., Fiocco, D., Rieu, A., Hols, P., Guzzo, J., and Spano, G. (2011). Inactivation of a small heat shock protein affects cell morphology and membrane fluidity in *Lactobacillus plantarum WCFS1*. Res. Microbiol. *162*, 419–425.

Collins, S.M., Surette, M., and Bercik, P. (2012). The interplay between the intestinal microbiota and the brain. Nat. Rev. Microbiol. *10*, 735–742.

Corcoran, B.M., Stanton, C., Fitzgerald, G., and Ross, R.P. (2008). Life under stress: The probiotic stress response and how it may be manipulated. Current Pharmaceutical Design *14*, 1382–1399.

Corcoran, B.M., Stanton, C., Fitzgerald, G.F., and Ross, R.P. (2005). Survival of probiotic lactobacilli in acidic environments is enhanced in the presence of metabolizable sugars. Appl. Environ. Microbiol. *71*, 3060–3067.

Corcoran, B.M., Stanton, C., Fitzgerald, G.F., and Ross, R.P. (2007). Growth of probiotic lactobacilli in the presence of oleic acid enhances subsequent survival in gastric juice. Microbiology-Sgm *153*, 291–299.

Cotter, P.D., and Hill, C. (2003). Surviving the acid test: Responses of Gram-positive bacteria to low pH. Microbiol. Mol. Biol. Rev. *67*, 429-+.

De Angelis, M., Bini, L., Pallini, V., Cocconcelli, P.S., and Gobbetti, M. (2001). The acid-stress response in *Lactobacillus sanfranciscensis* CB1. Microbiology *147*, 1863–1873.

De Angelis, M., and Gobbetti, M. (2004). Environmental stress responses in *Lactobacillus*: A review. Proteomics *4*, 106–122.

Delmas, F., Pierre, F., Coucheney, F., Divies, C., and Guzzo, J. (2001). Biochemical and physiological studies of the small heat shock protein Lo18 from the lactic acid bacterium *Oenococcus oeni*. J. Mol. Microbiol. Biotechnol. *3*, 601–610.

den Besten, H.M., Arvind, A., Gaballo, H.M., Moezelaar, R., Zwietering, M.H., and Abee, T. (2010). Short- and long-term biomarkers for bacterial robustness: a framework for quantifying correlations between cellular indicators and adaptive behavior. PLoS One *5*, e13746.

FAO/WHO (2001). Evaluation of health and nutritional properties of powder milk with live lactic acid bacteria. Report of FAO/WHO expert consultation 1–4 October.

Fiocco, D., Collins, M., Muscariello, L., Hols, P., Kleerebezem, M., Msadek, T., and Spano, G. (2009). The *Lactobacillus plantarum ftsH* gene is a novel member of the CtsR stress response regulon. J. Bacterol. *191*, 1688–1694.

Frees, D., and Ingmer, H. (1999). ClpP participates in the degradation of misfolded protein in *Lactococcus lactis*. Mol. Microbiol. *31*, 79–87.

Govers, M.J., Termont, D.S., Van Aken, G.A., and Van der Meer, R. (1994). Characterization of the adsorption of conjugated and unconjugated bile acids to insoluble, amorphous calcium phosphate. J. Lipid Res. *35*, 741–748.

Gross, G., van der Meulen, J., Snel, J., van der Meer, R., Kleerebezem, M., Niewold, T.A., Hulst, M.M., and Smits, M.A. (2008). Mannose-specific interaction of *Lactobacillus plantarum* with porcine jejunal epithelium. FEMS Immunol. Med. Microbiol. *54*, 215–223.

Guarner, F., Sanders, M.E., Gibson, G., Klaenhammer, T., Cabana, M., Scott, K., Reid, G., Delzenne, N.M., Fahey, G.C., and Hill, C. (2011). Probiotic and prebiotic claims in Europe: seeking a clear roadmap. Br. J. Nutr. *106*, 1765–1767.

Han, M.J., Yun, H., and Lee, S.Y. (2008). Microbial small heat shock proteins and their use in biotechnology. Biotechnol. Adv. *26*, 591–609.

Hartke, A., Frere, J., Boutibonnes, P., and Auffray, Y. (1997). Differential induction of the chaperonin GroEL and the Co-chaperonin GroES by heat, acid, and UV-irradiation in *Lactococcus lactis* subsp. *lactis*. Curr. Microbiol. *34*, 23–26.

Jacobsen, C.N., Rosenfeldt Nielsen, V., Hayford, A.E., Moller, P.L., Michaelsen, K.F., Paerregaard, A., Sandstrom, B., Tvede, M., and Jakobsen, M. (1999). Screening of probiotic activities of forty-seven strains of *Lactobacillus* spp. by *in vitro* techniques and evaluation of the colonization ability of five selected strains in humans. Appl. Environ. Microbiol. *65*, 4949–4956.

Kilstrup, M., Jacobsen, S., Hammer, K., and Vogensen, F.K. (1997). Induction of heat shock proteins DnaK, GroEL, and GroES by salt stress in *Lactococcus lactis*. Appl. Environ. Microbiol. *63*, 1826–1837.

Kim, J.E., Eom, H.J., Kim, Y., Ahn, J.E., Kim, J.H., and Han, N.S. (2012). Enhancing acid tolerance of *Leuconostoc mesenteroides* with glutathione. Biotechnol. Lett. *34*, 683–687.

Kim, W.S., Perl, L., Park, J.H., Tandianus, J.E., and Dunn, N.W. (2001). Assessment of stress response of the probiotic *Lactobacillus acidophilus*. Curr. Microbiol. *43*, 346–350.

Kovatcheva-Datchary, P., Egert, M., Maathuis, A., Rajilic-Stojanovic, M., de Graaf, A.A., Smidt, H., de Vos, W.M., and Venema, K. (2009). Linking phylogenetic identities of bacteria to starch fermentation in an *in vitro* model of the large intestine by RNA-based stable isotope probing. Environ. Microbiol. *11*, 914–926.

Lebeer, S., Vanderleyden, J., and De Keersmaecker, S.C. (2008). Genes and molecules of lactobacilli supporting probiotic action. Microbiol. Mol. Biol. Rev. *72*, 728–764.

Lebeer, S., Vanderleyden, J., and De Keersmaecker, S.C. (2010). Host interactions of probiotic bacterial surface molecules: comparison with commensals and pathogens. Nat. Rev. Microbiol. *8*, 171–184.

Lee, K., Pi, K., Kim, E.B., Rho, B.S., Kang, S.K., Lee, H.G., and Choi, Y.J. (2010). Glutathione-mediated response to acid stress in the probiotic bacterium, *Lactobacillus salivarius*. Biotechnology letters *32*, 969–972.

Lorca, G.L., Font de Valdez, G., and Ljungh, A. (2002). Characterization of the protein-synthesis dependent adaptive acid tolerance response in *Lactobacillus acidophilus*. J. Mol. Microbiol. Biotechnol. *4*, 525–532.

Marteau, P., Minekus, M., Havenaar, R., and Veld, J. (1997). Survival of lactic acid bacteria in a dynamic model of the stomach and small intestine: Validation and the effects of bile. Journal of Dairy Science *80*, 1031–1037.

Meijerink, M., van Hemert, S., Taverne, N., Wels, M., de Vos, P., Bron, P.A., Savelkoul, H.F., van Bilsen, J., Kleerebezem, M., and Wells, J.M. (2010). Identification of genetic loci in *Lactobacillus plantarum* that modulate the immune response of dendritic cells using comparative genome hybridization. PLoS One *5*, e10632.

Metchnikoff, E. (1907). Lactic acid as inhibiting intestinal putrefaction. In The Prolongation of Life: Optimistic Studies (W. Heinemann, London), pp. 161–183.

Mills, S., Stanton, C., Fitzgerald, G.F., and Ross, R.P. (2011). Enhancing the stress responses of probiotics for a lifestyle from gut to product and back again. Microb. Cell Fact. *10* (*Suppl. 1*), S19.

Molly, K., De Smet, I., Nollet, L., Vande Woestyne, M., and Verstraete, W. (1996). Effect of lactobacilli on the ecology of the gastro-intestinal microbiota cultured in the SHIME reactor. Microbial. Ecol. Health Dis. 9, 79–89.

Narberhaus, F. (2002). Alpha-crystallin-type heat shock proteins: socializing minichaperones in the context of a multichaperone network. Microbiol. Mol. Biol. Rev. *66*, 64–93.

Parente, E., Ciocia, F., Ricciardi, A., Zotta, T., Felis, G.E., and Torriani, S. (2010). Diversity of stress tolerance in *Lactobacillus plantarum*, *Lactobacillus pentosus* and *Lactobacillus paraplantarum*: A multivariate screening study. Int. J. Food Microbiol. *144*, 270–279.

Prasad, J., McJarrow, P., and Gopal, P. (2003). Heat and osmotic stress responses of probiotic *Lactobacillus rhamnosus* HN001 (DR20). in relation to viability after drying. Appl. Environ. Microbiol. *69*, 917–925.

Pretzer, G., Snel, J., Molenaar, D., Wiersma, A., Bron, P.A., Lambert, J., de Vos, W.M., van der Meer, R., Smits, M.A., and Kleerebezem, M. (2005). Biodiversity-based identification and functional characterization of the mannose-specific adhesin of *Lactobacillus plantarum*. J. Bacterol. *187*, 6128–6136.

Rauch, M., and Lynch, S.V. (2012). The potential for probiotic manipulation of the gastrointestinal microbiome. Curr. Opin. Biotechnol. *23*, 192–201.

Remus, D.M., Kleerebezem, M., and Bron, P.A. (2011). An intimate *tete-a-tete* – How probiotic lactobacilli communicate with the host. Eur. J. Pharmacol. *668*, S33-S42.

Ricciardi, A., Parente, E., Guidone, A., Ianniello, R.G., Zotta, T., Abu Sayem, S.M., and Varcamonti, M. (2012). Genotypic diversity of stress response in *Lactobacillus plantarum*, *Lactobacillus paraplantarum* and *Lactobacillus pentosus*. Int. J. Food Microbiol. *157*, 278–285.

Salotra, P., Singh, D.K., Seal, K.P., Krishna, N., Jaffe, H., and Bhatnagar, R. (1995). Expression of DnaK and GroEL homologues in *Leuconostoc esenteroides* in response to heat shock, cold shock or chemical stress. FEMS Microbiol. Lett. *131*, 57–62.

Seydlova, G., Halada, P., Fiser, R., Toman, O., Ulrych, A., and Svobodova, J. (2012). DnaK and GroEL chaperones are recruited to the *Bacillus subtilis* membrane after short-term ethanol stress. J. Appl. Microbiol. *112*, 765–774.

Snel, J., Marco, M.L., Kingma, F., Noordman, W.M., Rademaker, J., and Kleerebezem, M. (2011). Competitive selection of lactic acid bacteria that persist in the human oral cavity. Appl. Environ. Microbiol. *77*, 8445–8450.

Spano, G., and Massa, S. (2006). Environmental stress response in wine lactic acid bacteria: Beyond *Bacillus subtilis*. Crit. Rev. Microbiol. *32*, 77–86.

Tian, H., Tan, J., Zhang, L., Gu, X., Xu, W., Guo, X., and Luo, Y. (2012). Increase of stress resistance in *Lactococcus lactis* via a novel food-grade vector expressing a *shsp* gene from *Streptococcus thermophilus*. Brazilian J. Microbiol. *43*, 1157–1164.

van Bokhorst-van de Veen, H., Abee, T., Tempelaars, M., Bron, P.A., Kleerebezem, M., and Marco, M.L. (2011). Short- and long-term adaptation to ethanol stress and its cross-protective consequences in *Lactobacillus plantarum*. Appl. Environ. Microbiol. *77*, 5247–5256.

Van Bokhorst-van de Veen, H., Bongers, R.S., Wels, M., Bron, P.A., and Kleerebezem, M. (2013a). Transcriptome signatures of class I and III stress response deregulation in *Lactobacillus plantarum* reveal pleiotropic adaptation. Microb. Cell Fact. *12*, 112.

van Bokhorst-van de Veen, H., Lee, I.C., Marco, M.L., Wels, M., Bron, P.A., and Kleerebezem, M. (2012a). Modulation of *Lactobacillus plantarum* gastrointestinal robustness by fermentation conditions enables identification of bacterial robustness markers. PLoS One 7, e39053.

van Bokhorst-van de Veen, H., Smelt, M.J., Wels, M., van Hijum, S.A., de Vos, P., Kleerebezem, M., and Bron, P.A. (2013b). Genotypic adaptations associated with prolonged persistence of *Lactobacillus plantarum* in the murine digestive tract. Biotechnol. J. *8*, 895–904.

van Bokhorst-van de Veen, H., van Swam, I., Wels, M., Bron, P.A., and Kleerebezem, M. (2012b). Congruent strain specific intestinal persistence of *Lactobacillus plantarum* in an intestine-mimicking *in vitro* system and in human volunteers. PLoS One 7, e44588.

van de Guchte, M., Serror, P., Chervaux, C., Smokvina, T., Ehrlich, S.D., and Maguin, E. (2002). Stress responses in lactic acid bacteria. Antonie Van Leeuwenhoek *82*, 187–216.

van Hemert, S., Meijerink, M., Molenaar, D., Bron, P.A., de Vos, P., Kleerebezem, M., Wells, J.M., and Marco, M.L. (2010). Identification of *Lactobacillus plantarum* genes modulating the cytokine response of human peripheral blood mononuclear cells. BMC Microbiol. *10*, 293.

Varmanen, P., and Savijoki, K. (2011). Responses of lactic acid bacteria to heat stress. In Stress Responses of Lactic Acid Bacteria, E. Tsakalidou, and K. Papadimitriou, eds. (Springer US), pp. 55–66.

Wang, Y., Delettre, J., Corrieu, G., and Béal, C. (2011). Starvation induces physiological changes that act on the cryotolerance of *Lactobacillus acidophilus* RD758. Biotechnology Progress *27*, 342–350.

Watanabe, M., van der Veen, S., and Abee, T. (2012). Impact of respiration on resistance of *Lactobacillus plantarum* WCFS1 to acid stress. Appl. Environ. Microbiol. *78*, 4062–4064.

Weber, F.J., and de Bont, J.A. (1996). Adaptation mechanisms of microorganisms to the toxic effects of organic solvents on membranes. Biochim. Biophys. Acta *1286*, 225–245.

Wels, M., Francke, C., Kerkhoven, R., Kleerebezem, M., and Siezen, R.J. (2006). Predicting *cis*-acting elements of *Lactobacillus plantarum* by comparative genomics with different taxonomic subgroups. Nucleic Acids Res. *34*, 1947–1958.

Wels, M., Overmars, L., Francke, C., Kleerebezem, M., and Siezen, R.J. (2011). Reconstruction of the regulatory network of *Lactobacillus plantarum* WCFS1 on basis of correlated gene expression and conserved regulatory motifs. Microb. Biotechnol. *4*, 333–344.

Wu, C., Zhang, J., Du, G., and Chen, J. (2013a). Aspartate protects *Lactobacillus casei* against acid stress. Appl. Microbiol. Biotechnol. *97*, 4083–4093.

Wu, C., Zhang, J., Du, G., and Chen, J. (2013b). Heterologous expression of *Lactobacillus casei* RecO improved the multiple-stress tolerance and lactic acid production in *Lactococcus lactis* NZ9000 during salt stress. Bioresour. Technol. *143*, 238–241.

Wu, C., Zhang, J., Wang, M., Du, G., and Chen, J. (2012). *Lactobacillus casei* combats acid stress by maintaining cell membrane functionality. J. Ind Microbiol. Biotechnol. *39*, 1031–1039.

Zhang, J., Du, G.C., Zhang, Y., Liao, X.Y., Wang, M., Li, Y., and Chen, J. (2010a). Glutathione protects *Lactobacillus sanfranciscensis* against freeze–thawing, freeze-drying, and cold treatment. Appl. Environ. Microbiol. *76*, 2989–2996.

Zhang, J., Wu, C.D., Du, G.C., and Chen, J. (2012). Enhanced acid tolerance in *Lactobacillus casei* by adaptive evolution and compared stress response during acid stress. Biotechnol. Bioproc. E. *17*, 283–289.

Zhang, Y., Zhang, Y., Zhu, Y., Mao, S., and Li, Y. (2010b). Proteomic analyses to reveal the protective role of glutathione in resistance of *Lactococcus lactis* to osmotic stress. Appl. Environ. Microbiol. *76*, 3177–3186.

Zotta, T., Ricciardi, A., Ciocia, F., Rossano, R., and Parente, E. (2008). Diversity of stress responses in dairy thermophilic streptococci. Int. J. Food Microbiol. *124*, 34–42.

Zotta, T., Ricciardi, A., Guidone, A., Sacco, M., Muscariello, L., Mazzeo, M.F., Cacace, G., and Parente, E. (2012). Inactivation of *ccpA* and aeration affect growth, metabolite production and stress tolerance in *Lactobacillus plantarum* WCFS1. Int. J. Food Microbiol. *155*, 51–59.

Biology of Reactive Oxygen Species, Oxidative Stress, and Antioxidants in Lactic Acid Bacteria

14

Marta C. T. Leite, Bryan Troxell, Jose M. Bruno-Bárcena and Hosni M. Hassan

Abstract

Lactic acid bacteria (LAB) are important in food fermentations and in human health. Due to their physiological and ecological heterogeneity, they encounter a variety of stressors including acid, osmolality, heat, bile-salt, and oxygen. LAB are exposed to oxidative stress caused by the partially reduced reactive oxygen species (ROS) generated from endogenous sources as well as from the environment. Findings over the past fifty years have demonstrated that this group of organisms comprise a heterogeneous mixture of different genera, where some members of the group have the capacity to synthesis antioxidant enzymes like Mn-containing superoxide dismutase (MnSOD), non-haem catalases (i.e. Mn-containing catalase –MnKat), and haem catalases (when provided with exogenous source of haem). Furthermore, some members of the group can accumulate large intracellular concentrations of manganese to use in the detoxification of ROS. In this chapter, we discuss the natural defences against ROS in LAB as well as the technological practices used in the food and nutraceutical industries to protect LAB from oxidative stress and loss of viability during processing and storage.

Introduction

Lactic acid bacteria (LAB) are a diverse group of Gram-positive (low G+C) obligate fermenters that occupy several environmental niches from foodstuffs to plants and animals including humans. Due to their physiological and ecological heterogeneity, they evolved to endure a variety of stressors (i.e. acid, osmotic, cold, salt, bile-salt, oxygen, heat, starvation). As a result LAB possess remarkable systems to combat these environmental stressors and in many cases the stress response is mostly species/strain specific (Lin and Yen, 1999; Van de Guchte *et al.*, 2001). LAB are one of the most important groups of microorganisms used by the traditional and modern food industry. Organisms belonging to the LAB group have been identified as responsible for the biotransformation of sugars present in foods. LAB are widely manufactured for direct consumption (i.e. probiotics) or use as starter cultures for the production of fermented milks, cheeses, sourdough, sausages, vegetables, etc. Use of LAB in foods adds flavour, texture, prevents spoilage and contributes to human health (Smit *et al.*, 2005; Bron and Kleerebezem, 2011). Although prevention of food spoilage by LAB is generally attributed to acidification and accumulation of lactic acid during the fermentation, many LAB also produce other metabolic by-products like bacteriocins, hydrogen peroxide (H_2O_2), or acetate, that can inhibit its own growth as well as the growth of a wide range of bacteria including pathogens (e.g. *Listeria, Clostridium, Salmonella,* etc.) (Serrazanetti *et al.*, 2013; Rattanachaikunsopon and Phumkhachorn, 2010; Soomro *et al.*, 2002; Ouwehand, 1998).

An array of technological, nutritional, and health benefits are attributed to LAB. These include: production of commodity chemicals such as lactate, the synthesis of B-vitamins, improve lactose utilization in the gastrointestinal (GI) tract of humans and animals. They can also prevent the antagonistic effects of pathogens, reduce serum cholesterol levels, modulate the immune system, and moderate the carcinogenicity of some chemicals (Azcarate-Peril *et al.*, 2011). Owing to their nutritional and health benefits, there is a growing public demand for additional application of LAB in food products. Therefore, an increase in new food products are being developed to incorporate new and previously characterized LAB strains into a variety of fermented food products including yoghurt, cheeses, drinks, dietary supplements, etc. (Shah *et al.*, 2010; Corcoran *et al.*, 2008; Ong *et al.*, 2006; Liong and Shah, 2005; Gomes and Malcata, 1999).

Whether used as starter cultures or as probiotics, LAB are exposed to dynamic environmental fluctuations during food processing or in the GI tract that may influence their growth, survival, and functionality. Unfavourable environmental conditions encountered during food and industrial processing include temperature variations, osmolality, organic acid stress, desiccation, and oxidative stress (Bron and Kleerebezem, 2011; Mills *et al.*, 2011; Zhang and Li, 2013). These stressful conditions extend during storage of the food product and may affect the shelf-life and the number of viable cells in the final product. These challenges do not end at that point since, LAB must also survive the environment in the GI tract (i.e. low pH, hydrolytic and proteolytic enzymes, and bile-salts) in order to reach the right niche, proliferate, and provide health benefits (Kim *et al.*, 2008a; Li *et al.*, 2009;

Bron and Kleerebezem, 2011; Gaudreau *et al.*, 2013). In order to exert the beneficial health-promoting properties on the host, it is critical to use LAB strains with high viability and functionality, and utilize adequate production and recovery/storage methods during manufacturing. Currently for probiotic application, it is recommended that the LAB-containing products contain at least 10^7 viable cells per gram or millilitre; however, these numbers vary depending on the strain and the type of application (FAO/WHO, 2002; Corcoran *et al.*, 2008; Bron and Kleerebezem, 2011; Grzeskowiak *et al.*, 2011). Thus, surviving adverse conditions is essential for LAB to overcome the technological barriers and for the successful delivery of health promoting properties.

Although LAB contain systems to combat environmental stresses, during industrial production, these protective systems are not sufficient to overcome the encountered stresses (Zhang and Li, 2013; Mills *et al.*, 2011; Smid and Hugenholtz, 2010). Post-acidification, temperature fluctuations, dissolved O_2 concentration, and permeability through packing materials are among the environmental conditions affecting the viability of LAB in food products during storage (Dave and Shah, 1998; Shah *et al.*, 2010; Kailasapathy, 2002). Clearly, exposure to O_2 plays a major role in the survival of LAB during proliferation, food processing, and storage. While oxygen itself is unable to damage cells, exposure to O_2 results in the formation of the partially reduced reactive oxygen species (ROS): superoxide anion (O_2^-), hydrogen peroxide (H_2O_2), and the highly reactive hydroxyl radical (HO·). They collectively cause oxidative stress and contribute to the damage of cellular proteins, lipids, DNA, and ultimately lead to cell death. Thus, preventing oxidative stress by the selection of ROS tolerant strains or the application of techniques that reduce oxidative stress will ensure cell viability during storage and GI transit (Talwalkar and Kailasapathy, 2004; Mills *et al.*, 2011). Moreover, it has been suggested that LAB strains tolerant to oxidative stress may have cross protection against other stresses like drying, cold, and acid stresses (Zhang and Li, 2013; Zotta *et al.*, 2012; Zhang *et al.*, 2010b; Bruno-Barcena *et al.*, 2010). Some LAB have O_2-consuming enzymes such as NADH oxidases (Nox), pyruvate oxidase (Pox), and lactate oxidase (Lox). However, in LAB, the presence of important antioxidant enzymes such as manganese superoxide dismutase (MnSOD), manganese or haem catalases (MnKat or haem-Kat) is scarce. Indeed, the presence of antioxidant enzymes in LAB varies by species and even by strain. Thus, there is substantial heterogeneity in the presence of antioxidant enzymes in LAB, and that may explain why some strains are more sensitive to O_2 than others (van de Guchte *et al.*, 2006; Talwalkar and Kailasapathy, 2004; Vasiljevic and Shah, 2008, Gaudreau *et al.*, 2013; Serrazanetti *et al.*, 2013). In this review we focus on ROS, oxidative stress in LAB, and discuss the different mechanisms employed by this group of organisms to cope with oxygen toxicity.

Production and toxicity of reactive oxygen species

Oxygen (O) is the eighth element in the periodic table and the first in the chalcogen group, which consists of sulfur, selenium, tellurium, and polonium. Oxygen is one of the more abundant elements within the Earth's crust and atmosphere and its most prevalent form is the stable form, dioxygen (O_2) (Urey, 1952). O_2 is paramagnetic and has two unpaired electrons (e^-) with parallel spin resulting in spin restriction. During the reduction of O_2 to H_2O, the spin restriction of O_2 favours the univalent pathway that results in the formation of the toxic ROS (Fridovich, 1978; Hassan and Fridovich, 1980). Thus, the partial reduction of O_2 forms O_2^- through the univalent addition of an e^-, H_2O_2 is formed through the addition of $2e^-$ $2H^+$, and HO· is formed from the addition of $3e^-$ $3H^+$. Both O_2^- and HO· are radical forms of O_2, but H_2O_2 is not a radical because it does not contain an unpaired e^-. This property of H_2O_2 lengthens its half-life thereby promoting its diffusion from the site of generation to distal sites. Enzymes within the e^- transport chain (ETC) are notable sources of ROS produced within bacteria and the mitochondria of eukaryotic cells. Additional sources of ROS in cells are the autooxidation of reduced metabolites with low redox potential, as well as enzymes capable of donating e^- to O_2 (Hassan and Fridovich, 1980). Here, we focus on the formation of ROS by enzymes within the ETC and enzymes within metabolic pathways.

Complex-I of the ETC

Complex-I (NADH:ubiquinone oxidoreductase) within the ETC is perhaps the most studied example of the production of ROS in biology. Complex-I is present within the mitochondria and bacteria capable of aerobic respiration. Reactions generated within the tricarboxylic acid cycle (TCA or Krebs cycle) produce reducing power by converting nicotinamide dinucleotide (NAD^+) into the reduced form, NADH. NADH is the e^- and H^+ donor to the ETC and enters this pathway through the function of Complex-I. Complex-I is a large multiprotein complex, in *Escherichia coli* there are at least 13 proteins present, which transfers two e^- from NADH to a lipophilic molecule, ubiquinone; during this process, H^+ are transported into the periplasm, or inner membrane space of the mitochondria sustaining the proton motive force (PMF) necessary for the synthesis of ATP during respiration. Complex-I contains a flavin mononucleotide (FMN) cofactor, which becomes reduced during NADH oxidation, and numerous Fe–S clusters involved in the function of the NADH-ubiquinone reaction. O_2^- is produced by complex I through the univalent reduction of O_2 by the action of the reduced FMN moiety (Kussmaul and Hirst, 2006). The formation of O_2^- by complex-I is slow, but nevertheless represents a significant source of ROS within the cell (Murphy, 2009). The formation of O_2^- by complex-I is influenced by the ratio of NAD^+:NADH, i.e. raising the NAD^+:NADH ratio reduces the formation of O_2^- suggesting that reduction of FMN by NADH promotes O_2^- formation. This is supported by evidence that NAD^+ binds to the FMN thereby blocking the transfer of e^- to O_2 (Berrisford and Sazanov, 2009). These studies support the broad concept that the redox potential of the cell (NAD^+:NADH ratio) influences the propensity to generate endogenous ROS.

Oxidases within LAB

In spite of the lack of complex-I within LAB, numerous strains within this group produce substantial amounts of H_2O_2 during *in vitro* growth. For instance, *Lactobacillus delbrueckii, L. johnsonni,* and *L. gasseri* produce millimolar concentrations of H_2O_2 during aerobic growth (Marty-Teysset *et al.*, 2000; Pridmore *et al.*, 2008). Oxidases within LAB capable of reducing O_2 include Nox, Pox, and Lox. Nox enzymes that form either H_2O or H_2O_2 have been identified in LAB (Condon, 1987; Jansch *et al.*, 2011). Pox and Lox have been identified and both appear to play a role in the formation of H_2O_2 within LAB (Condon, 1987; Quatravaux *et al.*, 2006; Barre *et al.*, 2007). A similarity among these oxidases is the indication of a flavin-dependent cofactor that may be involved in the formation of H_2O_2. Thus, flavoproteins, as in complex-I and oxidases, are prominent enzymes within biology that contribute to the formation of H_2O_2, and are probably the major H_2O_2 generators within LAB during aerobic growth.

The devil within: endogenous production of H_2O_2 inhibits growth of LAB

When cultivated in the presence of O_2, certain LAB enter into stationary phase earlier and have reduced biomass, which can be prevented by growth in anaerobic conditions or by the addition of exogenous catalase to aerobically growing cultures (Hertzberger *et al.*, 2014). In the case of *L. johnsonnii,* recent work indicates that an NADH-dependent flavin reductase, encoded by two genes present in tandem, is responsible for nearly all of the H_2O_2 produced during *in vitro* growth (Hertzberger *et al.*, 2014). This indicates that the metabolism of LAB produce sufficient quantities of H_2O_2 to inhibit growth.

How does H_2O_2 inhibit cell growth? Over a century ago, H.G.H. Fenton observed the reaction between tartaric acid, H_2O_2, and ferrous iron (Fe^{2+}) salts (Fenton, 1894). The chemistry that occurs under this condition is known as Fenton reaction. Fenton reaction produces an extremely toxic compound HO· that can damage any macromolecular within the cell. Proteins are damaged by HO·, but can be repaired depending on the extent of amino acid damage. If amino acids are extensively damaged by HO·, then *de novo* synthesis of proteins is required to maintain metabolic pathways. Damage of proteins eventually reduces the production of ATP and lowers the overall energy within the cell. Perhaps the most well-known macromolecule that is damaged by HO· is DNA (Imlay and Linn, 1988). The phosphodiester bonds within DNA are broken by HO· and result in fragmented DNA. Although cells contain dedicated DNA repair systems, when ROS reach higher levels the repair systems are overwhelmed and extensive DNA fragmentation occur. In addition, HOv can damage lipid moieties within the plasma membrane, resulting in cell lysis. Clearly, H_2O_2 reacts with Fe^{2+} to produce HO· that ultimately destroys the backbone of DNA, damages proteins, and results in cell death (Fig. 14.1). O_2^- eventually dismutes into H_2O_2 and can alter the [Fe] pool, thereby promoting Fenton chemistry. Not surprisingly, many different cell types have mechanisms to reduce the half-life

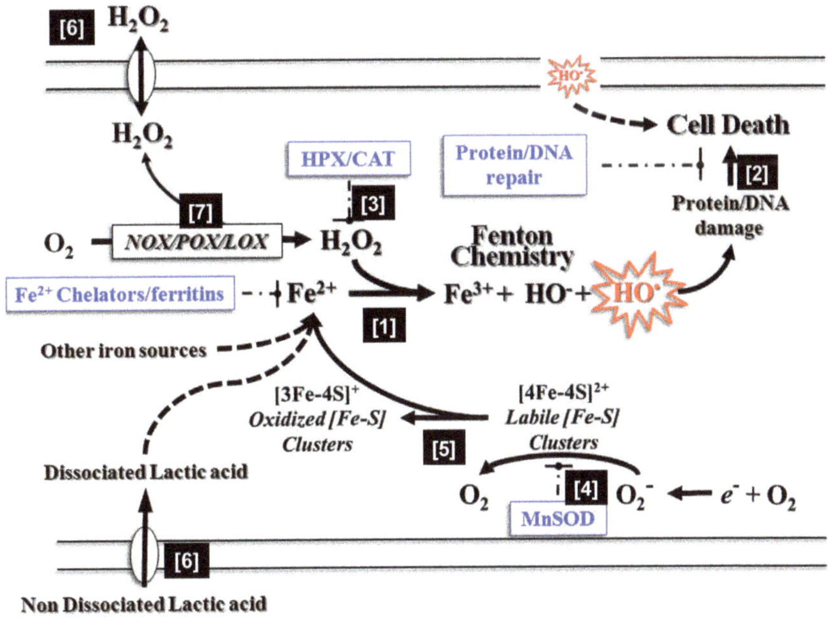

Figure 14.1 Schematic representation of oxidants and antioxidants in LAB-cells. The reaction between ferrous iron (Fe^{2+}) and H_2O_2 produces the highly reactive free radical, the hydroxyl radical (HO·) via Fenton chemistry [1]. HO· can damage proteins, DNA, and lipids present in the bacterial membrane, which ultimately results in cell death [2]. However, protective mechanisms are present within LAB, such as, protein and DNA repair systems that can repair the damage exerted by HO· (shown by a dashed line with a blunt end). Moreover, hydroperoxidases and catalases (HPX/CAT) [3] can reduce [H_2O_2] thereby reducing [HO·] within the cell (shown by a dashed line with a blunt end). Iron chelators or ferritin proteins within the cell can reduce the presence of Fe^{2+} thereby further reducing Fenton chemistry. Furthermore, the function of MnSOD [4] can reduce the steady-state concentration of O_2^- (shown by a dashed line with a blunt end), and prevent the release of Fe^{2+} from proteins containing iron–sulfur clusters [5] that can participate in Fenton chemistry. Exogenous factors, such as H_2O_2, and protonated lactic acid [6] that can increase [Fe^{2+}] can cause an increase in Fenton chemistry and ultimately cause cell death. In addition, endogenous factors such as NADH oxidase (Nox), pyruvate oxidase (Pox), and lactate oxidase (Lox) [7] can consume O_2 and produce H_2O_2 within the cell leading to oxidative stress.

of O_2^- resulting in the limitation of the Fenton reaction. Because there is no dedicated enzyme(s) that can eliminate HO·, strong selective pressures have evolved to limit its formation by the eliminating both O_2^- and H_2O_2 (Fridovich, 1995). As a group, some LAB have enzymes dedicated to the removal of both O_2^- and/or H_2O_2. In addition, certain species of LAB have unique methods for the non-enzymatic removal of O_2^-. Indeed, LAB represent a diverse group in the study of defences against ROS.

Antioxidant enzymes and scavenging strategies

Protection from O_2^-

Oxidases within LAB are capable of generating O_2^- (see above). O_2^- can be dismutated either spontaneously in a pH-dependent manner or enzymatically through the function of superoxide dismutase (SOD). SOD function with second order rate constants that are diffusion limiting (~2×10^9/M/s), which are several orders of magnitude faster than spontaneous dismutation; the second order rate constant for the spontaneous dismutation is ~10^7/M/s), M/s), at pH 5 and ~10^5/M/s), at pH 7 (Bielski, 1978). Rapid dismutation of O_2^- is important to inhibit the leaching of Fe^{3+} from iron–sulfur $[4Fe–4S]^{2+}$ cluster-containing proteins. O_2^- can serve as a reductant to liberate Fe^{3+} from proteins and increase the concentration of free Fe^{2+} within the cell. Thus, the inability to reduce the steady state level of O_2^- can increase the susceptibility of the cell to Fenton chemistry (Fig. 14.1). This is evident by the increased intracellular Fe^{2+} and protection of SOD against killing by H_2O_2 (Keyer and Imlay, 1996; Bruno-Barcena *et al.*, 2004). Metals play an important role in the defence against ROS. SODs require metal cofactors for enzymatic function, notably manganese, iron, and copper are essential for MnSODs, FeSODs, and CuZnSODs, respectively. LAB do not contain FeSODs or CuZnSODs; however, MnSODs have been found in several species of LAB (i.e. *L. lactis* and *Streptococcus thermophilus*). In addition to serving as a cofactor for SODs, Mn can serve as a potent scavenger of O_2^-. For example, *L. plantarum*, which lacks SOD, accumulates large quantities of Mn allowing for comparable O_2^- scavenging activities as SOD proficient organisms (Archibald and Fridovich, 1981a,b). Mn forms complexes with phosphates and other molecules within the cell and these complexes can act as O_2^- or H_2O_2 scavengers (Archibald Fridovich, 1982; Stadtman *et al.*, 1990; reviewed in Culotta and Daly, 2013). Therefore, *L. plantarum*, and perhaps other LAB, can serve as an important model system in the study of fundamental and novel aspects of ROS defence.

Protection from H_2O_2

Some oxidases favour the $2e^-/2H^+$ reduction of O_2 to form H_2O_2. Enzymes capable of degrading H_2O_2 (i.e. catalases and peroxidases) are scarce in LAB. However, a pseudocatalase, a manganese-containing enzyme (MnKat) has been identified in *L. plantarum* (Johnston and Delwiche, 1965; Kono and Fridovich, 1983); while *L. sakei* is reported to contain a haem- containing catalase (Haem-Kat) (Knauf *et al.*, 1992). Therefore, these species exhibit protection against H_2O_2; however many LAB do not contain catalases and may rely on alternative methods of protection against H_2O_2 toxicity. Redox reactions within the cell are critical for maintaining cellular activity and ATP synthesis. Typically, the cytoplasm of bacteria is kept in a reduced state, which favours protein function and metabolic activity. H_2O_2 can disrupt this reducing environment by oxidizing thiol groups of cysteines and forming disulfide bonds between adjacent thiol groups thereby inhibiting protein function. Formation of disulfide bonds is reversible and protein function can be restored following H_2O_2 exposure. Thioredoxins (TrxA) are small, ubiquitous proteins that repair disulfide bonds (Holmgren, 1985). Reducing power, in the form of NADH or NADPH, is utilized by thioredoxins reductases (TrxB) to repair oxidized TrxA, which allows for continued reduction of oxidized thiol groups. TrxB has been shown to be important in the oxidative stress defence for *L. plantarum* (Serrano *et al.*, 2007). Within *L. casei*, both TrxA and TrxB are important for growth under aerobic conditions (Serata *et al.*, 2012). These results demonstrate the importance of TrxA/TrxB, within some LAB, during growth conditions that promote the formation of H_2O_2.

Regulation of antioxidant enzymes in LAB

Very little is known about the regulation of SODs, catalases and other antioxidant enzymes in LAB. However, several studies demonstrated that MnSOD is not regulated at the level of transcription. Indeed, the activity of MnSOD was dependent on the intracellular concentration of the correct metal cofactor, Mn^{2+} (Andrus *et al.*, 2003; Bruno-Barcena *et al.*, 2004). Similarly the expression of active haem-Kat or Mn-Kat is dependent on the presence of the appropriate sources and concentrations of haem or manganese, respectively.

Importance of O_2 tolerance in LAB and its industrial applications

LAB are generally classified as oxygen-indifferent because they can tolerate O_2, but grow best under anaerobic conditions. There is a wide spectra among the different strains with regard to oxygen tolerance that range from those that require almost complete anaerobic conditions to those able to tolerate microaerophilic environments. Nevertheless, the presence of dissolved O_2 beyond the microaerophilic conditions is detrimental for the growth of many strains, and it is generally accepted that LABs could not use O_2 as the main terminal electron acceptor. However, some members of the group, e.g. *L. lactis* and *L. plantarum*, contain an incomplete respiratory chain capable of aerobic respiration when provided with an exogenous source of haem. Additionally, certain strains of *L. plantarum*, when grown in the presence of haem and vitamin K2, can reduce O_2 via an active partial respiratory chain composed of NADH dehydrogenase, menaquinone, and cytochrome oxidase (Brooijmans *et al.*, 2009; Duwat *et al.*, 2001; Gaudu *et al.*, 2002; Watanabe *et al.*, 2012; Corcoran *et al.*, 2008; Miyoshi *et al.*, 2003; Lechardeur *et al.*, 2011).

All steps of industrial production including the mixing and homogenization steps, drying, filling and packaging can result in the generation of ROS (Mills *et al.*, 2011; Serrazanetti *et al.*, 2013). Additionally, LAB may encounter oxidative stress during its transit through the GI since production of ROS can be triggered by stressors like bile salts (Corcoran *et al.*, 2008; Begley *et al.*, 2005; Alcantara and Zuniga, 2012). Considerable levels of oxygen get incorporated into milk during mixing steps and storage and especially in the first stages of fermentation; aeration can induce important changes in the process that affects sugar metabolism of LAB and/or inhibit their growth (Serrazanetti *et al.*, 2013). Some of the technological practices are designed to reduce oxidative stress and ROS, for example in the production yoghurt, the co-culturing of *S. thermophilus* and *L. delbrueckii* accomplishes that goal. Thus, high levels of H_2O_2 and ROS negatively influences *L. delbrueckii* cultures grown under aerobic or microaerophilic conditions and analysis of its genome sequence confirmed that it is poorly equipped with antioxidant enzymes (Fu *et al.*, 2013; Marty-Teysset *et al.*, 2000; Van de Guchte *et al.*, 2001; Van de Guchte *et al.*, 2006). On the other hand its partner, in yoghurt production, *S. thermophilus* produces a MnSOD that protects the cell against O_2^-, but it lacks catalase(s) (Chang and Hassan, 1997; Thibessard *et al.*, 2004). Although, *S. thermophilus* does not express a catalase, the presence of MnSOD can protect against H_2O_2 by preventing the Fe-catalysed generation of HO· (Fig. 14.1; Bruno-Barcena *et al.*, 2004). Despite the synergistic interaction between *S. thermophilus* and *L. delbrueckii* subsp. *bulgaricus*, some strain combinations were found to inhibit fermentation rate compared with single strain cultures. It has been reported that association in yoghurt of *L. delbrueckii* subsp. *bulgaricus* (isolated from commercial starter culture YoFlex L702) and *L. acidophilus* ATCC 700396 or *L. acidophilus* NCFM resulted in pronounced inhibition of growth of the probiotic strain (i.e. *L. acidophilus*) due to acidification and H_2O_2 generated by the starter culture (i.e. *L. delbrueckii*) (Ng *et al.*, 2011; Lankaputhra and Shah, 1996). This signifies the importance of investigating interactions between starter cultures to avoid exposure of the strains to inhibitory byproducts such as H_2O_2, leading to growth arrest and cell death during the fermentation process and storage. Exposure to O_2 during storage constitutes a technological challenge, especially on the functionality of probiotics. Once O_2 diffuses through the packaging material it will especially affect strain viability and activity (Talwalkar and Kailasapathy, 2004; Cruz *et al.*, 2012a,b; Granato *et al.*, 2010). To extend the shelf-life and to minimize oxidative stress and loss of viability of LAB during industrial applications, several approaches have been suggested. These include some of the following techniques.

Combination of starter cultures and probiotics

Coculturing of O_2-sensitive probiotics with high O_2-depleting starter strains (e.g. *S. thermophilus* or *L. lactis*) has been demonstrated to be a possible strategy to improve survival of the probiotic strains during fermentation and storage (Lourens-Hattingh and Viljeon, 2001; Talwalkar and Kalasapathy, 2004; Abe *et al.*, 2009; Odakami *et al.*, 2011). On the other hand, one should avoid the combination of O_2-sensitive probiotics with the starter culture *L. delbrueckii* subsp. *bulgaricus* because of its ability to produce high levels of H_2O_2, which causes oxidative stress and affects viability of the cocultured microorganisms (Ng *et al.*, 2011; Lankaputhra and Shah, 1996; Vinderola *et al.*, 2002).

Packaging systems

Applications of new packaging materials and developments in the use of a modified atmosphere have provided means for enhancing survival of O_2-sensitive LAB and bifidobacteria during storage. Evaluation of different types of packaging materials has been tested for yoghurts and fermented milks containing probiotics. Miller and coworkers (2002 and 2003) compared the effect of high-impact polystyrene packaging with an O_2-barrier material and an O_2 scavenging active packaging system on the dissolved O_2 levels of two types of yoghurt. The O_2-barrier packaging combined with the O_2 scavenging material was shown to be the most effective system, not only preventing O_2 diffusion, but also decreasing dissolved O_2 levels in yoghurt throughout storage. Thus, the application of an O_2 barrier with an O_2 scavenging systems can be used to protect O_2-sensitive strains and develop a more favourable environment during storage.

The addition of glucose oxidase (GOX), an O_2-consuming enzyme, in yoghurt containing probiotics has been shown as a potential option to decrease oxidative stress during refrigerated storage (Cruz *et al.*, 2010, 2011, 2012a, 2012b). The authors recently reported that addition of GOX in combination with a packaging system that reduces O_2 permeability resulted in reduced O_2 permeation in yoghurt containing *L. acidophilus* and *B. longum* and showed increased viability and product shelf-life during refrigerated storage (Cruz *et al.*, 2013). Interestingly, one of the end-products of GOX is H_2O_2 (Gibson *et al.*, 1964), which may suggest that oxygen itself is more important than H_2O_2 during prolonged storage. Indeed, this subject needs further studies.

Supplementation

Supplementation of growth media or food matrices with antioxidants can provide a useful strategy for maintaining low redox potential and protect the viability of LAB strains. Addition of ascorbic acid, an O_2 scavenger, allowed for a temporary reduction of O_2 levels and redox potential and improving survival of *L. acidophilus* in yoghurt. This may suggest a possible application of this food additive to protect probiotics from ROS toxicity (Dave and Shah, 1998). In support of this, improved survival of the probiotics *L. rhamnosus* HN001 and *L. paracasei* Lpc37 was achieved by providing high levels of antioxidants such as vitamin C alone or in combination with grape extract, fruit juice, or green tea extracts (Shah *et al.*, 2010).

Green tea extracts are rich in antioxidant compounds, among them, catechins, which act *in vitro* by scavenging ROS and reactive nitrogen species (RNS), or by chelating metal ions such as Fe (Wiseman *et al.*, 1997; Frei and Higdon, 2003). The effect of catechins on the growth of the O_2-sensitive *L. helveticus* has been evaluated. Supplementation of the culture medium with green tea extract greatly promoted the growth of *L. helveticus* under aerobic conditions. Green tea extract seems to reduce the redox potential of the culture medium resulting in lower production of ROS. Growing under aerobic conditions, Fourier transform infrared spectroscopy analysis (FTIR) also suggests that components of the green tea extract induce a change in the lipid composition in the phospholipid bilayer of *L. helveticus* limiting the negative effects of O_2 exposure (Gaudreau *et al.*, 2013).

Although metals, such as Fe^{2+} and Cu^{2+}, can promote the formation of HO·; Mn^{2+} is an important metal used by LAB to defend against ROS. Thus, addition of Mn^{2+} to the growth media promotes the growth of *L. plantarum* under aerobic conditions, leading to higher cell densities (Watanabe *et al.*, 2012). *L. plantarum* accumulates high intracellular concentrations of Mn^{2+} to scavenge O_2^- during aerobic growth (Archibald and Fridovich, 1981a,b, 1982). Thus, addition of Mn^{2+} to milk, meat, and vegetables undergoing fermentative processes is an easy strategy to improve performance of starter cultures (Watanabe *et al.*, 2012; Zhang and Li, 2013).

Glutathione (GSH) synthesis and accumulation is another potential anti-oxidative defence mechanism of LAB and its ability to grow under aerobic conditions. GSH is synthesized from three amino acids, glycine, cysteine and glutamate. GSH is an intracellular low molecular weight thiol which contributes towards stress management in cells by acting as e^- donor. It is an important antioxidant molecule that is highly efficient in detoxifying ROS and peroxides. GSH plays important roles in response to different types of stress including, osmotic, acid and oxidative stresses (Kim *et al.*, 2012; Zhang *et al.*, 2012; Zhang *et al.*, 2007). Supplementation of the LAB growth media with GSH has been investigated for its role in stress protection and as a media component. In the mice model, deletion of a key enzyme in the synthesis of GSH, Gclc, results in a profound change in the biomarkers associated with oxidative stress, which results in liver failure and death (Chen *et al.*, 2007). An important function of GSH is to scavenger HO· since this ROS is not eliminated by any enzymatic reaction (Pompella *et al.*, 2003). Overexpression of genes gshA and gshB from *E. coli*, involved in GSH synthesis, in *L. lactis* NZ9000, which is incapable of synthesis as well as uptake of GSH, resulted in increased resistance to acid and oxidative stress (Fu *et al.*, 2006, Zhang *et al.*, 2007). Even though genes involved in glutathione synthesis are not found in many LAB, an important component of glutathione–dependent oxido-reduction system, glutathione reductase (GR), are widely present in LAB genomes. It has been shown that two GshR found in *L. plantarum* playan important role against oxidative stress caused by bile salts (Hamon *et al.*, 2011; Zangh and Li, 2013). Knockout of *gshR* in *L. sanfranciscensis*, used in wheat sourdoughs production, resulted in a strong reduction in oxidative stress tolerance and inhibition of aerobic growth (Jansch *et al.*, 2007).

Most of GSH biological functions are related to the conversion of reduced glutathione to its oxidized form (GSSG) by the enzyme glutathione peroxidase (Gpx) that utilizes glutathione to scavenge H_2O_2 and defend cells against oxidative stress. The later reduction of GSSG by GR to regenerate GSH helps to maintain the GSH pool (Pophaly *et al.*, 2012; Zhang and Li, 2013). Glutathione can be naturally synthesized by some LAB such as *S. thermophilus* and *E. faecalis* but most of other LAB species are only able to accumulate glutathione from the medium (Janowiak and Griffith, 2005; Kim *et al.*, 2008a; Zhang *et al.*, 2009). Accumulation of glutathione in stationary cells of *L. lactis* subsp. *cremosis* offers increased resistance to H_2O_2. Additionally, the anti-oxidative activity of 11 strains of LAB was found to correlate with the cellular concentration of GSH (Li *et al.*, 2003; Yoon and Byun, 2004). Also, GSH seems to be important in overcoming additional stresses accompanied with exposure to bile-salts, cold storage and freeze drying (Pophaly *et al.*, 2012; Zhang and Li, 2013). *L. plantarum* WCFS1 showed an increased expression of glutathione reductase after exposure to bile stress (Bron *et al.*, 2004). *L. sanfranciscensis* supplemented with glutathione and later freeze dried showed higher survival rates when compared to cysteine-supplemented cells. Prevention of membrane fatty acid oxidation, maintenance of chain length of fatty acids, and protection against secondary stress conditions generated during the treatment were associated to this protective effect of glutathione (Zhang *et al.*, 2010b, 2010c, 2012).

Glutathione supplementation has also improved resistance towards osmotic and acid stress. For example, *L. salivarius* (isolated from pig faeces), *L. lactis* subsp. *cremosis* SK11, *L. lactis* subsp. *cremosis* NZ9000 and *L. mesentereroides* ATCC8293 were protected from acid challenge and *L. lactis* was able to survive up to 5 M NaCl after glutathione supplementation. Therefore, LAB are resistant to multiple stresses when glutathione was present in the growth media (Zhang *et al.*, 2007; Lee *et al.*, 2010; Zhang *et al.*, 2010a,b; Kim *et al*, 2012). Metabolism of glutathione by LAB also presents important technological applications, such as accelerating cheese ripening, improving sourdough fermentation as well as promoting growth of *Oenococcus oeni* stimulating malolactic fermentation in wine (Rauhut *et al.*, 2004; Jansch *et al.*, 2007; Ganzle *et al.*, 2008).

Overall, glutathione supplementation or coculturing of glutathione-producing strains can be a successful strategy to improve fitness and survival of LAB during harsh processing treatments as well as hostile conditions in the GI tract.

Encapsulation

Encapsulation is a physicochemical or mechanical process used to entrap a substance in a material in order to produce small particles. The encapsulated material is called active agent or core material and is surrounded by a matrix called coating or shell (Anal and Singh, 2007; Chen and Chen, 2007; Burgain *et al.*, 2011). Encapsulation technology has been applied to protect probiotic LAB from adverse environments and improve viability in both food products and the transit through the GI tract. Besides protecting cells from harsh conditions, encapsulation allows their targeted release in a metabolically active state at the appropriate sites in the digestive tract (Burgain *et al.*, 2011).

Several methods have been applied to encapsulate lactic acid and probiotics including extrusion, emulsion, spray drying, freeze drying and spray coating techniques (Krasaekoopt *et al.*, 2003). Among the carrier materials the most commonly used are alginate, kappa-carrageenan, starch, chitosan, gelatin, lipids, whey protein, xanthan and locust bean gum. The viability of encapsulated bacteria depends on the physicochemical properties of the capsules and recoating microcapsules further increases bacterial survival (Shah, 2000; Adhrikari *et al.*, 2000; Krasaekoopt *et al.*,

2003; Corona-Hernandez *et al.*, 2013). Microencapsulation is a promising technology to protect cells from industrial processing and for the efficient release of viable cells in the GI tract. Microencapsulation has been reported to protect and reduce loss of viability of probiotics by blocking the effects of a variety of stressors such as O_2, bile-salts, and high temperature, among others. Anoxic regions inside the capsules appears to provide a mechanism for the protective role of encapsulation of O_2-sensitive strains; however, few studies have evaluated the protective effect of encapsulation against O_2 toxicity (Talwalkar and Kailasapathy, 2004; Rodrigues *et al.*, 2011).

The protective role of microencapsulation against oxidative stress was evaluated in *L. acidophilus* 2409 encapsulated in calcium alginate and grown aerobically in reconstituted skim milk broth. The results showed significantly higher cell counts when encapsulated compared to free cells. The protective effect of encapsulation was also tested in yoghurt, but the results demonstrated that only 6 out of 12 strains tested exhibited higher survival values (Talwalkar and Kailasapathy, 2003). Thus, encapsulation can protect probiotics from ROS toxicity, but further optimizations are needed before its application in fermented dairy products.

Even though little is known about the protective role of encapsulation against the toxic effects of ROS, several studies have reported protection against exposure to bile-salts. *L. gasseri* encapsulated in chitosan-coated alginate microcapsules was resistant to simulated gastric conditions and bile salts solution resulting in improved survival when compared with free cells (Chavarri *et al.*, 2010). The formation of an insoluble complex between chitosan and bile salts takes place on the surface of chitosan-alginate microparticles thereby limiting the diffusion of bile salts into the matrix core and protecting the probiotics (Koo *et al.*, 2001; Yu *et al.*, 2001; Corona-Hernandez *et al.*, 2013). Survival rates of *L. plantarum* CRL1815 and *L. rhamnosus* ATCC53103 exposed to bile salts solution was also improved after encapsulation in gelatin combined with xanthan (Jimenez-Pranteda *et al.*, 2012).

Stress adaptation

An adaptive cellular response can be induced by exposing a strain to sublethal levels of stress that will improve its resistance towards potential higher levels of the same stressor. In this regard, *L. acidophilus* 2409 exposed to increasing concentrations of dissolved O_2 in yoghurt was adapted to oxidative stress and was capable of withstanding higher levels of dissolved O_2 even after 35 days of storage (Talwalkar and Kailasapathy, 2004). Investigation of the growth performance and oxidative stress adaptation of *L. plantarum* WCFS1 under respiratory and fermentative growth conditions demonstrated that cells harvested from exponential or stationary phases from aerobically growth cultures showed highest resistance to H_2O_2. Increased resistance of aerobically grown cultures seems to be related to the higher expression and activity of catalase and to the rapid consumption of O_2 by the active partial electron transport chain. Leakage of electrons from the electron transport chain intermediates and formation of ROS might also be related to the induction of oxidative stress resistance mechanisms in *L. plantarum* under aerobic growth (Watanabe *et al.*, 2012). For *L. lactis* MG1363, better survival during long-term storage was demonstrated when the cells were originally grown under respiratory conditions due to the reduced intracellular O_2 concentrations (Duwat *et al.*, 2001; Rezaiki *et al.*, 2004)

Genetic modification

Currently, there are available strategies for efficient heterologous gene expression and genetic manipulation of LAB strains. Through genetic manipulation of LAB, researchers are able to study the mechanisms for ROS tolerance (Andrus *et al.*, 2003; Bruno-Barcena *et al.*, 2004, 2005, 2010; Bron and Kleerebezem, 2011). To date, the use of genetically modified organisms in food is not allowed; therefore, the acquired physiological knowledge is indirectly employed to improve cell survival, cell tolerance, and in general cell robustness, e.g. to make the strains able to tolerate the overall stressful environments that these important organisms encounter either during manufacturing or in the GI tract.

Commonly, LAB used as starters in fermented food, such as *S. thermophilus* and *L. delbrueckii subsp. bulgaricus*, are tolerant to the conditions encountered during fermentation; however, these strains frequently battle with oxidative stress, which leads to growth inhibition and cell death (Fu, 2013; Imlay, 2003). Heterologous production of catalases and MnSOD has been widely studied to improve oxidative stress resistance of LAB strains used as starters and probiotic. As example, heterologous expression of the haem-dependent catalase gene, *katE*, from *L. brevis* ATCC 367 in the yoghurt starter bacteria *S. thermophilus* ST5 resulted in 52- and 143-fold enhanced oxidative stress resistance when exponential or stationary phases cells were exposed to 6 mM and 10 mM of H_2O_2, respectively. Expression of the *katE* in *S. thermophilus* was also beneficial to *L. delbrueckii subsp. bulgaricus* ATCC11842 increasing its survival values during co-culture (Fu *et al.*, 2013).

Overexpression of *katA* gene from *L. sakei* SR911 in *L. rhamnosus* AS 12466 or in the catalase-deficient *L. plantarum* TISTR850, markedly improved their resistance to oxidative stress (An *et al.*, 2010; Noonpakdee *et al.*, 2004). Heterologous expression of a Mn-Kat from *L. plantarum* ATCC 14431 in *L. casei* BL23 enabled the cells to better survive under aerated conditions and in presence of H_2O_2 (Rochat *et al.*, 2006, 2007). Likewise, the gene *sodA* from *S. thermophilus* encoding a MnSOD was cloned in a shuttle vector and expressed in strains of *L. reuteri* NCK938, *L. jonsonii* NCK89, *L. gasseri* NCK334 and *L. acidophilus* NCK56 providing a clear protection against ROS and H_2O_2 (Bruno-Barcena *et al.*, 2004). Later integration of *sodA* into the chromosome of *L. gasseri* confirmed previous results, and added information about shelf-life protection during storage (Bruno-Barcena *et al.*, 2005). The co-expression of catalase and superoxide dismutase genes demonstrated to improve protection against oxidative stress in *L. rhamnosus* AS1.2466 that showed a 400-fold higher survival when exposed to 10 mM of H_2O_2 than the cells expressing only the catalase gene (An *et al.*, 2010). Enhancing oxidative stress resistance in LAB may also result in more tolerance to other kinds of stress like acid and osmotic stresses. It has been demonstrated that Mn-SOD from *S. thermophilus* is not only important for oxidative stress resistance (Andrus *et al.*, 2003), but also for acid stress tolerance (Bruno-Barcena *et al.*, 2010). Furthermore, the knock-out of the *sodA* increased *S. thermophilus* sensitivity to lactic acid. The inclusion of iron chelators provided a partial protection against lactic acid supporting the evidence that acid stress triggers iron-mediated oxidative stress in Gram-positive and Gram-negative bacteria that can be relieved by MnSOD or the presence of iron chelators (Bruno- Barcena *et al.*, 2010) (Fig. 14.1).

Besides the antioxidant enzymes, catalase and MnSOD, the thiol/disulfide redox systems glutathione:glutaredoxin reductase and thioredoxin:thioredoxin reductase are important mechanisms to cope with oxidative stress in LAB. Like the glutathione:GR system, the thioredoxin system constitutes an important part of the bacterial oxidative defence system. This system help cells to maintain redox homeostasis and protect bacteria from the damage caused by ROS. In addition, the thioredoxin system controls thiol–disulfide bond balance, playing an essential role in the maintenance of protein structure and activity (Vido *et al.*, 2005; Serata *et al.*, 2012). The importance of the thioredoxin system in *L. plantarum* WCFS1 was investigated by overexpressing the *trxB1* gene, coding for a TrxR. The higher TrxR activity resulted in increased resistance towards oxidative stress induced by diamide or H_2O_2. The overexpression of the thioredoxin reductase gene also activates genes associated with DNA repair and stress response (Serrano *et al.*, 2007).

As antioxidative systems may contribute to overall resistance of LAB to ROS, identification of potential targets for genetic engineering may contribute in the design of strains more resistant to multiple environmental stresses, like acid and bile salts, thereby enhancing the robustness of industrially applied LAB.

Antioxidant properties of probiotic LAB

Oxidative stress is triggered by the imbalance between pro-oxidants and the antioxidant defence system within the cell. ROS and RNS damage cellular lipids, proteins, nucleic acids and carbohydrates (Fridovich, 1978, 1995; Halliwell and Gutterridge, 1999; Imlay, 2013). In the presence of O_2, the by-products of normal metabolism can participate in the spontaneous production of ROS. In addition, ROS can be generated by exogenous sources such as environmental pollutants, radiation, toxins, redox-cycling compounds, drugs, Fe imbalance, and symbiosis within the gut (Sies, 1997, 2014; Hassan and Fridovich, 1979). Although aerobic metabolism generates a substantial amount of energy for the cell, the accumulation of ROS can lead to cellular damage and death (Fridovich, 1995; Hassan; 1997; Sies, 1997, 2014; Halliwell and Gutterridge, 1999). In humans, the generation of ROS is associated with several diseases and disorders including premature ageing, atherosclerosis, myocardial infarction, Alzheimer's disease, cirrhosis, diabetes, metabolic syndrome, obesity, gastrointestinal diseases and cancer (Halliwell and Gutterridge, 1999; Stocker and Keaney, 2004; Kals *et al.*, 2006; Krzystek-Korpacka *et al.*, 2008; Tsukahara, 2007; Vincent *et al.*, 2007; Castellani *et al.*, 2008). As a result, there has been a strong selective pressure to reduce cellular oxidative stress and maintain cell homeostasis, and this selective pressure gave rise to the evolution of multiple mechanisms of cellular protection. Enzymatic antioxidants, such as superoxide dismutases (SODs), catalases (CATs), glutathione peroxidases (GPXs) and oxygenases, as well as non-enzymatic antioxidants, such as reduced glutathione, vitamin C and E, constitute an integrated antioxidant defence system. Surprisingly, the ability of the host to defend against ROS appears to be tightly linked to the constituents within the gastrointestinal microbiota. Specific strains with antioxidative properties influence the host's resistance to oxidative stress within the gut lumen, inside the host cells, and in the blood, protecting tissues and organs from the damage caused by ROS (Kullisaar *et al.*, 2012; Milkelsaar and Zilmer, 2009). Probiotic LAB have been known to have health benefits to the host (Metchnikoff, 1907; Dobrogosz, *et al.*, 2010). A limited number of naturally occurring and genetically modified probiotic LAB strains have been investigated for their antioxidant properties (Ebel *et al.*, 2014; Carroll *et al.*, 2007; LeBlanc *et al.*, 2011; Rochat *et al.*, 2007; Watterlot *et al.*, 2010; Han *et al.*, 2006; Li *et al.*, 2012; Zhang *et al.*, 2010a; Kullisaar *et al.*, 2002, 2011). Biomarkers of oxidative stress are used to evaluate damage caused by ROS like oxidized proteins (nitrate tyrosine, protein carbonyls), oxidized nucleic acid bases (8-oxo-2-deoxyguanosine), oxidized carbohydrates (glycated products), and oxidized lipids (malondialdehyde, isoprostanes, oxidized low-density lipoproteins). Unsaturated fatty acids are prone to lipid peroxidation during exposure to ROS. Malondialdehyde (MDA) is a mutagenic and carcinogenic reactive substance produced as end product of lipid peroxidation and its presence and accumulation is used as a marker of oxidative stress in biological systems and foods. Isoprostanes and oxidized low-density lipoproteins (oxLDL) are also used as oxidative stress biomarkers, the latter associated with the development of cardiovascular diseases (Gil *et al.*, 2006; Kullisaar *et al.*, 2012).

Promising results have been reported when naturally occurring probiotics were administered to animal disease models. The data of reduction in the levels of oxidative stress biomarkers and enhancement of the host's antioxidative defences have revealed signs of protection by preserving tissue integrity, and inhibition of disease progression (Amaretti *et al.*, 2013; Balakrishnan *et al.*, 2013; Li *et al.*, 2013; Kullisaar *et al.*, 2003, 2011; Zhang *et al.*, 2010a). Lin and Yen (1999) showed that *L. acidophilus* E and *B. longum* B6 scavenge MDA on cultured mammalian cells. Additionally, supplementation of *L. fermentum* (isolated from gut mucosa of healthy weaning piglets) to pigs' diet increased SOD and GPX enzymes activities in serum (Wang *et al.*, 2009).

The antioxidant potential of *L. casei* (Zhang) has been investigated, *in vivo* using the hyperlipidaemic Wistar rat as a model system, by measuring lipid peroxidation, activities of antioxidant enzymes, and biomarkers of liver oxidative stress. Thus, supplementing the diet with different doses of *L. casei* (Zhang) diminished the levels of MDA in serum and liver, which is indicative of its antioxidative properties. In addition, the activities of SOD and GPX, key enzymes that protect cells against oxidative injury, were increased in serum and liver samples. Liver function and hepatic cell integrity were improved by the probiotic as indicated by the lower concentrations of glutamic–pyruvic transaminase (GPT) and glutamic oxalacetic transaminase (GOT), indicators of liver function, in the blood. Moreover, total antioxidant capacity (T-AOC) in liver, as measured by the ferric reducing/

antioxidant power (FRAP) assay, showed that *L. casei* (Zhang) supplementation significantly increased T-AOC. These results suggest that *L. casei* (Zhang) possesses important antioxidant properties improving antioxidative indexes in hyperlipidaemic rats (Zhang *et al.*, 2010a).

Studies using *L. fermentum* ME-3 isolated from human showed that the strain possesses antimicrobial, antioxidative and antiatherogenic properties both *in vitro* and *in vivo* (Kullisaar *et al.*, 2002, 2003, 2010; Truusalu *et al.*, 2004; Songisepp *et al.*, 2005; Jarvenpaa *et al.*, 2007). It has also been demonstrated that when *L. fermentum* ME-3 was consumed with food products, it enhanced the systemic antioxidative activity of human serum, lengthened the lag phase of LDL oxidation, suppressed production of atherogenic oxLDL, and decreased the serum levels of oxidized glutathione, oxLDL and baseline conjugated dienes of low-density lipoprotein (BCD-LDL) (Kullisaar *et al.*, 2003, 2011; Mikelsaar *et al.*, 2007). Several clinical trials have demonstrated the beneficial properties of *L. fermentum* ME-3 in reducing the inflammatory and atherogenic factor LDL and the oxLDL fraction in the blood by extending its resistance to oxidation (Kullisaar *et al.*, 2003, 2011; Songisepp *et al.*, 2005; Mikelsaar *et al.*, 2008). It has been shown that *L. fermentum* ME-3 is able to reduce the concentration of host systemic oxidative stress markers in urine: such as isoprostanes and 8-oxo-2-deoxyguanosine (Kullissar *et al.*, 2003, 2011; Songisepp *et al.*, 2005). Administration of *L. fermentum* ME-3 reduces the postprandial elevation of triglycerides levels in blood and improves HDL bioquality, with lower levels of harmful oxidation products. Systemic antioxidative effect of *L. fermentum* ME-3 seems to be related to the alleviation of oxidative stress and inflammatory-related abnormalities in the GI tract, improving the bioquality of LDL and HDL particles (Kullisaar *et al.*, 2012). Some of the antioxidative properties of *L. fermentum* ME-3 are credited to its ability to produce and accumulate GSH, maintaining a high ratio GSH/GSSG (Kullisaar *et al.*, 2002). *L. fermentum* ME-3 has a complete GSH system, being able to synthesize, transport, and reduce GSH.

GSH has an important role in the maintenance of the integrity of the intestinal mucosa and low levels of GSH correlate with ulcerative colitis and Crohn's disease. GSH supplementation improves colonic damage, restores GSH levels, and decreases lipid peroxidation in a rat model of TNBS colitis (Loguercio *et al.*, 2003, Coskun *et al.*, 2010). Probiotic strains that produce or promote the intestinal release of GSH could have potential therapeutic applications. To this end, treatment with *L. fermentum* 5716, a probiotic strain capable of producing GSH and its precursor, ameliorated colonic inflammation in TNBS induced rat colitis, attributed to its ability to release GSH in the gut, and restored colonic GSH levels (Peran *et al.*, 2007). At the early stages of experimental acute pancreatitis GSH depletion correlates with the extent of pancreatic injury, but treatment with probiotics markedly reduced the severity of acinar cell injury through improved intestinal barrier function via mechanisms involving GSH biosynthesis. Using a well-established model of acute pancreatitis, the treatment with a multispecies probiotic containing *L. acidophilus* W70, *L. casei* W56, *L. salivarus* W24, *L. lactis* W58, *B. bifidum* W23, and *B. lactis* W52 enhanced de novo GSH synthesis and increased the systemic and pancreatic GSH content (Lutgendorff *et al.*, 2008, 2009; Spyropoulos *et al.*, 2011).

As shown above, only few naturally occurring LAB have the capacity to exert an anti-inflammatory response within the host. Therefore, scientists have attempted to construct genetically modified LAB expressing different antioxidant genes in order to enhance their anti-inflammatory impact. For instance, the delivery of SOD to the intestine by LAB has also been associated with an anti-inflammatory phenotype. *L. plantarum* and *L. lactis* genetically engineered to produce and release SOD were found to promote an anti-inflammatory response within a colitis rat model (Han *et al.*, 2006). In addition, the oral administration of *L. gasseri* engineered to produce a manganese-containing SOD reduced the severity of colitis in IL-10-deficient mice (Carroll *et al.*, 2007).

Exopolysaccharides (EPS) produced by probiotics are important molecules that may be used for reducing oxidative stress in the gastrointestinal tract. Various health benefits have been attributed to EPS including anti-tumour, anti-inflammation, anti-diabetes and modulation of factors within the innate immune system (Meng *et al.*, 2010, Akramiene *et al.*, 2007; Volman *et al.*, 2008; Mantovani *et al.*, 2008). The protective effect of an EPS isolated from *L. plantarum* LP6 against oxidative stress was evaluated by observing the viability of PC12 cells after exposure to H_2O_2. Viability of PC12 cells as well as some oxidative stress biomarkers were improved by EPS, resulting in the enhancement of CAT and SOD activities and reduction in the level of MDA (Li *et al.*, 2013). EPS isolated from *L. plantarum* C88, designated LPC-1, and EPS isolated from *L. helveticus* MB2-1 have demonstrated antioxidant properties and ROS scavenging ability. LPC-1 inhibited the formation of MDA and increased SOD activity and the total antioxidant capacity of H_2O_2-treated Caco-2 cells. EPS isolated from *L. helveticus* MB2-1 showed strong scavenging activities against the DPPH free radical, HO•, and O_2^- in addition to demonstrating the ability to chelate Fe^{2+} (Li *et al.*, 2014; Zhang *et al.*, 2013). Another study compared the effects of a high-EPS producing probiotic strain with a low-EPS producing probiotic strain of *L. delbrueckii* subsp. *bulgaricus* in a rat model of acetic acid-induced colitis. Both strains were able to attenuate the severity of experimental colitis, but the high-EPS producing strain demonstrated better results by lowering oxidative stress biomarkers (Sengül *et al.*, 2011). Thus, EPS isolated from LAB have demonstrated antioxidant properties scavenging ROS, up-regulating enzymatic and non-enzymatic antioxidative activities, as well as reducing lipid peroxidation.

Conclusion and future perspectives

Overall, LAB have promising applications towards preventing and treating diseases and disorders associated with oxidative stress. During bacterial infections, LABs may modulate the host and pathogen towards beneficial outcomes. In addition, there is evidence that LABs may modulate the host antiviral responses as well. These data suggest that future treatment of bacterial and viral infections may include LABs as a supplement to the standard treatments. In sub-Saharan Africa where the number of HIV/AIDS cases is reaching a catastrophic level, the use of *L. rhamnosus*

GR-1 may have a beneficial role when supplementing the anti-retroviral therapy (ART) regime. Although future experiments and clinical trials are required to discern the efficacy of LABs in treating bacterial, and viral infections, or perhaps certain cancers, the application of these probiotics and their antioxidant properties may one day become standard supplements to established treatments.

References

Abe, F., Tomita, S., Yaeshima, T., and Iwatsuki, K. (2009). Effect of production conditions on the stability of a human bifidobacterial species *Bifidobacterium longum* in yoghurt. Lett. Appl. Microbiol. *49*, 715–720.

Adhikari, K., Mustapha, A., Grun, I.V., and Fernando, L. (2000). Viability of microencapsulated bifidobacteria in set yoghurt during refrigerated storage. J. Dairy Sci. *83*, 1946–1951.

Akramiene, D., Kondrotas, A., Didziapetriene, J., and Kevelaitis, E. (2007). Effects of beta-glucans on the immune system. Medicina (Kaunas) *43*, 597–606.

Alcántara, C., and Zúñiga, M. (2012). Proteomic and transcriptomic analysis of the response to bile stress of *Lactobacillus casei* BL23. Microbiology. *158*, 1206–1218.

Amaretti A., di Nunzio M., Pompei, A., Raimondi, S., Rossi, M., and Bordoni, A. (2013). Antioxidant properties of potentially probiotic bacteria: *in vitro* and *in vivo* activities. Appl. Microbiol. Biotechnol. *97*, 809–817.

An, H.R., Zhou, H., Huang, Y., Wang, G.H., Luan, C.G., Mou, J., Luo, Y.B., and Hao, Y.L. (2010). High-level expression of haem-dependent catalase gene *katA* from *Lactobacillus sakei* protects *Lactobacillus rhamnosus* from oxidative stress. Mol. Biotechnol. *45*, 155–160.

Anal, A.K., and Singh, H. (2007). Recent advances in microencapsulation of probiotics for industrial applications and targeted delivery. Trends Food Sci. Technol. *18*, 240–251.

Andrus, J.M., Bowen, S.W., Klaenhammer, T.R., and Hassan, H.M. (2003). Molecular characterization and functional analysis of the manganese-containing superoxide dismutase gene (*sodA*) from *Streptococcus thermophilus* AO54. Arch. Biochem. Biophys. *420*,103–113.

Archibald, F.S., and Fridovich, I. (1982). Investigations of the state of the manganese in *Lactobacillus plantarum*. Arch. Biochem. Biophys. *215*, 589–596.

Archibald, F.S., and Fridovich, I. (1981a). Manganese and defences against oxygen toxicity in *Lactobacillus plantarum*. J. Bacteriol. *145*, 442–451.

Archibald, F.S., and Fridovich, I. (1981b). Manganese, superoxide dismutase, and oxygen tolerance in some lactic acid bacteria. J. Bacteriol. *146*, 928–936.

Azcárate-Peril, M.A., Sikes, M., and Bruno-Bárcena, J.M. (2011). The intestinal microbiota, gastrointestinal environment and colorectal cancer: a putative role for probiotics in prevention of colorectal cancer. Am. J. Physiol. Gastrointest. Liver Physiol. *301*, G401–24.

Balakrishnan, R., Kumar, C.S., Reddy, K.K., Rani, M.U., Srikanth, M.K., Kavitha, K., Boobaln, G., and Reddy, A.G. (2013). Antioxidant activity of coated probiotic *Lactobacillus casei* on chromium (VI) induced oxidative stress in rats. Proc. Natl. Acad. Sci., India Section B: Bio. Sci., 1–6. doi: 10.1007/s40011-013-0215-5

Barré, O., Mourlane, F., and Solioz, M. (2007). Copper induction of lactate oxidase of *Lactococcus lactis*: a novel stress response. J. Bacteriol. *189*, 5947–5954.

Begley, M., Gahan, C.G.M., and Hill, C. (2005). The interaction between bacteria and bile. FEMS Microbiol. Rev. *29*, 625–651.

Berrisford, J.M., and Sazanov, L.A. (2009). Structural basis for the mechanism of respiratory complex I. J. Biol. Chem. *284*, 29773–29783.

Bielski, B.H.J. (1978). Reevaluation of the spectral and kinetic properties of HO_2 and O_2^- free radicals. Photochem. Photobiol. *28*, 645–649.

Bron, P.A., and Kleerebezem, M. (2011). Engineering lactic acid bacteria for increased industrial functionality. Bioeng. Bugs 2, 80–87.

Bron, P.A., Marco, M., Hoffer, S.M., Van Mullekom, E., de Vos, W.M., and Kleerebezem, M. (2004). Genetic characterization of the bile salt response in *Lactobacillus plantarum* and analysis of responsive promoters *in vitro* and *in situ* in the gastrointestinal tract. J. Bacteriol. *186*, 7829–7835.

Brooijmans, R., Smit, B., Santos, F., Van Riel, J., De Vos, W.M., and Hugenholtz, J. (2009). Haem and menaquinone induced electron transport in lactic acid bacteria. Microb. Cell Fact. *8*, 28.

Bruno-Barcena, J.M., Azcarate-Peril, M.A., and Hassan, H.M. (2010). Role of antioxidant enzymes in bacterial resistance to organic acids. Appl. Environ. Microbiol. *76*, 2747–2753.

Bruno-Barcena, J.M., Andrus, J.M., Libby, S.L., Klaenhammer, T.R., and Hassan, H.M. (2004). Expression of a heterologous manganese superoxide dismutase gene in intestinal lactobacilli: Protection against the toxicity of hydrogen peroxide. Appl. Environ. Microbiol. *70*, 4702–4710.

Bruno-Barcena, J.M., Azcarate-Peril, M.A., Klaenhammer, T.R., and Hassan, H.M. (2005). Marker-free chromosomal integration of the manganese superoxide dismutase gene (*sodA*) from *Streptococcus thermophilus* into *Lactobacillus gasseri*. FEMS Microbiol. Lett. *246*, 91–101.

Burgain, J., Gaiani, C., Linder, M., and Scher, J. (2011). Encapsulation of probiotic living cells: From laboratory scale to industrial applications. J. Food Engin. *104*, 467–483.

Carroll, I.M., Andrus, J.M., Bruno-Barcena, J.M., Klaenhammer, T.R., Hassan, H.M., and Threadgill, D.S. (2007). The anti-inflammatory properties of *Lactobacillus gasseri* expressing manganese superoxide dismutase (MnSOD) using the interleukin 10-deficient mouse model of colitis. Am. J. Physiol. Gastrointest. Liver Phyiol. *293*, 729–739.

Castellani, R.J., Lee, H.G., Zhu, X., Perry, G., and Smith, M.A. (2008). Alzheimer disease pathology as a host response. J. Neuropathol. Exp. Neurol. *67*, 523–531.

Chang, S.K., and Hassan, H.M. (1997). Characterization of superoxide dismutase in *Streptococcus thermophilus*. Appl. Environ. Microbiol. *63*, 3732–3735.

Chávarri, M., Marañón, I., Ares, R., Ibáñez, F.C., Marzo, F., and Villarán, M.D.C. (2010). Microencapsulation of a probiotic and prebiotic in alginate–chitosan capsules improves survival in simulated gastro-intestinal conditions. Int. J. Food Microbiol. *142*, 185–189.

Chen, M.J., and Chen, K.N. (2007). Applications of probiotic encapsulation in dairy products. In: J.M. Lakkis (Ed.), Encapsulation and Controlled Release Technologies in Food Systems. Wiley-Blackwell, USA, pp. 83–107.

Chen, Y., Miller, M.L., Shen, D., Shertzer, H.G., Stringer, K.F., Wang, B., Schneider, S.N., Nebert, D.W., and Dalton, T.P. (2007). Hepatocyte-specific Gclc deletion leads to rapid onset of steatosis with mitocondrial injury and liver failure. Hepatology *45*, 1118–1128.

Condon, S. (1987). Responses of lactic acid bacteria to oxygen. FEMS Microbiol. Rev. *46*, 269–280.

Corona-Hernandez, R.I., Álvarez-Parrilla, E., Lizardi-Mendoza, J., Islas-Rubio, A.R., Rosa, L., and Wall-Medrano, A. (2013). Structural Stability and Viability of Microencapsulated Probiotic Bacteria: A Review. Com. Rev. Food Sci. Food Safety *12*, 614–628.

Corcoran, B.M., Stanton, C., Fitzgerald, G., and Ross, R.P. (2008). Life under stress: the probiotic stress response and how it may be Manipulated. Curr. Pharmac. Design *14*, 1382–1399.

Coskun, S., Aslim, B., and Yuksekgad, Z.N. (2010). Effect of two strains of *Lactobacillus delbrueckii* subsp. *bulgaricus* on nitric oxide generation and antioxidant status of rat intestine. Med. Chem. Res. *19*, 1082–1091.

Cruz, A.G., Castro, W.F., Faria, J.A.F., Bolini, H.M.A., Celeghini, R.M.S., Raices, R.S.L., Oliveira, C.A.F., Freitas, M.Q., Conte Junior, C.A., and Marsico, E.T. (2013). Stability of probiotic yoghurt added with glucose oxidase in plastic materials with different permeability oxygen rates during the refrigerated storage. Food Res. Intl. *51*, 723–728.

Cruz, A.G., Castro, W.F., Faria, J.A.F., Bogusz, S., Granato, D., Celeguini, R.M.S., Lima-Pallone, J., and Godo, H.T. (2012a). Glucose oxidase: A potential option to decrease the oxidative stress in stirred probiotic yoghurt. LWT—Food Sci. Tech. *47*, 512–515.

Cruz, A.G., Castro, W.F., Faria, J.A.F., Lollo, P.C.B., Amaya-Farfán, J., Freitas, M.Q., Rodrigues, D., Oliveira, C.A., and Godoy, H.T. (2012b). Probiotic yoghurts manufactured with increased glucose oxidase levels: postacidification, proteolytic patterns, survival of probiotic microorganisms, production of organic acid and aroma compounds. J. Dairy Sci. *95*, 2261–2269.

Cruz, A.G., Cadena, R.S., Faria, J.A.F., Oliveira, C.A.F., Cavalcanti, R.N., Bona, E., Bolini, H.M., and Silva, M.A.A.P. (2011). Consumer acceptability and purchase intent of probiotic yoghurt with added glucose oxidase using sensometrics, artificial neural networks and logistic regression. Int. J. Dairy Technol. *64*, 549–556.

Cruz, A.G., Faria, J.A.F., Walter, E.H.M., Andrade, R.R., Cavalcanti, R.N., Oliveira, C.A.F., and Granato, D. (2010). Processing optimization of probiotic yoghurt containing glucose oxidase using response surface methodology. J. Dairy Sci. *93*, 1058–1069.

Culotta, V.C., and Daly, M.J. (2013). Manganese complexes: diverse metabolic routes to oxidative stress resistance in prokaryotes and yeast. Antioxid. Redox. Signal. *19*, 933–944.

Dave, R.I., and Shah, N.P. (1998). Ingredient supplementation effects on viability of probiotic bacteria in yoghurt. J. Dairy Sci. *81*, 2804–2816.

Dobrogosz, W.J., Peacock, T.J., and Hassan, H.M. (2010). Evolution of the probiotic concept: from conception to validation and acceptance in medical science. Adv. Appl. Microbiol. *72*, 1–41.

Duwat, P., Sourice, S., Cesselin, B., Lamberet, G., Vido, K., Gaudu, P., LeLoir, Y., Violet, F., Loubiere, P., and Gruss, A. (2001). Respiration capacity of the fermenting bacterium *Lactococcus lactis* and its positive effects on growth and survival. J. Bacteriol. *183*, 4509–4516.

Ebel, B., Lemetais, G., Beney, L., Cachon, R., Sokol, H., Langella, P., and Gervais, P. (2014). Impact of Probiotics on Risk Factors for Cardiovascular Diseases. A Review, Crit. Rev. Food Sci. Nut. *54*, 175–189.

FAO/WHO. (2002). Guidelines for the evaluation of probiotics in food. London Ontario, Canada.

Fenton, H.J.H. (1894). Oxidation of tartaric acid in presence of iron. J. Chem. Soc. Trans. *65*, 899–911.

Frei, B., and Higdon, J.V. (2003). Antioxidant activity of tea polyphenols *in vivo*: Evidence from animal studies. J. Nutr. *133*, 3275S-3284S.

Fridovich, I. (1995). Superoxide Radical and superoxide dismutases. Annu. Rev. Biochem. *64*, 97–112.

Fridovich, I. (1978). Oxygen radicals, hydrogen peroxide, and oxygen toxicity. In Free Radicals in Biology, W.A. Pryor, ed. (New York, NY: Academic Press), pp. 239–277.

Fu, L.Y., Kong, J., Sun, Z.L., Zhang, L., Zhang, X., and Guo, T. (2013). Enhancing the oxidative resistance of yoghurt starter bacteria with heterologous catalase expression in *Streptococus thermophilus*. Int. Dairy J. *30*, 68–72.

Fu, R.Y., Bongers, R.S., Van Swam, I.I., Chen, J., Molenaar, D., Kleerebezem, M., Hugenholtz, J., and Li, Y. (2006). Introducing glutathione biosynthetic capability into *Lactococcus lactis* subsp. cremoris NZ9000 improves the oxidative-stress resistance of the host. Metabol. Eng. *8*, 662–671.

Gänzle, M.G., Loponen, J., and Gobbetti, M. (2008). Proteolysis in sourdough fermentations: mechanisms and potential for improved bread quality. Trends Food Sci. Tech. *19*, 513–521.

Gaudreau, H., Champagne, C.P., Remondetto, G.E., Bazinet, L., and Subirade, M. (2013). Effect of catechins on the growth of oxygen-sensitive probiotic bacteria. Food Res. Int. *53*, 751–757.

Gaudu, P., Vido, K., Cesselin, B., Kulakauskas, S., Tremblay, J., Rezaiki, L., Lamberet, G., Sourice, S., Duwat, P., and Gruss, A. (2002). Respiration capacity and consequences in *Lactococcus lactis*. Antonie Van Leeuwenhoek *82*, 263–269.

Gibson, Q.H., Swoboda, B.E.P., and Massey, V. (1964). Kinetics and mechanism of action of glucose oxidase. J. Biol. Chem. *239*, 3927–3934.

Gil, L., Siems, W., Mazurek, B., Gross, J., Schroeder, P., Voss, P., and Grune, T. (2006). Age-associated analysis of oxidative stress parameters in human plasma and erythrocytes. Free Radic. Res. *40*, 495–505.

Gomes A.M.P., and Malcata, F.X. (1999). *Bifidobacterium* spp. and *Lactobacillus acidophilus*: biological, biochemical, technological and therapeutical properties relevant for use as probiotics. Trends Food Sci. Technol. *10*, 139–157.

Granato, D., Branco, G.F., Cruz, A.G., Faria, J.A.F., and Shah, N.P. (2010). Probiotic dairy products as functional foods. Comprehensive Rev. Food Sci. Food Safety. *9*, 455–470.

Grzeskowiak, L., Isolauri, E., Salminen, S., and Gueimonde, M. (2011). Manufacturing process influences properties of probiotic bacteria. Br. J. Nutr. *105*, 887–894.

Halliwell, B., and Gutteridge, J.M.C. (1999). Free Radicals in Biology and Medicine, 3rd. edn. New York: Oxford University Press.

Hamon, E., Horvatovich, P., Izquierdo, E., Bringel, F., Marchioni, E., Aoude-Werner, D., and Ennahar, S. (2011). Comparative proteomic analysis of *Lactobacillus plantarum* for the identification of key proteins in bile tolerance. BMC Microbiol. *11*, 63.

Han, W., Mercenier, A., Ait-Belgnaoui, A., Pavan, S., Lamine, F., van Swam, I.I., Kleerebezem, M., Salvador-Cartier, C., Hisbergues, M., Bueno, L., Theodorou, V., and Fioramonti, J. (2006). Improvement of an experimental colitis in rats by lactic acid bacteria producing superoxide dismutase. Inflamm. Bowel. Dis. *12*, 1044–1052.

Hassan, H.M. (1997). Cytotoxcity of oxyradicals and the evolution of superoxide dismutases. In Oxygen, Gene Expression, and Cellular Function, L.B. Clerch and D.J. Massaro, eds., Vol. 105, Lung Biology in Health and Disease, C. Lenfant, Exec. Ed. (New York, NY: Marcel Dekker), pp. 27–47.

Hassan, H.M., and Fridovich, I. (1980). Superoxide dismutases: detoxification of a free radical. In Enzymatic Basis of Detoxification. Vol. I., W.B. Jakoby, ed. (New York, NY: Academic Press), pp. 311–332.

Hassan, H.M., and Fridovich, I. (1979). Intracellular production of superoxide radical and hydrogen peroxide by redox active compounds. Arch. Biochem. Biophys. *196*, 385–395.

Hertzberger, R., Arents, J., Dekker, H.L., Pridmore, R.D., Gysler, C., Kleerebezem, M., and Teixeira de Mattos, M.J. (2014). H_2O_2 production in species of the *Lactobacillus acidophilus* group, a central role for a novel NADH dependent flavin reductase. Appl. Environ. Microbiol. doi: 10.1128/AEM.04272-13.

Holmgren, A. (1985). Thioredoxin. Annu. Rev. Biochem. *54*, 237–271.

Imlay, J.A. (2013). The molecular mechanisms and physiological consequences of oxidative stress: lessons from a model bacterium. Nat. Revs. Microbiol. Doi:10.1038/nrmicro3032.

Imlay, J.A. (2003). Pathways of oxidative damage. Annu. Rev. Microbiol. *57*, 395–418.

Imlay, J.A., and Linn, S. (1988). Toxic DNA damage by hydrogen peroxide through the Fenton reaction *in vitro* and *in vivo*. Science *240*, 640–642.

Janowiak, B.E., and Griffith, O.W. (2005). Glutathione synthesis in *Streptococcus agalactiae*. One protein accounts for gamma-glutamylcysteine synthetase and glutathione synthetase activities. J. Biol. Chem. *280*, 11829–11839.

Jansch, A., Freiding, S., Behr, J., and Vogel, R.F. (2011). Contribution of the NADH-oxidase (Nox) to the aerobic life of *Lactobacillus sanfranciscensis* DSM20451(T). Food Microbiol. *28*, 29–37.

Jansch, A., Korakli, M., Vogel, R.F., and Gänzle, M.G. (2007). Glutathione reductase from *Lactobacillus sanfranciscensis* DSM20451T: contribution to oxygen tolerance and thiol exchange reactions in wheat sourdoughs. Appl. Environ. Microbiol. *73*, 4469–4476.

Jarvenpaa, S., Tahvonen, R.L., Ouwehand, A.C., Sandell, M., Jarvenpaa, E., and Salminen, S. (2007). A probiotic, *Lactobacillus fermentum* ME-3, has antioxidative capacity in soft cheese spreads with different fats. J. Dairy Sci. *90*, 3171–3177.

Jimenez-Pranteda, M.L., Poncelet, D., Nader-Macias, M.E., Arcos, A., Aguilera, M., Monteoliva-Sanchez, M., and Ramos-Comenzana, A. (2012). Stability of lactobacilli encapsulated in various microbial polymers. J. Biosci. Bioeng. *113*, 179–184.

Johnston, M.A., and Delwiche, E.A. (1965). Isolation and characterization of the cyanide-resistant and azide-resistant catalase of *Lactobacillus plantarum*. J. Bacteriol. *90*, 352–356.

Kailasapathy, K. (2002). Microencapsulation of probiotic bacteria: technology and potential applications. Curr. Issu. Intest. Microbiol. *3*, 39–48.

Kals, J., Kampus, P., Kals, M., Zilmer, K., Kullisaar, T., Teesalu, R., Pulges, A., and Zilmer, M. (2006). Impact of oxidative stress on arterial elasticity in patients with atherosclerosis. Am. J. Hypertens. *19*, 902–908.

Keyer, K., and Imlay, J.A. (1996). Superoxide accelerates DNA damage by elevating free-iron levels. Proc. Natl. Acad. U.S.A.. *26*, 13635–13640.

Kim, E.K., Cha, C.J., Cho, Y.J., Cho, Y.B., and Roe, J.H. (2008a). Synthesis of gamma-glutamylcysteine as a major low-molecular-weight thiol in lactic acid bacteria *Leuconostoc* spp. Biochem. Biophs. Res. Comm. *369*, 1047–1051.

Kim, S.J., Cho, S.Y., Kim, S.H., Song, O.J., Shin, I.S., Cha, D.S., and Park, H.J. (2008b). Effect of microencapsulation on viability and other characteristics in *Lactobacillus acidophilus* ATCC 43121. LWT – Food Sci. Technol. *41*, 493–500.

Kim, J.E., Eom, H.J., Kim, Y., Ahn, J.E., Kim, J.H., and Han, N.S. (2012). Enhancing acid tolerance of *Leuconostoc mesenteroides* with glutathione. Biotechnol. Lett. *34*, 683–687.

Knauf, H.J., Vogel, R.F., and Hammes, W.P. (1992). Cloning, sequence, and phenotypic expression of *kata*, which encodes the catalase of *Lactobacillus sake* LTH677. Appl. Environ. Microbiol. *58*, 832–839.

Kono, Y., and Fridovich, I. (1983). Isolation abd characterization of the pseudocatalase of *Lactobacillus plantarum*. J. Biol. Chem. *258*, 6015–6019.

Koo, S., Cho, Y., Huh, C., Baek, Y., and Park, J. (2001). Improvement of the stability of *Lactobacillus casei* YIT 9018 by microencapsulation using alginate and chitosan. J. Microbiol. Biotechnol. *11*, 376–383.

Krasaekoopt, W., Bhandari, B., and Deeth, H. (2003). Evaluation of encapsulation techniques of probiotics for yoghurt. Int. Dairy J. *13*, 3–13.

Krzystek-Korpacka, M., Neubauer, K., Berdowska, I., Boehm, D., Zielinski, B., Petryszyn, P., Terlecki, G., Paradowski, L., and Gamian, A. (2008). Enhanced formation of advanced oxidation protein products in IBD. Inflamm. Bowel Dis. *14*, 794–802.

Kullisaar, T., Songisepp, E., and Zilmer, M. (2012). Probiotics and Oxidative Stress, In, Oxidative Stress – Environmental Induction and Dietary Antioxidants, V. Lushchak (Ed.), ISBN: 978-953-51-0553-4, InTech Europe, 51000 Rijeka, Croatia. doi: 10.5772/33924.

Kullisaar, T., Shepetova, J., Zilmer, K., Songisepp, E., Rehema, A., Mikelsaar, M., and Zilmer, M. (2011). An antioxidant probiotic redudes postprandial lipeemia and oxidative stress. Cent. Eur. J. Biol. *6*, 32–40.

Kullisaar, T., Songisepp, E., Aunapuu, M., Kilk, K., Arend, A., Mikelsaar, M., Rehema, A., and Zilmer, M. (2010). Complete glutathione system in probiotic *L. fermentum* ME-3. Appl. Biochem. Microbiol. *46*, 527–531.

Kullisaar, T., Songisepp, E., Mikelsaar, M., Zilmer, K., Vihalemm, T., and Zilmer, M. (2003). Antioxidative probiotic fermented goats' milk decreases oxidative stress-mediated atherogenicity in human subjects, Br. J. Nutr. *90*, 449–456.

Kullisaar, T., Zilmer, M., Mikelsaar, M., Vihalemm, T., Annuk, H., Kairane, C., and Kilk, A. (2002). Two antioxidative lactobacilli strains as promising probiotics. Int. J. Food Microbiol. *72*, 215–224.

Kussmaul, L., and Hirst, J. (2006). The mechanism of superoxide production by NADH:ubiquinone oxidoreductase (complex I) from bovine heart mitochondria. Proc. Natl. Acad. Sci. U.S.A.. *103*, 7607–7612.

Lankaputhra, W.E.V., and Shah, N.P. (1996). A simple method for selective enumeration of *Lactobacillus acidophilus* in yoghurt supplemented with *L. acidophilus* and *Bifidobacterium* spp. Milchwissenschaft. *51*, 446–450.

LeBlanc, J.G., del Carmen, S., Miyoshi, A., Azevedo, V., Sesma, F., Langella, P., Bermudez-Humaran, L.G., Watterlot, L., Perdigon, G., and de Moreno de LeBlanc, A. (2011). Use of superoxide dismutase and catalase producing lactic acid bacteria in TNBS induced Crohn's disease in mice. J. Biotechnol. *151*, 287–293.

Lechardeur, D., Cesselin, B., Fernandez, A., Lamberet, G., Garrigues, C., Pedersen, M., Gaudu, P., and Gruss, A. (2011). Using haem as an energy boost for lactic acid bacteria. Curr. Opin. Biotechnol. *22*, 143–9.

Lee, K.B., Pi, K.B., Kim, E.B., Rho, B.S., Kang, S.K., Lee, H.G., and Choi, Y.J. (2010). Glutathione-mediated response to acid stress in the probiotic bacterium *Lactobacillus salivarius*. Biotechnol. Lett. *32*, 969–972.

Li, W., Cheng, X., Jiang, M., and Dong, M. (2014). Structural elucidation and antioxidant activities of exopolysaccharides from *L. helveticus* MB2-1. Carbohydrate Polymers. *102*, 351–359.

Li, J.Y., Jin, M.M., Meng, J., Gao, S.M., and Lu, R.R. (2013). Exopolysaccharide from *Lactobacillus plantarum* LP6: Antioxidation and the effect on oxidative stress. Carbohydrate Polymers. *98*, 1147–1152.

Li, S., Zhao, Y., Zhang, L., Zhang, X., Huang, L., Li, D., Niu, C., Yang, Z., and Wang, Q. (2012). Antioxidant activity of *Lactobacillus plantarum* strains isolated from traditional Chinese fermented foods. Food Chem. *135*, 1914–1919.

Li, X.Y., Chen, X.G., Cha, D.S., Park, H.J., and Liu, C.S. (2009). Microencapsulation of probiotic bacteria with alginate–gelatin and its properties. J. Microencapsulation *26*, 315–324.

Li, Y., Hugenholtz, J., Abee, T., and Molenaar, D. (2003). Glutathione protects *Lactococcus lactis* against oxidative stress. Appl. Environ. Microbiol. *69*, 5739–5745.

Lin, M.Y., and Yen, C.L. (1999). Antioxidative ability of lactic acid bacteria. J. Agric. Food Chem. *47*, 1460–1466.

Liong, M.T., and Shah, N.P. (2005). Acid and bile tolerance and cholesterol removal ability of lactobacilli strains. J. Dairy Sci. *88*, 55–56.

Loguercio, C., D'Argenio, G., Delle-Cave M., Cosenza, V., Della-Valle, N., Mazzacca, G., and Del Vecchio Blanco, C. (2003). Glutathione supplementation improves oxidative damage in experimental colitis. Dig. Liver Dis. *35*, 635–641.

Lourens-Hattingh, A., and Viljeon, C.B. (2001). Yoghurt as probiotic carrier food. Int. Dairy J. *11*, 1– 17.

Lutgendorff, F., Nijmeijer, R.M., Sandström, P.A., Trulsson, L.M., Magnusson, K.E., Timmerman, H.M., van Minnen, L.P., Rijkers, G.T., Gooszen, H.G., Akkermans, L.M., and Söderholm, J.D. (2009). Probiotics prevent intestinal barrier dysfunction in acute pancreatitis in rats via induction of ileal mucosal glutathione biosynthesis. PLoS One *4*, e4512.

Lutgendorff, F., Trulsson, L.M., van Minnen, L.P., Rijkers, G.T., Timmerman, H.M., Franzén, L.E., Gooszen, H.G., Akkermans, L.M., Söderholm, J.D., and Sandström, P.A. (2008). Probiotics enhance pancreatic glutathione biosynthesis and reduce oxidative stress in experimental acute pancreatitis. Am. J. Physiol. Gastrointest. Liver Physiol. *295*, G1111–G1121.

Mantovani, M.S., Bellini, M.F., Angeli, J.P., Oliveira, R.J., Silva, A.F., and Ribeiro, L.R. (2008). beta-Glucans in promoting health: prevention against mutation and cancer. Mut. Res. *658*, 154–161.

Marty-Teysset, C., de la Torre, F., and Garel, J. (2000). Increased production of hydrogen peroxide by *Lactobacillus delbrueckii* subsp. bulgaricus upon aeration: involvement of an NADH oxidase in oxidative stress. Appl. Environ. Microbiol. *66*, 262–267.

Meng, F., Liu, X., Jia, L., Song, Z., Deng, P., and Fan, K. (2010). Optimization for the production of exopolysaccharides from Morchella esculenta SO-02 in submerged culture and its antioxidant activities *in vitro*. Carbohydrate Polymers. *79*, 700–704.

Metchnikoff, E. (1907). The prolongation of life. Optimistic studies. Heinemann, London.

Mikelsaar, M., and Zilmer, M. (2009). *Lactobacillus fermentum* ME-3—an antimicrobial and antioxidative probiotic. Microb. Ecol. Health Dis. *21*, 1–27.

Mikelsaar, M., Hütt, P., Kullisaar, T., Zilmer, K., and Zilmer, M. (2008). Double benefit claims for antimicrobial and antioxidative probiotic. Microb. Ecol. Health Dis. *20*, 184–188.

Mikelsaar, M., Zilmer, M., Kullisaar, T., Annuk, H., and Songisepp, E. (2007). Strain of micro- organism *Lactobacillus fermentum* ME-3 as novel antimicrobial and antioxidative probiotic. USA Patent 7, 244, 424 B2. Approved 17 July 2007.

Miller, C.W., Nguyen, M.H., Rooney, M., and Kailasapathy, K. (2003). The control of dissolved oxygen content in probiotic yoghurts by alternative packaging materials. Packag. Technol. Sci. *16*, 61–67.

Miller, C.W., Nguyen, M.H., Rooney, M., and Kailasapathy, K. (2002). The influence of packing materials on the dissolved oxygen content of probiotic yoghurt. Packag. Technol. Sci. *15*, 133–8.

Mills, S., Stanton, C., Fitzgerald, G.F., and Ross, R.P. (2011). Enhancing the stress responses of probiotics for a lifestyle from gut to product and back again. Microb. Cell Fact. *10* (Suppl. 19), 1–15.

Miyoshi, A., Rochat, T., Gratadoux, J.J., Le Loir, Y., Oliveira, S.C., Langella, P., and Azevedo, V. (2003). Oxidative stress in *Lactococcus lactis*. Genet. Mol. Res. *2*, 348–359.

Murphy, M.P. (2009). How mitochondria produce reactive oxygen species. Biochem. J. *417*, 1–13.

Ng, E.W., Yeung, M., and Tong, P. (2011). Effect of starter cultures on the survival of *Lactobacillus acidophilus*. Int. J. Food Microbiol. *145*, 169–175,

Noonpakdee, W., Sitthimonchai, S., Panyim, S., and Lertsiri, S. (2004). Expression of the catalase gene *katA* in starter culture *Lactobacillus plantarum* TISTR850 tolerates oxidative stress and reduces lipid oxidation in fermented meat product. Int. J. Food Microbiol. *95*, 127–135.

Odamaki, T., Xiao, J.Z., Yonezawa, S., Yaeshima, T., and Iwatsuki, K. (2011). Improved viability of bifidobacteria in fermented milk by cocultivation with *Lactococcus lactis* subspecies *lactis*. J. Dairy Sci. *94*, 1112–1121.

Ong, L., Henriksson, A., and Shah, N.P. (2006). Development of probiotic Cheddar cheese containing *Lb. acidophilus*, *Lb. paracasei*, *Lb. casei* and *Bifidobacterium* spp. and the influence of these bacteria on proteolytic patterns and production of organic acid. Int. Dairy J. *16*, 446–456.

Ouwehand, A.C. (1998). Antimicrobial components from lactic acid bacteria. In: Salminen, S., and von Wright, A. (eds.). Lactic Acid Bacteria: Microbiology and Functional Aspects, 2nd edn. New York: Marcel Dekker Inc., pp. 139–160.

Peran, L., Camuesco, D., Comalada, M., Bailon, E., Henriksson, A., Xaus, J., Zarzuelo, A., and Galvez, J. (2007). A comparative study of the preventative effects exerted by three probiotics, *Bifidobacterium lactis*, *Lactobacillus casei* and *Lactobacillus acidophilus*, in the TNBS model of rat colitis. J. Appl. Microbiol. *103*, 836–844.

Pompella, A., Visvikis, A., Paolicchi, A., De Tata, V., and Casini, A.F. (2003). The changing faces of glutathione, a cellular protagonist. Biochem. Pharmacol. *66*, 1499–1503.

Pophaly, S.D., Singh, R., Pophaly, S.D., Kaushik, J.K., and Tomar, S.K. (2012). Current status and emerging role of glutathione in food grade lactic acid bacteria. Microb. Cell Fact. *11*, 114.

Pridmore, R.D., Pittet, A.C., Praplan, F., and Cavadini, C. (2008). Hydrogen peroxide production by *Lactobacillus johnsonii* NCC 533 and its role in anti-*Salmonella* activity. FEMS Microbiol. Lett. *283*, 210–215.

Quatravaux, S., Remize, F., Bryckaert, E., Colavizza, D., and Guzzo, J. (2006). Examination of *Lactobacillus plantarum* lactate metabolism side effects in relation to the modulation of aeration parameters. J. Appl. Microbiol. *101*, 903–912.

Rauhut, D., Gawron-Scibek, M., Beisert, B., Kondzior, M., Schwarz, R., Kürbel, H., Grossmann, M., and Krieger, S. (2004). Impact of S-containing amino acids and glutathione on growth of *Oenococcus oeni* and malolactic fermentation. In LES XVIe Entretiens Scientifiques Lallemand. Porto: pp. 33–38.

Rattanachaikunsopon, P., and Phumkhachorn, P. (2010). Lactic acid bacteria: their antimicrobial compounds and their uses in food production. Annals Biol. Res. *1*, 218–228.

Rezaiki, L., Cesselin, B., Yamamoto, Y., Vido, K., van West, E., Gaudu, P., and Gruss, A. (2004). Respiration metabolism reduces oxidative and acid stress to improve long-term survival of *Lactococcus lactis*. Mol. Microbiol. *53*, 1331–1342.

Rochat, T., Bermudez-Humaran, L., Gratadoux, J.J., Fourage, C., Hoebler, C., Corthier, G., and Langella, P. (2007). Anti-inflammatory effects of *Lactobacillus casei* BL23 producing or not a manganese-dependant catalase on DSS-induced colitis in mice. Microb. Cell Fact. *6*, 22.

Rochat, T., Gratadoux, J.J., Gruss, A., Corthier, G., Maguin, E., Langella, P., and van de Guchte, M. (2006). Production of a heterologous nonheme catalase by *Lactobacillus casei*: an efficient tool for removal of H_2O_2 and protection of *Lactobacillus bulgaricus* from oxidative stress in milk. Appl. Environ. Microbiol. 72, 5143–5149.

Rodrigues, D., Sousa, S., Rocha-Santos, T., Silva, J.P., Sousa Lobo, J.M., Costa, P., Amaral, P.H., Pintado, M.M., Gomes, A.M., Malcata, F.X., and Freitas, A.C. (2011). Influence of l-cysteine, oxygen and relative humidity upon survival throughout storage of probiotic bacteria in whey protein-based microcapsules. Int. Dairy J. *21*, 869–876.

Sengül, N., Isik, S., Aslím, B., Uçar, G., and Demirbag, A.E. (2011). The Effect of Exopolysaccharide-Producing Probiotic Strains on Gut Oxidative Damage in Experimental Colitis. Dig. Dis. Sci. *56*, 707–714.

Serata, M., Iino, T., Yasuda, E., and Sako, T. (2012). Roles of thioredoxin and thioredoxin reductase in the resistance to oxidative stress in *Lactobacillus casei*. Microbiol. *158*, 953–962.

Serrano, L.M., Molenaar, D., Wels, M., Teusink, B., Bron, P., de Vos, W., and Smid, E. (2007). Thioredoxin reductase is a key factor in the oxidative stress response of *Lactobacillus plantarum* WCFS1. Microb. Cell Fact. *6*, 29.

Serrazanetti, D.I., Gottardi, D., Montanari, C., and Gianotti, A. (2013). Dynamic Stresses of Lactic Acid Bacteria Associated to Fermentation Processes, Lactic Acid Bacteria – R & D for Food, Health and Livestock Purposes, J.M. Kongo (Ed.), InTech. doi: 10.5772/51049.

Shah, N.P. (2000). Probiotic bacteria: selective enumeration and survival in dairy foods. J. Dairy Sci. *83*, 894–907.

Shah, N.P., Ding, W.K., Fallourd, M.J., and Leyer, G. (2010). Improving the stability of probiotic bacteria in model fruit juices using vitamins and antioxidants. J. Food Sci. *75*, M278–282.

Sies, H. (2014). Role of metabolic H_2O_2 generation: Redox signalling and oxidative stress. J. Biol. Chem. (in press). doi.1074/jbc.R113.544635.

Sies, H. (1997). Oxidative stress: oxidants and antioxidants. Exp. Physiol. *82*, 291–295.

Smid, E.J., and Hugenholtz, J. (2010). Functional genomics for food fermentation processes. Annu. Rev. Food Sci. Technol. *1*, 497–519.

Smit, G., Smit, B.A., and Engels, W.J. (2005). Flavour formation by lactic acid bacteria and biochemical flavour profiling of cheese products. FEMS Microbiol. Rev. *29*, 591–610.

Songisepp, E., Kals, J., Kullisaar, T., Mändar, R., Hütt, P., Zilmer, M., and Mikelsaar, M. (2005). Evaluation of the functional efficacy of an antioxidative probiotic in healthy volunteers. Nutr. J. *4*, 22.

Soomro, A.H., Masud, T., and Anwaar, K. (2002). Role of lactic acid bacteria (LAB) in food preservation and human health – a review. Pakistan J. Nutr. *1*, 20–24.

Spyropoulos, B.G., Misiakos, E.P., Fotiadis, C., and Stoidis, C.N. (2011). Antioxidant properties of probiotics and their protective effects in the pathogenesis of radiation-induced enteritis and colitis. Dig. Dis. Sci. *56*, 285–294.

Stadtman, E.R., Berlett, B.S., and Chock, P.B. (1990). Manganese-depedent disproportionation of hydrogen peroxide in bicarbonate buffer. Proc. Natl. Acad. Sci. U.S.A. *87*, 384–388.

Stocker, R., and Keaney, J.F. (2004). Role of oxidative modifications in atherosclerosis. Physiol. Rev. *84*, 1381–1478.

Talwalkar A., and Kailasapathy, K. (2003). Effect of microencapsulation on oxygen toxicity in probiotic bacteria. Aust. J. Dairy Technol. *58(1)*, 36–39.

Talwalkar A., and Kailasapathy, K. (2004). A review of oxygen toxicity in probiotic yoghurts: influence on the survival of probiotic bacteria and protective techniques. Comp. Rev. Food Sci. Food Safety. 3, 117–124.

Thibessard, A., Borges, F., Fernandez, A., Gintz, B., Decaris, B., and Leblond-Bourget, N. (2004). Identification of *Streptococcus thermophilus* CNRZ368 genes involved in defense against superoxide stress. Appl. Environ. Microbiol. *70*, 2220–2229.

Truusalu, K., Naaber, P., Kullisaar, T., Tamm, H., Mikelsaar, R.H., Zilmer, K., Zilmer, M., and Mikelsaar. M. (2004). The influence of antibacterial and antioxidative probiotic lactobacilli on gut mucosa in mouse model of *Salmonella* infection. Microb. Ecol. Health Dis. *16*, 180- 187.

Tsukahara, H. (2007). Biomarkers for oxidative stress: clinical application in pediatric medicine. Curr. Med. Chem. *14*, 339–351.

Urey, H.C. (1952). On the early chemical history of the earth and the origin of life. Proc. Natl. Acad. Sci. U.S.A. *38*, 351–363.

van de Guchte, M., Ehrlich, S.D., and Maguin, E. (2001). Production of growth- inhibiting factors by *Lactobacillus delbrueckii*. J. Appl. Microbiol. *91*, 147–153.

van de Guchte, M., Penaud, S., Grimaldi, C., Barbe, V., Bryson, K., Nicolas, P., Robert, C., Oztas, S., Mangenot, S., Couloux, A., Loux, V., Dervyn, R., Bossy, R., Bolotin, A., Batto, J.M., Walunas, T., Gibrat, J.F., Bessières, P., Weissenbach, J., Ehrlich, S.D., and Maguin, E. (2006). The complete genome sequence of *Lactobacillus bulgaricus* reveals extensive and ongoing reductive evolution. Proc. Natal. Acad. Sci., U.S.A. *103*, 9274–9279.

Vasiljevic, T., and Shah, N.P. (2008). Probiotics – from Metchnikoff to bioactives. Int. Dairy J. *18*, 714–728.

Vincent, H.K., Innes, K.E., and Vincent, K.R. (2007). Oxidative stress and potential interventions to reduce oxidative stress in overweight and obesity. Diabetes Obes. Metab. *9*, 813–839.

Vinderola, C.G., Mocchiutti, P., and Reinheimer, J.A. (2002). Interaction among lactic acid starter and probiotic bacteria used for fermented dairy products. J. Dairy Sci. *85*, 721–729.

Vido, K., Diemer, H., Van Dorsselaer, A., Leize, E., Juillard, V., Gruss, A., and Gaudu, P. (2005). Roles of thioredoxin reductase during the aerobic life of *Lactococcus lactis*. J. Bacteriol. *187*, 601–610.

Volman, J.J., Ramakers, J.D., and Plat, J. (2008). Dietary modulation of immune function by beta-glucans. Physiol. Behav. *94*, 276–284.

Wang, A.N., Yi, X.W., Yu, H.F., Dong, B., and Qiao, S.Y. (2009). Free radical scavenging activity of *Lactobacillus fermentum in vitro* and its antioxidative effect on growing-finishing pigs. J. Appl. Microbiol. *107*, 1140–1148.

Watanabe, M., van der Veen, S., Nakajima, H., and Abee, T. (2012). Effect of respiration and manganese on oxidative stress resistance of *Lactobacillus plantarum* WCFS1. Microbiol. *158*, 293–300.

Watterlot, L., Rochat, T., Sokol, H., Cherbuy, C., Bouloufa, I., Lefèvre, F., Gratadoux, J.-J., Honvo-Hueto, E., Chilmonczyk, S., Blugeon, S., Corthier, G., Langella, P., and Berm'udez-Humar'an, L.G. (2010). Intragastric administration of a superoxide dismutase-producing recombinant *Lactobacillus casei* BL23 strain attenuates DSS colitis in mice. Int. J. Food Microbiol. *144*, 35–41.

Wiseman, S.A., Balentine, D.A., and Frei, B. (1997). Antioxidants in tea. Crit. Rev. Food Sci. Nut. *37*, 705–708.

Yoon, Y.H., and Byun, J.R. (2004). Occurrence of glutathione sulphydryl (GSH) and antioxidant activities in probiotic *Lactobacillus* spp. Asian-Aust J. Anim. Sci. *17*, 1582–1585.

Yu, W., Yim, T., Lee, K., and Heo, T. (2001). Effect of skim milk alginate beads on survival rate of bifidobacteria. Biotechnol. Biopros. Eng. *6*, 133–138.

Zhang, L., Lui, C., Zhao, Y., Zhang, X., Zeng, X., Yang, Z., and Li, S. (2013). Antioxidant activity of an exopolysaccharide isolated from L. *plantarum* C88. Int. J. Biol. Macromol. *54*, 270–275.

Zhang, Y., and Li, Y. (2013). Engineering the antioxidative properties of lactic acid bacteria for improving its robustness. Curr. Opin. Biotechnol. *24*, 142–147.

Zhang, Y., Li, Y., Chen, W., Du, G.C., and Chen, J. (2012). Glutathione improves the cold resistance of *Lactobacillus sanfranciscensis* by physiological regulation. Food Microbiol. *31*, 285–292.

Zhang, Y., Du, R., Wang, L., and Zhang, H. (2010a). The antioxidative effects of probiotic *Lactobacillus casei* Zhang on the hyperlipidemic rats. Eur. Food Res. Technol. *231*, 151–158.

Zhang, Y., Du, G.C., Zhang, Y., Liao, X.Y., Wang, M., Li, Y., and Chen, J. (2010b). Glutathione protects *Lactobacillus sanfranciscensis* against freeze–thawing, freeze–drying, and cold treatment. Appl. Environ. Microbiol. *76*, 2989–2996.

Zhang, Y., Zhu, Y., Mao, S., and Li, Y. (2010c). Proteomic analyses to reveal the protective role of glutathione in resistance of *Lactococcus lactis* to osmotic stress. Appl. Environ. Microbiol. *76*, 3177–3186.

Zhang, W.Y., Yu, D.L., Sun, Z.H., Airideng, C., Hu, S.N., Meng, H., and Zhang, H.P. (2009). Preliminary analysis of glutathione S-transferase homologue from *Lactobacillus casei* Zhang. Ann. Microbiol. *59*, 727–731.

Zhang, J., Fu, R.Y., Hugenholtz, J., Li, Y., and Chen, J. (2007). Glutathione protects *Lactococcus lactis* against acid stress. Appl. Environ. Microbiol. *73*, 5268–5275.

Zotta, T., Ricciardi, A., Guidone, A., Sacco, M., Muscariello, L., Mazzeo, M.F., Cacace, G., and Parente, E. (2012). Inactivation of ccpA and aeration affect growth, metabolite production and stress tolerance in *Lactobacillus plantarum* WCFS1. Int. J. Food Microbiol., *155*, 51–59.

Part III

Prebiotics

Functional Aspects of the Endogenous Microbiota that Benefit the Host

15

Koen Venema

Abstract

Over the past two or three decades we have learned a great deal about our gut microbiota. It is well-established that the microbiota plays a role in just about any disease and disorder that human mankind suffers from. Primarily this is due to the interaction of the microbiota with the host at various levels: (i) modulation of the host immune system, which has effects far beyond the local effects in the gut and may reach up to the brain, and (ii) production of healthy or toxic microbial metabolites that also may have systemic effects, either directly or after co-metabolism by the host. The composition, and hence the activity of the microbiota, is determined by food and medication. Certain food components, such as prebiotics, fibres or polyphenols, may lead to production of health-promoting metabolites, while, *for example*, excess protein may lead to the production or what are generally considered toxic metabolites. Antibiotics and other drugs may influence the (composition and) activity of the microbiota in such as way that the activity has deleterious effects on the host. This review discusses some of the functional aspects of the gut microbiota thought to be important for health of the host. This chapter will review recent insight in the functional role of the microbiota in health and disease.

Gut microbiota

In the not so distant past the colon was just considered a storage place for indigestible matter, prior to defecation. Researchers were unaware of the bloom of microbes present in our guts, let alone their effect on health. Nowadays, the gut microbiota is considered a forgotten organ (Baquero and Nombela, 2012; Beebe *et al.*, 2014), that shows a large metabolic capacity that may approach the metabolic capacity carried out by our own liver. However, in health the metabolism of the colonic microbiota occurs under strictly anaerobic conditions, while that of the liver occurs in the presence of oxygen. Hence, the metabolic reactions occurring in those two 'organs' are completely different, with fermentation occurring in the large intestine. The approximately 10^{14} cells of the microbiota strongly influence host metabolism and immune function, and through this human health.

Composition of the gut microbiota

Since the human intestinal microbiota was known to play a key role in several metabolic, nutritional, physiological, and immunological processes, recent years have seen a rapid development in the techniques for studying this previously overlooked organ (Ottman *et al.*, 2012). The first revolution into a better understanding of the gut microbiota came from improved molecular DNA methods, that allowed to study the composition of the microbiota using culture-independent methods based on the molecular signature present in the 16S ribosomal RNA gene. This has taught us that in general the gut microbiota is composed of some 1000+ species dominated by the phyla *Bacteroidetes* and *Firmicutes* (Ottman *et al.*, 2012), although not every individual carries all these thousands of species (Qin *et al.*, 2010), and there seems to be a core microbiome (Rajilic-Stojanovic *et al.*, 2012; Tap *et al.*, 2009). Based on results obtained in the MetaHIT consortium, the composition of the gut microbiota of different individuals has been tried to categorize into different classes, such as the enterotypes (Arumugam *et al.*, 2011), although there is no consensus about this concept. Ideas range from 2 or 3 enterotypes, or several dozen (Rajilic-Stojanovic *et al.*, 2012; Yong, 2012) to gradients of different communities between different well-balanced symbiotic consortia (Jeffery *et al.*, 2012). The latter may be more intuitive, since grouping the microbiota of individual subjects into enterotypes, based on the dominance of certain genera, does seem like an oversimplification of a very complex situation, also given the genetic differences of the host around the globe and different dietary habits (see below). The same MetaHIT consortium also tried to categorize the microbiota in terms of richness, where richness seemed to correlate with metabolic markers (Le Chatelier *et al.*, 2013). In general, richness declines if there is a deviation from health, and only a few bacterial species were thought to be sufficient to distinguish between individuals with high and low bacterial richness, and even between lean and obese participants. That there is a difference in microbiota composition between lean and obese individuals has been established by several groups (Duncan *et al.*, 2007; Qin *et al.*, 2012; Santacruz *et al.*, 2009; Turnbaugh *et al.*, 2006), however, it is striking to see that the differences described by these groups are rarely the same

(Venema, 2010). This leads one to believe that there is much more than the microbiota composition that defines obesity (discussed below). Similarly, differences in microbiota composition have been found between healthy individuals and those suffering from, for example, inflammatory bowel disease (IBD) (Hansen *et al.*, 2012; Shanahan, 2012; Sokol *et al.*, 2006), irritable bowel syndrome (IBS) (Bonfrate *et al.*, 2013; Jalanka-Tuovinen *et al.*, 2013; Ng *et al.*, 2013; Shanahan, 2012), cancer (Shanahan, 2012; Yang *et al.*, 2013), and even non-gut related disease or disorders like allergy (Stsepetova *et al.*, 2007) and autism (Kang *et al.*, 2013; Wang *et al.*, 2013), amongst others.

Although the increase in throughput of sequencing methods and the substantial drop in price allows us to gather millions or trillions of reads, metagenomics of the gut microbiota (Blottiere *et al.*, 2013) only reveals the potential for metabolic reactions to occur, without actually being able to show that these gene are expressed. Metatranscriptomics and metaproteomics study those genes that are expressed in bacteria (or a bacterial consortium) (Ottman *et al.*, 2012), but if consortia are studied, it is frequently unclear which bacterium the mRNA comes from and how the different bacteria in the consortium interact in the food-chain.

Activity of the gut microbiota

It becomes more and more evident that the activity of the gut microbiota is equally if not more important than its composition. Sure, composition matters, because some microbial genes may be present or absent in certain individuals, such as those for the conversion of isoflavone daidzin in soy into equol (Tsuji *et al.*, 2010), which is dependent on a *Slackia* species, which occurs only in approximately 40% of the Japanese population, with a tendency towards a higher incidence in elderly populations. Composition also matters in terms of the presence of Gram-negative species, and the production of lipopolysaccharide (LPS) by these bacteria. Translocation of LPS over the gut barrier leads to complications from continuous low grade inflammation to septic shock. However, not all LPS molecules produced by the different species are equal, and therefore not all are equally toxic (Cuthbertson, 2012). Therefore it matters which species are present.

However, generally, the major functional aspects of the microbiota are more or less species independent. Namely, there are only a limited number of biochemical routes from say glucose to say acetate. Therefore, there has to be enormous functional overlap between the several hundred species in our gut. Also, the microbiota strives to optimal use of the available substrates, going for an optimal thermodynamic system or entropy. This leads to a tremendous functional overlap between the microbiotas of several individuals even though their composition may be entirely different. This is for instance exemplified in a study where the microbiota of 10 different individuals was fed lactulose in an *in vitro* model of the colon. Even though the composition of these individual donors differed, the metabolic outcome was strikingly similar, with low amounts of butyrate being produced on this substrate by the microbiota of all 10 individuals (Venema *et al.*, 2003). Below some of the functional aspect of the microbiota related to health and disease are discussed.

Genetics drive the composition of the gut microbiota

As we will see below, diet is a major driver for changes in microbiota composition and activity. However, our own genetic make-up is another. In an elegant study that compared the microbiota of humanized mice with and without α1–2 fucosyltransferase 2 (Fut2) activity, mimicking human (non-)secretor type, differences were shown in microbiota composition, related to the fucosylation status of the host glycans (Kashyap *et al.*, 2013). The secretor status is defined by the expression of the ABH and Lewis histo-blood group antigens in the intestinal mucus and other secretions and is determined by α1–2 fucosyltransferase 2 enzyme. However, in this study, a glucose-rich plant polysaccharide-deficient diet eliminated the effect of Fut2-genotype, showing that differences in host genotype that affect the carbohydrate decoration of host glycans in the distal gut interact with diet to alter the composition and function of resident microbes in a diet-dependent manner. Another team showed that the blood group secretor status (encoded by Fut2) is strongly associated with the composition of bifidobacteria in the colon (Wacklin *et al.*, 2011). Bifidobacterial diversity, richness, and abundance were all significantly reduced in samples from the non-secretor individuals. The non-secretor individuals were rarely colonized by several genotypes related to *Bifidobacterium bifidum*, *B. adolescentis* and *B. catenulatum/pseudocatenulatum*. The authors concluded that expression of ABH and Lewis glycan epitopes in the mucosa appears to be one of the host genetic determinants for the composition of the intestinal microbiota. In another elegant study by the same team, blood group was shown to be a discriminating factor for gut microbiota composition. Specifically, the microbiota from the individuals harbouring the B blood-group antigen differed from the non-B antigen groups and showed higher diversity of the *Eubacterium rectale–Clostridium coccoides* and *Clostridium leptum*-groups (containing numerous butyrate-producing species) in comparison with other blood groups (Makivuokko *et al.*, 2012a). The same group together with us used faecal samples pooled from individuals with the same blood group to screen for high surviving bifidobacterial strains in an approach to develop 'personalized' probiotics (Makivuokko *et al.*, 2012b).

Dietary modulation of the gut microbiota

There are numerous examples of how diet shapes the composition of the microbiota. Most examples have been gathered over the past decades on the effects of non-digestible carbohydrates on microbiota composition, *e.g.* for inulin and fructo-oligosaccharides (FOS) (for a review see Meyer and Stasse-Wolthuis, 2009), galacto-oligosaccharides (GOS) (for a review see Oozeer *et al.*, 2013), or less defined extracts, such as a wheat bran extract containing arabinoxylan oligosaccharides (AXOS) (Francois *et al.*, 2013). However, also other dietary components are under investigation for their effects on the gut microbiota composition. Examples are green tea (Jin *et al.*, 2012), or polyphenols and polyphenol-rich dietary components (reviewed in (Etxeberria *et al.*, 2013)), and probiotics, either alone (Di Gioia *et al.*, 2014; Hemarajata and Versalovic, 2013; Ng *et al.*, 2013; Persborn *et al.*, 2013) or in combination with prebiotics (Ceapa *et al.*, 2013; De

Preter *et al.*, 2011; Steed *et al.*, 2008). Numerous other studies have been done in *in vitro* models of the colon, varying in degree of complexity (Venema *et al.*, 2009; Venema and van den Abbeele, 2013), but the discussion of these studies goes beyond the scope of this chapter. Differences in whole diets on microbiota composition are also addressed. For instance, differences between a vegetarian and omnivore diet (Matijasic *et al.*, 2013), or between a Mediterranean and Scandinavian diet (Tabernero *et al.*, 2011). An interesting (yet unpublished) study is the effect of roasting on microbiota composition, where we studied the effect of roasting of non-cultivated wild species of tubers (such as *Vigna frutescens*) collected by the Hadza community, a hunter-gatherer group from northern Tanzania, on microbiota composition of a European and Hadzanian microbiota (Schnorr *et al.*, unpublished results).

Apart from direct effects on the microbiota, dietary compounds can have indirect effects on host metabolism as well, e.g. FOS leading to a decrease in plasma LPS in obese women, thus lowering metabolic endotoxaemia (Dewulf *et al.*, 2013). Some of these effects are likely to be caused by the fermentative metabolites produced by the gut microbiota from these substrates and will be addressed below.

Effects of one dietary component affecting the metabolism of another – the example of the conversion of daidzin

Dietary isoflavones have received considerable attention because of their potential to prevent postmenopausal conditions. Daidzein, the major isoflavone in soy, is metabolized to equol by intestinal microbiota (Tsuji *et al.*, 2010). Only approximately one-third of human population harbours an intestinal microbiota capable of transforming daidzein into equol. While it is not entirely understood how bacterial species produce equol in the intestine, a few human intestinal species catalysing the conversion of daidzein to equol have been identified, including *Adlercreutzia equolifaciens*, *Slackia isoflavoniconvertens*, *Slackia equolifaciens*, and *Lactococcus garvieae* (Matthies *et al.*, 2012). The conversion is optimal in the presence of H_2 gas, acting as electron donor in this biotransformation (Decroos *et al.*, 2005). In an attempt to increase the production of the bioactive compound, addition of FOS led to the inhibition of equol production (Decroos *et al.*, 2005), or to no change at all (Tousen *et al.*, 2013). Others were more successful: lactulose increased the kinetics of equol production *in vitro* (Zheng *et al.*, 2014). The authors hypothesized that lactulose led to a chance in microbiota composition and this increased H_2 production. They speculated that 'hydrogen-producing prebiotics' might be a novel way to promote equol production. In mice, xylitol feeding led to increased equol production (Tamura *et al.*, 2013). There were also some changes observed in the composition of the gut microbiota upon xylitol feeding, with an increase in bifidobacteria and a reduction in *Bacteroides* species. The authors hypothesized that these changes could lead to the increase in equol production. Feeding of a single *Slackia isoflavoniconvertens* strain to gnotobiotic mice harbouring a simplified human microbiota led to the production of equol when daidzin was fed (Matthies *et al.*, 2012). This seems a perfect model system to evaluate conditions that could lead the increase of equol in equol producers. When this strain is fed to non-equol producers, it will give these individuals the benefit of equol as well.

This is just a single example of how our dietary intake may influence the metabolism of the gut microbiota, not only of the components that were present in that meal and that made it to the colon, but also of dietary components, ingested with subsequent meals, such as equol. Below I will discuss the effect of carbohydrates on protein fermentation.

Effect of drugs on gut microbiota

It is clear that drugs, such as antibiotics, have an effect on the composition and activity of the microbiota. Apart from killing the intended pathogen, antibiotics in addition completely disturb the microbiota balance. In some cases this may lead to severe side-effects such as antibiotic-induced diarrhoea and *Clostridium difficile* overgrowth. In the case of *C. difficile* overgrowth, more antibiotics are not always efficacious. When the infection becomes refractive to antibiotic treatment, the patients are at high risk of mortality. Faecal microbiota transplantation (FMT, also referred to as Repoopulate) has been shown to be very effective, with a ~90% cure rate (McCune *et al.*, 2013; Petrof *et al.*, 2013). Faecal donations from healthy volunteers are then distilled in the colon to combat the *C. difficile* infection. There is some debate about the safety of the use of complete faeces for this, as this may contain opportunistic pathogens, viruses, etc., even though it was collected from a healthy individual. New strategies that make use of a multi-strain cocktail of save gut microbes are being developed (Allen-Vercoe and Petrof, 2013). Another approach is to provide probiotics to restore the ecological balance, either during or after antibiotic administration. In a comparative study it was shown that the metabolic pattern under combined therapy with antibiotics plus probiotics had the most beneficial and consistent effect on intestinal metabolic profiles, compared to probiotics administration following antibiotics. With respect to composition, antibiotic therapy showed a decrease in several indigenous bacterial groups in the microbiota, and there was no significant recovery of these groups when the antibiotic therapy was followed by administration of probiotics. However, simultaneous application of antibiotics and probiotics had a stabilizing effect on the intestinal microbiota with increased bifidobacteria and lactobacilli (Rehman *et al.*, 2012). However, there is an absence or insufficiency of high-quality evidence to support routine use of probiotics to prevent antibiotic associated diarrhoea (Butler *et al.*, 2012).

There is increasing evidence that there are both direct and indirect effects of the human gut microbiota on xenobiotic metabolism and through this on human physiology. These will be discussed below.

Functional aspects of the gut microbiota in health and disease

Inherently, human metabolism is under tight homeostatic regulation. We constantly experience phenomena that want to perturb this balance, whether the initiator of this change is induced by the environment (i.e. lifestyle, diet), the microbiome, or genetics. There are a large number of publications that offer intriguing associations between the microbiota and disease, but mechanistic links are frequently absent. Microbially encoded functions that contribute to disease progression are likely to be common across distinct populations, which is why metagenomics has not yet provided conclusive answers. Instead, the readout of a chemical signature between different populations (healthy or diseased, lean or obese, *etc.*) may be a definitive way to determine this. Metabolites are a functional barometer for signifying and understanding these states irrespective of the composition of the microbiota. As we have seen above, there is tremendous functional overlap between different individuals, and hence even though the microbiota composition may be different, the chemical concoctions that they will produce upon fermenting a certain substrate will be very much the same.

For a long time now, the microbiota has been described as a functional, symbiotic organ that helps maintain health under normal, homeostatic conditions. The members of the microbiota have been thought to keep pathogens in check, by production of antimicrobial substances (short-chain fatty acids, lactate, bacteriocins), of taking up space (adhesion to the gut wall or mucus) or competing for nutrients, referred to as colonization resistance or pathogen exclusion. Even for this phenomena the mechanism of action has not been corroborated *in vivo*, basically due to the inaccessibility of the human colon. Nevertheless, most people believe that these mechanisms are in place in normal homeostasis.

Most recent functional investigations in the gut microbiota rely on metagenomics (Bienenstock *et al.*, 2013) and metabolomics approaches (Jacobs *et al.*, 2009; Marcobal *et al.*, 2013). That metagenomics, which describes the potential functionality of a microbial consortium, does not always provide straight forward answers is exemplified by recent reports that describe metagenomic signatures of type 2 diabetes in European females (Karlsson *et al.*, 2013) and in Chinese individuals (Qin *et al.*, 2012). Both studies successfully produced predictive metagenomic signatures, but the signatures were different and the metagenomic models were incongruent.

The overall microbial fermentative process is complex and requires the contribution of different functional groups of microorganisms linked in a trophic chain (de Graaf and Venema, 2008). Most of the functional groups of microorganisms appear to be present at rather similar levels in all healthy volunteers (Chassard *et al.*, 2008), indicating tremendous functional overlap between individuals and suggesting that optimal numbers of these various bacterial groups are crucial for efficient gut fermentation, as well as for host nutrition and health. However, below the tip of the iceberg there is also enormous metabolic diversity. This was already exemplified above with the example of daidzin to equol conversion, which takes place in approximately one third of the population. Another example is the capacity of the human microbiota to ferment cellulose. The population level of the cellulolytic community estimated using filter paper cellulose as substrate varied considerably depending on the subject, from undetectable (< log 3/g) to log 7.2/g faeces (Chassard *et al.*, 2008). This confirms results from more than 3 decades ago (Slavin *et al.*, 1981), where cellulose digestibility *in vivo* ranged from undetectable to 38% of the dose of purified cellulose provided. The cellulose-degrading microbial community of the human gut also varies according to the presence or absence of methanogens (Chassard *et al.*, 2010). In non-methane-excreting subjects the main cellulose-degrading bacteria belong essentially to *Bacteroidetes*, whereas this is members from the *Firmicutes* phylum in methane-excreting individuals.

Some of the major and minor metabolic capacities of the gut microbiota that are believed to be important for health and disease are discussed below.

Healthy microbial metabolites

The major metabolites produced by the gut microbiota are the so-called short-chain fatty acids (SCFA). They are produced primarily from carbohydrate fermentation, although protein fermentation does contribute to SCFA production as well. The major SCFA produced are acetate, propionate and butyrate, with smaller or undetectable amounts of valerate and caproate. The SCFA that has received the most attention so far is butyrate. Butyrate is the major fuel for colonocytes and as such considered to be very important for the health of the gut epithelium (Hamer *et al.*, 2008; Havenaar, 2011). One of the proposed beneficial effects of butyrate on human intestinal health is the prevention and inhibition of colon carcinogenesis, due to induction of differentiation and apoptosis of transformed cells (Hamer *et al.*, 2008, 2012). In addition, it has been shown to have anti-inflammatory properties (Hamer *et al.*, 2008, 2012). Even though there are other fuels that can be used by colonic epithelial cells, lack of butyrate has been considered to be detrimental for health. *For example*, butyrate concentrations are lower in inflammatory bowel disease (IBD) (Hamer *et al.*, 2008). Moreover, butyrate has anti-inflammatory effects. Inhibition of histone deacetylase activity, resulting in hyperacetylation of histones, and as a consequence suppression of NF-κB activation, is a likely mechanism. In addition, it has been proposed that butyrate reinforces the colonic barrier function, by increasing production of mucins and antimicrobial peptides, such as β-defensin or trefoil factors. It has also been shown that butyrate decreases intestinal epithelial permeability by increasing the expression of tight junction proteins (Van Immerseel *et al.*, 2010). This is why in the past there has been a great focus on butyrate production by the gut microbiota. Substrates have been selected that lead to an increase in the amount and ratio of butyrate using a healthy microbiota (Fassler *et al.*, 2006; Rose *et al.*, 2010) or a microbiota originating from IBD patients (Rose *et al.*, 2010). Members of the microbiota involved in the production of butyrate have attracted attention too. Examples are members of the Lachnospiraceae and Ruminococcaceae, such as *Eubacterium rectale*, *Roseburia intestinalis*, and *Faecalibacterium prausnitzii* (Louis and Flint, 2009). They are considered potential therapeutic probiotics for IBD (Van Immerseel *et al.*, 2010). The prevalence of *F. prausnitzii* is often decreased also in other conditions of intestinal dysbiosis (Miquel *et al.*, 2014; Miquel *et al.*, 2013). These include, besides IBD, irritable bowel syndrome (IBS) and coeliac disease.

Another SCFA that has attracted attention, although not yet as much as butyrate, is propionate. When taken up by the host it is largely extracted by the liver (Bloemen *et al.*, 2009), and is involved in gluconeogenesis, at least in ruminants. Whether this also is the case in humans is controversial (Al-Lahham *et al.*, 2010a). Propionate inhibits (acetate-induced) fatty acid synthesis in the liver in both animals and humans and is thought to be involved in cholesterol synthesis (Al-Lahham *et al.*, 2010a). Moreover it has been shown to inhibit inflammation and modulate adipokine production in adipose tissue (Al-Lahham *et al.*, 2010b, 2012). Inhibitory effects of propionate on free fatty acids metabolism and inflammation, suggest that it could be a potential therapeutic agent to improve insulin sensitivity. In line with the modulation of adipokines from adipose tissue (Al-Lahham *et al.*, 2012), in humans two studies demonstrated that dietary supplementation of propionate caused satiety (Liljeberg *et al.*, 1995; Ruijschop *et al.*, 2008), likely by the induction of satiety hormones such as PYY (Al-Lahham *et al.*, 2010a). Two G-protein-coupled receptors (GPRs), free fatty acid receptor 3 (FFAR3) and FFAR2, also known as GPR41 and GPR43, respectively, have been identified as receptors for SCFA (Brown *et al.*, 2003; Le Poul *et al.*, 2003; Nilsson *et al.*, 2003). The effects of propionate on inhibition of adipose tissue inflammation and modulation of adipokine expression was shown to be dependent on (one of these) GPRs (Al-Lahham *et al.*, 2012; Al-Lahham *et al.*, 2010b). Recently, evidence has shown that propionate may also be beneficial for colon carcinogenesis. It was shown that *in vitro*, propionate reduces cell growth of the immortalized murine bone marrow-derived pro-B-cell line BaF3, through a cAMP level-dependent pathway. Furthermore, the activation of GPR43 lessens the proliferation of BaF3 and other human cancer cell lines. *In vivo*, colonic production of propionate reduced the infiltration of BaF3 in the liver (Bindels *et al.*, 2012). Recently, it was shown that mice fed a high-fibre diet, leading to increased circulating levels of SCFA, were protected against allergic inflammation in the lung (Trompette *et al.*, 2014). Treatment of mice specifically with propionate led to allergic inflammation alterations. These effects were dependent on GPR41, but not GPR43. These results show that enteric SCFA can shape the immunological environment in peripheral tissues, likely through crosstalk with these tissues by circulating immunological effector molecules.

The distribution of the three different pathways used by bacteria for propionate formation was evaluated using genomic and metagenomic analysis of the human gut microbiota and by designing degenerate primer sets for these pathways (Reichardt *et al.*, 2014). The acrylate pathway is restricted to only a few human colonic species. The propanediol pathway was more abundant, occurring within the Lachnospiraceae in relatives of *Ruminococcus obeum* and in *Roseburia inulinivorans*. However, the dominant source of propionate was concluded to be the succinate pathway (Reichardt *et al.*, 2014). This insight helps in a better understanding of the microbial ecology of SCFA formation, and may allow modulation of propionate formation (or the other SCFA) by the human gut microbiota.

Although detrimental or toxic effects of metabolites are discussed in more detail below, the role of propionate in autism spectrum disorders (ASD) is discussed here. A subgroup of patients with ASD has been shown to have mitochondrial defects (Frye *et al.*, 2013; MacFabe *et al.*, 2011; Thomas *et al.*, 2012). Under normal circumstances SCFA are primarily metabolized in the liver. However, in this subgroup of ASD patients, higher than normal levels of SCFA can be present in the circulating blood, and can cross the gut–blood and blood–brain barriers. In a mouse model of ASD with mitochondrial defects, intracerebroventricularly infusion of propionate was shown to cause abnormal behavioural symptoms, including restrictive and/or repetitive interests and behaviours, impaired social behaviour, cognitive impairment, and convulsions, very similar to ASD in humans (Frye *et al.*, 2013; MacFabe *et al.*, 2011; Thomas *et al.*, 2012). In line with this, faecal samples of autistic disorder children showed higher levels of propionate than that of control children (De Angelis *et al.*, 2013). This shows that not in all cases propionate may be beneficial. Interestingly, these faecal samples also showed increased concentration of toxic proteolytic metabolites (see below).

Acetate is also gaining attention. As a central metabolite in human metabolism, it is involved in many biochemical pathways, including fatty acid and glucose metabolism. Its role in obesity, insulin resistance and type 2 diabetes is currently under investigation (Diamant *et al.*, 2011).

In an attempt to better understand SCFA production by the gut microbiota we have developed tools making use of stable isotopes (primarily ^{13}C) to trace the fate of substrates in the human colonic microbiota (reviewed in (de Graaf and Venema, 2008; Egert *et al.*, 2006)). Uniformly ^{13}C-labelled carbohydrates were provided to a human gut microbiota in a validated dynamic *in vitro* model of the colon (TIM-2), and the appearance of label in microbial metabolites and microbial biomass (16S rRNA) was determined over time. The appearance of label in the biomass indicated precisely which bacterial species were involved in fermentation of the substrates, and when coupled to the chemical analyses, also indicated which metabolites were produced by these microbes (de Graaf and Venema, 2008; Egert *et al.*, 2006). Using data from these ^{13}C-labelling experiments, gut microbiota metabolism could be modelled *in silico* (Binsl *et al.*, 2010; de Graaf *et al.*, 2010). In this way the food-chain occurring in the black box of the colon (or brown box if you like) is started to become unravelled.

Toxic microbial metabolites

Another major substrate that is fermented by the gut microbiota is protein. Most protein that makes it to the colon is incorporated into bacterial biomass, since most members of the microbiota prefer to ferment carbohydrate. However, in the distal gut, where fermentable carbohydrate is depleted, the microbiota will switch to protein fermentation. This leads to the production of all kind of putrefactive microbial metabolites, such as ammonia, branched-chain fatty acids (BCFA), phenolic compounds, H_2S and methanethiol, and nitrosamines and other biogenic amines, which are all thought to be harmful for the host (Cummings *et al.*, 1979; Hughes *et al.*, 2000; Russell *et al.*, 2011). In addition to host proteins, significant amounts of dietary protein (~ 6 g) reach the large intestine per day. When dietary protein is consumed in high amounts, even more dietary protein will reach the colon (Cummings *et al.*, 1979; Magee *et al.*, 2000), which results in increased fermentation and production of toxic microbial metabolites. These occur then in concentrations that have been implicated in increased colon-cancer risk. Overall, high protein intake, in particular of red meat, may increase risk of developing colorectal cancer (Gill and Rowland, 2002; Norat and Riboli, 2001). In a recent study, this was confirmed: high-protein diets resulted in increased proportions of BCFA and concentrations of phenylacetic acid and N-nitroso compounds (Russell *et al.*, 2011). A high-protein, low-carbohydrate diet also decreased the proportion of butyrate, which was shown to be concomitant with a reduction in the butyrate-producing *Roseburia*/*Eubacterium rectale* group of bacteria. In addition, this diet also greatly reduced concentrations of fibre-derived, antioxidant phenolic acids such as ferulic acid and its derivatives. In line with previous research, the authors concluded that long-term adherence to such high-protein diets may increase the risk of colonic disease.

Choline is an essential vitamin, with numerous roles in the body. It acts as a methyl donor in biochemical reactions, is a precursor for the biosynthesis of phospholipids, acetylcholine and of lipoproteins, and is involved in homocysteine reduction. It primarily however is a precursor for phosphatidylcholine (PC). The intestinal microbiota plays a role in the catabolism of choline. It is first degraded to trimethylamine (TMA), acetate and ethanol, after which TMA is further converted to trimethylamine-N-oxide (TMAO) by several members of the microbiota. It has been shown that microbial production of TMAO is associated with cardiovascular disease (Tang *et al.*, 2013). Similarly, intestinal microbiota metabolism of L-carnitine, a nutrient in red meat and central to fat metabolism, promotes atherosclerosis, but only in those patients with accompanying high levels of microbially derived plasma TMAO (Koeth *et al.*, 2013). This is an example of microbial metabolism of an essential nutrient that is detrimental to the host.

Other detrimental functionalities include the production of secondary bile acids and metabolism of cholesterol. Since there has not been a lot of progress in this over the past few years, these will not be discussed here in detail.

Metabolites from minor dietary ingredients

Although consumed in lower amounts, other dietary ingredients may lead to the production of bioactive metabolites. This has long been known, e.g. for glucosinolates and polyphenols. Sinigrin, a glucosinolate in *Brassica* vegetables, such as Brussels sprouts and broccoli, is a thioglucoside. Sinigrin is deglucosylated by the gut microbiota by glucosidase enzymes, upon which allyl isothiocyanate (AITC) is formed (Krul *et al.*, 2002). AITC is considered to be a strong antioxidant and anti-carcinogenic compound, and beneficial for the gut. The most frequently proposed cancer preventative mechanisms are modulation of the activities of phase I (cytochrome P450s) and phase II (glutathione-S-transferase, UDP-glucuronosyltransferase and quinone reductase) biotransformation enzymes, redox regulation, antiproliferation and induction of cell cycle arrest, and by increasing the rate of apoptosis in cancer cells. The induction of phase I and II enzymes may have broad consequences for the host as will be discussed below.

Polyphenols are present in many fruits and vegetables, in addition to tea and chocolate. They can occur as aglycones, but frequently are glycosylated. Only small quantities of the aglycones are absorbed by the small intestine, the rest makes it to the colon. Virtually all glycosylated polyphenols are unabsorbable and reach the colon unaltered. Here both the aglycone and the glycone (after deglycosylation) are metabolized by the gut microbiota into all kinds of more simple phenolic compounds (Bordonaro *et al.*, 2014; van Duynhoven *et al.*, 2013). At present the bioactivity of most of these more simple phenolic compounds are unknown, although some have been studied in detail, and have been shown to be beneficial (e.g. Gao *et al.*, 2006). It is thought that these metabolites from polyphenols are antioxidants. However, most, if not all, studies on the bioactivity of the phenolic metabolites for polyphenols have been performed on single metabolites, and it is unknown what the bioactivity is of a mixture of these, let alone whether they act synergistically or antagonistically with other metabolites produced by the gut microbiota. It is clear that much more data is required to understand the bioactivity of polyphenols and especially their microbial metabolites.

Recently gut bacteria that can digest gluten, a common component in the human diet that is capable of triggering coeliac disease pathogenesis in genetically predisposed individuals, have been identified (Caminero *et al.*, 2014). Mainly species from the genera *Lactobacillus*, *Streptococcus*, *Staphylococcus*, *Clostridium* and *Bifidobacterium* were capable of degrading gluten. Species of some of these genera are nowadays in use as probiotics, and these could offer promising new treatment modalities for coeliac disease.

An interesting new field is the gut–microbiota–brain axis. The microbiota has been shown to affect brain development and mood (Diaz Heijtz *et al.*, 2011; Dinan and Cryan, 2013; Messaoudi *et al.*, 2011), and to play a role in autism (Frye *et al.*, 2013; MacFabe *et al.*, 2011; Thomas *et al.*, 2012). While most of these effects are likely to be caused by immune mediators which are modulated by the gut microbiota and that signal with the brain, the gut microbiota is also known to produce several neuroactive compounds, such as melatonin and serotonin, which both are produced from the precursor tryptophan. The gut harbours more neurons than the spinal cord. Therefore, there is a tremendous potential for the gut microbiota to influence the nervous system. Other neuroactive compounds include histamine, γ-aminobutyric acid (GABA), dopamine, to name just a few (Forsythe *et al.*, 2012). Some (if not all) of these are also produced by the gut microbiota. Since most of these compounds are produced from precursors that are also present in the diet, it is clear that the combination of diet and gut microbiota activity also influences the gut–brain axis.

It is hypothesized that in irritable bowel syndrome (IBS) the modulation of the production of serotonin may be involved in reduced pain perception. We have shown that enemas of butyrate resulted in a significant reduction in pain, urge and discomfort scores in healthy individuals undergoing a barostat procedure (Vanhoutvin *et al.*, 2009). It was hypothesized that the decrease in visceral perception due to butyrate treatment could be via modulation of serotonin activity, although other mechanisms are not excluded. SCFA are known to stimulate intraluminal serotonin release in the rat colon, resulting in an increase in motility. Previous studies in humans have shown that an increased serotonin release leads to a decrease in visceral perception. Serotonin exerts its actions by activation of specific serotonin receptors, of which 5-HT_3 and 5-HT_4 are known to induce colonic contractions leading to an increased motility. Some evidence exists for a neuroprotective effect of 5-HT_4 receptor stimulation (Vanhoutvin *et al.*, 2009).

Gut microbes affect the host's response to drugs

Host health is an intricate interaction between genetics, environmental factors and gut microbiota. All of these aspects influence how we respond to dietary intake, either directly, or via the gut microbiota (Fig. 15.1). A similar schematic could be drawn for the interaction with drugs, where in the upper lower corner of the figure, diet would be replaced by medication. There are numerous examples of how gut microbes affect the efficacy of drugs. Some of these will be discussed here.

The effects of the gut microbiota on host responses to drugs may be subtle, such as the increased tolerance to metformin by a mixture of inulin, oat β-glucan and a blueberry pomace extract (Greenway *et al.*, 2014), where the mixture led to a decrease in the metformin induced diarrhoea. Other effects are more drastic and could lead to lower drug efficacy or even detrimental effects, such as increasing the toxicity of drugs. These can occur through direct metabolization of the drug by the microbiota, or by indirect effects, e.g. by changing the expression of host genes involved in xenobiotic metabolism (Haiser and Turnbaugh, 2013; Saad *et al.*, 2012).

An example of direct metabolism of an orally taken drug is the reduction of the cardiac drug digoxin. It is reduced by the microbiota to dihydrodigoxin, which is inactive. As for the conversion of daidzin to equol, dihydrodigoxin is not present in all individuals receiving digoxin therapy, since not all individual carry the bug, *Eggerthella lenta*, required for the reduction step (Dobkin *et al.*, 1982). Treatment with antibiotics in those that did have *E. lenta* prevented the conversion from taking place.

Another example of direct metabolism is the activation of a common colon cancer chemotherapeutic, irinotecan. It is deconjugated by bacterial β-glucuronidases, and by this it is reactivated in the large intestine. This causes severe diarrhoea (Wallace *et al.*, 2010). The way to circumvent this from occurring was the development of a highly specific inhibitor of β-glucuronidase, which prevented the reactivation (Wallace *et al.*, 2010). Rather than adding a broad-spectrum antibiotic, that would kill off most of the gut microbiota and cause dysbiosis and accompanying negative consequences on host health, the development of a specific inhibitor is an elegant manner to reduce side-effects of drug treatment.

An activity that so far has only been detected in several *Bacteroides* species is the dephosphorylation of sorivudine, an antiviral drug against for instance herpes simplex virus, to the metabolite (E)-5-(2-bromovinyl)uracil (BVU). When given in combination with anticancer drugs such as 5-fluorouracil (5-FU), 5-FU accumulates in blood and leads to an increased toxicity, because BVU prevents the degradation of 5-FU by the host. Treatment with antibiotics that inhibit intestinal anaerobes, including *Bacteroides*, was effective in reducing circulating concentrations of BVU (Nakayama *et al.*, 1997). However, such a triple drug regimen, sorivudine plus 5-FU plus antibiotics, is evidently not wanted in terms of side-effects, destroying the gut microbiota with its health benefits, or even cost-wise. Different strategies should be developed to prevent BVU from being produced by the gut microbiota, perhaps by developing a specific inhibitor for the phosphorylase of *Bacteroides*. Although that would still be triple therapy, it is disturbing the microbiota less than broad spectrum antibiotics.

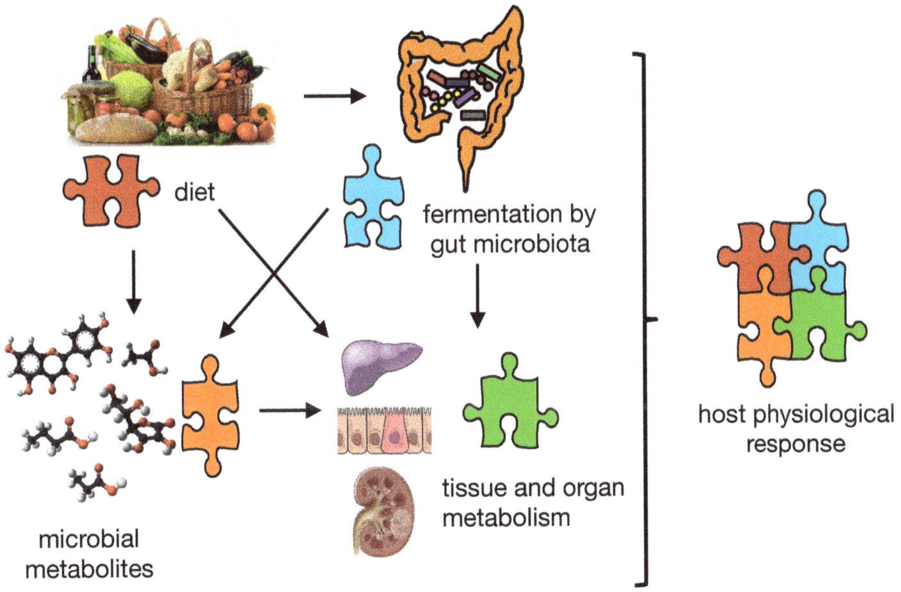

Figure 15.1 Interactions between diet, the host and gut microbiome at the metabolite level, driving host physiological response.

As discussed above, AITC led to phase I and phase II enzyme induction. It is these same enzyme inductions that could have an effect on drug-metabolism by the host, and hence it is expected that microbial metabolites of sinigrin and other glucosinolates with similar induction capacity influence drug efficacy. Most drugs are metabolized by the gut wall or the liver. This usually is a modification by phase I and phase II enzymes. These activities are employed by the host to detoxify the xenobiotics. Reactions include glucuronidation and sulfation. This leads to a decrease in the bioactive drug in the systemic circulation, lowering the efficacy. It is clear that dietary components or their microbial metabolites that induce phase I and phase II biotransformation enzymes lead to increased detoxification, thereby reducing efficacy even further. The phase I/II modified drugs are usually excreted via bile into the gastrointestinal tract. Due to them being transformed they are not taken up by the gut for a second time. However, the microbiota has glucuronidase and sulfatase activity, and may liberated the active drug again in the colon, where it could potentially be taken up again for a second (or third or more) bypass through the system, in the process of enterohepatic cycling. In this way, the microbiota may in fact increase the drug efficacy by allowing them to return to the system after being expelled from it by the liver. However, this only occurs when there is no further metabolization of the drugs by the microbiota, which, in light of their biochemical potential equalling or surpassing that of the liver, is unlikely.

In a similar manner it has been shown that feeding of probiotics or prebiotics influences the expression of the enzymes β-glucuronidase and β-glucosidase, two enzymes involved in deglycosylation (De Preter *et al.*, 2011). For example, prebiotics have been shown to increase intestinal bifidobacteria concentrations, which produce low levels of β-glucuronidase, and to suppress faecal activities of carcinogen metabolizing enzymes in humans and rats (Ballongue *et al.*, 1997; De Preter *et al.*, 2008; Hughes and Rowland, 2001).

Some pharmaceutical companies have made use of the gut microbiota activity by creating prodrugs which are activated by their metabolic activity. Examples of these are sulfasalazine and olsalazine, which are precursors of mesalazine (also known as 5-aminosalicylic acid; 5-ASA), an anti-inflammatory drug used in ulcerative colitis, a disease which occurs primarily in the colon. By activating the precursors through azoreductase activity of the gut microbiota, the active compound 5-ASA is released where it is required and far lower doses can be given to patients, limiting side-effects and toxicity of the drug compared to when it is given orally as the active compound (5-ASA).

Bacterial metabolites, primarily SCFA, have also been shown to affect epigenetic expression of host genes (Remely *et al.*, 2014b). Butyrate is well known for its epigenetic activity as a histone deacetylase inhibitor (HDACi), which may influence DNA-methylation and through this expression of genes (Hamer *et al.*, 2008). Modulation of histone acetylation by SCFA may also prevent obesity through anti-inflammatory effects (Duncan *et al.*, 2008). In a recent study, epigenetic methylation of the promoter region of FFAR3 was analysed and correlated to the diversity of the microbiota in lean and obese individuals and type 2 diabetic patients (Remely *et al.*, 2014b). Both diversity as well as the abundance of *F. prausnitzii* were significantly lower in obese individuals and type 2 diabetics. This was correlated to a significant lower methylation state of five CpGs in the promoter region of FFAR3. The authors concluded that butyrate may impact strategies for regenerative medicine by promoting epigenetic remodelling and the expression of associated genes. However, they did not measure butyrate concentrations and inferred this from the correlation between the lower abundance of butyrate-producing *F. prausnitzii* and the epigenetic activity of butyrate. It remains to be seen whether there is a direct link between butyrate and the observed effects. Nevertheless, in a similar study the same group showed that methylation of CpGs in the first exon of TLR4 and the promoter region of TLR2 were significantly lower in obese and type 2 diabetic patients, respectively (Remely *et al.*, 2014a). Again these changes were correlated to changes in microbiota composition. The lower methylation state would lead to increased gene expression, leading to increased inflammation. Epigenetic regulation of inflammatory reactions by dietary intervention, whether or not through the microbiota (Fig. 15.1), may thus be effective in the prevention of metabolic syndrome.

It is clear from the above that the gut microbiota is increasingly being recognized as a key site of metabolism for drugs and other xenobiotic compounds that are relevant to human health. The molecular complexity of the gut microbiota revealed by recent metagenomics studies has highlighted the need as well as the challenges for system-level modelling of xenobiotic metabolism in the gut (Klunemann *et al.*, 2014). Possible pathways through which the gut microbiota can modify xenobiotics and the available computational tools towards modelling complex xenometabolic processes have been recently reviewed (Klunemann *et al.*, 2014). Diverse computational tools and relevant experimental findings were combined into a unified perspective towards building holistic models of xenobiotic metabolism in the gut (Klunemann *et al.*, 2014). Pathway prediction tools included phase I/II biotransformation pathways on the side of the host, and species and whole community metabolic profiling on the side of the gut microbiota. This would already include a lot of the metabolic reactions from Fig. 15.1 that together define host response.

Future trends

The more we understand the functional role of the microbiota, the more we can make use of this knowledge to increase the beneficial effects of our gut symbionts. The actual number of metabolites produced by our microbiome is not known, partly due to the incomplete annotation of the microbial genome sequences and their functions. Likewise, among the hundreds of already identified gut microbial metabolites, the number of metabolites that have the ability to exert a physiological effect on the host is also unknown. By some this is assumed to be limited considering that many metabolites are present only transiently and in very low concentrations (Jacobs *et al.*, 2009). However, I wonder how transient this really is. It is clear that some specific metabolites, like daidzin from equol, will only be produced when daidzin is ingested (and provided that the microbiota in question is capable of modifying daidzin, as only ~ 30% of the population has a microbiota that can do this), which in Western countries is not frequently. Other precursors of microbial metabolites may be ingested at a more frequent basis, and may thus lead to consistent metabolite production, despite changes in dietary pattern from day to day. Alternatively, even though dietary patterns are not equal from day to day, precursors may be present in multiple dishes, or similar components may lead to fairly similar metabolites (e.g. the class of polyphenols), or to metabolites that have overlapping bioactivity. Certain metabolites may be highly bioactive already at low concentrations, or may have opposite activities at different concentrations (e.g. H_2S). Similarly, production of two metabolites may lead to synergisms, additive effects, neutralizing effects, or even antagonism. Since most of this info is not yet known, it is difficult to speculate about the importance of the transient nature or concentrations required that lead to bioactivity. It is my believe that studies that increase the insight in functionality of the gut microbiota, such as those using stable isotopes, are of utmost importance to decipher or define the functionalities that are of benefit to the host.

Nevertheless, the are some interesting perspectives with regards to microbiota in health and disease. Below I provide just a snap-shot of possibilities:

- Already long wished for, but not yet really established, are carbohydrates that ferment slowly. So slowly, that they would reach the distal colon and could perhaps prevent the putrefactive fermentation that normally occurs there. Prevention of production of the proteolytic fermentative metabolites would certain lead to a more healthier gut milieu.
- One may look for tailored prebiotics, such as the 'hydrogen producing prebiotics' for increased equol production, or prebiotics that influence methane production.
- Alternatively, prebiotics may be selected on the basis of a certain amount or ratio of SCFA, that are expected to have benefit for the host. *For example*, an increase in propionate may lead to more satiety through the induction of hormones like PYY.
- In contrast, low concentrations of propionate would be required for autistic disorder patients with metabolic defects in their mitochondria.
- While the exact role and mechanisms of the microbiota in obesity is still a matter of research and debate, it can be foreseen that certain dietary components may affect gut microbiota-induced obesity and its associated diseases such as type II diabetes. Also this may require screening dietary components for certain ratios or concentrations of SCFA.
- Similarly, one may look for prebiotic substances that negate the activity of the microbiota on, for example, choline, to prevent production of TMAO.

Screening of the proper combinations of dietary ingredients that show the desired effect can be done is sophisticated *in vitro* models (Venema and van den Abbeele, 2013). A combination of -omics tools (or a meta-omics platform) is required for proper understanding of gut microbiota functionality (Del Chierico *et al.*, 2014).

An interesting concept that is already known now for almost a decade is the fact that *B. thetaiotaomicron* induces expression of components of its own preferred meal in the gut. *B. thetaiotaomicron* was shown to adaptively forage host mucosal glycans when dietary polysaccharides are absent, by increasing expression of a fucosyl-transferase (Fut2) that decorates the gut wall with fucose. Subsequently, *B. thetaiotaomicron* used this fucose as a growth substrate (Hooper *et al.*, 1999). In knock-out (KO) mice that lack fucosyltransferase ($Fut2^{-/-}$), the *B. thetaiotaomicron* fucose-catabolic pathway was markedly down-regulated, whereas an operon responsive to terminal β-galactose, the precursor that accumulates in the KO mice, was significantly up-regulated (Kashyap *et al.*, 2013). This shows the intricate cross-talk between our gut microbes and ourselves. Similar sensors in the genome of the microorganism might explain why it has such a large repertoire of genes involved in acquiring and processing dietary polysaccharides (see below) and why it can grow on so many different dietary carbohydrates (Xu *et al.*, 2004).

Our diet contains a lot of carbohydrate structures from, for example, plants, that our own pancreatic enzymes cannot digest. Our gut bacteria play a pivotal role in the digestion of these dietary polysaccharides by producing a very large number of carbohydrate-active enzymes (El Kaoutari *et al.*, 2013; Koropatkin *et al.*, 2012). The genus *Bacteroides* alone already produces > 500 carbohydrate-active enzymes (Martens *et al.*, 2009; McNulty *et al.*, 2013), with as an example 503 genes in *Bacteroides cellulosilyticus* strain WH2, comprising 373 glycosylhydrolases, 23 pectate lyases, 28 carbohydrate esterases, and 84 glycosyltransferases (McNulty *et al.*, 2013), which are more carbohydrate active enzymes than any previously sequenced member of the *Bacteroidetes*. Hence it is expected that this genus will thrive on a large number of different carbohydrates, both of dietary and endogenous origin. Indeed, in a study where gnotobiotic mice were colonized with an artificial microbiota comprising 12 sequenced human gut bacterial species, *B. cellulosilyticus* WH2 was shown to be an adaptive forager that tailors its versatile carbohydrate utilization strategy to available dietary polysaccharides, with a strong emphasis on plant-derived xylans abundant in dietary staples like cereal grains. Using transcriptional profiling (RNA-Seq) two highly expressed, diet-specific polysaccharide utilization loci (PULs) were found, one with characteristics of xylan utilization systems, that were highly expressed when triggered by carbohydrates (McNulty *et al.*, 2013). Another tool used to look at carbohydrate active enzymes was a custom microarray that contained non-redundant DNA probes for more than 6500 genes encoding glycoside hydrolases and lyases selected from 174 reference genomes from distal gut bacteria. Examination of eight stool samples allowed the identification of a core of carbohydrate-active enzymes, containing 46 families of glycoside hydrolases and polysaccharide lyases, which suggests the functional stability of the gut microbiota despite large taxonomical variations between individuals (El Kaoutari *et al.*, 2013). Using comparative genomics a regulatory network for carbohydrate utilization *in B. thetaiotaomicron* was created (Ravcheev *et al.*, 2013). The inferred regulatory network contains 308 genes encoding polysaccharide and sugar catabolic enzymes, carbohydrate-binding and transport systems, and transcription factors. The analysed transcription factors control pathways for utilization of host and dietary glycans to monosaccharides and their further interconversions to intermediates of the central metabolism. The reconstructed regulatory network allowed to suggest and refine specific functional assignments for sugar catabolic enzymes and transporters, providing a substantial improvement to the existing metabolic models for *B. thetaiotaomicron* (Ravcheev *et al.*, 2013).

It is clear from the research described above that we need molecular tools like RNA-Seq or micro-arrays to grasp what is going on in our gut. It is also clear that we still have a long way to go before we will understand a little bit better what drives a stable and healthy microbiota community. Molecular approaches such as those described by Klunemann *et al.* (2014), Del Chierico *et al.* (2014) and Ravcheev *et al.* (2013), to develop integrative models for both microbe and host activity are required to make proper steps forward in our understanding of the role of the gut microbiota in health and disease. On the human side transporters and receptors such as those for the SCFA provide the molecular link between the gut microbiota and the host (Ganapathy *et al.*, 2013).

References

Al-Lahham, S., Roelofsen, H., Rezaee, F., Weening, D., Hoek, A., Vonk, R., and Venema, K. (2012). Propionic acid affects immune status and metabolism in adipose tissue from overweight subjects. Eur. J. Clin. Invest. *42*, 357–364.

Al-Lahham, S.H., Peppelenbosch, M.P., Roelofsen, H., Vonk, R.J., and Venema, K. (2010a). Biological effects of propionic acid in humans; metabolism, potential applications and underlying mechanisms. Biochim. Biophys. Acta *1801*, 1175–1183.

Al-Lahham, S.H., Roelofsen, H., Priebe, M., Weening, D., Dijkstra, M., Hoek, A., Rezaee, F., Venema, K., and Vonk, R.J. (2010b). Regulation of adipokine production in human adipose tissue by propionic acid. Eur. J. Clin. Invest. *40*, 401–407.

Allen-Vercoe, E., and Petrof, E.O. (2013). Artificial stool transplantation: progress towards a safer, more effective and acceptable alternative. Expert Rev. Gastroenterol. Hepatol. *7*, 291–293.

Arumugam, M., Raes, J., Pelletier, E., Le Paslier, D., Yamada, T., Mende, D.R., Fernandes, G.R., Tap, J., Bruls, T., Batto, J.M., *et al.* (2011). Enterotypes of the human gut microbiome. Nature *473*, 174–180.

Ballongue, J., Schumann, C., and Quignon, P. (1997). Effects of lactulose and lactitol on colonic microflora and enzymatic activity. Scand J. Gastroenterol. Suppl. *222*, 41–44.

Baquero, F., and Nombela, C. (2012). The microbiome as a human organ. Clin. Microbiol. Infect. *18* (Suppl. 4), 2–4.

Beebe, K., Sampey, B., Watkins, S.M., Milburn, M., and Eckhart, A.D. (2014). Understanding the apothecaries within: the necessity of a systematic approach for defining the chemical output of the human microbiome. Clin. Transl. Sci. *7*, 74–81.

Bienenstock, J., Gibson, G., Klaenhammer, T.R., Walker, W.A., and Neish, A.S. (2013). New insights into probiotic mechanisms: a harvest from functional and metagenomic studies. Gut Microbes *4*, 94–100.

Bindels, L.B., Porporato, P., Dewulf, E.M., Verrax, J., Neyrinck, A.M., Martin, J.C., Scott, K.P., Buc Calderon, P., Feron, O., Muccioli, G.G., *et al.* (2012). Gut microbiota-derived propionate reduces cancer cell proliferation in the liver. Br. J. Cancer *107*, 1337–1344.

Binsl, T.W., De Graaf, A.A., Venema, K., Heringa, J., Maathuis, A., De Waard, P., and Van Beek, J.H. (2010). Measuring non-steady-state metabolic fluxes in starch-converting faecal microbiota *in vitro*. Benef. Microbes *1*, 391–405.

Bloemen, J.G., Venema, K., van de Poll, M.C., Olde Damink, S.W., Buurman, W.A., and Dejong, C.H. (2009). Short chain fatty acids exchange across the gut and liver in humans measured at surgery. Clin. Nutr. *28*, 657–661.

Blottiere, H.M., de Vos, W.M., Ehrlich, S.D., and Dore, J. (2013). Human intestinal metagenomics: state of the art and future. Curr. Opin. Microbiol. *16*, 232–239.

Bonfrate, L., Tack, J., Grattagliano, I., Cuomo, R., and Portincasa, P. (2013). Microbiota in health and irritable bowel syndrome: current knowledge, perspectives and therapeutic options. Scand J. Gastroenterol. *48*, 995–1009.

Bordonaro, M., Venema, K., Putri, A.K., and Lazarova, D. (2014). Approaches that ascertain the role of dietary compounds in colonic cancer cells. World J. Gastrointest Oncol *6*, 1–10.

Brown, A.J., Goldsworthy, S.M., Barnes, A.A., Eilert, M.M., Tcheang, L., Daniels, D., Muir, A.I., Wigglesworth, M.J., Kinghorn, I., Fraser, N.J., *et al.* (2003). The Orphan G protein-coupled receptors GPR41 and GPR43 are activated by propionate and other short chain carboxylic acids. J. Biol. Chem. *278*, 11312–11319.

Butler, C.C., Duncan, D., and Hood, K. (2012). Does taking probiotics routinely with antibiotics prevent antibiotic associated diarrhoea? BMJ. *344*, e682.

Caminero, A., Herran, A.R., Nistal, E., Perez-Andres, J., Vaquero, L., Vivas, S., Ruiz de Morales, J.M., Albillos, S.M., and Casqueiro, J. (2014). Diversity of the cultivable human gut microbiome involved in gluten metabolism: isolation of microorganisms with potential interest for coeliac disease. FEMS Microbiol. Ecol.

Ceapa, C., Wopereis, H., Rezaiki, L., Kleerebezem, M., Knol, J., and Oozeer, R. (2013). Influence of fermented milk products, prebiotics and probiotics on microbiota composition and health. Best Pract. Res. Clin. Gastroenterol. 27, 139–155.

Chassard, C., Scott, K.P., Marquet, P., Martin, J.C., Del'homme, C., Dapoigny, M., Flint, H.J., and Bernalier-Donadille, A. (2008). Assessment of metabolic diversity within the intestinal microbiota from healthy humans using combined molecular and cultural approaches. FEMS Microbiol. Ecol. *66*, 496–504.

Chassard, C., Delmas, E., Robert, C., and Bernalier-Donadille, A. (2010). The cellulose-degrading microbial community of the human gut varies according to the presence or absence of methanogens. FEMS Microbiol. Ecol. *74*, 205–213.

Cummings, J.H., Hill, M.J., Bone, E.S., Branch, W.J., and Jenkins, D.J. (1979). The effect of meat protein and dietary fibre on colonic function and metabolism. II. Bacterial metabolites in feces and urine. Am. J. Clin. Nutr. *32*, 2094–2101.

Cuthbertson, L. (2012). Lipopolysaccharide biosynthesis. In Bacterial glycomics – current research, technology and applications, C.W. Reid, S.M. Twine, and A.N. Reid, eds. (Norfolk, UK, Caister Academic Press), pp. 1–25.

De Angelis, M., Piccolo, M., Vannini, L., Siragusa, S., De Giacomo, A., Serrazzanetti, D.I., Cristofori, F., Guerzoni, M.E., Gobbetti, M., and Francavilla, R. (2013). Faecal microbiota and metabolome of children with autism and pervasive developmental disorder not otherwise specified. PLoS One *8*, e76993.

de Graaf, A.A., Maathuis, A., de Waard, P., Deutz, N.E., Dijkema, C., de Vos, W.M., and Venema, K. (2010). Profiling human gut bacterial metabolism and its kinetics using [U-13C]glucose and NMR. NMR Biomed. *23*, 2–12.

de Graaf, A.A., and Venema, K. (2008). Gaining insight into microbial physiology in the large intestine: a special role for stable isotopes. Adv. Microb. Physiol. *53*, 73–168.

De Preter, V., Hamer, H.M., Windey, K., and Verbeke, K. (2011). The impact of pre- and/or probiotics on human colonic metabolism: does it affect human health? Mol. Nutr. Food Res. *55*, 46–57.

De Preter, V., Raemen, H., Cloetens, L., Houben, E., Rutgeerts, P., and Verbeke, K. (2008). Effect of dietary intervention with different pre- and probiotics on intestinal bacterial enzyme activities. Eur. J. Clin. Nutr. *62*, 225–231.

Decroos, K., Vanhemmens, S., Cattoir, S., Boon, N., and Verstraete, W. (2005). Isolation and characterization of an equol-producing mixed microbial culture from a human faecal sample and its activity under gastrointestinal conditions. Arch. Microbiol. *183*, 45–55.

Del Chierico, F., Gnani, D., Vernocchi, P., Petrucca, A., Alisi, A., Dallapiccola, B., Nobili, V., and Lorenza, P. (2014). Meta-omic platforms to assist in the understanding of NAFLD gut microbiota alterations: tools and applications. Int. J. Mol. Sci. *15*, 684–711.

Dewulf, E.M., Cani, P.D., Claus, S.P., Fuentes, S., Puylaert, P.G., Neyrinck, A.M., Bindels, L.B., de Vos, W.M., Gibson, G.R., Thissen, J.P., *et al.* (2013). Insight into the prebiotic concept: lessons from an exploratory, double blind intervention study with inulin-type fructans in obese women. Gut 62, 1112–1121.

Di Gioia, D., Aloisio, I., Mazzola, G., and Biavati, B. (2014). Bifidobacteria: their impact on gut microbiota composition and their applications as probiotics in infants. Appl. Microbiol. Biotechnol. 98, 563–577.

Diamant, M., Blaak, E.E., and de Vos, W.M. (2011). Do nutrient–gut-microbiota interactions play a role in human obesity, insulin resistance and type 2 diabetes? Obes. Rev. *12,* 272–281.

Diaz Heijtz, R., Wang, S., Anuar, F., Qian, Y., Bjorkholm, B., Samuelsson, A., Hibberd, M.L., Forssberg, H., and Pettersson, S. (2011). Normal gut microbiota modulates brain development and behavior. Proc. Natl. Acad. Sci. U.S.A. *108,* 3047–3052.

Dinan, T.G., and Cryan, J.F. (2013). Melancholic microbes: a link between gut microbiota and depression? Neurogastroenterol. Motil. 25, 713–719.

Dobkin, J.F., Saha, J.R., Butler, V.P., Jr., Neu, H.C., and Lindenbaum, J. (1982). Inactivation of digoxin by Eubacterium lentum, an anaerobe of the human gut flora. Trans. Assoc. Am. Physicians 95, 22–29.

Duncan, S.H., Belenguer, A., Holtrop, G., Johnstone, A.M., Flint, H.J., and Lobley, G.E. (2007). Reduced dietary intake of carbohydrates by obese subjects results in decreased concentrations of butyrate and butyrate-producing bacteria in feces. Appl. Environ. Microbiol. 73, 1073–1078.

Duncan, S.H., Lobley, G.E., Holtrop, G., Ince, J., Johnstone, A.M., Louis, P., and Flint, H.J. (2008). Human colonic microbiota associated with diet, obesity and weight loss. Int. J. Obes. (Lond.) 32, 1720–1724.

Egert, M., de Graaf, A.A., Smidt, H., de Vos, W.M., and Venema, K. (2006). Beyond diversity: functional microbiomics of the human colon. Trends Microbiol. *14,* 86–91.

El Kaoutari, A., Armougom, F., Leroy, Q., Vialettes, B., Million, M., Raoult, D., and Henrissat, B. (2013). Development and validation of a microarray for the investigation of the CAZymes encoded by the human gut microbiome. PLoS One *8,* e84033.

Etxeberria, U., Fernandez-Quintela, A., Milagro, F.I., Aguirre, L., Martinez, J.A., and Portillo, M.P. (2013). Impact of polyphenols and polyphenol-rich dietary sources on gut microbiota composition. J. Agric. Food Chem. *61,* 9517–9533.

Fassler, C., Arrigoni, E., Venema, K., Brouns, F., and Amado, R. (2006). *In vitro* fermentability of differently digested resistant starch preparations. Mol. Nutr. Food Res. *50,* 1220–1228.

Forsythe, P., Kunze, W.A., and Bienenstock, J. (2012). On communication between gut microbes and the brain. Curr. Opin. Gastroenterol. *28,* 557–562.

Francois, I.E., Lescroart, O., Veraverbeke, W.S., Marzorati, M., Possemiers, S., Hamer, H., Windey, K., Welling, G.W., Delcour, J.A., Courtin, C.M., *et al.* (2013). Effects of a wheat bran extract containing arabinoxylan oligosaccharides on gastrointestinal parameters in healthy preadolescent children: a double-blind, randomized, placebo-controlled, crossover trial. J. Pediatr. Gastroenterol. Nutr. *58,* 647–653.

Frye, R.E., Melnyk, S., and Macfabe, D.F. (2013). Unique acyl-carnitine profiles are potential biomarkers for acquired mitochondrial disease in autism spectrum disorder. Transl Psychiatry 3, e220.

Ganapathy, V., Thangaraju, M., Prasad, P.D., Martin, P.M., and Singh, N. (2013). Transporters and receptors for short-chain fatty acids as the molecular link between colonic bacteria and the host. Curr. Opin. Pharmacol. *13,* 869–874.

Gao, K., Xu, A., Krul, C., Venema, K., Liu, Y., Niu, Y., Lu, J., Bensoussan, L., Seeram, N.P., Heber, D., *et al.* (2006). Of the major phenolic acids formed during human microbial fermentation of tea, citrus, and soy flavonoid supplements, only 3,4-dihydroxyphenylacetic acid has antiproliferative activity. J. Nutr. *136,* 52–57.

Gill, C.I., and Rowland, I.R. (2002). Diet and cancer: assessing the risk. Br. J. Nutr. *88* (Suppl. 1), S73–87.

Greenway, F., Wang, S., and Heiman, M. (2014). A novel cobiotic containing a prebiotic and an antioxidant augments the glucose control and gastrointestinal tolerability of metformin: a case report. Benef. Microbes *5,* 29–32.

Haiser, H.J., and Turnbaugh, P.J. (2013). Developing a metagenomic view of xenobiotic metabolism. Pharmacol. Res. *69,* 21–31.

Hamer, H.M., De Preter, V., Windey, K., and Verbeke, K. (2012). Functional analysis of colonic bacterial metabolism: relevant to health? Am. J. Physiol. Gastrointest Liver Physiol. *302,* G1–9.

Hamer, H.M., Jonkers, D., Venema, K., Vanhoutvin, S., Troost, F.J., and Brummer, R.J. (2008). Review article: the role of butyrate on colonic function. Aliment. Pharmacol. Ther. *27,* 104–119.

Hansen, R., Russell, R.K., Reiff, C., Louis, P., McIntosh, F., Berry, S.H., Mukhopadhya, I., Bisset, W.M., Barclay, A.R., Bishop, J., *et al.* (2012). Microbiota of de-novo paediatric IBD: increased *Faecalibacterium prausnitzii* and reduced bacterial diversity in Crohn's but not in ulcerative colitis. Am. J. Gastroenterol. *107,* 1913–1922.

Havenaar, R. (2011). Intestinal health functions of colonic microbial metabolites: a review. Benef. Microbes *2,* 103–114.

Hemarajata, P., and Versalovic, J. (2013). Effects of probiotics on gut microbiota: mechanisms of intestinal immunomodulation and neuromodulation. Ther. Adv. Gastroenterol. *6,* 39–51.

Hooper, L.V., Xu, J., Falk, P.G., Midtvedt, T., and Gordon, J.I. (1999). A molecular sensor that allows a gut commensal to control its nutrient foundation in a competitive ecosystem. Proc. Natl. Acad. Sci. U.S.A. *96,* 9833–9838.

Hughes, R., Magee, E.A., and Bingham, S. (2000). Protein degradation in the large intestine: relevance to colorectal cancer. Curr. Issues Intest Microbiol. *1,* 51–58.

Hughes, R., and Rowland, I.R. (2001). Stimulation of apoptosis by two prebiotic chicory fructans in the rat colon. Carcinogenesis *22,* 43–47.

Jacobs, D.M., Gaudier, E., van Duynhoven, J., and Vaughan, E.E. (2009). Non-digestible food ingredients, colonic microbiota and the impact on gut health and immunity: a role for metabolomics. Curr. Drug Metab. *10,* 41–54.

Jalanka-Tuovinen, J., Salojarvi, J., Salonen, A., Immonen, O., Garsed, K., Kelly, F.M., Zaitoun, A., Palva, A., Spiller, R.C., and de Vos, W.M. (2013). Faecal microbiota composition and host–microbe cross-talk following gastroenteritis and in postinfectious irritable bowel syndrome. Gut.

Jeffery, I.B., Claesson, M.J., O'Toole, P.W., and Shanahan, F. (2012). Categorization of the gut microbiota: enterotypes or gradients? Nat. Rev. Microbiol. *10,* 591–592.

Jin, J.S., Touyama, M., Hisada, T., and Benno, Y. (2012). Effects of green tea consumption on human faecal microbiota with special reference to Bifidobacterium species. Microbiol. Immunol. *56,* 729–739.

Kang, D.W., Park, J.G., Ilhan, Z.E., Wallstrom, G., Labaer, J., Adams, J.B., and Krajmalnik-Brown, R. (2013). Reduced incidence of *Prevotella* and other fermenters in intestinal microflora of autistic children. PLoS One *8,* e68322.

Karlsson, F.H., Tremaroli, V., Nookaew, I., Bergstrom, G., Behre, C.J., Fagerberg, B., Nielsen, J., and Backhed, F. (2013). Gut metagenome in European women with normal, impaired and diabetic glucose control. Nature *498,* 99–103.

Kashyap, P.C., Marcobal, A., Ursell, L.K., Smits, S.A., Sonnenburg, E.D., Costello, E.K., Higginbottom, S.K., Domino, S.E., Holmes, S.P., Relman, D.A., *et al.* (2013). Genetically dictated change in host mucus carbohydrate landscape exerts a diet-dependent effect on the gut microbiota. Proc. Natl. Acad. Sci. U.S.A. *110,* 17059–17064.

Klunemann, M., Schmid, M., and Patil, K.R. (2014). Computational tools for modeling xenometabolism of the human gut microbiota. Trends Biotechnol. *32,* 157–165.

Koeth, R.A., Wang, Z., Levison, B.S., Buffa, J.A., Org, E., Sheehy, B.T., Britt, E.B., Fu, X., Wu, Y., Li, L., *et al.* (2013). Intestinal microbiota metabolism of l-carnitine, a nutrient in red meat, promotes atherosclerosis. Nat. Med. *19,* 576–585.

Koropatkin, N.M., Cameron, E.A., and Martens, E.C. (2012). How glycan metabolism shapes the human gut microbiota. Nat. Rev. Microbiol. *10,* 323–335.

Krul, C., Humblot, C., Philippe, C., Vermeulen, M., van Nuenen, M., Havenaar, R., and Rabot, S. (2002). Metabolism of sinigrin (2-propenyl glucosinolate) by the human colonic microflora in a dynamic *in vitro* large-intestinal model. Carcinogenesis 23, 1009–1016.

Le Chatelier, E., Nielsen, T., Qin, J., Prifti, E., Hildebrand, F., Falony, G., Almeida, M., Arumugam, M., Batto, J.M., Kennedy, S., *et al.* (2013). Richness of human gut microbiome correlates with metabolic markers. Nature *500,* 541–546.

Le Poul, E., Loison, C., Struyf, S., Springael, J.Y., Lannoy, V., Decobecq, M.E., Brezillon, S., Dupriez, V., Vassart, G., Van Damme, J., *et al.* (2003). Functional characterization of human receptors for short-chain fatty acids and their role in polymorphonuclear cell activation. J. Biol. Chem. *278*, 25481–25489.

Liljeberg, H.G., Lonner, C.H., and Bjorck, I.M. (1995). Sourdough fermentation or addition of organic acids or corresponding salts to bread improves nutritional properties of starch in healthy humans. J. Nutr. *125*, 1503–1511.

Louis, P., and Flint, H.J. (2009). Diversity, metabolism and microbial ecology of butyrate-producing bacteria from the human large intestine. FEMS Microbiol. Lett. *294*, 1–8.

MacFabe, D.F., Cain, N.E., Boon, F., Ossenkopp, K.P., and Cain, D.P. (2011). Effects of the enteric bacterial metabolic product propionic acid on object-directed behaviour, social behaviour, cognition, and neuroinflammation in adolescent rats: Relevance to autism spectrum disorder. Behav. Brain Res. *217*, 47–54.

Magee, E.A., Richardson, C.J., Hughes, R., and Cummings, J.H. (2000). Contribution of dietary protein to sulfide production in the large intestine: an *in vitro* and a controlled feeding study in humans. Am. J. Clin. Nutr. 72, 1488–1494.

Makivuokko, H., Lahtinen, S.J., Wacklin, P., Tuovinen, E., Tenkanen, H., Nikkila, J., Bjorklund, M., Aranko, K., Ouwehand, A.C., and Matto, J. (2012a). Association between the ABO blood group and the human intestinal microbiota composition. BMC Microbiol. *12*, 94.

Makivuokko, H., Wacklin, P., Koenen, M.E., Laamanen, K., Alakulppi, N., Venema, K., and Matto, J. (2012b). Isolation of bifidobacteria for blood group secretor status targeted personalized nutrition. Microb. Ecol. Health Dis. *23*, 18578.

Marcobal, A., Kashyap, P.C., Nelson, T.A., Aronov, P.A., Donia, M.S., Spormann, A., Fischbach, M.A., and Sonnenburg, J.L. (2013). A metabolomic view of how the human gut microbiota impacts the host metabolome using humanized and gnotobiotic mice. ISME J. 7, 1933–1943.

Martens, E.C., Koropatkin, N.M., Smith, T.J., and Gordon, J.I. (2009). Complex glycan catabolism by the human gut microbiota: the Bacteroidetes Sus-like paradigm. J. Biol. Chem. *284*, 24673–24677.

Matijasic, B.B., Obermajer, T., Lipoglavsek, L., Grabnar, I., Avgustin, G., and Rogelj, I. (2013). Association of dietary type with faecal microbiota in vegetarians and omnivores in Slovenia. Eur. J. Nutr. 53, 1051–1064.

Matthies, A., Loh, G., Blaut, M., and Braune, A. (2012). Daidzein and genistein are converted to equol and 5-hydroxy-equol by human intestinal Slackia isoflavoniconvertens in gnotobiotic rats. J. Nutr. *142*, 40–46.

McCune, V.L., Struthers, J.K., and Hawkey, P.M. (2013). Faecal transplantation for the treatment of *Clostridium difficile* infection: a review. Int. J. Antimicrob. Agents 43, 201–206.

McNulty, N.P., Wu, M., Erickson, A.R., Pan, C., Erickson, B.K., Martens, E.C., Pudlo, N.A., Muegge, B.D., Henrissat, B., Hettich, R.L., *et al.* (2013). Effects of diet on resource utilization by a model human gut microbiota containing *Bacteroides cellulosilyticus* WH2, a symbiont with an extensive glycobiome. PLoS Biol. *11*, e1001637.

Messaoudi, M., Violle, N., Bisson, J.F., Desor, D., Javelot, H., and Rougeot, C. (2011). Beneficial psychological effects of a probiotic formulation (*Lactobacillus* helveticus R0052 and *Bifidobacterium longum* R0175). in healthy human volunteers. Gut Microbes 2, 256–261.

Meyer, D., and Stasse-Wolthuis, M. (2009). The bifidogenic effect of inulin and oligofructose and its consequences for gut health. Eur. J. Clin. Nutr. *63*, 1277–1289.

Miquel, S., Mart, N.R., Bridonneau, C., Robert, V., Sokol, H., Bermudez-Humaran, L.G., Thomas, M., and Langella, P. (2014). Ecology and metabolism of the beneficial intestinal commensal bacterium. Gut Microbes 5.

Miquel, S., Martin, R., Rossi, O., Bermudez-Humaran, L.G., Chatel, J.M., Sokol, H., Thomas, M., Wells, J.M., and Langella, P. (2013). *Faecalibacterium prausnitzii* and human intestinal health. Curr. Opin. Microbiol. *16*, 255–261.

Nakayama, H., Kinouchi, T., Kataoka, K., Akimoto, S., Matsuda, Y., and Ohnishi, Y. (1997). Intestinal anaerobic bacteria hydrolyse sorivudine, producing the high blood concentration of 5-(E)-(2-bromovinyl)uracil that increases the level and toxicity of 5-fluorouracil. Pharmacogenetics *7*, 35–43.

Ng, S.C., Lam, E.F., Lam, T.T., Chan, Y., Law, W., Tse, P.C., Kamm, M.A., Sung, J.J., Chan, F.K., and Wu, J.C. (2013). Effect of probiotic bacteria on the intestinal microbiota in irritable bowel syndrome. J. Gastroenterol. Hepatol. *28*, 1624–1631.

Nilsson, N.E., Kotarsky, K., Owman, C., and Olde, B. (2003). Identification of a free fatty acid receptor, FFA2R, expressed on leucocytes and activated by short-chain fatty acids. Biochem. Biophys. Res. Commun *303*, 1047–1052.

Norat, T., and Riboli, E. (2001). Meat consumption and colorectal cancer: a review of epidemiologic evidence. Nutr. Rev. *59*, 37–47.

Oozeer, R., van Limpt, K., Ludwig, T., Ben Amor, K., Martin, R., Wind, R.D., Boehm, G., and Knol, J. (2013). Intestinal microbiology in early life: specific prebiotics can have similar functionalities as human-milk oligosaccharides. Am. J. Clin. Nutr. *98*, 561S–571S.

Ottman, N., Smidt, H., de Vos, W.M., and Belzer, C. (2012). The function of our microbiota: who is out there and what do they do? Front. Cell Infect. Microbiol. *2*, 104.

Persborn, M., Gerritsen, J., Wallon, C., Carlsson, A., Akkermans, L.M., and Soderholm, J.D. (2013). The effects of probiotics on barrier function and mucosal pouch microbiota during maintenance treatment for severe pouchitis in patients with ulcerative colitis. Aliment Pharmacol. Ther. *38*, 772–783.

Petrof, E.O., Claud, E.C., Gloor, G.B., and Allen-Vercoe, E. (2013). Microbial ecosystems therapeutics: a new paradigm in medicine? Benef. Microbes *4*, 53–65.

Qin, J., Li, R., Raes, J., Arumugam, M., Burgdorf, K.S., Manichanh, C., Nielsen, T., Pons, N., Levenez, F., Yamada, T., *et al.* (2010). A human gut microbial gene catalogue established by metagenomic sequencing. Nature *464*, 59–65.

Qin, J., Li, Y., Cai, Z., Li, S., Zhu, J., Zhang, F., Liang, S., Zhang, W., Guan, Y., Shen, D., *et al.* (2012). A metagenome-wide association study of gut microbiota in type 2 diabetes. Nature *490*, 55–60.

Rajilic-Stojanovic, M., Heilig, H.G., Tims, S., Zoetendal, E.G., and de Vos, W.M. (2012). Long-term monitoring of the human intestinal microbiota composition. Environ. Microbiol.

Ravcheev, D.A., Godzik, A., Osterman, A.L., and Rodionov, D.A. (2013). Polysaccharides utilization in human gut bacterium Bacteroides thetaiotaomicron: comparative genomics reconstruction of metabolic and regulatory networks. BMC Genomics *14*, 873.

Rehman, A., Heinsen, F.A., Koenen, M.E., Venema, K., Knecht, H., Hellmig, S., Schreiber, S., and Ott, S.J. (2012). Effects of probiotics and antibiotics on the intestinal homeostasis in a computer controlled model of the large intestine. BMC Microbiol. *12*, 47.

Reichardt, N., Duncan, S.H., Young, P., Belenguer, A., McWilliam Leitch, C., Scott, K.P., Flint, H.J., and Louis, P. (2014). Phylogenetic distribution of three pathways for propionate production within the human gut microbiota. ISME J.

Remely, M., Aumueller, E., Jahn, D., Hippe, B., Brath, H., and Haslberger, A.G. (2014a). Microbiota and epigenetic regulation of inflammatory mediators in type 2 diabetes and obesity. Benef. Microbes *5*, 33–43.

Remely, M., Aumueller, E., Merold, C., Dworzak, S., Hippe, B., Zanner, J., Pointner, A., Brath, H., and Haslberger, A.G. (2014b). Effects of short-chain fatty acid producing bacteria on epigenetic regulation of FFAR3 in type 2 diabetes and obesity. Gene *537*, 85–92.

Rose, D.J., Venema, K., Keshavarzian, A., and Hamaker, B.R. (2010). Starch-entrapped microspheres show a beneficial fermentation profile and decrease in potentially harmful bacteria during *in vitro* fermentation in faecal microbiota obtained from patients with inflammatory bowel disease. Br. J. Nutr. *103*, 1514–1524.

Ruijschop, R.M.A.J., Boelrijk, A.E.M., and te Giffel, M.C. (2008). Satiety effects of a dairy beverage fermented with propionic acid bacteria. Int. Dairy J. *18*, 945–950.

Russell, W.R., Gratz, S.W., Duncan, S.H., Holtrop, G., Ince, J., Scobbie, L., Duncan, G., Johnstone, A.M., Lobley, G.E., Wallace, R.J., *et al.* (2011). High-protein, reduced-carbohydrate weight-loss diets promote metabolite profiles likely to be detrimental to colonic health. Am. J. Clin. Nutr. *93*, 1062–1072.

Saad, R., Rizkallah, M.R., and Aziz, R.K. (2012). Gut Pharmacomicrobiomics: the tip of an iceberg of complex interactions between drugs and gut-associated microbes. Gut Pathog. *4*, 16.

Santacruz, A., Marcos, A., Warnberg, J., Marti, A., Martin-Matillas, M., Campoy, C., Moreno, L.A., Veiga, O., Redondo-Figuero, C., Garagorri, J.M., *et al.* (2009). Interplay between weight loss and gut microbiota composition in overweight adolescents. Obesity (Silver Spring) *17*, 1906–1915.

Shanahan, F. (2012). The colonic microbiota and colonic disease. Curr. Gastroenterol. Rep. *14*, 446–452.

Slavin, J.L., Brauer, P.M., and Marlett, J.A. (1981). Neutral detergent fibre, hemicellulose and cellulose digestibility in human subjects. J. Nutr. *111*, 287–297.

Sokol, H., Seksik, P., Rigottier-Gois, L., Lay, C., Lepage, P., Podglajen, I., Marteau, P., and Dore, J. (2006). Specificities of the fecal microbiota in inflammatory bowel disease. Inflamm. Bowel Dis. *12*, 106–111.

Steed, H., Macfarlane, G.T., and Macfarlane, S. (2008). Prebiotics, synbiotics and inflammatory bowel disease. Mol. Nutr. Food Res. *52*, 898–905.

Stsepetova, J., Sepp, E., Julge, K., Vaughan, E., Mikelsaar, M., and de Vos, W.M. (2007). Molecularly assessed shifts of *Bifidobacterium* subsp. and less diverse microbial communities are characteristic of 5-year-old allergic children. FEMS Immunol. Med. Microbiol. *51*, 260–269.

Tabernero, M., Venema, K., Maathuis, A.J., and Saura-Calixto, F.D. (2011). Metabolite production during *in vitro* colonic fermentation of dietary fibre: analysis and comparison of two European diets. J. Agric. Food Chem. *59*, 8968–8975.

Tamura, M., Hoshi, C., and Hori, S. (2013). Xylitol affects the intestinal microbiota and metabolism of daidzein in adult male mice. Int. J. Mol. Sci. *14*, 23993–24007.

Tang, W.H., Wang, Z., Levison, B.S., Koeth, R.A., Britt, E.B., Fu, X., Wu, Y., and Hazen, S.L. (2013). Intestinal microbial metabolism of phosphatidylcholine and cardiovascular risk. N. Engl. J. Med. *368*, 1575–1584.

Tap, J., Mondot, S., Levenez, F., Pelletier, E., Caron, C., Furet, J.P., Ugarte, E., Munoz-Tamayo, R., Paslier, D.L., Nalin, R., *et al.* (2009). Towards the human intestinal microbiota phylogenetic core. Environ. Microbiol. *11*, 2574–2584.

Thomas, R.H., Meeking, M.M., Mepham, J.R., Tichenoff, L., Possmayer, F., Liu, S., and MacFabe, D.F. (2012). The enteric bacterial metabolite propionic acid alters brain and plasma phospholipid molecular species: further development of a rodent model of autism spectrum disorders. J. Neuroinflammation 9, 153.

Tousen, Y., Uehara, M., Abe, F., Kimira, Y., and Ishimi, Y. (2013). Effects of short-term fructooligosaccharide intake on equol production in Japanese postmenopausal women consuming soy isoflavone supplements: a pilot study. Nutr. J. *12*, 127.

Trompette, A., Gollwitzer, E.S., Yadava, K., Sichelstiel, A.K., Sprenger, N., Ngom-Bru, C., Blanchard, C., Junt, T., Nicod, L.P., Harris, N.L., *et al.* (2014). Gut microbiota metabolism of dietary fibre influences allergic airway disease and hematopoiesis. Nat. Med. *20*, 159–166.

Tsuji, H., Moriyama, K., Nomoto, K., Miyanaga, N., and Akaza, H. (2010). Isolation and characterization of the equol-producing bacterium Slackia sp. strain NATTS. Arch. Microbiol. *192*, 279–287.

Turnbaugh, P.J., Ley, R.E., Mahowald, M.A., Magrini, V., Mardis, E.R., and Gordon, J.I. (2006). An obesity-associated gut microbiome with increased capacity for energy harvest. Nature *444*, 1027–1031.

van Duynhoven, J., Vaughan, E.E., van Dorsten, F., Gomez-Roldan, V., de Vos, R., Vervoort, J., van der Hooft, J.J., Roger, L., Draijer, R., and Jacobs, D.M. (2013). Interactions of black tea polyphenols with human gut microbiota: implications for gut and cardiovascular health. Am. J. Clin. Nutr. *98*, 1631S1641S.

Van Immerseel, F., Ducatelle, R., De Vos, M., Boon, N., Van De Wiele, T., Verbeke, K., Rutgeerts, P., Sas, B., Louis, P., and Flint, H.J. (2010). Butyric acid-producing anaerobic bacteria as a novel probiotic treatment approach for inflammatory bowel disease. J. Med. Microbiol. *59*, 141–143.

Vanhoutvin, S.A., Troost, F.J., Kilkens, T.O., Lindsey, P.J., Hamer, H.M., Jonkers, D.M., Venema, K., and Brummer, R.J. (2009). The effects of butyrate enemas on visceral perception in healthy volunteers. Neurogastroenterol. Motil. *21*, 952-e976.

Venema, K. (2010). Role of gut microbiota in the control of energy and carbohydrate metabolism. Curr. Opin. Clin. Nutr. Metab. Care *13*, 432–438.

Venema, K., Havenaar, R., and Minekus, M. (2009). Improving *in vitro* simulation of the GI tract – models that mimic the stomach and intestines. In Designing functional Foods – measuring and controlling food structure breakdown and nutrient absorption, J.D. McClements, and E.A. Decker, eds. (Cambridge, UK, Woodhead Publishing Ltd), pp. 314–229.

Venema, K., and van den Abbeele, P. (2013). Experimental models of the gut microbiome. Best Pract Res. Clin. Gastroenterol. *27*, 115–126.

Venema, K., van Nuenen, M.H.M.C., van den Heuvel, E.G., Pool, W., and van der Vossen, J.M.B.M. (2003). The effect of lactulose on the composition of the intestinal microbiota and short-chain fatty acid production in human volunteers and a computer-controlled model of the proximal large intestine. Microb. Ecol. Health Dis. *15*, 94–105.

Wacklin, P., Makivuokko, H., Alakulppi, N., Nikkila, J., Tenkanen, H., Rabina, J., Partanen, J., Aranko, K., and Matto, J. (2011). Secretor genotype (FUT2 gene) is strongly associated with the composition of Bifidobacteria in the human intestine. PLoS One *6*, e20113.

Wallace, B.D., Wang, H., Lane, K.T., Scott, J.E., Orans, J., Koo, J.S., Venkatesh, M., Jobin, C., Yeh, L.A., Mani, S., *et al.* (2010). Alleviating cancer drug toxicity by inhibiting a bacterial enzyme. Science *330*, 831–835.

Wang, L., Christophersen, C.T., Sorich, M.J., Gerber, J.P., Angley, M.T., and Conlon, M.A. (2013). Increased abundance of *Sutterella* spp. and *Ruminococcus torques* in faeces of children with autism spectrum disorder. Mol. Autism *4*, 42.

Xu, J., Chiang, H.C., Bjursell, M.K., and Gordon, J.I. (2004). Message from a human gut symbiont: sensitivity is a prerequisite for sharing. Trends Microbiol. *12*, 21–28.

Yang, T., Owen, J.L., Lightfoot, Y.L., Kladde, M.P., and Mohamadzadeh, M. (2013). Microbiota impact on the epigenetic regulation of colorectal cancer. Trends Mol. Med. *19*, 714–725.

Yong, E. (2012). Symbiotic gut microbes modulate human metabolic phenotypes. Nature. doi:10.1038/nature.2012.10276.

Zheng, W., Hou, Y., Su, Y., and Yao, W. (2014). Lactulose promotes equol production and changes the microbial community during *in vitro* fermentation of daidzein by faecal inocula of sows. Anaerobe *25*, 47–52.

Studying the Microbiota and Microbial Ecology of the GI Tract in the Omics Era: Tools for Stools

16

Kieran M. Tuohy, Francesca Fava and Nicola Segata

Abstract

The human gut microbiota is now recognized as an important contributor to host health and disease, with diet: microbe interactions playing critical roles in immune function, energy balance and even brain development and cognitive function. Importantly, these advances in the fields of physiology and nutrition have come at a time when methodologies for studying microbial communities are providing a comprehensive tool-kit of unparalleled resolution and breadth of coverage, allowing for the first time efficacious measurement of important ecological parameters within the gut microbiome. Over the past thirty years a diverse tool kit has been developed, including molecular methods both for targeted microbial quantitation and semi-quantitative but high resolution/broad spectrum methods capable of identifying theoretically all microorganisms within an ecosystem. These methods too allow us to measure both the metabolic potential and metabolic kinetic of the gut microbiota allowing researchers to shed new light on what was until very recently the black hole of microbiota ecological function within the gut. This chapter describes a collection of molecular genetic and metabolomic approaches used to study the human gut microbiota.

Introduction

It is becoming increasingly recognized that the interaction between diet and our intestinal microbiota plays an important role in human health and disease (Tuohy and Del Rio, 2014). As described elsewhere in this book, the gut microbiota within any given human comprises many hundreds of different microbial species, bacteria, viruses and fungi (yeast), which together make up an interdependent cross-kingdom microscopic community which combined encode a 'metagenome' many times more diverse than our own human genome. This microbiota mainly resides within the colon or large bowel, but also the small intestine, urogenital tract, skin, and oral cavities harbour ecosystems of considerable species richness and microbial load. Indeed, recent multidisciplinary studies have revealed the extent to which gut bacteria contribute to many different aspects of human physiology traditionally viewed as deriving exclusively from the genetic information encoded by the human genome (Koeth *et al.*, 2013; Panda *et al.*, 2014). A key role for the gut microbiota in regulating both metabolic and immune homeostasis has been identified, not only in the gut, but also systemically. Acting through dietary carbohydrate/fibre fermentation into short-chain fatty acids and the enterohepatic circulation of bile acids, the flux of metabolites derived from diet–microbe interactions in the gut impacts on satiety and energy intake, adipose tissue structure and function (including leptin and adipokine production), fat storage and thermogenesis, inflammation [including resolution of chronic inflammation and regulatory T-cells (Treg) function], macrophage cell signalling and reverse cholesterol transport. Most recently, a role of the gut microbiome in both brain development and brain function including mood and neurodegeneration have also been investigated (Shapiro *et al.*, 2014; Fava, 2014, Borre *et al.*, 2014). Importantly, such revolutionary insights from the fields of physiology and nutrition have come at a time when methodologies for studying microbial communities are providing a comprehensive tool-kit of unprecedented resolution and breadth of coverage, allowing for the first time efficacious measurement of important ecological parameters within the gut microbiome.

Methodologies for studying the human gut microbiota

Traditionally the majority of gut microorganisms have been considered as 'unculturable' with early molecular studies suggesting that up to 70% of bacteria identified using 16S rRNA gene sequencing had not been grown previously under laboratory conditions and therefore were, in fact, new to science (Suau *et al.*, 1999). Such observations appeared to confirm earlier assertions that microbial unculturability could account for the 'great plate count anomaly', whereby the majority of environmental bacteria visible under the microscope did not appear to be enumerated using classical nutrient agar plate counts (Staley *et al.*, 1985). However, pioneering work, particularly in Scotland and Holland, has questioned this perceived wisdom with the cultivation of strictly anaerobic, fastidious microorganisms from the intestinal tract, including dominant and prevalent commensal members of what may be emerging as a core human microbiome important for human health, including organisms such as *Faecalibacterium prausnitzii*, *Roseburia* spp. and *Akkermansia muciniphilia* (Duncan *et al.*, 2002; Derrien *et al.*, 2004; Flint et al, 2012). It may be therefore that rather than being 'unculturable' the majority of strictly anaerobic bacteria within the gut microbiota should be classed as 'not yet cultured' and await 'discovery' by those suitably skilled and willing to dedicate the time necessary to isolate fastidious, strictly anaerobic and often sub-dominant members of this microbial consortia. Indeed, recently it has been shown that for the 29 most abundant gut bacterial species in metagenomic libraries, representing about 50% of the total gut microbiota diversity within individuals, cultivated representatives exist, providing a powerful framework against which novel isolates may be compared both genotypically and phenotypically (Walker *et al.*, 2011). Growing microorganisms in pure culture under laboratory conditions is still important even in the post-genomics era, because it allows the phenotype of an organism to be fully studied and compared with its genetic make up. In addition, it is currently the only way to ascribe novel gene functions, and considering that many microbial gene functions in the vast majority of microorganisms present on earth are still without known function, traditional cultivation approaches will remain an important tool for the gut microbiologist going forward. None the less, the complexity of the gut microbiota in terms of species richness, the diversity of their metabolic activities and interactions with the host and the importance of studying host–microbe interactions *in situ*, necessitate powerful culture independent technologies. Over the past thirty years a diverse tool kit has been developed, including molecular methods for both targeted microbial quantitative methods and semi-quantitative but high-resolution/broad-spectrum methods capable of identifying theoretically all microorganisms within an ecosystem. These methods measure microorganisms in environmental samples either directly (e.g. via microscopy) or indirectly upon isolation of reporter molecules like DNA, RNA, proteins or the metabolites flowing through community metabolic pathways. Together this tool-kit (represented schematically in Fig. 16.1) allows the investigator to ask three fundamental questions of microbial communities: (i) Who's there? (ii) Do your populations grow or shrink under environmental pressure? (iii) And what is your function within the ecosystem, or in this case, within the human–microbiota super-organism (Inglis *et al.*, 2012; de Bruijn 2011a,b).

Who's there? – describing microbial community species make-up

Determining the species richness of a microbial community or how many different species are present within that community, is a first step in describing a microbial community. It can be used to monitor the stability or 'health' of an ecosystem over time or how the community responds to different environmental challenges, and when combined with species evenness, or the relative abundance of different species within that community (discussed below), it is an important component of microbial community

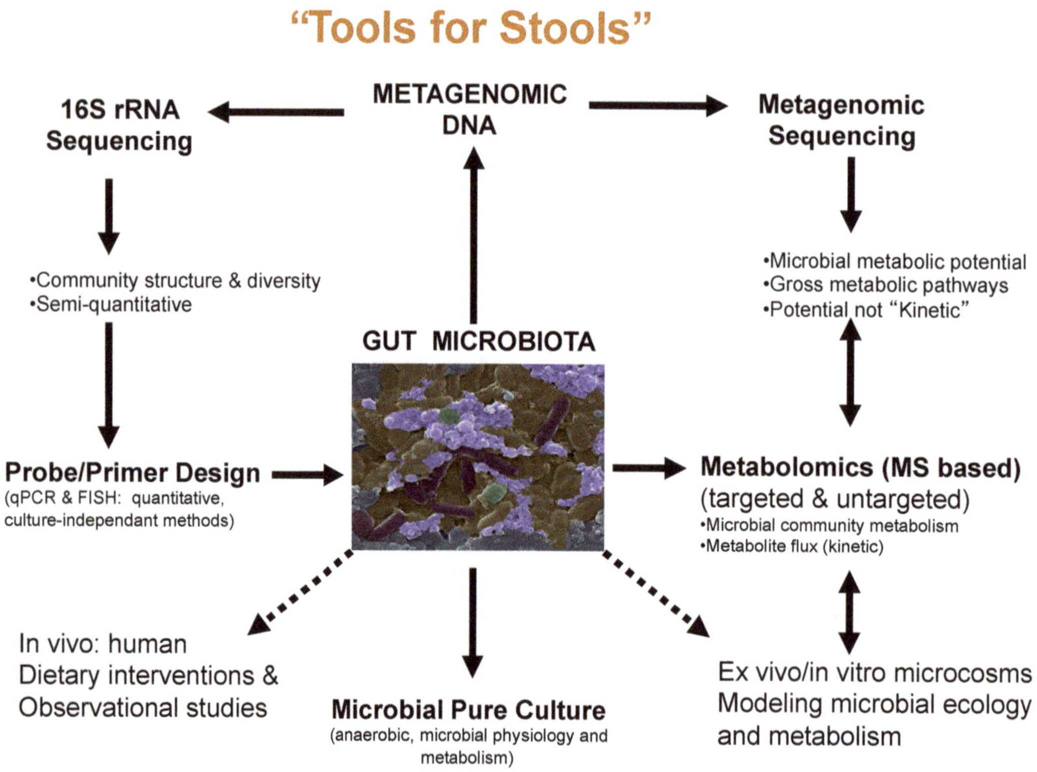

Figure 16.1 A schematic representing the application of traditional microbiological culture, culture independent molecular microbiology tools and metabolic profiling using metabolomics to study the ecology of the human gut microbiota directly *in vivo* and also indirectly, using ex vivo samples/microcosms or *in vitro* model systems.

diversity. Moreover, determining the phylogenetic composition of a community begins the process of inferring ecological function or rather describing the ecological roles or niche of different members within that community. DNA fingerprinting techniques, for example temperature or DENATURING GRADIENT GEL electrophoresis (T/DGGE) and AUTOMATED rRNA intergenic spacer analysis (ARISA) are commonly used approaches to measure microbial species richness (Inglis *et al.*, 2012; McCartney 2002). They are relatively rapid, universal in terms of microbial coverage, and importantly, are cheap, at least compared to more high-resolution next-generation sequencing (NGS) technologies. However, as the cost of high-throughput sequencing is becoming much more competitive, these techniques are being largely overtaken by metagenomics/metataxonomics (Ventura *et al.*, 2009). They are still though a useful tool, especially when used in combination with other more phylogenetically descriptive approaches and when employed with full knowledge of their relative limitations. Targeting the 16S rRNA gene, a universal chronometer within bacteria, T/DGGE profiles are based on the differential melting of double-stranded 16S rRNA gene fragment amplicons across a temperature or denaturing gradient. Double-stranded DNA amplicons of different DNA sequence will migrate at different rates across these gradients depending on their melting temperature governed largely by their GC content. The resulting band patterns on the electrophoresis gel, or 'fingerprints' provide an estimate of species richness, as each band theoretically corresponds to a single 16S rRNA gene fragment amplicon and thus species. This information can then be used with traditional ecological measures such as the Shannon–Weaver index (Gafan *et al.*, 2005). T/DGGE has a number of limitations. Firstly, used to describe complex microbial communities such as the human gut microbiota, T/DGGE band profiles typically only represent the most abundant species, maybe the top 10%, because of their log difference relative abundance compared to subdominant species. Secondly, although each DNA band in the T/DGGE band profile theoretically corresponds to a single DNA amplicon and therefore species or phylotype, DNA amplicons sometimes co-migrate because they have similar melting temperatures even with different DNA sequence. This can be true for both closely and very distantly related species of bacteria. Finally, because both temperature and denaturing gradient gel electrophoresis, are gel electrophoresis based approaches, heterogeneity occurs between gels introducing experimental error and making it difficult to combine or compare results from different gel runs, although with sufficient internal standards and specialized software, inter-gel comparisons and multi-gel datasets are possible. Finally, T/DGGE is not phylogenetically descriptive. Bands can be cut from the gel and their DNA sequenced, but this process is time consuming, limiting the number of bands which can be identified in this manner. Similarly, the size of the 16S rRNA fragment amplified using T/DGGE primer sets, because of the need for efficient amplicon separation on the gel and long GC clamps, is small, usually abut 200bp, giving low phylogenetic resolution, typically identification to the genus level at best.

Another commonly employed finger-printing technique is ARISA, or Automated rRNA Intergenic Spacer Analysis. This approach is based on the length of interspacer regions between the 16S rRNA gene and 23S rRNA gene in bacteria, which appears to be different between different bacterial phylotypes. Typically, different species have different length of this interspacer region, allowing the number of different phylotypes/species to be determined by the number of different length spacer regions amplified from a microbial community. The technique uses capillary electrophoresis rendering it more reproducible than gel based fingerprinting techniques, and is automated and rapid. However, it gives very little phylogenetic information, as the length of 16S–23S rRNA interspacer region does not appear to be evolutionarily conserved, meaning that different bacteria with very different phylogenetic backgrounds can give the same inter-spacer length (Popa *et al.*, 2009).

Recent advances in sequencing technology have made high-though put sequencing both affordable and readily available. Similarly, bioinformatic tools aimed at the end user or microbiologist rather than specialized bioinformaticians, have opened up next generation sequencing to the vast majority of laboratories working on microbial ecology. The 454-pyrosequencing technology of Roche, allows up to 700bp fragment sequencing and when combined with bar-coded primer sets targeting variable regions of the 16S rRNA gene, provides an extremely useful tool to determine species make-up of a microbial community. Depending on depth of sequencing, this approach has the potential to identify the great majority of the bacteria present in an environmental sample and the gastrointestinal tract has been targeted with large scale studies (Segata *et al.*, 2012a). Primer choice and variable region of the 16S rRNA gene chosen for analysis both impact on the data output. For example, Huse *et al.* (2012), found considerable variation in species richness and phylogenetic makeup when comparing metataxonomic data generated using primers targeting the V1–V3 region of the 16S rRNA gene compared to data derived from V3–V5 metagenomic datasets from the same gut microbiota samples. Our own work has found similar results with primer choice apparently having a dramatic effect on representation of different phylogenetic groups within metataxonomic datasets derived from the same faecal samples. However, in general high-through put sequencing of amplified community 16S rRNA gene fragments, or 'metataxonomics' as it has been dubbed, provides thus far unrivalled characterization of the species make up of microbial communities both in terms of species richness and the identity of these species. Observations of different primer sets giving different results from the same sample highlight the importance to recognizing the limitations of metataxonomics and of confirming observations made using these datasets with other parallel methods such as targeted culture independent microbial quantification using either FISH or qPCR as described in Box 16.1.

Shotgun metagenomics is a PCR-free approach which employs high-throughput NGS methods to provide a quantitative snapshot of, ideally, all genomes present in the community. Initially limited to large research consortia due to its costs (Qin *et al.*, 2010, The Human Microbiome Consortium 2012), as the price of sequencing continues to decrease, shotgun metagenomics is now an affordable tool in many laboratories. The experimental

Box 16.1 A cautionary note when using PCR-based methods for assessing microbial diversity

T/DGGE, ARISA and metataxonomics using bar coded 16S rRNA targeted primer sets are all are based on amplification of segments of the 16S rRNA gene using PCR and therefore, are subject to PCR bias (Berry *et al.*, 2011; Sim *et al.*, 2012). Mixed or complex consortia of microorganisms will possess a pool of 16S rRNA species with variable ratios of GC and AT rich DNA. Since the GC content of any piece of double stranded DNA will largely determine its melting temperature, 16S rRNA gene fragment amplicons from mixed microbial consortia with bacteria possessing different GC content are likely to have different melting temperatures and therefore display differential efficiency during PCR. PCR will therefore select against amplicons with high GC content and enrich for amplicons with more AT-rich sequences. This bias is inherent in any PCR based method applied to mixed DNA species and in the case of PCR based molecular methods used to measure species richness within complex microbial communities introduces a bias which must be recognized and appreciated to correctly interpret the data (Rajendhran and Gunasekaran, 2011). This situation is especially important when the high GC bacteria are sub-dominant species within the consortia. A good example are the bifidobacteria which have a GC content in the region of 60% and relative abundance lower than 2% within the adult human gut microbiota. This compares to clostridia which have a GC content of 25–55% and a much higher relative abundance (over 40% of the adult gut microbiota). Bifidobacteria are often underrepresented in many metataxonomic studies of the human gut microbiota compared to other components of the microbiota for both biases in GC content and primer selection, and other regions of the ribosomal machinery are more suitable to assess the diversity and abundance of such genera (Milani *et al.*, 2014).

protocol basically consists only of a DNA extraction and a NGS library preparation step, thus avoiding the bias-prone 16S rRNA gene amplification step. However, advanced bioinformatics tools are crucial to mine shotgun metagenomic datasets (Segata *et al.*, 2013a) and obtain species or sub-species level taxonomic profiles (as well as reconstructing the metabolic potential of the community as discussed later on). In addition, metagenomic assembly (Boisvert *et al.*, 2012; Namiki *et al.*, 2012) followed by mapping against the reconstructed genomes and genome characterization using specific analysis tools (Seemann 2014; Segata *et al.*, 2013b) potentially provides strain-level taxonomic characterizations with single nucleotide polymorphism (SNP)-level resolution thus overcoming a major limitation of early metataxonomic studies. However, several current limitations of metagenomic assembly including high computational costs, low coverage for many members of the microbiome, and trans-species conserved regions, make it possible to investigate only the most abundant microorganisms in a microbial community. Assembly-free taxonomic tools are in fact popular choices for the analysis of gut shotgun metagenomes as they are able to profile very low-abundance organisms as well. These include tools for species-level taxonomic profiling based on hundreds of species-specific markers (Segata *et al.*, 2012b), universal markers (Sunagawa *et al.*, 2013), or exact k-mer matching (Wood and Salzberg 2014). Assembly-free approaches exploit the collection of available sequenced reference genomes (currently approaching 20,000 bacterial genomes) and, although the richness of available reference genomes is still a crucial limitation in the analysis of environmental microbiomes, for gut-associated microbial communities the coverage of reference information is becoming sufficiently comprehensive. This is mainly due to new cultivation-free genome sequencing tools [e.g. by single-cell sequencing (Rinke *et al.*, 2013) or mouse monocolonization (Sczesnak *et al.*, 2011)] and as the result of large-scale projects specifically targeting previously unsequenced species (Fodor *et al.*, 2012).

Do your populations grow or shrink under environmental pressures? Culture-independent enumeration of microorganisms within the gut microbiota

A number of methods have been developed to enumerate microorganisms in environmental samples without the need for microbial cultivation. These methods, designed to overcome the 'great plate count anomaly', can be divided into direct methods, which enumerate microorganisms directly within the natural consortia usually by microscopic visualization of microorganisms labelled with reporter molecules, or indirect methods, which usually rely on extraction of nucleic acids followed by molecular assessment of the relative abundance of different phylogenetic groupings (Tuohy and McCartney, 2006). The most commonly employed direct, culture independent molecular method used to accurately enumerate microorganisms within environmental samples is fluorescence *in situ* hybridization (FISH) using labelled oligonucleotide probes targeting 16S rRNA. This method accurately quantifies microbes at various phylogenetic levels from species to phylum, depending on probe design (Amann *et al.*, 2001). Environmental samples collected directly from the natural consortia under study are fixed in paraformaldehyde, permeabilized using SDS and then hybridized with fluorescently labelled oligonucleotide probes designed to target different phylogenetically relevant regions of the 16S rRNA gene. The probe attaches onto the 16S rRNA itself, within ribosomes, increasing signal intensity under fluorescent microscopy, and the labelled fluorescent bacteria are counted directly under the microscope. A limitation of FISH therefore, is that probes are only available for bacteria which have already been cultivated, as probe design and validation require empirical confirmation of probe accessibility and attachment onto the target 16S rRNA within the ribosome. FISH is accurate and reproducible, especially using modern computerized microscopic counting or flow cytometry, with precision determined by probe design, usually targeting variable regions of the 16S rRNA and is currently the 'gold standard' for culture independent direct microbial enumeration in environmental samples (Jansen *et al.*, 1999; Rochet *et al.*, 2004). Another advantage of FISH is that it targets rRNA on the ribosome itself rather than the 16S rRNA gene, which means that the signal intensity is directly correlated with the number of ribosomes in a cell and therefore cellular metabolic activity. Bacteria enumerated using 16S rRNA targeted FISH therefore represent the metabolically active portion of cells within a microbial community (Rosselló-Mora *et al.*, 1999). In addition, FISH has recently been coupled with high-resolution secondary ion mass spectrometry (NanoSIMS) and RAMAN microscopy allowing simultaneous phylogenetic marking of target microorganisms and determination of the ability to metabolize isotopically labelled substrates *in situ* within environmental samples (Berry *et al.*, 2013; Stecher *et al.*, 2013).

Of the indirect sequencing-free culture independent methods for enumerating microorganisms from environmental samples, quantitative PCR (qPCR) is probably the most commonly employed. qPCR using either Taqman or Syber Green technologies, quantifies the copy number of target sequences present in a mixture of DNA targets extracted from the environmental sample. The 16S rRNA gene is commonly the target, with primers covering variable regions depending on depth of phylogenetic classification required, again, typically phylum to species level. The method is reliant on DNA extraction efficiency from the environmental sample, can be impacted by PCR inhibitors and, in the absence of prior knowledge from a cultured target microorganism, makes assumptions about the number of copies of 16S rRNA gene present in a target microorganism or group of microorganisms. Alternatively, data can be expressed as numbers of 16S rRNA copies rather than related to numbers of target microorganisms per gram of sample, or as a relative abundance or percentage of total 16S rRNA gene copies in the sample (Furet *et al.*, 2009; Ahmed *et al.*, 2007). Although it is generally considered less accurate than FISH, it is commonly employed because it is perceived to be more rapid. It is also useful for enumerating previously uncultured bacteria for which FISH probes cannot be effectively validated.

What is your function within the ecosystem? Measuring microbial metabolic potential and metabolic kinetic

The age of 'genomics', at least in terms of human health and disease, has passed and we now realize that disease risk is determined by more than the sum of our 'genetically encoded parts'. Environmental factors are now recognized to contribute significantly to chronic disease risk at the population level outside a limited number of genetic diseases. Moreover, as discussed elsewhere in this book, the gut microbiota is emerging as significant player in human health and disease, linked through different axes to diverse host physiological functions especially in the liver, adipose tissue, muscle and brain (Joyce *et al.*, 2014; Geurts *et al.*, 2014; O'Mahony *et al.*, 2014). In this post genomics era it has become apparent that gut microbial activities are important determinants of chronic disease risk. The post-genomics era has endowed us with a number of meta'omic tools (Segata *et al.*, 2013a) capable, for the first time, of measuring both the microbial metabolic potential (locked within the genes of the gut metagenome) and the metabolic kinetics (the flux of metabolites flowing through pathways encoded by the gut metagenome) (Tuohy *et al.*, 2009). Of these, metagenomics (Furrie, 2006), metatranscriptomics (Berry *et al.*, 2012) and metabolomics (Yap *et al.*, 2008) are the most advanced and are now routinely applied when studying the gut microbiota, how it changes with disease, diet and drugs to mediate host health.

As already introduced in the context of taxonomic profiling, shotgun metagenomics aims at recovering a full snapshot of the genetic material in the community. Assembling the genomes in the community *de novo* is also in this case the ideal scenario, as standard genome-based functional analyses can be then employed on the reconstructed contigs, but as discussed above, reference-free tools can capture only a fraction of organisms. Functionally characterized protein databases such as the Kyoto Encyclopaedia of Genes and Genomes (KEGG), and the Carbohydrate-Active enZYmes database (CAZY) for enzymes related to carbohydrate metabolism can be used as a reference for profiling the overall metabolic potential of the community (Huson *et al.*, 2011). As with all functional assays, this operation partially suffers from the lack of annotations for the majority of the bacterial genes (with the partial exception of few model organisms) and from the converting microbial functions into discrete sequence-similarity based bins (e.g. KEGG orthologous clusters and modules). Advanced tools also exist for estimating the community's metabolic potential by post processing the single gene/protein family hits into functional pathways and metabolic models (Abubucker *et al.*, 2012). Recent giant steps in understanding the metabolic potential of the gut microbiome in health and disease were based on shotgun metagenomic sequencing and include timely publicly funded projects around the world such as the Human Microbiome Project and MetaHIT (The Human Microbiome Project Consortium, 2012; Qin *et al.*, 2010) as well as Chinese (Qin *et al.*, 2012) and European (Karlsson *et al.*, 2013) cohorts investigating the role of the gut microbiota in diabetes.

Metatranscriptomic RNA sequencing approaches interrogate directly the active portion of the microbiome's genetic repertoire. Although community RNA sequencing requires additional care in order to avoid technical biases, such as removal of human RNA, and minimizing the fraction of ribosomal RNA in the sample, preliminary successful metatranscriptomic analyses of the human gut microbiota have been performed (Booijink *et al.*, 2010; David *et al.*, 2014; Gosalbes *et al.*, 2011). The analysis challenge of distinguishing variations in transcriptional levels with respect to variations in underlying gene/genome abundances is hard to tackle given the lack of robust universal housekeeping genes, making it usually necessary to have coupled metagenomic data on the same sample for normalizing transcripts with gene abundances (Franzosa *et al.*, 2014). Transcript identification and quantification can typically occur by *de novo* transcript sequence assembly or by mapping against functionally characterized gene/protein families. However, the large transcript abundance dynamic range makes assembly unfeasible for non-highly expressed genes while the mapping approach suffers from organismal aspecificity. Metatranscriptomics is thus a promising tool for unravelling the transcriptional activity of the gut microbiome, but robust analysis pipelines are still to be developed.

With advances in analytical chemical techniques and chemometrics, metabolomics has emerged as a powerful tool for measuring the profile of metabolites present in biological matrices. Metabolomics uses data generated using NMR or 'hyphenated' MS strategies and multivariate statistics to profile metabolites, or fluxes in metabolites, present in biological samples and was the first of the 'omics' technologies to fully appreciate the contribution of the gut microbiota to mammalian host physiology (Brindle *et al*, 2003; Nicholson and Lindon, 2008; Xie *et al.*, 2013). As applied to human health and nutrition, metabolomics (or sometimes called metabonomics), aims to monitor all detectable metabolites in a biofluid (e.g. blood, urine, faecal water) and identify molecules which change in concentration over time or in response to disease or an outside influence such as dietary change or drug/xenobiotic/pollutant exposure (Nicholson and Lindon 2008). Unlike the other established 'omics' technologies, metabolomics therefore, rather than trying to measure metabolic potential as encoded by the nucleic acids of an organism, endeavours to measure the metabolic flux or metabolic kinetics of an organism and how this changes under experimental conditions. Metabolomics techniques can either be applied in a targeted manner, targeting specific pathways or class of compounds in a highly accurate and quantitative manner, or a broader and more inclusive semi-quantitative untargeted (holistic) manner. Targeted metabolomics has the advantage of accurate quantification of selected known compounds allowing subtle changes in the flux of metabolites through pathways to be measured, while untargeted holistic metabolomics allows a more global view of metabolic flux through a biological system. Both are now being applied to study the importance of the gut microbiota and have greatly contributed to the emergence of nutritional metabolomics, which is shedding new light not only on which nutrients impact on human health and disease but is demonstrating that many compounds of dietary origin which regulate human physiology, are actually derived from co-metabolic pathways of host and intestinal microbiota (Martin *et al.*, 2010; Heinzmann *et al.*, 2012; Swann and Claus, 2014; Gasperotti *et al.*, 2014).

Other post-gemonics approaches are also being developed, including proteomics or metaproteomics, but currently lack wide spread application in studies of gut microbiota because of the complexity of the data produced or their lack of reproducibility. Metagenomics, including shotgun metagenomics, metatranscriptomics and metabolomics on the other hand, have by and large overcome early problems with automated chemical or genetic functional annotation mainly thanks to advances in bioinformatics, and are currently being applied to a range of human biofluids (Martin *et al.*, 2012; Blottière *et al.*, 2013; Delmont *et al.*, 2013; De Filippo *et al.*, 2012; Stanstrup *et al.*, 2013).

State of play: the colonic microbiome in all its complexity

The colon is the main site of microbial colonization within the gut and home to many thousands of different microbial strains with microbial numbers reaching up to 10^{11} cells/g contents. Most studies use the faecal microbiota as a readily accessible surrogate for the colonic microbiota because of the technical and ethical implications of directly sampling intestinal microbial communities. Although it is known that the gut microbiota differs both along the length of the gastrointestinal tract and across the gut transverse section, from mucosa to luminal contents, for the most part the faecal microbiota allows a good approximation of the colonic microbiota composition and is the only viable option for larger studies measuring the microbial ecology of the human gut. In health, climax microbial populations are attained in the colon and this richness of microbial diversity is mainly due to the slowing of intestinal transit in this region of the gut, allowing sizable microbial populations to grow on food components which escape digestion in the upper gut. Many of these non-digestible foods, such as plant fibres and polyphenols are complex molecules which are difficult to break down even by consortia of microorganisms working together. The pH of the proximal colon is moderately acidic, about 5.5, due to the actions of microbial fermentation, while pH increases distally through the transverse colon (typically 6.2) and distal colon (typically 6.8). The pH in different regions of the colon can itself regulate microbial metabolic activities, with studies showing that different species are involved in carbohydrate fermentation in the proximal colon and distal colon (Walker *et al.*, 2005). There is a considerable degree of cross-feeding and cooperation between microorganisms within the colonic microbiota particularly in the breakdown of complex plant materials, and in health, competition for ecological niche is fierce, endowing a remarkable degree of homeostatic control and ecological stability upon microbiota architecture (Veilleux and Rowland, 1981).

The recent application of post-genomics approaches to study the ecology of the intestinal microbiota has led to considerable advances in our understanding both of the structure and function of the gut microbiota. Now numerous high-through put metataxonomic studies show that the gut microbiota is dominated by a relatively small number of bacterial phyla (Firmicutes, Bacteroidetes and Actinobacteria) with other bacteria occurring as sub-dominant but prevalent community members (Proteobacteria and Verrucomicrobia) (Eckburg *et al.*, 2005; Durbán *et al.*, 2011; Codling *et al.*, 2010). On the question of 'culturability' recent studies are showing that, at least for the dominant members of the gut microbiota, most species have cultured representatives making it much easier to assign putative ecological function. Walker *et al.* (2011), studying the faecal microbiota of overweight men, found 29 different phylotypes dominated with relative abundance of more than 1%. All of these phylotypes or species had closely related strains which had been previously cultured. They also found however, that phylotypes which were sub-dominant, were less likely to have closely related strains which had previously been cultivated indicating that the 'great plate count anomaly', at least as concerns the human gut microbiota, may reflect the lack of efficacious selective isolation and growth conditions for sub-dominant members of complex microbial communities, rather than any inherent 'unculturability' of gut microorganisms. Having said that however, it has to be remembered that the vast majority of gut bacteria are strict anaerobes, highly sensitive to oxygen and some are fastidious, likely requiring very specific growth substrates and co-factors for growth in pure culture.

We derive our intestinal microbiota both from our mothers and our environment, with diet early in the post-natal period being particularly important for acquisition of gut microorganisms. Although a complex and relatively stable microbiota emerges by the age of two years, this microbiota does respond to dietary change, especially in the relative abundance of key bacterial groups or species, e.g. the bifidobacteria and Prevotella, and in metabolic activity. Once this climax microbial community is reached, it displays a remarkable degree of self-regulation and homeostasis, with microorganisms ingested with food or probiotics, for example, finding it difficult to occupy ecological niche already occupied by autochthonous strains. This community stability persists until either old age or disease disturb homeostatic control mechanisms, especially the hosts immune system and mucosal integrity, often leading to reduced microbial diversity both in terms of species richness and evenness of species abundance (Claesson *et al.*, 2012; Mondot *et al.*, 2013). Thus the ecological principles of community structure hold strong for the gut microbiota and interestingly are mirrored in health and disease of the host, with many chronic diseases now shown to present with reduced gut microbiota diversity. Using TGGE, Zoetendal *et al.* (1998) showed that not only is the gut microbiota stable over extended periods of time in terms of species composition, but that each individual harbours a unique collection of microorganisms. Inter-individual differences were even found between identical twins, although within families, twins showed the greatest degree of relatedness in terms of microbiota species composition. Similarity measures decreased as the degree of relatedness decreased, highlighting the important role of both host genes and environment in determining gut microbiota composition (Zoetendal *et al.*, 2001). Combining culture independent molecular microbiology (DGGE) with measurement of gut microbiota metabolic output via NMR based metabonomics, Li *et al.* (2008) related species of gut bacteria with profiles of metabolites in urine, suggesting correlations between important gut bacteria such as *Faecalibacterium pruasnitzii*, bifidobacteria, and even previously uncultured bacterial species with urinary metabolites. They confirmed the individual nature of microbiota makeup and that microbial relatedness reflects the degree of genetic relatedness between hosts and that both microbiota composition and faecal metabolome differed between Chinese and American study groups, mainly due to metabolites of host–microbiota co-metabolic pathways.

Much concern has arisen recently in the need to define a 'core' gut microbiome, made up of species characteristic of the human gut microbiota. The identification of highly prevalent species common to most humans such as *Faecalibacterium prausnitzii* and certain members of the Lachnospiraceae and Bacteroidetes for example, would argue that a core human gut microbiota exists, but also since each of us harbours a unique collection of microorganisms, considerable metabolic redundancy must also exist even between distantly related gut bacteria if different strains or even species are capable of fulfilling the same specific ecological niche in different hosts. Using NGS data Tap *et al.* (2009), found that the majority of faecal operational taxonomic units (OTUs) within a given subject were unique to that subject with only about 2.1% present in more than half of the 17 subjects studied. However, this 2.1% of more prevalent OTU's accounted for 35.8% of total sequences, and was made up of 66 different phylotypes mainly *Faecalibacterium, Ruminococcus, Eubacterium, Dorea, Bacteroides, Alistipes* and *Bifidobacterium*. Salonen *et al.* (2012) using a gut microbiota specific 16S rRNA targeted phylogenetic microarray (the HITchip) showed that in a hundred subjects, the size of the core microbiota depended on the depth of analysis, – the more samples analysed, the more bacteria joined the core microbiota. Similarly, the% prevalence used to define membership of the core microbiota was also a critical determinant of core microbiota size. Surprisingly however, they also found that biological parameters, most notably disease (inflammatory bowel disease in this case), also appeared to determine the size of the core microbiota. As discussed above, this illustrates the apparent link between reduced microbiota diversity and ecosystem fitness or health, and thus host health (Mondot *et al.*, 2013). Both Turnbaugh *et al.* (2010) and later Sekelja *et al.* (2011) described a core human microbiome comprising *Lachnospiraceae*, which contains many high butyrate-producing species including *Eubacterium eligens, Roseburia, Dorea* and the ruminococci. This core group appears to have closely co-evolved alongside animals all through evolution, and are commonly represented within gut metagenomes of invertebrates, vertebrates and mammals. Arumugam *et al.* (2011) sequenced 22 human metagenomes and upon analysis proposed the existence of three enterotypes characterized by *Bacteroides, Prevotella* and *Ruminococcus* (which includes *Blautia* and unclassified *Lachnospiraceae*). Their enterotypes were defined both by species composition and co-occurrence, and functional metabolic genes. Although useful for describing a core human microbiome, such groupings appear to be effected by limits and thresholds set by bioinformatic/statistical tests (Koren *et al.*, 2013) and Huse *et al.* (2012) suggested that the enterotype model although fitting most metagenomic datasets well at the genus level, appears to unravel at lower taxonomic levels.

In summary, a small core microbiota is likely to be acquired through vertical transfer from our mothers, and indeed recent studies are confirming close association between microbiota of mother and child. Indeed, transfer of specific microorganisms, including strictly anaerobic species, both through breast milk and even before birth has been demonstrated (Donnet-Hughes *et al.*, 2010). Acquisition after that of a metabolically versatile, auxiliary microbiota capable of completing the gut microbiota and occupying more specialized ecological niche in the complex gut microbiota food web is likely to be affected largely through environmental exposures, through strains we encounter during early childhood, from the foods we eat, especially fresh low processed or fermented foods bearing diverse food associated microbiota themselves, from co-habitation with family and even companion animals (Dietert *et al.*, 2014). Drug or pollutant exposure too is likely to have an impact on which species of bacteria take up residence within an individual's gut, especially exposure to the selective pressure of antibiotic therapy. Of course, which species and which metabolic capabilities they encode will have a dramatic effect on the ecosystem as a whole both in terms of diversity of metabolites products and the relative efficiencies of different metabolic pathways. A good example of this is the complex interaction between sulfate-reducing bacteria and methanogens within the gut, which appear to compete for short-chain fatty acids at once altering the profile of small molecules produced in the colon and impacting on microbiota fermentation efficiency as a whole through relief of end product inhibition caused by H_2 (Gibson *et al*, 1988; Carbonero *et al.*, 2012).

Conclusions

The omics era has endowed us with a range of culture independent microbiological techniques, broad spectrum and quantitative analytical chemical methodologies and user friendly bioinformatic pipelines which when combined and employed with full appreciation of their relative advantages and shortcomings, can characterize both the composition and activity of the human gut microbiota. Initial studies of an observational nature have described both the phylogenetic and metabolic architecture of the gut microbiome in health and disease, from different body sites and also in diverse populations from around the world. However, we are now moving into a new age of gut microbiome research, past the observational and into the hypothesis driven experimental research phase, with the necessary physiological insight into microbiota driven metabolic and immune pathways and equipped with appropriate 'tools for stools' allowing us to measure the microbial ecology of the gut microbiome as it responds to different environmental stimuli.

References

Abubucker S., Segata, N., Goll, J., Schubert, A.M., Izard, J., Cantarel, B.L., Rodriguez-Mueller B., Zucker, J., Thiagarajan, M., *et al.* (2012). Metabolic reconstruction for metagenomic data and its application to the human microbiome. PLoS Comput. Biol. *8*, e1002358.

Ahmed, S., Macfarlane, G.T., Fite, A., McBain, A.J., Gilbert, P., and Macfarlane, S. (2007). Mucosa-associated bacterial diversity in relation to human terminal ileum and colonic biopsy samples. Appl. Environ. Microbiol. 73, 7435–7442.

Amann, R., Fuchs, B.M., and Behrens, S. (2001). The identification of microorganisms by fluorescence *in situ* hybridization. Curr. Opin. Biotechnol. *12*, 231–236.

Arumugam, M., Raes, J., Pelletier, E., Le Paslier, D., Yamada, T., Mende, D.R., Fernandes, G.R., Tap, J., Bruls, T., Batto, J.M., *et al.* Enterotypes of the human gut microbiome. Nature 473, 174–180.

Berry, D., Ben Mahfoudh, K., Wagner, M., and Loy, A. (2011). Barcoded primers used in multiplex amplicon pyrosequencing bias amplification. Appl. Environ. Microbiol. *77*, 7846–7849.

Berry, D., Schwab, C., Milinovich, G., Reichert, J., Ben Mahfoudh, K., Decker, T., Engel, M., Hai, B., Hainzl, E., Heider, S., Kenner, L., Müller, M., Rauch, I., Strobl, B., Wagner, M., Schleper, C., Urich, T., and Loy, A. (2012). Phylotype-level 16S rRNA analysis reveals new bacterial indicators of health state in acute murine colitis. ISME J. *6*, 2091–2106.

Berry. D., Stecher, B., Schintlmeister, A., Reichert, J., Brugiroux, S., Wild, B., Wanek, W., Richter, A., Rauch, I., Decker, T., Loy, A., and Wagner, M. (2013). Host-compound foraging by intestinal microbiota revealed by single-cell stable isotope probing. Proc. Natl. Acad. Sci. U.S.A. *110*, 4720–4725.

Blottière, H.M., de Vos, W.M., Ehrlich, S.D., and Doré, J. (2013). Human intestinal metagenomics: state of the art and future. Curr. Opin. Microbiol. *16*, 232–239.

Boisvert, S., Raymond, F., Godzaridis, E., Laviolette, F., and Corbeil, J. (2012). Ray Meta: scalable de novo metagenome assembly and profiling. Genome Biol. *13*, R122.

Booijink, C.C., Boekhorst, J., Zoetendal, E.G., Smidt, H., Kleerebezem, M., and de Vos, W.M. (2010). Metatranscriptome analysis of the human faecal microbiota reveals subject-specific expression profiles, with genes encoding proteins involved in carbohydrate metabolism being dominantly expressed. Appl. Environ. Microbiol. *76*, 5533–5540.

Borre. Y.E., Moloney, R.D., Clarke, G., Dinan, T.G., and Cryan, J.F. (2014). The impact of microbiota on brain and behavior: mechanisms & therapeutic potential. Adv. Exp. Med. Biol. *817*, 373–403.

Brindle, J.T., Nicholson, J.K., Schofield, P.M., Grainger, D.J., and Holmes, E. (2003). Application of chemometrics to 1H NMR spectroscopic data to investigate a relationship between human serum metabolic profiles and hypertension. Analyst *128*, 32–36.

Carbonero, F., Benefiel, A.C., and Gaskins, H.R. (2012). Contributions of the microbial hydrogen economy to colonic homeostasis. Nat. Rev. Gastroenterol. Hepatol. *9*, 504–518.

Claesson, M.J., Jeffery, I.B., Conde, S., Power, S.E., O'Connor, E.M., Cusack, S., Harris, H.M., Coakley, M., Lakshminarayanan, B., O'Sullivan, O., *et al.* (2012). Gut microbiota composition correlates with diet and health in the elderly. Nature *488*, 178–184.

Codling, C., O'Mahony, L., Shanahan, F., Quigley, E.M., and Marchesi, J.R. (2010). A molecular analysis of facal and mucosal bacterial communities in irritable bowel syndrome. Dig. Dis. Sci. *55*, 392–397.

David, L.A., Mauric,e C.F., Carmody, R.N., Gootenberg, D.B., Button, J.E., Wolfe, B.E., Ling, A.V., Devlin, A.S., Varma, Y., Fischbach, M.A., *et al.* (2014). Diet rapidly and reproducibly alters the human gut microbiome. Nature *505*, 559–563.

de Bruijn, F.J. (2011). Handbook of Molecular Microbial Ecology I: Metagenomics and Complementary Approaches (Ed. Frans J. de Bruijn), Wiley-Blackwell, Oxford, UK.

de Bruijn, F.J. (2011). Handbook of Molecular Microbial Ecology II: Metagenomics in Different Habitats (Ed. Frans J. de Bruijn), Wiley-Blackwell, Oxford, UK.

De Filippo, C., Ramazzotti, M., Fontana, P., and Cavalieri, D. (2012). Bioinformatic approaches for functional annotation and pathway inference in metagenomics data. Brief. Bioinform. *13*, 696–710.

Derrien, M., Vaughan, E.E., Plugge, C.M., de Vos, W.M. (2004). *Akkermansia muciniphila* gen. nov., sp. nov., a human intestinal mucin-degrading bacterium. Int. J. Syst. Evol. Microbiol. *54*, 1469–1476.

Dietert, R.R. (2014). The microbiome in early life: self-completion and microbiota protection as health priorities. Birth Defects Res. B. Dev. Reprod. Toxicol. *101*, 333–340.

Donnet-Hughes, A., Perez, P.F., Doré, J., Leclerc, M., Levenez, F., Benyacoub, J., Serrant, P., Segura-Roggero, I., and Schiffrin, E.J. (2010). Potential role of the intestinal microbiota of the mother in neonatal immune education. Proc. Nutr. Soc. *69*, 407–415.

Duncan, S.H., Barcenilla, A., Stewart, C.S., Pryde, S.E., and Flint, H.J. (2002). Acetate utilization and butyryl coenzyme A (CoA): acetate-CoA transferase in butyrate-producing bacteria from the human large intestine. Appl. Envirol. Microbiol. *68*, 5186–5190.

Durbán, A., Abellán, J.J., Jiménez-Hernández, N., Ponce, M., Ponce, J., Sala, T., D'Auria, G., Latorre, A., and Moya, A. (2011). Assessing gut microbial diversity from feces and rectal mucosa. Microb. Ecol. *61*, 123–133.

Eckburg, P.B., Bik, E.M., Bernstein, C.N., Purdom, E., Dethlefsen, L., Sargent, M., Gill, S.R., Nelson, K.E., and Relman, D.A. (2005). Diversity of the human intestinal microbial flora. Science *308*, 1635–1638.

Fava, F. (2014). Gut Microbiota Immune System Crosstalk: Implications for Metabolic Disease. In, 'Diet:Microbe interactions in the gut effects on human health and disease' (Eds. Kieran Tuohy and Daniele Del Rio), Academic Press, Elsevier, USA, pp. 127–137.

Flint, H.J., Scott, K.P., Louis, P., and Duncan, S.H. (2012). The role of the gut microbiota in nutrition and health. Nat. Rev. Gastroenterol. Hepatol. *9*, 577–589.

Furrie, E. (2006). A molecular revolution in the study of intestinal microflora. Gut *55*, 141–143.

Furet, J.P., Firmesse, O., Gourmelon, M., Bridonneau, C., Tap, J., Mondot, S., Doré, J., and Corthier, G. (2009). Comparative assessment of human and farm animal faecal microbiota using real-time quantitative PCR. FEMS Microbiol. Ecol. *68*, 351–362.

Fodor, A.A., DeSantis, T.Z., Wylie, K.M., Badger, J.H., Ye, Y., Hepburn, T., Hu, P., Sodergren, E., Liolios, K., Huot-Creasy H., Birren, B.W., and Earl, A.M. (2012). The 'most wanted' taxa from the human microbiome for whole genome sequencing. PLoS One 7, e41294.

Franzosa, E.A., Morgan, X.C., Segata, N., Waldron, L., Reyes, J., Earl, A.M., Giannoukos, G., Boylan, M.R., Ciulla, D., Gevers, D., *et al.* (2014). Relating the metatranscriptome and metagenome of the human gut. Proc. Natl. Acad. Sci. U.S.A. *111*, E2329–2338.

Gafan, G.P., Lucas, V.S., Roberts, G.J., Petrie, A., Wilson, M., and Spratt, D.A. (2005). Statistical analyses of complex denaturing gradient gel electrophoresis profiles. J. Clin. Microbiol. *43*, 3971–3978.

Gasperotti, M., Masuero, D., Guella, G., Mattivi, F., and Vrhovsek, U. (2014). Development of a targeted method for twenty-three metabolites related to polyphenol gut microbial metabolism in biological samples, using SPE and UHPLC-ESI-MS/MS. Talanta. *128*, 221–230.

Geurts, L., Neyrinck, A.M., Delzenne, N.M., Knauf, C., and Cani, P.D. (2014). Gut microbiota controls adipose tissue expansion, gut barrier and glucose metabolism: novel insights into molecular targets and interventions using prebiotics. Benef. Microbes. *5*, 3–17.

Gibson, G.R., Cummings, J.H., and Macfarlane, G.T. (1988). Competition for hydrogen between sulphate-reducing bacteria and methanogenic bacteria from the human large intestine. J. Appl. Bacteriol. *65*, 241–7.

Gosalbes, M.J., Durbán, A., Pignatelli, M., Abellan, J.J., Jiménez-Hernández, N., Pérez-Cobas, A.E., Latorre, A., and Moya, A. (2011). Metatranscriptomic approach to analyze the functional human gut microbiota. PLoS One *6*, e17447.

Heinzmann, S.S., Merrifield, C.A., Rezzi, S., Kochhar, S., Lindon, J.C., Holmes, E., and Nicholson, J.K. (2012). Stability and robustness of human metabolic phenotypes in response to sequential food challenges. J. Proteome Res. *11*, 643–655.

Huse, S.M., Ye, Y., Zhou, Y., and Fodor, A.A. (2012). A core human microbiome as viewed through 16S rRNA sequence clusters. PLoS One 7*(6)*, e34242.

Huson, D.H., Mitra, S., Ruscheweyh, H.J., Weber, N., and Schuster, S.C. (2011). Integrative analysis of environmental sequences using MEGAN4. Genome Res. *21*, 1552–1560.

Inglis, G.D., Thomas, M.C., Thomas, D.K., Kalmokoff, M.L., Brooks, S.P., and Selinger, L.B. (2012). Molecular methods to measure intestinal bacteria: a review. J. AOAC. Int. *95*, 5–23.

Jansen, G.J., Wildeboer-Veloo, A.C., Tonk, R.H., Franks, A.H., and Welling, G.W. (1999). Development and validation of an automated, microscopy-based method for enumeration of groups of intestinal bacteria. J. Microbiol. Methods *37*, 215–221.

Joyce, S.A., MacSharry, J., Casey, P.G., Kinsella, M., Murphy, E.F., Shanahan, F., Hill, C., and Gahan, C.G. (2014). Regulation of host weight gain and lipid metabolism by bacterial bile acid modification in the gut. Proc. Natl. Acad. Sci. U.S.A. *111*, 7421–7426.

Karlsson F.H., Tremaroli, V., Nookaew, I., Bergström, G., Behre, C.J., Fagerberg, B., Nielsen, J., and Bäckhed, F. (2013). Gut metagenome in European women with normal, impaired and diabetic glucose control. Nature *498*, 99–103.

Koeth. R.A., Wang, Z., Levison, B.S., Buffa, J.A., Org, E., Sheehy, B.T., Britt, E.B., Fu, X., Wu, Y., Li, L., Smith, J.D., *et al.* Intestinal microbiota metabolism of l-carnitine, a nutrient in red meat, promotes atherosclerosis. Nat. Med. *19*, 576–585.

Koren, O., Knights, D., Gonzalez, A., Waldron, L., Segata, N., Knight, R., Huttenhower, C., and Ley, R.E. (2013). A guide to enterotypes across the human body: meta-analysis of microbial community structures in human microbiome datasets. PLoS Comput. Biol. *9(1)*, e1002863.

Li, M., Wang, B., Zhang, M., Rantalainen, M., Wang, S., Zhou, H., Zhang, Y., Shen, J., Pang, X., Zhang, M., *et al.* (2008). Symbiotic gut microbes modulate human metabolic phenotypes. Proc. Natl. Acad. Sci. U.S.A. *105*, 2117–2122.

Martin, F.P., Collino, S., Rezzi, S., and Kochhar, S. (2012). Metabolomic applications to decipher gut microbial metabolic influence in health and disease. Front. Physiol. *3*, 113.

Martin, F.P., Sprenger, N., Montoliu, I., Rezzi, S., Kochhar, S., and Nicholson, J.K. (2010). Dietary modulation of gut functional ecology studied by fecal metabonomics. J. Proteome Res. *9*, 5284–5295.

McCartney, A.L. (2002). Application of molecular biological methods for studying probiotics and the gut flora. Br. J. Nutr. *88* (Suppl. 1), S29–37.

Milani C., Lugli, G.A., Turroni, F., Mancabelli, L., Duranti, S., Viappiani, A., Mangifesta, M., Segata, N., van Sinderen D., and Ventura, M. (2014). Evaluation of bifidobacterial community composition in the human gut by means of a targeted amplicon sequencing (ITS) protocol. FEMS microbiol. Ecol. 2014 Aug 13. doi: 10.1111/1574-6941.12410. [Epub ahead of print]

Mondot, S., de Wouters, T., Doré, J., and Lepage, P. (2013). The human gut microbiome and its dysfunctions. Dig. Dis. *31*, 278–285.

Namiki,T., Hachiya, T., Tanaka, H., and Sakakibara, Y. (2012). MetaVelvet: an extension of Velvet assembler to de novo metagenome assembly from short sequence reads. Nucl. Acids Res. *40*, e155.

Nicholson, J.K., and Lindon, J.C. (2008). Systems biology: metabonomics. Nature *455*, 1054–1056.

O'Mahony, S.M., Clarke, G., Borre, Y.E., Dinan, T.G., and Cryan, J.F. (2014). Serotonin, tryptophan metabolism and the brain–gut-microbiome axis. Behav. Brain Res. Jul 29. pii: S0166-4328(14)00476-8.

Panda, S., Guarner, F., and Manichanh, C. (2014). Structure and functions of the gut microbiome. Endocr. Metab. Immune Disord. Drug Targets. Jul 14. [Epub ahead of print]

Popa, R., Popa, R., Mashall, M.J., Nguyen, H., Tebo, B.M., and Brauer, S. (2009). Limitations and benefits of ARISA intra-genomic diversity fingerprinting. J. Microbiol. Methods *78*, 111–118.

Qin, J., Li, Y., Cai, Z., Li, S., Zhu, J., Zhang, F., Liang, S., Zhang, W., Guan, Y., Shen, D., *et al.* (2012). A metagenome–wide association study of gut microbiota in type 2 diabetes. Nature *490*, 55–60.

Qin, J., Li, R., Raes, J., Arumugam, M., Burgdorf, K.S., Manichanh, C., Nielsen, T., Pons, N., Levenez, F., Yamada, T., *et al.* (2010). A human gut microbial gene catalogue established by metagenomic sequencing. Nature *464*, 59–65.

Rajendhran, J., and Gunasekaran, P. (2011). Microbial phylogeny and diversity: small subunit ribosomal RNA sequence analysis and beyond. Microbiol. Res. *166*, 99–110.

Rinke, C., Schwientek, P., Sczyrba, A., Ivanova, N.N., Anderson, I.J., Cheng, J.F., *et al.* (2013). Insights into the phylogeny and coding potential of microbial dark matter. Nature *499*, 431–437.

Rochet, V., Rigottier-Gois, L., Rabot, S., and Doré, J. (2004). Validation of fluorescent *in situ* hybridization combined with flow cytometry for assessing interindividual variation in the composition of human fecal microflora during long-term storage of samples. J. Microbiol. Methods *59*, 263–270.

Rosselló-Mora, R., Thamdrup, B., Schäfer, H., Weller, R., and Amann, R. (1999). The response of the microbial community of marine sediments to organic carbon input under anaerobic conditions. Syst. Appl. Microbiol. *22*, 237–248.

Salonen, A., Salojärvi, J., Lahti, L., and de Vos, W.M. (2012). The adult intestinal core microbiota is determined by analysis depth and health status. Clin. Microbiol. Infect. *18* (Suppl. 4), 16–20.

Sczesnak, A., Segata, N., Qin, X., Gevers, D., Petrosino, J.F., Huttenhower, C., Littman, D.R., and Ivanov, I.I. (2011). The genome of th17 cell-inducing segmented filamentous bacteria reveals extensive auxotrophy and adaptations to the intestinal environment. Cell Host Microbe *10*, 260–272.

Seemann, T. (2014). Prokka: rapid prokaryotic genome annotation. Bioinformatics. *30*, 2068–2069.

Segata, N., Bornigen, D., Morgan, X.C., and Huttenhower, C. (2013b). PhyloPhlAn is a new method for improved phylogenetic and taxonomic placement of microbes. Nat. Commun. *4*, 2304.

Segata, N., Boernigen, D., Tickle, T.L., Morgan, X.C., Garrett, W.S., and Huttenhower, C. (2013a). Computational meta'omics for microbial community studies. Mol. Syst. Biol. *9*, 666.

Segata N., Haake, S.K., Mannon, P., Lemon, K.P., Waldron, L., Gevers, D., Huttenhower, C., and Izard, J. (2012a). Composition of the adult digestive tract bacterial microbiome based on seven mouth surfaces, tonsils, throat and stool samples. Genome Biol. *13*, R42.

Segata. N., Waldron. L., Ballarini, A., Narasimhan, V., Jousson, O., and Huttenhower, C. (2012b). Metagenomic microbial community profiling using unique clade-specific marker genes. Nat. Methods *9*, 811–814.

Sekelja, M., Berget, I., Næs, T., and Rudi, K. (2011). Unveiling an abundant core microbiota in the human adult colon by a phylogroup-independent searching approach. ISME. J. *5*, 519–531.

Sim, K., Cox, M.J., Wopereis, H., Martin, R., Knol, J., Li, M.S., Cookson, W.O., Moffatt, M.F., and Kroll, J.S. (2012). Improved detection of bifidobacteria with optimized 16S rRNA-gene based pyrosequencing. PLoS One *7(3)*, e32543.

Shapiro, H., Thaiss, C.A., Levy, M., and Elinav, E. (2014). The cross talk between microbiota and the immune system: metabolites take center stage. Curr. Opin. Immunol. *30C*, 54–62.

Staley, J.T., and Konopka, A. (1985). Measurement of *in situ* activities of nonphotosynthetic microorganisms in aquatic and terrestrial habitats. Annu. Rev. Microbiol. *39*, 321–346.

Stecher, B., Berry, D., and Loy, A. (2013). Colonization resistance and microbial ecophysiology: using gnotobiotic mouse models and single-cell technology to explore the intestinal jungle. FEMS Microbiol. Rev. *37*, 793–829.

Stanstrup, J., Gerlich, M., Dragsted, L.O., and Neumann, S. (2013). Metabolite profiling and beyond: approaches for the rapid processing and annotation of human blood serum mass spectrometry data. Anal. Bioanal. Chem. *405*, 5037–5048.

Suau, A., Bonnet, R., Sutren, M., Godon, J.J., Gibson, G.R., Collins, M.D., and Doré, J. (1999). Direct analysis of genes encoding 16S rRNA from complex communities reveals many novel molecular species within the human gut. Appl. Environ. Microbiol. *65*, 4799–4807.

Sunagawa S., Mende, D.R., Zeller, G., Izquierdo-Carrasco F., Berger, S.A., Kultima, J.R., Coelho, L.P., Arumugam, M., Tap, J., Nielsen, H.B. *et al.* (2013). Metagenomic species profiling using universal phylogenetic marker genes. Nat. Methods *10*, 1196–1199.

Swann, J.R., and Claus, S.P. (2014). Nutrimetabonomics: nutritional applications of metabolic profiling. Sci. Prog. *97*, 41–47.

Tap, J., Mondot, S., Levenez, F., Pelletier, E., Caron, C., Furet, J.P., Ugarte, E., Muñoz-Tamayo, R., Paslier, D.L., Nalin, R., Dore, J., and Leclerc, M. (2009). Towards the human intestinal microbiota phylogenetic core. Environ. Microbiol. *11*, 2574–2584.

The Human Microbiome Project Consortium. (2012). Structure, function and diversity of the healthy human microbiome. Nature *486*, 207–214.

The Human Microbiome Consortium (2012). A framework for human microbiome research. Nature *486*, 215–221.

Tuohy, K., and Del Rio, D. (2014). Diet:Microbe Interactions in the Gut – Effects on Human Health and Disease (Elsevier Academic Press, USA).

Tuohy K.M., Gougoulias, C., Qing Shen, Walton, G., Fava, F., and Priya Ramnani, P. (2009). Studying the human gut microbiota in the trans-Omics era – focus on metagenomics and metabonomics. Curr. Pharmaceut. Design *15*, 1415–1427.

Tuohy, K.M., and McCartney, A.L. (2006). Molecular microbial ecology of the human gut. In: Gibson, G.R., and Rastall, R.A. (eds), Prebiotics: Development and Application. Chichester: John Wiley & Sons Ltd., 135–155.

Turnbaugh, P.J., Quince, C., Faith, J.J., McHardy, A.C., Yatsunenko, T., Niazi, F., Affourtit, J., Egholm, M., Henrissat, B., Knight, R., and Gordon, J.I. (2010). Organismal, genetic, and transcriptional variation in the deeply sequenced gut microbiomes of identical twins. Proc. Natl. Acad. Sci. U.S.A. *107*, 7503–7508.

Veilleux, B.G., and Rowland, I.R. (1981). Simulation of the rat intestinal ecosystem using a two-stage continuous culture system. J. Gen. Microbiol. *123*, 103.

Ventura, M., Turroni, F., Canchaya, C., Vaughan, E.E., O'Toole. P.W., and van Sinderen, D. (2009). Microbial diversity in the human intestine and novel insights from metagenomics. Fron.t Biosci. (Landmark Ed.) *14*, 3214–3221.

Walker, A.W., Duncan, S.H., McWilliam Leitch, E.C., Child, M.W., and Flint, H.J. (2005). pH and peptide supply can radically alter bacterial populations and short-chain fatty acid ratios within microbial communities from the human colon. Appl. Environ. Microbiol. *71*, 3692–3700.

Walker, A.W., Ince, J., Duncan, S.H., Webster, L.M., Holtrop, G., Ze, X., Brown, D., Stares, M.D., Scott, P., Bergerat, A., Louis, P., McIntosh, F., Johnstone, A.M., Lobley, G.E., Parkhill, J., and Flint, H.J. (2011). Dominant and diet-responsive groups of bacteria within the human colonic microbiota. ISME J. *5*, 220–230.

Wood, D.E., and Salzberg, S.L. (2014). Kraken: ultrafast metagenomic sequence classification using exact alignments. Genome biology 15, R46.

Xie, G., Zhang, S., Zheng, X., and Jia, W. (2013). Metabolomics approaches for characterizing metabolic interactions between host and its commensal microbes. Electrophoresis *34*, 2787–2798.

Yap, I.K., Li, J.V., Saric,J., Martin, F.P., Davies, H., Wang, Y., Wilson, I.D., Nicholson, J.K., Utzinger, J., Marchesi, J.R., and Holmes, E. (2008). Metabonomic and microbiological analysis of the dynamic effect of vancomycin-induced gut microbiota modification in the mouse. J. Proteome Res. *7*, 3718–3728.

Zoetendal, E.G., Akkermans, A.D., De Vos, W.M. (1998). Temperature gradient gel electrophoresis analysis of 16S rRNA from human faecal samples reveals stable and host-specific communities of active bacteria. Appl. Environ. Microbiol. *64*, 3854–3859.

Zoetendal, E., Akkermans, A., Akkermans-Van Vleit, W., De Visser, J., and De Vos, W. (2001). The host genotype affects the bacterial community in the human gastrointestinal tract. Microbial Ecol. Health Dis. *13*, 129–134.

Metagenomics of the Gut Microbiota as a Tool for Discovery of New Probiotics and Prebiotics

17

Hyun Ju You, Jiyeon Si and GwangPyo Ko

Abstract

The human gut comprises diverse microbiota which put substantial impact on various human diseases. The gut microbiota is linked to the maintaining gastrointestinal homeostasis and the association between metabolic and inflammatory diseases such as obesity, type 2 diabetes, low-grade chronic inflammation, and inflammatory bowel disease. Consequentially, the growing interest in gut microbiota has resulted in the development of functional foods such as probiotics and prebiotics, which have been reduce the risk of disease and improve host health. Recent advances in nucleotide sequencing technology widen the profile of the previously unknown microbial communities. International collaborative projects such as the MetaHIT initiative and the Human Microbiome Project (HMP) revealed the diversity of the gut microbiota and their metabolic functions and interactions in the host using this new technique. Current advancement of these tools eases analysis of the massive datasets, yet challenges remain for a complete construction of the microbiome. Here we discuss the development of metagenomics along with the challenges and future perspectives for metagenomic studies to find new probiotics and prebiotics.

The gut microbiota as a hidden controller of human health

Since the Ukrainian microbiologist and immunologist, Elie Metchnikoff, suggested a scientific rationale for the use of probiotics in modifying and normalizing the intestinal microbiota (Mechnikov, 1908), significant advances have been made in the selection and application of probiotic strains that benefit health. In the past decade, Metchnikoff's hypothesis regarding replacing harmful microbes with useful bacteria restores the imbalance in gut ecology and improves host health has been brought to life, together with accumulating studies on the human microbiome, most of which utilize metagenomic analysis.

The human body and its symbiotic microbiota generate a 'superorganism', which has co-evolved to establish an equilibrium that facilitates optimum nutrient processing, protection from pathogens, detoxification of xenobiotics, and regulation of metabolism and homeostasis (Eberl, 2010). These mutualistic symbionts inhabit defined niches on nutrient-rich body surfaces, including the entire gastrointestinal tract, vagina, skin, oral and nasal cavity, pharynx, oesophagus, and urogenital tract. Among the 'normal' human microbiota, the gut microbiota is referred to as a forgotten organ, having a distinct community structure that contributes to important biological and metabolic functions (O'Hara and Shanahan, 2006). The total number of microbes colonizing the gut is ~10^{14}, which is 10-fold greater than the number of human cells in the body (Sekirov *et al.*, 2010). In addition, the collective genomes of the gut microbes (*i.e.* the gut microbiome) encode 3 million genes, exceeding the human genome by more than two orders of magnitude (Human Microbiome Project, 2012a, b). These gene products provide a wide range of biochemical and metabolic functions that complement host development and physiology (Sommer and Backhed, 2013).

The composition of the gut microbiota is unique for each individual (Lozupone *et al.*, 2012). Host physiology affects the bacterial communities by providing variable conditions in each ecological niche. Factors such as age, gender, genotype and health status impact the colonization of individual hosts (Roth and James, 1988). Although its precise structure is not known due to its diversity and dynamic nature, the intestinal microbiota consists of ~500–1000 species of several bacterial phyla (Sommer and Backhed, 2013). The most abundant phyla in the human gut are Firmicutes (35–80%) and Bacteroidetes (17–60%), but other bacterial phylotypes belong to the phyla Proteobacteria, Verrumicrobia, Actinobacteria, Fusobacteria and Cyanobacteria (Human Microbiome Project, 2012b; Qin *et al.*, 2010) are also present. At the genus level, the major constituents of the human gut microbiota are obligate anaerobes of the genera *Bacteroides, Eubacterium, Clostridium, Ruminococcus, Peptococcus, Peptostreptococcus, Bifidobacterium* and *Fusobacterium*, as well as less-predominant facultative anaerobes, such as *Escherichia, Enterobacter, Enterococcus, Klebsiella, Lactobacillus* and *Proteus* (Shen *et al.*, 2013). In 2012, Lozupone *et al.*, reported that there was no ubiquitous 'core' microbiome at the species level (Lozupone *et al.*, 2012), as previously suggested core species represent less than 1% of the total microbial phylotypes in some individuals. The species composition of the gut microbial community shows a marked variability; however, the profiles of functional genes are quite similar across healthy individuals (Human Microbiome Project, 2012b; Qin *et al.*, 2010). This functional redundancy reflects the evolutionary convergence of unrelated taxa, as well as the link between community structure and function.

The concept of 'enterotypes' could be useful for understanding of the structure–function relationships of the gut microbiota. Shotgun metagenomes of stool samples from 39 healthy individuals were analysed for both phylogenetic and functional signals (Burcelin *et al.*, 2013). The data indicated that microbial communities were segregated into three distinct clusters dominated by *Bacteroides*, *Prevotella*, or *Ruminococcus* and/or other Firmicutes genera, which were referred to as 'enterotypes' by the authors. Each enterotype was based on co-occurring associations of specific genera and species that regulate the composition and function of other secondary species within the host gut. Although other studies have not reported the same pattern (Lozupone *et al.*, 2012; Wu *et al.*, 2011; Yatsunenko *et al.*, 2012), at least in part, the long-term diet pattern affects the enterotype composition. For instance, the *Bacteroides* enterotype was associated with higher dietary consumption of protein and saturated fats, whereas the *Prevotella* enterotype has been associated with high ingestion of carbohydrates and simple sugars, but low intake of protein (Wu *et al.*, 2011).

The gut is a dynamic habitat that undergoes constant and rapid changes in microbial composition due to variations in host diet, lifestyle, hygiene, or use of antibiotics. Unlike the host genome, the microbiome can fluctuate rapidly as a result of modifications in either community structure or individual microbial genomes in accordance with these diverse changes in extrinsic/intrinsic factors. As a crucial environmental factor, dietary change affects the homeostasis of the gut microbiota to a great extent. During co-evolution with humans over millennia, the gut microbiota adapted to the hunter–gatherer diet (Ley *et al.*, 2008). This type of diet was composed of mainly non-digestible carbohydrates, fibre, and plant polysaccharides, digestion of which was assisted by the tiny inhabitants in the gut. It allowed the host to obtain additional energy, which would otherwise be lost. For the last 50 years, the human gut microbiota has been faced a new challenge: adapting to Western-style diets. The high fat and high sugar diet has gradually impaired the fermentative functions of the gut bacteria. The end-products of bacterial fermentation, short-chain fatty acids (SCFA), play an important role in energy metabolism, lipid digestion, and modulation of the inflammatory response (Cook and Sellin, 1998). Reduced fermentation disrupts gut homeostasis and may cause chronic diseases, including obesity, inflammatory bowel disease, cardiovascular disease, irritable bowel syndrome, and cancer (Bomba *et al.*, 2002; Jenkins *et al.*, 1999; Roediger, 1980). The growing interest in colonic health has resulted in the development of functional foods such as probiotics and prebiotics, which have been reduce the risk of disease and improve host health.

At present, molecular profiling of microbes that inhabit the gut is used to demonstrate the changes in the community due to either genetic or environmental factors. Traditionally, microbial functions have been investigated on culture plates under artificial laboratory conditions. Of the gut microbiota, approximately 70–80% cannot be grown in pure culture, limiting knowledge to readily cultivable bacteria (Eckburg *et al.*, 2005; Suau *et al.*, 1999). Recent advances in molecular techniques enabled the determination of microbial composition using the 16S rRNA gene. Additionally, further development of genomic approaches has opened up a new era of functional studies of the complex society within the gut. However, further research is required to reveal the complex host–microbe interactions and their contribution to maintaining host physiology. Another demanding task is to identify the mechanisms underlying the health-promoting effects of specific prebiotics in terms of modification of the gut microbiota. In this chapter, we describe the analysis pipeline used in studies of the human microbiome and review current knowledge and challenges of the metagenomic approach to prebiotics research.

The preventative role of prebiotics in metabolic diseases

Unlike the stomach, which contains its gastric acid, the intestines provide neutral conditions, harbouring a prosperous microbial community (Wang *et al.*, 2005). It is covered by the intestinal mucosa, a major location for host–microbe interactions (O'Hara and Shanahan, 2006). To date, many researchers have investigated the role of the gut microbiota in maintaining gastrointestinal homeostasis in the mucosal layer and the association between disturbances in the gut microbial community and various diseases (Backhed *et al.*, 2004; Cani *et al.*, 2007; Cani and Delzenne, 2011; Greenblum *et al.*, 2012; Ley *et al.*, 2005, 2006; Ridaura *et al.*, 2013; Schele *et al.*, 2013). For the last decade, such studies have focused on the development of metabolic and inflammatory diseases such as obesity, type 2 diabetes, low-grade chronic inflammation, and inflammatory bowel disease.

Obesity is caused by an imbalance between energy intake and expenditure and is associated with various metabolic disturbances, including glucose intolerance, type 2 diabetes, hypertension, dyslipidaemia, fibrinolysis disorders, cardiovascular diseases, and non-alcoholic fatty liver disease (NAFLD). The gut microbiota can affect host signalling systems related to hunger/satiety (Schele *et al.*, 2013) and the intestinal transit time of food (Wichmann *et al.*, 2013). In 2005, Gordon and colleagues reported the first evidence that lean and obese mice have different gut microbial communities (Ley *et al.*, 2005). The gut microbiota of leptin-deficient mice, which typically weigh three times as much as normal mice, contained ~50% fewer *Bacteroidetes* and 50% more *Firmicutes* as compared with normal mice. Because the same pattern was identified in humans (Ley *et al.*, 2006), these studies represent the first report of an association between the gut microbiota and obesity.

Subsequent research using germ-free mice provided mechanistic evidence that the gut microbiota could modulate host fat metabolism and energy homeostasis. Despite eating less food before the transfer, germ-free mice colonized with gut microbes from conventional mice showed a 60% increase in body fat over 2 weeks (Backhed *et al.*, 2004). Moreover, the germ-free mouse colonized by the gut microbiota of an obese mouse showed twofold body fat gain compared to mice colonized with the gut microbiota from a normal-weight mouse. Moreover, transfer of an obese microbiome from human donors to germ-free mice resulted in greater weight-gain, compared with transfer of the microbiome of a lean person (Ridaura *et al.*, 2013). However, these findings are hardly proof of 'transplantable obesity'. The germ-free mice given an obesity-associated microbiota did not become obese themselves (DeWeerdt, 2014). This suggests that the combined effect of host genome, microbiome, and environmental factors could lead to the end-point pathological status of obesity. To identify the host factors controlling the gut microbial community, Benson *et al.*, examined faecal samples from 645 intercrossed mice and analysed the influence of host genotype on variations in microbial composition (Benson *et al.*, 2010). Quantitative trait loci (QTL) analysis, using 530 single nucleotide polymorphisms, showed that 18 QTL had significant linkage to the relative abundances of specific microbial taxa. Host genetics could explain a significant proportion of the variation in microbial composition. Based on their results, the authors suggested that the microbial composition of the gut could be understood as a complex, polygenic trait influenced by combinations of host genomic loci and environmental factors.

Although host genetics is a key driver of the gut microbial community structure, a variety of exogenous factors has a marked influence on the microbial composition and associated diseases. In obesity, diet is a major factor in shaping the microbial composition and the associated pathological process. Dietary intervention causes marked changes in the levels of gut microbes. The gut microbiota can metabolize ingested dietary material as well as products of host metabolism; in turn, microbial metabolites can be further converted by the host. Indeed, co-metabolism of the host and symbionts makes difficult demonstration of the causal effect of specific microbial taxa in terms of the development of obesity.

Prebiotics, defined as indigestible carbohydrate such as oligofructose and inulin, are one of the strategies used to manipulate the gut microbiota. Recently, Zhao *et al.* (2014) conducted a clinical trial of a dietary regimen including whole grains, traditional Chinese medicinal foods, and prebiotics in severely obese subjects. After 9 weeks, the 93 participants achieved an average weight loss of ~6 kg, in addition to improvements in markers of metabolic health such as insulin sensitivity, lipid profiles, and blood pressure. A diet enriched with specific non-digestible carbohydrates has been reported to result in greater intestinal SCFA production and thereby alleviation of body weight gain, fat mass development, and obesity and diabetes severity (Cani and Delzenne, 2011).

Obesity and its related metabolic disorders are closely associated with systemic low-grade inflammation. Changes in the gut microbiota induced by a high-fat diet and obesity are implicated in impaired gut barrier function and obesity-related low-grade inflammation, due to mechanisms involving bacteria-derived lipopolysaccharide (LPS) (Cani *et al.*, 2007). Cani and colleagues demonstrated that prebiotics alleviated the inflammation and insulin resistance associated with obesity. This was associated with improved gut barrier function due to increased levels of gut hormones, such as glucagon-like peptide 1 and 2 (GLP-1 and GLP-2); the endocannabinoid system was also involved (Cani *et al.*, 2006; Cani *et al.*, 2009; Muccioli *et al.*, 2010). Genetic obese-, diet-induced obese-, and diabetic mice showed restoration of glucose and lipid metabolism as well as gut barrier function following oral administration of oligofructose for 8 weeks. The oligofructose-induced changes in the gut microbiota were associated with increased enteroendocrine L-cell number, and improved leptin sensitivity and glucose homeostasis (Everard *et al.*, 2011).

Structural analysis of the gut microbial community using 16S rRNA gene pyrosequencing revealed that prebiotic treatment affected markedly the levels of several bacterial taxa from the phylum to the phylotype level. The gut microbiome could thus represent a novel target that affect host metabolism in obesity and type 2 diabetes. In 2013, Dewulf *et al.*, reported that obese females who took a prebiotic supplement every day for 3 months showed a slight decrease in fat mass and a reduction in blood levels of pro-inflammatory markers (Dewulf *et al.*, 2013). However, additional studies are required to elucidate the precise mechanisms underlying the roles of prebiotics in metabolic diseases in terms of effects on the gut microbiota. Identification of bacteria at the species level is necessary to demonstrate causality in humans. It is well known that prebiotic interventions with inulin and galactooligosaccharides promote the growth of the genus *Bifidobacterium*, whose presence is generally associated with beneficial health effects. In addition, inulin-type fructans have been shown to increase levels of *Faecalibacterium prausnitzii*, which modulates inflammation and diabetes in obese individuals (Ramirez-Farias *et al.*, 2009).

Feeding of genetic- or diet-induced obese mice with prebiotics increased the abundance of *Akkermansia muciniphila* by ~100-fold in the gut; this was correlated with an improved metabolic status including reduced fat-mass gain, metabolic endotoxaemia, adipose tissue inflammation, and insulin resistance (Everard *et al.*, 2011, 2013). This effect could be caused by mechanisms associated with the restoration of intestinal mucus production and improvement of gut barrier function. Oral supplementation of *A. muciniphila* counteracted the decrease of the antimicrobial peptide regenerating islet-derived 3-gamma (Reg3γ) in a high-fat-diet-induced obesity model, e.g. Reg3γ could reinforce the mucus barrier and associated immunological function (Everard *et al.*, 2014).

The development of metagenomics from genomics

Based on great discoveries of the 19th century, the investigation of microbes and their functions was conducted using pure culture methods. However, it has long been known that only 1% of all microbes can be cultivated in pure culture. Theoretically, there are reportedly 160 taxa/ml in the ocean, 6400–38,000 taxa/g in soil, and 70 taxa/ml in acid mine drainage. In soil, less than 1000 taxa are thought to be potentially culturable (Curtis *et al.*, 2002). In the case of the human gut, roughly 80% of the phylotypes represent uncultured bacteria (Margulies *et al.*, 2005). The term 'the great plate-count anomaly' was coined to explain this problem (Staley and Konopka, 1985). The general method of microbial cultivation is to use artificial media composed of commercially available compounds, however, this methods cannot take account of microbes interact with other microbes to survive. An alternative method of culturing bacteria is the diffusion chamber method, which imitates the environmental niche (Pavelic *et al.*, 1998). When the appropriate medium was provided, this method was reported to increase the recovery of isolates, including rarely cultivated taxa, from pond water (Bollmann *et al.*, 2007). Recently, *in vitro* gut model systems, such as the Simulator of the Human Intestinal Microbial Ecosystem (SHIME) or the TNO *in vitro* model of the colon (TIM-2), have been developed to maintain the community structure of faecal samples and dynamically monitor microbial processes at the site of metabolic activity. Since 1999, the effects of various probiotic strains (Decroos *et al.*, 2006; Gmeiner *et al.*, 2000; Sivieri *et al.*, 2013) and prebiotic materials such as whey retentate, starch, lactulose, arabinogalactan, and fructo-oligosaccharides and galacto-oligosaccharides (Decroos *et al.*, 2006; Gmeiner *et al.*, 2000; Meddah *et al.*, 2001; Sivieri *et al.*, 2014; Terpend *et al.*, 2013; Kovatcheva-Datchary *et al.*, 2009; Maathuis *et al.*, 2012; Rose *et al.*, 2010; Venema and van den Abbeele, 2013; Venema *et al.*, 2003), on changes in microbial communities have been investigated using an *in vitro* gut ecology simulation systems.

Despite much effort, studies of the human microbiome still lack means to reproduce accurately the *in vivo* situation. Thus, it is nearly impossible to study microbial composition and functions without genetically analysing their DNA using molecular methods. The term 'metagenomics' was coined in 1998 and has been applied to investigation of the natural world of microbes (Handelsman *et al.*, 1998). This work began with cloning extrinsic DNA derived from faecal samples into a bacterial plasmids (Schmidt *et al.*, 1991). Before this, the Human Genome Project (HGP) spurred the development of sequencing technologies and bioinformatic tools (Lander *et al.*, 2001). The initial step in this large-scale sequencing project was the discovery of DNA as the hereditary material and development of genetic recombination (Collins *et al.*, 2003). The project produced an unprecedented quantity of complex genomic data. Deciphering of the genetic code provided information regarding heritable diseases and links between the host genotype and associated diseases (Desiere, 2004).

The HGP led to the opening of the 'post-genomics era' and other international collaborative projects, such as the MetaHIT initiative and the Human Microbiome Project (HMP), which aimed to characterize the microbial communities of human body sites and understand the interactions between the microbiota and the host. The Human Microbiome Project was feasible for two reasons. First, the technological advances in sequencing methods facilitated determination of microbial functions in a culture-independent manner. Second, the recognition of the bacterial 16S rRNA gene was critical because this ubiquitous gene enabled evaluation of the phylogeny of communities and provided the molecular tools necessary to classify all members to the appropriate taxonomic level (from kingdom to species).

Next-generation sequencing (NGS)

Despite the concern regarding the tremendous amount of microbes residing in the human body, information about these inhabitants was limited for many years by the insufficiency of relevant technology and reference genomes (Scholz *et al.*, 2012a). The recent advent of next-generation sequencing (NGS) yielded high-throughput sequences at reduced cost and time. The cost of sequencing DNA nucleotides decreased from $10,000 to $0.10 US dollars per Mb between 2001 and 2011 as a result of NGS techniques (Sboner *et al.*, 2011); thus this innovative technology provided the groundwork for metagenomics. By directly isolating bacterial DNA from a sample, this approach removed the bias from cultivability and so expanded the bacterial phylogeny (Tuohy *et al.*, 2009). This advance was enabled by 16S rRNA gene-based methods, which identify all bacterial taxa regardless of their variability, enabling analysis of species richness and relative abundance within a community (Tuohy *et al.*, 2009). The first investigation of communities using 16S rRNA gene-based methods was performed on 286 environmental samples; bacterial DNA was amplified using barcoded primers so that the massive number of samples could be run in a single sequence run (Hamady *et al.*, 2008).

Prior to the advent of NGS technologies, the identification of bacterial phylogeny relied on Sanger sequencing, phylochip, FISH (fluorescence *in situ* hybridization), T-RFLP (terminal restriction fragment length polymorphism), or DGGE (denaturing gradient gel electrophoresis) methods (Nocker *et al.*, 2007). In the case of bacterial community 'fingerprinting' techniques, the amplified 16S rRNA genes (amplicons) are analysed by electrophoresis or hybridization. Cloning and sequencing with an automated capillary sequencer produces long reads (~800 bp) of the 16S rRNA gene, thereby identifying bacteria with high phylogenetic resolution (Young and Schmidt, 2004). Until 2006, Sanger sequencing of cloned sample DNA was the major approach used for 16S-based community analysis and early phase metagenomics studies (Temperton and Giovannoni, 2012). Although these methods can be used to determine the bacterial members and community shift, the depth of the analytic methods is rather shallow. For a successful metagenomic analysis, the minimum amount of unamplified starting material is ~1 ng, which corresponds to roughly 100 human diploid genomes (Linnarsson, 2010). Therefore, the development of new NGS technologies or refinement of current instruments is required to improve throughput and cost efficiency (Pushkarev *et al.*, 2009).

The leading platforms in the field of NGS include the Roche 454 FLX Titanium (Margulies *et al.*, 2005) and the Illumina Genome Analyser (GA) II (Ng and Kirkness, 2010), which has been upgraded recently to the HiSeq platform. The principle of NGS involves a clonal amplification to directly generate clusters of DNA templates and transformation of the signal produced when synthesizing DNA. The Roche 454 FLX Titanium is based on a pyrosequencing method (Mardis, 2008). In pyrosequencing, the incorporation of a nucleotide releases pyrophosphate, which subsequently undergoes conversion by luciferase to generate ATP and light. The intensity of the light emitted as each nucleotide is incorporated is recorded. The Illumina platform utilizes the same approach (sequencing-by-synthesis) but detects the four nucleotides as four different colours. Each nucleotide is labelled with base-specific fluorescence, thereby producing distinct signals. The incorporation of nucleotides is a unique reaction as the 3′-OH group is chemically blocked. After an imaging step, the fluorescence and the blocking group are chemically washed out to prepare for the next incorporation.

Compared to the Illumina system, Roche 454 pyrosequencing generates two- to three-fold longer read lengths, but with lower throughput. While the Illumina affords deep-coverage sequencing data, assembling the short reads is problematic. These technologies have different pros and cons, and the general characteristics and comparative analyses of these platforms have been reported (Bragg and Tyson, 2014; Cho and Blaser, 2012; Kim *et al.*, 2013; Linnarsson, 2010; Scholz *et al.*, 2012b; Shi *et al.*, 2010; see also Table 17.1).

Information limit for rich data

The continued reduction in sequencing costs lowered the barrier of entry to metagenomics, resulting in the accumulation of datasets (Desai *et al.*, 2012). The massive datasets generated by metagenomics promoted public accessibility for investigation (Temperton and Giovannoni, 2012). The analysis of metagenomic datasets requires biological understanding and more importantly, computational skills to develop up-to-date bioinformatic tools (Scholz *et al.*, 2012a). Additionally, the large datasets require ever-larger space to be stored. Sequencing has become easier to access, but dealing with its products remains a challenge. This has induced data sharing through open public databases, which accelerated the development of relevant bioinformatic tools (Table 17.2).

The ultimate goal of microbiome analysis is to reconstruct, *all the genomes*, and decipher their functional roles in the human body using a metagenomic approach. However, this is not feasible owing to the limitation of NGS technologies (*e.g.* short read-based analysis) and the computational complexity of alignment, assembly, and annotation of the sequence reads. Although many noble algorithms have been developed, the use of NGS platforms necessitates securing at least several hundred gigabytes of data storage per sample, and several problems with short read assembly still remain. The algorithms require a crippling amount of memory to reconstruct the genomes found within a metagenome, and low-abundance genomes in a sample are sometimes not assembled (Scholz *et al.*, 2012b). To achieve functional and taxonomic classification of NGS data, read-based computational algorithms must overcome long analysis time (due to the number of reads) and high error rates (due to the short read length).

There are two general types of read-based metagenome analysis methods: amplicon sequencing (16S rRNA gene analysis or

Table 17.1 NGS platform and expected throughputs (adapted from Scholz *et al.*, 2012b)

Platform	Run time	Read length (bp)	Throughput per run
Roche			
454 GS junior	10 hours	400	35 Mb
454 GS junior+	18 hours	700	70 Mb
GS FLX Titanium	10 hours	600	450 Mb
GS FLX Titanium XL+	23 hours	1000	700 Mb
Illumina			
MiSeq	5–65 hours	2×300	14.9 Gb
NextSeq 500	12–30 hours	2×150	120 Gb
HiSeq 2500	<1–6 days	2×125	1002 Gb
HiSeq X	<3 days	2×150	1800 Gb
Life Technologies			
SOLiD 5500xl	6 days	2×60	95 Gb
SOLiD 5500xl Wildfire	10 days	2×50	240 Gb
SOLiD 5500	6 days	2×60	48 Gb
SOLiD 5500 Wildfire	10 days	2×50	120 Gb
Ion torrent			
PGM 314	2–4 hours	400	100 Mb
PGM 316	3–5 hours	400	1 Gb
PGM 318	4–7 hours	400	2 Gb
PI	2–4 hours	200	10 Gb
Pacific Bioscience			
RS II: P4-C2	3 h	5500	275 Mb
RS II: P5-C3	3 h	8500	375 Mb

Table 17.2 Sequence and metadata browsers for microbiome analysis

Resource	Type	Features	Website	Reference
HMPDACC	DB	Central repository for all HMP data	http://www.hmpdacc.org/	Human Microbiome Project (2012a,b)
MetaHIT	Consortium	Metagenomics of the human intestinal tract	http://www.metahit.eu/	Human Microbiome Project (2012a), Le Chatelier *et al.* (2013), Qin *et al.* (2010)
IHMC	Consortium	International Human Microbiome Consortium	http://www.human-microbiome.org/	
ENA	DB	EMBL-EBI, European Nucleotide Archive	http://www.ebi.ac.uk/ena/	
NCBI SRA	DB	NCBI, next-generation Sequence Read Archive	http://www.ncbi.nlm.nih.gov/sra/	
DDBJ SRA	DB	DDBJ, next-generation Sequence Read Archive	http://www.ddbj.nig.ac.jp	
INSDC	Initiative	International Nucleotide Sequence Database Collaboration (ENA, NCBI, and DDBJ)	http://www.insdc.org/	
Trace Archive	DB	Permanent repository of DNA sequence chromatograms (traces), base calls, and quality estimates for single-pass reads from various large-scale sequencing projects	NCBI, http://www.ncbi.nlm.nih.gov/Traces/trace.cgi DDBJ, http://trace.ddbj.nig.ac.jp/dta/	
GenBank	DB	Annotated sequences database, NCBI	http://www.ncbi.nlm.nih.gov/genbank/	
BioSample	DB	Submitter-supplied descriptive information and metadata	http://www.ncbi.nlm.nih.gov/biosample http://trace.ddbj.nig.ac.jp/biosample/	
BioProject	DB	Collection of biological data, provide users a single place to find links to the diverse data types generated for the project.	http://www.ncbi.nlm.nih.gov/bioproject http://trace.ddbj.nig.ac.jp/bioproject/	
VMC	Consortium	Vaginal Microbiome Consortium	http://vmc.vcu.edu/resources	Fettweis *et al.* (2012)
HOMD	DB	Human Oral Microbiome Database	http://www.homd.org/	Chen *et al.* (2010)

gene-targeted metagenomics, community profiling) and whole-sample sequencing (whole-genome shotgun metagenomics, functional profiling). Amplicon sequencing involves operational taxonomic unit (OTU)-based or homology-based approaches, and shotgun metagenomics requires additional analytic processes, including binning, assembly, and annotation (Fig. 17.1).

Structure and function of the healthy human microbiome (learning from HMP data)

In 2008, the NIH launched a $157 million, 5-year human microbiome project, which investigated a total of 4,788 specimens representing body habitats, such as the gut, vagina, skin, mouth and respiratory (nasal) tract, from 242 healthy adults (129 males, 113 females) (Zarco *et al.*, 2012). The body habitats analysed by the HMP provide distinct niches for host–microbe interactions (Human Microbiome Project, 2012a). Phylogenetic analysis of community 'structure' based on 16SrRNA gene analysis revealed the primary patterns at each body site. The diversity and abundance of each habitat's signature microbes varied widely even among healthy subjects, and indicated strong niche specialization (Human Microbiome Project, 2012b). Oral and stool communities showed especially diverse community composition, whereas vaginal sites harboured relatively simple communities (Fig. 17.2). Saliva had among the highest median alpha diversities of OTUs, but one of the lowest beta diversities, indicating that the populations are similar. In contrast, the skin showed the highest beta diversity but was intermediate in alpha diversity. The vagina had the lowest alpha diversity, with low beta diversity at the genus level but high among OTUs due to the presence of distinct *Lactobacillus* spp.

Regarding functional analysis of the microbial community, the metagenomic carriage of metabolic pathways was stable among individuals despite variation in community structure and ethnic/racial background (Human Microbiome Project, 2012b). Indeed, no taxa were universally present at all body sites and in all individuals. However, unlike microbial taxa, several pathways were ubiquitous among individuals and body sites. The most abundant of these 'core' pathways include the ribosome and translational machinery, nucleotide charging, ATP synthesis, and glycolysis, as expected. The relative abundances of pathways were more constant and uniformly diverse than those of organisms. This study used both marker gene and metagenomic data to assess the ecology of microbial metabolic and functional pathways in these communities. The data can be downloaded from the HMP Data Analysis and Coordination Center browser (Table 17.2).

16S rRNA-based community-level phylogenetic analysis

As eluded to above, DNA-based microbiome studies can be divided into two large categories. Targeted amplicon studies select marker genes that are universal to the entire community (Kuczynski *et al.*, 2012) and are used to investigate microbial composition and diversity within the community (structure). Whole-genome studies are employed to reveal overall microbial functions, but exclude information about the individual microbial taxa (function). Microbial community studies are based mainly on the ribosomal small subunit gene, 16S rRNA (Kuczynski *et al.*,

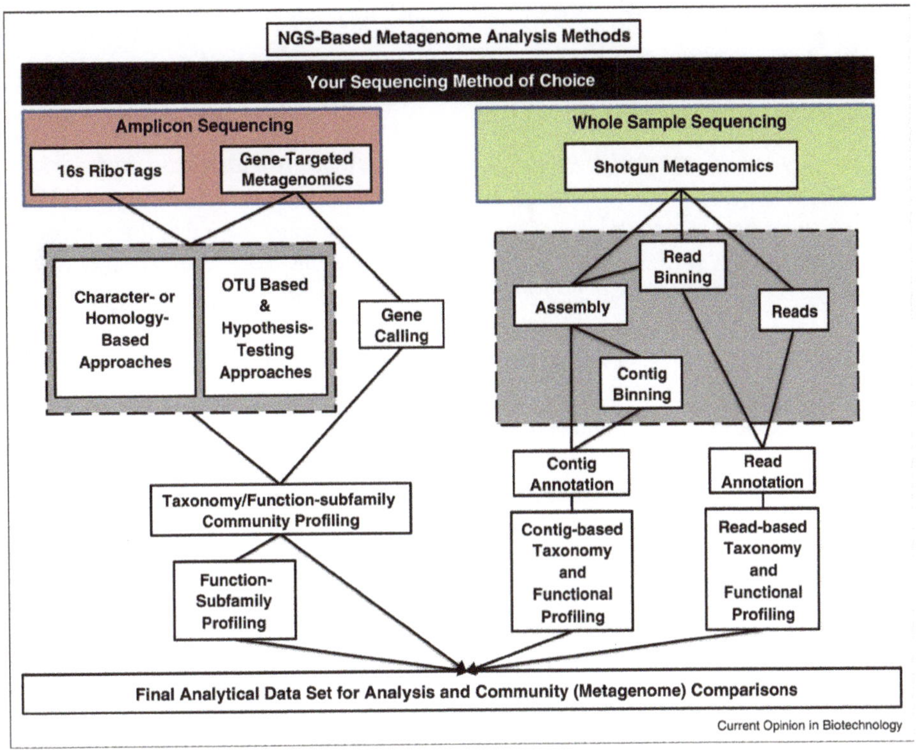

Figure 17.1 Analysis of metagenomic data from amplicon sequencing or whole-sample shotgun metagenome sequencing (Scholz *et al.*, 2012b).

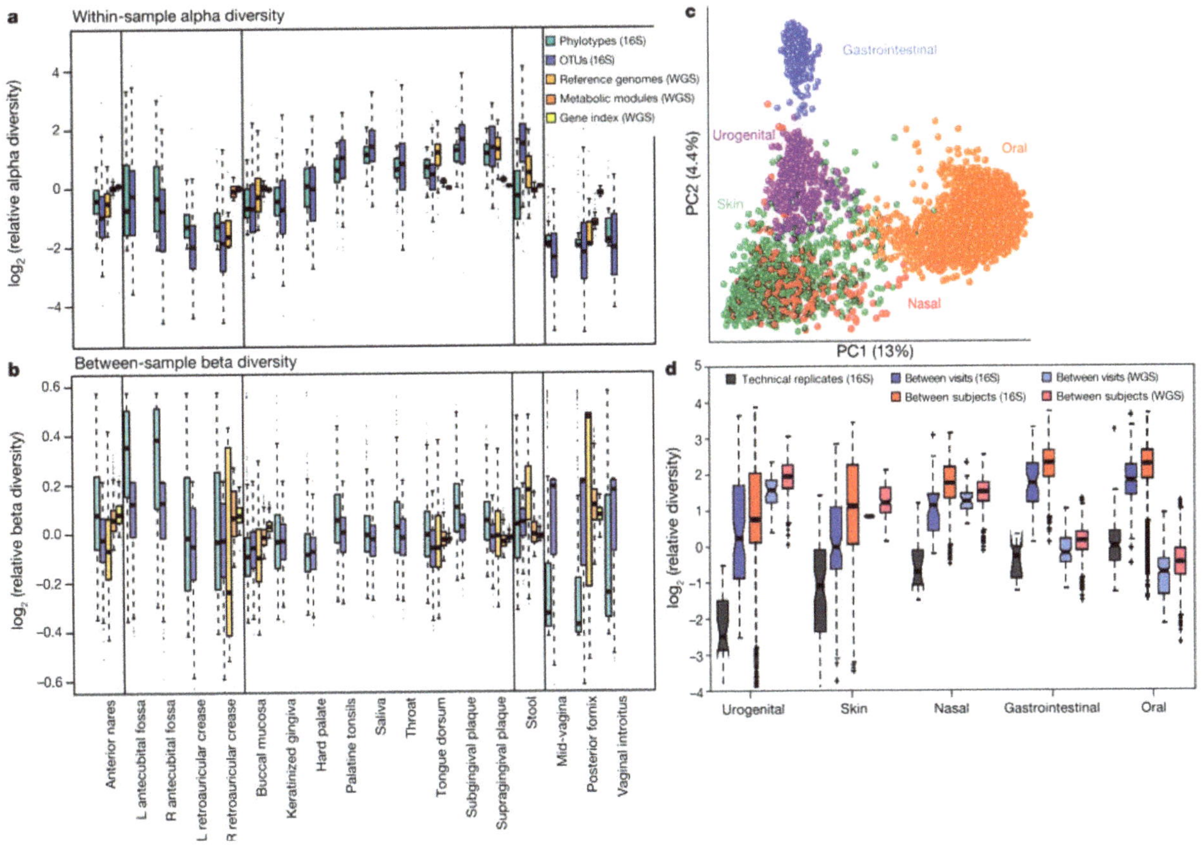

Figure 17.2 Diversity of the human microbiome (from HMP data (Human Microbiome Project, 2012b)).

2012). This conserved gene contains short hypervariable regions. One advantage of the ribosomal subunit is that it evolves slowly and thus, found in all bacteria. Whereas other marker genes can identify only a limited numbers of taxa, 16S rRNA allows bacterial taxa to be distinguished by means of its hypervariable regions. Another advantage of the ribosomal gene is that databases of reference sequences and taxonomies have been established, e.g. greengenes (DeSantis *et al.*, 2006), SILVA (Pruesse *et al.*, 2007), and the Ribosomal Database Project (RDP) (Cole *et al.*, 2009). DNA sequences are clustered into groups based on sequence similarity and compared with the reference sequences in the database. These sequence clusters are defined as OTUs (Wooley *et al.*, 2010). An OTU is a proxy for taxonomic classification and its size can be changed by altering the sequence similarity threshold. Taxonomic assignment proceeds according to the defined OTUs using the reference sequences. However, the analysis of a community based on the 16S rRNA gene has limitation with regard to the identification of community members and their abundance due to the biological techniques. Although it covers a broad spectrum of microbes, species richness changes depending on the variable (V) region amplified by the specific primer; *i.e.* V6 and a combination of V1 and V2 overestimate richness, whereas V3, V7, and a combination of V7 and V8 underestimate richness (Youssef *et al.*, 2009). Moreover, the short lengths of the hypervariable regions reduce the resolution for defining the end branch of the phylogenetic tree.

Methods for analysing 16S rRNA gene sequencing data have been developed for many years and various have been reviewed elsewhere (Caporaso *et al.*, 2010; Degnan and Ochman, 2012; Hamady and Knight, 2009; Jumpstart Consortium Human Microbiome Project Data Generation Working, 2012; Kozich *et al.*, 2013; Kuczynski *et al.*, 2011; Navas-Molina *et al.*, 2013). Quality checks and data-cleaning steps along the analysis flow are very important to ensure the validity of the results. QIIME and Mothur are popular analysis tools for 16S rRNA gene-based microbial sequencing data, although they contain many necessary pre- and post-analysis steps. They can trim, denoise, screen, align, and analyse sequences, as well as generate visual displays of the results. For evaluation of community structure using 16S-based NGS analysis, several factors can represent changes in microbial composition both between and within samples diversity. For example, UniFrac is a β-diversity measure that uses phylogenetic information to compare environmental samples (Lozupone *et al.*, 2006). UniFrac can be coupled with standard multivariate statistical techniques, including principal coordinates analysis (PCoA, see also Fig 17.2), and then be used to identify factors that account for the differences among microbial communities. UniFrac measures the difference between two metagenomic samples as the amount of evolutionary history, which is measured as the fraction of branch length in a phylogenetic tree that leads to descendants of one sample or the other, but not both (Lozupone and Knight, 2005). There is also a weighted version that directly accounts for differences in relative abundances that produce different but complementary results to the unweighted version (Lozupone *et al.*, 2011). UniFrac can be used to test whether the phylogenetic lineages between samples are significantly different, or to cluster many samples using multivariate statistical techniques.

Functional metagenomics

Termed 'superorganism,' the human body is intertwined with trillions of microbes. Defining the microbial genome would expand our knowledge of microbial biotransformation. Metagenomics is the study of the genomes of the microbial community members using DNA isolated directly from environmental samples (DeLong, 2002; Handelsman, 2004; Riesenfeld *et al.*, 2004; Rodríguez-Valera, 2004). Metagenomics is a rapidly growing field that seeks to characterize the metabolic features and composition of the microbial community. The first metagenomic study characterized on samples from an abandoned mine site in Iron Mountain (Tyson *et al.*, 2004). The acid drain of the mine created a simple microbial biofilm community. Due to its simplicity, the results provided near-complete genomes of key organisms from 78 million bases of Sanger sequencing reads. In 2000, bacteriorhodopsin was discovered through metagenomic analysis of seawater (Béja *et al.*, 2000). The research team cloned 130-kb DNA fragments into BAC libraries and identified a mechanism for ATP generation from light, which was thought to occur only in hypersaline ponds. This is regarded as one of the most successful metagenomic investigations. Another landmark study of metagenomes analysed surface water from the Sargasso Sea in 2004; 148 novel bacterial phylotypes were identified, including 1800 genome species (Venter *et al.*, 2004). Moreover, 782 novel rhodopsin-like photoreceptors were identified from 1.2 million novel genes.

In addition to environmental metagenome studies, attempts to reveal metabolic functions of the human gut microbiota have continued. Apart from identifying bacterial compositions within the human gut, metagenomic approaches have also been applied to investigate the functional and metabolic potential of all bacterial genes present (Tuohy *et al.*, 2009). Most analyses of the human gut microbiota used faecal samples. An initial gut metagenome study of two healthy adults was performed by comparing the human faecal microbiome and previously sequenced microbial genomes (Gill *et al.*, 2006). The data confirmed that the human gut microbiome is enriched with genes associated with carbohydrate, amino acid, and xenobiotic metabolism, energy production, methanogenesis, and the biosynthesis of vitamins and isoprenoids. Particularly, many functions not represented in our own human genome were found, e.g. the metabolism of polysaccharides present in dietary plants, such as xylans, pectins, arabino-containing carbohydrates and fructans. They were related mainly to carbohydrate fermentation, thereby emphasizing on the role of the human gut microbiota in energy harvesting. Metabolic activities could thus differ among individuals depending on their microbiome suggesting the nutritional and energy values from food intake could be a relative concept (Turnbaugh and Gordon, 2008). Furthermore, connections between the human gut microbiome and chronic diseases have been investigated in attempt to explain the functional alterations caused by such diseases (Erickson *et al.*, 2012; Greenblum *et al.*, 2012). For example, metagenomic analysis of faecal samples of patients with Crohn's disease has indicated impaired bacterial carbohydrate metabolism and depletion of some SCFA production pathways (Erickson *et al.*, 2012). Altered enzymes levels have been associated with obesity and inflammatory bowel disease (Greenblum *et al.*, 2012). These results could be utilized to identify potential biomarkers or risk indicators of obesity, malnutrition, and dietary recommendations (Prentice *et al.*, 2008).

Metagenomic analysis has enabled examination of gut microbes with greater detail by analysing not only the microorganisms present in the gut, but also their genes. In 2013, the MetaHIT Consortium, a European initiative to investigate the associations between the gut microbiome and chronic diseases, analysed the microbiomes of 169 obese and 123 non-obese individuals (Le Chatelier *et al.*, 2013). They reported that people could be classified into two groups differed in terms of the diversity of microbial genes in their guts. Those having a low diversity showed a tendency to have more body fat and increased biomarkers of poor metabolic health, as compared with individuals with a more diverse microbiome. Knight and colleagues calculated that microbial genes could sort lean from obese individuals with 90% accuracy. The authors commented that 'microbial genes are a much better readout of whether you're likely to be obese or not than human genes are' (DeWeerdt, 2014).

Metagenome analysis workflow

Sequencing approaches for whole genomes are unlike those for ribosomal genes. Previous metagenomic sequencing involved randomly sheared DNA cloned into a bacterial plasmid and sequencing by the capillary method (*e.g.* Sanger sequencing) (Gilbert and Dupont, 2011), which has been replaced by high-throughput NGS recently. An initial step following sequencing is assembly of the datasets. Subsequently, gene calling is carried out mostly at the protein level (Raes *et al.*, 2007b). The prepared products are further analysed in terms of microbial diversity, functional composition, and comparative metagenomics (Fig. 17.1 and Table 17.3).

Sample preparation

The preparation of samples for NGS-based metagenomics requires either a large amount of starting material or the amplification of DNA from smaller samples. Increased sample volumes can blur the distinction between individual cells, and certain types of amplification can introduce bias. In the general protocol, the starting material for NGS analysis is double-stranded DNA, such as isolated genomic DNA. This should be fragmented, polished, size-selected, adapter-ligated, purified, and quantified to be processed for a sequencing. When working with low concentrations of DNA, low-adsorbing plasticware should be used. Because standard polypropylene microfuge tubes can adsorb as much as 100 ng of DNA, it is recommended that polyallomer tubes be used, with the addition of detergent (e.g. 0.02% Tween-20) in each reaction step (Linnarsson, 2010).

DNA fragmentation is generally achieved by mechanical force, such as nebulization using a disposable nebulizer driven by pressurized air or by the Covaris AFA ultrasound device. Sometimes enzymatic digestion such as fragmentase (New England Biolabs) or random transposon insertion such as Nextera (Epicenter) can be used. Nextera has the advantage of introducing adapter sequences simultaneously with fragmentation, although insertion bias remains a concern. The choice of fragmentation method dictates the amount of starting material required. For example, the Covaris instrument generally produces a narrower fragment distribution compared to that of nebulization, leading to more than fourfold better recovery (Quail *et al.*, 2008). Physical methods are prone to cross-contamination and it is difficult to process large numbers of samples. Relatively simple enzymatic digestion can allow preparation of large numbers of samples in parallel, but batch-to-batch reproducibility is a concern.

Following fragmentation, adapters must be ligated. Each fragment has distinct upstream and downstream adaptors. For example, the Illumina GA platform uses a single Y-shaped adapter consisting of A-B (or B-A) adapter sequences, whereas the FLX system uses double-stranded A and B adapters. In this step, two fragments can be accidently joined together, forming a chimera, which generates misleading paired-end reads. To avoid this problem, 200-bp fragments are selected from a gel prior to ligation with 100-bp adapters, and 300-bp fragments are selected again after ligation.

When the starting material does not contain several billion input molecules of DNA, amplification is required. The final quality control and quantification is a crucial step. If the sample concentration is too low, much of the capacity of the instrument is wasted and the cost efficiency is reduced. If too high, most detection sites will overlap or be non-clonal. An intercalating fluorescent dye, such as SYBR or PicoGreen, can be used to quantify double-stranded DNA specifically, and is recommended for initial quantification of sequencing libraries. To achieve the maximum throughput and most accurate quantification, it is necessary to estimate the specific fragments ligated with the correct adapter sequence. Simple quantitative PCR is used for this purpose. The general protocol for the preparation of short paired-end libraries from genomic dsDNA was well documented in the mini-review by Linnason (2010).

Assembly

For reconstruction of the raw reads, there are two methods of completing genome reconstruction (Scholz *et al.*, 2012a). The first is to assemble the fragments into longer contigs, which together become scaffolds (Temperton and Giovannoni, 2012). This is less complicated when using long Sanger reads, whereas reduced read lengths are more problematic in the assembly of repetitive sequences, multistrain, and multispecies mixtures. The assembly of short reads requires an immense amount of memory, posing additional difficulty in the reconstruction. Successful assembly is also affected by the abundance of the genome because highly abundant genomes are assembled more effectively. This creates another limitation in whole-genome sequencing (WGS), in that only a significant fraction of the community is assembled. To overcome this limitation, the library can undergo filtration and

Table 17.3 Resources available for metagenomic analysis

Type	Resource	Features	Website	Reference
Reference sequence DB	Greengenes	Quality-controlled microbial DNA data and tools	http://greengenes.lbl.gov/cgi-bin/nph-index.cgi	DeSantis *et al.* (2006)
	SILVA	Quality checked and regularly updated datasets	http://www.arb-silva.de/	Pruesse *et al.* (2007), Quast *et al.* (2013)
	RDP	Online data analysis, rRNA derived phylogenetic trees, aligned and annotated rRNA sequences	http://rdp.cme.msu.edu/ https://github.com/rdpstaff	Cole *et al.* ((2009)
Annotation and analysis tools	MG-RAST (MetaGenomics Rapid Annotaton using Subsystem)	Annotation, taxonomic and functional assignment, pathway reconstruction single platform (depositing, locating, analysing, visualizing, data sharing)	http://metagenomics.nmpdr.org	Glass *et al.* (2010)
	IMG-M (Integrated Microbial Genomics/Metagenomics)		http://img.jgi.doe.gov	Markowitz *et al.* (2012, 2014)
	MetaRep (METAgenomics REPorts)	High-performance comparative metagenomics	http://www.jcvi.org/metarep/ http://www.jcvi.org/hmp-metarep/	Goll *et al.* (2010, 2012)
	CAMERA (Community cyberinfrastructure from Advanced Micobial Ecology Research and Analysis)	Data analysis pipeline, single platform	http://camera.calit2.net	Sun *et al.* (2011b)
	MEGAN	Metagenome analysis software	http://www-ab.informatik.uni-tuebingen.de/software/megan	Huson *et al.* (2007), Huson and Weber (2013), Mitra *et al.* (2011)
	DIYA	Bacterial annotation pipeline	https://github.com/bioteam/DIYA	Stewart *et al.* (2009)
	RATT	Transfer annotation from reference genome onto assembly	http://ratt.sourceforge.net/ (Now, part of PAGIT, http://www.sanger.ac.uk/resources/software/pagit/	Otto *et al.* (2011)
	VMGAP	Virus-specific annotation pipeline		Lorenzi *et al.* (2011)
	CloVR	Virtual machine, automated sequence analysis from desktop, cloud computing.	http://clovr.org	Angiuoli *et al.* (2011), White *et al.* (2013)
	Galaxy	Web-based data analysis pipeline	https://usegalaxy.org/	Blankenberg *et al.* (2010)
	HUMAnN	Direct read alignment against protein DB, without prior assembly	http://huttenhower.sph.harvard.edu/humann	Abubucker *et al.* (2012)
	Ergatis	Genome annotation and comparative analysis	http://ergatis.sourceforge.net/index.html	Hemmerich *et al.* (2010), Orvis *et al.* (2010b)
	MetaPathways	Direct read alignment against protein DB, without prior assembly	http://hallam.microbiology.ubc.ca/MetaPathways/	Konwar *et al.* (2013)
Assembly tools	RaY	De novo metagenome assembly and profiling	http://denovoassembler.sourceforge.net/	Boisvert *et al.* (2012)
	Velvet	De novo assembly, short read alignment, Kmer DeBruijin graph traversal-based method	http://www.ebi.ac.uk/~zerbino/velvet/	Namiki *et al.* ((2012), Powell and Seemann (2013), Zerbino (2010), Zerbino *et al.* (2009)
	SOAPdenovo	Short read assembly, Kmer DeBruijin graph traversal-based method	http://soap.genomics.org.cn/soapdenovo.html	Luo *et al.* (2012)
	Newbler	De novo assembler, designed for 454 platform	http://454.com/products/analysis-software/index.asp	Miller *et al.* (2010)
	ABySS	Kmer DeBruijin graph traversal-based method	http://www.bcgsc.ca/platform/bioinfo/software/abyss	Simpson *et al.* (2009)
	ALLPATHs	Kmer DeBruijin graph traversal-based method	http://www.broadinstitute.org/software/allpaths-lg/blog/	Butler *et al.* (2008), Maccallum *et al.* (2009)
	Genovo	Generative probabilistic model of read generation from environmental samples	http://cs.stanford.edu/group/genovo/	Laserson *et al.* (2011)

Table 17.3 Continued

Type	Resource	Features	Website	Reference
	CLCbio	Kmer DeBruijin graph traversal-based method	http://clcbio.com	
	Meta-IDBA	Kmer DeBruijin graph traversal-based method	http://i.cs.hku.hk/~alse/hkubrg/projects/metaidba/	Peng *et al.* (2011)
	MetaVelvet	Kmer DeBruijin graph traversal-based method	http://metavelvet.dna.bio.keio.ac.jp/	Namiki *et al.* (2012)
	SMASHCommunity	Assembly and gene calling, data exploration, visualization	http://www.bork.embl.de/software/smash/	Arumugam *et al.* (2010)
Mapping/ Alignment tools	BWA	Mapping low-divergent sequences, based on Burrows-Wheeler aligners	http://bio-bwa.sourceforge.net/	Li and Durbin (2009)
	Bowtie	Memory-efficient short read aligner, based on Burrows-Wheeler index	http://bowtie-bio.sourceforge.net/index.shtml	Langmead *et al.* (2009)
	Novoalign	Single-ended and paired-end reads, Illumina GA platform, Needleman-Wunsch algorithm	http://www.novocraft.com/main/index.php	
	SOAP	Alignment of short oligonucleotides	http://soap.genomics.org.cn/soap1/	Li *et al.* (2008, 2009)
	MrFAST	Mapping NGS reads, prediction of copy-number variation of duplicated segments and genes	http://mrfast.sourceforge.net/	Alkan *et al.* (2009)
	CloudBurst	Parallel read-mapping algorithm, SNP discovery, genotyping	http://sourceforge.net/projects/cloudburst-bio/	Nguyen *et al.* (2011), Schatz (2009)
	BFAST	Mapping of short reads to reference genomes	http://sourceforge.net/projects/bfast/	Homer *et al.* (2009)
	MUMer	Rapid genome alignment	http://mummer.sourceforge.net/	Kurtz *et al.* (2004)
	BLAST (Basic Local Alignment Search Tool)	Local similarity between sequences	http://blast.ncbi.nlm.nih.gov/Blast.cgi	
	MAQ (Mapping and Assembly with Quality)	Based on Burrows–Wheeler aligners	http://maq.sourceforge.net/	Kumps *et al.* (2010)
Reference DB for functional metagenomics	KEGG (Kyoto Encyclopedia of Genes and Genomes)	Integrated database (16 main DB), broadly categorized systems, genomic, chemical, health information	http://www.genome.jp/kegg/	
	SEED	Comparative genomics environment, curation of genomic data, by annotator across many genes	http://www.theseed.org	
	CAZy (Carbohydrate Active enZyme database)	Genomic, structural, and biochemical information on CAZymes	http://www.cazy.org/	
	PFAM	Collection of protein families using multiple sequence alignments and hidden Markov models	http://pfam.xfam.org/	
	Phylofacts	Structural phylogenomic encyclopaedia for protein functional and structural classification	http://phylogenomics.berkeley.edu/phylofacts/	
	Uniprot	Protein sequence and functional information	http://www.uniprot.org/	
	Mascot	Search engine that uses mass spectrometry data to identify proteins from primary sequences	http://www.matrixscience.com/home.html	
16S based analysis tools	MOTHUR	16S-focused, quality-control, taxonomic profiling, data exploration, statistical analysis, visualization	http://www.mothur.org/wiki/Main_Page	Schloss *et al.* (2009)
	QIIME (Quantitative Insights Into Microbial Ecology)		http://qiime.org/	Caporaso *et al.* (2010), Kuczynski *et al.* (2011), Navas-Molina *et al.* (2013)
	VAMPS (Visualization and Analysis of Microbial Populations)		http://vamps.mbl.edu/	Huse *et al.* (2014)

Table 17.3 Continued

Type	Resource	Features	Website	Reference
Data exploration, statistical analysis and visualization tools	bioBakery	Metagenome-focused, taxonomic and functional profiling, data exploration, statistical analysis, visualization	http://huttenhower.sph.harvard.edu/biobakery	
	MetaPhlAn	Clade-specific marker genes, assign reads	http://huttenhower.sph.harvard.edu/metaphlan	Segata *et al.* (2012)
	PhymmBL	Classification method using interpolated Markov models	http://www.cbcb.umd.edu/software/phymm/	Brady and Salzberg (2009, 2011),
	STAMP (Statistical Analysis of Metagenomic Profiles)	Data exploration, quality control, taxonomic profiling, data exploration, statistical analysis, visualization	http://kiwi.cs.dal.ca/Software/STAMP	Parks and Beiko (2010)
	LEfSe		http://huttenhower.sph.harvard.edu/galaxy/	Segata *et al.* ((2011)
	MaAsLin		http://huttenhower.sph.harvard.edu/galaxy/	
	MetaStats		http://metastats.cbcb.umd.edu/	Foster (2003)
	PICRUSt	Determination of virtual metagenome by using 16S data set	http://huttenhower.sph.harvard.edu/galaxy/	
	microPITA	Targeted selection of samples	http://huttenhower.sph.harvard.edu/galaxy/	Tickle *et al.* (2013)

normalization to enrich the low-abundance species (Dale *et al.*, 2005; Hallam *et al.*, 2004).

The second assembly method is read-based reconstruction. It has an advantage over contig-based assembly due to its shorter analysis time and lower error rate (Scholz *et al.*, 2012a). Whereas the contigs (and scaffolds) are created as the result of linking short reads, read-based assembly reconstructs the community using reference databases. Despite the extremely rapid analysis time, only those genomes found in the database can be used. The analysis time varies depending on the computational resources and alignment methods used. For example, it is estimated that the Burrow-Wheeler aligner performs the assembly of millions of reads within hours (Langmead *et al.*, 2009), whereas BLAST is sensitive enough to identify distant homologues, but takes weeks (Altschul *et al.*, 1997).

Currently, assembly programs for NGS data operate based on Kmer DeBruijn graph traversal-based methods (de Bruijin graph) (Liu *et al.*, 2011; Miller *et al.*, 2010). The Kmer approach is an advanced form of traditional overlap-based assembly strategies to reduce the long analysis time arising from intense calculation (Scholz *et al.*, 2012a). The number of calculations required is calculated by N^2, where N is the number of reads in an assembly. A disproportional community composition results in genomes with different abundances, hindering complete coverage of the microbiome. This is not problematic when analysing single genomes, but is still extremely challenging in the case of complex populations (Xie *et al.*, 2010). For extended coverage, investment in computational memory and time is required. Some assemblers seek to resolve this problem by binning genomes according to their abundance in the population, separating short-reads by Kmer frequency and/or collapsing the graph (Scholz *et al.*, 2012a). Another approach is to generate discrete groups with reads more amenable to assembly.

Gene calling

Gene calling is performed to identify the genes in the isolated microbial genome of a certain isolate and to understand the biochemistry, physiology and ecology of the isolate (Mavromatis *et al.*, 2007; Pati *et al.*, 2010). This process also suffers from fragmented genes due to short read lengths, sequencing errors that reduce sequence quality, and chimeras generated from metagenomic assembly in addition to phylogenetic diversity (Mavromatis *et al.*, 2007; Raes *et al.*, 2007b). The most popular *ab initio* methods identify protein-coding, intergenic, and boundary regions by statistical methods. The variation in gene model predictions increases depending on the degree of complexity of the microbial community, and other factors such as G+C content and pseudogenes. Evidence-based methods using a reference databases may reduce the uncertainty of the predicted gene model; however, they cannot identify genes containing chimeric or shuffled sequences. To alleviate the gene prediction problem, heuristically estimated codon models and postprocessing steps – such as overlap removal, translation initiation site adjustment and frameshift detection – have been adopted in some studies (Noguchi *et al.*, 2006; Tech and Meinicke, 2006). Therefore, metagenome gene prediction requires further development to enable precise construction of the genome of a complex community.

Annotation

Annotation is the final step in evaluating community functional potential. Briefly, annotation is the process of attaching biological information to sequences (Stein, 2001). The quality of gene annotation is dependent on gene calling because functions can be assigned to properly constructed genes. As is the case for processes of assembly and gene calling, short-length NGS products impede the annotation of genes. Of the 69 million protein sequences in the current (2014_06) release of the UniProt database, 1.33% has been confirmed either at the protein or transcript level (http://www.ebi.ac.uk/uniprot/TrEMBLstats). Compared with the statistics 2 years ago (~3% from 20.6 million sequences), the amounts of sequencing data has increased considerably, but the rate of discovery of putative protein has not (Temperton and Giovannoni, 2012). This has led researchers to choose automated annotation over manual curation, in which humans make the decisions. Automated annotation involving homology transfer from similar sequences with known function increased the number of annotations of putative proteins exponentially. Manual curation may improve upon the automated output, but ultimately experimental verification of the qualified genome structure is required. And even then, there are numerous examples where manual annotation/prediction of function have been wrong based on experimental results. For example, annotation of the ligand for sugar transporters is hardly ever correct, due to the fact that single amino acid changes may determine the specificity of the transporter. These single amino acid changes cannot be taken into account by automatic or manual annotation. It requires experimental proof, but that is impossible for all annotated genes.

Currently, there is no centralized method for annotating the diverse range of genomes, and a variety of platforms exist (Scholz *et al.*, 2012a). Annotation can be performed in-house using the available gene calling and functional annotation workflows included in web-based workflow systems (Orvis *et al.*, 2010a). There are also online annotation services available (Markowitz *et al.*, 2008; Meyer *et al.*, 2008). In addition to various platforms, different strategies are utilized perform annotation. The reference-based approach compares ORFs against databases, such as NCBI Clusters of Orthologous Groups (COGs), TIGR funccats, STRING extended COGs, SEED, and KEGG (Raes *et al.*, 2007c). However, the success rate depends on the sensitivity of the database, which is estimated to be 25–50% of proteins per published metagenome. Profile-based homology searches protein modules from domain databases (Bateman *et al.*, 2004; Letunic *et al.*, 2006; Mulder *et al.*, 2007) and is useful in identifying functionally distinct families within known superfamilies. Neighbourhood-based approaches are practical for annotating novel proteins from known processes.

Comparative metagenomics

The direct sequencing of environmental samples – including those from humans – has advanced our understanding of microbial communities. In addition, the application of high-throughput shotgun sequencing has provided global views that were not obtainable using older sequencing methods (Tringe *et al.*, 2005). Moreover, comparisons with other communities can further deepen the investigation; e.g. to examine whether the revealed compositions and functions are distinctive features of the community (Fraser *et al.*, 2000). The comparison of samples from the same environmental conditions can indicate environmental effects on the community (Raes *et al.*, 2007b); *e.g.* the functional variation of a marine microbiota was influenced by water pressure and light reduction when sampled from gradually increasing depths (DeLong *et al.*, 2006). Comparisons of different environments can reveal the general relationship between metagenome properties and the habitat from which they were derived (Raes *et al.*, 2007b). As an example, GC content varies depending on the sampling site, being low in surface sea water and high in farm soil (Foerstner *et al.*, 2005). This was confirmed in a larger-scale study, the Global Ocean Survey (GOS), in which the GC content increased with sampling closer to land (Rusch *et al.*, 2007). Furthermore, comparative analysis of multiple metagenomes provides information on the evolution of microbial species, e.g. the surface seawater microbiome, which had a low GC content, showed a rapidly evolving small genome, whereas soil showed the opposite relationship (Raes *et al.*, 2007a).

Recent metagenomic comparisons of the human gut microbiota have provided a picture of the development of the gut microbial community (Kurokawa *et al.*, 2007; Vaishampayan *et al.*, 2010). An analysis of faecal samples from subjects of various age and both genders demonstrated that the gut microbiota from unweaned infants was simple, comprising few microbial species, and exhibited high inter-individual variation. Conversely, the gut microbiota from adults and weaned children was complex, but with functional redundancy independent of age and gender. The number of genes in gut microbiomes differed between adults and children: 237 gene families were found in adults and 136 families in children. Moreover, the degree of overlap was small. Comparing the gut microbiota and environmental samples from seawater and soil, the human gut microbiome showed distinctive patterns enriched for the category 'Carbohydrate transport and metabolism' and an under-represented for 'Lipid transport and metabolism.' Interestingly, the functional profile was differed markedly according to the weaning state. Adults and weaned children exhibited an over-representation of 'Defence mechanisms' and an under-representation of 'Cell motility', 'Secondary metabolites biosynthesis, transport and catabolism' and 'Post-translational modification, protein turnover, chaperones.' The gut microbial functionality in infants showed large variability. Comparative metagenomics of the gut microbiota between mother and infant showed the progression of the infant's gut microbiota to a fully functional grown-up state (Vaishampayan *et al.*, 2010). The infant's gut harboured maternal *Bacteroides* phylotypes until 1 month of age; however, the microbial composition changed to that in which *Bacteroides* and *Bifidobacterium* predominated at 11 months after birth. The adult-like gut microbiota composition is thought to be established when the infant begins to intake solid foods. Moreover, the analysis of phylogenetic composition presented clusters of mother with infants at 11 months, but separating infants at 1 month.

Challenges in current metagenomics and future perspectives

Rapid progress in the development of sequencing and analytical tools over past decades has afforded tremendous biological insights into the microbial communities. Despite these efforts, challenges remain in terms of the prediction of gene function and metagenomic investigations. Initial challenges due to sequencing throughput in genome analysis were overcome by the advent of high-throughput NGS, which allowed metagenomic approaches by reducing the price, but at the cost of reduced sequencing length (Temperton and Giovannoni, 2012). Short-length sequencing technologies were first used for the purpose of resequencing the human genome. Having the base template to compare to, long reads were not necessary but this was not the case in. In complex communities comprising multiple strains, short read lengths were insufficient for genome assembly, gene prediction, and functional analysis. Data analysis has replaced sequencing as the limiting factor in metagenomic studies. The development of innovative sequencing techniques and novel algorithms for assembly and analysis would be helpful to reduce the gap between real versus computationally constructed genomes.

Inexpensive sequencing has contributed to generation of massive sequencing data, but whether the data produced represents the entire community is unclear (Desai *et al.*, 2012). For example, a hypothetical calculation estimates that 6,000 sequencing runs are required to achieve one fold coverage of 1 g of soil, at a cost of $267 million. However, this is merely hypothetical as the whole community composition has not been fully identified and the current analysis presented the data as relative abundances (Tyson *et al.*, 2004). Current metagenomic technologies are not capable of determining the depth of coverage required for an objective community. Even if this could be calculated, tools with which to interpret the complexity would need to be developed in parallel. Furthermore, most studies investigating host–microbe interactions in gut ecology and the role of the gut microbiota in the development of related diseases were carried out using faecal samples due to the difficulty of obtaining samples from the various regions of the intestine. However, the gut microbiota of faecal samples is not necessarily representative of the small and large intestine.

Metagenomics is constantly advancing in terms of tools and analysis pipelines. Analysing metagenomic data and discovering associations between the gut microbiome and various disease states are the first steps towards unravelling the mechanisms by which changes in microbial communities can disrupt gastrointestinal homeostasis. Now, it is not sufficient to find that people have different microbial communities that 'presumably' lead to their individual phenotypic difference. Over the last decade, many association studies in humans have investigated the relationship between diet, the gut microbiome, and body weight. The HMP represented an intriguing health-related association study; however, the discovery of associations only be rarely translated into direct (food or pharmaceutical) interventions. Results from this line of research have not established whether the changes are a cause or effect of obesity. Therefore, additional experimental confirmation is required. Results from animal models, especially using germ-free or gnotobiotic mice, can offer much evidence regarding causality and the precise interactions between the host and the intestinal microbiome that lead to gastrointestinal disorders. However, there are many questions to be answered, as many as have been answered to date. For example, questions regarding vertical transmission need to be extended to other family members and the degree of functional redundancy of the microbiota in different niches should be compared more deeply (Turnbaugh and Gordon, 2008). Those animal models should provide information regarding the co-evolution and interaction between the host and the microbiome by excluding the confounding factors from human life (e.g. diet, disease, environmental exposures) (Turnbaugh and Gordon, 2008). Indeed, the correlation between obesity and the human gut microbiota was verified by the increased energy harvest capacity in the obese mouse (Turnbaugh *et al.*, 2006). Ultimately, the results of animal models can be applied to humans to gain further insights into our physiology and develop methods of manipulating the gut microbiome to achieve beneficial effects. However, laboratory mice live tightly controlled lives, whereas humans are constantly surrounded by dynamic environmental changes; moreover, humans are more genetically diverse than laboratory mice. To date, the effectiveness of diet and/or microbiota-based weight interventions for obesity has varied markedly among individuals. Thus the results from microbiome studies of laboratory animals must be translated into novel preventative strategies and therapies for a variety of gastrointestinal disorders.

Microbial cultivation is another suggested method of compensating for the uncertainty of metagenomic analysis. The provenance of metagenomics was the impossibility of cultivation of the entire microbial community. Ironically, the limitations of metagenomic analysis may necessitate the return to cultivation techniques. Consider the case of SAR11 Alphaproteobacteria, the most abundant aerobic and heterotrophic marine bacterial taxon first discovered in the Sargasso Sea (Giovannoni *et al.*, 1990; Morris *et al.*, 2002; Rappé and Giovannoni, 2003). Metagenomic analysis of the sea surface water identified a high frequency of genes for C1 metabolism, which was confirmed by axenic culture of the bacteria in the laboratory (Sun *et al.*, 2011a). Not only can microbial culture complement the metagenomes but it is also needed for verification. SAR11 provides another example. The metagenomically discovered bacteriorhodopsin from surface seawater was expected to allow growth of SARl1 when exposing cultures to light. However, bacteriorhodopsin is activated under conditions of carbon starvation, which was reported a decade after the metagenome study. Although metagenomic approaches have provided great insight into the gut microbiome, further studies are required to determine how 'specific' changes in the gut microbial community can prevent and reverse the metabolic and immunological alterations associated with diverse disease states.

References

Abubucker, S., Segata, N., Goll, J., Schubert, A.M., Izard, J., Cantarel, B.L., Rodriguez-Mueller, B., Zucker, J., Thiagarajan, M., Henrissat, B., *et al.* (2012). Metabolic reconstruction for metagenomic data and its application to the human microbiome. PLoS Comput Biol. *8*, e1002358.

Alkan, C., Kidd, J.M., Marques-Bonet, T., Aksay, G., Antonacci, F., Hormozdiari, F., Kitzman, J.O., Baker, C., Malig, M., Mutlu, O., *et al.* (2009). Personalized copy number and segmental duplication maps using next-generation sequencing. Nat. Genet. *41*, 1061–1067.

Altschul, S.F., Madden, T.L., SchŠffer, A.A., Zhang, J., Zhang, Z., Miller, W., and Lipman, D.J. (1997). Gapped BLAST and PSI-BLAST: a new generation of protein database search programs. Nucleic Acids Res. *25*, 3389–3402.

Angiuoli, S.V., Matalka, M., Gussman, A., Galens, K., Vangala, M., Riley, D.R., Arze, C., White, J.R., White, O., and Fricke, W.F. (2011). CloVR: a virtual machine for automated and portable sequence analysis from the desktop using cloud computing. BMC Bioinformatics *12*, 356.

Arumugam, M., Harrington, E.D., Foerstner, K.U., Raes, J., and Bork, P. (2010). SmashCommunity: a metagenomic annotation and analysis tool. Bioinformatics *26*, 2977–2978.

Béja, O., Aravind, L., Koonin, E.V., Suzuki, M.T., Hadd, A., Nguyen, L.P., Jovanovich, S.B., Gates, C.M., Feldman, R.A., and Spudich, J.L. (2000). Bacterial rhodopsin: evidence for a new type of phototrophy in the sea. Science *289*, 1902–1906.

Backhed, F., Ding, H., Wang, T., Hooper, L.V., Koh, G.Y., Nagy, A., Semenkovich, C.F., and Gordon, J.I. (2004). The gut microbiota as an environmental factor that regulates fat storage. Proc. Natl. Acad. Sci. U.S.A. *101*, 15718–15723.

Bateman, A., Coin, L., Durbin, R., Finn, R.D., Hollich, V., Griffiths-Jones, S., Khanna, A., Marshall, M., Moxon, S., and Sonnhammer, E.L. (2004). The Pfam protein families database. Nucleic Acids Res. *32*, D138-D141.

Benson, A.K., Kelly, S.A., Legge, R., Ma, F., Low, S.J., Kim, J., Zhang, M., Oh, P.L., Nehrenberg, D., Hua, K., *et al.* (2010). Individuality in gut microbiota composition is a complex polygenic trait shaped by multiple environmental and host genetic factors. Proc. Natl. Acad. Sci. U.S.A. *107*, 18933–18938.

Blankenberg, D., Von Kuster, G., Coraor, N., Ananda, G., Lazarus, R., Mangan, M., Nekrutenko, A., and Taylor, J. (2010). Galaxy: a web-based genome analysis tool for experimentalists. Curr. Protoc Mol. Biol. *Chapter 19*, Unit 19 10 11–21.

Boisvert, S., Raymond, F., Godzaridis, E., Laviolette, F., and Corbeil, J. (2012). Ray Meta: scalable de novo metagenome assembly and profiling. Genome Biol. *13*, R122.

Bollmann, A., Lewis, K., and Epstein, S.S. (2007). Incubation of environmental samples in a diffusion chamber increases the diversity of recovered isolates. Appl. Environ. Microbiol. *73*, 6386–6390.

Bomba, A., Nemcova, R., Gancarcikova, S., Herich, R., Guba, P., and Mudronova, D. (2002). Improvement of the probiotic effect of micro-organisms by their combination with maltodextrins, fructo-oligosaccharides and polyunsaturated fatty acids. Br. J. Nutr. *88*, S95-S99.

Brady, A., and Salzberg, S. (2011). PhymmBL expanded: confidence scores, custom databases, parallelization and more. Nat. Methods *8*, 367.

Brady, A., and Salzberg, S.L. (2009). Phymm and PhymmBL: metagenomic phylogenetic classification with interpolated Markov models. Nat. Methods *6*, 673–676.

Bragg, L., and Tyson, G.W. (2014). Metagenomics using next-generation sequencing. Methods Mol. Biol. *1096*, 183–201.

Burcelin, R., Serino, M., Chabo, C., Garidou, L., Pomie, C., Courtney, M., Amar, J., and Bouloumie, A. (2013). Metagenome and metabolism: the tissue microbiota hypothesis. Diabetes Obes Metab. *15*, 61–70.

Butler, J., MacCallum, I., Kleber, M., Shlyakhter, I.A., Belmonte, M.K., Lander, E.S., Nusbaum, C., and Jaffe, D.B. (2008). ALLPATHS: de novo assembly of whole-genome shotgun microreads. Genome Res. *18*, 810–820.

Cani, P.D., Amar, J., Iglesias, M.A., Poggi, M., Knauf, C., Bastelica, D., Neyrinck, A.M., Fava, F., Tuohy, K.M., Chabo, C., *et al.* (2007). Metabolic endotoxemia initiates obesity and insulin resistance. Diabetes *56*, 1761–1772.

Cani, P.D., and Delzenne, N.M. (2011). The gut microbiome as therapeutic target. Pharmacol. Ther. *130*, 202–212.

Cani, P.D., Knauf, C., Iglesias, M.A., Drucker, D.J., Delzenne, N.M., and Burcelin, R. (2006). Improvement of glucose tolerance and hepatic insulin sensitivity by oligofructose requires a functional glucagon-like peptide 1 receptor. Diabetes *55*, 1484–1490.

Cani, P.D., Possemiers, S., Van de Wiele, T., Guiot, Y., Everard, A., Rottier, O., Geurts, L., Naslain, D., Neyrinck, A., Lambert, D.M., *et al.* (2009). Changes in gut microbiota control inflammation in obese mice through a mechanism involving GLP-2-driven improvement of gut permeability. Gut *58*, 1091–1103.

Caporaso, J.G., Kuczynski, J., Stombaugh, J., Bittinger, K., Bushman, F.D., Costello, E.K., Fierer, N., Pena, A.G., Goodrich, J.K., Gordon, J.I., *et al.* (2010). QIIME allows analysis of high-throughput community sequencing data. Nat. Methods *7*, 335–336.

Chen, T., Yu, W.-H., Izard, J., Baranova, O.V., Lakshmanan, A., and Dewhirst, F.E. (2010). The Human Oral Microbiome Database: a web accessible resource for investigating oral microbe taxonomic and genomic information. Database *2010*, baq013.

Cho, I., and Blaser, M.J. (2012). The Human Microbiome: at the interface of health and disease. Nat. Rev. Genet. *13*, 260–270.

Cole, J.R., Wang, Q., Cardenas, E., Fish, J., Chai, B., Farris, R.J., Kulam-Syed-Mohideen, A., McGarrell, D.M., Marsh, T., and Garrity, G.M. (2009). The Ribosomal Database Project: improved alignments and new tools for rRNA analysis. Nucleic Acids Res. *37*, D141–D145.

Collins, F.S., Green, E.D., Guttmacher, A.E., and Guyer, M.S. (2003). A vision for the future of genomics research. Nature *422*, 835–847.

Cook, S., and Sellin, J. (1998). Review article: short-chain fatty acids in health and disease. Aliment. Pharmacol. Ther. *12*, 499–507.

Curtis, T.P., Sloan, W.T., and Scannell, J.W. (2002). From the Cover: Estimating prokaryotic diversity and its limits. Proc. Natl. Acad. Sci. U.S.A. *99*, 10494–10499.

Dale, C., Dunbar, H., Moran, N.A., and Ochman, H. (2005). Extracting single genomes from heterogenous DNA samples: A test case with Carsonella ruddii, the bacterial symbiont of psyllids (Insecta). J. Insect Sci. *5*, 3.

Decroos, K., Eeckhaut, E., Possemiers, S., and Verstraete, W. (2006). Administration of equol-producing bacteria alters the equol production status in the Simulator of the Gastrointestinal Microbial Ecosystem (SHIME). J. Nutr. *136*, 946–952.

Degnan, P.H., and Ochman, H. (2012). Illumina-based analysis of microbial community diversity. ISME J. *6*, 183–194.

DeLong, E.F. (2002). Microbial population genomics and ecology. Curr. Opin. Microbiol. *5*, 520–524.

DeLong, E.F., Preston, C.M., Mincer, T., Rich, V., Hallam, S.J., Frigaard, N.-U., Martinez, A., Sullivan, M.B., Edwards, R., and Brito, B.R. (2006). Community genomics among stratified microbial assemblages in the ocean's interior. Science *311*, 496–503.

Desai, N., Antonopoulos, D., Gilbert, J.A., Glass, E.M., and Meyer, F. (2012). From genomics to metagenomics. Curr. Opin. Biotechnol. *23*, 72–76.

DeSantis, T.Z., Hugenholtz, P., Larsen, N., Rojas, M., Brodie, E.L., Keller, K., Huber, T., Dalevi, D., Hu, P., and Andersen, G.L. (2006). Greengenes, a chimera-checked 16S rRNA gene database and workbench compatible with ARB. Appl. Environ. Microbiol. *72*, 5069–5072.

Desiere, F. (2004). Towards a systems biology understanding of human health: Interplay between genotype, environment and nutrition. Biotechnol. Annu. Rev. *10*, 51–84.

DeWeerdt, S. (2014). Microbiome: A complicated relationship status. Nature *508*, S61–63.

Dewulf, E.M., Cani, P.D., Claus, S.P., Fuentes, S., Puylaert, P.G., Neyrinck, A.M., Bindels, L.B., de Vos, W.M., Gibson, G.R., Thissen, J.P., *et al.* (2013). Insight into the prebiotic concept: lessons from an exploratory, double blind intervention study with inulin-type fructans in obese women. Gut *62*, 1112–1121.

Eberl, G. (2010). A new vision of immunity: homeostasis of the superorganism. Mucosal Immunol. 3, 450–460.

Eckburg, P.B., Bik, E.M., Bernstein, C.N., Purdom, E., Dethlefsen, L., Sargent, M., Gill, S.R., Nelson, K.E., and Relman, D.A. (2005). Diversity of the human intestinal microbial flora. Science *308*, 1635–1638.

Erickson, A.R., Cantarel, B.L., Lamendella, R., Darzi, Y., Mongodin, E.F., Pan, C., Shah, M., Halfvarson, J., Tysk, C., and Henrissat, B. (2012). Integrated metagenomics/metaproteomics reveals human host-microbiota signatures of Crohn's disease. PLoS One 7, e49138.

Everard, A., Belzer, C., Geurts, L., Ouwerkerk, J.P., Druart, C., Bindels, L.B., Guiot, Y., Derrien, M., Muccioli, G.G., Delzenne, N.M., *et al.* (2013). Cross-talk between Akkermansia muciniphila and intestinal epithelium controls diet-induced obesity. Proc. Natl. Acad. Sci. U.S.A. *110*, 9066–9071.

Everard, A., Lazarevic, V., Derrien, M., Girard, M., Muccioli, G.G., Neyrinck, A.M., Possemiers, S., Van Holle, A., Francois, P., de Vos, W.M., *et al.* (2011). Responses of gut microbiota and glucose and lipid metabolism to prebiotics in genetic obese and diet-induced leptin-resistant mice. Diabetes *60*, 2775–2786.

Everard, A., Lazarevic, V., Gaïa, N., Johansson, M., StŒhlman, M., Backhed, F., Delzenne, N.M., Schrenzel, J., François, P., and Cani, P.D. (2014). Microbiome of prebiotic-treated mice reveals novel targets involved in host response during obesity. ISME J. *8*, 2116–2130.

Fettweis, J.M., Serrano, M.G., Girerd, P.H., Jefferson, K.K., and Buck, G.A. (2012). A new era of the vaginal microbiome: advances using next-generation sequencing. Chem. Biodivers. *9*, 965–976.

Foerstner, K.U., Von Mering, C., Hooper, S.D., and Bork, P. (2005). Environments shape the nucleotide composition of genomes. EMBO reports *6*, 1208–1213.

Foster, E.K. (2003). METASTATS: behavioural science statistics for Microsoft Windows and the HP49G programmable calculator. Behav. Res. Methods Instrum Comput 35, 325–328.

Fraser, C.M., Eisen, J., Fleischmann, R.D., Ketchum, K.A., and Peterson, S. (2000). Comparative genomics and understanding of microbial biology. Emerg. Infect. Dis. *6*, 505.

Gilbert, J.A., and Dupont, C.L. (2011). Microbial metagenomics: beyond the genome. Ann. Rev. Mar. Sci. 3, 347–371.

Gill, S.R., Pop, M., DeBoy, R.T., Eckburg, P.B., Turnbaugh, P.J., Samuel, B.S., Gordon, J.I., Relman, D.A., Fraser-Liggett, C.M., and Nelson, K.E. (2006). Metagenomic analysis of the human distal gut microbiome. Science *312*, 1355–1359.

Giovannoni, S.J., Britschgi, T.B., Moyer, C.L., and Field, K.G. (1990). Genetic diversity in Sargasso Sea bacterioplankton. Nature *345*, 60–63.

Glass, E.M., Wilkening, J., Wilke, A., Antonopoulos, D., and Meyer, F. (2010). Using the metagenomics RAST server (MG-RAST) for analysing shotgun metagenomes. Cold Spring Harb Protoc *2010*, pdb prot5368.

Gmeiner, M., Kneifel, W., Kulbe, K.D., Wouters, R., De Boever, P., Nollet, L., and Verstraete, W. (2000). Influence of a synbiotic mixture consisting of *Lactobacillus* acidophilus 74–2 and a fructooligosaccharide preparation on the microbial ecology sustained in a simulation of the human intestinal microbial ecosystem (SHIME reactor). Appl. Microbiol. Biotechnol. 53, 219–223.

Goll, J., Rusch, D.B., Tanenbaum, D.M., Thiagarajan, M., Li, K., Methe, B.A., and Yooseph, S. (2010). METAREP: JCVI metagenomics reports – an open source tool for high-performance comparative metagenomics. Bioinformatics 26, 2631–2632.

Goll, J., Thiagarajan, M., Abubucker, S., Huttenhower, C., Yooseph, S., and Methe, B.A. (2012). A case study for large-scale human microbiome analysis using JCVI's metagenomics reports (METAREP). PLoS One 7, e29044.

Greenblum, S., Turnbaugh, P.J., and Borenstein, E. (2012). Metagenomic systems biology of the human gut microbiome reveals topological shifts associated with obesity and inflammatory bowel disease. Proc. Natl. Acad. Sci. U.S.A. *109*, 594–599.

Hallam, S.J., Putnam, N., Preston, C.M., Detter, J.C., Rokhsar, D., Richardson, P.M., and DeLong, E.F. (2004). Reverse methanogenesis: testing the hypothesis with environmental genomics. Science *305*, 1457–1462.

Hamady, M., and Knight, R. (2009). Microbial community profiling for human microbiome projects: Tools, techniques, and challenges. Genome Res. *19*, 1141–1152.

Hamady, M., Walker, J.J., Harris, J.K., Gold, N.J., and Knight, R. (2008). Error-correcting barcoded primers for pyrosequencing hundreds of samples in multiplex. Nat. Methods *5*, 235–237.

Handelsman, J. (2004). Metagenomics: application of genomics to uncultured microorganisms. Microbiol. Mol. Biol. Rev. *68*, 669–685.

Handelsman, J., Rondon, M.R., Brady, S.F., Clardy, J., and Goodman, R.M. (1998). Molecular biological access to the chemistry of unknown soil microbes: a new frontier for natural products. Chem. Biol. *5*, R245–249.

Hemmerich, C., Buechlein, A., Podicheti, R., Revanna, K.V., and Dong, Q. (2010). An Ergatis-based prokaryotic genome annotation web server. Bioinformatics *26*, 1122–1124.

Homer, N., Merriman, B., and Nelson, S.F. (2009). BFAST: an alignment tool for large scale genome resequencing. PLoS One *4*, e7767.

Human Microbiome Project, C. (2012a). A framework for human microbiome research. Nature *486*, 215–221.

Human Microbiome Project, C. (2012b). Structure, function and diversity of the healthy human microbiome. Nature *486*, 207–214.

Huse, S.M., Mark Welch, D.B., Voorhis, A., Shipunova, A., Morrison, H.G., Eren, A.M., and Sogin, M.L. (2014). VAMPS: a website for visualization and analysis of microbial population structures. BMC Bioinformatics *15*, 41.

Huson, D.H., Auch, A.F., Qi, J., and Schuster, S.C. (2007). MEGAN analysis of metagenomic data. Genome Res. *17*, 377–386.

Huson, D.H., and Weber, N. (2013). Microbial community analysis using MEGAN. Methods Enzymol. *531*, 465–485.

Jenkins, D.J., Kendall, C.W., and Vuksan, V. (1999). Inulin, oligofructose and intestinal function. J. Nutr. *129*, 1431S-1433s.

Jumpstart Consortium Human Microbiome Project Data Generation Working, G. (2012). Evaluation of 16S rDNA-based community profiling for human microbiome research. PLoS One 7, e39315.

Kim, M., Lee, K.H., Yoon, S.W., Kim, B.S., Chun, J., and Yi, H. (2013). Analytical tools and databases for metagenomics in the next-generation sequencing era. Genomics Inform *11*, 102–113.

Konwar, K.M., Hanson, N.W., Page, A.P., and Hallam, S.J. (2013). MetaPathways: a modular pipeline for constructing pathway/genome databases from environmental sequence information. BMC Bioinformatics *14*, 202.

Kovatcheva-Datchary, P., Egert, M., Maathuis, A., Rajilic-Stojanovic, M., de Graaf, A.A., Smidt, H., de Vos, W.M., and Venema, K. (2009). Linking phylogenetic identities of bacteria to starch fermentation in an *in vitro* model of the large intestine by RNA-based stable isotope probing. Environ. Microbiol. *11*, 914–926.

Kozich, J.J., Westcott, S.L., Baxter, N.T., Highlander, S.K., and Schloss, P.D. (2013). Development of a dual-index sequencing strategy and curation pipeline for analyzing amplicon sequence data on the MiSeq Illumina sequencing platform. Appl. Environ. Microbiol. 79, 5112–5120.

Kuczynski, J., Lauber, C.L., Walters, W.A., Parfrey, L.W., Clemente, J.C., Gevers, D., and Knight, R. (2012). Experimental and analytical tools for studying the human microbiome. Nat. Rev. Genet. *13*, 47–58.

Kuczynski, J., Stombaugh, J., Walters, W.A., Gonzalez, A., Caporaso, J.G., and Knight, R. (2011). Using QIIME to analyze 16S rRNA gene sequences from microbial communities. Curr. Protoc Bioinformatics *Chapter 10*, Unit 10 17.

Kumps, C., Van Roy, N., Heyrman, L., Goossens, D., Del-Favero, J., Noguera, R., Vandesompele, J., Speleman, F., and De Preter, K. (2010). Multiplex Amplicon Quantification (MAQ), a fast and efficient method for the simultaneous detection of copy number alterations in neuroblastoma. BMC Genomics *11*, 298.

Kurokawa, K., Itoh, T., Kuwahara, T., Oshima, K., Toh, H., Toyoda, A., Takami, H., Morita, H., Sharma, V.K., and Srivastava, T.P. (2007). Comparative metagenomics revealed commonly enriched gene sets in human gut microbiomes. DNA Res. *14*, 169–181.

Kurtz, S., Phillippy, A., Delcher, A.L., Smoot, M., Shumway, M., Antonescu, C., and Salzberg, S.L. (2004). Versatile and open software for comparing large genomes. Genome Biol. *5*, R12.

Lander, E.S., Linton, L.M., Birren, B., Nusbaum, C., Zody, M.C., Baldwin, J., Devon, K., Dewar, K., Doyle, M., FitzHugh, W., *et al.* (2001). Initial sequencing and analysis of the human genome. Nature *409*, 860–921.

Langmead, B., Trapnell, C., Pop, M., and Salzberg, S.L. (2009). Ultrafast and memory-efficient alignment of short DNA sequences to the human genome. Genome Biol. *10*, R25.

Laserson, J., Jojic, V., and Koller, D. (2011). Genovo: de novo assembly for metagenomes. J. Comput. Biol. *18*, 429–443.

Le Chatelier, E., Nielsen, T., Qin, J., Prifti, E., Hildebrand, F., Falony, G., Almeida, M., Arumugam, M., Batto, J.M., Kennedy, S., *et al.* (2013). Richness of human gut microbiome correlates with metabolic markers. Nature *500*, 541–546.

Letunic, I., Copley, R.R., Pils, B., Pinkert, S., Schultz, J., and Bork, P. (2006). SMART 5: domains in the context of genomes and networks. Nucleic Acids Res. *34*, D257-D260.

Ley, R.E., Backhed, F., Turnbaugh, P., Lozupone, C.A., Knight, R.D., and Gordon, J.I. (2005). Obesity alters gut microbial ecology. Proc. Natl. Acad. Sci. U.S.A. *102*, 11070–11075.

Ley, R.E., Hamady, M., Lozupone, C., Turnbaugh, P.J., Ramey, R.R., Bircher, J.S., Schlegel, M.L., Tucker, T.A., Schrenzel, M.D., and Knight, R. (2008). Evolution of mammals and their gut microbes. Science *320*, 1647–1651.

Ley, R.E., Turnbaugh, P.J., Klein, S., and Gordon, J.I. (2006). Microbial ecology: human gut microbes associated with obesity. Nature *444*, 1022–1023.

Li, H., and Durbin, R. (2009). Fast and accurate short read alignment with Burrows-Wheeler transform. Bioinformatics 25, 1754–1760.

Li, R., Li, Y., Kristiansen, K., and Wang, J. (2008). SOAP: short oligonucleotide alignment program. Bioinformatics 24, 713–714.

Li, R., Yu, C., Li, Y., Lam, T.W., Yiu, S.M., Kristiansen, K., and Wang, J. (2009). SOAP2: an improved ultrafast tool for short read alignment. Bioinformatics 25, 1966–1967.

Linnarsson, S. (2010). Recent advances in DNA sequencing methods – general principles of sample preparation. Exp. Cell Res. *316*, 1339–1343.

Liu, Y., Schmidt, B., and Maskell, D.L. (2011). Parallelized short read assembly of large genomes using de Bruijn graphs. BMC Bioinformatics *12*, 354.

Lorenzi, H.A., Hoover, J., Inman, J., Safford, T., Murphy, S., Kagan, L., and Williamson, S.J. (2011). TheViral MetaGenome Annotation Pipeline(VMGAP): an automated tool for the functional annotation of viral Metagenomic shotgun sequencing data. Stand Genomic Sci. *4*, 418–429.

Lozupone, C., Hamady, M., and Knight, R. (2006). UniFrac – an online tool for comparing microbial community diversity in a phylogenetic context. BMC Bioinformatics 7, 371.

Lozupone, C., and Knight, R. (2005). UniFrac: a new phylogenetic method for comparing microbial communities. Appl. Environ. Microbiol. *71*, 8228–8235.

Lozupone, C., Lladser, M.E., Knights, D., Stombaugh, J., and Knight, R. (2011). UniFrac: an effective distance metric for microbial community comparison. ISME J. 5, 169–172.

Lozupone, C.A., Stombaugh, J.I., Gordon, J.I., Jansson, J.K., and Knight, R. (2012). Diversity, stability and resilience of the human gut microbiota. Nature *489*, 220–230.

Luo, R., Liu, B., Xie, Y., Li, Z., Huang, W., Yuan, J., He, G., Chen, Y., Pan, Q., Liu, Y., *et al.* (2012). SOAPdenovo2: an empirically improved memory-efficient short-read de novo assembler. Gigascience *1*, 18.

Maathuis, A.J., van den Heuvel, E.G., Schoterman, M.H., and Venema, K. (2012). Galacto-oligosaccharides have prebiotic activity in a dynamic *in vitro* colon model using a (13)C-labelling technique. J. Nutr. *142*, 1205–1212.

Maccallum, I., Przybylski, D., Gnerre, S., Burton, J., Shlyakhter, I., Gnirke, A., Malek, J., McKernan, K., Ranade, S., Shea, T.P., *et al.* (2009). ALLPATHS 2: small genomes assembled accurately and with high continuity from short paired reads. Genome Biol. *10*, R103.

Mardis, E.R. (2008). Next-generation DNA sequencing methods. Annu. Rev. Genomics Hum. Genet. 9, 387–402.

Margulies, M., Egholm, M., Altman, W.E., Attiya, S., Bader, J.S., Bemben, L.A., Berka, J., Braverman, M.S., Chen, Y.-J., and Chen, Z. (2005). Genome sequencing in microfabricated high-density picolitre reactors. Nature *437*, 376–380.

Markowitz, V.M., Chen, I.M., Chu, K., Szeto, E., Palaniappan, K., Grechkin, Y., Ratner, A., Jacob, B., Pati, A., Huntemann, M., *et al.* (2012). IMG/M: the integrated metagenome data management and comparative analysis system. Nucleic Acids Res. *40*, D123–129.

Markowitz, V.M., Chen, I.M., Chu, K., Szeto, E., Palaniappan, K., Pillay, M., Ratner, A., Huang, J., Pagani, I., Tringe, S., *et al.* (2014). IMG/M 4 version of the integrated metagenome comparative analysis system. Nucleic Acids Res. *42*, D568–573.

Markowitz, V.M., Ivanova, N.N., Szeto, E., Palaniappan, K., Chu, K., Dalevi, D., Chen, I.M., Grechkin, Y., Dubchak, I., Anderson, I., *et al.* (2008). IMG/M: a data management and analysis system for metagenomes. Nucleic Acids Res. *36*, D534–538.

Mavromatis, K., Ivanova, N., Barry, K., Shapiro, H., Goltsman, E., McHardy, A.C., Rigoutsos, I., Salamov, A., Korzeniewski, F., and Land, M. (2007). Use of simulated data sets to evaluate the fidelity of metagenomic processing methods. Nat. Methods *4*, 495–500.

Mechnikov, I. (1908). Optimistic studies. By Il'ia Il'ich Mechnikov (New York, Putman's Sons).

Meddah, A.T., Yazourh, A., Desmet, I., Risbourg, B., Verstraete, W., and Romond, M.B. (2001). The regulatory effects of whey retentate from bifidobacteria fermented milk on the microbiota of the Simulator of the Human Intestinal Microbial Ecosystem (SHIME). J. Appl. Microbiol. *91*, 1110–1117.

Meyer, F., Paarmann, D., D'Souza, M., Olson, R., Glass, E.M., Kubal, M., Paczian, T., Rodriguez, A., Stevens, R., Wilke, A., *et al.* (2008). The metagenomics RAST server – a public resource for the automatic phylogenetic and functional analysis of metagenomes. BMC Bioinformatics 9, 386.

Miller, J.R., Koren, S., and Sutton, G. (2010). Assembly algorithms for next-generation sequencing data. Genomics 95, 315–327.

Mitra, S., Stark, M., and Huson, D.H. (2011). Analysis of 16S rRNA environmental sequences using MEGAN. BMC Genomics *12 Suppl. 3*, S17.

Morris, R.M., Rappe, M.S., Connon, S.A., Vergin, K.L., Siebold, W.A., Carlson, C.A., and Giovannoni, S.J. (2002). SAR11 clade dominates ocean surface bacterioplankton communities. Nature *420*, 806–810.

Muccioli, G.G., Naslain, D., Backhed, F., Reigstad, C.S., Lambert, D.M., Delzenne, N.M., and Cani, P.D. (2010). The endocannabinoid system links gut microbiota to adipogenesis. Mol. Syst Biol. *6*, 392.

Mulder, N.J., Apweiler, R., Attwood, T.K., Bairoch, A., Bateman, A., Binns, D., Bork, P., Buillard, V., Cerutti, L., and Copley, R. (2007). New developments in the InterPro database. Nucleic Acids Res. 35, D224–D228.

Namiki, T., Hachiya, T., Tanaka, H., and Sakakibara, Y. (2012). MetaVelvet: an extension of Velvet assembler to de novo metagenome assembly from short sequence reads. Nucleic Acids Res. *40*, e155.

Navas-Molina, J.A., Peralta-Sanchez, J.M., Gonzalez, A., McMurdie, P.J., Vazquez-Baeza, Y., Xu, Z., Ursell, L.K., Lauber, C., Zhou, H., Song, S.J., *et al.* (2013). Advancing our understanding of the human microbiome using QIIME. Methods Enzymol. *531*, 371–444.

Ng, P.C., and Kirkness, E.F. (2010). Whole genome sequencing. In Genetic Variation (Springer), pp. 215–226.

Nguyen, T., Shi, W., and Ruden, D. (2011). CloudAligner: A fast and full-featured MapReduce based tool for sequence mapping. BMC Res. Notes *4*, 171.

Nocker, A., Burr, M., and Camper, A.K. (2007). Genotypic microbial community profiling: a critical technical review. Microb. Ecol. *54*, 276–289.

Noguchi, H., Park, J., and Takagi, T. (2006). MetaGene: prokaryotic gene finding from environmental genome shotgun sequences. Nucleic Acids Res. *34*, 5623–5630.

O'Hara, A.M., and Shanahan, F. (2006). The gut flora as a forgotten organ. EMBO reports 7, 688–693.

Orvis, J., Crabtree, J., Galens, K., Gussman, A., Inman, J.M., Lee, E., Nampally, S., Riley, D., Sundaram, J.P., and Felix, V. (2010a). Ergatis: a web interface and scalable software system for bioinformatics workflows. Bioinformatics *26*, 1488–1492.

Orvis, J., Crabtree, J., Galens, K., Gussman, A., Inman, J.M., Lee, E., Nampally, S., Riley, D., Sundaram, J.P., Felix, V., *et al.* (2010b). Ergatis: a web interface and scalable software system for bioinformatics workflows. Bioinformatics *26*, 1488–1492.

Otto, T.D., Dillon, G.P., Degrave, W.S., and Berriman, M. (2011). RATT: Rapid Annotation Transfer Tool. Nucleic Acids Res. *39*, e57.

Parks, D.H., and Beiko, R.G. (2010). Identifying biologically relevant differences between metagenomic communities. Bioinformatics *26*, 715–721.

Pati, A., Ivanova, N.N., Mikhailova, N., Ovchinnikova, G., Hooper, S.D., Lykidis, A., and Kyrpides, N.C. (2010). GenePRIMP: a gene prediction improvement pipeline for prokaryotic genomes. Nat. Methods 7, 455–457.

Pavelic, P., Ragusa, S.R., Flower, R.L., Rinck-Pfeiffer, S.M., and Dillon, P.J. (1998). Diffusion chamber method for *in situ* measurement of pathogen inactivation in groundwater. Water Res. 32, 1144–1150.

Peng, Y., Leung, H.C., Yiu, S.M., and Chin, F.Y. (2011). Meta-IDBA: a de Novo assembler for metagenomic data. Bioinformatics *27*, i94–101.

Powell, D.R., and Seemann, T. (2013). VAGUE: a graphical user interface for the Velvet assembler. Bioinformatics *29*, 264–265.

Prentice, A.M., Gershwin, M.E., Schaible, U.E., Keusch, G.T., Victora, C.G., and Gordon, J.I. (2008). New challenges in studying nutrition–disease interactions in the developing world. J. Clin. Invest. *118*, 1322–1329.

Pruesse, E., Quast, C., Knittel, K., Fuchs, B.M., Ludwig, W., Peplies, J., and Glšckner, F.O. (2007). SILVA: a comprehensive online resource for quality checked and aligned ribosomal RNA sequence data compatible with ARB. Nucleic Acids Res. 35, 7188–7196.

Pushkarev, D., Neff, N.F., and Quake, S.R. (2009). Single-molecule sequencing of an individual human genome. Nat. Biotechnol. *27*, 847–850.

Qin, J., Li, R., Raes, J., Arumugam, M., Burgdorf, K.S., Manichanh, C., Nielsen, T., Pons, N., Levenez, F., Yamada, T., *et al.* (2010). A human gut microbial gene catalogue established by metagenomic sequencing. Nature *464*, 59–65.

Quail, M.A., Kozarewa, I., Smith, F., Scally, A., Stephens, P.J., Durbin, R., Swerdlow, H., and Turner, D.J. (2008). A large genome centre's improvements to the Illumina sequencing system. Nat. Methods *5*, 1005–1010.

Quast, C., Pruesse, E., Yilmaz, P., Gerken, J., Schweer, T., Yarza, P., Peplies, J., and Glockner, F.O. (2013). The SILVA ribosomal RNA gene database project: improved data processing and web-based tools. Nucleic Acids Res. *41*, D590–596.

Raes, J., Foerstner, K.U., and Bork, P. (2007a). Get the most out of your metagenome: computational analysis of environmental sequence data. Curr. Opin. Microbiol. *10*, 490–498.

Raes, J., Foerstner, K.U., and Bork, P. (2007b). Get the most out of your metagenome: computational analysis of environmental sequence data. Curr. Opin. Microbiol. *10*, 490–498.

Raes, J., Harrington, E.D., Singh, A.H., and Bork, P. (2007c). Protein function space: viewing the limits or limited by our view? Curr. Opin. Struct Biol. *17*, 362–369.

Ramirez-Farias, C., Slezak, K., Fuller, Z., Duncan, A., Holtrop, G., and Louis, P. (2009). Effect of inulin on the human gut microbiota: stimulation of *Bifidobacterium adolescentis* and *Faecalibacterium prausnitzii*. Br. J. Nutr. *101*, 541–550.

Rappé, M.S., and Giovannoni, S.J. (2003). The uncultured microbial majority. Annu. Rev. Microbiol. *57*, 369–394.

Ridaura, V.K., Faith, J.J., Rey, F.E., Cheng, J., Duncan, A.E., Kau, A.L., Griffin, N.W., Lombard, V., Henrissat, B., Bain, J.R., *et al.* (2013). Gut microbiota from twins discordant for obesity modulate metabolism in mice. Science *341*, 1241214.

Riesenfeld, C.S., Schloss, P.D., and Handelsman, J. (2004). Metagenomics: genomic analysis of microbial communities. Annu. Rev. Genet. *38*, 525–552.

Rodŕuez-Valera, F. (2004). Environmental genomics, the big picture? FEMS Microbiol. Lett. *231*, 153–158.

Roediger, W. (1980). Role of anaerobic bacteria in the metabolic welfare of the colonic mucosa in man. Gut *21*, 793–798.

Rose, D.J., Venema, K., Keshavarzian, A., and Hamaker, B.R. (2010). Starch-entrapped microspheres show a beneficial fermentation profile and decrease in potentially harmful bacteria during *in vitro* fermentation in faecal microbiota obtained from patients with inflammatory bowel disease. Br. J. Nutr. *103*, 1514–1524.

Roth, R.R., and James, W.D. (1988). Microbial ecology of the skin. Annu. Rev. Microbiol. *42*, 441–464.

Rusch, D.B., Halpern, A.L., Sutton, G., Heidelberg, K.B., Williamson, S., Yooseph, S., Wu, D., Eisen, J.A., Hoffman, J.M., and Remington, K. (2007). The Sorcerer II global ocean sampling expedition: northwest Atlantic through eastern tropical Pacific. PLoS Biol. *5*, e77.

Sboner, A., Mu, X.J., Greenbaum, D., Auerbach, R.K., and Gerstein, M.B. (2011). The real cost of sequencing: higher than you think! Genome Biol. *12*, 125.

Schatz, M.C. (2009). CloudBurst: highly sensitive read mapping with MapReduce. Bioinformatics 25, 1363–1369.

Schele, E., Grahnemo, L., Anesten, F., Hallen, A., Backhed, F., and Jansson, J.O. (2013). The gut microbiota reduces leptin sensitivity and the expression of the obesity-suppressing neuropeptides proglucagon (Gcg) and brain-derived neurotrophic factor (Bdnf) in the central nervous system. Endocrinology *154*, 3643–3651.

Schloss, P.D., Westcott, S.L., Ryabin, T., Hall, J.R., Hartmann, M., Hollister, E.B., Lesniewski, R.A., Oakley, B.B., Parks, D.H., Robinson, C.J., *et al.* (2009). Introducing mothur: open-source, platform-independent, community-supported software for describing and comparing microbial communities. Appl. Environ. Microbiol. *75*, 7537–7541.

Schmidt, T.M., DeLong, E.F., and Pace, N.R. (1991). Analysis of a marine picoplankton community by 16S rRNA gene cloning and sequencing. J. Bacterol. *173*, 4371–4378.

Scholz, M.B., Lo, C.-C., and Chain, P.S. (2012a). Next generation sequencing and bioinformatic bottlenecks: the current state of metagenomic data analysis. Curr. Opin. Biotechnol. 23, 9–15.

Scholz, M.B., Lo, C.C., and Chain, P.S. (2012b). Next generation sequencing and bioinformatic bottlenecks: the current state of metagenomic data analysis. Curr. Opin. Biotechnol. 23, 9–15.

Segata, N., Izard, J., Waldron, L., Gevers, D., Miropolsky, L., Garrett, W.S., and Huttenhower, C. (2011). Metagenomic biomarker discovery and explanation. Genome Biol. *12*, R60.

Segata, N., Waldron, L., Ballarini, A., Narasimhan, V., Jousson, O., and Huttenhower, C. (2012). Metagenomic microbial community profiling using unique clade-specific marker genes. Nat. Methods *9*, 811–814.

Sekirov, I., Russell, S.L., Antunes, L.C.M., and Finlay, B.B. (2010). Gut microbiota in health and disease. Physiol. Rev. *90*, 859–904.

Shen, J., Obin, M.S., and Zhao, L. (2013). The gut microbiota, obesity and insulin resistance. Mol. Aspects Med. *34*, 39–58.

Shi, W.B., Syrenne, R., Sun, J.Z., and Yuan, J.S. (2010). Molecular approaches to study the insect gut symbiotic microbiota at the 'omics' age. Insect Sci. *17*, 199–219.

Simpson, J.T., Wong, K., Jackman, S.D., Schein, J.E., Jones, S.J., and Birol, I. (2009). ABySS: a parallel assembler for short read sequence data. Genome Res. *19*, 1117–1123.

Sivieri, K., Morales, M.L., Saad, S.M., Adorno, M.A., Sakamoto, I.K., and Rossi, E.A. (2014). Prebiotic Effect of Fructooligosaccharide in the Simulator of the Human Intestinal Microbial Ecosystem (SHIME Model). J. Med. Food. *17*, 894–901.

Sivieri, K., Villarreal, M.L., Adorno, M.A., Sakamoto, I.K., Saad, S.M., and Rossi, E.A. (2013). *Lactobacillus* acidophilus CRL 1014 improved 'gut health' in the SHIME(R) reactor. BMC Gastroenterol. *13*, 100.

Sommer, F., and Backhed, F. (2013). The gut microbiota – masters of host development and physiology. Nat. Rev. Microbiol. *11*, 227–238.

Staley, J.T., and Konopka, A. (1985). Measurement of *in situ* activities of nonphotosynthetic microorganisms in aquatic and terrestrial habitats. Annu. Rev. Microbiol. 39, 321–346.

Stein, L. (2001). Genome annotation: from sequence to biology. Nat. Rev. Genet. 2, 493–503.

Stewart, A.C., Osborne, B., and Read, T.D. (2009). DIYA: a bacterial annotation pipeline for any genomics lab. Bioinformatics 25, 962–963.

Suau, A., Bonnet, R., Sutren, M., Godon, J.-J., Gibson, G.R., Collins, M.D., and Doré, J. (1999). Direct analysis of genes encoding 16S rRNA from complex communities reveals many novel molecular species within the human gut. Appl. Environ. Microbiol. *65*, 4799–4807.

Sun, J., Steindler, L., Thrash, J.C., Halsey, K.H., Smith, D.P., Carter, A.E., Landry, Z.C., and Giovannoni, S.J. (2011a). One carbon metabolism in SAR11 pelagic marine bacteria. PLoS One *6*, e23973.

Sun, S., Chen, J., Li, W., Altintas, I., Lin, A., Peltier, S., Stocks, K., Allen, E.E., Ellisman, M., Grethe, J., *et al.* (2011b). Community cyberinfrastructure for Advanced Microbial Ecology Research and Analysis: the CAMERA resource. Nucleic Acids Res. 39, D546–551.

Tech, M., and Meinicke, P. (2006). An unsupervised classification scheme for improving predictions of prokaryotic TIS. BMC Bioinformatics *7*, 121.

Temperton, B., and Giovannoni, S.J. (2012). Metagenomics: microbial diversity through a scratched lens. Curr. Opin. Microbiol. *15*, 605–612.

Terpend, K., Possemiers, S., Daguet, D., and Marzorati, M. (2013). Arabinogalactan and fructo-oligosaccharides have a different fermentation profile in the Simulator of the Human Intestinal Microbial Ecosystem (SHIME (R)). Environ. Microbiol. Rep *5*, 595–603.

Tickle, T.L., Segata, N., Waldron, L., Weingart, U., and Huttenhower, C. (2013). Two-stage microbial community experimental design. ISME J. *7*, 2330–2339.

Tringe, S.G., Von Mering, C., Kobayashi, A., Salamov, A.A., Chen, K., Chang, H.W., Podar, M., Short, J.M., Mathur, E.J., and Detter, J.C. (2005). Comparative metagenomics of microbial communities. Science *308*, 554–557.

Tuohy, K.M., Gougoulias, C., Shen, Q., Walton, G., Fava, F., and Ramnani, P. (2009). Studying the human gut microbiota in the trans-omics era-focus on metagenomics and metabonomics. Curr. Pharm. Des *15*, 1415–1427.

Turnbaugh, P.J., and Gordon, J.I. (2008). An invitation to the marriage of metagenomics and metabolomics. Cell *134*, 708–713.

Turnbaugh, P.J., Ley, R.E., Mahowald, M.A., Magrini, V., Mardis, E.R., and Gordon, J.I. (2006). An obesity-associated gut microbiome with increased capacity for energy harvest. Nature *444*, 1027–1031.

Tyson, G.W., Chapman, J., Hugenholtz, P., Allen, E.E., Ram, R.J., Richardson, P.M., Solovyev, V.V., Rubin, E.M., Rokhsar, D.S., and Banfield, J.F. (2004). Community structure and metabolism through reconstruction of microbial genomes from the environment. Nature *428*, 37–43.

Vaishampayan, P.A., Kuehl, J.V., Froula, J.L., Morgan, J.L., Ochman, H., and Francino, M.P. (2010). Comparative metagenomics and population dynamics of the gut microbiota in mother and infant. Genome Biol. Evol. *2*, 53–66.

Venema, K., and van den Abbeele, P. (2013). Experimental models of the gut microbiome. Best Pract Res. Clin. Gastroenterol. *27*, 115–126.

Venema, K., van Nuenen, M.H., van den Heuvel, E.G., Pool, W., and van der Vossen, J.M. (2003). The effect of lactulose on the composition of the intestinal microbiota and short-chain fatty acid production in human volunteers and a computer-controlled model of the proximal large intestine. Microb. Ecol. Health Dis. *15*, 94–105.

Venter, J.C., Remington, K., Heidelberg, J.F., Halpern, A.L., Rusch, D., Eisen, J.A., Wu, D., Paulsen, I., Nelson, K.E., and Nelson, W. (2004). Environmental genome shotgun sequencing of the Sargasso Sea. Science *304*, 66–74.

Wang, M., Ahrné, S., Jeppsson, B., and Molin, G. (2005). Comparison of bacterial diversity along the human intestinal tract by direct cloning and sequencing of 16S rRNA genes. FEMS Microbiol. Ecol. *54*, 219–231.

White, J.R., Maddox, C., White, O., Angiuoli, S.V., and Fricke, W.F. (2013). CloVR-ITS: Automated internal transcribed spacer amplicon sequence analysis pipeline for the characterization of fungal microbiota. Microbiome *1*, 6.

Wichmann, A., Allahyar, A., Greiner, T.U., Plovier, H., Lunden, G.O., Larsson, T., Drucker, D.J., Delzenne, N.M., Cani, P.D., and Backhed, F. (2013). Microbial modulation of energy availability in the colon regulates intestinal transit. Cell Host Microbe *14*, 582–590.

Wooley, J.C., Godzik, A., and Friedberg, I. (2010). A primer on metagenomics. PLoS Comput Biol. *6*, e1000667.

Wu, G.D., Chen, J., Hoffmann, C., Bittinger, K., Chen, Y.Y., Keilbaugh, S.A., Bewtra, M., Knights, D., Walters, W.A., Knight, R., *et al.* (2011). Linking long-term dietary patterns with gut microbial enterotypes. Science 334, 105–108.

Xiao, S., Fei, N., Pang, X., Shen, J., Wang, L., Zhang, B., Zhang, M., Zhang, X., Zhang, C., Li, M., *et al.* (2014). A gut microbiota-targeted dietary intervention for amelioration of chronic inflammation underlying metabolic syndrome. FEMS Microbiol. Ecol. *87*, 357–367.

Xie, G., Chain, P., Lo, C.C., Liu, K.L., Gans, J., Merritt, J., and Qi, F. (2010). Community and gene composition of a human dental plaque microbiota obtained by metagenomic sequencing. Mol. Oral Microbiol. *25*, 391–405.

Yatsunenko, T., Rey, F.E., Manary, M.J., Trehan, I., Dominguez-Bello, M.G., Contreras, M., Magris, M., Hidalgo, G., Baldassano, R.N., Anokhin, A.P., *et al.* (2012). Human gut microbiome viewed across age and geography. Nature *486*, 222–227.

Young, V.B., and Schmidt, T.M. (2004). Antibiotic-associated diarrhoea accompanied by large-scale alterations in the composition of the faecal microbiota. J. Clin. Microbiol. *42*, 1203–1206.

Youssef, N., Sheik, C.S., Krumholz, L.R., Najar, F.Z., Roe, B.A., and Elshahed, M.S. (2009). Comparison of species richness estimates obtained using nearly complete fragments and simulated pyrosequencing-generated fragments in 16S rRNA gene-based environmental surveys. Appl. Environ. Microbiol. *75*, 5227–5236.

Zarco, M., Vess, T., and Ginsburg, G. (2012). The oral microbiome in health and disease and the potential impact on personalized dental medicine. Oral Dis. *18*, 109–120.

Zerbino, D.R. (2010). Using the Velvet de novo assembler for short-read sequencing technologies. Curr. Protoc Bioinformatics *Chapter 11*, Unit 11 15.

Zerbino, D.R., McEwen, G.K., Margulies, E.H., and Birney, E. (2009). Pebble and rock band: heuristic resolution of repeats and scaffolding in the velvet short-read de novo assembler. PLoS One *4*, e8407.

Emerging Applications of Established Prebiotics: Promises Galore

18

Seema Patel

Abstract

Prebiotics are non-digestible food ingredients that invigorate and boost proliferation of health-promoting bacteria in the gastro-intestinal tract, while inhibiting enteric pathogens. The term prebiotics so far encompasses fermentable carbohydrates and oligosaccharides. Prebiotics were catapulted to prominence after their stimulatory role towards bifidobacteria was recognized. They are currently being implicated as sweetener, fat replacer, starter culture formulation, for bacteriocin induction as well as for gut health maintenance, laxative and colitis prevention. However, their functionality has now ramified to multiple significant domains. The unconventional applications have been met with moderate to excellent success *viz.* cancer inhibition, immune-augmentation, anti-allergic, diabetes control, cholesterol reduction, cardiovascular risk mitigation, obesity management, sepsis prevention, neural protection, vaginal health restoration, bone density increment, guard for kidney ailments, addressing dermatological issues, as well as feed additive and antibiotic substitute in aquaculture and veterinary care. The versatile usage of prebiotics has motivated intense research in this field. The exploration of new biological roles for established prebiotics, generation of new compounds from unconventional sources (even extending beyond carbohydrates) and upgrading of existing technology have been identified as promising. This chapter presents in a nutshell, the known prebiotics and their conventional as well as evolving implications for enrichment of food, drugs, cosmetics and use in the veterinary sector.

Introduction

Prebiotics are generally defined as non-digestible poly- and oligosaccharides, which promote the growth of beneficial bacteria in the colon, while holding back the proliferation of pathogens, namely *Salmonella sp., Escherichia coli, Campylobacter jejuni* etc. The term 'prebiotic' was coined by Gibson and Roberfroid (1995). Functional prebiotics are recognized by their resistance to gastric acidity, mammalian gastrointestinal (GI) enzymes and GI absorption; fermentation by intestinal microbiota and selective stimulation of beneficial bacteria (Gibson *et al.*, 2004). There exists an array of prebiotics with diverse origin and unique chemical attributes. The standard prebiotics with widespread utility are inulin, fructooligosaccharides (FOS), galactooligosaccharides (GOS), glucooligosaccharides, isomaltooligosaccharides (IMO), xylooligosaccharides (XOS), arabinoxylooligosaccharides (AXOS), mannanoligosaccharides (MOS), human milk oligosaccharides (HMO), pectin-derived acidic oligosaccharides, lactulose, polydextrose and lactitol (Stowell 2007; Sabater-Molina *et al.*, 2009; Xu *et al.*, 2009; Femia *et al.*, 2010). Mannitol, maltodextrin, raffinose and sorbitol are also prebiotics with proven healthy properties (Mandal *et al.*, 2009; Yeo and Liong, 2010; Vamanu and Vamanu, 2010). The fermentability of dietary fibres, namely bran β-glucan, gum and pectin into short-chain fatty acids (SCFAs), namely acetate, propionate and butyrate, suggest their potential prebiotic applications as well (Vaidya and Sheth, 2010; Lin *et al.*, 2011), although the selective stimulation of beneficial bacteria in the colon has not been shown for these. The interesting therapeutic spectrum of functional prebiotics makes it a quintessential area of biomedical science. To facilitate wider and faster dissemination of recent information on new strides of prebiotics, the key established and emerging applications of prebiotics have been discussed below (Fig. 18.1). It is crucial to know that not all prebiotics have similar bioactivity and hence not all applications are feasible for established (or new) prebiotics.

Established roles of prebiotics

The role of prebiotics in protecting gut health and enriching food has been appreciated since decades. The recent developments in these functions are discussed briefly before moving on to the emerging roles.

As food additive, starter culture, bacteriocin inducer

More and more findings supporting the useful role of prebiotics in pasta, bakery and allied food processing are coming forth. Soluble fibres of inulin influences dough and bread quality (Hager *et al.*, 2011). Its supplementation has been evidenced to lower the generation time of *Streptococcus thermophilus* and

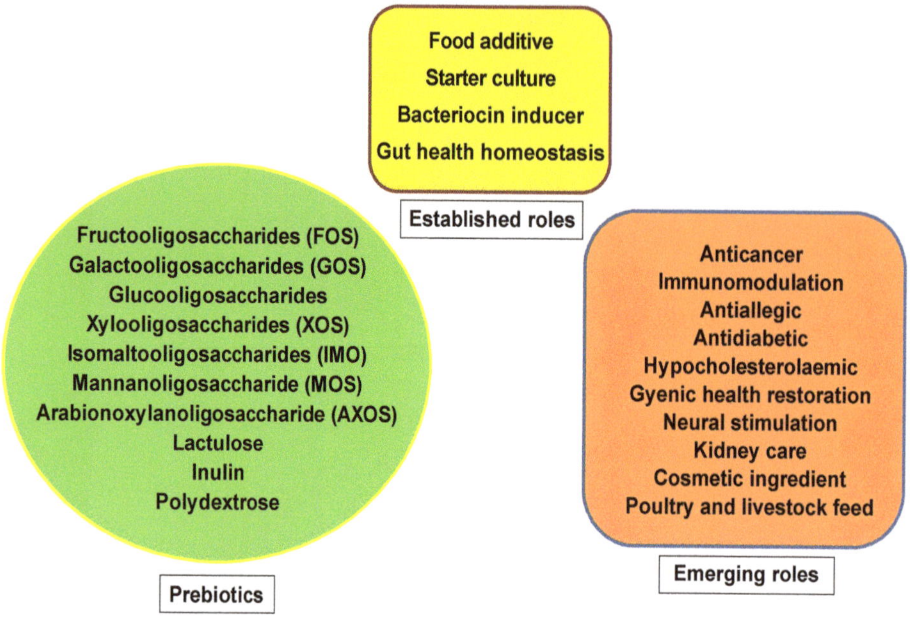

Figure 18.1 Established and emerging applications of prebiotics.

Lactobacillus acidophilus, thus improving the quality of yoghurt (Oliveira *et al.*, 2011). Rodrigues *et al.* (2012) investigated the effect of FOS and inulin (50:50) on the free fatty acid profile of cheese and reported an increase in the conjugated linoleic acid content (known to have several health benefits). Herfel *et al.* (2011) evaluated the prebiotic effect of polydextrose enrichment in cow milk-based infant formula for piglets. Results demonstrated that 18 days feeding increased ileal lactobacilli population and enhanced propionic and lactic acid concentrations of the ileum. MacDonald *et al.* (2011) reported that a diet designed for infants having phenylketonuria with prebiotic GOS/FOS oligosaccharide blend of 0.8 g/100 ml, maintained serum phenylalanine levels, enhanced the levels of bifidobacteria and lowered stool pH. The supplementation of soy milk with FOS and maltodextrin increased the α-galactosidase activity of *Lactobacillus* and Bifidobacterium strains, leading to enhanced hydrolysis and utilization of fructose and glucose, promoting their growth (Yeo and Liong 2010). The addition of dry spirulina to milk (6 mg/ml) stimulated the growth of *Lactococcus lactis* by about 27% (de Caire *et al.*, 2000). Long-chain and short-chain inulin combined in different proportions when added (7.5 g/100 g) to low-fat custards improved its rheological and sensory properties (Tárrega *et al.*, 2010). The addition of lactulose in skim milk increased the probiotics counts (particularly *Bifidobacterium lactis*), the acidification rate, concurrently reducing the fermentation time (Oliveira *et al.*, 2011). Functional *petit-suisse* cheese, made with the combination of prebiotics (inulin, oligofructose, honey) and probiotics (*L. acidophilus* La-5 and *B. lactis* BL04) was found to be more health-friendly, as fast fermentation and high lactic acid production, along with increased growth rates of bifidobacteria and lactobacilli was observed (Cardarelli *et al.*, 2007).

Ice creams supplemented with *Lactobacillus casei* and 2.5% inulin showed superior nutritional and sensory properties (Criscio *et al.*, 2010). Rößle *et al.* (2010) developed synbiotic fresh-cut apple wedges by applying *Lactobacillus rhamnosus* GG along with oligofructose and inulin. Fructan analysis showed that all prebiotics remained stable over the 14 days storage period and an intake of 100 g of apple supplied 2–3 g of prebiotics. Different dehydrated prebiotic fibres, namely oat bran, β-glucan and green banana flour, provided substratum for adherence and trehalose acted as a cell-protectant prolonging the viability of *Lactobacillus casei*. Oat bran added to a dairy fruit beverage was well-accepted by consumers (Guergoletto *et al.*, 2010). Apple purees enriched with two commercial FOS prebiotics showed stability for 30 days during storage at 4°C (Keenan *et al.*, 2011). *Lactobacillus plantarum*, *Bifidobacterium* spp. and *Bifidobacterium lactis* were found in buckwheat diet, confirming its candidature as a prebiotic product (Préstamo *et al.*, 2003). Lyophilized inulin syrup made from Jerusalem artichoke tuber was used as a prebiotic ingredient in the small-scale manufacture of wafer crackers (Hempel *et al.*, 2006). The impact of incorporating arabinoxylan oligosaccharides (AXOS) on the quality of a sugar-snap cookie was investigated. Replacement of 30% initial sucrose level by AXOS resulted in cookies with comparable size, though colour turned darker (Pareyt *et al.*, 2011). Peach-flavoured yoghurt containing the prebiotic was accepted by consumers (Gonzalez *et al.*, 2011). Allgeyer *et al.* (2010) studied the popularity of dairy products fortified with polydextrose, which proved to be an acceptable vehicle to deliver the prebiotic health effects at the end of the 30-day storage period.

Bacteriocins are antimicrobial by nature and possess the ability to replace chemical preservatives. Sorbitol has been reported to positively influence bacteriocin production from *Pediococcus acidilactici* LAB 5 (Mandal *et al.*, 2009). Vamanu and Vamanu (2010) confirmed the induction by prebiotics, namely inulin from chicory and dahlia, raffinose and lactulose on the synthesis of bacteriocins from *Lactobacillus paracasei* CMGB16.

Gut health maintenance, colitis and constipation prevention

A healthy gut is vital for deterring many metabolic syndromes. In this regard, certain prebiotics play a very effective role. Necrotizing enterocolitis (NEC) is a major cause of morbidity and mortality in premature infants. Prebiotic administration has been evidenced to manipulate the intestinal bacterial profile and accelerate the growth of commensal bacteria, eventually decreasing the incidence of NEC (Stenger *et al.*, 2010). Prebiotic supplemented formula increase stool colony counts of bifidobacteria and lactobacilli in preterm neonates without adversely affecting weight gain (Srinivasjois *et al.*, 2009). FOS consumption has been reported to increase faecal bolus and the frequency of depositions, sparing the misery of constipation (Sabater-Molina *et al.*, 2009). Passariello *et al.* (2011) evaluated the efficacy of zinc and prebiotics (FOS and XOS)-fortified hypotonic oral rehydration solution (ORS) for the treatment of diarrhoea in children. The combination reduced diarrhoea duration by stimulating water and electrolyte absorption across gut mucosa and eliminating pathogens. Partty *et al.* (2013) evaluated the impact of synbiotics on infants' well-being, crying, growth and normal microbiota. A mix of *Lactobacillus rhamnosus* GG, GOS and polydextrose administered for first two months of life resulted in less fussing, crying and lowered *Clostridium histolyticum* in the stool.

Emerging roles

Keeping track of the nascent, promising roles of prebiotics is the core of this topic. This is a rapidly evolving area for certain prebiotics, with wide-ranging possibilities being reported.

Anticancer agent (evidences so far restricted to animal experiments)

Prebiotics are rapidly gaining credence for their cancer prevention ability. Though the modes of action vary widely, competitive exclusion of pathogens, augmentation of advantageous commensals, degradation of carcinogenic compounds, binding of secondary bile acids, creation of an acidic milieu, and indirect effects on DNA protection have been recognized as important mechanisms. The consumption of arabinoxylan-oligosaccharides (AXOS) has been reported to reduce the occurrence of preneoplastic lesions in the colon of rats treated with the carcinogen 1,2-dimethylhydrazine (Femia *et al.*, 2010). In another rat model, the prebiotic epilactose inhibited the conversion of primary bile acids to secondary bile acids, the promoters of colon cancer (Watanabe *et al.*, 2008). Butyrate has been evidenced to suppress expression of transcription factor NF-κB in the human colon adenocarcinoma cell line HT-29, whereas acetate is known to increase the peripheral blood antibody production and natural killer activity in cancer patients (Macfarlane *et al.*, 2008). Gourineni *et al.* (2011) evaluated the chemopreventive effects of prebiotics fructans and isoflavones-rich soybean meal in mice. The combination significantly reduced incidence of azoxymethane-induced colon tumours. Komiyama *et al.* (2011) evaluated germinated barley foodstuff at a 10% dose in chronic and subacute colitis mice models. It lowered the caecal succinate content and significantly attenuated mucosal damage. It led to minimized risk of colon cancer as no dysplasia was observed. Itoh *et al.* (2012) evaluated anticancer effect of prebiotic brown rice extract fermented with *Aspergillus oryzae* on human colorectal HCT116 cells. Reduced viability of the tumour cells was attributed to exertion of oxidative damage, significant elevation of Bax protein, activation of caspase-3 in the cells, eventually resulting in apoptosis. Daily ingestion of the prebiotic diet for prevention of gut carcinogenesis was advocated. Bindels *et al.* (2012) reported that fermentation of inulin into propionate can counteract malignant cell proliferation in the liver tissue, as determined *in vitro* as well as *in vivo*. BaF3 cell growth was inhibited by propionate, through a cAMP level-dependent pathway. Also, the role of free fatty acid receptor 2 (GPR43), the receptor for propionate, in proliferation was elucidated. In mice transplanted with Bcr-Abl-transfected BaF3 cells, the mechanism was deciphered to be due to reduction in hepatic BaF3 cell infiltration and lowered inflammation. Hijova *et al.* (2013) conducted a 28-week study on anticancer potency of inulin in rats and showed that it nullified the adverse effect of dimethylhydrazine. It decreased the numbers of cyclooxygenase-2 (COX-2) and nuclear factor-kappaB (NFκB)-positive cells in the tunica mucosae and tela submucosae of the colon. The expression of IL-2, tumour necrosis factor (TNF-α) and IL-10 was also diminished. It showed that dietary intake of inulin prevents preneoplastic changes and inflammation that promote colon cancer development. Most of the *in vivo* trials have been studied on colon cancer. Effects of prebiotics should be investigated against other forms of malignant cancers.

Immune potentiator and anti-allergic

Prebiotics have been evidenced to reinforce the epithelial barrier and bolster gut mucosal immunity (Stoidis *et al.*, 2011). Prebiotics have a positive influence on the gut-associated lymphoid tissues (GALT) (Janardhana *et al.*, 2009). Inulin, FOS, MOS and AXOS consumption are capable of modulating the immune parameters in GALT, secondary lymphoid tissues and peripheral circulation (Bodera, 2008). Grüber *et al.* (2010) assessed the effect of supplementation of an infant formula with prebiotic containing neutral oligosaccharides and pectin-derived acidic oligosaccharides, on the occurrence of atopic dermatitis and obtained positive result. Vos *et al.* (2010) investigated the immunomodulatory effect of supplementation of a mixture of GOS, FOS and pectin-derived acidic oligosaccharides and reported positive modulation (Delgado *et al.*, 2011). Stam *et al.* (2011) hypothesized from a randomized study that the above prebiotic mixture promoted T helper 1 (Th1) and regulatory T-cells (Treg)-dependent immune responses and induced down-regulation of IgE-mediated allergic responses. Additionally, the same prebiotic administration did not interfere with the desired vaccine-specific serum antibody responses in healthy term infants. Kubota *et al.* (2014) hypothesized that the consumption of prebiotics by pregnant and lactating women is capable of preventing the onset of allergic disorders in their children by increasing the concentrations of immunoactive agents such as cytokines in breast milk. The subjects who consumed FOS (4 g/twice daily), had higher mRNA expression and protein level of cytokine IL-27 (linked with immunoregulatory function) in their breast milk than those in the control group. This intervention could be exploited to address allergic disorders in children. Moreno-Vilet *et al.* (2014) reported that agave fructans increased the levels of CD69 expression, proliferative activity and nitric oxide production in monocytes when administered with the probiotic *Lactobacillus* casei strain SACCO BGP93. It was reported that fructans are involved in the activation and selective differentiation of cells of the immune system through interactions with probiotics. Sagar *et al.* (2014) reported that a combination of *Bifidobacterium breve* with non-digestible oligosaccharides has strong anti-inflammatory effects in an asthmatic mouse model. Suppression of pulmonary airway inflammation; reduction in T-cell activation and mast cell degranulation; modulated expression of pattern recognition receptors, cytokines and transcription factors and induction of a Treg response were assumed to be the mode of action.

Antidiabetic

Significant evidences exist corroborating the glycaemic control ability of certain prebiotics. Everard *et al.* (2011) reported that in obese mice prebiotics (oligofructose) improve glucose homeostasis, which could be a key strategy in diabetes therapy. *d-chiro*-Inositol, found in buckwheat contributed to the improvement of insulin resistance by potentiating the action of insulin (Takahama *et al.*, 2011). Bomhof *et al.* (2014) studied the effect of oligofructose and *Bifidobacterium animalis* subsp. *lactis* BB-12 on gut microbiota and host metabolism administered to obese rats for 8 weeks. The combination improved glycaemia, while oligofructose could individually reduce insulin levels. Moroti *et al.* (2012) evaluated the effect of the consumption of a symbiotic shake containing *Lactobacillus acidophilus*, *Bifidobacterium bifidum* (Fortitech strains) and FOS on glycaemia and cholesterol levels in elderly people. A daily dose of 200 ml symbiotic shake containing 10^8 cfu/ml each of the probiotics along with 2g FOS was administered for 30 days. Analysis of blood samples showed a significant increase in high-density lipoprotein (HDL) cholesterol and a significant reduction in fasting glycaemia.

Cholesterol, cardiovascular, liver disease risk and obesity lowering

Synbiotic treatment of *Pediococcus acidilactici* LAB 5 with sorbitol for 1 month lowers the plasma cholesterol level of experimental mice (Mandal *et al.*, 2009). The intake of whole grain with resistant carbohydrates consistently has been associated with reduced risk of vascular issues (Harris and Kris-Etherton, 2010). The increase in plasma ferulic acid level following the ingestion of enzyme-treated prebiotic durum wheat consumption was reported and correlated to reduced risk of cardiovascular disease (Napolitano *et al.*, 2009). Cani *et al.* (2009) conducted a randomized controlled trial for two weeks and reported the lowered hunger sensation after a prebiotic (oligofructose) supplementation to be the outcome of increased postprandial plasma gut hormones (glucagon-like peptide 1 and peptide YY) concentration. Hess *et al.* (2011) suggested that satiety-induction of fermentable dietary fibres can be exploited to combat the global epidemic proportions of obesity. Wong *et al.* (2010) studied the effect of soy-fortified prebiotic (oligofructose-enriched inulin) in hyperlipidaemic adults. Greater reduction in LDL and increase in HDL cholesterol was reported as the result of dietary intervention. The role of prebiotics in potentiating the efficacy of soy foods in improvement of serum cholesterol profile has been validated. Buckwheat diet decreased the level of cholesterols in serum by enhancing their removal through faeces, improved diabetes by inhibiting the absorption of sugars and enhanced growth of beneficial intestinal bacteria (Takahama *et al.*, 2011). Nilsson *et al.* (2013) reported that a meal of brown bean (rich in GOS) modulated appetite regulatory hormones, within a time frame of 11–14 h. Suppression of hunger hormone and increment in satiety hormone was witnessed. Pachikian *et al.* (2013) reported that FOS is capable of attenuating n-3 PUFA depletion-caused hepatic steatosis, by changing gene expression in the liver. Decreased hepatic micro-RNA33 expression and increased colonic glucagon-like peptide 1 production was reported.

Management of sepsis and neural impairment

A convincing number of studies endorse the role of prebiotics in favourably modulating brain networking via gut microbiota (Savignac *et al.*, 2013). Nosocomial sepsis in preterm infants results in neuro-developmental impairment (Nair and Soraisham, 2013). Tackling this requires prevention of sepsis in the preterm infants in the neonatal intensive care unit (ICU), but this remains a major challenge. The immature intestinal epithelium of preterm infants is prone to penetration by bacteria and yeast. Modulation of the gut microbiota by prebiotics can help avert the risk (Nair and Soraisham, 2013). In critically ill patients (major abdominal surgery, trauma and ICU patients) synbiotic therapy has been shown to significantly reduce septic complications (Shimizu *et al.*, 2013). Savignac *et al.* (2013) reported that oral administration of FOS and GOS to rats increase hippocampal brain-neurotropic factor and N-methyl-D-aspartate-selective glutamate receptor1 (NR1) subunit expression compared to controls. GOS also increased hippocampal N-methyl-D-aspartate-selective glutamate receptor2A (NR2A) subunits and frontal cortex NR1 and d-serine. Involvement of gut hormones in imparting the significant anxiolytic effects is deduced (Savignac *et al.*, 2013). Further studies must be focussed on implication of prebiotics in addressing brain disorders.

Restoration of vaginal health and promotion of bone health

The female urogenital tract is susceptible to various infections. Depletion of lactobacilli in the vagina results in bacterial vaginosis, where the normal microbiota is replaced by several bacterial pathogens, usually *Gardnerella vaginalis* and obligate anaerobes. The infection results in complications such as premature labour, low birth weight and increased risk of HIV. Prebiotics when administered along with probiotics show the ability to restore vaginal ecosystem by increasing lactobacilli density. In post-menopausal women, anaerobic pathogens tend to dominate the vaginal microbiota. Pliszczak *et al.* (2011) designed a vaginal bio-adhesive delivery system based on pectinate-hyaluronic acid microparticles for probiotics and prebiotics encapsulation. Pliszczak *et al.* (2012) further investigated the influence of hyaluronic acid and FOS on gelation for the vaginal delivery system. The results showed mucoadhesive enhancing properties of the two components.

Rodrigues *et al.* (2012b) evaluated the effect of yacon flour and *Bifidobacterium longum* on the modulation of bone health of rats. Sacrificing the animals followed by tibia composition analysis revealed significant increase in calcium, phosphorus and magnesium content. Highest increase in fracture strength was observed in the synbiotic diet-fed group. Total anaerobe and weight of the caecum improved. The increased mineral level in bone might delay the onset of degenerative disease. Yang *et al.* (2013) reported that the water extract of an orchid *Anoectochilus formosanus* (200–400 mg/kg daily) prevents bone loss, associated with ovarian hormone deficiency in the ovariectomized rats. The mechanism was decoded to be due to enhanced number of faecal bifidobacteria, accompanied with decrease in caecal pH and increased caecal weight and higher SCFA. Whisner *et al.* (2013) also evaluated the role of prebiotics in maintaining bone density. In adolescents who consumed GOS fortified-smoothie drinks for a period of 3 weeks, significant improvements in calcium absorption was observed. The elevated absorption of the mineral was hypothesized to be mediated by changes in the levels of bifidobacteria.

Alleviation of kidney and bladder issues

Uraemic toxins (e.g. ammonia, indoxyl sulphate and p-cresyl sulfate) accelerate the progression of chronic kidney disease, rendering it essential to alleviate the toxin overload of the serum. The modulation of gut bacteria by prebiotics or probiotics is considered to be effective in reducing the production of indole in the intestine, which could lead to reduced serum levels of toxin (Niwa, 2013). Renal failure might be averted by administering prebiotics.

HMO acts as receptor analogues to inhibit the adhesion of pathogens on the epithelial surface of infant intestine and interact directly with immune cells (Boehm and Stahl 2007). Lin *et al.* (2014) investigated the role of HMO in protection of human bladder epithelial cells infected with the pathogen. It came forth that the oligosaccharide shields bladder epithelial cells from deleterious cytotoxic and proinflammatory effects, leading to reduced risk of invasion and adherence of the pathogen.

For skin health and as cosmetic ingredient

Skin health is compromised (reduced skin hydration, disrupted keratinization) by excess phenol generation in gut. It was demonstrated that prebiotics when applied on the skin can effectively decrease *Propionibacterium acnes* population and cure acnes, as met with a significant 91% success in human trial (Bockmühl *et al.*, 2007). Miyazaki *et al.* (2014) conducted a double-blind, placebo-controlled trial and demonstrated that the daily intake of *Bifidobacterium breve* strain Yakult and GOS reduces serum total phenol levels (a causative factor of skin issues). It leads to the prevention of skin dryness and disruption of keratinization in healthy adult women. Capitanio *et al.* (2012) reported that seaweed-derived oligosaccharides complexed with 0.1% zinc pyrrolidone when used topically for 8 weeks produces a significant reduction in acne. A number of prebiotics-enriched cosmetic products are expected to be formulated in near future, when more evidence of their benefit on skin health will be produced.

In poultry, fishery, pig and cattle feed

Prebiotics have proven very useful in pisci-, poultry and livestock rearing as well. Elevated defence against pathogens, stronger immunity and better growth performance are some of the observed benefits. In terms of mode of action, treatment of MQ-NCSU chicken macrophage cell line with prebiotic β-1,4-mannobiose dose-dependently increased phagocytic activity of macrophages towards *Salmonella* strains (Ibuki *et al.*, 2011). Prebiotic XOS enhanced the growth performance and digestive enzyme activities of carps (Xu *et al.*, 2009). Ahmdifar *et al.* (2011) investigated the effects of inulin ingestion on haematological and biochemical parameters in juvenile great sturgeons. A dose-dependent increase in enzymes and white blood cell counts was reported. Holstein heifer calves supplemented with a product, which contained fermentation products of *Lactobacillus gasseri* OLL2716 and *Propionibacterium freudenreichi* ET-3, had more lactobacilli in their faeces (Heinrichs *et al.*, 2009). Brown algae *Ascophyllum*, when incorporated in feed material, improved the gut microbiota of weaned piglets, by inhibiting their small intestinal and hindgut microbiota, especially *E. coli* (Dierick *et al.*, 2010). Oral administration of chicory root for 16 days to pigs increased the mRNA and protein expression of cytochrome P450 1A2 and 2A, the enzymes having crucial role in metabolism of drugs and endogenous compounds (Rasmussen *et al.*, 2011). Mechanistically, a product, formulated with a non-living yeast cell walls or MOS, acted as an anti-adhesive for Shiga toxin-producing *Escherichia coli* O157:H7 colonization and a mycotoxin (found in contaminated feed) *in vitro*. Also, its consumption improved milk production and feed conversion efficiency in dairy cattle (Baines *et al.*, 2011). The results imply that prebiotics should be implicated in feed and supplements of a wider range of animals.

Novel potential prebiotics production from unconventional sources and their optimization

Recent times have seen the production of potential prebiotics from diverse sources, ranging from agro-wastes, dairy effluents and industrial refuge to novel sources as seaweeds, cereal brans and mushrooms. The list is vast and increasing. To name a few, dairy whey permeate, soybean whey (a by-product of tofu manufacturing) (Tenorio *et al.*, 2010; Golowczyc *et al.*, 2013), soy sauce lees (Yang *et al.*, 2011a), mung bean, Bengal gram husk and wheat bran, triticale bran (Madhukumar and Muralikrishna, 2010; Wang *et al.*, 2010; Agil and Hosseinian, 2012), spent osmotic sugar solution (the residue left after osmotic dehydration of fruits and vegetables) (Aachary and Prapulla, 2009), barley husks, spent grains (Gullón *et al.*, 2011a), konjac glucomannan hydrolysate (Connolly *et al.*, 2010), apple pomace (Gullón *et al.*, 2011b), bergamot peel (Mandalari *et al.* (2007), *Lentinula edodes* stipe, *Pleurotus eryngii* base, and *Flammulina velutipes* base

(Synytsya *et al.*, 2009; Chou *et al.*, 2013), hazelnut and almond skin (Montella *et al.*, 2013; Liu *et al.*, 2014).

Some novel sources have been identified lately, namely yacon (Ojansivu *et al.*, 2011), yam (Huang *et al.*, 2012), lupin kernel fibre (Smith *et al.*, 2006), blueberry (Molan *et al.*, 2009a), Indian mulberry (Yang *et al.*, 2011b), dragon fruits (Wichienchot *et al.*, 2010), glycated pea proteins (Dominika *et al.*, 2011), selenium-containing green tea (Molan *et al.*, 2009b), seaweeds (*Gelidium*, *Ulva*) (O'Sullivan *et al.*, 2010; Ramnani *et al.*, 2012; Gupta and Abu-Ghannam, 2011), burdock (Li *et al.*, 2008), Agave (*A. tequilana*, *A. angustifolia*, *A. salmiana*) (Ávila-Fernández *et al.*, 2011; Allsopp *et al.*, 2013, Moreno-Vilet, 2014). Endemic plant sources have also been identified that boast of functional prebiotics, e.g. Australian and New Zealand plants *Arthropodium cirratum* and *Cordyline australis* (Vidanarachchi *et al.*, 2009). The bioactive components of these sources have not been identified yet, and the clinical proof of these compounds being indeed prebiotic, according to the current definition, remains to be established.

However, the generation from all sources is not economically viable. Discovery of cost-effective approaches for the conversion and recovery of prebiotics are required. The prevalent current technology for the production of prebiotic oligosaccharides is acid, enzymatic, microwave hydrolysis of polysaccharides sourced from plants or microbes (de Paula *et al.*, 2008; Ávila-Fernández *et al.*, 2011) or synthesis by transglycosylation reactions employing microbial enzymes (Chen *et al.*, 2011, Nimpiboon *et al.*, 2011). Also, cloning and expressing genes coding for enzymes catalysing conversion of polysaccharides into oligosaccharides has received some success, namely glucosidase gene from *Thermoanaerobacter ethanolicus* expressed in *E. coli*, and β-hexosyltransferase gene from *Sporobolomyces singularis* expressed in yeast *Pichia pastoris* (Wang *et al.*, 2009; Dagher *et al.*, 2013). Also, enzyme immobilization on various matrices has resulted in higher yield of prebiotic oligosaccharides (Quang *et al.*, 2011). Roasting, baking and boiling increased the resistant starch content of processed cereals and its products (Vaidya and Sheth, 2010).

Human milk oligosaccharides (HMO) are increasingly being recognized as vital prebiotics for infant well-being. However, its complexity in terms of structural and functional diversity restricts its inclusion in dietary ingredients. Now, enzymatic production of some of its components is being carried out with moderate success. Monomers N-acetylglucosamine and fucose are identified as important factors. Separation and concentration of HMO-like oligosaccharides from bovine milk is also being carried out (Barile and Rastall, 2013). Baumgärtner *et al.* (2014) engineered an *E. coli* capable of producing lacto-N-tetraose (a component of HMO) by chromosomal integration of the genes lgtA and wbgO, encoding β-1,3-N-acetylglucosaminyltransferase and β-1,3-galactosyltransferase.

Future directions and concluding remarks

Though limited by concrete validations, there are several areas which are expected to be benefited by ingestion of certain prebiotics. Investigations in those areas should be expedited. However, not all probiotics show efficiency in all disorders, and hence the choice of prebiotic to use is critical for success. Neonatal health, early life immune insufficiency, metabolic syndromes, female genital tract infections, geriatric degenerative problems, mutagen-induced cancers and post-operative complications might be tackled by providing the right (existing or novel) prebiotics. It is essential to recognize that, according to the prebiotic definition, all therapeutic benefits stem from manipulating the gut microbiota, safeguarding the gut integrity against invaders and alleviating toxic metabolites. If harnessed properly, prebiotics-based dietary intervention can ensure longer, healthier life.

As outlined above, the discussed applications in no way limit the utility range of prebiotics in the nutraceutical, pharmaceutical and cosmetic sector. The implication possibilities are immense, waiting to be unravelled. To wrap up this topic, it could be safely stated that prebiotics act as defenders of gut security. They mediate the host–microbe cross-talk at the epithelial mucosa. The incorporation of prebiotics in the diet seems to be an effective way to tackle myriad metabolic syndromes, namely cancer, immune efficiency, hypersensitivity, obesity, cardiovascular, hypertension, diabetes, gynaecological diseases, psychological disorders, renal ailments, skin complications, although how the change in gut microbiota leads to these benefits is hardly understood. Expanding the existing array of prebiotics can be a robust step towards this goal. Tailored prebiotics with specific functional properties should be the next endeavour to target the crux of health issues. The reported clinical significance can be further validated and underlying mechanism decoded through adequate number of well-designed, randomized, double blind, placebo-controlled human trials. Efforts must be directed towards exposing the hidden potential of prebiotics to ensure nutritious food and to improve human and animal health.

References

Aachary, A.A., and Prapulla, S.G. (2009). Value addition to spent osmotic sugar (SOS) by enzymatic conversion to fructooligosaccharides (FOS), a low calorie prebiotic. Innovat. Food Sci. Emerg. Tech. *10*, 284–288.

Agil, R., and Hosseinian, F. (2012). Dual functionality of triticale as a novel dietary source of prebiotics with antioxidant activity in fermented dairy products. Plant Foods Hum. Nutr. *67*, 88–93.

Ahmdifar, E., Akrami, R., Ghelichi, A., and Zarejabad, A.M. (2011). Effects of different dietary prebiotic inulin levels on blood serum enzymes, hematologic, and biochemical parameters of great sturgeon (*Huso huso*) juveniles. Comp. Clin. Pathol. *20*, 447–451.

Allgeyer, L.C., Miller, M.J., and Lee, S.-Y. (2010). Sensory and microbiological quality of yoghurt drinks with prebiotics and probiotics. Dairy Sci. *93*, 4471–4479.

Allsopp, P., Possemiers, S., Campbell, D., Oyarzabal, I.S., Gill, C., and Rowland, I. (2013). An exploratory study into the putative prebiotic activity of fructans isolated from *Agave angustifolia* and the associated anticancer activity. Anaerobe 22, 38–44.

Angioloni, A., and Collar, C. (2011). Physicochemical and nutritional properties of reduced-caloric density high fibre breads. LWT-Food Sci. Technol. *44*, 747–758.

Ávila-Fernández, A., Galicia-Lagunas, N., Rodríguez-Alegría, M.E., Olvera, C., and López-Munguía, A. (2011). Production of functional oligosaccharides through limited acid hydrolysis of agave fructans. Food Chem. *129*, 380–386.

Baines, D., Erb, S., Lowe, R., Turkington, K., Sabau, E., Kuldau, G., Juba, J., Masson, L., Mazza, A., and Roberts, R. (2011). A prebiotic, Celmanax ™, decreases *Escherichia coli* O157:H7 colonization of bovine cells and feed-associated cytotoxicity *in vitro*. BMC Res. Notes *4*, 110.

Barile, D., and Rastall, A. (2013). Human milk and related oligosaccharides as prebiotics. Curr. Opin. Biotechnol. 2, 214–219.

Baumgärtner, F., Conrad, J., Sprenger, G.A., and Albermann, C. (2014). Synthesis of the human milk oligosaccharide lacto-N-tetraose in metabolically engineered, plasmid-free *E. coli*. ChemBioChem *15*, 1896–1900.

Bindels, L.B., Porporato, P., Dewulf, E.M., Verrax, J., Neyrinck, A.M., Martin, J.C., Scott, K.P., Buc Calderon, P., Feron, O., Muccioli, G.G., *et al.* (2012). Gut microbiota-derived propionate cancer cell proliferation in the liver. Br. J. Cancer *107*, 1337–1344.

Bockmühl, D., Jassoy, C., Nieveler, S., Scholtyssek, R., Wadle, A., and Waldmann-Laue, M. (2007). Prebiotic cosmetics: an alternative to antibacterial products. Int. J. Cosmet. Sci. *29*, 63–64.

Bodera, P. (2008). Influence of prebiotics on the human immune system (GALT). Rec. Pat. Inflamm. Allergy Drug Discover. 2, 149–153.

Boehm, G., and Stahl, B. (2007). Oligosaccharides from milk. J. Nutr. *137*, 847–849.

Bomhof, M.R., Saha, D.C., Reid, D.T., Paul, H.A., and Reimer, R.A. (2014). Combined effects of oligofructose and *Bifidobacterium animalis* on gut microbiota and glycemia in obese rats. Obesity *22*, 763–767.

Cani, P.D., Lecourt, E., Dewulf, E.M., Sohet, F.M., Pachikian, B.D., Naslain, D., De Backer, F., Neyrinck, A.M., and Delzenne, N.M. (2009). Gut microbiota fermentation of prebiotics increases satietogenic and incretin gut peptide production with consequences for appetite sensation and glucose response after a meal. Am. J. Clin. Nutr. *90*, 1236–1243.

Capitanio, B., Sinagra, J.L., Weller, R.B., Brown, C., and Berardesca, E. (2012). Randomized controlled study of a cosmetic treatment for mild acne. Clin. Exp. Dermatol. 37, 346–349.

Cardarelli, H.R., Saad, S.M.I., Gibson, G.R., and Vulevic, J. (2007). Functional *petit-suisse* cheese: measure of the prebiotic effect. Anaerobe *13*, 200–207.

Chen, J., Chen, X., Xu, X., Ning, Y., Jin, Z., and Tian, Y. (2011). Biochemical characterization of an intracellular ^{6}G-fructofuranosidase from *Xanthophyllomyces dendrorhous* and its use in production of neo-fructooligosaccharides (neo-FOSs). Bioresour. Technol. *102*, 1715–1721.

Chou, W.T., Sheih, I.C., and Fang, T.J. (2013). The applications of polysaccharides from various mushroom wastes as prebiotics in different systems. J. Food Sci. 78, 1041–1048.

Connolly, M.L., Lovegrove, J.A., and Tuohy, K.M. (2010). Konjac glucomannan hydrolysate beneficially modulates bacterial composition and activity within the faecal microbiota. J. Functional Foods 2, 219–224.

Criscio, T.D., Fratianni, A., Mignogna, R., Cinquanta, L., Coppola, R., Sorrentino, E., and Panfili, G. (2010). Production of functional probiotic, prebiotic, and synbiotic ice creams. J. Dairy Sci. *93*, 4555–4564.

Dagher, S.F., Azcarate-Peril, M.A., and Bruno-Barcena, J.M. (2013). Heterologous expression of a bioactive β-hexosyltransferase, an enzyme producer of prebiotics, from *Sporobolomyces singularis*. Appl. Environ. Microbiol. *79*, 1241–1249.

de Caire, G.Z., Parada, J.L., Zaccaro, M.C., and de Cano, M.M.S. (2000). Effect of *Spirulina platensis* biomass on the growth of lactic acid bacteria in milk. World J. Microbiol. Biotechnol. *16*, 563–565.

Delgado, G.T.C., Tamashiro, W.M.D.S.C., Junior, M.R.M., Moreno, Y.M.F., and Pastore, G.M. (2011). The putative effects of prebiotics as immunomodulatory agents. Food Res. Int. *4*, 3167–3173.

de Paula, V.C., Pinheiro, I.O., Lopes, C.E., and Calazans, G.M.T. (2008). Microwave-assisted hydrolysis of Zymomonas mobilis levan envisaging oligofructan production. Bioresour. Technol. *99*, 2466–2470.

Dierick, N., Ovyn, A., De Smet, S. (2010). *In vitro* assessment of the effect of intact marine brown macro-algae *Ascophyllum nodosum* of the gut flora piglets. Livestock Sci. *133*, 154–156.

Dominika, S., Arjan, N., Karyn, R., and Henryk, K. (2011). The study on the impact of glycated pea proteins on human intestinal bacteria. Int. J. Food Microbiol. *145*, 267–272.

Everard, A., Lazarevic, V., Derrien, M., Girard, M., Muccioli, G.M., Neyrinck, A.M., Possemiers, S., Holle, A.V., Francois, P., de Vos W.M., *et al.* (2011). Responses of gut microbiota and glucose and lipid metabolism to prebiotics in genetic obese and diet-induced leptin-resistant mice. Diabetes *11*, 2775–2786.

Femia, A.P., Salvadori, M., Broekaert, W.F., Francois, I.E.J.A., Delcour, J.A., Chritophe, M.C., and Giovanna, C. (2010). Arabinoxylan-oligosaccharides (AXOS) reduce preneoplastic lesions in the colon of rats treated with 1,2-dimethylhydrazine (DMH). Eur. J. Nutr. *49*, 127–132.

Gibson, G.R., and Roberfroid, M.B. (1995). Dietary modulation of the human colonic microbiota: introducing the concept of prebiotics. J. Nutr. *125*, 1401–1412.

Gibson, G.R., Probert, H.M., Loo, J.V., Rastall, R.A., and Roberfroid, M.B. (2004). Dietary modulation of the human colonic microbiota: updating the concept of prebiotics. Nutr. Res. Rev. *17*, 259–275.

Golowczyc, M., Vera, C., Santos, M., Guerrero, C., Carasi, P., Illanes, A., Gomez-Zavaglia, A., and Tymczyszyn, E. (2013). Use of whey permeate containing *in situ* synthesized galacto-oligosaccharides for the growth and preservation of *Lactobacillus plantarum*. J. Dairy Res. *80*, 374–381.

Gonzalez, N.J., Adhikari, K., and Sancho-Madriz, M.F. (2011). Sensory characteristics of peach-flavoured yoghurt drinks containing prebiotics and synbiotics. LWT – Food Sci. Technol. *44*, 158–163.

Gourineni, V.P., Verghese, M., Boateng, J., Shckelford, L., Bhat, N.K., and Walker, L.T. (2011). Combinational effects of prebiotics and soybean against azoxymethane-induced colon cancer *in vivo*. J. Nutr. Metab. ID 868197. doi: 10.1155/2011/868197.

Grüber, C., van Stuijvenberg, M., Mosca, F., Moro, G., Chirico, G., Braegger, C.P., Riedler, J., Boehm, G., and Wahn, U. (2010). Reduced occurrence of early atopic dermatitis because of immunoactive prebiotics among low-atopy-risk infants. J. Allergy Clin. Immunol. *126*, 791–797.

Guergoletto, K.B., Magnani, M., Martin, J.S., and Andrade, C.G.T.D., and Garcia, S. (2010). Survival of *Lactobacillus casei* (LC-1). adhered to prebiotic vegetal fibres. Innov. Food Sci. Emerg. Techn. *11*, 415–421.

Gullón, P., González-Muñoz, M.J., and Parajó, J.C. (2011a). Manufacture and prebiotic potential of oligosaccharides derived from industrial solid wastes. Bioresour. Technol. *102*, 6112–6119.

Gullón, B., Gullón, P., Sanz, Y., Alonso, J.L., and Parajó, J.C. (2011b). Prebiotic potential of a refined product containing pectic oligosaccharides. LWT-Food Sci. Technol. *44*, 1687–1696.

Gupta, S., and Abu-Ghannam, N. (2011). Bioactive potential and possible health effects of edible brown seaweeds. Trends Food Sci. Technol. *22*, 315–326.

Hager, A.-S., Ryan, L.A.M., Schwab, C., Ganzle, M.G., O'Doherty, J.V., and Arendt, E.K. (2011). Influence of the soluble fibres inulin and oat glucan on quality of dough and bread. Eur. Food Res. Technol. *232*, 405–413.

Harris, K.A., and Kris-Etherton, P.M. (2010). Effects of whole grains on coronary heart disease risk. Curr. Atheroscler. Rep. *12*, 368–376.

Heinrichs, A.J., Jones, C.M., Elizondo-Salazar, J.A., and Terrill, S.J. (2009). Effects of a prebiotic supplement on health of neonatal dairy calves. Livestock Sci. *125*, 149–154.

Hempel, S., Jacob, A., and Rohm, H. (2006). Influence of inulin modification and flour type on the sensory quality of prebiotic wafer crackers. Eur. Food Res. Technol. *224*, 335–341.

Herfel, T.M., Jacobi, S.K., Lin, X., Fellner, V., Walker, D.C., Jouni, Z.E., and Odle, J. (2011). Polydextrose enrichment of infant formula demonstrates prebiotic characteristics by altering intestinal microbiota, organic acid concentrations, and cytokine expression in suckling piglets. J. Nutr. *141*, 2139–2145.

Hess, J.R., Birkett, A.M., Thomas, W., and Slavin, J.L. (2011). Effects of short-chain fructooligosaccharides on satiety responses in healthy men and women. Appetite *56*, 128–134.

Hijova, E., Szabadosova, V., Stofilova, J., and Hrckova, G. (2013). Chemopreventive and metabolic effects of inulin on colon cance development. J. Vet. Sci. *14*, 387–393.

Huang, C.-H., Cheng, J.-Y., Deng, M.-C., Chou, C.-H., and Jan, T.-R. (2012). Prebiotic effect of diosgenin, an immunoactive steroidal sapogenin of the Chinese yam. Food Chem. *132*, 428–432.

Huerta, L.M., Vera, C., Guerrero, C., Wilson, L., and Illanes, A. (2011). Synthesis of galacto-oligosaccharides at very high lactose concentrations with immobilized β-galactosidases from *Aspergillus oryzae*. Proc. Biochem. *46*, 245–252.

Ibuki, M., Kovacs-Nolan, J., Fukui, K., Kanatani, H., and Mine, Y. (2011). β 1–4 mannobiose enhances Salmonella-killing activity and activates innate immune responses in chicken macrophages. Vet. Immunol. Immunopath. *139*, 289–295.

Itoh, M., Nishibori, N., Sagara, T., Horie, Y., Motojima, A., and Morita, K. (2012). Extract of fermented brown rice induces apoptosis of human colorectal tumour cells by activating mitochondrial pathway. Phytother. Res. *26*, 1661–1666.

Janardhana, V., Broadway, M.M., Bruce, M.P., Lowenthal, J.W., Geier, M.S., Hughes, R.J., and Bean, A.G. (2009). Prebiotics modulate immune responses in the gut-associated lymphoid tissue of chickens. J. Nutr. *139*, 1404–1409.

Keenan, D.F., Brunton, N., Butler, F., Wouters, R., and Gormley, R. (2011). Evaluation of thermal and high hydrostatic pressure processed apple purees enriched with prebiotic inclusions. Innov. Food Sci. Emerg. Technol. *12*, 261–268.

Komiyama, Y., Mitsuyama, K., Masuda, J., Yamasaki, H., Takedatsu, H., Andoh, A., Tsuruta, O., Fukuda, M., and Kanauchi, O. (2011). Prebiotic treatment in experimental colitis reduces the risk of colitic cancer. J. Gastroenterol. Hepatol. *26*, 1298–1308.

Kubota, T., Shimojo, N., Nonaka, K., Yamashita, M., Ohara, O., Igoshi, Y., Ozawa, N., Nakano, T., Morita, Y., Inoue, Y., *et al.* (2014). Prebiotic consumption in pregnant and lactating women increases IL-27 expression in human milk. Br. J. Nutr. *111*, 625–632.

Li, D., Kim, J.M., Jin, Z., and Zhou, J. (2008). Prebiotic effectiveness of inulin extracted from edible burdock. Anaerobe *14*, 29–34.

Lin, B., Gong, J., Wang, Q., Cui, S., Yu, H., and Huang, B. (2011). In-vitro assessment of the effects of dietary fibres on microbial fermentation and communities from large intestinal digesta of pigs. Food Hydrocol. *25*, 180–188.

Lin, A.E., Autran, C.A., Espanola, S.D., Bode, L., and Nizet, V. (2014). Human milk oligosaccharides protect bladder epithelial cells against uropathogenic *Escherichia coli* invasion and cytotoxicity. J. Infect. Dis. *209*, 389–398.

Liu, Z., Lin, X., Huang, G., Zhang, W., Rao, P., and Ni, L. (2014). Prebiotic effects of almonds and almond skins on intestinal microbiota in healthy adult humans. Anaerobe *26*, 1–6.

MacDonald, A., Cochrane, B., Wopereis, H., and Loveridge, N. (2011). Specific prebiotics in a formula for infants with phenylketonuaria. Mol. Genet. Metab. *104*, 55–59.

Macfarlane, G.T., Steed, H., and Macfarlane, S. (2008). Bacterial metabolism and health-related effects of galacto-oligosaccharides and other prebiotics. J. Appl. Microbiol. *104*, 305–344.

Madhukumar, M.S., and Muralikrishna, G. (2010). Structural characterization and determination of prebiotic activity of purified xylooligosaccharides obtained from Bengal gram husk (*Cicer arietinum* L.). and wheat bran (*Triticum aestivum*). Food Chem. *118*, 215–222.

Mandal, V., Sen, S.K., and Mandal, N.C. (2009). Effect of prebiotics on bacteriocin production and cholesterol lowering activity of *Pediococcus acidilactici* LAB 5. World J. Microbiol. Biotechnol. *25*, 1837–1841.

Mandalari, G., Palop, C.N., Tuohy, K., Gibson, G.R., Bennett, R.N., Waldron, K.W., Bisignano, G., Narbad, A., and Faulds, C.B. (2007). *In vitro* evaluation of the prebiotic activity of a pectic oligosaccharide-rich extract enzymatically derived from bergamot peel. Appl. Microbiol. Biotechnol. 73, 1173–1179.

Mitsou, E.K., Panopoulou, N., Turunen, K., Spiliotis, V., and Kyriacou, A. (2010). Prebiotic potential of barley derived β-glucan at low intake levels: A randomized, double-blinded placebo-controlled clinical study. Food Res. Int. *43*, 1086–1092.

Miyazaki, K., Masuoka, N., Kano, M., and Iizuka, R. (2014). Bifidobacterium fermented milk and galacto-oligosaccharides lead to improved skin health by decreasing phenols production by gut microbiota. Benef. Microbes *5*, 121–128.

Molan, A.L., Lila, M.A., Mawson, J., and De, S. (2009a). *In vitro* and *in vivo* evaluation of the prebiotic activity of water-soluble blueberry extracts. World J. Microbiol. Biotechnol. *25*, 1243–1249.

Molan, A.L., Flanagan, J., Wei, W., and Moughan, P.J. (2009b). Selenium-containing green tea has higher antioxidant and prebiotic activities than regular green tea. Food Chem. *114*, 829–835.

Molan, A.L., Liu, Z., and Kruger, M. (2010). The ability of blackcurrant to positively modulate key markers of gastrointestinal function in rats. World J. Microbiol. Biotechnol. *26*, 1735–1743.

Montella, R., Coisson, J.D., Travaglia, F., Locatelli, M., Bordiga, M., Meyrand, M., Barile, D., and Arlorio, M. (2013). Identification and characterization of water and alkali soluble oligosaccharides from hazelnut skin (*Corylus avellana* L.). Food Chem. *140*, 717–725.

Moreno-Vilet, L., Garcia-Hernandez, M.H., Delgado-Portales, R.E., Corral-Fernandez, N.E., Cortez-Espinosa, N., Ruiz-Cabrera M.A., and Portales-Perez, D.P. (2014). *In vitro* assessment of agave fructans (Agave salmiana) as prebiotics and immune system activators. Int. J. Biol. Macromol. *63*, 181–187.

Moroti, C., Souza Magri, L.F., de Rezende Costa, M., Cavallini, D.C., and Sivieri, K. (2012). Effect of the consumption of a new symbiotic shake on glycemia and cholesterol levels in elderly people with type 2 diabetes mellitus. Lipids Health Dis. *11*, 29. doi: 10.1186/1476-511X-11-29.

Nair, V., and Soraisham, A.S. (2013). Probiotics and prebiotics: role in prevention of nosocomial sepsis in preterm infants. Int. J. Pediatr. ID 874726. doi: 10.1155/2013/874726.

Napolitano, A., Costabile, A., Martin-Pelaez, S., Vitaglione, P., Klinder, A., Gibson, G.R., and Fogliano, V. (2009). Potential prebiotic activity of oligosaccharides obtained by enzymatic conversion of durum wheat insoluble dietary fibre into soluble dietary fibre. Nutr. Metab. Cardiovasc. Dis. *19*, 283–290.

Nilsson, A., Johansson, E., Ekstrom, L., and Bjorck, I. (2013). Effects of a brown beans evening meal on metabolic risk markers and appetite regulating hormones at a subsequent standardized breakfast: a randomized cross-over study. PLoS One *8*, e59985.

Nimpiboon, P., Nakapong, S., Pichyangkura, R., Ito, K., and Pongsawasdi, P. (2011). Synthesis of a novel prebiotic trisaccharide by a type I α-glucosidase from *B. licheniformis* strain TH4-2. Proc. Biochem. *46*, 448–457.

Niwa, T. (2013). Targeting protein-bound uraemic toxins in chronic kidney disease. Expert Opin. Ther. Targets *17*, 1287–1301.

Ojansivu, I., Ferreira, C.L., and Salminen, S. (2011). Yacon, a new source of prebiotic oligosaccharides with a history of safe use. Trends Food Sci. Technol. *22*, 40–46.

Oliveira, R.P.D.S., Florence, A.C.R., Perego, P., Oliveira, M.N.D., and Converti, A. (2011). Use of lactulose as prebiotic and its influence on the growth, acidification profile and viable counts of different probiotics in fermented skim milk. Int. J. Microbiol. *145*, 22–27.

Oliveira, R.P.D., Perego, P., De Oliveira, M.N., and Converti, A. (2011). Effect of inulin as a prebiotic to improve growth and counts of a probiotic cocktail in fermented skim milk. LWT – Food Sci. Technol. *44*, 520–523.

O'Sullivan, L., Murphy, B., McLoughlin, P., Duggan, P., Lawlor, P.G., Hughes, H., and Gardiner, G.E. (2010). Prebiotics from marine macroalgae for human and animal health applications. Mar. Drugs *8*, 2038–2064.

Pachikian, B.D., Essaghir, A., Demoulin, J.B., Catry, E., Neyrinck, A.M., Dewulf, E.M., Sohet, F.M., Portois, L., Clerbaux, L.A., Carpentier, Y.A., *et al.* (2013). Prebiotic approach alleviates hepatic steatosis: implication of fatty acid oxidative and cholesterol synthesis pathways. Mol. Nutr. Food Res. *57*, 347–359.

Pareyt, B., Goovaerts, M., Broekaert, W.F., and Delcour, J.A. (2011). Arabinoxylan oligosaccharides (AXOS) as a potential sucrose replacer in sugar-snap cookies. LWT – Food Sci. Technol. *44*, 725–728.

Partty, A., Luoto, R., Kalliomaki, M., Salminen, S., and Isolauri, E. (2013). Effects of early prebiotic and probiotic supplementation on development of gut microbiota and fussing and crying in preterm infants: a randomized, double-blind, placebo-controlled trial. J. Pediatr. *163*, 1272–1277.

Passariello, A., Terrin, G., De Marco, G., Cecere, G., Ruotolo, S., Marino, A., Cosenza, L., Tardi, M., Nocerino, R., and Canani, R.B. (2011). Efficacy of a new hypotonic oral rehydration solution containing zinc and prebiotics in the treatment of childhood acute diarrhoea: a randomized controlled trial. J. Pediatr. *158*, 288–292.

Pliszczak, D., Bourgeois, S., Bordes, C., Valour, J.P., Mazoyer, M.A., Orecchioni, A.M., Nakache, E., and Lantéri, P. (2011). Improvement of an encapsulation process for the preparation of pro- and prebiotics-loaded bioadhesisve microparticles by using experimental design. Eur. J. Pharm. Sci. *44*, 83–92.

Pliczczak, D., Bordes, C., Bourgeois, S., Marote, P., Zahouani, H., Tupin, S., Mattei, C.P., and Lanteri, P. (2012). Mucoadhesion evaluation of polysaccharide gels for vaginal application by using rheological and indentation measurements. Colloids Surf B Biointerfaces 92, 168–174.

Préstamo, G., Pedrazuela, A., Peñas, E., Lasunción, M.A., and Arroyo, G. (2003). Role of buckwheat diet on rats as prebiotic and healthy food. Nutr. Res. 23, 803–814.

Quang, D., Nguyen, Q.D., Judit, M., Rezessy-Szabó, J.M., Czukor, B., and Hoschke, A. (2011). Continuous production of oligofructose syrup from Jerusalem artichoke juice by immobilized endo-inulinase. Proc. Biochem., *46*, 298–303.

Ramnani, P., Chitarrari, R., Tuohy, K., Grant, J., Hotchkiss, S., Philp, K., Campbell, R., Gill, C., and Rowland, I. (2011). *In vitro* fermentation and prebiotic potential of novel low molecular weight polysaccharides derived from agar and alginate seaweeds. Anaerobe *18*, 1–6.

Rasmussen, M.K., Zamaratskaia, G., and Ekstrand, B. (2011). *In vivo* effect of dried chicory root (*Cichorium intybus* L.). on xenobiota metabolizing cytochrome P450 enzymes in porcine liver. Toxicol. Lett. *200*, 88–91.

Rößle, C., Brunton, N., Gormley, R.T., Ross, P.R., and Butler, F. (2010). Development of potentially synbiotic fresh-cut apple slices. J. Funct. Foods *2*, 245–254.

Rodrigues, D., Rocha-Santos, T.A.P., Gomes, A.M., Goodfellow, B.J., and Freitas, A.C. (2012a). Lipolysis in probiotic and symbiotic cheese: The influence of probiotic bacteria, prebiotic compounds and ripening time on free fatty acid profiles. Food Chem. *131*, 1414–1421.

Rodrigues, F.C., Castro, A.S., Rodrigues, V.C., Fernandes, S.A., Fontes, E.A., de Oliveira, T.T., Martino, H.S., and de Luces Fortes Ferreira, C.L. (2012b). Yacon flour and *Bifidobacterium longum* modulate bone health in rats. J. Med. Food *15*, 664–670.

Sabater-Molina, M., Larque, E., Torrella, F., and Zamora, S. (2009). Dietary fructooligosaccharides and potential benefits on health. J. Physiol. Biochem. *65*, 315–328.

Sagar, A., Vos, A.P., Morgan, M.E., Garssen, J., Georgiou, N.A., Boon, L., Kraneveld, A.D., and Folkerts, G. (2014). The combination of*Bifidobacterium breve* with non-digestible oligosaccharides suppresses airway inflammation in a murine model for chronic asthma. Biochimia et Biophysica Acta (BBA)- Mol. Basis. Dis. *1842*, 573–583.

Savignac, H.M., Corona, G., Mills, H., Chen, L., Spencer, J.P., Tzortzis, G., and Burnet, P.W. (2013). Prebiotic feeding elevates central brain derived neurotrophic factor, N-methyl-d-aspartate receptor subunits and d-serine. Neurochem. Int. *63*, 756–764.

Shimizu, K., Ogura, H., Asahara, T., Nomoto, K., Morotomi, M., Tasaki, O., Matsushima, A., Kuwagata, Y., Shimazu, T., and Sugimoto, H. (2013). Probiotic/synbiotic therapy for treating critically ill patients from a gut microbiota perspective. Dig. Dis. Sci. *58*, 23–32.

Smith, S.C., Choy, R., Johnson, S.K., Hall, R.S., and Wileboer, A.C.M. (2006). Lupin kernel fibre consumption modifies faecal microbiota in healthy men as determined by rRNA gene fluorescent *in situ* hybridization. Eur. J. Nutr. *45*, 335–341.

Srinivasjois, R., Rao, S., and Patole, S. (2009). Prebiotic supplementation of formula in preterm neonates: A systematic review and meta-analysis of randomized controlled trials. Clin. Nutr. *28*, 237–242.

Stam, J., van Stuijvenberg, M., Garssen, J., Knipping, K., and Sauer, P.J.J. (2011). A mixture of three prebiotics does not affect vaccine specific antibody responses in healthy term infants in the first year of life. Vaccine *29*, 7766–7772.

Stenger, M.R., Reber, K.M., Giannone, P.J., and Nankervis, C.A. (2010). Probiotics and prebiotics for the prevention of necrotizing enterocolitis. Curr. Infect. Dis. Rep. *13*, 13–20.

Stoidis, C.N., Misiakos, E.P., Patapis, P., Fotiadis, C.I., and Spyropoulos, B.G. (2011). Potential benefits of pro- and prebiotics on intestinal mucosal immunity and intestinal barrier in short bowel syndrome. Nutr. Res. Rev. *24*, 21–30.

Synytsya, A., Míčková, K., Synytsya, A., Jablonský, I., Spěváček, J., Erban, V., Kovářiková, E., and Čopíková, J. (2009). Glucans from fruit bodies of cultivated mushrooms *Pleurotus ostreatus* and *Pleurotus eryngii*: Structure and potential prebiotic activity. Carbohydr. Polym. *76*, *548–556*.

Takahama, U., Tanaka, M., and Hirota, S. (2011). Flour and breads and their fortification in health and disease prevention. In Buckwheat Flour and Bread, V.R. Preedy, R.R. Watson and V.B. Patel, eds. (Academic Press, London, UK), pp. 141–151.

Tárrega, A., Rocafull, A., and Costell, E. (2010). Effect of blends of short and long chain inulin on the rheological and sensory properties of prebiotic low-fat custards. LWT – Food Sci. Technol. *43*, 556–562.

Tenorio, M.D., Espinosa-Martos, I., and Prestamo, G. (2010). Soyabean whey enhance mineral balance and caecal fermentation in rats. Eur. J. Nutr. *49*, 155–163.

Uchida, M., Mogami, O., andMatsueda, K. (2007). Characteristic of milk whey culture with *Propionibacterium freudenreichii ET-3* and its application to the inflammatory bowel disease therapy. Inflammopharmacology *15*, 105–108.

Vaidya, R.H., and Sheth, M.K. (2010). Processing and storage of Indian cereal and cereal products alters its resistant starch content. J. Food Sci. Technol. *48*, 622–627.

Vamanu, E., and Vamanu, A. (2010). The influence of prebiotics on bacteriocin synthesis using the strain *Lactobacillus paracasei* CMGB16. Afr. J. Microbiol. Res. *4*, 534–537.

Vidanarachchi, J.K., Iji, P.A., Mikkelsen, L.L., Sims, I., and Choct, M. (2009). Isolation and characterization of water-soluble prebiotic compounds from Australian and New Zealand plants. Carbohydr. Polym. *77*, 670–676.

Vos, A.P., Knol, J., Stahl, B., M'Rabet, L., and Garssen, J. (2010). Specific prebiotic oligosaccharides modulate the early phase of a murine vaccination response. Int. Immunopharmacol. *10*, 619–625.

Wang, Y.-H., Jiang, Y., Duan, Z.-Y., Shao, W.-L., and Li, H.-Z. (2009). Expression and characterization of a glucosidase from *Thermoanaerobacter ethanolicus* JW200 with potential for industrial application. Biologia *64*, 1053–1057.

Wang, S.-L., Liang, Y.-C., and Liang, T.-W. (2010). Purification and characterization of a novel alkali-stable amylase from *Chryseobacterium taeanense* TKU001 and application in antioxidant and prebiotic. Proc. Biochem. *46*, 745–750.

Watanabe, J., Nishimukai, M., Taguchi, H., Senoura, T., Hamada, S., Matsui, H., Yamamoto, T., Wasaki, J., Hara, H., and Ito, S. (2008). Prebiotic properties of epilactose. J. Dairy Sci. *91*, 4518–4526.

Wichienchot, S., Jatupornpipat, M., and Rastall, R.A. (2010). Oligosaccharides of pitaya (dragon fruit) flesh and their prebiotic properties. Food Chem. *120*, 850–857.

Whisner, C.M., Martin, B.R., Schoterman, M.H., Nakatsu, C.H., McCabe L.D., Wastney, M.E., van den Heuvel E.G., and Weaver, C.M. (2013). Galacto-oligosaccharides increase calcium absorption and gut bifidobacteria in young girls: a double–blind cross-over trial. Br. J. Nutr. *110*, 1292–1303.

Wong, J.M.W., Kendall, C.W.C., de Souza, R., Emam, A., Marchie, A., Vidgen, E., Holmes, C., and Jenkins, D.J.A. (2010). The effect on the blood lipid profile of soy foods combined with a prebiotic: a randomized controlled trial. Metabolism 59, 1331–1340.

Xu, B., Wang, Y., Li, J., and Lin, Q. (2009). Effect of prebiotic xylooligosaccharides on growth performances and digestive enzyme activities of allogynogenetic crucian carp (*Carassius auratus gibelio*). Fish Physiol. Biochem. 35, 351–357.

Yang, B., Prasad, K.N., Xie, H., Lin, S., and Jiang, Y. (2011a). Structural characteristics of oligosaccharides from soy sauce lees and their potential prebiotic effect on lactic acid bacteria. Food Chem. *126*, 590–594.

Yang, Z., Hu, J., and Zhao, M. (2011b). Isolation and quantitative determination of inulin-type oligosaccharides in roots of *Morinda officinalis*. Carbohydr. Polym. *83*, 1997–2004.

Yang, L.C., Wu, J.B., Lu, T.J., and Lin, W.C. (2013). The prebiotic effect of *Anoectochilus formosanus* and its consequences on bone health. Br. J. Nutr. *109*, 1779–1788.

Yeo, S.-K., and Liong, M.-T. (2010). Effect of prebiotics on viability and growth characteristics of probiotics in soymilk. J. Sci. Food Agric. *90*, 267–275.

Prebiotics: Technological Aspects and Human Health

19

Vanessa Rios de Souza, Camila Carvalho Menezes, Luciana Rodrigues Cunha, Patricia Aparecida Pimenta Pereira and Uelinton Manoel Pinto

Abstract

Prebiotics are known as selectively fermented ingredients that allow modifications in the composition and/or activity of the gastrointestinal microbiota conferring benefits to the host wellbeing. They are commonly described as short-chain carbohydrates containing between two and sixty monosaccharides, which are not digested by human or animal enzymes making them selectively fermented by specific beneficial colon bacteria. Besides promoting the intestinal microbiota, they also prevent the incidence of gastrointestinal infections, modulate the immune system, increase the bioavailability of minerals, regulate metabolic disorders related to obesity and diabetes and decrease the risk of cancer. In addition to the functional characteristics, the use of prebiotics by the food industry has been prompted by their favourable technological aspects. In this chapter we present the main functional and technological aspects of inulin, fructo-oligosaccharides (FOS), lactulose, galactoligosaccharides (GOS), resistant starch (RS), soybean oligosaccharides (SOS), xylooligosaccharides (XOS) and isomaltooligosaccharides (IMO).

Introduction

Since consumers are more aware of the relationships between diet and health, the demand for foods that promote health beyond providing basic nutrition, increases constantly. In the past recent years, consumers have been increasingly confronted with so-called functional food products among which probiotics and prebiotics constitute the most popular examples (De Preter *et al.*, 2011).

The most well accepted definition of prebiotic is 'a selectively fermented ingredient that allows specific changes, both in the composition and/or activity in the gastrointestinal microbiota that confers benefits upon host wellbeing and health' (Gibson *et al.*, 2004; Roberfroid, 2007). According to Saad *et al.* (2012) all prebiotics currently described are short-chain carbohydrates with a degree of polymerization (DP) between two and sixty monosaccharides, and are thought to be non-digestible by human, or animal digestive enzymes, and are selectively fermented by specific genera of colon bacteria (Cummings and Macfarlane, 2002).

According to Wang (2009) the effect of a prebiotic is, essentially, indirect because it selectively feeds one or a limited number of microorganisms, thus causing a selective modification of the host's intestinal (especially colonic) microbiota. It is not the prebiotic by itself but rather the changes induced in the microbiota composition that is responsible for its health benefits. Indeed, the most important bacterial genera targeted for selective stimulation are bifidobacteria and lactobacilli (Teitelbaum and Walker, 2002).

The health benefits associated with the intake of prebiotics are modulation of the native intestinal microbiota and prevention of adhesion and colonization by pathogens (Asahara *et al.*, 2001; Bomba *et al.*, 2002; Cummings and Macfarlane, 2002); prevention of the incidence or improvement in the severity and duration of gastrointestinal infections (Macfarlane *et al.*, 2006; Lomax and Calder, 2009; Drakoularakou *et al.*, 2010); modulation of the immune system (Hosono *et al.*, 2003; Seifert and Watzl, 2007; Lomax and Calder, 2009); increase in the bioavailability of minerals, particularly calcium (Bosscher *et al.*, 2006; Scholz-Ahrens and Schrezenmeir, 2007; Demigné *et al.*, 2008; Roberfroid *et al.*, 2010); regulation of metabolic disorders related to obesity (Cani *et al.*, 2006; Delzenne *et al.*, 2005; Urias-Silvas *et al.*, 2008; Cani *et al.*, 2009; Parnell and Reimer, 2009, Peters *et al.*, 2009, Roberfroid *et al.*, 2010); and risk reduction of cancer (Taper and Roberfroid, 2002, Verghese *et al.*, 2002; Liong, 2009).

Prebiotics show both important technological characteristics and interesting nutritional properties (Chow, 2002; Huebner *et al.*, 2007). In food formulations, they can significantly improve sensory characteristics, upgrading both taste and mouthfeel. Aside from having the function of fibre and prebiotic in itself, prebiotics have been successfully used in foods based on their following properties: sugar and fat replacement, texture improvement, body and mouthfeel, moisture retention, heat resistance, foam stabilization, stability to food processing conditions, and others (Wang, 2009).

In relation to the nutritional properties, for prebiotics to serve as functional food ingredients, they must be chemically stable to food processing treatments such as heat, low pH, and Maillard reaction conditions. Stability to processing conditions is essential if prebiotics are to confer beneficial effects for the host. That is, a prebiotic would no longer provide selective stimulation of beneficial microorganisms if it was degraded to its component mono- and disaccharides or chemically altered so that it was unavailable for bacterial metabolism. In contrast, processing may cause a chemical alteration of the prebiotic that may result in increased selective enrichment of beneficial colonic bacteria (Huebner *et al.*, 2008).

Most prebiotics and prebiotic candidates identified today are non-digestible oligosaccharides (Wang, 2009). They are obtained either by extraction from plants (e.g. chicory inulin), possibly followed by an enzymatic hydrolysis (e.g. oligofructose from inulin) or by synthesis (by trans-glycosylation reactions) from mono- or disaccharides such as sucrose (fructooligosaccharides) or lactose (trans-galactosylated oligosaccharides or galactooligosaccharides) (Crittenden and Playne, 1996).

The currently established prebiotics with an impact on human health are inulin, fructooligosaccharides (FOS), lactulose, galactoligosaccharides (GOS), isomaltooligosaccharides (IMO), xylooligosaccharides (XOS), resistant starch (RS) and soybean oligosaccharides (SOS) (Charalampopoulos and Rastall, 2012). The most common prebiotics (inulin, FOS, lactulose and GOS) according to Charalampopoulos and Rastall (2012) have *in vivo* evidence from studies that show their beneficial effects, while others have limited number of data.

When incorporating prebiotics in food, one must take into account the particular characteristics of each prebiotic. Among these characteristics, the fermentation rate, the type of affected microbiota, the recommended dose for a prebiotic effect and the maximum dose that does not induce adverse effects based on scientific studies and by law, must be considered. One must also take into account its stability and its technological features, which are directly related to the physicochemical, rheological and sensory attributes of a particular product where the prebiotic would be added.

Prebiotics

Inulin

Inulin, from the fructans class, is a fructose polymer, typically with a terminal glucose unit (Roberfroid *et al.*, 1993; Roberfroid and Slavin, 2000). The β-D-fructofuranosyl units of inulin and FOS are joined by β(1,2) linkages, which are not hydrolysed in the upper human gastrointestinal tract (Stewart *et al.*, 2008; Quemener *et al.*, 1997). Its DP varies from 2 to 60 units, with an average of 12 units (Izzo and Niness, 2001; Roberfroid, 2007).

Powder inulin is white, amorphous and hygroscopic, having a molecular mass of approximately 1600 Da. Its solubility in water depends on the temperature in such a way that at 10°C about 6% is soluble while at 90°C 35% is soluble. Therefore, the remaining crystals may be responsible for insolubility or recrystallization after cooling a dispersed solution. The dispersion capacity can be improved by adding sugars and starch to the mix. Inulin's ability to bind water is on a ratio of 2:1, while the formation of an inulin gel is dependent on its concentration, DP, temperature, presence of hydrocolloids and mono and divalent cations. As inulin has practically no odour or flavour, it contributes little to the sweet taste, especially because commercial products have low concentrations of mono and disaccharides (Silva, 1996). However, there are studies that show a positive influence on the sweet taste (Franck, 2002; Tungland and Meyer, 2002; Tárrega and Costell, 2006; Tárrega *et al.*, 2010; Rocafull and Costell, 2010).

The enzymatic or acid hydrolysis of inulin produces linear oligomers structurally called GFn and Fm, where G stands for glucose, F for fructose and n or m for the number of fructofuranosyl units obtained by hydrolysis. The n and m values vary between 2 and 9 (De Bruyn *et al.*, 1992; Roberfroid, 1993). The oligomers GFn and Fm have similar physicochemical properties. However, Fm oligomers have reducing power while GFn do not, even though both have a terminal reducing fructose group. Fructose oligomers (the fraction with a low DP) are called fructo-sugars, fructooligosaccharides (FOS) or simply oligofructoses (Roberfroid, 1993). The fructans are present in a great variety of plants and some bacteria. Like starch, inulin acts as a storage carbohydrate in many plants and undergoes hydrolysis to molecules with a lower DP, besides working as an osmoregulator. This facilitates plant survival especially in regions of cold weather (Carpita *et al.*, 1989). The concentration of inulin in plants varies according to the plant variety, time from harvest to consumption and to the storage conditions (Rutherford and Whittle, 1982; Suzuki and Cutclife, 1989).

According to Meyer *et al.* (2011) inulin can be found amongst others in onions (1–5% on a fresh weight basis), garlic (4–12%), banana (0.2%) and chicory roots (15–20%). Most of the inulin and oligofructose commercially available on the industrial food ingredient market nowadays is either synthesized from sucrose or extracted from chicory roots (Niness, 1999).

Inulin can be obtained after harvest, selection and proper washing through a series of processes starting with grinding, followed by aqueous extraction, filtration, centrifugation, concentration, purification, precipitation and drying (Lopez-Molina *et al.*, 2005). The processes vary according to specific conditions of each step. López-Molina *et al.* (2005) for instance obtained inulin from artichoke using ultrafiltration in the filtration and concentration steps. The saccharides were then precipitated by cooling at low temperatures which varied from −24°C to 16°C. Franck (2002), on the other hand, developed a process to produce inulin extracted from chicory roots, combining hot water extraction, purification in ionic exchange resin and drying by atomization. Carvalho *et al.* (1998) extracted inulin from *Vernonia herbácea* roots in a 80% boiling ethanol solution, followed by immersion in water for many times and then concentrating, freezing and thawing and finally centrifuging in order to obtain fructans with high molecular weight.

In order to make inulin-based dietary fibre claims inulin should be added in amounts that range from 3 to 6 g/100 g or 100 ml of the product (European Parliament, 2006) and amounts of 3–8 g per portion have to be added in order to assure a bifidogenic effect (Coussement, 1999; Meyer and Stasse-Wolthuis, 2009). According to Bouhnik *et al.* (2007) stimulation of bifidobacteria with relatively high doses of inulin ranging from 8 to 40 g/day have been repeatedly demonstrated in humans, and recent *in vitro* data indicate that inulin has a clear prebiotic effect at a concentration corresponding to a human daily dose of only 5 g (Van deWiele *et al.*, 2004). Bouhnik *et al.* (2007) also investigated the effect of low-dose inulin consumption concluding that even at doses as low as 2.5 g twice a day; inulin could exert a prebiotic effect by stimulating bifidobacteria growth. According to Flamm *et al.* (2001) up to 20 g/day of inulin may be consumed without adverse gastrointestinal effects such as diarrhoea and flatulence.

Inulin, as well as FOS, have been suggested to be less stable than other oligosaccharides at low pH conditions and high temperatures, especially the combination of the two. In very acidic conditions, the β (2–1) bonds between the fructose units can be partially hydrolysed (Bosscher, 2009). It has moderate solubility in water (10% soluble at room temperature), which is convenient for its incorporation into liquid food systems, has a neutral taste and it is slightly sweet (less than 10% of the sweet power of sucrose) (Franck, 2008). Dry heating of inulin from chicory or Jerusalem artichoke for up to 60 min at temperatures between 135°C and 195°C, resulted in significant degradation ranging from 20% to 100% and led to the formation of new products, likely to be di-D-fructose dianhydrides (Bohm *et al.*, 2005; Bohm *et al.*, 2006). But although the heated inulin had been chemically altered, when added to a mixed faecal culture it showed significantly greater stimulation of the growth of bifidobacteria and lactobacilli, and a significant decrease in the growth of potential pathogens. Glibowski and Bukowska (2011) showed that the chemical stability of inulin at low pH decreased with an increase in heating time and temperature, so the inulin application in food systems may have limited use in acidic products especially when heated above 60°C during the production process. According to these authors in products at pH ≥ 5, the degradation of this fructan does not occur even at thermal processing.

The differences in chain length between inulin and oligofructose account for their distinctly different functional attributes. Owing to its longer chain length, inulin is less soluble than oligofructose and has the ability to form inulin microcrystals when sheared in water or milk. These crystals are not discretely perceptible in the mouth, but they interact to form a smooth creamy texture and provide a fat-like mouthfeel (Niness, 1999).

The short-chain fraction, oligofructose, is much more soluble and sweeter than native and long-chain inulin, and can contribute to improve mouthfeel because its properties are closely related to those of other sugars (Meyer *et al.*, 2011).

The extensive use of inulin in the food industry is based on its nutritional and technological properties. The technological use of inulin is based on its properties as a fat replacer and texture modifier and also as a sugar replacer (Meyer *et al.*, 2011), especially in combination with high intensity sweeteners, such as Stevia, where inulin masks the negative off-flavour (Guggisberg *et al.*, 2011). It is used primarily in low fat dairy products, including fermented milks, yoghurts, dairy desserts, cheeses and ice cream, as a fat replacer (Kusuma *et al.*, 2009; Buriti *et al.*, 2010; Meyer *et al.*, 2011; Rinaldoni *et al.*, 2012) and in baked products as a texture modifier (Angioloni *et al.*, 2008; Hager *et al.*, 2011). Owing to the bond type β(2→1) between the chains, inulin and oligofructoses are not digested, and because of this feature, their use is allowed as food fibres, meaning that they are resistant to human enzymatic digestion (Trowel and Burkitt, 1986; Lee and Prosky, 1995; Hoebregs, 1997). The physiological definition of a food fibre is related to an improvement of the intestinal function. Generally, the accepted physiological effects of fibre include, besides its effects on intestinal function, an improvement in the blood lipid parameters with decreased caloric value (Roberfroid, 1993), where a drop in the plasma levels of triacylglycerols and cholesterol is observed in hypercholesterolaemic patients (Fiordaliso *et al.*, 1995).

Studies have shown that the intake of inulin and oligofructoses protect against the development of cancer and the growth of tumours (Roberfroid, 1991; Gibson *et al.*, 1995; Meyer and Stasse-Wolthuis, 2009; Verma and Shukla, 2013; Wu *et al.*, 2014). Inulin and oligofructoses are known to selectively stimulate the growth of bifidobacteria and to keep population levels of Bacteroides, Clostridia and coliform bacteria low (Gibson and Roberfroid, 1995; Gibson *et al.*, 1995, Guigoz *et al.*, 2002). Additional health beneficial effects due to the consumption of inulin and oligofructoses have been shown with respect to increased absorption of minerals, especially calcium (Delzenne *et al.*, 1995; Ohta *et al.*, 1994).

Fructooligosaccharides (FOS)

Fructooligosaccharides (FOS) consist of molecules of sucrose in which one or two units of fructoses are attached by β-(2–1) bonds to the molecule of fructose within sucrose (Thamer and Penna, 2005). Depending on the place of enzymatic cleavage of inulin, FOS may contain a terminal glucose unit (Niness, 1999).

The fructooligosaccharides (FOS) are naturally found in many foods, with a wide distribution among those of vegetal origin. These oligosaccharides have fructose as the main monomeric unit at different degrees of polymerization, and can actually be industrially produced by means of extraction, microbial fermentations or through enzymatic action. Usually, 10% of the chains have DP of 2 (F2) and 5 (GF4). The partial hydrolysis of these chains produces oligofructoses which are a mixture of molecules of glucopyranosyl-(fructofuranosyl)n-1-fructose (GFn) and fructopyranosyl-(fructofuranosyl)n-1-fructose (Fn) in which the DP varies between 2 and 7, with an average of 4. When these fructans have DP of 10, they can also be called fructooligosaccharides (FOS) (Roberfroid, 2005). The nomenclature FOS is only given to oligomers of fructose, which are commonly known as 1-kestose (GF2), nistose (GF3) and fructofuranosyl nistose (GF4) in which the fructosyl units (F) are linked at position β-2,1 to sucrose, which differs them from other oligomers (Yun, 1996).

The most common sources of FOS are roots of chicory, Jerusalem artichoke, asparagus, and members of the onion family (Wack and Blaschek, 2006). The industrial processes for FOS production is through the enzymatic synthesis from sucrose by microbial enzymes with transfructosylating activity (β-fructofuranosidases, also designated as fructosyltransferases) (Mussatto and Teixeira, 2010) and through the controlled enzymatic hydrolysis of the polysaccharide inulin (inulin oligofructose), which can be extracted from chicory roots, for example (Passos and Park, 2003; Crittenden and Playne, 1996).

The bifidogenic effect of oligofructose, as inulin and in general all oligosaccharides, is now well established through various studies, not only in adult participants but also in other age groups (Coussement, 1999; Marteau, 2001; Manning and Gibson, 2004; Bruzzese *et al.*, 2006; Venter, 2007; Meyer and Stasse-Wolthuis, 2009). According to Crittenden and Playne (1996) most oligosaccharides have been demonstrated to increase bifidobacteria numbers in the colon at doses of < 15 g/day.

Rivero-Urgell and Santamaria-Orleans (2001) reported that the fructooligosacharide intake required for having a bifidogenic effect ranges from 2 to 10 g/day in adults. Macfarlane and Cummings (1999) observed that the intake of 8 to 15g/day of fructooligosacharides increased by 10 times the population of bifidobacteria in faeces, while reducing the counts of clostridia and enterobacteria. Human studies with FOS as prebiotics have shown that a daily dose of 4–20 g significantly increased lactobacilli and bifidobacteria populations in the gut (Rycroft *et al.*, 1999). However, in individuals with an already high level of bifidobacteria it is difficult to increase levels even further.

Typically, the oligosaccharides are well tolerated. However, in some instances an increase in flatulence has been reported, usually associated with higher levels of ingestion, i.e. an excess of 40 g/day. At this level, the oligosaccharides may also induce diarrhoea (Spiegel *et al.*, 1994; Carabin and Flamm, 1999; Sako *et al.*,1999).

According to Charalampopoulos and Rastall (2012) FOS, like inulin, have been suggested to be less stable than other oligosaccharides at conditions of low pH and high temperatures, especially the combination of both. Huebner *et al.* (2008) determined the effect of processing conditions on the prebiotic activity of commercial prebiotics using a prebiotic activity assay; the results showed that only heating at low pH caused a significant reduction in prebiotic activity, with one of the FOS products being the least stable. In buffered solutions, FOS have been found to be degraded at increased temperature and decreased pH (L'Homme *et al.*, 2003). Several other studies show the effect of different conditions on FOS stability. For instance, it has been reported that heating a FOS solution at pH 3.5 at 145°C for 10 s resulted in the hydrolysis of approximately 10% of FOS (Torres *et al.*, 2010). It has also been shown that the exposure of a FOS solution to high temperatures (85°C for 30 min) resulted in a significant reduction in its prebiotic activity (Huebner *et al.*, 2008). Moderate to high levels of degradation (10–30%) have been found at temperatures between 60°C and 70°C, and pH 2.0 and 2.5 (Wang, 2009). When using fruit juices as the food matrix, it was concluded that FOS were susceptible to hydrolysis during pasteurization, and could even be degraded by over 80%, depending on the conditions (Klewicki, 2007). In another study it was shown that at pH 2, 63% and 38% of FOS were hydrolysed after 56 days of storage at 37°C and 18 weeks of storage at 4°C, respectively (Courtin *et al.*, 2009).

In very acidic conditions, the β-(2–1) bonds between the fructose units can be partially hydrolysed (Bosscher, 2009). The lower stability of FOS at acidic pH, especially at high temperatures (> 70°C), could be explained by the fact that the fragility of the fructose-fructose and fructose-glucose C–O osidic bonds involved in the glycosidic linkages in FOS increases as the temperature increases (Matusek *et al.*, 2009).

FOS have good solubility (up to 85% soluble at room temperature). They are fairly sweet (30–35% compared with sucrose) and have similar technological properties to sucrose and glucose syrups; as a result, they are frequently used as sugar replacements (Bosscher, 2009). They have been already applied in a variety of dairy products, as they are the ideal ingredients to give bulk with fewer calories and increase the functional value without compromising the taste and mouthfeel of the products (Franck, 2008).

Due to its characteristics, FOS have applications in many areas. It has one third the sweetener power of sucrose but it is not caloric. Besides, FOS are more soluble than sucrose, they do not crystalize, nor precipitate and do not leave a dry or sand sensation in the mouth (Yun, 1996). FOS can be used in formulations of dairy ice-cream and deserts which have in the label the terms 'low sugar', 'no sugar added' or 'reduced calories'; in formulations destined to diabetic persons; in functional products that promote additional nutritional effects in the areas of prebiotics, symbiotics, dietetic fibres; in yoghurts promoting a symbiotic effect (besides the intrinsic probiotic effect of yoghurts); in cookies and bakery products replacing carbohydrates and reducing the amount of sugar; in cereal bars; juices and nectars, sauces, among others (Yun, 1996; Bornet *et al.*, 2002; Sangeetha *et al.*, 2005). They can also be used in foods destined to animals, with the same prebiotic effects and also as feed additives for hogs and domestic birds (Passos *et al.*, 2003).

There are several studies that prove the many beneficial effects of prebiotics in the body. FOS have shown outstanding effects promoting gastrointestinal health, inhibiting the development of pre-cancerous wounds in the intestines, promoting the growth of diverse intestinal bacteria (especially the strict anaerobes such as bacteroides, eubacteria, bifidobacteria and clostridia, which constitute 99% of the intestinal microbiota), decreasing the pH of the colon, eliminating putrefactive bacteria, increasing the intestinal permeability after infections, improving the resistance to *Salmonella* Enteritidis colonization, having beneficial effects in cases of intestinal inflammation and increasing absorption of some minerals such as calcium (Manning and Gibson, 2004; Swennen *et al.*, 2006; Mussatto *et al.*, 2007; Búrigo *et al.*, 2007; Venter, 2007). Besides, FOS provide benefits to the immunological function (Hosono *et al.*, 2003; Suzuki and Hara, 2004; Nakamura *et al.*, 2004).

Lactulose

Lactulose (4-O-β-D-galactopyranosyl-D-fructose) is a synthetic disaccharide composed of fructose and galactose by β-1,4 glycosidic bonds. This oligosaccharide does not occur naturally and it is produced by the isomerization of lactose by rearranging the glucose residues to fructose (Ruttloff *et al.*, 1967). Many compounds such as sodium hydroxide, sulphites, phosphates, calcium carbonate, alkaline and enzymes can be used as catalysts. Kim and Oh (2012) reported the synthesis of lactulose from lactose using the enzyme cellobiose 2-epimerase from *Caldicellulosiruptor saccharolyticus*, which represents a significant step towards a large scale enzymatic production of this compound. However, the industrial process most commonly used is through isomerization by means of alkaline treatment (Olano and Corzo, 2009; Panesar and Kumari, 2011).

Even though lactulose is absent in raw milk, it can be formed during thermal processing because milk treated at high temperatures favours the isomerization process of lactose to lactulose. Usually the concentration of lactulose varies from 0.3 g/l to 1.6 g/l in UHT and conventionally sterilized milk (autoclaved). The formation of this disaccharide is not favoured in milk due to its neutral pH (close to pH 6.8) which does not offer optimal conditions for isomerization (Olano *et al.*, 1989; Mendoza *et al.*, 2005). Some approaches such as an increase in the temperature during processing or the time of exposure did not result in higher amounts of lactulose in milk, as it has been shown that this sugar is degraded by heating. According to Berg and Van Boekel (1994) lactulose formation induced by heating is, at the same time, decreased by its decomposition to galactose and formic acid. In the development of low-calorie guava preserves with lactulose, Menezes *et al.* (2012) reported an average degradation of 18% of lactulose indicating that the production conditions at high temperatures for long time are unfavourable for disaccharide stability.

Even though lactulose may be degraded by heating, it has important properties that contribute to its use in the food industry. It is usually sweeter and more soluble than lactose, which makes it suitable as a sweetener for diabetics, as a sugar substitute in baking products, beverages, infant milk powders, yoghurts, dairy desserts and in various liquid and dried food preparations which are routinely manufactured for the elders (Tamura *et al.*, 1993; Crittenden and Playne, 1996; Strohmaier, 1998; Gänzle, 2011). Besides, lactulose also has flavour enhancing properties, favourable browning behaviour, excellent solubility in water (Panesar and Kumari, 2011) and has shown to improve the survival of probiotic bacteria in yoghurt (Tabatabaie and Mortazavi, 2008).

In the pharmaceutical industry, lactulose is used as an effective drug against different diseases like acute and chronic constipation (Mizota *et al.*, 1987; Tamura *et al.*, 1993; Foster *et al.*, 2010), treatment of hepatic encephalopathy (Gänzle, 2011), complication of liver disease, tumour prevention, improved immunological response, maintenance of blood glucose and insulin level (Schumann, 2002). Lactulose is also applied in a wide variety of foods as a bifidogenic factor or as a functional ingredient for intestinal regulation. Studies in healthy adults proved that lactulose is an effective food-grade prebiotic, showing that consumption of 10 g per day of lactulose resulted in a statistically significant increase in the number of bifidobacteria, whereas the clostridia population decreased, for which reason they are referred to as bifidogenic factors in nutrition (Nagendra *et al.*, 1995; Voragen, 1998; Marteau, 2001; Seki *et al.*, 2007). Oliveira *et al.* (2011) concluded that addition of lactulose in skim milk increased the counts of all probiotics (*Lactobacillus* and *Bifidobacterium*), particularly *B. lactis* (bifidogenic effect). Concerning infant formula, the incorporation of 0.5% lactulose is considered adequate to stimulate the bifidobacterial microbiota to the extent observed in breast-fed babies, while the presence of lactulose at 1% level in the formula may also provide a partial laxative effect (Nagendra *et al.*, 1995).

Galactooligosaccharides (GOS)

Galactooligosaccharides are composed of oligosaccharides with a chain of 3–8 galactose units with an end-capped glucose produced from lactose by the transgalactosylation activity of the enzyme β-galactosidase (Anthony *et al.*, 2006; Martins and Burket, 2009). The oligosaccharides can be linear or branched with multiple different β-glycosidic linkages (Coulier *et al.*, 2009).

The formation of GOS occurs from a substrate rich in lactose, notably milk, whey or a mixture of both (Burkert and Martins, 2009). These oligosaccharides are generally found in human milk (5 to 8 g/l), bovine milk (0.03 to 0.06 g/l), goat milk (from 0.25 to 0.30 g/l) and ovine milk (0.02 to 0.04 g/l) (Martinez-Perez *et al.*, 2006). During manufacture of functional fermented milks, oligosaccharides are produced in variable amounts mainly depending on the strains used (Joung and Cheong, 2001; Lamoureux *et al.*, 2002; Yadav *et al.*, 2007). Villaluenga *et al.* (2008) determined the GOS content of 14 commercially available fermented milk and found 0.36–0.58% of GOS in commercial yoghurts containing bifidobacteria, 0.29–0.44% of GOS in ready-to-drink yoghurts containing *L. casei* and 0.22–0.25% of GOS in traditional yoghurts.

GOS are industrially produced by transgalactosidation reactions using β-galactosidases (Kimura *et al.*, 1995; Sako *et al.*, 1999) from lactose, leading to a final product comprising of oligosaccharides with a DP of up to 9, and high amounts of mono- and disaccharides. In this sense, the removal of mono- and disaccharide fractions from a GOS mixture is necessary to both evaluate their functional properties, e.g. prebiotic activity, and to determine their structures. On the other hand, the removal of these carbohydrates could help to expand the applications of the purified GOS in the food and pharmaceutical industries. Among these applications could be the inclusion of purified GOS as ingredients in (i) diabetic foods, which could allow diabetic people to utilize the beneficial properties of GOS, free from carbohydrates that increase the level of postprandial glucose (ii) specialized foods for individuals that have intolerance to lactose and (iii) low calorie foods with a reduction in some mono- and disaccharides (Hernández *et al.*, 2009).

According to Rycroft *et al.* (1999), human studies with GOS have shown that daily doses of 4–20 g significantly increased lactobacilli and bifidobacteria populations in the gut. According to Mussatto and Mancilla (2007), some authors have suggested that an intake of 10 g/day of GOS is sufficient to cause a bifidogenic effect. However, when the subjects' initial number of indigenous bifidobacteria are low, which is often the case in middle-aged and elderly people, a daily intake of 2.5 g/day is enough to lead to an increase in the faecal bifidobacteria population (Sako *et al.*, 1999; Gibson *et al.*, 2004).

Anthony *et al.* (2006) investigated the safety of Vivinal® GOS syrup at 2500 or 5000 mg/kg body weight per day in rats, and there was no toxicological effect. Davis *et al.* (2010) studied the effect of different doses of GOS on faecal microbiota in healthy adults, focusing on bifidobacteria. This study showed that a high purity GOS administered in a cooking product at doses of 5 g or more was bifidogenic, while a dose of 2.5 g showed no significant effect. According to Carabin and Flamm (1999) at higher levels of ingestion, i.e. levels above 40 g/day, GOS and other oligosaccharides may induce diarrhoea. Sako and collaborators (1999) and Spiegel *et al.* (1994) reported that GOS consumption higher than 20 g/day caused diarrhoea.

Playne and Crittenden (1996) reported that GOS were stable in water-based solutions during heating at 100°C and pH 3 for 10 min; at pH 2, only 5% of GOS was degraded and at 37°C and pH 2 they were stable during storage for several months. A high GOS stability in various fruit juices, with pH ranging from 2.7 to 4.1, at a range of pasteurization conditions (88–95°C for a varying number of minutes) was also demonstrated; in most cases, more than 99% of GOS endured pasteurization. The likely reason for the high stability of GOS is the presence of β-linkages, which are very stable to hydrolysis, although other factors, such as the sugar residues present, the ring form and the anomeric configuration can play a role as well (Klewicki, 2007).

Due to the stability of GOS to acidic conditions and high temperatures they can be potentially added to a variety of acid or heated foods, such as yoghurts, fermented milks, buttermilk, and pasteurized fruit juices (Torres *et al.*, 2010; Sangwan *et al.*, 2011). They are also suitable ingredients for use in bread and baked products, owing to their high moisture retaining capacity (Torres *et al.*, 2010).

They are, as other oligosaccharides, very soluble in water, mildly sweet (30–35% compared to sucrose) and can be used to improve the textural properties of dairy products, such as yoghurts (Sangwan *et al.*, 2011). However, according to Charalampopoulos and Rastall (2012), there are few commercial examples of GOS-containing food products, and very little information on the impact of GOS on the physicochemical properties of the food carrier.

Galactooligosaccharides are considered bifidogenic factors because they serve as substrate for bifidobacteria, stimulating their metabolism and growth in the human intestine (Roberfroid, 2007). The potential of GOS mixtures to stimulate bifidobacteria has been studied extensively, both *in vivo* and *in vitro* (Van Laere *et al.*, 2000; Macfarlane *et al.*, 2008; Barboza *et al.*, 2009). Davis *et al.* (2011) showed that consumption of GOS resulted in a highly specific enrichment of bifidobacteria in adult subjects.

Resistant starch (RS)

Chemically, starches are polysaccharides composed of α-D-glucopyranosyl units linked together with α-D-(1–4) and/or α-D-(1–6) linkages, and comprise two molecular types: amylose, the straight-chain polyglucan comprising approximately 1000 α-D-(1–4) linked glucoses; and amylopectin, the branched glucan, comprising approximately 4000 glucose units with branches occurring as α-D-(1–6) linkages, in addition to the α-D-(1–4) bonds (Haralampu, 2000; Sharma *et al.*, 2008).

Resistant starch (RS) refers to the portion of starch and starch products that resist α-amylase digestion as they pass through the gastrointestinal tract. Resistant starch comprises such a diverse range of compounds that is commonly grouped into 4 major categories, usually referred to as RS1–4. Currently, RS are better defined according to their physical and chemical characteristics (Nugent, 2005). According to Simsek and El (2012), the term RS1 is used to define the physically inaccessible starch locked within the cell walls of grains, seeds and bulbs. The RS2 encompasses the ungelatinized native granular starch with type B crystallinity. The RS3 refers to indigestible starch which is formed after treatment with water and heat, also known as retrograded starch. The RS4 comprises a group of starches which have been chemically modified by esters, phosphates and ethers, as well as cross linked starches, all with characteristics of being undigested in the human small intestines (Nugent, 2005).

The amount of RS in foods is related to the type and amount of starch (crystal and granular structure, amylose–amylopectin ratio, the chain length of amylose, and linearization of amylopectin), along with the processing, cooking, and storage conditions of food (heat and moisture conditions) (Sajilata *et al.*, 2006; Fuentes-Zaragoza *et al.*, 2010). Fuentes-Zaragoza and collaborators (2010) reported that the main sources of RS1 are the grains and seeds; RS2 are green bananas, raw potatoes and some legumes; RS3 are products with prolonged and/or repeated moist heat treatment and RS4 are foods in which modified starches have been used.

Resistant starch was first recognized in the 1980s when it was found to be resistant to enzymatic hydrolysis (Englyst et at., 1992), and it has been defined as 'the sum of starch and products of starch degradation not absorbed in the small intestine of healthy individuals' (Asp, 1992). Resistant starch is not digested in the small intestine and, therefore, has many of the physiological health benefits associated with dietary fibres, which puts it as a functional fibre (Sajilata *et al.*, 2006). Resistant starch acts as a prebiotic component by promoting the growth and activity of probiotic bacteria and can interact with other prebiotic dietary fibres such as β-glucans (Topping *et al.*, 2003; Brown, 2004; Goldring, 2004). These prebiotics are naturally found in cereal grains, seeds and in heated starch or starch-containing foods (Charalampopoulos *et al.*, 2002). According to Fuentes-Zaragoza *et al.* (2010) unripe banana is considered the richest RS non-processed food. Several studies have suggested that consumption of unripe bananas confers beneficial effects for human health, a fact often associated with its high resistant starch content, which ranges between 47% and 57%. The recommended dietary fibre intake is 25 g/day for woman and 38 g/day for men (Institute of Medicine, 2005). According to Kendall *et al.* (2004), RS doses of 20–30 g/day are needed in order to observe physiological effects. Most foods have RS contents of 3 g or less per serving (Murphy *et al.*, 2008). The Commonwealth Scientific and Industrial Research Organization (CSIRO) has recommended that the total intake of resistant starch should be around 20 g a day based on a study by Baghurst *et al.* (2001).

Besides its prebiotic properties, RS has also shown promising results as ingredients for the manufacture of bakery products such as bread, muffins, and breakfast cereals (Fuentes-Zaragoza *et al.*, 2010). Its physical properties, particularly its low water-holding capacity, make it a functional ingredient that provides good handling and improves texture in the final product (Baixauli *et al.*, 2008). Beyond the beneficial physiological effects, as a food ingredient, RS has desirable physicochemical properties such as white appearance, bland flavour, swelling, gel formation and increased viscosity (Simsek and El, 2012). According to Haralampu (2000), among the four types of RS, RS3 is of particular interest because of its thermal stability, which allows it to be stable in most normal cooking operations, and enables its use as an ingredient in a wide variety of conventional foods.

Aigster *et al.* (2011) concluded that RS-supplemented products were acceptable to consumers; however, incorporation of high levels of RS resulted in changes in sensory characteristics; furthermore RS levels increased after 4 weeks' storage.

Soybean oligosaccharides (SOS)

Soybean oligosaccharides are water-soluble carbohydrates of low molecular weight, mainly represented by raffinose and stachyose (Manzanares and Graff, 1998). Raffinose is a trisaccharide made of a galactose molecule linked by a α-(1–6) bond to glucose from sucrose (Gal-Glu-Fru). This oligosaccharide is also known as gossypose, melitose, and melitreose. Stachyose is a tetrasaccharide composed of a galactose molecule linked by α-(1–6) bonds to the terminal galactose from raffinose (Gal-Gal-Glu-Fru) (Kim *et al.*, 2003). It is also known as galactosyl-raffinose, lupeose, and β-galactan. In these oligosaccharides, sucrose is the basal unit with additional α-D-galactopyranosyl (α-d-Gal) always linked to the left side of sucrose on the carbon 6 of the molecule (Dey, 1985; Silva *et al.*, 1992).

These oligosaccharides are present in large quantities in soybean, ranging from 100 to 150 g/kg of soy (De Reu *et al.*, 1997). Its quantities vary according to the degree of maturation and cultivation conditions such as the composition of the soil and the temperature (Sakla *et al.*, 1988). A study by Espinosa and Rupieres (2006) has shown that soybeans with higher degrees of maturation have higher levels of these oligosaccharides.

Unlike other oligosaccharides, SOS can be extracted directly from soybeans and there is no need for enzymatic processes (Crittenden and Playne, 1996). The extraction is performed in water or in ethanolic solutions of 70–90% (Quigly *et al.*, 1999). The extract has no colour and a sweet taste similar to sucrose, corresponding to 70% sweetener capacity compared to sucrose on a dry weight basis (Massai, 1988). As a result of this property, they can be used commercially as sugar replacers acting as sweeteners in low-calorie foods, since they are not metabolized by human digestive enzymes, and also added to juices and soda for better sensory properties. Besides the sweet characteristic, SOS have humectant properties, protecting food against water loss (Crittenden and Playne, 1996), are thermostable (Francis *et al.*, 2001) and do not have chelating properties, that is, they do not bind minerals, facilitating their addition to foods (Crittenden and Playne, 1996).

Besides having chemical and technological properties which make them good alternatives as ingredients in the processing of foods, their biological and functional properties have shown promising results due to their synergistic effects with probiotic microorganisms, antioxidant and anti-diabetic activities and the prevention of cardiac arrests. Zhow and collaborators (2012) evaluated the fermentation characteristics of SOS *in vitro* and observed a significant increase in the population of *Lactobacillus* and *Bifidobacterium* and a drop in the counts of *E. coli* and *Streptococcus* among the fermenting microorganisms inoculated with 2% SOS in comparison with the glucose control. Hara *et al.* (1997) showed that the daily intake of 1–2 g of SOS promoted a substantial increase in the numbers of bifidobacteria in faeces of individuals who had low levels of these bacteria in the intestine. According to Hayakawa *et al.* (1990), raffinose and stachyose were fermented by bifidobacteria under *in vitro* conditions. They also observed a significant reduction in the population of *Clostridium* in faeces of 6 individuals who were fed with 10 g/day of SOS for a period of 3 weeks. Similar results were observed by Wada *et al.* (1992) in a study in which seven volunteers were fed 0.6 g of SOS/day for 3 weeks and by Benno *et al.* (1987), who worked with seven volunteers who took 15 g of raffinose/day for 4 weeks.

According to Fei *et al.* (2014), women suffering from gestational diabetes mellitus can benefit from the concomitant treatment of insulin and the consumption of 10g of SOS/day, since a significant reduction in the oxidative stress and an alleviation in the resistance to insulin in these women were observed, when compared with those that did not consume SOS, indicating that these oligosaccharides can function as important agents controlling complications of gestational diabetes mellitus. Chen *et al.* (2010) verified a reduction in the oxidative stress and an improvement in the levels of glucose and lipids in blood of rats treated with 150, 300 and 450 mg of SOS/kg of body weight for 45 days for all doses. Similar results were found by Zhao and Yang (2007). An increase in the antioxidant enzymatic activities of CAT, SOD and GPx in livers of chicken fed with SOS for six weeks was also observed (Huang *et al.*, 2006). Deng *et al.* (2007) reported that the intake of SOS decreased the incidence of fat in livers of fishes. The antioxidant mechanism of action of SOS remains to be elucidated. According to Chen *et al.* (2010) this activity could be related to the redox potential of phenolic polysaccharides formed between phenolic compounds of high molecular weight (that are not removed in the extraction process) and the oligosaccharides. These compounds would have an important role in the absorption and neutralization of free radicals, chelating singlet oxygen and/or triplet oxygen or decomposing peroxide.

The consumption of 150 and 300 mg of SOS by hypercholesterolaemic rats reverted the plasmatic levels of lactic acid, nitric oxide (NO), endothelial vascular factor and the levels of Na^{+}-K^{+}-ATPase, Ca^{2+}-ATPase and Mg^{2+}-ATPase, indicating that SOS can have beneficial effects against some coronary diseases (Chen *et al.*, 2010).

Besides the activities mentioned above, the consumption of SOS has also shown promising results in the prevention of chronic diseases such as cancer, osteoporosis, atherosclerosis and the symptoms of menopause (Espinosa-Martosy and Ruperez, 2006; Mateos-Aparicio *et al.*, 2008; Chen *et al.*, 2010).

Despite the numerous benefits, the consumption of SOS should be moderate as high doses are associated with flatulence and intestinal discomfort. The discomfort is associated with the lack of the α-galactosidase enzyme which is necessary to hydrolyse the α-1,6 bonds in these sugars. When ingested, these oligosaccharides are not metabolized in the small intestine of humans and monogastric animals, reaching the colon where they are anaerobically fermented, producing gases such as hydrogen, carbon dioxide and methane, as well as organic acids. These substances, besides causing flatulence, can cause diarrhoea and abdominal pain (Sánchez-Mata *et al.*, 1998). Tomomatsu (1994) suggests that the daily doses of SOS should be limited to 2 g.

Xylooligosaccharides (XOS)

Xylooligosaccharides (XOS) are oligomers of xylose units with varying proportions and types of side-groups such as acetyl, uronic acids and phenolics. The composition depends on the vegetal origin and extraction process (Kabel *et al.*, 2002). The structures of XOS vary in the DP, with monomeric units and types of bonds according to the xylan sources. Usually, XOS are oligosaccharides formed by xylose residues linked by β-(1–4) bonds (Aachary and Prapulla, 2008) with DP varying from 2 to 10 units, also known as xylobiose, xylotriose and so on and so forth. Xylobiose is the xylo-oligosaccharide most used in the food industry (Vazquez *et al.*, 2000). Besides containing xylose residue, these oligosaccharides are normally found associated with other secondary groups such as α-d-glucopyranosyl uronic acid or its derivative 4-O-methyl, acetyl, or residues of arabinofuranosyl. The presence of these secondary groups results in branched XOS with different biological properties (Aachary and Prapulla, 2011).

Xylooligosaccharides are naturally found in fruits, vegetables, milk, honey, besides forest and agricultural residues (Vázquez *et al.*, 2000). The industrial production of XOS is obtained through lignocellulosic materials (LCMs). The LCMs are constituted by three main polymers: lignan (phenolic structure), cellulose (linear polymer composed of glucose units linked by β-1–4 bonds) and hemicellulose which is a heteropolysaccharide composed of several monosaccharides, including xylose, arabinose, mannose and ramnose (Garrote *et al.*, 1999).

Depending on the nature of the lignocellulosic material, polymers of xylose (xylan), arabinose (arabinan) or mannose (mannan) may account for a substantial part of hemicellulose, which can be substituted via ether or ester bonds, for example with α-d-glucopyranosyl uronic acid or its 4-O-methyl derivative, acetyl groups and acids (Ebringerova and Heize 2000; Garrote *et al.*, 2007). The typical materials used in the production of XOS are rich in xylan with some ether heterocyclic chains that should be hydrolysed in order to generate degraded compounds of long chain length (Eneysaya *et al.*, 2007). These compounds can be produced in the following ways: through the treatment with native enzymes contained in LCMs or through chemical fractioning of LCMs from isolated or solubilized xylan, with the enzymatic addition of these polymers to XOS or through the degradation of xylan to XOS by water vapour or diluted solutions of mineral acids (Vázquez *et al.*, 2000). In order to obtain a prebiotic effect, studies have shown that one should consume 4 g of XOS per day (Chung *et al.*, 2007; Otieno and Ahring, 2012).

Xylooligosaccharides are stable in a broad range of pH values varying from 2.5 to 8.0 and also up to 100°C, which is highly advantageous compared with other non-digestible oligosaccharides such as fructooligosaccharides (FOS). For instance, Wang and collaborators (2009) reported a degradation of only 3% of XOS under pasteurization conditions (30 min, 60–100°C, at pH 2). The reported water activity of xylobiose is higher than xylose, but almost the same as glucose. It has pleasing odour and low caloric input, which allows its use in anti-obesity diets (Bhat, 1998). They have interesting physicochemical characteristics as they are moderately sweet, stable at a large pH range and temperatures and have adequate sensory characteristics for incorporation in many foods (Wang *et al.*, 2009). In the processing of foods, XOS present advantages related to inulin due to its thermal and acid resistance which allows its use in low pH juices and carbonated drinks, besides being able to intensify the taste of these drinks (Vázquez *et al.*, 2001).

Mumtaz *et al.* (2008) evaluated the physicochemical and sensorial aspects of yoghurt enriched with XOS and compared those with yoghurts containing gelatin (0.4% w/w). Their data showed that pH, acidity and total solids were significantly affected by the addition of XOS with differences detected sensorially when XOS was applied up to 3%.

The use of XOS as an ingredient in functional foods is based on their benefits to human health, selective metabolism towards bifidobacteria, increased production of short-chain fatty acids and a reduction in the sizes of stomach ulcers (Parajó *et al.*, 2004). Katapodis *et al.* (2003) have also shown that XOS present antioxidant properties due to the presence of phenolic groups, besides preventing anaemia and atherosclerosis. Other studies have shown anti-allergic effects, cytotoxicity against lymphoblastic leukaemia and apoptosis induction in cancerous cells (Ando *et al.*, 2004; Lecerf *et al.*, 2012, Prescilla *et al.*, 2012).

Okazaki *et al.* (1990a,b) used a mixture of xylose, xylobiose, xylotriose and other saccharides as carbon sources for *in vitro* fermentations with *Bifidobacterium adolescentis* and showed a noticeable capacity of this bacterium to use both xylobiose and xylotriose. According to Hopkings and collaborators (1998), working with commercial sources of XOS, the capacity of bifidobacteria to grow on XOS depends on the strain. Kajihara *et al.* (2000) evaluated the effects of XOS on the levels of ammonia in blood samples and the intestinal microbiota of patients suffering from liver cirrhosis and observed that the oral administration of 3g/day of XOS during 2 weeks resulted in a reduction of ammonia in the blood, while other parameters were not affected. They also showed that XOS promoted the growth of intestinal bifidobacteria, while suppressing *Bacteriodes*, with no complications such as diarrhoea or flatulence being observed. In a study by Tateyama *et al.* (2005), it was shown that XOS (4.2 g/day for 4 weeks) alleviated constipation problems in pregnant women.

Isomaltooligosaccharides (IMO)

Isomaltooligosaccharides are glucosyl-saccharides containing one or more glycosidic α-(1,6) bonds with or without α-(1,4) bonds. The commercially available IMO are mainly composed of isomaltose (IG2), panose (P) and isomaltotriose (IG3) (Takaku, 1988).

In 1974 an enzyme named isomaltodextranase that was able to produce IMO was reported for the first time. This was an exo-dextranase produced and exported to the extracellular milieu by a bacteria isolated from soil which was later identified as *Arthrobacter globiformis* T6. Then, other microorganisms were also identified as able to produce α-glycosidases with strong transglycosylation activity such as *Aspergillus niger* (Pazur and French, 1952), *Aureobasidium pullulans* (Yun, Lee and Song, 1994), *Aspergillus carbonarius* (Duan *et al*, 1994), *Aspergillus nidulans* (Kato et *al*., 2002), *Bacillus stearothermophilus* (Mala et *al*., 1999) and *Xanthophyllomyces dendrorhous* (Fernández-Arrojo et *al*., 2007). Some IMO can also be produced by the treatment of sucrose or dextran with dextranase (Goulas et *al*., 2004; Aslan and Tanriseven, 2007). These saccharides also appear naturally in several fermented foods and honey (Komoto et *al*., 1988; Deepa et *al*., 2008).

The enzyme that produces IMO by transglycosylation of maltose is also called transglycosidase because α-D-glycosidase has transglycosylation activity and is the key enzyme in the production of IMO (Chen et *al*., 2001). The conventional method for the production of IMO from starch begins with the breakage of the α-(1,4) bonds and α-D-glyco-oligosaccharides using α-amylase, pululanase, and β-amylase (Kuriki et *al*., 1993), then convert to α-(1,6)-oligosaccharides linked to α-D-glycosidase (Takaku, 1988).

Studies employing α-D-gluco-oligosaccharides of known structure have demonstrated that isomaltodextranase recognizes the α-isomaltosyl residue and catalyses the cleavage of whatever α-linkage attaches that disaccharide to a glucose residue. A high concentration of water in the system is undesirable as water acts as a competitive acceptor leading to maltose hydrolysis (Pazur and French, 1952). After the synthesis of IMO, monosaccharides and/or disaccharides can be removed by means of chromatographic methods, nanofiltration (Goulas et *al*., 2002) or through fermentation by yeasts (Crittenden and Playne, 2002; Yoon, Mukerjea and Robyt, 2003).

Isomaltooligosaccharides are used as food ingredients in Asia, but their use has increased in western countries (Seibel and Buchholz 2010; Goffin et *al*., 2011). These oligosaccharides with α-(1–6) glycosidic bonds like isomaltose, panose and isomaltotriose are found in fermented foods such as 'miso', soy sauce and sake (Kohmoto et *al*., 1992). They are used in several foods and drinks as sweeteners (Nakakuki, 2005), having 40% the sweetener power of sucrose and a calorie value of 2.8 to 3.2 kcal/g (Kaneko, Yokoyama and Suzuki, 1995). According to Kohmoto et *al*. (1992) the calorie content of a commercial preparation containing IMO was 75% lower than the same one prepared with maltose.

Since the eighties, IMO have attracted attention and been used as food additives due to their unique characteristics (Chen et *al*., 2001). They can improve constipation as well as other fibres (Wang *et al*, 2001) and promote the growth of bifidobacteria and reduce the incidence of dental caries (Miyake et *al*., 1985; Machida et *al*., 1988; Chen et *al*., 2001; Rycroft et *al*., 2001). According to Want et *al*. (2001), IMO are used in the treatment of chronic constipation and hyperlipidaemia due to complications in haemodialysis. These effects were observed in patients with a minimum dosage of 8 g per day (Kohmoto et *al*., 1991).

Isomaltooligosaccharides are hardly digested in the upper gastrointestinal tract (Kohmoto et *al*., 1992; Oku and Nakamura, 2003). The ingestion of IMO containing isomaltose (9.0%), panose (21.8%), isomaltose (2%) and isomaltotetraose (7.4%) significantly increased the number of faecal bifidobacteria in humans (Chen et *al*., 2001; Yen et *al*., 2011; Ketabi et *al*., 2011). However, according to Kaneko et *al*. (1995) part of isomaltose is hydrolysed by isomaltase in the jejunum, which suggests that only residual IMO is used by bifidobacteria in the colon. Therefore, there is a concern when classifying IMO as prebiotics with only those with high DP, exemplified by isomaltotetraose and larger oligomers, being truly considered as non-digestible (Debnam et *al*., 1998; Ketabi et *al*., 2011). Despite these reports, scientific data concerning the influence of these oligomers on the intestinal microbiota, especially using cultivation independent methods, are still limited (Yen et *al*., 2010; Ketabi et *al*., 2011).

According to Kaneko et *al*. (1994), IMO are tolerated at higher doses in comparison with other non-digestible oligosaccharides in the diet. This can be considered an advantage since a great deal of the laxative effects happens at higher doses.

Conclusion

When creating functional products with added prebiotics, it is important to determine what is the most suitable prebiotic for the particular food product being developed. One should take into account the particular characteristics of the prebiotics, including their nutritional and technological properties such as stability to processing conditions, minimum and maximum concentrations that lead to the desirable effects, as well as the adverse effects at the concentration range under use. The health benefits associated with the positive technological applications are the main reasons for the widespread use of prebiotics in today's human diet.

Future perspectives

The study of prebiotics would benefit from more in depth clinical studies in order to prove the beneficial effects on human health. In addition, microbiome alterations due to the use of prebiotics should be considered in order to evaluate the specific changes in all the intestinal groups of microorganisms. A great deal of work remains to be done to better understand the behaviour of prebiotics on food matrices, especially considering their interactions with other components. Other studies should also be performed to evaluate the synergistic or antagonistic effects of the combinations of two or more prebiotics, since the combination of two or more prebiotics may lead to better characteristics than the individual ones. Countries should also improve their laws in order to protect consumers from false claims and advertisements.

References

Aachary, A.A., and Prapulla, S.G. (2008). Corncob-induced endo-1,4-β-d-xylanase of Aspergillus oryzae MTCC 5154: production and characterization of xylobiose from glucuronoxylan. J. Agric. Food Chem. 56, 3981–3988.

Aigster, A., Duncan, S.E., Conforti, F.D., and Barbeau, W.E. (2011). Physicochemical properties and sensory attributes of resistant starch-supplemented granola bars and cereals. Food Sci. Technol. 44, 2159–2165.

Ando, H., Ohba, H., Sasaki, T., Takamine, K., Kamino, Y., and Moriwaki, S. (2004). Hot-compressed-water decomposed products from bamboo manifest a selective cytoxicity against acute lymphoblastic leukaemia cells. Toxicol. *in vitro 18*, 765–771.

Angioloni, A., and Collar, C. (2008). Functional response of diluted doughmatrixes in high-fibre systems: a viscometric and rheological approach. Food Res. Int. *41*, 803–812.

Anthony J.C., Merriman, T.N., and Heimbach, J.T. (2006). 90-Day oral (gavage) study in rats with galactooligosaccharides syrup. Food Chem. Toxicol. *44*, 819–826.

Asahara, T., Nomoto, K., Shimizu, K., Watanuki, M., and Tanaka, R. (2001). Increased resistance of mice to *Salmonella enterica* serovar Typhimurium infection by synbiotic administration of Bifidobacteria and transgalactosylated oligosaccharides. J. Appl. Microbiol. *91*, 985–996.

Aslan, Y., and Tanriseven, A. (2007). Immobilization of *Penicillium lilacinum* dextranase to produce isomaltooligosaccharides from dextran. Biochem. Eng. J. *34*, 8–12.

Asp, N.G. (1992). Preface: resistant starch, proceedings of the 2nd plenary meeting of EURESTA: European Flair Concerted Action No. 11 on physiological implications of the consumption of resistant starch in man. Eur. J. Clin. Nutr. *46*, S2-S1.

Baghurst, K.I., Baghurst, P.A., and Record, S.J. (2001). Dietary fibre, nonstarchpolysaccharide and resistant starch intakes in Australia. In Handbook of dietary fibre in human health, G.A. Spiller, ed. (Boca Raton: CRC Press), pp. 583–591.

Baixauli, R., Salvador, A., Martinez-Cervera, S., and Fiszman, S.M. (2008). Distinctivesensory features introduced by resistant starch in baked products. Food Sci. Technol. *41*, 1927–1933.

Barboza, M., Sela, D.A., Pirim, C., Locascio, R.G., Freeman, S.L., German, J.B., Mills, D.A., and Lebrilla, C.B. (2009). Glycoprofiling bifidobacterial consumption of galacto-oligosaccharides by mass spectrometry reveals strain-specific, preferential consumption of glycans. Appl. Environ. Microbiol. *75*, 7319–7325.

Barclay, T., Ginic-Markovic, M., Cooper, P., and Petrovsky, N. (2010). Inulin – a versatile polysaccharide with multiple pharmaceutical and food chemical uses. J. Excipients and Food Chem. *1*, 27–50.

Bayarri, S., and Costell, C.R. (2010). Comparing λ-carrageenan and an inulin blend as fat replacers in carboxymethyl cellulose dairy desserts. Rheological and sensory aspects. Food Hydrocolloid. *24*, 578–587.

Benno, Y., Endo, K., Shiragami, N., Sayama, K., and Mitsuoka, T. (1987). Effect of raffinose on human faecal microflora. Bifidobacteria Microflora. *6*, 59–63.

Berg, H.E., and Van Boekel, M.A.J.S. (1994). Degradation of lactose during heating of milk. 1. Reaction pathways. Neth Milk Dairy J. *48*, 157–175.

Bhat, M.K. (1998). Oligosaccharides as Functional Food Ingredients and their Role in Improving the Nutritional Quality of Human Food and Health. *Recent Res. Dev. Agric. Food Chem. 2*, 787–802.

Bohm, A., Kaiser, I., Trebstein, A., and Henle, T. (2005). Heat-induced degradation of inulin. Eur. Food Res. Technol. *220*, 466–471.

Bohm, A., Kleessen, B., and Henle, T. (2006). Effect of dry heated inulin on selected intestinal bacteria. Eur. Food Res. Technol. *222*, 737–740.

Bomba, A., Nemcova, R., Gancarcikova, S., Herich, R., Guba, P., and Mudronova, D. (2002). Improvement of the probiotic effect of microorganisms by their combination with maltodextrins, fructo-oligosaccharides and polyunsaturated fatty acids. Br. J. Nutr. *88*, 95–99.

Bornet, F.R., Brounsl, J.F., Tahiro, Y., and Duvillier, V. (2002). Nutritional aspects of short-chain fructooligosaccharides: natural occurrence, chemistry, physiology and health implications. Digest Liver Dis. *Suppl. 21*, S111–20.

Bosscher, D. (2009). Fructan prebiotics derived from inulin. In Prebiotics and Probiotics Science and Technology, D. Charalampopoulos and R.A. Rastall, eds. (New York, USA: Springer), pp. 163–206.

Bosscher, D., Van Loo, J., and Franck, A. (2006). Inulin and oligofructose as functionalingredients to improve bone mineralization. Int. Dairy J.*16*, 1092–1097.

Bouhnik, Y., Raskine, L., Champion, K., Andrieux, C., and Penven, S. (2007). Prolonged administration of low-dose inulin stimulates the growth of bifidobacteria in humans. Nutr. Res. *27*, 187– 193.

Brown, I.L. (2004). Applications and uses of resistant starch. J. Assoc. Off. Anal. Chem. *87*, 727–732.

Bruzzese, E., Volpicelli, M., Salvini, F., Bisceglia, M., Lionetti, P., Cinquetti, M., Iacono, G., and Guarino, A. (2006). Early administration of GOS/FOS prevents intestinal and respiratory infections in infants. J. Pediatr. Gastroenterol. Nutr. *42*, PN2–18.

Búrigo, T., Fagundes, R.L.M., Trindade, E.B.S.M., and Vasconcelo, H.C.F.F. (2007). Efeito bifidogênico do frutooligossacarídeo na microbiota intestinal de pacientes com neoplasia hematológica. Rev. Nutr. *20*, 491–497.

Buriti, F.C.A., Castro, I.A., and Saad, S.M.I. (2010). Effects of refrigeration, free zingand replacement of milk fat by inulin and whey protein concentrate on texture profile and sensory acceptance of synbiotic guava mousses. Food Chem. *123*, 1190–1197.

Cani, P.D., Knauf, C., Iglesias, M.A., Drucker, D.J., Delzenne, N.M., and Burcelin, R. (2006). Improvement of glucose tolerance and hepatic insulin sensitivity by oligofructose requires a functional glucagon-like peptide 1 receptor. Diabetes. *55*, 1484–1490.

Cani, P.D., Lecourt, E., Dewulf, E.M., Sohet, F.M., Pachikian, B.D., Naslain, D., De Backer, F., Neyrinck, A.M., and Delzenne, N.M. (2009). Gut microbiota fermentation of prebiotics increases satietogenic and incretin gut peptide production with consequences for appetite sensation and glucose response after a meal. Am. J. Clin. Nutr. *90*, 1236–1243.

Carabin, I.G., and Flamm, W.G. (1999). Evaluation of safety of inulin and oligofructose as dietary fiber. Regul. Toxicol. Pharmacol. *30*, 268–282.

Carpita, N.C., Kamabus, J., and Housley, T.L. (1989). Linkage structure of fructans and fructan oligomers from *Triticum astivum* and *Fistuca arundinaceae* leaveas. J. Plant Physiol. *134*, 162–168.

Carvalho, M.A.M., Pinto, M.M., and Figueiredo-Ribeiro, R.C.L. (1998). Inulin production by *Vernonia herbacea* as influenced by mineral fertilization and time of harvest. Rev. Bras. Bot. *21*, 275–280.

Charalampopoulos, D., and Rastall, R. (2012). Prebiotics in foods. Curr. Opin. Biotechnol. *23*, 187–191.

Charalampopoulos, D., Wang, S., Pandiella, S.S., and Webb, C. (2002). Application of cereals and cereal components in functional foods: a review. Int. J. Food Microbiol. *79*, 131–141.

Chen, H.L., Lu, Y.H., Lin, J.J., andKo, L.Y. (2001). Effects of isomalto-oligosaccharides on bowel functions and indicators of nutritional status in constipated elderly men. J. Am. Coll. Nutr. *20*, 44–49.

Chen, H., Li-Jun, L., Jian-Jun, Z., Bo, X., and Rui, L. (2010). Effect of soybean oligosaccharides on blood lipid, glucose levels and antioxidant enzymes activity in high fat rats. Food Chem. *119*, 1633–1636.

Chow, J.M. (2002). Probiotics and prebiotics: A brief overview. J. Renal Nutr. *12*, 76–86.

Chung, Y.C., Hsu, C.K., Ko, C.Y., and Chan, Y.C. (2007). Dietary intake of xylooligosaccharides improves the intestinal microbiota, faecal moisture, and pH value in the elderly. Nutr. Res. *27*, 756–761.

Coulier, L., Timmermans, J., Bas, R., Van Den Dool, R., Haaksman, I., Klarenbeek, B., Slaghek, T., and Van Dongen, W. (2009). In-depth characterization of prebiotic galacto-oligosaccharides by a combination of analytical techniques. J. Agric. Food Chem. *57*, 8488–8495.

Courtin, C.M., Swennen, K., Verjans, P., and Delcour, J.A. (2009). Heat and pHstability of prebiotic arabinoxylooligosaccharides, xylooligosaccharides and fructooligosaccharides. Food Chem. *112*, 831–837.

Coussement, P.A.A. (1999). Inulin and oligofructose: safe intakes and legal status. J. Nutr. *129*, 1412–1417.

Crittenden, R.G., and Playne, M.J. (1996). Production, properties and applications of food-grade oligosaccharides. Trends Food Sci. Technol. *7*, 353–361.

Crittenden, R.G., and Playne, M.J. (2002). Purification of food-grade oligosaccharides using immobilized cells of *Zymomonas mobilis*. Appl. Microbiol. Biotechnol. *58*, 297–302.

Cummings, J.H., and Macfarlane, G.T. (2002). Gastrointestinal effect of prebiotics. Br. J. Nutr. *87*, 145–151.

Davis, L.M. G., Martínez, I., Walter, J., and Hutkins, R. (2010). A dose dependent impact f prebiotic galactooligosaccharides on the intestinal microbiota of healthy adults. Int. J. Food Microbiol. *144*, 285–292.

Davis, L.M., Martinez, I., Walter, J., Goin, C.,and Hutkins, R.W. (2011). Barcoded pyrosequencing reveals that consumption of galactooligosaccharides results in a highly specific bifidogenic response in humans. PLoS One. *6*, e25200.

De Bruyn, A., Alvarez, A.P., Sandra, P., and De Leenheer, L. (1992). Isolation and identification of *O*-β-d-fructofuranosyl-(2 →

1)-O-β-d-fructofuranosyl-(2 → 1)-d-fructose, a product of the enzymic hydrolysis of the inulin from *Chicorium intybus*. Carbohydr. Res. *235*, 303–308.

De Preter, V., Hamer, H.M., Windey, K., and Verbeke, K. (2011). The impact of pre- and/or probiotics on human colonic metabolism: Does it affect human health? Mol. Nutr. Food Res. *55*, 46–57.

De Reu, J.C., Linssen, V.A.J.M., Rombouts, F.M., and Nout, M.J.R. (1997). Consistency, polysaccharidase activities and non-starch polysaccharides content of soya beans during tempe fermentation. J. Sci. Food Agric. *73*, 357–363.

Debnam, E.S., Denholm, E.E., and Grimble, G.K. (1998). Acute and chronic exposure of rat intestinal mucosa to dextran promotes SGLT1-mediated glucose transport. Eur. J. Clin. Invest. *28*, 651–658.

Deepa, G., Vasudeva, S., and Akhilender, K. (2008). Nutrient composition and physicochemical properties of Indian medicinal rice Njavara. Food Chem. *106*, 165–171.

Delzenne, N.M., Cani, P.D., Daubioul, C., and Neyrinck, A.M. (2005). Impact of inulinand oligofructose on gastrointestinal peptides. Br. J. Nutr. 93(*Suppl. 1*), S157-SS16.

Delzenne, N., Aertssens, J., Verplaetse, H., Roccaro, M., and Roberfroid, M. (1995). Effect of fermentable fructooligosaccharides on mineral, nitrogen and energy digestive balance in the rat. Life Sci. *57*, 1579–1587.

Demigné, C., Jacobs, H., Moundras, C., Davicco, M., Horcajada, M., Bernalier, A., and Coxam, V. (2008). Comparison of native or reformulated chicory fructans, or nonpurifiedchicory, on rat cecal fermentation and mineral metabolism. Eur. J. Nutr. *47*, 366–374.

Deng, J.M., Mai, K.S., Ai, Q.H., and Zhang, W.B. (2007). Effects of soybean oligosaccharides on lipid metabolism of Japanese flounder (Paralichthys olivaceus Temminck et Schlegel) fed animal or plant protein source-based diets. Front. Agric. China. *1*, 315–323.

Dey, P.M. (1985). d-Galactose containing oligosaccharides. In Biochemistry of storage carbohydrates in green plants. Dey, P.M., Dixon, R.A (London, UK: Academic Press), pp. 53–129.

Drakoularakou, A., Tzortzis, G., Rastall, R.A., and Gibson, G.R. (2010). A doubleblind,placebo-controlled, randomized human studyassessing the capacity of a novel galacto-oligosaccharidemixture in reducing travellers' diarrhoea. Eur. J. Clin. Nutr. *64*, 146–152.

Duan, K.J., Sheu, D.C., Lin, M.T., and Hsueh, H.C. (1994). Reaction mechanism of isomaltooligosaccharides synthesis by α-glucosidase from *Aspergillus carbonarius*. Biotechnol. Lett. *16*,1151–1156.

Ebringerová, A., and Heinze, T. (2000). Xylan and xylan derivativesbiopolymers with valuable properties, 1. Naturally occurring xylans structures, isolation procedures properties. Macromol. Rapid Commun. *21*, 542–556.

Eneysaya, E.V., Ivanen, D.R., Bobrov, K.S., Isaeva-Ivanova, L.S., Shabalin, K.A., Savel'ev, A.N., Golubev, A.M., and Kulminskaya, A.A.(2007). Biochemical and kinetic analysis of the GH3 family β-xylosidase from *Aspergillus awamori* X-100. Arch. Biochem. Biophys. *457*, 225- 234.

Englyst, H.N., Kingman, S.M., and Cummings, J.H. (1992). Classification and measurement of nutritionally important starch fractions. Eur. J. Clin. Nutr. *46*, S33-S50.

Espinosa-Martosy, I., and Rupérez, P. (2006). Soybean oligosaccharides. Potential as new ingredients in functional food. Nutr. Hosp. *21*, 92–96.

European Parliament. (2006). Regulation (EC) No 1924/2006 of the European parliament and of the council of 20 December 2006 on nutrition and health claims made on foods. OJEU, 9–25.

Fei, B., Ling, L., Hua, C., and Ren, S. (2014). Effects of soybean oligosaccharides on antioxidant enzyme activities and insulin resistance in pregnant women with gestational diabetes mellitus. Food Chem. *158*, 429–432.

Fernàndez-Arrojo, L., Marìn, D., De Segura, A.G., Linde, D., Alcalde, M., and Gutiérrez-Alonso, P. (2007). Transformation of maltose into prebiotic isomaltooligosaccharides by a novel α-glucosidase from *Xanthophyllomyces dendrorhous*. Process. Biochem. *42*, 1530–1536.

Fiordaliso, M., Kok, N., Desager, J.P., Goethals, F., Deboyser, D., Roberfroid, M., and Delzenne, N. (1995). Dietary oligofructose lowers triglycerides, phospholipids and cholesterol in serum and very low density lipoproteins in rats. Lipids. *30*, 163–167.

Flamm, G., Glinsmann, W., Kritchevsky, D., Prosky, L., and Roberfroid, M. (2001). Inulin and oligofructose as dietary fiber: a review of the evidence. Crit. Rev. Food Sci. Nutr. *41*, 353–362.

Foster, K.J., Lin, S., and Turck, C.J. (2010). Current and emerging strategies for treating hepatic encephalopathy. Crit. Care Nurs. Clin. North Am. *22*, 341–350.

Franck, A. (2002). Technological functionality of inulin and oligofructose. Br. J. Nutr. *87*,S287–S291.

Franck, A. (2008). Food applications of prebiotics. In Hanbook of Prebiotics, G.R., Gibson, and M.B., Roberfroid, eds. (CRC Press, Florida), pp. 437–448.

Fuentes-Zaragoza, E., Riquelme-Navarrete, M.J., Sánchez-Zapata, E., and PérezÁlvarez, J.A. (2010). Resistant starch as functional ingredient: A review. Food Res. Int. *43*, 931–942.

Gänzle, M.G. (2011). Encyclopedia of Dairy Sciences, Second Edition (San Diego, USA: Academic Press).

Garrote, G., Domínguez, H., and Parajó, J.C. (1999). Mild autohydrolysis: an environmentally friendly technology for xylooligosaccharide production from wood. J. Chem. Technol. Biotechnol. *74*, 1101–1109.

Garrote, G., Falqué, E., Domínguez, H., and Parajó, J.C. (2007). Autohydrolysis of agricultural residues: study of reaction byproducts. Bioresource Tecnol. *98*, 1951–1957.

Gibson, G.R., Beatty, E.R., Wang, X., and Cummings, J.H. (1995). Selective stimulation of bifidobacteria in the human colon by oligofructose and inulin. Gastroenterology. *108*, 975–982.

Gibson, G.R., Probert, H.M., Van Loo, J.A.E., and Roberfroid, M.B. (2004). Dietary modulation of the human colonic microbiota: updating the concept of prebiotics. Nutr. Res. Rev. *17*, 257–259.

Gibson, G.R., and Roberfroid, M.B. (1995). Dietary modulation of the human colonic microbiota – introducing the concept of prebiotics. J. Nutr. *125*, 1401–1412.

Glibowski, P., and Bukowska, A. (2011). The effect of pH, temperature and heating time on inulin chemical stability. Acta Sci. Po. Technolo. Aliment. *10*, 189–196.

Goldring, J.M. (2004). Resistant starch: Safe intakes and legal status. J. Assoc. Off. Anal. Chem. *87*, 733–739.

Goulas, A.K., Kapasakalidis, P.G., Sinclair, H.R., Rastall, R.A., and Grandison, A.S. (2002). Purification of oligosaccharides by nanofiltration. J. Membr. Sci. *209*, 321–335.

Guggisberg, D., Piccinali, P., and Schreier, K. (2011). Effects of sugar substitution with Stevia, Actilight™ and Stevia combinations or Palatinose™ on rheological and sensory characteristics of low-fat and whole milk set yoghurt. Int. Dairy J. *21*, 636–644.

Hager, A.-S., Ryan, L.A.M., Schwab, C., Gaenzle, M.G., O'Doherty, J.V., and Arendt, E.K. (2011). Influence of the soluble fibers inulin and oat betaglucanon quality of dough and bread. Eur. Food Res. Technol. *232*, 405–413.

Hara, T., Ikeda, N., Hatsumi, K., Watabe, J., Lino, H., and Mitsuoka, T. (1997). Effects os small amount ingestion of soybean oligosaccharides on bowel habits and fecal flora of volunteers. Jap. J. Nutr. *55*, 79–84.

Haralampu, S.G. (2000). Resistant starch: A review of the physical properties and biological impact of RS3. Carbohyd. Polym. *41*, 285–292.

Hata, Y., Yamamoto, M., and Nakajima, K. (1991). Effects of soybean oligosaccharides on human digestive organs: estimation of fifty per cent effective dose and maximum non-effective dose based on diarrhoea. J. Clin. Biochem. Nutr. *10*, 135–144.

Hayakawa, K., Mizutani, J., Wada, K., Masai, T., Yoshihara, I., and Mitsuoka, T. (1990). Effects of Soybean Oligosaccharides on Human Faecal Flora. Microb. Ecol. Health Dis. 3, 293–303.

Hernández, O., Ruiz-Matute, A.I., Olano, A., Moreno, F.J., and Sanz, M.L. (2009). Comparison of fractionation techniques to obtain prebiotic galactooligosaccharides. Int. Dairy J. *19*, 531–536.

Hoebregs, H. (1997). Fructans in foods and food products, ion-exchange chromatographic method: collaborative study. J. AOAC Int. *80*, 1029–1037.

Hopkins, M., Cummings, J., and Mac Farlane, G. (1998). Inter-species differences in maximum specific growth rates and cell yields of bifidobacteria cultured on oligosaccharides and other simple carbohydrate sources. J. Appl. Microbiol. *85*, 381–386.

Hosono, A., Ozawa, A., Kato, R., Ohnishi, Y., Nakanishi, Y., Kimura, T., and Nakamura, R. (2003). Dietary fructooligosaccharides induce immunoregulation of intestinal IgA secretion by murine Peyer's patch cells. Biosci. Biotechnol. Biochem. *67*, 758–764.

Huang, L., Zhang, Z., and Chen, P. (2006). Effect of soybean-oligosaccharides on antioxidation and the immunity of broliers. J. Agric. Univ. HeBei. *29*, 88–94.

Huebner, J., Wehling, R.L., and Hutkins, R.W. (2007). Functional activity of commercial prebiotics. Int. Dairy J. *17*, 770–775.

Huebner, J., Wehling, R.L., Parkhurst, A., and Hutkins, R.W. (2008). Effect of processing conditions on the prebiotic activity of commercial prebiotics. Int. Dairy J. *18*, 287–293.

Institute of Medicine (2005). Dietary reference intakes for energy, carbohydrate, fiber, fat, fatty acids, cholesterol, protein, and amino acids. (Washington, USA: National Academy Press).

Izzo, M., and Niness, K. (2001). Formulating nutrition bars with inulin and oligofructose. Cereal Food. World. *46*, 102–106.

Joung, H.K., and Cheong, W.J. (2001). Simultaneous quantitative determination of mono-, di-, tri-, tetra-, and pentasaccharides in yoghurt products by a simple HPLCsystem with a refractive index detector. Bull. Korean Chem. Soc. *22*, 123–126.

Kabel, M.A., Carvalheiro, F., Garrote, G., Avgerinos, E., Koukios, E., Parajó, J.C., Gírio, F.M., Schols, A.G.J., and Voragen, A.G.J. (2002). Hydrothermally treated xylan rich by-products yields different classes of xylo-oligosaccharides. Carbohyd. Polym. *50*, 47–56.

Kajihara, M., Kato, S., Konishi, M., Yamagishi, Y., Horie, Y., and Ishii, H. (2000). Xylooligosaccharide decreases blood ammonia levels in patients with liver cirrhosis. Am. J. Gastroenterol. *95*, 2514.

Kaneko, T., Yokoyama, A., and Suzuki, M. (1995). Digestibility characteristics of isomaltooligosaccharides in comparison with several saccharides using the rat jejunum loop method. Biosci. Biotechnol. Biochem. *59*, 1190–1194.

Katapodis, P., Vardakou, M., Kalogeris, E., Kekos, D., Macris, B.J., and Christakopoulos, P. (2003). Enzymatic production of a feruloylated oligosaccharide with antioxidant activity from wheat flour arabinoxylan. Eur. J. Nutr. *42*, 55–60.

Kato, N., Suyama, S., Shirokane, M., Kato, M., Kobayashi, T., and Tsukagoshi, N. (2002). Novel α-glucosidase from *Aspergillus nidulans* with strong transglycosylation activity. Appl. Environ. Microbiol. *68*,1250–1256.

Kendall,W.C., Emam, A., Agustin, S.A., and Jenkins, J.A. (2004). Resistant starches and health. J. Assoc. Off. Anal. Chem. *87*, 769–774.

Ketabi, A., Dieleman, L.A., and Ganzle, M.G. (2011). Influence of isomalto-oligosaccharides on intestinal microbiota in rats. J. App. Microbiol. *110*, 1297–1306.

Kim, Y.S., and Oh, D.K. (2012). Lactulose production from lactose as a single substrate by a thermostable cellobiose 2-epimerase from *Caldicellulosiruptor saccharolyticus*. Biorsource Technol. *104*, 668–672.

Kimura, K., Matsumoto, K., Ishihara, C., Harada, K., and Miyagi, A. (1995). Structure determination of galacto-oligosaccharides by pyridylamination and NMR spectroscopy. Carbohydr. Res. *270*, 33–42.

Klewicki, R. (2007). The stability of gal-polyols and oligosaccharides during pasteurization at a low pH. LWT Food Sci. Technol. *40*, 1259–1265.

Kohmoto, T., Fukui, F., Takaku, H., and Mitsuoka, T. (1991). Dose-response test of isomaltooligosaccharides for increasing faecal bifidobacteria. Agric. Biol. Chem. *55*, 2157–2159.

Kohmoto, T., Tsuji, K., Kaneko, T., Shiota, M., Fukui, F., Takaku, H., Nakagawa, Y., Ichikawa, T., and Kobayashi, S. (1992). Metabolism of 13 C-isomaltooligosaccharides in healthy men. Biosci. Biotechnol. Biochem. *56*, 937–940.

Kuriki, T., Yanase, M., Takata, H., Takesada, Y., Imanaka, T., and Okada, S. (1993). A new way of producing isomalto-oligosaccharide syrup by using the transglycosylation reaction of neopullulanase. Appl. Environ. Microbiol. *59*, 953–959.

Kusuma, G.D., Paseephol, T., and Sherkat, F. (2009). Prebiotic and rheological effects of Jerusalem artichoke inulin in low-fat yoghurt. Aust. J. Dairy Tecnol. *64*, 159–163.

L'Homme, C., Arbelot, M., Puigserver, A., and Biagini, A. (2003). Kinetics of hydrolysis of fructooligosaccharides in mineral-buffered aqueous solutions: Influence of pH and temperature. J. Agric. Food Chem. *51*, 224–228.

Lamoureux, L., Roy, D., and Gauthier, S.F. (2002). Production of oligosaccharides in yoghurt containing Bifidobacteria and yoghurt cultures. J. Dairy Sci. *85*, 1058–1069.

Lecerf, J.M., Dépeint, F., Clere, E., Dugenet, Y., Niamba, C.N., Rhazi, L., Cayzeele, A., Abdelnour, G., Jaruga, A., Younes, H., *et al.* (2012). Xylo-oligosaccharide (XOS) in combination with inulin modulates both the intestinal environment and immune status in healthy subjects, while XOS alone only shows prebiotic properties. Brit. J. Nutr. *108*, 1847–1858.

Lee, S., and Prosky, L. (1995). International survey on dietary fibre: definition, analysis and reference material. J. AOAC Int. *78*, 22–36.

Liong M.-T. (2008). Roles of probiotics and prebiotics in colon cancer prevention: postulated mechanisms and *in vivo* evidence. Int. J. Mol. Sci. *9*, 854–863.

Lomax, A.R., and Calder, P.C. (2009). Prebiotics, immune function, infection and inflammation: a review of the evidence. Br. J. Nutr. *101*, 633–658.

López-Molina, D., Navarro-Martínez, M.D., Rojas, M.F., Hiner, A.N., Chazarra, S., and Rodríguez-López, J.N. (2005). Molecular properties and prebiotic effect of inulin obtained from artichoke (*Cynara scolymus L.*). Hytochemistry. *66*, 1476–1484.

Macfarlane, G.T., and Cummings, J.H. (1999). Probiotics and prebiotics: can regulating the activities of intestinal bacteria benefit health? Br. Med. J. *18*, 999–1003.

Macfarlane, G.T., Steed, H., and Macfarlane, S. (2008). Bacterialmetabolism and health-related effects of galacto-oligosaccharides andother prebiotics. J. Appl. Microbiol. *104*, 305–344.

Machida, Y., Fukui, F., and Komoto, T. (1988). Promoting the proliferation of intestinal bifidobacteria. US Patent 4782045.

Mala, S., Dvorakova, H., Hrabal, R., and Kralova, B. (1999). Towards regioselective synthesis of oligosaccharides by use of alpha-glucosidases with different substrate-specificity. Carbohydr. Res. *322*, 209–218.

Manning, T.S., and Gibson, G.R. (2004). Prebiotics. Best Practice & Research Clinical. Gastroenterology. *18*, 287–298.

Manzanares, P., Graaff, L.H., and Visser, J. (1998). Characterization of galactosidases from *Aspergillus niger*: purification of a novel α-galactosidase activity. Enzyme Microb. Technol. *22*, 383–390.

Marteau, P. (2001). Prebiotics and probiotics for gastrointestinal health. Clin. Nutr. *20*, 41–45.

Martínez-Ferez, A., Rudloff, S., Guadix, A., Henkel, C.A., Pohlentz, G., Boza, J.J., Guadix, E.M., and Kunz, C. (2006). Goat's milk as natural source lactose-derived oligosaccharides: Isolation by membrane technology. Int. Dairy J. *16*, 173–181.

Martins, A.R., and Burket, C.A.V. (2009). Galacto-oligossacarídeos (GOS) e seus efeitos prebióticos e bifidogênicos. Braz. J. Food Technol. *12*, 230–240.

Mateos-Aparicio, I., Redondo-Cuenca, A., Villanueva-Suárez, M.J., and Zapata-Revilla, M.A. (2008). Soybean, a promising health source. Nutr. Hosp. *23*, 305–312.

Matusek, A., Meresz, P., Le, T.K.D., and Oersi, F. (2009). Effect of temperature and pH on the degradation of fructo-oligosaccharides. Eur. Food Res. Technol. *228*, 355–365.

Mendoza, M.R., Olano, A., and Villamiel, M. (2005). Chemical indicators of heat treatment in fortified and special milks. J. Agric. Food Chem. *53*, 2995–2299.

Menezes, C.C., Carneiro, J.D.S., Borges, S.V., Silva, V.S.N., Brigagão, M.R.P.L., and Azevedo, L. (2012). Development of low-calorie guava preserves with prebiotics and evaluation of their effects on carcinogenesis biomarkers in rats. Food Chem. Toxicol. *50*, 3719–3724.

Meyer, D., and Stasse-Wolthuis, M. (2009). The bifidogenic effect of inulin and oligofructose and its consequences for gut health. Eur. J. Clin. Nutr. *63*, 1277–1289.

Meyer, D., Bayarri, S., Tárrega, A., and Costell, E. (2011). Inulin as texture modifier in dairy products. Food Hydrocolloids. *25*, 1881–1890.

Miyake, T., Yoshida, M., and Takeuchi, K. (1985). Imparting low-or anti-cariogenic property to orally-usable products. US Patent 4518581.

Mizota, T., Tamura, Y., Tomita, M., and Okonogi, S. (1987). Lactulose as a sugar with physiological significance. IDF Bull. *212*, 69–76.

Mumtaz, S., Rehman, S.U., Huma, N., Jamil, A., and Nawaz, H. (2008). Xylooligosaccharide enriched yoghurt: physicochemical and sensoryevaluation. Pak. J. Nutri. *7*, 566–569.

Murphy, M.M., Douglass, J.S., and Birkett, A. (2008). Resistant starch intakes in the United States. J. Am. Diet. Assoc. *108*, 67–78.

Mussatto, S.I., and Mancilha, I.M. (2007). Non-digestible oligosaccharides: *A review*. Carbohyd. Polym. *68*, 587–597.

Mussatto, S.I., and Teixeira, J.A. (2010). Increase in the fructooligosaccharides yield and productivity by solid-state fermentation with *Aspergillus japonicus* using agro-industrial residues as support and nutrient source. Bioch. Eng. *J. 53*, 154–157.

Nagendra, R., Mahadevamma, S., Baskaran, V., and Rao, S.V. (1995). Shelf-life of spray-dried infant formula supplemented with lactulose. J. Food Process. Preserv. *19*, 303–315.

Nakakuki, T. (2005). Present status and future prospects of functional oligosaccharide development in Japan. J. Appl. Glycosci. *52*, 267–271.

Nakamura, Y., Nosaka, S., Suzuki, M., Nagafuchi, S., Takahashi, T., Yajima, T., Takenouchi-Ohkubo, N., Iwase, T., and Moro, I. (2004). Dietary fructooligosaccharides up-regulate immunoglobulin A response and polymeric immunoglobulin receptor expression in intestines of infant mice. Clin. Exp. Immunol. *137*, 52–58.

Niness, K.R. (1999). Inulin and oligofructose: what are they? J. Nutr. *129*, 1402S–1406S.

Nugent, A.P. (2005). Health properties of resistant starch. BNF, Nutr. Bull. *30*, 27–54.

Ohta, A., Ohtuki, M., Takizawa, T., Inaba, H., Adachi, T., and Kimura, S. (1994). Effects of fructooligosaccharides on the absorption of Mg and Ca by cectomized rats. Int. J. Vitam. Nutr. Res. *64*, 316–323.

Ohta, A., Motohashi, Y., Ohtsuki, M., Hirayama, M., Adachi, T., and Sakuma, K. (1997). Dietary fructooligosaccharides change the concentration of calbindin-D9k differently in the mucosa of the small and large intestine of rats. J. Nutr. *128*, 934–939.

Okazaki, M., Fujikawa, S., and Mastumoto, N. (1990a). Effects of xylooligosaccharides on growth of Bifidobacteria. Bifidobact. Microfl. *9*, 77–86.

Okazaki, M., Fujikawa, S., and Matsumoto, N. (1990b). Effects of xylooligosaccharide on growth of bifidobacteria. J. Japan Soc. Nutr. Food Sci. *43*, 395–401.

Olano, A., and Corzo, N. (2009). Lactulose as food ingredient. J. Sci. Food Agric. *89*, 1987–1990.

Olano, A., Calvo, M.M., and Corzo, N. (1989). Changes in the carbohydrate fraction of milk during heating processes. Food Chem. *31*, 259–265.

Otieno, D.O., and Ahring, B.K. (2012). A thermochemical pretreatment process to produce xylooligosaccharides (XOS), arabinooligosaccharides (AOS) and mannooligosaccharides (MOS) from lignocellulosic biomasses. Bioresour. Technol. *112*, 285–292.

Panesar, P.S., and Kumari, S. (2011). Lactulose: Production, purification and potential applications. Biotechnol. Adv. *29*, 940–948.

Parajó, J.C., Garrote, G., Cruz, J.M., and Dominguez, H. (2004). Production of xylooligosaccharides by autohydrolysis of lignocellulosic materials. Trends Food Sci. Technol. *15*, 115–120.

Parnell, J.A., and Reimer, R.A. (2009). Weight loss during oligofructose supplementation is associated with decreased ghrelin and increased peptide YY in overweight and obese adults. Am. J. Clin. Nutr. *89*, 1751–1759.

Pazur, J.H., and French, D. (1952). The action of transglucosidase of Aspergillus oryzae on maltose. J. Biol. Chem. *196*, 265–272.

Peters, H.P.F., Boers, H.M., Haddeman, E., Melnikov, S.M., and Qvyjt, F. (2009). No effect of added beta-glucan or of fructooligosaccharide on appetite or energy intake. Am. J. Clin. Nutr. *89*, 58–63.

Playne, M.J., and Crittenden, R. (1996). Commercially available oligosaccharides. Bull. Int. Dairy Federation. *313*, 10–22.

Prescilla, V.J., Rocio, R.A.M., Johan, G., and Leon, M.J.K. (2012). Difficulties in describing allergic disease modulation by pre-, pro- and Synbiotics. Curr. Pharm. Design. *18*, 2369–2374.

Quemener, B., Thibault, J.F., and Coussement, P. (1997). Integration of inulin determination in the AOAC method for measurement of total dietary fibre. Int. J. Biol. Macromol. *21*, 175–178.

Rinaldoni, A.N., Campderrós, M.E., and Padilla, A.P. (2012). Physico-chemical and sensory properties of yoghurt from ultrafiltreted soy milk concentrate added with inulin. Food Sci. Technol. *45*, 142–147.

Rivero-Urgell, M., and Santamaria-Orleans, A. (2001). Oligosaccharides: Application in infant food. Early Hum. Dev. *65*, S43–S52.

Roberfroid, M. (2007). Prebiotics: the concept revisited. J. Nutr. *137*, 830–837.

Roberfroid, M.B. (1991). Dietary modulation of experimental neoplastic development: role of fat and fibre content and caloric intake. Mutat. Res. *259*, 351–362.

Roberfroid, M.B. (1993). Dietary fibre, inulin and oligofructose: a review comparing their physiological effects. Crit. Rev. Food Sci. Nutr. *33*, 103–148.

Roberfroid, M.B. (2007). Inulin-type fructans: functional food ingredients. J. Nutr. *137*, 2493S-2502S.

Roberfroid, M.B. (2005). Introducing inulin-type fructans. Br. J. Nutr. *93*, S13–S25.

Roberfroid, M., and Slavin, J. (2000). Nondigestible oligosaccharides. Crit. Rev. Food Sci. Nutr. *40*, 461–480.

Roberfroid, M., Gibson, G.R., and Delzenne, N. (1993). The biochemistry ofoligofructose, a nondigestible fibre: an approach to calculate its caloric value. Nutr. Rev. *51*, 137–146.

Roberfroid, M., Gibson, G.R., Hoyles, L., McCartney, A.L., Rastall, R., Rowland, I., Wolvers, D., Watzl, B., Szajewska, H., Stahl, B., *et al.* (2010). Prebiotic effects: metabolic and health benefits. Br. J. Nutr. *104* (Suppl. 2), S1–63.

Rutherford, P.P., and Whittle, R. (1982). The carbohydrate composition of onions during long term cold storage. J. Hortic. Sci. *57*, 349–356.

Ruttloff, H., Taufel, A., Krause, W., Haenel, H., and Taufel, K. (1967). Enzymatic hydrolysis of galacto-oligosaccharides in the human and animal intestine with particular regard to *L. bifidus*: behavior of lactulose in the intestine. Nahrung. *11*, 39–46.

Rycroft, C.E., Fooks, L.J., and Gibson, G.R. (1999). Methods for assessing the potential of prebiotics and probiotics. Curr. Opin. Clin. Nutr. Metab. Care. *2*, 481–484.

Saad, N., Delattre, C., Urdaci, M., Schmitter, J.M., and Bressollier, P. (2013). An overview of the last advances in probiotic and prebiotic field. Food Sci. Technol. *50*, 1–6.

Sabater-Molina, M., Larqué, R., Torella, F., and Zamora, S. (2009). Dietary fructooligosaccharides and potencial benefits on health. J. Physiol. Biochem. *65*, 315–328.

Sajilata, M.G., Singhal, R.S., and Kulkarni, P.R. (2006). Resistant starch – A review. Compr. Rev. Food. Sci. Food Saf. *5*, 1–17.

Sakla, A.B., Ghali, Y., El-Farra, A., and Rizk, L.F. (1988). The effect of environmental conditions on the chemical composition of soybean seeds: relationship between the protein, oil, carbohydrate and trypsin inhibitor content. Food Chem. *29*, 221–231.

Sako, T., Matsumoto, K., and Tanaka, R. (1999). Recent progress on research and applications of non-digestible galacto-oligosaccharides. Int. Dairy J. *9*, 69–80.

Sánchez-Mata, M.C., Peñuela-Teruel, M.J., Cámara-Hurtado, M., Díez-Marqués, C., and Torija-Isasa, M.E. (1998). Determination of mono-, di-, and oligosaccharides in legumes by high-performance liquid chromatography using an amino-bonded silica column. J. Agric. Food Chem. *9*, 3648–3652.

Sangeetha, P.T., Ramesha, M.N., and Prapulla, S.G. (2005). Recent trends in the microbial production, analysis and application of Fructooligosaccharides. Trends Food Sci. Technol. *16*, 442–457.

Sangwan, V., Tomar, S.K., Singh, R.R.B., Singh, A.K., and Ali, B. (2011). Galactooligosaccharides: novel components of designer foods. J. Food Sci. *76*, R103–R111.

Scholz-Ahrens, K.E., and Schrezenmeir, J. (2007). Inulin and oligofructose and mineral metabolism: the evidence from animal trials. J. Nutr. *137(11 Suppl.)*, 2513S-2523S.

Seibel, J., and Buchholz, K. (2010). Tools in oligosaccharide synthesis: current research and application. Adv. Carb. Chem. Biochem. *63*, 101–138.

Seifert, S., and Watzl, B. (2007). Inulin and oligofructose: review of experimental data on immune modulation. J. Nutr. *137*, 2563S-2567S.

Seki, N., Hamano, H., Liyama, Y., Asano, Y., Kokubo, S., Yamauchi, K., Tamura, Y., Uenishi, K., and Kudou, H. (2007). Effect of lactulose on calcium and magnesium absorption: a study using stable isotopes in adult men. J. Nutr. Sci. Vitam. *53*, 5–12.

Sharma, A., Yadav, B.S., and Ritika (2008). Resistant starch: Physiological roles and food applications. Food Res. Int. *24*, 193–234.

Silva, H.C., Braga, G.L., Bianchi, M.L.P., and Lopes, D.C.O. (1992). Oligossacarídeos da família da rafinose e flatulência. Cad. Nut. *4*, 48–60.

Silva, R.F. (1996). Use of inulin as a natural texture modifier. Cereal Foods World *41*, 792–794.

Simsek, S., and El, S.N. (2012). Production of resistant starch from taro (*Colocasia esculenta L.* Schott) corn and determination of its effects on health by *in vitro* methods. Carbohyd. Polym. *90, 1204–1209.*

Spiegel, J.E., Rose, R., Karabell, P., Frankos, V.H., and Schmitt, D.F. (1994). Safety and benefits of fructooligosaccharides as food ingredients. Food Technol. *48*, 85–89.

Stewart, M.L., Timm, D.A., and Slavin, J.L. (2008). Fructooligosaccharides exhibit more rapid fermentation than long-chain inulin in an *in vitro* fermentation system. Nutr. Res. *28*, 329–334.

Strohmaier, W. (1998). Lactulose: status of health-related applications. Int. Dairy Federation. *9804*, 262–271.

Suzuki, M., and Cutclife, J.A. (1989). Fructan in onion bulbs in relation to storage life. Can. J. Plant Sci. *69*, 1327–1333.

Suzuki, T., and Hara, H. (2004). Various Nondigestible Saccharides Open a Paracellular Calcium Transport Pathway with the Induction of Intracellular Calcium Signalling in Human Intestinal Caco-2 Cells. J. Nutr. *134*, 1935–1941.

Swennen, K., Courtin, C.M., and Delcour, J.A. (2006). Non-digestible Oligosaccharides with Prebiotic Properties. Crit. Rev. Food Sci. Nutr. *46*, 459–470.

Tabatabaie, F., and Mortazavi, A. (2008). Influence of lactulose on the survival of probiotic strains in yoghurt. World Appl. Sci. J.*3*, 88–90.

Takaku, H. (1988). Anomalously linked oligosaccharides mixture ('Alo mixture'). In Handbook of amylase and related enzymes. Edited by the Amylase Research Society of Japan. Pergamon Press, Oxford. 215–217.

Tamura, Y., Mizota, T., Shimamura, S., and Tomita, M. (1993). Lactulose and its application to the food and pharmaceutical industries. Bull. Int. Dairy Fed. *289*, 43–53.

Taper, H.S., and Roberfroid, M.B. (2002). Inulin/Oligofructose and anticancer therapy. Br. J. Nutr. *87*, 283–286.

Tárrega, A., and Costell, E. (2006). Effect of composition on the rheological behaviour and sensory properties of semisolid dairy dessert. Food Hydrocolloid. *20*, 914–922.

Tárrega, A., Rocafull, A., and Costell, E. (2010). Effect of blends of short and long-chain inulin on the rheological and sensory properties of prebiotic low-fat custards. Food Sci. Technol. *43*, 556–562.

Tateyama, I., Hashii, K., Johno, I., Iino, T., Hirai, K., Suwa, Y., and Kiso, Y. (2005). Effect of xylooligosaccharide intake on severe constipation in pregnant women. J. Nutr. Sci. Vitaminol. *51*,445–448.

Teitelbaum, J.E., and Walker, W.A. (2002). Nutritional impact of pre- and probiotics as protective gastrointestinal organisms. Annu. Rev. Nutr. *22*, 107–138.

Thamer, K.G., and Penna, A.L.B. (2005). Efeito do teor de soro, açúcar e de frutooligossacarídeos sobre a população de bactérias lácticas probióticas em bebidas fermentadas. Rev. Bras. Cienc. Farm. *41*, 393–400.

Tomomatsu, H. (1994). Health Effects of oligosaccharides. Food Technol. *10*, 61–65.

Topping, D.L., Fukushima, M., and Bird, A.R. (2003). Resistant starch as a prebiotic and synbiotic: state of the art. Proc. Nutr. Soc. *62*, 171–176.

Torres, D.P.M., Gonçalves, M.D.F., Teixeira, J.A., and Rodrigues, L.R. (2010). Galacto-oligosaccharides: production, properties, applications, and significance as prebiotics. Compr. Rev. Food. Sci. Food Saf. *9*, 438–454.

Trowel, H., and Burkitt, D. (1986). Physiological role of dietary fibre: a ten years review. J. Dent. Child. *53*, 444–447.

Tungland, B.C., and Meyer, D. (2002). Non digestible oligo-and polysaccharides (dietary fiber): their physiology and role in human health and food. Compr Rev. Food Sci. F. *1*, 73–92.

Urias-Silvas, J.E., Cani, P.D., Delmee, E., Neyrinck, A., López, M.G., and Delzenne, N.M. (2008). Physiological effects of dietary fructans extracted from *Agave tequilana* Gto. and *Dasylirion spp*. Br. J. Nutr. *99*, 254–261.

Van deWiele, T., Boon, N., Possemiers, S., Jacobs, H., and Verstraete, W. (2004). Prebiotic effects of chicory inulin in the simulator of the human intestinal microbial ecosystem. FEMS Microbiol. Ecol. *51*, 143–154.

Van Laere, K.M., Abee, T., Schols, H.A., Beldman, G., and Voragen, A.G. (2000). Characterization of a novel β-galactosidase from Bifidobacterium adolescentis DSM 20083 active towards transgalactooligosaccharides. Appl. Environ. Microbiol. *66*, 1379–1384.

Vazquez, M.J., Alonso, J.L., Domınguez, H., and Parajo, J.C. (2000). Xylooligosaccharides: manufacture and applications. Trends Food Sci. Technol. *11*, 387–393.

Vázquez, M.J., Garrote, G., Alonso, J.L., Domínguez, H., and Parajó, J.C. (2005). Refining of autohydrolysis liquors for manufacturing xylooligosaccharides: evaluation of operational strategies. *Bioresour. Technol. 96*, 889–896.

Venter, S.C. (2007). Prebiotics: an update. JFECS. 35, 17- 25.

Verghese, M., Rao, D.R., Chawan, C.B., Williams, L.L., and Shackelford, L. (2002). Dietary inulin suppresses azoxymethane-induced aberrant crypt foci and colon tumours at the promotion stage in young Fisher 344 rats. J. Nutr. *132*, 2809–2813.

Verma, A., and Shukla, G. (2013). Administration of prebiotic inulin suppresses 1,2 dimethyhydrazine dihydrochloride induced procarcinogenic biomarkers faecal enzymes and preneoplastic lesion in early colon carcinogenesis in Sprague Dawley rats. J. Funct Foods. *5*, 991–996.

Villaluenga, C.M., Cobas, A.C., Corzo, N., and Olano, A. (2008). Study of galactooligosaccharide composition in commercial fermented milks. J. Food Compos. Anal. *21*, 540–544.

Voragen, A.G.J. (1998). Technological aspects of functional food related carbohydrates. Trends Food Sci. Technol. *9*, 328–335.

Wack, M., and Blaschek, W. (2006). Determination of the structure and degree of polymerization of fructans from *Echinacea purpurea* roots. Carbohydr. Res. *341*, 1147–1153.

Wada, K., Watabe, J., Mizutani, J., Tomoda, M., and Saitoh, Y. (1992). Effects of soybean oligosaccharides in a beverage on human faecal flora and metabolites. J. Agric. Chem. Soc. *66*, 127–135.

Wang, H., Lim, P., Kao, M., Chan, E., Lin, L., and Wang, N. (2001). Use of isomalto-oligosaccharide in the treatment of lipid profiles and constipation in hemodialysis patients. J. Ren. Nutr. *11*, 73–79.

Wang, J., Sun, B., Cao, Y., and Tian, Y. (2009). Enzymatic preparation of wheat bran xylooligosaccharides and their stability during pasteurization and autoclave sterilization at low pH. Carbohy. Polym. *77*, 816–821.

Wang, Y. (2009). Prebiotics: Present and future in food science and technology. Food Res. Int. *42*, 8–12.

Wu, W.T., Yang, L.C., and Chen, H.L. (2014). Effects of konjac glucomannan, inulin and cellulose on acute colonic responses to genotoxic azoxymethane. Food Chem. *155*, 304–310.

Yadav, H., Jain, S., and Sinha, P.R. (2007). Formation of oligosaccharides in skim milk fermented with mixed dahi cultures, *Lactococcus lactis subsp.* diacetylactis and probiotic strains of lactobacilli. J. Dairy Res. *74*, 1–6.

Yen, C.-H., Tseng, Y.-H., Kuo, Y.-W., Lee, M.-C., and Chen, H.-L. (2011). Long-term supplementation of isomalto- oligosaccharides improved colonic microflora profile, bowel function, and blood cholesterol levels in constipated elderly people – a placebo-controlled trial. Nutrition. *27*, 445–450.

Yoon, S.-H., Mukerjea, R., and Robyt, J.F. (2003). Specificity of yeast (*Saccharomyces cerevisiae*) in removing carbohydrates by fermentation. Carbohydr. Res. *338*, 1127–1132.

Yun, J.W. (1996). Fructooligosaccharides – Occurrence, preparation and applications. Enzyme Microb. Tech. *19*, 107–117.

Yun, J.W., Lee, M.G., and Song, S.K. (1994). Continuous production of isomalto-oligosaccharides from maltose syrup by immobilized cells of permeabilized *Aureobacterium pullulans*. Biotechnol. Lett. *16*, 1145–1150.

Zhao, Y.J., and Yang, Y.H. (2007). Random double blind test on effect of soybean oligosaccharides in regulation of blood lipid in patients with hyperlipidemia. Prog. Mod. Biomed. 7, 1526–1535.

New and Tailored Prebiotics – Established Applications

20

Shanthi G. Parkar, Paul A. Blatchford, Caroline C. Kim, Douglas I. Rosendale and Juliet Ansell

Abstract

'Prebiotics' have been recognized for nearly 20 years, and have recently been defined as selectively fermented ingredients that that modulate the gut microbial composition and/or activity to deliver benefits to human health, through a range of host and microbial derived mediators. Here, we discuss the emerging range of biomolecules with potential prebiotic properties beyond the well-characterized galacto- and fructo-oligosaccharides. These include carbohydrate-based foods that provide accessible glycoside substrates to the gut microbiome, polyphenols and glucosinolates, which comprise both fermentable sugar moieties and/or alternative phytochemical substrates. We examined the microbial and molecular responses to potential prebiotics that validate their benefits to host health in terms of the ability of the prebiotics to alter the gut microbiome and the molecular responses that mediate the link between prebiotics and established host benefits including improved gastrointestinal function.

Introduction

A considerable number of human clinical trials have now demonstrated an association between our resident microbiota and our physiology and metabolism. Diet-induced changes to the microbiota have been associated with protective effects in diverse health issues such as infant colic, cancer, bone mineral density, type 2 diabetes and metabolic syndrome. This range of benefits is a consequence of the multiple interactions between the hugely complex mix of bacteria, immune cells, and our nervous system, all contributing to our body's metabolic homeostasis. Some scientific studies have shown an association between specific bacteria and disease, e.g. higher numbers of *Pseudomonas aeruginosa* have been observed in people suffering from irritable bowel syndrome (IBS) (Kerckhoffs *et al.*, 2011) or lower numbers of *Faecalibacterium prausnitzii* found in individuals with Crohn's disease (Sokol *et al.*, 2008). However, whether such associations are the cause or the effect is uncertain since pathology typically develops over a lifetime, and the very nature of our large bowel means that we are unable to see or measure what is going on *in situ*. In addition, many general fibre benefits (such as faecal bulking, laxation, bile acid binding) co-exist with selective prebiotic effects, and unlike traditional nutritional deficiencies where there is a clear linkage with disease (for example vitamin C deficiency and scurvy), for gut health there is no straight linkage between a dietary component and a health endpoint. In fact, there is no generally accepted definition of gut health, just a growing body of evidence that only serves to demonstrate its complexity. Since the pathologies we are talking about usually develop over long periods of time, it is largely epidemiological evidence that supports the notion that diet can drive better health outcomes through microbially mediated events.

With the advent of 'next generation' molecular and bioinformatic technologies, researchers have been able to study whole population ecology, genome wide metagenomic and metabolomic changes. This allows us to move beyond a 'bottom-up' or selected approach of measuring the impact of a food on a particular species, for instance bifidobacteria or lactobacilli, to a wider 'top-down' approach of looking for patterns between the vast array of possible genera of bacteria and diet, lifestyle plus in some instances disease pathology. Through this approach, previously obscure bacterial species and their abilities to generate by-products that are part of a co-metabolism pathway with comparatively better known 'beneficial' bacteria are coming to the fore as plausible mechanisms of gut health benefits.

The original concept of prebiotics was first developed and published by Gibson and Roberfroid (1995) as a food that resists digestion, absorption and adsorption; is fermented by the resident microbiota; is able to stimulate the growth/activity of one or a few bacteria; and is generally accepted to have a benefit to host health. Since this initial notion was put forward, there has been considerable research effort in the area and even the formation of a scientific association dedicated to prebiotic and probiotic research. The International Scientific Association for Probiotics and Prebiotics (ISAPP), recently collectively redefined a prebiotic as: 'a selectively fermented ingredient that results in specific changes in the composition and/or activity of the gastrointestinal microbiota, thus conferring benefit(s) upon host health' (Gibson *et al.*, 2010).

The fructooligosaccharides (FOS) and galactooligosaccharides (GOS) are examples of confirmed prebiotics, although not accepted by regulatory bodies such as the European Food Safety Authority (EFSA). However, there is now substantial scientific evidence to back up the efficacy of a range of foods and food ingredients that fulfil the ISAPP definition stated above. This ISAPP definition is intentionally inclusive, so whilst specific modulation of the microbiota is a requirement of prebiotics, the host consequences of this may be wide-ranging. Of paramount importance for defining a prebiotic is that it modulates the growth of selected members of the microbiota directly or indirectly. Examples of direct modulation are increased growth of beneficial organisms by primary utilisation of the substrate or as a result of secondary cross-feeding upon microbial breakdown products or inhibition of pathogens. Microbes are indirectly modulated by provision of ecological niches which may be occupied or exploited by members of the microbiota. The bacterial changes are often accompanied by physicochemical properties of some prebiotics, which are generally recognized amongst dietary fibres as increasing bulk and water holding capacity, decreasing gut transit time and improving laxation. Increased bulk is also a result of increased microbial mass – a commonly recognized and accepted definition of prebiosis.

Consequences of increased microbial growth include increased occupation of important ecological niches close to the epithelia (commonly referred to as 'epithelial adhesion'), and the generation of microbial metabolites which in turn provide a range of health benefits. Foremost amongst these metabolites are the short-chain fatty acids or SCFAs (acetate, propionate, butyrate, etc.), organic acids (lactate, succinate), biotransformed dietary phytochemicals, bacteriocins and microbially synthesised vitamins. These metabolites, and the bacteria that produce them, in turn modulate local and systemic host responses. Local responses include enhanced mucosal integrity by modulating the mucus physical barrier (mucin synthesis and resistance to degradation) and epithelial integrity (e.g. tight junctions); antimicrobial protein (defensin) secretion; IgA; and immunomodulation of the underlying gut immune system (e.g. dendritic cell sampling, tolerance, toll-like receptors; Magrone and Jirillo, 2013). Systemic responses include altered nutrient uptake, energy deposition and utilization, e.g. reduction in energy intake and markers of insulin resistance (Parnell and Reimer, 2009; Vulevic *et al.*, 2013), increased satiety (Cani *et al.*, 2009) and improved bone mineral density (Abrams *et al.*, 2007; Holloway *et al.*, 2007).

All of these prebiotic effects will be individually discussed in the context of the newly emerging prebiotics: our focus will, however, be on an integral part of the definition, i.e. selective modulation of microbial growth. However, some of these potential new prebiotics have been reported more on the consequences of microbial modification than they do on actual impact on microbial abundances themselves. If these health benefits are validated in human supplementation trials and proven to occur as a consequence of selective fermentation of the substrate then they fulfil the established definition of prebiotic.

New and tailored prebiotics

We discuss here new prebiotics as biomolecules, recently identified or tailored for specialised microbial selection. For the microbiota to selectively respond to prebiotics, they must possess the equipment required to degrade them.

While carbohydrates are a major part of our diet, humans have evolved to digest only a narrow range of plant glycans efficiently, with only about 17% of their glycoside hydrolase (GH) capacity possessing any digestive properties (El Kaoutari *et al.*, 2013). The majority of the enzymes that extract energy from cell wall carbohydrates are of microbial origin, as identified in the databases of carbohydrate active enzymes, http://www.cazy.org/Genomes.html. Our resident gut microbiota has access to many substrate sources in the gastrointestinal tract, including plant derived substances (such as cellulose, hemicellulose, pectin, and starches), endogenous mucin and secondary metabolites from cross-feeding bacteria. These microbes contribute to the vast glycosidic capacity in the colon by first depolymerising the complex glycan structures with polysaccharide lyases (PL) before further hydrolysis with GH. There are a number of GHs with specificity for different substrates, depending on their ability to hydrolyse the glycosidic bond between glycosyl-glycosyl or glycosyl-non glycosyl units.

Traditionally, the production of prebiotics has been driven by availability and cost of the substrate or starting ingredient such as waste from industrial processes or abundant natural resources. This may be complemented or preceded by subjecting candidate sources to simple screening processes. Thus far, there has not been a substantive body of work performed to actively manufacture prebiotics to cause a proliferation in certain desirable bacterial groups. Various methods exist for producing prebiotics including extraction from the original source, transglycosylation or transgalactosylation, isomerization, condensation and hydrolysis (Giese *et al.*, 2010). The resulting end product generated from these methods all have a bearing on the specificity and effectiveness of the prebiotic product. For example, hydrolysis of lactose using different bifidobacterial β-galactosidases yielded GOS with different linkages and structure (Rabiu *et al.*, 2001). When fermented, using the same strains of the source β-galactosidase, the species that produced the GOS generally fermented it best. Thus, the molecular features of the prebiotic produced using these methods have microbiological and physiological relevance. The size, structure, degree of polymerization, glycosidic linkages and monosaccharide composition will ultimately determine which bacteria can utilise the molecule as a substrate for growth.

Different bacteria have different arsenals of enzymes, carbohydrate binding and uptake systems, and different flexibility with regards to how many carbohydrates they can access. The greater the diversity or versatility of degradative capabilities the gut bacteria possess, the greater are the gastrointestinal health benefits. For example, *Bacteroides thetaiotaomicron* is an extremely versatile saccharolytic gut microbe that can express genes from different polysaccharide utilization loci (PULs) to exploit different structures that become available (e.g. starch, fructans, mucin, etc.) (Koropatkin *et al.*, 2012). Similarly, *Ruminococcus bicirculans* is a recently discovered gut commensal with a unique dual chromosome architecture to enable rapid adaptation to dietary fluctuations (Wegmann *et al.*, 2013).

The new prebiotics that we discuss here typically possess linkages which specific microbial groups in the gut have the capacity to cleave, to access the released carbon sources. While we mainly consider glycosidic linkages between different sugars, or between sugars and other phytochemicals of varying structure and bioactivity, in some cases we believe that non-carbohydrate compounds also fulfil the established prebiotic criteria by resulting in specific changes in growth and/or activity of members of the microbiota. From a health perspective, it is desirable that prebiotics reach the distal colon to promote proliferation of bacteria for generation of beneficial metabolites and detoxification of accumulated toxic or pathological metabolites (Medani *et al.*, 2011; Windey *et al.*, 2012). Plant cell walls obtained though consumption of fresh whole fruits, vegetables or cereals are naturally delivered to this site as their physical structure is digested slowly, thereby delivering pectic polysaccharides and indigestible non-starch polysaccharides to the colon. For man-made prebiotics, this may be achieved by modulating the glycosidic linkages, the chain length, or by using a combinatorial mix of complex bioactives, so that the bacteria will take longer to break down the prebiotic into monosaccharides and metabolites. These novel prebiotics are summarised in Table 20.1 and explained in detail in the subsequent sections.

Saccharide-based prebiotics

Di-, tri-, and oligosaccharides

Sugar polymers joined by glycosidic linkages that are not hydrolysed or are not absorbed by human pancreatic or intestinal glycoside hydrolases are naturally pushed towards the colon for digestion by microbial enzymes. We discuss below some of the widely accessible sources of these molecules.

Human milk

As the role of intestinal bacteria in regulating host physiology is inarguably established, one can appreciate the vulnerability of the neonate born with a near sterile gut and who has to rapidly develop a healthy microbiome that can regulate development and immune responses to protect the infant (Dai and Walker, 1999; Fanaro *et al.*, 2003). In a study using rat pups, the type of bacteria that colonized the newborn gut was shown to determine metabolic predisposition, obesity was associated with an increased lactobacilli count and decreased Bacteroidetes (Sefcikova *et al.*, 2011). Thus, one cannot underestimate the nutritional importance of the first feed of the neonate. The mother's milk is a source of probiotic bacteria including the commonly associated lactic acid bacteria and bifidobacteria. With the advancement of sequencing technologies, it has become apparent that milk also harbours low levels of obligate anaerobes such as *Bacteroides* and butyrogenic *Roseburia* and *Faecalibacterium* (Jost *et al.*, 2013). Potentially all these inhabitants may have the opportunity to establish in the gut and influence the developing microbiome by directly or indirectly metabolizing human milk oligosaccharide (HMO) molecules to facilitate their growth.

Human breast milk is a rich source of glycoconjugates and oligosaccharides that modulate the colonization of the initially germ free neonatal gut (Dai and Walker, 1999). Breast-fed infants have a developing microbiota dominant in bifidobacteria and lactobacilli (Harmsen *et al.*, 2000; Turroni *et al.*, 2012), which are primary targets for prebiotic supplementation (Gibson and Roberfroid, 1995). This most certainly has come about through the co-evolution of bifidobacteria with HMOs. The oligosaccharides are the third most abundant chemical species in human milk; with concentrations of up to 18.5 g/l and upwards of 200 HMO compositions have been observed with a length of between 3 and 32 sugar residues (Miller *et al.*, 1994; Ninonuevo *et al.*, 2006; Sela, 2011).

HMOs invariably consist of a lactose core among glucose and galactose molecules with associated moieties of fucose,

Table 20.1 New and tailored prebiotics

Category	Macrostructure	Structure	Common sources
Saccharide-based	Di-, tri- and oligosaccharides; sugar alcohols	e.g. Human milk oligosaccharides	Milk
	Storage polysaccharides	Resistant starch	Fruit, vegetables, cereals, algae
	Structural polysaccharides (plant)	Pectin, hemicellulose, cellulose	Fruit, vegetables, cereals
	Structural polysaccharides (marine)	Alginate, carageenan, fucoidan	seaweed
	Synthetic polysaccharides	polydextrose	
	Microbial exopolysaccharides		*Bifidobacterium* spp., lactic acid bacteria
Phytochemicals	Polyphenols	Phenolic acids, flavonoids, lignans, stilbenes	Fruit, vegetables, cereals, algae
	Glucosinolates	thioglucoside-N-hydroxysulfates	Cruciferous vegetables
'Next generation' prebiotics	Fatty acids	*Trans* fatty acids, conjugated linoleic acid	Oils from fish, seeds, cereal bran
	Glycopeptides	Disaccharide-containing glycopeptide	Synthetic

N-acetylglucosamine and N-acetylneuraminic acid (Barile and Rastall, 2013; Kunz *et al.*, 2000). They also have both alpha-glycosidic and beta-glycosidic linkages which further enhance their intricacy (Barile and Rastall, 2013). Probiotic bacteria, such as bifidobacteria and lactobacilli, have specially adapted enzyme machinery that enables them to utilize human glycans such as HMOs (and mucin) as substrates for growth (Turroni *et al.*, 2012). For example, in some strains of bifidobacteria, glycosidases and carbohydrate transporter encoding genes make up over 10% of their genetic repertoire (Sela and Mills, 2010). Bifidobacteria possess a range of glycosyl hydrolases that include fucosidases, β-galactosidases, sialidases and a β-hexosaminidase, which can hydrolyse some of the unique bonds, encountered in HMOs and facilitates subsequent catabolism and growth (Ventura *et al.*, 2014). This bestows a competitive advantage on bifidobacteria over other bacterial groups.

One important aspect of prebiotic effectiveness is the ability of the molecule to persist intact through to the colon, resisting hydrolysis and absorption in the upper gut (Gibson and Roberfroid, 1995). This is proven to be fulfilled by HMOs, as shown in experiments with brush border, salivary and pancreatic digestive enzymes (Gnoth *et al.*, 2000). Another of the key precepts of the prebiotic definition (that is often the hardest to fulfil) is that candidate molecules must induce selective fermentation (Roberfroid, 2007). That is the targeted proliferation of one or a small number of bacterial species or groups rather than the non-specific stimulation of bacterial groups devoid of beneficial health effects. For HMOs in particular, bifidobacteria are an obvious target since they constitute at least 80% of the neonatal gut microbiota, as recently found using 454 Titanium FLX sequencing (Turroni *et al.*, 2012). When selecting candidate HMO derived prebiotics, one must assess their impact on many species to determine the broader fermentation profile of the molecule. A comparative *in vitro* study on 16 different strains of gut bacteria found that various bifidobacteria and bacteroides strains were able to metabolize HMOs, whereas clostridia, enterococci, eubacteria and *Escherichia coli* were not (Marcobal *et al.*, 2010). Another study by the same group found that *Bifidobacterium infantis* could utilize certain HMO compositions by up-regulating mucus degradation genes, thereby gaining a growth advantage over competing bacteria (Marcobal *et al.*, 2011). In addition to the direct supply of substrate, HMOs can also act as anti-adhesion modulators through competition with epithelial receptors for diarrheal pathogens such as *E. coli*, *Vibrio cholerae* and *Salmonella fyris* (Coppa *et al.*, 2006; Marcobal *et al.*, 2010). HMOs were found to prevent the adhesion of enteropathogenic *E. coli* to cultured epithelial cells and reduced colonization by the pathogens in suckling mice, providing a direct barrier against infection (Manthey *et al.*, 2014). Furthermore, enrichment with HMOs and milk enzymes such as lysozyme prevents the growth of potentially pathogenic clostridia so that the bifidobacteria are selectively enriched (Rockova *et al.*, 2011). There is also evidence to suggest that HMOs can also directly interact with the intestinal epithelia, causing physiological and immunological effects. Gut contractility and motility were directly diminished by interaction with fucosylated HMO species, showing their potential use as therapeutic agents in gut disorders (Bienenstock *et al.*, 2013). An *ex vivo* study found that incubation of HMOs with porcine peripheral blood mononuclear cells (PBMCs) induced an increase in production of immunomodulatory molecules, such as IL-10 (Comstock *et al.*, 2014). An *in vitro* study using HMO and bovine derived oligosaccharides found they altered the expression of many cytokines, chemokines and cell surface receptors on HT-29 colonic epithelial cells (Lane *et al.*, 2013). This suggests that HMOs may interact and communicate with the naïve infant immune system playing a potential role in its maturation in the first weeks postpartum. Commercially formulated HMOs may be able to deliver these properties and therefore act as prebiotics. A potential niche for the use of HMOs is in the supplementation of infant milk formula, but they can also be considered for broader use in adults or those suffering from gastrointestinal or immune dysfunction (Coulet *et al.*, 2014).

A high level of complexity exists in the unique and varied HMO structure which renders the production of HMOs by chemical or enzymatic means extremely difficult. Nevertheless, attempts have been made to reproduce their unique characteristics, most commonly by the enzymatic condensation of lactose dimers. Commercially viable manufacture protocols of HMO-like components (from other milk sources) and further validation of benefits in *in vivo* human clinical trials are necessary to utilize the prebiotic potential of HMOs.

Dairy milk

Cow's milk is the major dairy product consumed worldwide, and selective fractionation of the bovine milk glycome is being considered to identify prebiotic saccharides similar to the HMO glycome (Zivkovic and Barile, 2011). One such molecule is epilactose, which is present in small quantities in heat-treated milk (Mu *et al.*, 2013b). The product is now being harvested by enzymatic conversion of milk lactose using a microbial cocktail rich in cellobiose-2-epimerase (Krewinkel *et al.*, 2014), which could potentially add value to many dairy products.

Synthetic saccharides

Microbial glycoside hydrolases have been employed to synthesize new small molecule prebiotics such as lactulose, lactosucrose from lactose-derived sugars and GOS from lactulose, lactulosucrose and 2-α-glucosyl-lactose (Díez-Municio *et al.*, 2014).

The disaccharide lactulose (galactosyl $\beta(1\to4)$ fructose) may be microbially synthesized by the isomerization of lactose. Its prebiotic potential lies in its resistance to human digestive enzymes coupled with stimulation of bifidobacteria and inhibition of potential pathogens (Ballongue *et al.*, 1997; Candela *et al.*, 2010). Physiological benefits included reduction of pro-carcinogenic enzymes such as azoreductase and metabolites such as phenol and cresol (Ballongue *et al.*, 1997).

Lactosucrose is an artificial sweetener consisting of galactose, glucose and fructose, manufactured by enzymatic transglycosylation of lactose and sucrose (Mu *et al.*, 2013a). The β-(1–4) glycosidic linkage between the galactose and glucose is not cleaved by human intestinal enzymes. It has been found to have a prebiotic effect in rats, specifically promoting microbial metabolism and decreasing triggers for intestinal inflammation, such as intestinal pathogens and luminal ammonia (Zhou *et al.*, 2014).

Fruits and vegetables

Fruit and vegetables are a particularly rich source of mono-, di- and oligosaccharides, in particular fructooligosaccharides, the raffinose-family of oligosaccharides, and prebiotic sugar alcohols. Fruits such as blueberry, pear, watermelon, and nectarine and vegetables such as garlic, spring garlic, leek, white onion, and scallion are rich in oligosaccharides (Jovanovic-Malinovska *et al.*, 2013). These oligosaccharides are largely uncharacterised in terms of their chemical structure and prebiotic potential. Most fruits also contain low amounts of fructooligosaccharides which are well established for their prebiotic potential.

The efficiency of a prebiotic depends on its glycosidic backbone: plants offer a huge source of diverse polysaccharides. Heteropolysaccharides from plant cell walls such as xylans and pectins would appear to be ideal substrates to generate di- and oligo-saccharides with prebiotic potential (Yoo *et al.*, 2012). Various isomalto-oligosaccharides and xylo-oligosaccharides are now being examined and may provide alternative prebiotics due to their ability to promote probiotic bacteria under *in vitro* colonic fermentation conditions (Candela *et al.*, 2010).

Sugar alcohols

Low doses of 10 g/day of the artificial sweetener, lactitol (4-beta-D-galactopyranosyl-D-glucitol) were trialled in 75 adults, with minimal adverse effects, whilst promoting significant increase in bifidobacteria and concentrations of faecal propionate and butyrate (Finney *et al.*, 2007). SCFAs in the colon, especially propionate and butyrate have been linked with hormonal changes that control appetite mainly related to hormones such as glucagon-like peptide 1, peptide YY (PYY) and leptin (Hosseini *et al.*, 2011; Tarini and Wolever, 2010). In rats, acute ingestion of lactitol was found to significantly increase the acute postprandial PYY response (Reimer *et al.*, 2012). In one study, long-term intake decreased weight gain in growing rats and attenuated the postprandial decrease in circulating concentrations of PYY (Gee and Johnson, 2005).

Sugar alcohols are commonly found in fruit and vegetable fibres. An example is the 5-carbon polyalcohol xylitol. It is commonly used in dental products such as chewing gum due to its anti-cariogenic properties. In the Simulator of Human Intestinal Microbial Ecosystem (SHIME) model, it did not stimulate bifidobacteria over control levels, but butyrate (a beneficial SCFA) concentrations were found to be nearly 10 times those of the control levels. In combination with polydextrose (which mainly increased the SCFAs, acetate and propionate), production of butyrate was sustained in all the 4 vessels simulating the different sections of the colon (Makelainen *et al.*, 2007). Both xylitol and polydextrose were found to be non bifidogenic indicating that pathways for the beneficial SCFAs may be through selective stimulation of the mature cross feeding microbiome rich in butyrogenic bacteria (El Aidy *et al.*, 2013). This may involve different metabolic pathways of cross feeding, including utilization of primary metabolites such as acetate and lactate or partial breakdown of complex and poorly broken down substrates (Belenguer *et al.*, 2006).

Synthetic oligosaccharides

Oligosaccharides have been commercially synthesized for many years, but are only now being evaluated for their prebiotic properties (Putaala *et al.*, 2011). Polydextrose is a synthetic highly branched polymer of glucose linked with small quantities of sorbitol and citric acid. All possible glycosidic bonds are present, α and β 1–2, 1–3, 1–4 and 1–6, among which the 1–6 bond predominates. This compound has an average degree of polymerization (DP) of 12 and an average molecular weight of 200 g/mol (Konar *et al.*, 2014). This randomly linked glucose polymer is indigestible in the small intestine and slowly degraded by bacteria. In an *in vitro* model, polydextrose was found to increase total SCFAs especially in the segment representative of the end of the descending colon and sigmoid/rectum (Makelainen *et al.*, 2007). There was a three-fold increase in acetate and propionate, indicating its ability to sustain the release of beneficial metabolites throughout the distal gut. It is becoming more frequently used in infant formulas blended with established prebiotics as it is associated with increases in beneficial bacteria and decreases problems associated with colic in infants (Ashley *et al.*, 2012; Herfel *et al.*, 2010; Partty *et al.*, 2013). It has also been assessed in an *in vivo* double blind, placebo controlled, crossover human clinical study where a reduction in faecal water genotoxicity was observed and other anti-cancer effects demonstrated (Costabile *et al.*, 2012).

Storage polysaccharides – resistant starch

While resistant starch (RS) is not a new food, its use as a prebiotic has gained momentum only recently, hence its inclusion here. RS is defined as 'the sum of starch and starch degradation products that reach the human large intestine' (Englyst *et al.*, 1996), i.e. are not absorbed in the small intestine of healthy individuals. RS is a low viscous linear molecule of α–1,4-D-glucan with α-1,6 branch points, essentially derived from amylose (Fuentes-Zaragoza *et al.*, 2011). It is resistant to amylase digestion in the small intestine and this is influenced by its encapsulation within the cell wall, by microarchitecture of the starch polymer, and by the crystalline structure of the starch formed due to retrogradation depending on source and on chemical or heat treatment (Homayouni *et al.*, 2014).

The five types of resistant starch (which may be native, processed or manufactured) are outlined in Table 20.2. RS arrives in the large bowel either encapsulated within broken down grains and seeds, as a soluble jelly like matrix or as an insoluble polysaccharide. The solid entities provide substrates that the bacteria can adhere to. Primary contact is a prerequisite for gut bacteria to utilize the starch via their vast repertoire of glycosidases (Flint *et al.*, 2008). Adhesion structures such as the Type IV pili in *Ruminococcus bicirculans* (Wegmann *et al.*, 2013) are essential for these organisms to initiate degradation of resistant dietary energy sources reaching the colon (Ze *et al.*, 2013).

Fibre from oats, wheat, potatoes, rice, carrots, peas, citrus fruits and apples were tested for their ability to retain *Lactobacillus acidophilus* by adhesive interactions, RS rich-wheat and oat fibres

Table 20.2 Categories of resistant starch

Type	Nature	Example	Digestion in small intestine
1	Physically inaccessible starch protected within cell walls and food matrices.	Whole grains and seeds	Slow rate
2	Granular native starch, with highly crystalline structure	The starch granules in raw potato and green banana. Also includes processed starch granules, e.g. high-amylose corn starch (HACS) such as Hi maize R260	Very slow rate due to limited accessibility to digestive enzymes
3	Retrograded non-granular starch, heat stable	Cooked and cooled potato or bread, retrograded HACS	Slow rate
4	Chemically modified starch and resistant dextrins	Esterification of RS3 with acetate, propionate, butyrate or phosphate	Resist hydrolysis
5	Water-insoluble linear polyalpha-1,4-D-glucan which was not subject to degradation by α-amylases	Resistant maltodextrins, amylose–lipid complexed starch	Slow rate

Adapted from Brouns *et al.*, 2002; Fuentes-Zaragoza *et al.*, 2011; Homayouni *et al.*, 2014.

(but not potato fibre) were found to retain more than 60% of the bacteria whilst citrus and apple fibre retained only 5–20% (Holko and Hrabe, 2012). This ability of bacteria to adhere to their substrate may be an important factor in determining their efficiency as probiotic bacteria, especially since the abundance of lactobacilli and bifidobacteria may constitute as little as 1.8% and 4.9%, respectively, of the total gut microbiota (Louis *et al.*, 2007). In the case of bifidobacteria, the bacterial membrane proteins involved in adhesion appears to be specific for the α-1,4-linked glucose sugars comprising starch, although this adhesion was inhibited by proteases and the lowered pH of the stomach (Crittenden *et al.*, 2001; Ghalia *et al.*, 2013), which means that at least some of the bacteria may be lost in gastric transit due to inactivation of the bacterial membrane proteins.

RS is known to enhance butyrate production *in vivo*, resulting in butyrate values twice that achieved with wheat bran and over four times that achieved with pectin (Fuentes-Zaragoza *et al.*, 2011). When dietary supplementation with RS was compared with a non-starch polysaccharide diet using 46 healthy adults, RS was found to cause a significant increase in faecal butyrate compared to an entry diet, inter-individual and gender based variability in faecal butyrate concentrations notwithstanding (McOrist *et al.*, 2011). RS appears to be well tolerated even at moderately high dietary intakes of 20–25 g/day (Klosterbuer *et al.*, 2013), perhaps up to 33 g/day without unpleasant side effects such as bloating (Martinez *et al.*, 2010). Fermentation of RS to enhance butyrate production is particularly relevant in the distal segments of the large bowel where there are maximum concentrations of toxic metabolites, making the tissues most vulnerable to disease (Brouns *et al.*, 2002). It is here where butyrate can best exert its protective effects (Topping *et al.*, 2001).

RS is a long polyglucan polymer with differing ratios of amylose and amylopectin, the latter being more susceptible to digestion by gut amylases as its branched structure increases the surface area of the substrate accessible to the enzyme. In fact, the higher the amylose content the greater its resistance to digestion and thus the greater the formation of resistant starch (Singh *et al.*, 2010): yet the presence of a small amount of amylopectin within the RS microarchitecture that reaches the colon may aid microbial utilization of RS in the colon. High amylose starch (HAS) is a RS2 substrate with up to 70% amylase, a microarchitecture that influences the extent of butyrate formation. HAS treated to high temperature at different moisture conditions was measured for fermentability by faecal bacteria in an *in vitro* colonic model. The greatest butyrate production was from oven-heated native starch with an increased amylose but with a residual amylopectin fraction. Starches cooked with moisture lacked the residual amylopectin, and produced lower butyrate levels (Zhou *et al.*, 2013).

RS fermentation clearly modulates microbial populations in *in vitro* and in animal trials, and emerging evidence supports its prebiotic effects in humans.

It appears that dietary interventions at an early age are more likely to result in permanent beneficial microbial changes compared to later in life when the gut composition is well established and more resistant to change. Weanling rats fed with a RS2-rich diet (corn starch) were found to benefit from a bloom in gut microbial bifidobacteria and Bacteroidetes providing opportunities for long lasting microbial changes due to dietary interventions at an early age. There were changes in intestinal physiology that favoured absorptive capacity (colonic crypts) and gene expression changes to enhance anti-oxidant, anti-inflammatory and detoxification effects (Young *et al.*, 2012). Similarly, a RS2-rich diet was found to stimulate total bacteria, especially lactobacilli and bifidobacteria and reduce the adiposity associated with obesity, which was modelled in the rats by performing ovariectomies (Keenan et al., 2013).

A trial with overweight adults found that a diet rich in RS3 favoured *Eubacterium rectale* and *Ruminococcus bromii*, however these bacteria decreased when the volunteers switched over to a low carbohydrate diet (Walker *et al.*, 2011). Ruminococci increased to over 25% of total bacteria, which is consistent with their metabolic capacity to utilize starch (Walker *et al.*, 2011) and this may be attributed partly to their ability to colonize particulate RS (Leitch *et al.*, 2007). Conversely, two of the volunteers had very poor digestibility of RS corresponding with low ruminococci in their faecal microbiota (Walker *et al.*, 2011). Microbial adaptation to a RS rich diet was recently confirmed in a trial in pigs, whose proximal colon was directly investigated for changes in microbial abundance and fermentation byproducts. There was an increase in primary feeders such as *Parabacteroides*, *Prevotella* and *Mitsuokella multiacida*. These generate organic acids such as acetate, lactate and succinate that feed back into the system to enhance propionate and butyrate production. This was confirmed by a significant increase in butyrogenic bacteria such as *F. prausnitzii* and *Megasphaera elsdenii* (Haenen *et al.*, 2013). Overall RS diet shifted the colonic microbial balance to a lower ratio of Firmicutes to Bacteroidetes than was found with digestible starch. There is ongoing debate over the link between obesity and an increase in Firmicutes: Bacteroidetes ratio given the reverse of this ratio appears to be associated with a lean microbial phenotype (Ley *et al.*, 2006). Furthermore, RS diet favoured a healthier microbial balance evidenced by a decrease in potential pathogens such as *Leptospira*, *E. coli*, *Pseudomonas* spp., *Actinobacillus* spp., and *Pasteurella* spp. This is not the first time that changes in gut major microbial populations have been associated with changes in body weight (Delzenne and Cani, 2011). For example, an increase in Firmicutes and a concomitant suppression of Bacteroidetes was demonstrated in the gut microbiome of diet-induced obese mice. The obese gut microbiome also showed a drastic decrease in the metabolically versatile (beneficial) bifidobacteria indicating a possible loss of metabolic functions in the microbial metabolome. Obesity has been associated with low-grade systemic inflammation: gut microbiota and their metabolites – especially butyrate – have been suggested to attenuate many of these inflammatory responses by enhancing gut barrier integrity (Brahe *et al.*, 2013).

RS is a much preferred substrate for gut bacteria over protein. An abundance of RS substrate has been found to enhance SCFA production with nearly a two-fold increase in butyrate and concomitant decrease in branched chain fatty acids (BCFAs) which are produced from protein sources (Haenen *et al.*, 2013). With the digestible starch diet, the carbon source was exhausted much earlier along the gastrointestinal tract and caecum, so that the bacteria then switched to undigested or endogenous protein sources in the colon resulting in harmful BCFAs (Haenen *et al.*, 2013). An RS rich diet helps to prolong the 'shelf-life' or retention of non-starch polysaccharides (NSP) that are usually consumed with RS. In a study in pigs, a diet rich in NSP such as potato, wheat and barley was found to be utilized by the bacteria mainly in the caecum, however addition of tapioca starch shifted the fermentation of NSP further along the gastrointestinal tract to the more distal sections of the large bowel. Concurrently, there was a three-fold increase in protein content in the distal part of the colon and this was related to an increase in soluble sugars, starch degrading enzymes and a higher microbial mass (Jonathan *et al.*, 2013). While RS diets encourage microbial diversity and functional abundance, individual responses to the type of substrate can vary. For example, in a comparison between RS2 and RS4 diets, RS4 was found to be more effective in stimulating Actinobacteria and Bacteroidetes, while decreasing Firmicutes. The species level diversity was modulated in that RS4 caused significant increases in *Bifidobacterium adolescentis* and *Parabacteroides distasonis*, while RS2 significantly raised the proportions of *Ruminococcus bromii* and *Eubacterium rectale* when compared to RS4 (Martinez *et al.*, 2010). In a recent study with 86 participants, RS4-enriched flour was found to lower the body fat and waist circumference in healthy adults. In a subset with metabolic syndrome, this diet improved the blood lipid profile with a significant decrease in total cholesterol, a change important for metabolic regulation and cardiovascular wellness (Nichenametla *et al.*, 2014). The microbial groups that were actively involved in the breakdown of RS were tracked using ^{13}C-labelled starch as a substrate in an *in vitro* intestinal model incubated with faecal bacteria. The isotope was predominantly enriched in populations of *R. bromii* indicating their role as primary RS degraders, with associated increases in secondary degraders such as *E. rectale* and *Prevotella* spp. (Kovatcheva-Datchary *et al.*, 2009).

Whole foods rich in resistant starch include dietary staples such as potatoes, sweet potatoes and cassava. A nutritional trial conducted for 90 days using 59 adults with moderately raised serum glucose and cholesterol levels studied the role of RS on parameters of metabolic function such as blood glucose, haemoglobin A1C, and blood lipids. Consumption of RS rich foods (controlled to a serve of 50 grams of available carbohydrates) increased the beneficial HDL-C (high density lipoprotein cholesterol) in the participants compared to a control (bread) diet (Trinidad *et al.*, 2013). While the gut microbial parameters were not explored in the Trinidad *et al.* (2013) study, there is abundant evidence demonstrating the prebiotic effects of RS rich diets in modulating gut microecology and enhancing probiotic bacteria (Fuentes-Zaragoza *et al.*, 2011). In a rat study, a diet rich in high amylose corn starch (HACS) (RS2) increased microbial proliferation along with plasma measures of satiety hormones such as peptide YY and GLP-1 (Keenan *et al.*, 2006). This is supported by findings from the previously discussed pig study, where the RS diet up-regulate metabolic pathways such as the tricarboxylic acid cycle and lipid oxidation resulting in the lowering of cholesterol. This may be explained by an increase in expression of the lipid regulating signal peroxisome proliferator-activated receptor g, which has previously been shown to be selectively up-regulated by butyrate (Wächtershäuser *et al.*, 2000).

RS rich diets have been recognized to protect against gastrointestinal diseases including colorectal cancer due to beneficial effects of metabolites such as butyrate (Brouns *et al.*, 2002; Le Leu *et al.*, 2005). Butyrate *per se* may not explain all the beneficial effects, as at least some of the benefits may actually arise from elimination of toxic metabolites that accumulate in the distal colon. For example, the reduction of hydrogen sulfide production which occurs by inhibiting sulfate-reducing bacteria (SRB). Instead the excess hydrogen may be channelled to the hydrogenotrophs such as methanogenic archaea as demonstrated in the following example. Reduced susceptibility of native Africans to sporadic colorectal cancer has been correlated with their RS-rich diet, low consumption of animal products and greater microbial production of colonic methane. This adaptation to methane production correlates with low abundance of SRB that generate hydrogen sulfide, which has been linked with inflammatory bowel disease (IBD) (Zinkevich and Beech, 2000). The faecal samples from the native Africans showed greater diversity in their archaeal genotype and higher breath methane. Africans on a 'Western diet' (high in calories, and rich in animal protein and fat) showed greater individual variability in their faecal archaea, which may be linked to the variable availability of the starch substrates in the colon (Nava *et al.*, 2012).

Considering the ability of RS to reach the colon and manipulate its acidic metabolites, namely acetate, propionate and butyrate, RS acylated with these acids was trialled for their effects on the gut in rats (Annison *et al.*, 2003). Butyrylated starch was found to cause the greatest increase in SCFAs in rat caecum and colon. This finding has recently been corroborated in humans implying that the butyrate was selectively released in the colon, mediated mainly by organisms such as *Parabacteroides distasonis*, implying the prebiotic potential of these novel starches (Clarke *et al.*, 2011).

Structural polysaccharides – plant cell walls

The most widely accepted definition of dietary fibre refers to polysaccharides, oligosaccharides, and other edible plant substances that resist digestion and absorption in the human gastrointestinal tract. Plant cell walls from common vegetables, fruits, whole grain cereals, and legumes are major sources of such dietary fibres, being a complex polysaccharide matrix, which is formed by extensive interactions between three major components of the cell wall polysaccharides, namely cellulose, hemicelluloses, and pectin. Although the consumption of plant fibres has been associated with the improvement of gastrointestinal well-being and the prevention of health-threatening diseases (e.g. colon cancer, IBD, and Crohn's disease) the prebiotic potential of the individual plant cell wall polysaccharides has been largely overlooked in the light of more promising and better-studied prebiotics, such as inulin and fructo-oligosaccharides. Emerging evidence suggests that the polysaccharide components of the plant cell wall, particularly pectin, demonstrate significant prebiotic effects, hinting that these polysaccharides may potentially provide prebiotic benefits.

Cellulose

Cellulose is a tightly packed aggregate of linear polymers of 1,4-linked β-D-glycopyranose. The polymer aggregates form long microfibrils (of 7000–12,000 glucose molecules/units) varying in length, width, and in the degree of order, depending on the plant source (Simoneit *et al.*, 1999).

Hemicellulose

Hemicelluloses are a collective term for plant cell wall polysaccharides that are neither cellulose nor pectin, characterized by having β-1,4-linked backbones of glucose, mannose, and xylose (Scheller and Ulvskov, 2010). The structural components of hemicelluloses include xyloglucans, xylans, mannans, glucomannans, β-1,3-linked glucans, and β-1,4-linked glucans. Xyloglucan is the most abundant form of sugar in hemicelluloses, consisting of a backbone of 1,4-linked β-D-glucose residues with short side chains containing xylose, galactose, and terminal fucose. Unlike cellulose, hemicelluloses are branched, and form polymers that are shorter in length (about 200 sugar molecules/unit) (Simoneit *et al.*, 1999).

Pectin

Pectin is a major non-cellulosic component of plant cell wall polysaccharides, comprising approximately one third of the cell wall dry weight of vascular plants (Sriamornsak, 2003). The pectic polysaccharides have a linear backbone structure containing galacturonic acid (GalA) and rhamnose (Rha) units joined in alternating chains via α-1,4-glycosidic linkages, accompanied by side chains containing neutral sugars such as arabinose, galactose and xylose. The major structural components of the pectin backbone includes homogalacturonan (HG), rhamnogalacturonan-I (RG-I) and rhamnogalacturonan-II (RG-II). HG, which constitutes the majority of uronic acid content in pectin-containing fruits and dietary plants, is a homogenous polymer of α-1,4-linked-D-galacturonic acid (Bush *et al.*, 2001). Approximately half of the galacturonic acid residues present in HG are either methyl-esterified at C-6 or acetylated at O-2 and O-3 (Ridley *et al.*, 2001). GalA residues that are neither methyl-esterified nor acetylated carry a negative charge, enabling formation of a characteristic gel by chelating Ca^{2+} ions. The degree of methyl-esterification and acetylation is dependent on plant developmental stage and influences enabling cell growth and ripening of fruits. HG forms successive connections to RG-I and RG-II that carry most of the side chains found on pectin. The RG-I and RG-II regions of pectin are compositionally heterogeneous, containing diverse neutral sugars. Rhamnose residues in RG-I are branched by homoarabinans, homogalactans, and arabinogalactan. Some arabinan and galactan side chains can be substituted with ferulic acid side chains, which can dimerize to strengthen the pectin network. The RG-II class is less abundant with a higher degree of structural complexity than RG-I. RG-II carries four groups of side chains, designated A to D, that contain at least 12 glycosyl residues covalently linked together by more than 20 different types of glycosidic linkages (O'Neill *et al.*, 2004). RG-II contains sugar residues that are rarely found in plants such as D-apiose, L-aceric acid, 2-keto-3-deoxy-D-manno-octulosonic acid and 2-O-methyl L-fucose. RG-II plays a major role in structuring the pectic network by forming dimers through borate ester cross-links. The degree of dimerization is subject to change during various plant developmental stages, enabling dynamic structural modifications that facilitate cell growth.

Cross-linkage between the polysaccharide components of the cell wall

Due to the difficulty of extracting an intact cell wall complex using conventional extraction methods, how these structural polymers cross-link to each other to form a sturdy cell wall structure is not well understood. For the last two decades, plant cell wall polysaccharides have been envisaged as a series of cellulose microfibrils consisting of a highly organized crystalline domain of cellulose linked to less organized, and more easily accessible amorphous regions of cellulose that are embedded into a highly cross-linked matrix of hemicelluloses and pectin. However, recent studies of cellulose chemical structure revealed that the 'crystalline' and 'amorphous' regions of cellulose microfibrils may actually be structural distortions caused by the use of harsh chemical means to extract cellulose components from the plant cell wall (Wei *et al.*, 2009). Electron microscopic structural analysis of plant cell wall cellulose indicated that cellulose is cross-linked by the constituents of hemicelluloses, such as xyloglucans (Somerville *et al.*, 2004). Owing to the steric restraint caused by its side chain positions, only part of the xyloglucan backbone of hemicelluloses can form hydrogen bonds along the surface of cellulose microfibrils (Hayashi *et al.*, 1987; Vincken *et al.*, 1994). Although no direct linkage between cellulose and pectin was initially believed to exist, growing evidence suggests that the neutral sugars of pectin side chains such as arabinan, galactan, and arabinogalactan, can bind to cellulose, as shown in *in vitro* binding assays using pectin and cellulose isolated from the cell walls of potato and apple (Zykwinska *et al.*, 2005). The exact mechanism of cross-linking is yet to be elucidated: hydrogen bond formation between the arabinan and galactan side chains of pectin and cellulose microfibrils is suspected, similar to the hemicelluloses–cellulose interaction. The strength of pectin–cellulose interaction was found to be weaker than that of hemicelluloses–cellulose interaction, despite the abundance of arabinan and galactan side chains on pectin (Zykwinska *et al.*, 2005). This may suggest that not all pectin side chains are capable of bonding to cellulose, possibly because the steric interference and bulkiness of pectin side chains discourage all side chains from engaging in hydrogen bond formation with cellulose. An ample amount of literature supports the existence of direct linkage between xyloglucan of hemicelluloses and RG-I of pectin, although the nature of linkage is still unknown. Researchers found alkaline-stable connections between xyloglucan and pectin, allowing both types of polysaccharides to be isolated together during the alkaline extraction of the plant cell wall (Scheller and Ulvskov, 2010).

Prebiotic potential of plant cell wall polysaccharides

Cellulose

The hard matrix of the cellulose-hemicellulose can efficiently resist degradation and fermentation by gut microorganisms, reaching the lower intestine with a substantial amount of fibre still intact or only partially degraded. The potential prebiotic effects of cellulose consumption have been summarized in Table 20.3. In the lower intestine, the water-holding capacity of these fermentable fibres increases the faecal bulk and stimulates colonic transit (Monro *et al.*, 2012). Monro and colleagues found that among various prebiotic ingredients tested (barley β-glucan, inulin, and cellulose), the consumption of cellulose caused maximum faecal bulking in rats, largely due to the high survival of dry matter through the gastrointestinal tract. These data are consistent with results obtained from cellulose-fed humans (Cummings, 1993; Eastwood *et al.*, 1973; Spiller *et al.*, 1980). The accelerated transit of colonic fermenta has a particular role in alleviating constipation, a common symptom of IBS. However, the consumption of insoluble fibres was only marginally effective at managing many IBS symptoms, suggesting that bulking agents such as cellulose should be used as a supplementary means to relieve minor symptoms of IBD (Zuckerman, 2006). An extra physicochemical effect of dietary fibre may lie in its capacity to bind to bile acids, which in turn may serve to reduce serum cholesterol by preventing the reabsorption of bile acid from the gut, and thus back into the liver (Kahlon and Chow, 2000). However, no obvious increases in bile acid production were observed after the consumption of cellulose fibres by human subjects (Eastwood *et al.*, 1973; Schneeman and Gallaher, 1980). This may be because an overall increase in the amount of digesta diluted the intestinal concentration of bile acid or it may be that cellulose has weaker binding affinity. *In vitro* studies comparing the bile acid binding capacity of dietary fibres suggest that cellulose binding to bile acid

Table 20.3 The physiological effects of cellulose/hemicelluloses/pectin consumption

Host response	Cellulose	Hemicellulose	Pectin
Faecal bulking	Significantly increase faecal bulking (Cummings, 1993; Eastwood *et al.*, 1973; Monro *et al.*, 2010; Spiller *et al.*, 1980)	Increase faecal bulking (Eastwood *et al.*, 1973; Holloway and Greig, 1984)	Unaffected (Dongowski *et al.*, 2000; Sanaka *et al.*, 2007; Spiller *et al.*, 1980)
Organic acid production	Acetate (Vince *et al.*, 1990)	Acetate/propionate (Zampa *et al.*, 2004)	Acetate/propionate/butyrate (Dongowski *et al.*, 2000; Licht *et al.*, 2010; Manderson *et al.*, 2005; Parkar *et al.*, 2012)
Lactobacillus/ Bifidobacterium	Unaffected (Paturi *et al.*, 2010); Significant increase in bifidobacteria (Campbell and Fahey Jr, 1997; Morishita and Konishi, 1994)	Both stimulated (Zampa *et al.*, 2004)	Both stimulated (De Vuyst and Leroy, 2011; Holck *et al.*, 2011; Licht *et al.*, 2010; Mandalari, 2006; Manderson *et al.*, 2005; Olano-Martin *et al.*, 2002; Onumpai *et al.*, 2011; Parkar *et al.*, 2012)
Immunomodulation	Unconfirmed	Unconfirmed	Inhibit pathogen adhesion to gut cells (Parkar *et al.*, 2010); Stimulate T-cells in lymph nodes (Lim *et al.*, 1997; Schley and Field, 2002); Induce apoptosis in colon cancer cells (Olano-Martin *et al.*, 2003a); Protect colonocytes from E. coli toxin (Olano-Martin *et al.*, 2003b)
Cholesterol control	No changes in bile acid production (Eastwood *et al.*, 1973; Kahlon and Chow, 2000; Schneeman and Gallaher, 1980)	High bile acid binding capacity (Kahlon and Chow, 2000)	Decrease serum/liver cholesterol (Arjmandi *et al.*, 1992; GarciaDiez *et al.*, 1996; Jenkins *et al.*, 1975; Kay and Truswell, 1977; Palmer and Dixon, 1966)

was almost 15 times weaker than the bile acid-binding strength of wheat and rice bran that are richer in hemicelluloses than cellulose (Kahlon and Chow, 2000). Dietary supplementation with cellulose, followed by a subsequent *in vitro* faecal fermentation using cellulose as a substrate showed no apparent increase in the amount of total organic acid when compared to the negative control (Vince *et al.*, 1990). The absence of notable changes in the organic acid profile may simply indicate that the gut bacteria were only able to partially ferment much of the cellulose which in this study was highly crystallized and chemically refined. A study by Mikkelsen *et al.* (2011) showed that *in vitro* porcine faecal fermentation using cotton wool generated approximately half the amount of total organic acids produced from the faecal fermentation of bacterial cellulose composites, indicating highly refined crystalline cellulose materials do not accurately reflect the fermentation of dietary cellulose occurring in the intestine. Despite the fact that cellulose is extremely difficult to hydrolyse enzymatically due to the present of resilient glycosidic bonds and the close association with other polysaccharides (Medie *et al.*, 2012), a very high total gut digestibility of 70% of cellulose has been observed, suggesting an extensive degradation and fermentation of natural cellulose does occur during its passage through the human intestine (Flint *et al.*, 2012; Robert and Bernalier-Donadille, 2003). The study by Mikkelsen *et al.* (Mikkelsen *et al.*, 2011) also showed that the fermentation of cellulose, regardless of the source types, led to the production of a moderate amount of acetate and low butyrate and propionate. This result was corroborated in a rat study upon dietary supplementation with cellulose, where moderate but significantly less acetate was produced compared to broccoli-supplemented diet, while no notable changes in butyrate and propionate productions occurred (Paturi *et al.*, 2010). A progressive reduction in propionate production upon dietary supplementation with gradually increasing cellulose was also observed in a similar rat study (Morishita and Konishi, 1994). Propionate, a SCFA normally produced at a late stage of fermentation, can still be generated in significant amounts from readily fermentable carbohydrates when they are able to be completely fermented within a relatively short period of time. In the human intestine where the transit of undigested material occurs within 72 hours, a complete fermentation of cellulose would be rather difficult. As a result, compared to soluble carbohydrates, insoluble cellulose is unlikely to be a major precursor to propionate production in the human gut.

A major role of cellulose during cell wall degradation is in providing increased surface areas for bacterial attachment. Fibre bulking effects of cellulose also creates micro-environments that are distinctively different from the intestinal surroundings (e.g. mucosa), providing gut bacteria with gradients of substrates (e.g. SCFAs) within the bulk that can mediate the growth of a large variety of cross-feeders (Sakata *et al.*, 2003). Bulking also leads to a prolonged detainment of the end products of fermentation in the lower intestine, increasing the retention time allowing more opportunity for nutrients to be absorbed across the intestinal barrier. Ability of bacteria to attach to the substrate surface is crucial for efficient digestion and fermentation of recalcitrant plant materials. Given the enormous structural diversity of plant fibres, there are a large variety of bacterial genomes enriched with different genes encoding an array of enzymes that target specific carbohydrate structure for degradation. However, only a handful of bacterial species are equipped with systems to break down recalcitrant fibres; these 'keystone species', as Flint termed them, modulate the microbial attacks on fermentable plant materials by initiating breakdown of the hard matrix of cellulose and hemicelluloses, exposing the more easily accessible parts of the cell wall to microbial degradation, and subsequently release of previously inaccessible substrates for further fermentation by secondary feeders (Flint *et al.*, 2012). These degraders of recalcitrant fibres have a large impact on determining inter-individual variations in physiological outcomes, as they stimulate primary- and secondary-level increases in the diversity of microorganisms that are efficient at extracting energy from plant source polysaccharides that are considered difficult substrates to use. Indeed, a majority of cellulolytic bacteria that have been isolated from humans are Gram-positive anaerobes which belong to the members of the genus *Ruminococcus*, *Clostridium*, *Roseburia*, and *Eubacterium* (Chassard *et al.*, 2010; Khouvine, 1923; Montgomery, 1988; Robert and Bernalier-Donadille, 2003; Wedekind *et al.*, 1988). The consumption of cellulose by rats was shown to significantly increase the abundance of the *Bacteroides–Prevotella–Porphyromonas* group compared with either a high-fat diet or a broccoli-based diet, although probiotic strains such as *Bifidobacterium* sp. and *Lactobacillus* sp. were largely unaffected (Paturi *et al.*, 2010). This contrasts with other studies that found that increasing the proportion of cellulose in the diet of rats significantly increased the abundance of bifidobacteria in the large intestine (Morishita *et al.*, 1994; Campbell *et al.*, 1997). The concurrent increase in the abundance of both Gram-positive anaerobes and bifidobacteria may indicate that cellulose fermentation does not directly stimulate the growth of bifidobacteria, but rather that the end-products which assimilate from cellulose degradation by Gram-positive anaerobes can be salvaged by bifidobacteria, enabling their growth (Morishita *et al.*, 1994).

While not a major component of the microbiome, methanogenic archaea such as *Methanobrevibacter smithii* and *Methanosphaera stadtmanae* play an important role in the utilization of excess hydrogen to form methane (Dridi *et al.*, 2009). The presence of methanogens indicates that the hydrogen gas substrate is not being used by SRB for the production of toxic gaseous metabolites such as hydrogen sulfide (Sahakian *et al.*, 2010). Presence of gut methanogens appears to influence the microbial adaptations to extraction of energy from the substrate. *Bacteroides* and butyrogenic *Ruminococcus* group of cellulose degraders tend to dominate the gut in methane non-excreting individuals. In individuals with methanogens, the microbial diversity shifted towards the Firmicutes, with the cellulose degraders from *Ruminococcus*, *Enterococcus* and *Roseburia* spp. (Chassard *et al.*, 2010) with the excess hydrogen being channelled to methane production.

Overall, dietary supplementation of cellulose appears to be particularly effective at increasing the amount of faecal bulk and improving laxation. Owing to the technical and ethical difficulties of measuring intestinal SCFA concentration in human subjects, animal models (mostly rats) have been used to study the effects of cellulose consumption on the production of major SCFAs, including acetate, butyrate, and propionate. However, these results do not provide conclusive evidence that cellulose can be used to manipulate SCFA concentrations in humans; human trials would be required to support such a claim. Furthermore, cellulose showed only limited efficiency as a prebiotic in that no direct linkage between cellulose consumption and the stimulation of probiotic bacterial growth has been reported in human cases.

Hemicellulose

Hemicellulose is extensively cross-linked to cellulose and pectic components of the cell wall, making it difficult to extract them in a pure form. As a result, few studies have been carried out to elucidate the prebiotic potential of hemicelluloses as an independent prebiotic ingredient (Table 20.3). However, cereal bran, which is very high in hemicelluloses and low in cellulose and pectin, was shown to exhibit a reasonable faecal bulking effect comparable to common fruits and other vegetable sources (Eastwood *et al.*, 1973). When hemicellulose fractions were alkaline-extracted from various plant sources, dry hemicellulose of wheat bran showed the highest water holding capacity among all vegetables and fruits (bean, cabbage, lettuce, onion, peach, pear, pumpkin, tomato) that were tested for their faecal bulking efficiency (Holloway and Greig, 1984). Although xylan-rich hemicelluloses have been examined for their possible candidacy as an emerging prebiotic, there is insufficient scientific evidence to support the claim (Gibson *et al.*, 2004). Cereal and germinated barley were shown to have some degrees of therapeutic and preventative effects on chronic gastrointestinal conditions such as ulcerative colitis (UC) and IBS, although clinical evidence is scant and it is not clear whether the observed gut health benefits accrue from eating whole foods, or from the presence of hemicelluloses *per se* (Bamba *et al.*, 2002; Fujimori *et al.*, 2007; Zuckerman, 2006). Wheat bran and rice bran showed a high binding capacity to bile acid *in vitro* (Kahlon and Chow, 2000), suggesting that a regular consumption of cereal bran may help lowering serum cholesterol. *In vitro* faecal fermentation of xylo-oligosaccharides showed a significant increase in *Lactobacillus*, which may have generated sufficient concentrations of lactate to provide a substrate for secondary feeders, as propionate production peaked at a very late stage of the fermentation (Zampa *et al.*, 2004). Propionate production has been associated with concomitant increases in lactobacilli and bifidobacteria indicating the convergence of metabolic pathways to utilize lactate production through cross-feeding of secondary metabolites (Chaia *et al.*, 1994; Chung *et al.*, 1999; Kaneko *et al.*, 1994). The situation is clearly complex. For example, xylan in hemicelluloses is known to be fermented by various butyrogens such as *F. prausnitzii* and the *Roseburia intestinalis/Eubacterium rectale* group that convert acetate and lactate to butyrate (Chassard and Bernalier-Donadille, 2006; Duncan *et al.*, 2002). However, no apparent butyrogenic effects were observed during the faecal fermentation of xylo-oligosaccharides, indicating that a preferential conversion from lactate and acetate to propionate, rather than to butyrate, may have taken place (Morrison *et al.*, 2006; Zampa *et al.*, 2004).

Pectin

Pectin is the initial polysaccharide exposed to the digestive system following plant consumption, and its removal exposes other polysaccharides such as xylans, mannans, and cellulose to microbial degradation in the large intestine. Pectin's potential to boost the gastrointestinal health and immune functions has been demonstrated in a number of studies, making it a possible candidate for the next-generation of prebiotics (Table 20.3). Contrary to the belief that the majority of pectin is completely fermented during the passage through gastrointestinal tract, a considerable amount of pectin from guar gum was observed to be excreted in faeces of rats, contributing to the increased faecal mass (Nyman and Asp, 1982). The degree of methyl esterification moderate how and which pectin is fermented by the gut bacteria; highly methylated pectin efficiently resists fermentation and can be excreted out in the faeces (Dongowski *et al.*, 2000; Nyman and Asp, 1982). However, the total faecal weight decreased compared to that from the baseline diet for a group of human subjects who consumed a controlled diet supplemented with high methoxyl citrus pectin (Spiller *et al.*, 1980). This result is corroborated by another human study where individuals ate a controlled diet supplemented with high methoxyl pectin produced stool weight comparable to those with low fibre intake (Cummings *et al.*, 1978). These results suggest that regardless of the degree of methyl modification, pectin is efficiently fermented in the human gut, and its contribution to faecal bulking is minimal. Pectin was however found to increase transit time through the gastrointestinal tract, prolonging the retention of nutrients in the small gut to allow increased absorption across the intestinal barrier (Sanaka *et al.*, 2007). In rats, pectin was shown to significantly reduce the concentration of serum and liver cholesterol and to increase faecal bile acid excretion, indicating a possible hypocholesterolaemic effect of pectin by mechanisms that include the increase in bile secretion and the sequestering of phospholipids (Arjmandi *et al.*, 1992; GarciaDiez *et al.*, 1996). A major SCFA product of soluble fibre fermentation, propionate, was also shown to control the accumulation of serum and liver cholesterol, suggesting that pectin might indirectly contribute to controlling the liver lipid production through propionate production (Chen *et al.*, 1984). Similar cholesterol-lowering effects were observed in human subjects who received a diet supplemented with citrus-pectin; a significant reduction in serum lipids level, as well as an increased excretion of cholesterol in faeces was observed (Jenkins *et al.*, 1975; Kay and Truswell, 1977; Palmer and Dixon, 1966). Pectin has shown an apparent butyrogenic effect; *in vitro* faecal fermentation using orange peel pectin as the substrate stimulated the production of major SCFAs, including lactate, acetate, propionate, and butyrate with butyrate concentration being considerably higher than that from FOS, a common prebiotic (Manderson *et al.*, 2005). Similar SCFA profiles were observed during *in vitro* faecal fermentation using citrus pectin (Dongowski *et al.*, 2000), apple pectin (Licht *et al.*, 2010), and fruits (e.g. kiwi fruit) that are known to be rich in pectin (Parkar *et al.*, 2012). The butyrogenic effect of pectin is corroborated by the significant stimulation of bifidobacteria that cross-feeds with butyrate producers to generate butyrate (De Vuyst and Leroy, 2011; Holck *et al.*, 2011; Licht *et al.*, 2010;

Mandalari, 2006; Manderson *et al.*, 2005; Olano-Martin *et al.*, 2002; Onumpai *et al.*, 2011; Parkar *et al.*, 2012). *Lactobacillus* has also shown to be stimulated by pectin supplementation, although to a lesser extent than bifidobacteria (Holko and Hrabe, 2012; Licht *et al.*, 2010; Mandalari, 2006; Manderson *et al.*, 2005; Onumpai *et al.*, 2011; Parkar *et al.*, 2012).

Pectin also appears to provide immunological benefits. Kiwi fruit pectin was shown to inhibit pathogenic adhesion of *Salmonella enterica* serovar Typhimurium to Caco-2 cells, increasing the competitiveness of cell attachment by probiotic strains of *Lactobacillus* and *Bifidobacterium* (Parkar *et al.*, 2010). Feeding rats with 5% pectin increased the number of CD4+ and CD8+ T-cells in lymph nodes and lymphocytes, suggesting that there are immunomodulatory effects of pectic oligosaccharides complementing their gastrointestinal benefits (Eiwegger *et al.*, 2010; Lim *et al.*, 1997; Schley and Field, 2002). Pectic oligosaccharides induced apoptosis in human colonic adenocarcinoma cells (Olano-Martin *et al.*, 2003a), and in addition protected colonocytes against the destructive *E. coli* verocytotoxins, possibly providing important protective mechanisms against colon cancer (Olano-Martin *et al.*, 2003b). Whilst microbial changes were not directly demonstrated in these latter studies, one may speculate that at least some of these benefits are derived by their microbial modification after reaching the colon.

Plant cell wall polysaccharide composite: Perhaps the major difficulty in correlating cell wall polysaccharides with prebiotic benefits has been the complex interactions between these molecules in the plant cell and the great structural diversity in the cell wall of different plants, even within the same crop at differing physiological maturity. The composite cell wall in kiwifruit is well characterized (Carnachan *et al.*, 2012). Kiwi fruit contains about 2–3% (g/100 g fresh weight) cell wall non-starch polysaccharides, comprising 60% insoluble fractions, of which 67% is cellulose and the remaining 33% is hemicellulose. About 40% of cell wall polysaccharide is soluble pectin, 90% of which is the true pectin, rhamnogalacturonan (Carnachan *et al.*, 2012). As described above, each of these fractions are largely unchanged during gastrointestinal transit until they undergo microbial fermentation in the colon. It may be anticipated that the complex substrates are first attacked by the carbohydrate-degrading enzymes of the primary feeders, followed by a bloom of secondary feeders that make use of released oligosaccharides or fermentation by-products from the primary feeders. Thus while classical prebiosis implies selection of specific groups of bacteria (Gibson *et al.*, 2010), some of the benefits may be derived from increased functionalities resulting from diet induced divergence of microbial groups and subsequent microbial adaptations to complex substrates through coupling of diverse metabolic pathways.

As a case study to demonstrate this: fresh whole green kiwi fruit were subjected to simulated gastric and ileal digestion, and the remaining cell wall polysaccharides after dialysis of the digesta were fermented for 48 hours with faecal bacteria obtained from 10 healthy donors. The 48-hour kiwifruit fermenta showed an increase in *Bacteroides-Prevotella-Porphyromonas* group, *Lactobacillus* and *Bifidobacterium* spp. compared with the control (Parkar *et al.*, 2012). The complexity of the changes in microbiota was then analysed in detail for three individuals using 16S rRNA pyrosequencing (Rosendale *et al.*, 2012). Kiwi fruit polysaccharides stimulated production of beneficial metabolites in all three individuals. The baseline microbial ecology of each individual influenced the relative microbial abundances and diversity as well as the enzymatic adaptations required to break down the complex substrates and reach similar endpoints. Of note were immediate and lasting reduction in the average abundance of Proteobacteria phyla; Bacillaceae and Clostridiaceae (Firmicutes phyla); and Prevotellaceae (Bacteroidetes phyla). While some members of Firmicutes phyla (Ruminococcaceae) decreased after 24 hours, others (Lachnospiraceae and *Veillonella*) increased. Correlating these data with the emerging patterns of the carbohydrate utilizing enzymes and metabolites generated, one may appreciate how changes in microbial diversity drives polysaccharide utilization irrespective of the starting point. Thus, the abundant Bacteroidaceae with their glycoside hydrolases may initiate and drive utilization of soluble pectin (Martens *et al.*, 2011) which is further enhanced by the bloom in Lachnospiraceae as they continue the catabolic processes (Scott *et al.*, 2014). After 48 hours, the change in available carbohydrates drives a change in microbial diversity with resurgence in Ruminococcaceae, which have now been described as keystone species for degradation of leftover cellulose (Ze *et al.*, 2013). There were sustained changes in the Bifidobacteriaceae that are known to adapt to diverse glycan substrates to provide probiotic benefits (Turroni *et al.*, 2010). In addition to modulating microbial diversity and coupling metabolic pathways, kiwi fruit fermentation influenced proliferation and even adhesion of a sample probiotic organism *Bifidobacterium longum* ATCC 15707 to human intestinal epithelial cells (Parkar *et al.*, 2012). While this was a simplified model, the demonstrated effects of kiwifruit on adhesion of key bacteria were exciting as it is recognized that adhesion to the gut wall is the first step in microbial colonization (Bavington and Page, 2005). These adhesion effects may have been due to fermentation byproducts including pectin oligosaccharide fractions. Previously it has been demonstrated that kiwifruit pectin rich in galactose, arabinose and galacturonic acid enhanced the adhesion of *Lactobacillus rhamnosus* to intestinal cells *in vitro*, but inhibited the adhesion of *S. enterica* sv. Typhimurium (Parkar *et al.*, 2010).

Structural polysaccharides – marine

Whilst marine plants and algae (seaweed) have been exploited for thousands of years, there is increasing interest in incorporating properties of seaweed in food, textile and cosmetic applications. Seaweed are macroalgae, broadly classified into brown (Phaeophyceae), red (Rhodophyceae) or green (Chlorophyceae) algae on the basis of their specific pigments, the nature of their chlorophyll, the presence or absence of flagella, their cell wall and intercellular mucilage chemistry, and the nature of their photosynthetically derived storage carbohydrates. Here we focus on the gut microbial fermentability of seaweed polysaccharides (Table 20.4) which may have potential prebiotic effects.

The fermentability of 'total dietary fibre' extracts from brown seaweeds *Himanthalia elongate*, *Undaria pinnatifida* and *Laminaria digitata* (particle sizes 200–500 um) were studied using sugar

Table 20.4 Examples of commonly characterized edible seaweeds, and their carbohydrates. (modified from Bocanegra *et al.*, 2009).

Division/common name, example species	Cell wall structural carbohydrate	Storage carbohydrate	Intercellular mucilage
Chlorophyta/green algae e.g. *Enteromorpha* spp., *Ulva* spp.	Cellulose (β1–4 glucose), glucuronans (β1–4 glucuronic acid), xylans (β1–4 xylose), mannans (β1–4 mannose)	Starch (α1–4 and α1–6 glucose) inside chloroplasts	Sulfated polysaccharides, e.g. ulvan (4-glucuronic acid –α1–4-rhamnose, rhamnose-3-O-sulfates and glucuronic acid-2-O-sulfates, some xylose and glucose)
Phaeophyta/brown algae e.g. *Laminaria* spp., *Himanthalia fusiform, Undaria pinnafitida* (Wakame)	Cellulose (β1–4 glucose), chitin (β1–4 N-acetylglucosamine)	Laminarans (β1–3, β1–6 glucans associated with mannitol residues)	Alginates (β1–4 D-mannuronic acid, α1–4 guluronic acid polymers), sulfated polysaccharides, e.g. fucoidans (α1–2, 1–3 or 1–4 fucose, 2-O-, or 4-O-sulfates, also galactose, mannose, xylose, glucose, glucuronic acid and acetate)
Rhodophyta/red algae, e.g. *Porphyra tenera* (Nori), *P. haitanensis, P. capensis*	Cellulose (β1–4 glucose), xylans (β1–4 xylose), mannans (β1–4 mannose)	Floridean starch (structurally similar to starch granules but lacking amylose) outside chloroplasts	Sulfated polysaccharides, e.g. agar (agarose: repeating agarobiose (galactose-α1–4–3,6-anhydrogalactose)) and agaropectin: sulfated and pyruvated D- and L-galactose); carageenans (alternating α1–3 and β1–4 galactose and 3,6-anhydrogalactose, 2-O-, 4-O- and 6-O-sulfates); and porphyrins (β3–4, α2–6 or α1–4 galactose, 6-O-sulfates)

beet fibre as a positive control (Michel *et al.*, 1996). Substrate disappearance, SCFA production and gas production were monitored during the course of the fermentation. Most of each of the total algal fibre extracts were utilized, but conversion to SCFA was incomplete, in contrast to the sugar beet reference fibre. Refined fibre preparations from seaweeds such as laminarans, alginates, mannuronates and fucans were also compared in this study and their fermentability was in the same order. That is, laminaran β-glucans were the most readily and extensively metabolized to SCFA, while fucans appeared to be poorly or not fermented at all. The relative fermentability of the total dietary fibre extracts was similar, although *L. digitata* (which contained the highest sugar dry weight composition and high relative laminaran composition) tended to be more readily fermented.

Interestingly, the microbial response to alginate favoured acetate and propionate production whilst the response to laminarans strongly favoured the more reduced acids propionate and butyrate. Thus substrates appeared to select for different functional groups of bacteria, with alginate possibly enhancing the numbers of propionate-producing bacteria such as Bacteroides. In contrast, the laminaran β-glucans were highly fermented, and strongly selected for butyrate and propionate-producing populations. This was consistent with the response shown to β-glucans from other sources, such as barley, which was found to increase Bacteroides-Prevotella group numbers and enhance propionate production (Hughes *et al.*, 2008).

Red (*P. tenera*) and brown algae (*U. pinnatifida*) were compared for specific microbial responses in an *in vivo* study (Goni *et al.*, 2001). Total fibre extracts from both sources were poorly fermented in the caecum of rats fed either substrate compared with control (lactulose-fed) animals. However, when caecal microbiota were incubated with the same substrate *ex vivo*, up-regulated fermentation was observed for the agar, carageenan and porphyran-rich red algae, whereas the alginate- and fucoidan-rich brown algae did not show further signs of fermentability (Goni *et al.*, 2001). These data indicate that with some substrates, microbial adaptation can occur. The sulfated polymers found in red algae are predominantly galactans, whilst the non-laminaran, sulfated polymers found in brown algae consist of a wider variety of fucose- and uronate–rich complex polysaccharides, the former perhaps more readily desulfated and depolymerized than the latter. Seaweed polysaccharides are similar to the gastrointestinal mucin that intestinal bacteria are regularly exposed to: both are predominantly sulfated and galactose rich (Gibson *et al.*, 1990). Consequently, one can speculate that Bacteroidetes phylum members, that are known mucin-degraders and possess high sulfatase activities (Wright *et al.*, 2000) may be selectively manipulated in response to these seaweed substrates. However, an increased availability of free sulfate in the colon may favour populations of dissimilatory sulfate-reducing bacteria (SRB) such as *Desulfovibrio* (Gibson *et al.*, 1993), which have been implicated in inflammatory bowel processes (Zinkevich and Beech, 2000). Nevertheless, there may be positive outcomes, such as lowering methane production via SRB outcompeting of methanogens for necessary H_2.

The idea that a selective response to seaweed carbohydrates is possible was indicated by comparatively recent work showing that richness and diversity (Shannon) indices decreased upon dietary supplementation of rats with brown seaweed alginate in a dose-responsive manner (Janczyk, 2010). DGGE profiling showed a reduction in bands corresponding to butyrogenic groups such as *Clostridium* and *Eubacterium* spp., including *E. rectale*. Of course, given the poorly fermented nature of sulfated seaweed carbohydrates, the selective microbial response might be the consequence of survival of the least poorly adapted, rather than a bloom in favoured organisms. However, given the known health benefits of some *Bacteroides* (e.g. *B. thetaiotaomicron* (Martens *et al.*, 2011)), and the ongoing debate on the consequences of an increased Bacteroidetes: Firmicutes ratio in weight management and metabolic disease (Duncan *et al.*, 2008; Ley *et al.*, 2006) further research would seem warranted to understand the prebiotic potential of these seaweed polysaccharides.

Microbial exopolysaccharides

Probiotic bacteria when ingested are known to provide gastrointestinal protection and promote health benefits (Rijkers *et al.*, 2011). At least some of these benefits may be attributed to bacterial components or secretions. Many strains of lactic acid bacteria

(LAB) secrete extracellular polysaccharide either associated with bacterial cell wall or liberated into the medium. These exopolysaccharides (EPS) may provide functional benefits to human health such as cholesterol-lowering, immunomodulating and prebiotic effects. EPS also offer the normal benefits associated with typical polysaccharides as stabilizer, viscosifier, emulsifier or gelling agent, thereby providing the product with natural thickness and acceptable flavour and sensory attributes (Behare *et al.*, 2009).

EPS isolated from two Bifidobacterium strains, *B. longum* subsp. *longum* IPLA E44 and *B. animalis* subsp. *lactis* IPLA R1, were found to resist digestion by gastrointestinal enzymes *in vitro* (Salazar *et al.*, 2009). Their effect on bacteria in pH-controlled faecal batch cultures was compared with inulin. The EPS from the R1 strain stimulated growth of *Lactobacillus/Enterococcus*, *Bifidobacterium* and *Atopobium* in a similar way to that seen with inulin with increases in acetate, and lactate concentrations. As cultures aged, there was a build-up of propionate due to selective increases in *Clostridium* cluster IX with EPS R1 while *Bacteroides* increased irrespective of substrates due to a nonspecific stimulation by the glycans. Further trials are needed to establish their prebiotic role.

Phytochemicals

Polyphenols

Polyphenols are small bioactive phytochemicals widely present in fruits, vegetables, cereals and beverages (http://www.phenol-explorer.eu/ offers a comprehensive database of polyphenols in foods). Polyphenols serve useful antibacterial and antioxidant purposes in plants and these roles may have also made them useful to humans after consumption. The two major classes of dietary polyphenols are the phenolic acids and the flavonoids (Manach, 2004). Plant-derived polyphenols generally occur in a conjugated form (glycosylated, acylated, esterified, polymerized) (Hervert-Hernandez and Goni, 2011). The most common conjugates are the glycosides comprising about 80% of all polyphenolics in plant tissues, while aglycones are less common in nature (Perez-Jimenez *et al.*, 2010). The glycosides are carried to the colon and then metabolized by microbial glycosidases while the aglycones may be absorbed through the small intestinal membrane by diffusion or active transport proteins (Manach, 2004; Manach *et al.*, 2005). Polyphenols are consumed as a part of a food matrix, whether whole food or processed food, and are a major component of the undigested material that reaches the colon (Hervert-Hernandez and Goni, 2011). Polyphenols bound to the fibre matrix are released into the lumen after microbial breakdown of the fibre microarchitecture (Palafox-Carlos *et al.*, 2011).

Polyphenols have evolved in plants to provide antibacterial protection, and thus can modulate bacterial growth and survival. The effect of polyphenols on intestinal bacteria with further gastrointestinal and systemic benefits are summarized in Table 20.5. Our own work and work by others shows that polyphenols can modulate gut microbes by inhibiting pathogens and stimulating probiotic bacteria. Out of a range of phenolic acids and flavonoids studied, two flavonoids, naringenin and quercetin were inhibitory to representative gut bacteria such as *L. rhamnosus*, *E. coli*, *Staphylococcus aureus*, *S. enterica* sv. Typhimurium at physiologically relevant colonic concentrations of 125 to 250 μg/ml. The Gram-positive *Staphylococcus* was found to be more sensitive to the polyphenols than *Lactobacillus* which may be considered to be more 'adapted' to the gut environment (Parkar *et al.*, 2008). The antimicrobial effects of three flavonoids were later studied with intestinal bacteria from species of *Bacteroides*, *Lactobacillus*, *Enterococcus*, *Bifidobacterium*, *Ruminococcus* and *Escherichia* and similar inhibitory effects were noted. The aglycone flavonoids, naringenin, hesperetin and quercetin were more active than the glycosylated forms, naringin, hesperidin, and rutin respectively (Duda-Chodak, 2012). The glycosylated forms while less toxic to the bacteria are more susceptible to bacterial metabolism as the microorganisms in the gut have developed a diverse array of glycosidases to adapt to the undigested plant foods that reach the colon (Ley *et al.*, 2008). This is especially true of the *Bacteroides* genus. In comparison, dietary catechols such as those present in tea (caffeic acid, catechin and chlorogenic acid) appear to modulate a complex bacterial growth depending on the intestinal environment. For example, caffeic acid, epicatechin, catechin and 3-O-methyl-gallic acid are largely antimicrobial, especially towards potential clostridial pathogens, with different effects on the rate of growth of other gut bacteria such as *Bacteroides*, *Bifidobacterium* and *Lactobacillus* species (Lee *et al.*, 2006). While catechin, caffeic and cholorogenic acid had only moderately inhibitory effects on *E. coli* and *S. enterica* sv. Typhimurium in one study (Parkar *et al.*, 2008), these catechols stimulated the growth of enteropathogenic *E. coli* and *S. enterica* in an iron-deficient basal growth medium. The mechanism was suggested to be an up-regulation of bacterial siderophores, causing an uptake of iron from sequestrating proteins such as lactoferrin and transferrin (Freestone *et al.*, 2007). Caffeic acid esters may also have a protective effect on cultured intestinal cells implying anti-cancer benefits (Bordonaro *et al.*, 2014). Indeed, apple polyphenol metabolites generated by faecal bacteria were found to inhibit histone deacetylases in intestinal cell systems and thus may provide protection in synergy with butyrate to inhibit the growth of cancer cells (Bordonaro *et al.*, 2014).

Berryfruit such as blackcurrants, boysenberries, cranberries and raspberries are rich in polyphenolic bioactives such as phenolic acids, anthocyanidins, ellagitannins which contribute to the antimicrobial and protective effects in the gut (Cavanagh, 2003; Puupponen-Pimia *et al.*, 2005; Werlein *et al.*, 2005). Berryfruit are also rich in phenolic derivatives of hydroxycinnamic and hydroxybenzoic acids which are known to be antimicrobial (Alakomi *et al.*, 2007). Blackcurrant anthocyanins such as cyanidin and delphinidin or their glycosides were found to have an inhibitory effect on *S. enterica* sv. Typhimurium in micromolar amounts, and these concentrations may be achieved in the colon with an average intake of blackcurrant juice (Parkar *et al.*, 2014). Berry phenolics such as cyanidin glucoside and gallic acid were found to inhibit *E. coli* with weakening of the outer membrane (Lacombe *et al.*, 2010). The glycoside conjugates may also have prebiotic effects, with stimulatory effects on *Lactobacillus rhamnosus* (Parkar *et al.*, 2014). Lactobacilli have been reported to metabolize delphinidin glucoside (Avila *et al.*, 2009) implying that blackcurrant delphinidins may support the growth of the bacteria and that lactobacilli

Table 20.5 Categories of phytochemicals with potential prebiotic effects

Phytochemical	Types	Example	Microbial metabolites	*In vivo* effects of the microbial activation
Polyphenol				
Phenolic acids (account for one third of dietary polyphenols)	Hydroxybenzoic acid	Protocatechuic acid	3-Methoxy-4-hydroxybenzoic (vanillic) acid	Antibacterial and antioxidant (Proestos *et al.*, 2006); vanillic acid is also anti-inflammatory (Kim *et al.*, 2010)
	Hydroxyl cinnamic acid	Chlorogenic acid, ferulic acid	Caffeoylquinic acid lactones, hydroxycinnamic acids, urinary excretion of dihydrocaffeic acid-3-O-sulfate and feruloylglycine (Stalmach *et al.*, 2009)	Inhibit potential pathogens (Aziz *et al.*, 1998); reduced intestinal mucosal inflammation and oxidative damage in animal models of inflammatory bowel disease (Tuohy *et al.*, 2012)
Flavonoids (two thirds of dietary polyphenols)	Flavonols	Quercetin, kaempferol, myricetin	2-(3,4-Dihydroxyphenyl)acetic acid, 2,3-(3-hydroxyphenyl)acetic acid, 3,4-dihydroxybenzoic acid, phloroglucinol, acetate, butyrate, 3-(3,4-dihydroxyphenyl) propionic acid, 3-(3-hydroxyphenyl)propionic acid	Inhibit potential pathogens (Aziz *et al.*, 1998); Antioxidant (Olthof *et al.*, 2003)
	Flavones	Apigenin, luteolin	3-(4-hydroxyphenyl)propionate, phloroglucinol	Antioxidant, anti-inflammatory (Tuohy *et al.*, 2012)
	Isoflavones	Daidzein, genistein, glycetin	Equol, O-demethylangolensin	Phytoestrogen with protective effects on heart and bone, possible colon cancer protection (Tuohy *et al.*, 2012)
	Flavanones	Rutin, naringenin, hesperetin, eriodictyol	Quercetin, glucose, SCFAs, ethanol, 3-hydroxyphenylacetic acid, 3,4-dihydroxybenzoic acid, 4-hydroxybenzoic acid, 3-(3-hydroxyphenyl)propionic acid, 3,4-dihydroxyphenylacetic acid	Antioxidant (Olthof *et al.*, 2003), anti-platelet aggregation activity (Kim *et al.*, 1998)
	Monomeric flavanols	Epicatechin, catechin	4-hydroxyphenylacetic acid, 3-(3-hydroxyphenyl)propionic acid, 5-(3,4-dihydroxyphenyl)-γ-valeric acid (−)-5-(3′,4′-dihydroxyphenyl)-γ-valerolactone	Antimicrobial/antimycotic activity *in vitro*; antimicrobial activity against Gram-negative enterobacteria via outer membrane destabilization (Tuohy *et al.*, 2012)
	Polymeric flavanols	Proanthocyanidins, condensed tannins	3,4-Dihydroxyphenylpropionic acid, 3-hydroxyphenylpropionic acid, 3,4-dihydroxyphenylacetic acid	Reduced inflammatory response from lipopolysaccharide-stimulated blood lymphocytes (Tuohy *et al.*, 2012)
	Anthocyanidin	Cyanidin, delphinidin, pelargonidin	Protocatechuic acid	Antioxidant, important for heart health (Williamson and Clifford, 2010)
Lignans		Secoisolariciresinol diglucoside	Enterolactone, enterodiol	Phytoestrogen important for heart and bone health; possible colon cancer protectants (Tuohy *et al.*, 2012)
Stilbene		Resveratrol	Resveratrol glucuronides and sulfates	Enhanced microbial changes, antioxidant, important for heart health (Queipo-Ortuno *et al.*, 2012)
Glucosinolates				
		Glucoraphanin	Sulforaphane (Bheemreddy and Jeffery, 2007)	Beneficial changes in faecal microbial metabolism (Mullaney, 2014)

Adapted from Manach (2004), Moco *et al.*, (2012), Tuohy *et al.* (2012), Williamson and Clifford (2010).

may have an initiator role in activating the anthocyanin. Strains of lactobacilli and bifidobacteria were also found to metabolize and grow on grape seed polyphenols rich in catechins and oligomeric polyphenols (Tabasco *et al.*, 2011). Similarly, selected strains of bifidobacteria were found to activate the deglycosylation of secoisolariciresinol diglucoside, which is essential for further microbial biotransformation in the colon (Roncaglia *et al.*, 2011).

A diverse range of polyphenols are metabolized by the intestinal bacteria to generate phenolic acid metabolites such as 3-(4-OH phenyl)propionic acid, 3-phenylpropionic acid and 4-OH phenylacetic acid. These metabolites were shown to be inhibitory towards pathogenic species such as *Salmonella* and *Escherichia* including *E. coli* 0157:H7, while they were found to be less inhibitory on common gut inhabitants of the *Clostridium*, *Bifidobacterium* and *Lactobacillus* species (Lee *et al.*, 2006). We also found that the metabolites enhanced the growth of beneficial *B. longum* (Parkar *et al.*, 2013). The mechanism of this inhibitory effect towards Gram-negative pathogens has been suggested to be the destabilization of the bacterial outer membrane (Alakomi *et al.*, 2007). Thus the antimicrobial efficiency of a polyphenol may vary with the gut environment, the complexity of microbes in the gut, the nature of the polyphenol itself and the types of metabolites that it generates (Fogliano *et al.*, 2011; Hervert-Hernandez and Goni, 2011; Lee *et al.*, 2006). Polyphenols have also influenced the adhesion of bacteria to cultured intestinal epithelial cells, with the glycosylated phloridzin (phloretin 2′-O-glucose) having a role in both inhibiting the adhesion of salmonellae and enhancing the adhesion of lactobacilli to the gut cells in culture (Parkar *et al.*, 2008). Since, as previously noted, bacterial adhesion is the first step in their colonization to the gut wall for further beneficial effects, selective enhancement of adhesion by probiotic bacteria will assist in reducing opportunities for adhesion by pathogens such as salmonella (Lee *et al.*, 2003; Todoriki *et al.*, 2001).

The prebiotic potential of polyphenols to enhance the growth of beneficial bacteria has been demonstrated using models simulating the intestine, as well as *in vivo* trials with animals and humans.

One of the major roles of prebiotic foods is their selective utilization by bacteria to form beneficial short-chain fatty acid (SCFA) metabolites. While polyphenols may not be of sufficient mass (compared with fibre) in the colon to be major substrates for SCFA production, they may influence the SCFA profile indirectly by selectively inhibiting or enhancing the growth and colonization of certain bacterial groups. For example, a mixture of pomegranate polyphenols rich in gallic acid, ellagic acid and glucose units was investigated for its effect on faecal microbiota in an *in vitro* culture. There was enhanced growth of total bacteria, *Bifidobacterium* spp. and *Lactobacillus* spp., with concomitant increases in acetate, propionate and butyrate. There were no changes in the *Clostridium coccoides-Eubacterium rectale* group and the *Clostridium histolyticum* group (Bialonska *et al.*, 2010). Take for example, the metabolic breakdown of a typical polyphenol such as rutin (Fig. 20.1). As shown in Fig. 20.1a, this common polyphenol reaches the colon, either free or as conjugates or bound to fibre with H bonds and van der Waals interactions (Palafox-Carlos *et al.*, 2011). The first step is glycosidic cleavage of the rutin from the fibre, followed by a series of hydrolysis steps catalysed by microbial enzymes (Fig. 20.1b). The polyphenol is first hydrolysed from the rutinose sugar to release quercetin by, for example, *Bacteroides* spp. equipped with β-glycosidases (Bokkenheuser *et al.*, 1987). Then the phenolic A ring is broken down to phloroglucinol, while the B ring is metabolized into hydroxylated forms of phenylacetate or phenylpropionate. Several organisms of the *Clostridia* class including *Eubacterium ramulus* and *Eubacterium oxidoreducens* have been associated with the fermentation of the aglycone quercetin, leading to the formation of taxifolin, 3,4-dihydroxyphenylacetic acid, acetate and butyrate (Moco *et al.*, 2012). Polyphenol glycosides such as rutin may thus serve to initiate or activate propionate producing commensals such as *Bacteroides* species which are well equipped with a diverse range of glycosidases to break down glycosidic linkages that abound in the polysaccharide rich dietary fibre that reaches the colon (Dabek *et al.*, 2008; Macfarlane *et al.*, 1998; Macy *et al.*, 1978). *In vitro* incubation with pure rutin was found to increase beneficial bacteria such as *Bifidobacterium* spp., and there was an increase in the ratio of propionate to butyrate in comparison to fermentation with no carbohydrate control, inulin or the aglycone, quercetin (Parkar *et al.*, 2013). Thus polyphenolic glycosides may have a novel role in stimulating metabolites such as propionate, a metabolic regulator of metabolic disorders and obesity (Arora *et al.*, 2011).

Polyphenol-rich foods that have been explored for their prebiotic benefits include tea, coffee, cocoa and coloured fruits such as grapes, and berry fruit. Cocoa flavonols rich in oligomeric procyanidins have been shown to reach the colon to a large extent (75% or more) in an *in vitro* model of the gastric and ileal digestion of water insoluble fractions of cocoa. These fractions were predicted to reach the colon in their free polyphenolic form or bound to hemicellulose and pectin-rich cocoa fibre. In the same model, incubation with faecal bacteria resulted in a favourable balance of bacteria with a significant increase in lactobacilli and bifidobacteria, and with an increase in butyrate production (Fogliano *et al.*, 2011). The major phenolic metabolites produced were 3-hydroxyphenylpropionic acid (3-HPP), 3-hydroxyphenylacetic acid (3-HPA) and 3,4-dihydroxybenzoic acid (3,4-DHBA) which reached maximum concentration in the *in vitro* gut section mimicking the terminal colon. These metabolites could therefore provide protection to the colonic mucosa (Jenner *et al.*, 2005) due to their antioxidant (Duenas *et al.*, 2011), anti-inflammatory, and antiproliferative properties (Gao *et al.*, 2006). The association between fibre and bound polyphenols could therefore deliver a rich prebiotic cocktail to the colonic bacteria and modulate the microbial balance. Perhaps not surprisingly then, there is increasing interest in how combinations of prebiotics could be delivered with the assistance of food processing technology; for example, novel food processing technology is driving the development of novel prebiotic combinations such as encapsulation of cocoa polyphenols in a matrix of gastric resistant high-amylose maize starch (Vitaglione *et al.*, 2013).

A diet rich in polyphenols has been prescribed for weight loss and restoring metabolic and immunological profile, including restoring blood glucose and lipid profile (Rastmanesh, 2011). Plant based diets rich in polyphenols that change the microbial balance in the gut have been mooted as a more effective option than solely increasing intake of selective probiotic bacteria. The effect of red wine polyphenols, including resveratrol, was studied in 10 volunteers for 4 weeks. There was a significant increase in beneficial bacteria such as *Bifidobacterium*, *Bacteroides uniformis*, *Eggerthella lenta*, and *Blautia coccoides-Eubacterium rectale* groups coupled with metabolic improvement in terms of blood pressure and blood lipid profile, with significant correlations between blood cholesterol, the inflammatory C-reactive protein concentrations and faecal bifidobacterial numbers (Queipo-Ortuno *et al.*, 2012).

Tea polyphenols such as flavonoids, catechins and gallic acid were found to be poorly absorbed in the small intestine. In a trial with healthy adults, a single dose of tea polyphenolic extract was found to be metabolized by colonic bacteria with the release of the polyphenolic phase II conjugates in systemic circulation within 2 hours (van Duynhoven *et al.*, 2014). Microbial degradation patterns of wine and the tea polyphenolic extracts have also been investigated in an *in vitro* gastrointestinal model, which is much easier to use to demonstrate changes in SCFAs and polyphenolic

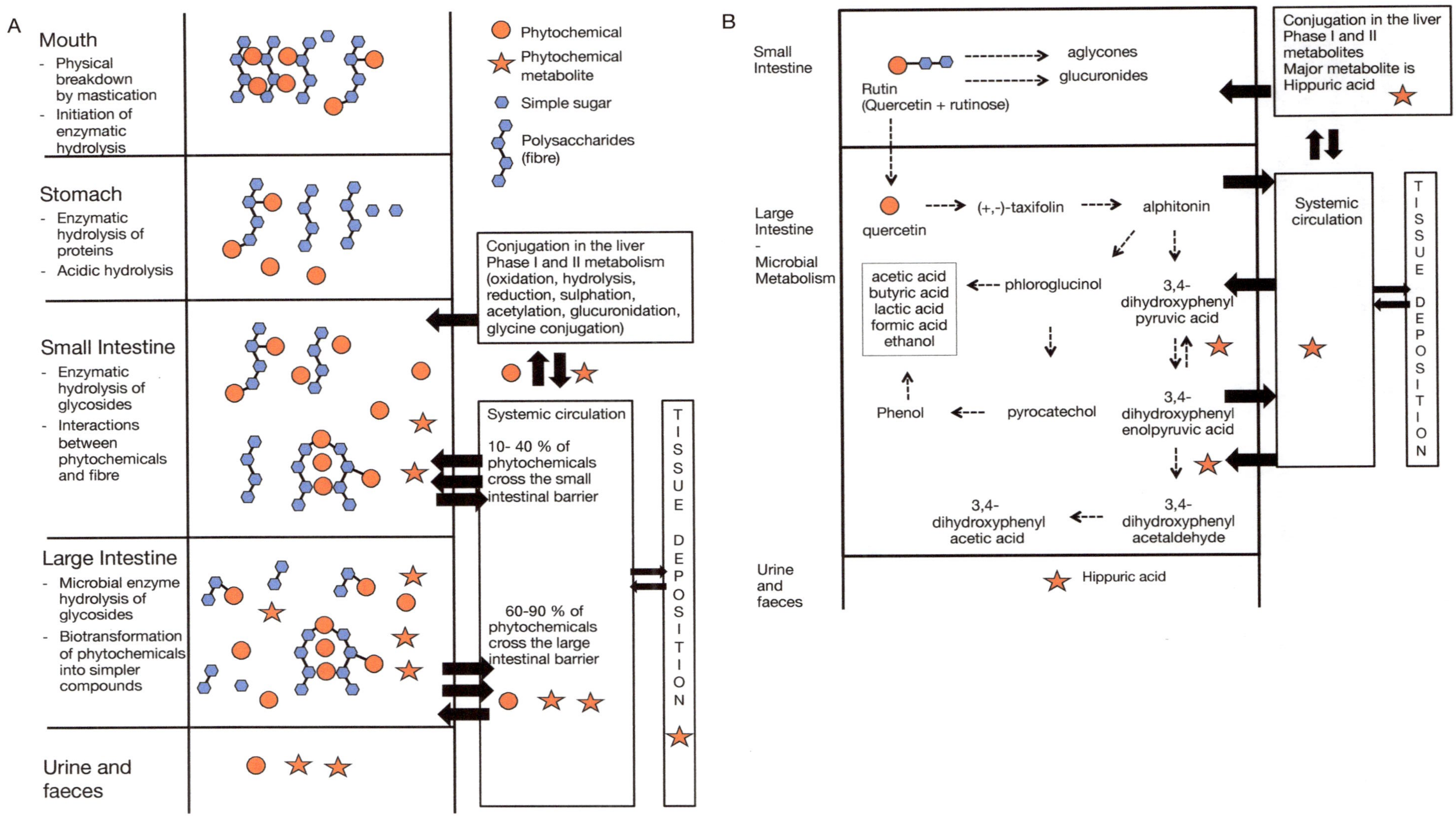

Figure 20.1 Schematic fate of a fibre-bound generic phytochemical throughout the gastrointestinal tract (A) with example of the gastrointestinal fate of rutin and breakdown products (B). The figures have been modified from (Palafox-Carlos *et al.*, 2011) (A) and (Moco *et al.*, 2012) (B). Reprinted (adapted) with permission. Copyright © 2010 Institute of Food Technologists® and © 2012 American Chemical Society.

metabolites in the various segments of the colon. Continuous feeding with polyphenol extracts appeared to modulate SCFA profiles, with an increase in the propionate–butyrate molar ratio (van Dorsten *et al.*, 2012). The magnitude of change in the gut microenvironment will depend on the intake of total dietary polyphenols, the nature of the polyphenols consumed, the variation in the gut microbiota in the individual (van Duynhoven *et al.*, 2011), and the matrix in which the bioactive polyphenols are consumed. Given that polyphenols appear to be capable of reaching the colon and modulating microecology and metabolism, polyphenols may potentially be included in functional foods or nutraceuticals as novel prebiotics.

Glucosinolates

Cruciferous vegetable-derived β-thioglucoside-N-hydroxysulfates (glucosinolates, e.g. glucoraphanin) are renowned for the bioactivity of their aglycone breakdown products, such as isothiocyanates (e.g. sulforaphane from glucoraphanin). The biological activities of isothiocyanates and other byproducts, such as host phase II antioxidant enzyme induction (Cohen *et al.*, 2000), and the epidemiological evidence supporting the consequent beneficial effects (Das *et al.*, 2000) can be read elsewhere, and are outside the scope of this chapter. Here our primary interest is the potential prebiotic activity conferred during the process of removal of the sulfur-linked sugar moiety from the intact glucosinolate by the gut microbiota.

Normally, hydrolysis of the glucose moiety occurs as a result of a sequestered plant thioglucosidase commonly known as a myrosinase, but if this myrosinase is inactivated through storage, processing or cooking (Bheemreddy and Jeffery, 2007; Zhu *et al.*, 2010), then hydrolysis by the gut microbiota occurs (Lai *et al.*, 2010; Mullaney *et al.*, 2013a,b, 2014). It is possible that this glucosinolate substrate could stimulate growth and/or activity of gastrointestinal bacteria associated with health and well-being. This has yet to be proven, but it is known that different members of the microbiota catabolize glucosinolates differently (Mullaney *et al.*, 2013b). Microbial exposure to glucosinolates may not necessarily result in prominent differences in microbial makeup or identification of distinct keystone species (Li *et al.*, 2010). Nevertheless, *ex vivo* supplementation of microbiota with glucosinolates, in the presence of added probiotic bacteria, does result in marked changes in SCFA profiles. This did not occur upon probiotic supplementation alone, indicative of a change in the activity of members of the microbiota due to the glucosinolates (Mullaney, 2014). To our knowledge there are no other published reports of animal or human clinical trials detailing prebiotic effects of glucosinolate consumption so the potential prebiotic (or synbiotic) potential of these thioglucoside-N-hydroxysulfates remains to be seen.

Next-generation prebiotics

As the definition of a prebiotic includes resisting digestion, this review would not be complete without consideration of nutrients or ingredients that do this by being part of a complex food matrix such as microencapsulation.

Other macronutrients

Lipid moieties may manipulate microbial ecology either through direct stimulation of probiotic bacteria or by inhibition of pathogens, which is an indirect prebiotic effect. For example, *trans* fatty acids selectively promote the growth of lactobacilli, with oleic acid being stimulatory and palmitoleic acid inhibitory (Endo *et al.*, 2006). Polyunsaturated fatty acids (PUFAs) such as α-linolenic, eicosapentaenoic and docosahexaenoic acids strongly inhibit *Lactobacillus* growth. In contrast, octadecenoic acids such as oleic and elaidic acid substantively promoted the growth of *Lactobacillus gasseri*, largely due to their incorporation into membrane lipids (Endo *et al.*, 2006). A trial diet rich in saturated fatty acids was found to decrease Bacteroidetes and Lachnospiraceae in mice guts, while linoleic acid suppressed Porphyromonadaceae of the Bacteroidetes phyla (Liu *et al.*, 2012).

Microbially produced fatty acids such as conjugated linoleic acids (CLA) have also been found to be inhibitory to the growth of laboratory cultures of pathogenic bacteria such as *Bacillus cereus*, *Pseudomonas aeruginosa*, *S. enterica* sv. Typhimurium, *Vibrio parahaemolyticus*, and *Proteus mirabilis*. The Gram-positive bacteria were more sensitive to cell disruption due to peroxidative breakdown of cell membranes (Byeon *et al.*, 2009). While PUFAs may in themselves not conform with the conventional definition of prebiotics, their combination with classical prebiotics such as inulin has been used to deliver novel prebiotic benefits. The bifidogenic effect of inulin combined with CLA, a well-known bifidobacterial metabolite, was associated with resetting parameters of metabolic dysfunction such as serum cholesterol in obese patients (Druart *et al.*, 2014). This combinatorial approach may hold promise in restoring metabolism and controlling nutrient uptake. A nutritional mixture of bovine milk oligosaccharides, *L. rhamnosus* NCC4007, arachidonic and docosahexaenoic acid was studied for its effects on growth of intrauterine growth-restricted (IUGR) rats. Short-term intervention with this nutritional mixture promoted catch-up growth and normalized excessive adiposity in IUGR rats (Castaneda-Gutierrez *et al.*, 2014).

Glycopeptide-based prebiotics

Glycopeptides have not been considered as prebiotics, although they can act upon bacteria and therefore have the potential to alter the composition of the microbiota. However, selective hydrolysis utilizing bacterial enzymes may help to generate prebiotic glycopeptides with tailored specificity to the producing organism. For example, an endo-α-*N*-acetylgalactosaminidase from *B. longum* was able to catalyse the production of core 1 disaccharide-containing glycopeptides (Ashida *et al.*, 2010). These may have specificity to bifidobacteria and cause their selective proliferation if delivered in high enough doses. It has still to be demonstrated that these molecules have prebiotic efficacy *in vitro* and *in vivo* and if they come through this test, can be produced in a commercially viable manner.

Microencapsulated prebiotics

Microencapsulation of prebiotic fibres with potential probiotic bacteria has been trialled to deliver synbiotic combinations (Ying *et al.*, 2007). Mucilage polysaccharides from kiwifruit were combined with *L. rhamnosus* and then further encapsulated in alginate formulations to ensure the survival of the bacteria in simulated gastrointestinal conditions (Ying *et al.*, 2007). Such microencapsulated formulations may improve the survival of bacteria and their delivery along with prebiotic substrates to the colon.

Future trends

Future prebiotic consumption/trends seem likely to incorporate/use more fresh/whole plant material containing or enriched for prebiotic components such as many of the polysaccharide- or phytochemical-based ingredients we have specified here. This is because humans have adapted to a plant-rich diet and these fruits, vegetables, and cereals are our richest source of dietary carbohydrates, as well as polyphenolic components that may serve as novel functional prebiotic substrates for our gut microbiota. Keeping this microbial population healthy by catering to different classes of beneficial bacteria, stimulating specific groups, inhibiting potential pathogens and/or enhancing the extent of cross feeding and functional diversity to achieve a healthy metabolome, may perhaps be optimized by consuming whole foods. It is a well established fact that consuming a balanced diet rich in fruits, vegetables and whole grain cereals provides a stable microbiome favourable to humans (Tuohy *et al.*, 2012).

While saccharides and phytochemical ingredients appear to show potential as novel emerging prebiotics, how they are best packaged and delivered is still being investigated. Ultimately it may well be the packaging of these bioactives in a whole plant food that will provide holistic benefits ideally suited to the host. Thus, adjustments to the volumes and proportions of fruits, vegetables and cereals consumed as part of an everyday diet may provide the simplest and most effective mechanism to deliver a composite mixture of microbial nutrients that sustain prebiotic processes through the gastrointestinal tract.

Conclusion

Here we have reviewed a variety of fruit-, vegetable-, cereal, algal- and milk-derived, fractionated or synthesized compounds, predominantly carbohydrates and phytochemicals, which we regard as possessing the potential to broadly fulfil the current definition of prebiotics. This highlights opportunities to examine new ingredients as functional foods in future trials for prebiotics. We believe that a larger selection of prebiotics, beyond the classically recognized FOS and GOS will allow selective manipulation of the gut microbiome to confer novel prebiotic benefits.

Acknowledgements

We thank John Monro, Guna Paturi and Cath Kingston for critical review of manuscript and Tony Corbett for his help with the graphics. We acknowledge funding from Foods for Health at Different Life Stages Programme (MBIE contract #C11X1312).

References

Abrams, S.A., Griffin, I.J., Hawthorne, K.M., and Ellis, K.J. (2007). Effect of prebiotic supplementation and calcium intake on body mass index. J. Pediatr. *151*, 293–298.

Alakomi, H.L., Puupponen-Pimia, R., Aura, A.M., Helander, I.M., Nohynek, L., Oksman-Caldentey, K.M., and Saarela, M. (2007). Weakening of *Salmonella* with selected microbial metabolites of berry-derived phenolic compounds and organic acids. J. Agric. Food Chem. *55*, 3905–3912.

Annison, G., Illman, R.J., and Topping, D.L. (2003). Acetylated, propionylated or butyrylated starches raise large bowel short-chain fatty acids preferentially when fed to rats. J. Nutr. *133*, 3523–3528.

Arjmandi, B.H., Ahn, J., Nathani, S., and Reeves, R.D. (1992). Dietary soluble fiber and cholesterol affect serum-cholesterol concentration, hepatic portal venous short-chain fatty-acid concentrations and fecal sterol excretion in rats. J. Nutr. *122*, 246–253.

Arora, T., Sharma, R., and Frost, G. (2011). Propionate. Anti-obesity and satiety enhancing factor? Appetite *56*, 511–515.

Ashida, H., Ozawa, H., Fujita, K., Suzuki, S., and Yamamoto, K. (2010). Syntheses of mucin-type O-glycopeptides and oligosaccharides using transglycosylation and reverse-hydrolysis activities of *Bifidobacterium* endo-alpha-N-acetylgalactosaminidase. Glycoconjugate J. *27*, 125–132.

Ashley, C., Johnston, W.H., Harris, C.L., Stolz, S.I., Wampler, J.L., and Berseth, C.L. (2012). Growth and tolerance of infants fed formula supplemented with polydextrose (PDX) and/or galactooligosaccharides (GOS): double-blind, randomized, controlled trial. Nutr. J. *11*, 11–38.

Avila, M., Hidalgo, M., Sanchez-Moreno, C., Pelaez, C., Requena, T., and de Pascual-Teresa, S. (2009). Bioconversion of anthocyanin glycosides by *Bifidobacteria* and *Lactobacillus*. Food Res. Int. *42*, 1453–1461.

Aziz, N.H., Farag, S.E., Mousa, L.A., and Abo-Zaid, M.A. (1998). Comparative antibacterial and antifungal effects of some phenolic compounds. Microbios *93*, 43–54.

Ballongue, J., Schumann, C., and Quignon, P. (1997). Effects of lactulose and lactitol on colonic microflora and enzymatic activity. Scand J. Gastroenterol. *32*, 41–44.

Bamba, T., Kanauchi, O., Andoh, A., and Fujiyama, Y. (2002). A new prebiotic from germinated barley for nutraceutical treatment of ulcerative colitis. J. Gastroenterol. Hepatol. *17*, 818–824.

Barile, D., and Rastall, R.A. (2013). Human milk and related oligosaccharides as prebiotics. Curr. Opin. Biotechnol. *24*, 214–219.

Bavington, C., and Page, C. (2005). Stopping bacterial adhesion: a novel approach to treating infections. Thematic Review Series *72*, 335–334.

Behare, P.V., Singh, R., Kumar, M., Prajapati, J.B., and Singh, R.P. (2009). Exopolysaccharides of lactic acid bacteria: A review. J. Food Sci. Tech. Mys. *46*, 1–11.

Belenguer, A., Duncan, S.H., Calder, A.G., Holtrop, G., Louis, P., Lobley, G.E., and Flint, H.J. (2006). Two routes of metabolic cross-feeding between *Bifidobacterium adolescentis* and butyrate-producing anaerobes from the human gut. Appl. Environ. Microbiol. *72*, 3593–3599.

Bheemreddy, R.M., and Jeffery, E.H. (2007). The metabolic fate of purified glucoraphanin in F344 rats. J. Agric. Food Chem. *55*, 2861–2866.

Bialonska, D., Ramnani, P., Kasimsetty, S.G., Muntha, K.R., Gibson, G.R., and Ferreira, D. (2010). The influence of pomegranate by-product and punicalagins on selected groups of human intestinal microbiota. Int. J. Food Microbiol. *140*, 175–182.

Bienenstock, J., Buck, R.H., Linke, H., Forsythe, P., Stanisz, A.M., and Kunze, W.A. (2013). Fucosylated but not sialylated milk oligosaccharides diminish colon motor contractions. PLoS One *8*, e76236.

Bocanegra, A., Bastida, S., Benedi, J., Rodenas, S., and Sanchez-Muniz, F.J. (2009). Characteristics and nutritional and cardiovascular-health properties of seaweeds. J. Med. Food *12*, 236–258.

Bokkenheuser, V.D., Shackleton, C.H.L., and Winter, J. (1987). Hydrolysis of dietary flavonoid glycosides by strains of intestinal bacteroides from humans. Biochem. J. *248*, 953–956.

Bordonaro, M., Venema, K., Putri, A.K., and Lazarova, D. (2014). Approaches that ascertain the role of dietary compounds in colonic cancer cells. World J. Gastrointest. Onco. *6*, 1–10.

Brahe, L.K., Astrup, A., and Larsen, L.H. (2013). Is butyrate the link between diet, intestinal microbiota and obesity-related metabolic diseases? Obesity Rev. *14*, 950–959.

Brouns, F., Kettlitz, B., and Arrigoni, E. (2002). Resistant starch and 'the butyrate revolution'. Trends Food Sci. Technol. *13*, 251–261.

Bush, M.S., Marry, M., Huxham, I.M., Jarvis, M.C., and McCann, M.C. (2001). Developmental regulation of pectic epitopes during potato tuberisation. Planta *213*, 869–880.

Byeon, J.I., Song, H.S., Oh, T.W., Kim, Y.S., Choi, B.D., Kim, H.C., Kim, J.O., Shim, K.H., and Ha, Y.L. (2009). Growth Inhibition of Foodborne and Pathogenic Bacteria by Conjugated Linoleic Acid. J. Agric. Food Chem. *57*, 3164–3172.

Campbell, J.M., and Fahey Jr, G.C. (1997). Psyllium and methylcellulose fermentation properties in relation to insoluble and soluble fibre standards. Nutr. Res. *17*, 619–629.

Candela, M., Maccaferri, S., Turroni, S., Carnevali, P., and Brigidi, P. (2010). Functional intestinal microbiome, new frontiers in prebiotic design. Int. J. Food Microbiol. *140*, 93–101.

Cani, P.D., Lecourt, E., Dewulf, E.M., Sohet, F.M., Pachikian, B.D., Naslain, D., De Backer, F., Neyrinck, A.M., and Delzenne, N.M. (2009). Gut microbiota fermentation of prebiotics increases satietogenic and incretin gut peptide production with consequences for appetite sensation and glucose response after a meal. Am. J. Clin. Nutr. *90*, 1236–1243.

Carnachan, S.M., Bootten, T.J., Mishra, S., Monro, J.A., and Sims, I.M. (2012). Effects of simulated digestion *in vitro* on cell wall polysaccharides from kiwifruit (Actinidia spp.). Food Chem. *133*, 132–139.

Castaneda-Gutierrez, E., Moser, M., Garcia-Rodenas, C., Raymond, F., Mansourian, R., Rubio-Aliaga, I., Viguet-Carrin, S., Metairon, S., Ammon-Zufferey, C., Avanti-Nigro, O., *et al.* (2014). Effect of a mixture of bovine milk oligosaccharides, *Lactobacillus rhamnosus* NCC4007 and long-chain polyunsaturated fatty acids on catch-up growth of intra-uterine growth-restricted rats. Acta Physiol. *210*, 161–173.

Cavanagh, H., Hipwell, M., and Wilkinson, J.M. (2003). Antibacterial activity of berry fruits used for culinary purposes. J. Med. Food *6*, 57–61.

Chaia, A.P., Desaad, A.M.S., Deruizholgado, A.P., and Oliver, G. (1994). Influence of propionate on growth and fermentative activity of lactobacilli. J. Appl. Bacteriol. *77*, 134–139.

Chassard, C., and Bernalier-Donadille, A. (2006). H2 and acetate transfers during xylan fermentation between a butyrate-producing xylanolytic species and hydrogenotrophic microorganisms from the human gut. FEMS Microbiol. Lett. *254*, 116–122.

Chassard, C., Delmas, E., Robert, C., and Bernalier-Donadille, A. (2010). The cellulose-degrading microbial community of the human gut varies according to the presence or absence of methanogens. FEMS Microbiol. Ecol. *74*, 205–213.

Chen, W.J.L., Anderson, J.W., and Jennings, D. (1984). Propionate may mediate the hypocholesterolemic effects of certain soluble plant fibres in cholesterol-fed rats. Proc. Soc. Exp. Biol. Med. *175*, 215–218.

Chung, H.S., Kim, Y.B., Chun, S.L., and Ji, G.E. (1999). Screening and selection of acid and bile resistant bifidobacteria. Int. J. Food Microbiol. *47*, 25–32.

Clarke, J.M., Topping, D.L., Christophersen, C.T., Bird, A.R., Lange, K., Saunders, I., and Cobiac, L. (2011). Butyrate esterified to starch is released in the human gastrointestinal tract. Am. J. Clin. Nutr. *94*, 1276–1283.

Cohen, J.H., Kristal, A.R., and Stanford, J.L. (2000). Fruit and vegetable intakes and prostate cancer risk. J. Natl Cancr Inst *92*, 61–68.

Comstock, S.S., Wang, M., Hester, S.N., Li, M., and Donovan, S.M. (2014). Select human milk oligosaccharides directly modulate peripheral blood mononuclear cells isolated from 10-d-old pigs. Br. J. Nutr. *111*, 819–828.

Coppa, G.V., Zampini, L., Galeazzi, T., Facinelli, B., Ferrante, L., Capretti, R., and Orazio, G. (2006). Human milk oligosaccharides inhibit the adhesion to Caco-2 cells of diarrheal pathogens: *Escherichia coli, Vibrio choleraee*, and *Salmonella fyris*. Pediatr. Res. *59*, 377–382.

Costabile, A., Fava, F., Roytio, H., Forssten, S.D., Olli, K., Klievink, J., Rowland, I.R., Ouwehand, A.C., Rastall, R.A., Gibson, G.R., *et al.* (2012). Impact of polydextrose on the faecal microbiota: a double-blind, crossover, placebo-controlled feeding study in healthy human subjects. Br. J. Nutr. *108*, 471–481.

Coulet, M., Phothirath, P., Allais, L., and Schilter, B. (2014). Pre-clinical safety evaluation of the synthetic human milk, nature-identical, oligosaccharide 2′-O-Fucosyllactose (2 FL). Regul. Toxicol. Pharmacol. *68*, 59–69.

Crittenden, R., Laitila, A., Forssell, P., Matto, J., Saarela, M., Mattila-Sandholm, T., and Myllarinen, P. (2001). Adhesion of bifidobacteria to granular starch and its implications in probiotic technologies. Appl. Environ. Microbiol. *67*, 3469–3475.

Cummings, J.H. (1993). The effect of dietary fibrer on fecal weight and composition. Dietary Fiber in Human Nutrition, 2nd edn (edited by GA Spiller), 263349.

Cummings, J.H., Branch, W., Jenkins, D.J., Southgate, D.A., Houston, H., and James, W.P. (1978). Colonic response to dietary fibre from carrot, cabbage, apple, bran. Lancet *1*, 5–9.

Dabek, M., McCrae, S.I., Stevens, V.J., Duncan, S.H., and Louis, P. (2008). Distribution of beta-glucosidase and beta-glucuronidase activity and of beta-glucuronidase gene gus in human colonic bacteria. FEMS Microbiol. Ecol. *66*, 487–495.

Dai, D., and Walker, W.A. (1999). Protective nutrients and bacterial colonization in the immature human gut. Adv. Pediatr. *46*, 353–382.

Das, S., Tyagi, A.K., and Kaur, H. (2000). Cancer modulation by glucosinolates: A review. Curr. Sci. *79*, 1665–1671.

De Vuyst, L., and Leroy, F. (2011). Cross-feeding between bifidobacteria and butyrate-producing colon bacteria explains bifdobacterial competitiveness, butyrate production, and gas production. Int. J. Food Microbiol. *149*, 73–80.

Delzenne, N.M., and Cani, P.D. (2011). Interaction between obesity and the gut microbiota: Relevance in nutrition. Annu. Rev. Nutr. *31*, 15–31.

D'ez-Municio, M., Herrero, M., Olano, A., and Moreno, F.J. (2014). Synthesis of novel bioactive lactose-derived oligosaccharides by microbial glycoside hydrolases. Microbial Biotech. *7*, 315–331.

Dongowski, G., Lorenz, A., and Anger, H. (2000). Degradation of pectins with different degrees of esterification by *Bacteroides thetaiotaomicron* isolated from human gut flora. Appl. Environ. Microbiol. *66*, 1321–1327.

Dridi, B., Henry, M., El Khechine, A., Raoult, D., and Drancourt, M. (2009). High prevalence of *Methanobrevibacter smithii* and *Methanosphaera stadtmanae* detected in the human gut using an improved DNA detection protocol. PLoS One *4*, e7063.

Druart, C., Dewulf, E.M., Cani, P.D., Neyrinck, A.M., Thissen, J.-P., and Delzenne, N.M. (2014). Gut microbial metabolites of polyunsaturated fatty acids correlate with specific fecal bacteria and serum markers of metabolic syndrome in obese women. Lipids *49*, 397–402.

Duda-Chodak, A. (2012). The inhibitory effect of polyphenols on human gut microbiota. J. Physiol. Pharmacol. *63*, 497–503.

Duenas, M., Surco-Laos, F., Gonzalez-Manzano, S., Gonzalez-Paramas, A.M., and Santos-Buelga, C. (2011). Antioxidant properties of major metabolites of quercetin. Eur. Food Res. Technol. *232*, 103–111.

Duncan, S.H., Hold, G.L., Barcenilla, A., Stewart, C.S., and Flint, H.J. (2002). Roseburia intestinalis sp nov., a novel saccharolytic, butyrate-producing bacterium from human faeces. Int. J. Syst Evol. Microbiol. *52*, 1615–1620.

Duncan, S.H., Lobley, G.E., Holtrop, G., Ince, J., Johnstone, A.M., Louis, P., and Flint, H.J. (2008). Human colonic microbiota associated with diet, obesity and weight loss. Int. J. Obes *32*, 1720–1724.

Eastwood, M.A., Kirkpatrick, J.R., Mitchell, W.D., Bone, A., and Hamilton, T. (1973). Effects of dietary supplements of wheat bran and cellulose on faeces and bowel function. Br. Med. J. *4*, 392–394.

Eiwegger, T., Stahl, B., Haidl, P., Schmitt, J., Boehm, G., Dehlink, E., Urbanek, R., and Szepfalusi, Z. (2010). Prebiotic oligosaccharides: *In vitro* evidence for gastrointestinal epithelial transfer and immunomodulatory properties. Pediatr. Allergy Immunol. *21*, 1179–1188.

El Aidy, S., Van den Abbeele, P., Van de Wiele, T., Louis, P., and Kleerebezem, M. (2013). Intestinal colonization: How key microbial players become established in this dynamic process. Bioessays *35*, 913–923.

El Kaoutari, A., Armougom, F., Gordon, J.I., Raoult, D., and Henrissat, B. (2013). The abundance and variety of carbohydrate-active enzymes in the human gut microbiota. Nature Rev. Microbiol. *11*, 497–504.

Endo, Y., Kamisada, S., Fujimoto, K., and Saito, T. (2006). Trans fatty acids promote the growth of some *Lactobacillus* strains. J. Gen Appl. Microbiol. *52*, 29–35.

Englyst, H.N., Kingman, S.M., Hudson, G.J., and Cummings, J.H. (1996). Measurement of resistant starch *in vitro* and *in vivo*. Br. J. Nutr. *75*, 749–755.

Fanaro, S., Chierici, R., Guerrini, P., and Vigi, V. (2003). Intestinal microflora in early infancy: composition and development. Acta Paediatr *92*, 48–55.

Finney, M., Smullen, J., Foster, H.A., Brokx, S., and Storey, D.M. (2007). Effects of low doses of lactitol on faecal microflora, pH, short-chain fatty acids and gastrointestinal symptomology. Eur. J. Nutr. *46*, 307–314.

Flint, H.J., Bayer, E.A., Rincon, M.T., Lamed, R., and White, B.A. (2008). Polysaccharide utilization by gut bacteria: potential for new insights from genomic analysis. Nat. Rev. Microbiology *6*, 121–131.

Flint, H.J., Scott, K.P., Duncan, S.H., Louis, P., and Forano, E. (2012). Microbial degradation of complex carbohydrates in the gut. Gut Microbes 3, 289–306.

Fogliano, V., Corollaro, M.L., Vitaglione, P., Napolitano, A., Ferracane, R., Travaglia, F., Arlorio, M., Costabile, A., Klinder, A., and Gibson, G. (2011). *In vitro* bioaccessibility and gut biotransformation of polyphenols present in the water-insoluble cocoa fraction. Mol. Nutr. Food Res. *55*, S44–S55.

Freestone, P.P.E., Walton, N.J., Haigh, R.D., and Lyte, M. (2007). Influence of dietary catechols on the growth of enteropathogenic bacteria. Int. J. Food Microbiol. *119*, 159–169.

Fuentes-Zaragoza, E., Sanchez-Zapata, E., Sendra, E., Sayas, E., Navarro, C., Fernandez-Lopez, J., and Perez-Alvarez, J.A. (2011). Resistant starch as prebiotic: A review. Starch-Starke *63*, 406–415.

Fujimori, S., Tatsuguchi, A., Gudis, K., Kishida, T., Mitsui, K., Ehara, A., Kobayashi, T., Sekita, Y., Seo, T., and Sakamoto, C. (2007). High dose probiotic and prebiotic cotherapy for remission induction of active Crohn's disease. J. Gastroenterol. Hepatol. *22*, 1199–1204.

Gao, K., Xu, A., Krul, C., Venema, K., Liu, Y., Niu, Y., Lu, J., Bensoussan, L., Seeram, N.P., Heber, D., *et al.* (2006). Of the major phenolic acids formed during human microbial fermentation of tea, citrus, and soy flavonoid supplements, only 3,4-dihydroxyphenylacetic acid has antiproliferative activity. J. Nutr. *136*, 52–57.

GarciaDiez, F., GarciaMediavilla, V., Bayon, J.E., and Gonzalez-Gallego, J. (1996). Pectin feeding influences faecal bile acid excretion, hepatic bile acid and cholesterol synthesis and serum cholesterol in rats. J. Nutr. *126*, 1766–1771.

Gee, J.M., and Johnson, I.T. (2005). Dietary lactitol fermentation increases circulating peptide YY and glucagon-like peptide-1 in rats and humans. Nutrition *21*, 1036–1043.

Ghalia, S., Zaiton, H., and Maryam, A. (2013). Adhesion of probiotic bacteria to resistant rice starch. Am. J. Appl. Sci. *10*, 313–321.

Gibson, G.R., Macfarlane, S., and Cummings, J.H. (1990). The fermentability of polysaccharides by mixed human faecal bacteria in relation to their suitability as bulk-forming laxatives. Lett. Appl. Microbiol. *11*, 251–254.

Gibson, G.R., Macfarlane, S., and Macfarlane, G.T. (1993). Metabolic interactions involving sulfate-reducing and methanogenic bacteria in the human large intestine. FEMS Microbiol. Ecol. *12*, 117–125.

Gibson, G.R., Probert, H.M., Van Loo, J., Rastall, R.A., and Roberfroid, M.B. (2004). Dietary modulation of the human colonic microbiota: updating the concept of prebiotics. Nutr. Res. Rev. *17*, 259–275.

Gibson, G.R., and Roberfroid, M.B. (1995). Dietary modulation of the human colonic microbiota – introducing the concept of prebiotics. J. Nutr. *125*, 1401–1412.

Gibson, G.R., Scott, K.P., Rastall, R.A., Tuohy, K.M., Hotchkiss, A., Dubert-Ferrandon, A., Gareau, M., Murphy, E.F., Saulnier, D., Loh, G., *et al.* (2010). Dietary prebiotics: current status and new definition. Food Science and Technology Bulletin: Functional Foods 7, 1–19.

Giese, E.C., Barbosa, A.M., and Dekker, R.F.H. (2010). Pathways to bioactive oligosaccharides: Biological functions and potential applications. In Handbook of Carbohydrate Polymers: Development, Properties and Applications, R. Ito, and Y. Matsuo, eds. (Nova Science Publishers, Incorporated), pp. 279–309.

Gnoth, M.J., Kunz, C., Kinne-Saffran, E., and Rudloff, S. (2000). Human milk oligosaccharides are minimally digested *in vitro*. J. Nutr. *130*, 3014–3020.

Goni, I., Gudiel-Urbano, M., Bravo, L., and Saura-Calixto, F. (2001). Dietary modulation of bacterial fermentative capacity by edible seaweeds in rats. J. Agric. Food Chem. *49*, 2663–2668.

Haenen, D., Zhang, J., da Silva, C.S., Bosch, G., van der Meer, I.M., van Arkel, J., van den Borne, J.J., Gutiérrez, O.P., Smidt, H., and Kemp, B. (2013). A diet high in resistant starch modulates microbiota composition, SCFA concentrations, and gene expression in pig intestine. J. Nutr. *143*, 274–283.

Harmsen, H.J.M., Wildeboer-Veloo, A.C.M., Raangs, G.C., Wagendorp, A.A., Klijn, N., Bindels, J.G., and Welling, G.W. (2000). Analysis of intestinal flora development in breast-fed and formula-fed infants by using molecular identification and detection methods. J. Pediatr. Gastroenterol. Nutr. *30*, 61–67.

Hayashi, T., Marsden, M.P.F., and Delmer, D.P. (1987). Xyloglucan–cellulose interactions *in vitro* and *in vivo*. Plant Physiol. *83*, 384–389.

Herfel, T.M., Jacobi, S., Lin, X., Walker, D.C., Jouni, Z.E., and Odle, J. (2010). Polydextrose enrichment of infant formula acts as a prebiotic by increasing ileal lactobacilli and lactic acid concentration and decreasing pH. FASEB J. *24 (Meeting Abstract Supplement)*, 327.7.

Hervert-Hernandez, D., and Goni, I. (2011). Dietary polyphenols and human gut microbiota: A review. Food Rev. Int. 27, 154–169.

Holck, J., Lorentzen, A., Vigsnaes, L.K., Licht, T.R., Mikkelsen, J.D., and Meyer, A.S. (2011). Feruloylated and nonferuloylated arabino-oligosaccharides from sugar beet pectin selectively stimulate the growth of *Bifidobacterium* spp. in human faecal *in vitro* fermentations. J. Agric. Food Chem. *59*, 6511–6519.

Holko, I., and Hrabe, J. (2012). The adhesion of *Lactobacillus acidophilus* to dietary fibre. Int. J. Probiotics and Prebiotics 7, 165–168.

Holloway, L., Moynihan, S., Abrams, S.A., Kent, K., Hsu, A.R., and Friedlander, A.L. (2007). Effects of oligofructose-enriched inulin on intestinal absorption of calcium and magnesium and bone turnover markers in postmenopausal women. Br. J. Nutr. *97*, 365–372.

Holloway, W.D., and Greig, R.I. (1984). Water holding capacity of hemicelluloses from fruits, vegetables and wheat bran. J. Food Sci. *49*, 1632–1633.

Homayouni, A., Amini, A., Keshtiban, A.K., Mortazavian, A.M., Esazadeh, K., and Pourmoradian, S. (2014). Resistant starch in food industry: A changing outlook for consumer and producer. Starch-Starke *66*, 102–114.

Hosseini, E., Grootaert, C., Verstraete, W., and Van de Wiele, T. (2011). Propionate as a health-promoting microbial metabolite in the human gut. Nutr. Rev. *69*, 245–258.

Hughes, S.A., Shewry, P.R., Gibson, G.R., McCleary, B.V., and Rastall, R.A. (2008). *In vitro* fermentation of oat and barley derived beta-glucans by human faecal microbiota. FEMS Microbiol. Ecol. *64*, 482–493.

Janczyk, P., Pieper, R., Wolf, C., Freyer, G., and Souffrant, W.B. (2010). Alginate fed as a supplement to rats affects growth performance, crude protein digestibility, and caecal bacterial community. Archiva Zootechnica *13*, 5–18.

Jenkins, D.J., Newton, C., Leeds, A.R., and Cummings, J.H. (1975). Effect of pectin, guar gum, and wheat fibre on serum-cholesterol. Lancet *1*, 1116–1117.

Jenner, A.M., Rafter, J., and Halliwell, B. (2005). Human fecal water content of phenolics: The extent of colonic exposure to aromatic compounds. Free Radical Biol. Med. *38*, 763–772.

Jonathan, M.C., Haenen, D., Souza da Silva, C., Bosch, G., Schols, H.A., and Gruppen, H. (2013). Influence of a diet rich in resistant starch on the degradation of non-starch polysaccharides in the large intestine of pigs. Carbohydr. Polym. *93*, 232–239.

Jost, T., Lacroix, C., Braegger, C., and Chassard, C. (2013). Assessment of bacterial diversity in breast milk using culture-dependent and culture-independent approaches. Br. J. Nutr. *110*, 1253–1262.

Jovanovic-Malinovska, R., Kuzmanova, S., and Winkelhausen, E. (2013). Oligosaccharide profile in fruits and vegetables as sources of prebiotics and functional foods. Int. J. Food Prop *17*, 949–965.

Kahlon, T.S., and Chow, F.I. (2000). *In vitro* binding of bile acids by rice bran, oat bran, wheat bran, and corn bran. Cereal Chem. *77*, 518–521.

Kaneko, T., Mori, H., Iwata, M., and Meguro, S. (1994). Growth stimulator for bifidobacteria produced by *Propionibacterium freudenreichii* and several intestinal bacteria. J. Dairy Sci. *77*, 393–404.

Kay, R.M., and Truswell, A.S. (1977). Effect of citrus pectin on blood lipids and fecal steroid excretion in man. Am. J. Clin. Nutr. *30*, 171–175.

Keenan, M.J., Janes, M., Robert, J., Martin, R.J., Raggio, A.M., McCutcheon, K.L., Pelkman, C., Tulley, R., Goita, M.F., Durham, H.A., *et al.* (2013). Resistant starch from high amylose maize (HAM-RS2). reduces body fat and increases gut bacteria in ovariectomized (OVX) rats. Obesity *21*, 981–984.

Keenan, M.J., Zhou, J., McCutcheon, K.L., Raggio, A.M., Bateman, H.G., Todd, E., Jones, C.K., Tulley, R.T., Melton, S., Martin, R.J., *et al.* (2006).

Effects of resistant starch, a non-digestible fermentable fibre, on reducing body fat. Obesity *14*, 1523–1534.

Kerckhoffs, A.P.M., Ben-Amor, K., Samsom, M., van der Rest, M.E., de Vogel, J., Knol, J., and Akkermans, L.M.A. (2011). Molecular analysis of faecal and duodenal samples reveals significantly higher prevalence and numbers of *Pseudomonas aeruginosa* in irritable bowel syndrome. J. Med. Microbiol. *60*, 236–245.

Khouvine, Y. (1923). Digestion de la cellulose par la flore intestinale d l'homme B. cellutlosae dissolvens, n. sp.[in French]. Ann. Inst. Pasteur (Paris) *37*, 711–752.

Kim, D.H., Jung, E.A., Sohng, I.S., Han, J.A., Kim, T.H., and Han, M.J. (1998). Intestinal bacterial metabolism of flavonoids and its relation to some biological activities. Arch. Pharmacol. Res. *21*, 17–23.

Kim, S.-J., Kim, M.-C., Um, J.-Y., and Hong, S.-H. (2010). The beneficial effect of vanillic acid on ulcerative colitis. Molecules *15*, 7208–7217.

Klosterbuer, A.S., Hullar, M., Li, F., Traylor, E., Lampe, J.W., Thomas, W., and Slavin, J.L. (2013). Gastrointestinal effects of resistant starch, soluble maize fibre and pullulan in healthy adults. Br. J. Nutr. *110*, 1068–1074.

Konar, N., Ozhan, B., Artik, N., and Poyrazoglu, E.S. (2014). Using polydextrose as a prebiotic substance in milk chocolate: effects of process parameters on physical and rheological properties. CyTA – Journal of Food *12*, 150–159.

Koropatkin, N.M., Cameron, E.A., and Martens, E.C. (2012). How glycan metabolism shapes the human gut microbiota. Nature Rev. Microbiol. *10*, 323–335.

Kovatcheva-Datchary, P., Egert, M., Maathuis, A., Rajilic-Stojanovic, M., de Graaf, A.A., Smidt, H., de Vos, W.M., and Venema, K. (2009). Linking phylogenetic identities of bacteria to starch fermentation in an *in vitro* model of the large intestine by RNA-based stable isotope probing. Environ. Microbiol. *11*, 914–926.

Krewinkel, M., Gosch, M., Rentschler, E., and Fischer, L. (2014). Epilactose production by 2 cellobiose 2-epimerases in natural milk. J. Dairy Sci. *97*, 155–161.

Kunz, C., Rudloff, S., Baier, W., Klein, N., and Strobel, S. (2000). Oligosaccharides in human milk: Structural, functional, and metabolic aspects. Annu. Rev. Nutr. *20*, 699–722.

Lacombe, A., Wu, V.C.H., Tyler, S., and Edwards, K. (2010). Antimicrobial action of the American cranberry constituents; phenolics, anthocyanins, and organic acids, against *Escherichia coli* O157:H7. Int. J. Food Microbiol. *139*, 102–107.

Lai, R.-H., Miller, M.J., and Jeffery, E. (2010). Glucoraphanin hydrolysis by microbiota in the rat caecum results in sulforaphane absorption. Food & Function *1*, 161–166.

Lane, J.A., O'Callaghan, J., Carrington, S.D., and Hickey, R.M. (2013). Transcriptional response of HT-29 intestinal epithelial cells to human and bovine milk oligosaccharides. Br. J. Nutr. *110*, 2127–2137.

Le Leu, R.K., Brown, I.L., Hu, Y., Bird, A.R., Jackson, M., Esterman, A., and Young, G.P. (2005). A synbiotic combination of resistant starch and *Bifidobacterium lactis* facilitates apoptotic deletion of carcinogen-damaged cells in rat colon. J. Nutr. *135*, 996–1001.

Lee, H.C., Jenner, A.M., Low, C.S., and Lee, Y.K. (2006). Effect of tea phenolics and their aromatic faecal bacterial metabolites on intestinal microbiota. Res. Microbiol. *157*, 876–884.

Lee, Y.-K., Puong, K.-Y., Ouwehand, A.C., and Salminen, S. (2003). Displacement of bacterial pathogens from mucus and Caco-2 cell surface by lactobacilli. J. Med. Microbiol. *52*, 925–930.

Leitch, E.C.M., Walker, A.W., Duncan, S.H., Holtrop, G., and Flint, H.J. (2007). Selective colonization of insoluble substrates by human faecal bacteria. Environ. Microbiol. *9*, 667–679.

Ley, R.E., Hamady, M., Lozupone, C., Turnbaugh, P.J., Ramey, R.R., Bircher, J.S., Schlegel, M.L., Tucker, T.A., Schrenzel, M.D., Knight, R., *et al.* (2008). Evolution of mammals and their gut microbes. Science *320*, 1647–1651.

Ley, R.E., Turnbaugh, P.J., Klein, S., and Gordon, J.I. (2006). Human gut microbes associated with obesity. Nature *444*, 1022–1023.

Li, F., Hullar, M.A.J., Beresford, S.A.A., and Lampe, J.W. (2010). Exploring the *in vivo* and *in vitro* metabolism of glucosinolates by human gut microbiota. FASEB J. *24* (*Meeting Abstract Supplement*), lb371.

Licht, T.R., Hansen, M., Bergstrom, A., Poulsen, M., Krath, B.N., Markowski, J., Dragsted, L.O., and Wilcks, A. (2010). Effects of apples and specific apple components on the caecal environment of conventional rats: role of apple pectin. BMC Microbiol. *10*, 13.

Lim, B.O., Yamada, K., Nonaka, M., Kuramoto, Y., Hung, P., and Sugano, M. (1997). Dietary fibres modulate indices of intestinal immune function in rats. J. Nutr. *127*, 663–667.

Liu, T., Hougen, H., Vollmer, A.C., and Hiebert, S.M. (2012). Gut bacteria profiles of Mus musculus at the phylum and family levels are influenced by saturation of dietary fatty acids. Anaerobe *18*, 331–337.

Louis, P., Scott, K.P., Duncan, S.H., and Flint, H.J. (2007). Understanding the effects of diet on bacterial metabolism in the large intestine. J. Appl. Microbiol. *102*, 1197–1208.

Macfarlane, S., Quigley, M.E., Hopkins, M.J., Newton, D.F., and Macfarlane, G.T. (1998). Polysaccharide degradation by human intestinal bacteria during growth under multi-substrate limiting conditions in a three-stage continuous culture system. FEMS Microbiol. Ecol. *26*, 231–243.

Macy, J.M., Ljungdahl, L.G., and Gottschalk, G. (1978). Pathway of succinate and propionate formation in *Bacteroides fragilis*. J. Bacterol. *134*, 84–91.

Magrone, T., and Jirillo, E. (2013). The interplay between the gut immune system and microbiota in health and disease: Nutraceutical intervention for restoring intestinal homeostasis. Curr. Pharm. Des *19*, 1329–1342.

Makelainen, H.S., Makivuokko, H.A., Salminen, S.J., Rautonen, N.E., and Ouwehand, A.C. (2007). The effects of polydextrose and xylitol on microbial community and activity in a 4-stage colon simulator. J. Food Sci. *72*, M153-M159.

Manach, C., Scalbert, Augustin, Morand, Christine, Remesy, Christian, Jimenez, Liliana (2004). Polyphenols: food sources and bioavailability. Am. J. Clin. Nutr. *79*, 727–747.

Manach, C., Williamson, G., Morand, C., Scalbert, A., and Remesy, C. (2005). Bioavailability and bioefficacy of polyphenols in humans. I. Review of 97 bioavailability studies. Am. J. Clin. Nutr. *81*, 230S--242S.

Mandalari, G., Nueno Palop C., Tuohy, K., Gibson, G.R., Bennett, R.N., Waldron, K.W., Bisignano, G., Narbad, A., and Faulds, C.B. (2006). *In vitro* evaluation of the prebiotic activity of a pectic oligosaccharide-rich extract enzymatically derived from bergamot peel. Appl. Microbiol. Biotechnol. 73, 173–179.

Manderson, K., Pinart, M., Tuohy, K.M., Grace, W.E., Hotchkiss, A.T., Widmer, W., Yadhav, M.P., Gibson, G.R., and Rastall, R.A. (2005). *In vitro* determination of prebiotic properties of oligosaccharides derived from an orange juice manufacturing by-product stream. Appl. Environ. Microbiol. *71*, 8383–8389.

Manthey, C.F., Autran, C.A., Eckmann, L., and Bode, L. (2014). Human milk oligosaccharides protect against enteropathogenic *Escherichia coli* attachment *in vitro* and EPEC colonization in suckling mice. J. Pediatr. Gastroenterol. Nutr. *58*, 167–170.

Marcobal, A., Barboza, M., Froehlich, J.W., Block, D.E., German, J.B., Lebrilla, C.B., and Mills, D.A. (2010). Consumption of human milk oligosaccharides by gut-related microbes. J. Agric. Food Chem. *58*, 5334–5340.

Marcobal, A., Barboza, M., Sonnenburg, E.D., Pudlo, N., Martens, E.C., Desai, P., Lebrilla, C.B., Weimer, B.C., Mills, D.A., German, J.B., *et al.* (2011). *Bacteroides* in the infant gut consume milk oligosaccharides via mucus-utilization pathways. Cell Host & Microbe *10*, 507–514.

Martens, E.C., Lowe, E.C., Chiang, H., Pudlo, N.A., Wu, M., McNulty, N.P., Abbott, D.W., Henrissat, B., Gilbert, H.J., Bolam, D.N., *et al.* (2011). Recognition and degradation of plant cell wall polysaccharides by two human gut symbionts. PLoS Biol. *9*, e1001221.

Martinez, I., Kim, J., Duffy, P.R., Schlegel, V.L., and Walter, J. (2010). Resistant starches types 2 and 4 have differential effects on the composition of the fecal microbiota in human subjects. PLoS One *5*, e15046.

McOrist, A.L., Miller, R.B., Bird, A.R., Keogh, J.B., Noakes, M., Topping, D.L., and Conlon, M.A. (2011). Faecal butyrate levels vary widely among individuals but are usually increased by a diet high in resistant starch. J. Nutr. *141*, 883–889.

Medani, M., Collins, D., Docherty, N.G., Baird, A.W., O'Connell, P.R., and Winter, D.C. (2011). Emerging Role of Hydrogen Sulfide in Colonic Physiology and Pathophysiology. Inflamm. Bowel Dis. *17*, 1620–1625.

Medie, F.M., Davies, G.J., Drancourt, M., and Henrissat, B. (2012). Genome analyses highlight the different biological roles of cellulases. Nature Rev. Microbiol. *10*, 227-U.

Michel, C., Lahaye, M., Bonnet, C., Mabeau, S., and Barry, J.L. (1996). *In vitro* fermentation by human faecal bacteria of total and purified dietary fibres from brown seaweeds. Br. J. Nutr. *75*, 263–280.

Mikkelsen, D., Gidley, M.J., and Williams, B.A. (2011). *In vitro* fermentation of bacterial cellulose composites as model dietary fibres. J. Agric. Food Chem. *59*, 4025–4032.

Miller, J.B., Bull, S., Miller, J., and McVeagh, P. (1994). The oligosaccharide composition of human-milk – temporal and individual variations in monosaccharide components. J. Pediatr. Gastroenterol. Nutr. *19*, 371–376.

Moco, S., Martin, F.-P.J., and Rezzi, S. (2012). Metabolomics view on gut microbiome modulation by polyphenol-rich foods. J. Proteome Res. *11*, 4781–4790.

Monro, J.A., Paturi, G., Butts, C.A., Young, W., De Guzman, C.E., McLachlan, A., Roy, N.C., and Ansell, J. (2012). Prebiotic effects of fermentable carbohydrate polymers may be modulated by faecal bulking of non-fermentable polysaccharides in the large bowel of rats. Int. J. Food Sci. Technol. *47*, 968–976.

Monro, J.A., Wallace, A., Mishra, S., Eady, S., Willis, J.A., Scott, R.S., and Hedderley, D. (2010). Relative glycaemic impact of customarily consumed portions of eighty-three foods measured by digesting *in vitro* and adjusting for food mass and apparent glucose disposal. Br. J. Nutr. *104*, 407–417.

Montgomery, L. (1988). Isolation of human colonic fibrolytic bacteria. Lett. Appl. Microbiol. *6*, 55–57.

Morishita, Y., and Konishi, Y. (1994). Effects of high dietary cellulose on the large intestinal microflora and short-chain fatty-acids in rats. Lett. Appl. Microbiol. *19*, 433–435.

Morrison, D.J., Mackay, W.G., Edwards, C.A., Preston, T., Dodson, B., and Weaver, L.T. (2006). Butyrate production from oligofructose fermentation by the human faecal flora: what is the contribution of extracellular acetate and lactate? Br. J. Nutr. *96*, 570–577.

Mu, W., Chen, Q., Wang, X., Zhang, T., and Jiang, B. (2013a). Current studies on physiological functions and biological production of lactosucrose. Appl. Microbiol. Biotechnol. *97*, 7073–7080.

Mu, W., Li, Q., Fan, C., Zhou, C., and Jiang, B. (2013b). Recent advances on physiological functions and biotechnological production of epilactose. Appl. Microbiol. Biotechnol. *97*, 1821–1827.

Mullaney, J.A., Ansell, J., Kelly, W.J., and Heyes, J.A. (2013a). Biotransformation of glucosinolates from a bacterial perspective. CAB Reviews *8*, 1–15.

Mullaney, J.A., Kelly, W.J., McGhie, T.K., Ansell, J., and Heyes, J.A. (2013b). Lactic acid bacteria convert glucosinolates to nitriles efficiently yet differently from Enterobacteriaceae. J. Agric. Food Chem. *61*, 3039–3046.

Mullaney, J.A., Kelly, W.J., McGhie, T.K., Heyes, J.A., and Ansell, J. (2015). Dietary glucosinolates and beneficial bacteria independently elevate quinine reductase in the bladder of Sprague–Dawley rats with host oxidation thus determining the end products of glucosinolate metabolism. in press.

Nava, G.M., Carbonero, F., Ou, J., Benefiel, A.C., O'Keefe, S.J., and Gaskins, H.R. (2012). Hydrogenotrophic microbiota distinguish native Africans from African and European Americans. Environ. Microbiol. Rep *4*, 307–315.

Nichenametla, S.N., Weidauer, L.A., Wey, H.E., Beare, T.M., Specker, B.L., and Dey, M. (2014). Resistant starch type 4-enriched diet lowered blood cholesterols and improved body composition in a double blind controlled cross-over intervention. Mol. Nutr. Food Res. *58*, 1365–1369.

Ninonuevo, M.R., Park, Y., Yin, H., Zhang, J., Ward, R.E., Clowers, B.H., German, J.B., Freeman, S.L., Killeen, K., Grimm, R., *et al.* (2006). A strategy for annotating the human milk glycome. J. Agric. Food Chem. *54*, 7471–7480.

Nyman, M., and Asp, N.G. (1982). Fermentation of dietary fibre components in the rat intestinal tract. Br. J. Nutr. *47*, 357–366.

O'Neill, M.A., Ishii, T., Albersheim, P., and Darvill, A.G. (2004). Rhamnogalacturonan II: structure and function of a borate cross-linked cell wall pectic polysaccharide. Annu. Rev. Plant Biol. *55*, 109–139.

Olano-Martin, E., Gibson, G.R., and Rastall, R.A. (2002). Comparison of the *in vitro* bifidogenic properties of pectins and pectic-oligosaccharides. J. Appl. Microbiol. *93*, 505–511.

Olano-Martin, E., Rimbach, G.H., Gibson, G.R., and Rastall, R.A. (2003a). Pectin and pectic-oligosaccharides induce apoptosis in *in vitro* human colonic adenocarcinoma cells. Anticancer Res. *23*, 341–346.

Olano-Martin, E., Williams, M.R., Gibson, G.R., and Rastall, R.A. (2003b). Pectins and pectic-oligosaccharides inhibit *Escherichia coli* O157: H7 Shiga toxin as directed towards the human colonic cell line HT29. FEMS Microbiol. Lett. *218*, 101–105.

Olthof, M.R., Hollman, P.C.H., Buijsman, M.N.C.P., van Amelsvoort, J.M.M., and Katan, M.B. (2003). Chlorogenic acid, quercetin-3-rutinoside and black tea phenols are extensively metabolized in humans. J. Nutr. *133*, 1806–1814.

Onumpai, C., Kolida, S., Bonnin, E., and Rastall, R.A. (2011). Microbial utilization and selectivity of pectin fractions with various structures. Appl. Environ. Microbiol. *77*, 5747–5754.

Palafox-Carlos, H., Ayala-Zavala, J.F., and Gonzalez-Aguilar, G.A. (2011). The role of dietary fibre in the bioaccessibility and bioavailability of fruit and vegetable antioxidants. J. Food Sci. *76*, R6-R15.

Palmer, G.H., and Dixon, D.G. (1966). Effect of pectin dose on serum cholesterol levels. Am. J. Clin. Nutr. *18*, 437–442.

Parkar, S., Rosendale, D., Paturi, G., Herath, T., Stoklosinski, H., Phipps, J., Hedderley, D., and Ansell, J. (2012). *In vitro* utilization of gold and green kiwifruit oligosaccharides by human gut microbial populations. Plant Foods Hum Nutr. *67*, 200–207.

Parkar, S.G., Redgate, E.L., McGhie, T.K., and Hurst, R.D. (2014). *In vitro* studies of modulation of pathogenic and probiotic bacterial proliferation and adhesion to intestinal cells by blackcurrant juices. Journal of Functional Foods *8*, 35–44.

Parkar, S.G., Redgate, E.L., Wibisono, R., Luo, X., Koh, E.T.H., and Schršder, R. (2010). Gut health benefits of kiwifruit pectins: Comparison with commercial functional polysaccharides. Journal of Functional Foods 2, 210–218.

Parkar, S.G., Stevenson, D.E., and Skinner, M.A. (2008). The potential influence of fruit polyphenols on colonic microflora and human gut health. Int. J. Food Microbiol. *124*, 295–298.

Parkar, S.G., Trower, T.M., and Stevenson, D.E. (2013). Fecal microbial metabolism of polyphenols and its effects on human gut microbiota. Anaerobe 23, 12–19.

Parnell, J.A., and Reimer, R.A. (2009). Weight loss during oligofructose supplementation is associated with decreased ghrelin and increased peptide YY in overweight and obese adults. Am. J. Clin. Nutr. *89*, 1751–1759.

Partty, A., Luoto, R., Kalliomaki, M., Salminen, S., and Isolauri, E. (2013). Effects of early prebiotic and probiotic supplementation on development of gut microbiota and fussing and crying in preterm infants: A randomized, double-blind, placebo-controlled trial. J. Pediatr. *163*, 1272–1277.

Paturi, G., Butts, C., Monro, J., Nones, K., Martell, S., Butler, R., and Sutherland, J. (2010). Cecal and colonic responses in rats fed 5 or 30% corn oil diets containing either 7.5% broccoli dietary fiber or microcrystalline cellulose. J. Agric. Food Chem. *58*, 6510–6515.

Perez-Jimenez, J., Neveu, V., Vos, F., and Scalbert, A. (2010). Systematic analysis of the content of 502 polyphenols in 452 foods and beverages: An application of the phenol-explorer database. J. Agric. Food Chem. *58*, 4959–4969.

Proestos, C., Boziaris, I.S., Nychas, G.J.E., and Komaitis, M. (2006). Analysis of flavonoids and phenolic acids in Greek aromatic plants: Investigation of their antioxidant capacity and antimicrobial activity. Food Chem. *95*, 664–671.

Putaala, H., Makivuokko, H., Tiihonen, K., and Rautonen, N. (2011). Simulated colon fibre metabolome regulates genes involved in cell cycle, apoptosis, and energy metabolism in human colon cancer cells. Mol. Cell Biochem. *357*, 235–245.

Puupponen-Pimia, R., Nohynek, L., Hartmann-Schmidlin, S., Kahkonen, M., Heinonen, M., Maatta-Riihinen, K., and Oksman-Caldentey, K.M. (2005). Berry phenolics selectively inhibit the growth of intestinal pathogens. J. Appl. Microbiol. *98*, 991–1000.

Queipo-Ortuno, M.I., Boto-Ordonez, M., Murri, M., Gomez-Zumaquero, J.M., Clemente-Postigo, M., Estruch, R., Cardona Diaz, F., Andres-Lacueva, C., and Tinahones, F.J. (2012). Influence of red wine polyphenols and ethanol on the gut microbiota ecology and biochemical biomarkers. Am. J. Clin. Nutr. *95*, 1323–1334.

Rabiu, B.A., Jay, A.J., Gibson, G.R., and Rastall, R.A. (2001). Synthesis and fermentation properties of novel galacto-oligosaccharides by beta-galactosidases from Bifidobacterium species. Appl. Environ. Microbiol. *67*, 2526–2530.

Rastmanesh, R. (2011). High polyphenol, low probiotic diet for weight loss because of intestinal microbiota interaction. Chem-Biol. Interact *189*, 1–8.

Reimer, R.A., Maurer, A.D., Eller, L.K., Hallam, M.C., Shaykhutdinov, R., Vogel, H.J., and Weljie, A.M. (2012). Satiety hormone and metabolomic response to an intermittent high energy diet differs in rats consuming long-term diets high in protein or prebiotic fiber. J. Proteome Res. *11*, 4065–4074.

Ridley, B.L., O'Neill, M.A., and Mohnen, D.A. (2001). Pectins: structure, biosynthesis, and oligogalacturonide-related signaling. Phytochemistry *57*, 929–967.

Rijkers, G.T., de Vos, W.M., Brummer, R.-J., Morelli, L., Corthier, G., and Marteau, P. (2011). Health benefits and health claims of probiotics: bridging science and marketing. Br. J. Nutr. *106*, 1291–1296.

Roberfroid, M. (2007). Prebiotics: The concept revisited. J. Nutr. *137*, 830S–837S.

Robert, C., and Bernalier-Donadille, A. (2003). The cellulolytic microflora of the human colon: evidence of microcrystalline cellulose-degrading bacteria in methane-excreting subjects. FEMS Microbiol. Ecol. *46*, 81–89.

Rockova, S., Rada, V., Marsik, P., Vlkova, E., Bunesova, V., Sklenar, J., and Splichal, I. (2011). Growth of bifidobacteria and clostridia on human and cow milk saccharides. Anaerobe *17*, 223–225.

Roncaglia, L., Amaretti, A., Raimondi, S., Leonardi, A., and Rossi, M. (2011). Role of bifidobacteria in the activation of the lignan secoisolariciresinol diglucoside. Appl. Microbiol. Biotechnol. 92, 159–168.

Rosendale, D.I., Blatchford, P.A., Sims, I.M., Parkar, S.G., Carnachan, S.M., Hedderley, D., and Ansell, J. (2012). Characterizing kiwifruit carbohydrate utilization *in vitro* and its consequences for human faecal microbiota. J. Proteome Res. *11*, 5863–5875.

Sahakian, A., Jee, S.-R., and Pimentel, M. (2010). Methane and the gastrointestinal tract. Dig. Dis. Sci. *55*, 2135–2143.

Sakata, T., Kojima, T., Fujieda, M., Takahashi, M., and Michibata, T. (2003). Influences of probiotic bacteria on organic acid production by pig caecal bacteria *in vitro*. Proc. Nur Soc *62*, 73–80.

Salazar, N., Ruas-Madiedo, P., Kolida, S., Collins, M., Rastall, R., Gibson, G., and de Los Reyes-Gavilan, C.G. (2009). Exopolysaccharides produced by *Bifidobacterium longum* IPLA E44 and *Bifidobacterium animalis* subsp. *lactis* IPLA R1 modify the composition and metabolic activity of human faecal microbiota in pH-controlled batch cultures. Int. J. Food Microbiol. *135*, 260–267.

Sanaka, M., Yamamoto, T., Anjiki, H., Nagasawa, K., and Kuyama, Y. (2007). Effects of agar and pectin on gastric emptying and post-prandial glycaemic profiles in healthy human volunteers. Clin. Exp. Pharmacol. Physiol. 34, 1151–1155.

Scheller, H.V., and Ulvskov, P. (2010). Hemicelluloses. Annual Review of Plant Biology, Vol 61 *61*, 263–289.

Schley, P.D., and Field, C.J. (2002). The immune-enhancing effects of dietary fibres and prebiotics. Br. J. Nutr. *87*, S221-S230.

Schneeman, B.O., and Gallaher, D. (1980). Changes in small intestinal digestive enzyme activity and bile acids with dietary cellulose in rats. J. Nutr. *110*, 584–590.

Scott, K.P., Martin, J.C., Duncan, S.H., and Flint, H.J. (2014). Prebiotic stimulation of human colonic butyrate-producing bacteria and bifidobacteria, *in vitro*. FEMS Microbiol. Ecol. *87*, 30–40.

Sefcikova, Z., Bujnakova, D., Racek, L., Kmet, V., and Mozes, S. (2011). Developmental changes in gut microbiota and enzyme activity predict obesity risk in rats arising from reduced nests. Physiol. Res. *60*, 337–346.

Sela, D.A. (2011). Bifidobacterial utilization of human milk oligosaccharides. Int. J. Food Microbiol. *149*, 58–64.

Sela, D.A., and Mills, D.A. (2010). Nursing our microbiota: molecular linkages between bifidobacteria and milk oligosaccharides. Trends Microbiol. *18*, 298–307.

Simoneit, B.R.T., Schauer, J.J., Nolte, C.G., Oros, D.R., Elias, V.O., Fraser, M.P., Rogge, W.F., and Cass, G.R. (1999). Levoglucosan, a tracer for cellulose in biomass burning and atmospheric particles. Atmos. Environ. 33, 173–182.

Singh, J., Dartois, A., and Kaur, L. (2010). Starch digestibility in food matrix: a review. Trends Food Sci. Technol. *21*, 168–180.

Sokol, H., Pigneur, B., Watterlot, L., Lakhdari, O., Bermudez-Humaran, L.G., Gratadoux, J.J., Blugeon, S., Bridonneau, C., Furet, J.P., Corthier, G., *et al.* (2008). *Faecalibacterium prausnitzii* is an anti-inflammatory commensal bacterium identified by gut microbiota analysis of Crohn disease patients. Proc. Natl. Acad. Sci. U.S.A. *105*, 16731–16736.

Somerville, C., Bauer, S., Brininstool, G., Facette, M., Hamann, T., Milne, J., Osborne, E., Paredez, A., Persson, S., Raab, T., *et al.* (2004). Towards a systems approach to understanding plant-cell walls. Science *306*, 2206–2211.

Spiller, G.A., Chernoff, M.C., Hill, R.A., Gates, J.E., Nassar, J.J., and Shipley, E.A. (1980). Effect of purified cellulose, pectin, and a low-residue diet on fecal volatile fatty acids, transit time, and fecal weight in humans. Am. J. Clin. Nutr. *33*, 754–759.

Sriamornsak, P. (2003). Chemistry of pectin and its pharmaceutical uses: a review. Silpakorn University Journal of Social Sciences, Humanities, and Arts *3*, 206–228.

Stalmach, A., Mullen, W., Barron, D., Uchida, K., Yokota, T., Cavin, C., Steiling, H., Williamson, G., and Crozier, A. (2009). Metabolite profiling of hydroxycinnamate derivatives in plasma and urine after the ingestion of coffee by humans: Identification of biomarkers of coffee consumption. Drug Metab. Disposition 37, 1749–1758.

Tabasco, R., Sanchez-Patan, F., Monagas, M., Bartolome, B., Moreno-Arribas, M.V., Pelaez, C., and Requena, T. (2011). Effect of grape polyphenols on lactic acid bacteria and bifidobacteria growth: Resistance and metabolism. Food Microbiol. *28*, 1345–1352.

Tarini, J., and Wolever, T.M.S. (2010). The fermentable fibre inulin increases postprandial serum short-chain fatty acids and reduces free-fatty acids and ghrelin in healthy subjects. Appl. Physiol. Nutr. Metab. *35*, 9–16.

Todoriki, K., Mukai, T., Sato, S., and Toba, T. (2001). Inhibition of adhesion of food-borne pathogens to Caco-2 cells by *Lactobacillus* strains. J. Appl. Microbiol. *91*, 154–159.

Topping, D., and Clifton, P.M. (2001). Short-chain fatty acids and human colonic function: Roles of resistant starch and nonstarch polysaccharides. Physiol. Rev. *81*, 1031–1064.

Trinidad, T.P., Sagum, R.S., Mallillin, A.C., Borlagdan, M.S., Leon, M.P.d. and Aviles, T.F. (2013). Sweet potato and cassava can modify cholesterol profile in humans with moderately raised serum cholesterol levels. Food and Nutrition Sciences *4*, 491–495.

Tuohy, K.M., Conterno, L., Gasperotti, M., and Viola, R. (2012). Up-regulating the human intestinal microbiome using whole plant foods, polyphenols, and/or fiber. J. Agric. Food Chem. *60*, 8776–8782.

Turroni, F., Bottacini, F., Foroni, E., Mulder, I., Kim, J.H., Zomer, A., Sanchez, B., Bidossi, A., Ferrarini, A., Giubellini, V., *et al.* (2010). Genome analysis of *Bifidobacterium bifidum* PRL2010 reveals metabolic pathways for host-derived glycan foraging. Proc. Natl. Acad. Sci. U.S.A. *107*, 19514–19519.

Turroni, F., Peano, C., Pass, D.A., Foroni, E., Severgnini, M., Claesson, M.J., Kerr, C., Hourihane, J., Murray, D., Fuligni, F., *et al.* (2012). Diversity of bifidobacteria within the infant gut microbiota. PLoS One 7, e36957.

van Dorsten, F.A., Peters, S., Gross, G., Gomez-Roldan, V., Klinkenberg, M., de Vos, R.C., Vaughan, E.E., van Duynhoven, J.P., Possemiers, S., van de Wiele, T., *et al.* (2012). Gut microbial metabolism of polyphenols from black tea and red wine/grape juice is source-specific and colon-region dependent. J. Agric. Food Chem. *60*, 11331–11342.

van Duynhoven, J., van der Hooft, J.J.J., van Dorsten, F.A., Peters, S., Foltz, M., Gomez-Roldan, V., Vervoort, J., De Vos, R.C., and Jacobs, D.M. (2014). Rapid and sustained systemic circulation of conjugated gut microbial catabolites after single-dose black tea extract consumption. J. Proteome Res. *13*, 2668–2678.

van Duynhoven, J., Vaughan, E.E., Jacobs, D.M., Kemperman, R.A., van Velzen, E.J.J., Gross, G., Roger, L.C., Possemiers, S., Smilde, A.K., Dore, J., *et al.* (2011). Metabolic fate of polyphenols in the human superorganism. Proc. Natl. Acad. Sci. U.S.A. *108*, 4531–4538.

Ventura, M., Turroni, F., Lugli, G.A., and van Sinderen, D. (2014). Bifidobacteria and humans: our special friends, from ecological to genomics perspectives. J. Sci. Food Agric. *94*, 163–168.

Vince, A.J., Mcneil, N.I., Wager, J.D., and Wrong, O.M. (1990). The effect of lactulose, pectin, arabinogalactan and cellulose on the production of organic-acids and metabolism of ammonia by intestinal bacteria in a fecal incubation system. Br. J. Nutr. *63*, 17–26.

Vincken, J.P., Beldman, G., and Voragen, A.G.J. (1994). The effect of xyloglucans on the degradation of cell-wall-embedded cellulose by the combined action of cellobiohydrolase and endoglucanases from Trichoderma-Viride. Plant Physiol. *104*, 99–107.

Vitaglione, P., Lumaga, R.B., Ferracane, R., Sellitto, S., Ramon Morello, J., Reguant Miranda, J., Shimoni, E., and Fogliano, V. (2013). Human bioavailability of flavanols and phenolic acids from cocoa-nut creams enriched with free or microencapsulated cocoa polyphenols. Br. J. Nutr. *109*, 1832–1843.

Vulevic, J., Juric, A., Tzortzis, G., and Gibson, G.R. (2013). A mixture of trans-galactooligosaccharides reduces markers of metabolic syndrome and modulates the fecal microbiota and immune function of overweight adults. J. Nutr. *143*, 324–331.

Wächtershäuser, A., Loitsch, S.M., and Stein, J. (2000). PPAR-[gamma] is selectively upregulated in Caco-2 cells by butyrate. Biochem. Biophys. Res. Commun *272*, 380–385.

Walker, A.W., Ince, J., Duncan, S.H., Webster, L.M., Holtrop, G., Ze, X., Brown, D., Stares, M.D., Scott, P., Bergerat, A., *et al.* (2011). Dominant and diet-responsive groups of bacteria within the human colonic microbiota. ISME J. *5*, 220–230.

Wedekind, K.J., Mansfield, H.R., and Montgomery, L. (1988). Enumeration and isolation of cellulolytic and hemicellulolytic bacteria from human feces. Appl. Environ. Microbiol. *54*, 1530–1535.

Wegmann, U., Louis, P., Goesmann, A., Henrissat, B., Duncan, S.H., and Flint, H.J. (2013). Complete genome of a new Firmicutes species belonging to the dominant human colonic microbiota ('Ruminococcus bicirculans'). reveals two chromosomes and a selective capacity to utilize plant glucans. Environ. Microbiol. *16*, 2879–2890.

Wei, H., Xu, Q., Taylor, L.E., 2nd, Baker, J.O., Tucker, M.P., and Ding, S.Y. (2009). Natural paradigms of plant cell wall degradation. Curr. Opin. Biotechnol. *20*, 330–338.

Werlein, H.D., Kutemeyer, C., Schatton, G., Hubbermann, E.M., and Schwarz, K. (2005). Influence of elderberry and blackcurrant concentrates on the growth of microorganisms. Food Control *16*, 729–733.

Williamson, G., and Clifford, M.N. (2010). Colonic metabolites of berry polyphenols: the missing link to biological activity? Br. J. Nutr. *104*, S48-S66.

Windey, K., De Preter, V., and Verbeke, K. (2012). Relevance of protein fermentation to gut health. Mol. Nutr. Food Res. *56*, 184–196.

Wright, D.P., Knight, C.G., Parkar, S.G., Christie, D.L., and Roberton, A.M. (2000). Cloning of a mucin-desulfating sulfatase gene from *Prevotella* strain RS2 and its expression using a *Bacteroides* recombinant system. J. Bacterol. *182*, 3002–3007.

Ying, D.Y., Parkar, S., Luo, X.X., Seelye, R., Sharpe, J.C., Barker, D., Saunders, J., Pereira, R., and Schršder, R. (2007). Microencapsulation of probiotics using kiwifruit polysaccharide and alginate chitosan. Acta Hortic. *753*, 801–808.

Yoo, H.-D., Kim, D., Paek, S.-H., and Oh, S.-E. (2012). Plant cell wall polysaccharides as potential resources for the development of novel prebiotics. Biomol Ther. *20*, 371–379.

Young, W., Roy, N.C., Lee, J., Lawley, B., Otter, D., Henderson, G., McCann, M.J., and Tannock, G.W. (2012). Changes in bowel microbiota induced by feeding weanlings resistant starch stimulate transcriptomic and physiological responses. Appl. Environ. Microbiol. *78*, 6656–6664.

Zampa, A., Silvi, S., Fabiani, R., Morozzi, G., Orpianesi, C., and Cresci, A. (2004). Effects of different digestible carbohydrates on bile acid metabolism and SCFA production by human gut micro-flora grown in an *in vitro* semi-continuous culture. Anaerobe *10*, 19–26.

Ze, X., Le Mougen, F., Duncan, S.H., Louis, P., and Flint, H.J. (2013). Some are more equal than others: the role of 'keystone' species in the degradation of recalcitrant substrates. Gut Microbes *4*, 236–240.

Zhou, X., Ruan, Z., Huang, X., Zhou, Y., Liu, S., and Yin, Y. (2014). The prebiotic lactosucrose modulates gut metabolites and microbiota in intestinal inflammatory rats. Food Sci. Biotechnol. *23*, 157–163.

Zhou, Z., Cao, X., and Zhou, J.Y.H. (2013). Effect of resistant starch structure on short-chain fatty acids production by human gut microbiota fermentation *in vitro*. Starch-Starke *65*, 509–516.

Zhu, N., Soendergaard, M., Jeffery, E.H., and Lai, R.-H. (2010). The impact of loss of myrosinase on the bioactivity of broccoli products in f344 rats. J. Agric. Food Chem. *58*, 1558–1563.

Zinkevich, V., and Beech, I.B. (2000). Screening of sulfate-reducing bacteria in colonoscopy samples from healthy and colitic human gut mucosa. FEMS Microbiol. Ecol. *34*, 147–155.

Zivkovic, A.M., and Barile, D. (2011). Bovine milk as a source of functional oligosaccharides for improving human health. Adv. Nutr: An Int. Rev. J. *2*, 284–289.

Zuckerman, M.J. (2006). The role of fibre in the treatment of irritable bowel syndrome: therapeutic recommendations. J. Clin. Gastroenterol. *40*, 104–108.

Zykwinska, A.W., Ralet, M.C.J., Garnier, C.D., and Thibault, J.F.J. (2005). Evidence for *in vitro* binding of pectin side chains to cellulose. Plant Physiol. *139*, 397–407.

Immunomodulating Effects of Prebiotics and Fibres

21

Hanne Frøkiær, Stine B. Metzdorff and Koen Venema

Abstract

The beneficial effects of fibres have long been attributed to their bulking effect, reducing the transit time of chyme through the gastrointestinal (GI) tract, thereby, for example, reducing the contact of potential toxic compounds with the gastrointestinal epithelium and preventing constipation. Lately however, it has been recognized that fibres may have prebiotics effects by modulating the activity of the gut microbiota, and even direct effects on the immune system. This chapter highlights the effects of fibres on the immune system. Mostly direct interactions with the host are discussed, although effects of microbial fermentation products [the so-called short-chain fatty acids (SCFA)] on the immune system is also briefly mentioned.

Introduction

Fibres from food, also referred to as dietary fibres, are composed of complex carbohydrates and lignin that are indigestible within the small intestine by the mammalian digestive enzymes (Turner and Lupton, 2011). As this definition implies, fibres constitute a highly diverse group of molecules that differs in regard to molecular composition, molecular size, and solubility, all factors that influence the putative immunomodulating effects of fibres. Most fibres originate from plants, but as microorganisms and fungi also produce complex indigestible carbohydrates, often with structures similar to fibres of plant origins, most researches also include fibres from such sources in the definition of dietary fibres. Although fibres are not degradable by the mammalian digestive enzymes, many bacteria express one or more enzymes and transporters that can degrade and take up the indigestible fibres, respectively. When the bacteria degrade the fibres, they grow in numbers and the increased number of one or some specific bacterial strains may influence barrier function, haematopoiesis and gut immunity. In addition, bacteria may metabolize the fibres resulting in production of short-chain fatty acids (SCFA), such as acetate, propionate and butyrate. These SCFA hold potent immune modulating activities, thus through fermentation dietary fibres may exert indirect immune modulatory effects. In particular many beneficial bacteria hold transporters and carbohydrate degrading enzymes. When fibres give rise to increase in bacterial numbers and/or metabolic activity of beneficial bacteria, thus exerting potential effects on the host health, the fibres are termed prebiotics. Hence, just as fibres with direct immune modulating activities, prebiotics represent a subpopulation of fibres and importantly, these two groups do only to some extend overlap.

In the following the immunomodulating effects of fibres are dealt with in two main sections; first the evidence for direct immune modulating effects of fibres is described (Fig. 21.1), then indirect effects through the increase in the number of beneficial bacteria and through increased production of SCFA is evaluated (Fig. 21.2). Finally, the evidence for immunomodulatory effects *in vivo* is described.

Direct effects of fibres on immune cells

Mammal cells, in particular cells of the immune system, express various receptors on their surface that recognize different carbohydrate structures, and some of these bind to the carbohydrate structures of fibres. Especially phagocytic cells such as neutrophils, macrophages and dendritic cells display a number of carbohydrate receptors with different and overlapping carbohydrate specificity (Table 21.1). Essentially, all the carbohydrate receptors recognize carbohydrate structures present on the surface of various microorganisms, including bacteria, viruses, yeasts and fungi, but the same or similar structures can be found in algae, plants, mushrooms and mosses, the most used sources for dietary fibres. As we will see in the following, the individual receptor may bind to various fibre types albeit not always with the same affinity. Most of the receptors belong to the C-type lectin receptor (CLR) family defined as proteins that have one or more domains with conserved residue motifs that are homologous to carbohydrate recognition domains. As the definition is based on the primary protein structure and not the functionality of the receptor, some CLRs contain domains that are not able to bind carbohydrate structures (Zelensky and Gready, 2004). All the receptors described here, however, are functional carbohydrate binding receptors. These receptors can be divided into two groups according to whether they display calcium-dependent binding or not.

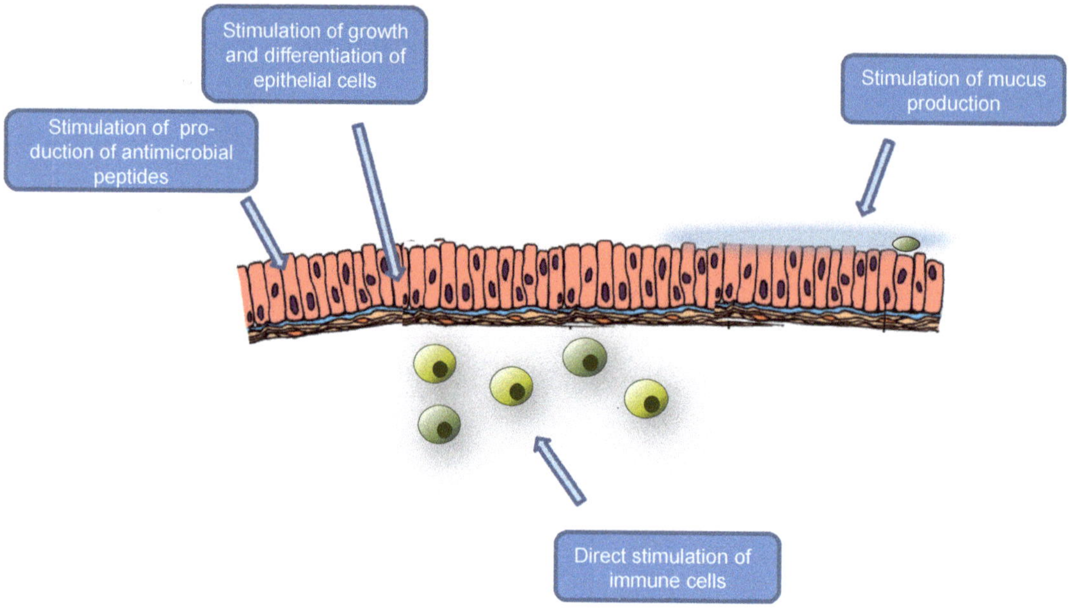

Figure 21.1 Schematic overview of putative direct mechanisms of prebiotics involved in beneficial as well as adverse effect on the intestinal epithelium: prebiotics themselves may stimulate growth and differentiation of epithelial cells, induce mucus and antimicrobial peptide production, and stimulate immune cells. In addition some prebiotics are suggested to act directly on the intestinal cells by binding to specific receptors or to act as irritants on the epithelial cells, which might lead to a higher permeability and risk of bacterial translocation from the intestine.

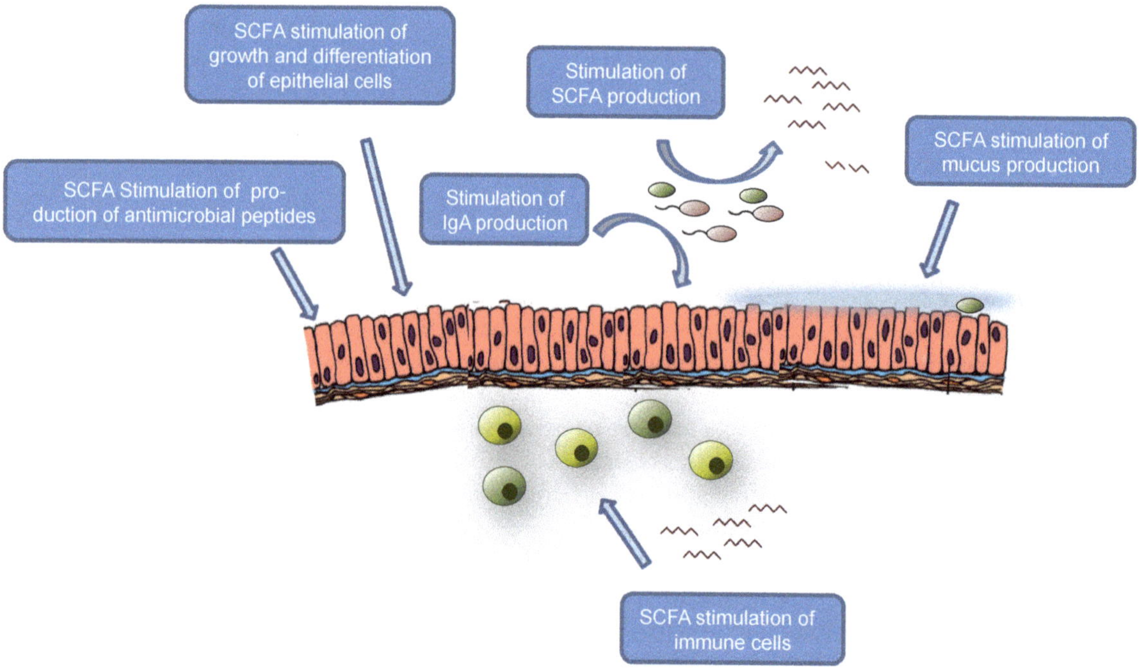

Figure 21.2 Schematic overview of putative indirect mechanisms of prebiotics involved in beneficial as well as adverse effect on the intestinal epithelium: prebiotics may stimulate bacterial growth leading to an increase in the number of beneficial bacteria and/or to the production short-chain fatty acids (SCFA). A higher number of beneficial bacteria or their metabolites may induce secretory IgA and mucus production, up-regulate genes involved in tight junction formation between epithelial cells, and stimulate immune cells. SCFA may stimulate growth and differentiation of epithelial cells, regulate tight junction formation between epithelial cells, and regulate proinflammatory immune cells.

The Mannose Receptor (MR), Dendritic Cell-Specific Intercellular adhesion molecule-3-Grabbing Non-integrin (DC-SIGN), macrophage galactose-specific lectin (MGL) and langerin all represent C-type lectins displaying calcium-dependent binding (Blixt *et al.*, 2004; Largent *et al.*, 1984), while Dectin-1 shows calcium-independent carbohydrate binding (Brown, 2006). Also the complement receptor (CR) 3 has carbohydrate binding activity. CR3 plays a key role in phagocytosis of complement coated

Table 21.1 Carbohydrate binding lectin receptors in the immune system, their presence on different immune cells and their specificity

	Receptor	Specificity	Cell types
Ca^{2+}-dependent	Mannose receptor (MR)	Mannose, mannan, chitin, fucose	Macrophages, dendritic cells, nonvascular endothelium
	DC-SIGN	Mannose, mannan, glucans β, fucose	Dendritic cells, macrophages
	Macrophage galactose specific lectin (MGL)	Galactose	Macrophages
	Langerin	Mannose	Dendritic cells, Langerhans cells
	Mincle	Mannose	Macrophages
Ca^{2+}-independent	Dectin 2	α-Glucans, mannose, mannan	Neutrophils, monocytes, macrophages, dendritic cells
	Dectin 1	β-Glucans	Neutrophils, monocytes, macrophages, dendritic cells

Based on Chen and Seviour (2007), Drummond *et al.* (2011), Geijtenbeek and Gringhuis (2009), Plato *et al.* (2013), Thornton *et al.* (1996).

microorganisms (opsonization), but may also bind directly to β-glucan structures (Ross *et al.*, 1985). Scavenger receptors, which comprise a group of promiscuous receptors binding a vast variety of macromolecules of both endogenous and microbial origin, have also been shown to bind carbohydrate structures present in fibres (Canton *et al.*, 2013). Still new carbohydrate receptors are identified, however, here only a number of carbohydrate receptors which have been related to carbohydrates found in fibres will be discussed.

Carbohydrate binding receptors and their binding specificity

These carbohydrate binding receptors expressed by cells of the immune system, in particular the myeloid cells, are especially adapted to recognize microbial carbohydrate structures, such as β-glucans within the yeast cell wall and mannosylated cell wall components on bacteria and yeast. The action and specificity of the carbohydrate receptors are described in several reviews (Geijtenbeek and Gringhuis, 2009; Sancho and Reis e Sousa, 2012) and only the most fibre relevant receptors will be described in details here. One important example of a fibre relevant receptor is the DC-associated C-type lectin 1 (Dectin-1) receptor, which binds to various β-glucans that contain a (1–3)-β-d-linked backbone, but with several folds higher affinity to β-(1–3),(1–6) β-glucans (laminarin) compared with β-(1–3),(1–4) β-glucans (Adams *et al.*, 2008; Brown and Gordon, 2001). Whereas the former structure is present in cell walls from brown algae (Allaby, 2006), the latter structures are found in plant fibres, where in particular barley and oat are rich sources (Åman and Graham, 1987). The dependency on the structure for a proper binding to the Dectin-1 receptor is reflected in the immune modulating effect of the β-glucans as demonstrated by their capacity to modulate the LPS stimulated cytokine response in dendritic cells (Wismar *et al.*, 2010). Although the structure as such is important for the immune modulatory effect, the length of the polymers may represent the major determinant of the immune modulatory activity, due to the fact that the longer the polymers the less soluble they are (Adams *et al.*, 2008; Mikkelsen *et al.*, 2013). Such polymers may also attain different aggregate/gel-like structures in solution that in turn may affect the cells differently.

Other receptors than Dectin-1 may also bind β-glucan. These include the integrin receptor CR3, scavenger receptors, and lactosylceramide (Brown and Gordon, 2003), but none of these other receptors have been clearly demonstrated to play a major role in mediation of the biological response to β-glucans. Common for the β-glucan binding receptors are their presence on different types of myeloid leucocytes, including neutrophils, monocytes, macrophages, and dendritic cells. These cells have in common that they belong to the innate immune system, hold phagocytic activity, and recognize conserved structures on microorganisms, so called microbial-associated molecular patterns (MAMPs). Microbially derived carbohydrates comprise one group of such MAMPs.

A number of MAMPs contains mannose in their structure, either as terminal carbohydrate units or as multiple mannose units linked together (Gringhuis *et al.*, 2009; Martinez-Pomares, 2012). The Mannose receptor recognizes mannose, both as a terminal sugar in complex carbohydrates and as mannose multimers, such as mannan (Martinez-Pomares, 2012; Taylor *et al.*, 2005). The ligands are present on a range of bacteria, fungi, virus-infected cells, and parasites. Other receptors that recognize mannose residues include langerin and Dectin-2. Dectin-2 and the mannose receptor also have affinity for α-glucans and chitin, respectively (Levitz, 2010).

The receptors hold different binding specificities but they also give rise to distinct effects in the antigen presenting cells (Feriotti *et al.*, 2013). One important effect of the carbohydrate–receptor interaction is induction of phagocytosis of microorganisms by binding to multiple carbohydrate receptors at the same time. This has been demonstrated for many of the receptors including the mannose receptor, Dectin-1, Dectin-2 and DC-SIGN (Kerrigan and Brown, 2009). Phagocytosis is a requirement for induction of a strong cellular immune response of importance for an efficient clearance of bacteria, fungi as well as viruses. Depending on the specific preparation, plant carbohydrates may hold single or multiple ligands that can ligate to one or many receptors on an antigen presenting cell, respectively, in turn giving rise to different effects in the cell. Thus, depending on the structure, such plant-derived carbohydrates may inhibit the binding of bacteria(l structures) to the PPRs and thus, modulate the immune response. Multiple receptors have been implicated in recognition of e.g. yeasts or zymosan. These include CR3 and dectin-1, which bind cell wall β-glucans; mannose receptor (MR), DC-SIGN, Toll-like receptors (TLRs) and galectin-3, which bind mannans or other mannose-containing compounds (Jozefowski *et al.*, 2012). However, involvement of some of these receptors in yeast recognition remains controversial. For instance, Brown, Taylor *et al.* (Brown *et al.*, 2002; Taylor *et al.*, 2002) reported that opsonin-independent zymosan binding was unaffected by genetic CR3 deficiency or a blocking monoclonal antibody against CR3, suggesting that CR3 was not the receptor for β-glucan recognition. Also mannose receptor-deficient macrophages showed normal phagocytic uptake of *C. albicans*. Moreover, resistance to *C. albicans* infection was unimpaired in mannose receptor-deficient mice (Lee *et al.*, 2003). Another class of C-type lectins that recognize microbial components are scavenger receptors (Pluddemann *et al.*, 2006). There are 8 classes and not all have been characterized in full detail. They bind to modified glucans (acetylated, phosphorylated) as well as to unmodified molecules. Apart from being involved in direct recognition and phagocytic clearance of bacteria, different scavenger receptors collaborate with other classes of PRRs and either increase the repertoire of innate immune recognition or modulate the biological responses of host cells to bacteria (Pluddemann *et al.*, 2006).

When binding to carbohydrate receptors on the surface of cells, fibres may initiate signal transduction to initiate either an immune stimulatory signal or an inhibiting signal. The induced intracellular signalling processes are complex and most likely not fully understood. Binding of fibres to some of these receptors including dectin-2, initiates signalling pathways through immunoreceptor tyrosine-based activation motif (ITAM)-containing adaptor molecules, such as Fc receptor γ-chain (FcRg) or Src kinases and SYK, leading to induction of pro-inflammatory cytokines such as IL-6 and TNF-α (Sato *et al.*, 2006; Yamasaki *et al.*, 2009). Conversely, binding of fibres to dectin-1, DC-SIGN and several other receptors induces signalling through protein kinases or phosphatases, but has not hitherto been shown to induce pro-inflammatory gene expression by the binding. Rather the binding of fibres to these inhibitors initiates a signal transduction that may affect signalling through Toll like receptors (TLRs), thus modulating the TLR-ligand initiated inflammatory signal. There are many speculations regarding the intracellular pathways involved and also partly elucidated intracellular pathways that to varying degree support the suggested interactions between carbohydrate receptor and TLR signalling (Geijtenbeek and Gringhuis, 2009). A popular theory regarding the modulating activity of fibres is that the fibre binds simultaneously with the binding of a microorganism or just a single TLR-ligand of that microorganisms, and that the simultaneous binding leads to intracellular interaction of the pathways leading to immune modulating of the TLR induced response. For instance, it has been shown that several structurally different non-starch polysaccharides can modulate LPS-induced cytokine production in dendritic cells in a structure-dependent manner (Wismar *et al.*, 2011).

Also, as most experiments are performed with carbohydrate macromolecules isolated from microorganisms the possibility exists that such fibre preparations contains impurities of microbial origin which may hold TLR-stimulating activities or by other means may interfere with signalling. Hence, whether a fibre holds immune stimulatory or inhibitory properties should be determined by extremely pure, preferably synthetic fibres.

Importantly, most if not all carbohydrate receptors seem to be involved in phagocytosis which is in good accordance with their presence on phagocytic cells as dendritic cells, monocytes, macrophages and neutrophils. Which receptors are present on the different phagocytic cell types, is not described very comprehensively. This hampers our understanding of how complex carbohydrates interact with and affect the different cell types. Many fibres from fungi, including mushrooms and yeast contain mannose units, either as terminal sugars or as series of many α–linked mannose units (Kery *et al.*, 1992). Such fibres may bind to different mannose specific receptors including DC-SIGN, Dectin-2 and MR, and accordingly the signal induced may be determined by the specific mannose-binding receptors present on a given cell. Beta-glucan stimulation through Dectin-1 and mannan stimulation through Dectin-2 have both been shown to induce NF-κB through Syk kinase and CARD9 signalling (Drummond *et al.*, 2011). Interestingly, while Dectin-1 has an ITAM-motif in its cytoplasmic domain, Dectin-2 and the other receptor of the Dectin-2 family, Mincle, have no known intracellular signalling regions, but may associate with the ITAM motif of the FcRg chain and thus, indirectly induce Syk activation (Sato *et al.*, 2006).

Apart from binding mannose, the mannose receptor also is capable of binding terminating fucose and N-acetyl glucosamine residues in a calcium-dependent manner (Taylor *et al.*, 2005). The MR lacks a potential signalling motif in the cytoplasmic tail and hence, requires collaboration with other PRRs to induce e.g. NF-κB activation (Gazi and Martinez-Pomares, 2009). The dependency for collaboration with other PRRs was nicely demonstrated by Tachado *et al.* (2007). Using non-phagocytic human embryonic kidney cells transfected with cDNA encoding for human TLR2 or human MR cDNA alone, no cytokine production was measured in response to the fungus *Pneumocystis*. Conversely, when co-transfecting the two receptors, cytokine production was induced (Tachado et al., 2007).

DC-SIGN also recognizes mannose, but with strong preference for high-mannose oligosaccharides, in a calcium-dependent manner. Moreover, as MR, it also recognizes fucose-containing ligands (Feinberg *et al.*, 2001). Upon mannan ligation, DC-SIGN signalling involves the RAF proto-oncogene serine/threonine-proteins kinase (Raf-1) (Gringhuis *et al.*, 2007). Phosphorylation of Raf-1 results in modulation of the phosphorylation and acetylation of NF-κB, which in turn induces transcription of the genes encoding proinflammatory cytokines such as IL-12 and IL-6. Upon fucose stimulation, however, the signalling was found to be independent of the Raf-1 signalling cascade and to suppress pro-inflammatory cytokine production, while increasing the IL-10 production (Gringhuis *et al.*, 2009). Whether this discrepancy actually is due to dual signalling pathways that are initiated by distinct ligands or alternatively, the different responses are caused by different experimental approaches or different sources of ligands, remains to be clarified.

As one would expect from the paragraphs above, knockout gene studies indicate some degree of redundancy of the different carbohydrate receptors. Whether this is an immunological safety net or a means to increase the capacity to distinguish between different carbohydrate structures, and hence microorganisms, by the collaboration of two or more receptors with distinct specificity, is not clear. Collaboration between different PRRs has been described. One example is the collaboration between Dectin-1 and TLR2 (Bauer *et al.*, 2009; Goodridge and Underhill, 2008; O'Neill, 2008). Stimulation with Dectin-1 and TLR2 agonists has synergistic effect on TNF, IL-6 and IL-23 production. Of note, such synergistic effects have also been shown for Dectin-1 in collaboration with TLR-1, TLR-4, TLR-5, TLR-7 and TLR-9 (Bonfim *et al.*, 2009; Dennehy *et al.*, 2008). While these studies demonstrate the collaboration between different receptors, they also underline an obstacle for elucidating specific mechanisms behind the action of specific structures; the fact that by far most of the fibres investigated originated from microorganisms and accordingly the likelihood of the presence of contaminants representing other carbohydrate structures as well as other ligands for TLRs exits. The presence of such PRR ligands may result in induction of other cellular pathways initiating other cytokines and accordingly other immune responses, clouding the effects of the fibre alone. As an example, curdlan a (1–3)-β-glucan from the Gram-negative soil bacterium *Alcaligene faecalis,* is often used to assess the immunomodulating effect of (1–3)-β-glucans. However, many preparations of curdlan are not pure and contain varying amounts of lipopolysaccharide (LPS), which may induce additional stimulatory pathways in the immune cells. To this end, we showed that a preparation of commercially available curdlan contained an appreciable amount of LPS (Wismar *et al.*, 2010). Likewise, some of the reported work on dectin-1 is at least partly based on the use of zymosan from yeast (Kerrigan and Brown 2010). Again, apart from an abundance of β-glucans, zymosan has been shown to contain mannan and TLR2 ligands, which may give rise to confusing results (Ikeda *et al.*, 2008; Jawhara *et al.*, 2012). Moreover, mannan has been shown to be contaminated with β-glucans (Brown *et al.*, 2002). While the use of knock out mouse strains, has clearly demonstrated the indispensability of dectin-1 for induction of Th1 and Th17 responses (LeibundGut-Landmann *et al.*, 2007, 2008; Osorio *et al.*, 2008), and that β-glucan is the ligand for dectin-1 was shown by transfected cell lines (Brown and Gordon, 2001), it remains to be demonstrated that β-glucan *per se* is capable of this induction.

Keeping in mind that the fibres by binding to carbohydrate structures may induce phagocytosis, it might also be this induced phagocytosis that modulates the TLR response, e.g. by enhancing the uptake of microorganisms or microbial compounds. It has been shown for some TLRs, that signalling from endosomes instead of from the plasma membrane may change the signalling and accordingly which cytokines that are being produced through TLR signalling. This compartmentalized segregation in signalling was demonstrated for TLR4 showing that ligand binding to and signalling through TLR4 present in the plasma membrane of antigen presenting cells induces signalling through the adaptor protein MyD88, while when signalling through TLR4 takes place from endosomal compartments, signalling takes place through the adaptor protein TRIF. The employment of different adaptor proteins according to the localization of the receptor results in activation of distinct signalling pathways and accordingly different cytokine are produced (Barton and Kagan, 2009). Even though this compartment dependent signalling has not been demonstrated for other TLRs or PRRs, it cannot be excluded that some fibres due to their capacity to induce phagocytosis act by facilitating the endocytosis of TLR ligands and thus might redirect the TLR signalling to other pathways.

Even preparations of plant derived fibres are likely to contain other macromolecules with putative immune stimulatory/modulatory properties. For instance they may contain LPS (from Gram-negatives) or LTA (from Gram-positives) or other immunostimulatory compounds of microbial origin simply owing to the fact that the fibres may serve as nutrients for the microorganisms. To be completely sure that any immunomodulatory effect demonstrated can be ascribed to the fibres it is advisable to test for endotoxin content and if possible for other contaminants. We tested a variety of fibre samples used to assess their immunomodulatory effects in dendritic cells and found that out of the 28 non-starch polysaccharides (NSP) tested, several contained endotoxins (Wismar *et al.*, 2011). Due to the nature of different NSP it was not possible to purify for unknown microbial TLR-ligand contaminants. Therefore, addition of a high amount of LPS to all stimuli was carried out to level out these differences in contamination. A diverse group of NSP was capable to suppress LPS-induced IL-12p70 production. Among those, the group comprising differently structured β-glucans and the galactomannan guar gum were at the same time able to augment the LPS-induced IL-10 production. Only curdlan gave rise to detectable amounts of IL-12p70 in the absence of LPS-stimulation (Wismar *et al.*, 2011).

When assessing the immunomodulating properties of fibres it is important to relate to a number of factors that may influence the experimental outcome. These comprise physicochemical parameters such as molecular mass, molecular fine structure, solubility, polymer conformation in solution, the presence of fibre particles and the experimental system used to assess the immunomodulating effect. The molecular fine structure, mass, solubility and polymer conformation are highly interrelated parameters and this makes the structure–functional relationships difficult to establish (Bohm and Kulicke, 1999; Tosh *et al.*, 2004). As an example, solubilized β-glucan has been shown to comprise a quite heterogeneous group of polysaccharide conformations, due to variation in degree of polymerization that affects the aggregation of the β-glucan molecules, which in turn may affects the solubility of the overall β-glucan preparation (Håkansson *et al.*, 2012). The solubility versus the particulate nature of a β-glucan preparation may also be a determinant for an immunomodulatory property. Even inert particles that are not being recognized by phagocytes may induce phagocytic activity. Accordingly, immunomodulating fibres may attain phagocytosis-inducing activity if they form insoluble particles in solution, independently of their receptor specific immunomodulating activity. This was demonstrated by comparison of a variety of dried β-glucan samples prepared from barley that were either fully solubilized or just suspended in an aqueous solution. When added together with the probiotic bacterium *Lactobacillus acidophilus* NCFM to dendritic cells, the two different preparations exhibited opposite immunomodulating properties: while the fully dissolved preparations increased the *L. acidophilus* NCFM induced production of the Th1 inducing cytokine IL-12, the dispensed preparation diminished the induced production of IL-12 but increased the production of the anti-inflammatory cytokine IL-10 (Mikkelsen *et al.*, 2014). Also, only dispensed β-glucan preparations induced a cytokine response (IL-10) response in the dendritic cells when added alone. These disparate effects of two preparations of the same sample highlight the complexity of the fibre structures and functionality and indicate that solubility is a key determinant of the immunomodulatory properties of carbohydrate fibres.

Last but not least the experimental system used may determine the outcome of the testing. When testing the immunomodulating properties of a series of β-glucan molecules, this was done by adding the β-glucan samples alone or together with either LPS from *E. coli* or whole Gram-positive *L. acidophilus* NCFM bacteria to stimulate dendritic cells (Mikkelsen *et al.*, 2014). These two stimuli are known to induce distinct intracellular pathways in the cells, despite that both give rise to IL-12 production of importance for NK cell activation and activation of a cellular immunity (Abbas *et al.*, 2012). While dissolved β-glucan samples inhibited the IL-12 production in LPS stimulated cells, the same samples enhanced the IL-12 induced production in *L. acidophilus* NCFM stimulated cells. Hence, the (microbiological) context of the interaction of fibres with immune cells is very important: it matters what is present in the close environment of the fibres and immune cells.

Taken together, the most well documented role of the receptors binding carbohydrate structures present in a variety of fibres is their role in phagocytosis. The receptors are in particular present on myeloid leucocytes such as monocytes, neutrophils and dendritic cells (Table 21.1). Their cooperative or antagonizing action when recognizing (different ligands on one and the same) fibre precludes a clear mode of action from being established. Similarly, due to numerous small changes in structure (e.g. in β-glucans) a clear-cut structure function relationship is still years if not decades away. The work is hampered by numerous factors such as solubility, length and heterogeneity of fibre preparations as well as overlapping specificity of many fibres and receptors.

The bioavailability of fibres

One intrusive question that emerges in relation to whether orally administered fibres hold the capacity to directly act as immunomodulators is if or to which extent such carbohydrate molecules will get into contact with cells of the immune system. To get on close hold of most leukocytes in the body, the fibres have to traverse the epithelial barrier, either to get into the circulation or to get in contact with the vast number of leukocytes present just below the epithelial barrier. Perhaps due to the lack of simple, specific and sensitive assays for quantification of the carbohydrate structures when studying the bioavailability of orally ingested fibres, only few studies have assessed the absorbance of fibres from the gastrointestinal tract. In most of the reported studies of fibre absorption, the methods to detect the presence of polysaccharides in tissue and blood have involved labelling of the carbohydrate with isotopes or fluorescence (Rice *et al.*, 2005; Sandvik *et al.*, 2007; Vetvicka *et al.*, 2007; Zhu *et al.*, 2011), but also detection by the use of specific antibodies was used (Hoshi *et al.*, 2008; Sakurai *et al.*, 1996). By the use of covalent labelling of the carbohydrate fibres, a 0.5–5% bioavailability in rats was estimated. As the methods used do not distinguish between intact or degraded macromolecules it cannot be concluded as to which extend the fibres had retained their bioactivity. In a study using ELISA based on antibodies specific to α-d-glucan separated from the edible mushroom *Tricholoma matsutake* a peak concentration in blood on 160 ng/ml was determined 24 hours after administration of a single dose of 450 mg/kg (roughly corresponding to 13 mg) to mice (Hoshi *et al.*, 2008). In addition, the available antibodies allowed detection of the fibre molecules in various tissue by immunostaining and showed the presence of the fibre in the gut mucosal layer, gut associated lymph node as well as the spleen during a 32 hour period after administration of the fibre. Even though the documentation of absorption of significant amounts of fibres upon oral intake is still not very solid, the existing studies indicate some absorption and hence, a possible *in vivo* direct immunomodulating effect of fibres showing such effects in *in vitro* assays is thus not unlikely.

Other examples of dietary fibres that (potentially) interact with the immune system of the host

Several dietary polysaccharides/fibres have been studied in *in vitro* assays, animal experiments and human volunteers for their immunmodulatory capacity (Ramberg *et al.*, 2010). It goes beyond the scope of this chapter to discuss these all in detail, but a selection of studies is briefly addressed. Apart from studies with glucans (both β and α) (Ramberg *et al.*, 2010), pectin has been studied as well, for its anti-inflammatory and cancer repressing properties (Hao *et al.*, 2013). Citrus pectin was treated with heat or pH-treated. The pH treatment at both high and low pH results in the degradation of the pectin into smaller fragments and the cleavage of the side chains from the backbones. Heat treatment leads to degradation in smaller fragments only. Heat-treated citrus pectin was shown to have anti-tumour activity *in vitro* and in a mouse model, whereas pH treated pectin was inactive. This indicates that the (degree of) methylation of pectin may be important for its activity. The anti-tumour activity of heat-treated pectin was not correlated with its immunological activities, or its inhibitory effect on galectin-3 (Hao *et al.*, 2013). In the light of the discovery that some pectic polysaccharides could bind to galectin-3, a galectin-3-targeting anti-tumour mechanism has been proposed. Several other pectins (swallow root pectin, Hemidesmus pectin, black cumin pectin, Andrographis pectin and Okra pectin) have been shown to interact with galectin-3 and inhibit galectin-3 mediated cancer cell processes (Hao *et al.*, 2013). The pH-treated pectin had higher immunomodulating activities than the heat-treated pectin, in addition to over 4000-fold higher galectin-3-inhibiting activities. It is noteworthy that the pH-treated pectin had no anti-tumour activity. Thus, anti-tumour activity of pectin on the one side and immunomodulation and galectin-3 inhibition on the other are likely brought about by different structural features on the pectin molecule.

Using HEK293 cells transfected with individual toll-like receptor, THP-1 cells and human PBMCs it was shown that fructooligosaccharides stimulate the immune system through TLRs, primarily TLR2 (Vogt *et al.*, 2013). The effect was dependent on the chain-length (or degree of polymerization; DP), assessed by cytokine production in human. β(2–1)-fructans with a low DP induced a regulatory cytokine balance compared to those with a higher DP as measured by IL-10/IL-12 ratios. Interestingly, the same fructooligosaccharides (FOS) also showed a DP-dependent effect on barrier function (transepithelial electrical resistance) in T84 monolayers, with again the shorter DP showing highest activity (Vogt *et al.*, 2014). Also this effect was TLR2-dependent as blocking of TLR2 led to abrogation of the activity of the FOS.

In addition, reports have described effects of galactomannan (guar gum), glucomannan and various heteroglycans (Ramberg *et al.*, 2010), as well as a mixture of galactooligosaccharides and fructooligosaccharides (van de Pol *et al.*, 2011; van der Aa *et al.*, 2011). In man only glucans (Kirmaz *et al.*, 2005; Koray *et al.*, 2009), fucoidan (a sulphated polysaccharide from brown algae) (Irhimeh *et al.*, 2007), and arabinogalactans (Riede *et al.*, 2013; Udani, 2013; Udani *et al.*, 2010) have been studied for their immunostimulatory effects.

Indirect effects of fibres – prebiotic effects

Indirect effects of fibres are caused by the action on bacteria that can utilize the carbohydrate and either, due to the higher number of such bacteria and an according change in microbial composition, stimulate the immune system differently and/or, by the fermentation of the carbohydrate, produce short-chain fatty acids (SCFA) and other metabolites that *per se* may hold immunomodulatory effects (Hamer *et al.*, 2008). Fibres that support the growth and metabolisms of bacteria exerting beneficial effects on the host health are called prebiotics. The term prebiotics was originally introduced as non-digestible food ingredients that beneficially affect the host by selectively stimulating the growth or activity of one or a limited number of bacteria already residing in the colon (Gibson and Roberfroid, 1995). Many of the endpoints associated with beneficial effects of prebiotics are diseases such as inflammatory bowel diseases, allergies, type II diabetes and cardiovascular diseases (Ouwehand *et al.*, 2005), which are all recognized to be influenced by imbalances in the immune system, e.g. a chronic inflammatory state sometimes combined with a hyper reactive local or systemic immune system. Common for the diseases are also that the gut microbiota is believed to play a major role in the development and, in some cases, in the 'maintenance' of the disease. This is addressed in more detail elsewhere in this book.

The full nature of prebiotic effects can only be addressed in *in vivo* models even though one may gain some information about the potential prebiotic properties of a fibre by *in vitro* growth of specific bacteria or the production of SCFA by bacteria growing on a substrate containing the fibre as the only carbohydrate source. Whether an increase in specific bacteria in the gastrointestinal (GI) tract will give rise to specific beneficial effects on the immune system may depend on the composition of the microbiota, the location in the gut, and specific immunological and physiological conditions in the gut. Fig. 21.2 depicts suggested indirect effects, so-called prebiotic effects of fibres.

Effects of prebiotics through increasing the number of (certain) bacterial groups within the microbiota

Prebiotics can have effects on different bodily locations: Even though the definition originally states that the effect of the prebiotics should affect bacterial growth and metabolism in the colon (and this is primarily addressed elsewhere in this book), increasing awareness that also the growth and metabolisms of microorganisms present in the small intestine may be affected by some prebiotics emerges. Even though the concentration of microorganisms here is several orders of magnitude lower than in the colon, and the flow through this part of the gastrointestinal tract is much faster than in the colon, thereby limiting the time available for the fermentation process, the upper part of the GI tract may be of major importance regarding the interaction of the bacteria with the immune system. Throughout the small intestine the mucosa is scattered with isolated lymphoid follicles that together with Peyer's patches (PP) serve as gate keepers, allowing a limited number of microorganisms and macromolecules to get access to the immune cells underlying the mucosal epithelium (Brandtzaeg, 2013). This is an important mechanism for the immune system to establish a protective barrier which prevents invasion of pathogens. Despite the fact that direct interactions of prebiotics with the epithelial layer has been shown *in vitro* as discussed above (Vogt *et al.*, 2014), it is highly likely that through modifying the microbiota composition in the upper GI tract, a community which is much more prone to compositional changes, prebiotics have an indirect effect on the immune systems as well, since these microbes interact with the cells of the underlying gastrointestinal associated lymphoid tissue (GALT).

Effects on gut infections

The microbial composition in the GI tract may affect the virulence of pathogenic bacteria, by providing conditions that either support or inhibit growth of the pathogen or by affecting the physical conditions of the microbiota, e.g. changes in the production of mucus. Hence, addition of a new source of nutrient, the prebiotic, that can only be exploited by some bacteria, creates changes in the microbial composition which in turn may affect how the microbiota affects the virulence of pathogens in the gut. *Salmonella* in mice fed various prebiotics showed that certain of the prebiotics (fructooligosaccharides and xylooligosaccharides) led to an increased translocation of salmonella in the gut (Petersen *et al.*, 2009). In a guinea pig model for *Listeria* infection, other prebiotics had a similar effect (Ebersbach *et al.*, 2010). This effect is likely to be associated with the ability of the pathogen to grow on the prebiotic. Other changes include production and secretion of IgA, production of antimicrobial peptides, and increased production of mucus (Lievin-Le Moal and Servin, 2006). Such properties of the epithelial layer represent key effector parts of the mucosal immune system, which may be modulated by the microbiota (Brandtzaeg, 2013).

Effects on production of antimicrobial peptides from enterocytes and Paneth cells

Antimicrobial peptides comprise a diverse group of low molecular proteins which by interaction with the bacterial membrane kill Gram-positive or Gram-negative bacteria. Production of some antimicrobial peptides takes place in the enterocytes, but the majority of the secreted antimicrobial peptides originates from the Paneth cells, granule-rich epithelial cells located at the base of small intestinal crypts (Bevins and Salzman, 2011). The Paneth cells are mainly confined to the small intestine in healthy humans, reflecting the importance of protecting this large area against microbial invasion. Some bacteria have been demonstrated to induce enhanced production of antimicrobial peptides from enterocytes *in vitro*. Bacteria shown to enhance production of antimicrobial peptides include bifodbacteria, lactobacilli and the probiotic *E. coli* Nissle 1917 strain, but not several other (commensal) *E. coli* strains (Schlee *et al.*, 2008; Wehkamp *et al.*, 2004; Bloch *et al.*, unpublished data). Notably, Paneth cells do not alter the secretion of antimicrobial peptides due to direct microbial stimulation. Rather, their secretion is affected by cytokines, in particular IL-22, the induction of which is determined by the gut microbiota (Hooper and Macpherson, 2010).

Hence, as the microbiota, through its presence as well as the amount of cells and the specific microbiota composition in the crypt of the epithelium of small intestine, may play a key role in the secretion of antimicrobial peptides, carbohydrate fibres that stimulate the growth of specific bacteria may represent an important means to maintain/establish a strong barrier towards invasive pathogens.

Microorganisms in the gastrointestinal tract may induce increased secretion of mucus

Mucus consists of a number of heavily glycosylated proteins called mucins of which mucins 1, 3, and 4 are membrane proteins present on the apical surface of the enterocytes. Mucin 2 (MUC-2) is secreted from the goblet cells located in the villi, and by interaction with the other mucins, Muc-2 forms a thick viscous gel/layer on top of the epithelial cells that separates microorganisms from the mucosa and accordingly protects against invasion of pathogens through the epithelial surface (Bergstrom *et al.*, 2010). Goblet cells are secretory cells derived from the stem cells located in the crypt bottom. They contain granules with densely packed MUC2 polymers that are released into the lumen upon stimulation. The importance of MUC2 in protection against pathogenic invasion can be demonstrated by the MUC2 knock out mouse or mice defective in the inflammasome proteins NLRP6, ASC or caspase1. Such mice have a dysfunctional gut microbiota and are prone to intestinal infection by pathogens, such as Prevotellaceae and TM7 bacteria (e.g. *Citrobacter rodentium* (Gill *et al.*, 2011; van der Sluis *et al.*, 2006, 2008; Wlodarska *et al.*, 2011)). Goblet cells of germ-free mice are fewer in number and smaller in size than in conventionally raised mice. This signifies a role of the microbiota in the development and maintenance of goblet cells and MUC2 production. For several microorganisms it has been shown that they increase the secretion of MUC2 and/or MUC3, e.g. *Clostridium tyrobutyricum* DSM 2637 (Hudcovic *et al.*, 2012), *Lactobacillus plantarum* 299v (Dykstra *et al.*, 2011; Mack *et al.*, 1999), *Lactobacillus rhamnosus* R0011, and *Bifidobacterium bifidum* R0071 (Dykstra *et al.*, 2011). Some of these strains are used as probiotics. Selective stimulation of these probiotic strains by prebiotics would likely enhance their functionality.

Effects of the specificity and amount of IgA secreted

Dimerized immunoglobulin A (IgA) molecules with specificity towards the microorganisms present in the gastrointestinal tract are secreted from the mucus into the lumen of the gut (van der Waaij *et al.*, 1996). Here the IgA binds to the bacteria and thereby facilitates their exclusion from the epithelial surface and accordingly, their translocation and access into the circulation. The amount of IgA secreted into the gut lumen strongly depends on the presence and diversity of the microbiota in the gut. Germ free mice and newborn babies are almost devoid of IgA in the faeces and the IgA concentration increases along with the postnatal colonization of the gut. Recently, several prebiotics/fibres have been shown to increase IgA, such as xylooligosaccharides (increased in saliva, but very likely then also in the gut, although this was not tested) (Childs *et al.*, 2014), galactooligosaccharides (Vulevic *et al.*, 2013), a galactooligosaccharide/inulin mixture (Gourbeyre *et al.*, 2013) fructo-oligosaccharides from yacon flour (Velez *et al.*, 2013) and β-1,4-mannobiose (Kovacs-Nolan *et al.*, 2013). The galactooligosaccharides/inulin mix also increased IgA in mother's milk of mice (Gourbeyre *et al.*, 2012), protecting the suckling offspring against pathogens. However, there were also several other prebiotics that showed no effect of faecal IgA (Lecerf *et al.*, 2012; Lomax *et al.*, 2012; Scalabrin *et al.*, 2012), despite showing effects on microbiota composition.

Effect on homeostasis in the gut – anti-inflammatory effects

The gut microbiota also plays an important role for homeostasis of the gut mucosal immune system and establishment of tolerance towards the commensal bacteria and food proteins. This has become evident from studies with germ free mice where the number of leukocytes in the lamina propria and epithelium is low compared with conventionally reared mice and in which tolerance towards food proteins cannot become established. The mechanisms involved in homeostasis and tolerance establishment are complex and only beginning to become understood and beyond the scope of this chapter. However, it may be noted that dendritic cells and macrophages present in Peyer's patches and lamina propria acquires specific phenotypes that stimulate the development of regulatory T-cells and IgA producing B-cells, that is dependent on the presence of the microbiota (Rescigno, 2014). 'Inhibit' type macrophages (called M1) can rapidly kill pathogens, and are thus the primary host defence, while 'Heal' type macrophages (M2) routinely repair and maintain tissue integrity (Mills and Ley, 2014). Intestinal macrophages undergo activation in response to environmental signals, including microbial products, and develop a M1 or M2 phenotype depending on the activation signals. It has been shown that activation of Dectin-1 induces M1 macrophages (Loures *et al.*, 2014).

Fibres that can be exploited as nutrients by a subgroup of the gut microbiota, may increase this group in proportion, which in turn may affect both the mucosal defence mechanisms outlined above as well as the maintenance of homeostasis in the gut. Depending on the specific structure of the fibre, different bacteria may thrive on it. Hence the mucosal barrier may be strengthened or weakened according to such changes. For instance, a multistrain probiotic product induced a mixed proinflammatory and anti-inflammatory phenotype in polarized and unpolarized macrophages (Isidro *et al.*, 2014), depending on the initial phenotype of the macrophages. At present it is unclear which probiotic component(s) in the mixture is/are responsible for this activity, but if this/these strain(s) could be selectively induced or repressed by feeding prebiotics, the polarization of macrophages could perhaps be steered. That this is feasible is exemplified by the fact that acidic oligosaccharides derived from pectin stimulated M1 macrophage activation through modulation of the microbiota (Bernard *et al.*, 2014).

Effects of prebiotics through increased production of microbial metabolites

Apart from the direct action/interaction of the microorganisms with the mucosa, which may influence the innate barrier functions as well as the mucosal homeostasis, microorganisms that grow on complex carbohydrate structures are recognized to produce various low-weight molecular compounds, such as short-chain fatty acids (SCFA; mainly acetate, propionate, and butyrate). These microbially produced SCFA may in turn affect the microbial composition simply by changing the pH in the gut. But perhaps most importantly, they may influence signalling in the host. Two major SCFA signalling mechanisms have been identified, inhibition of histone deacetylases (HDACs) and activation of G-protein-coupled receptors (GPCRs) (Tan *et al.*, 2014).

SCFA bind to two G-protein coupled receptors, the GPR41 (a.k.a. FFAR3) and GPR43 (FFAR2), with different affinities for the individual SCFA for both receptors. Binding of the ligand induces a signal transduction in the cell which, depending on the cell type, induces various responses. The GPR41 is primarily expressed in adipose cells, but also in the intestine, pancreas, lung and spleen (Brown *et al.*, 2003). The GPR43 is highly expressed in bone marrow and in white adipose tissue (Cox *et al.*, 2009; Kimura *et al.*, 2013). The high GPR43 expression in bone marrow reflects the expression in granulocytes and monocytes (Cox *et al.*, 2009). The SCFA produced by the gut microbiota act in the intestine as well as at other sites in the body. The preferred energy substrate for the gut epithelial cells is butyrate, and little butyrate escapes to the circulation (Hamer *et al.*, 2008). Instead, acetate and propionate are to a large extend absorbed into the circulation and may affect systemic cells expressing the SCFA receptors, such as adipose tissue and blood neutrophils. The overlap between in receptor affinity for the SCFA and the lack of synthetic small molecule ligands that display selectivity between these two receptors has, until recently, hindered efforts to resolve their individual functions. However, the importance of the two receptors and SCFA has been demonstrated by the use of knock-out mice. It has been shown that SCFA through GPR1-signalling are involved in the secretion of satiety hormones, GLP-1 and PYY. Both GPR 41 and GPR 43 are involved in energy metabolism and homeostasis. The SCFA have been shown to have an anti-inflammatory effect through ligation with (one of the two) GPCRs (Maslowski *et al.*, 2009; Masui *et al.*, 2013; Voltolini *et al.*, 2012), in addition to induction of regulatory T-cells (T-regs) in a GPR43-dependent manner in mice (Smith *et al.*, 2013).

Another mechanism through which SCFA, and in particular butyrate, modulate gene expression is through inhibition of HDACs (Licciardi *et al.*, 2011; Steliou *et al.*, 2012; Vinolo *et al.*, 2011). Regulation of HDAC activity may be critical for establishing immune-suppressive responses through T-reg cell activation, but also for controlling intestinal epithelial cells homeostatic responses to the luminal environment (Asselin and Gendron, 2014).

SCFA have been shown to be beneficial when given as enemas in inflammatory bowel disease (IBD), although not in all studies. Sometimes only butyrate was tested, sometimes mixtures of SCFA (Hamer *et al.*, 2008). The inconsistent results could be explained due to the fact that the use of enemas is often hampered by a low compliance rate and a short and discontinuous exposure of the colon mucosa to the SCFA. Alternatively, the inflammatory state of the volunteer might have clouded the effect of the SCFA. It has been shown for instance that butyrate enemas have a beneficial effect on pain perception during colonic extension in healthy individuals, but not in irritable bowel syndrome individuals, that have a low grade inflammation (Vanhoutvin *et al.*, 2009). Similarly, effects of butyrate enemas on inflammation and antioxidant status in the colonic mucosa of patients with ulcerative colitis in remission depended on the level of inflammation (Hamer *et al.*, 2010).

In should be mentioned that conversion of prebiotics into SCFA can also have adverse effects, especially in the case of prebiotics that are fermented fast (such as lactulose, short-chain fructooligosaccharides (FOS), or lactose in lactose-maldigesters) or are given at high doses. Fast fermentation leads to (transient) accumulation of SCFA and lactate in the proximal colon. This has been shown in rats to lead to increased translocation of *Salmonella* (Bovee-Oudenhoven *et al.*, 2003). The supposed mechanism pertains to epithelial stress or perhaps even damage due to the high concentrations of lactate and SCFA (accompanied with a low pH), allowing the pathogen to translocate. A similar mechanism is thought to be involved in lactose-intolerance where the colonic microbiota plays a role (He *et al.*, 2008). Accumulation of metabolites (e.g. due to reduced uptake by the epithelium) is thought to cause diarrhoea in these individuals. Not only lactose causes diarrhoea at high doses, this is also true for lactulose and FOS. Strikingly, for all these three carbohydrates it has been shown that a gradual increase in intake over the course of several weeks may adapt the microbiota and/or the colonic epithelium to cope with the higher doses of substrate and and/or metabolites (He *et al.*, 2008).

A too high production of SCFA may also negatively impact other processes in the body. Some of these are discussed elsewhere in this book. Of note is that studies have shown that a high concentration of propionate in the brain leads to repetitive behaviour similar as that observed in autism (MacFabe *et al.*, 2011). Since the liver is generally very well capable of clearing the SCFA from the portal vein (Bloemen *et al.*, 2010; Bloemen *et al.*, 2009), other metabolic disorders, e.g. linked to mitochondrial metabolism may be affected as well in autistic spectrum disorders (MacFabe *et al.*, 2011).

Adverse effects of fibres

Using *in vitro* platforms, fibres have been shown to be capable of immune stimulation as well as attenuation of the immune response after triggering with inflammatory ligands. In fact some fibres can have this dual (opposite) effect (Mikkelsen *et al.*, 2013) and (unpublished results), depending on the conditions. It seems therefore that the context or the environmental milieu, such as inflammatory state or presence of other ligands etc. may determine the outcome of the effect of fibres. Moreover, apart from an interaction with the immune system, fibres may have other capabilities themselves, such as binding to receptors on pathogenic (or beneficial) bacteria. The fibres then act as a decoy for receptors on epithelial or immune cells, and in such a manner can prevent the interaction of pathogens (or commensals or probiotics) with the host. Moreover, they can affect the immune system indirectly after fermentation by the microbiota. From this it is clear that depending on various other factors, the effects of fibres on the immune system may lead to adverse effects. For instance, if a fibre attenuates the immune response while a pathogen has translocated across the epithelium, the pathogen may not be cleared properly, because of a lack of phagocytosis and a lack of cellular immune response due to the attenuating effect of the fibre.

As yet, this is mostly hypothetical, as very few studies have tested the effects of fibres on the immune system. Nevertheless, there are certainly indications of the possibility of adverse effects *in vitro*. What makes matters more complex is that fibres, certainly the ones that are composed of multiple different monosaccharides and glycosidic linkages, may be recognized by different receptors, as they contain more than one ligand within the same molecule. Similarly, different fibres may be recognized by the same C-type lectin or other receptors. This will influence the overall outcome of the interaction of fibres with the host. Therefore, unless and until we gain better insight in mechanistic actions of fibres in a physiological setting, i.e. using complex experimental set-ups that are relevant for the *in vivo* situation, we should remain careful in using fibres for immunomodulatory purposes. Nevertheless, there have been some successful and interesting results as well, as discussed in the next paragraph.

In vivo studies – any evidence for an *in vivo* effect?

Despite the fact that many fibres through the binding to microbial PRRs have been demonstrated to hold immunomodulating effects *in vitro*, many questions regarding the actual effect *in vivo* upon either injection or oral administration/ingestion of the fibres remain to be answered before final conclusions on a direct immunomodulatory effect can be made. First of all, the question of whether ingested fibres in the gastrointestinal tract will be absorbed into circulation to any significant extent or in any other way may interact with receptors of cells of the immune system or other related cells remains to be determined. The few studies that indicate transfer across the GI epithelium (Hoshi *et al.*, 2008; Rice *et al.*, 2005; Sakurai *et al.*, 1996; Sandvik *et al.*, 2007; Vetvicka *et al.*, 2007; Zhu *et al.*, 2011) clearly need to be expanded. Also, fascinating studies that show that macrophages (formerly thought to be dendritic cells) can extend dendrites into the lumen of the GI tract and in this way can sample particles (including probiotics and fibres) indicate that there may in fact be a direct effect on immune cells (Rescigno, 2014).

A number of animal experiments have been performed in which the fibres have been injected either directly into the blood stream, intramuscularly or fed orally. From these experiments some evidence for a direct immune modulating effect of the fibres may be suggested, as can be evidenced below (discussed only for the oral studies).

Anti-infection activities of fibres

The direct immune modulating/stimulating activity of fibres has been speculated to impact the resistance towards microbial infections. As discussed briefly above, the fibres may act as decoys for the epithelial receptors, resulting in adhesion of microbes to the fibres molecules rather than to the epithelial cells. In this manner, the fibres preclude interaction with the host. This has been described, for instance, for mannan-derived oligosaccharides (MOS) and enterotoxic *E. coli* (ETEC) strains in pigs (e.g. Castillo *et al.*, 2008). There has been a long debate about whether or not MOS would be active *in vivo*, as the *E. coli* receptor for binding to mannan may not be expressed at 39°C, the body-temperature of pigs. But as evidenced by amongst others Castillo *et al.*, MOS is efficacious in reducing ETEC infection *in vivo*. However, not every MOS preparation may be equally effective. Alternatively, fibres may influence the immune system, leading to a reduction in infection (Rice *et al.*, 2005).

Anti-tumour activity of fibres

Perhaps the most investigated indirect immune modulatory activity of fibres hitherto is the growth inhibiting effects on tumours. In particular the β-glucans have been investigated for the antitumour activity (reviewed in (Ramberg *et al.*, 2010)), but also κ-carageenan has been investigated. Whereas β-glucans have been shown to have a beneficial effect, carrageenans have been associated with induction of ulcerative colitis and colon cancer. Carrageenans are seaweed extracts comprising high molecular weight sulphated polygalactosides. They are used in foods as thickening and gelling agents. When degraded to lower molecular weight forms, they have been shown to induce ulcerative colitis and colon cancer in laboratory animals (Pintauro and Gilbert, 1990; Tobacman, 2001). The sulphate groups are hypothesized to play a role in their detrimental effects.

The effects of β-glucans on tumour growth and survival depend on the type of tumour and have shown most efficient when the fibre treatment is combined with injection of antibodies against the tumour cells, e.g. (Modak *et al.*, 2005). Although the mechanism of action is not known, different mechanisms of action have been suggested. Among those are that the fibres promote antibody dependent cellular cytotoxicity, which facilitates the elimination of antibody coated cancer cells by macrophages and other phagocytic cells (Ross *et al.*, 1999) and that the fibres stimulate haematopoiesis hereby leading to a faster renewal of leucocytes during chemotherapy and radiation treatment.

Effects of fibres on immune cell activity and cytokine production

Apart from the immune-mediated effects in cancer, fibres also stimulate an increase in number and in activity of immune cells. For instance, lentinan has been shown to increase T-cell numbers, and the ration of helper cells over suppressor cells (Hanaue *et al.*, 1989). Also, laminarin has been shown to increase Dectin-1 expression on macrophages in the gut-associate lymphoid tissue (Rice *et al.*, 2005). In terms of activity of immune cells, β-glucans have been shown to increase plasma immunoglobulins (Chan *et al.*, 2007; Yun *et al.*, 2003), neutrophil respiratory burst (Murphy *et al.*, 2007), and cytokine production (Oh *et al.*, 2006; Yap and Ng, 2003). Other fibres, e.g. pectin, have also been shown to affect cytokine levels in plasma (Lim *et al.*, 2003). The references provided are just a few examples of numerous reports published in this field.

A few studies in man have also been carried out. Some with clear results, others with non-significant trends. For instance, arabinogalactan resulted in a trend towards increased white blood cells and oxidative burst activity (Nantz *et al.*, 2001), but showed a clear increase in IgG subtypes (Ramberg *et al.*, 2010). Other examples with β-glucans have been discussed above.

Conclusions and future perspectives

It is clear that fibres are interesting dietary ingredients with various functionalities. These range from normalizing stool frequencies to immunomodulatory and anti-tumour activities. Some microbial and dietary fibres have been studied in great detail already, such as fungal or grain-derived β-glucans and various pectins. However, due to great structural diversity amongst both the β-glucans and the various pectins, structure function relationships are still (eagerly) awaited. Unravelling these structure function relationships will perhaps be only half a decade away with the development of reporter cells that carry specific carbohydrate-recognition receptors (e.g. (Vogt *et al.*, 2013)), and the possibility to carefully modify existing bioactive polymers to decipher required oligosaccharide moieties and/or glycosidic linkages. Yet, complex carbohydrates, such as pectins, may carry many different structures that interact with multiple different lectins. Therefore, cells with single receptors are unlikely to tell the whole picture. Here, the ability to create knock out lines of rodents will help in our understanding of the role of specific receptors. This together should lead to a full understanding of the bioactivity of these important dietary ingredients. In this chapter, the focus has been on the direct and indirect immunomodulatory effects of fibres. Research so far has established that fibres may indeed affect the immune system both directly and indirectly. However, more knowledge is required regarding e.g. absorption of fibres from the GI tract and the relative impact of direct and indirect action, particularly regarding *in vivo* effects, to obtain a full understanding of how fibres affect our immune system.

References

Abbas, A.K., Lichtman, A.H., and Pillai, S., eds. (2012). Cellular and Molecular Immunology, 7th edn (Philadelphia, Elsevier/Saunders).

Adams, E.L., Rice, P.J., Graves, B., Ensley, H.E., Yu, H., Brown, G.D., Gordon, S., Monteiro, M.A., Papp-Szabo, E., Lowman, D.W., *et al.* (2008). Differential high–affinity interaction of dectin-1 with natural or synthetic glucans is dependent upon primary structure and is influenced by polymer chain length and side-chain branching. J. Pharmacol. Exp. Ther. *325*, 115–123.

Allaby, M., ed. (2006). A Dictionary of Plant Sciences, 2nd edn (Oxford, Oxford University Press).

Åman, P., and Graham, H. (1987). Analysis of total and insoluble mixed-linked (1–3),(1–4)-beta-d-glucans in barley and oats. J. Agric. Food Chem. *35*, 704–709.

Asselin, C., and Gendron, F.P. (2014). Shuttling of information between the mucosal and luminal environment drives intestinal homeostasis. FEBS Lett. *588*, 4148–4157.

Barton, G.M., and Kagan, J.C. (2009). A cell biological view of Toll-like receptor function: regulation through compartmentalization. Nat. Rev. Immunol. *9*, 535–542.

Bauer, S., Muller, T., and Hamm, S. (2009). Pattern recognition by Toll-like receptors. Adv. Exp. Med. Biol. *653*, 15–34.

Bergstrom, K.S., Kissoon-Singh, V., Gibson, D.L., Ma, C., Montero, M., Sham, H.P., Ryz, N., Huang, T., Velcich, A., Finlay, B.B., *et al.* (2010). Muc2 protects against lethal infectious colitis by disassociating pathogenic and commensal bacteria from the colonic mucosa. PLoS Pathog. *6*, e1000902.

Bernard, H., Desseyn, J.L., Bartke, N., Kleinjans, L., Stahl, B., Belzer, C., Knol, J., Gottrand, F., and Husson, M.O. (2015). Dietary pectin-derived acidic oligosaccharides improve the pulmonary bacterial clearance of *Pseudomonas aeruginosa* lung infection in mice by modulating intestinal microbiota and immunity. J. Infect. Dis. *211*, 156–165.

Bevins, C.L., and Salzman, N.H. (2011). Paneth cells, antimicrobial peptides and maintenance of intestinal homeostasis. Nat. Rev. Microbiol. *9*, 356–368.

Blixt, O., Head, S., Mondala, T., Scanlan, C., Huflejt, M.E., Alvarez, R., Bryan, M.C., Fazio, F., Calarese, D., Stevens, J., *et al.* (2004). Printed covalent glycan array for ligand profiling of diverse glycan binding proteins. Proc. Natl. Acad. Sci. U.S.A. *101*, 17033–17038.

Bloemen, J.G., Olde Damink, S.W., Venema, K., Buurman, W.A., Jalan, R., and Dejong, C.H. (2010). Short chain fatty acids exchange: Is the cirrhotic, dysfunctional liver still able to clear them? Clin. Nutr. *29*, 365–369.

Bloemen, J.G., Venema, K., van de Poll, M.C., Olde Damink, S.W., Buurman, W.A., and Dejong, C.H. (2009). Short chain fatty acids exchange across the gut and liver in humans measured at surgery. Clin. Nutr. *28*, 657–661.

Bohm, N., and Kulicke, W.-M. (1999). Rheological studies of barley (1–3)(1–4)-beta-glucan in concentrated solution: investigation of the viscoelastic flow behaviour into the sol-state. Carbohydr Res. *315*, 293–301.

Bonfim, C.V., Mamoni, R.L., and Blotta, M.H. (2009). TLR-2, TLR-4 and dectin-1 expression in human monocytes and neutrophils stimulated by Paracoccidioides brasiliensis. Med. Mycol 47, 722–733.

Bovee-Oudenhoven, I.M., ten Bruggencate, S.J., Lettink-Wissink, M.L., and van der Meer, R. (2003). Dietary fructo-oligosaccharides and lactulose inhibit intestinal colonization but stimulate translocation of salmonella in rats. Gut 52, 1572–1578.

Brandtzaeg, P. (2013). Gate-keeper function of the intestinal epithelium. Benef. Microbes 4, 67–82.

Brown, A.J., Goldsworthy, S.M., Barnes, A.A., Eilert, M.M., Tcheang, L., Daniels, D., Muir, A.I., Wigglesworth, M.J., Kinghorn, I., Fraser, N.J., *et al.* (2003). The Orphan G protein-coupled receptors GPR41 and GPR43 are activated by propionate and other short chain carboxylic acids. J. Biol. Chem. *278*, 11312–11319.

Brown, G.D. (2006). Dectin-1: a signalling non-TLR pattern-recognition receptor. Nat. Rev. Immunol. *6*, 33–43.

Brown, G.D., and Gordon, S. (2001). Immune recognition. A new receptor for beta-glucans. Nature *413*, 36–37.

Brown, G.D., and Gordon, S. (2003). Fungal beta-glucans and mammalian immunity. Immunity *19*, 311–315.

Brown, G.D., Taylor, P.R., Reid, D.M., Willment, J.A., Williams, D.L., Martinez-Pomares, L., Wong, S.Y., and Gordon, S. (2002). Dectin-1 is a major beta-glucan receptor on macrophages. J. Exp. Med. *196*, 407–412.

Canton, J., Neculai, D., and Grinstein, S. (2013). Scavenger receptors in homeostasis and immunity. Nat. Rev. Immunol. *13*, 621–634.

Castillo, M., Martin-Orue, S.M., Taylor-Pickard, J.A., Perez, J.F., and Gasa, J. (2008). Use of mannanoligosaccharides and zinc chelate as growth promoters and diarrhea preventative in weaning pigs: Effects on microbiota and gut function. J. Anim Sci. *86*, 94–101.

Chan, Y., Chang, T., Chan, C.H., Yeh, Y.C., Chen, C.W., Shieh, B., and Li, C. (2007). Immunomodulatory effects of *Agaricus blazei* Murill in Balb/cByJ. mice. J. Microbiol. Immunol. Infect. *40*, 201–208.

Chen, J., and Seviour, R. (2007). Medicinal importance of fungal beta-(1 – >3). (1 – >6)-glucans. Mycol Res. *111*, 635–652.

Childs, C.E., Roytio, H., Alhoniemi, E., Fekete, A.A., Forssten, S.D., Hudjec, N., Lim, Y.N., Steger, C.J., Yaqoob, P., Tuohy, K.M., *et al.* (2014). Xylo-oligosaccharides alone or in synbiotic combination with *Bifidobacterium animalis* subsp. *lactis* induce bifidogenesis and modulate markers of immune function in healthy adults: a double-blind, placebo-controlled, randomized, factorial cross-over study. Br. J. Nutr., *111*, 1945–1956.

Cox, M.A., Jackson, J., Stanton, M., Rojas-Triana, A., Bober, L., Laverty, M., Yang, X., Zhu, F., Liu, J., Wang, S., *et al.* (2009). Short-chain fatty acids act as antiinflammatory mediators by regulating prostaglandin E(2). and cytokines. World J. Gastroenterol. *15*, 5549–5557.

Dennehy, K.M., Ferwerda, G., Faro-Trindade, I., Pyz, E., Willment, J.A., Taylor, P.R., Kerrigan, A., Tsoni, S.V., Gordon, S., Meyer-Wentrup, F., *et al.* (2008). Syk kinase is required for collaborative cytokine production induced through Dectin-1 and Toll-like receptors. Eur. J. Immunol. *38*, 500–506.

Drummond, R.A., Saijo, S., Iwakura, Y., and Brown, G.D. (2011). The role of Syk/CARD9 coupled C-type lectins in antifungal immunity. Eur. J. Immunol. *41*, 276–281.

Dykstra, N.S., Hyde, L., Adawi, D., Kulik, D., Ahrne, S., Molin, G., Jeppsson, B., Mackenzie, A., and Mack, D.R. (2011). Pulse probiotic administration induces repeated small intestinal Muc3 expression in rats. Pediatr. Res. *69*, 206–211.

Ebersbach, T., Jorgensen, J.B., Heegaard, P.M., Lahtinen, S.J., Ouwehand, A.C., Poulsen, M., Frokiaer, H., and Licht, T.R. (2010). Certain dietary carbohydrates promote *Listeria* infection in a guinea pig model, while others prevent it. Int. J. Food Microbiol. *140*, 218–224.

Feinberg, H., Mitchell, D.A., Drickamer, K., and Weis, W.I. (2001). Structural basis for selective recognition of oligosaccharides by DC-SIGN and DC-SIGNR. Science *294*, 2163–2166.

Feriotti, C., Loures, F.V., Frank de Araujo, E., da Costa, T.A., and Calich, V.L. (2013). Mannosyl-recognizing receptors induce an M1-like phenotype in macrophages of susceptible mice but an M2-like phenotype in mice resistant to a fungal infection. PLoS One *8*, e54845.

Gazi, U., and Martinez-Pomares, L. (2009). Influence of the mannose receptor in host immune responses. Immunobiology *214*, 554–561.

Geijtenbeek, T.B., and Gringhuis, S.I. (2009). Signalling through C-type lectin receptors: shaping immune responses. Nat. Rev. Immunol. *9*, 465–479.

Gibson, G.R., and Roberfroid, M.B. (1995). Dietary modulation of the human colonic microbiota: introducing the concept of prebiotics. J. Nutr. *125*, 1401–1412.

Gill, N., Wlodarska, M., and Finlay, B.B. (2011). Roadblocks in the gut: barriers to enteric infection. Cell Microbiol. *13*, 660–669.

Goodridge, H.S., and Underhill, D.M. (2008). Fungal Recognition by TLR2 and Dectin-1. Handb. Exp. Pharmacol. 87–109.

Gourbeyre, P., Desbuards, N., Gremy, G., Le Gall, S., Champ, M., Denery-Papini, S., and Bodinier, M. (2012). Exposure to a galactooligosaccharides/inulin prebiotic mix at different developmental time points differentially modulates immune responses in mice. J. Agric. Food Chem. *60*, 11942–11951.

Gourbeyre, P., Desbuards, N., Gremy, G., Tranquet, O., Champ, M., Denery-Papini, S., and Bodinier, M. (2013). Perinatal and postweaning exposure to galactooligosaccharides/inulin prebiotics induced biomarkers linked to tolerance mechanism in a mouse model of strong allergic sensitization. J. Agric. Food Chem. *61*, 6311–6320.

Gringhuis, S.I., den Dunnen, J., Litjens, M., van der Vlist, M., and Geijtenbeek, T.B. (2009). Carbohydrate-specific signaling through the DC-SIGN signalosome tailors immunity to *Mycobacterium tuberculosis*, HIV-1 and Helicobacter pylori. Nat. Immunol. *10*, 1081–1088.

Gringhuis, S.I., den Dunnen, J., Litjens, M., van Het Hof, B., van Kooyk, Y., and Geijtenbeek, T.B. (2007). C-type lectin DC-SIGN modulates Toll-like receptor signaling via Raf-1 kinase-dependent acetylation of transcription factor NF-kappaB. Immunity *26*, 605–616.

Håkansson, A., Ulmius, M., and Nilsson, L. (2012). Asymmetrical flow field-flow fractionation enables the characterization of molecular and supramolecular properties of cereal β-glucan dispersions. Carbohydrate Polymers *87*, 518–523.

Hamer, H.M., Jonkers, D., Venema, K., Vanhoutvin, S., Troost, F.J., and Brummer, R.J. (2008). Review article: the role of butyrate on colonic function. Aliment Pharmacol. Ther. *27*, 104–119.

Hamer, H.M., Jonkers, D.M., Vanhoutvin, S.A., Troost, F.J., Rijkers, G., de Bruine, A., Bast, A., Venema, K., and Brummer, R.J. (2010). Effect of butyrate enemas on inflammation and antioxidant status in the colonic mucosa of patients with ulcerative colitis in remission. Clin. Nutr. *29*, 738–744.

Hanaue, H., Tokuda, Y., Machimura, T., Kamijoh, A., Kondo, Y., Ogoshi, K., Makuuchi, H., Nakasaki, H., Tajima, T., Mitomi, T., *et al.* (1989). Effects of oral lentinan on T-cell subsets in peripheral venous blood. Clin. Ther. *11*, 614–622.

Hao, M., Yuan, X., Cheng, H., Xue, H., Zhang, T., Zhou, Y., and Tai, G. (2013). Comparative studies on the anti-tumor activities of high temperature- and pH-modified citrus pectins. Food Funct. *4*, 960–971.

He, T., Venema, K., Priebe, M.G., Welling, G.W., Brummer, R.J., and Vonk, R.J. (2008). The role of colonic metabolism in lactose intolerance. Eur. J. Clin. Invest. *38*, 541–547.

Hooper, L.V., and Macpherson, A.J. (2010). Immune adaptations that maintain homeostasis with the intestinal microbiota. Nat. Rev. Immunol. *10*, 159–169.

Hoshi, H., Iijima, H., Ishihara, Y., Yasuhara, T., and Matsunaga, K. (2008). Absorption and tissue distribution of an immunomodulatory alpha-d-glucan after oral administration of *Tricholoma matsutake*. J. Agric. Food Chem. *56*, 7715–7720.

Hudcovic, T., Kolinska, J., Klepetar, J., Stepankova, R., Rezanka, T., Srutkova, D., Schwarzer, M., Erban, V., Du, Z., Wells, J.M., *et al.* (2012). Protective effect of Clostridium tyrobutyricum in acute dextran sodium sulphate-induced colitis: differential regulation of tumour necrosis factor-alpha and interleukin-18 in BALB/c and severe combined immunodeficiency mice. Clin. Exp. Immunol. *167*, 356–365.

Ikeda, Y., Adachi, Y., Ishii, T., Miura, N., Tamura, H., and Ohno, N. (2008). Dissociation of Toll-like receptor 2-mediated innate immune response to Zymosan by organic solvent-treatment without loss of Dectin-1 reactivity. Biol. Pharm. Bull. *31*, 13–18.

Irhimeh, M.R., Fitton, J.H., and Lowenthal, R.M. (2007). Fucoidan ingestion increases the expression of CXCR4 on human CD34+ cells. Exp. Hematol. *35*, 989–994.

Isidro, R.A., Bonilla, F.J., Pagan, H., Cruz, M.L., Lopez, P., Godoy, L., Hernandez, S., Loucil-Alicea, R.Y., Rivera-Amill, V., Yamamura, Y., *et al.* (2014). The probiotic mixture VSL#3 Alters the morphology and secretion profile of both polarized and unpolarized human macrophages in a polarization-dependent manner. J. Clin. Cell Immunol. *5*, 227.

Jawhara, S., Habib, K., Maggiotto, F., Pignede, G., Vandekerckove, P., Maes, E., Dubuquoy, L., Fontaine, T., Guerardel, Y., and Poulain, D. (2012). Modulation of intestinal inflammation by yeasts and cell wall extracts: strain dependence and unexpected anti-inflammatory role of glucan fractions. PLoS One *7*, e40648.

Jozefowski, S., Yang, Z., Marcinkiewicz, J., and Kobzik, L. (2012). Scavenger receptors and beta-glucan receptors participate in the recognition of yeasts by murine macrophages. Inflamm. Res. *61*, 113–126.

Kerrigan, A.M., and Brown, G.D. (2009). C-type lectins and phagocytosis. Immunobiology *214*, 562–575.

Kery, V., Krepinsky, J.J., Warren, C.D., Capek, P., and Stahl, P.D. (1992). Ligand recognition by purified human mannose receptor. Arch Biochem. Biophys. *298*, 49–55.

Kimura, I., Ozawa, K., Inoue, D., Imamura, T., Kimura, K., Maeda, T., Terasawa, K., Kashihara, D., Hirano, K., Tani, T., *et al.* (2013). The gut microbiota suppresses insulin-mediated fat accumulation via the short-chain fatty acid receptor GPR43. Nat. Commun *4*, 1829.

Kirmaz, C., Bayrak, P., Yilmaz, O., and Yuksel, H. (2005). Effects of glucan treatment on the Th1/Th2 balance in patients with allergic rhinitis: a double-blind placebo-controlled study. Eur. Cytokine Netw. *16*, 128–134.

Koray, M., Ak, G., Kurklu, E., Tanyeri, H., Aydin, F., Oguz, F.S., Temurhan, S., Ciltci, H., Carin, M., Onal, A.E., *et al.* (2009). The effect of beta-glucan on recurrent aphthous stomatitis. J. Altern. Complement Med. *15*, 111–112.

Kovacs-Nolan, J., Kanatani, H., Nakamura, A., Ibuki, M., and Mine, Y. (2013). beta-1,4-mannobiose stimulates innate immune responses and induces TLR4-dependent activation of mouse macrophages but reduces severity of inflammation during endotoxemia in mice. J. Nutr. *143*, 384–391.

Largent, B.L., Walton, K.M., Hoppe, C.A., Lee, Y.C., and Schnaar, R.L. (1984). Carbohydrate-specific adhesion of alveolar macrophages to mannose-derivatized surfaces. J. Biol. Chem. *259*, 1764–1769.

Lecerf, J.M., Depeint, F., Clerc, E., Dugenet, Y., Niamba, C.N., Rhazi, L., Cayzeele, A., Abdelnour, G., Jaruga, A., Younes, H., *et al.* (2012). Xylo-oligosaccharide (XOS) in combination with inulin modulates both the intestinal environment and immune status in healthy subjects, while XOS alone only shows prebiotic properties. Br. J. Nutr. *108*, 1847–1858.

Lee, S.J., Zheng, N.Y., Clavijo, M., and Nussenzweig, M.C. (2003). Normal host defense during systemic candidiasis in mannose receptor-deficient mice. Infect. Immun. *71*, 437–445.

LeibundGut-Landmann, S., Gross, O., Robinson, M.J., Osorio, F., Slack, E.C., Tsoni, S.V., Schweighoffer, E., Tybulewicz, V., Brown, G.D., Ruland, J., *et al.* (2007). Syk- and CARD9-dependent coupling of innate immunity to the induction of T helper cells that produce interleukin 17. Nat. Immunol. *8*, 630–638.

Leibundgut-Landmann, S., Osorio, F., Brown, G.D., and Reis e Sousa, C. (2008). Stimulation of dendritic cells via the dectin-1/Syk pathway allows priming of cytotoxic T-cell responses. Blood *112*, 4971–4980.

Levitz, S.M. (2010). Innate recognition of fungal cell walls. PLoS Pathog. *6*, e1000758.

Licciardi, P.V., Ververis, K., and Karagiannis, T.C. (2011). Histone deacetylase inhibition and dietary short-chain Fatty acids. ISRN Allergy *2011*, 869647.

Lievin-Le Moal, V., and Servin, A.L. (2006). The front line of enteric host defense against unwelcome intrusion of harmful microorganisms: mucins, antimicrobial peptides, and microbiota. Clin. Microbiol. Rev. *19*, 315–337.

Lim, B.O., Lee, S.H., Park, D.K., and Choue, R.W. (2003). Effect of dietary pectin on the production of immunoglobulins and cytokines by mesenteric lymph node lymphocytes in mouse colitis induced with dextran sulfate sodium. Biosci. Biotechnol. Biochem. *67*, 1706–1712.

Lomax, A.R., Cheung, L.V., Tuohy, K.M., Noakes, P.S., Miles, E.A., and Calder, P.C. (2012). beta2-1 Fructans have a bifidogenic effect in healthy middle-aged human subjects but do not alter immune responses examined in the absence of an *in vivo* immune challenge: results from a randomised controlled trial. Br. J. Nutr. *108*, 1818–1828.

Loures, F.V., Araujo, E.F., Feriotti, C., Bazan, S.B., Costa, T.A., Brown, G.D., and Calich, V.L. (2014). Dectin-1 Induces M1 Macrophages and Prominent Expansion of CD8+IL-17+ Cells in Pulmonary Paracoccidioidomycosis. J. Infect. Dis. *210*, 762–773.

MacFabe, D.F., Cain, N.E., Boon, F., Ossenkopp, K.P., and Cain, D.P. (2011). Effects of the enteric bacterial metabolic product propionic acid on object-directed behavior, social behavior, cognition, and neuroinflammation in adolescent rats: Relevance to autism spectrum disorder. Behav. Brain Res. *217*, 47–54.

Mack, D.R., Michail, S., Wei, S., McDougall, L., and Hollingsworth, M.A. (1999). Probiotics inhibit enteropathogenic E. coli adherence *in vitro* by inducing intestinal mucin gene expression. Am. J. Physiol. *276*, G941–950.

Martinez-Pomares, L. (2012). The mannose receptor. J. Leukoc. Biol. *92*, 1177–1186.

Maslowski, K.M., Vieira, A.T., Ng, A., Kranich, J., Sierro, F., Yu, D., Schilter, H.C., Rolph, M.S., Mackay, F., Artis, D., *et al.* (2009). Regulation of inflammatory responses by gut microbiota and chemoattractant receptor GPR43. Nature *461*, 1282–1286.

Masui, R., Sasaki, M., Funaki, Y., Ogasawara, N., Mizuno, M., Iida, A., Izawa, S., Kondo, Y., Ito, Y., Tamura, Y., *et al.* (2013). G protein-coupled receptor 43 moderates gut inflammation through cytokine regulation from mononuclear cells. Inflamm. Bowel Dis. *19*, 2848–2856.

Mikkelsen, M.S., Jespersen, B.M., Larsen, F.H., Blennow, A., and Engelsen, S.B. (2013). Molecular structure of large-scale extracted beta-glucan from barley and oat: Identification of a significantly changed block structure in a high beta-glucan barley mutant. Food Chem. *136*, 130–138.

Mikkelsen, M.S., Jespersen, B.M., Mehlsen, A., Engelsen, S.B., and Frøkiær, H. (2014). Cereal β-glucan immune modulating activity depends on the polymer fine structure Food Research International *62*, 829–836.

Mills, C.D., and Ley, K. (2014). M1 and M2 macrophages: the chicken and the egg of immunity. J. Innate Immun. *6*, 716–726.

Modak, S., Koehne, G., Vickers, A., O'Reilly, R.J., and Cheung, N.K. (2005). Rituximab therapy of lymphoma is enhanced by orally administered (1→3),(1→4)-D-β-glucan. Leuk. Res. *29*, 679–683.

Murphy, E.A., Davis, J.M., Brown, A.S., Carmichael, M.D., Ghaffar, A., and Mayer, E.P. (2007). Oat beta-glucan effects on neutrophil respiratory burst activity following exercise. Med. Sci. Sports Exerc *39*, 639–644.

Nantz, M.P., Parker, A.R., McGill, C., and Percival, S.S. (2001). Evaluation of arabinogalactan's effect on human immunity. FASEB J. *15*, A633.

O'Neill, L.A. (2008). When signaling pathways collide: positive and negative regulation of toll-like receptor signal transduction. Immunity *29*, 12–20.

Oh, G.S., Lee, M.S., Pae, H.O., Kwon, J., Lee, S.S., Jeong, J.G., Shin, M.K., Kwon, T.O., and Chung, H.T. (2006). Effects of oral administration of Phellinus linteus on the production of Th1- and Th2-type cytokines in mice. Immunopharmacol. Immunotoxicol. *28*, 281–293.

Osorio, F., LeibundGut-Landmann, S., Lochner, M., Lahl, K., Sparwasser, T., Eberl, G., and Reis e Sousa, C. (2008). DC activated via dectin-1 convert Treg into IL-17 producers. Eur. J. Immunol. *38*, 3274–3281.

Ouwehand, A.C., Derrien, M., de Vos, W., Tiihonen, K., and Rautonen, N. (2005). Prebiotics and other microbial substrates for gut functionality. Curr. Opin. Biotechnol. *16*, 212–217.

Petersen, A., Heegaard, P.M., Pedersen, A.L., Andersen, J.B., Sorensen, R.B., Frokiaer, H., Lahtinen, S.J., Ouwehand, A.C., Poulsen, M., and Licht, T.R. (2009). Some putative prebiotics increase the severity of *Salmonella enterica* serovar Typhimurium infection in mice. BMC Microbiol. *9*, 245.

Pintauro, S.J., and Gilbert, S.W. (1990). The effects of carrageenan on drug-metabolizing enzyme system activities in the guinea pig. Food Chem. Toxicol. *28*, 807–811.

Plato, A., Willment, J.A., and Brown, G.D. (2013). C-type lectin-like receptors of the dectin-1 cluster: ligands and signaling pathways. Int. Rev. Immunol. *32*, 134–156.

Pluddemann, A., Mukhopadhyay, S., and Gordon, S. (2006). The interaction of macrophage receptors with bacterial ligands. Expert Rev. Mol. Med. *8*, 1–25.

Ramberg, J.E., Nelson, E.D., and Sinnott, R.A. (2010). Immunomodulatory dietary polysaccharides: a systematic review of the literature. Nutr. J. *9*, 54.

Rescigno, M. (2014). Dendritic cell-epithelial cell crosstalk in the gut. Immunol. Rev. *260*, 118–128.

Rice, P.J., Adams, E.L., Ozment-Skelton, T., Gonzalez, A.J., Goldman, M.P., Lockhart, B.E., Barker, L.A., Breuel, K.F., Deponti, W.K., Kalbfleisch, J.H., *et al.* (2005). Oral delivery and gastrointestinal absorption of soluble glucans stimulate increased resistance to infectious challenge. J. Pharmacol. Exp. Ther. *314*, 1079–1086.

Riede, L., Grube, B., and Gruenwald, J. (2013). Larch arabinogalactan effects on reducing incidence of upper respiratory infections. Curr. Med. Res. Opin. *29*, 251–258.

Ross, G.D., Cain, J.A., and Lachmann, P.J. (1985). Membrane complement receptor type three (CR3). has lectin-like properties analogous to bovine conglutinin as functions as a receptor for zymosan and rabbit erythrocytes as well as a receptor for iC3b. J. Immunol. *134*, 3307–3315.

Ross, G.D., Vetvicka, V., Yan, J., Xia, Y., and Vetvickova, J. (1999). Therapeutic intervention with complement and beta-glucan in cancer. Immunopharmacology *42*, 61–74.

Sakurai, M.H., Matsumoto, T., Kiyohara, H., and Yamada, H. (1996). Detection and tissue distribution of anti-ulcer pectic polysaccharides from Bupleurum falcatum by polyclonal antibody. Planta Med. *62*, 341–346.

Sancho, D., and Reis e Sousa, C. (2012). Signaling by myeloid C-type lectin receptors in immunity and homeostasis. Annu. Rev. Immunol. *30*, 491–529.

Sandvik, A., Wang, Y.Y., Morton, H.C., Aasen, A.O., Wang, J.E., and Johansen, F.E. (2007). Oral and systemic administration of beta-glucan protects against lipopolysaccharide-induced shock and organ injury in rats. Clin. Exp. Immunol. *148*, 168–177.

Sato, K., Yang, X.L., Yudate, T., Chung, J.S., Wu, J., Luby-Phelps, K., Kimberly, R.P., Underhill, D., Cruz, P.D., Jr. and Ariizumi, K. (2006). Dectin-2 is a pattern recognition receptor for fungi that couples with the Fc receptor gamma chain to induce innate immune responses. J. Biol. Chem. *281*, 38854–38866.

Scalabrin, D.M., Mitmesser, S.H., Welling, G.W., Harris, C.L., Marunycz, J.D., Walker, D.C., Bos, N.A., Tolkko, S., Salminen, S., and Vanderhoof, J.A. (2012). New prebiotic blend of polydextrose and galacto-oligosaccharides has a bifidogenic effect in young infants. J. Pediatr. Gastroenterol. Nutr. *54*, 343–352.

Schlee, M., Harder, J., Koten, B., Stange, E.F., Wehkamp, J., and Fellermann, K. (2008). Probiotic lactobacilli and VSL#3 induce enterocyte beta-defensin 2. Clin. Exp. Immunol. *151*, 528–535.

Smith, P.M., Howitt, M.R., Panikov, N., Michaud, M., Gallini, C.A., Bohlooly, Y.M., Glickman, J.N., and Garrett, W.S. (2013). The microbial metabolites, short-chain fatty acids, regulate colonic Treg cell homeostasis. Science *341*, 569–573.

Steliou, K., Boosalis, M.S., Perrine, S.P., Sangerman, J., and Faller, D.V. (2012). Butyrate histone deacetylase inhibitors. Biores. Open Access *1*, 192–198.

Tachado, S.D., Zhang, J., Zhu, J., Patel, N., Cushion, M., and Koziel, H. (2007). Pneumocystis-mediated IL-8 release by macrophages requires coexpression of mannose receptors and TLR2. J. Leukoc. Biol. *81*, 205–211.

Tan, J., McKenzie, C., Potamitis, M., Thorburn, A.N., Mackay, C.R., and Macia, L. (2014). The role of short-chain fatty acids in health and disease. Adv. Immunol. *121*, 91–119.

Taylor, P.R., Brown, G.D., Reid, D.M., Willment, J.A., Martinez-Pomares, L., Gordon, S., and Wong, S.Y. (2002). The beta-glucan receptor, dectin-1, is predominantly expressed on the surface of cells of the monocyte/macrophage and neutrophil lineages. J. Immunol. *169*, 3876–3882.

Taylor, P.R., Gordon, S., and Martinez-Pomares, L. (2005). The mannose receptor: linking homeostasis and immunity through sugar recognition. Trends Immunol. *26*, 104–110.

Thornton, B.P., Vetvicka, V., Pitman, M., Goldman, R.C., and Ross, G.D. (1996). Analysis of the sugar specificity and molecular location of the beta-glucan-binding lectin site of complement receptor type 3 (CD11b/CD18). J. Immunol. *156*, 1235–1246.

Tobacman, J.K. (2001). Review of harmful gastrointestinal effects of carrageenan in animal experiments. Environ. Health Perspect. *109*, 983–994.

Tosh, S.M., Wood, P.J., Wang, Q., and Wesz, J. (2004). Structural characteristics and rheological properties of partially hydrolysed oat β-glucan: the effect of molecular weight and hydrolysis method. Carbohydrate Polymers 55, 425–436.

Turner, N.D., and Lupton, J.R. (2011). Dietary fiber. Adv. Nutr. 2, 151–152.

Udani, J.K. (2013). Immunomodulatory effects of ResistAid: A randomized, double-blind, placebo-controlled, multidose study. J. Am. Coll. Nutr. 32, 331–338.

Udani, J.K., Singh, B.B., Barrett, M.L., and Singh, V.J. (2010). Proprietary arabinogalactan extract increases antibody response to the pneumonia vaccine: a randomized, double-blind, placebo-controlled, pilot study in healthy volunteers. Nutr. J. 9, 32.

van de Pol, M.A., Lutter, R., Smids, B.S., Weersink, E.J., and van der Zee, J.S. (2011). Synbiotics reduce allergen-induced T-helper 2 response and improve peak expiratory flow in allergic asthmatics. Allergy *66*, 39–47.

van der Aa, L.B., van Aalderen, W.M., Heymans, H.S., Henk Sillevis Smitt, J., Nauta, A.J., Knippels, L.M., Ben Amor, K., and Sprikkelman, A.B. (2011). Synbiotics prevent asthma-like symptoms in infants with atopic dermatitis. Allergy *66*, 170–177.

van der Sluis, M., Bouma, J., Vincent, A., Velcich, A., Carraway, K.L., Buller, H.A., Einerhand, A.W., van Goudoever, J.B., Van Seuningen, I., and Renes, I.B. (2008). Combined defects in epithelial and immunoregulatory factors exacerbate the pathogenesis of inflammation: mucin 2-interleukin 10-deficient mice. Lab Invest. *88*, 634–642.

Van der Sluis, M., De Koning, B.A., De Bruijn, A.C., Velcich, A., Meijerink, J.P., Van Goudoever, J.B., Buller, H.A., Dekker, J., Van Seuningen, I., Renes, I.B., *et al.* (2006). Muc2-deficient mice spontaneously develop colitis, indicating that MUC2 is critical for colonic protection. Gastroenterology *131*, 117–129.

van der Waaij, L.A., Limburg, P.C., Mesander, G., and van der Waaij, D. (1996). *In vivo* IgA coating of anaerobic bacteria in human faeces. Gut *38*, 348–354.

Vanhoutvin, S.A., Troost, F.J., Kilkens, T.O., Lindsey, P.J., Hamer, H.M., Jonkers, D.M., Venema, K., and Brummer, R.J. (2009). The effects of butyrate enemas on visceral perception in healthy volunteers. Neurogastroenterol. Motil. *21*, 952-e76.

Velez, E., Castillo, N., Meson, O., Grau, A., Bibas Bonet, M.E., and Perdigon, G. (2013). Study of the effect exerted by fructo-oligosaccharides from yacon (Smallanthus sonchifolius) root flour in an intestinal infection model with Salmonella Typhimurium. Br. J. Nutr. *109*, 1971–1979.

Vetvicka, V., Dvorak, B., Vetvickova, J., Richter, J., Krizan, J., Sima, P., and Yvin, J.C. (2007). Orally administered marine (1 – >3)-beta-d-glucan Phycarine stimulates both humoral and cellular immunity. Int. J. Biol. Macromol *40*, 291–298.

Vinolo, M.A., Rodrigues, H.G., Nachbar, R.T., and Curi, R. (2011). Regulation of inflammation by short-chain fatty acids. Nutrients *3*, 858–876.

Vogt, L., Ramasamy, U., Meyer, D., Pullens, G., Venema, K., Faas, M.M., Schols, H.A., and de Vos, P. (2013). Immune modulation by different types of beta2 – >1-fructans is toll-like receptor dependent. PLoS One *8*, e68367.

Vogt, L.M., Meyer, D., Pullens, G., Faas, M.M., Venema, K., Ramasamy, U., Schols, H.A., and de Vos, P. (2014). Toll-like receptor 2 activation by beta2 ® 1-fructans protects barrier function of T84 human intestinal epithelial cells in a chain length-dependent manner. J. Nutr. *144*, 1002–1008.

Voltolini, C., Battersby, S., Etherington, S.L., Petraglia, F., Norman, J.E., and Jabbour, H.N. (2012). A novel antiinflammatory role for the short-chain fatty acids in human labour. Endocrinology *153*, 395–403.

Vulevic, J., Juric, A., Tzortzis, G., and Gibson, G.R. (2013). A mixture of trans-galactooligosaccharides reduces markers of metabolic syndrome and modulates the faecal microbiota and immune function of overweight adults. J. Nutr. *143*, 324–331.

Wehkamp, J., Harder, J., Wehkamp, K., Wehkamp-von Meissner, B., Schlee, M., Enders, C., Sonnenborn, U., Nuding, S., Bengmark, S., Fellermann, K., *et al.* (2004). NF-kappaB- and AP-1-mediated induction of human beta defensin-2 in intestinal epithelial cells by *Escherichia coli* Nissle 1917: a novel effect of a probiotic bacterium. Infect. Immun. *72*, 5750–5758.

Wismar, R., Brix, S., Frokiaer, H., and Laerke, H.N. (2010). Dietary fibres as immunoregulatory compounds in health and disease. Ann. N.Y. Acad. Sci. *1190*, 70–85.

Wismar, R., Brix, S., Laerke, H.N., and Frokiaer, H. (2011). Comparative analysis of a large panel of non-starch polysaccharides reveals structures with selective regulatory properties in dendritic cells. Mol. Nutr. Food Res. *55*, 443–454.

Wlodarska, M., Willing, B., Keeney, K.M., Menendez, A., Bergstrom, K.S., Gill, N., Russell, S.L., Vallance, B.A., and Finlay, B.B. (2011). Antibiotic treatment alters the colonic mucus layer and predisposes the host to exacerbated Citrobacter rodentium-induced colitis. Infect. Immun. *79*, 1536–1545.

Yamasaki, S., Matsumoto, M., Takeuchi, O., Matsuzawa, T., Ishikawa, E., Sakuma, M., Tateno, H., Uno, J., Hirabayashi, J., Mikami, Y., *et al.* (2009).

C-type lectin Mincle is an activating receptor for pathogenic fungus, Malassezia. Proc. Natl. Acad. Sci. U.S.A. *106*, 1897–1902.

Yap, A.T., and Ng, M.-L. (2003). Immunopotentiating properties of lentinan (1–3)-b-Dglucan extracted from culinary-medicinal Shiitake mushroom. Int. J. Medicinal Mushrooms 5, 19–39.

Yun, C.H., Estrada, A., Van Kessel, A., Park, B.C., and Laarveld, B. (2003). Beta-glucan, extracted from oat, enhances disease resistance against bacterial and parasitic infections. FEMS Immunol. Med. Microbiol. 35, 67–75.

Zelensky, A.N., and Gready, J.E. (2004). C-type lectin-like domains in Fugu rubripes. BMC Genomics 5, 51.

Zhu, W., Gu, B., Miao, J., Lu, J., and Zou, S. (2011). Dectin1 activation of beta-(1–3)/(1–6)-d-glucan produces an anti-mastitis effect in rats. Inflamm. Res. *60*, 937–945.

Prebiotics Beyond Fibres

22

Delphine M. Saulnier and Michael Blaut

Abstract

The research on the potential prebiotic properties of polyphenols, human milk oligosaccharides (HMOs), polypeptides and polyols – bioactive compounds not being strictly defined as fibres – has increased at a rapid pace. A selective stimulation of members of the microbiota traditionally associated with health benefits, such as bifidobacteria and lactobacilli, and an inhibition of potentially pathogenic species, has been demonstrated *in vitro* for most of these compounds. Evidence for health benefits associated with microbiota changes have started to emerge for polyphenols and HMOs in animal studies. An improvement of biomarkers for type 2 diabetes accompanied by an increase in bifidobacteria and lactobacilli occurred in a limited number of healthy subjects in response to the intake of polyphenols. Well-designed double-blind placebo controlled nutritional intervention studies should be conducted to confirm or refute a causal link between changes in microbiota composition and an improvement of type 2 diabetes or obesity. Here we present and discuss our perception of the future trends that we foresee in this field, and provide some recommendations for future research.

Introduction: scope and definitions

Epidemiological studies have indicated that diet has a major impact on human health. A diet high in dietary fibre has been associated with a decreased risk of non-communicable chronic diseases such type 2 diabetes, cardiovascular diseases (Cho *et al.*, 2013) or colon cancer (Bingham *et al.*, 2003). The human gastrointestinal (GI) tract contains an estimated 10^{14} microbial cells, primarily inhabiting the colon, and representing at least 160 species in a given individual (Qin *et al.*, 2010). Compositional profile and activity of the gut microbiota are being scrutinized, as they are intricately linked to host wellbeing. Most members of the bacterial community are commensals, meaning that they are neither advantageous nor harmful to the host. However, some community members promote health while others cause disease. Bacterial colonization begins preferentially at birth, even though recent work suggested that colonization already starts before birth (DiGiulio *et al.*, 2008). A variety of factors including diet are of considerable importance for the development of the microbiota later in life. Dietary fibre is considered the main substrate of intestinal bacteria. However, other non-digestible, non-nutritive dietary compounds such as polyphenols and glucosinolates may also undergo degradation by intestinal bacteria.

The definition of dietary fibre has been a matter of debate for several decades. Dietary fibre originally has been defined as 'that portion of food which is derived from cellular walls of plants which are digested very poorly by human beings' (Trowell, 1972). The recognition that polysaccharides added to food, notably hydrocolloids, could have similar effects as plant cell wall components led to a new definition to include 'polysaccharides and lignin that are not digested in the human small intestine' (Trowell *et al.*, 1976). The most current definition has been adopted in June 2009 by the Codex Executive Committee (http://www.codexalimentarius.net/download/report/710/al32_26e.pdf). Dietary fibres are defined herein as 'carbohydrate polymers with ten or more monomeric units, which are not hydrolysed by the endogenous enzymes in the small intestine of humans and belong to the following categories:

- edible carbohydrate polymers naturally occurring in the food as consumed;
- carbohydrate polymers, which have been obtained from food raw material by physical, enzymatic or chemical means and which have been shown to have a physiological effect of benefit to health as demonstrated by generally accepted scientific evidence to competent authorities;
- synthetic carbohydrate polymers which have been shown to have a physiological effect of benefit to health as demonstrated by generally accepted scientific evidence to competent authorities.

The decision on whether to include carbohydrates compounds from 3 to 9 monomeric units has been left to national authorities. The European Union also adopted an almost identical definition in October, 2008 for the term 'fibre', amending Annex II of Council Directive 90/496/EEC on nutrition labelling based on work by the Codex as well as by the European Food Safety Authority (http://eurlex.europa.eu/LexUriServ/LexUriServ.do?uri=OJ:L:2008:285:0009:0012:en:pdf). Beneficial physiological effects of fibres have to be demonstrated before addition of natural or synthetic fibres to foods. The main types of dietary fibre that are often considered include:

- Non-starch polysaccharides (NSP):
  - cellulose, hemicelluloses, pectins, hydrocolloids (i.e. gums, mucilages, glucans)
- Resistant oligosaccharides:
  - fructooligosaccharides (FOS), galactooligosaccharides (GOS), other resistant oligosaccharides
- Resistant starch:
  - physically enclosed starch, some types of raw starch granules, retrograded amylose, chemically and/or physically modified starches
- Lignin associated with the dietary fibre polysaccharides.

Prebiotics, 'non-digestible food ingredients that beneficially affect the host by selectively stimulating the growth and/or activity of one or a limited number of bacteria that can improve the host health' (Reid *et al.*, 2003), are an alternative (or adjunct) approach to probiotics and target indigenous beneficial bacteria already established in the gut. The health benefits of well-established prebiotics such as fructans (fructooligosaccharides (FOS), inulin), galactooligosaccharides (GOS), and lactulose are well documented [see Roberfroid *et al.* (2010) for an extensive review and also elsewhere in this book]. Historically, *Bifidobacterium* spp. and *Lactobacillus* spp. have been the preferred targets of prebiotics. Nowadays, it is suggested that the stimulation of other intestinal microbes, such as *Faecalibacterium prausnitzii* or other butyrate-producing bacteria, or in general bacteria that confer a specific health benefit on the host may be the targets of prebiotics. However, knowledge about the microbes mediating health effects is still scarce.

Health benefits of prebiotics and dietary fibres are often intimately intertwined. Both are non-digestible, and have often the same benefits, including improvement of intestinal functions (stool bulking, stool consistency, stool regularity), or increased satiety. For a long time, prebiotic research has focused on the assessment of a variety of fibres as potential prebiotics. However, not all fibres are prebiotics. Conversely all potential or well-established prebiotics such are lactulose do not fulfil the fibre definition. Research on potential prebiotic properties of other bioactive compounds, which are not defined as fibres such as polyphenols, polyols, human milk oligosaccharides (HMOs), or polypeptides, has increased at a rapid pace.

In this chapter, we review the evidence for health benefits and changes in gut microbiota in response to bioactive compounds which are not strictly defined as fibres, but owing to their effects, might be classified as prebiotics in the future. Overall, most of the studies have focused on the modulation of microbiota composition, often targeting a small proportion of bacterial groups encountered in the gastrointestinal tract. Very few of these studies have evaluated concomitant health benefits to the host and mechanisms of action. In the last part of this chapter, we summarize the main limitations that impede research and product development of these potential prebiotics. We conclude by giving our view of future trends and recommendations for future research.

Emerging prebiotics beyond fibres

Polyphenols

Polyphenols are secondary plant metabolites ubiquitously distributed in higher plants (Kemperman *et al.*, 2010). There is increasing evidence that polyphenols, their degradational intermediates, or their end products, can modulate the composition and the activity of the gut microbiota, and thereby contribute to host health (Neyrinck *et al.*, 2013; Tzounis *et al.*, 2011). Polyphenols are characterized by the presence of multiple phenol units, whose numbers and characteristics determine the physical, chemical, and biological (metabolic, toxic, therapeutic, etc.) properties. These phytochemicals comprise a wide range of molecules, ranging from low-molecular weight phenolic polymers to complex proanthocyanidins. The most important food sources are commodities widely consumed in large quantities such as fruit and vegetables, green tea, black tea, red wine, coffee, chocolate, olives, and extra virgin olive oil (D'Archivio *et al.*, 2010). Some polyphenols are specific to certain food items (isoflavones in soy, phloridzin in apples, flavanones in citrus fruit) whereas others, such as quercetin, are found in all plant products such as fruit, vegetables, cereals, leguminous plants, tea, and wine.

A high inter-individual variability exists in the metabolism of polyphenols, including their degradation and absorption (Bolca *et al.*, 2013). Gastrointestinal microbiota composition has been suggested to be one, if not the most important reason for individual differences in polyphenol bioavailability, conversion and absorption in humans (Crozier *et al.*, 2009). Polyphenol metabolism usually starts with the hydrolysis of polymeric, glycosylated and/or esterified native compounds by brush border enzymes and/or microbial enzymes (Kemperman *et al.*, 2010). Selective bacterial stimulation seems to be achieved by both inhibition of potential pathogens and stimulation of selected strains of bifidobacteria and lactobacilli. The products of degradation, in particular the short-chain fatty acid (SCFA) profiles, distinguish polyphenols from FOS or GOS. The microbial enzyme repertoire for polyphenol degradation does not seem as redundant as the one observed for well-established prebiotics such as FOS or GOS either. It usually requires multiple ensuing enzymatic reactions that often cannot be accomplished by a sole microorganism, suggesting a need for interactions among different members of the gut microbiota. An improvement of biomarkers for type 2 diabetes associated with the consumption of dietary polyphenols and concomitant changes in microbiota have been demonstrated in animal models of obesity associated with hypercholesterolaemia and inflammatory disorders (Neyrinck *et al.*, 2013).

Selective antimicrobial activity against pathogens (Selective inhibition)

In contrast to FOS or GOS, whose presence in the gastrointestinal directly stimulates microbes able to degrade and utilize monomers of the oligosaccharides, the selective stimulation of intestinal bacteria by polyphenols rather seems to be based on their potential to selectively inhibit the growth and/or the activity of pathogens in the gastrointestinal tract. This effect has been demonstrated *in vitro* for a variety of polyphenols derived from tea, cocoa or coffee. Inhibition of growth of *Clostridium perfringens*, or *Clostridium difficile* was inhibited by several tea-derived polyphenols while bifidobacteria and lactobacilli levels were less affected (Lee *et al.*, 2006). Flavonoid aglycones, but not their glycosides, were able to inhibit growth of some intestinal bacteria considered potentially detrimental to the host (Duda-Chodak, 2012). Selective inhibition rather than selective stimulation would therefore be one of the mechanisms for the changes observed in microbiota composition, allowing beneficial microbes to occupy ecological niches vacated by pathogens in the gastrointestinal tract.

Selective stimulation of bifidobacteria and lactobacilli

In contrast to fructans, which are considered to be preferentially bifidogenic, and in human studies rarely lead to increased *Lactobacillus* counts, some polyphenols seem additionally able to stimulate the growth of *Lactobacillus* species (Bialonska *et al.*, 2010; Chen *et al.*, 2012). A possible explanation for this observation is that some *Lactobacillus* species contribute directly to the degradation of some polyphenolic compounds. *Lactobacillus plantarum* IFPL935 significantly increased the concentration of the initial microbial ring-fission products derived from catechins and procyanidins, namely diphenylpropanol, and 4-hydroxy-5-(3′-hydroxyphenyl)valerate respectively (Barroso *et al.*, 2013). Lactobacilli were increased *in vitro* following the addition of polyphenols from pomegranate (Bialonska *et al.*, 2010), or with the addition of tea flavins (polyphenol formed during the oxidation of flavan-3-ol in tea leaves) (Chen *et al.*, 2012). *L. plantarum* IFPL935 metabolized grape seed extract fractions rich in catechins and epicatechins (Sanchez-Patan *et al.*, 2012) (Table 22.1) by means of a galloyl-esterase, decarboxylase and benzyl alcohol dehydrogenase (Tabasco *et al.*, 2011).

Growth stimulation of lactobacilli by polyphenols has also been demonstrated in rats with green tea powder (Axling *et al.*, 2012), and in healthy humans with cocoa-derived flavanols containing catechins and epicatechins (Tzounis *et al.*, 2011). A bifidogenic effect has also been observed both *in vitro* and *in vivo* in a few studies in response to diverse polyphenols including pomegranate byproducts, blanched almonds, or flavanols (Bialonska *et al.*, 2010; Mandalari *et al.*, 2010; Neyrinck *et al.*, 2013; Tzounis *et al.*, 2008). It has been suggested that some *Bifidobacterium* species can rather be stimulated by the degradational products of polyphenols than by the plant compound itself (Parkar *et al.*, 2008).

Microbial-derived phenolic metabolites

Anti-inflammatory, anti-proliferative, antiglycative (=counteraction of protein glycation), neuroprotective or photoprotective activities have been demonstrated in different cell lines for some microbial derived phenolic metabolites (Gao *et al.*, 2006; Gimenez-Bastida *et al.*, 2012; Gonzalez-Sarrias *et al.*, 2010; Henning *et al.*, 2013; Poquet *et al.*, 2008; Seeram *et al.*, 2007). In particular, potential benefits of microbial derived phenolic metabolites have been determined *in vitro* for 3,4-dihydroxyphenylacetic acid (one of the major phenolic acid formed during human microbial fermentation of tea, citrus and soy isoflavonoid supplements), hydroxy-dibenzopyranone derivative urolithins (derived from ellagic acid)), and dihydrocaffeic acid (Gao *et al.*, 2006;

Table 22.1 Effect of aerobic/anaerobic conditions on the catabolism of a grape seed extract (0.5 mg/ml) during growth of Lactobacillus plantarum IFPL 935 in ZMB1 medium containing 0.4% glucose (after (Sanchez-Patan *et al.*, 2012)). Total phenolic content of the grape seed extract was 750 mg of gallic acid equiv./g, as measured by the Folin–Ciocalteu reagent. Compounds present in this extract included gallic acid (8.00 mg/g), monomeric flavan-3-ols ((+)-catechin (136 mg/g) and (–)-epicatechin (135 mg/g)), procyanidin dimers (B1 (21.0 mg/g), B2 (42.4 mg/g), B3 (20.9 mg/g), and B4 (21.2 mg/g)) and trimmers (C1 (3.37 mg/g) and T2 (2.04 mg/g)) (–)-epicatechin-3-O-gallate (19.1 mg/g) and procyanidin gallates (B1–3-O-gallate (0.13 mg/g), B2–3-O-gallate (3.27 mg/g), and B2–3′-O-gallate (2.23 mg/g)), all giving a total content of 414 mg/g

	Residual concentration (mg/ml) after incubation for			
	7.5 h		24 h	
Phenolic compound	Aerobic	Anaerobic	Aerobic	Anaerobic
(+) Catechin	40.0 ± 2.1	38.3 ± 1.8	30.3 ± 1.3	33.4 ± 2.0
(-) Epicatechin	40.2 ± 1.7	40.3 ± 1.6	29.8 ± 2.5	33.4 ± 1.9
Epicatechin-3-*O*-gallate	1.53 ± 0.14	1.31 ± 0.15	nd	nd
Procyanidin B1–3-*O*-gallate	0.048 ± 0,011	0.060 ± 0.011	nd	nd
Procyanidin B2–3-*O*-gallate	0.113 ± 0.008	*0.055 ± 0.006	nd	nd
Procyanidin B3–3-*O*-gallate	0.303 ± 0.011	0.257 ± 0.014	0.023 ± 0.002	0.027 ± 0.001
Gallic acid	19.6 ± 1.3	20.0 ± 0.6	13.7 ± 0.4	*15.7 ± 0.6
Diphenyl-propane-2-ol	nd	nd	5.34 ± 0.07	*3.45 ± 0.21
Pyrogallol	nd	nd	7.66 ± 0.93	6.58 ± 0.71
Catechol/pyrocatechol	nd	nd	0.313 ± 0.014	*0.425 ± 0.014

*Significant differences from the mean values corresponding to the anaerobic condition ($p<0.05$).
nd, not detected.

Gimenez-Bastida *et al.*, 2012; Gonzalez-Sarrias *et al.*, 2010; Henning *et al.*, 2013; Poquet *et al.*, 2008; Seeram *et al.*, 2007). Further investigations are needed to confirm the health benefits of these metabolites *in vivo*.

SCFA production

Polyphenols may stimulate the *in vitro* production of SCFA (Parkar *et al.*, 2013). An increase in propionate was observed in blood following the ingestion of a combination of blueberry husk and a mix of probiotic lactobacilli, suggesting a potential contribution of the latter in the degradation of phenolic compounds or their intermediates (Branning *et al.*, 2009). Considering the potential hypophagic as well as hypocholesterolaemic effects of propionate in humans (Arora *et al.*, 2011), stimulation of propionate production by polyphenols may be of particular relevance for nutritional interventions in obese subjects and type 2 diabetes patients.

Health benefits of polyphenols associated with changes in microbiota: application to obesity and type 2 diabetes

A few studies support a negative association between a low number of bifidobacteria in faecal samples and obesity (Kalliomaki *et al.*, 2008; Wu *et al.*, 2010). Recent animal studies showed an improvement of symptoms related to obesity and metabolic syndrome, concomitant with changes in the composition of the intestinal microbiota, confirming a role of polyphenols as potential prebiotic (Dao *et al.*, 2011; Neyrinck *et al.*, 2013). An increase in bifidobacteria after oral administration of polyphenol-rich pomegranate peel extracts to obese mice suffering from hypercholesterolaemia and inflammatory disorders, correlated with a reduction of total cholesterol and LDL cholesterol (Neyrinck *et al.*, 2013). Additionally, a down-regulation of the expression of key inflammatory genes in the colon [interleukin 6 (IL-6)] and visceral tissues (IL-1b) was observed in the pomegranate group. These results corroborate previous findings obtained in fermentation experiments, which showed that pomegranate byproducts selectively stimulated bifidobacteria in faecal slurries (Bialonska *et al.*, 2010). Another polyphenol extract (resveratrol) exerted an anti-diabetic effect, counteracted the adverse effects induced by a high-fat diet in mice and restored the microbiota composition so that it became similar to that observed before dietary intervention (Dao *et al.*, 2011). Whether the changes in microbiota directly caused the observed improvement of health parameters is unknown.

These promising studies in animals have not yet been complemented by nutritional interventions in obese subjects or type 2 diabetes patients. However, few studies with a limited number of healthy subjects have been performed (Queipo-Ortuno *et al.*, 2012; Tzounis *et al.*, 2011). After a 4-week intake by 22 healthy volunteers of a high versus low-cocoa flavanol diet an improvement of plasma C-reactive protein (CRP), a risk factor for type 2 diabetes and cardiovascular diseases, was associated with an increase in bifidobacteria and lactobacilli, and a decrease in clostridia (Tzounis *et al.*, 2011). Changes in microbiota composition were previously monitored using batch cultures inoculated with faecal slurries from the targeted population group (Tzounis *et al.*, 2008). An increase in bifidobacteria and lactobacilli was associated with a decrease in cholesterol and CRP, and a decrease in HDL cholesterol and CRP, respectively, in a small pilot study with 10 healthy volunteers after ingestion of red wine polyphenols (Queipo-Ortuno *et al.*, 2012). *Bacteroides, Prevotella, Eggerthella lenta, Enterococcus, Blautia, Eubacterium rectale* group increased during the intervention. Since none of these studies carried out an extensive survey of the microbiota, it is likely that microbial groups other than those that proliferated may have contributed to polyphenol degradation and to the health benefits observed in response to polyphenol intake. Identification of host–microbe signalling pathways that contribute to an improved metabolite profile and reduced inflammation in obesity and type 2 diabetes would be key for the design of nutritional intervention studies.

Milk oligosaccharides

Human milk oligosaccharides (HMOs) are a heterogeneous mix of soluble glycans, representing the third main component of human milk after lactose and fats (Sela and Mills, 2010). They are present at concentration of >4 g/l in human breast milk, with higher concentrations present in colostrum. They contain *N*-acetylglucosamine with a degree of polymerization >4, and in which D-glucose, D-galactose, L-fucose and *N*-acetylneuraminic acid (also known as sialic acid) residues are incorporated in an assortment of isomeric configurations (Castanys-Munoz *et al.*, 2013).

Approximately 200 distinct HMOs have been identified in pooled breast milk samples (Newburg *et al.*, 2005). The concentrations and composition of HMOs depend on the lactation phase and the genetic background of the mother (Blank *et al.*, 2012). The simple addition of one of the monosaccharides fucose or sialic acid to lactose renders the resulting trisaccharides resistant to digestion by human enzymes (Engfer *et al.*, 2000). Only intestinal bacteria that possess the necessary glycosyl hydrolases are able to cleave and utilize these monosaccharides (Marcobal *et al.*, 2011). HMOs containing fucose and sialic acid share structural motifs with glycans occurring on the infant intestinal epithelium and are known to serve as receptors for pathogens. Such HMOs can therefore be used as decoy, in that they subvert the pathogen away from binding to the natural receptor site on the cell wall (Morrow *et al.*, 2005; Newburg *et al.*, 2005). HMOs from sources other than human milk have been investigated for possible prebiotic effects. However, oligosaccharides in bovine milk for instance differ greatly from those in human breast milk. Oligosaccharides in bovine colostrum contain HMOs whose structures are identical to neutral and acidic HMOs, including 3¢-sialyllactose (3SL) and 6¢-sialyllactose (6SL), which are found in human breast milk (Barile *et al.*, 2010). 3SL and 6SL are the main oligosaccharides present in bovine milk, but at considerably lower concentrations than in human breast milk (Kelly *et al.*, 2013). Some of these compounds, such as 3SL are commercially available in nearly pure forms. However, high costs for their production have so far limited research on 3SL and 6SL to *in vitro* studies or animal trials.

Selective stimulation of bifidobacteria, which are found in high abundance in the infant gastrointestinal tract, has been reported (Garrido *et al.*, 2013). New processes and technologies for HMO production would without doubt facilitate assessing their benefits in humans, in particular in the field of infant nutrition. In this paragraph we will focus on the effects of 3SL, 6SL, and 2¢-fucosyllactose (2FL) which probably have the best chances to be added to the list of proven prebiotics.

Selective stimulation of bacterial groups associated with health

The intestinal microbiota of breast-fed infants is usually dominated by *bifidobacteria* during the first months of life. Therefore, promotion of bifidobacterial growth is one of the preferred targets of HMOs. *Bifidobacterium* species have shown strain differences in their ability to degrade HMOs. Strains typically found in infants, *Bifidobacterium longum* subsp. *infantis* (*B. infantis*) are better HMO degraders than strains more often encountered in adults such as *B. longum* subsp. *longum* or *B. breve* (Ruiz-Moyano *et al.*, 2013). Whole-genome sequencing of bacteria has advanced our understanding of strain differences with respect to HMO degradation (Sela *et al.*, 2011; Sela and Mills, 2010). *B. infantis*, which efficiently utilizes various low mass HMOs, possesses a large gene cluster and other loci dedicated to HMO metabolism. Examples include four glycosyl hydrolases (sialidase, fucosidase, galactosidase, and hexaminase) required for the cleavage of HMOs into their constituent monosaccharides, and a large number of oligosaccharide transport related genes (LoCascio *et al.*, 2010; Sela *et al.*, 2008, 2012; Sela and Mills, 2010) (Fig. 22.1).

It has been speculated that the various catabolic strategies used for HMO degradation in different *Bifidobacterium* species are the results of various evolutionary mechanisms to outcompete other microorganisms because of a more efficient utilization of a complex substrate (Sela and Mills, 2010). Other bacterial species have the ability to at least partially degrade HMOs in pure culture. Using 3SL, 6SL, 2FL, lacto-*N*-neotetraose (LNnT), or *N*-acetylglucosamine as sole carbon source, several *Lactobacillus* species were capable of degrading at least some of these substrates, *L. plantarum* and *L. acidophilus* being the most versatile species for HMO degradation (Schwab and Ganzle, 2011). Three putative fucosidase genes in the *Lactobacillus casei* BL23 genome were cloned and the genes products purified (Rodriguez-Diaz *et al.*, 2011). The proteins varied in their ability to hydrolyse 2FL, 3-, 4-, and 6-fucosyl-GlcNAc indicating a possible role in the utilization of fucosylated oligosaccharides present in human milk. Alpha-l-fucosidases from *L. casei* BL23 acted preferentially on short- rather than long-chain fucosyloligosaccharides. Genes encoding α-l-fucosidases were found to be exclusively present in the *L. casei*/*Lactobacillus paracasei*/*Lactobacillus rhamnosus* group (Rodriguez-Diaz *et al.*, 2012). Beside bidifobacteria and lactobacilli, several *Bacteroides* species commonly found in the infant intestinal microbiota were able to at least partially degrade most of the HMOs tested in single culture, including 3SL, 6SL, 2FL, 3¢-fucosyllactose (3FL), and lactodifucotetraose (LTDF) (Yu *et al.*, 2013). Other infant-derived bacterial species such as *Escherichia coli*, *Enterococcus faecalis*, *Staphylococcus epidermidis*, *Clostridium leptum* or *Clostridium perfringens* were unable to grow on these substrates. Competition for HMOs in complex communities *in vitro* or *in vivo* may still lead to the stimulation of a limited number of bacteria. Although *B. infantis* and *Bacteroides thetaiotaomicron* were able to degrade LNnT in monoculture, feeding of LNnT to gnotobiotic mice di-associated with these two species favoured the growth of *B. infantis* over *B. thetaiotaomicron* (Marcobal *et al.*, 2011). Selective degradation of 6SL, 2FL, and LNnT by the infant intestinal microbiota has also been demonstrated in batch culture inoculated with faecal slurries obtained from formula-fed and breast-fed infants (Vester Boler *et al.*, 2013). 2FL and LNnT were more rapidly degraded than 6SL by the gut microbiota, and led to higher lactate production. Another mechanism that could explain a selective stimulation of bifidobacteria by HMOs *in vivo*, beside the preferential utilization of HMO as a substrate, relates to the bifidobacterial binding to human intestinal cells mediated by HMOs (Chichlowski *et al.*, 2012).

Selective pathogen inhibition

In contrast to well established prebiotic fructans and GOS, for which anti-pathogenic activity has been reported for a limited number of pathogens, there are a large number of *in vitro*, *ex vivo* and *in vivo* studies demonstrating growth inhibition of a variety of infective agents, including viruses, bacteria, and fungi by HMO (Table 22.2). Anti-pathogenic activity of HMO is thought to be due primarily to the inhibition of pathogen binding to cell receptors, so called 'decoy activity' (Newburg *et al.*, 2005). Fucosylated HMOs, whose production depends on maternal enzymes encoded by the genes associated with the expression of the Lewis blood group system, have important anti-infective activity. Their presence is significantly associated with a decreased risk of infectious diseases in breast-fed infants (Newburg *et al.*, 2005).

Applications: HMO supplements in infant formula

Because of the beneficial effects of HMOs in newborns and infants, addition of HMOs to formula has been on the agenda of the infant nutrition industry for several years. So far, the cost for their production has been far too high to carry out large clinical studies but also to add them to formula. However, the development of new methods for the production of the major HMOs present in human breast milk, such as 2FL, using for instance a bioengineered *Escherichia coli* strain, warrants further research on the effect of HMOs in infants (Baumgartner *et al.*, 2013). Safety and dose response studies are needed before oligosaccharides such as 2FL, 3SL, 6SL or LNnT, can be incorporated into infant formula. So far, such studies have only been performed with a limited number of HMOs (Coulet *et al.*, 2013a; Coulet *et al.*, 2013b). Several human studies advertised the effects of selected HMOs added to infant formula. Such studies can be found in the publicly accessible registry of clinical studies (http://clinicaltrials.gov/; 'human milk oligosaccharides'). Pre-term infants are a population at risk for necrotizing enterocolitis, which renders them a preferential study target. A pilot study performed in a limited number of premature infants who received either formula enriched with

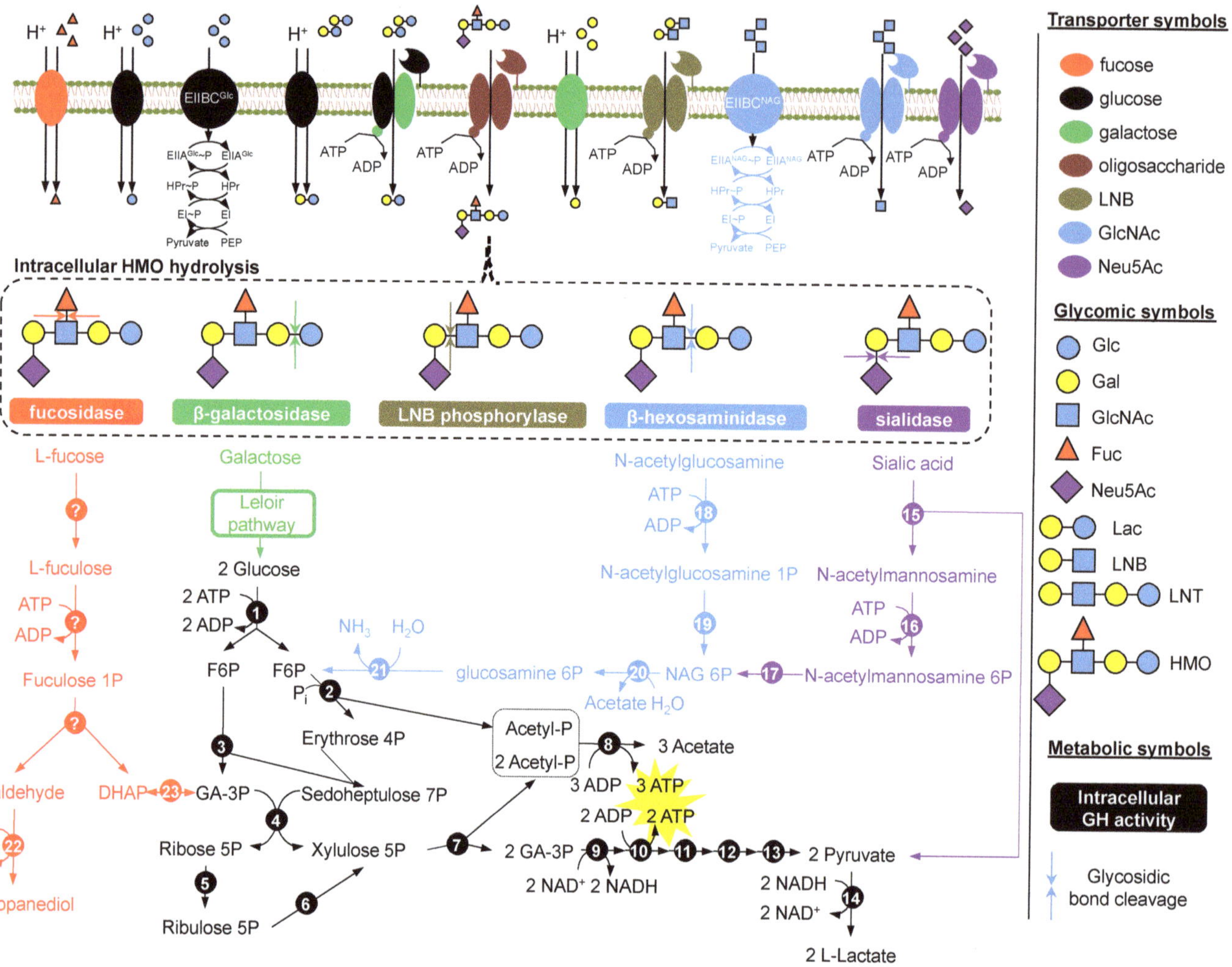

Figure 22.1 B. longum subsp. infantis metabolism of human milk oligosaccharides and derivatives. Carbohydrates transverse the cell membrane by ABC transporters (HMO, lactose, LNB, N-acetylglucosamine, and sialic acid), major facilitator superfamily permeases (fucose, glucose, galactose, and lactose), and PTS (glucose and N-acetylglucosamine). Intracellular glycosyl hydrolases process sugar polymers into constituent monosaccharides which are further converted before entering the catabolic pathways. The central fermentative pathway (Bifidobacterium shunt) is denoted in black with stoichiometric coefficients representing utilization of two hexoses. HPr: phosphocarrier protein; PEP: phosphoenol pyruvate; PTS, phosphotransferase system; P_i, phosphate; G6P, glucose 6-phosphate; F6P, fructose 6-phosphate; GA-3P, glyceraldehyde 3-phosphate; DHAP, dihydroxyacetone phosphate; EIIBC Glc, Enzyme IIBC Glucose transporter; EIIBC NAc, Enzyme IIBC N-acetylglucosamine transporter; GH, glycosyl hydrolase; Glc, glucose; Gal, galactose; GlcNAc, N-acetylglucosamine; Fuc, fucose; Neu5Ac, sialic acid; Lac, lactose; LNB, lacto-N-biose; LNT, lacto-N-tetraose; HMO, human milk oligosaccharide. 1, glucose-6-phosphate isomerase; 2, fructose-6-phosphate phosphoketolase; 3, transaldolase; 4, transketolase; 5, ribose 5-phosphate isomerase; 6, ribulose 5-phosphate epimerase; 7, xylulose-5-phosphate phosphoketolase; 8, acetate kinase; 9, glyceraldehyde-3-phosphate dehydrogenase; 10, phosphoglycerate kinase; 11, phosphoglycerate mutase; 12, enolase; 13, pyruvate kinase; 14, lactate dehydrogenase; 15, N-acetylneuraminate lyase; 16, N-acetylmannosamine kinase; 17, N-acetylmannosamine-6-phosphate 2-epimerase; 18, N-acetylhexosamine 1-kinase; 19, phosphoglucosamine mutase; 20, N-acetylglucosamine-6-phosphate deacetylase; 21, glucosamine-6-phosphate isomerase; 22, lactaldehyde reductase; 23, triose-phosphate isomerase. To be reprinted with permission from reference (Sela *et al.*, 2008). Copyright (2008) National Academy of Sciences, U.S.A.

GOS (F+GOS) or HMO (F+HMO) or own mother's milk enriched with human milk fortifier (H+H) or bovine milk fortifiers (H+B) has however been disappointing, because it revealed neither a health benefit, nor a stimulation of beneficial microbes (Underwood *et al.*, 2014). Absence of bifidobacteria at the beginning of the intervention was probably the main reason for the failure to stimulate bifidobacterial growth. Instead, an increase in clostridia was observed in the F+GOS and F+HMO groups, and an increase in Proteobacteria in the H+H and F+HMO groups. The result of this pilot study reveals the need to consider the microbiota composition at the beginning of the intervention, and suggests that the use of a synbiotic (i.e. HMO and an infant-derived *Bifidobacterium* strain) might have been better.

Glycosylated proteins and peptides

Proteins derived from mammalian milk have been also been studied for their potential prebiotic activities. Evidence for a prebiotic potential has been obtained for glycomacropeptide (GMP), alpha-lactalbumin (α-lac), and lactoferrin in *in vitro* studies (Liepke *et al.*, 2002) as well as in human studies. However, the

Table 22.2 Pathogen inhibition by select HMOs in *in vitro*, ex vivo, and *in vivo* studies. From Chichlowski *et al.*, 2011

Pathogen	HMO tested	Reference
Norwalk virus	Fucosylated oligosaccharides	Ruvoen-Clouet *et al.* (2006)
Campylobacter jejuni		Morrow *et al.* (2004), Ruiz-Palacios *et al.* (2003), Weichert *et al.* (2013)
Vibrio cholerae		Ruiz-Palacios *et al.* (2003)
Salmonella enterica serovar Fyris		Weichert *et al.* (2013)
Escherichia coli (heat-stable enterotoxin)		Newburg *et al.* (2004)
Pseudomonas aeruginosa		Weichert *et al.* (2013)
Streptococcus pneumoniae	Sialyllactose	Leach *et al.* (2005)
Cholera toxin		Idota *et al.* (1995)
Escherichia coli		Virkola *et al.* (1993)
Pseudomonas aeruginosa		Devaraj *et al.* (1994)
Aspergillus fumigates conidia		Bouchara *et al.* (1997)
Helicobacter pylori		Simon *et al.* (1997)
Influenza virus		Matrosovich *et al.* (1993)
Reovirus		(Iskarpatyoti *et al.* (2012)
Polyomavirus		Stehle *et al.* (1994)
HIV-1	Human milk oligosaccharides	Hong *et al.* (2009)
Streptococcus pneumoniae	lacto-N-neotetraose and sialylated derivatives	Andersson *et al.* (1986), Idanpaan-Heikkila *et al.* (1997)
Enteropathogenic *Escherichia coli* (EPEC)		Cravioto *et al.* (1991), Manthey *et al.* (2014)
Haemophilius influenzae		Idanpaan-Heikkila *et al.* (1997)

efficacy of these compounds in humans needs to be demonstrated in larger clinical studies.

Glycomacropeptide

Glycomacropeptide (GMP) is a derivative of κ-casein. *N*-Acetylneuraminic acid (NeuNAc) is one of the sugars associated with bovine GMP and the reported biological and functional properties of this dairy peptide. An early study indicated that GMP stimulates several *Bifidobacterium* and *Lactobacillus* species *in vitro* in single culture (Idota *et al.*, 1994). Additionally, GMP inhibited *in vitro* the adhesion of bacterial pathogens, viruses, and toxins to intestinal cells by a decoy mechanism (Bruck *et al.*, 2006a; Kawakami, 1997) similar to the decoy activity of HMO. Further investigations on the effects of GMP on the microbiota were performed using batch cultures and a two-stage continuous culture model mimicking the large intestine inoculated with faecal microbiota from healthy infants (Bruck *et al.*, 2002). GMP supported the growth of bifidobacteria and lactobacilli, while the number of other bacterial taxa, including *E. coli* and members of the *Clostridium histolyticum* group decreased, as assessed by fluorescent *in situ* hybridization (FISH) with a limited sets of probes targeting the 16S ribosomal RNA (Bruck *et al.*, 2002). Using again a two-stage continuous GI tract model, GMP was tested for its prebiotic effect in response to a challenge with enteropathogenic *E. coli* (EPEC) and *Salmonella enterica* serotype Typhimurium at the end of the fermentation (Bruck *et al.*, 2003a). Reduction of *Salmonella* Typhimurium was significantly decreased with GMP, while the levels of bifidobacteria and lactobacilli remained unchanged. Modulation of the gut microbiota and protection against EPEC infection by GMP was subsequently tested in infant macaques fed from birth till 5 months of age with either breast milk, or with a formula either enriched with lactose, GMP or α-lac (Bruck *et al.*, 2003b). At the end of the trial, the animals were challenged with EPEC, and the diarrhoea index was recorded. In comparison to the group fed the lactose-enriched formula, the group receiving the GMP-enriched formula experienced fewer diarrhoeas. The effect was attributed to the attachment of EPEC to the *N*-acetylneuraminic moiety of GMP. No specific bacterial groups were stimulated or inhibited by GMP as measured monthly with a limited set of FISH probes. Ultimately, the effect of a GMP-supplemented formula was tested in a randomized double-blinded placebo-controlled study in healthy Swedish infants from 6 weeks till 6 months of age (Bruck *et al.*, 2006b). The effects of the intake of a standard formula, a formula enriched in GMP, and breast-milk intake on gastrointestinal symptoms and microbiota modulation were compared. *Bifidobacterium* species were predominant in all groups and no significant differences were observed among the groups at 2 weeks, 2 months and 6 months. The small size of the study groups (17–31 infants per group) did not allow drawing any firm conclusion regarding the beneficial effects of GMP on gut microbiota composition. As the beneficial effects of GMP have previously been related to the prevention of infection by pathogens, a nutritional intervention in an infant population in which EPEC or *Salmonella* Typhimurium are frequently encountered might be a better setting to demonstrate the beneficial effect of GMP. Since only a fraction of the microbiota was monitored in this trial, possible effects may have remained undiscovered. Alternatively, GMP may be digested in the upper GI tract and never reach the colon where they should employ their bifidogenic activity.

Alpha-lactalbumin

Alpha-lactalbumin (α-lac) accounts for up to 28% of the human breast milk protein, and contains a high percentage (64.8%) of essential amino acids (Heine *et al.*, 1991). It has a similar structure as c-type lysozyme, and just like the latter is able to inhibit pathogens (Bruck *et al.*, 2006a; Pihlanto-Leppala *et al.*, 1999). Although the mechanisms underlying the antimicrobial activity of α-lac have not yet been unravelled, it has been speculated that it affects the integrity of the cytoplasmic membrane, just like lactoferrin-derived antibacterial peptides do (see below). Alpha-lac purified from cow's milk has often been studied as a potential prebiotic, both *in vitro* and *in vivo*, and in parallel with GMP (Bruck *et al.*, 2002, 2003a; Bruck *et al.*, 2006a; Bruck *et al.*, 2003b; Bruck *et al.*, 2006b). Stimulation of bifidobacterial growth by α-lac has also been demonstrated in pure culture experiments (Petschow and Talbott, 1991). Anti-microbial activity and modulation of the microbiota in response to α-lac supplementation were overall superior to those observed in response to GMP.

A recent multicentre study investigated the effects of an α-lac-enriched formula in 73 infants (Roze *et al.*, 2012). Unfortunately in this study, the formula was not only supplemented with α-lac but also enriched with prebiotics and probiotics, rendering it difficult to attribute the beneficial effects observed (reduction of atopic dermatitis at 6 months) and change in microbiota composition (increase in bifidobacteria) solely to α-lac.

Lactoferrin

Human lactoferrin (hLF) is one of the most abundant *N*-glycosylated proteins in human breast milk and colostrum, and its derived peptides (i.e. lactoferricin) display broad anti-inflammatory and anti-microbial activities (Barboza *et al.*, 2012; Gonzalez-Chavez *et al.*, 2009). Resistance of hLF to proteolysis may reflect an evolutionary development to permit its persistence in the infant gut (Brines and Brock, 1983). Prebiotic effects of hLF and bovine lactoferrin (bLF) have been suggested to be due to the selective inhibition of pathogens (Nibbering *et al.*, 2001), and their ability to promote bifidobacteria (Petschow and Talbott, 1991; Petschow *et al.*, 1999; Rahman *et al.*, 2009a,b). Screening for endo β-*N*-acetylglucosaminidase activity in 76 bifidobacterial isolates identified in *B. longum* subsp. *infantis* (40%), *B. breve* (36%) and *B. longum* subsp. *longum* (21%) as carriers of this enzyme (Garrido *et al.*, 2012). EndoB1–1, a novel *N*-acetyl glucosaminidase from *B. infantis* ATCC 15695 was constitutively expressed in this strain (Garrido *et al.*, 2012). Incubation of EndoB1-1 with either hLF or bLF led to the induction of genes responsible for HMO degradation, suggesting concerted expression of genes required for the utilization of these N-glycans.

The effect of bovine transgenic milk containing recombinant hLF was tested in piglets fed from day 7 until day 28 after birth and compared to two other groups (bovine milk only, or a mixture of bovine milk and bovine transgenic milk) (Hu *et al.*, 2012). Increase in body mass at the end of the study was higher in the group fed the recombinant hLF. This group also displayed a higher microbial diversity, compared with other groups. Bifidobacteria, enumerated using selective agar, were moderately higher in the ileum of pigs fed milk that contained recombinant hLF. Lactobacilli increased to a larger extent (1 to 2 logs) in the piglets receiving the recombinant hLF compared with the control group, in all parts of the GI tract (duodenum, jejunum, ileum, and colon). In contrast, pathogens such as *Salmonella* were decreased in the colon of piglets receiving the recombinant hLF. It is a downside of this study that the composition of the milk preparations was not reported. Therefore, it cannot be excluded that milk components other than recombinant hLF accounted for the observed effects.

Sugar alcohols (polyols) and disaccharides

Lactulose

Lactulose (4-O-β-d-Galactopyranosyl-β-D-fructofuranose) is a synthetic, non-digestible disaccharide that belongs to the list of accepted prebiotics (Roberfroid *et al.*, 2010). Extensive *in vitro* and *in vivo* studies have been performed for more than twenty years. These studies have demonstrated that lactulose selectively stimulates bifidobacteria and lactobacilli, increases mineral absorption, and reduces pre-neoplastic lesions and DNA damage in the colon of carcinogen-treated rodents (Ballongue *et al.*, 1997; Challa *et al.*, 1997; De Preter *et al.*, 2006; Rowland *et al.*, 1996; Terada *et al.*, 1992; Van den Heuvel *et al.*, 1999; Vanhoutte *et al.*, 2006; Venema *et al.*, 2003b). In the most recent clinical studies, a probiotic was often given together with lactulose, impeding a clear attribution of the observed effects observed to the probiotic or the prebiotic. Metagenomic screening for bacterial genes involved in the degradation of lactulose, as found in ileal mucosa of healthy adult volunteers, identified *Faecalibacterium prausnitzii*, *Streptococcus thermophilus*, and an unidentified *Clostridium* species (Cecchini *et al.*, 2013) that are potentially stimulated by lactulose.

Lactitol

Lactitol (4-O-α-D-galactopyranosyl-D-glucitol) is a sugar alcohol derived from lactose, which purportedly stimulates bacterial groups both *in vitro* and *in vivo* (Bjorklund *et al.*, 2012; Finney *et al.*, 2007; Makelainen *et al.*, 2010; Makivuokko *et al.*, 2010; Ouwehand *et al.*, 2009; Saulnier *et al.*, 2008). Intake of 10 g of lactitol increased bidfidobacteria, and led to a significant increase of faecal propionate and butyrate in 75 healthy adults (Finney *et al.*, 2007). However, evidence for a prebiotic activity of lactitol is not as strong as for lactulose. Health benefits attributed to lactitol are only supported by a few clinical studies in healthy adults and senior individuals (Ouwehand *et al.*, 2009). Genes and proteins involved in lactitol utilization have been characterized in the probiotic *L. acidophilus* NCFM (Andersen *et al.*, 2011; Andersen *et al.*, 2012; Majumder *et al.*, 2011). An increase in the mRNA levels of the permease encoded by *lacS* and β-galactosidase encoded by *lacZ* during growth on lactitol compared with growth on glucose indicate that these genes are likely involved in lactitol utilization. This was confirmed by proteome analysis which showed an increased expression of the small subunit of the β-galactosidase as well as of galactokinase, galactose-1-phosphate uridylyltransferase and UDP-glucose-4-epimerase (Majumder *et al.*, 2011).

Sorbitol and isomalt

Sorbitol is a polyolic monosaccharide that is only partially (25%) absorbed in the upper gastrointestinal tract (Holub *et al.*, 2009). Evidence for a prebiotic activity of sorbitol is scarce. Oral administration of sorbitol to rats increased the number of lactobacilli including *L. reuteri* (Venema *et al.*, 2003a). The operon encoding the enzyme required for the utilization of sorbitol was characterized in *L. casei* (Alcantara *et al.*, 2008). It codes for a sorbitol-6-phosphate dehydrogenase, regulators, and a phosphoenolpyruvate-dependent sorbitol phosphotransferase system.

Isomalt is a mixed disaccharide composed of 1-O-α-D-glucopyranosyl-D-mannitol and 6-O-α-D-glucopyranosyl-D-sorbitol, and considered to be poorly absorbed (0–14%) in the upper gastrointestinal tract (Holub *et al.*, 2009). Administration of a large dose of isomalt (30 g/day) or sucrose in conjunction with a controlled Western-type diet to 19 adult volunteers in a double-blind, placebo-controlled, cross-over trial led to a modest (0.3 log/g faeces), but significant increase of bifidobacteria after 4 weeks of isomalt treatment (Muthaiyan *et al.*, 2012). Activity of β-glucosidase, which catalyses the release toxic aglycons from xenobiotics, was 40% lower in subjects who obtained isomalt than in those who obtained sucrose. Fermentation of isomalt in batch cultures inoculated with faecal slurries resulted in the formation of butyrate. Several *Bifidobacterium* strains degraded isomalt completely when this substrate was added as the sole carbon source. Improvement of markers of risk for cardiovascular disease was not observed in the same study. This was probably due to the low number of hyperlipidaemic subjects enrolled ($n=6$) (Reid *et al.*, 2003). Administration of the same isomalt dose over a 12-week intervention to 31 type 2 diabetes patients led to a reduction of glycosylated haemoglobin, fructosamine, fasting blood glucose, insulin, proinsulin, C-peptide, insulin resistance, and oxidized LDL (Holub *et al.*, 2009). Microbiota composition was not monitored in this study.

Other polyols and disaccharides

Investigations of the potential prebiotic activities of other polyols and disaccharides are limited to *in vitro* studies only. Moreno *et al.*, primarily investigated the influence of the disaccharide structure on the selectivity of a wide range of disaccharides using micro-scale batch culture fermentations inoculated with faecal slurries (Moreno *et al.*, 2008). Although this study only monitored changes of a limited number of bacterial groups using FISH, the authors of this work used these data to define a 'prebiotic index score'. Using a limited number of faecal slurries from different individuals (n=3) for fermentation experiments, the authors observed that disaccharides with 1–2, 1–4, and 1–6 linkages generally resulted in higher prebiotic scores. Repetition of such experiments using pyrosequencing and metabolomics could lead to a better predictive understanding of how specific molecular units undergo fermentation by specific community members.

Limitations

Although specific polyphenols, HMOs, peptides and polyols have shown promising effects in the studies mentioned above, there are some important limitations. We have summarized the most important ones in Table 22.3.

Future trends

The preliminary data obtained with select polyphenols, HMOs, peptides and polyols as potential prebiotic warrant further studies to better determine how changes in microbiota composition and activity are linked to health benefits. Future studies should address this point, especially when considering current or upcoming regulations in Europe and in the USA, which require demonstration of health claims to obtain approval from the European Union Safety Authority (EFSA) for new dietary ingredients. For this purpose, the impact of novel prebiotics on well accepted biomarkers of disease risks should be assessed.

Table 22.3 Current study limitations of the prebiotic properties of bioactive compounds not being strictly defined as fibres

Limitations
The majority of potential new prebiotics have been investigated in Western populations consuming Western diets
The majority of potential new prebiotics have been investigated in the healthy population
Technical challenges have to be overcome for the synthesis of some of these compounds for the assessment of their effects in large clinical trials (i.e. HMO)
The impact of the selectivity of the microbiota has often been limited to lactobacilli and bifidobacteria
There are important technical challenges, and costs associated with process development for some of these bioactive compounds (i.e. HMOs)
A stringent regulatory environment may limit research and innovation on potential new prebiotics. Prebiotic compounds introduced as dietary supplements that are not already component of food may require regulatory approval as novel foods
Very few studies performed with new potential prebiotics have made the link between changes in microbiota and associated health benefits
The impact of the food matrix has so far not been considered
The dose response of potential new prebiotic has so far been hardly studied
The importance of microbe–microbe interactions, including cross-feeding, for the degradation of these compounds is largely unknown
The acceptance of new prebiotics other than fibres produced in transgenic animals (i.e. recombinant human lactoferrin and alpha lactalbumin) by the public is uncertain

The industry should be able to address some technical challenges for the synthesis of novel prebiotics, by using for instance reverse enzymatic engineering. (Gimenez-Bastida *et al.*, 2012; Moreno *et al.*, 2008; Muthaiyan *et al.*, 2012; Rodriguez-Diaz *et al.*, 2012). Novel prebiotics may results from the combination of different substrates (i.e. glycated peptides derived from GOS (Moreno *et al.*, 2008)). Application of existing prebiotics or novel prebiotics could be broadened and further investigated. This may be particularly relevant for HMOs, which should be assessed not only in healthy infants, but also in subjects suffering from gastro-intestinal disorders or elderly.

Future studies should also better consider the response variability to bioactive compounds that may depend on specific metabotypes or starting microbiota composition, in order to better define which population group would benefit most from prebiotic administration. These studies should also identify responders and non-responders in order to be able to predict which individuals would benefit the most from prebiotic supplementation. For example children from mothers that lack the fucosyl-transferase in their mammary glands are unlikely to carry bifidobacterial species that can degrade fucosylated HMO. Hence it would not be beneficial to feed these children these fucosylated prebiotics, unless they are accompanied with a probiotic that can ferment them.

There is a need to better understand how the chemical structure of bioactive compounds classified as prebiotics affect microbiota composition and activity. To investigate possible cross-feeding among microbial community members, multiple samples should be analysed at multiple time-points both *in vitro* and *in vivo*. The use of multiple omics technologies, such as metatranscriptomics, proteomics, metabolomics, in combination with well-defined markers of disease risk as clinical endpoints should help to get a better understanding of host–microbiota interactions and mechanisms associated with health benefits. The use of nutrikinetics in conjunction with metabolic modelling is a relatively new approach to associate degradational products of polyphenols with specific gut phylotypes (van Velzen *et al.*, 2009). The health benefit of novel prebiotics should be confirmed in large double-blind crossover placebo-controlled studies. They should help elucidate the mechanisms underlying changes in microbiota composition/activity and health benefits and confirm whether there is a causal link between these associations.

References

Alcantara, C., Sarmiento-Rubiano, L.A., Monedero, V., Deutscher, J., Perez-Martinez, G., and Yebra, M.J. (2008). Regulation of *Lactobacillus* casei sorbitol utilization genes requires DNA-binding transcriptional activator GutR and the conserved protein GutM. Appl. Environ. Microbiol. *74*, 5731–5740.

Andersen, J.M., Barrangou, R., Abou Hachem, M., Lahtinen, S., Goh, Y.J., Svensson, B., and Klaenhammer, T.R. (2011). Transcriptional and functional analysis of galactooligosaccharide uptake by lacS in *Lactobacillus* acidophilus. Proc. Natl. Acad. Sci. U.S.A. *108*, 17785–17790.

Andersen, J.M., Barrangou, R., Hachem, M.A., Lahtinen, S.J., Goh, Y.J., Svensson, B., and Klaenhammer, T.R. (2012). Transcriptional analysis of prebiotic uptake and catabolism by *Lactobacillus* acidophilus NCFM. PLoS One 7, e44409.

Andersson, B., Porras, O., Hanson, L.A., Lagergard, T., and Svanborg-Eden, C. (1986). Inhibition of attachment of Streptococcus pneumoniae and Haemophilus influenzae by human milk and receptor oligosaccharides. J. Infect. Dis. *153*, 232–237.

Arora, T., Sharma, R., and Frost, G. (2011). Propionate. Anti-obesity and satiety enhancing factor? Appetite *56*, 511–515.

Axling, U., Olsson, C., Xu, J., Fernandez, C., Larsson, S., Strom, K., Ahrne, S., Holm, C., Molin, G., and Berger, K. (2012). Green tea powder and *Lactobacillus plantarum* affect gut microbiota, lipid metabolism and inflammation in high-fat fed C57BL/6J mice. Nutrition & Metabolism *9*, 105.

Ballongue, J., Schumann, C., and Quignon, P. (1997). Effects of lactulose and lactitol on colonic microflora and enzymatic activity. Scand. J. Gastroenterol. Supplement *222*, 41–44.

Barboza, M., Pinzon, J., Wickramasinghe, S., Froehlich, J.W., Moeller, I., Smilowitz, J.T., Ruhaak, L.R., Huang, J., Lonnerdal, B., German, J.B., *et al.* (2012). Glycosylation of human milk lactoferrin exhibits dynamic changes during early lactation enhancing its role in pathogenic bacteria–host interactions. Mol. Cell. Proteomics *11*, M111 015248.

Barile, D., Marotta, M., Chu, C., Mehra, R., Grimm, R., Lebrilla, C.B., and German, J.B. (2010). Neutral and acidic oligosaccharides in Holstein-Friesian colostrum during the first 3 days of lactation measured by high performance liquid chromatography on a microfluidic chip and time-of-flight mass spectrometry. J. Dairy Sci. 93, 3940–3949.

Barroso, E., Sanchez-Patan, F., Martin-Alvarez, P.J., Bartolome, B., Moreno-Arribas, M.V., Pelaez, C., Requena, T., van de Wiele, T., and Martinez-Cuesta, M.C. (2013). *Lactobacillus plantarum* IFPL935 Favors the initial metabolism of red wine polyphenols when added to a colonic microbiota. J. Agric. Food. Chem. *61*, 10163–10172.

Baumgartner, F., Seitz, L., Sprenger, G.A., and Albermann, C. (2013). Construction of *Escherichia coli* strains with chromosomally integrated expression cassettes for the synthesis of 2′-fucosyllactose. Microb. Cell. Fact. *12*, 40.

Bialonska, D., Ramnani, P., Kasimsetty, S.G., Muntha, K.R., Gibson, G.R., and Ferreira, D. (2010). The influence of pomegranate by-product and punicalagins on selected groups of human intestinal microbiota. International journal of food microbiology *140*, 175–182.

Bingham, S.A., Day, N.E., Luben, R., Ferrari, P., Slimani, N., Norat, T., Clavel-Chapelon, F., Kesse, E., Nieters, A., Boeing, H., *et al.* (2003). Dietary fibre in food and protection against colorectal cancer in the European Prospective Investigation into Cancer and Nutrition (EPIC): an observational study. Lancet *361*, 1496–1501.

Bjorklund, M., Ouwehand, A.C., Forssten, S.D., Nikkila, J., Tiihonen, K., Rautonen, N., and Lahtinen, S.J. (2012). Gut microbiota of healthy elderly NSAID users is selectively modified with the administration of *Lactobacillus* acidophilus NCFM and lactitol. Age *34*, 987–999.

Blank, D., Dotz, V., Geyer, R., and Kunz, C. (2012). Human milk oligosaccharides and Lewis blood group: individual high-throughput sample profiling to enhance conclusions from functional studies. Advanc. Nutr. *3*, 440S–449S.

Bolca, S., Van de Wiele, T., and Possemiers, S. (2013). Gut metabotypes govern health effects of dietary polyphenols. Current opinion in biotechnology *24*, 220–225.

Bouchara, J.P., Sanchez, M., Chevailler, A., Marot-Leblond, A., Lissitzky, J.C., Tronchin, G., and Chabasse, D. (1997). Sialic acid-dependent recognition of laminin and fibrinogen by *Aspergillus fumigatus* conidia. Infect. Immun. *65*, 2717–2724.

Branning, C., Hakansson, A., Ahrne, S., Jeppsson, B., Molin, G., and Nyman, M. (2009). Blueberry husks and multi-strain probiotics affect colonic fermentation in rats. Br. J. Nutr. *101*, 859–870.

Brines, R.D., and Brock, J.H. (1983). The effect of trypsin and chymotrypsin on the *in vitro* antimicrobial and iron-binding properties of lactoferrin in human milk and bovine colostrum. Unusual resistance of human apolactoferrin to proteolytic digestion. Biochim. Biophys. Acta *759*, 229–235.

Bruck, W.M., Graverholt, G., and Gibson, G.R. (2002). Use of batch culture and a two-stage continuous culture system to study the effect of supplemental alpha-lactalbumin and glycomacropeptide on mixed populations of human gut bacteria. FEMS Microbiol. Ecol. *41*, 231–237.

Bruck, W.M., Graverholt, G., and Gibson, G.R. (2003a). A two-stage continuous culture system to study the effect of supplemental alpha-lactalbumin and glycomacropeptide on mixed cultures of human gut bacteria challenged with enteropathogenic *Escherichia coli* and Salmonella serotype Typhimurium. J. Appl. Microbiol. 95, 44–53.

Bruck, W.M., Kelleher, S.L., Gibson, G.R., Graverholt, G., and Lonnerdal, B.L. (2006a). The effects of alpha-lactalbumin and glycomacropeptide on the association of CaCo-2 cells by enteropathogenic *Escherichia coli, Salmonella typhimurium* and *Shigella flexneri*. FEMS Microbiol. Lett. 259, 158–162.

Bruck, W.M., Kelleher, S.L., Gibson, G.R., Nielsen, K.E., Chatterton, D.E., and Lonnerdal, B. (2003b). rRNA probes used to quantify the effects of glycomacropeptide and alpha-lactalbumin supplementation on the predominant groups of intestinal bacteria of infant rhesus monkeys challenged with enteropathogenic *Escherichia coli*. J. Pediatr. Gastroenterol. Nutr. 37, 273–280.

Bruck, W.M., Redgrave, M., Tuohy, K.M., Lonnerdal, B., Graverholt, G., Hernell, O., and Gibson, G.R. (2006b). Effects of bovine alpha-lactalbumin and casein glycomacropeptide-enriched infant formulae on faecal microbiota in healthy term infants. J. Pediatr. Gastroenterol. Nutr. 43, 673–679.

Castanys-Munoz, E., Martin, M.J., and Prieto, P.A. (2013). 2′-fucosyllactose: an abundant, genetically determined soluble glycan present in human milk. Nutr. Rev. *71*, 773–789.

Cecchini, D.A., Laville, E., Laguerre, S., Robe, P., Leclerc, M., Dore, J., Henrissat, B., Remaud-Simeon, M., Monsan, P., and Potocki-Veronese, G. (2013). Functional metagenomics reveals novel pathways of prebiotic breakdown by human gut bacteria. PLoS One *8*, e72766.

Challa, A., Rao, D.R., Chawan, C.B., and Shackelford, L. (1997). *Bifidobacterium longum* and lactulose suppress azoxymethane-induced colonic aberrant crypt foci in rats. Carcinogenesis *18*, 517–521.

Chen, H., Hayek, S., Rivera Guzman, J., Gillitt, N.D., Ibrahim, S.A., Jobin, C., and Sang, S. (2012). The microbiota is essential for the generation of black tea theaflavins-derived metabolites. PLoS One 7, e51001.

Chichlowski, M., De Lartigue, G., German, J.B., Raybould, H.E., and Mills, D.A. (2012). Bifidobacteria isolated from infants and cultured on human milk oligosaccharides affect intestinal epithelial function. J. Pediatr. Gastroenterol. Nutr. 55, 321–327.

Chichlowski, M., German, J.B., Lebrilla, C.B., and Mills, D.A. (2011). The influence of milk oligosaccharides on microbiota of infants: opportunities for formulas. Annual review of food science and technology *2*, 331–351.

Cho, S.S., Qi, L., Fahey, G.C., Jr. and Klurfeld, D.M. (2013). Consumption of cereal fibre, mixtures of whole grains and bran, and whole grains and risk reduction in type 2 diabetes, obesity, and cardiovascular disease. Am. J. Clin. Nutr. 98, 594–619.

Coulet, M., Phothirath, P., Allais, L., and Schilter, B. (2013a). Pre-clinical safety evaluation of the synthetic human milk, nature-identical, oligosaccharide 2′-O-Fucosyllactose (2′FL). Regul. Toxicol. Pharmacol. *68*, 59–69.

Coulet, M., Phothirath, P., Constable, A., Marsden, E., and Schilter, B. (2013b). Pre-clinical safety assessment of the synthetic human milk, nature-identical, oligosaccharide Lacto-N-neotetraose (LNnT). Food Chem. Toxicol. *62*, 528–537.

Cravioto, A., Tello, A., Villafan, H., Ruiz, J., del Vedovo, S., and Neeser, J.R. (1991). Inhibition of localized adhesion of enteropathogenic *Escherichia coli* to HEp-2 cells by immunoglobulin and oligosaccharide fractions of human colostrum and breast milk. J. Infect. Dis. *163*, 1247–1255.

Crozier, A., Jaganath, I.B., and Clifford, M.N. (2009). Dietary phenolics: chemistry, bioavailability and effects on health. Natural Product Reports *26*, 1001–1043.

D'Archivio, M., Filesi, C., Vari, R., Scazzocchio, B., and Masella, R. (2010). Bioavailability of the polyphenols: status and controversies. Int. J. Mol. Sci. *11*, 1321–1342.

Dao, T.M., Waget, A., Klopp, P., Serino, M., Vachoux, C., Pechere, L., Drucker, D.J., Champion, S., Barthelemy, S., Barra, Y., *et al.* (2011). Resveratrol increases glucose induced GLP-1 secretion in mice: a mechanism which contributes to the glycaemic control. PLoS One *6*, e20700.

De Preter, V., Vanhoutte, T., Huys, G., Swings, J., Rutgeerts, P., and Verbeke, K. (2006). Effect of lactulose and *Saccharomyces boulardii* administration on the colonic urea-nitrogen metabolism and the bifidobacteria concentration in healthy human subjects. Aliment. Pharmacol. Ther. *23*, 963–974.

Devaraj, N., Sheykhnazari, M., Warren, W.S., and Bhavanandan, V.P. (1994). Differential binding of *Pseudomonas aeruginosa* to normal and cystic fibrosis tracheobronchial mucins. Glycobiology *4*, 307–316.

DiGiulio, D.B., Romero, R., Amogan, H.P., Kusanovic, J.P., Bik, E.M., Gotsch, F., Kim, C.J., Erez, O., Edwin, S., and Relman, D.A. (2008). Microbial prevalence, diversity and abundance in amniotic fluid during preterm labor: a molecular and culture-based investigation. PLoS One 3, e3056.

Duda-Chodak, A. (2012). The inhibitory effect of polyphenols on human gut microbiota. Journal of physiology and pharmacology: an official journal of the Polish Physiological Society *63*, 497–503.

Engfer, M.B., Stahl, B., Finke, B., Sawatzki, G., and Daniel, H. (2000). Human milk oligosaccharides are resistant to enzymatic hydrolysis in the upper gastrointestinal tract. Am. J. Clin. Nutr. *71*, 1589–1596.

Finney, M., Smullen, J., Foster, H.A., Brokx, S., and Storey, D.M. (2007). Effects of low doses of lactitol on faecal microflora, pH, short-chain fatty acids and gastrointestinal symptomology. European Journal of Nutrition *46*, 307–314.

Gao, K., Xu, A., Krul, C., Venema, K., Liu, Y., Niu, Y., Lu, J., Bensoussan, L., Seeram, N.P., Heber, D., *et al.* (2006). Of the major phenolic acids formed during human microbial fermentation of tea, citrus, and soy flavonoid supplements, only 3,4-dihydroxyphenylacetic acid has antiproliferative activity. J. Nutr. *136*, 52–57.

Garrido, D., Dallas, D.C., and Mills, D.A. (2013). Consumption of human milk glycoconjugates by infant-associated bifidobacteria: mechanisms and implications. Microbiology *159*, 649–664.

Garrido, D., Nwosu, C., Ruiz-Moyano, S., Aldredge, D., German, J.B., Lebrilla, C.B., and Mills, D.A. (2012). Endo-beta-N-acetylglucosaminidases from infant gut-associated bifidobacteria release complex N-glycans from human milk glycoproteins. Molecular & cellular proteomics: MCP *11*, 775–785.

Gimenez-Bastida, J.A., Larrosa, M., Gonzalez-Sarrias, A., Tomas-Barberan, F., Espin, J.C., and Garcia-Conesa, M.T. (2012). Intestinal ellagitannin metabolites ameliorate cytokine-induced inflammation and associated molecular markers in human colon fibroblasts. J. Agric. Food. Chem. *60*, 8866–8876.

Gonzalez-Chavez, S.A., Arevalo-Gallegos, S., and Rascon-Cruz, Q. (2009). Lactoferrin: structure, function and applications. International journal of antimicrobial agents *33*, 301 e301–308.

Gonzalez-Sarrias, A., Larrosa, M., Tomas-Barberan, F.A., Dolara, P., and Espin, J.C. (2010). NF-kappaB-dependent anti-inflammatory activity of urolithins, gut microbiota ellagic acid-derived metabolites, in human colonic fibroblasts. Br. J. Nutr. *104*, 503–512.

Heine, W.E., Klein, P.D., and Reeds, P.J. (1991). The importance of alpha-lactalbumin in infant nutrition. J. Nutr. *121*, 277–283.

Henning, S.M., Wang, P., Abgaryan, N., Vicinanza, R., de Oliveira, D.M., Zhang, Y., Lee, R.P., Carpenter, C.L., Aronson, W.J., and Heber, D. (2013). Phenolic acid concentrations in plasma and urine from men consuming green or black tea and potential chemopreventive properties for colon cancer. Mol. Nutr. Food res. *57*, 483–493.

Holub, I., Gostner, A., Hessdorfer, S., Theis, S., Bender, G., Willinger, B., Schauber, J., Melcher, R., Allolio, B., and Scheppach, W. (2009). Improved metabolic control after 12-week dietary intervention with low glycaemic isomalt in patients with type 2 diabetes mellitus. Hormone and metabolic research = Hormon- und Stoffwechselforschung = Hormones et metabolisme *41*, 886–892.

Hong, P., Ninonuevo, M.R., Lee, B., Lebrilla, C., and Bode, L. (2009). Human milk oligosaccharides reduce HIV-1-gp120 binding to dendritic cell-specific ICAM3-grabbing non-integrin (DC-SIGN). Br. J. Nutr. *101*, 482–486.

Hu, W., Zhao, J., Wang, J., Yu, T., Wang, J., and Li, N. (2012). Transgenic milk containing recombinant human lactoferrin modulates the intestinal flora in piglets. Biochemistry and cell biology = Biochimie et biologie cellulaire *90*, 485–496.

Idanpaan-Heikkila, I., Simon, P.M., Zopf, D., Vullo, T., Cahill, P., Sokol, K., and Tuomanen, E. (1997). Oligosaccharides interfere with the establishment and progression of experimental pneumococcal pneumonia. J. Infect. Dis. *176*, 704–712.

Idota, T., Kawakami, H., Murakami, Y., and Sugawara, M. (1995). Inhibition of cholera toxin by human milk fractions and sialyllactose. Biosci. Biotechnol. Biochem. *59*, 417–419.

Idota, T., Kawakami, H., and Nakajima, I. (1994). Growth promoting effect of N-Acetylneuraminic acid-containing substance on bifidobacteria. Biosci. Biotechnol. Biochem. *58*, 1720–1722.

Iskarpatyoti, J.A., Morse, E.A., McClung, R.P., Ikizler, M., Wetzel, J.D., Contractor, N., and Dermody, T.S. (2012). Serotype-specific differences in inhibition of reovirus infectivity by human-milk glycans are determined by viral attachment protein sigma1. Virology *433*, 489–497.

Kalliomaki, M., Collado, M.C., Salminen, S., and Isolauri, E. (2008). Early differences in fecal microbiota composition in children may predict overweight. Am. J. Clin. Nutr. *87*, 534–538.

Kawakami, H. (1997). Biological signi¢cance of sialic acid-containing substances in milk and their application. Recent Res. Dev. Agric. Biol. Chem. *1*, 193–207.

Kelly, V., Davis, S., Berry, S., Melis, J., Spelman, R., Snell, R., Lehnert, K., and Palmer, D. (2013). Rapid, quantitative analysis of 3′- and 6′-sialyllactose in milk by flow-injection analysis-mass spectrometry: Screening of milks for naturally elevated sialyllactose concentration. J. Dairy Sci. *96*, 7684–7691.

Kemperman, R.A., Bolca, S., Roger, L.C., and Vaughan, E.E. (2010). Novel approaches for analysing gut microbes and dietary polyphenols: challenges and opportunities. Microbiology *156*, 3224–3231.

Leach, J.L., Garber, S.A., Marcon, A.A., and Prieto, P.A. (2005). *In vitro* and *in vivo* effects of soluble, monovalent globotriose on bacterial attachment and colonization. Antimicrob. Agents Chemother. *49*, 3842–3846.

Lee, H.C., Jenner, A.M., Low, C.S., and Lee, Y.K. (2006). Effect of tea phenolics and their aromatic fecal bacterial metabolites on intestinal microbiota. Res. Microbiol. *157*, 876–884.

Liepke, C., Adermann, K., Raida, M., Magert, H.J., Forssmann, W.G., and Zucht, H.D. (2002). Human milk provides peptides highly stimulating the growth of bifidobacteria. Eur. J. Biochem./FEBS *269*, 712–718.

LoCascio, R.G., Desai, P., Sela, D.A., Weimer, B., and Mills, D.A. (2010). Broad conservation of milk utilization genes in *Bifidobacterium longum* subsp. infantis as revealed by comparative genomic hybridization. Appl. Environ. Microbiol. *76*, 7373–7381.

Majumder, A., Sultan, A., Jersie-Christensen, R.R., Ejby, M., Schmidt, B.G., Lahtinen, S.J., Jacobsen, S., and Svensson, B. (2011). Proteome reference map of *Lactobacillus* acidophilus NCFM and quantitative proteomics towards understanding the prebiotic action of lactitol. Proteomics *11*, 3470–3481.

Makelainen, H., Saarinen, M., Stowell, J., Rautonen, N., and Ouwehand, A.C. (2010). Xylo-oligosaccharides and lactitol promote the growth of Bifidobacterium lactis and *Lactobacillus* species in pure cultures. Benef. Microbes *1*, 139–148.

Makivuokko, H., Forssten, S., Saarinen, M., Ouwehand, A., and Rautonen, N. (2010). Synbiotic effects of lactitol and *Lactobacillus* acidophilus NCFM in a semi-continuous colon fermentation model. Beneficial microbes *1*, 131–137.

Mandalari, G., Faulks, R.M., Bisignano, C., Waldron, K.W., Narbad, A., and Wickham, M.S. (2010). *In vitro* evaluation of the prebiotic properties of almond skins (Amygdalus communis L.). FEMS microbiology letters *304*, 116–122.

Manthey, C.F., Autran, C.A., Eckmann, L., and Bode, L. (2014). Human milk oligosaccharides protect against enteropathogenic E. coli (EPEC) attachment *in vitro* and EPEC colonization in suckling mice. J. Pediatr. Gastroenterol. Nutr. *58*, 167–170.

Marcobal, A., Barboza, M., Sonnenburg, E.D., Pudlo, N., Martens, E.C., Desai, P., Lebrilla, C.B., Weimer, B.C., Mills, D.A., German, J.B., *et al.* (2011). Bacteroides in the infant gut consume milk oligosaccharides via mucus-utilization pathways. Cell host & microbe *10*, 507–514.

Matrosovich, M.N., Gambaryan, A.S., Tuzikov, A.B., Byramova, N.E., Mochalova, L.V., Golbraikh, A.A., Shenderovich, M.D., Finne, J., and Bovin, N.V. (1993). Probing of the receptor-binding sites of the H1 and H3 influenza A and influenza B virus hemagglutinins by synthetic and natural sialosides. Virology *196*, 111–121.

Moreno, F.J., Quintanilla-Lopez, J.E., Lebron-Aguilar, R., Olano, A., and Sanz, M.L. (2008). Mass spectrometric characterization of glycated beta-lactoglobulin peptides derived from galacto-oligosaccharides surviving the *in vitro* gastrointestinal digestion. Journal of the American Society for Mass Spectrometry *19*, 927–937.

Morrow, A.L., Ruiz-Palacios, G.M., Altaye, M., Jiang, X., Guerrero, M.L., Meinzen-Derr, J.K., Farkas, T., Chaturvedi, P., Pickering, L.K., and Newburg, D.S. (2004). Human milk oligosaccharides are associated with protection against diarrhea in breast-fed infants. The Journal of pediatrics *145*, 297–303.

Morrow, A.L., Ruiz-Palacios, G.M., Jiang, X., and Newburg, D.S. (2005). Human-milk glycans that inhibit pathogen binding protect breast-feeding infants against infectious diarrhea. J. Nutr. *135*, 1304–1307.

Muthaiyan, A., Hernandez-Hernandez, O., Moreno, F.J., Sanz, M.L., and Ricke, S.C. (2012). Hydrolyzed caseinomacropeptide conjugated galactooligosaccharides support the growth and enhance the bile tolerance in *Lactobacillus* strains. J. Agric. Food Chem. *60*, 6839–6845.

Newburg, D.S., Ruiz-Palacios, G.M., Altaye, M., Chaturvedi, P., Guerrero, M.L., Meinzen-Derr, J.K., and Morrow, A.L. (2004). Human milk alphal,2-linked fucosylated oligosaccharides decrease risk of diarrhea due to stable toxin of E. coli in breastfed infants. Advan. Exp. Med. Biol. *554*, 457–461.

Newburg, D.S., Ruiz-Palacios, G.M., and Morrow, A.L. (2005). Human milk glycans protect infants against enteric pathogens. Annual review of nutrition *25*, 37–58.

Neyrinck, A.M., Van Hee, V.F., Bindels, L.B., De Backer, F., Cani, P.D., and Delzenne, N.M. (2013). Polyphenol-rich extract of pomegranate peel alleviates tissue inflammation and hypercholesterolaemia in high-fat diet-induced obese mice: potential implication of the gut microbiota. Br. J. Nutr. *109*, 802–809.

Nibbering, P.H., Ravensbergen, E., Welling, M.M., van Berkel, L.A., van Berkel, P.H., Pauwels, E.K., and Nuijens, J.H. (2001). Human lactoferrin and peptides derived from its N terminus are highly effective against infections with antibiotic-resistant bacteria. Infect. Immun. *69*, 1469–1476.

Ouwehand, A.C., Tiihonen, K., Saarinen, M., Putaala, H., and Rautonen, N. (2009). Influence of a combination of *Lactobacillus* acidophilus NCFM and lactitol on healthy elderly: intestinal and immune parameters. Br. J. Nutr. *101*, 367–375.

Parkar, S.G., Stevenson, D.E., and Skinner, M.A. (2008). The potential influence of fruit polyphenols on colonic microflora and human gut health. International journal of food microbiology *124*, 295–298.

Parkar, S.G., Trower, T.M., and Stevenson, D.E. (2013). Fecal microbial metabolism of polyphenols and its effects on human gut microbiota. Anaerobe *23*, 12–19.

Petschow, B.W., and Talbott, R.D. (1991). Response of bifidobacterium species to growth promoters in human and cow milk. Pediatr. Res. *29*, 208–213.

Petschow, B.W., Talbott, R.D., and Batema, R.P. (1999). Ability of lactoferrin to promote the growth of Bifidobacterium spp. *in vitro* is independent of receptor binding capacity and iron saturation level. J. Med. Microbiol. *48*, 541–549.

Pihlanto-Leppala, A., Marnila, P., Hubert, L., Rokka, T., Korhonen, H.J., and Karp, M. (1999). The effect of alpha-lactalbumin and beta-lactoglobulin hydrolysates on the metabolic activity of *Escherichia coli* JM103. Journal of applied microbiology *87*, 540–545.

Poquet, L., Clifford, M.N., and Williamson, G. (2008). Effect of dihydrocaffeic acid on UV irradiation of human keratinocyte HaCaT-cells. Archives of biochemistry and biophysics *476*, 196–204.

Qin, J., Li, R., Raes, J., Arumugam, M., Burgdorf, K.S., Manichanh, C., Nielsen, T., Pons, N., Levenez, F., Yamada, T., *et al.* (2010). A human gut microbial gene catalogue established by metagenomic sequencing. Nature *464*, 59–65.

Queipo-Ortuno, M.I., Boto-Ordonez, M., Murri, M., Gomez-Zumaquero, J.M., Clemente-Postigo, M., Estruch, R., Cardona Diaz, F., Andres-Lacueva, C., and Tinahones, F.J. (2012). Influence of red wine polyphenols and ethanol on the gut microbiota ecology and biochemical biomarkers. Am. J. Clin. Nutr. *95*, 1323–1334.

Rahman, M., Kim, W.S., Kumura, H., and Shimazaki, K. (2009a). Bovine lactoferrin region responsible for binding to bifidobacterial cell surface proteins. Biotechnology letters *31*, 863–868.

Rahman, M.M., Kim, W.S., Ito, T., Kumura, H., and Shimazaki, K. (2009b). Growth promotion and cell binding ability of bovine lactoferrin to *Bifidobacterium longum*. Anaerobe *15*, 133–137.

Reid, G., Jass, J., Sebulsky, M.T., and McCormick, J.K. (2003). Potential uses of probiotics in clinical practice. Clin. Microbiol. Rev. *16*, 658–672.

Roberfroid, M., Gibson, G.R., Hoyles, L., McCartney, A.L., Rastall, R., Rowland, I., Wolvers, D., Watzl, B., Szajewska, H., Stahl, B., *et al.* (2010).

Prebiotic effects: metabolic and health benefits. Br. J. Nutr. *104* (Suppl. 2), S1–63.

Rodriguez-Diaz, J., Monedero, V., and Yebra, M.J. (2011). Utilization of natural fucosylated oligosaccharides by three novel alpha-l-fucosidases from a probiotic *Lactobacillus* casei strain. Appl. Environ. Microbiol. *77*, 703–705.

Rodriguez-Diaz, J., Rubio-del-Campo, A., and Yebra, M.J. (2012). *Lactobacillus casei* ferments the N-Acetylglucosamine moiety of fucosyl-alpha-1,3-N-acetylglucosamine and excretes l-fucose. Appl. Environ. Microbiol. *78*, 4613–4619.

Rowland, I.R., Bearne, C.A., Fischer, R., and Pool-Zobel, B.L. (1996). The effect of lactulose on DNA damage induced by DMH in the colon of human flora-associated rats. Nutr. Cancer *26*, 37–47.

Roze, J.C., Barbarot, S., Butel, M.J., Kapel, N., Waligora-Dupriet, A.J., De Montgolfier, I., Leblanc, M., Godon, N., Soulaines, P., Darmaun, D., *et al.* (2012). An alpha-lactalbumin-enriched and symbiotic-supplemented v. a standard infant formula: a multicentre, double-blind, randomised trial. Br. J. Nutr. *107*, 1616–1622.

Ruiz-Moyano, S., Totten, S.M., Garrido, D.A., Smilowitz, J.T., German, J.B., Lebrilla, C.B., and Mills, D.A. (2013). Variation in consumption of human milk oligosaccharides by infant gut-associated strains of *Bifidobacterium breve*. Appl. Environ. Microbiol. *79*, 6040–6049.

Ruiz-Palacios, G.M., Cervantes, L.E., Ramos, P., Chavez-Munguia, B., and Newburg, D.S. (2003). *Campylobacter jejuni* binds intestinal H(O) antigen (Fuc alpha 1, 2Gal beta 1, 4GlcNAc), and fucosyloligosaccharides of human milk inhibit its binding and infection. J. Biol. Chem. *278*, 14112–14120.

Ruvoen-Clouet, N., Mas, E., Marionneau, S., Guillon, P., Lombardo, D., and Le Pendu, J. (2006). Bile-salt-stimulated lipase and mucins from milk of 'secretor' mothers inhibit the binding of Norwalk virus capsids to their carbohydrate ligands. Biochem. J. *393*, 627–634.

Sanchez-Patan, F., Tabasco, R., Monagas, M., Requena, T., Pelaez, C., Moreno-Arribas, M.V., and Bartolome, B. (2012). Capability of *Lactobacillus plantarum* IFPL935 to catabolize flavan-3-ol compounds and complex phenolic extracts. J. Agric. Food. Chem. *60*, 7142–7151.

Saulnier, D.M., Gibson, G.R., and Kolida, S. (2008). *In vitro* effects of selected synbiotics on the human faecal microbiota composition. FEMS Microbiol. Ecol. *66*, 516–527.

Schwab, C., and Ganzle, M. (2011). Lactic acid bacteria fermentation of human milk oligosaccharide components, human milk oligosaccharides and galactooligosaccharides. FEMS microbiology letters *315*, 141–148.

Seeram, N.P., Aronson, W.J., Zhang, Y., Henning, S.M., Moro, A., Lee, R.P., Sartippour, M., Harris, D.M., Rettig, M., Suchard, M.A., *et al.* (2007). Pomegranate ellagitannin-derived metabolites inhibit prostate cancer growth and localize to the mouse prostate gland. J. Agric. Food. Chem. *55*, 7732–7737.

Sela, D.A., Chapman, J., Adeuya, A., Kim, J.H., Chen, F., Whitehead, T.R., Lapidus, A., Rokhsar, D.S., Lebrilla, C.B., German, J.B., *et al.* (2008). The genome sequence of *Bifidobacterium longum* subsp. infantis reveals adaptations for milk utilization within the infant microbiome. Proc. Natl. Acad. Sci. U.S.A. *105*, 18964–18969.

Sela, D.A., Garrido, D., Lerno, L., Wu, S., Tan, K., Eom, H.J., Joachimiak, A., Lebrilla, C.B., and Mills, D.A. (2012). *Bifidobacterium longum* subsp. infantis ATCC 15697 alpha-fucosidases are active on fucosylated human milk oligosaccharides. Appl. Environ. Microbiol. *78*, 795–803.

Sela, D.A., Li, Y., Lerno, L., Wu, S., Marcobal, A.M., German, J.B., Chen, X., Lebrilla, C.B., and Mills, D.A. (2011). An infant-associated bacterial commensal utilizes breast milk sialyloligosaccharides. J. Biol. Chem. *286*, 11909–11918.

Sela, D.A., and Mills, D.A. (2010). Nursing our microbiota: molecular linkages between bifidobacteria and milk oligosaccharides. Trends in microbiology *18*, 298–307.

Simon, P.M., Goode, P.L., Mobasseri, A., and Zopf, D. (1997). Inhibition of Helicobacter pylori binding to gastrointestinal epithelial cells by sialic acid-containing oligosaccharides. Infect. Immun. *65*, 750–757.

Stehle, T., Yan, Y., Benjamin, T.L., and Harrison, S.C. (1994). Structure of murine polyomavirus complexed with an oligosaccharide receptor fragment. Nature *369*, 160–163.

Tabasco, R., Sanchez-Patan, F., Monagas, M., Bartolome, B., Victoria Moreno-Arribas, M., Pelaez, C., and Requena, T. (2011). Effect of grape polyphenols on lactic acid bacteria and bifidobacteria growth: resistance and metabolism. Food microbiology *28*, 1345–1352.

Terada, A., Hara, H., Kataoka, M., and Mitsuoka, T. (1992). Effect of lactulose on the composition and metabolic activity of the human faecal flora. Microb. Ecol. Health Dis. *92*, 43–50.

Trowell, H. (1972). Fibre: a natural hypocholesteremic agent. Am. J. Clin. Nutr. *25*, 464–465.

Trowell, H., Southgate, D.A., Wolever, T.M., Leeds, A.R., Gassull, M.A., and Jenkins, D.J. (1976). Letter: Dietary fibre redefined. Lancet *1*, 967.

Tzounis, X., Rodriguez-Mateos, A., Vulevic, J., Gibson, G.R., Kwik-Uribe, C., and Spencer, J.P. (2011). Prebiotic evaluation of cocoa-derived flavanols in healthy humans by using a randomized, controlled, double-blind, crossover intervention study. Am. J. Clin. Nutr. *93*, 62–72.

Tzounis, X., Vulevic, J., Kuhnle, G.G., George, T., Leonczak, J., Gibson, G.R., Kwik-Uribe, C., and Spencer, J.P. (2008). Flavanol monomer-induced changes to the human faecal microflora. Br. J. Nutr. *99*, 782–792.

Underwood, M.A., Kalanetra, K.M., Bokulich, N.A., Mirmiran, M., Barile, D., Tancredi, D.J., German, J.B., Lebrilla, C.B., and Mills, D.A. (2014). Prebiotic oligosaccharides in premature infants. J. Pediatr. Gastroenterol. Nutr. *58*, 352–360.

Van den Heuvel, E.G., Muijs, T., Van Dokkum, W., and Schaafsma, G. (1999). Lactulose stimulates calcium absorption in postmenopausal women. Journal of bone and mineral research: the official journal of the American Society for Bone and Mineral Research *14*, 1211–1216.

van Velzen, E.J., Westerhuis, J.A., van Duynhoven, J.P., van Dorsten, F.A., Grun, C.H., Jacobs, D.M., Duchateau, G.S., Vis, D.J., and Smilde, A.K. (2009). Phenotyping tea consumers by nutrikinetic analysis of polyphenolic end-metabolites. Journal of proteome research *8*, 3317–3330.

Vanhoutte, T., De Preter, V., De Brandt, E., Verbeke, K., Swings, J., and Huys, G. (2006). Molecular monitoring of the fecal microbiota of healthy human subjects during administration of lactulose and Saccharomyces boulardii. Appl. Environ. Microbiol. *72*, 5990–5997.

Venema, K., van Nuenen, M.H.M.C., van den Heuvel, E.G., Wietske, P., and van der Vossen, J.M.B.M. (2003a). The effect of lactulose on the composition of the intestinal microbiota and short-chain fatty acid production in human volunteers and a computercontrolled model of the proximal large intestine. Microbial Ecology in Health and Disease *15*, 94–105.

Venema, K., van Nuenen, M.H.M.C., van den Heuvel, E.G., Wietske, P., and van der Vossen, J.M.B.M. (2003b). The effect of lactulose on the composition of the intestinal microbiota and short-chain fatty acid production in human volunteers and a computercontrolled model of the proximal large intestine. Microb. Ecol. Health Dis. *15*, 94–105.

Vester Boler, B.M., Rossoni Serao, M.C., Faber, T.A., Bauer, L.L., Chow, J., Murphy, M.R., and Fahey, G.C., Jr. (2013). *In vitro* fermentation characteristics of select nondigestible oligosaccharides by infant fecal inocula. J. Agric. Food. Chem. *61*, 2109–2119.

Virkola, R., Parkkinen, J., Hacker, J., and Korhonen, T.K. (1993). Sialyloligosaccharide chains of laminin as an extracellular matrix target for S fimbriae of *Escherichia coli*. Infect. Immun. *61*, 4480–4484.

Weichert, S., Jennewein, S., Hufner, E., Weiss, C., Borkowski, J., Putze, J., and Schroten, H. (2013). Bioengineered 2′-fucosyllactose and 3-fucosyllactose inhibit the adhesion of *Pseudomonas aeruginosa* and enteric pathogens to human intestinal and respiratory cell lines. Nutrition Research *33*, 831–838.

Wu, X., Ma, C., Han, L., Nawaz, M., Gao, F., Zhang, X., Yu, P., Zhao, C., Li, L., Zhou, A., *et al.* (2010). Molecular characterization of the faecal microbiota in patients with type II diabetes. Curr. Microbiol. *61*, 69–78.

Yu, Z.T., Chen, C., and Newburg, D.S. (2013). Utilization of major fucosylated and sialylated human milk oligosaccharides by isolated human gut microbes. Glycobiology *23*, 1281–1292.

Synbiotics – More than just the Sum of Pro- and Prebiotics?

23

Koen Venema

Abstract

This is a comprehensive but critical overview of the studies performed with synbiotics (products containing both probiotics and prebiotics) up to date. Most of the studies reviewed did not have the desired controls, *i.e.* arms with only the probiotic or only the prebiotic. This makes it impossible to judge whether the synbiotic (that is the combination of both probiotics and prebiotics) was required for the observed effect, or whether one of the components alone would have done the trick as well. Studies in adults, children and elderly, with and without disorders or diseases are highlighted. Only human trials are discussed, although a few *in vitro* studies and studies in experimental animals are reviewed to highlight potential mechanisms of action.

Probiotics and prebiotics

It probably is not necessary in a book devoted to probiotics and prebiotics to define the phrases, but for the sake of being able to define synbiotics it will be done here for completion.

Probiotics have been defined by an expert panel commissioned by the Food and Agriculture Organization of the United Nations (FAO) and the World Health Organization (WHO) as 'Live microorganisms which when administered in adequate amounts confer a health benefit on the host' (FAO/WHO, 2002). This definition is preferred over historical definition as it includes both intestinal and other forms of probiotics, such as vaginal or topical probiotics. Although dead cells and cell components may also exert some health-promoting physiological effects, this definition requires that the term 'probiotic' only be applied to live microbes having a substantiated beneficial effect (Reid *et al.*, 2003). Thus, microbes administered alive are considered probiotics regardless of their ability to survive intestinal transit. As such, strains or species that do not survive intestinal transit, such as *S. thermophilus*, could be considered probiotic. As already stated, although a preparation of non-viable (dead) bacteria usually also can mediate a physiological benefit (Adams, 2010), dead cells are not considered to be 'probiotics' under the present definition. Such preparations have recently been referred to as postbiotics. Postbiotics have been defined as nonviable bacterial products or metabolic byproducts from probiotic microorganisms that have biological activity in the host (Tsilingiri and Rescigno, 2013). For some physiological benefit it is clear that the cells need to be active. However, for other benefits, e.g. lactose intolerance, it is hypothesized that lysis of the cells within the gastrointestinal (GI) tract is required to increase digestion of lactose in lactose-intolerant individuals (Venema, 2012). New trends that may soon be reality are the development of new therapeutic strategies, such as the development of phagebiotics, psychobiotics, and (genetically modified) pharmabiotics (Eeckhaut *et al.*, 2014; Shanahan *et al.*, 2012).

A prebiotic was originally defined as 'a nondigestible food ingredient that beneficially affects the host by selectively stimulating the growth and/or activity of one or a limited number of bacteria in the colon, and thus improves host health' (Gibson and Roberfroid, 1995). This definition was later refined to include, besides the GI tract, other areas of the host that may benefit from selective targeting of particular microorganisms (Gibson *et al.*, 2004). A more recently adopted definition by the International Scientific Association of Probiotics and Prebiotics (ISAPP) is: 'a selectively fermented ingredient that results in specific changes in the composition and/or activity of the gastrointestinal microbiota, thus conferring benefit(s) upon host health' (Blatchford *et al.*, 2014), in which the phrase 'gastrointestinal' re-appeared. I suppose the jury is still out on non-GI prebiotics. To be classified as a prebiotic, a dietary compound historically had to fulfil three criteria: (1) resistance to gastric acidity, hydrolysis by mammalian enzymes and gastrointestinal absorption, (2) fermentation by intestinal microbiota, and (3) selective stimulation of the growth and/or activity of intestinal bacteria associated with health and wellbeing. Given the fact that other areas of the host besides the GI tract are now also targeted, these criteria are not all applicable any more. In fact, they are not applicable at all any more, unless the phrase 'intestinal' is deleted from criteria 2 and 3. However, most prebiotics developed up to date are geared towards the GI tract.

Most prebiotics that have been developed so far are carbohydrates, or fibres. Though most prebiotics (so far) are fibres, not all fibres are prebiotics (Slavin, 2013). Although the concept of prebiotics is relatively new, foods high in prebiotics have been consumed since prehistoric times: archaeological evidence from dry cave deposits in the northern Chihuahuan Desert, that straddles the US–Mexico border, show intensive utilization of desert plants that were high in prebiotic fructans (Slavin, 2013). Dietary intake of fructans was estimated to be about 135 g/day for the typical adult male hunter–forager. The fructans inulin and fructo-oligosaccharides (FOS) occur naturally in foods such as leeks, asparagus, chicory, Jerusalem artichokes, garlic, and onions. Nowadays, consumption in typical US and European diets has been estimated to be only several grams per day (Slavin, 2013). Despite the possibility of ingesting high quantities of inulin (135 g/day for hunter-foragers), in our current time inulin and FOS consumption in large quantities is associated with impaired gastrointestinal tolerance (Grabitske and Slavin, 2009; Lied *et al.*, 2011). Other prebiotic fibres (e.g. wheat dextrin, polydextrose) exhibit higher GI tolerability (up to 30–45 g per day) (Pasman *et al.*, 2006). Examples of well-established prebiotics are inulin and FOS, galacto-oligosaccharides (GOS; a.k.a. transgalacto-oligosaccharides or TOS) and lactulose. Less well studied prebiotics include lactitol, isomalto-oligosaccharides (IMO), gluco-oligosaccharides (Gl-OS), lactosucrose, xylo-oligosaccharides (XOS), arabinoxylo-oligosaccharides (AXOS), resistant starch, raffinose (Gal-Glc-Fru) and stachyose (Gal-Gal-Glc-Fru) from soy, and (acidic) pectic oligosaccharides, although there is accumulating evidence for these.

Inulin and FOS are composed of linear chains of β (2,1) linked fructose units, with or without a terminal glucose. GOS are a mixture of oligosaccharides derived from lactose by enzymatic transglycosylation. The mixture generally consists of oligosaccharides from tri- to pentasaccharides with β (1,6), β (1,3) and β (1,4) linkages, which make it refractive to digestion. Lactulose is manufactured by the isomerization of lactose to generate the disaccharide galactosyl- β (1,4)-fructose. Lactitol is a sugar alcohol and is the hydrogenated form of lactose. IMO are manufactured from starch, which is first hydrolysed by the combined action of α-amylase and pullulanase, and the resultant malto-oligosaccharides are acted upon by α-glucosidase to create IMO. IMO is partly digestible by pancreatic enzymes, but is digested slowly. Depending on the individual enzyme activity and food transit, variable amount of IMO may reach the colon. Gl-OS are synthesized by the action of the enzyme dextran-sucrase on sucrose in the presence of maltose. Due to the presence of glycosidic bonds uncharacteristic of starch, it is not hydrolysed by pancreatic α-amylase. Lactosucrose is produced from a mixture of lactose and sucrose using the enzyme β-fructofuranosidase. XOS and AXOS are manufactured by the enzymatic hydrolysis of (arabino)xylan from various cereal sources.

A large part of the adult world population cannot digest lactose (Venema, 2012), as the expression of the lactase enzyme in the gut wall is decreased with increasing age. This phenomenon is called lactase deficiency (no lactase) or hypolactasia (some remaining activity) (Venema, 2012). If these individuals would ingest lactose (Gal-Glc), the disaccharide would make it to the colon. As such it can be considered a prebiotic. However, lactose ingestion is very frequently associated with gastrointestinal complaints. This lactose intolerance is thought to occur due to fermentation of lactose by the gut microbiota (He *et al.*, 2008). There is a lot of anecdotal evidence and scientific data that slowly increasing the dose of lactose would diminish the intolerance. Adaptation to long term lactose ingestion in lactose maldigesters has been thought to be related to adaptation of the colonic microbiota and colonic function (He *et al.*, 2008; Hertzler and Savaiano, 1996). The mechanism underlying lactose intolerance is hypothesized to be the very quick fermentation of the disaccharide by the microbiota, likely leading to accumulation of microbial metabolites and gas, which lead to the experience of bloating, rumbling and diarrhoea (He *et al.*, 2008). Adaptation of the microbiota, or adaptation of the colonic epithelium to cope with the increased metabolites is thought to relieve the symptoms. A similar mechanism may also be in play for oligosaccharides such as inulin and FOS, which also have been shown to cause GI complaints after high doses (see above). Experiments in validated *in vitro* models of the colon have indicated that these oligosaccharides (degree of polymerization (DP) ranging from 3 to 9) are also very quickly fermented (van Nuenen *et al.*, 2003), and thus might also lead to accumulation of microbial metabolites and gas, with consequent GI complaints. A very-long chain inulin (average DP >55, with a maximum DP of ~75) extracted from globe artichoke had pronounced prebiotic effects in human subjects, but was well tolerated (Costabile *et al.*, 2010). Besides DP also the biochemical structure of the oligosaccharides determines (speed of) fermentability. GOS contains many different molecules with different DP and glycosidic linkages (Coulier *et al.*, 2009), which cannot all be degraded by the same microbial species. Similarly, fermentation of linear α-1,6 dextrans and dextrans with α-1,2 branching by the human microbiota was different (Sarbini *et al.*, 2011). In addition, a branched inulin from Agave shows different properties than the linear inulin from, for example, chicory: fermentation of the Agave fructan was slower (Koenen, Cruz Rubio and Venema, personal communication). This may be (part of) the reason why the hunter-foragers in the Chihuahuan Desert could ingest higher quantities of these fructans, as the Agava fructans are highly branched (López and Urías-Silvas, 2007). It has been shown that different strains of bifidobacteria fall into one of four clusters with reference to their metabolism of inulin and shorter chain FOS, or the monosaccharide FOS. Some strains only metabolized fructose; other strains preferred fructose but could also metabolize FOS, but with decreasing affinity as molecular weight (or DP) increased. Some strains preferred FOS with little ability to grow on the monosaccharide or inulin, while the last cluster had equal affinity to grow on fructose and FOS and have some ability to grow on inulin (Falony *et al.*, 2009). Similarly, structurally different wheat-derived arabinoxylo-oligosaccharides with different DP showed different fermentation characteristics (Van Craeyveld *et al.*, 2008). This shows that the structure, but also the dose defines the beneficial properties of the prebiotic that is ingested. Whilst we do not have a clear and comprehensive understanding of the molecular mechanisms involved in the bacterial fermentation of complex carbohydrates in the gut, it is clear that these processes will involve the action of glycosylhydrolases and transport systems to take up breakdown products. This is discussed in more detail in Chapter 15.

Synbiotics

Prebiotics function complementary to, and possibly synergistically with, probiotics. Gibson and Roberfroid first defined synbiotics as 'a mixture of probiotics and prebiotics that beneficially affects the host by improving the survival and implantation of live microbial dietary supplements (read: probiotics) in the gastrointestinal tract, by selectively stimulating the growth and/or by activating the metabolism of one or a limited number of health-promoting bacteria, and thus improving host welfare' (Gibson and Roberfroid, 1995). Searching PubMed for the occurrence of the phrase 'synbiotic' or 'synbiotics' in titles of papers gives 244 hits (26 March 2014). Dropping the requirement for the phrase to specifically occur in the title, the number of hits increased to 505. Similar numbers for probiotics and prebiotics are 12.266 and 3791 hits, respectively. This indicates that there is relatively less research focused on the combination of probiotics and prebiotics (synbiotics) compared to the individual components, although undoubtedly some papers that do not use the phrase synbiotic(s) but do use the combination of probiotics and prebiotics will have been missed in this simple search.

The prebiotics present in synbiotic mixtures are generally considered to have one of two possible functions. First and foremost they act as prebiotics, in the sense that they stimulate the numbers and activity of beneficial microbes endogenous to the (gut of the) host, independent of the activity of the probiotic (complementary approach). Secondly, they are considered a 'lunch-box' for the probiotics and should increase their survival while the beneficial microbes travel through the hostile environment of the GI tract (synergistic approach) (Kolida and Gibson, 2011). However, the lunch-box function of prebiotics assumes that the probiotics are able to use the prebiotics as a substrate, and this is certainly not always true, as some products combine probiotic lactobacilli with a bifidogenic prebiotic. In this case the probiotic will not benefit from the ability to use the prebiotic as a substrate for growth. Not all currently commercially available probiotics for instance are capable of fermenting FOS or GOS. Indications on the potential growth of a probiotic on a selected substrate can be obtained through a genomic scan of the glycosidase spectra of the strain (assuming that the genome sequence is available), which may allow for a rational selection of the potential prebiotics that this particular strain can use (Kolida and Gibson, 2011). A second assumption is that the presence of a fermentable substrate indeed does increase the survival of the probiotic. Although there is evidence that this is the case (e.g. Martinez *et al.*, 2011), we also have some yet unpublished data that this depends on the dose of prebiotics added. At relatively low doses it appears that survival of the probiotic strain may in fact be lower than that without the presence of the prebiotic. We hypothesize that this is caused by the following: Most probiotics when tested for survival have been cultured until the stationary phase of growth before they are harvested. When grown *in situ* in a product, this harvesting is simply distribution over individual packages and shipping out for retailing. When the strains are to be incorporated into tablets, capsules or other formulations, they are usually freeze- or spray-dried. When entering the stationary growth phase, bacteria express stress proteins that help them cope this phase of low substrate availability, etc. This has been generally shown in lactic acid bacteria and is also true for probiotics (Mills *et al.*, 2011). These stress-proteins protect the cells against stresses they can encounter in the production pipeline, such as low or high temperature, low pH, high pressure, etc. When cells that have expressed these stress proteins encounter the hostile environment in the GI tract, with low gastric acidity and high bile concentrations in the small intestine, the cells are usually cross-protected to a certain extent by the stress proteins. However, this is never a full protection, and some cells will die when travelling through the GI tract. This has been nicely shown using a dynamic *in vitro* model of the stomach and small intestine (TIM-1), that accurately mimics the dynamic parameters occurring in the upper GI tract, such as gastric acidity, bile and pancreatic enzyme concentrations, and food transit (Marteau *et al.*, 1997). These models are excellent tools to study the mechanisms underlying optimal survival of probiotics in the GI tract. For instance, by changing growth conditions of the probiotics, the effect of protective stress proteins on their survival can be established. Also, some strains are more sensitive to acid, others are more sensitive to bile. When this is known, specific measures can be taken to increase survival of these strains, e.g. by screening for optimal growth conditions, or by developing a protective coating for strains that are sensitive to gastric acid, whether with or without a prebiotic (e.g. Chaluvadi *et al.*, 2012; Chavarri *et al.*, 2010; Eiberger *et al.*, 2014; Riaz and Masud, 2013).

Now what happens when these probiotic strains which have expressed (increased) stress proteins are allowed to have access to a prebiotic during GI transit? When the probiotics start to feed on the prebiotic they leave the stationary phase and enter the logarithmic growth phase. In this log-phase there is no need to express the stress proteins, and hence cells are less able to cope with the stresses encountered in the GI tract. Hence survival drops. However, one could argue that since the prebiotic is a substrate for growth, one would get cell multiplication as well. One would be right. However, there appears to be a delicate balance between the dose of prebiotics added that destroys cross-protection by the stress proteins (and leads to lower survival) and the dose that allows for sufficient growth to occur such that the number of viable cells increases. This increase in viable cells then appears to be increased survival, but in reality is the sum of increased death due to loss of protection by stress proteins and increased numbers of viable cells due to growth. As researchers, we have no way of distinguishing this. From this, two things follow: (a) there should be a dose of prebiotics that exactly balances increased death and increased cell numbers and that for the researcher does not seem to affect survival; and (b) apparent increased survival can be obtained by choosing (usually be trial and error) the proper dose of the prebiotic. The TIM-1 system referred to above is a perfect tool to study this under standardized conditions, as we will see in the section on potential mechanisms of synbiotics.

Not all studies that investigated a synbiotic have compared the effects with those of the single ingredients, *i.e.* the probiotic or prebiotic alone (de Vrese and Schrezenmeir, 2008). Moreover, not all studies show that the combination of probiotics and prebiotics was more effective than the single ingredients alone, at least not on the parameters that were studied. Not all studies evaluated the effect of the synbiotic on microbiota composition and activity, but rather focused on parameters such as immunomodulation (by the probiotic). Below some recent examples of synbiotics are discussed. Both the definition of probiotics and prebiotics include 'benefit for the host', implying that clinical data is required for a product to be able to carry the label probiotic or prebiotic (and I do not mean regulatory label here, because according to EFSA none of the current products fulfil the requirements). This also implies that for synbiotics there should be clinical data. This is why the examples mostly focus on clinical data, although *in vitro* or animal data is discussed in the last section, where it relates to deciphering the underlying mechanisms.

Recent examples of synbiotics

Above two different approaches were discussed for synbiotics: the synergistic and complementary approach. Each approach has different implications. The complementary approach targets host well-being separately with a probiotic and a prebiotic. Because of this, both components need to be administered in an efficacious dose. The relatively high doses of prebiotics (usually more than 6 g/day) required to mediate an effect on the gut microbiota will, in most cases, exclude the option of encapsulation. With a synergistic approach, the synbiotic is perceived as a single product, whereby the primary role of the prebiotic is to improve the survivability and colonization of the probiotic. The implication is that the dose of the prebiotic required to accomplish this may be as such a smaller dose than for the complementary approach. In addition, since the prebiotic is intended to increase the probiotic survival, also a lower dose of probiotics can be used. Very few of the studies described to date however have addressed dose. Several of the studies described below should strictly spoken have been omitted from this review, because the bacterial strains used cannot be considered probiotics (or were not even mentioned at all). Nevertheless, I have included the studies here, to show what kind of results can be obtained.

Healthy adults

Two early studies investigated the effect of GOS and/or *B. breve* 4006 (Tanaka *et al.*, 1983) and lactulose and/or *B. longum* BB536 (Tomoda *et al.*, 1991) on the microbiota composition. The study of Tanaka *et al.*, provided 3 or 10 g of GOS with or without *B. breve* 4006 to volunteers. The GOS increased both the indigenous bifidobacteria, as well as the added *B. breve* strain (Tanaka *et al.*, 1983). *B. breve* was increased by GOS in a dose-dependent manner. Although this is a clear example of a synergistic effect, it is unclear whether the strain in question can be considered a probiotic. Tomoda *et al.*, studied the lactulose-induced increase in bifidobacteria in volunteers. They also showed an increase in indigenous and probiotic bifidobacteria, in this case *B. longum* BB526. There were some interesting changes within the genus *Bifidobacterium*, but those go beyond the scope of this review. The amount of lactulose provided was 0.65 g/serving (Tomoda *et al.*, 1991). Apart from a bifidogenic effect by feeding lactulose, also a decrease in faecal ammonia was observed, indicating reduced putrefactive fermentation. Lactulose nowadays is frequently used as a laxative and administered as a standard treatment for chronic hepatic encephalopathy (Havenaar and Van Dokkum, 2001). Hepatic encephalopathy is a challenging clinical complication of liver dysfunction with a wide spectrum of neuropsychiatric abnormalities that range from mild disturbances in cognitive function and consciousness to coma and death caused by excessive amounts of ammonia. Fermentation of lactulose leads to incorporation of ammonia (as a source of nitrogen for the microbiota) into biomass of the microbiota, thereby reducing the ammonia load on the liver. It is hypothesized that the synergistic effect of BB536 and lactulose leads to a healthier luminal milieu in the colon of the volunteers (Tomoda *et al.*, 1991). In the two studies described above, the authors made sure that the bifidobacterial strain used could in fact use the prebiotic that was chosen. In more recent studies this is not always done, and arbitrary combinations of commercially available probiotics and prebiotics have been selected. Morelli *et al.* (2003) studied a three-strain probiotic mixture with inulin provided in a sachet. The probiotic mixture contained the strains *L. paracasei* B21060, *L. paracasei* B21070 and *L. gasseri* B21090. Only of the first strain (B21060) data is available of its probiotic action. The subjects showed an increase in *L. paracasei*. *L. gasseri* could not be identified. Since there were no individual probiotic or prebiotic treatments, the synergistic effect of the inulin is unclear. A small increase in bifidobacteria was seen as well. Vanhoutte and colleagues studied *Saccharomyces boulardii* Hansen CBS 5926 and/or lactulose (De Preter *et al.*, 2006; Vanhoutte *et al.*, 2006). The lactulose was shown to increase bifidobacteria and reduce urinary ammonia secretion, while the probiotic yeast did not display any detectable changes on ammonia excretion, nor did it influence the bifidobacterial levels, leaving one to wonder why the combination was chosen. The same group also studied the effect of *L. casei* Shirota or *B. breve* Yakult and FOS (De Preter *et al.*, 2007). Faecal and urinary excretion of ammonia and p-cresol were analysed as indicators of proteolytic colonic fermentation. The prebiotic alone mediated significant increases in faecal bifidobacteria concentrations and decreases in urinary p-cresol and ammonia. There was no additive effect observed with either synbiotic formulation. In a later study this group turned to metabolite profiling of faecal samples (De Preter *et al.*, 2011). In nine volunteers, volatile metabolites were measured before, during and after a 4-week synbiotic treatment, composed of *L. casei* Shirota and FOS-enriched inulin. Only three volatile organic compounds, acetate, dimethyl trisulfide and ethyl benzene, were significantly affected. The acetate levels increased, whereas the dimethyl trisulfide levels decreased during and after the intervention. For ethyl benzene a decrease was only observed during the synbiotic intervention period. No changes in the composition of the predominant microbiota was observed. Although the single components were not tested, the effects are interesting as they show an increase in saccharolytic activity and a decrease in proteolytic activity. The examples above are some cases where it is unclear whether one of the components in the synbiotic mixture has an effect or not, or what the benefit of the addition of one of the components to the synbiotic mixture was. When both individual components and mixture were tested, the results imply that there was neither synergy nor an additive effect between the probiotic and the prebiotic components of these synbiotic formulations.

Gut microbiota metabolites in adults were also assessed in two other studies. Given the developments in the analytical technology, global metabolite analysis seems to be become a trend, and in my view a must, to study microbial metabolites produced under different circumstances, as these metabolites are responsible for a great deal of the cross-talk with the host (see also Chapter 15). The first study investigated the effect of a synbiotic composed of *B. longum* and *L. acidophilus* with FOS on metabolites in faecal samples of 16 healthy volunteers using NMR (Ndagijimana *et al.*, 2009). Despite the fact that the metabolite profiles of faecal samples before and after the consumption were characterized by significant inter-individual variability, more than 150 molecules in the profile changed, with amino acids and SCFA being those most significantly affected by the treatment. In particular, a significant general increase of SCFA content was observed in parallel with a significant decrease in both the aromatic (phenylalanine and tyrosine) and aliphatic amino acids content. Unfortunately, this study did not indicate which microbial strains were used, so it is unclear whether these are true probiotics and whether the product indeed could be considered a synbiotic. The same group studied the effect of another synbiotic on metabolites in faecal samples from 20 volunteers (Vitali *et al.*, 2010). FOS was combined with the strains *L. helveticus* Bar13 and *B. longum* Bar33. The strains were selected on the basis of *in vitro* an animal experiment, and no prior clinical data is available, so these strains cannot be considered probiotics. Nevertheless, the product showed an effect on faecal metabolites, this time analysed using GC-MS, rather than NMR. Only 500 mg of FOS was given, and no changes in microbiota composition occurred, as could be expected with this low amount of FOS. Despite showing no changes in microbiota composition, the content of the SCFA acetate and valerate, several ketones, carbon disulfide and methyl acetate was significantly increased by the synbiotic food consumption (Vitali *et al.*, 2010). The increase in some of the ketones was discussed in terms of health. Ooi *et al.* (2010) studied the effect of a synbiotic on the fatty acid composition and morphology of red blood cells. Hypercholesterolaemia causes morphological abnormalities in red blood cells (RBC) leading to their decreased life span. Supplementation of the synbiotic (*L. acidophilus* CHO-220 and inulin) reduced the cholesterol–phospholipids ratio of the RBC membrane by 47% over the 12-week period of the study. In addition, supplementation of the synbiotic reduced the concentration of saturated fatty acids (SFA), increased unsaturated fatty acids (UFA), and increased the ratio of UFA:SFA (Ooi *et al.*, 2010). Like with many studies it is unclear whether the strain used was a true probiotic. The authors only stated that it was a human-derived strain. Childs *et al.*, studied the effect of XOS and/or *B. lactis* Bi-07 on immune function in healthy adults (Childs *et al.*, 2014). XOS alone or in combination with Bi-07 increased the amounts of bifidobacteria in the stool. The individual supplementations showed distinct effects on immune parameters. *For example*, XOS showed an increased fasting plasma HDL content, lower expression of CD16/56 on natural killer T-cells and lower IL-10 secretion, while Bi-07 lowered lipopolysaccharide-stimulated IL-4 secretion in whole-blood cultures and salivary IgA content, and increased IL-6 secretion. The synbiotic did not show additive or synergistic effects, but rather distinct effects, such as a reduction of expression of CD19 on B-cells. Nevertheless, the authors concluded that the 'synbiotic may confer further benefits' (Childs *et al.*, 2014), although to me it is unclear how that can be concluded.

Three studies have been done in lactating women. The first studied the effect of synbiotics on total antioxidant capacity of breast milk (Nikniaz *et al.*, 2013a). The synbiotic mixture contained 394 mg FOS and the following freeze-dried strains: *L. casei* PXN 37, *L. rhamnosus* PXN 54, *Streptococcus thermophilus* PXN 66, *B. breve* PXN 25, *L. acidophilus* PXN 35, *B. longum* PXN 30, and *L. bulgaricus* PXN 39. It is unclear whether these strains can be considered true probiotics. The synbiotic significantly increased the total antioxidant capacity of breast milk (Nikniaz *et al.*, 2013a). In a follow-up publication the same group described the effect of the same synbiotic on immune parameters of breast milk (Nikniaz *et al.*, 2013b). The breast milk IgA and TGF-β2 levels increased significantly in the supplemented group. Interestingly, the incidence of diarrhoea in infants decreased significantly in the supplemented group while no significant changes were observed in the placebo group after the experimental period. A third study by this group investigated the effect of intake of the same synbiotic on energy intake of lactating mothers, and BMI and growth of their children (Ostadrahimi *et al.*, 2013). Synbiotic supplementation resulted in a slight increase in maternal weight and BMI. Also, the weight gain of the infants in the synbiotic group was significantly higher than the placebo group (Ostadrahimi *et al.*, 2013). The authors concluded that synbiotics may prevent weight loss in lactating mothers and result in weight gain in infants. They advised that further experiments are required to study these effects in undernourished lactating mothers and their infants.

Allergic adults

The effect of a synbiotic on allergic responses in adults with established allergic asthma and house dust mite allergy was tested by van de Pol *et al.* (2011). The synbiotic consisted of a mixture of 8 g of the prebiotics GOS and FOS (9:1) and the probiotic *B. breve* M-16V. Compared to the placebo, bronchial inflammation did not differ. However, treatment significantly increased both the morning and evening peak expiratory flow, a measure of lung function. In addition, the increase in serum IL-5 after allergen challenge was significantly inhibited by the synbiotic, as was *ex vivo* allergen-induced Th2-cytokine production by PBMCs (van de Pol *et al.*, 2011).

Elderly

Bartosch *et al.* (2005) used a synbiotic comprising *B. bifidum* BB-02 and *B. lactis* BL-01 together with inulin in elderly volunteers. The synbiotic increased bifidobacterial counts in the elderly, but the probiotics or inulin were not tested alone. Ouwehand *et al.*, used a combination of lactitol and the probiotic *L. acidophilus* NCFM. A significant increase in faecal *L. acidophilus* NCFM levels was observed in the synbiotic group, as well as an increase in bifidobacteria (Ouwehand *et al.*, 2009). Also this study lacked an arm with the prebiotic or probiotic alone. Given that so many of disorders are age-related in prevalence, a unifying hypothesis, called 'inflamm-ageing', has emerged, which proposes that a common immunological factor could be responsible for an age-related rise in inflammatory disorders (Quigley, 2013). It is believed that synbiotics could help modify some of these immunological factors. Macfarlane *et al.* (2013) addressed this in an intervention study where they administered a synbiotic (an unknown strain of *B. longum* and inulin) to subjects aged between 65 and 90 and demonstrated a significant reduction in the important pro-inflammatory cytokine TNF-α. The study also showed an increase in the numbers of bifidobacteria and an associated increase in butyrate production. Butyrate has been shown to have anti-inflammatory activity (Hamer *et al.*, 2008), so this may have contributed to the lowering of TNF-α.

Children

The safety aspects in clinical trials with probiotics and synbiotics in young infants (0–2 years of age) was recently reviewed (van den Nieuwboer *et al.*, 2014). Almost all studies concluded that none of the adverse effects were related to the study product; the study products are generally well tolerated. The authors stated that inconsistent, imprecise and potentially incomplete reporting as well as the variation in probiotic strains, dosages, administration regimes, study populations and reported outcomes, greatly limited the ability to generalize the conclusions. Nevertheless, the authors concluded that probiotic administration (also in synbiotics) to infants between 0 and 24 months is safe (van den Nieuwboer *et al.*, 2014). Only 9 of the 57 clinical trials evaluated in the period 2008–2013 used synbiotics. In a study with 4 probiotics strains and GOS, pregnant mothers carrying infants at high risk for allergy received the mixture of 4 probiotic species (*L. rhamnosus* GG and *L. rhamnosus* LC705, *B. breve* Bb99, and *Propionibacterium freudenreichii* subsp. *shermanii* JS) or a placebo for 4 weeks before delivery. Their infants received the same probiotics with 0.8 g of GOS, or a placebo, daily for 6 months after birth (Kukkonen *et al.*, 2008). During the 6-month intervention, antibiotics were prescribed less often in the synbiotic group than in the placebo group. Throughout the follow-up period, respiratory infections occurred less frequently in the synbiotic group, indicating that feeding synbiotics to newborn infants seemed to increase resistance to respiratory infections during the first 2 years of life (Kukkonen *et al.*, 2008). The same study was also used to study allergy of the infants, but those results will be discussed below in the section on children prone to allergy. A synbiotic mixture of *L. plantarum* ATCC 20195 cells and 150 mg of FOS was given to children born by Caesarean section (Panigrahi *et al.*, 2008). The intention was to study *L. plantarum* implantation in the gut of these children. Naturally, the strain could be isolated from children who were supplemented with the microbe, but the effect of FOS could not be assessed as there was no probiotic-only control. Interestingly, the infants stayed colonized for several months after administration was stopped (Panigrahi *et al.*, 2008). Chouraqui *et al.* (2008) evaluated three formulas containing different mixtures of the probiotics *B. longum* BL999, *L. rhamnosus* LPR, and *L. paracasei ST11* as well as a mixture of probiotics and the prebiotics GOS and FOS (in a ratio of 9:1) for their safety and effect on the incidence of diarrhoea in infants. There was no difference in weight gain between the groups. The only difference observed was an increase in stool frequency in those groups that were supplemented with the prebiotic. It is likely that this is attributable to the prebiotic, although not all controls were taken along in the study, e.g. the BL999 + ST11 was not tested alone, but only in the presence of the prebiotic mixture (Chouraqui *et al.*, 2008). In the BL999 + ST11 + GOS/FOS group the frequency of liquid stools was increased, which in children is unwanted. Underwood *et al.*, tested the effect of two different synbiotic preparations, one composed of *L. rhamnosus* GG plus inulin, the other 3 unidentified strains of *B. longum*, *B. bifidum*, and *B. infantis* together with *L. acidophilus* DDS®-1 plus inulin (Underwood *et al.*, 2009). Not unsurprisingly, infants receiving the product with the bifidobacterial strains were more likely to become colonized with bifidobacteria. No significant differences in weight gain or stool SCFA content were detected. No control groups were included. Tolerance and safety of *L. paracasei* subsp. *paracasei* CRL-431 in combination with *B. animalis* subsp. *lactis* Bb-12 in a GOS-containing infant formula was tested by Vlieger *et al.* (2009). There were no safety issues with the synbiotic combination. The CRL-431 strain had previously been shown to have positive effects in the treatment of diarrhoea in children. It is unclear why Bb-12 was added. Whereas the above studies that assessed safety and tolerance did not show significant differences in weight gain, Firmansyah *et al.* (2011) reported that the use of *B. longum* BL999, *L. rhamnosus* LPR, and prebiotics (inulin and FOS) led to a significant increase in weight gain in Indonesian children after a 12 month supplementation compared to control milk. Apart from the control milk, no groups with either probiotics or prebiotics alone were included. The gain in the synbiotics group resulted in a change in 'z-score weight-for-age' closer to the WHO Child Growth Standard. Sazawal *et al.* (2010) evaluated the efficacy of adding GOS and *B. lactis* HN019 to milk, in preventing diarrhoea, respiratory infections and severe illnesses, in children aged 1–4 years. There was a trend of a synbiotic effect on diarrhoea ($P = 0.08$). Significant reductions were observed in the synbiotic group in the incidence of dysentery episodes, the incidence of pneumonia and severe acute lower respiratory infection. Compared to placebo children in the synbiotic group less days with severe illness and high fever occurred (Sazawal *et al.*, 2010). The effect of *L. paracasei* B21060 plus arabinogalactan and XOS in children with acute diarrhoea was tested by Passariello *et al.* (Passariello *et al.*, 2012). Compared to placebo, the resolution rate of diarrhoea after 3 days was significantly higher in the synbiotic group. Children in the synbiotic group showed a significant reduction of almost 20 h in the duration of diarrhoea, and reductions in daily stool outputs and stool consistency. The need for adjunctive medication and hospitalization was also reduced. The adjunctive medication in fact was only used in the placebo group and included anti-diarrheal and anti-nausea drugs and surprisingly: probiotics (Passariello *et al.*, 2012). How the use of probiotics (no further details were given as to which probiotics) influenced the outcome of the trial is unclear. No groups with only the probiotic or only the prebiotic were included. To evaluate the effect of a growing-up milk containing synbiotics on immune function and child growth, Xuan *et al.* (2013) conducted a cluster randomized, multicentre, double-blind, placebo controlled clinical trial in children between 18 and 36 months of age in Vietnam. Eligible children from several kindergartens were randomly assigned to receive an infant formula containing *L. paracasei* NCC2461 and *B. longum* NCC3001 together with inulin and FOS, or a control milk, respectively, for 5 months. The synbiotic formula was also fortified with vitamins (A, C, and E), minerals (zinc and selenium), and docosahexaenoic acid. Blood IgA level and growth parameters were increased in the synbiotic group. Compared with the control group, there was also a trend of decreased vitamin A deficiency. It is highly likely that this is due to the vitamin A fortification, rather than the synbiotic (Xuan *et al.*, 2013). As for many of the other studies, no probiotic- or prebiotic-only groups were taken along. This also holds true for the study by Safavi *et al.* (2013), who investigated the effects of 8 weeks of supplementation with synbiotics (composed of unidentified strains of *L. casei*, *L. rhamnosus*, *S. thermophilus*, *B. breve*, *L. acidophilus*, *B. longum*, and *L. bulgaricus* along with FOS) on some cardio-metabolic risk factors in overweight children. Seventy overweight children from 6 to 18 y with a body mass index (BMI) equal to or higher than the age- and gender-specific 85th percentile of generally accepted normal weight were included in the study. Over the 8 weeks there was a significant loss of weight in the synbiotic group, as well as reduction in waist circumference, but a higher waist-to-hip ratio. Moreover, the synbiotic group had significant decreases in serum triglycerides, and total- and low density lipoprotein-cholesterol levels (Safavi *et al.*, 2013). In the same children, Kelishadi *et al.* (2014) showed that, compared with the placebo group, the synbiotic group showed significant decreases in plasma TNF-α and IL-6, as well as a significant increase in adiponectin. These changes were no longer significant after adjustment for BMI, indicating that the benefit was primarily for those children losing weight.

Allergic children

In the study with the four probiotic strains and GOS referred to above (Kukkonen *et al.*, 2008), the same team also looked at the effect of probiotics and GOS on the cumulative incidence of allergic diseases (food allergy, eczema, asthma, and allergic rhinitis) and IgE sensitization in infants at high risk for allergy (Kukkonen *et al.*, 2007). Synbiotic treatment did not show effects on the cumulative incidence of all allergic diseases and only tended to reduce IgE-associated atopic. However, it did reduce eczema and especially atopic eczema significantly. Lactobacilli and bifidobacteria more frequently colonized the guts of supplemented infants (Kukkonen *et al.*, 2007). In a follow-up study the same team investigated the effect of the same four-strain probiotic + GOS on faecal calprotectin, α_1-antitrypsin, TNF-α, and IgA, and on haemoglobin (Hb) (Kuitunen *et al.*, 2009). The authors concluded that the synbiotic used caused a gut mucosal inflammation and an unexpected transient decreased Hb value during intervention. The authors assessed the clinical significance of the small and transient difference in Hb to be small, but the supplementation period was relatively short. The authors warned that during more intensive and longer interventions Hb concentrations should be monitored. It is unclear whether the effects observed in fact have a beneficial effect on these children.

The same synbiotic (*B. breve* M-16V and a GOS/FOS mixture) that was used by van de Pol *et al.* in allergic adults (van de Pol *et al.*, 2011), was also used by van der Aa and colleagues in a number of studies in infants. In a first report van der Aa *et al.*, showed that this mixture did not have an effect on the development of atopic dermatitis (AD) in children (van der Aa *et al.*, 2010), although in a subgroup of infants with IgE-associated AD, SCORAD score improvement was significantly greater in the synbiotic than in the placebo group. In a second publication this group showed that the same synbiotic reduced the prevalence of 'frequent wheezing' and 'wheezing and/or noisy breathing apart from colds' in the synbiotic group approximately 2-fold and 10-fold, respectively. Significantly less children in the synbiotic group had started to use asthma medication after baseline (van der Aa *et al.*, 2011). In a third study, *ex vivo* cytokine production by PBMCs and circulating regulatory T (T_{reg}) cell percentage in infants with atopic dermatitis were studied (van der Aa *et al.*, 2012). The authors showed that the synbiotic mixture had no detectable effect on plasma levels of the analysed atopic disease markers, *ex vivo* cytokine production nor on circulating T_{reg} cell percentage, although they were able to show down-regulation of IL-12 production in *ex vivo* egg- and peanut-stimulated PBMCs. The authors concluded that their results did not support the use of this synbiotic in clinical practice. In a more recent study, the mixture of prebiotics was changed from GOS:FOS (9:1) to FOS: pectin-derived acidic oligosaccharides (85:15). The probiotic strain used was still the same: *B. breve* M-16V (Harvey *et al.*, 2014). This synbiotic mixture was used in two studies both reported in the same paper. In the first, the safety and tolerance was studied of the synbiotics in an amino acid–based formula (AAF) in healthy children, and no changes were observed to control regarding growth parameters and tolerance. The second study aimed to investigate the safety of feeding the synbiotic in AAF on hypoallergenicity in 30 infants and children with established IgE–mediated cow's milk allergy. No allergic reactions were detected after feeding (Harvey *et al.*, 2014). However, no comparison was made with AAF alone, and therefore the role of the synbiotic is unclear. No doubt however, future studies will address the functionality of this synbiotic mixture. Rozé *et al.* (2012) evaluated the safety, tolerance and preventative effect on atopic dermatitis of an experimental α-lactalbumin-enriched and synbiotic-supplemented infant formula in 97 non-breastfed babies. The experimental infant formula contained *L. rhamnosus* LCS-742 and *B. longum* subsp. *infantis* M63 with 0.4 g GOS and 0.02 g FOS per 100 ml of formula. Growth was similar in both groups at 6 months of age. At 1 month infants in the synbiotic group exhibited less crying or agitation, and more quiet behaviour. At 6 months, atopic dermatitis was less frequently observed in the synbiotic group. Moreover, in the control group a decrease of faecal secretory IgA concentration between 1 and 6 months was observed. This decrease was significantly associated with atopic dermatitis and negatively correlated to the level of colonization by bifidobacteria (Roze *et al.*, 2012).

Synbiotics in colon cancer

Only 3 *in vivo* human study on the protective effect of synbiotics against colon cancer have been published. Rafter *et al.*, aimed to investigate the effect of a synbiotic preparation, containing *L. rhamnosus* GG and *B. lactis* Bb-12 encapsulated in enteric coated tablets combined with 12 g inulin. Two groups were targeted: polypectomized and colon cancer patients. Following synbiotic ingestion, significant increases in faecal lactobacilli and bifidobacteria, and a decrease in *Clostridium perfringens* were observed in both patient groups. In polypectomized patients, the synbiotic improved the epithelial barrier function and reduced the ability of faecal water to induce colonic cell necrosis, and it prevented increases in IL-2 by PBMCs. In cancer patients, synbiotic ingestion increased the production of IFNγ (Rafter *et al.*, 2007). The results of this study are encouraging; however, they do not offer enough insight into the mechanism of synbiotic action, as groups with only the probiotics or prebiotic were not taken along. Nevertheless there are ample indications of the possible mechanism of synbiotics in colon cancer from animal studies. These are addressed in the section on potential mode of action. Reddy *et al.*, aimed to study whether it is possible to modulate gut microbiota and preserve intestinal barrier function during elective colorectal surgery by using combinations of the antibiotic neomycin, synbiotics and/or mechanical bowel preparation (MBP) (Reddy *et al.*, 2007). The primary endpoint of the study was prevalence of *Enterobacteriaceae* in the gut microbiota. There was a significant decrease in *Enterobacteriaceae* in the group receiving synbiotics + antibiotic + MBP compared with the other groups. The synbiotic group also had a significantly lower incidence of translocation after bowel mobilization. There were no significant differences between the groups in intestinal permeability, inflammatory response or septic morbidity (Reddy *et al.*, 2007). Wortheley *et al.* (2009) described the effect of the combination of *B. lactis* LAFTI B94 with resistant starch (RS) or either component alone on luminal, inflammatory, epigenetic, and epithelial biomarkers of colorectal cancer in a cross-over trial in 20 individuals. Despite the proper set-up of the study to show synergy between the probiotic and prebiotic, unfortunately the only difference observed were changes in the faecal microbiota. Treatment with either component alone or together did not significantly alter any other faecal, serum, or epithelial variables. The microbiota was assessed using DGGE. The major difference was an increase in the proportion of patients with *Lachnospiraceae* spp. after the synbiotic treatment. In a recent meta-analysis He *et al.* (2013) concluded that perioperatively use of probiotics or synbiotics as prophylaxis in patients undergoing colorectal resection improved clinical outcomes. Only the study of Reddy *et al.* (2007) was included in the analysis, as the others were not aimed at perioperative patients. The authors concluded, however, that it is not yet a preventative strategy, and that the best probiotic species and the optimal dose still needs to be investigated.

Inflammatory bowel disease (IBD)

The efficacy of synbiotics in IBD has also been investigated, with two studies in ulcerative colitis (UC) and two studies in Crohn's disease (CD). in UC, Furrie *et al.*, first screened a number of *Bifidobacterium* isolates, isolated from healthy colonic mucosa, faeces, or obtained from culture collections. Besides their ability to withstand oxygen, acid, and bile, their adhesion to epithelial cells, their ability to survive freeze drying and long-term storage, and their ability to metabolize FOS as an energy source was also determined (Furrie *et al.*, 2005). Moreover, the ability of the microbial strains to reduce production of IL-1α was investigated in the HT29 epithelial cell line. A *B. longum* isolate from healthy rectal mucosa (no strain number provided) was selected for a clinical study together with inulin in ulcerative colitis (UC) patients. Clinical status was scored and rectal biopsies were collected before and after treatment, and transcription levels of epithelium related immune markers were measured. Sigmoidoscopy scores were reduced in the synbiotic group compared with placebo, although it should be said that scores at the start of the study were already higher in this group compared to placebo. The mRNA levels for several human beta defensins were significantly reduced in the synbiotic group after treatment. Also TNF-α and IL-1α expression was significantly reduced. Biopsies in the test group had reduced inflammation and regeneration of epithelial tissue (Furrie *et al.*, 2005). This is one of the few studies that followed a rational for strain selection that would complement the selected prebiotic as well as target the disease. The strain was specifically selected for its ability to down-regulate *in vitro* production of cytokines that have been involved in the active state of UC. It would have been interesting to see what effect the probiotic and prebiotic alone would have been. This was unfortunately not tested. Fujimore *et al.* (2009) investigated the effect of an unknown strain of *B. longum* and/or psyllium in UC patients. A quality of life (QoL) score was assessed, as well as several blood parameters. Overall QoL was only significantly improved in the synbiotic group. Significant individual scores that improved were emotional function for the probiotic group, bowel function for the prebiotic group and systemic and social functions for the synbiotic group. C-reactive protein decreased significantly only with synbiotic therapy (Fujimori *et al.*, 2009). The synbiotic group clearly experienced greater quality of life changes, but it remains to be seen whether this is a synergistic effect of the combination of the probiotic and prebiotic (which some would argue is not an established prebiotic at all), as the synbiotic group had different QoL outcomes compared to the other groups.

In CD, a multicentre, randomized, placebo-controlled study investigating the efficacy of a synbiotic (composed of *Pediococcus pentosaceus* 5-33:3, *Leuconostoc mesenteroides* 77:1, *L. casei* spp. *paracasei* F19, *L. plantarum* 2362, and β-glucan, resistant starch, inulin, and pectin) in 30 ileal resection CD patients over a period of 24 months failed to note an effect on remission or disease scores compared to placebo (Chermesh *et al.*, 2007). Fujimori *et al.* (2007) examined the effect of a synbiotic (*B. breve*, *L. casei*, and *B. longum* with psyllium fibre) in a small open label study of 10 active CD patients. Eight patients with moderately active CD lowered their CDAI scores by over 70 points and six achieved remission. Several inflammatory parameters were lowered in these patients, as well as the occurrence of diarrhoea and abdominal pain. Strain numbers for the probiotics used were not provided, and unfortunately there was no control group. Steed *et al.*, investigated the same synbiotic that was used beneficially in UC patients by Furrie *et al.* (2005) as described above. The mixture composed of an unknown strain of *B. longum* and inulin was provided for 6 months to individuals with a Crohn's disease activity index (CDAI) score between 150 and 450 aged between 18 and 79 years (Steed *et al.*, 2010). The main outcome of the study was to show a reduction in mucosal TNF-α in patients receiving the synbiotic. Secondary outcome was numbers of patients in remission. Significant improvements in clinical outcomes occurred with synbiotic consumption, with reductions in both CDAI and histological scores. As primary outcome, TNF-α expression in synbiotic patients at 3 months was significantly reduced, although not at 6 months (Steed *et al.*, 2010). Of note, TNF-α expression in biopsies from inflamed tissue was significantly higher than in non-inflamed tissue.

Infections

Schrezenmeier *et al.* (2004) studied the effect of a synbiotic (composed of *L. acidophilus* NCFM and *B. animalis* subsp. *lactis* Bi-07 with FOS) in young children aged 1–6 years with acute bacterial infections. The group with synbiotic had lower relapse of bacterial infection after 14 days of treatment compared to probiotics only and placebo, but the differences were not statistically significant. The synbiotic increased stool lactobacilli and bifidobacteria more than the probiotics alone. The synbiotic group showed a significantly greater weight gain compared to placebo (Schrezenmeir *et al.*, 2004). The efficacy of synbiotic supplementation in the prevention of common winter diseases in children was studied by Cazzola *et al.* (2010) The synbiotic was composed of *L. helveticus* R0052, *B. infantis* R0033, *B. bifidum* R0071, and FOS (Cazzola *et al.*, 2010). The relative risk of at least one illness episode was 25% lower in the synbiotic group compared to placebo, with the main differences in the number of children who suffered from at least one ear, nose and throat, respiratory tract or gastrointestinal disorder. The number of lost school days due to sickness was also significantly lower in the synbiotic group (Cazzola *et al.*, 2010). Vandenplas and De Hert (2012) reported on behalf of the Probiotical Study Group the effect of a synbiotic in acute infectious gastroenteritis. The synbiotic was formulated in a sachet and composed of *L. rhamnosus* PHA113, *L. acidophilus* PHA 121, *B. infantis* PHA211, *B. lactis* PHA222 and *S. thermophilus* PHA311, with 20 mg of FOS. Children with acute infectious gastroenteritis were included. They were all treated with an oral rehydration solution and half of the children received the synbiotic supplement. Median duration of diarrhoea was significantly shorter by 1 day in the synbiotic. Significantly more concomitant medication (antibiotics, antipyretics, antiemetics) was prescribed in the placebo group. The authors calculated that although use of the synbiotic supplementation itself increased immediate costs, add-on medication and extra consultations were reduced in this group, resulting in an overall reduction of health care cost of 25% in the synbiotic group (Vandenplas and De Hert, 2012).

Surgical and intensive care unit patients

In a study with patients scheduled to undergo elective laparotomy, the efficacy of a synbiotic in preventing postoperative complications was evaluated (Anderson *et al.*, 2004). The synbiotic was composed of *L. acidophilus* La5, *B. lactis* Bb-12, and contained the yoghurt cultures *L. bulgaricus* and *S. thermophilus* together with FOS. Treatment was initiated one to two weeks preoperatively and continued until patients were discharged from the hospital. In this study no difference was observed between the synbiotic and placebo groups with regard to bacterial translocation and colonization, systemic inflammation, and septic complications (Anderson *et al.*, 2004). In another study with the same synbiotic reported by Jain *et al.*, the effects of the synbiotic on critically ill patients were also limited (Jain *et al.*, 2004). There were no significant differences between the groups in terms of intestinal permeability, septic complications or mortality. Only a lower incidence of potentially pathogenic bacteria and lower total number of microorganisms in their nasogastric aspirates was observed in the synbiotic group (Jain *et al.*, 2004). Another study investigating the effect of a synbiotic composed of four strains and four fibres/prebiotics in patients scheduled for liver transplantation. The synbiotic was composed of *P. pentosaceus* 5-33:3, *Leuc. mesenteroides* 77:1, *L. paracasei* subsp. *paracasei* F19 and *L. plantarum* 2362 plus β-glucan, inulin, pectin and resistant starch, as used by Chermesh *et al.*, in UC patients (Chermesh *et al.*, 2007). Patients were divided in two treatment groups, one receiving synbiotic and the other receiving only the fibre components of the synbiotic. Interestingly, in the group receiving the full synbiotic formula, postoperative infections were significantly lower compared to the fibre only group. Also, in the synbiotic group, the required antibiotic treatment duration was significantly lower (Rayes *et al.*, 2005). In a later study, the same group evaluated the efficacy of this synbiotic against bacterial infection rates following pylorus-preserving pancreato-duodenectomy (Rayes *et al.*, 2007). The synbiotic or placebo (only the four fibres) treatment was administered one day preoperatively and was continued for eight days postoperatively. The results were in agreement with the liver transplantation study with bacterial infections being significantly lower in the synbiotic group compared with the fibre group (Rayes *et al.*, 2007). Despite these promising results, the authors in a review of available data were quite cautious about the use of synbiotics in clinical trials. They stated that some probiotics in fact may be harmful (Rayes *et al.*, 2009). The same 4 strain/4 fibre mixture was also effective in reducing endotoxaemia and blood ammonia levels, besides increasing faecal *Lactobacillus* amounts in patients with cirrhosis suffering from minimal hepatic encephalopathy (MHE) (Liu *et al.*, 2004). The gut microbiota of the MHE patients were characterized by an overgrowth of *Escherichia coli* and staphylococcal species, which was normalized by the synbiotic. MHE reversal was observed in 50% of the synbiotic treated patients with a significant reduction in endotoxaemia (Liu *et al.*, 2004). More recently, a study assessed the clinical efficacy of *B. longum* W11 plus FOS in the treatment of MHE. After 90 days of treatment, fasting ammonia serum levels were significantly decreased in the synbiotic group. In addition, several neuropsychological markers were improved that can be affected by the encephalopathy, such as abstract reasoning, tactile performance, tactile-visual and spatial memory, rhythm perception and memory, etc. (Malaguarnera *et al.*, 2007).

Three studies with the four-strain/four-fibre synbiotic (*P. pentosaceus* 5–33:3, *Leuc. mesenteroides* 77:1, *L. paracasei* subsp. *paracasei* F19 and *L. plantarum* 2362 plus β-glucan, inulin, pectin and resistant starch) have been performed in trauma patients. In one study synbiotic therapy resulted in significantly lower infection rate, severe sepsis rate, and a reduced number of days in the intensive care unit and on mechanical ventilation. Mortality was also lower in the synbiotic group, but this difference was not statistically significant (Kotzampassi *et al.*, 2006). The same team studied the same synbiotic to control inflammation and infection in patients with multiple injuries (Giamarellos-Bourboulis *et al.*, 2009). The synbiotics decreased significantly the risk for sepsis by bloodstream infections, by both decreasing the occurrence of sepsis, as well as the time to progression to primary bacteraemia. Within the different causes of bacterial infection, the synbiotic decreased the development of ventilator-associated pneumonia with *Acinetobacter baumannii*. Synbiotic treatment was accompanied by a reduction of white blood cell counts and LPS and C-reactive protein (CRP) levels, indicating diminished bacterial translocation (Giamarellos-Bourboulis *et al.*, 2009). In another study, Spindler-Vesel *et al.* (2007) used the same synbiotic cocktail in a study on multiple injured patients. Patients either received an enteral formula enriched with glutamine, fermentable fibre, a peptide diet, or a standard enteral formula plus the synbiotic. Mortality was very low and comparable in the four groups. Significantly fewer patients in the synbiotic group developed infections than in the other three groups by a factor of approximately 2. This effect was accompanied by a decrease of the lactulose–mannitol index as a parameter for intestinal permeability in the synbiotic group (Spindler-Vesel *et al.*, 2007).

In patients with biliary cancer undergoing hepato-biliary resection the impact of synbiotic consisting of *L. casei* Shirota, *B. breve* Yakult, and GOS on the clinical course was investigated (Kanazawa *et al.*, 2005). Patients received either enteral nutrition alone, or in combination with the synbiotic postoperatively for 14 days. In the synbiotic group, bacterial infections were significantly lower compared to the group with only enteral feeding. In addition, a significant reduction of pathogenic bacteria and an increase of organic acids in the faeces were observed (Kanazawa *et al.*, 2005). In another study on liver resection for bile duct carcinoma, one group of patients received a synbiotic only postoperatively, another group received the synbiotic both preoperatively and postoperatively (Sugawara *et al.*, 2006). The synbiotic was the same as used by Kanazawa *et al.* (2005), but at a higher dose. The preoperative treatment was administered orally and delivered daily. The postoperative treatment was delivered parenterally. Pre- plus postoperative treatment with synbiotics resulted in a significantly lower bacterial infection rate compared to only postoperative treatment. Also white blood cell counts, and CRP levels were significantly lower compared to the group receiving only the postoperative treatment (Sugawara *et al.*, 2006). Unfortunately no placebo was taken along. In an earlier case-report, this same synbiotic (*L. casei* Shirota, *B. breve* Yakult, and GOS) was shown to the effective in improving the intestinal barrier function in a girl with short bowel syndrome (Kanamori *et al.*, 2001). Prior to staring the synbiotic regimen, the patient experienced frequent high fever attacks (14 in 2 years) and metabolic acidosis (8 times in 2 years). The microbiota of the patient showed severe dysbiosis prior to synbiotic treatment. After the synbiotic treatment an initial fever attack occurred but after that for another 2 years both fever attacks and metabolic acidosis did not occur any more. Whereas the patient gained less than 1.5 kg of weight in the 2 years prior to the synbiotic treatment, she gained approximately 7 kg in the subsequent years. Moreover, intestinal motility was gradually restored (Kanamori *et al.*, 2001). The same group also published a case-report in which the same synbiotic was used in a critically ill 9-month-old girl with type IV laryngotracheo-oesophageal cleft (Kanamori *et al.*, 2002). The patient had four episodes of severe pneumonia in 10 months prior to the synbiotic treatment, with only one episode after. Body weight gain in the 9 months prior to synbiotics was only approximately 1 kg, and 3 kg in the 10 months after. Prior to the synbiotic treatment she did not have normal bowel movements and enemas were required daily, while bowel movements resumed soon after the commencement of synbiotic therapy (Kanamori *et al.*, 2002). While these case reports obviously miss proper controls and placebo treatments, they are promising and encourage further studies with this synbiotic mixture. One such follow-up study was recently reported. Yokoyama *et al.*, determined the perioperative effect of *L. casei* Shirota, *B. breve* Yakult, and GOS on bacterial translocation after oesophagectomy (Yokoyama *et al.*, 2014). Mesenteric lymph nodes (MLN) and blood were studied for bacterial translocation. Bacteria were more frequently detected in MLN samples in the control group than in the synbiotic group. In addition, bacteraemia was less prevalent in the synbiotic group. This was accompanied by a lower neutrophil count on postoperative days 1, 2 and 7 in the synbiotics group (Yokoyama *et al.*, 2014).

Others

In irritable bowel syndrome, Dughera *et al.* (2007) investigated the effect of a synbiotic product consisting of *B. longum* W11 and FOS in an open label, prospective, uncontrolled multicentre study on 129 constipated IBS subjects. A significant increase in stool frequency was reported as well as alleviation of pain and bloating as compared with baseline samples. However, this was an open label study in the absence of a placebo, and although the results are encouraging, further RCTs (randomized controlled trials) are required to confirm efficacy.

Gonzalez-Hernandez *et al.*, tested the efficacy of synbiotic therapy on microbial translocation and inflammation and immunological status in HIV-infected patients (Gonzalez-Hernandez *et al.*, 2012). The synbiotic used was fructans from agave with *L. rhamnosus* HN001 and *B. lactis* Bi-07. The fructans are composed of linear and branched FOS. Groups with only the prebiotic or only the probiotic were also taken along. The synbiotic group showed a significant reduction in bacterial DNA concentrations in plasma in these HIV-infected patients. Moreover, CD4+ T-cells count increased and the level of IL- 6 decreased significantly in this group. The probiotic and synbiotic groups, but not the prebiotic group, demonstrated a decrease in total bacterial load in faeces. Only the probiotic group exhibited a significant increase of *Bifidobacterium* in faeces of the HIV-patients (Gonzalez-Hernandez *et al.*, 2012).

Takahashi *et al.*, reported a case report on the efficacy of *B. breve* Yakult, *L. casei* Shirota and GOS to prevention D-lactate acidosis in short bowel syndrome (Takahashi *et al.*, 2013). The patient had recurrent episodes of neurological dysfunction owing to D-lactate acidosis. Fasting was required to keep D-lactate levels low. Antibiotic treatment was not capable of eliminating D-lactate producers. After synbiotic treatment serum D-lactate levels declined to below the level required for onset of symptoms. However, faecal levels of D-lactate remained very high. Prior to synbiotic treatment the frequency of defecation was low, but this increased after the intake of the synbiotic. The authors concluded that, although faecal concentrations of D-lactate were high, the synbiotic increased intestinal peristalsis, thereby increasing the frequency of defecation resulting in prevention of excessive absorption of D-lactate.

Potential mode of action

One of the first studies to investigate the potential application of synbiotics against colon cancer was by Rowland and coworkers (Rowland *et al.*, 1998). They evaluated the effect of *B. longum* 25 with or without inulin in male Sprague–Dawley rats that were treated with azoxymethane so as to induce aberrant crypt foci (ACF) formation. The probiotic was prescreened for its *in vitro* ability to ferment inulin HP prior to the *in vivo* study. Rats ingested a control diet, probiotic only, prebiotic only, and the synbiotic diet. *B. longum* 25 and inulin both reduced small ACF formation, caecal ammonia concentrations, and β-glucuronidase activity when tested individually compared to the control diet. When used in combination, they mediated a significantly higher inhibition of both small and large size ACF formation. Both inulin and the synbiotic caused significant increases in β-glucosidase activities; however, *B. longum* alone did not have an effect (Rowland *et al.*, 1998). The inclusion in this study of proper controls gives valuable insight into the behaviour of this synbiotic. Lowered ammonia and β-glucuronidase activity have been correlated to a lower genotoxicity of faecal water. More animal studies have shown the beneficial effects of synbiotics on tumour-preventative efficacy and faecal water genotoxicity, e.g. a synbiotic composed of *L. rhamnosus* GG, *B. lactis* Bb-12 and inulin (Klinder *et al.*, 2004), which has later been used by Rafter *et al.* (2007) in a clinical trial as discussed above.

The pathogenesis of inflammatory bowel disease (IBD) is thought to involve different factors, including the microbiota. The intestinal mucosal immune system has the formidable task of remaining unresponsive to the vast indigenous bacterial population (~10^{14} cells), and yet be able to discriminate and eliminate pathogenic microorganisms to control infection. In IBD evidence has been found for a disturbed balance between pro-inflammatory and anti-inflammatory cytokines. Osman *et al.*, studied the effect of a synbiotic in the DSS-induced mouse model of IBD (Osman *et al.*, 2006). They tested numerous markers for potential mechanisms, such as gut microbiota composition, bacterial translocation, SCFA, and inflammatory markers, such as myeloperoxidase (MPO) and several cytokines, as well as antioxidant status by measuring malondialdehyde (MDA). They used two *B. infantis* strains (DSM 15158 or DSM 15159) alone or in combination with FOS. FOS was also tested alone. Disease activity index decreased in all groups compared to control. Also, SCFA increased and bacterial translocation decreased significantly in all groups compared to placebo. IL-1β was reduced in all groups, except for *B. infantis* DSM 15158. Moreover, colonic MPO activity decreased significantly in all groups compared to the control. MDA was only reduced in the groups receiving *B. infantis* DSM 15159, with and without FOS. This was a well-designed study comparing the effect of each of the synbiotic constituents alone and in combination and investigating the effect of strain specificity against the disease. Some unexpected effects were observed. *For example*, translocation to the liver was reduced by FOS and below the detection limit for *B. infantis* DSM 15158, yet it was not significantly different (if at all, a little increased) when these were combined together. All groups reduced disease activity index. However, the authors speculated that the observed anti-inflammatory effect may have been reached through different mechanisms. These include a direct effect by the probiotics, or indirect effects through SCFA, which have been shown to be anti-inflammatory, or influencing the antioxidant status in the gut (Osman *et al.*, 2006).

Schouten *et al.* (2009) used a mixture of GOS and FOS (9:1) alone and/or *B. breve* M-16V to mechanistically investigate how the synbiotic (components) influences allergy to cow's milk. Mice were orally sensitized with whey. Antigen-induced anaphylaxis was scored, as well as acute allergic skin response by measuring ear swelling. In mice fed the synbiotic mixture, both the allergic skin response and the anaphylactic reaction were strongly reduced compared with whey-sensitized mice fed the control diet. However, the prebiotic or probiotic alone were significantly less effective in reducing the allergic skin response than the synbiotic diet and neither reduced the anaphylactic reaction. This indicated that both components in the synbiotic are required to get the desired effect. The molecular mechanism behind this effect was unravelled by the same team in a recent publication (de Kivit *et al.*, 2012). Based on the hypothesis that GOS/FOS may interfere with mast cell degranulation and may induce Th1- and T_{reg}-cell polarization, and the observation that galectin-9 was shown to neutralize IgE and to induce T_{reg} type immune responses, the authors investigated the effect of the synbiotic on galectin-9 induction, and through this on suppression of allergic symptoms. This was done in both mice and man (infants). Galectin-9 expression by intestinal epithelial cells and serum galectin-9 levels were increased both in mice and humans after the synbiotic and this correlated with reduced acute allergic skin reaction and mast cell degranulation. In mice, the mechanism was studied in more detail. Galectin-9 in serum was only induced in whey-sensitized mice, not in sham-sensitized mice. Moreover, the synbiotic was more effective in inducing galectin-9 than the individual components. The synbiotic enhanced Th1- and T_{reg}-cell differentiation in MLN and in PBMC cultures of mice exposed to galectin-9. Based on these results, the mechanisms by which the synbiotic suppresses IgE-mediated allergic reactions was hypothesized to be the following: During allergic sensitization, a Th2-type effector response is raised against an allergen (in this case whey), which results in IgE production by B-cells. Upon repeated exposure to the allergen, IgE-primed mast cells degranulate, resulting in clinical symptoms. Supplementation of the synbiotic, but not the individual components, to the diet 2 weeks before and during sensitization induces galectin-9 secretion by the intestinal epithelium driving Th1 and T_{reg} cell development. During the challenge phase, galectin-9 induced by the synbiotic directly inhibits mast cell degranulation, leading to a reduction in symptoms (de Kivit *et al.*, 2012).

D'Souza *et al.* (2010) studied the effect of a synbiotic on gene expression in the gut of the host, in this case neonatal rats. Rats received an infant formula supplemented with GOS/FOS, *Saccharomyces boulardii* lyo, or both. Samples from the terminal ileum were collected for histological, biochemical, and molecular analyses. They hypothesized that the synbiotics would influence the anti-oxidant status in the gut thereby reducing inflammation, and assessed this by measuring some of the genes involved in anti-oxidant status. These were constitutively expressed nitric oxide synthase (NOS) and inducible NOS (iNOS) and caveolin-1, which is involved in regulating the functions of the different NOS isoforms, and the genes for superoxide dismutase (SOD), catalase and glutathione peroxidase (GPX), regulating oxidative stress. Primarily, supplementation up-regulated the genes that regulate oxidative stress. Overall, the effect of the synbiotic was more beneficial than that of the individual components, and the effect of the prebiotic was greater than that of the probiotic, although this varied more at the level of gene expression of the genes regulating oxidative stress. The findings suggest synbiotics are effective anti-oxidants and may be beneficial for combating the adverse effects of reactive oxygen species in inflammation (D'Souza *et al.*, 2010).

We have investigated the effect of prebiotics on the survival and activity of probiotics in TNO's validated, dynamic *in vitro* models of the gastrointestinal tract (nick-named TIM). The TIM-1 system simulates the stomach and small intestine (Minekus *et al.*, 1995), whereas TIM-2 mimics the colon (Minekus *et al.*, 1999). TIM-1 has been validated with respect to survival of probiotics (Marteau *et al.*, 1997), and is currently frequently used to predict survival of probiotic strains in a clinical trial, and to investigate mechanistically what influences survival of probiotics. We also used it to study survival of *L. amylovorus* DSM 16698 (Martinez *et al.*, 2011). The microorganism was introduced in the system alone or with prebiotic GOS and/or the probiotic *B. animalis* subsp. *lactis* Bb-12 (Bb-12). Interestingly, co-administration of *L. amylovorus* DSM 16698 with Bb-12 increased survival of *L. amylovorus* DSM 16698 by approximately a factor 2. Addition of GOS alone increased survival even slightly more. The highest survival was seen with both Bb-12 and GOS added to the *L. amylovorus* DSM 16698 culture. The mechanism of how the two strains influence each other's survival is unclear, but we have seen this phenomenon before (unpublished results). With the particular dose of GOS used in these experiments, an increase in survival of approximately 4-fold was observed. Both *L. amylovorus* and Bb-12 were observed to promptly degrade GOS compounds in as measured by High Performance Anion Exchange Chromatography of samples obtained from TIM-1, indicating metabolic activity. It is clear that the dose used served as a lunch-box for both strains, which resulted in a cumulative increase of viable counts due to increased survival and/or growth (Martinez *et al.*, 2011). The combination of *L. amylovorus* DSM 16698 and GOS was also tested in TIM-2, to study its effect on colonic metabolism and microbiota composition. As *L. amylovorus* DSM 16698 is a pig isolate, this was tested in both a pig and human microbiota (Martinez *et al.*, 2013). Interestingly, clear differences were obtained between the pig and human microbiota, reinforcing the notion that differences in microbiota composition between target host organisms need to be considered when animal data are extrapolated to humans, as is often done with pre- and probiotic intervention studies. Amongst the effects observed, high performance anion exchange chromatography revealed that GOS degradation was more pronounced in TIM-2 inoculated with a pig inoculum compared to a human inoculum. GOS stimulated the growth of lactobacilli and bifidobacteria in faecal material from both origins. The cumulative production of SCFA and ammonia was higher in the pig compared to the human microbiota. However, lactate accumulation was higher in the human microbiota and increased after co-administration with GOS (Martinez *et al.*, 2013). Although *in vitro* models do have some limitations, they are excellent tools to get mechanistic insight in the potential mode of action of synbiotics.

Future trends

In the studies reviewed here not all microbial isolates used could be considered probiotics, as the clinical evidence for them was lacking prior to their use in the studies presented here. In some cases it might even be argued that the fibres used are not prebiotics. Nevertheless, I have decided to include these studies to illustrate effects of the combination of the two.

To date, most studies investigating the efficacy of synbiotic administration have used multi-strain probiotics, and sometimes even multiple prebiotics (or fibres). In most cases it is unclear why the combinations of these strains and prebiotic(s) have been chosen, and whether or not the multiple strains interact or even interfere with one another. Nevertheless, some of the mixtures show efficacy in multiple clinical trials, for instance the mixture composed of *P. pentosaceus* 5–33:3, *Leuc. mesenteroides* 77:1, *L. paracasei* subsp. *paracasei* F19 and *L. plantarum* 2362 plus β-glucan, inulin, pectin and resistant starch, in surgical and UC patients. Also, the GOS/FOS mixture used in many studies has been optimized in pre-clinical trials, and has been shown to be effective in a number of clinical trials.

What is largely lacking from most human trials is the underlying mechanism by which the synbiotic works. In the majority of studies reviewed here, there were no control groups with only the probiotic(s) or only the prebiotic(s), which makes it essentially impossible to decipher whether both components are required for the observed effect(s). Moreover, it is unclear whether the effect of the synbiotic is due to the lunch-box phenomenon, or separate beneficial effects of the probiotic and prebiotic. This detail is required to develop novel functional products that use probiotics and/or prebiotics. If only one of the components is active, adding the other is just adding to the price, but nothing else. Therefore, it is highly recommended to perform studies towards the underlying mechanisms, something that the EFSA is also advocating.

Given the multitude of diseases and disorders synbiotics have been tested in, I believe that the future for synbiotics is bright. However, it is essential to separate the wheat from the chaff, and only continue with those compounds that are truly expected to give a benefit.

References

Adams, C.A. (2010). The probiotic paradox: live and dead cells are biological response modifiers. Nutr. Res. Rev. *23*, 37–46.

Anderson, A.D., McNaught, C.E., Jain, P.K., and MacFie, J. (2004). Randomized clinical trial of synbiotic therapy in elective surgical patients. Gut 53, 241–245.

Bartosch, S., Woodmansey, E.J., Paterson, J.C., McMurdo, M.E., and Macfarlane, G.T. (2005). Microbiological effects of consuming a synbiotic containing *Bifidobacterium bifidum, Bifidobacterium lactis,* and oligofructose in elderly persons, determined by real-time polymerase chain reaction and counting of viable bacteria. Clin. Infect. Dis. *40*, 28–37.

Blatchford, P., Ansell, J., de Godoy, M.R.C., Fahey, G., Garcia-Mazcorro, J.F., Gibson, G.R., Goh, Y.J., Hotchkiss, A.T., Hutkins, R., Lacroix, C., *et al.* (2014). Prebiotic mechanisms, functions and applications. Int. J. Probiotics Prebiotics *in press*.

Cazzola, M., Pham-Thi, N., Kerihuel, J.C., Durand, H., and Bohbot, S. (2010). Efficacy of a synbiotic supplementation in the prevention of common winter diseases in children: a randomized, double-blind, placebo-controlled pilot study. Ther. Adv. Respir. Dis. *4*, 271–278.

Chaluvadi, S., Hotchkiss, A.T., Jr., Call, J.E., Luchansky, J.B., Phillips, J.G., Liu, L., and Yam, K.L. (2012). Protection of probiotic bacteria in a synbiotic matrix following aerobic storage at 4 degrees C. Benef. Microbes 3, 175–187.

Chavarri, M., Maranon, I., Ares, R., Ibanez, F.C., Marzo, F., and Villaran Mdel, C. (2010). Microencapsulation of a probiotic and prebiotic in alginate-chitosan capsules improves survival in simulated gastro-intestinal conditions. Int. J. Food Microbiol. *142*, 185–189.

Chermesh, I., Tamir, A., Reshef, R., Chowers, Y., Suissa, A., Katz, D., Gelber, M., Halpern, Z., Bengmark, S., and Eliakim, R. (2007). Failure of Synbiotic 2000 to prevent postoperative recurrence of Crohn's disease. Dig. Dis. Sci. 52, 385–389.

Childs, C.E., Roytio, H., Alhoniemi, E., Fekete, A.A., Forssten, S.D., Hudjec, N., Lim, Y.N., Steger, C.J., Yaqoob, P., Tuohy, K.M., *et al.* (2014). Xylo-oligosaccharides alone or in synbiotic combination with *Bifidobacterium animalis* subsp. *lactis* induce bifidogenesis and modulate markers of immune function in healthy adults: a double-blind, placebo-controlled, randomized, factorial cross-over study. Br. J. Nutr., 1–12.

Chouraqui, J.P., Grathwohl, D., Labaune, J.M., Hascoet, J.M., de Montgolfier, I., Leclaire, M., Giarre, M., and Steenhout, P. (2008). Assessment of the safety, tolerance, and protective effect against diarrhoea of infant formulas containing mixtures of probiotics or probiotics and prebiotics in a randomized controlled trial. Am. J. Clin. Nutr. *87*, 1365–1373.

Costabile, A., Kolida, S., Klinder, A., Gietl, E., Bauerlein, M., Frohberg, C., Landschutze, V., and Gibson, G.R. (2010). A double-blind, placebo-controlled, cross-over study to establish the bifidogenic effect of a very-long-chain inulin extracted from globe artichoke (*Cynara scolymus*) in healthy human subjects. Br. J. Nutr. *104*, 1007–1017.

Coulier, L., Timmermans, J., Bas, R., Van Den Dool, R., Haaksman, I., Klarenbeek, B., Slaghek, T., and Van Dongen, W. (2009). In-depth characterization of prebiotic galacto-oligosaccharides by a combination of analytical techniques. J. Agric. Food Chem. 57, 8488–8495.

D'Souza, A., Fordjour, L., Ahmad, A., Cai, C., Kumar, D., Valencia, G., Aranda, J.V., and Beharry, K.D. (2010). Effects of probiotics, prebiotics, and synbiotics on messenger RNA expression of caveolin-1, NOS, and genes regulating oxidative stress in the terminal ileum of formula-fed neonatal rats. Pediatr. Res. *67*, 526–531.

de Kivit, S., Saeland, E., Kraneveld, A.D., van de Kant, H.J., Schouten, B., van Esch, B.C., Knol, J., Sprikkelman, A.B., van der Aa, L.B., Knippels, L.M., *et al.* (2012). Galectin-9 induced by dietary synbiotics is involved in suppression of allergic symptoms in mice and humans. Allergy *67*, 343–352.

De Preter, V., Ghebretinsae, A.H., Abrahantes, J.C., Windey, K., Rutgeerts, P., and Verbeke, K. (2011). Impact of the synbiotic combination of *Lactobacillus casei* shirota and oligofructose-enriched inulin on the faecal volatile metabolite profile in healthy subjects. Mol. Nutr. Food Res. *55*, 714–722.

De Preter, V., Vanhoutte, T., Huys, G., Swings, J., De Vuyst, L., Rutgeerts, P., and Verbeke, K. (2007). Effects of *Lactobacillus casei* Shirota, *Bifidobacterium breve*, and oligofructose-enriched inulin on colonic nitrogen-protein metabolism in healthy humans. Am. J. Physiol. Gastrointest Liver Physiol. *292*, G358–368.

De Preter, V., Vanhoutte, T., Huys, G., Swings, J., Rutgeerts, P., and Verbeke, K. (2006). Effect of lactulose and *Saccharomyces boulardii* administration on the colonic urea-nitrogen metabolism and the bifidobacteria concentration in healthy human subjects. Aliment. Pharmacol. Ther. *23*, 963–974.

de Vrese, M., and Schrezenmeir, J. (2008). Probiotics, prebiotics, and synbiotics. Adv. Biochem. Eng Biotechnol. *111*, 1–66.

Dughera, L., Elia, C., Navino, M., and Cisaro, F. (2007). Effects of symbiotic preparations on constipated irritable bowel syndrome symptoms. Acta Biomed 78, 111–116.

Eeckhaut, V., Ducatelle, R., Sas, B., Vermeire, S., and Van Immerseel, F. (2014). Progress towards butyrate-producing pharmabiotics: *Butyricicoccus pullicaecorum* capsule and efficacy in TNBS models in comparison with therapeutics. Gut 63, 367.

Eiberger, I., Fuerst, T., Maathuis, A.J.H., and Venema, K. (2014). Use of a dynamic, computer-controlled *in vitro* model of the stomach and small intestine (TIM-1). as a tool in the development of probiotic products, to study survival of the bacteria and their delivery to the site of action. submitted.

Falony, G., Lazidou, K., Verschaeren, A., Weckx, S., Maes, D., and De Vuyst, L. (2009). *In vitro* kinetic analysis of fermentation of prebiotic inulin-type fructans by *Bifidobacterium* species reveals four different phenotypes. Appl. Environ. Microbiol. 75, 454–461.

FAO/WHO (2002). Guidelines for the Evaluation of Probiotics in Food: Joint FAO/WHO Working Group meeting, London Ontario, Canada, 30 April-1 May 2002.

Firmansyah, A., Dwipoerwantoro, P.G., Kadim, M., Alatas, S., Conus, N., Lestarina, L., Bouisset, F., and Steenhout, P. (2011). Improved growth of toddlers fed a milk containing synbiotics. Asia Pac J. Clin. Nutr. *20*, 69–76.

Fujimori, S., Gudis, K., Mitsui, K., Seo, T., Yonezawa, M., Tanaka, S., Tatsuguchi, A., and Sakamoto, C. (2009). A randomized controlled trial on the efficacy of synbiotic versus probiotic or prebiotic treatment to improve the quality of life in patients with ulcerative colitis. Nutrition 25, 520–525.

Fujimori, S., Tatsuguchi, A., Gudis, K., Kishida, T., Mitsui, K., Ehara, A., Kobayashi, T., Sekita, Y., Seo, T., and Sakamoto, C. (2007). High dose probiotic and prebiotic cotherapy for remission induction of active Crohn's disease. J. Gastroenterol. Hepatol. 22, 1199–1204.

Furrie, E., Macfarlane, S., Kennedy, A., Cummings, J.H., Walsh, S.V., O'Neil D.A., and Macfarlane, G.T. (2005). Synbiotic therapy (*Bifidobacterium longum*/Synergy 1). initiates resolution of inflammation in patients with active ulcerative colitis: a randomized controlled pilot trial. Gut *54*, 242–249.

Giamarellos-Bourboulis, E.J., Bengmark, S., Kanellakopoulou, K., and Kotzampassi, K. (2009). Pro- and synbiotics to control inflammation and infection in patients with multiple injuries. J. Trauma *67*, 815–821.

Gibson, G.R., Probert, H.M., Loo, J.V., Rastall, R.A., and Roberfroid, M.B. (2004). Dietary modulation of the human colonic microbiota: updating the concept of prebiotics. Nutr. Res. Rev. *17*, 259–275.

Gibson, G.R., and Roberfroid, M.B. (1995). Dietary modulation of the human colonic microbiota: introducing the concept of prebiotics. J. Nutr. *125*, 1401–1412.

Gonzalez-Hernandez, L.A., Jave-Suarez, L.F., Fafutis-Morris, M., Montes-Salcedo, K.E., Valle-Gutierrez, L.G., Campos-Loza, A.E., Enciso-Gomez, L.F., and Andrade-Villanueva, J.F. (2012). Synbiotic therapy decreases microbial translocation and inflammation and improves immunological status in HIV-infected patients: a double-blind randomized controlled pilot trial. Nutr. J. *11*, 90.

Grabitske, H.A., and Slavin, J.L. (2009). Gastrointestinal effects of low-digestible carbohydrates. Crit. Rev. Food Sci. Nutr. *49*, 327–360.

Hamer, H.M., Jonkers, D., Venema, K., Vanhoutvin, S., Troost, F.J., and Brummer, R.J. (2008). Review article: the role of butyrate on colonic function. Aliment. Pharmacol. Ther. *27*, 104–119.

Harvey, B.M., Langford, J.E., Harthoorn, L.F., Gillman, S.A., Green, T.D., Schwartz, R.H., and Burks, A.W. (2014). Effects on growth and tolerance and hypoallergenicity of an amino acid-based formula with synbiotics. Pediatr. Res. *75*, 343–351.

Havenaar, R., and Van Dokkum, W. (2001). Lactulose. A review on effects, clinical results, and safety aspects in relation to its influence on the colonic environment. In Handbook of Dietary Fibre, S. Sungsoo Cho, and M.L. Dreher, eds. (New York, Marcel Dekker), pp. 179–193.

He, D., Wang, H.Y., Feng, J.Y., Zhang, M.M., Zhou, Y., and Wu, X.T. (2013). Use of pro-/synbiotics as prophylaxis in patients undergoing colorectal resection for cancer: a meta-analysis of randomized controlled trials. Clin. Res. Hepatol. Gastroenterol. *37*, 406–415.

He, T., Venema, K., Priebe, M.G., Welling, G.W., Brummer, R.J., and Vonk, R.J. (2008). The role of colonic metabolism in lactose intolerance. Eur. J. Clin. Invest. *38*, 541–547.

Hertzler, S.R., and Savaiano, D.A. (1996). Colonic adaptation to daily lactose feeding in lactose maldigesters reduces lactose intolerance. Am. J. Clin. Nutr. *64*, 232–236.

Jain, P.K., McNaught, C.E., Anderson, A.D., MacFie, J., and Mitchell, C.J. (2004). Influence of synbiotic containing *Lactobacillus acidophilus* La5, *Bifidobacterium lactis* Bb 12, Streptococcus thermophilus, *Lactobacillus bulgaricus* and oligofructose on gut barrier function and sepsis in critically ill patients: a randomized controlled trial. Clin. Nutr. *23*, 467–475.

Kanamori, Y., Hashizume, K., Sugiyama, M., Morotomi, M., and Yuki, N. (2001). Combination therapy with *Bifidobacterium breve*, *Lactobacillus casei*, and galactooligosaccharides dramatically improved the intestinal function in a girl with short bowel syndrome: a novel synbiotics therapy for intestinal failure. Dig. Dis. Sci. *46*, 2010–2016.

Kanamori, Y., Hashizume, K., Sugiyama, M., Mortomi, M., Yuki, N., and Tanaka, R. (2002). A novel synbiotic therapy dramatically improved the intestinal function of a paediatric patient with laryngotracheo-esophageal cleft (LTEC) in the intensive care unit. Clin. Nutr. *21*, 527–530.

Kanazawa, H., Nagino, M., Kamiya, S., Komatsu, S., Mayumi, T., Takagi, K., Asahara, T., Nomoto, K., Tanaka, R., and Nimura, Y. (2005). Synbiotics reduce postoperative infectious complications: a randomized controlled trial in biliary cancer patients undergoing hepatectomy. Langenbecks Aarch Surg *390*, 104–113.

Kelishadi, R., Farajian, S., Safavi, M., Mirlohi, M., and Hashemipour, M. (2014). A randomized triple-masked controlled trial on the effects of synbiotics on inflammation markers in overweight children. J. Pediatr. (Rio J) *90*, 161–168.

Klinder, A., Forster, A., Caderni, G., Femia, A.P., and Pool-Zobel, B.L. (2004). Faecal water genotoxicity is predictive of tumour-preventative activities by inulin-like oligofructoses, probiotics (*Lactobacillus rhamnosus* and *Bifidobacterium lactis*), and their synbiotic combination. Nutr. Cancer *49*, 144–155.

Kolida, S., and Gibson, G.R. (2011). Synbiotics in health and disease. Annu. Rev. Food Sci. Technol. *2*, 373–393.

Kotzampassi, K., Giamarellos-Bourboulis, E.J., Voudouris, A., Kazamias, P., and Eleftheriadis, E. (2006). Benefits of a synbiotic formula (Synbiotic 2000Forte) in critically Ill trauma patients: early results of a randomized controlled trial. World J. Surg *30*, 1848–1855.

Kuitunen, M., Kukkonen, K., and Savilahti, E. (2009). Pro- and prebiotic supplementation induces a transient reduction in haemoglobin concentration in infants. J. Pediatr. Gastroenterol. Nutr. *49*, 626–630.

Kukkonen, K., Savilahti, E., Haahtela, T., Juntunen-Backman, K., Korpela, R., Poussa, T., Tuure, T., and Kuitunen, M. (2007). Probiotics and prebiotic galacto-oligosaccharides in the prevention of allergic diseases: a randomized, double-blind, placebo-controlled trial. J. Allergy Clin. Immunol. *119*, 192–198.

Kukkonen, K., Savilahti, E., Haahtela, T., Juntunen-Backman, K., Korpela, R., Poussa, T., Tuure, T., and Kuitunen, M. (2008). Long-term safety and impact on infection rates of postnatal probiotic and prebiotic (synbiotic) treatment: randomized, double-blind, placebo-controlled trial. Pediatrics *122*, 8–12.

Lied, G.A., Lillestol, K., Lind, R., Valeur, J., Morken, M.H., Vaali, K., Gregersen, K., Florvaag, E., Tangen, T., and Berstad, A. (2011). Perceived food hypersensitivity: a review of 10 years of interdisciplinary research at a reference centre. Scand J. Gastroenterol. *46*, 1169–1178.

Liu, Q., Duan, Z.P., Ha, D.K., Bengmark, S., Kurtovic, J., and Riordan, S.M. (2004). Synbiotic modulation of gut flora: effect on minimal hepatic encephalopathy in patients with cirrhosis. Hepatology *39*, 1441–1449.

López, M.G., and Urías-Silvas, J.E. (2007). Agave fructans as prebiotics. In Recent Advances in Fructooligosaccharides Research, S. Norio, B. Noureddine, and O. Shuichi, eds. (Kerala, India, Research Signpost), pp. 297–310.

Macfarlane, S., Cleary, S., Bahrami, B., Reynolds, N., and Macfarlane, G.T. (2013). Synbiotic consumption changes the metabolism and composition of the gut microbiota in older people and modifies inflammatory processes: a randomized, double-blind, placebo-controlled crossover study. Aliment. Pharmacol. Ther. *38*, 804–816.

Malaguarnera, M., Greco, F., Barone, G., Gargante, M.P., and Toscano, M.A. (2007). *Bifidobacterium longum* with fructo-oligosaccharide (FOS) treatment in minimal hepatic encephalopathy: a randomized, double-blind, placebo-controlled study. Dig. Dis. Sci. *52*, 3259–3265.

Marteau, P., Minekus, M., Havenaar, R., and Huis in't Veld, J.H. (1997). Survival of lactic acid bacteria in a dynamic model of the stomach and small intestine: validation and the effects of bile. J. Dairy Sci. *80*, 1031–1037.

Martinez, R.C., Aynaou, A.E., Albrecht, S., Schols, H.A., De Martinis, E.C., Zoetendal, E.G., Venema, K., Saad, S.M., and Smidt, H. (2011). *In vitro* evaluation of gastrointestinal survival of *Lactobacillus* amylovorus DSM 16698 alone and combined with galactooligosaccharides, milk and/or *Bifidobacterium animalis* subsp. *lactis* Bb-12. Int. J. Food Microbiol. *149*, 152–158.

Martinez, R.C., Cardarelli, H.R., Borst, W., Albrecht, S., Schols, H., Gutierrez, O.P., Maathuis, A.J., de Melo Franco, B.D., De Martinis, E.C., Zoetendal, E.G., *et al.* (2013). Effect of galactooligosaccharides and *Bifidobacterium animalis* Bb-12 on growth of *Lactobacillus amylovorus* DSM 16698, microbial community structure, and metabolite production in an *in vitro* colonic model set up with human or pig microbiota. FEMS Microbiol. Ecol. *84*, 110–123.

Mills, S., Stanton, C., Fitzgerald, G.F., and Ross, R.P. (2011). Enhancing the stress responses of probiotics for a lifestyle from gut to product and back again. Microb. Cell Fact. *10 Suppl. 1*, S19.

Minekus, M., Marteau, P., Havenaar, R., and Huis in't Veld, J.H. (1995). A multi compartmental dynamic computer-controlled model simulating the stomach and small intestine. Alternatives to Laboratory Animals (ATLA) 23, 197–209.

Minekus, M., Smeets-Peeters, M., Bernalier, A., Marol-Bonnin, S., Havenaar, R., Marteau, P., Alric, M., Fonty, G., and Huis in't Veld, J.H. (1999). A computer-controlled system to simulate conditions of the large intestine with peristaltic mixing, water absorption and absorption of fermentation products. Appl. Microbiol. Biotechnol. *53*, 108–114.

Morelli, L., Zonenschain, D., Callegari, M.L., Grossi, E., Maisano, F., and Fusillo, M. (2003). Assessment of a new synbiotic preparation in healthy volunteers: survival, persistence of probiotic strains and its effect on the indigenous flora. Nutr. J. *2*, 11.

Ndagijimana, M., Laghi, L., Vitali, B., Placucci, G., Brigidi, P., and Guerzoni, M.E. (2009). Effect of a synbiotic food consumption on human gut metabolic profiles evaluated by (1)H Nuclear Magnetic Resonance spectroscopy. Int. J. Food Microbiol. *134*, 147–153.

Nikniaz, L., Mahdavi, R., Ostadrahimi, A., Hejazi, M.A., and Vatankhah, A.M. (2013a). Effects of synbiotic supplementation on total antioxidant capacity of human breastmilk. Breastfeed Med. *8*, 217–222.

Nikniaz, L., Ostadrahimi, A., Mahdavi, R., Hejazi, M.A., and Salekdeh, G.H. (2013b). Effects of synbiotic supplementation on breast milk levels of IgA, TGF-beta1, and TGF-beta2. J. Hum Lact *29*, 591–596.

Ooi, L.G., Bhat, R., Rosma, A., Yuen, K.H., and Liong, M.T. (2010). A synbiotic containing *Lactobacillus* gasseri [corrected] CHO-220 and inulin improves irregularity of red blood cells. J. Dairy Sci. *93*, 4535–4544.

Osman, N., Adawi, D., Molin, G., Ahrne, S., Berggren, A., and Jeppsson, B. (2006). *Bifidobacterium infantis* strains with and without a combination of oligofructose and inulin (OFI) attenuate inflammation in DSS-induced colitis in rats. BMC Gastroenterol. *6*, 31.

Ostadrahimi, A., Nikniaz, L., Mahdavi, R., Hejazi, M.A., and Nikniaz, Z. (2013). Effects of synbiotic supplementation on lactating mothers' energy intake and BMI, and infants' growth. Int. J. Food Sci. Nutr. *64*, 711–714.

Ouwehand, A.C., Tiihonen, K., Saarinen, M., Putaala, H., and Rautonen, N. (2009). Influence of a combination of *Lactobacillus acidophilus* NCFM and lactitol on healthy elderly: intestinal and immune parameters. Br. J. Nutr. *101*, 367–375.

Panigrahi, P., Parida, S., Pradhan, L., Mohapatra, S.S., Misra, P.R., Johnson, J.A., Chaudhry, R., Taylor, S., Hansen, N.I., and Gewolb, I.H. (2008). Long-term colonization of a *Lactobacillus plantarum* synbiotic preparation in the neonatal gut. J. Pediatr. Gastroenterol. Nutr. *47*, 45–53.

Pasman, W., Wils, D., Saniez, M.H., and Kardinaal, A. (2006). Long-term gastrointestinal tolerance of NUTRIOSE FB in healthy men. Eur. J. Clin. Nutr. *60*, 1024–1034.

Passariello, A., Terrin, G., Cecere, G., Micillo, M., De Marco, G., Di Costanzo, M., Cosenza, L., Leone, L., Nocerino, R., and Canani, R.B. (2012). Randomized clinical trial: efficacy of a new synbiotic formulation containing *Lactobacillus paracasei* B21060 plus arabinogalactan and xilooligosaccharides in children with acute diarrhoea. Aliment. Pharmacol. Ther. *35*, 782–788.

Quigley, E.M. (2013). Commentary: synbiotics and gut microbiota in older people – a microbial guide to healthy ageing. Aliment. Pharmacol. Ther. *38*, 1141–1142.

Rafter, J., Bennett, M., Caderni, G., Clune, Y., Hughes, R., Karlsson, P.C., Klinder, A., O'Riordan, M., O'Sullivan, G.C., Pool-Zobel, B., *et al.* (2007). Dietary synbiotics reduce cancer risk factors in polypectomized and colon cancer patients. Am. J. Clin. Nutr. *85*, 488–496.

Rayes, N., Seehofer, D., and Neuhaus, P. (2009). Prebiotics, probiotics, synbiotics in surgery – are they only trendy, truly effective or even dangerous? Langenbecks Aarch Surg *394*, 547–555.

Rayes, N., Seehofer, D., Theruvath, T., Mogl, M., Langrehr, J.M., Nussler, N.C., Bengmark, S., and Neuhaus, P. (2007). Effect of enteral nutrition and synbiotics on bacterial infection rates after pylorus-preserving pancreatoduodenectomy: a randomized, double-blind trial. Ann. Surg. *246*, 36–41.

Rayes, N., Seehofer, D., Theruvath, T., Schiller, R.A., Langrehr, J.M., Jonas, S., Bengmark, S., and Neuhaus, P. (2005). Supply of pre- and probiotics reduces bacterial infection rates after liver transplantation – a randomized, double-blind trial. Am. J. Transplant *5*, 125–130.

Reddy, B.S., Macfie, J., Gatt, M., Larsen, C.N., Jensen, S.S., and Leser, T.D. (2007). Randomized clinical trial of effect of synbiotics, neomycin and mechanical bowel preparation on intestinal barrier function in patients undergoing colectomy. Br. J. Surg *94*, 546–554.

Reid, G., Sanders, M.E., Gaskins, H.R., Gibson, G.R., Mercenier, A., Rastall, R., Roberfroid, M., Rowland, I., Cherbut, C., and Klaenhammer, T.R. (2003). New scientific paradigms for probiotics and prebiotics. J. Clin. Gastroenterol. *37*, 105–118.

Riaz, Q.U., and Masud, T. (2013). Recent trends and applications of encapsulating materials for probiotic stability. Crit. Rev. Food Sci. Nutr. *53*, 231–244.

Rowland, I.R., Rumney, C.J., Coutts, J.T., and Lievense, L.C. (1998). Effect of *Bifidobacterium longum* and inulin on gut bacterial metabolism and carcinogen-induced aberrant crypt foci in rats. Carcinogenesis *19*, 281–285.

Roze, J.C., Barbarot, S., Butel, M.J., Kapel, N., Waligora-Dupriet, A.J., De Montgolfier, I., Leblanc, M., Godon, N., Soulaines, P., Darmaun, D., *et al.* (2012). An alpha-lactalbumin-enriched and symbiotic-supplemented v. a standard infant formula: a multicentre, double-blind, randomized trial. Br. J. Nutr. *107*, 1616–1622.

Safavi, M., Farajian, S., Kelishadi, R., Mirlohi, M., and Hashemipour, M. (2013). The effects of synbiotic supplementation on some cardio-metabolic risk factors in overweight and obese children: a randomized triple-masked controlled trial. Int. J. Food Sci. Nutr. *64*, 687–693.

Sarbini, S.R., Kolida, S., Naeye, T., Einerhand, A., Brison, Y., Remaud-Simeon, M., Monsan, P., Gibson, G.R., and Rastall, R.A. (2011). *In vitro* fermentation of linear and alpha-1,2-branched dextrans by the human faecal microbiota. Appl. Environ. Microbiol. *77*, 5307–5315.

Sazawal, S., Dhingra, U., Hiremath, G., Sarkar, A., Dhingra, P., Dutta, A., Verma, P., Menon, V.P., and Black, R.E. (2010). Prebiotic and probiotic fortified milk in prevention of morbidities among children: community-based, randomized, double-blind, controlled trial. PLoS One *5*, e12164.

Schouten, B., van Esch, B.C., Hofman, G.A., van Doorn, S.A., Knol, J., Nauta, A.J., Garssen, J., Willemsen, L.E., and Knippels, L.M. (2009). Cow milk allergy symptoms are reduced in mice fed dietary synbiotics during oral sensitization with whey. J. Nutr. *139*, 1398–1403.

Schrezenmeir, J., Heller, K., McCue, M., Llamas, C., Lam, W., Burow, H., Kindling-Rohracker, M., Fischer, W., Sengespeik, H.C., Comer, G.M., *et al.* (2004). Benefits of oral supplementation with and without synbiotics in young children with acute bacterial infections. Clin. Pediatr. (Phila.) *43*, 239–249.

Shanahan, F., Dinan, T.G., Ross, P., and Hill, C. (2012). Probiotics in transition. Clin. Gastroenterol. Hepatol. *10*, 1220–1224.

Slavin, J. (2013). Fibre and prebiotics: mechanisms and health benefits. Nutrients *5*, 1417–1435.

Spindler-Vesel, A., Bengmark, S., Vovk, I., Cerovic, O., and Kompan, L. (2007). Synbiotics, prebiotics, glutamine, or peptide in early enteral nutrition: a randomized study in trauma patients. JPEN J. Parenter Enteral Nutr. *31*, 119–126.

Steed, H., Macfarlane, G.T., Blackett, K.L., Bahrami, B., Reynolds, N., Walsh, S.V., Cummings, J.H., and Macfarlane, S. (2010). Clinical trial: the microbiological and immunological effects of synbiotic consumption – a randomized double-blind placebo-controlled study in active Crohn's disease. Aliment. Pharmacol. Ther. *32*, 872–883.

Sugawara, G., Nagino, M., Nishio, H., Ebata, T., Takagi, K., Asahara, T., Nomoto, K., and Nimura, Y. (2006). Perioperative synbiotic treatment to prevent postoperative infectious complications in biliary cancer surgery: a randomized controlled trial. Ann. Surg. *244*, 706–714.

Takahashi, K., Terashima, H., Kohno, K., and Ohkohchi, N. (2013). A stand-alone synbiotic treatment for the prevention of d-lactic acidosis in short bowel syndrome. Int. Surg *98*, 110–113.

Tanaka, R., Takayama, H., Morotomi, M., Kuroshima, T., Ueyama, S., Matsumoto, K., Kuroda, A., and Mutai, M. (1983). Effects of administration of TOS and *Bifidobacterium breve* 4006 on the human faecal flora. Bifidobact Microflora 2, 17–24.

Tomoda, T., Nakano, Y., and T., K. (1991). Effect of yoghurt and yoghurt supplemented with *Bifidobacterium* and/or lactulose in healthy persons: a comparative study. Bifidobact Microflora *10*, 123–130.

Tsilingiri, K., and Rescigno, M. (2013). Postbiotics: what else? Benef. Microbes *4*, 101–107.

Underwood, M.A., Salzman, N.H., Bennett, S.H., Barman, M., Mills, D.A., Marcobal, A., Tancredi, D.J., Bevins, C.L., and Sherman, M.P. (2009). A randomized placebo-controlled comparison of 2 prebiotic/probiotic combinations in preterm infants: impact on weight gain, intestinal microbiota, and faecal short-chain fatty acids. J. Pediatr. Gastroenterol. Nutr. *48*, 216–225.

Van Craeyveld, V., Swennen, K., Dornez, E., Van de Wiele, T., Marzorati, M., Verstraete, W., Delaedt, Y., Onagbesan, O., Decuypere, E., Buyse, J., *et al.* (2008). Structurally different wheat-derived arabinoxylooligosaccharides have different prebiotic and fermentation properties in rats. J. Nutr. *138*, 2348–2355.

van de Pol, M.A., Lutter, R., Smids, B.S., Weersink, E.J., and van der Zee, J.S. (2011). Synbiotics reduce allergen-induced T-helper 2 response and improve peak expiratory flow in allergic asthmatics. Allergy *66*, 39–47.

van den Nieuwboer, M., Claassen, E., Morelli, L., Guarner, F., and Brummer, R.J. (2014). Probiotic and synbiotic safety in infants under two years of age. Benef. Microbes *5*, 45–60.

van der Aa, L.B., Heymans, H.S., van Aalderen, W.M., Sillevis Smitt, J.H., Knol, J., Ben Amor, K., Goossens, D.A., and Sprikkelman, A.B. (2010). Effect of a new synbiotic mixture on atopic dermatitis in infants: a randomized-controlled trial. Clin. Exp. Allergy *40*, 795–804.

van der Aa, L.B., Lutter, R., Heymans, H.S., Smids, B.S., Dekker, T., van Aalderen, W.M., Sillevis Smitt, J.H., Knippels, L.M., Garssen, J., Nauta, A.J., *et al.* (2012). No detectable beneficial systemic immunomodulatory effects of a specific synbiotic mixture in infants with atopic dermatitis. Clin. Exp. Allergy *42*, 531–539.

van der Aa, L.B., van Aalderen, W.M., Heymans, H.S., Henk Sillevis Smitt, J., Nauta, A.J., Knippels, L.M., Ben Amor, K., and Sprikkelman, A.B.

(2011). Synbiotics prevent asthma-like symptoms in infants with atopic dermatitis. Allergy *66*, 170–177.

van Nuenen, M.H.M.C., Meyer, P.D., and Venema, K. (2003). The effect of various inulins and *Clostridium difficile* on the metabolic activity of the human colonic microbiota *in vitro*. Microb. Ecol. Health Dis. *15*, 137–144.

Vandenplas, Y., and De Hert, S. (2012). Cost/benefit of synbiotics in acute infectious gastroenteritis: spend to save. Benef. Microbes 3, 189–194.

Vanhoutte, T., De Preter, V., De Brandt, E., Verbeke, K., Swings, J., and Huys, G. (2006). Molecular monitoring of the faecal microbiota of healthy human subjects during administration of lactulose and *Saccharomyces boulardii*. Appl. Environ. Microbiol. *72*, 5990–5997.

Venema, K. (2012). Intestinal fermentation of lactose and prebiotic lactose derivatives, including human milk oligosaccharides. Int. Dairy J. *22*, 123–140.

Vitali, B., Ndagijimana, M., Cruciani, F., Carnevali, P., Candela, M., Guerzoni, M.E., and Brigidi, P. (2010). Impact of a synbiotic food on the gut microbial ecology and metabolic profiles. BMC Microbiol. *10*, 4.

Vlieger, A.M., Robroch, A., van Buuren, S., Kiers, J., Rijkers, G., Benninga, M.A., and te Biesebeke, R. (2009). Tolerance and safety of *Lactobacillus paracasei* subsp. *paracasei* in combination with *Bifidobacterium animalis* subsp. *lactis* in a prebiotic-containing infant formula: a randomized controlled trial. Br. J. Nutr. *102*, 869–875.

Worthley, D.L., Le Leu, R.K., Whitehall, V.L., Conlon, M., Christophersen, C., Belobrajdic, D., Mallitt, K.A., Hu, Y., Irahara, N., Ogino, S., *et al.* (2009). A human, double-blind, placebo-controlled, crossover trial of prebiotic, probiotic, and synbiotic supplementation: effects on luminal, inflammatory, epigenetic, and epithelial biomarkers of colorectal cancer. Am. J. Clin. Nutr. *90*, 578–586.

Xuan, N.N., Wang, D., Grathwohl, D., Lan, P.N., Kim, H.V., Goyer, A., and Benyacoub, J. (2013). Effect of a Growing-up Milk Containing Synbiotics on Immune Function and Growth in Children: A Cluster Randomized, Multicenter, Double-blind, Placebo Controlled Study. Clin. Med. Insights Pediatr. 7, 49–56.

Yokoyama, Y., Nishigaki, E., Abe, T., Fukaya, M., Asahara, T., Nomoto, K., and Nagino, M. (2014). Randomized clinical trial of the effect of perioperative synbiotics versus no synbiotics on bacterial translocation after oesophagectomy. Br. J. Surg *101*, 189–199.

Part IV

Clinical and Medical Aspects of Pro- and Prebiotics

Pro- and Prebiotics: the Role of Gut Microbiota in Obesity

24

Marc R. Bomhof and Raylene A. Reimer

Abstract

Obesity is a multifactorial disease that is widespread and continuing to increase in prevalence worldwide. The composition of the gut microbiota is altered in obesity and is associated with impaired gut barrier, enhanced proinflammatory response and metabolic disturbances. Manipulation of the gut microbiota as a therapeutic intervention in obesity and other chronic diseases is of major interest. Diet plays a major role in shaping the composition of the gut microbiota and prebiotics and probiotics have received much attention in this regard. This chapter examines the evidence for the effect of prebiotics and probiotics on obesity with special attention given to body weight and adiposity, appetite regulation, inflammation and gut barrier integrity and glucose and lipid metabolism.

Introduction

Overweight (BMI ≥ 25) and obesity (BMI ≥ 30) are defined as 'abnormal or excessive fat accumulation that may impair health' (WHO). Whereas obesity was once considered a problem that was limited to developed nations, obesity has seen a dramatic increase worldwide. Obesity rates have nearly doubled over the past 35 years and today more than 10% of the adult population is classified as obese (WHO) and nearly 40 million preschool children are considered overweight (de Onis *et al.*, 2010). Excess body fat predisposes individuals to a host of chronic health conditions including type 2 diabetes, hypertension, cardiovascular disease, non-alcoholic fatty liver disease, and some types of cancer (Mokdad *et al.*, 2003; Tilg *et al.*, 2009), making obesity one of the leading causes of death worldwide (WHO).

The increasing incidence of obesity around the world reflects the vulnerability of the human biological system to alterations in the physical and nutritional environment. While the human body is physiologically adept at dealing with periodic food scarcity, an environment in which energy is ubiquitous threatens overall metabolic homeostasis. Body weight is ultimately defined by energy balance, where calibration between 'energy in' and 'energy out' dictates whether weight is lost, gained, or stable. Given the limited success of countless regimes that promote weight loss through caloric restriction and/or increased physical activity, it is clear that simply trying to manipulate energy intake and expenditure through diet and exercise has limited efficacy for long term weight loss. Over the past few decades, extensive research has demonstrated that the human body is very efficient at defending against sustained weight loss through the regulation of physiological processes that control hunger, energy expenditure, and metabolic efficiency (Rosenbaum and Leibel, 2010). Given the powerful influence of these control mechanisms, it is now acknowledged that obesity is not just a lifestyle choice but rather a complex disease that is driven by biological, environmental, and psychological factors (Sharma and Padwal, 2010). Targeting the underlying physiological factors that affect energy balance, rather than the behaviours and activities that result from these factors is perhaps the best approach to developing effective treatments and prevention strategies for obesity.

In recent years it has been demonstrated that the gut microbiota plays a prominent role in the regulation of host energy balance and metabolism (Ley *et al.*, 2006). Gut microbiota has both the ability to affect 'energy in' as well as the metabolic efficiency by which energy substrates are processed. Considered an environmental factor for the body, the gut microbiota offers a component of the human biological system that can be tailored to fit the demands and needs of the human body. Given these qualities, gut microbiota has become a very promising target in the restoration of energy homeostasis and metabolic health.

The gut microbiota is the product of thousands of millennia of co-evolution between bacteria and host that has led to a complex ectosymbiotic relationship yielding benefits for both bacteria and host (Neish, 2009). Microbes benefit from the delivery of an abundant supply of energy in the digestive tract, while the host benefits from the competitive exclusion of opportunistic and pathogenic organisms and the adoption of various metabolic and digestive functions mediated by bacteria (Fukuda *et al.*, 2011; Neish, 2009). It is estimated that there are approximately one hundred trillion microbes that inhabit the gut, outnumbering the ten trillion eukaryotic cells that comprise the human body by a factor of ten (Petschow *et al.*, 2013; Qin *et al.*, 2010). The collection of genes that are coded in the makeup of all the microbiota, referred to as the microbiome, exceeds that of the human genome by ~150 times (Qin *et al.*, 2010). While enhancing our ability to culture more of the gut microbiota is currently a new frontier in microbiota research (Cani, 2013), the advent of sequencing technologies, utilizing the highly conserved, intra-species 16S ribosomal RNA (rRNA) gene region as a taxonomic marker, has allowed for a more extensive characterization of the various phylotypes (including the non-cultured variation) that dwell within the gut. It has only been in recent years that much of the diversity of the microbiota has been identified, due to the fact that approximately 70% of the organisms that lived in the gut could not be cultured via traditional culture techniques. Altogether, an estimated 1150 species or phylotypes have been identified (Qin *et al.*, 2010). Firmicutes and Bacteroidetes are recognized as the most prominent phyla representing ~90% of the bacteria, whereas Proteobacteria (e.g. genus *Escherichia*), Actinobacteria (e.g. genus *Bifidobacterium*), Fusobacteria, and Verrucomicrobia are found in lower abundance (Eckburg *et al.*, 2005). Eckburg and colleagues, by harvesting faecal samples from various locations within the colon, established that there is spatial variability in bacterial phylotypes (Eckburg *et al.*, 2005). Proximal versus distal and mucosal-associated versus luminal domains contain unique communities of bacteria, varying according to phylotype make-up and complexity (Booijink *et al.*, 2010; Sommer and Backhed, 2013). The duodenum/stomach, jejunum/ileum, and colon contain 10^1–10^3, 10^4–10^7, and 10^{11}–10^{12} colony forming units (cfu)/ml, respectively (Sekirov *et al.*, 2010). Approximately 70% of the gut microbiota associated with the body resides in the colon (Sekirov *et al.*, 2010). Altogether, the total mass of the gut microbiota is approximately 1.5 kg (Neish, 2009). It has been demonstrated that despite interpersonal phylotype diversity, a core set of genes exists in all microbiomes. Referred to as functional redundancy, a core genome ensures that key processes are not affected by changes in diversity (Sommer and Backhed, 2013; Turnbaugh *et al.*, 2009).

Given the vast metabolic capacity of the gut microbiota and the close proximity that the gut microbiota occupies to the intestinal epithelium, gut microbiota has been described as a 'metabolic organ' within the host providing critical physiological functions (O'Hara and Shanahan, 2006). The vast genetic potential of the microbiota has led some to draw parallels between the gut microbiota and the liver in regards to metabolic capacity (Sommer and Backhed, 2013). Known functions of the gut microbiota include immune system development, epithelial barrier function, protection against pathogens through production of antimicrobial factors such as bacteriocins and organic acids, synthesis of vitamin K and B, dehydroxylation and deconjugation of bile acids, enterocyte differentiation, xenobiotic metabolism, regulation of intestinal mobility, and cation absorption (Hsiao *et al.*, 2008; Neish, 2009; O'Hara and Shanahan, 2006; Ridlon *et al.*, 2006). Another prominent function of the gut microbiota, and perhaps the most relevant to a discussion on energy balance, is fermentation of non-digestible substrate which leads to a range of metabolic by-products that provide energy to the host. Plant polysaccharides are a diverse group of carbohydrates comprised of unique sugars, such as xylan, pectin, and arabinose, each with unique glycosidic bonds that cannot be digested by human enzymes. Anaerobic microbes, however, contain various glycoside hydrolases, which permit fermentation of these complex carbohydrates. This process produces a range of metabolic by-products, including the short-chain fatty acids acetate, propionate, and butyrate, which are absorbed via passive diffusion and by mono-carboxylic acid transporters along the epithelium (Roberfroid *et al.*, 2010). Butyrate is primarily used as fuel for colonocytes, whereas propionate and acetate are delivered via the portal vein to the liver. Propionate is a gluconeogenic substrate utilized by hepatocytes and intestinal cells while acetate is metabolized by other tissues (Roberfroid *et al.*, 2010). Proteins that enter the colon are fermented by proteolytic bacteria and yield by-products such as isobutyrate and isovalerate. The by-products of protein fermentation are reported to be more toxic to the host and may be involved in pathophysiological conditions (Roberfroid *et al.*, 2010). Cud-chewing animals are able to harness ~70% of their energy requirements from plant food because they have an abundance of cellulolytic bacteria (Bergman, 1990; Flint, 1997). Although humans do not rely on bacterial derived energy to the same extent, it is estimated that humans acquire ~10% of total daily energy from gut fermentation (Bergman, 1990). Additionally, through production of metabolites, gut microbiota is reputed to affect the host's ability to store and metabolize nutrients by influencing host-gene expression (Tremaroli *et al.*, 2010). Genes that are involved in satiety and lipogenesis are notable examples of host-processes that can be altered through microbiota activity (Delzenne and Cani, 2011). With a vast array of microbial metabolic potential at the disposal of the host, researchers are only beginning to understand the complex interactions between microbial metabolism and human physiology.

Early neonatal influences play a prominent role in sculpting gut microbiota. Although the digestive tract was historically thought to be sterile at birth, evidence that bacteria are present in meconium of neonates demonstrates that colonization may be initiated prior to birth (Jimenez *et al.*, 2008). Prominent influences on early microbiota development include mode of delivery (vaginal versus caesarean-section), feeding type (breastfed versus formula fed), gestational age, and antibiotic use (Azad *et al.*, 2013; Penders *et al.*, 2006). It is theorized that a 'healthy' microbiota develops from early exposure to the bacteria that are present in the birth canal, and subsequent 'feeding' of these bacteria by substrates that are present in breast milk. Deviations from an optimal bacteria profile may increase susceptibility to disease states in later life (Penders *et al.*, 2006). During the weaning period, gut microbiota undergoes significant change and transitions into an adult-like gut microbiota (Reinhardt *et al.*, 2009). Across the lifespan, gut microbiota are highly influenced by diet and the host innate and adaptive immune system (McFall-Ngai, 2007; Neish, 2009). Altogether, the combination of these variables contributes to the development of a very complex and individualized network of gut microbiota. Over time each human being develops a distinctively unique microbial fingerprint within the intestine that appears very stable over time (Costello *et al.*, 2009; Turnbaugh *et al.*, 2009).

Gut microbiota and energy balance

Given that the gut microbiota can unlock energy supplies that are inaccessible to the host based on their mammalian enzymatic capacity, Gordon and colleagues employed a germ free (GF) mouse model to study the role of gut microbiota on host metabolism. Although GF does not represent a physiological state, the model offers important insight into the mechanisms by which gut microbiota influences whole body energy homeostasis. Formative work by Backhed *et al.* (Backhed *et al.*, 2004) revealed that gut microbiota, regardless of food intake, can affect body composition and body weight. Their first study in 2004 demonstrated that GF mice, relative to conventionally raised (CONV-R) mice, had 42% less body fat despite consuming approximately 30% more rodent chow than their CONV-R counterparts (Backhed *et al.*, 2004). Interestingly, based on decreased oxygen consumption measurements, GF mice did not display an elevated metabolic rate (Backhed *et al.*, 2004). Furthermore, when GF animals were colonized with the caecal matter from CONV-R mice, referred to as CONV-D mice, body fat increased by 57%. Alongside increased adiposity, it was also demonstrated that CONV-D mice had elevated plasma glucose levels during insulin and glucose tolerance tests, suggesting that gut microbiota decreased insulin sensitivity (Backhed *et al.*, 2004). This finding coincided with elevated intestinal saccharide uptake, measured via 2-deoxyglucose gavage, in CONV-D animals. Additionally, CONV-D mice had increased liver triglyceride (TAG) levels (Backhed *et al.*, 2004). In a second study completed in 2007, GF and CONV-D mice were fed a Western diet (WD). In this study, it was revealed that gut microbiota increases body weight, in addition to affecting adiposity (Backhed *et al.*, 2007). Calculations showed that the mice had similar faecal energy excretion, suggesting that underlying mechanisms beyond energy harvest were mediating the reduced body weight (Backhed *et al.*, 2007). Overall, these studies reveal that an absence of gut microbiota protects against increased adiposity, liver steatosis, and insulin resistance. In a somewhat contradictory study, Fleissner *et al.* (2010) demonstrated that GF mice are not necessarily protected against weight gain (Fleissner *et al.*, 2010). The authors used three different diets in their study: one low fat diet and two high-fat (HF) diets – a standard HF diet and WD, which was enriched in hydrogenated oils and sucrose and similar to the WD diet used in the Gordon *et al.*, studies. GF mice fed the HF diet had increased weight relative to conventional mice. Interestingly, the GF mice that received the WD had decreased adiposity relative to the GF mice fed standard HF diet (Fleissner *et al.*, 2010).

Based on evidence from the GF model, it appears that microbiota influence adiposity by means and mechanisms that go beyond simple energy extraction. One mechanism explaining the decreased adiposity in GF animals was an up-regulation of fasting-induced adipocyte factor (FIAF), also known as angiopoietin-like protein 4 (Angptl4), in the intestinal epithelium. FIAF is an inhibitor of the enzyme lipoprotein lipase (LPL) which is responsible for cleaving fatty acids from lipoproteins in circulation for uptake by other tissues (Backhed *et al.*, 2004). No changes in FIAF expression were observed in white adipose tissue and liver thereby suggesting that intestinal microbes were responsible for the observed change in intestinal FIAF. To confirm the proposed link between gut microbiota and FIAF-mediated changes in adiposity, GF FIAF$^{-/-}$ knockout (KO) mice were studied. Without the FIAF gene, the reductions in adiposity seen with GF were absent. Furthermore, the GF FIAF$^{-/-}$ KO mice had a 67% increase in LPL activity compared to GF wild-type mice. As further confirmation, plasma TAG concentrations remained elevated longer in response to an olive oil gavage in GF animals, suggesting greater LPL suppression relative to CONV-D animals (Backhed *et al.*, 2007). In the absence of markers of adipogenesis and lipogenesis (Backhed *et al.*, 2004), the increased adiposity is consistent with hypertrophied adipose tissue secondary to elevated LPL activity in CONV-D mice. LPL activity was found to be elevated in both epididymal and heart tissue in CONV-D animals (Backhed *et al.*, 2004), demonstrating the FIAF regulation at the level of the gut affects multiple tissues.

Apart from FIAF, gut microbiota alters the expression and modification of numerous adiposity-associated gene and proteins. CONV-D mice had increased expression of the long-term satiety hormone leptin, which is one possible explanation for why food intake was reduced in the CONV-D group (Backhed *et al.*, 2004). In the liver, associated with the 2.3 fold increase in liver TAGs, there was an increase in the expression of carbohydrate response element binding protein (CHREBP) and sterol response element binding protein-1 (SREBP-1), both transcriptional factors that regulate fat synthesis. As verification, two genes that are regulated by these transcriptional factors, acetyl-CoA carboxylase (Acc1) and fatty acid synthase (FAS) were also elevated in GF mice. Additionally, it was found that compared to CONV-D mice, GF mice had increased levels of phosphorylated AMP-activated protein kinase (AMPK), an enzyme commonly known to be the fuel gauge of the cell, in skeletal muscle and liver tissue (Backhed *et al.*, 2007). When activated, AMPK limits lipogenesis and enhances fatty acid oxidation. The gene expression pattern observed in the gastrocnemius muscle of GF animals was consistent with increased fat oxidation. Furthermore, in the liver, increased P-AMPK was associated with decreased glycogen synthase activity alongside depressed glycogen storage (Backhed *et al.*, 2007). It is clear from these data that gut microbiota plays a role in regulating host metabolism by affecting various physiological processes involved in energy partitioning and adiposity.

It is well established that gut microbiota play a critical role in maintaining host health. In recent years, it has been observed that gut microbiota differs between the obese and lean state (Ley *et al.*, 2006). It is theorized that in the context of obesity, the symbiotic relationship between gut microbiota and host shifts to a dysbiotic state, such that the microbiota become pathogenic or cause harm to the host (Cani and Delzenne, 2009). The first studies that looked at the profile of gut microbiota in the lean versus obese state observed a shift in the relative abundance of the taxa present, with a profile characterized by increased Firmicutes and decreased Bacteroidetes in obesity (Ley *et al.*, 2005, 2006). Ley *et al.* (2005) found that genetically obese *ob/ob* mice had a 50% reduction in Bacteroidetes and a proportional increase in Firmicutes. A subsequent study revealed that human obesity was also associated with increased Firmicutes and decreased Bacteroidetes

(Ley *et al.*, 2006). Interestingly, weight loss in obese individuals was associated with an increase in the overall proportion of Bacteroidetes (Ley *et al.*, 2006). A study by Nadal *et al.* (2009) using fluorescence *in situ* hybridization (FISH), confirmed that *Bacteroides-Prevotella* was increased in obese adolescents after 4 kg weight loss.

While other studies have confirmed the decreased Bacteroidetes to Firmicutes ratio in obesity (Everard *et al.*, 2011; Fleissner *et al.*, 2010; Turnbaugh *et al.*, 2009), some studies do not show this characteristic shift. Vrieze *et al.* (2012) found that obese individuals had increased Bacteroidetes and lower *Clostridium* cluster XIVa relative to lean subjects. Schwiertz *et al.* (2010) found that overweight and obesity favoured an increase in Bacteroidetes relative to Firmicutes. A study by Duncan *et al.* (2008) using FISH, did not find a relationship between the major bacterial phyla and BMI or weight loss. The authors did not rule out that more detailed analysis at the community level could yield insight into species differences in lean and obese individuals. Some comorbid conditions of obesity, such as type 2 diabetes also appear to be associated with phylum level compositional changes in gut microbiota (Larsen *et al.*, 2010).

Another prominent genus of bacteria that is affected in obesity is *Bifidobacterium*, which belongs to the phyla Actinobacteria (Cani *et al.*, 2007a). Cani *et al.* (2007a), feeding a HF diet to mice to induce obesity, observed that HF feeding led to a decrease in bifidobacteria. Further demonstrating the association between obesity and decreased bifidobacteria, a prospective human study found that young children with low counts of bifidobacteria and high counts of *Staphylococcus aureus*, as measured by FISH, were at increased risk for excess weight gain by age 7 (Kalliomaki *et al.*, 2008).

Apart from altered abundance of particular microbial taxa, it has also been observed that aberrations from the core-microbiome are associated with obesity. In a study of monozygotic and dizygotic twins concordant for leanness or obesity, Turnbaugh *et al.* (2009) observed that the microbiomes of obese versus lean individuals, although containing a core set of microbial functions, showed alterations in bacterial genes and metabolic pathways. For instance, an obese microbiome was enriched in genes involved in carbohydrate, lipid, and protein metabolism. A notable example was an increase in enzymes involved in the phosphotransferase system, which mediates carbohydrate metabolism. Interestingly, it was observed that many of the enriched genes in obesity belonged to the Actinobacteria and Firmicutes phyla. It is hypothesized that deviations from the core-microbiome may predispose to altered metabolic and physiological states, such as obesity (Turnbaugh *et al.*, 2009).

Alterations in the functional capacity of the microbiome are likely an important part of increased energy extraction in obesity. Research by Turnbaugh and colleagues established in *ob/ob* mice that an obesity-associated gut microbiota was more efficient at extracting energy from food (Turnbaugh *et al.*, 2006). The obese microbiome was found to be enriched for genes that facilitate breakdown of dietary saccharides, including glucosidases and galactosidases. Furthermore, SCFA concentrations were increased in the caecum. Confirming that these SCFA were absorbed, bomb calorimetry experiments demonstrated that *ob/ob* mice had decreased energy in their faecal matter. Additionally, a recent human study found that bacterial groups commonly found in obesity are associated with increased energy extraction from stool (Jumpertz *et al.*, 2011). Jumpertz *et al.*, in a highly controlled inpatient, randomized, cross-over study, found that a 3400 kcal diet, balanced for macronutrients (24% protein, 16% fat, 60% carbohydrate), promoted an increase in Firmicutes and decrease in Bacteroidetes, relative to a 2400 kcal diet. This bacterial shift was accompanied by a 150 kcal increase in energy extraction (Jumpertz *et al.*, 2011). In another study, Schwiertz *et al.* (2010), studying lean and obese subjects, elucidated that overweight and obese individuals had increased concentrations of faecal SCFA relative to lean. It was observed that the proportion of the SCFA pool that was represented by propionate was increased in obese individuals. Energy extraction appears to be a prominent factor in the link between a dysbiotic gut microbiota and obesity.

One of the current topics of debate is the precise aetiology of the characteristic microbial shift in obesity and whether or not a dysbiotic microbiota is a cause or consequence of metabolic disease. Diet is recognized as one of the primary factors that affects the gut microbiota. Whereas diets were historically filled with whole grains and plant based carbohydrates in an agrarian-type society, today many foods are higher in fat, sugar, and energy. To study whether a switch to a HF diet is responsible for an altered microbiota, Hildebrandt *et al.*, used RELMβ$^{-/-}$ KO mice, a mouse model known for its resistance to developing obesity (Hildebrandt *et al.*, 2009). When RELMβ$^{-/-}$ KO mice were switched from a lean to a HF diet, they did not develop obesity, unlike their wild-type counterparts (Hildebrandt *et al.*, 2009). Despite the weight differences between the wild-type and KO mice, both groups of mice developed a dysbiotic microbiota profile, with decreased Bacteroidetes and increased Firmicutes and Proteobacteria on the HF diet (Hildebrandt *et al.*, 2009). Similarly, a study by Ravussin *et al.* (2012) set out to determine if diet or body weight was primarily responsible for gut microbiota shifts. To study this, four different groups of mice were compared: (1) diet-induced obese (DIO), 20% weight reduced on a HF diet (60% of calories from fat); (2) DIO mice *ad libitum* on HF diet; (3) Lean control fed *ad libitum* on a lean diet (10% calories from fat); (4) lean control, 20% weight reduced on a lean diet. After the weight reduction in the DIO mice, body composition and body weight was equal to the mice fed the lean control diet *ad libitum*. Based on 16S rRNA gene sequencing, the authors determined that diet, rather than weight reduction, had the greatest impact on bacterial composition (Ravussin *et al.*, 2012). Altogether, these studies provide strong evidence that diet, rather than body weight, plays a significant role in regulating the microbiota environment.

Alternatively, there is evidence that a dysbiotic profile of microbiota is not diet related but the result of genotype or metabolic disease itself. As evidence of this, *ob/ob* mice, a mouse model that develops obesity due to a lack of leptin production, developed dysbiosis despite eating the same polysaccharide-rich diet as wild-type animals (Ley *et al.*, 2005). Furthermore, in a study by Geurts *et al.* (2011), *db/db* mice, which are deficient for the leptin receptor, were found to develop an altered microbial profile, irrespective of diet. The *db/db* genotype was found to have a phyla level decrease in Bacteroidetes and an increase in Firmicutes, Proteobacteria, and Fibrobacteres. Overall, this evidence suggests that there is a specific host-level manipulation of the gut microbiota, such that the microbiota is specifically tailored to meet the demands and needs of the host, as dictated by host genotype. The precise mechanisms by which the host genotype manipulates the gut microbiota are not completed understood. One hypothesis is that the microbiota is regulated by the immune system. However, given that the changes in bacteria are phyla wide, and not just species specific, the pattern of microbiota shift does not correlate with typical immune-mediated bacterial change (Turnbaugh *et al.*, 2006).

Despite the proposed role of host genotype affecting the gut microbiota, there is also evidence that the gut microbiota is the cause of host phenotype. When adult GF C57BL/6J mice were inoculated with bacteria from either an obese or lean host, the phenotype that developed in recipient mice was found to match that of the donor mice (Turnbaugh *et al.*, 2006). Furthermore, in another transplantation experiment, it was found that while body weight might not be affected by microbiota, some of the metabolic complications of obesity such as diabetes or non-alcoholic fatty liver disease (NAFLD) can be transmitted. Le Roy *et al.* (2013) selected the microbiota from two C57BL/6J donor mice based on their response to a HF diet. The responder was found to have developed hyperglycaemia and inflammation whereas the non-responder did not develop these complications in response to a HF diet. After inoculating GF mice with microbiota from either the responder or non-responder, it was found that mice receiving the responder microbiota developed fasting hyperglycaemia and insulinaemia and hepatic macrovesicular steatosis (Le Roy *et al.*, 2013). Although mice that received the non-responder microbiota still developed the same level of obesity as the mice receiving the responder bacteria, the additional metabolic complications did not develop. Altogether, this evidence suggests that gut microbiota play a causative role in metabolic disease. While animal models can be used to assess the causal role of gut microbiota in obesity, it is not yet clear whether gut microbiota plays a causal role in human disease (Petschow *et al.*, 2013).

In addition to diet and host genotype influence on gut microbiota in obesity, additional environmental influences such as antibiotics are also thought to play a critical role. In the agricultural industry, sub-therapeutic doses of antibiotics have been used for decades to facilitate weight gain and increase fat mass in animals (Feighner and Dashkevicz, 1987). Despite the effectiveness of antibiotics for increasing adiposity, the precise mechanisms mediating these changes were unknown. With these data, Martin and colleagues hypothesized that modern day exposure to antibiotics is potentially contributing to obesity in humans (Cho *et al.*, 2012). To study whether this could be true, Cho *et al.* employed an animal model in which young mice were administered subtherapeutic levels of antibiotics. The results from the study indicated that low doses of antibiotics did indeed lead to increased adiposity. Furthermore, the antibiotics induced both phyla-level and gene-level changes in the microbiome, such that the capacity of the microbiome to generate SCFAs was increased. This study provides evidence that the indiscriminate use of antibiotics may have detrimental effects on host-health and metabolism by disturbing the equilibrium of health-promoting bacteria in the gut. Previous studies in humans have identified that the antibiotic ciprofloxacin rapidly alters the gut microbiota with approximately one third of the bacterial taxa populations affected (Dethlefsen *et al.*, 2008; Dethlefsen and Relman, 2011). While the pattern of microbial shift has some inter-individual similarity, the changes in gut microbiota with ciprofloxacin varied amongst individuals (Dethlefsen and Relman, 2011). Furthermore, even though some taxa returned to baseline levels within a week, ciprofloxacin was demonstrated to alter baseline state, suggesting that antibiotics may lead to long-term changes in gut microbiota.

Although a causal link between a dysbiotic gut microbiota and obesity has not been established, there are several mechanisms that have been proposed to explain the association between dysbiosis and the pathogenesis of obesity. The underlying physiological processes that are involved in the development of obesity that are directly affected by the gut microbiota include satiety hormone secretion, gut permeability, endocannabinoid tone, adipose tissue metabolism, and bile acid signalling. To some extent all of these mechanisms can be manipulated and restored through the actions of prebiotics and probiotics. The following sections will review evidence for the effectiveness of prebiotics and probiotics in the restoration of gut microbiota and metabolic health that has been established to date.

Prebiotics

A prebiotic is by definition a non-digestible food ingredient that elicits 'the selective stimulation of growth and/or activity (ies) of one or a limited number of microbial genus (era)/species in the gut microbiota that confer (s) health benefits to the host' (Roberfroid *et al.*, 2010). While dietary fibre is classified as a non-digestible food ingredient, not all fibre is considered a prebiotic, owing to the fact that not all dietary fibres can be selectively fermented by one or a few members of the microbiota and thereby confer health benefits. Saccharolytic bacteria, which are able to break down and ferment complex oligosaccharides containing glucosyl, fuctosyl, galactosyl, and xylosyl moieties (Delzenne, 2003), are generally considered to be beneficial due to the metabolic by-products that they produce (Brownawell *et al.*, 2012). Proteolytic bacteria, which utilize undigested proteins as a source of energy, are considered to be more pathogenic to the host. Therefore, food ingredients that contain the nondigestible oligosaccharide substrates that match the enzymatic glycosidic hydrolase capacity of health promoting bacteria are primary prebiotic candidates. The main bacterial genera that are targeted by prebiotics are *Bifidobacterium* and *Lactobacillus*, which are both known for health promoting qualities (Roberfroid *et al.*, 2010). To date, two chemical groups that have been identified for their prebiotic effects include inulin-type fructans (ITF) and galacto-oligosaccharides (GOS) (Roberfroid *et al.*, 2010). ITF and GOS are known for their ability to stimulate bifidobacteria and to some extent lactobacilli (Roberfroid *et al.*, 2010), at the expense of other harmful bacteria (Macfarlane *et al.*, 2006).

Prebiotics are characterized by their degree of polymerization (DP). ITFs are composed of linear chains of β2–1 fructosyl-fructose linkages. An ITF with a DP of 10–60 is classified as inulin and a DP of 2–9 is fructo-oligosaccharide (FOS) or oligofructose (OFS) (Roberfroid *et al.*, 2010). GOS is composed of β1–6, β1–3, and β1–4 galactosyl-galactose linkages and has a DP of 2–8 (Roberfroid *et al.*, 2010). The shorter ITF polymers have been found to be highly fermentable in the caecal-colon (Coussement, 1999; Roberfroid and Delzenne, 1998). Inulin is found naturally in plant foods, including onions, leeks, garlic, asparagus, bananas, Jerusalem artichokes, chicory root, wheat, oats, and soybeans (Coussement, 1999; Delzenne, 2003). ITFs, due to their reputed health benefits and because intake from food sources is relatively low, are being increasingly incorporated as functional ingredients in a variety of foods. The fermentation of prebiotics is approximately 25–35% as efficient as metabolism of conventional starch and yields approximately 2kcal/gram, or 8.4KJ/g (Roberfroid, 1999). Although considered to be safe for consumption, some of the side effects of prebiotic include infrequent bloating, flatulence, and soft stools.

A substantial amount of research has been conducted in both animals and humans that highlight the antiobesogenic potential of prebiotics. OFS supplementation in animal models consistently reduces energy intake, body weight, and adipose tissue mass and improves metabolic parameters such as blood lipids and insulin resistance (Cani *et al.*, 2007a,b; Everard *et al.*, 2011; Parnell and Reimer, 2010). Human studies have yielded similar results, showing that prebiotics help to facilitate weight loss, satiety, and improved glucose tolerance (Cani *et al.*, 2009a; Kellow *et al.*, 2013; Parnell and Reimer, 2009). There are several mechanisms by which prebiotics are believed to mediate these effects.

Upon consumption of nutrients the gastrointestinal tract secretes hormones and gut peptides that communicate satiety centrally. In obesity, the production and secretion of satiety hormones in the basal and/or nutrient-stimulated state is impaired (Batterham *et al.*, 2003; Naslund and Hellstrom, 1998; Ranganath *et al.*, 1996; Vilsboll *et al.*, 2001). Prebiotics on the other hand can augment the secretion of key satiety hormones and thereby enhance satiety. Several key gut derived appetite signals include glucagon like peptide-1 (GLP-1) and peptide YY (PYY); both produced in enteroendocrine L-cells and known to be anorexigenic. Ghrelin, chiefly produced in the stomach, is known for its orexigenic effects. Generated from post-translational modification of the precursor proglucagon, GLP-1, an incretin hormone, delays gastric emptying, reduces food intake, increases β-cell proliferation, enhances insulin secretion, and promotes glucose homeostasis (Burcelin *et al.*, 2007; Cani *et al.*, 2007b). Cani *et al.* (2004, 2007b)demonstrated that OFS supplementation increased both serum GLP-1 and proglucagon mRNA expression in the proximal colon of rats and this was associated with a twofold increase in GLP-1 producing L-cells in the proximal colon. Additionally, OFS supplementation resulted in a 2.5-fold increase in colonic proglucagon mRNA levels compared to HF fed mice (Cani *et al.*, 2007a). Everard *et al.* (2011) showed that prebiotic supplementation increased both L-cell number and proglucagon expression and subsequent GLP-1 secretion in *ob/ob* mice. Concomitantly, adipose tissue mass was decreased and glucose control and leptin sensitivity were improved.

Animal studies that demonstrate enhanced satiety hormone response in prebiotic fed animals is consistent with the research in human trials. Cani *et al.*, in a double-blind, parallel group, placebo controlled trial showed that 16g of prebiotics/day (oligofructose-enriched inulin, Orafti Synergy 1 for 2 weeks increased fermentation, as indicated by a hydrogen breath test, as well as decreased hunger in response to a meal tolerance test. Prebiotic treatment in healthy individuals increased plasma PYY and GLP-1 secretion in addition to decreased postprandial glucose (Cani *et al.*, 2009a). These findings were supported by Parnell *et al.*, in a randomized, double-blind placebo controlled trial in overweight and obese individuals (Parnell and Reimer, 2009). Supplementation with 21g/day of OFS for 12 weeks resulted in increased PYY levels and blunted ghrelin response compared to placebo. Self-reported caloric intake was reduced and glycaemic control improved with the OFS supplementation (Parnell and Reimer, 2009). Unlike the study by Cani *et al.*, however, no changes in GLP-1 concentrations were observed (Parnell and Reimer, 2009). Another single-blind, cross-over study examined the effect of 8 g OFS consumed at breakfast and dinner for two weeks in healthy adults (Cani *et al.*, 2006). At the end of the experimental phase, energy intake and satiety were measured upon consumption of a free-access buffet breakfast, lunch, and dinner meal. Using a visual analogue scale (VAS), it was demonstrated that OFS versus placebo increased satiety during breakfast and dinner (Cani *et al.*, 2006). At dinner, OFS also reduced hunger and prospective food consumption. Combined energy intake at breakfast and lunch was decreased by 5% with OFS (Cani *et al.*, 2006). In a similar cross-over study, Verhoef *et al.* (2011) measured the effect of 2 weeks' supplementation of 16 g, 10 g, or 0 g OFS on satiety and energy intake in healthy human subjects. In response to a standardized, free-access meal, 16 g OFS produced the most significant reduction in energy intake (~11%). This effect corresponded with increased PYY and GLP-1 area under the curve relative to the lower 10 g dose of OFS (Verhoef *et al.*, 2011). The authors suggest that a 16 g/day dose of OFS is more effective at decreasing energy intake than 10 g/day.

These longer term prebiotic human studies demonstrated a clear effect on satiety. However, short term studies with prebiotics have not always shown the same effects. Using subjective ratings, Hess *et al.*, assessed the effect of short-chain FOS (scFOS) in 20 healthy subjects using a double-blind, crossover study. Each subject underwent three different days of testing. On each day, the subjects received a dose of either 0, 10, or 16 g scFOS, with half of the dose consumed in the morning and the other half consumed 2 hours before dinner. The subjective ratings failed to show that scFOS enhanced acute satiety or decreased hunger, despite hydrogen breath tests confirming that prebiotic fermentation occurred 240 min post fibre consumption (Hess *et al.*, 2011). These findings were supported in a study by Peters *et al.*, in which the satiating effects of FOS and beta-glucan were studied in 21 healthy individuals. Three meal replacement bars, each containing 8g FOS, consumed over the course of 1.5 days did not elicit an increase in satiety relative to control treatments (Peters *et al.*,

2009). The authors speculate that a longer term trial with FOS may be required to elicit satiety-enhancing effects.

The mechanisms by which prebiotics elicit these changes in gut satiety hormones are not completely understood. Cani *et al.* proposed that precursor cells may be differentiated into GLP-1 producing L-cells via expression of neurogenin 3 and NeuroD in response to the prebiotic fermentation products (Cani *et al.*, 2007b). Recent work has confirmed that metabolic by-products of prebiotic fermentation play a vital role in GLP-1 secretion. Tolhurst *et al.* (2012) using mixed colonic cultures *in vitro*, found that SCFAs led to increased secretion of GLP-1. SCFA also increased the expression of the G-protein-coupled receptors, GPR43 and GPR41 in enteroendocrine L-cells. The known role of SCFA in the activation of GPR41 and GPR43 (Le Poul *et al.*, 2003) was confirmed by evidence that SCFA increased cytosolic calcium concentrations in the intestinal L-cells, a response consistent with the GPR43-Gq signal transduction pathway (Tolhurst *et al.*, 2012). Furthermore, in a FFAR2$^{-/-}$ KO and FFAR3$^{-/-}$ KO mouse model, it was observed that GLP-1 secretion was blunted, demonstrating that FFAR is vital for GLP-1 release (Tolhurst *et al.*, 2012). Consistent with these findings, butyrate has also been found to stimulate GLP-1 secretion from enteroendocrine L-cells in cell culture (Yadav *et al.*, 2013). Altogether, these data provide strong evidence for the role of SCFA in the differentiation of enteroendocrine L-cells and associated GLP-1 secretion.

Like GLP-1, PYY appears to be under a similar SCFA-induced control mechanism (Miyauchi *et al.*, 2010; Samuel *et al.*, 2008). Samuel *et al.* (2008) employed a GPR41$^{-/-}$ KO mouse model to study the effects of these receptors on host metabolism. GPR41$^{-/-}$ KO mice, relative to wild-type, had decreased secretion of PYY. Additionally, it was identified that in the absence of PYY, gastric motility was increased, leading to reduced energy extraction from the diet. These effects were reported to elicit a reduction in body weight and adiposity. To confirm this signalling mechanism, when GPR41$^{-/-}$ KO mice and wild-type mice were rendered GF, the differences in gut hormone secretion and anthropometrics disappeared, confirming that fermentative by-products are critical for PYY signalling.

Interestingly, the same G-protein coupled receptors (GPR41 and GPR43) that are found on enteroendocrine L-cells are also expressed in adipocytes (Miyauchi *et al.*, 2010). It has been demonstrated by Ge *et al.* (2008) that activation of these receptors with the SCFA acetate inhibits lipolysis in both cell culture and *in vivo*. This mechanism was confirmed using a GPR43$^{-/-}$ KO mouse model, which showed that in the absence of the receptor, the decreased lipolytic effect was abolished (Ge *et al.*, 2008). Given the known effects of prebiotic fermentation on GPR43 in the gut, Dewulf *et al.* (2010) conducted a study to assess the effects of prebiotics on GPR43 and associated metabolism in white adipose tissue. ITF treatment significantly increased bifidobacteria and decreased *Roseburia* spp. and *Clostridium cluster* XIVa. Alongside decreased adipocyte size, expression of GPR43 was reduced by ITF, which corresponded with decreased markers of adipogenesis. Furthermore, it was observed that basal lipolysis was increased with ITF, which is somewhat contradictory to the findings by Ge *et al.* (2008). However, it was observed that ITF improved insulin-induced lipolysis inhibition, demonstrating that ITF improved adipocyte insulin sensitivity. Overall, there appears to be clear impacts of prebiotic fermentation and associated SCFA production on adipose tissue, however the precise mechanisms mediating these processes requires further investigation.

Altogether, the evidence points to the production of SCFA as the primary mechanism by which prebiotics mediate the beneficial effects on satiety and adipose tissue metabolism. Paradoxically, the generation of these same molecules has been cited as the mechanism by which microbiota contributes to weight gain. To explore this dichotomy, De Vadder *et al.* (2014) examined the effects of FOS and SCFA on intestinal gluconeogenesis (IGN), which is a mechanism that has been linked to beneficial effects on glucose control and energy homeostasis. The researchers observed that mice given FOS or SCFA had decreased weight gain and improved glucose and insulin tolerance (De Vadder *et al.*, 2014). Highlighting the role of IGN in mediating these improvements, it was observed that propionate and butyrate both stimulated intestinal gluconeogenesis through independent mechanisms. While butyrate activated IGN via a cAMP dependent mechanism, it is believed that propionate increases IGN via activation of FFAR3 in the portal vein which activates peripheral neural signalling pathways. Coinciding with these effects, increased neural activity was observed in the dorsal vagal complex (DVC), an area of the brain associated with regulation of energy homeostasis (De Vadder *et al.*, 2014). Confirming the necessity of the IGN mechanism for mediating the benefits of prebiotic, it was observed that the improvements in weight, adiposity, and glycaemia were abrogated when IGN function was disrupted using glucose 6-phosphatase (G6pc)$^{-/-}$ KO mice. Overall, this study provides novel mechanistic insight into the ability of SCFA, produced through the fermentation of prebiotics, to mediate beneficial, antiobesogenic effects on the host (De Vadder *et al.*, 2014).

A study by Respondek *et al.* (2008) using obese dogs examined the effect of prebiotics in the mediation of insulin sensitivity in obesity. Obese beagle dogs were provided with a diet containing 1% (wt:wt dry diet) scFOS for a period of 6 weeks. Hyperinsulinaemic–euglycaemic clamp studies confirmed that insulin sensitivity was improved in obese dogs that received scFOS (Respondek *et al.*, 2008). scFOS increased expression of uncoupling protein-2 (UCP-2) and tended to increase carnitine palmitoyltransferase-1 (CPT-1) in adipose tissue, suggesting that these proteins are involved in prebiotic-induced improvements in insulin sensitivity. Furthermore, highlighting the importance of gut microbiota in mediating insulin sensitivity, a study by Vrieze *et al.* (2012) found that faecal transplants from lean human donors improved insulin sensitivity in obese human males. Autologous 'from self' faecal transplants failed to elicit the same improvements. Analysis of microbiota revealed that increased insulin sensitivity was associated with an increase in butyrate-producing bacteria.

Associated with a 'dysbiotic' shift in the bacteria profile, HF feeding has been linked with impaired gut integrity and increased circulatory levels of the pro-inflammatory molecule lipopolysaccharide (LPS) (Cani *et al.*, 2007c, 2008). LPS, a cell-wall component of Gram-negative bacteria that can cross the epithelium and initiate a low grade inflammatory state that is 10–50 times less than would be experienced during septic shock, is thought to be involved in the development of insulin resistance and diabetes in obesity (Cani *et al.*, 2007c). In circulation, LPS binds to receptors Cluster of Differentiation 14 (CD14) and Toll-like receptors (TLRs) on innate immune cells (mature monocytes, macrophages, neutrophils), which triggers the release of pro-inflammatory cytokines, such as tumour necrosis factor (TNF)-α, Interleukin (IL)- 6, and IL-1, all of which are associated with the propagation of insulin resistance (Cani *et al.*, 2007c; Lu *et al.*, 2008). A HF diet is believed to increase the number of Gram-negative bacteria in the gut and impair gut permeability, thus allowing increased LPS translocation into circulation (Cani *et al.*, 2008). Furthermore, one of the hallmarks of a HF diet is a decrease in the proportion of the gut microbiota represented by the health promoting *Bifidobacterium* species.

Given that prebiotics elicit a potent 'bifidogenic' effect, Cani *et al.* investigated whether increasing bifidobacteria in the colon would ameliorate endotoxaemia associated with HF feeding (Cani *et al.*, 2007a). As hypothesized, supplementing HF fed mice with OFS for 14 weeks restored bifidobacteria. Measuring plasma levels of IL-1α, IL-1β, and IL-6 and adipose tissue derived proinflammatory cytokines, IL-1, TNF-α, and plasminogen activator inhibitory type 1 (PAI-1), bifidobacteria was found to be inversely associated with metabolic endotoxaemia. Furthermore, OFS supplementation improved glucose control and glucose-induced insulin secretion (Cani *et al.*, 2007a). In additional studies, obese *ob/ob* mice treated with a prebiotic had reduced plasma LPS concentrations and oxidative stress compared to control (Cani *et al.*, 2009b; Everard *et al.*, 2011). To confirm the link between LPS and metabolic endotoxaemia, Cani *et al.* used broad spectrum antibiotics to manipulate the microbiota and decrease the supply of LPS in the gut. The researchers found that treating HF-fed and *ob/ob* mice with antibiotics reduced metabolic endotoxaemia associated with LPS and down-regulated visceral adipose tissue PAI-1, IL-1, and TNF-α expression (Cani *et al.*, 2008). Altogether, this evidence highlights the potential for prebiotics to mitigate inflammation and maintain glucose homeostasis and insulin sensitivity.

With evidence that prebiotics improved endotoxaemia, Cani *et al.* completed a range of experiments that looked at how microbiota affects intestinal permeability. Genetically obese *ob/ob* mice were supplemented with OFS, which resulted in significant increases in lactobacilli, bifidobacteria, and *Clostridium coccoides–Eubacterium rectale* (Cani *et al.*, 2009b). More recent evidence also shows that OFS elicits increases in Bacteroidetes and decreased Firmicutes in the same mouse model (Everard *et al.*, 2011). *Ob/ob* mice supplemented with a prebiotic had decreased intestinal permeability relative to control, determined by measuring the amount of DX-4000-FITC that passed from the gut into circulation (Cani *et al.*, 2009b). Having previously identified that HF feeding disrupts intestinal integrity via reduced expression of epithelial tight junction proteins zona occludens-1 (ZO-1) and occludin (Cani *et al.*, 2008), Cani *et al.* demonstrated that mRNA of the tight junction proteins ZO-1 and occludin in the jejunum were up-regulated in response to prebiotic. As evidence of the intimate link between metabolic endotoxaemia and gut barrier function, plasma DX-4000-FITC concentrations were found to be positively correlated with portal plasma LPS concentrations (Cani *et al.*, 2009b).

GLP-2, a 33 amino acid peptide that is co-secreted with GLP-1 from the L-cells in the intestine, is thought to play a pivotal role in mediating intestinal adaptation, epithelial cell proliferation, and maintenance of gut permeability (Cani *et al.*, 2009b; Musso *et al.*, 2010). With an established link between endotoxaemia, intestinal permeability, gut microbiota, and GLP-1, Cani *et al.*, hypothesized that GLP-2 might play a critical role in mediating beneficial prebiotic effects. As predicted, *ob/ob* mice fed OFS had increased portal plasma GLP-2 relative to control diet. Blocking GLP-2 action with a GLP-2 antagonist blunted the prebiotic-mediated improvements on metabolic endotoxaemia and gut integrity (Cani *et al.*, 2009b). To further probe these mechanisms, a pharmacological infusion of GLP-2 was administered to mice which produced similar improvements in gut permeability and endotoxaemia (Cani *et al.*, 2009b). Similarly, Hadjiyanni *et al.* (2009) demonstrated that treatment with a GLP-2 analogue decreased jejunal transepithelial permeability in non-obese diabetic mice.

In obesity there appears to be a clear link between disturbances in the gut microbiota and impaired intestinal permeability leading to systemic LPS-induced endotoxaemia. Based on evidence that obesity is also characterized by increased endocannabinoid (eCB) tone (Scherer and Buettner, 2009), which is in part regulated by LPS, Muccioli *et al.*, proposed a link between gut microbiota and eCBs in the pathogenesis of obesity (Muccioli *et al.*, 2010). The increased eCB tone in obesity is associated with elevated levels of eCBs, which are small active lipid mediators, as well as increased eCB G-protein receptor expression. One of these receptors, the eCB CB_1 receptor, plays an important role in energy regulation (Scherer and Buettner, 2009). Using prebiotics to correct disturbances in gut microbiota, it was observed that prebiotics reduced colonic eCB CB_1 mRNA expression as well as concentrations of Anandamide (AEA), an endogenous CB_1 ligand (Muccioli *et al.*, 2010). Expression of fatty acid amide hydrolase (FAAH), the enzyme responsible for catabolizing AEA, was increased with prebiotic (Muccioli *et al.*, 2010). These changes were associated with reduced intestinal permeability indicated by lower plasma LPS levels. Blocking CB_1 signalling was also found to improve gut permeability, thus demonstrating that the eCB system is an important mediator of intestinal permeability. Moving away from the gut, it was demonstrated that adipose tissue in obesity also had increased eCB tone, as evidenced by increased mRNA expression of CB_1 and N-acylphosphatidylethanolamine phospholipase D, an enzyme involved in AEA synthesis, as well as decreased FAAH (Muccioli *et al.*, 2010). OFS supplementation reduced CB_1 expression in adipose tissue which corresponded with increased adipogenesis, as indicated by elevated markers of adipocyte differentiation factors and lipogenesis. In contrast, LPS can significantly reduce adipogenesis which appears to have consequences for adipose tissue endocrine function as well. Evidence shows that both the eCB system and endotoxaemia affect the apelinergic system, which alters the secretion of the adipokine apelin which plays a role in glucose homeostasis (Geurts *et al.*, 2011). Given the prominent regulatory function of adipose tissue in mediating whole body metabolism, the eCB system, as regulated by the gut microbiota, is identified as a significant variable in the pathogenesis of obesity.

The precise bacteria that are mediating the beneficial effects on intestinal permeability, endocannabinoid tone, and LPS are not completely understood. The mucin-degrading bacteria *Akkermansia muciniphila* represents approximately 3–5% of the gut microbial community. Given its inverse association with body weight, Everard *et al.*, investigated the role of this specific organism in mediating the host metabolism using *ob/ob* and DIO mice (Everard *et al.*, 2013). Both groups of obese mice were found to have low levels of *A. muciniphila*. With OFS supplementation, levels of *A. muciniphila* were restored alongside reductions in body weight and adiposity. Additionally, OFS reduced LPS, a marker of endotoxaemia, and macrophage infiltration into adipose tissue (Everard *et al.*, 2013). To elucidate further the role of *A. muciniphila* in mediating host metabolism, live *A. muciniphila* was provided via gavage to obese mice. The results of live *A. muciniphila* treatment included improved body weight and composition, decreased hepatic expression of glucose 6 phosphatase (a gluconeogenic enzyme), and increased markers of adipocyte differentiation and lipid oxidation (Everard *et al.*, 2013). Moreover, *A. muciniphila* colonization increased levels of endocannabinoids in the ileum which was speculated to mediate overall improvements in gut permeability and endotoxaemia (Everard *et al.*, 2013). Altogether, this study provides valuable insight into a specific organism that alters the endocannabinoid system and adipose tissue metabolism in favour of a lean phenotype.

Animal studies have established a link between disturbances in the gut microbiota and low grade endotoxaemia in obesity. Several human studies have sought to determine whether prebiotic supplements mediate the same improvements in inflammation, gut integrity, and adiposity as seen in animal studies. Deghan *et al.* (2014) conducted a randomized control trial in women with type 2 diabetes evaluating the effect of 10g inulin/day versus maltodextrin on inflammation and endotoxaemia. Supplementation with inulin decreased energy intake, fasting blood glucose, HbA1c, fasting insulin, and HOMA-IR relative to the placebo. Furthermore, at the end of the 8 week trial, plasma concentrations of high-sensitive C-reactive protein, TNF-α, and LPS were reduced. In another double-blind trial Lecerf *et al.*, randomized sixty healthy participants to receive either xylo-oligosaccharide (XOS), XOS + inulin, or control for a period of four weeks (Lecerf *et al.*, 2012). Supplementation with a combination of XOS and inulin increased faecal concentrations of SCFA with specific increases in propionate and butyrate. Furthermore, individuals that received XOS + inulin had decreased plasma LPS (Lecerf *et al.*, 2012). In a randomized, cross-over trial, Russo *et al.*, provided twenty healthy males with inulin-enriched pasta (~11g inulin/day) or placebo for a period of 5 weeks (Russo *et al.*, 2012). The washout period between control and treatment was 8 weeks. Intestinal barrier function was determined by analysing lactulose-mannitol excretion in urine. Inulin decreased lactulose in the urine, indicating improved intestinal permeability. Moreover, this change was associated with enhanced fasting GLP-2, a gut trophic hormone, and decreased plasma zonulin, a marker positively associated with paracellular intestinal permeability.

A prebiotic galacto–oligosaccharide mixture [(Bi2muno (B-GOS)], was used by Vulevic *et al.* (2013) in a double-blind, randomized, placebo controlled, crossover study in 45 overweight adults to assess the effect of this prebiotic on parameters of metabolic syndrome, gut microbiota profiles, and immune function. Consumption of 5.5 g/day B-GOS increased bifidobacteria and led to decreases in *Bacteroides* spp. and *C. histolyticum*. Although there were no changes in plasma cytokines with the prebiotic, B-GOS increased faecal secretory sIgA and lowered both faecal calprotectin (marker of intestinal inflammation) and plasma C-reactive protein. Additionally B-GOS decreased insulin, total cholesterol (TC), TAG, and TC:high density lipoprotein (HDL)-C although no changes were noted in glucose, HDL-C, or LDL-C (Vulevic *et al.*, 2013).

In another double-blind, placebo controlled trial, 30 obese women were treated with a 16 g/day (8 g twice daily) mixture of inulin and OFS for a period of 3 months (Dewulf *et al.*, 2013). Using HITChip analysis, it was identified that ITF treatment resulted in an increase in Firmicutes and Actinobacteria and a decrease in Bacteroidetes. At the genus level, ITF led to increased *bifidobacterium* and *Faecalibacterium prausnitzii*. Furthermore, qPCR confirmed the increase in *bifidobacterium* and also demonstrated an increase in lactobacilli. In support of the animal studies which showed the prebiotic supplementation decreased endotoxaemia, it was demonstrated that bifidobacteria and *F. prausnitzii* were inversely correlated with serum LPS. ITF improved post-OGTT glycaemia but no changes in body weight, HbA1c, fasting glycaemia and insulinaemia, post-OGTT insulinaemia, HOMA index, adiponectinaemia, cholesterol, or TAG were identified (Dewulf *et al.*, 2013). Metabolomic analysis of plasma and urine was also performed, revealing positive correlations between decreased *Bacteroides intestinalis*, *Bacteroides vulgatus*, and *Propionibacterium* and suppressed plasma lactate and phosphatidylcholine with ITF. Overall, despite some of the subtle metabolic effects of prebiotic treatment, several of the measured parameters in regards to inflammation and endotoxaemia parallel observations in animal models.

One of the other mechanisms by which microbiota are believed to elicit metabolic change in the host is via the signalling actions of bile acids (Turnbaugh, 2012). Primary bile acids are synthesized in the liver through enzymatic modification of cholesterol and conjugated to either taurine or glycine, which serves to enhance hydrophilicity of the amphipathic molecule (Lefebvre *et al.*, 2009). In the intestines bile acids are subject to bacterial modification, with the primary modifications being deconjugation, oxidation, and dehydroxylation (Ridlon *et al.*, 2006). Additionally, free bile acids, by disrupting phospholipid membranes, elicit bactericidal effects on gut microbiota (Kurdi *et al.*, 2006). Bile acids are important endogenous signalling molecules for numerous receptors such as farnesoid X-receptor (FXR), G-protein coupled receptor TGR5, and pregnane X receptor (Watanabe *et al.*, 2006). Through these receptors bile acids are able to regulate whole body energy homeostasis through the modulation of lipid metabolism, satiety, and to some extent glucose metabolism (Lefebvre *et al.*, 2009; Thomas *et al.*, 2009; Watanabe *et al.*, 2006). Currently one of the proposed mechanisms by which bariatric surgery is thought to mediate beneficial metabolic improvements is via alterations to bile acid metabolism and signalling (Myronovych *et al.*, 2014).

Several animal studies have assessed the effects of prebiotic supplementation on bile acid metabolism. Respondek *et al.*, completed faecal metabolomic analysis on human microbiota inoculated C57BL/6J mice fed a 10% scFOS HF diet. Metabolomic analysis revealed that bile acids were one of the primary metabolites affected by prebiotic supplementation (Respondek *et al.*, 2013). In another study, male Wistar rats supplemented with scFOS for 2 weeks were found to have altered hepatic FXR and PPARα expression (Fukasawa *et al.*, 2010). Moreover, healthy humans that consumed 10 g/day scFOS for three months had significantly altered bile acids, with an overall increase in primary bile acids (Boutron-Ruault *et al.*, 2005). Van Meer *et al.* (2008) fed rats a mixture of GOS and FOS for 3 weeks to elucidate the potential effects of the prebiotics on bile acid metabolism, citing implications for fat absorption. Although supplementation yielded significant increases in both lactobacilli and bifidobacteria, no changes were observed in bile salt composition or bile flow. Overall, given the complex interaction between bile acids and microbiota and the effects that bile acid can elicit on host, targeting bile acid metabolism with prebiotics offers potential for halting the pathogenesis of obesity, although further research in this area is required.

Overall, an abundance of animal research and a growing list of human clinical trials have demonstrated beneficial anthropometric and metabolic effects with prebiotics. Kellow *et al.* (2013) conducted a systematic review of human trials to evaluate the metabolic outcomes associated with prebiotic supplementation. A total of 26 randomized control trials were deemed eligible for inclusion, with the sum of the participants from all studies totalling 831. The duration of the prebiotic trials included in the analysis spanned 2 days to 28 weeks. Meta-analysis provided confirmation of enhanced self-reported satiety with prebiotics relative to placebo (Kellow *et al.*, 2013). Although, multiple trials demonstrated increased GLP-1 and PYY as well as decreases in energy intake with prebiotics, the reviewed evidence did not support these claims. The meta-analysis did indicate that prebiotics significantly reduced postprandial glucose and insulin (Kellow *et al.*, 2013). Decreases in fasting glucose, fasting insulin, and insulin resistance were not supported by the data. Furthermore, there was insufficient evidence for the role of prebiotics in the reduction of total cholesterol, LDL, TAG, and inflammatory markers. The meta-analysis did not support findings for decreased CRP with prebiotic supplementation. Overall, based on the available evidence, the authors suggest that prebiotics are beneficial for improving satiety and decreasing postprandial glucose and insulin concentrations (Kellow *et al.*, 2013).

Probiotics

Probiotics are defined as 'live micro-organisms which when administered in adequate amounts confer a health benefit on the host' (FAO/WHO, 2001). Nobel prize winner Elie Metchnikoff (1845–1916) first conceptualized the idea of probiotics after hypothesizing that intestinal bacteria were causing 'autointoxication' which accelerated ageing by releasing toxins into the vascular and neural system (Tannock, 2004). Metchnikoff inferred that the consumption of fermented milk was responsible for prolonging the health and lives of Bulgarian peasants, and therefore speculated that 'implanting' a lactic acid producing bacteria would kill certain harmful proteolytic bacteria and contribute to improved health and longevity. Today probiotics are consumed in yoghurt, yoghurt drinks, kefir, and dietary supplements for their purported benefits on amongst others, intestinal health. Although not all studies are consistent in their findings, some animal and human studies have identified that probiotics elicit antiobesogenic effects. In changing the gut microbiota, probiotics have the ability to change the metabolic function of the gut microbiome, enhance gut barrier function, and influence host satiety and metabolic function through various endocrine and signalling mechanisms. In particular *Bifidobacterium* spp. and *Lactobacillus* spp. have the potential to affect energy homeostasis, adiposity, lipidaemia, glycaemia, and body weight.

Several observations suggest that probiotics elicit a positive influence on blood lipids which are commonly elevated in obesity and the metabolic syndrome. A 1998 study by Taranto *et al.*, identified that supplementation with *L. reuteri* CRL 1098 (10^4 cells/day) for 7 days in Swiss albino hypercholesterolaemic mice reduced serum cholesterol and TAG by approximately 40% and increased the ratio of HDL to LDL by 20% relative to control mice (Taranto *et al.*, 1998). A randomized, double-blind study by Bukowska *et al.* (1998) observed a 7% reduction in total cholesterol and 10% reduction LDL with 1×10^{10} cfu/day *L. plantarum* 299v consumed for a period of 6 weeks. No changes were identified in TAG, HDL, or BMI. In another randomized, double-blind, cross-over study, a fermented milk beverage (375ml daily) containing *L. acidophilus* DN 112.053 and DN 112.096 (10^7–10^8 cfu/ml) and 2.5% FOS was compared to standard yoghurt beverage in healthy adult males (Schaafsma *et al.*, 1998). Improvements in total cholesterol, LDL, and the LDL/HDL ratio were observed with the probiotic beverage. Of note, given that the beverage contained a prebiotic, it is not clear whether or not the probiotic, the prebiotic, or the combination elicited the beneficial effects.

Although there is promising evidence for improved lipid profiles with probiotics, not all demonstrate improved lipidaemia. Greany *et al.* (2008), in a randomized, single-blinded, placebo-controlled trial, assessed the effects of 2 months of 3×10^9 cfu/day *L. acidophilus* DDS-1 and *B. longum* UABL-14 supplementation in 55 normocholesterolaemic individuals. Post-treatment, no changes in total cholesterol, HDL, LDL, or TAG were observed. In another study it was determined that supplementation with 1.8×10^{11} cfu/day *L. acidophilus* LA-1for six weeks in hypercholesterolaemic individuals did not elicit improvements in serum lipids or anthropometrics (Lewis and Burmeister, 2005).

As an explanation for improved lipidaemia with probiotics, it is theorized that probiotic-induced changes in gut microbiota limit the uptake of lipid substrates by the host (Wang *et al.*, 2009). Wang *et al.*, using *L. plantarum* MA2 that was isolated from traditional Chinese kefir, provided HF fed male Sprague–Dawley rats with 10^{11} cells/day *L. plantarum* MA2 daily for five weeks. Probiotic reduced serum cholesterol, TAG, and LDL (Wang *et al.*, 2009). These findings corresponded with increased faecal cholesterol, TAG, and propionate, thus supporting the idea that probiotics inhibit absorption of lipids in the gut. In support of this mechanism, a study by Hamad *et al.* (2009) found that *L. gasseri* SBT2055 blocked absorption of fat in the intestinal tract (Hamad *et al.*, 2009). When fed to lean and obese Zucker rats for 4 weeks, *L. gasseri* SBT2055 significantly reduced total serum cholesterol as well as hepatic cholesterol. Furthermore, the probiotic led to reductions in mesenteric adipose weight, adipocyte size, and serum leptin. It was confirmed that the probiotic increased faecal fatty acids and cholesterol. To assess whether the actions of *L. gasseri* SBT2055 limited uptake of fat by the host a second experiment was conducted in Sprague Dawley rats using thoracic cannulas. A 4 hour collection of lymph demonstrated that rats given the probiotic had reduced transport of TAG and phospholipid and tended to have reduced transport of cholesterol (Hamad *et al.*, 2009). The authors speculate that lactobacilli might elicit these effects by binding intestinal lipids.

In addition to the proposed mechanism by which probiotics inhibit lipid absorption, it is believed that probiotics might affect blood lipids via bile acid metabolism. Martin *et al.*, using a human baby flora (HBF) mouse model which was derived by inoculating GF animals with microbiota from a newborn infant, assessed the metabolic effects of the probiotics *L. paracasei* NCC2461and *L. rhamnosus* NCC4007 (Martin *et al.*, 2008). Probiotics elicited reductions in LDL and VLDL as well as increased plasma TAG. Faecal matter had decreased concentrations of the SCFA acetate and butyrate as well as marked changes in unconjugated bile acids and tauro-conjugated bile acids, with an overall decrease in faecal excretion of bile acids. Furthermore, the ileum was found to have increased concentrations of the bile acids ursodeoxycholic acid (UDCA) and chenodeoxycholic acid (CDCA). Altogether, given this evidence, the authors speculate that some of the observed metabolic effects in lipid metabolism may be mediated by gut microbiota – induced changes in enterohepatic circulation and bile acid signalling (Martin *et al.*, 2008). The purported effects of *Lactobacillus* species on hepatic metabolism are corroborated by a study in which C57BL/6J mice fed a high fat/cholesterol diet were provided with either *L. plantarum* KY1032 (1×10^{10} cfu/day), *L. curvatus* HY7601 (1×10^{10} cfu/day), or a combination of the two probiotics for a period of nine weeks (Yoo *et al.*, 2013). Yoo *et al.*, observed that the combination of both probiotics reduced adipose tissue mass, plasma cholesterol, liver steatosis, and liver cholesterol (Yoo *et al.*, 2013). These changes were associated with decreased expression of genes responsible for fatty acid synthesis as well as genes mediating fatty acid oxidation in the liver. Conversely, it has also been observed that HF induced weight gain, inflammation, and hepatic steatosis is positively correlated with endogenous *L. gasseri* and *L. taiwanesis* (Zeng *et al.*, 2013). According to the authors of the study, many *Lactobacillus* species are resistant to the bactericidal effects of bile acids and therefore increased *L. gasseri* and *L. taiwanesis* abundance may reflect their resistance to elevated concentrations of bile acids associated with a HF diet. Consequently, the bile acids that are secreted in response to a HF diet are subject to modification by *Lactobacillus* species, which could potentially have implications in the development of liver steatosis (Zeng *et al.*, 2013).

To assess whether a bile acid modifying probiotic could improve lipid profiles in hypercholesterolaemic humans, Jones *et al.* (2012a,b) conducted two studies using microencapsulated bile salt hydrolase (BSH)-active *L. reuteri* NCIMB 30242. In the first study, the probiotic was administered in a yoghurt beverage (5×10^{10} cfu twice daily for 6 weeks) (Jones *et al.*, 2012a), whereas in the second study the probiotic was provided in capsule form (2.9×10^{9} cfu twice daily for 9 weeks) (Jones *et al.*, 2012b). In these double-blind, placebo-controlled, randomized studies, each with more than 100 participants, it was demonstrated that both modes of supplementation resulted in reduced total cholesterol, LDL, as well as apoB-100. No changes in HDL or TAG were observed (Jones *et al.*, 2012a,b). *L. reuteri* NCIMB 30242 increased plasma deconjugated bile acids. Additionally, it was observed that the probiotic reduced plasma non-cholesterol sterols, suggesting an inhibition of sterol absorption at the level of the gut (Jones *et al.*, 2012b). Overall, the authors purport a beneficial effect of *L. reuteri* NCIMB 30242 for improvement of hypercholesterolaemia in humans.

In addition to improved lipid profiles with probiotics, a number of studies demonstrate that probiotics from the genus *Lactobacillus* and *Bifidobacterium* can elicit effects on adiposity and body weight. The effect of probiotic on body weight, however, appears to be species and strain specific. Noting that probiotics from various *Lactobacillus* species have been used for decades in the agricultural industry to enhance growth of farm animals, Million *et al.*, conducted a meta-analysis of clinical trials and experimental models on the effect of *Lactobacillus* spp. probiotics on body weight (Million *et al.*, 2012). A total of 17 randomized control trials, 51 farm animal studies, and 14 experimental models were included in the meta-analysis. Although the extent to which conclusions can be made about species-level effects on body weight is limited, given that body weight changes are differentially affected by bacterial strains, the authors identified that *L. acidophilus* was associated with increased weight gain whereas *L. gasseri* decreased body weight in both humans and animals (Million *et al.*, 2012). *L. ingluviei* or *L. fermentum* were both associated with weight gain in animals. Additionally, *L. plantarum* was noted to have an anti-obesity effect in animals. Interestingly, it has also been observed that some endogenous *Lactobacillus* species are correlated with impaired glucose control. Karlsson *et al.* (2013), in a study of European women, identified that while *L. gasseri* was not correlated with BMI, there was a positive correlation between *L. gasseri* and fasting glucose and haemoglobin A1c. Overall, it appears that different *Lactobacillus* strains elicit differential effects on body weight and host metabolism.

Given the variable effects of *Lactobacillus* spp. on body weight, Drissi *et al.* (2014) investigated the complete genomes of 13 different *Lactobacillus* strains (previously identified to affect weight gain) to identify the candidate genes that might explain the differential effects of *Lactobacillus* species on body weight. Marked differences were observed between the genomes associated with weight gain versus weight protection. Species that predispose to weight gain were associated with increased bacteriocin production and an increased ability to generate fructose and glucose from sucrose (Drissi *et al.*, 2014). Weight-protective *Lactobacillus* spp. were noted to have increased gene content for glucose permease, which is an enzyme that facilitates transport of glucose across a membrane where it can be broken down through glycolytic pathways. Weight protective species also have an enhanced gene capacity to catabolize fructose, inhibit oxidative stress, and synthesize acetate (Drissi *et al.*, 2014). In addition to differences at the probiotic genome level, Van Baarlen *et al.* (2011), in a double-blind, placebo-controlled study identified that *L. acidophilus* Lafti-L10, *L. casei* CRL-431, and *L. rhamnosus* GG elicit variable transcriptional response in duodenal cells of healthy humans. Potentially, these two mechanisms involving both the genome of the probiotic and the associated transcriptional response in the host could explain the differential weight patterns observed with probiotics of different *Lactobacillus* strains.

A variable response on body weight is also characteristic of *Bifidobacterium* spp. Yin *et al.*, compared the effects of four different endogenous *Bifidobacterium* strains (L66-5, L75-4, M13-4 and FS31-12) on obese male Sprague–Dawley rats. Of the four strains tested, M13–4 resulted in weight gain relative to control, L66–5 decreased body weight gain, and L75-4 and FS31-12 remained weight neutral (Yin *et al.*, 2010). Despite the differential effects on body weight, the authors noted that all four strains tended to lower serum and liver cholesterol. Chen *et al.* (2012) assessed the effects of *B. adolescentis* (unknown strain) on body weight, adiposity, and metabolic health in HF fed male Wistar rats. Twelve weeks of supplementation decreased body and adipose tissue weight gain. Furthermore, *B. adolescentis* mitigated HF diet induced increases in liver steatosis and insulin resistance. Kondo *et al.* (2010) demonstrated that *B. breve* B-3 dose dependently decreased body and epididymal fat weight gain in C57BL/6J mice fed a HF diet. Additionally, fasting glucose and insulin levels were improved, which may relate to the finding that *B. breve* B-3 increased expression of adiponectin in epididymal adipose tissue. Altogether, there appears to be unique strain specific responses on body weight and adiposity with different probiotics.

Evidence for antiobesogenic properties of a probiotic in an animal model does not always equate to benefits in human studies. In a study conducted by Yun *et al.*, *db/db* mice were fed 2×10^{10} cfu/day *L. gasseri* BNR17, a probiotic isolated from breast milk, for 12 weeks (Yun *et al.*, 2009). Although no differences were observed in body weight, *L. gasseri* BNR17 improved fasting and postprandial glucose levels. A subsequent study by Kang *et al.*, identified that *L. gasseri* BNR17 reduced body weight gain and adipose tissue mass in male Sprague–Dawley rats fed a high carbohydrate diet (Kang *et al.*, 2010). These changes were associated with increased expression of genes involved in fat oxidation and a decrease in expression of genes involved in fatty acid synthesis (Kang *et al.*, 2013). Furthermore, *L. gasseri* BNR17 was associated with both an increase in GLUT-4 transporter and a decrease in plasma leptin. With convincing evidence for the antiobesogenic properties of *L. gasseri* BNR17, Jung *et al.*, conducted a study evaluating the effectiveness of *L. gasseri* BNR17 for treating obesity in humans (Jung *et al.*, 2013). In the randomized, double-blind trial 57 individuals with obesity received either 6×10^{10} cfu/day *L. gasseri* BNR17 or placebo for a period of 12 weeks. At the end of the study, subjects that consumed *L. gasseri* BNR17 failed to achieve a significant reduction in body weight or plasma lipids relative to the control group (Jung *et al.*, 2013).

Kadooka *et al.* (2010) conducted a double-blind, randomized control trial in 87 overweight individuals (BMI 24.2–30.7). Subjects were randomized to receive either 1×10^{11} *L. gasseri* SBT 2055 or placebo consumed in a fermented milk beverage for 12 weeks. Abdominal visceral and subcutaneous fat, as measured with computed tomography, was significantly reduced by 4.6% with *L. gasseri* SBT 2055 treatment. Additionally, body weight decreased by 1.4% and BMI by 1.5% relative to baseline in the treatment group. Furthermore, it was observed that adiponectin was increased significantly with *L. gasseri* SBT 2055. A similar study conducted by the same research team tested the effectiveness of lower doses of *L. gasseri* SBT 2055 (Kadooka *et al.*, 2013). Japanese individuals with a high waist circumference were assigned to one of three groups for 12 weeks: (1) 2×10^9 cfu/day; (2) 2×10^8 cfu/day; (3) 0 cfu/day control. Both doses of *L. gasseri* SBT 2055 significantly reduced abdominal visceral fat areas by ~8%. Significant improvements in BMI and waist circumference were also observed (Kadooka *et al.*, 2013). The authors noted that improvements in adiposity upon cessation of probiotic consumption trended back towards baseline, suggesting that continued intake of probiotics is important for sustainable results. Given that *L. gasseri* SBT2055 has been demonstrated to decrease lipid absorption in the intestine in an animal model, Ogawa *et al.* (2014) studied the effect of *L. gasseri* SBT2055 supplementation in a group of hypertriglyceridaemic individuals. Using a single-blinded, cross-over study participants underwent 4 weeks on control and 4 weeks of *L. gasseri* SBT2055 treatment with a 4-week washout period in between. Relative to control, it was observed that *L. gasseri* SBT2055 significantly reduced postprandial serum non-esterified fatty acids (NEFA) and TAG (Ogawa *et al.*, 2014). Altogether, this combination of studies highlights the antiobesogenic effects of *L. gasseri* SBT2055.

Antiobesogenic properties have also been observed with the probiotic *L. rhamnosus* GG (ATCC53103) and *B. animalis* subsp. *lactis* BB-12. Ilmonen *et al.* (2011) studied the effect of probiotic supplementation on anthropometrics and body weight in pregnant women. A total of 256 pregnant women were randomly assigned to either diet counselling or control in their first trimester of pregnancy. Women who were assigned to receive dietary counselling were further randomized to receive either *L. rhamnosus* GG, and *B. animalis* subsp. *lactis* BB-12 (1×10^{10} cfu/day) or control for the duration of the pregnancy and up to six months post-partum, granted the mother continued to breastfeed. Overall, the probiotic supplements were well tolerated, with only a small per cent of participants experience minor gastrointestinal symptoms which diminished over the course of the trial. At 6 months post-partum dietary counselling in combination with a probiotic supplement lowered maternal waist circumference measurements and decreased the risk of central adiposity (waist circumference ≥80cm). The researchers determined that in order to prevent one person from developing central adiposity, 4 people needed to be treated with the dietary counselling and probiotic supplements (Ilmonen *et al.*, 2011).

In a study by Luoto *et al.* (2010) the effects of perinatal probiotic supplementation on child growth and development was examined. A total of 159 women were randomized to receive either placebo or 1×10^{10} cfu/day *L. rhamnosus* GG for 4 weeks prior to delivery and for 6 month post partum. Anthropometric measurements were taken on the children at 3, 6, 12, and 24 months and at 4, 7, and 10 years of age. The authors observed that in the early phases of life (fetal period until 24–48 months) children whose mothers consumed *L. rhamnosus* GG had a slightly restricted weight gain, particularly in individuals who later went on to develop overweight (Luoto *et al.*, 2010).

Furthermore in a somewhat serendipitous discovery, Woodard *et al.* (2009) found that probiotics promote sustainable weight loss after Roux en Y gastric bypass surgery. In the study, 44 patients were randomized to either a control or probiotic. Post-surgery, the experimental group consumed 2.4×10^9 cfu/day *Lactobacillus*. Unfortunately, the authors did not specify which species and strain of *Lactobacillus* were provided to patients. Relative to control, probiotic administration reduced bacterial overgrowth and significantly increased weight loss at 3 months and 6 months post-operatively (Woodard *et al.*, 2009). The authors noted that increased weight loss was an unexpected benefit. As an interesting note, the authors speculate that individuals who have undergone gastric bypass surgery may be an ideal group to study the use of probiotics as the survival of probiotics is more prominent in individuals without a gastric pouch (Woodard *et al.*, 2009).

A clinical study conducted by Gobel *et al.* (2012) using the probiotic *L. salivarium* Ls-33 failed to show improvements in body weight, adiposity, and other metabolic parameters in adolescents with obesity. In the double-blind, placebo-controlled trial, 50 adolescents (12–15 years, BMI ≥ 30) were randomized to receive either 1×10^{10} cfu/day *L. salivarium* Ls-33 or placebo for 12 weeks. Upon completion of the study, no changes were observed in inflammatory markers, fasting glucose and insulin, serum lipids, and anthropometrics. Although 89% of the treatment subjects had detectable levels of *L. salivarius* in their faecal matter, *L. salivarium* Ls-33 failed to elicit any beneficial effects on parameters of metabolic syndrome (Gobel *et al.*, 2012).

A study by Andreasen *et al.* (2010) evaluated the potential of the probiotic *L. acidophilus* NCFM for improving insulin sensitivity. In the randomized, double-blind, control trial, male subjects with either type 2 diabetes, impaired glucose tolerance, or normal glucose tolerance underwent a baseline and post-intervention hyperinsulinaemic–euglycaemic clamp. Participants received either *L. acidophilus* NCFM or placebo for a period of 4 weeks. It was observed that *L. acidophilus* NCFM preserved insulin sensitivity relative to the group that received the placebo. In addition to the insulin sensitivity measurements, the systemic inflammatory response was also assessed in participants using a 0.3 ng/kg LPS injection. *L. acidophilus* NCFM did not affect the systemic inflammatory response (Andreasen *et al.*, 2010). The authors suggest that the probiotic *L. acidophilus* NCFM may be useful for preventing aberrations in insulin sensitivity.

While some studies show significant benefit with probiotics on body weight, adiposity, and glucose control, the precise mechanisms mediating these changes are not completely understood. A variety of mechanisms have been proposed, including synthesis of SCFA, production of conjugated linolenic acid, altered tissue gene expression, changes in cytokine and satiety hormone secretions, as well as modulation of gut permeability and associated low grade inflammation.

In a study by Yadav *et al.* (2013), the probiotic VSL#3, a combination of *B. breve, B. longum, B. infantis, L. acidophilus, L. plantarum, L. paracasei, L. bulgaricus,* and *Streptococcus thermophilus*, was administered to both HF fed C57J/B6 mice, DIO mice, as well as *ob/ob* mice for approximately 8 weeks. It was demonstrated that VSL#3 decreased body weight gain, fat mass, adipocyte size, liver steatosis, as well as improved glucose and insulin tolerance, inflammation, and adipokine secretion from adipose tissue (Yadav *et al.*, 2013). Furthermore, VSL#3 decreased food intake in HF fed animals, which was associated with increased expression of agouti-related protein and neuropeptide Y and reduced expression of proopiomelanocortin in the hypothalamus, suggesting a role of VSL#3 in decreasing central hunger cues. It was observed that GLP-1 was significantly increased with VSL#3 treatment. Additionally, VSL#3 increased faecal and plasma concentrations of the SCFA butyrate (Yadav *et al.*, 2013). A subsequent *in vitro* experiment using human L-cells demonstrated that butyrate stimulates GLP-1 secretion (Yadav *et al.*, 2013). The authors purport that VSL#3-induced SCFA production and consequent GLP-1 release is a probable mechanism regulating anthropometric and metabolic improvements observed with probiotic treatment.

Previous studies have identified that gut microbiota affects adipose tissue mass by regulating the activity of lipoprotein lipase. Given the established link between the increased expression of Angiopoietin-like protein 4 (Angptl4) and decreased activity of LPL in GF mice (Backhed *et al.*, 2004), Aronsson *et al.* (2010) investigated the potential role of Angptl4 expression in the mediation of *L. paracasei* subsp. *paracasei* F19-induced changes in host adiposity. Mice supplemented with *L. paracasei* F19 gained significantly less body fat and were found to have an increased VLDL TAG load, although no changes in cholesterol levels were noted. It was observed that *L. paracasei* F19 induced levels of circulating Angptl4, suggesting that increased TAG was secondary to decreased LPL activity. Co-culture experiments with colonic cell-lines confirmed that *L. paracasei* F19 has the ability to stimulate Angptl4 expression. GF mice monocolonized with *L. paracasei* F19 also demonstrated enhanced circulating levels of Angptl4, thus providing confirmation for the potential role of Angptl4 expression in mediating probiotic-induced changes in adiposity. A study by (Fak and Backhed, 2012), however, did not observe changes in the expression of ileal Angptl4 with the probiotic *Lactobacillus reuteri* ATCC PTA 4659, despite the fact that body, adipose, and liver mass in Apoe$^{-/-}$ mice were significantly reduced. Instead, it was observed that hepatic expression of carnitine palmitoyltransferase 1a (Cpt1a) was increased, suggesting that β-oxidation in the liver was a mediating factor.

It has also been proposed that probiotics mediate improved metabolic effects on the host through the synthesis of conjugated linoleic acid (CLA). CLA has been proposed to reduce adiposity via suppression of appetite, increasing energy expenditure, inhibiting lipogenesis, increasing lipolysis, and increasing adipocyte apoptosis (Kennedy *et al.*, 2010). Lee *et al.*, conducted a study in DIO mice using the probiotic *L. rhamnosus* PL60, which is a human derived strain known for its ability to synthesize trans10, cis12 CLA (Lee *et al.*, 2006). *L. rhamnosus* PL60 supplementation elicited reductions in white adipose tissue and a down-regulation of TNF-α and FAS mRNA and an up-regulation of UCP-2. Although CLAs have previously been linked to increased liver steatosis, it was found that *L. rhamnosus* PL60 ameliorated liver steatosis (Lee *et al.*, 2006). Overall, results suggest that CLA might be responsible for mediating some of the antiobesogenic effects or probiotics. A study by Takemura *et al.* (2010), however, despite seeing reductions in adipocyte size and a trend towards reduced adipose tissue mass, serum cholesterol, and leptin with *L. plantarum* strain No. 14, did not detect alterations in plasma CLA concentrations. It has been identified that *Bifidobacterium* species have the ability to convert linoleic acid into CLA and alpha-linolenic acid into conjugated linolenic acid (CLNA), which is another fatty acid associated with health benefits (Gorissen *et al.*, 2010). Gorissen *et al.* (2010) confirmed that *B. breve* (LMG 11040, 11084, 11613, and 13194), *B. bifidum* LMG 10645, *and B. pseudolongum* subsp. *pseudolongum* LMG 11595 were able to synthesize both CLA and CLNA. Potentially, synthesis of CLA is a mechanism by which *Bifidobacterium* spp. stimulate health-promoting effects.

It is believed that maintenance of gut permeability and barrier function is another mechanism by which probiotics contribute to improved energy homeostasis. As demonstrated by Cani *et al.* (2007c), impaired intestinal permeability is associated with increased endotoxaemia due to translocation of LPS into circulation. Given that the development of NAFLD is linked with impaired intestinal permeability, Ritze *et al.* (2014) assessed the effects of *L. rhamnosus* GG (5.2×10^7 cfu/g of body weight per day) in female C57BL/6 mice fed a high-fructose diet. In addition to mitigating inflammation and steatosis in the liver, *L. rhamnosus* GG prevented reductions in the tight junction proteins occludin and claudin-1 and decreased portal plasma LPS concentrations by a factor of 3 relative to the control group. In another study, Naito *et al.*, found that the probiotic *L. casei* Shirota improved insulin resistance and glucose tolerance in DIO C57BL/6J mice (Naito *et al.*, 2011). Although *L. casei* Shirota did not elicit a decrease in adipose tissue mass, probiotic treatment significantly decreased plasma concentrations of LPS-binding protein, evidence that *L. casei* Shirota decreased endotoxaemia. It has also been demonstrated in DIO Sprague Dawley rats that the probiotic *B. animalis* subsp. *lactis* BB-12 stimulates the secretion of the gut trophic hormone GLP-2 (Bomhof *et al.*, 2014), a hormone associated with increased tight junction integrity and reduced intestinal permeability (Cani *et al.*, 2009b). The increase in GLP-2 was associated with improved glycaemia and lower insulin, however, the result was not corroborated by decreased intestinal permeability or inflammatory markers. There is also evidence that the probiotic VSL#3 maintains intestinal barrier function through the production of mucin, which provides a protective barrier against pathogens. Daily supplementation with 3×10^9 cfu/day VSL#3 in male Wistar rats increased luminal mucin by 60% which was associated with up-regulation of MUC2 expression in goblet cells (Caballero-Franco *et al.*, 2007). Overall, these studies highlight the potential for probiotics to protect against the pathogenesis of obesity by regulating gut permeability and barrier function.

Altogether, there is an abundance of animal and human studies that demonstrate the influence of specific probiotics on the regulation of energy balance, body weight, and metabolic health. The mixed results associated with various species and strains of probiotic in both animal models and clinical studies emphasize that the metabolic effects are strain specific and have the potential to contribute to weight loss or weight gain. Given the current evidence, it is difficult to conclude which probiotic supplements are most beneficial for managing obesity. Defining an optimal probiotic for prevention and treatment of obesity and associated metabolic disease will require further research and a better understanding of the mechanisms by which individual probiotics mediate their metabolic effects.

Future perspectives

Although much has been learned in recent years about the impact of gut microbiota on host physiology, there still exists a significant portion of the metabolic potential contained within the microbiome that has yet to be discovered or harnessed. As the science and technology advances within the microbiota field and new discoveries are made into the unique function of individual bacteria and the interactions between microbe, diet, and host, the opportunities for the development of new prebiotic and probiotic strategies in the management of obesity are diverse and numerous. Although the ultimate cure for obesity will certainly require a multifactorial approach given the complex nature of the disease, the optimization of gut microbiota for improved energy balance and metabolic health using minimally invasive prebiotic and probiotic strategies offers great potential for assisting individuals worldwide in the pursuit to prevent and treat this exigent disease.

References

Andreasen, A.S., Larsen, N., Pedersen-Skovsgaard, T., Berg, R.M., Moller, K., Svendsen, K.D., Jakobsen, M., and Pedersen, B.K. (2010). Effects of *Lactobacillus* acidophilus NCFM on insulin sensitivity and the systemic inflammatory response in human subjects. Br. J. Nutr. *104*, 1831–1838.

Aronsson, L., Huang, Y., Parini, P., Korach-Andre, M., Hakansson, J., Gustafsson, J.A., Pettersson, S., Arulampalam, V., and Rafter, J. (2010). Decreased fat storage by *Lactobacillus paracasei* is associated with increased levels of angiopoietin-like 4 protein (ANGPTL4). PLoS One *5*, e13087.

Azad, M.B., Konya, T., Maughan, H., Guttman, D.S., Field, C.J., Chari, R.S., Sears, M.R., Becker, A.B., Scott, J.A., Kozyrskyj, A.L., and CHILD Study Investigators. (2013). Gut microbiota of healthy Canadian infants:

profiles by mode of delivery and infant diet at 4 months. CMAJ. *185*, 385–394.

Backhed, F., Ding, H., Wang, T., Hooper, L.V., Koh, G.Y., Nagy, A., Semenkovich, C.F., and Gordon, J.I. (2004). The gut microbiota as an environmental factor that regulates fat storage. Proc. Natl. Acad. Sci. U.S.A. *101*, 15718–15723.

Backhed, F., Manchester, J.K., Semenkovich, C.F., and Gordon, J.I. (2007). Mechanisms underlying the resistance to diet-induced obesity in germ-free mice. Proc. Natl. Acad. Sci. U.S.A. *104*, 979–984.

Batterham, R.L., Cohen, M.A., Ellis, S.M., Le Roux, C.W., Withers, D.J., Frost, G.S., Ghatei, M.A., and Bloom, S.R. (2003). Inhibition of food intake in obese subjects by peptide YY3-36. N. Engl. J. Med. *349*, 941–948.

Bergman, E.N. (1990). Energy contributions of volatile fatty acids from the gastrointestinal tract in various species. Physiol. Rev. *70*, 567–590.

Bomhof, M.R., Saha, D.C., Reid, D.T., Paul, H.A., and Reimer, R.A. (2014). Combined effects of oligofructose and Bifidobacterium animalis on gut microbiota and glycaemia in obese rats. Obesity (Silver Spring) *22*, 763–771.

Booijink, C.C., El-Aidy, S., Rajilic-Stojanovic, M., Heilig, H.G., Troost, F.J., Smidt, H., Kleerebezem, M., De Vos, W.M., and Zoetendal, E.G. (2010). High temporal and inter-individual variation detected in the human ileal microbiota. Environ. Microbiol. *12*, 3213–3227.

Boutron-Ruault, M.C., Marteau, P., Lavergne-Slove, A., Myara, A., Gerhardt, M.F., Franchisseur, C., Bornet, F., and Eripolyp Study Group. (2005). Effects of a 3-mo consumption of short-chain fructo-oligosaccharides on parameters of colorectal carcinogenesis in patients with or without small or large colorectal adenomas. Nutr. Cancer *53*, 160–168.

Brownawell, A.M., Caers, W., Gibson, G.R., Kendall, C.W., Lewis, K.D., Ringel, Y., and Slavin, J.L. (2012). Prebiotics and the health benefits of fibre: current regulatory status, future research, and goals. J. Nutr. *142*, 962–974.

Bukowska, H., Pieczul-Mroz, J., Jastrzebska, M., Chelstowski, K., and Naruszewicz, M. (1998). Decrease in fibrinogen and LDL-cholesterol levels upon supplementation of diet with *Lactobacillus plantarum* in subjects with moderately elevated cholesterol. Atherosclerosis *137*, 437–438.

Burcelin, R., Cani, P.D., and Knauf, C. (2007). Glucagon-like peptide-1 and energy homeostasis. J. Nutr. *137*, 2534S-2538S.

Caballero-Franco, C., Keller, K., De Simone, C., and Chadee, K. (2007). The VSL#3 probiotic formula induces mucin gene expression and secretion in colonic epithelial cells. Am. J. Physiol. Gastrointest. Liver Physiol. *292*, G315–22.

Cani, P.D. (2013). Gut microbiota and obesity: lessons from the microbiome. Brief Funct. Genomics *12*, 381–387.

Cani, P.D., and Delzenne, N.M. (2009). Interplay between obesity and associated metabolic disorders: new insights into the gut microbiota. Current Opinion in Pharmacology *9*, 737–743.

Cani, P.D., Dewever, C., and Delzenne, N.M. (2004). Inulin-type fructans modulate gastrointestinal peptides involved in appetite regulation (glucagon-like peptide-1 and ghrelin) in rats. Br. J. Nutr. *92*, 521–526.

Cani, P.D., Joly, E., Horsmans, Y., and Delzenne, N.M. (2006). Oligofructose promotes satiety in healthy human: a pilot study. Eur. J. Clin. Nutr. *60*, 567–572.

Cani, P.D., Neyrinck, A.M., Fava, F., Knauf, C., Burcelin, R.G., Tuohy, K.M., Gibson, G.R., and Delzenne, N.M. (2007a). Selective increases of bifidobacteria in gut microflora improve high-fat-diet-induced diabetes in mice through a mechanism associated with endotoxaemia. Diabetologia *50*, 2374–2383.

Cani, P.D., Hoste, S., Guiot, Y., and Delzenne, N.M. (2007b). Dietary non-digestible carbohydrates promote L-cell differentiation in the proximal colon of rats. Br. J. Nutr. *98*, 32–37.

Cani, P.D., Amar, J., Iglesias, M.A., Poggi, M., Knauf, C., Bastelica, D., Neyrinck, A.M., Fava, F., Tuohy, K.M., Chabo, C., *et al.* (2007c). Metabolic endotoxemia initiates obesity and insulin resistance. Diabetes *56*, 1761–1772.

Cani, P.D., Bibiloni, R., Knauf, C., Waget, A., Neyrinck, A.M., Delzenne, N.M., and Burcelin, R. (2008). Changes in gut microbiota control metabolic endotoxemia-induced inflammation in high-fat diet-induced obesity and diabetes in mice. Diabetes *57*, 1470–1481.

Cani, P.D., Lecourt, E., Dewulf, E.M., Sohet, F.M., Pachikian, B.D., Naslain, D., De Backer, F., Neyrinck, A.M., and Delzenne, N.M. (2009a). Gut microbiota fermentation of prebiotics increases satietogenic and incretin gut peptide production with consequences for appetite sensation and glucose response after a meal. Am. J. Clin. Nutr. *90*, 1236–1243.

Cani, P.D., Possemiers, S., Van de Wiele, T., Guiot, Y., Everard, A., Rottier, O., Geurts, L., Naslain, D., Neyrinck, A., Lambert, D.M., Muccioli, G.G., and Delzenne, N.M. (2009b). Changes in gut microbiota control inflammation in obese mice through a mechanism involving GLP-2-driven improvement of gut permeability. Gut *58*, 1091–1103.

Chen, J., Wang, R., Li, X.F., and Wang, R.L. (2012). *Bifidobacterium adolescentis* supplementation ameliorates visceral fat accumulation and insulin sensitivity in an experimental model of the metabolic syndrome. Br. J. Nutr. *107*, 1429–1434.

Cho, I., Yamanishi, S., Cox, L., Methe, B.A., Zavadil, J., Li, K., Gao, Z., Mahana, D., Raju, K., Teitler, I., *et al.* (2012). Antibiotics in early life alter the murine colonic microbiome and adiposity. Nature *488*, 621–626.

Costello, E.K., Lauber, C.L., Hamady, M., Fierer, N., Gordon, J.I., and Knight, R. (2009). Bacterial community variation in human body habitats across space and time. Science *326*, 1694–1697.

Coussement, P.A. (1999). Inulin and oligofructose: safe intakes and legal status. J. Nutr. *129*, 1412S-7S.

de Onis, M., Blossner, M., and Borghi, E. (2010). Global prevalence and trends of overweight and obesity among preschool children. Am. J. Clin. Nutr. *92*, 1257–1264.

De Vadder, F., Kovatcheva-Datchary, P., Goncalves, D., Vinera, J., Zitoun, C., Duchampt, A., Backhed, F., and Mithieux, G. (2014). Microbiota-Generated Metabolites Promote Metabolic Benefits via Gut-Brain Neural Circuits. Cell *156*, 84–96.

Dehghan, P., Gargari, B.P., Jafar-Abadi, M.A., and Aliasgharzadeh, A. (2014). Inulin controls inflammation and metabolic endotoxemia in women with type 2 diabetes mellitus: a randomized-controlled clinical trial. Int. J. Food Sci. Nutr. *65*, 117–123.

Delzenne, N.M. (2003). Oligosaccharides: state of the art. Proc. Nutr. Soc. *62*, 177–182.

Delzenne, N.M., and Cani, P.D. (2011). Interaction between obesity and the gut microbiota: relevance in nutrition. Annu. Rev. Nutr. *31*, 15–31.

Dethlefsen, L., Huse, S., Sogin, M.L., and Relman, D.A. (2008). The pervasive effects of an antibiotic on the human gut microbiota, as revealed by deep 16S rRNA sequencing. PLoS Biol. *6*, e280.

Dethlefsen, L., and Relman, D.A. (2011). Incomplete recovery and individualized responses of the human distal gut microbiota to repeated antibiotic perturbation. Proc. Natl. Acad. Sci. U.S.A. *108 Suppl. 1*, 4554–4561.

Dewulf, E.M., Cani, P.D., Claus, S.P., Fuentes, S., Puylaert, P.G., Neyrinck, A.M., Bindels, L.B., de Vos, W.M., Gibson, G.R., Thissen, J.P., and Delzenne, N.M. (2013). Insight into the prebiotic concept: lessons from an exploratory, double blind intervention study with inulin-type fructans in obese women. Gut *62*, 1112–1121.

Dewulf, E.M., Cani, P.D., Neyrinck, A.M., Possemiers, S., Holle, A.V., Muccioli, G.G., Deldicque, L., Bindels, L.B., Pachikian, B.D., Sohet, F.M., *et al.* (2010). Inulin-type fructans with prebiotic properties counteract GPR43 overexpression and PPARgamma-related adipogenesis in the white adipose tissue of high-fat diet-fed mice. J. Nutr. Biochem.

Drissi, F., Merhej, V., Angelakis, E., El Kaoutari, A., Carriere, F., Henrissat, B., and Raoult, D. (2014). Comparative genomics analysis of *Lactobacillus* species associated with weight gain or weight protection. Nutr. Diabetes *4*, e109.

Duncan, S.H., Lobley, G.E., Holtrop, G., Ince, J., Johnstone, A.M., Louis, P., and Flint, H.J. (2008). Human colonic microbiota associated with diet, obesity and weight loss. Int. Obes. J. (Lond.) 32, 1720–1724.

Eckburg, P.B., Bik, E.M., Bernstein, C.N., Purdom, E., Dethlefsen, L., Sargent, M., Gill, S.R., Nelson, K.E., and Relman, D.A. (2005). Diversity of the Human Intestinal Microbial Flora. Science *308*, 1635–1638.

Everard, A., Belzer, C., Geurts, L., Ouwerkerk, J.P., Druart, C., Bindels, L.B., Guiot, Y., Derrien, M., Muccioli, G.G., Delzenne, N.M., de Vos, W.M., and Cani, P.D. (2013). Cross-talk between Akkermansia muciniphila and intestinal epithelium controls diet-induced obesity. Proc. Natl. Acad. Sci. U.S.A. *110*, 9066–9071.

Everard, A., Lazarevic, V., Derrien, M., Girard, M., Muccioli, G.G., Neyrinck, A.M., Possemiers, S., Van Holle, A., Francois, P., de Vos, W.M., *et al.* (2011). Responses of gut microbiota and glucose and lipid metabolism to prebiotics in genetic obese and diet-induced leptin-resistant mice. Diabetes *60*, 2775–2786.

Fak, F., and Backhed, F. (2012). *Lactobacillus reuteri* prevents diet-induced obesity, but not atherosclerosis, in a strain dependent fashion in Apoe−/− mice. PLoS One 7, e46837.

FAO/WHO. (2001). Health and Nutritional Properties of Probiotics in Food including Powder Milk with Live Lactic Acid Bacteria. Expert consultation on evaluation of health and nutritional properties of probiotics in food including powder milk with live lactic acid bacteria. http://www.who.int/foodsafety/publications/fs_management/en/probiotics.pdf.

Feighner, S.D., and Dashkevicz, M.P. (1987). Subtherapeutic levels of antibiotics in poultry feeds and their effects on weight gain, feed efficiency, and bacterial cholyltaurine hydrolase activity. Appl. Environ. Microbiol. *53*, 331–336.

Fleissner, C.K., Huebel, N., Abd El-Bary, M.M., Loh, G., Klaus, S., and Blaut, M. (2010). Absence of intestinal microbiota does not protect mice from diet-induced obesity. Br. J. Nutr. *104*, 919–929.

Flint, H.J. (1997). The rumen microbial ecosystem – some recent developments. Trends Microbiol. *5*, 483–488.

Fukasawa, T., Kamei, A., Watanabe, Y., Koga, J., and Abe, K. (2010). Short-chain fructooligosaccharide regulates hepatic peroxisome proliferator-activated receptor alpha and farnesoid X receptor target gene expression in rats. J. Agric. Food Chem. *58*, 7007–7012.

Fukuda, S., Toh, H., Hase, K., Oshima, K., Nakanishi, Y., Yoshimura, K., Tobe, T., Clarke, J.M., Topping, D.L., Suzuki, T., *et al.* (2011). Bifidobacteria can protect from enteropathogenic infection through production of acetate. Nature *469*, 543–547.

Ge, H., Li, X., Weiszmann, J., Wang, P., Baribault, H., Chen, J.L., Tian, H., and Li, Y. (2008). Activation of G protein-coupled receptor 43 in adipocytes leads to inhibition of lipolysis and suppression of plasma free fatty acids. Endocrinology *149*, 4519–4526.

Geurts, L., Lazarevic, V., Derrien, M., Everard, A., Van Roye, M., Knauf, C., Valet, P., Girard, M., Muccioli, G.G., Francois, P., *et al.* (2011). Altered gut microbiota and endocannabinoid system tone in obese and diabetic leptin-resistant mice: impact on apelin regulation in adipose tissue. Front. Microbiol. 2, 149.

Gobel, R.J., Larsen, N., Jakobsen, M., Molgaard, C., and Michaelsen, K.F. (2012). Probiotics to adolescents with obesity: effects on inflammation and metabolic syndrome. J. Pediatr. Gastroenterol. Nutr. *55*, 673–678.

Gorissen, L., Raes, K., Weckx, S., Dannenberger, D., Leroy, F., De Vuyst, L., and De Smet, S. (2010). Production of conjugated linoleic acid and conjugated linolenic acid isomers by Bifidobacterium species. Appl. Microbiol. Biotechnol. *87*, 2257–2266.

Greany, K.A., Bonorden, M.J., Hamilton-Reeves, J.M., McMullen, M.H., Wangen, K.E., Phipps, W.R., Feirtag, J., Thomas, W., and Kurzer, M.S. (2008). Probiotic capsules do not lower plasma lipids in young women and men. Eur. J. Clin. Nutr. *62*, 232–237.

Hadjiyanni, I., Li, K.K., and Drucker, D.J. (2009). Glucagon-like peptide-2 reduces intestinal permeability but does not modify the onset of type 1 diabetes in the nonobese diabetic mouse. Endocrinology *150*, 592–599.

Hamad, E.M., Sato, M., Uzu, K., Yoshida, T., Higashi, S., Kawakami, H., Kadooka, Y., Matsuyama, H., Abd El-Gawad, I.A., and Imaizumi, K. (2009). Milk fermented by *Lactobacillus* gasseri SBT2055 influences adipocyte size via inhibition of dietary fat absorption in Zucker rats. Br. J. Nutr. *101*, 716–724.

Hess, J.R., Birkett, A.M., Thomas, W., and Slavin, J.L. (2011). Effects of short-chain fructooligosaccharides on satiety responses in healthy men and women. Appetite *56*, 128–134.

Hildebrandt, M.A., Hoffmann, C., Sherrill-Mix, S.A., Keilbaugh, S.A., Hamady, M., Chen, Y.Y., Knight, R., Ahima, R.S., Bushman, F., and Wu, G.D. (2009). High-fat diet determines the composition of the murine gut microbiome independently of obesity. Gastroenterology *137*, 1716–24. e1–2.

Hsiao, W.W.L., Metz, C., Singh, D.P., and Roth, J. (2008). The Microbes of the Intestine: An Introduction to Their Metabolic and Signalling Capabilities. Endocrinology & Metabolism Clinics of North America *37*, 857–871.

Ilmonen, J., Isolauri, E., Poussa, T., and Laitinen, K. (2011). Impact of dietary counselling and probiotic intervention on maternal anthropometric measurements during and after pregnancy: a randomized placebo-controlled trial. Clin. Nutr. *30*, 156–164.

Jimenez, E., Marin, M.L., Martin, R., Odriozola, J.M., Olivares, M., Xaus, J., Fernandez, L., and Rodriguez, J.M. (2008). Is meconium from healthy newborns actually sterile? Res. Microbiol. *159*, 187–193.

Jones, M.L., Martoni, C.J., Parent, M., and Prakash, S. (2012a). Cholesterol-lowering efficacy of a microencapsulated bile salt hydrolase-active *Lactobacillus reuteri* NCIMB 30242 yoghurt formulation in hypercholesterolaemic adults. Br. J. Nutr. *107*, 1505–1513.

Jones, M.L., Martoni, C.J., and Prakash, S. (2012b). Cholesterol lowering and inhibition of sterol absorption by *Lactobacillus reuteri* NCIMB 30242: a randomized controlled trial. Eur. J. Clin. Nutr. *66*, 1234–1241.

Jumpertz, R., Le, D.S., Turnbaugh, P.J., Trinidad, C., Bogardus, C., Gordon, J.I., and Krakoff, J. (2011). Energy-balance studies reveal associations between gut microbes, caloric load, and nutrient absorption in humans. Am. J. Clin. Nutr. *94*, 58–65.

Jung, S.P., Lee, K.M., Kang, J.H., Yun, S.I., Park, H.O., Moon, Y., and Kim, J.Y. (2013). Effect of *Lactobacillus* gasseri BNR17 on Overweight and Obese Adults: A Randomized, Double-Blind Clinical Trial. Korean J. Fam. Med. *34*, 80–89.

Kadooka, Y., Sato, M., Imaizumi, K., Ogawa, A., Ikuyama, K., Akai, Y., Okano, M., Kagoshima, M., and Tsuchida, T. (2010). Regulation of abdominal adiposity by probiotics (*Lactobacillus* gasseri SBT2055). in adults with obese tendencies in a randomized controlled trial. Eur. J. Clin. Nutr. *64*, 636–643.

Kadooka, Y., Sato, M., Ogawa, A., Miyoshi, M., Uenishi, H., Ogawa, H., Ikuyama, K., Kagoshima, M., and Tsuchida, T. (2013). Effect of *Lactobacillus* gasseri SBT2055 in fermented milk on abdominal adiposity in adults in a randomized controlled trial. Br. J. Nutr. *110*, 1696–1703.

Kalliomaki, M., Carmen Collado, M., Salminen, S., and Isolauri, E. (2008). Early differences in fecal microbiota composition in children may predict overweight. Am. J. Clin. Nutr. *87*, 534–538.

Kang, J.H., Yun, S.I., and Park, H.O. (2010). Effects of *Lactobacillus* gasseri BNR17 on body weight and adipose tissue mass in diet-induced overweight rats. J. Microbiol. *48*, 712–714.

Kang, J.H., Yun, S.I., Park, M.H., Park, J.H., Jeong, S.Y., and Park, H.O. (2013). Anti-obesity effect of *Lactobacillus* gasseri BNR17 in high-sucrose diet-induced obese mice. PLoS One *8*, e54617.

Karlsson, F.H., Tremaroli, V., Nookaew, I., Bergstrom, G., Behre, C.J., Fagerberg, B., Nielsen, J., and Backhed, F. (2013). Gut metagenome in European women with normal, impaired and diabetic glucose control. Nature *498*, 99–103.

Kellow, N.J., Coughlan, M.T., and Reid, C.M. (2013). Metabolic benefits of dietary prebiotics in human subjects: a systematic review of randomized controlled trials. Br. J. Nutr. 1–15.

Kennedy, A., Martinez, K., Schmidt, S., Mandrup, S., LaPoint, K., and McIntosh, M. (2010). Antiobesity mechanisms of action of conjugated linoleic acid. J. Nutr. Biochem. *21*, 171–179.

Kondo, S., Xiao, J.Z., Satoh, T., Odamaki, T., Takahashi, S., Sugahara, H., Yaeshima, T., Iwatsuki, K., Kamei, A., and Abe, K. (2010). Antiobesity effects of *Bifidobacterium breve* strain B-3 supplementation in a mouse model with high-fat diet-induced obesity. Biosci. Biotechnol. Biochem. *74*, 1656–1661.

Kurdi, P., Kawanishi, K., Mizutani, K., and Yokota, A. (2006). Mechanism of growth inhibition by free bile acids in lactobacilli and bifidobacteria. J. Bacteriol. *188*, 1979–1986.

Larsen, N., Vogensen, F.K., van den Berg, F.W., Nielsen, D.S., Andreasen, A.S., Pedersen, B.K., Al-Soud, W.A., Sorensen, S.J., Hansen, L.H., and Jakobsen, M. (2010). Gut microbiota in human adults with type 2 diabetes differs from non-diabetic adults. PLoS One *5*, e9085.

Le Poul, E., Loison, C., Struyf, S., Springael, J.Y., Lannoy, V., Decobecq, M.E., Brezillon, S., Dupriez, V., Vassart, G., Van Damme, J., Parmentier, M., and Detheux, M. (2003). Functional characterization of human receptors for short-chain fatty acids and their role in polymorphonuclear cell activation. J. Biol. Chem. *278*, 25481–25489.

Le Roy, T., Llopis, M., Lepage, P., Bruneau, A., Rabot, S., Bevilacqua, C., Martin, P., Philippe, C., Walker, F., Bado, A., *et al.* (2013). Intestinal microbiota determines development of non-alcoholic fatty liver disease in mice. Gut *62*, 1787–1794.

Lecerf, J.M., Depeint, F., Clerc, E., Dugenet, Y., Niamba, C.N., Rhazi, L., Cayzeele, A., Abdelnour, G., Jaruga, A., Younes, H., *et al.* (2012).

Xylo-oligosaccharide (XOS) in combination with inulin modulates both the intestinal environment and immune status in healthy subjects, while XOS alone only shows prebiotic properties. Br. J. Nutr. *108*, 1847–1858.

Lee, H., Park, J., Seok, S., Baek, M., Kim, D., Lee, K., Paek, K., Lee, Y., and Park, J. (2006). Human originated bacteria, *Lactobacillus rhamnosus* PL60, produce conjugated linoleic acid and show anti-obesity effects in diet-induced obese mice. Biochimica Et Biophysica Acta (BBA) – Molecular and Cell Biology of Lipids *1761*, 736–744.

Lefebvre, P., Cariou, B., Lien, F., Kuipers, F., and Staels, B. (2009). Role of bile acids and bile acid receptors in metabolic regulation. Physiol. Rev. *89*, 147–191.

Lewis, S.J., and Burmeister, S. (2005). A double-blind placebo-controlled study of the effects of *Lactobacillus* acidophilus on plasma lipids. Eur. J. Clin. Nutr. *59*, 776–780.

Ley, R.E., Backhed, F., Turnbaugh, P., Lozupone, C.A., Knight, R.D., and Gordon, J.I. (2005). Obesity alters gut microbial ecology. Proc. Natl. Acad. Sci. U.S.A. *102*, 11070–11075.

Ley, R.E., Turnbaugh, P.J., Klein, S., and Gordon, J.I. (2006). Microbial ecology: human gut microbes associated with obesity. Nature *444*, 1022–1023.

Lu, Y.C., Yeh, W.C., and Ohashi, P.S. (2008). LPS/TLR4 signal transduction pathway. Cytokine *42*, 145–151.

Luoto, R., Kalliomaki, M., Laitinen, K., and Isolauri, E. (2010). The impact of perinatal probiotic intervention on the development of overweight and obesity: follow-up study from birth to 10 years. Int. J. Obes. (Lond.) *34*, 1531–1537.

Macfarlane, S., Macfarlane, G.T., and Cummings, J.H. (2006). Review article: prebiotics in the gastrointestinal tract. Aliment. Pharmacol. Ther. *24*, 701–714.

Martin, F.P., Wang, Y., Sprenger, N., Yap, I.K., Lundstedt, T., Lek, P., Rezzi, S., Ramadan, Z., van Bladeren, P., Fay, L.B., *et al.* (2008). Probiotic modulation of symbiotic gut microbial-host metabolic interactions in a humanized microbiome mouse model. Mol. Syst. Biol. *4*, 157.

McFall-Ngai, M. (2007). Adaptive immunity: care for the community. Nature *445*, 153.

Million, M., Angelakis, E., Paul, M., Armougom, F., Leibovici, L., and Raoult, D. (2012). Comparative meta-analysis of the effect of *Lactobacillus* species on weight gain in humans and animals. Microb. Pathog. *53*, 100–108.

Miyauchi, S., Hirasawa, A., Ichimura, A., Hara, T., and Tsujimoto, G. (2010). New frontiers in gut nutrient sensor research: free fatty acid sensing in the gastrointestinal tract. J. Pharmacol. Sci. *112*, 19–24.

Mokdad, A.H., Ford, E.S., Bowman, B.A., Dietz, W.H., Vinicor, F., Bales, V.S., and Marks, J.S. (2003). Prevalence of obesity, diabetes, and obesity-related health risk factors, 2001. JAMA *289*, 76–79.

Muccioli, G.G., Naslain, D., Backhed, F., Reigstad, C.S., Lambert, D.M., Delzenne, N.M., and Cani, P.D. (2010). The endocannabinoid system links gut microbiota to adipogenesis. Mol. Syst. Biol. *6*, 392.

Musso, G., Gambino, R., and Cassader, M. (2010). Obesity, diabetes, and gut microbiota: the hygiene hypothesis expanded? Diabetes Care *33*, 2277–2284.

Myronovych, A., Kirby, M., Ryan, K.K., Zhang, W., Jha, P., Setchell, K.D., Dexheimer, P.J., Aronow, B., Seeley, R.J., and Kohli, R. (2014). Vertical sleeve gastrectomy reduces hepatic steatosis while increasing serum bile acids in a weight-loss-independent manner. Obesity (Silver Spring) *22*, 390–400.

Nadal, I., Santacruz, A., Marcos, A., Warnberg, J., Garagorri, M., Moreno, L.A., Martin-Matillas, M., Campoy, C., Marti, A., Moleres, A., *et al.* (2009). Shifts in clostridia, bacteroides and immunoglobulin-coating fecal bacteria associated with weight loss in obese adolescents. Int. J. Obes. (Lond.) 33, 758–767.

Naito, E., Yoshida, Y., Makino, K., Kounoshi, Y., Kunihiro, S., Takahashi, R., Matsuzaki, T., Miyazaki, K., and Ishikawa, F. (2011). Beneficial effect of oral administration of *Lactobacillus* casei strain Shirota on insulin resistance in diet-induced obesity mice. J. Appl. Microbiol. *110*, 650–657.

Naslund, E., and Hellstrom, P.M. (1998). Glucagon-like peptide-1 in the pathogenesis of obesity. Drug News. Perspect. *11*, 92–97.

Neish, A.S. (2009). Microbes in gastrointestinal health and disease. Gastroenterology *136*, 65–80.

Ogawa, A., Kadooka, Y., Kato, K., Shirouchi, B., and Sato, M. (2014). *Lactobacillus* gasseri SBT2055 reduces postprandial and fasting serum non-esterified fatty acid levels in Japanese hypertriacylglycerolemic subjects. Lipids Health. Dis. *13*, 36.

O'Hara, A.M., and Shanahan, F. (2006). The gut flora as a forgotten organ. EMBO Rep. *7*, 688–693.

Parnell, J.A., and Reimer, R.A. (2010). Effect of prebiotic fibre supplementation on hepatic gene expression and serum lipids: a dose–response study in JCR:LA-cp rats. Br. J. Nutr. *103*, 1577–1584.

Parnell, J.A., and Reimer, R.A. (2009). Weight loss during oligofructose supplementation is associated with decreased ghrelin and increased peptide YY in overweight and obese adults. Am. J. Clin. Nutr. *89*, 1751–1759.

Penders, J., Thijs, C., Vink, C., Stelma, F.F., Snijders, B., Kummeling, I., van den Brandt, P.A., and Stobberingh, E.E. (2006). Factors influencing the composition of the intestinal microbiota in early infancy. Pediatrics *118*, 511–521.

Peters, H.P., Boers, H.M., Haddeman, E., Melnikov, S.M., and Qvyjt, F. (2009). No effect of added beta-glucan or of fructooligosaccharide on appetite or energy intake. Am. J. Clin. Nutr. *89*, 58–63.

Petschow, B., Dore, J., Hibberd, P., Dinan, T., Reid, G., Blaser, M., Cani, P.D., Degnan, F.H., Foster, J., Gibson, G., *et al.* (2013). Probiotics, prebiotics, and the host microbiome: the science of translation. Ann. N. Y. Acad. Sci. *1306*, 1–17.

Qin, J., Li, R., Raes, J., Arumugam, M., Burgdorf, K.S., Manichanh, C., Nielsen, T., Pons, N., Levenez, F., Yamada, T., *et al.* (2010). A human gut microbial gene catalogue established by metagenomic sequencing. Nature *464*, 59–65.

Ranganath, L.R., Beety, J.M., Morgan, L.M., Wright, J.W., Howland, R., and Marks, V. (1996). Attenuated GLP-1 secretion in obesity: cause or consequence? Gut *38*, 916–919.

Ravussin, Y., Koren, O., Spor, A., LeDuc, C., Gutman, R., Stombaugh, J., Knight, R., Ley, R.E., and Leibel, R.L. (2012). Responses of gut microbiota to diet composition and weight loss in lean and obese mice. Obesity (Silver Spring) *20*, 738–747.

Reinhardt, C., Reigstad, C.S., and Backhed, F. (2009). Intestinal microbiota during infancy and its implications for obesity. J. Pediatr. Gastroenterol. Nutr. *48*, 249–256.

Respondek, F., Gerard, P., Bossis, M., Boschat, L., Bruneau, A., Rabot, S., Wagner, A., and Martin, J.C. (2013). Short-chain fructo-oligosaccharides modulate intestinal microbiota and metabolic parameters of humanized gnotobiotic diet induced obesity mice. PLoS One *8*, e71026.

Respondek, F., Swanson, K.S., Belsito, K.R., Vester, B.M., Wagner, A., Istasse, L., and Diez, M. (2008). Short-chain fructooligosaccharides influence insulin sensitivity and gene expression of fat tissue in obese dogs. J. Nutr. *138*, 1712–1718.

Ridlon, J.M., Kang, D.J., and Hylemon, P.B. (2006). Bile salt biotransformations by human intestinal bacteria. J. Lipid Res. *47*, 241–259.

Ritze, Y., Bardos, G., Claus, A., Ehrmann, V., Bergheim, I., Schwiertz, A., and Bischoff, S.C. (2014). *Lactobacillus rhamnosus* GG Protects against Non-Alcoholic Fatty Liver Disease in Mice. PLoS One *9*, e80169.

Roberfroid, M., Gibson, G.R., Hoyles, L., McCartney, A.L., Rastall, R., Rowland, I., Wolvers, D., Watzl, B., Szajewska, H., Stahl, B., *et al.* (2010). Prebiotic effects: metabolic and health benefits. Br. J. Nutr. *104 Suppl. 2*, S1–63.

Roberfroid, M.B. (1999). Caloric value of inulin and oligofructose. J. Nutr. *129*, 1436S-7S.

Roberfroid, M.B., and Delzenne, N.M. (1998). Dietary fructans. Annu. Rev. Nutr. *18*, 117–143.

Rosenbaum, M., and Leibel, R.L. (2010). Adaptive thermogenesis in humans. Int. J. Obes. (Lond.) *34* (Suppl. 1), S47–55.

Russo, F., Linsalata, M., Clemente, C., Chiloiro, M., Orlando, A., Marconi, E., Chimienti, G., and Riezzo, G. (2012). Inulin-enriched pasta improves intestinal permeability and modifies the circulating levels of zonulin and glucagon-like peptide 2 in healthy young volunteers. Nutr. Res. *32*, 940–946.

Samuel, B.S., Shaito, A., Motoike, T., Rey, F.E., Backhed, F., Manchester, J.K., Hammer, R.E., Williams, S.C., Crowley, J., Yanagisawa, M., and Gordon, J.I. (2008). Effects of the gut microbiota on host adiposity are modulated by the short-chain fatty-acid binding G protein-coupled receptor, Gpr41. Proc. Natl. Acad. Sci. U.S.A. *105*, 16767–16772.

Schaafsma, G., Meuling, W.J., van Dokkum, W., and Bouley, C. (1998). Effects of a milk product, fermented by *Lactobacillus acidophilus* and with fructo-oligosaccharides added, on blood lipids in male volunteers. Eur. J. Clin. Nutr. *52*, 436–440.

Scherer, T., and Buettner, C. (2009). The dysregulation of the endocannabinoid system in diabesity-a tricky problem. J. Mol. Med. (Berl.). *87*, 663–668.

Schwiertz, A., Taras, D., Schafer, K., Beijer, S., Bos, N.A., Donus, C., and Hardt, P.D. (2010). Microbiota and SCFA in lean and overweight healthy subjects. Obesity (Silver Spring) *18*, 190–195.

Sekirov, I., Russell, S.L., Antunes, L.C., and Finlay, B.B. (2010). Gut microbiota in health and disease. Physiol. Rev. *90*, 859–904.

Sharma, A.M., and Padwal, R. (2010). Obesity is a sign – over-eating is a symptom: an aetiological framework for the assessment and management of obesity. Obes. Rev. *11*, 362–370.

Sommer, F., and Backhed, F. (2013). The gut microbiota – masters of host development and physiology. Nat. Rev. Microbiol. *11*, 227–238.

Takemura, N., Okubo, T., and Sonoyama, K. (2010). *Lactobacillus plantarum* strain No. 14 reduces adipocyte size in mice fed high-fat diet. Exp. Biol. Med. *235*, 849–856.

Tannock, G.W. (2004). A special fondness for lactobacilli. Appl. Environ. Microbiol. *70*, 3189–3194.

Taranto, M.P., Medici, M., Perdigon, G., Ruiz Holgado, A.P., and Valdez, G.F. (1998). Evidence for Hypocholesterolemic Effect of *Lactobacillus reuteri* in Hypercholesterolemic Mice. J. Dairy Sci. *81*, 2336–2340.

Thomas, C., Gioiello, A., Noriega, L., Strehle, A., Oury, J., Rizzo, G., Macchiarulo, A., Yamamoto, H., Mataki, C., Pruzanski, M., *et al.* (2009). TGR5-mediated bile acid sensing controls glucose homeostasis. Cell. Metab. *10*, 167–177.

Tilg, H., Moschen, A.R., and Kaser, A. (2009). Obesity and the microbiota. Gastroenterology *136*, 1476–1483.

Tolhurst, G., Heffron, H., Lam, Y.S., Parker, H.E., Habib, A.M., Diakogiannaki, E., Cameron, J., Grosse, J., Reimann, F., and Gribble, F.M. (2012). Short-chain fatty acids stimulate glucagon-like peptide-1 secretion via the G-protein-coupled receptor FFAR2. Diabetes *61*, 364–371.

Tremaroli, V., Kovatcheva-Datchary, P., and Backhed, F. (2010). A role for the gut microbiota in energy harvesting? Gut *59*, 1589–1590.

Turnbaugh, P.J. (2012). Microbiology: fat, bile and gut microbes. Nature *487*, 47–48.

Turnbaugh, P.J., Hamady, M., Yatsunenko, T., Cantarel, B.L., Duncan, A., Ley, R.E., Sogin, M.L., Jones, W.J., Roe, B.A., Affourtit, J.P., *et al.* (2009). A core gut microbiome in obese and lean twins. Nature *457*, 480–484.

Turnbaugh, P.J., Ley, R.E., Mahowald, M.A., Magrini, V., Mardis, E.R., and Gordon, J.I. (2006). An obesity-associated gut microbiome with increased capacity for energy harvest. Nature *444*, 1027–1031.

van Baarlen, P., Troost, F., van der Meer, C., Hooiveld, G., Boekschoten, M., Brummer, R.J., and Kleerebezem, M. (2011). Human mucosal *in vivo* transcriptome responses to three lactobacilli indicate how probiotics may modulate human cellular pathways. Proc. Natl. Acad. Sci. U.S.A. *108 Suppl. 1*, 4562–4569.

van Meer, H., Boehm, G., Stellaard, F., Vriesema, A., Knol, J., Havinga, R., Sauer, P.J., and Verkade, H.J. (2008). Prebiotic oligosaccharides and the enterohepatic circulation of bile salts in rats. Am. J. Physiol. Gastrointest. Liver Physiol. *294*, G540–7.

Verhoef, S.P., Meyer, D., and Westerterp, K.R. (2011). Effects of oligofructose on appetite profile, glucagon-like peptide 1 and peptide YY3–36 concentrations and energy intake. Br. J. Nutr. *106*, 1757–1762.

Vilsboll, T., Krarup, T., Deacon, C.F., Madsbad, S., and Holst, J.J. (2001). Reduced postprandial concentrations of intact biologically active glucagon-like peptide 1 in type 2 diabetic patients. Diabetes *50*, 609–613.

Vrieze, A., Van Nood, E., Holleman, F., Salojarvi, J., Kootte, R.S., Bartelsman, J.F., Dallinga-Thie, G.M., Ackermans, M.T., Serlie, M.J., Oozeer, R., *et al.* (2012). Transfer of intestinal microbiota from lean donors increases insulin sensitivity in individuals with metabolic syndrome. Gastroenterology *143*, 913–6.e7.

Vulevic, J., Juric, A., Tzortzis, G., and Gibson, G.R. (2013). A mixture of trans-galactooligosaccharides reduces markers of metabolic syndrome and modulates the faecal microbiota and immune function of overweight adults. J. Nutr. *143*, 324–331.

Wang, Y., Xu, N., Xi, A., Ahmed, Z., Zhang, B., and Bai, X. (2009). Effects of *Lactobacillus plantarum* MA2 isolated from Tibet kefir on lipid metabolism and intestinal microflora of rats fed on high-cholesterol diet. Appl. Microbiol. Biotechnol. *84*, 341–347.

Watanabe, M., Houten, S.M., Mataki, C., Christoffolete, M.A., Kim, B.W., Sato, H., Messaddeq, N., Harney, J.W., Ezaki, O., Kodama, T., *et al.* (2006). Bile acids induce energy expenditure by promoting intracellular thyroid hormone activation. Nature *439*, 484–489.

WHO. World Health Organization Fact Sheet – Obesity and Overweight. http://www.who.int/mediacentre/factsheets/fs311/en/index.html.

Woodard, G.A., Encarnacion, B., Downey, J.R., Peraza, J., Chong, K., Hernandez-Boussard, T., and Morton, J.M. (2009). Probiotics improve outcomes after Roux-en-Y gastric bypass surgery: a prospective randomized trial. J. Gastrointest. Surg. *13*, 1198–1204.

Yadav, H., Lee, J.H., Lloyd, J., Walter, P., and Rane, S.G. (2013). Beneficial metabolic effects of a probiotic via butyrate-induced GLP-1 hormone secretion. J. Biol. Chem. *288*, 25088–25097.

Yin, Y.-., Yu, Q.-., Fu, N., Liu, X.-. and Lu, F.-. (2010). Effects of four Bifidobacteria on obesity in high-fat diet induced rats. World Journal of Gastroenterology *16*, 3394–3401.

Yoo, S.R., Kim, Y.J., Park, D.Y., Jung, U.J., Jeon, S.M., Ahn, Y.T., Huh, C.S., McGregor, R., and Choi, M.S. (2013). Probiotics *L. plantarum* and L. curvatus in combination alter hepatic lipid metabolism and suppress diet-induced obesity. Obesity (Silver Spring) *21*, 2571–2578.

Yun, S.I., Park, H.O., and Kang, J.H. (2009). Effect of *Lactobacillus* gasseri BNR17 on blood glucose levels and body weight in a mouse model of type 2 diabetes. J. Appl. Microbiol. *107*, 1681–1686.

Zeng, H., Liu, J., Jackson, M.I., Zhao, F.Q., Yan, L., and Combs, G.F.,Jr. (2013). Fatty liver accompanies an increase in *lactobacillus* species in the hind gut of C57BL/6 mice fed a high-fat diet. J. Nutr. *143*, 627–631.

The Role of the Gut Microbiota in Brain Function

25

Julia König, John-Peter Ganda Mall, Ignacio Rangel, Hanna Edebol and Robert-Jan Brummer

Abstract

The intestinal microbiota forms a complex ecosystem that has an important impact on our health, and an increasing number of disorders are associated with disturbances in this ecosystem. There is growing evidence that even brain function can be affected by an aberrant gut microbiota and that the bidirectional signalling along the microbe–gut–brain axis plays a significant role in the well-being of the gut as well as the brain. In this chapter, mechanistic pathways explaining how the gut microbiota can affect brain function are described. In addition, its role is elucidated in relation to disorders such as anxiety, depression, autism spectrum disorder, Parkinson's disease and Alzheimer's disease, and possible beneficial action of prebiotics and probiotics is discussed.

Introduction to microbe–gut–brain interactions

Food compounds, their digestive products and microbiota are the main intraluminal constituents of the human intestine. Nutrition- and specifically microbe–gut interactions have a substantial and clinically relevant impact on the development of the immune system and affect intestinal barrier function throughout life with consequences for e.g. the resistance to pathogenic bacteria, gut inflammation and abdominal complaints. This has major implications for the prevention and therapy of a variety of 'modern' diseases like irritable bowel syndrome (IBS), inflammatory bowel disease (IBD) and other non-intestinal disorders like allergy, type-1 diabetes and obesity and its associated metabolic disorders (OAMD), for all of which the incidence has dramatically increased during the last decades.

In addition to the role of nutrition and microbiota in primary intestinal disorders as well as the role of the brain in affecting gut function, there is increasing evidence that certain aspects of brain function are affected by gut function, which in turn is affected by its nutritional and microbial composition. This bidirectional relationship between gut and brain function is strengthened by findings of an aberrant gut microbial composition in IBS, as well as in major depression and autism spectrum disorder.

It has become evident that even behaviour can be affected by the intestinal microbiota (Bravo *et al.*, 2012; Cryan and Dinan, 2012). This has been nicely demonstrated by a study of Bercik *et al.* (2011), in which the gut microbiota of mice belonging to the timid and less explorative strain BALB/c was exchanged with the microbiota of highly explorative NIH Swiss mice. This resulted in a more explorative behaviour of the BALB/c mice, similar to that of the donor mice. Brain activity in healthy women was modulated after they consumed a fermented milk product (containing *Bifidobacterium animalis* subsp. *lactis* CNCM I-2494, *Streptococcus thermophiles* CNCM I-1630, *Lactobacillus bulgaricus* CNCM I-1632 and I-1519, and *Lactobacillus lactis* subsp. *lactis* CNCM I-1631). Its ingestion reduced task-related brain responses and altered resting–state networks, thus successfully demonstrating an effect on gut–brain communication in humans (Tillisch *et al.*, 2013).

The integrated model of gut–brain signalling has five main components: (1) the gut microbiota and its metabolites, (2) the intestinal epithelium providing the barrier between the luminal contents and the intestine, (3) the enteric nervous system (ENS) whose sensory neurons, interneurons and motor neurons regulate blood flow, peristalsis and secretion of water and electrolytes, (4) the intermediary metabolism including protein, carbohydrate and fat metabolism, and (5) the brain (see Fig. 25.1).

Important mediators of the bidirectional signalling between the gut and the brain include serotonin (5-HT) and other mono-aminergic, opioid and endocannabinoid compounds, the autonomic nervous system, the hypothalamic–pituitary axis (HPA), gut hormones, cytokines, and other gut-derived metabolic signalling molecules such as growth factors (Kilkens *et al.*, 2004a,b, 2005; Labus *et al.*, 2011).

Mechanistic pathways

The gut microbiota can affect our brain function through various mechanisms, comprising direct action via microbe–host interaction or indirect action mediated by microbial metabolites.

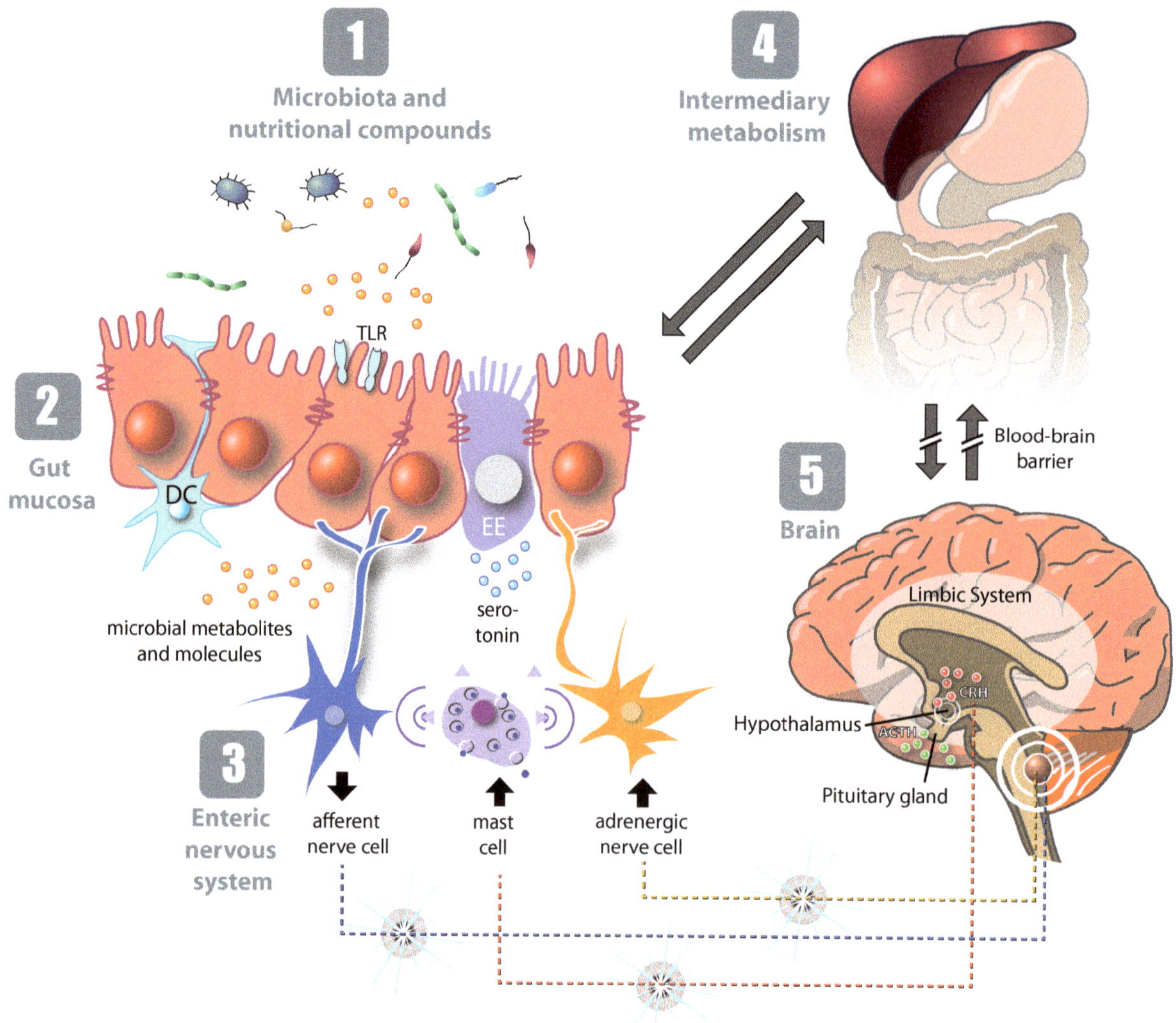

Figure 25.1 Schematic model of gut–brain signalling representing five components, with a central role of host–microbe interaction and intestinal barrier function.

Direct effects of gut microbiota on the nervous system and vice versa

Some bacteria are able to directly activate the vagus nerve, as demonstrated by a study from Perez-Burgos *et al.* (Perez-Burgos *et al.*, 2013). Luminal infusion of *Lactobacillus rhamnosus* (JB-1) into the jejunum of mice led to activation of vagal afferents, shown by an increase in their firing frequency. The strain *Lactobacillus salivarius* UCC4331, on the other hand, had no effect. In vagotomised mice, the neuronal effects of the probiotic were abolished (Bravo *et al.*, 2011; Perez-Burgos *et al.*, 2013). The administration of *Lactobacillus reuteri* ATCC 23272 enhanced the excitability of enteric neurons in the colon of rats (Kunze *et al.*, 2009).

It has been shown that several intestinal-derived *Lactobacillus* and *Bifidobacterium* strains are able to produce the neurotransmitter γ-aminobutyric acid (GABA) from glutamate in culture (Barrett *et al.*, 2012; Higuchi *et al.*, 1997). Other *in vitro* studies demonstrated that the stress hormone noradrenaline stimulated the growth of pathogenic bacteria (Freestone *et al.*, 1999; Freestone *et al.*, 2000; O'Donnell *et al.*, 2006), suggesting a direct interaction between the gut microbiota and the nervous system.

Metabolic pathways

The metabolism of the gut microbiota is closely connected to the metabolic processes of the host, and many of the physiological effects of the microbiota can be ascribed to their fermentation end products of complex carbohydrates and proteins reaching the colon. These include mainly the short-chain fatty acids (SCFAs) acetate, propionate, and butyrate, branched-chain fatty acids (isovalerate, isobutyrate), hydrogen, carbon dioxide, methane, ammonia and hydrogen sulfide (Blaut and Clavel, 2007).

Acetate is the most abundantly produced SCFA in the colon and its production is a common feature of most gut microbiota members. Propionate formation is mainly restricted to Bacteroidetes, *Clostridium* cluster IX (*Veillonella*), *Clostridium* cluster XI (*Megasphaera*), and Actinobacteria (*Propionibacterium*). Butyrate production is generally restricted to some *Clostridium* clusters (IV and XIVa) from the Firmicutes phylum. An important butyrate producer is *Faecalibacterium prausnitzii*. It is one of the most abundant bacteria in faecal samples of healthy adults, and its prevalence is altered in many intestinal disorders (Miquel *et al.*, 2013).

The complex network of metabolic processes taking place among the different members of the microbiota affects the availability of certain molecules. For example, it was shown that *Eubacterium hallii* and *Anaerostipes caccae* strains were able to produce butyrate using lactate, which in turn was synthetized by a *B. adolescentis* strain from starch (Duncan *et al.*, 2004). This nutritional interdependence between two or several species is known as metabolic cross-feeding. Loss of certain strains can disturb this interaction and change the abundance of certain SCFAs, and subsequently modify their interaction with the host.

Butyrate is probably the most extensively studied SCFA, due to its pleiotropic effects on metabolism, immune function and epithelial barrier (Hamer *et al.*, 2008). It is an important energy source for colon epithelial cells and has protective effects on colon mucosa function, for example via modulating oxidative stress (Hamer *et al.*, 2009).

The physiological effects of SCFAs seem to be closely associated with the activation of the G-protein coupled receptors GPR41 and GPR43 (Tazoe *et al.*, 2008). These receptors have been detected in immune cells such as macrophages, dendritic cells and mast cells (Karaki *et al.*, 2006, 2008), and it is hypothesized that they could play a role in the indirect effect of the gut microbiota on host health via their metabolites (Tazoe *et al.*, 2008).

There is increasing evidence that SCFAs and other microbial metabolic compounds do not only have nutritional effects, but are also involved in the functioning of the nervous system, either by direct contact with the enteric nervous system or via the vagus nerve, and maybe even directly on the central nervous system after passing the blood brain barrier. It was shown in rats that SCFAs could stimulate colonic motility via the release of serotonin from enterochromaffin cells, which in turn stimulated receptors on the vagal sensory fibres (Fukumoto *et al.*, 2003).

Studies in mice have shown that administration of specific probiotic strains which produce conjugated linoleic acid affected brain fatty acid composition (Wall *et al.*, 2012). Feeding of n-3 polyunsaturated fatty acids to juvenile pigs influenced the fat composition of their brains (Dullemeijer *et al.*, 2008). It is tempting to hypothesize that also in humans, nutrition itself or its microbial fermentation products could have a direct effect on the brain composition and thus brain function.

Gut microbiota, the intestinal barrier and brain function

An important mediator/connector in the bidirectional signalling between the gut microbiota and the brain is the intestinal barrier. A dysfunctional barrier is involved in various disorders of brain function, and studies suggest that it can be directly affected by the enteric nervous system (ENS) (Cheadle *et al.*, 2013; Neunlist *et al.*, 2003).

Formation of the intestinal barrier

The intestinal barrier is composed of the mucus layer, the epithelial layer with its tight junction proteins and the lamina propria. Also the lumen itself provides protection from pathogens as it contains, depending on the localization along the gastrointestinal tract, gastric acid or biliary and pancreatic fluids that can degrade bacteria and antigens (Peterson *et al.*, 1989; Rubinstein *et al.*, 1985; Sung *et al.*, 1993). The commensal bacteria can inhibit the colonization of pathogens by competitive adhesion, through pH modification and production of bacteriocins (antimicrobial substances) (Corr *et al.*, 2009; Davila *et al.*, 2013).

The mucus layer is produced by the so-called goblet cells. It shields the epithelium from physical and chemical damage and protects against the large abundance of bacteria residing in the colon, making it especially important to that part of the intestine (Johansson *et al.*, 2011). The sticky property of the mucus is gained by glycosylation of the mucus-forming protein MUC2. The small intestine contains only a low number of bacteria and thus only a single porous layer of mucus to allow a more efficient absorption of nutrients. The mucus in the colon consists of two layers: one dense form adjacent to the epithelium and one loose form in contact with the lumen (Atuma *et al.*, 2001). The MUC2 proteins in the outer layer of the colon mucus have attachment sites to which bacteria can adhere, thus inhibiting them from reaching the epithelium and keeping the inner layer mostly devoid of bacteria (Johansson *et al.*, 2008, 2011; Martens *et al.*, 2009). The outer mucus layer also serves as an energy source for bacteria due to its large abundance of carbohydrates (Martens *et al.*, 2009). The inner dense layer has the consistency of a gel and contains enzymes, antimicrobial peptides and immunoglobulin IgA, which act together as a defence for the epithelium (Sellers and Morton, 2014).

The lamina propria is made up of loose connective tissue supporting the overlying epithelium. It contains capillaries, lymphatic vessels, and a large amount of immune cells. Both cells of the innate immune system (such as macrophages, dendritic cells and mast cells) and the adaptive immune system (T-cells and IgA producing B-cells) contribute to maintain immune homeostasis (Farache *et al.*, 2013).

Tight junctions

While many cells of the epithelial layer can produce factors that aid in mucosal defence, one of the most important features of the intestinal barrier function is its ability to regulate the paracellular permeability through its tight junction proteins. Tight junctions are multiple protein complexes consisting of transmembrane and cytosolic proteins that respond to various external stimuli and selectively regulate paracellular permeability of ions, solutes and water, and thus also maintain cell polarity (Suzuki, 2013).

The four integral transmembrane proteins claudin, occludin, junctional adhesion molecule and tricellulin have been identified so far. The extracellular domains of the proteins form a link and a barrier with adjacent cells, while the intracellular domains are attached to cytosolic scaffold proteins. These scaffold proteins anchor the transmembrane proteins to the actin cytoskeleton, which enables tight junction regulation through cytoskeletal contraction (Fanning *et al.*, 1998; Furuse *et al.*, 1994, 1998; Itoh *et al.*, 1999). Phosphorylation of the myosin light chain, a subunit of the motor protein myosin, by myosin light chain kinase (MLCK) leads to contraction of the cytoskeleton and thus to opening of the barrier and to increased permeability (Madara *et al.*, 1987; Turner *et al.*, 1997).

Gut microbiota and tight junctions

One of the mechanisms how bacteria can affect intestinal barrier function is through modulation of tight junction proteins. Several *in vitro* and *in vivo* animal studies have shown an effect of probiotics on expression or localization changes of tight junction proteins (Madsen, 2012). The probiotic strain *Lactobacillus plantarum* MB452 increased tight junction integrity in Caco-2 cells as measured by increased gene expression of occludin, ZO-1 and ZO-2 and transepithelial electrical resistance (Anderson *et al.*, 2010). Further strains shown to improve epithelial barrier function using similar models include *Streptococcus thermophilus* ATCC19258) and *Lactobacillus acidophilus* (TCC4356)(Resta-Lenert and Barrett, 2003) as well as *Lactobacillus rhamnosus* CGMCC 1.3724 and *Bifidobacterium lactis* CNCM I-3446 (Mathias *et al.*, 2010).

An effect of probiotics on barrier function has also been shown in humans. Administration of the strain *L. plantarum* WCFS1 into the duodenum of healthy volunteers via a feeding catheter resulted in increased localization of ZO-1 and occludin into the tight junctions of the intestinal epithelium. Further cell culture experiments suggested TLR-2 signalling as an important regulator of these processes (Karczewski *et al.*, 2010).

As will be discussed later, also bacterial metabolites play an important role in the mechanisms of action of gut microbiota. The bacterial fermentation end product butyrate increased assembly of tight junction proteins and transepithelial resistance in Caco-2 cells (Peng *et al.*, 2009) and reduced bacterial translocation in stressed epithelial cells (Lewis *et al.*, 2010).

The beneficial effects of probiotic strains on barrier function are probably also mediated by indirect effects, such as protection against oxidative stress or cytokine-induced damage, up-regulation of heat-shock proteins and anti-apoptotic effects (Madsen, 2012). It has been shown that bacteria do not necessarily need to be intact, but that also bacterial DNA, released factors, cell wall components and metabolites can be effective (Madsen, 2012).

The gut–brain axis and intestinal barrier function

The so-called enteric glial cells (EGCs) as part of the ENS span the gastrointestinal wall and are in close proximity to the intestinal epithelial cells. This enables them to actively transmit and receive signals between the epithelium and the ENS through neurotransmitters (de Jonge, 2013).

In an experiment with transgenic mice, the ablation of EGCs resulted in impaired mucosal barrier function leading to inflammation. EGC-derived GSNO (S-nitrosoglutathione) attenuated this inflammation through increased expression of ZO-1 and occludin (Savidge *et al.*, 2007). Another study used a cytokine mixture (TNF-α, IFN-γ and IL-1β) to induce tight junction disruption in Caco-2 cells, which was accompanied by an altered localization of ZO-1 and occludin away from the cell contact points. Co-culture with EGCs, or GSNO alone, restored the localization of ZO-1 and occludin to their original location and normalized barrier function (Cheadle *et al.*, 2013). The same study also showed that the cytokine mixture increased MLCK activity, leading to increased phosphorylation of the myosin light chain followed by opening of the tight junctions through enhanced contraction of the actin cytoskeleton.

Increased intestinal permeability and gut inflammation are associated with neuroinflammation. The vagal afferent and efferent nerves mediate bi-directional information between the gut and the brain and can also contribute to immune homeostasis. It has been found that vagal efferent fibres are able to inhibit the release of pro-inflammatory cytokines and protect against inflammation through the cholinergic anti-inflammatory pathway (Ji *et al.*, 2014). Cultured vagal afferent nerves are activated by LPS and TNF-alpha, a mechanism suggested to play a role in the detection of pathogenic bacterial colonization or gut inflammation by the nervous system (Riley *et al.*, 2013). Systemic injection of LPS has been shown to increase IL-1β protein levels in the dorsal vagal complex as well as in different parts of the brain (Hansen *et al.*, 2000).

Various disorders of brain function related to gut microbiota

An increasing number of disorders of brain function are being associated with the gut ecosystem. Especially a disturbed composition and/or an impaired intestinal barrier function seem to play a major role.

Anxiety, stress and depression

An essential component of the communication between the brain and the gut is the hypothalamic-pituitary-axis (HPA-axis). Any substantial level of stress, including psychological stress, activates this HPA-axis, which starts with the release of corticotrophin-releasing hormone (CRH). CRH increases intestinal permeability via mast cell activation (Wallon *et al.*, 2008). This increased intestinal permeability can result in immune activation and in lipopolysaccharide (LPS) translocation into the blood (endotoxaemia), as it was demonstrated in rats exposed to acute psychological stress (Ait-Belgnaoui *et al.*, 2012). These rats also showed increased mRNA expression of pro-inflammatory cytokines in the hypothalamus.

Nowadays it is widely accepted that depression is an inflammatory state. Studies in mice have shown that a systemic and CNS-associated increase in pro-inflammatory cytokines can lead to depressive-like behaviour (Kubera *et al.*, 2011). Also in humans, patients with major depression have been found to have higher levels of hormones associated with HPA activation, including CRH (Heim *et al.*, 2008). In addition, significantly increased serum levels of IgM and IgA towards LPS have been detected in chronic depressed patients compared to healthy controls (Maes *et al.*, 2012). This LPS originated from Gram-negative bacteria belonging to the normal commensal gut microbiota, hence showing that there is an increased permeability in the gut of depressed patients facilitating bacterial translocation (Maes *et al.*, 2012). In an earlier study, the infusion of low dose *Salmonella abortus equi* endotoxins increased anxiety and depressive mood in healthy participants, possibly as a result of increased cytokine release (mainly TNF-α and IL-6) (Reichenberg *et al.*, 2001).

The relationship between inflammatory cytokines and decreased mood is supported by the fact that depression is a common side effect of chronic hepatitis treatment with INF-α (Udina *et al.*, 2012). In addition, depression is more prevalent in IBD patients compared to healthy subjects, and shows a correlation with the proinflammatory cytokine response (Filipovic and Filipovic, 2014; Goodhand *et al.*, 2012).

It is an ongoing debate whether depression is caused by elevated inflammatory activity or if this inflammatory activity is a result of depression. It is plausible that some kind of trauma or chronic stress leads to overactivation of the HPA axis, resulting in increased intestinal permeability towards bacteria and/or LPS. This could lead to a vicious circle of increased immune activation and inflammatory activity and further translocation of bacteria and/or LPS, as several cytokines are known to damage the intestinal barrier. For instance, IL-1β and TNF-α both increase the activity of MLCK and thus cause opening of the tight junctions (Al-Sadi *et al.*, 2013; Ye *et al.*, 2006).

The use of probiotics could be a novel therapeutic approach for the treatment of depression. In the study investigating rats exposed to acute psychological stress by Ait-Belgnaoui *et al.* (2012), mentioned above, oral administration of the probiotic *Lactobacillus farciminiis* CIP 103136 resulted in reduced HPA axis response to stress through decrease in corticosterone and CRH expression. This probiotic strain also decreased colonic hyperpermeability by blocking the activity of MLCK and thus prevented LPS translocation (Ait-Belgnaoui *et al.*, 2006). Administration of *L. rhamnosus* JB-1 reduced anxiety and depression-related behaviour in mice by modulating GABA receptor expression in the brain and lowering corticosterone levels (Bravo *et al.*, 2012). These effects were vagal nerve dependent.

There are also a few human studies that showed promising results of probiotics regarding mood disorders. Administration of the strains *Lactobacillus helveticus* R0052 and *Bifidobacterium longum* R0175 showed beneficial effects on psychological distress (shown by lower scores in the Hopkins symptom checklist-90 scale and the hospital and depression scale) and reduced cortisol levels compared to the placebo group in a double-blind, placebo-controlled randomized trial including 55 healthy human volunteers (30 days of administration) (Messaoudi *et al.*, 2011). In addition, consumption of a milk drink containing *Lactobacillus casei* Shirota improved mood in subjects with initially poor mood, while it had no effect when considering the entire study population (n=132) (Benton *et al.*, 2007). This could be explained by a generally high baseline mood in which further improvements were unlikely. The same strain led to a decrease in anxiety symptoms in patients with chronic fatigue syndrome, however, an association with enhanced bowel function and/or reduced abdominal complaints was not assessed (Rao *et al.*, 2009).

These studies support the important link between gut and brain function and the search for alternative treatment modalities, primarily directed to the intestine, for brain function disorders.

Autism spectrum disorder

There is a consensus that children with autism spectrum disorder (ASD) have a higher likelihood to suffer from gastrointestinal problems, in spite of the patients' decreased capability to express their discomfort. Furthermore, varying interpretations of their gastrointestinal discomfort hampers quantification (Adams *et al.*, 2011). Although the pathophysiology of the gastrointestinal discomfort is still unclear, ASD patients, and their first-degree relatives, were reported to have an increased intestinal permeability (de Magistris *et al.*, 2010).

Several studies have analysed the gut microbiota in children with ASD. Apart from one study (Gondalia *et al.*, 2012), an altered composition has been found both in faecal samples (Adams *et al.*, 2011; Finegold *et al.*, 2010; Kang *et al.*, 2013; Parracho *et al.*, 2005; Wang *et al.*, 2011) as well as in the ileal and caecal mucosa (Williams *et al.*, 2011; Williams *et al.*, 2012) compared to healthy control children. Due to different examination methods and small patient numbers, the results are rather inconsistent, and no specific microbial species could be linked to ASD so far. This inconsistency might be the reason why no studies have been published yet investigating potential treatments aiming at correcting aberrations in the intestinal ecosystem. A possible therapeutic option might be the administration of probiotics, which ideally would improve both gastrointestinal discomfort as well as autistic behaviour. However, even though many ASD patients take probiotics (personal communication, Robert J. Brummer, School of Health and Medical Sciences, Faculty of Medicine and Health, Örebro University, Örebro, Sweden) and animal studies have shown promising initial results, no clinical intervention trials have been published yet.

Additional evidence for the important role of the gut microbiota in ASD comes from a small pilot study investigating the effect of vancomycin, an antibiotic compound that is not absorbed in the intestine (Sandler *et al.*, 2000). Treatment with vancomycin for 8 weeks improved autistic behaviour in 8 out of 10 treated children. Unfortunately, this effect did not persist after ending the treatment. Faecal microbiota transplantation, or alternatively, the application of a mixture of selected beneficial bacteria strains (so-called synthetic stool) could be another treatment option worth to investigate in the future, but should be based on a proper hypothesis.

There is also some evidence that bacterial metabolites play a role in the pathophysiology of ASD. Propionate, a microbially produced short-chain fatty acid, induces autistic-like behaviour in mice (Macfabe, 2012). In humans, however, lower levels of propionate and other organic acids in faecal samples of children with ASD compared to healthy controls were found (Adams *et al.*, 2011). The authors of the latter study commented that an interpretation is difficult as low levels could also be due to increased absorption in the intestine caused by increased permeability or prolonged transit time because of constipation rather than less fermentation by the gut microbiota.

Parkinson's disease

Parkinson's disease is the second most common neurodegenerative disease and is characterized by region-specific loss of dopaminergic neuromelanin-containing neurons in the substantia nigra of the brain (Braak *et al.*, 2003; Lees *et al.*, 2009). Three distinct neuronal inclusions, termed Lewy bodies, pale bodies and Lewy neurites, mainly cause the neuronal degeneration. The main component of the Lewy bodies is the pre-synaptic protein α-synuclein, which is prone to form aggregates causing damage to the neurons (Braak *et al.*, 2003; Lees *et al.*, 2009).

It is generally known that Parkinson's disease patients suffer from gastrointestinal discomfort. Previously, these symptoms were seen as a consequence of intestinal motility disorder and thus as a result of the disease in the central nervous system itself. More recently, however, it was shown that nearly half of all Parkinson's disease patients report long-term constipation prior to onset of signs of motor impairment, suggesting a link between early gastrointestinal problems and later onset of Parkinson's disease (Abbott *et al.*, 2001; Kaye *et al.*, 2006; Petrovitch *et al.*, 2009; Savica *et al.*, 2009).

Forsyth *et al.* (2011) reported that Parkinson's disease patients ($n=9$) had a significantly increased intestinal permeability, as demonstrated by increased urinary sucralose values (marker for total intestinal permeability). These values correlated with increased immunostaining for *Escherichia coli* in sigmoid colon biopsies, suggesting bacterial translocation into the epithelium and lamina propria. In addition, only the colon biopsies of Parkinson's disease patients showed positive staining for α-synuclein, which correlated with staining for nitrotyrosine, a marker of oxidative stress. The authors speculated that intestinal α-synuclein aggregates could play a major role in the pathogenesis of Parkinson's disease.

Future studies might elucidate further links between gut and brain function in Parkinson's disease, and should preferably comprise investigations of the gut microbiota. Pro- and/or prebiotics that strengthen the epithelial barrier function might be beneficial to these patients and should perhaps be considered in the clinic.

Alzheimer's disease

Amyloid protein precursor (APP) is a pivotal protein in the formation of the plaque-inducing amyloid-β protein, which causes neuronal degeneration in the brain of Alzheimer's disease patients. Its expression was increased by the cytokines IL-1β and TNF-α *in vitro* (Buxbaum *et al.*, 1992; Haass *et al.*, 2012; Lahiri *et al.*, 2003). Also systemic administration of LPS in mice has been shown to increase APP expression and caused neuronal cell death in the hippocampus and the cerebral cortex of mice brains through the formation of amyloid-β protein (Lee *et al.*, 2008).

Zhang *et al.* detected significantly increased plasma levels of LPS in Alzheimer's disease patients ($n=18$) compared with healthy controls ($n=21$) (Zhang *et al.*, 2009), which might suggest that Alzheimer's disease patients suffer from elevated intestinal permeability deteriorating neuroinflammation. Pro- and/or prebiotics that have an anti-inflammatory effect or reduce bacterial and/or LPS translocation might be beneficial for these patients. This should be addressed in preclinical studies and clinical trials.

Conclusion

There is increasing evidence for bidirectional signalling along the microbe–gut–brain axis, which opens up novel treatment modalities not only for primary intestinal disorders but also for disorders of brain function such as depression, Alzheimer's disease, Parkinson's disease, and autism's spectrum disorders, and possibly for attention deficit hyperactivity disorder, although there are no data on the latter yet. These treatment modalities are directed at altering the intestinal ecosystem and hence could be realized both through primarily nutritional or bacterial modulation, i.e. using prebiotics and probiotics.

Future trends/perspectives

Based on the characterization of the intestinal ecosystem and host response more individualized treatments can be developed. These could not only be applied for the prevention and treatment of disorders but also for improving resilience of gut–brain function, e.g. better resistance/handling of naturally occurring daily life stressors.

Novel treatments including specific narrow-spectrum antibiotics or faecal microbial transplantation could be applied for more severe and irreversible aberrations of the gut ecosystem.

References

Abbott, R.D., Petrovitch, H., White, L.R., Masaki, K.H., Tanner, C.M., Curb, J.D., Grandinetti, A., Blanchette, P.L., Popper, J.S., and Ross, G.W. (2001). Frequency of bowel movements and the future risk of Parkinson's disease. Neurology *57*, 456–462.

Adams, J.B., Johansen, L.J., Powell, L.D., Quig, D., and Rubin, R.A. (2011). Gastrointestinal flora and gastrointestinal status in children with autism – comparisons to typical children and correlation with autism severity. BMC Gastroenterol. *11*, 22–230X-11–22.

Ait-Belgnaoui, A., Durand, H., Cartier, C., Chaumaz, G., Eutamene, H., Ferrier, L., Houdeau, E., Fioramonti, J., Bueno, L., and Theodorou, V. (2012). Prevention of gut leakiness by a probiotic treatment leads to attenuated HPA response to an acute psychological stress in rats. Psychoneuroendocrinology *37*, 1885–1895.

Ait-Belgnaoui, A., Han, W., Lamine, F., Eutamene, H., Fioramonti, J., Bueno, L., and Theodorou, V. (2006). *Lactobacillus* farciminis treatment suppresses stress induced visceral hypersensitivity: a possible action through interaction with epithelial cell cytoskeleton contraction. Gut *55*, 1090–1094.

Al-Sadi, R., Guo, S., Ye, D., Dokladny, K., Alhmoud, T., Ereifej, L., Said, H.M., and Ma, T.Y. (2013). Mechanism of IL-1beta modulation of intestinal epithelial barrier involves p38 kinase and activating transcription factor-2 activation. J. Immunol. *190*, 6596–6606.

Anderson, R.C., Cookson, A.L., McNabb, W.C., Park, Z., McCann, M.J., Kelly, W.J., and Roy, N.C. (2010). *Lactobacillus plantarum* MB452 enhances the function of the intestinal barrier by increasing the expression levels of genes involved in tight junction formation. BMC Microbiol. *10*, 316–2180–10–316.

Atuma, C., Strugala, V., Allen, A., and Holm, L. (2001). The adherent gastrointestinal mucus gel layer: thickness and physical state *in vivo*. Am. J. Physiol. Gastrointest. Liver Physiol. *280*, G922–9.

Barrett, E., Ross, R.P., O'Toole, P.W., Fitzgerald, G.F., and Stanton, C. (2012). gamma-Aminobutyric acid production by culturable bacteria from the human intestine. J. Appl. Microbiol. *113*, 411–417.

Benton, D., Williams, C., and Brown, A. (2007). Impact of consuming a milk drink containing a probiotic on mood and cognition. Eur. J. Clin. Nutr. *61*, 355–361.

Bercik, P., Denou, E., Collins, J., Jackson, W., Lu, J., Jury, J., Deng, Y., Blennerhassett, P., Macri, J., McCoy, K.D., Verdu, E.F., and Collins, S.M. (2011). The intestinal microbiota affect central levels of brain-derived neurotropic factor and behavior in mice. Gastroenterology *141*, 599–609, 609.e1–3.

Blaut, M., and Clavel, T. (2007). Metabolic diversity of the intestinal microbiota: implications for health and disease. J. Nutr. *137*, 751S-5S.

Braak, H., Del Tredici, K., Rub, U., de Vos, R.A., Jansen Steur, E.N., and Braak, E. (2003). Staging of brain pathology related to sporadic Parkinson's disease. Neurobiol. Aging *24*, 197–211.

Bravo, J.A., Forsythe, P., Chew, M.V., Escaravage, E., Savignac, H.M., Dinan, T.G., Bienenstock, J., and Cryan, J.F. (2011). Ingestion of *Lactobacillus* strain regulates emotional behavior and central GABA receptor expression in a mouse via the vagus nerve. Proc. Natl. Acad. Sci. U.S.A. *108*, 16050–16055.

Bravo, J.A., Julio-Pieper, M., Forsythe, P., Kunze, W., Dinan, T.G., Bienenstock, J., and Cryan, J.F. (2012). Communication between gastrointestinal bacteria and the nervous system. Curr. Opin. Pharmacol. *12*, 667–672.

Buxbaum, J.D., Oishi, M., Chen, H.I., Pinkas-Kramarski, R., Jaffe, E.A., Gandy, S.E., and Greengard, P. (1992). Cholinergic agonists and interleukin 1 regulate processing and secretion of the Alzheimer beta/A4 amyloid protein precursor. Proc. Natl. Acad. Sci. U.S.A. *89*, 10075–10078.

Cheadle, G.A., Costantini, T.W., Lopez, N., Bansal, V., Eliceiri, B.P., and Coimbra, R. (2013). Enteric glia cells attenuate cytomix-induced intestinal epithelial barrier breakdown. PLoS One *8*, e69042.

Corr, S.C., Hill, C., and Gahan, C.G. (2009). Understanding the mechanisms by which probiotics inhibit gastrointestinal pathogens. Adv. Food Nutr. Res. *56*, 1–15.

Cryan, J.F., and Dinan, T.G. (2012). Mind-altering microorganisms: the impact of the gut microbiota on brain and behaviour. Nat. Rev. Neurosci. *13*, 701–712.

Davila, A.M., Blachier, F., Gotteland, M., Andriamihaja, M., Benetti, P.H., Sanz, Y., and Tome, D. (2013). Intestinal luminal nitrogen metabolism: role of the gut microbiota and consequences for the host. Pharmacol. Res. *69*, 114–126.

de Jonge, W.J. (2013). The Gut's Little Brain in Control of Intestinal Immunity. ISRN Gastroenterol. *2013*, 630159.

de Magistris, L., Familiari, V., Pascotto, A., Sapone, A., Frolli, A., Iardino, P., Carteni, M., De Rosa, M., Francavilla, R., Riegler, G., Militerni, R., and Bravaccio, C. (2010). Alterations of the intestinal barrier in patients with autism spectrum disorders and in their first-degree relatives. J. Pediatr. Gastroenterol. Nutr. *51*, 418–424.

Dullemeijer, C., Zock, P.L., Coronel, R., Den Ruijter, H.M., Katan, M.B., Brummer, R.J., Kok, F.J., Beekman, J., and Brouwer, I.A. (2008). Differences in fatty acid composition between cerebral brain lobes in juvenile pigs after fish oil feeding. Br. J. Nutr. *100*, 794–800.

Duncan, S.H., Louis, P., and Flint, H.J. (2004). Lactate-utilizing bacteria, isolated from human faeces, that produce butyrate as a major fermentation product. Appl. Environ. Microbiol. *70*, 5810–5817.

Fanning, A.S., Jameson, B.J., Jesaitis, L.A., and Anderson, J.M. (1998). The tight junction protein ZO-1 establishes a link between the transmembrane protein occludin and the actin cytoskeleton. J. Biol. Chem. *273*, 29745–29753.

Farache, J., Zigmond, E., Shakhar, G., and Jung, S. (2013). Contributions of dendritic cells and macrophages to intestinal homeostasis and immune defense. Immunol. Cell Biol. *91*, 232–239.

Filipovic, B.R., and Filipovic, B.F. (2014). Psychiatric comorbidity in the treatment of patients with inflammatory bowel disease. World J. Gastroenterol. *20*, 3552–3563.

Finegold, S.M., Dowd, S.E., Gontcharova, V., Liu, C., Henley, K.E., Wolcott, R.D., Youn, E., Summanen, P.H., Granpeesheh, D., Dixon, D., *et al.* (2010). Pyrosequencing study of fecal microflora of autistic and control children. Anaerobe *16*, 444–453.

Forsyth, C.B., Shannon, K.M., Kordower, J.H., Voigt, R.M., Shaikh, M., Jaglin, J.A., Estes, J.D., Dodiya, H.B., and Keshavarzian, A. (2011). Increased intestinal permeability correlates with sigmoid mucosa alpha-synuclein staining and endotoxin exposure markers in early Parkinson's disease. PLoS One *6*, e28032.

Freestone, P.P., Haigh, R.D., Williams, P.H., and Lyte, M. (1999). Stimulation of bacterial growth by heat-stable, norepinephrine-induced autoinducers. FEMS Microbiol. Lett. *172*, 53–60.

Freestone, P.P., Lyte, M., Neal, C.P., Maggs, A.F., Haigh, R.D., and Williams, P.H. (2000). The mammalian neuroendocrine hormone norepinephrinesupplies iron for bacterial growth in the presence of transferrin or lactoferrin. J. Bacteriol. *182*, 6091–6098.

Fukumoto, S., Tatewaki, M., Yamada, T., Fujimiya, M., Mantyh, C., Voss, M., Eubanks, S., Harris, M., Pappas, T.N., and Takahashi, T. (2003). Short-chain fatty acids stimulate colonic transit via intraluminal 5-HT release in rats. Am. J. Physiol. Regul. Integr. Comp. Physiol. *284*, R1269–76.

Furuse, M., Itoh, M., Hirase, T., Nagafuchi, A., Yonemura, S., Tsukita, S., and Tsukita, S. (1994). Direct association of occludin with ZO-1 and its possible involvement in the localization of occludin at tight junctions. J. Cell Biol. *127*, 1617–1626.

Furuse, M., Sasaki, H., Fujimoto, K., and Tsukita, S. (1998). A single gene product, claudin-1 or -2, reconstitutes tight junction strands and recruits occludin in fibroblasts. J. Cell Biol. *143*, 391–401.

Gondalia, S.V., Palombo, E.A., Knowles, S.R., Cox, S.B., Meyer, D., and Austin, D.W. (2012). Molecular characterization of gastrointestinal microbiota of children with autism (with and without gastrointestinal dysfunction) and their neurotypical siblings. Autism Res. *5*, 419–427.

Goodhand, J.R., Wahed, M., Mawdsley, J.E., Farmer, A.D., Aziz, Q., and Rampton, D.S. (2012). Mood disorders in inflammatory bowel disease: relation to diagnosis, disease activity, perceived stress, and other factors. Inflamm. Bowel Dis. *18*, 2301–2309.

Haass, C., Kaether, C., Thinakaran, G., and Sisodia, S. (2012). Trafficking and proteolytic processing of APP. Cold Spring Harb Perspect. Med. *2*, a006270.

Hamer, H.M., Jonkers, D., Venema, K., Vanhoutvin, S., Troost, F.J., and Brummer, R.J. (2008). Review article: the role of butyrate on colonic function. Aliment. Pharmacol. Ther. *27*, 104–119.

Hamer, H.M., Jonkers, D.M., Bast, A., Vanhoutvin, S.A., Fischer, M.A., Kodde, A., Troost, F.J., Venema, K., and Brummer, R.J. (2009). Butyrate modulates oxidative stress in the colonic mucosa of healthy humans. Clin. Nutr. *28*, 88–93.

Hansen, M.K., Nguyen, K.T., Goehler, L.E., Gaykema, R.P., Fleshner, M., Maier, S.F., and Watkins, L.R. (2000). Effects of vagotomy on lipopolysaccharide-induced brain interleukin-1beta protein in rats. Auton. Neurosci. *85*, 119–126.

Heim, C., Mletzko, T., Purselle, D., Musselman, D.L., and Nemeroff, C.B. (2008). The dexamethasone/corticotropin-releasing factor test in men with major depression: role of childhood trauma. Biol. Psychiatry *63*, 398–405.

Higuchi, T., Hayashi, H., and Abe, K. (1997). Exchange of glutamate and gamma-aminobutyrate in a *Lactobacillus* strain. J. Bacteriol. *179*, 3362–3364.

Itoh, M., Furuse, M., Morita, K., Kubota, K., Saitou, M., and Tsukita, S. (1999). Direct binding of three tight junction-associated MAGUKs, ZO-1, ZO-2, and ZO-3, with the COOH termini of claudins. J. Cell Biol. *147*, 1351–1363.

Ji, H., Rabbi, M.F., Labis, B., Pavlov, V.A., Tracey, K.J., and Ghia, J.E. (2014). Central cholinergic activation of a vagus nerve-to-spleen circuit alleviates experimental colitis. Mucosal Immunol. *7*, 335–347.

Johansson, M.E., Larsson, J.M., and Hansson, G.C. (2011). The two mucus layers of colon are organized by the MUC2 mucin, whereas the outer layer is a legislator of host–microbial interactions. Proc. Natl. Acad. Sci. U.S.A. *108* (Suppl. 1), 4659–4665.

Johansson, M.E., Phillipson, M., Petersson, J., Velcich, A., Holm, L., and Hansson, G.C. (2008). The inner of the two Muc2 mucin-dependent mucus layers in colon is devoid of bacteria. Proc. Natl. Acad. Sci. U.S.A. *105*, 15064–15069.

Kang, D.W., Park, J.G., Ilhan, Z.E., Wallstrom, G., Labaer, J., Adams, J.B., and Krajmalnik-Brown, R. (2013). Reduced incidence of Prevotella and other fermenters in intestinal microflora of autistic children. PLoS One *8*, e68322.

Karaki, S., Mitsui, R., Hayashi, H., Kato, I., Sugiya, H., Iwanaga, T., Furness, J.B., and Kuwahara, A. (2006). Short-chain fatty acid receptor, GPR43, is expressed by enteroendocrine cells and mucosal mast cells in rat intestine. Cell Tissue Res. *324*, 353–360.

Karaki, S., Tazoe, H., Hayashi, H., Kashiwabara, H., Tooyama, K., Suzuki, Y., and Kuwahara, A. (2008). Expression of the short-chain fatty acid receptor, GPR43, in the human colon. J. Mol. Histol. *39*, 135–142.

Karczewski, J., Troost, F.J., Konings, I., Dekker, J., Kleerebezem, M., Brummer, R.J., and Wells, J.M. (2010). Regulation of human epithelial tight junction proteins by *Lactobacillus plantarum in vivo* and protective effects on the epithelial barrier. Am. J. Physiol. Gastrointest. Liver Physiol. *298*, G851–9.

Kaye, J., Gage, H., Kimber, A., Storey, L., and Trend, P. (2006). Excess burden of constipation in Parkinson's disease: a pilot study. Mov. Disord. *21*, 1270–1273.

Kilkens, T.O., Honig, A., Fekkes, D., and Brummer, R.J. (2005). The effects of an acute serotonergic challenge on brain-gut responses in irritable bowel syndrome patients and controls. Aliment. Pharmacol. Ther. *22*, 865–874.

Kilkens, T.O., Honig, A., Maes, M., Lousberg, R., and Brummer, R.J. (2004a). Fatty acid profile and affective dysregulation in irritable bowel syndrome. Lipids *39*, 425–431.

Kilkens, T.O., Honig, A., van Nieuwenhoven, M.A., Riedel, W.J., and Brummer, R.J. (2004b). Acute tryptophan depletion affects brain-gut responses in irritable bowel syndrome patients and controls. Gut *53*, 1794–1800.

Kubera, M., Obuchowicz, E., Goehler, L., Brzeszcz, J., and Maes, M. (2011). In animal models, psychosocial stress-induced (neuro)inflammation, apoptosis and reduced neurogenesis are associated to the onset of depression. Prog. Neuropsychopharmacol. Biol. Psychiatry *35*, 744–759.

Kunze, W.A., Mao, Y.K., Wang, B., Huizinga, J.D., Ma, X., Forsythe, P., and Bienenstock, J. (2009). *Lactobacillus reuteri* enhances excitability of colonic AH neurons by inhibiting calcium-dependent potassium channel opening. J. Cell. Mol. Med. *13*, 2261–2270.

Labus, J.S., Mayer, E.A., Jarcho, J., Kilpatrick, L.A., Kilkens, T.O., Evers, E.A., Backes, W.H., Brummer, R.J., and van Nieuwenhoven, M.A. (2011). Acute tryptophan depletion alters the effective connectivity of emotional arousal circuitry during visceral stimuli in healthy women. Gut *60*, 1196–1203.

Lahiri, D.K., Chen, D., Vivien, D., Ge, Y.W., Greig, N.H., and Rogers, J.T. (2003). Role of cytokines in the gene expression of amyloid beta-protein precursor: identification of a 5'-UTR-binding nuclear factor and its implications in Alzheimer's disease. J. Alzheimers Dis. *5*, 81–90.

Lee, J.W., Lee, Y.K., Yuk, D.Y., Choi, D.Y., Ban, S.B., Oh, K.W., and Hong, J.T. (2008). Neuro-inflammation induced by lipopolysaccharide causes cognitive impairment through enhancement of beta-amyloid generation. J. Neuroinflammation *5*, 37–2094–5–37.

Lees, A.J., Hardy, J., and Revesz, T. (2009). Parkinson's disease. Lancet *373*, 2055–2066.

Lewis, K., Lutgendorff, F., Phan, V., Soderholm, J.D., Sherman, P.M., and McKay, D.M. (2010). Enhanced translocation of bacteria across metabolically stressed epithelia is reduced by butyrate. Inflamm. Bowel Dis. *16*, 1138–1148.

Macfabe, D.F. (2012). Short-chain fatty acid fermentation products of the gut microbiome: implications in autism spectrum disorders. Microb. Ecol. Health Dis. 23, 10.3402/mehd.v23i0.19260. eCollection 2012.

Madara, J.L., Moore, R., and Carlson, S. (1987). Alteration of intestinal tight junction structure and permeability by cytoskeletal contraction. Am. J. Physiol. 253, C854–61.

Madsen, K. (2012). Enhancement of epithelial barrier function by probiotics. J. Epithelial Biol. Pharmacol. *5*, 55–59.

Maes, M., Kubera, M., Leunis, J.C., and Berk, M. (2012). Increased IgA and IgM responses against gut commensals in chronic depression: further evidence for increased bacterial translocation or leaky gut. J. Affect. Disord. *141*, 55–62.

Martens, E.C., Roth, R., Heuser, J.E., and Gordon, J.I. (2009). Coordinate regulation of glycan degradation and polysaccharide capsule biosynthesis by a prominent human gut symbiont. J. Biol. Chem. *284*, 18445–18457.

Mathias, A., Duc, M., Favre, L., Benyacoub, J., Blum, S., and Corthesy, B. (2010). Potentiation of polarized intestinal Caco-2 cell responsiveness to probiotics complexed with secretory IgA. J. Biol. Chem. *285*, 33906–33913.

Messaoudi, M., Lalonde, R., Violle, N., Javelot, H., Desor, D., Nejdi, A., Bisson, J.F., Rougeot, C., Pichelin, M., Cazaubiel, M., and Cazaubiel, J.M. (2011). Assessment of psychotropic-like properties of a probiotic formulation (*Lactobacillus* helveticus R0052 and *Bifidobacterium longum* R0175). in rats and human subjects. Br. J. Nutr. *105*, 755–764.

Miquel, S., Martin, R., Rossi, O., Bermudez-Humaran, L.G., Chatel, J.M., Sokol, H., Thomas, M., Wells, J.M., and Langella, P. (2013). *Faecalibacterium prausnitzii* and human intestinal health. Curr. Opin. Microbiol. *16*, 255–261.

Neunlist, M., Toumi, F., Oreschkova, T., Denis, M., Leborgne, J., Laboisse, C.L., Galmiche, J.P., and Jarry, A. (2003). Human ENS regulates the intestinal epithelial barrier permeability and a tight junction-associated protein ZO-1 via VIPergic pathways. Am. J. Physiol. Gastrointest. Liver Physiol. *285*, G1028–36.

O'Donnell, P.M., Aviles, H., Lyte, M., and Sonnenfeld, G. (2006). Enhancement of *in vitro* growth of pathogenic bacteria by norepinephrine: importance of inoculum density and role of transferrin. Appl. Environ. Microbiol. 72, 5097–5099.

Parracho, H.M., Bingham, M.O., Gibson, G.R., and McCartney, A.L. (2005). Differences between the gut microflora of children with autistic spectrum disorders and that of healthy children. J. Med. Microbiol. *54*, 987–991.

Peng, L., Li, Z.R., Green, R.S., Holzman, I.R., and Lin, J. (2009). Butyrate enhances the intestinal barrier by facilitating tight junction assembly via activation of AMP-activated protein kinase in Caco-2 cell monolayers. J. Nutr. *139*, 1619–1625.

Perez-Burgos, A., Wang, B., Mao, Y.K., Mistry, B., McVey Neufeld, K.A., Bienenstock, J., and Kunze, W. (2013). Psychoactive bacteria *Lactobacillus rhamnosus* (JB-1). elicits rapid frequency facilitation in vagal afferents. Am. J. Physiol. Gastrointest. Liver Physiol. *304*, G211–20.

Peterson, W.L., Mackowiak, P.A., Barnett, C.C., Marling-Cason, M., and Haley, M.L. (1989). The human gastric bactericidal barrier: mechanisms of action, relative antibacterial activity, and dietary influences. J. Infect. Dis. *159*, 979–983.

Petrovitch, H., Abbott, R.D., Ross, G.W., Nelson, J., Masaki, K.H., Tanner, C.M., Launer, L.J., and White, L.R. (2009). Bowel movement frequency in late-life and substantia nigra neuron density at death. Mov. Disord. *24*, 371–376.

Rao, A.V., Bested, A.C., Beaulne, T.M., Katzman, M.A., Iorio, C., Berardi, J.M., and Logan, A.C. (2009). A randomized, double-blind, placebo-controlled pilot study of a probiotic in emotional symptoms of chronic fatigue syndrome. Gut Pathog. *1*, 6–4749–1–6.

Reichenberg, A., Yirmiya, R., Schuld, A., Kraus, T., Haack, M., Morag, A., and Pollmacher, T. (2001). Cytokine-associated emotional and cognitive disturbances in humans. Arch. Gen. Psychiatry *58*, 445–452.

Resta-Lenert, S., and Barrett, K.E. (2003). Live probiotics protect intestinal epithelial cells from the effects of infection with enteroinvasive *Escherichia coli* (EIEC). Gut *52*, 988–997.

Riley, T.P., Neal-McKinney, J.M., Buelow, D.R., Konkel, M.E., and Simasko, S.M. (2013). Capsaicin-sensitive vagal afferent neurons contribute to the detection of pathogenic bacterial colonization in the gut. J. Neuroimmunol. *257*, 36–45.

Rubinstein, E., Mark, Z., Haspel, J., Ben-Ari, G., Dreznik, Z., Mirelman, D., and Tadmor, A. (1985). Antibacterial activity of the pancreatic fluid. Gastroenterology *88*, 927–932.

Sandler, R.H., Finegold, S.M., Bolte, E.R., Buchanan, C.P., Maxwell, A.P., Vaisanen, M.L., Nelson, M.N., and Wexler, H.M. (2000). Short-term benefit from oral vancomycin treatment of regressive-onset autism. J. Child Neurol. *15*, 429–435.

Savica, R., Carlin, J.M., Grossardt, B.R., Bower, J.H., Ahlskog, J.E., Maraganore, D.M., Bharucha, A.E., and Rocca, W.A. (2009). Medical records documentation of constipation preceding Parkinson disease: A case–control study. Neurology *73*, 1752–1758.

Savidge, T.C., Newman, P., Pothoulakis, C., Ruhl, A., Neunlist, M., Bourreille, A., Hurst, R., and Sofroniew, M.V. (2007). Enteric glia regulate intestinal barrier function and inflammation via release of S-nitrosoglutathione. Gastroenterology *132*, 1344–1358.

Sellers, R.S., and Morton, D. (2014). The colon: from banal to brilliant. Toxicol. Pathol. *42*, 67–81.

Sung, J.Y., Shaffer, E.A., and Costerton, J.W. (1993). Antibacterial activity of bile salts against common biliary pathogens. Effects of hydrophobicity of the molecule and in the presence of phospholipids. Dig. Dis. Sci. *38*, 2104–2112.

Suzuki, T. (2013). Regulation of intestinal epithelial permeability by tight junctions. Cell. Mol. Life Sci. *70*, 631–659.

Tazoe, H., Otomo, Y., Kaji, I., Tanaka, R., Karaki, S.I., and Kuwahara, A. (2008). Roles of short-chain fatty acids receptors, GPR41 and GPR43 on colonic functions. J. Physiol. Pharmacol. *59* (Suppl. 2), 251–262.

Tillisch, K., Labus, J., Kilpatrick, L., Jiang, Z., Stains, J., Ebrat, B., Guyonnet, D., Legrain-Raspaud, S., Trotin, B., Naliboff, B., and Mayer, E.A. (2013). Consumption of fermented milk product with probiotic modulates brain activity. Gastroenterology *144*, 1394–401, 1401.e1–4.

Turner, J.R., Rill, B.K., Carlson, S.L., Carnes, D., Kerner, R., Mrsny, R.J., and Madara, J.L. (1997). Physiological regulation of epithelial tight junctions is associated with myosin light-chain phosphorylation. Am. J. Physiol. 273, C1378–85.

Udina, M., Castellvi, P., Moreno-Espana, J., Navines, R., Valdes, M., Forns, X., Langohr, K., Sola, R., Vieta, E., and Martin-Santos, R. (2012). Interferon-induced depression in chronic hepatitis C: a systematic review and meta-analysis. J. Clin. Psychiatry 73, 1128–1138.

Wall, R., Marques, T.M., O'Sullivan, O., Ross, R.P., Shanahan, F., Quigley, E.M., Dinan, T.G., Kiely, B., Fitzgerald, G.F., Cotter, P.D., Fouhy, F., and Stanton, C. (2012). Contrasting effects of *Bifidobacterium breve* NCIMB 702258 and *Bifidobacterium breve* DPC 6330 on the composition of murine brain fatty acids and gut microbiota. Am. J. Clin. Nutr. 95, 1278–1287.

Wallon, C., Yang, P.C., Keita, A.V., Ericson, A.C., McKay, D.M., Sherman, P.M., Perdue, M.H., and Soderholm, J.D. (2008). Corticotropin-releasing hormone (CRH) regulates macromolecular permeability via mast cells in normal human colonic biopsies *in vitro*. Gut 57, 50–58.

Wang, L., Christophersen, C.T., Sorich, M.J., Gerber, J.P., Angley, M.T., and Conlon, M.A. (2011). Low relative abundances of the mucolytic bacterium *Akkermansia muciniphila* and *Bifidobacterium* spp. in feces of children with autism. Appl. Environ. Microbiol. 77, 6718–6721.

Williams, B.L., Hornig, M., Buie, T., Bauman, M.L., Cho Paik, M., Wick, I., Bennett, A., Jabado, O., Hirschberg, D.L., and Lipkin, W.I. (2011). Impaired carbohydrate digestion and transport and mucosal dysbiosis in the intestines of children with autism and gastrointestinal disturbances. PLoS One 6, e24585.

Williams, B.L., Hornig, M., Parekh, T., and Lipkin, W.I. (2012). Application of novel PCR-based methods for detection, quantitation, and phylogenetic characterization of Sutterella species in intestinal biopsy samples from children with autism and gastrointestinal disturbances. Mbio 3, 10.1128/mBio.00261–11. Print 2012.

Ye, D., Ma, I., and Ma, T.Y. (2006). Molecular mechanism of tumor necrosis factor-alpha modulation of intestinal epithelial tight junction barrier. Am. J. Physiol. Gastrointest. Liver Physiol. *290*, G496–504.

Zhang, R., Miller, R.G., Gascon, R., Champion, S., Katz, J., Lancero, M., Narvaez, A., Honrada, R., Ruvalcaba, D., and McGrath, M.S. (2009). Circulating endotoxin and systemic immune activation in sporadic amyotrophic lateral sclerosis (sALS). J. Neuroimmunol. *206*, 121–124.

Infant Development, Currently the Main Applications of Probiotics and Prebiotics?

26

Giuseppe Mazzola, Irene Aloisio and Diana Di Gioia

Abstract

Research on probiotics and prebiotics for use in infants is very active and results on their efficacy to prevent and combat several diseases are at present available. Bifidobacteria and lactobacilli are considered beneficial bacteria for the gut, the former being the predominant group of healthy breast-fed newborns.

One of the major area of probiotic research in children has been the treatment and prevention of diarrhoea. Moreover, a large number of infant pathologies, both enteric (infantile colics, necrotizing enterocolitis, coeliac disease) and not strictly enteric (allergies, obesity, neurological disease) have revealed promising preventative and therapeutic effects of probiotics, although these applications need additional experimental evidences. Recent studies have shown that probiotic strain characteristics are crucial to reach a targeted therapeutic effect.

One of the major aspect affecting the gut microbial composition of breast-fed neonates is the presence of oligosaccharides in breast milk. These molecules exert a prebiotic effect which is crucial for the development of a healthy gut microbiota. Research studies have been focused on the selection of fibres possessing a prebiotic role similar to human milk oligosaccharides. Galactooligosaccharides and fructoligosaccharides are abundantly used in infant formula, frequently as mixtures of the two molecules. Several studies have shown that the capability of stimulating beneficial bacteria and of shaping the gut microbiota is similar to that of breast milk. On the contrary, studies regarding the use of prebiotics in infants for the prevention of allergies showed contradictory results.

Therefore, it is possible to conclude that children are a very important target, if not the main one, for probiotic and prebiotic administration and the European industry is aware of that.

Introduction

Probiotics and prebiotics constitute a central growing market for the food/pharmaceutical industry and for the baby food industry as well. The industrial significance of these health promoting bacteria and molecules has driven a lot of research studies aimed at understanding their functionality and activity. The children associated market is of great relevance, because infants are very susceptible to diseases and non-pharmaceutical treatments are particularly searched for this age group.

Probiotic and prebiotic research is moving forward on two fronts: basic science, i.e. laboratory studies planned to elucidate the mechanisms through which these supplements exert their activity, and clinical trials to evaluate the safety and efficacy of probiotics in various medical conditions. Many early clinical trials of probiotics/prebiotics had methodological limitations, and definitive clinical evidence to support using specific probiotic strains for specific health purposes is sometimes lacking. Nevertheless, there is preliminary evidence for several uses of probiotics and prebiotics, and more studies are under way. This chapter is focused on the use of probiotics and prebiotics in infants for the treatment of infant diseases and, in the case of prebiotics, will describe attempts aimed at making formula milks as close as possible to human milk to overcome limitations connected to the lack of feeding with maternal milk. The basis of probiotics/prebiotics use is that the health status of the gut in infants is extremely important for the well-being of the whole organism in the successive stages of life. The definition of an equilibrated gut microbiota composition in the early stages of life and, therefore, the understanding of how colonization occurs in newborns, are of crucial importance.

Gut microbiota acquisition in infants

Three different successive phases in the composition of the intestinal microbiota can be described almost in all individuals: (i) the birth and the whole period of liquid diet; (ii) the weaning time up to adulthood, when the introduction of a solid diet determines profound changes giving rise to a stable community; (iii) the elderly period, when further strong changes in the microbiota occur deriving from both physiological modification of the gastrointestinal tract and an unbalanced diet (Claesson *et al.*, 2011). Studies on gnotobiotic mice have illustrated the essential role of the gastrointestinal microbiota in normal gut development and it is argued that the microbial diversity of the human gut is the result of co-evolution between microbial communities and their hosts (Ley *et al.*, 2006). It is well assessed that the intestinal microbiota of the newborn is a complex ecosystem composed of numerous genera, species and strains of bacteria, performing various unique activities affecting colonic and systemic physiology and that its establishment begins soon after birth. Birth brings about an immediate end to the sterility of the fetus environment and, within a few hours, bacteria start to appear in the faeces. The first microbial population the vaginally born newborn comes into contact with are the maternal intestinal and vaginal microbiota; successively, the newborn is exposed to microbes from the environment. Still, the first microbial colonization of the infant gastrointestinal tract is a remarkable episode in the human lifecycle and it is well known that several factors, in addition to the maternal microbiota, influence the gut colonization by microorganisms, including the mode of delivery, the type of infant feeding, the perinatal newborn circumstances (hospitalization and chemotherapy treatments) (Biasucci *et al.*, 2010; Di Gioia *et al.*, 2014).

The first bacteria healthy newborns come in contact with are facultative anaerobes (*Staphylococcus*, *Enterobacteriaceae* and *Streptococcus* spp.), because of the positive reduction/oxidation potential of the neonates' intestine at birth. These bacteria consume oxygen paving the way to strict anaerobes such as members of the *Bifidobacterium*, *Bacteroides* and *Clostridium* genera (Biasucci *et al.*, 2010). Epidemiological studies performed in the last decade focused on the determination of factors affecting gut composition (e.g. the KOALA study, Penders *et al.*, 2006) and have shown that anaerobic colonization may be delayed in caesarean section delivered infants. Additionally, an increased incidence of *Clostridium perfringens* and *Clostridium difficile* in caesarean section newborns is reported in relation to the hospital environment. Another factor influencing composition of the intestinal microbiota in neonates is the type of feeding (Penders *et al.*, 2006). In full term breastfed neonates *Bacteroides* spp. and bifidobacteria can appear 4 days after birth and after 1 week they their counts increase rapidly to constitute 80–90% of the total bacteria, whereas the microbiota of formula-fed infants is more complex, with *Bifidobacterium* spp., enterobacteria and *Streptococcus* spp. in similar proportion. Another notable difference is that formula fed infants have much higher counts of *Clostridium* spp. than breast-fed infants (Penders *et al.*, 2006). An important difference is the relative buffering capacity of the two feedings. Breast milk has poor buffering capacity, compared with formula milk, and this leads to market differences in the colon pH of breast and formula fed infants: 5.1 and 6.5, respectively. The low pH in breast-fed infants promotes the growth of bifidobacteria and lactobacilli, but is inhibitory to many other bacteria (Tham *et al.*, 2011). Moreover, human milk oligosaccharides, as described in a subsequent section of this chapter, are prebiotic agents that selectively encourage the growth of beneficial (probiotic) organisms (Coppa and Gabrielli, 2008). Both adults and neonates are regularly exposed to microorganisms via the diet, but with different effects. The microorganisms entering newborns through breast milk are more likely to colonize than those entering healthy adults possessing a stable climax community. Evidence about the isolation and identification of commensal or potential probiotic bacteria, including bifidobacteria, from milk of healthy women is not conclusive. Even though authors are aware that human milk is difficult to sample and microbial contamination can never be totally discarded, some studies have demonstrated the presence of live bifidobacteria in human milk (Martin *et al.*, 2003; Solis *et al.*, 2010).

In contrast to full term neonates, little information is available concerning the composition of the microbiota in premature infants. The inter-individual variability in these subjects is very high and many parameters, such as antibiotic use and diet, may lead to different ecosystems in these children, which could explain discrepancies between studies. In particular, preterm newborns often need parental feeding, due to the immaturity of their intestines, they often need respiratory support, they are vulnerable for infections and often require antibiotic treatment. One of the most significant differences between preterm and full term infants microbiota is the colonization of bifidobacteria that are not frequently identified in the first month of life of premature newborns (Westerbeek *et al.*, 2006). This alteration has been linked to the increased risk of severe gastrointestinal disorders such as necrotizing enterocolitis which affects predominantly premature and low weight newborns (Lin *et al.*, 2008).

The gut microbiota of infants, due to change of diet, becomes more stable and homogenous after weaning and it gradually gets closer to the typical adult microbiota (Magne *et al.*, 2006; Koenig *et al.*, 2011). A large-scale longitudinal study on development and change in the composition of gut microbiota during the process of weaning was carried out by Fallani *et al.* (2010). Within this study faecal samples of 605 infants (from five European countries), approximately 4 and 6 weeks after the introduction of first solid foods, were collected in order to investigate the association with determining factors such as mode of delivery, previous feeding practices, age of weaning and the impact of possible antibiotic treatment. The samples were analysed by fluorescence *in situ* hybridization combined with flow cytometry using a panel of 10 rRNA targeted group- and species-specific oligonucleotide probes.

After 1 month of weaning, bifidobacteria and *Bacteroides* continue to represent the predominant groups. However, the number of detectable *Bifidobacterium* species decreased after weaning together with counts of *C. perfringens* and *C. difficile*, while other strictly anaerobic clostridia increased.

Association between gut microbiota composition and health status

Because of immature intestinal immune function, the newborn is susceptible to intestinal and systemic infections. An immature intestinal epithelial barrier may predispose infants and children to intestinal inflammatory diseases, such as infectious enteritis, inflammatory bowel disease, and necrotizing enterocolitis. Therefore, the development of an healthy microbiota during the postnatal period is critical for the establishment of normal physiology of the intestinal tract. Moreover, a better understanding of the factors that regulate gut barrier maturation may yield insight into strategies to prevent these intestinal diseases. The microbiota is in close contact with the intestinal mucosa and epithelial surface which is, after the respiratory area, the largest surface of the body, occupying approximately 250–400 m^2 (Nataro, 2005). Some anatomical and physiological aspects of the host organism are directly linked to the presence and activity of the resident microorganisms, such as formation of the intestinal walls, production of organic acids and vitamins, stimulation of immune system, etc. The presence of abundant bifidobacteria and lactobacilli may provide some protection against incoming enteric pathogens. The benefits exerted by the strains belonging to *Bifidobacterium* and *Lactobacillus* genera have been widely studied in the last ten years (Laux *et al.*, 2005; Jankowska *et al.*, 2008; Cencic and Langerholc, 2010; Montier *et al.*, 2012). They are able to compete for nutrient with enteric pathogens, to strongly adhere to the intestinal mucosa and to stimulate the development of both humoral and cellular mucosal immune system. These and many other features make the majority of them excellent probiotic candidates for paediatric use.

Beneficial bacteria in the newborn gut

Bifidobacteria

The intestinal microbiota of breast-fed newborns is predominantly composed of bifidobacteria. This was already established in several studies performed in the 1980s (Biavati *et al.*, 1984) with the use of traditional agar plating technique. The most represented species in both breast-fed and formula-fed infants were found to be *Bifidobacterium infantis* (at present classified as *B. longum* subsp. *infantis*), *B. breve*, *B. longum* (at present classified as *B. longum* subsp. *longum*) and *B. bifidum*. *B. pseudocatenulatum* and *B. catenulatum* were also present although in a lower number in both type of newborns, whereas *B. dentium* was found only in breast-fed neonates. In the last 20 years the development of molecular techniques has completely revolutionized the way of performing microbial ecology studies. Temperature gradient gel electrophoresis (TGGE) studies of DNA amplified from infant faeces collected during the breast feeding period confirmed that bifidobacteria were the predominant group, precisely *B. infantis*, *B. longum* and *B. breve*, whereas in the post-weaning period a decrease in the *Bifidobacterium* population was observed (Magne *et al.*, 2006). Real-time PCR analyses of targeted *Bifidobacterium* species indicated that the number of species in breast-fed newborn was initially higher. *B. animalis* and *B. dentium* were not detectable in samples irrespective of the type of feeding, whereas *B. longum* subsp. *infantis*, *B. breve*, *B. bifidum* and *B. longum* subsp. *longum* were detected in all samples, *B. longum* subsp. *infantis* being the major species found. *B. adolescentis*, which is most commonly found in adults, was present only in formula-fed infants (Klaassens *et al.*, 2009). Further insights in the infant gut microbiota were recently carried out using next generation sequencing targeted on two hypervariable regions of the 16SrRNA gene by Turroni *et al.* (2012). This study elucidated that the most abundant class in infant faecal samples was *Bifidobacteriales*, being present at 80.6%. The predominant bifidobacterial species detected were *B. longum* and *B. bifidum* at 56.2% and 10.7%, respectively. Regarding the inter-individual variability in the infant gut microbiota, the statistical analyses performed revealed a large conservation of members of the *Actinobacteria* with a high proportion (ranging from 21.7% to 90.6%) belonging to the *Bifidobacteriaceae* family. In particular, *B. breve* was always detected with an average of 5.5% of total reads. In contrast, *B. adolescentis* was detected in a relatively high average percentage (3.4%), but it was only present in about 2% of the subjects.

Lactobacilli

Together with *Bifidobacterium* and *Bacteroides*, the *Lactobacillus* genus is one of the major component of the breast-fed newborn intestinal microbiota. However, the distribution of the different *Lactobacillus* species in infants has not been studied in as much detail as in the case of bifidobacteria. Satokari *et al.* (2002) carried out the analysis of *Lactobacillus* diversity in breast-fed and bottle-fed newborns using denaturing gradient gel electrophoresis (DGGE) approach. The analysis revealed that the predominant *Lactobacillus* population consisted of one or two dominant species and *L. acidophilus* resulted to be the most common species irrespective of the type of feeding. A further study has been focused on the distribution of *Lactobacillus* species in faeces of breast-fed newborns and newborns receiving standard formula milk or standard formula milk supplemented with galactooligosaccharides and fructooligosaccharides (Haarman and Knol, 2005) using real time PCR with primers and probe sets designed on the intergenic spacer of 16S–23S rRNA gene. This approach allowed to have a more detailed analysis of *Lactobacillus* species because sequences are less conserved than 16S rRNA gene sequence. The *Lactobacillus* species distribution in breast-fed was mainly composed of *L. acidophilus*, *L. casei and L. paracasei*. The same distribution was also found in newborns receiving prebiotic supplementation. On the other hand supplementation with standard formula resulted in a different *Lactobacillus* distribution with more *L. delbruekii* and *L. reuteri* and less *L. paracasei* and *L. acidophilus*. *L. fermentum*, *L. plantarum*, and *L. rhamnosus* were also present at a very low percentage at the beginning of the intervention period and these species seemed to disappear completely during the intervention in all feeding groups. Changes in the distribution of *Lactobacillus* species has been also demonstrated by Salminen *et al.* (2014). The authors carried out a double-blind, randomized prospective study of 21- to 30–day-old healthy and vaginally born infants showing that *L. paracasei* and *L. rhamnosus* increased in newborn fed with formula milk supplemented with galactooligosaccharides.

Probiotics and their paediatric use

One of the primary areas of probiotic research in children has been the treatment and prevention of bacterial acute diarrhoea and antibiotic associated diarrhoea (Wiegering *et al.*, 2011). Moreover, a large number of infant pathologies, both enteric and not strictly enteric, have revealed promising preventative and therapeutic effects of probiotics, although these applications need additional experimental evidence (Chen and Walker, 2011; Taibi and Comelli, 2014). Therefore, research is currently going on for the use of probiotics in necrotizing enterocolitis, infantile colics and coeliac disease, as described below and briefly summarized in Table 26.1. It is likely that, in the future, probiotics will become a consolidated therapy for these diseases. In addition, a number of pathologies, which are not strictly gut dysfunctions, have in recent years been correlated to alterations of the gut microbiota, such as allergy, obesity, neurological and psychiatric diseases (Table 26.2). However, not all the probiotic strains drive a targeted therapeutic effect, thus, there is a need for rational selection of specific probiotic strains, matched for precise clinical indications. This implies that there is also a need to have a better insight in the effects that a specific probiotic strain can have in patient cohorts as well as in healthy populations. Furthermore, several studies demonstrated that the combination of more strains, also belonging to different genera (mainly *Lactobacillus* and *Bifidobacterium*), may exhibit a synergistic effect and research is therefore aimed at finding proper combinations.

Treatment and prevention of acute diarrhoea and antibiotic-associated diarrhoea

Infectious gastroenteritis is one of the leading causes of morbidity especially in newborns and children under 5 years of age (Wiegering *et al.*, 2011). Although gastroenteritis-associated mortality is rare in Western Europe, an increased incidence has been noticed in some national registers over recent years. The most common causes of infant gastroenteritis are viruses (e.g. rotavirus, adenovirus and norovirus) and bacteria (e.g. *Salmonella* spp., *Shigella* spp., *Campylobacter* spp., *C. difficile*, *Klebsiella pneumoniae*, *Enterobacter cloacae*, *E. coli*), but parasite infections are also reported. The vast majority of the published trials show a statistically significant benefit and clinical benefit of a few, well-identified probiotic strains. The effect is strain-dependent and dose-dependent. Positive outcomes in the treatment and prevention of diarrhoea with probiotics date back in the '90s and the most studied strain is *L. rhamnosus* GG (LGG). LGG was particularly effective in the preventative therapy of diarrhoea, whereas seriously sick children received less benefits from the treatment (Arvola *et al.*, 1999).

The targeted effects of probiotics on viral diarrhoea have been the subject of recent studies trying to elucidate the mechanisms of diarrhoea symptoms relieve. In particular, in the study

Table 26.1 Overview of potential probiotic applications and studies performed in the prevention and treatment of gastrointestinal diseases

Pathology	Probiotic microorganism	Reported effect(s)	Reference
Acute diarrhoea	*Lactobacillus rhamnosus* GG	Reduced risk of diarrhoea	Szajewska *et al.* (2006)
	B. animalis subsp. *lactis*, *L. acidophilus*, *L. rhamnosus* strains	Decreased duration of diarrhoea	Grandy *et al.* (2010)
	B. longum subsp. *infantis* CECT7210	Increased inhibition of rotavirus virulence (*in vitro* study)	Munoz *et al.* (2011)
	B. animalis subsp. *lactis* plus *B. longum* subsp. *infantis*, *Streptoccoccus thermophilus*, *Lactobacillus rhamnosus*, *Lactobacillus acidophilus* strains	Decreased duration of diarrhoea	Vandelplas and Hert (2011)
	L. reuteri DSM 17938	Reduced incidence of nosocomial diarrhoea	Wanke and Szajewska (2012)
	B. breve DSMZ24706, *B. breve* DSMZ24707, *B. breve* DSMZ24708, *B. longum* DSMZ24709	Increased inhibition of diarrhoea pathogens (*in vitro* study)	Aloisio *et al.* (2012)
	B. longum subsp. *infantis* 35624	Reduced *Salmonella* spp. infection (*in vitro* study)	Sydmonds *et al.* (2012)
	L. ruminis SPM0211	Decreased duration of viral diarrhoea	Kang *et al.* (2013)
Infantile colics	*B. breve* DSMZ24706, *B. breve* DSMZ24707, *B. breve* DSMZ24708, *B. longum* DSMZ24709	Antimicrobial activity against gas forming coliforms (*in vitro* study); clinical trial is on going	Aloisio *et al.* (2012)
	L. reuteri DSM 17938	Reduced gas colics symptoms	Savino *et al.* (2010)
	L. reuteri DSM 17938	Reduced time of crying	Szajewska *et al.* (2013)
	L. reuteri DSM 17938	No effect on treatment of infant colics	Sung *et al.* (2014)
Necrotizing enterocolitis	*B. bifidum* strain in Infloran product	Reduced incidence of NEC and death	Lin *et al.* (2008)
	B. breve strain (Yakult preparation)	Increased intestinal motility	Braga *et al.* (2011)
	L. reuteri DSM 17938	Reduced frequency of sepsis	Oncel *et al.* (2013)
Coeliac disease	VLS#3 product (*B. longum* subsp. *longum*, *B. breve*, *B. longum* subsp. *infantis*, *S. thermophilus*, *L. plantarum*, *L. casei* *L. acidophilus*, *L. delbrueckii*)	Decreased toxicity of gluten (murine model and human cell lines)	De Angelis *et al.* (2006)
	VLS#3 product	Decreased toxicity of gluten during food processing	Kaur *et al.* (2002)
	B. animalis subsp. *lactis* and *L. lactis*	Reduced damage induced by gliadin (human cell lines)	Lindfors *et al.* (2008)

Table 26.2 Overview of potential probiotic applications and studies performed in infant diseases not directly related to the gastrointestinal tract. The obesity and neurology disease sections only report animal studies considering that conclusive applications in humans are not available yet

Pathology	Probiotic microorganism	Reported effect(s)	Reference
Allergies	*B. breve* BB99, *L. rhamnosus* GG, *L. rhamnosus* LC705, *P. freudenreichii* subsp. *shermanii* DSM 7076	Reduced eczema caused by cow's milk allergy	Viljanen *et al.* (2005)
	B. longum BL999, *L. rhamnosus* LPR	No effect on prevention of eczema	Soh *et al.* (2009)
	B. breve BB99, *L. rhamnosus* GG, *L. rhamnosus* LC705, *P. freudenreichii* subsp. *shermanii* DSM 7076	Decreased incidence of IgE-associated allergy	Kuitunen *et al.* (2009)
	L. johnsonii EM1	Reduced perennial allergic rhinitis	Lue at al. (2012)
Obesity	*Bifidobacterium* spp. strains	Increased glucose-tolerance and decreased proinflammatory cytokines (murine model)	Cani *et al.* (2007)
	VLS#3 product	Increased hepatic natural killer T-cell (murine model)	Ma *et al.* (2008)
	Bifidobacterium spp. strains	Reduced serum and liver triglyceride (murine model)	Yin *et al.* (2010)
	Mixture of three *Bifidobacterium* strains	Reduced serum cholesterol level (murine model)	An *et al.* (2011)
Neurology and psychiatry	*B. infantis* 35624	Reduced hyperactivity (murine model)	Sudo *et al.* (2004)
	B. infantis 35624	Increased level of plasma tryptophan (murine model)	Desbonnet *et al.* (2008)
	L. helveticus R0052, *B. longum* R0175	Reduced anxiety (murine model)	Messaoudi *et al.* (2010)
	B. longum NCC3001	Decreased infection- induced behavioural changes (murine model)	Bercik *et al.* (2011)

of Munoz *et al.* (2011) a *B. longum* subsp. *infantis* CECT7210 strain, isolated from infant faeces, was selected for the capability of inhibiting *in vitro* rotavirus replication and of protecting cells from virus infection. It was then tested *in vivo* in mice showing protection against rotavirus infection. A strong antiviral activity has been also obtained with *L. ruminis* SPM0211 *in vitro* (Kang *et al.*, 2013). Moreover, mixtures of probiotics showed good results in the reduction of duration and the severity of the acute rotavirus diarrhoea. A multiple species product composed of a *L. acidophilus*, a *L. rhamnosus* and a *B. longum* strain (strain numbers not reported), administered to infants aged 1–23 months, was able to reduce the duration of the disease (Grandy *et al.*, 2010). Similar results have been obtained testing the efficacy of a synbiotic product containing *Streptococcus thermophilus, L. rhamnosus, L. acidophilus, B. animalis* subsp. *lactis, B. infantis* strains (strain numbers not reported) and fructooligosaccharides in children with acute diarrhoea (Vandenplas and De Hert, 2011). The primary end-points were duration of diarrhoea and the number of children who had a normalized stool consistency. A further study was aimed at evaluating the efficacy of administering L. reuteri DSM 17938 for the prevention of nosocomial diarrhoea. In hospitalized children, the administration of L. reuteri DSM 17938 compared with placebo treatment had no effect on the overall incidence of nosocomial diarrhoea, mainly caused by rotavirus infection (Wanke and Szajewska, 2012). Therefore, the studies on viral associated diarrhoea still lack conclusive evidence. Alternatively, there is a strain-dependent effect and not all strains are equally effective, or even the dose chosen may have been wrong.

Some probiotic strains also resulted in effective reduction of the risk of antibiotic-associated diarrhoea in newborns and children. A clinical trial, performed with 766 infants, indicated that treatment with the probiotic strain *L. rhamnosus* GG, compared to placebo, reduced the risk of diarrhoea from 28.5% to 11.9% (Szajewska *et al.*, 2006). A number of study have focused on the administration of probiotic in the treatment of diarrhoea caused by bacteria, mainly *Clostridium* spp. *Bifidobacterium* strains, derived from infant faeces and mainly belonging to the *B. breve* species, were capable of inhibiting the growth of pathogens causing infectious diarrhoea of bacterial origins in infants, making them potential probiotic candidates for a formulation aimed at the prevention or the cure of bacterial diarrhoea (Aloisio *et al.*, 2012). Furthermore, a recent study demonstrated the efficacy of *B. longum* subsp. *infantis* 35624 in inhibiting *Salmonella* in the mouse gut thus preventing its pathogenic effects, including the reduction of pathogen-induced inflammatory response (Symonds *et al.*, 2012).

Treatment and prevention of necrotizing enterocolitis (NEC)

NEC is the most common gastrointestinal emergency in the neonatal intensive care unit and a major cause of morbidity in preterm infants. It is characterized by a gastrointestinal dysfunction progressing to pneumatosis intestinalis, systemic shock, and rapid death in severe cases. The most common risk factors are considered to be prematurity, enteral feeding and the occurring of bacterial colonization, which often operate in synergy (Claud and Walker, 2001). There is strong evidence that the initial bacterial colonization after birth plays a pivotal role in the development of NEC. As already evidenced, preterm newborns show a different colonization pattern with respect to full term newborns and more pathogenic microorganisms such as enterobacteria and enterococci remain predominant until the 20th day of life. For this reason it has been suggested that a major aetiological factor for NEC is a different microbiota, particularly as NEC usually occurs after 8–9 days postpartum when anaerobic bacteria start to colonize the gut (Mai *et al.*, 2011). This study also pointed out that correction of the abnormal microbiota composition may be a strategy to prevent NEC. Several authors have reported systematic reviews of randomized and controlled trials of probiotic supplementation in preterm infants (Deshpande *et al.*, 2007, 2010; Barclay *et al.*, 2007; Stockman, 2009), showing that probiotic supplementation was able to reduce the incidence of NEC in neonates. The action attributed to probiotics is species specific (Verdu, 2009). Administration of several strains of *B. bifidum*, *B. breve* and *B. animalis* subsp. *lactis* to preterm and low-birthweight infants showed evident clinical benefits for the treatment of NEC (Lin *et al.*, 2008; Khailova *et al.*, 2009; Braga *et al.*, 2011; Underwood *et al.*, 2012). The mechanisms by which probiotics operate are thought to be: competition against the colonization of potential pathogenic microorganism, immunomodulation, nutritional competition and improved intestinal motility. Treatment with *B. breve* associated with *L. casei* (Yakult LB product) in preterm infants underlined a positive correlation between improved intestinal motility and NEC (Braga *et al.*, 2011). In particular, infants in the probiotics group achieved full enteral feeding faster than did those in the control group and the survival curve, determined with the Kaplan–Meier method to evaluate the complete transition time of orogastric feeding tube to breastfeeding, showed a shorter time in the probiotics group.

On the contrary, *L. reuteri* DSM 17938 does not seem to affect the overall rates of NEC and/or death in preterm infants followed up in the neonatal intensive care unit, although significant reductions were observed in the frequency of proven sepsis, rates of feeding intolerance and duration of hospital stay (Oncel *et al.*, 2014). Because of the complexity of the pathogenesis of NEC, the administration of different *Bifidobacterium* and *Lactobacillus* strains may benefit by different actions on the host and, therefore, multistrain probiotic preparations are likely more effective than single-strain ones (Deshpande *et al.*, 2011).

Prevention of infantile colics: are probiotics a feasible option?

About 10–30% of infants are affected by infantile colics in the first months of life. The manifestations of colics are excessive, inconsolable crying, frequently accompanied by flushing of the face, meteorism, drawing-up of the legs and passing of gas. Microbiota of infants affected by colics was in fact found to be richer in gas-forming coliforms with respect to non-colicky ones (Savino *et al.*, 2009). Regarding possible therapies, the study of Savino *et al.* (2007) examined, for the first time, the modulation of intestinal microbiota of infants suffering from colics by administering a probiotic strain. A cohort of breastfed colicky infants was randomly assigned to treatment with a strain of *L. reuteri* ATCC 55730 and simethicone, an orally administered anti-foaming agent traditionally used to reduce discomfort or pain caused by excessive gas. This study evidenced that infants treated with *L. reuteri* had a significant reduction in crying compared to infants treated with simethicone. The positive effect of probiotic administration on the reduction of gas colic symptoms and on the modulation of intestinal microbiota was confirmed in a successive trial with the strain *L. reuteri* DSM 17938 (Savino *et al.*, 2010). Moreover, the efficacy of the administration of the latter strain in reducing the daily average time of crying time was also confirmed by Szajewska *et al.* (2013). In addition, research is going on with the aim of obtaining new probiotic strains for possible use in colics treatment, such as other *Lactobacillus* strains (Savino *et al.*, 2011) or *Bifidobacterium* strains (Aloisio *et al.*, 2012). The efficacy of probiotic supplementation on the reduction of crying time and successful treatment of infantile colic was the object of a recent systematic review (Anabrees *et al.*, 2013). The final conclusion was that, although *L. reuteri* may be effective as a treatment strategy for crying in exclusively breastfed infants with colics, the evidence supporting probiotic use for the treatment of infant colic or crying in formula-fed infants remains unresolved. Larger rigorously designed studies are necessary to draw more definitive conclusions. Recently, a large clinical trial on the efficacy of *L. reuteri* DSM 17938 in reducing the symptoms of infant colics obtained different results from previous trials (Sung *et al.*, 2014). *L. reuteri* DSM 17938 administration to both breastfed and formula fed newborns did not reduce crying or fussing. The efficacy of DSM 17938 and other probiotic strains should therefore be further explored, in particular considering that dose and duration of administration of DSM 17938 are similar in these studies.

Coeliac disease (CD): *in vitro* studies need to be supported by clinical trials

CD is a chronic inflammatory disorder of the small intestinal mucosa induced by the ingestion of wheat gluten, or other similar proteins found in barley and rye. CD involves genetic and environmental factors in predisposed individuals (Green and Cellier, 2007). CD is characterized by two types of immune response to gluten-derived peptides: an adaptive Th_1-dependent immune response, within the intestinal mucosa, and an innate immune response. Both responses lead to the release of proinflammatory cytokines like IFN-γ and IL-15 (Londei *et al.*, 2005) and result in a consequent inflammation and intestinal tissue remodelling (Meresse *et al.*, 2009). At present, the only effective treatment for the disease is a strict life-long gluten-free diet.

Palma *et al.* (2012) evaluated the gut microbial colonization during the first 4 months of life in breast-fed and formula fed healthy full-term infants with a genetic CD risk. They demonstrated that the milk-feeding type and the HLA-DQ genes (the major genetic risk factor for CD) influence the gut bacterial colonization. In particular, the reduction of *Bifidobacterium* spp. such as *B. longum* was associated with an increased CD risk. Until now relatively few studies considered the efficacy of the administration of probiotic strains in children affected by coeliac disease. De Angelis *et al.* (2006) showed the capacity of the probiotic preparation VLS#3, a mixture of lactic acid bacteria and bifidobacteria to decrease the toxicity of gluten during food processing. Furthermore VLS#3 has been shown to colonize the intestine and increase epithelial barrier function in the host (Kaur *et al.*, 2002) and it has the potential of being used to modify and improve gliadin degradation in the gastrointestinal tract after ingestion. Moreover, a recent study investigated the capability of probiotic bacteria (*L. fermentum* and *B. lactis* strains; strain numbers not reported) to inhibit direct toxic effects induced by gliadin on epithelial cells (Lindfors *et al.*, 2008). *B. lactis* was more efficient than *L. fermentum*, perhaps because the peptidase repertoire of *B. lactis* was more efficient than that of *L. fermentum* in breaking up gliadin into small harmless peptide products.

Therefore, although several studies have addressed the ability of probiotic bacteria to detoxify gliadin after an extensive incubation period and their role in suppressing proinflammatory effects *in vitro*, to our knowledge no studies have investigated whether different live probiotic bacteria can inhibit gliadin-induced toxic effects directly on *in vivo* epithelial cells.

Prevention of allergies

The incidence of allergic disorders such as atopic dermatitis, rhinitis and asthma has increased strikingly in developed countries over the past two decades. One of the most reliable hypotheses of this increase is a relative lack of microbial stimulation and a failure of immunoregulation due to low exposure of the infantile gut immune system to harmless microorganisms associated with the environment, the so-called hygiene hypothesis (Cabana *et al.*, 2007). The manifestations of allergic disease are age dependent. Infants commonly present symptoms and signs of atopic eczema, gastrointestinal symptoms and recurrent wheezing. Asthma and rhinoconjunctivitis become prevalent in later childhood. Sensitization to food allergens usually occurs in the first 2–3 years of life, followed by indoor allergens (e.g. house dust mite and pets) and, subsequently, outdoor allergens (e.g. rye and timothy grass) (Halken, 2004). Genetic susceptibility plays a large role in the development of food allergy, in fact the risk of development of allergy increases substantially with a positive family history of allergic diseases. The prevalence of allergic disease in childhood is 7% to 8% for food allergy, 15% to 20% for atopic eczema, and 31% to 34% for asthma or recurrent wheezing (Halken, 2004). Food hypersensitivities affect approximately 6% of infants less than 3 years of age (Osterballe *et al.*, 2005). Although the exact aetiology of allergic diseases remain ambiguous, epidemiological data have shown that atopic children show recurrent differences in the gut microbiota composition with respect to healthy children, with higher levels of clostridia and lower levels of bifidobacteria, although it is unclear whether this is cause or consequence. In addition to lower numbers of bifidobacteria, infants suffering from atopic disease harbour a peculiar pattern of bifidobacteria comprising adult-like strains, such as *B. adolescentis*, compared with healthy infants with a typical infant pattern (Ouwehand *et al.*, 2001). Some other studies have also shown that early colonization with pathogenic bacteria is more likely to occur in children who go on to develop allergy; in contrast, lactobacilli and bifidobacteria are found more commonly in the composition of the gut microbiota of non-allergic children and this seems to correlate with protection against atopy (Kalliomäki *et al.*, 2001; Özdemir, 2009). Therefore, the possibility of using probiotics to prevent the development of allergic disease has a sound scientific background.

A probiotic mixture, containing *B. breve* Bb99 strain, the LGG strain, *L rhamnosus* LC705 in addition to *Propionibacterium freudenreichii* subsp. *shermanii* DSM 7076, was administered to mothers during the last month of pregnancy and their infants received it from birth until age 6 months. The treatment resulted in a decreased incidence of IgE-associated allergy, such as atopic diseases, in Caesarean-delivered children until the age of 5, compared to placebo (Kuitunen *et al.*, 2009). The same probiotic mixture was also found to be effective in the treatment of eczema in infants with proven cow's milk allergy (Viljanen *et al.*, 2005). Conversely, commercially available cow's milk formula supplemented with *B. longum* BL999 and *L. rhamnosus* LPR administered in the first 6 months of life to Asian infants at risk of allergic disease showed no effect on prevention of eczema or allergen sensitization in the first year of life (Soh *et al.*, 2009). The administration of *L. johnsonii* EM1 administered to 7- to 12 year-old children was found to be more effective against allergic rhinitis than antihistamine treatment with levocetirizine (Lue *et al.*, 2012). Other studies have been focused on the effects of probiotics in the treatment of food allergy, but conclusive effects have not been evidenced yet (Boyle and Tang, 2006). Some preliminary positive results have been obtained with LGG supplementation, but no experimentation is available on bifidobacteria.

Obesity: a correlation with abnormal gut microbiota

Nowadays, obesity prevalence is increasing especially among children and adolescents and it can be considered a worldwide epidemic. Recently, obesity has been associated with structural alterations in the gut microbiota, suggesting potential causality between specific microbial taxa and this disorder. In particular, studies have focused not only on individual bacterial species but also on the whole microbial community (Tennyson and Friedman, 2008). However, controversial data make it clear that the connection between the microbiota composition and excess weight is very complex. See other contributions in this book for more details. The explanations for the ability of the gut microorganisms to affect obesity development include an improvement of the energy harvest from the diet, influence on lipase activity, a decrease of lipopolysaccharide inflammation that is related to fat induced uptake and the control of endotoxaemia and insulin resistance (Blaut and Bischoff, 2010).

Million *et al.* (2012) found that bifidobacteria could be associated to the lean status, on the contrary *Lactobacillus* spp., *Staphylococcus aureus* and *Escherichia coli* have been associated to overweight status. Members of the genus *Bifidobacterium* were shown to be higher in number in children who remained normal weight at 7 years than in children developing overweight (Kalliomäki *et al.*, 2008), allowing the authors to conclude that an aberrant compositional development of the gut microbiota precedes overweight and this may offer new possibilities for preventative and therapeutic applications of bifidobacteria in weight management. Furthermore, other studies reported a decrease of total bifidobacteria in faeces of obese patients (Schwiertz at al., 2010; Collado *et al.*, 2008). A recent study analysed the faecal concentration of the main intestinal microbial groups in obese, overweight, lean and anorexic subjects. A positive correlation was found between certain *Lactobacillus* species, in particular *L. reuteri*, and obesity. On the contrary, *B. animalis* has been associated with a lower body mass index (Million *et al.*, 2013). Several meta-analyses, in the past years, have shown that breastfeeding is associated with a reduced likelihood of overweight or obesity in childhood and that the duration of breastfeeding is inversely associated with the risk of overweight (Owen *et al.*, 2005; Harder *et al.*, 2005). Several results mainly performed on murine models (listed in Table 26.2) suggested that some strains of bifidobacteria and lactobacilli have effect on the obese status reducing serum total cholesterol, decreasing proinflammatory cytokines and increasing glucose tolerance (Cani *et al.*, 2007; Ma *et al.*, 2008; Yin *et al.*, 2010; An *et al.*, 2011). However, the *in vivo* administration of probiotics in obese children or adults needs to be further investigated.

Treatment of neurological and psychiatric disease

The gut and the brain are highly integrated and communicate in a bidirectional manner and this connection is usually called 'the gut–brain axis' (Rhee *et al.*, 2009). An example of this relation is that psychiatric disorders frequently coexist with common pathological gastrointestinal conditions, such as irritable bowel syndrome, which is quite common in adults. The gastrointestinal tract is a site of interaction between microorganisms, immune cells, and the neuronal network. In this respect, beneficial microbes such as lactobacilli and bifidobacteria seem to be particularly sensitive to signals from the central nervous system, taking into consideration that stressful conditions, including emotional stress, are very often accompanied by a decrease of these organisms in the gastrointestinal tract. Neurochemical molecules can affect the microbiota composition changing the gut motility and increasing the acidity. Moreover, bacteria themselves can influence the endocrine system by the production of several biologically active peptides, nitric oxide, melatonin, gamma-aminobutyric acid and serotonin (Collins and Bercik, 2009). Pro-inflammatory cytokine such as IL-4 and IFNγ are implicated in a range of psychiatric disorders including depression, and multiple studies in animals and humans have shown that manipulation of the gut microbial composition influences systemic cytokine levels (e.g. Cryan and O'Mahoni, 2011). A decrease in the desirable gastrointestinal tract bacteria will lead to deterioration in gastrointestinal, neuroendocrine and immune relationships and, ultimately, disease. Therefore, studies focusing on the impact of enteric microbiota on the central nervous system are essential to the understanding of the influence of the gut microbiota on this system.

A possible approach to study the microbiota–gut–brain axis is the use of germ-free mice (Table 26.2). Research performed with acute stressed germ free mice, showing hyperactivity of the body major stress response system (the hypothalamic–pituitary–adrenal axis), have evidenced that the stress response was normalized by administration of a *B. infantis* strain (strain number not reported) (Sudo *et al.*, 2004). Bercik *et al.* (2011) showed that infection-induced behavioural changes in mice could be reversed by *B. longum* NCC3001 administration. *B. infantis* 35624 has been shown in Sprague–Dawley rats to induce an increase in levels of plasma tryptophan, a precursor of serotonin which is a key neurotransmitter within the gut–brain axis possessing antidepressant properties (Desbonnet *et al.*, 2008).

The research in the neurology and psychiatry sector has scarcely reached the point of intervention studies targeted to humans and, in particular, to infants. However, probiotics may offer a potential therapeutic that could beneficially alter the gut–brain axis and modify aberrant behaviours in infant related to altered immune inflammatory outputs such as autism spectrum disorders. One of the few intervention studies performed has assessed the effect of a combination of *L. helveticus* R0052 and *B. longum* R0175 in both human subjects and rats showing that these probiotics reduced anxiety in animals and had beneficial psychological effects with a decrease in serum cortisol in human adults (Messaoudi *et al.*, 2011). Moreover a study on human neonates showed that the pattern of electrical activity in the brain is less complex in neonates born by Caesarean section than in age matched neonates born by vaginal delivery. These results raise the possibility that different colonization patterns influence early postnatal brain development and also have longer term consequences (Kim *et al.*, 2003). The aim for future research in this field is to definitely clarify the effects of the gut microbiota on several brain-related functions in order to identify the microbial species that are critical for the development of a healthy phenotype and those that may have negative impacts on behaviour, mood and emotion in humans. This will pose the basis for targeted probiotic intervention trials, also early in life, in infants.

Prebiotics and their paediatric use

Prebiotics are non-digestible food ingredients that beneficially affects the host by selectively stimulating the growth and/or activity of one or a limited number of bacterial species in the colon, that can improve the host health (Roberfroid, 2007). They are low molecular weight carbohydrates, mainly oligosaccharides (i.e. composed of three to ten monomers linked together), currently mostly constituted of fructose, glucose or galactose. Prebiotics resist gastric acidity, are not hydrolysed by human enzymes and are not absorbed in the upper gastro-intestinal tract. They reach the colon virtually intact and stimulate the colonic microbiota, resulting in growth and fermentation products influencing the host immunity, metabolism and mineral absorption (Roberfroid *et al.*, 2010). The indigestibility of these molecules in the upper gastro-intestinal tract has allowed their definition as non-digestible oligosaccharides (NDOs). It is therefore true that prebiotics allow specific changes both in the composition and/or activity of the intestinal microbiota, which results in nutritional benefits (Saulnier *et al.*, 2009) and positive aspects to human health, as already evidenced in Chapters 21–23 of this book. Beneficial bacteria play useful roles in aspects of nutrition and prevention of disease. The most important of these species belongs to the genera *Lactobacillus* and *Bifidobacterium*. NDOs entering the colon are fermented selectively by members of these genera capable of producing glycolytic enzymes, so they are hydrolysed into mono- or disaccharides which can be transported into the cell where they

Table 26.3 Properties of non-digestible oligosaccharides mainly used in infants (see following section for studies regarding their use)

Name	Structure	Method of manufacture	Degree of polymerization (DP)
Galactooligosaccharides	(Gal)n-Glu	Synthesis from lactose by β-galactosidase	2–8
Fructooligosaccharides (FOS)	(Fru)n-Glu	Synthesis from sucrose or hydrolysis of inulin	2–10 (DP from 2 to 5: short-chain FOS; DP higher than 5: long-chain FOS)
Polydextrose	(Glu)n-Sor	Synthesis from glucose	10–12
Lactulose	Gal-Fru	Isomerization of lactose.	2
Acidic pectic oligosaccharides	Mixture of linear oligomers and small polymers of GalA with a 50% methylation degree	Enzymatic digestion of food grade pectin	n.a.

Gal, galactose; Glu, glucose; Fru, fructose; Sor, sorbitol; GalA, galacturonic acid.

can be metabolized into short-chain fatty acids (SCFAs). SCFAs are responsible for a large array of beneficial effects (Roberfroid *et al.*, 2010). The ability of bifidobacteria to utilize a large variety of oligosaccharides makes them able to adapt and compete in an environment with changing nutritional conditions. For this reason bifidobacteria are considered one of the most efficient groups at utilizing NDOs allowing them to proliferate with respect to other species. The particularly important role that bifidobacteria exert in the infant gut makes the use of prebiotics very important in infant nutrition, in particular for those infants who are not fed by maternal milk. The main characteristics of the most common NDOs used in infants are presented in Table 26.3.

Prebiotic activity of human milk oligosaccharides

The composition of the intestinal microbiota of breast-fed neonates previously outlined is in part due to the presence of complex molecules possessing prebiotic effects in human milk. The most abundant molecules with this role are oligosaccharides, referred to as human milk oligosaccharides (HMOs). HMOs are resistant to digestive processes and thereby reach the colon, where they can be fermented by intestinal bacteria, suggesting they have a particular prebiotic role (Arslanoglu *et al.*, 2007). Because of the intensive interaction between the intestinal microbiota and the epithelium as well as the intestinal immune cells, this prebiotic effect is crucial for the expansion and education of the immune system early in life (Schouten *et al.*, 2011). HMOs are synthesized in the mammary gland by the action of specific glycosyltransferases by the sequential addition of monosaccharide units to the lactose molecule. With few exceptions, all known HMOs have a lactose core and are elongated via linkage to one or more units of galactose and N-acetylglucosamine and can be decorated with several fucose and sialic acid residues. Because of the possibilities for different backbone length, different glycosidic bonds and decoration with different combinations of the basic building blocks, many HMO structures exist (Pfenninger *et al.*, 2002). A simple example of HMO configuration is depicted in Fig. 26.1. Over the years the prebiotic effect of HMOs has been confirmed. *In vitro* fermentation studies clearly demonstrated that the bifidogenic effect of maternal milk is mainly due to the 'non protein fraction' and that HMOs have a pivotal role in stimulating the selective development of bifidobacteria (Ward *et al.*, 2006). In this study, it has been demonstrated that *B. infantis* is able to use complex carbohydrates such as HMOs as the sole source of carbon and energy.

Recent studies focused on the molecular mechanisms underlying the role of human milk on the development of intestinal bifidobacterial community. The identification of genes expressed by *B. breve* strains, upon HMO stimulation, represented preliminary insight in the understanding of the molecular mechanisms governing the initial stages of bacterial colonization in newborns (Turroni *et al.*, 2012). Human milk can also be the direct source of beneficial bacteria: some papers reported the isolation of bifidobacteria from human milk (Martin *et al.*, 2003), although it is also possible that bifidobacteria are introduced into human milk through newborn-mother contacts. Another characteristic substance of human milk is lactoferrin, a glycoprotein member of the

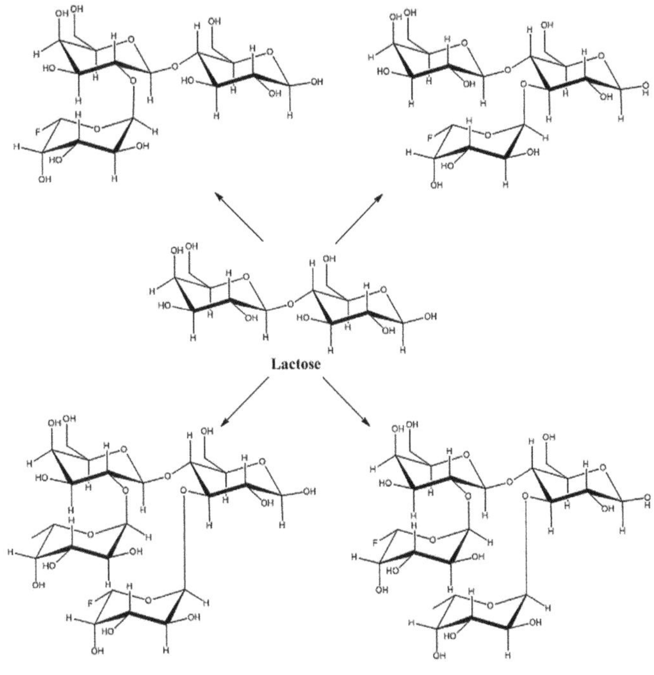

Figure 26.1 Examples of simple chemical structures of some human milk oligosaccharides. F represents fucose residues bound to lactose backbone.

transferrin family. In human milk it is the most abundant protein, on the contrary it is present only in traces in cow's milk. A small percentage of lactoferrin (about 6% to 10%) is estimated not to be digested by breast-fed infants, it could consequently reach the colon and play a role as a prebiotic. The availability of bovine lactoferrin has made it possible to add lactoferrin to infant formulas and to study the effect of feeding such formulas to infants. Recent studies have found that human milk lactoferrin appears to exert a prebiotic effect but addition of bovine lactoferrin in formula milk has little effect on the newborn faecal microbiota (Coppa *et al.*, 2006).

Other groups of substances studied for their possible prebiotic role are nucleotides. Human milk contains high concentrations of preformed nucleotides, whereas cow's milk is devoid of such compounds. Some studies have also suggested a prebiotic role for lactose as it has been demonstrated that lactose reaching the colon stimulates the growth of bifidobacteria, although the amount of ingested lactose reaching a neonate's colon is very low, ~10% of the ingested dose (Szilagyi, 2002). Is it also true that a certain amount of lactose could be metabolized by bifidobacteria (Parche *et al.*, 2006). In particular, studies have demonstrated that *B. longum* exhibits a preferential metabolic pathway for the use of lactose. In addition, bifidobacteria possess several homologous genes encoding enzymes which are involved in the metabolism and transport of numerous sugars.

Formula milk enriched with prebiotics: are they comparable to human milk?

Cow's milk, which is commonly used in the preparation of infant milk formulas, and human milk have significant differences. HMOs are one of the most important components in human milk, in contrast, these oligosaccharides are present only in small amounts in cow's milk. Within the complex mechanisms that regulate the development of the intestinal microbiota, the ability to utilize complex carbohydrates is believed to exert an important influence on the development of specific bacterial strains over others; in the gastrointestinal tract of breast-fed neonates, the relationship between HMOs and the development of bifidobacteria represents a typical example of this situation. For this reason, prebiotic oligosaccharides, in particular galactooligosaccharides (GOS) and fructoligosaccharides (FOS), are abundantly used in infant formula (Coppa *et al.*, 2006; Boehm and Stahl, 2007). The role of these molecules in newborns as an additive to cow's milk started to be studied from the beginning of the twenty-first century and, since then, several achievements have been reached. These are described in detail below and briefly summarized in Table 26.4.

A preliminary study (Boehm *et al.*, 2002) checked the prebiotic capacity of an oligosaccharide mixture consisting of a 9:1 mixture of short chain (sc)-GOS (derived from lactose) and long chain (lc)- FOS (high-molecular-weight fraction of inulin extracted from chicory roots) on pre-term infants. The mixture was designed to mimic the molecular size distribution of HMO and was used at a concentration of 1% (w/v), similar to the oligosaccharide content of human milk, to enrich the composition of a formula-milk. The result of this study showed that infants fed with this formula had a bifidobacteria content that was in the upper range of the values found in infants fed human breast milk. The same oligosaccharide-supplemented formula milk was assessed in healthy term infants, by using two different concentrations of oligosaccharides (0.4% and 0.8%, w/v) (Moro *et al.*, 2002). After a 28-day feeding period, the number of bifidobacteria was significantly increased in both groups receiving supplemented formulas with respect to the control group (fed with a formula milk without prebiotics) and the effect was dose dependent. The number of lactobacilli also increased significantly in both groups fed the supplemented formulas, but there was no dose dependent effect. Supplementation had a significant dose dependent influence on stool consistency, which was softer in prebiotic supplemented newborns, whereas it had no influence on the incidence of side effects (crying, regurgitation, vomiting) or growth. A successive study using the same enriched formula milk with GOS-FOS at 0.8% (w/v) showed that not only the number of beneficial bacteria was increased, but also the number of potential pathogens was reduced in newborns fed prebiotic-supplemented milk compared to the control group (Boehm *et al.*, 2004). Moreover, the amount of bifidobacteria was similar to that typical of breast-fed infants. *In vitro*, the SCFAs produced on the

Table 26.4 Positive effects upon administration of formula milk enriched with prebiotics

Prebiotics added to formula milk	Total prebiotic concentration in the formula milk	Detected positive effects	Reference
scGOS/lcFOS 9:1	0.4% and 0.8% (w/v)	Increase in bifidobacteria and lactobacilli content; softer stools; no side effects	Moro *et al.* (2002)
scGOS/lcFOS, 9:1	0.8% (w/v)	Pathogen reduction in stools	Boehm *et al.* (2004)
scGOS/lcFOS, 9:1	0.6% (w/v)	Metabolic activity of the microbiota similar to that of breast-fed infants; increased saccharolytic activity in the colon with respect to proteolytic activity	Bakker-Zierikzee *et al.* (2005)
scGOS/lcFOS, 9:1	0.8% (w/v)	Fewer infectious episodes; fewer number of infections requiring antibiotics; reduced incidence of infections	Arslanoglu *et al.* (2007)
PDX/GOS, 1:1	0.4% (w/v)	Consistency of stools similar to breast-fed infants	Ziegler *et al.* (2007)
PDX/GOS/LOS, 0.50:0.33:0.17	0.8% (w/v)	Consistency of stools similar to breast-fed infants	Ziegler *et al.* (2007)
PDX/GOS, 1:1	0.4% (w/v)	Gut microbiota composition similar to breast-fed infants	Salminen *et al.* (2014)

PDX, polydextrose; scGOS, short-chain galactooligosaccharides; lcFOS, long-chain fructooligosaccharides.

GOS–FOS mixture were similar to those produced on HMO. In the clinical trials, the pattern of faecal SCFAs of infants fed the oligosaccharide mixture was similar to that of breast-fed infants but was significantly different from that of a group of infants fed with an unsupplemented formula.

Bakker-Zierikzee *et al.* (2005) compared the effects of two infant formulas, one containing a mixture of GOS and FOS 9:1 (0.6% w/v) and the other containing the probiotic strain *B. animalis* Bb-12 (6.0×10^{10} cfu/l infant formula), on the composition and metabolic activity of the intestinal bacteria after 4 months of administration. The control group received a non-supplemented standard formula and a group of breast-fed infants was also included as a reference group. Compared with the groups fed Bb-12 and standard formula, the GOS/FOS formula group showed higher faecal acetate ratio and lactate concentration, similar to that of breast-fed infants. The differences in the SCFAs observed in the GOS/FOS group are consistent with a shift from a more proteolytic/putrefactive to a more saccharolytic colon physiology, which can be considered a health benefit for the infant. Differences in bifidobacteria counts between the GOS/FOS, Bb-12 and the standard groups were not statistically significant, even though bifidobacteria were administered.

A mixture of sc-GOS and lc-FOS (0.8% w/v) has been shown to have prebiotic and immunomodulatory effects comparable to HMO in healthy term infants with a parental history of atopy (Arslanoglu *et al.*, 2007). Although these oligosaccharides are not identical to HMO, studies in preterm and term infants have shown that a formula supplementation with this prebiotic scGOS/lcFOS mixture results in an intestinal microbiota similar to that found in breast-fed infants (Knol *et al.*, 2005; Boehm *et al.*, 2002). Infants supplemented with the scGOS/lcFOS formula milk had fewer infectious episodes, in particular respiratory infections, and fewer number of infections requiring antibiotics during the first 6 month of life.

Other studies explored the use of different prebiotic mixtures. Ziegler *et al.* (2007) studied the effect of different combinations of polydextrose (PDX), GOS, and lactulose (LOS), on the overall growth and tolerance in healthy term infants up to 120 days of age. Beside the control group fed with a standard formula, the other two groups were fed with a control formula supplemented with 0.4% (w/v) of a prebiotic blend containing PDX and GOS, 1:1 or with a control formula supplemented with 0.8% of a prebiotic blend containing PDX, GOS, and LOS, 0.50:0.33:0.17. There were no statistically significant differences among the 3 formula groups for body weight growth rate or length growth rate. The supplemented formula groups had looser stools, more similar to those of breast-fed infants, than the control group. However, a slight increase in the eczema frequency was observed when newborns were fed with the prebiotic blend containing PDX and GOS compared to placebo. Unfortunately, the microbial composition in the stool was not studied in this work.

In a double-blind, randomized, prospective study, healthy term infants vaginally born and exclusively formula-fed received a standard cow's milk-based formula or the same formula added with PDX and GOS (1:1 ratio), 0.4% (w/v). A reference breast-fed group was included. The study allowed to conclude that modifying formula feeding by adding prebiotics may bring the gut microbiota closer to that of breast-fed infants (Salminen *et al.*, 2014).

Prebiotics in infants for prevention of allergy

As already pointed out in section 24.5.4, food allergy and allergic disease represent a substantial health problem that is increasing in children. Among other factors, an altered microbial exposure in the gastrointestinal tract may be partly responsible for the increase of allergic diseases in populations with a western lifestyle. The gastrointestinal microbiota modulates mucosal physiology, barrier function and systemic immunological and inflammatory responses. The efficiency of this gastrointestinal barrier is reduced in the newborn period and it can make newborns more susceptible to allergies. The composition of the intestinal microbiota is different in infants with atopic eczema, in particular a reduced bifidobacteria content has been shown in infants with eczema and atopic sensitization (Osborn and Sinn, 2013). Moreover, the difference in microbial composition may precede the development of eczema. The recognition of the importance of intestinal microbiota in allergies has led to the development of strategies aimed at manipulating bacterial colonization in formula fed infants, including the use of prebiotics.

Among the several studies present in the literature, Osborn and Sinn (2013) considered four of them as the most rigorous from a scientific point of view in the field of prebiotic use in infants to prevent allergies. The outcomes are briefly summarized in Table 26.5. In the study of Moro *et al.* (2006) infants possessing an high risk of eczema development were fed with an hydrolysed whey protein formula supplemented with a mixture of GOS and FOS (0.8% w/v) versus the same formula with added maltodextrin at the same concentration. The incidence of development of the dermatitis was reduced in the GOS-FOS administered group and it was associated with a significantly higher number of bifidobacteria compared with controls; conversely, no significant difference in lactobacilli counts was detected. This study showed for the first time a beneficial effect of prebiotics on the development of atopic dermatitis in a high risk population of infants. Although the mechanism of this effect was not fully elucidated, it appeared likely that oligosaccharides modulate postnatal immune development by altering bowel microbiota and had a potential role in primary allergy prevention during infancy. On the other hand, a different achievement was reported in the study of Ziegler *et al.* (2007), already described in the previous section. Although the authors reported a slight increase in the eczema frequency when newborns were fed with the prebiotic blend containing PDX and GOS, this study did not enrol infants with a high risk of development of eczema. Gruber *et al.* (2010) recruited infants with low atopy risk before the age of 8 weeks to receive a regular cow's milk formula with added GOS and FOS (9:1) and acidic pectic oligosaccharides (OS) (total 0.8%) versus a control group who received cow's milk based formula without added oligosaccharides. The main outcome of this study was that formula supplementation with the specific mixture of oligosaccharides was effective as primary prevention of atopic dermatitis. The authors speculated that the effect persisted beyond the first

Table 26.5 Detected effects on the risk of allergies upon administration of formula milk enriched with prebiotics

Target of the study	Prebiotics used	Detected effects	Reference
Infants with high risk of eczema development	GOS/FOS 9:1 (0.8% w/v)	Reduction in the incidence of dermatitis	Moro *et al.* (2006)
Infants with no specified risk of eczema development	PDX/GOS, 1:1 (0.4% w/v)	Slight increase in the eczema frequency	Ziegler *et al.* (2007)
Infants with low atopy risk	GOS/FOS (9:1) plus acidic oligosaccharides (0.8% w/v)	Positive effect in the prevention of atopic dermatitis; the preventative effect persisted beyond the first birthday; a reduced incidence of respiratory allergy later in life	Gruber *et al.* (2010)
Preterm infants	Mixture of neutral and acidic oligosaccharides	No reduction in the incidence of allergic diseases during the first year of life	Westerbeek *et al.* (2010); a follow-up of this work can be found in Westerbeek *et al.* (2013)

PDX, polydextrose; scGOS, short-chain galactooligosaccharides; lcFOS, long-chain fructooligosaccharides.

birthday and might even result in a reduced incidence of respiratory allergy later in life.

Other studies were aimed at determining the effect of short-term enteral supplementation of GOS/FOS and acidic pectic oligosaccharides during the neonatal period in preterm infants on the incidence of allergic and infectious diseases during the first year of life (Westerbeek *et al.*, 2010 and 2013). A group of newborns received enteral neutral and acidic oligosaccharides supplementation or placebo (maltodextrin) between day 3 and 30 of life. It was concluded that short-term enteral supplementation of a prebiotic mixture of neutral and acidic oligosaccharides during the neonatal period in preterm infants did not decrease the incidence of allergic and infectious diseases during the first year of life. Moreover, enteral supplementation of the prebiotic mixture did not significantly reduce the risk of serious infectious morbidity in preterm infants, but there was a trend towards a lower incidence of serious infectious morbidity, especially for infections with endogenous bacteria (Westerbeek *et al.*, 2010). This was ascribed to an increase of postnatal intestinal colonization (Westerbeek *et al.*, 2013).

Conclusion and future trends

A critical examination of the results present in the literature allows to conclude that the research on probiotics and prebiotics is currently very active and a lot of results on their efficacy are at present available. This intense research activity has been stimulated by the evidence, acquired mainly in the last twenty years and supported by recent findings obtained via next generation molecular techniques, that the correlation between the microbiota composition and sickness exists for several diseases both concerning the gut system and those not strictly enteric. Research has also allowed to conclude that the efficacy of probiotics for the treatment and prevention of a target disease is strain specific and that not only the capability of colonizing the gut is important, but also the production of anti-inflammatory molecules and the stimulation of the gut immune system and the systemic immune system. Gaining new experimental results is particularly important for the research field concerning children, because they represent an interesting target for industry and because non pharmaceutical treatments are particularly looked for.

Extremely relevant are the results obtained to make bottle nutrition more similar to breast feeding with prebiotic supplementation. These studies are all in agreement in stating that prebiotic enriched formula makes the gut microbial composition of formula fed infants more similar to that of breast-fed newborns. On the contrary, the research on the potentiality of prebiotics to stimulate the beneficial colonic microbiota in case of disease is still sparse and mainly focused on allergic diseases. It is foreseeable that the opportunity of using prebiotics to prevent and reduce the symptoms of diseases such as coeliac disease, obesity and neurological upset is explored in future research activities, also considering that no adverse effects have been associated with prebiotic administration in newborn and infants.

Therefore, children are a very important target, if not the main one, for probiotic and prebiotic administration and the European industry is aware of that.

References

Aloisio, I., Santini, C., Biavati, B., Dinelli, G., Cencič, A., Chingwaru, W., Mogna, L., and Di Gioia, D. (2012). Characterization of *Bifidobacterium* spp. strains for the treatment of enteric disorders in newborns. Appl. Microbiol. Biotechnol. *96*, 1561–1576.

An, H.M., Park, S.Y., Lee, D.K., Kim, J.R., Cha, M.K., Lee, S.W., Lim, H.T., Kim, K.J., and Ha, N.J. (2011). Antiobesity and lipid-lowering effects of *Bifidobacterium* spp. in high fat diet-induced obese rats. Lipids Health Dis. *10*, 116.

Anabrees, J., Indrio, F., Paes, B., and AlFaleh, K. (2013). Probiotics for infantile colic: a systematic review. BMC Pediatr. *13*, 186–195.

Arslanoglu, S., Moro, G.E., and Boehm, G. (2007). Early supplementation of prebiotic oligosaccharides protects formula-fed infants against infections during the first 6 months of life. J. Nutr. *137*, 2420–2424.

Arvola, T., Laiho, K., Torkkeli, S., Mykkanen, H., Salminen, S., Maunula, L., and Isolauri, E. (1999). Prophylactic *Lactobacillus* GG reduces antibiotic-associated diarrhoea in children with respiratory infections: a randomized study. Pediatrics *104*, e64.

Bakker-Zierikzee, A.M., Alles, M.S., Knol, J., Kok, F.J., Tolboom, J.J.M., and Bindels, J.G. (2005). Effects of infant formula containing a mixture of galacto- and fructo-oligosaccharides or viable *Bifidobacterium animalis* on the intestinal microflora during the first 4 months of life. Br. J. Nutr. *94*, 783–790.

Barclay, A.R., Stenson, B., Simpson, J.H., Weaver, L.T., and Wilson, D.C. (2007). Probiotics for necrotizing enterocolitis: a systematic review. J. Pediatr. Gastroenterol. Nutr. *45*, 569–576.

Bercik, P., Park, A.J., Sinclair, D., Khoshdel, A., Lu, J., Huang, X., Deng, Y., Blennerhassett, P.A., Fahnestock, M., Moine, D., *et al.* (2011). The anxiolytic effect of *Bifidobacterium longum* NCC3001 involves vagal pathways for gut–brain communication. Neurogastroenterol. Motil. *23*, 1132–1139.

Biasucci, G., Rubini, M., Riboni, S., Morelli, L., Bessi, E., and Retetangos, C. (2010). Mode of delivery affects the bacterial community in the newborn gut. Early Hum. Dev. *86 Suppl. 1*, 13–15.

Biavati, B., Castagnoli, P., Crociani, F., and Trovatelli, L.D. (1984). Species of the *Bifidobacterium* in the faeces of infants. Microbiologica 7, 341–345.

Blaut, M., and Bischoff, S.C. (2010). Probiotics and obesity. Ann. Nutr. Metab. *57* (Suppl.), 20–23.

Boehm, G., Lidestri, M., Casetta, P., Jelinek, J., Negretti, F., Stahl, B., and Marini, A. (2002). Supplementation of a bovine milk formula with an oligosaccharide mixture increases counts of faecal bifidobacteria in preterm infants. Arch. Dis. Child. Fetal Neonatal Ed. *86*, F178–81.

Boehm, G., Jelinek, J., Stahl, B., van Laere, K., Knol, J., Fanaro, S., Moro, G., and Vigi, V. (2004). Prebiotics in infant formulas. J. Clin. Gastroenterol. *38*, S76–9.

Boehm, G., and Stahl, B. (2007). Oligosaccharides from milk. J. Nutr. *137*, 847S-9S.

Boyle, R.J., and Tang, M.L.K. (2006). The role of probiotics in the management of allergic disease. Clin. Exp. Allergy *36*, 568–576.

Braga, T.D., da Silva, G.A.P., de Lira, P.I.C., and de Carvalho Lima, M. (2011). Efficacy of *Bifidobacterium breve* and *Lactobacillus casei* oral supplementation on necrotizing enterocolitis in very-low-birth-weight preterm infants: a double-blind, randomized, controlled trial. Am. J. Clin. Nutr. *93*, 81–86.

Cabana, M.D., McKean, M., Wong, A.R., Chao, C., and Caughey, A.B. (2007). Examining the hygiene hypothesis: the trial of infant probiotic supplementation. Paediatr. Perinat. Epidemiol. *21* (Suppl. 3), 23–28.

Cani, P.D., Neyrinck, A.M., Fava, F., Knauf, C., Burcelin, R.G., Tuohy, K.M., Gibson, G.R., and Delzenne, N.M. (2007). Selective increases of bifidobacteria in gut microflora improve high-fat-diet-induced diabetes in mice through a mechanism associated with endotoxaemia. Diabetologia *50*, 2374–2383.

Cencic, A., and Langerholc, T. (2010). Functional cell models of the gut and their applications in food microbiology-a review. Int. J. Food Microbiol. *141* (Suppl.), S4–14.

Chen, C.-C., and Walker, W.A. (2011). Clinical applications of probiotics in gastrointestinal disorders in children. Nat. Med. J. India. *24*, 153–160.

Claesson, M.J., Cusack, S., O'Sullivan, O., Greene-Diniz, R., de Weerd, H., Flannery, E., Marchesi, J.R., Falush, D., Dinan, T., Fitzgerald, G., *et al.* (2011). Composition, variability, and temporal stability of the intestinal microbiota of the elderly. Proc. Natl. Acad. Sci. U.S.A. *108* (Suppl.), 4586–4591.

Claud, E.C., and Walker, W.A. (2001). Hypothesis: inappropriate colonization of the premature intestine can cause neonatal necrotizing enterocolitis. FASEB J. *15*, 1398–1403.

Collado, M.C., Donat, E., Ribes-Koninckx, C., Calabuig, M., and Sanz, Y. (2008). Imbalances in faecal and duodenal *Bifidobacterium* species composition in active and non-active coeliac disease. BMC Microbiol. *8*, 232.

Collins, S.M., and Bercik, P. (2009). The relationship between intestinal microbiota and the central nervous system in normal gastrointestinal function and disease. Gastroenterology *136*, 2003–2014.

Coppa, G.V., Zampini, L., Galeazzi, T., and Gabrielli, O. (2006). Prebiotics in human milk: a review. Dig. Liver Dis. *38 Suppl. 2*, S291–4.

Coppa, G., and Gabrielli, O. (2008). Human milk oligosaccharides as prebiotics. In Therapeutic microbiology: probiotics and related strategies, J. Versalovic, and M. Wilson, eds. (Washington DC, USA: American Society for Microbiology Press), pp. 131–146.

Cryan, J.F., and O'Mahony, S.M. (2011). The microbiome–gut–brain axis: from bowel to behavior. Neurogastroenterol. Motil. *23*, 187–192.

De Angelis, M., Rizzello, C.G., Fasano, A., Clemente, M.G., De Simone, C., Silano, M., De Vincenzi, M., Losito, I., and Gobbetti, M. (2006). VSL#3 probiotic preparation has the capacity to hydrolyze gliadin polypeptides responsible for Coeliac Sprue. Biochim. Biophys. Acta *1762*, 80–93.

Desbonnet, L., Garrett, L., Clarke, G., Bienenstock, J., and Dinan, T.G. (2008). The probiotic *Bifidobacteria infantis*: An assessment of potential antidepressant properties in the rat. J. Psychiatr. Res. *43*, 164–174.

Deshpande, G., Rao, S., Patole, S., and Edward, K. (2007). Probiotics for prevention of necrotizing enterocolitis in preterm neonates with very low birthweight: a systematic review of randomized controlled trials. Lancent *369*, 1614–1620.

Deshpande, G., Rao, S., Patole, S., and Bulsara, M. (2010). Updated meta-analysis of probiotics for preventing necrotizing enterocolitis in preterm neonates. Pediatrics *125*, 921–930.

Deshpande, G.C., Rao, S.C., Keil, A.D., and Patole, S.K. (2011). Evidence-based guidelines for use of probiotics in preterm neonates. BMC Med. *9*, 92.

Di Gioia, D., Aloisio, I., Mazzola, G., and Biavati, B. (2014). Bifidobacteria: their impact on gut microbiota composition and their applications as probiotics in infants. Appl. Microbiol. Biotechnol. *98*, 563–577.

Fallani, M., Young, D., Scott, J., Norin, E., Amarri, S., Adam, R., Aguilera, M., Khanna, S., Gil, A., Edwards, C.A., *et al.* (2010). Intestinal microbiota of 6-week-old infants across Europe: geographic influence beyond delivery mode, breast-feeding, and antibiotics. J. Pediatr. Gastroenterol. Nutr. *51*, 77–84.

Grandy, G., Medina, M., Soria, R., Terán, C.G., and Araya, M. (2010). Probiotics in the treatment of acute rotavirus diarrhoea. A randomized, double-blind, controlled trial using two different probiotic preparations in Bolivian children. BMC Infect. Dis. *10*, 253.

Green, P.H.R., and Cellier, C. (2007). Celiac disease. N. Engl. J. Med. *357*, 1731–1743.

Grüber, C., van Stuijvenberg, M., Mosca, F., Moro, G., Chirico, G., Braegger, C.P., Riedler, J., Boehm, G., and Wahn, U. (2010). Reduced occurrence of early atopic dermatitis because of immunoactive prebiotics among low-atopy-risk infants. J. Allergy Clin. Immunol. *126*, 791–797.

Haarman, M., and Knol, J. (2005). Quantitative real-time PCR assays to identify and quantify faecal *Bifidobacterium* species in infants receiving a prebiotic infant formula. Appl. Environ. Microbiol. *71*, 2318–2324.

Halken, S. (2004). Prevention of allergic disease in childhood: clinical and epidemiological aspects of primary and secondary allergy prevention. Pediatr. Allergy Immunol. *15 Suppl. 16*, 4–5.

Harder, T., Bergmann, R., Kallischnigg, G., and Plagemann, A. (2005). Duration of breastfeeding and risk of overweight: a meta-analysis. Am. J. Epidemiol. *162*, 397–403.

Jankowska, A., Laubitz, D., Antushevich, H., Zabielski, R., and Grzesiuk, E. (2008). Competition of *Lactobacillus paracasei* with *Salmonella enterica* for adhesion to Caco-2 cells. J. Biomed. Biotechnol. *59*, 795–810.

Kalliomäki, M., Kirjavainen, P., Eerola, E., Kero, P., Salminen, S., and Isolauri, E. (2001). Distinct patterns of neonatal gut microflora in infants in whom atopy was and was not developing. J. Allergy Clin. Immunol. *107*, 129–134.

Kalliomaki, M., Carmen Collado, M., Salminen, S., and Isolauri, E. (2008). Early differences in faecal microbiota composition in children may predict overweight. Am. J. Clin. Nutr. *87*, 534–538.

Kang, C., Gayen, S., Wang, W., Severin, R., Chen, A.S., Lim, H.A., Chia, C.S.B., Schüller, A., Doan, D.N.P., Poulsen, A., *et al.* (2013). Exploring the binding of peptidic West Nile virus NS2B-NS3 protease inhibitors by NMR. Antiviral Res. *97*, 137–144.

Kaur, I.P., Chopra, K., and Saini, A. (2002). Probiotics: potential pharmaceutical applications. Eur. J. Pharm. Sci. *15*, 1–9.

Khailova, L., Dvorak, K., Arganbright, K.M., Halpern, M.D., Kinouchi, T., Yajima, M., and Dvorak, B. (2009). *Bifidobacterium bifidum* improves intestinal integrity in a rat model of necrotizing enterocolitis. Am. J. Physiol. Gastrointest. Liver Physiol. *1297*, 940–949.

Kim, H.-R., Jung, K.-Y., Kim, S.-Y., Ko, K.-O., Lee, Y.-M., and Kim, J.-M. (2003). Delivery modes and neonatal EEG: spatial pattern analysis. Early Hum. Dev. *75*, 35–53.

Klaassens, E.S., Boesten, R.J., Haarman, M., Knol, J., Schuren, F.H., Vaughan, E.E., and de Vos, W.M. (2009). Mixed-species genomic microarray analysis of fecal samples reveals differential transcriptional responses of bifidobacteria in breast- and formula-fed infants. Appl. Environ. Microbiol. *75*, 2668–2676.

Knol, J., Scholtens, P., Kafka, C., Steenbakkers, J., Gro, S., Helm, K., Klarczyk, M., Schöpfer, H., Böckler, H.-M., and Wells, J. (2005). Colon microflora in infants fed formula with galacto- and fructo-oligosaccharides: more like breast-fed infants. J. Pediatr. Gastroenterol. Nutr. *40*, 36–42.

Koenig, J.E., Spor, A., Scalfone, N., Fricker, A.D., Stombaugh, J., Knight, R., Angenent, L.T., and Ley, R.E. (2011). Succession of microbial consortia in the developing infant gut microbiome. Proc. Natl. Acad. Sci. U.S.A. *108 Suppl*, 4578–4585.

Kuitunen, M., Kukkonen, K., Juntunen-Backman, K., Korpela, R., Poussa, T., Tuure, T., Haahtela, T., and Savilahti, E. (2009). Probiotics prevent IgE-associated allergy until age 5 years in cesarean-delivered children but not in the total cohort. J. Allergy Clin. Immunol. *123*, 335–341.

Laux, D., Cohen, P., and Coneay, T. (2005). Role of the mucus layer in bacterial colonization of the intestine. In Colonization of the mucosa surfaces, J. Nataro, P. Cohen, H. Mobley, and J. Weiser eds. (Washington DC, USA: American Society for Microbiology Press), pp 199–212.

Ley, R.E., Peterson, D.A., and Gordon, J.I. (2006). Ecological and evolutionary forces shaping microbial diversity in the human intestine. Cell *124*, 837–848.

Lin, H.-C., Hsu, C.-H., Chen, H.-L., Chung, M.-Y., Hsu, J.-F., Lien, R., Tsao, L.-Y., Chen, C.-H., and Su, B.-H. (2008). Oral probiotics prevent necrotizing enterocolitis in very low birth weight preterm infants: a multicenter, randomized, controlled trial. Pediatrics *122*, 693–700.

Lindfors, K., Blomqvist, T., Juuti-Uusitalo, K., Stenman, S., Venäläinen, J., Mäki, M., and Kaukinen, K. (2008). Live probiotic *Bifidobacterium lactis* bacteria inhibit the toxic effects induced by wheat gliadin in epithelial cell culture. Clin. Exp. Immunol. *152*, 552–558.

Londei, M., Ciacci, C., Ricciardelli, I., Vacca, L., Quaratino, S., and Maiuri, L. (2005). Gliadin as a stimulator of innate responses in celiac disease. Mol. Immunol. *42*, 913–918.

Lue, K.-H., Sun, H.-L., Lu, K.-H., Ku, M.-S., Sheu, J.-N., Chan, C.-H., and Wang, Y.-H. (2012). A trial of adding *Lactobacillus johnsonii* EM1 to levocetirizine for treatment of perennial allergic rhinitis in children aged 7–12 years. Int. J. Pediatr. Otorhinolaryngol. *76*, 994–1001.

Ma, X., Hua, J., and Li, Z. (2008). Probiotics improve high fat diet-induced hepatic steatosis and insulin resistance by increasing hepatic NKT-cells. J. Hepatol. *49*, 821–830.

Magne, F., Hachelaf, W., Suau, A., Boudraa, G., Mangin, I., Touhami, M., Bouziane-Nedjadi, K., and Pochart, P. (2006). A longitudinal study of infant faecal microbiota during weaning. FEMS Microbiol. Ecol. *58*, 563–571.

Mai, V., Young, C.M., Ukhanova, M., Wang, X., Sun, Y., Casella, G., Theriaque, D., Li, N., Sharma, R., Hudak, M., *et al.* (2011). Fecal microbiota in premature infants prior to necrotizing enterocolitis. PLoS One *6*, e20647.

Martín, R., Langa, S., Reviriego, C., Jimínez, E., Marín, M.L., Xaus, J., Fernández, L., and Rodríguez, J.M. (2003). Human milk is a source of lactic acid bacteria for the infant gut. J. Pediatr. *143*, 754–758.

Meresse, B., Ripoche, J., Heyman, M., and Cerf-Bensussan, N. (2009). Celiac disease: from oral tolerance to intestinal inflammation, autoimmunity and lymphomagenesis. Mucosal Immunol. *2*, 8–23.

Messaoudi, M., Lalonde, R., Violle, N., Javelot, H., Desor, D., Nejdi, A., Bisson, J.-F., Rougeot, C., Pichelin, M., Cazaubiel, M., *et al.* (2011). Assessment of psychotropic-like properties of a probiotic formulation (*Lactobacillus helveticus* R0052 and *Bifidobacterium longum* R0175). in rats and human subjects. Br. J. Nutr. *105*, 755–764.

Million, M., Maraninchi, M., Henry, M., Armougom, F., Richet, H., Carrieri, P., Valero, R., Raccah, D., Vialettes, B., and Raoult, D. (2012). Obesity-associated gut microbiota is enriched in *Lactobacillus reuteri* and depleted in *Bifidobacterium animalis* and *Methanobrevibacter smithii*. Int. J. Obes. (Lond.) *36*, 817–825.

Million, M., Angelakis, E., Maraninchi, M., Henry, M., Giorgi, R., Valero, R., Vialettes, B., and Raoult, D. (2013). Correlation between body mass index and gut concentrations of *Lactobacillus reuteri, Bifidobacterium animalis, Methanobrevibacter smithii* and *Escherichia coli*. Int. J. Obes. (Lond.) *37*, 1460–1466.

Montier, Y., Lorentz, A., Krämer, S., Sellge, G., Schock, M., Bauer, M., Schuppan, D., and Bischoff, S.C. (2012). Central role of IL-6 and MMP-1 for cross talk between human intestinal mast cells and human intestinal fibroblasts. Immunobiology *217*, 912–919.

Moreno Munoz, J.A., Chenoll, E., Casinos, B., Bataller, E., Ramon, D., Genoves, S., Montava, R., Ribes, J.M., Buesa, J., Fabrega, J., *et al.* (2011). Novel probiotic *Bifidobacterium longum* subsp. *infantis* CECT 7210 strain active against rotavirus infections. Appl. Environ. Microbiol. *77*, 8775–8783.

Moro, G., Minoli, I., Mosca, M., Fanaro, S., Jelinek, J., Stahl, B., and Boehm, G. (2002). Dosage-related bifidogenic effects of galacto- and fructooligosaccharides in formula-fed term infants. J. Pediatr. Gastroenterol. Nutr. *34*, 291–295.

Moro, G., Arslanoglu, S., Stahl, B., Jelinek, J., Wahn, U., and Boehm, G. (2006). A mixture of prebiotic oligosaccharides reduces the incidence of atopic dermatitis during the first six months of age. Arch. Dis. Child. *91*, 814–819.

Nataro, J. (2005). Interactions of the commensal flora with the human gastrointestinal tract. In Colonization of the mucosa surfaces, J. Nataro, P. Cohen, H. Mobley, and J. Weiser eds. (Washington DC, USA: American Society for Microbiology Press), pp 179–186.

Oncel, M.Y., Sari, F.N., Arayici, S., Guzoglu, N., Erdeve, O., Uras, N., Oguz, S.S., and Dilmen, U. (2014). *Lactobacillus reuteri* for the prevention of necrotizing enterocolitis in very low birthweight infants: a randomized controlled trial. Arch. Dis. Child. Fetal Neonatal Ed. *99*, F110–5.

Osborn, D.A., and Sinn, J.K.H. (2013). Prebiotics in infants for prevention of allergy. Cochrane Database Syst. Rev. *3*, CD006474.

Osterballe, M., Hansen, T.K., Mortz, C.G., Høst, A., and Bindslev-Jensen, C. (2005). The prevalence of food hypersensitivity in an unselected population of children and adults. Pediatr. Allergy Immunol. *16*, 567–573.

Ouwehand, A.C., Isolauri, E., He, F., Hashimoto, H., Benno, Y., and Salminen, S. (2001). Differences in *Bifidobacterium* flora composition in allergic and healthy infants. J. Allergy Clin. Immunol. *108*, 144–145.

Owen, C.G., Martin, R.M., Whincup, P.H., Smith, G.D., and Cook, D.G. (2005). Effect of infant feeding on the risk of obesity across the life course: a quantitative review of published evidence. Pediatrics *115*, 1367–1377.

Özdemir, Ö. (2009). Gut flora development in infancy and its effect on immune system. J. Pediatr. Inf. 3, 202–203.

Palma, G. De, Capilla, A., Nova, E., Castillejo, G., Varea, V., Pozo, T., Garrote, J.A., Polanco, I., López, A., Ribes-Koninckx, C., *et al.* (2012). Influence of milk-feeding type and genetic risk of developing coeliac disease on intestinal microbiota of infants: the PROFICEL study. PLoS One *7*, e30791.

Parche, S., Beleut, M., Rezzonico, E., Jacobs, D., Arigoni, F., Titgemeyer, F., and Jankovic, I. (2006). Lactose-over-glucose preference in *Bifidobacterium longum* NCC2705: glcP, encoding a glucose transporter, is subject to lactose repression. J. Bacteriol. *188*, 1260–1265.

Penders, J., Thijs, C., van den Brandt, P. a, Kummeling, I., Snijders, B., Stelma, F., Adams, H., van Ree, R., and Stobberingh, E.E. (2006). Gut microbiota composition and development of atopic manifestations in infancy: the KOALA Birth Cohort Study. Gut *56*, 661–667.

Pfenninger, A., Karas, M., Finke, B., and Stahl, B. (2002). Structural analysis of underivatized neutral human milk oligosaccharides in the negative ion mode by nano-electrospray MS(n) (part 1: methodology). J. Am. Soc. Mass Spectrom. *13*, 1331–1340.

Rhee, S.H., Pothoulakis, C., and Mayer, E.A. (2009). Principles and clinical implications of the brain-gut-enteric microbiota axis. Nat. Rev. Gastroenterol. Hepatol. *6*, 306–314.

Roberfroid, M. (2007). Prebiotics: the concept revisited. J. Nutr. *137*, 830S-7S.

Roberfroid, M., Gibson, G.R., Hoyles, L., McCartney, A.L., Rastall, R., Rowland, I., Wolvers, D., Watzl, B., Szajewska, H., Stahl, B., *et al.* (2010). Prebiotic effects: metabolic and health benefits. Br. J. Nutr. *104* (Suppl.), S1–63.

Salminen, S., Isolauri, E., Endo, A., and Scalabrin, D. (2014). Early gut colonization with lactobacilli in vaginally born infants fed breast milk or infant formula (637.9). FASEB J. *28*, 637–639.

Satokari, R.M., Vaughan, E.E., Favier, C.F., Doré, J., Edwards, C., and Vos, W.M. de (2002). Diversity of *Bifidobacterium* and *Lactobacillus* spp. in breast-fed and formula-fed infants as assessed by 16S rDNA sequence differences. Microb. Ecol. Health Dis. *14*, 97–105.

Saulnier, D.M.A., Spinler, J.K., Gibson, G.R., and Versalovic, J. (2009). Mechanisms of probiosis and prebiosis: considerations for enhanced functional foods. Curr. Opin. Biotechnol. *20*, 135–141.

Savino, F., Pelle, E., Palumeri, E., Oggero, R., and Miniero, R. (2007). *Lactobacillus reuteri* (American Type Culture Collection Strain 55730). versus simethicone in the treatment of infantile colic: a prospective randomized study. Pediatrics *119*, 124–130.

Savino, F., Cordisco, L., Tarasco, V., Calabrese, R., Palumeri, E., and Matteuzzi, D. (2009). Molecular identification of coliform bacteria from colicky breastfed infants. Acta Paediatr. *98*, 1582–1588.

Savino, F., Cordisco, L., Tarasco, V., Palumeri, E., Calabrese, R., Oggero, R., Roos, S., and Matteuzzi, D. (2010). *Lactobacillus reuteri* DSM 17938 in infantile colic: a randomized, double-blind, placebo-controlled trial. Pediatrics *126*, 526–533.

Savino, F., Cordisco, L., Tarasco, V., Locatelli, E., Di Gioia, D., Oggero, R., and Matteuzzi, D. (2011). Antagonistic effect of *Lactobacillus* strains against gas-producing coliforms isolated from colicky infants. BMC Microbiol. *11*, 157.

Schouten, B., Van Esch, B.C.A.M., Kormelink, T.G., Moro, G.E., Arslanoglu, S., Boehm, G., Knippels, L.M.J., Redegeld, F.A., Willemsen, L.E.M., and Garssen, J. (2011). Non-digestible oligosaccharides reduce immunoglobulin free light-chain concentrations in infants at risk for allergy. Pediatr. Allergy Immunol. *22*, 537–542.

Schwiertz, A., Taras, D., Schäfer, K., Beijer, S., Bos, N. a, Donus, C., and Hardt, P.D. (2010). Microbiota and SCFA in lean and overweight healthy subjects. Obesity (Silver Spring) *18*, 190–195.

Soh, S.E., Aw, M., Gerez, I., Chong, Y.S., Rauff, M., Ng, Y.P.M., Wong, H.B., Pai, N., Lee, B.W., and Shek, L.P.-C. (2009). Probiotic supplementation in the first 6 months of life in at risk Asian infants-effects on eczema and atopic sensitization at the age of 1 year. Clin. Exp. Allergy *39*, 571–578.

Solís, G., de Los Reyes-Gavilan, C.G., Fernández, N., Margolles, a, and Gueimonde, M. (2010). Establishment and development of lactic acid bacteria and bifidobacteria microbiota in breast-milk and the infant gut. Anaerobe *16*, 307–310.

Stockman, J.A. (2009). Newborn: probiotics for prevention of necrotizing enterocolitis in preterm neonates with very low birthweight: a systematic review of randomized controlled trials. In G. Deshpande, S. Rao, and S. Patole eds. (Philadelphia PA, USA: Yearbook of Pediatrics), pp 441–443.

Sudo, N., Chida, Y., Aiba, Y., Sonoda, J., Oyama, N., Yu, X.-N., Kubo, C., and Koga, Y. (2004). Postnatal microbial colonization programs the hypothalamic-pituitary-adrenal system for stress response in mice. J. Physiol. *558*, 263–275.

Sung, V., Hiscock, H., Tang, M.L.K., Mensah, F.K., Nation, M.L., Satzke, C., Heine, R.G., Stock, A., Barr, R.G., and Wake, M. (2014). Treating infant colic with the probiotic *Lactobacillus reuteri*: double blind, placebo controlled randomized trial. BMJ. *348*, g2107.

Symonds, E.L., O'Mahony, C., Lapthorne, S., O'Mahony, D., Mac Sharry, J., O'Mahony, L., and Shanahan, F. (2012). *Bifidobacterium infantis* 35624 protects against salmonella-induced reductions in digestive enzyme activity in mice by attenuation of the host inflammatory response. Clin. Transl. Gastroenterol. 3, e15.

Szajewska, H., Ruszczyński, M., and Radzikowski, A. (2006). Probiotics in the prevention of antibiotic-associated diarrhoea in children: a meta-analysis of randomized controlled trials. J. Pediatr. *149*, 367–372.

Szajewska, H., Gyrczuk, E., and Horvath, A. (2013). *Lactobacillus reuteri* DSM 17938 for the management of infantile colic in breastfed infants: a randomized, double-blind, placebo-controlled trial. J. Pediatr. *162*, 257–262.

Szilagyi, A. (2002). Review article: lactose-a potential prebiotic. Aliment. Pharmacol. Ther. *16*, 1591–1602.

Taibi, A., and Comelli, E.M. (2014). Practical approaches to probiotics use. Appl. Physiol. Nutr. Metab. 1–7.

Tennyson, C. a, and Friedman, G. (2008). Microecology, obesity, and probiotics. Curr. Opin. Endocrinol. Diabetes. Obes. *15*, 422–427.

Tham, C.S.-C., Peh, K.-K., Bhat, R., and Liong, M.-T. (2011). Probiotic properties of bifidobacteria and lactobacilli isolated from local dairy products. Ann. Microbiol. *62*, 1079–1087.

Turroni, F., Peano, C., Pass, D. a, Foroni, E., Severgnini, M., Claesson, M.J., Kerr, C., Hourihane, J., Murray, D., Fuligni, F., *et al.* (2012). Diversity of bifidobacteria within the infant gut microbiota. PLoS One 7, e36957.

Underwood, M.A., Kananurak, A., Coursodon, C.F., Adkins-Reick, C.K., Chu, H., Bennett, S.H., Wehkamp, J., Castillo, P.A., Leonard, B.C., Tancredi, D.J., *et al.* (2012). *Bifidobacterium bifidum* in a rat model of necrotizing enterocolitis: antimicrobial peptide and protein responses. Pediatr. Res. *71*, 546–551.

Vandenplas, Y., and De Hert, S.G. (2011). Randomized clinical trial: the synbiotic food supplement Probiotical vs. placebo for acute gastroenteritis in children. Aliment. Pharmacol. Ther. *34*, 862–867.

Verdu, E.F. (2009). Probiotics effects on gastrointestinal function: beyond the gut? Neurogastroenterol. Motil. *21*, 477–480.

Viljanen, M., Savilahti, E., Haahtela, T., Juntunen-Backman, K., Korpela, R., Poussa, T., Tuure, T., and Kuitunen, M. (2005). Probiotics in the treatment of atopic eczema/dermatitis syndrome in infants: a double-blind placebo-controlled trial. Allergy *60*, 494–500.

Wanke, M., and Szajewska, H. (2012). Lack of an effect of *Lactobacillus reuteri* DSM 17938 in preventing nosocomial diarrhoea in children: a randomized, double-blind, placebo-controlled trial. J. Pediatr. *161*, 40–43.

Ward, R.E., Niñonuevo, M., Mills, D.A., Lebrilla, C.B., and German, J.B. (2006). *In vitro* fermentation of breast milk oligosaccharides by *Bifidobacterium infantis* and *Lactobacillus gasseri*. Appl. Environ. Microbiol. 72, 4497–4499.

Westerbeek, E.A.M., van den Berg, A., Lafeber, H.N., Knol, J., Fetter, W.P.F., and van Elburg, R.M. (2006). The intestinal bacterial colonization in preterm infants: a review of the literature. Clin. Nutr. *25*, 361–368.

Westerbeek, E.A., van den Berg, J.P., Lafeber, H.N., Fetter, W.P., Boehm, G., Twisk, J.W., and van Elburg, R.M. (2010). Neutral and acidic oligosaccharides in preterm infants: a randomized, double-blind, placebo-controlled trial. Am. J. Clin. Nutr. *91*, 679–686.

Westerbeek, E.A.M., Slump, R.A., Lafeber, H.N., Knol, J., Georgi, G., Fetter, W.P.F., and van Elburg, R.M. (2013). The effect of enteral supplementation of specific neutral and acidic oligosaccharides on the faecal microbiota and intestinal microenvironment in preterm infants. Eur. J. Clin. Microbiol. Infect. Dis. *32*, 269–276.

Wiegering, V., Kaiser, J., Tappe, D., Weissbrich, B., Morbach, H., and Girschick, H.J. (2011). Gastroenteritis in childhood: a retrospective study of 650 hospitalized paediatric patients. Int. J. Infect. Dis. *15*, 401–407.

Yin, Y.-N., Yu, Q.-F., Fu, N., Liu, X.-W., and Lu, F.-G. (2010). Effects of four Bifidobacteria on obesity in high-fat diet induced rats. World J. Gastroenterol. *16*, 3394–3401.

Ziegler, E., Vanderhoof, J.A., Petschow, B., Mitmesser, S.H., Stolz, S.I., Harris, C.L., and Berseth, C.L. (2007). Term infants fed formula supplemented with selected blends of prebiotics grow normally and have soft stools similar to those reported for breast-fed infants. J. Pediatr. Gastroenterol. Nutr. *44*, 359–364.

Pro- and Prebiotics in Management of Patients with Irritable Bowel Syndrome

27

Ratnakar Shukla, Ujjala Ghoshal and Uday C. Ghoshal

Abstract

Recently, several studies suggested that altered quantity and quality in gut microbiota, called dysbiosis, are important in the pathogenesis of irritable bowel syndrome (IBS). It is therefore worthwhile to evaluate whether therapeutic manipulation of the gut microbiota using probiotics would be effective and safe in management of these patients. Several studies showed that probiotics are effective in manipulation of altered gut microbiota and in improving the global IBS symptoms. Probiotics reduce visceral hypersensitivity, improve colonic transit time, enhance the intestinal epithelial barrier, modulate immune response via production of anti-inflammatory and regulatory cytokines, and thus, inhibit the inflammation. Prebiotics, which are sometimes added to probiotics (symbiotics) improve the fermentation pattern, stool consistency, abdominal pain and bloating and flatulence in patients with IBS. Several meta-analyses have confirmed the efficacy of probiotics in management of IBS. Hence, it can be concluded that probiotics are useful in treatment of IBS, particularly diarrhoea predominant subtypes, and most studies showed that abdominal pain, bloating and flatulence are the symptoms, which were relieved most.

Introduction

Irritable bowel syndrome (IBS) is an enigmatic disorder, which is characterized by abdominal discomfort or pain, altered stool form and passage, bloating or distension and flatulence (Dai *et al.*, 2013; Lee and Bak, 2011). There are several possible pathophysiological mechanisms contributing to the development of IBS, such as altered gastrointestinal motility, visceral hypersensitivity, genetic factors, abnormal brain–gut axis and dysregulated gastro-intestinal neuro-peptides and hormones. Initially, IBS was believed to be related largely to psychological factors (Bixquert Jimenez, 2009; Ghoshal *et al.*, 2014). However, in the recent times, organic factors including alternation in normal gut microbiota (dysbiosis) have been incriminated in the pathogenesis of this disorder (Spiller, 2008). The evidence in favour of this hypothesis include: studies showing quantitative increase in gut microbes (such as small intestinal bacterial overgrowth) and qualitative change in faecal microbiota among patients with IBS compared to controls, and trials showing manipulation of gut microbiota using probiotics and antibiotics to be effective in treatment of the condition. Several studies reported dysbiosis in patients with IBS such as reduction in the number of anaerobic bacteria mainly *Bifidobacterium*, *Bacteroides*, *Lactobacillus* and increase in number of *Enterobacteriaceae* and *Clostridium* species (Irvine *et al.*, 2008; Nobaek *et al.*, 2000; Spiller, 2008). Recent reports have shown that different sub-types of IBS have different microbiota and patients with diarrhoea-predominant IBS more often have altered gut microbiota than those with other sub-types of IBS (Malinen *et al.*, 2005). For example, reduced levels of *Lactobacillus* and *Bifidobacterium* species in diarrhoea predominant (IBS-D) and increased level of *Veillonella* species and *Ruminococcus bromii* in constipation predominant (IBS-C) have been reported (Tack *et al.*, 2012). Furthermore, low-grade inflammation has also been reported in a subgroup of patients with IBS (Brint *et al.*, 2011).

A symbiotic relationship exists between complex microbiota in the gut and the host. Disturbance of this mutual interaction between gut microbiota and the host may lead to low grade inflammation due to abnormal cytokine response, which may cause IBS (Ghoshal *et al.*, 2014). Gut-associated bacteria can be classified into two types: luminal bacteria and mucosa-associated bacteria. Luminal microbiota in IBS is hypothesized to be responsible for abdominal bloating, and flatulence due to fermentation of carbohydrates while mucosal microbiota is more likely to be associated with host immune response leading to inflammation in the gastrointestinal tract (Tack *et al.*, 2012).

Since altered gut microbiota is an important factor in a proportion of patients with IBS, its manipulation using antibiotics, probiotics and prebiotics may have potential role in management of these patients (Bixquert Jimenez, 2009). Recently, in a large randomized controlled trial, rifaximin was found to relieve symptoms such as bloating, abdominal pain, and loose or watery stools among non-constipated patients with IBS compared with placebo (Pimentel *et al.*, 2011). Recent evidence suggests that several probiotic strains also have the ability to reduce abdominal bloating, pain or discomfort and improve the quality of life in patients with IBS (Bixquert Jimenez, 2009; Floch, 2003; Ng *et al.*, 2013). In this chapter, we review the current literature on probiotics and prebiotics in treatment of patients with IBS.

Probiotics

Probiotics are live microorganisms, which alleviate gut-dysbiosis when administered in sufficient quantity (Fuller, 1989; Schrezenmeir and de Vrese, 2001). Furthermore, some probiotic strains inhibit production of cytokines and chemokines and sustain the integrity of intestinal epithelial cells. They also play a pivotal role in fermentation and maintenance of the homeostatic condition of intestinal microbiota by reducing the growth of pathogenic bacteria (Ghoshal *et al.*, 2012; Neish, 2009; Stanton *et al.*, 2005; Zhang *et al.*, 2008). Recent studies have shown that lactic acid bacteria have ability to act against the harmful microbes and are helpful in relieving several gastrointestinal ailments (Chang, 2014; Kailasapathy and Chin, 2000; Zeng *et al.*, 2008). Probiotics also induce production of lactic acid and bacteriocins, which potentiate the gut commensal bacteria via reducing the growth of invasive bacteria (Ghoshal *et al.*, 2012; Pessione, 2012). Some probiotic bacteria stimulate the production of butyric acid. Butyric acid has the ability to increase the number of enterocytes and lead to the neutralization of the activity of dietary carcinogens (Kailasapathy and Chin, 2000; Lupton, 2004). Thus, emerging evidence based on studies including randomized controlled trials suggests that probiotics play an important role in several functional gastrointestinal disorders like IBS, inflammatory bowel disease (ulcerative colitis and Crohn's disease), lactose intolerance and small intestinal bacterial overgrowth (SIBO) (Grover *et al.*, 2012; Vyas and Ranganathan, 2012).

Probiotic can be classified according to the genus, species and strain. It is important to understand the properties at these three levels for therapeutic purposes (Sanders, 2008; Vieira *et al.*, 2013). The efficacy and health benefits of each probiotic strain should not only be evaluated *in vitro* but should also be studied in various gastrointestinal disorders (Kailasapathy and Chin, 2000). Probiotics should have 10^6 colony forming units/ml (cfu/ml) until the date of expiry as it exerts biological activity when the organisms are viable in sufficient number (Kailasapathy and Chin, 2000; Tamime *et al.*, 1995). The therapeutic dose of probiotics is proposed to be 10^8–10^9 cells (Kailasapathy and Chin, 2000). Most common bacterial species used in probiotics are *Lactobacillus* species (*Lactobacillus acidophilus, L. delbrueckii, L. rhamnosus, Lactobacillus casei, Lactobacillus casei, L. plantarum, L. helveticus, L. salivarius, L. reuteri*), *Bifidobacterium* species (*B. longum, B. lactis, B. breve, B. animalis*), *Streptococcus thermophilus, Enterococcus faecium, E. faecalis, Saccharomyces boulardii* (Ghoshal *et al.*, 2012; Hardy *et al.*, 2011; Kailasapathy and Chin, 2000; Vieira *et al.*, 2013) as well as non-pathogenic bacilli such as *E. coli* Nissle 1917 (Collins and Gibson, 1999). Non pathogenic *E. coli* produces anti-microbial peptides such as microsin S and bacteriocin, which inhibits growth of pathogenic bacteria (Dai *et al.*, 2013). Nowadays, probiotics are available in the market in several forms such as powder, liquid, gel, paste, granules and most commonly in the form of capsules or sachets, besides being present in a variety of (fermented) foods (Collins and Gibson, 1999). Effective probiotics must have the following characteristics: alive, reproducing and maintaining themselves, stable while passage through the acidic stomach content, metabolically active inside the gut, not degraded by hydrolytic enzymes and bile salts in the small intestine, non-pathogenic, genetically stable, should be of human origin (Dai *et al.*, 2013; Ghoshal *et al.*, 2012; Steer *et al.*, 2000). Though in several Western countries, probiotics are used as food additives (Gardiner *et al.*, 1998, 1999; Stanton *et al.*, 2001), therapeutic potential of these agents in treatment of several gastrointestinal disorders including irritable bowel syndrome (IBS) is rapidly emerging.

Therapeutic implications

Recent studies have shown that probiotic bacteria help in restoring gut homeostasis by its ability to manipulate gut microbiota (Grajek *et al.*, 2005). They have the ability to alter the immune response in the gut and inhibit growth of pathogenic bacteria in gut of patients with IBS. Probiotics may also alleviate the symptoms by its action on lactose malabsorption (Dai *et al.*, 2013; Kumar *et al.*, 2012; Sharma *et al.*, 2013), which is common among patients with IBS (Ghoshal *et al.*, 2009). Since probiotics are known to be effective in treatment of acute infectious diarrhoea, travellers' diarrhoea and antibiotic-associated diarrhoea, these are likely to be useful in prevention and treatment of post-infectious IBS (PI-IBS); however, so far there is no study on this issue. Probiotics may be used to relieve abdominal distension or bloating in a subset of patients with IBS. Since dysbiosis is more common among patients with non-constipating IBS, probiotics may be preferred among these patients (Bixquert Jimenez, 2009).

Mechanistically, *L. helveticus* MTCC 5463 strain is known to prevent de-conjugation of bile acids *in vitro* and exert antimicrobial activity against bad bugs such as *Bacillus cereus, Staphylococcus aureus, Pseudomonas aeruginosa, Salmonella enterica* serovar *typhi, Escherichia coli* (Grover *et al.*, 2012; Prajapati *et al.*, 2011). Similarly, the actions of *Lactobacillus plantarum* 299v include: (i) binding ability to mucosal epithelium and tolerate acidic pH, which are needed for successful colonization (ii) ability to reduce methotrexate-induced enterocolitis in rats (iii) change intraluminal fermentation in the gut lumen (iv) alteration of expression of mucin-producing gene in cell culture (v) efficacy in treatment of SIBO, which is not uncommon among patients with IBS (Ghoshal and Srivastava, 2014). The efficacy of all probiotics, however, is not equal. Therefore, the rationality of the use of multi-species probiotics in different indications is considered. VSL # 3 is such a multi-species probiotics, which is used widely in several gastrointestinal disorder including IBS. Quite a few randomized controlled trials documented efficacy of probiotics in the treatment of patients with IBS (Table 27.1) (Cappello *et al.*, 2013; Ducrotte *et al.*, 2012; Farup *et al.*, 2012; Francavilla *et al.*, 2010; Grossi *et al.*, 2010; Guandalini *et al.*, 2010; Guglielmetti *et al.*, 2011; Ki Cha *et al.*, 2012; Kim *et al.*, 2003, 2005; Kruis *et al.*, 2012; Min *et al.*, 2012). Therefore, several experts suggest that probiotics may be used in treatment of IBS (Dai *et al.*, 2013; Hosseini *et al.*, 2012a,b; Whelan and Quigley, 2013). Previous studies have shown that *Lactobacillus* and *Bifidobacterium* led to global improvement in IBS symptoms (Ki Cha *et al.*, 2012; Ortiz-Lucas

Table 27.1 Randomized controlled study of probiotics in patients with irritable bowel syndrome

Strain number	Name of probiotic/probiotic strain	Dose/day	Sample size	Duration	Clinical outcome	Reference
1	*L. plantarum* MF 1298	1 capsule of 10^{10} cfu	16	4 weeks	Significant improvement in symptoms of IBS	Farup *et al.* (2012)
2	*L. plantarum* 299v (DSM 9843)	1 capsule of 10^{10} cfu	214	4 weeks	Significant improvement in abdominal pain and bloating in IBS fulfilling Rome III criteria	Ducrotte *et al.* (2012)
3	Composite yoghurt containing high dose *B. animalis* subsp. *lactis* Bb-12, and acacia dietary fibre, together with two classic yoghurt starter cultures, *Streptococcus thermophilus* and *L. acidophilus*	Twice (*B. animalis* subsp. *lactis* Bb-12 $\geq 10^{11}$ cfu, *S. thermophilus* $\geq 3 \times 10^9$ cfu and *L. acidophilus* $\geq 10^9$ cfu/bottle	130	8 weeks	Significant relief in overall symptoms of IBS	Min *et al.* (2012)
4	*Escherichia coli Nissle* 1917 (EcN-1917)	MUTAFLOR® (with $2.5 - 25 \times 10^9$ cfu *E. coli Nissle* 1917) once daily for 4 days and thereafter two capsules MUTAFLOR® once daily throughout the trial	60	12 weeks	Meteorism and nausea showed significant improvements only with EcN	Kruis *et al.* (2012)
5	Probiotic mixture contains *Lactobacillus acidophilus* KCTC 11906 BP, *L. plantarum* KCTC 11867BP, *L. rhamnosus* KCTC 11868BP, *B. breve* KCTC 11858BP, *B. lactis* KCTC 11903BP, *B. longum* KCTC 11860BP and *Streptococcus thermophilus* KCTC 11870BP	One capsule (1.0×10^{10}cells/d) twice a day	50	8 weeks	Ameliorate over-all symptoms of IBS and also associated with stabilization of faecal micro-biota. However, no significant effect was seen with individual symptoms	Ki Cha *et al.* (2012)
6	Symbiotic mixture (Probinul*) contains *L. plantarum*, *L. casei* subsp. *rhamnosus*, *L. gasseri*, *L. acidophilus*, *L. salivarus*, *B. infantis*, *B. longum*, *L. sporogenes* and *Streptococcus thermophilus*, prebiotic inulin (2.2 g) and 1.3 g of tapioca-resistant starch.	Twice	64	4 weeks	Reduce the severity of flatulence in IBS patients; lack of adverse events but failed to relieve symptoms globally	Cappello *et al.* (2013)
7	VSL#3* (*Lactobacillus casei*, *L. plantarum*, *L. acidophilus*, *L. delbrueckii*, *Bifidobacterium longum*, *B. breve*, *B. infantis* and *Streptococcus thermophilus*)	Once	59	6 weeks	Significantly improved the quality of life in patients with IBS	Guandalini *et al.* (2010)
8	Flortec (*L. paracasei* B-21060) and FlorVis (*Lactobacillus rhamnosus* GG)	Twice	174	10 days	Flortech was statistically superior over FlorVis	Grossi *et al.* (2010)
9	*Lactobacillus rhamnosus* GG	Twice (3×10^9 cfu)	141	8 weeks	Significantly reduced the frequency and severity of abdominal pain in children with IBS	Francavilla *et al.* (2010)
10.	*Bifidobacterium bifidum* MIMBb75	Once	122	4 weeks	Significantly reduced global IBS symptoms and improved quality of life	Guglielmetti *et al.* (2011)
11	VSL#3	Twice	25	10 weeks	Abdominal bloating was significantly reduced but not other parameters (bowel dysfunction, abdominal pain, flatulence or urgency) compared with placebo in IBS-D	Kim *et al.* (2003)
12	VSL#3	Twice	48	8 weeks	Significantly improved flatulence and colonic transit time compared to placebo group in patients with IBS	Kim *et al.* (2005)

*Strain numbers have not been provided by the authors.

et al., 2013). However, it is important to note that the results of the different studies cannot be generalized as these used different probiotic bacterial strains, formulation and dosages and had different study design (Dai *et al.*, 2013).

Quite a few meta-analyses also showed efficacy of probiotics in treatment of patients with IBS. In a meta-analysis of 20 clinical trials (1404 study subjects) on 23 probiotics, global IBS symptoms [relative risk (RR) 0.77; 95% confidence interval (CI)

0.62–0.94) and abdominal pain (RR 0.78; 95% CI 0.69–0.88) were found to improve with probiotics compared to placebo (McFarland and Dublin, 2008). Similar findings were reported by Nikfar *et al*, in which 8 randomized placebos controlled clinical trials were included; treatment with probiotic was found to be associated with improvement in the IBS symptoms (RR 0.82; 95% CI 0.71–0.93; $P=0.0042$) (Nikfar *et al.*, 2008). In another meta-analysis on 10 randomized controlled trials, treatment with *Bifidobacterium breve, B. longum* or *Lactobacillus acidophilus* were found to be associated with improvement in abdominal pain; *B. breve, B. infantis, L. casei* or *L. plantarum* were associated with improvement in abdominal distension; *B. breve, B. infantis, L. casei, L. plantarum, B. longum, L. acidophilus, L. bulgaricus* and *Streptococcus salivarius subsp.* were associated with improvement in flatulence. Thus, the above evidence may suggest that certain probiotics may be used in treatment of IBS, particularly to relieve some specific symptoms such as abdominal pain, bloating, flatulence (Ortiz-Lucas *et al.*, 2013). However, since probiotic effects are strain-dependent, the authors of the meta-analysis should have provided the strain numbers, rather than lump all strain of e.g. *B. longum* together (at least 4 different strains were tested, all in multispecies products). Moreover, practically all product were multistrain products, and no doubt there will have been interactions between the different species within such a product, and hence their conclusion is too generalized.

Mechanism of action of probiotics

Molecular mechanisms of action of probiotics have not yet been elucidated clearly. Experimental studies showed that though some actions of probiotics are common, others are significantly different for different strains (O'Hara and Shanahan, 2007). The interaction between host, gut microbiota and probiotic bacteria is quite complex (Dai *et al.*, 2013; O'Hara and Shanahan, 2007). In IBS, probiotic bacteria broadly work through the following mechanisms after adherence to intestinal epithelial cell: (a) inhibition of adherence of pathogenic bacteria by producing the antimicrobial substances (b) potentiation of integrity of epithelial barrier, and (c) reduction of intestinal inflammation by activation of regulatory cytokines and chemokines (Candela *et al.*, 2008; Niedzielin *et al.*, 2001; O'Hara and Shanahan, 2007). Each of these mechanisms of action will be briefly discussed in the subsequent sections.

Prevention of invasion of intestinal epithelium by pathogens

Emerging evidence suggests that adhesive strain of probiotic bacteria (e.g. *L. plantarum* 299v, *L. acidophilus* ATCC4356, and *Streptococcus thermophilus* ATCC19258) inhibit ligands liberated by pathogens that damage epithelial barriers (Swidsinski *et al.*, 2002). Several species of *Lactobacillus* (*L. rhamnosus, L. plantarum, L. paracasei*) and *Bifidobacterium longum* reduce the attachment of enterotoxigenic *E. coli* to the intestinal epithelium (strain 0157:87) (Gopal *et al.*, 2001). Recent studies revealed that combination of specific types of probiotic bacterial strains have synergistic adhesive effects, thereby enhance the efficiency of other useful bacteria for adhesion and inhibit the penetration of invasive microbes (Ouwehand *et al.*, 2000).

Reinforcement of epithelial barrier, altering the local environment

Mucosal epithelial barrier is the first line of defence against pathogens causing various diseases (Caballero-Franco *et al.*, 2007; Saulnier *et al.*, 2009). Beneficial and pathogenic bacteria compete with one another for shelter and foods on the large surface of the gut epithelium (Gross *et al.*, 2008; Hardy *et al.*, 2011). The thick layer of mucus, epithelium, glycocalyx, adhesive binding proteins, which are enhanced by colonization by probiotic bacteria, inhibit the attachment by pathogenic bacteria (Bermudez-Brito *et al.*, 2012). A leaky gut epithelium has been reported in patients with IBS including SIBO and food-borne infections, which are known to pre-dispose to PI-IBS (Dunlop *et al.*, 2006; Ghoshal *et al.*, 2012; Lee and Bak, 2011; Lin and Pimentel, 2005); an altered gut microbiota is also known to cause leaky gut (Lee and Bak, 2011). Abnormal small intestinal permeability is well-known among patients with IBS, particularly in diarrhoea-predominant (IBS-D) disease and PI-IBS (Dunlop *et al.*, 2006; Lee and Bak, 2011). Reduced expression of tight junction proteins in small bowel and colon have been reported in patients with IBS-D and PI-IBS (Barbara *et al.*, 2012). Dunlop *et al.*, reported that permeability of small bowel was higher in patients with IBS-D without infection than IBS-D with infection (Zeng *et al.*, 2008). Expression of E-cadherin, which regulates the paracellular permeability, was also reduced in patients with IBS-D and alternating IBS (Piche, 2014). In a randomized single-blind placebo-controlled study, small bowel permeability and overall IBS symptoms scores improved in probiotic fermented milk (*Streptococcus thermophilus, Lactobacillus bulgaricus, Lactobacillus acidophilus* and *Bifidobacterium longum,* strain numbers not provided) treatment over 4 weeks compared to milk beverages without any probiotic though colonic permeability was unchanged at the end of duration (Zeng *et al.*, 2008). In a recent study on mouse model of PI-IBS, though *Lactobacillus acidophilus* HB56003 and *Bifidobacterium longum* HB55020 improved the intestinal permeability, visceral hypersensitivity and inflammation, *Streptococcus faecalis* HB62001 did not (Wang *et al.*, 2014).

Immune response mediated by probiotics

Like commensal organisms, probiotic bacteria also enhance the immune protection locally as well as systemically (O'Mahony *et al.*, 2005). Increased numbers of inflammatory cells and entero-endocrine cells have been reported in mucosal biopsies of patients with PI-IBS, which is associated with persistent mild inflammation (Andresen and Baumgart, 2006). The inflammatory hypothesis for IBS is also supported by increased levels of Th1-mediated pro-inflammatory cytokines (TNF-α, IL-1β, IL-12, IL-6) and reduced level of anti-inflammatory cytokines (IL-10) (Barbara *et al.*, 2012; Hughes *et al.*, 2013). In Fig. 27.1, a summary of the interaction of probiotics and the host microbiota and modulation of the immune system is shown. In one study, effects of two probiotic bacterial strain, namely *Lactobacillus salivarius* UCC 4331 and *Bifidobacterium infantis* 35624 were studied on IBS symptoms and the ratio of anti-inflammatory to pro-inflammatory cytokines (IL-10/IL-12) levels. *Bifidobacterium infantis* 35624 improved all the symptoms of IBS significantly

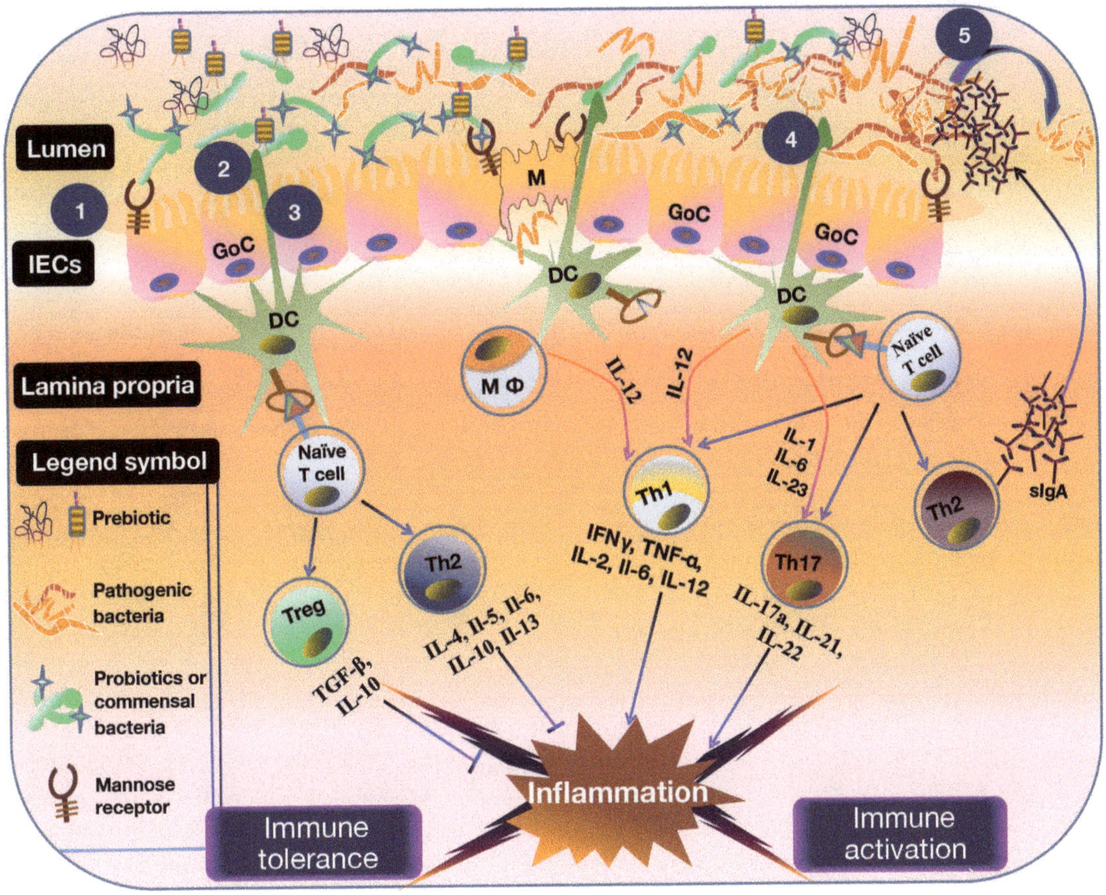

Figure 27.1 Modulation of immune response by probiotics and prebiotics. Stimulation of immune response is represented. Prebiotics act as adjuvant and facilitate the colonization of probiotics or commensal bacteria which compete with the pathogenic bacteria for adherence with mannose receptors. Probiotic induces the expression of mucin, mucus secretion by goblet cell (GoC) and Th2 derived sIgA that inhibits the attachment of disease causing pathogens. Interaction of pathogens or pathogen derived ligands interact with TLR present on antigen presenting cell (DC or MΦ) induce adaptive immune response. IL-12 which is produced by both dendritic cell and macrophage induces Th1 cell. Activated DC stimulates Th17 cell via secreting IL-1, IL-6, IL-23 cytokines. On immune activation, Th1 cell secretes IFNγ, TNF-α, IL-2, Il-6, IL-12 and Th17 cell secretes IL-17a, IL-21, IL-22 cytokines that lead to cause intestinal inflammation. Whilst interaction of probiotic or commensal bacteria with TLR induces Treg and Th2 cells. Th2 and Treg cells stimulate the production of anti-inflammatory (IL-4, Il-5, Il-6, IL-10, Il-13) and regulatory cytokines (TGF-β, IL-10) that mediate tolerogenic immune response. (1) Mucus layer; (2) interaction of probiotics or commensals with toll like receptor on dendritic cell (DC) or macrophage cells (MΦ) induces the regulatory immune response; (3) toll like receptor (TLR); (4) interaction of pathogenic bacteria with TLRs induces the proinflammatory immune response; (5) Th2 stimulated secretory immunoglobulin A (sIgA) inhibits the pathogenic bacteria. IECs: intestinal epithelial cells; GoC: goblet cell.

except stool frequency and stool consistency compared to placebo. After 8 weeks of treatment, *Bifidobacterium infantis* 35624 increased the level of anti-inflammatory cytokines and normalized the ratio of IL-10/IL-12. *Bifidobacterium infantis* 35624 was found to be more effective than *Lactobacillus salivarius* UCC 4331 (Andresen and Baumgart, 2006; O'Mahony *et al.*, 2005). In a murine model of PI-IBS, administration of *Lactobacillus paracasei* NCC2461 attenuated muscle hypercontractility via reduction in the level of transforming growth factor-1, cyclooxygenase-2 and prostaglandin E_2 in muscle. Thus, *Lactobacillus paracasei* NCC2461 may be useful in treating PI-IBS (Verdu *et al.*, 2004; Verdu and Collins, 2005). Increased populations of T-cells and mast cells were reported in small and large intestine in patients with IBS than healthy controls. Mediators released by these immunogenic cells may impact on enteric nervous system, tight junction proteins and also affect the integrity of intestinal epithelial cells (Barbara *et al.*, 2012). One study reported that administration of *Lactobacillus rhamnosus* GR-1 and *Lactobacillus reuteri* RC-14 augmented T regulatory cell population in patients with IBS; regulatory T-cells confer protective immune response (Borchers *et al.*, 2009). Another study reported that multi-species probiotic (VSL#3) improved Th1-mediated immune response in murine colitis by producing regulatory cytokines such as TGF-β and IL-10 and maintained the cytokine balance (Di Giacinto *et al.*, 2005).

Prebiotics and their role as therapeutic agents

Gibson and Roberfroid proposed the term prebiotic (Gibson *et al.*, 2004; Gibson and Roberfroid, 1995). Prebiotics are defined as non digestible carbohydrates or dietary fibres that increase the numbers or activity of beneficial members of the endogenous microbiota and in that way improve health. Prebiotics act as adjuvant for probiotic bacteria or already existing bacteria in the gut, which confer health benefits (Gibson and Roberfroid, 1995; Vieira *et al.*, 2013). For instance, galacto-oligosaccharides enhanced proliferation of beneficial bacteria such as lactic acid bacteria and improved stool consistency, flatulence and abdominal bloating in patients with IBS (Silk *et al.*, 2009). Inulin or inulin type fructans and trans-galacto-oligosaccharides are only two non-digestible oligosaccharides that fulfil the prebiotic classification criteria proposed by Gibson *et al.* (2004) (de Vrese and Marteau, 2007; Gibson *et al.*, 2004).

Impact of prebiotics in patients with IBS

Data on prebiotics in treatment of IBS are scanty. A few studies reported beneficial effects of prebiotics in IBS patients. A dietary mixture consisting of inulin, L-tryptophan, angelica, vegetal charcoal, vitamins and probiotics, improved abdominal pain, abdominal distension and bowel movement among 37 patients with IBS during a 6-month period (Astegiano *et al.*, 2006). In a randomized double blind study, 5 g of short-chain fructooligosaccharides improved quality of life and reduced the gastrointestinal symptoms among 105 patients with IBS over a period of 6 weeks compared with placebo (Paineau *et al.*, 2008).

Previous studies suggested that prebiotics might be helpful in the alleviation of IBS symptoms (Bixquert Jimenez, 2009). Silk *et al.*, tested efficacy of low and higher dose of trans-galactooligosaccharide (B-GOS) on faecal microbiota and symptoms in patients with IBS. In this study, 3.5 g/day of prebiotic, 7 g/day of prebiotic or placebo were given to 44 patients with IBS over 12 weeks. Both the dosage of B-GOS resulted in higher counts of *Bifidobacteria*. A dose of 3.5 g/day of B-GOS significantly improved stool consistency, flatulence and abdominal bloating, total symptoms score, subjective global assessment while 7 g/day of B-GOS significantly improved the subjective global assessment and anxiety only (Silk *et al.*, 2009).

Impaired fermentation of prebiotics results in production of propionate and acetate, which might worsen IBS symptoms (Fung and Szilagyi, 2012). Some prebiotics are fermented rapidly in proximal gut and enhance the production of gas, which may cause abdominal bloating and distension (Konig and Brummer, 2014). Moreover, exaggerated visceral hypersensitivity among patients with IBS might result in increased perception of gas-related symptoms (Symons *et al.*, 1992). Therefore, prebiotics that are given to IBS patients, should be slowly fermented throughout the entire colon so that colonic gas could be uniformly distributed (Konig and Brummer, 2014).

In a multicentre, prospective, randomized, double-blind, placebo-controlled, parallel-group study, a daily dose of 20 grams of fructooligosaccharides (FOS) powder or placebo was given to patients with IBS (52 in the FOS group and 46 in the placebo group) over 12 weeks. IBS symptoms significantly improved in the FOS-group compared to placebo after 4 and 6 weeks. In the FOS-group, IBS symptoms improved in 58% and worsened in 8%; in placebo group, IBS symptoms were improved in 65% and worsened in 13% (Olesen and Gudmand-Hoyer, 2000). However, differences were not statistically significant.

Lactulose has been considered as an important synthetic prebiotic that increases the number of beneficial bacteria, stool frequency and stool water content in patients with IBS-C (Spiller, 2008). Partially hydrolysed guar gum (PHGG) also acts a prebiotic and is helpful in improving the quality of life in patients with IBS. In a clinical trial, PHGG alleviated the symptoms of IBS in both IBS-C and IBS-D during and after treatment. Furthermore, it enhanced the counts of *Lactobacillus* and *Bifidobacterium* species and increased the levels of SCFAs in colon (Giannini *et al.*, 2006).

Dietary components play a major role in alleviating the IBS symptoms. Shepherd *et al.* reported that supplementation of low fermentable oligo-, di-, mono-saccharides and polyols (FODMAPs) diet improved 74% symptoms in patients with IBS and fructose malabsorption (Murray *et al.*, 2012). Therefore, low FODMAP diet should be recommended to IBS patients (Major and Spiller, 2014). Diets having high fibre content may lead to abdominal bloating, abdominal pain, flatulence in patients with IBS (Major and Spiller, 2014). For example administration of bran produces extensive amount of gases in patients with IBS even though it is beneficial in constipation predominant patients (Macfarlane *et al.*, 2006; Snook and Shepherd, 1994). Likewise, fructose and sorbitol are good substrates for colonic bacteria but they trap excessive amount of water, which tend to cause diarrhoea (Symons *et al.*, 1992). Some prebiotics like oligofructoses, inulin and galactose-oligosaccharides increase water content of stool and hence, may be useful in patients with IBS-C. Although galactose-oligosaccharides have been shown to cause abdominal bloating and flatulence in healthy volunteers (Spiller, 2008), these symptoms have been shown to be alleviated among patients with IBS (Silk *et al.*, 2009). In practice, therefore, it is important to use only moderate doses of these prebiotics to avoid a possible increase in abdominal bloating and flatulence and possibly to combine these with probiotics.

The combination of a probiotic and a prebiotic is termed synbiotic. This combination may have synergistic effects in alleviating IBS symptoms. Tsuchiya *et al.*, studied the effect of a novel synbiotic (SCM-III: composed of 3 probiotic species, namely *Lactobacillus acidophilus*, *L. helveticus*, and a *Bifidobacterium* species (strain numbers not specified) in a vitamin- and phytoextracts-enriched medium) on IBS symptoms; in this study, 10 ml of SCM-III thrice a day or the same dose of heat inactivated synbiotic (pseudo-placebo, as only the probiotic is inactivated during heat-treatment, and not the prebiotic) were given to 68 patients with IBS for 12 weeks. After 6 weeks of treatment, abdominal pain, bloating and bowel habits were improved in 80% of the patients. Furthermore, SCM-III increased the population of *Lactobacillus*, *Bifidobacterium* species and eubacteria. Nevertheless, this study had several limitations including pseudo-randomization, single blinded design and small sample size (Tsuchiya *et al.*, 2004). In another open labelled controlled trial, probinul consisting of *L. plantarum*, *L. casei* subsp. *rhamnosus*, *L. gasseri*, *L. acidophilus*, *L. salivarius*, *L. sporogenes* (as this is a non-existing species this is likely *Bacillus coagulans*), *B. infantis*, *B. longum*, and *S. thermophilus* (strain numbers not specified), prebiotic (inulin: 2 gram) and 1.3 grams of tapioca-resistant starch were given to 26 patients with IBS for a duration of six months. Probinul improved flatulence in patients with IBS (Bucci *et al.*, 2014). Additional RCTs are needed to understand the synergistic role of synbiotics in several gut associated diseases including IBS. Immune-modulatory effect of SCFAs (Saulnier *et al.*, 2009) and adjuvant nature of prebiotics (Vieira *et al.*, 2013) when combined with probiotics could be potential therapeutic agent in IBS.

Conclusion and future perspectives

We conclude that probiotics are useful in treatment of IBS, particularly non-constipating sub-types; most studies showed that abdominal pain, bloating and flatulence are the symptoms, which were relieved most. More studies are, however, needed to evaluate the most useful strains, dose, duration of probiotics and potential role of addition of prebiotics (synbiotics) in management of IBS.

References

Andresen, V., and Baumgart, D.C. (2006). Role of probiotics in the treatment of irritable bowel syndrome: potential mechanisms and current clinical evidence. International Journal of Probiotics and Prebiotics *1*, 11.

Astegiano, M., Pellicano, R., Terzi, E., Simondi, D., and Rizzetto, M. (2006). Treatment of irritable bowel syndrome. A case control experience. Minerva Gastroenterol. Dietol 52, 359–363.

Barbara, G., Cremon, C., Carini, G., Bellacosa, L., Zecchi, L., De Giorgio, R., Corinaldesi, R., and Stanghellini, V. (2011). The immune system in irritable bowel syndrome. J. Neurogastroenterol. Motil. *17*, 349–359.

Barbara, G., Zecchi, L., Barbaro, R., Cremon, C., Bellacosa, L., Marcellini, M., De Giorgio, R., Corinaldesi, R., and Stanghellini, V. (2012). Mucosal permeability and immune activation as potential therapeutic targets of probiotics in irritable bowel syndrome. J. Clin. Gastroenterol. *46 Suppl*, S52–55.

Bermudez-Brito, M., Plaza-Diaz, J., Munoz-Quezada, S., Gomez-Llorente, C., and Gil, A. (2012). Probiotic mechanisms of action. Ann. Nutr. Metab. *61*, 160–174.

Bixquert Jimenez, M. (2009). Treatment of irritable bowel syndrome with probiotics. An etiopathogenic approach at last? Rev. Esp. Enferm. Dig. *101*, 553–564.

Borchers, A.T., Selmi, C., Meyers, F.J., Keen, C.L., and Gershwin, M.E. (2009). Probiotics and immunity. J. Gastroenterol. *44*, 26–46.

Brint, E.K., MacSharry, J., Fanning, A., Shanahan, F., and Quigley, E.M. (2011). Differential expression of toll-like receptors in patients with irritable bowel syndrome. Am. J. Gastroenterol. *106*, 329–336.

Bucci, C., Tremolaterra, F., Gallotta, S., Fortunato, A., Cappello, C., Ciacci, C., and Iovino, P. (2014). A pilot study on the effect of a symbiotic mixture in irritable bowel syndrome: an open-label, partially controlled, 6-month extension of a previously published trial. Tech. Coloproctol. *18*, 345–353.

Caballero-Franco, C., Keller, K., De Simone, C., and Chadee, K. (2007). The VSL#3 probiotic formula induces mucin gene expression and secretion in colonic epithelial cells. Am. J. Physiol. Gastrointest Liver Physiol. *292*, G315–322.

Candela, M., Perna, F., Carnevali, P., Vitali, B., Ciati, R., Gionchetti, P., Rizzello, F., Campieri, M., and Brigidi, P. (2008). Interaction of probiotic *Lactobacillus* and Bifidobacterium strains with human intestinal epithelial cells: adhesion properties, competition against enteropathogens and modulation of IL-8 production. Int. J. Food Microbiol. *125*, 286–292.

Cappello, C., Tremolaterra, F., Pascariello, A., Ciacci, C., and Iovino, P. (2013). A randomised clinical trial (RCT) of a symbiotic mixture in patients with irritable bowel syndrome (IBS): effects on symptoms, colonic transit and quality of life. Int. J. Colorectal Dis. *28*, 349–358.

Chang, F.Y. (2014). Irritable bowel syndrome: The evolution of multidimensional looking and multidisciplinary treatments. World J. Gastroenterol. *20*, 2499–2514.

Collins, M.D., and Gibson, G.R. (1999). Probiotics, prebiotics, and synbiotics: approaches for modulating the microbial ecology of the gut. Am. J. Clin. Nutr. *69*, 1052S-1057S.

Dai, C., Zheng, C.Q., Jiang, M., Ma, X.Y., and Jiang, L.J. (2013). Probiotics and irritable bowel syndrome. World J. Gastroenterol. *19*, 5973–5980.

de Vrese, M., and Marteau, P.R. (2007). Probiotics and prebiotics: effects on diarrhea. J. Nutr. *137*, 803S-811S.

Di Giacinto, C., Marinaro, M., Sanchez, M., Strober, W., and Boirivant, M. (2005). Probiotics ameliorate recurrent Th1-mediated murine colitis by inducing IL-10 and IL-10-dependent TGF-beta-bearing regulatory cells. J. Immunol. *174*, 3237–3246.

Ducrotte, P., Sawant, P., and Jayanthi, V. (2012). Clinical trial: *Lactobacillus plantarum* 299v (DSM 9843). improves symptoms of irritable bowel syndrome. World J. Gastroenterol. *18*, 4012–4018.

Dunlop, S.P., Hebden, J., Campbell, E., Naesdal, J., Olbe, L., Perkins, A.C., and Spiller, R.C. (2006). Abnormal intestinal permeability in subgroups of diarrhea-predominant irritable bowel syndromes. Am. J. Gastroenterol. *101*, 1288–1294.

Farup, P.G., Jacobsen, M., Ligaarden, S.C., and Rudi, K. (2012). Probiotics, symptoms, and gut microbiota: what are the relations? A randomized controlled trial in subjects with irritable bowel syndrome. Gastroenterol. Res. Pract. 214102.

Floch, M.H. (2003). Probiotics, Irritable Bowel Syndrome, and Inflammatory Bowel Disease. Curr. Treat Options Gastroenterol. *6*, 283–288.

Francavilla, R., Miniello, V., Magista, A.M., De Canio, A., Bucci, N., Gagliardi, F., Lionetti, E., Castellaneta, S., Polimeno, L., Peccarisi, L., *et al.* (2010). A randomized controlled trial of *Lactobacillus* GG in children with functional abdominal pain. Pediatrics *126*, e1445–1452.

Fuller, R. (1989). Probiotics in man and animals. J. Appl. Bacteriol *66*, 365–378.

Fung, Q.M., and Szilagyi, A. (2012). Carbohydrate Elimination or Adaptation Diet for Symptoms of Intestinal Discomfort in IBD: Rationales for 'Gibsons' Conundrum'. Int. J. Inflam *2012*, 493717.

Gardiner, G., Ross, R.P., Collins, J.K., Fitzgerald, G., and Stanton, C. (1998). Development of a probiotic cheddar cheese containing human-derived *Lactobacillus paracasei* strains. Appl. Environ. Microbiol. *64*, 2192–2199.

Gardiner, G., Stanton, C., Lynch, P.B., Collins, J.K., Fitzgerald, G., and Ross, R.P. (1999). Evaluation of cheddar cheese as a food carrier for delivery of a probiotic strain to the gastrointestinal tract. J. Dairy Sci. *82*, 1379–1387.

Ghoshal, U.C., and Srivastava, D. (2014). Irritable bowel syndrome and small intestinal bacterial overgrowth: meaningful association or unnecessary hype. World J. Gastroenterol. *20*, 2482–2491.

Ghoshal, U.C., Kumar, S., Chourasia, D., and Misra, A. (2009). Lactose hydrogen breath test versus lactose tolerance test in the tropics: does positive lactose tolerance test reflect more severe lactose malabsorption? Trop Gastroenterol. *30*, 86–90.

Ghoshal, U.C., Shukla, R., Ghoshal, U., Gwee, K.A., Ng, S.C., and Quigley, E.M. (2012). The gut microbiota and irritable bowel syndrome: friend or foe? Int. J. Inflam. 151085.

Ghoshal, U., Shukla, R., Agarwal, V., Dhole, T.N., and Ghoshal, U.C. (2014). Su2048 fecal microbiota in patients with irritable bowel syndrome compared to healthy controls by real-time PCR: an evidence of dysbiosis. Gastroenterology *146*, S-532.

Giannini, E.G., Mansi, C., Dulbecco, P., and Savarino, V. (2006). Role of partially hydrolyzed guar gum in the treatment of irritable bowel syndrome. Nutrition *22*, 334–342.

Gibson, G.R., and Roberfroid, M.B. (1995). Dietary modulation of the human colonic microbiota: introducing the concept of prebiotics. J. Nutr. *125*, 1401–1412.

Gibson, G.R., Probert, H.M., Loo, J.V., Rastall, R.A., and Roberfroid, M.B. (2004). Dietary modulation of the human colonic microbiota: updating the concept of prebiotics. Nutr. Res. Rev. *17*, 259–275.

Gopal, P.K., Prasad, J., Smart, J., and Gill, H.S. (2001). *In vitro* adherence properties of *Lactobacillus rhamnosus* DR20 and Bifidobacterium lactis DR10 strains and their antagonistic activity against an enterotoxigenic *Escherichia coli*. Int. J. Food Microbiol. *67*, 207–216.

Grajek, W., Olejnik, A., and Sip, A. (2005). Probiotics, prebiotics and antioxidants as functional foods. Acta Biochim Pol *52*, 665–671.

Gross, G., van der Meulen, J., Snel, J., van der Meer, R., Kleerebezem, M., Niewold, T.A., Hulst, M.M., and Smits, M.A. (2008). Mannose-specific interaction of *Lactobacillus plantarum* with porcine jejunal epithelium. FEMS Immunol. Med. Microbiol. *54*, 215–223.

Grossi, E., Buresta, R., Abbiati, R., and Cerutti, R. (2010). Clinical trial on the efficacy of a new symbiotic formulation, Flortec, in patients with acute diarrhea: a multicenter, randomized study in primary care. J. Clin. Gastroenterol. *44 Suppl. 1*, S35–41.

Grover, S., Rashmi, H.M., Srivastava, A.K., and Batish, V.K. (2012). Probiotics for human health -new innovations and emerging trends. Gut Pathog. *4*, 15.

Guandalini, S., Magazzu, G., Chiaro, A., La Balestra, V., Di Nardo, G., Gopalan, S., Sibal, A., Romano, C., Canani, R.B., Lionetti, P., *et al.* (2010). VSL#3 improves symptoms in children with irritable bowel syndrome: a multicenter, randomized, placebo-controlled, double-blind, crossover study. J. Pediatr. Gastroenterol. Nutr. *51*, 24–30.

Guglielmetti, S., Mora, D., Gschwender, M., and Popp, K. (2011). Randomised clinical trial: Bifidobacterium bifidum MIMBb75 significantly alleviates irritable bowel syndrome and improves quality of life – a double-blind, placebo-controlled study. Aliment. Pharmacol. Ther. *33*, 1123–1132.

Hardy, H., Harris, J., Lyon, E., Beal, J., and Foey, A.D. (2011). Probiotics, prebiotics and immunomodulation of gut mucosal defences: homeostasis and immunopathology. Nutrients *5*, 1869–1912.

Hosseini, A., Nikfar, S., and Abdollahi, M. (2012a). Are probiotics effective in management of irritable bowel syndrome? Aarch Med. Sci. *8*, 403–405.

Hosseini, A., Nikfar, S., and Abdollahi, M. (2012b). Probiotics use to treat irritable bowel syndrome. Expert Opin. Biol. Ther. *12*, 1323–1334.

Hughes, P.A., Zola, H., Penttila, I.A., Blackshaw, L.A., Andrews, J.M., and Krumbiegel, D. (2013). Immune activation in irritable bowel syndrome: can neuroimmune interactions explain symptoms? Am. J. Gastroenterol. *108*, 1066–1074.

Irvine, E.J., Kim, J., Alders, G.L., and Ching, E. (2008). S1852 IBS Patients Have a Poorer Quality Diet and Exercise Less Than Organic GI Disease Patients or Normal Controls. Gastroenterology *134*, A-284.

Kailasapathy, K., and Chin, J. (2000). Survival and therapeutic potential of probiotic organisms with reference to *Lactobacillus* acidophilus and Bifidobacterium spp. Immunol. Cell Biol. *78*, 80–88.

Ki Cha, B., Mun Jung, S., Hwan Choi, C., Song, I.D., Woong Lee, H., Joon Kim, H., Hyuk, J., Kyung Chang, S., Kim, K., Chung, W.S., *et al.* (2012). The effect of a multispecies probiotic mixture on the symptoms and fecal microbiota in diarrhea-dominant irritable bowel syndrome: a randomized, double-blind, placebo-controlled trial. J. Clin. Gastroenterol. *46*, 220–227.

Kim, H.J., Camilleri, M., McKinzie, S., Lempke, M.B., Burton, D.D., Thomforde, G.M., and Zinsmeister, A.R. (2003). A randomized controlled trial of a probiotic, VSL#3, on gut transit and symptoms in diarrhoea-predominant irritable bowel syndrome. Aliment. Pharmacol. Ther. *17*, 895–904.

Kim, H.J., Vazquez Roque, M.I., Camilleri, M., Stephens, D., Burton, D.D., Baxter, K., Thomforde, G., and Zinsmeister, A.R. (2005). A randomized controlled trial of a probiotic combination VSL# 3 and placebo in irritable bowel syndrome with bloating. Neurogastroenterol. Motil. *17*, 687–696.

Konig, J., and Brummer, R.J. (2014). Alteration of the intestinal microbiota as a cause of and a potential therapeutic option in irritable bowel syndrome. Benef. Microbes, 1–15.

Kruis, W., Chrubasik, S., Boehm, S., Stange, C., and Schulze, J. (2012). A double-blind placebo-controlled trial to study therapeutic effects of probiotic *Escherichia coli* Nissle 1917 in subgroups of patients with irritable bowel syndrome. Int. J. Colorectal Dis. *27*, 467–474.

Kumar, M., Kumar, R., Poovai, P.D., and Kalaichelvan, P.T. (2012). Probiotics and multitude of health benefits. J. Res. Bio 2, 102–113.

Lee, B.J., and Bak, Y.T. (2011). Irritable bowel syndrome, gut microbiota and probiotics. J. Neurogastroenterol. Motil. *17*, 252–266.

Lin, H.C., and Pimentel, M. (2005). Bacterial concepts in irritable bowel syndrome. Rev. Gastroenterol. Disord. 5 (Suppl. 3), S3–9.

Lupton, J.R. (2004). Microbial degradation products influence colon cancer risk: the butyrate controversy. J. Nutr. *134*, 479–482.

Macfarlane, S., Macfarlane, G.T., and Cummings, J.H. (2006). Review article: prebiotics in the gastrointestinal tract. Aliment. Pharmacol. Ther. *24*, 701–714.

Major, G., and Spiller, R. (2014). Irritable bowel syndrome, inflammatory bowel disease and the microbiome. Curr. Opin. Endocrinol. Diabetes Obes. *21*, 15–21.

Malinen, E., Rinttila, T., Kajander, K., Matto, J., Kassinen, A., Krogius, L., Saarela, M., Korpela, R., and Palva, A. (2005). Analysis of the fecal microbiota of irritable bowel syndrome patients and healthy controls with real-time PCR. Am. J. Gastroenterol. *100*, 373–382.

McFarland, L.V., and Dublin, S. (2008). Meta-analysis of probiotics for the treatment of irritable bowel syndrome. World J. Gastroenterol. *14*, 2650–2661.

Min, Y.W., Park, S.U., Jang, Y.S., Kim, Y.H., Rhee, P.L., Ko, S.H., Joo, N., Kim, S.I., Kim, C.H., and Chang, D.K. (2012). Effect of composite yoghurt enriched with acacia fiber and Bifidobacterium lactis. World J. Gastroenterol. *18*, 4563–4569.

Murray, K., Wilkinson-Smith, V., Hoad, C., Costigan, C., Cox, E., Lam, C., Marciani, L., Gowland, P., and Spiller, R.C. (2012). Differential effects of FODMAPs (fermentable oligo-, di-, mono-saccharides and polyols) on small and large intestinal contents in healthy subjects shown by MRI. Am. J. Gastroenterol. *109*, 110–119.

Neish, A.S. (2009). Microbes in gastrointestinal health and disease. Gastroenterology *136*, 65–80.

Ng, S.C., Lam, E.F., Lam, T.T., Chan, Y., Law, W., Tse, P.C., Kamm, M.A., Sung, J.J., Chan, F.K., and Wu, J.C. (2013). Effect of probiotic bacteria on the intestinal microbiota in irritable bowel syndrome. J. Gastroenterol. Hepatol. *28*, 1624–1631.

Niedzielin, K., Kordecki, H., and Birkenfeld, B. (2001). A controlled, double-blind, randomized study on the efficacy of *Lactobacillus plantarum* 299V in patients with irritable bowel syndrome. Eur. J. Gastroenterol. Hepatol. *13*, 1143–1147.

Nikfar, S., Rahimi, R., Rahimi, F., Derakhshani, S., and Abdollahi, M. (2008). Efficacy of probiotics in irritable bowel syndrome: a meta-analysis of randomized, controlled trials. Dis. Colon Rectum *51*, 1775–1780.

Nobaek, S., Johansson, M.L., Molin, G., Ahrne, S., and Jeppsson, B. (2000). Alteration of intestinal microflora is associated with reduction in abdominal bloating and pain in patients with irritable bowel syndrome. Am. J. Gastroenterol. 95, 1231–1238.

O'Hara, A.M., and Shanahan, F. (2007). Mechanisms of action of probiotics in intestinal diseases. ScientificWorldJournal 7, 31–46.

O'Mahony, L., McCarthy, J., Kelly, P., Hurley, G., Luo, F., Chen, K., O'Sullivan, G.C., Kiely, B., Collins, J.K., Shanahan, F., *et al.* (2005). *Lactobacillus* and bifidobacterium in irritable bowel syndrome: symptom responses and relationship to cytokine profiles. Gastroenterology *128*, 541–551.

Olesen, M., and Gudmand-Hoyer, E. (2000). Efficacy, safety, and tolerability of fructooligosaccharides in the treatment of irritable bowel syndrome. Am. J. Clin. Nutr. *72*, 1570–1575.

Ortiz-Lucas, M., Tobias, A., Saz, P., and Sebastian, J.J. (2013). Effect of probiotic species on irritable bowel syndrome symptoms: A bring up to date meta-analysis. Rev. Esp. Enferm. Dig. *105*, 19–36.

Ouwehand, A.C., Isolauri, E., Kirjavainen, P.V., Tolkko, S., and Salminen, S.J. (2000). The mucus binding of Bifidobacterium lactis Bb_{12} is enhanced in the presence of *Lactobacillus* GG and Lact. delbrueckii subsp. bulgaricus. Lett. Appl. Microbiol. *30*, 10–13.

Paineau, D., Payen, F., Panserieu, S., Coulombier, G., Sobaszek, A., Lartigau, I., Brabet, M., Galmiche, J.P., Tripodi, D., Sacher-Huvelin, S., *et al.* (2008). The effects of regular consumption of short-chain fructo-oligosaccharides on digestive comfort of subjects with minor functional bowel disorders. Br. J. Nutr. *99*, 311–318.

Pessione, E. (2012). Lactic acid bacteria contribution to gut microbiota complexity: lights and shadows. Front. Cell. Infect. Microbiol. *2*, 86.

Piche, T. (2014). Tight junctions and IBS – the link between epithelial permeability, low-grade inflammation, and symptom generation? Neurogastroenterol. Motil. *26*, 296–302.

Pimentel, M., Lembo, A., Chey, W.D., Zakko, S., Ringel, Y., Yu, J., Mareya, S.M., Shaw, A.L., Bortey, E., and Forbes, W.P. (2011). Rifaximin therapy for patients with irritable bowel syndrome without constipation. N. Engl. J. Med. *364*, 22–32.

Prajapati, J.B., Khedkar, C.D., Chitra, J., Suja, S., Mishra, V., Sreeja, V., Patel, R.K., Ahir, V.B., Bhatt, V.D., Sajnani, M.R., *et al.* (2011). Whole-genome shotgun sequencing of an Indian-origin *Lactobacillus* helveticus strain, MTCC 5463, with probiotic potential. J. Bacterol. *193*, 4282–4283.

Sanders, M.E. (2008). Probiotics: definition, sources, selection, and uses. Clin. Infect. Dis. *46* (Suppl. 2), S58–61; discussion S144–151.

Saulnier, D.M., Spinler, J.K., Gibson, G.R., and Versalovic, J. (2009). Mechanisms of probiosis and prebiosis: considerations for enhanced functional foods. Curr. Opin. Biotechnol. *20*, 135–141.

Schrezenmeir, J., and de Vrese, M. (2001). Probiotics, prebiotics, and synbiotics – approaching a definition. Am. J. Clin. Nutr. *73*, 361S-364S.

Sharma, S., Arora, M., and Baldi, A. (2013). Probiotics in India: current status and future prospects. PharmAspire *4*, 68–79.

Silk, D.B., Davis, A., Vulevic, J., Tzortzis, G., and Gibson, G.R. (2009). Clinical trial: the effects of a trans-galactooligosaccharide prebiotic on faecal microbiota and symptoms in irritable bowel syndrome. Aliment. Pharmacol. Ther. *29*, 508–518.

Snook, J., and Shepherd, H.A. (1994). Bran supplementation in the treatment of irritable bowel syndrome. Aliment. Pharmacol. Ther. *8*, 511–514.

Spiller, R. (2008). Review article: probiotics and prebiotics in irritable bowel syndrome. Aliment. Pharmacol. Ther. *28*, 385–396.

Stanton, C., Gardiner, G., Meehan, H., Collins, K., Fitzgerald, G., Lynch, P.B., and Ross, R.P. (2001). Market potential for probiotics. Am. J. Clin. Nutr. 73, 476S-483S.

Stanton, C., Ross, R.P., Fitzgerald, G.F., and Van Sinderen, D. (2005). Fermented functional foods based on probiotics and their biogenic metabolites. Curr. Opin. Biotechnol. *16*, 198–203.

Steer, T., Carpenter, H., Tuohy, K., and Gibson, G.R. (2000). Perspectives on the role of the human gut microbiota and its modulation by pro- and prebiotics. Nutr. Res. Rev. *13*, 229–254.

Swidsinski, A., Ladhoff, A., Pernthaler, A., Swidsinski, S., Loening-Baucke, V., Ortner, M., Weber, J., Hoffmann, U., Schreiber, S., Dietel, M., *et al.* (2002). Mucosal flora in inflammatory bowel disease. Gastroenterology *122*, 44–54.

Symons, P., Jones, M.P., and Kellow, J.E. (1992). Symptom provocation in irritable bowel syndrome. Effects of differing doses of fructose-sorbitol. Scand J. Gastroenterol. 27, 940–944.

Tack, J., Simren, M., and Krammer, H. (2012). Probiotics in bloating distention and irritable bowel syndrome – from clinical evidence to guidelines. Eur. Gastroenterol. Hepatol. Rev. *8*, 72–76.

Tamime, A.Y., Marshall, V.M., and Robinson, R.K. (1995). Microbiological and technological aspects of milks fermented by bifidobacteria. J. Dairy Res. *62*, 151–187.

Tsuchiya, J., Barreto, R., Okura, R., Kawakita, S., Fesce, E., and Marotta, F. (2004). Single-blind follow-up study on the effectiveness of a symbiotic preparation in irritable bowel syndrome. Chin J. Dig. Dis. *5*, 169–174.

Verdu, E.F., Bercik, P., Bergonzelli, G.E., Huang, X.X., Blennerhasset, P., Rochat, F., Fiaux, M., Mansourian, R., Corthesy-Theulaz, I., and Collins, S.M. (2004). *Lactobacillus paracasei* normalizes muscle hypercontractility in a murine model of postinfective gut dysfunction. Gastroenterology *127*, 826–837.

Verdu, E.F., and Collins, S.M. (2005). Irritable bowel syndrome and probiotics: from rationale to clinical use. Curr. Opin. Gastroenterol. *21*, 697–701.

Vieira, A.T., Teixeira, M.M., and Martins, F.S. (2013). The Role of Probiotics and Prebiotics in Inducing Gut Immunity. Front Immunol. *4*, 445.

Vyas, U., and Ranganathan, N. (2012). Probiotics, prebiotics, and synbiotics: gut and beyond. Gastroenterol. Res. Pract. 872716.

Wang, H., Gong, J., Wang, W., Long, Y., Fu, X., Fu, Y., Qian, W., and Hou, X. (2014). Are There Any Different Effects of Bifidobacterium, *Lactobacillus* and Streptococcus on Intestinal Sensation, Barrier Function and Intestinal Immunity in PI-IBS Mouse Model? PLoS One *9*, e90153.

Whelan, K., and Quigley, E.M. (2013). Probiotics in the management of irritable bowel syndrome and inflammatory bowel disease. Curr. Opin. Gastroenterol. *29*, 184–189.

Zeng, J., Li, Y.Q., Zuo, X.L., Zhen, Y.B., Yang, J., and Liu, C.H. (2008). Clinical trial: effect of active lactic acid bacteria on mucosal barrier function in patients with diarrhoea-predominant irritable bowel syndrome. Aliment. Pharmacol. Ther. *28*, 994–1002.

Zhang, Z., Liu, C., and Guo, X. (2008). [Safety of probiotics – a review]. Wei Sheng Wu Xue Bao *48*, 257–261.

Pro- and Prebiotics for Oral Health

28

Svante Twetman, Mette Rose Jørgensen and Mette Kirstine Keller

Abstract

Recent insights in the importance of a natural microbial balance in the oral biofilm for maintaining health have brought a new concept to dentistry; to prevent ecologically disruptions rather than merely treating its consequences by restoring cavities. Consequently, an emerging interest in the use of probiotic bacteria for oral health is evident, albeit the local and systemic mechanisms of actions are still largely unknown. This chapter provides a brief background on the role of the oral biofilm in health and disease and examines the evidence for probiotic therapy within dentistry, based on clinical trials. Several studies have displayed an antagonistic role of probiotic lactobacilli and bifidobacteria against salivary *S. mutans* and four studies have shown reduced caries incidence and root caries arrest. Other trials have reported beneficial effects on endpoints related to gingivitis and periodontitis such as plaque index, gingival index, probing depth, subgingival microbiota and pro-inflammatory cytokine levels in gingival crevicular fluid. No adverse effects have been reported but further research is needed to confirm findings and strengthen the evidence before clinical recommendations can be advocated.

Introduction

The Human Microbiome Project has provided insights that human biofilms have co-evolved with mankind and play an important role for health and well-being. New pivotal molecular biological techniques, such as DNA-hybridization, 16S rDNA sequencing and microarrays, have opened up possibilities to study earlier uncultivable species and enabled researchers to study complex biofilm compositions and their functions. Today, the Human Oral Microbiome Database contains over 600 taxa and 13 phyla, of which only around 40% are cultivable with traditional agar-based methods (Dewhirst *et al.*, 2010). The current understanding is that the oral microbiota does not play a passive role but contributes actively to the maintenance of oral health; an oral biofilm can even be directly protective for certain conditions, such as enamel erosion. Thus, dentistry is facing a truly new paradigm to control and manage the various oral diseases. The aim of this chapter is to shortly review the background for this new concept and examine the evidence for using prebiotics and probiotic bacteria in the prevention, alleviation, control and management of some common dental diseases. This chapter has a clear focus on clinical trials in humans with endpoints of potential relevance for patients in everyday practice.

Replacement therapy versus bacteriotherapy

The term replacement therapy is commonly used for the event when 'a harmless effector strain is implanted in the host's microflora to maintain or restore a natural microbiome by interference and/or inhibition of other microorganisms' (Hillman and Socransky, 1987). The concept of bacteriotherapy is however even wider since also systemic effects via immune-modulation are suggested. The hypothesis behind bacteriotherapy rests on the concept of bacterial interference, i.e. using harmless bacteria to displace pathogenic organisms. Common examples of harmless effector strains are probiotic bacteria which are defined as 'live microorganisms which when administered in adequate amounts confer a health benefit on the host' (WHO, 2001). For clinical practice, there are two principal options; the use of natural commensal strains or to apply genetically modified bacteria in which certain properties have been added or deleted. The first successful attempt of bacteriotherapy was described by the French physician Tissier in 1908 who treated diarrhoea with colon irrigation (stool transplants with bifidobacteria). Since then, and even today, faecal transplantation with a bacterial microbiota from a healthy donor is an option in the management of diarrhoea caused by antibiotic resistant strains of *Clostridium difficile*. This approach of combating bacterial infections has also long been used in veterinary medicine. To understand the rationale of bacteriotherapy in the dental context, it is necessary to further elaborate on oral biofilms and the new paradigm of preventing oral diseases.

Oral biofilms in health and disease

Biofilms are found all over the body and the dental plaque is probably the body's most easily accessible biofilm. A biofilm is an

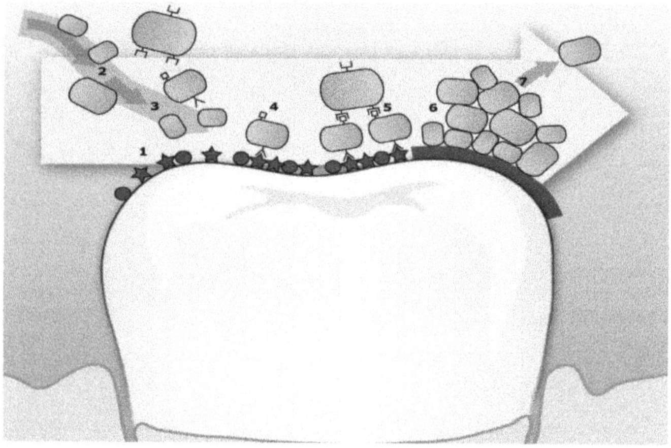

Figure 28.1. Schematic overview of pellicle formation, adhesion and colonization according to Hasslöf (2013), adapted after Marsh *et al.* (2009). 1. Pellicle formation. 2. Bacteria are transported passively to the tooth surface. 3. Bacteria may be held reversibly by weak electrostatic forces of attraction. 4. Attachment becomes irreversible mediated by adhesins on the bacteria and receptors in the pellicle. 5. Secondary colonizers attach to primary colonizers (co-adhesion). 6. Growth of bacteria results in biofilm maturation 7. De-attachment may occur.

aggregate of microorganisms in which cells adhere to each other on a surface that is protected. The different steps in the oral biofilm formation are shown in Fig. 28.1. The mature and complex biofilm is a community characterized by advanced inter-bacterial functions such as co-aggregation, metabolic exchange, cell-to-cell communication (quorum sensing) and exchange of genetic material.

The exact composition of the oral biofilm varies from one subject to the other and by location in the oral cavity (smooth surface, fissures, proximity to a saliva gland, etc.). Most of all, biofilms are influenced by genetic factors, diet, environment and stress. The newborn baby has a sterile mouth but already after a few hours, the oral cavity is colonized by the first streptococci. The mother is the primary source and the 'adult' biofilm is formed during the first 2–3 years of life and environmental conditions have a great influence on its microbial composition. A child's mouth is like an 'open window'; the sequence and timing of exposure to various microbes dictates the establishment and composition of the oral biofilm. For example, infants delivered with Caesarean section have a reduced diversity and fewer 'healthy' and protective bacteria in their saliva compared to those delivered vaginally (Nelun Barfod *et al.*, 2011; Lif Holgerson *et al.*, 2011). In the long run, this may even have a negative impact on their caries prevalence later in life (Barfod *et al.*, 2012). According to the 'hygiene hypothesis' (Strachan, 2000), children exposed to a diversity of friendly bacteria have a lower risk of developing asthma and ectopic eczema which has led to recommendations of baby drops, baby formulas and porridge containing probiotic bacteria. A recent finding of oral interest was that children using pacifiers 'cleaned' in their parent's mouth exhibited less eczema compared to those children who were provided with sterile pacifiers (Hesselmar *et al.*, 2013). The two groups could also be clearly distinguished based on the pattern of microbes in their saliva at 4 months of age. The examples above illustrate the importance of establishing a healthy biofilm in childhood.

A stable and diverse biofilm is associated with health and it is important to emphasize that the oral microbiota actively contributes to the maintenance of oral and dental health (Devine and Marsh, 2009). As stated above, biofilms can even be protective under certain conditions. When a stable biofilm is stressed, however, the balance between beneficial and pathogenic bacteria is disturbed and the diversity is reduced. The major reason for biofilm stress resulting in enamel demineralization is frequent and excessive sugar, and especially refined sucrose intake (Takahashi and Nyvad, 2011). Low-pH conditions in the biofilm are extended due to bacterial acid production and the biofilm adapts to this new environment through an overgrowth of aciduric and acidogenic bacteria. Similarly, when the biofilm stability is challenged by inflammatory proteins produced by pathogenic bacteria, the growth of proteolytic phenotypes is favoured, resulting in gingivitis and eventually, in periodontal disease. In both cases, the normal stability and balance between beneficial and hostile bacteria is lost and the diversity is significantly reduced. This current understanding is described as the 'ecological plaque hypothesis' (Marsh, 2009).

An important consequence of this ecological theory is that caries, the world's most common dental disease, is no longer regarded as a 'classical' infectious disease caused by specific pathogens, i.e. *S. mutans*. In older days, during the unspecific plaque hypothesis, the preventative mantra was to remove all plaque (biofilm) mechanically through meticulous tooth brushing at least two times a day. During the era of the specific plaque hypothesis, the general focus was to eliminate not all, but specific pathogens with antimicrobial treatment, for example chlorhexidine (Loesche, 1976). Today, with the current ecological view, treatment should attempt to control rather than eliminate the plaque flora (Marsh, 2004). Thus, the use of probiotics fits very well with the paradigm shift in oral microbiology; away from treating dental diseases by targeting specific oral pathogens towards an ecological and microbial-community-based approach.

Mechanisms of action

The mechanisms of probiotic action in the oral cavity are yet not known in detail. However, it is clear that both a local as well as a systemic event is occurring. The local events taking place in the oral cavity are summarized in Fig. 28.2, namely co-aggregation, competitive exclusion, bacteriocin production and competition for nutrients (Reid *et al.*, 2011). With caries influenced by the supragingival microbiota, the early studies concentrated on utilizing bacteria that were known to compete with, or that expressed bacteriocins or bacteriocin-like inhibitory substances against, caries-associated bacteria. On the other hand, gingivitis and periodontitis are inflammatory diseases mainly mediated by Gram-negative subgingival bacteria. Therefore, the probiotic effect is rather systemic through stimulation of the immune system since the direct access to periodontal pockets through local oral administration is limited. In this context, an interesting clinical observation was recently reported; volunteers exhibited

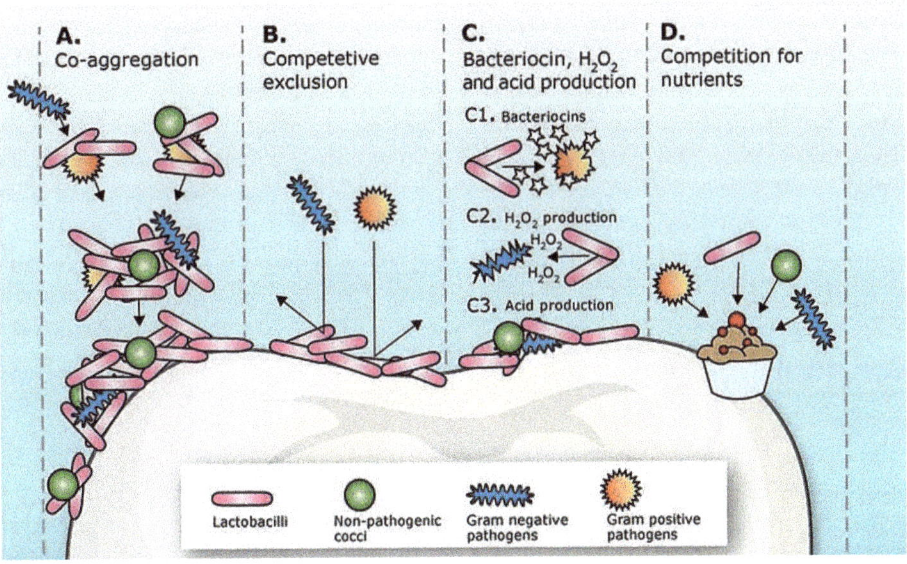

Figure 28.2 Proposed mechanisms of local effects on microorganisms by probiotic lactobacilli (with permission from Hasslöf, 2013).

increased levels of salivary IgA after 12 weeks ingestion of *L. reuteri* PTA5289 (Ericson *et al.*, 2013), thereby supporting the assumption that intake of beneficial bacteria might influence the adaptive immune response of the host.

The main probiotic strains used in oral science belong to the lactobacilli and bifidobacteria genera. Both are acidogenic and frequently occurring in the guts while they are rarely identified in high counts in the healthy oral biofilm or in saliva. From an efficiency point of view, there is no evidence to suggest lactobacilli or bifidobacteria over the other. Other genera with suggested probiotic properties are *Streptococcus* (i.e. *S. salivarius*) and *Proprionibacterium* (Wescombe et al., 2012).

Vehicles, dose and safety

Live 'friendly' bacteria must be taken on a regular, more or less daily basis in order to be detectable in plaque, saliva or stools. Thus, probiotic bacteria are not permanently installed in the human oral cavity in adults. Plateau levels, considered as therapeutic, are obtained in saliva after around 4 days of ingestion and lost again within a week after termination in most cases. It has been suggested that the chance of permanent installation increases following exposure early in life (Twetman and Stecksén-Blicks, 2008) and there are case reports describing subjects that seem to be permanent carriers of probiotic lactobacilli (Yli-Knuuttila *et al.*, 2006). It is important to stress that a permanent colonization is not a pre-requisite for probiotic action. For example, Toivainen *et al.* (2014) showed that short-term consumption of *L. rhamnosus* GG and *Bifidobacterium animalis* subsp. *lactis* BB-12 improved the periodontal status of otherwise healthy volunteers without affecting the profile of the oral microbiota.

The common vehicles for administration are dairy products (milk, yoghurt, sour cream) or chewing gums, tablets, capsules and lozenges. In some countries, food-items with probiotic additives are marketed; examples are cheese, bread, energy bars and fruit drinks. For infants, probiotic drops combined with D-vitamin are available. For topical applications in oral care, special oils, gels and toothpastes have been developed for self-care and professional use. Although there seems to be a dose–response relationship for oral effects (Twetman *et al.*, 2009), this area is yet almost unexplored. Nevertheless, common dose recommendations are 1–2 decilitres of yoghurt/milk per day with each millilitre containing $1 \times 10^{8-9}$ cfu of live bacteria or 1–2 tablets per day (approx. 1×10^9 cfu per tablet). The recommended dose for infants is five drops per day (2×10^7 cfu/drop).

Probiotic supplements in dentistry are from a regulatory point of view classified as food additives and labelled 'generally regarded as safe' (GRAS). This means that there are no reasons to fear overconsumption or adverse effects. It has, however, been suggested to handle critically ill patients and fragile elderly with some caution (Cannon *et al.*, 2005). The use of live bacteria has currently been discussed on the basis that there are some commercial products, claimed to be 'probiotic', that actually are based on heat-killed lactobacilli or lyophilized cell-walls of probiotic bacteria (Holtz *et al.*, 2013). Furthermore, in the rapidly growing interest in self-controlled health, a wide range of probiotic products have been marketed directly to the consumer with none or very limited research. However, it is the responsibility of the clinician to advocate safe probiotic products supported by sound background research and controlled clinical trials.

Prebiotics and oral health

Prebiotics are non-digestible food ingredients that affect the host by stimulating the growth and activity of selected beneficial bacteria, such as oligosaccharides, inulin and lactulose (de Vrese and Schrezenmeir, 2008). Although investigated in general medicine, the concept seems not yet to have entered the oral arena. A PubMed search in December 2013 did not identify any published literature of clinical relevance. Nevertheless, some interesting and indirect findings based on surrogate endpoints are worth mentioning. One clinical trial has compared the microbial diversity in saliva in babies that were breast fed or bottle fed (Holgerson *et al.*, 2013). The human milk is rich of oligosaccharides and the breast fed babies exhibited a greater diversity and more health-associated bacteria than their formula-fed peers. At three months of age, 25% of the breast-feed children were harbouring beneficial lactobacilli compared to zero per cent among the bottle-fed babies (Holgerson *et al.*, 2013). Another recent intervention study has linked the intake of fibre-rich diet with an increased salivary secretion rate in children (Laine *et al.*, 2014). An increased saliva secretion and buffer capacity is negatively associated with caries activity, because, besides from mechanical cleaning, the saliva contains biofilm-modulating agents such as electrolytes, antibacterial enzymes, immunoglobulins and amino acids (Dodds *et al.*, 2005). One of these amino acids, the semi-essential arginine, is of particular interest in the oral health context since it may act in a 'prebiotic' manner. The 'healthy' microbiota has arginolytic properties in the oral biofilm, converting arginine to ammonium, and thereby counteracting the low-pH stress in the oral biofilm (Cummins, 2013; Wolff *et al.*, 2013). The concept has been evaluated as an adjunct to fluoride and calcium in recent clinical trials in children and adults with active root caries lesions. Thus, arginine-containing toothpastes have displayed a 20% enhanced caries prevention in children (Kraivaphan *et al.*, 2013), improved remineralization of early caries lesions (Yin *et al.*, 2013) and a superior efficacy in arresting active root caries in adults compared to matched positive control dentifrices containing fluoride alone (Souza *et al.*, 2013). However, all studies were conducted in countries with generally high caries prevalence and the results need therefore to be confirmed in communities with a lower caries burden and better general access to dental care.

Probiotics and oral health

Probiotic bacteria have been employed for a number of oral conditions including caries, gingivitis, periodontal conditions, halitosis and dry mouth (Table 28.1). The number of publications identified through PubMed, based on the search term 'probiotics' AND 'oral health', was in December 2013 close to 350. However, the majority was *in vitro* laboratory studies and interestingly, a large number of narrative reviews, indicating that probiotics is a 'hot topic'. The scientific quality of the clinical trials was mixed; several studies applied an adequate randomized controlled research design while the risk of bias and confounders was moderate to high in others. Among the common limitations were small study groups, short-term interventions and a use of intermediate end-points. On the other hand, the research design and performance was fairly homogenous in many studies. However, a certain publication bias cannot be ruled out; projects on 'new' topics describing positive results are 'selling' and more likely published while studies with less good or questionable outcomes either are not submitted or even rejected. Furthermore, the high number of repetitive narrative reviews, all citing a limited number of original studies, might contribute to an over-report of enthusiastic messages concerning probiotics.

Dental caries

In Table 28.2, 26 clinical trials with caries-related endpoints are summarized. A quick glance revealed a significant inhibitory effect on salivary mutans streptococci levels in 19 out the 26 studies (73%). It was, however, not possible to quantify the magnitude of the effect because in most cases, simplified cultivation techniques with categorized bacterial scores were used rather than conventional counts of colony forming units. Another obvious limitation was that follow-up information on possible sustained mutans streptococci reduction was largely lacking. Interestingly, the mode of the delivery of the probiotic strains did not seem to significantly influence the suppressive effect, nor were any obvious differences between strains of probiotic lactobacilli or bifidobacteria revealed.

One might ask why a reduction of the surrogate endpoint *S. mutans* is of clinical importance. High counts of salivary mutans streptococci are considered as a biomarker of an acid-stressed 'cariogenic' oral environment and therefore, the results may indicate neutralized biofilm conditions and a reduced caries risk. With two exceptions, no significant increase in lactobacilli counts was recorded and most important, no side effects were reported. Two recent clinical trials have confirmed that probiotic *L. reuteri* supplements (strains SD2112, DSM17938 and PTA 5289) exerted no negative influence on biofilm acidogenicity (Marttinen *et al.*, 2011; Keller and Twetman, 2012). This was an important finding since concerns have been postulated that an increased lactobacilli exposure could increase the progression of existing cavitated

Table 28.1 Examples of probiotic bacteria investigated with respect to oral health

Species	Vehicle	Potential effects
Lactobacillus brevis	Lozenges	*S. mutans* antagonist, acid production, mucositis
Lactobacillus rhamnosus	Milk, yoghurt, cheese	*S. mutans* inhibition, caries reduction
Lactobacillus reuteri	Lozenge, chewing gum	Plaque/gingivitis reduction, anti-inflammatory
Lactobacillus paracasei	Dietary supplement	*S. mutans* inhibition, caries reduction in rats
Lactobacillus plantarum	Fruit drink	Hampers caries-associated bacteria
Lactobacillus salivarius	Tablet	Gingivitis/parodontitis reduction
Bifidobacterium lactis	Yoghurt, ice-cream	*S. mutans* reduction in saliva
Streptococcus salivarius	Lozenge	Halitosis relief
Weissella cibaria	Rinse	Halitosis relief, plaque reduction

Table 28.2 Clinical trials with caries-related microbial endpoints

First author, year	Design[a]	*n*; age (years)	Vehicle, time[b]	Strain	Major outcome[c]
Näse, 2001	RCT	594; 1–6	Milk, 7m	*L. rhamnosus* GG	MS↘
Ahola, 2002	RCT	74; 18–35	Cheese, 3w	*Lactobacilli* mix	MS↘
Nikawa, 2004	Cross-over	40; 20	Milk, 2w	*L. reuteri* 55730	MS↘
Montalto, 2004	RCT	35; 23–37	Capsules, 45d	*Lactobacilli* mix MS	→, LB↗
Caglar, 2005	cross-over	21; 21–24	Yoghurt, 2w	*Bifidobacterium* DN-173010	MS↘
Caglar, 2006	RCT	120; 21–25	Tablets, 2w	*L. reuteri* ATCC55730	MS↘
Caglar, 2007	RCT	80; 21–24	Gum, 3w	*L. reuteri* 55730+PTA5289	MS↘
Caglar, 2008	RCT	20; 20	Lozenge, 10d	*L. reuteri* 55730+PTA5289	MS↘
Caglar, 2008	Cross-over	40; 20–24	Ice cream, 10d	*B. lactis* Bb-12	MS↘
Cildir, 2009	Cross-over	24; 12–16	Yoghurt, 2w	*B. animalis* ssp. *lactis* DN173010	MS↘
Stecksén-Blicks, 2009	CRCT	174; 1–5	Milk, 21m	*L. rhamnosus* LB21	MS→
Singh, 2011	Cross-over	40; 12–14	Ice-cream, 10d	*B. lactis* Bb-12/*L. acidophilus*	MS↘
Lexner, 2010	RCT	20; 12–15	Milk, 2w	*L. rhamnosus* LB21	MS→
Jindal, 2011	RCT	150; 7–14	Powder, 14d	Darolac[d]	MS↘
Marttinen, 2011	Cross-over	13; 20–30	Tablets, 2w	*L. rhamnosus* GG/*reuteri* SD2112	MS→; LB↗
Chuang, 2011	RCT	70; 20–26	Tablets, 2w	*L. plantarum* GMNL-33	MS↘
Cildir, 2011	RCT	19[e]; 4–12	Drop, 25d	*L. reuteri* DSM17938+PTA5289	MS, LB→
Glavina, 2012	NC	25; 6–10	Yoghurt, 2w	*L. rhamnosus* GG	MS↘, LB→
Juneja, 2012	RCT	40; 12–15	Milk, 9w	*L. rhamnosus* hct70	MS↘
Sudhir, 2012	RCT	40; 10–12	Curd; 3w	*L. acidophilus* (Nestlé curd)	MS↘
Taipale, 2012	RCT	106; 2m-2	Pacifier/2yr	*B. lactis* Bb-12	MS→
Burton, 2013	RCT	100; child	?, 3m	*S. salivarius* M18	Plaque score ↘
Campus, 2013	RCT	191; 6–8	Tablets, 6w	*L. brevis* CD2	MS↘
Holz, 2013	RCT	60; adults	Candies, 1.5d	*L. paracasei* DSMZ16671	MS↘
Chinappa, 2013	CCT	40; 12–14	Ice-cream, 7d	*B. lactis* Bb-12/*L. acidophilus*	MS↘
Cannon, 2013	CCT	60; 6–12	Tablets, 30d	PerioBalance/EvoraKids[f]	MS↘, LB↘

[a]RCT, randomized controlled trial; CRCT, cluster randomized controlled trial.
[b]Time of intervention, in days (d), weeks (w) or months (m).
[c]MS, mutans streptococci; LB, lactobacilli; ↘=decreased; →=unchanged; ↗=increased.
[d]*L. rhamnosus*, *B. longum*, *Saccharomyces cerevisae*.
[e]CLP, cleft, lip, palate patients.
[f]PerioBalance=*L. reuteri* DSM17938+PTA5289; EvoraKids=*Streptococcus uberis* KJ2, *Streptococcus oralis* KJ3, *Streptococcus rattus* JH1 45.

caries lesions in susceptible patients. Based on the data presented above and two recent systematic reviews (Cagetti *et al.*, 2013; Laleman *et al.*, 2014), it can safely be concluded that probiotics play a role as antagonistic agent on mutans streptococci and acidogenic/aciduric bacteria that contributes to the caries process without any evidently harmful effects.

A recent study investigated if probiotic supplements (*L. reuteri* DSN17938 and PTA5289) were able to delay mutans streptococci re-growth after full mouth disinfection with chlorhexidine gluconate, a commonly used oral antiseptic. This was not the case in comparison with a placebo control group (Keller *et al.*, 2011). Further analysis of the samples of the test group disclosed however that subjects with detectable DNA from the administrated *L. reuteri* DSN17938 strain showed almost no mutans streptococci re-growth in contrast to those with no presence of the strain specific DNA (Fig. 28.3; Romani Vestman *et al.*, 2013). This illustrates the important fact that the response to probiotic therapy might be strain-dependent and highly individual.

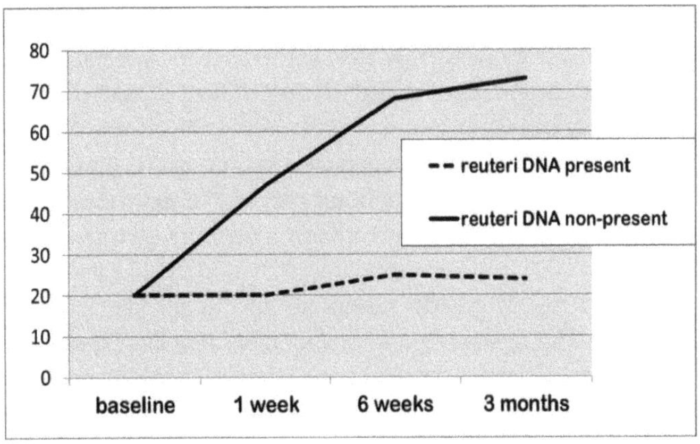

Figure 28.3 Regrowth of salivary mutans streptococci (MS) after full mouth disinfection with chlorhexidine. MS levels (cfu) differed significantly depending on the presence or absence of L. reuteri DSM 17938 detected by PCR (data from Romani Vestman *et al.*, 2013).

Table 28.3 Clinical trials with clinical caries as endpoint

First author, year	Design[a]	N; age (years)	Vehicle, time[b] (dropout rate)	Strain	Caries, test/ctr	PF[c] (%)
Näse, 2001	RCT	594; 1–6	Milk, 7m (24%)	*L. rhamnosus* GG	15%/19%[d]	21%
		164; 3–4	Milk, 7m	*L. rhamnosus* GG	10%/23%[d]	56%
Stecksén-Blicks, 2009	CRCT	174; 1–5	Milk, 21m (25%)	*L. rhamnosus* LB21 + 2.5 ppm fluoride	Δdmfs 0.4/1.6	75%
Petersson, 2011	RCT	200; 56–84	Milk, 15m (38%)	*L. rhamnosus* LB21 + 5.0 ppm fluoride	rev 68%/24%[e]	65%
				L. rhamnosus LB21	rev 54%/24%[e]	55%
Taipele, 2013	RCT	106; 4	Pacifier, 8m (11%)	*B. animalis* BB-12	CF[f] 69%/64%	5% (NS)
Hasslöf, 2013	RCT	180; 6	Cereals, 12m (9%)	*L. paracasei* F19	CF 80%/74%	6% (NS)
Stensson, 2013	RCT	113; 9	Drops, 12m (40%)	*L. reuteri* ATCC	CF 82%/58%	29%
Keller, 2014	RCT	36; 12–17	Lozenges, 3m	*L. reuteri* ATCC	QLF[g]	NS

[a]RCT, randomized controlled trial; CRCT, cluster randomized controlled trial.
[b]Intervention time in months (m).
[c]PF=prevented fraction; NS=not statistically significant.
[d]Caries prevalence (%).
[e]Root caries reversals (%).
[f]CF, caries-free.
[g]QLF, quantitative light fluorescence; NS, no significant differences between probiotic and placebo groups.

The seven clinical trials with caries incidence or root caries arrest as endpoints are listed in Table 28.3. A calculated mean value of the finding indicates 39% effectiveness, expressed as prevented fraction, in caries prevention and control. The first three studies utilized milk as vehicle for probiotic 'cocktail', with or without small amounts of fluoride. The pioneering Finnish study by Näse and co-workers in 2001 was the first major trial to demonstrate a caries-preventative effect in preschool children, 3–4 years of age, during 7 months consumption of milk supplemented with *L. rhamnosus* GG. In addition to an improved dental health, the children of the test group displayed better general health with fewer infections and a reduced need of antibiotics (Hatakka *et al.*, 2001). A randomized controlled trial in Sweden adopted the same concept and demonstrated an impressive 75% caries reduction over 21 months in preschool children when the probiotic milk drink (*L. rhamnosus* LB21) was supplemented with 2.5 ppm fluoride (Stecksen-Blicks *et al.*, 2009). A drawback however, was that the effects of the probiotic bacteria and fluoride could not be separated from each other. Also in this trial, the general health among the test children was improved with fewer doctor visits and significantly less prescription of antibiotics. Collectively, the two studies suggested a certain metabolic 'domino effect', illustrating that the oral cavity is a significant part of the gastro-intestinal tract. The third milk-study showed that probiotic supplements (*L. rhamnosus* LB21), alone or in adjunct to fluoride, could arrest and reverse existing root caries lesions in older adults (Petersson *et al.*, 2011). This is important information because life-expectancy is increasing globally and the ageing population maintains their own teeth to a greater extent. This means an increasing number of natural teeth, vulnerable for dental diseases in an ageing population that needs to be cared for. The effect of probiotic lozenges (*L. reuteri* DSM17928 and PTA5289) on early non-cavitated caries lesions in caries-susceptible adolescents was recently evaluated with quantitative light fluorescence (QLF), a device that can unveil small alterations in the mineral content of enamel (Keller *et al.*, 2014). After 3 months of ingestion, no significant differences in remineralization was found in comparison with placebo, albeit a significant decrease in fluorescence over time was noted in the probiotic test group.

A recent trial from Finland has tested an interesting concept (Taipale *et al.*, 2013). Probiotic lozenges containing *Bifidobacterium animalis* subsp. *lactis* BB-12 were administrated to infants with the help of a specially designed pacifier during the babies first year of life. The long-term impact on caries development was modest and non-significant at the age of 4-years which may be attributed to an uncertain compliance. Similar findings were displayed in a dental follow-up study of children receiving probiotic gruel (*L. paracasei* F19) during infancy (4–12 months). Although the test children showed less eczema at the medical follow-ups, the effect on caries at 6-years of age was promising but statistically non-significant (Hasslöf *et al.*, 2013). It should be noted that the general level of caries was very low in the two latter studies and that the number of children lost to follow-up was high in the study by Hasslöf *et al.* (2013). Contrasting and highly interesting findings have recently been presented by Stensson *et al.* (2013). The authors conducted a follow-up study in 9-year-old children who were given five probiotic drops per day containing *L. reuteri* ATCC 55730 from birth to 12 months of age. The primary outcome of the placebo-controlled trial was the prevalence of IgA-mediated eczema at the age of 2, but 60% of the children were available for a dental examination 7 years later. The children of the probiotic group displayed significantly lower caries prevalence (initial and manifest in primary teeth) in comparison with the control group (18% vs. 42%). Also the frequency of proximal caries lesions was significantly lower as shown in Fig. 28.4. These long-term results do reinforce the concept that probiotics may be most important and effective for caries prevention when applied early in life. In summary, the effect of probiotics on the development and arrestment of caries lesions seems promising, but further

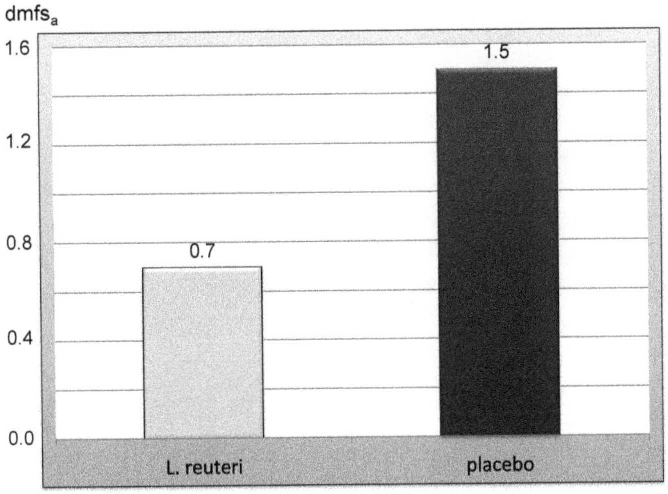

Figure 28.4 Mean proximal caries frequency (decayed, missed, filled surfaces; $dmfs_a$) in 9-year-olds given probiotic drops containing L. reuteri ATCC55730 (5/day) or placebo during their first year of life (data from Stensson *et al.*, 2013).

high-quality research is needed to build up evidence and justify clinical recommendations.

Gingivitis and periodontitis

Gingivitis is the most common form of periodontal disease and is in its most common form an inflammatory response to bacterial biofilms adherent to tooth surfaces. Periodontitis is a more advanced gum infection associated with an overactive immune system that damages the soft tissue and bone that supports the tooth. With periodontitis, the alveolar bone around the teeth is progressively lost, eventually leading to deep periodontal pockets and tooth loss. Twenty-two clinical trials with endpoints related to plaque amount, gingival condition, probing depth and a subgingival microbiota associated with periodontal diseases are compiled in Table 28.4. All except two studies reported beneficial effects of selected endpoints which may indicate that probiotic technologies have a prosperous future in periodontics. The majority of the trials was performed in adults with *L. reuteri* administrated in tablets or lozenges and displayed reduced amounts of plaque linked

Table 28.4 Clinical trials with endpoints related to gingivitis and periodontitis

First author, year	Design[a]	N; age	Vehicle, time[b]	Strain	Major outcome[c]
Kang, 2006	Cross-over	72; adults	Rinse, 1d	*W. cibaria* CSM1	OHI-S↘
Krasse, 2006	RCT	59; adults	Gums, 2w	*L. reuteri* LR1	PI, GI↘
Riccia, 2007	RCT	29; adults	Lozenges, 4d	*L. brevis* CD2	PI, GI↘, cytokines↘
Shimauchi, 2008	RCT	66; adults	Tablets, 8 w	*L. salivarius* WB21	PI, GI, BOP↘
Twetman, 2009	RCT	38; adults	Gums, 4 w	*L. reuteri*[d]	BOP→; TNF-a ↘
Mayanagi, 2009	RCT	66, adults	Tablets	*L. salivarius* WB21	Periopath↘
Staab, 2009	RCT	50; adults	Milk drink	*L. casei* Shirota	PI, GI↗
Tsubura, 2009	RCT	54; adults	Mouth rinse	*B. subtilis* E-300	GI↘, BOP→
Zahradnik, 2009	CCT	20; adults	Mouth rinse	ProBiora[e]	Periopath↘
Harini, 2010	RCT	45; child	Rinse, 2w	Unknown	PI, GI↘[e]
Sinkiewicz, 2010	RCT	23; adults[f]	Diet, 12w	*L. reuteri*[d]	LB↗, PI↘
Iwamoto, 2010	NC	20; adults	Tablet, 4w	*L. salivarius* WB21	BOP→
Vivekananda, 2010	RCT	30; adults[f]	Srp[g]+tablets, 6w	*L. reuteri*[d]	PI, GI, BOP, PD↘
Kleinhans, 2012	RCT	45; preg	Lozenges, 3m	*L. reuteri*[d]	PI, GI↘
Slawik, 2011	Exp. ging.	28; adults	Milk drink; 2w	*L. casei* Shirota	BOP↘
Iniesta, 2012	RCT	40; adults	Tablets, 8w	*L. reuteri*[d]	Periopath↘
Vicario, 2013	RCT	20; adults	Tablets, 4w	*L. reuteri*[d]	PI, BOP, PD↘
Burton, 2013	RCT	100; child	Rinse, 3m	*S. salivarius* M18	PI↘
Hallström, 2013	Exp. ging.	40; adults	Tablets, 3w	*L. reuteri*[d]	PI, GI→
Shah, 2013	NC	30; adults[d]	Srp[e] +tablets, 2w	*L. brevis* CD2	PI, GI, PD↘
Karuppaiah, 2013	RCT	216;14–17	Curd, 4w	Unknown	PI↘
Teughels, 2013	RCT	30[d]; >35[d]	12d, lozenges	*L. reuteri*[d]	PD, BOP↘[h]

[a]RCT, randomized controlled trial; crossover, randomized crossover trial; CCT, controlled clinical trial; exp. ging, experimental gingivitis study; NC, non-controlled, open label.
[b]Intervention time in days (d), weeks (w) or months (m).
[c]OHI-S, oral hygiene index; PI, plaque index; GI, gingival index; BOP, bleeding on probing; PD, pocket probing depth; ↘, decreased; →, unchanged; ↗, increased.
[d]L. reuteri ATCC 55730 and ATCC PTA 5289.
[e]*S. uberis* KJ2, *S. oralis* KJ3 and *S. rattus* JH145.
[f]Subjects with periodontitis.
[g]SRP, scaling and root planning.
[h]No differences compared to placebo.

with an improved gingival bleeding index which was immediately recognized by the patients. One study showed a decreased concentration of the pro-inflammatory cytokine TNF-α in the gingival crevicular fluid following daily intake of *L. reuteri* ATCC55730 and PTA5289 (Twetman *et al.*, 2009). This finding has been regarded as a proof of concept concerning the systemic influence on the oral cavity achieved via probiotic therapy. Another support for this theory was provided by Iniesta *et al.* (2012), who demonstrated significant reductions of *P. gingivalis* in subgingival samples after orally administrated tablets containing *L. reuteri* (DSM17938 and PTA5289). However, due to inconsistent study designs and reporting of endpoints, there is yet no meta-analysis or systematic review available on the impact of probiotic supplements on gingivitis and dental plaque reduction.

Three recent studies have combined the use of probiotics with scaling and root planing (SRP) on patients with chronic periodontitis which is regarded as the established best clinical practice for treatment of periodontal diseases. In the first study, the combined use was more successful to reduce pocket depth than either SRP or probiotic therapy (*L. reuteri* DSM17928 and PTA5289) alone (Vivekananda *et al.*, 2010). In the second study (Shah *et al.*, 2013), the effects of probiotic tablets (*L. brevis* CD2) were compared with doxacycline medication during 14 days after SRP in young adult patients with aggressive periodontitis. Both alternatives had a similar reducing effect on plaque index, gingival index and periodontal pocket probing depths after 2 months. The findings that the use of probiotics was an equal treatment option to antibiotics were of special interest because the prescription of systemic antibiotics in dentistry can be reduced. In the long run, this will contribute to lowering the emerging risk of developing drug-resistant microorganisms. The third study, conducted in Turkey, confirmed that oral administration of lozenges containing *L. reuteri* (DSM17938 and ATCC PTA5289) could be a useful adjunct to SRP in the management of chronic periodontitis (Teughels *et al.*, 2013). The combined therapy resulted in gain of attachment in terms of reduced periodontal pockets, as well as significantly reduced counts of *Porphyromonas gingivalis* (Teughels *et al.*, 2013). However, as pointed out by Gupta (2011), more research and larger long-term studies are needed before evidence-based treatment guidelines for managing gingivitis and periodontal diseases can be released. There is still a pertinent lack of information on the effect of topical applications. Teughels and co-workers (2007) have demonstrated that subgingival applications of a mixture of beneficial oral bacteria including *S. sanguinis*, *S. salivarius* and *S. mitis* could significantly suppress re-colonization of periodontal pathogens (*P. gingivalis* and *P. intermedia*) in Beagle dogs. There are ongoing projects in humans and if similar findings are verified, new breakthrough therapeutic strategies in periodontics can be foreseen.

Mucositis

The former mentioned precaution to avoid probiotic therapy in immune-suppressed and critically ill patients has recently been challenged. Oral mucositis is a frequent side-effect appearing in patients with head- and neck cancer treated with radiation and chemotherapy. The condition is very painful and hinders normal eating and oral hygiene procedures. In a randomized controlled trial, 200 adult patients were treated with probiotic lozenges containing *L. brevis* CD2 or placebo (six tablets per day) during and after the acute phases of cancer treatment (Sharma *et al.*, 2012). The prevalence of mucositis was significantly reduced in the test group compared with the placebo group (7% vs. 28%) and also the severity of the condition was significantly improved (Fig. 28.5). No similar reports are yet available for children and it should be underlined that the daily dose of probiotic bacteria was three times higher than in the previously mentioned trials with caries and periodontal endpoints.

Candidosis

Overgrowth of yeasts, especially *Candida albicans*, is common in elderly and generally regarded as a sign of a disturbed balance of the oral environment. Oral candida, impairing life quality and wellbeing, is often seen in ageing persons with xerostomia (dry mouth) and as a consequence of polypharmacy. The effect of probiotic lactobacilli (*L. rhamnosus* GG) on oral candida has been evaluated in elderly by Hattaka and co-workers (2007). They were able to demonstrate a decreased prevalence of candida after 16 weeks' consumption of probiotic cheese (50 g per day) and no side effects occurred. In fact, also the saliva secretion rate and buffering capacity was improved in the test group but the long-term effects of the intervention were not evaluated. A recent study has evaluated the effect of locally administrated probiotics, a powder mix of *B. longum*, *L. bulgaricus* and *S. thermophilus* (Shangri Pharmaceutical Co. Ltd., Neimenggu China) on candida-associated stomatitis (Li *et al.*, 2013). The findings showed that the probiotic adjunct improved clinical conditions and reduced the prevalence of *Candida* spp. when compared with conventional antifungal therapy (sodium bicarbonate plus 2% nystatin). Ongoing studies in Scandinavia concerning candida in lichen planus patients and in disabled elderly nursing home residents will further increase

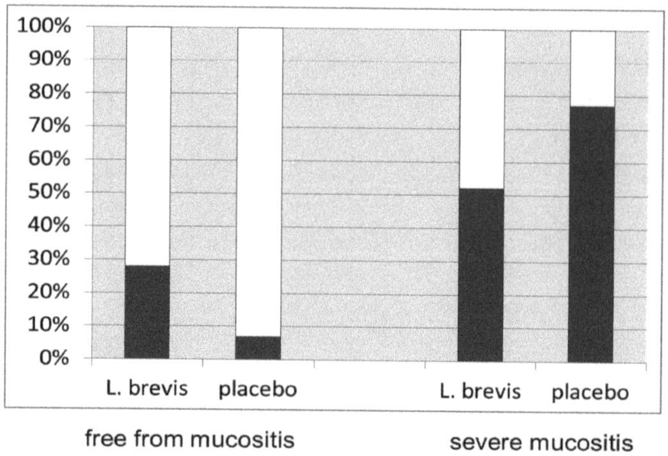

Figure 28.5 Prevalence and severity of radiation- and chemotherapy-induced oral mucositis (black bars) in patients with head and neck cancer after intake of probiotic lozenges (L. brevis CD2) or placebo during, and for one week after completion of anti-cancer therapy (data from Sharma *et al.*, 2012). Values in the figure denote per cent.

the knowledge around these issues and eventually form the ground for clinical recommendations.

Halitosis (bad breath)

Bad breath is a stigma that may lead to social problems and a reduced quality of life. One of the common reasons is the production of volatile sulphur compounds by commensal oral bacteria such as *Fusobacterium nucleatum*. In this context, probiotic therapy has been suggested to relieve subjective symptoms in addition to meticulous oral hygiene and tongue scraping. The theory behind this is that probiotic bacteria can co-aggregate effectively with *F. nucleatum* and suppress Gram-positive bacteria involved in halitosis. Preliminary reports from studies with bacteriocin-producing *S. salivarius* K12, *Weissella cibaria* CSM1 and *L. salivarius* WB21 have suggested a reduction of the volatile sulphur compound levels and biofilm formation (Burton *et al.*, 2006; Kang *et al.*, 2006; Iwamoto *et al.*, 2010). Furthermore, the use of lozenges containing two strains of *L. reuteri* (DSM 17938 and ATCC PTA5289) has been demonstrated to improve self-perceived bad breath among young adults as judged by an organoleptic panel (Keller *et al.*, 2012). The possible effect on severe and long-lasting halitosis cases remains to be evaluated.

Future trends and some unanswered questions

Currently, several studies are exploring the role of the oral microbiome in health and disease, thereby extending the knowledge within the research area. Thus, in our opinion, the concept of probiotics will continue to gain research attention within dentistry, especially as a link between oral medicine and behaviour and lifestyle associated general diseases such as the metabolic syndrome. The probiotic concept is also attractive for informed consumers and along with an increasing patient awareness, new products and brands will be offered over the counter. While the use of natural beneficial bacteria seems more or less non-controversial, the market for genetically modified strains may be questioned, at least for early introduction in babies and infants. One crucial issue is to define and find the optimal strains for the various dental diseases. So far, dentistry has adopted strains from gastrointestinal or urogenital use which may be suboptimal in the oral environment. Consequently, the search and mapping of effective probiotic candidates for oral health is intensive (Lang *et al.*, 2010; Bosch *et al.*, 2012; Güngör *et al.*, 2013). There are experimental *in vitro* indications that orally derived probiotic strains are to be preferred but very few have been tested in clinical settings. For example, it has been shown that lactic acid bacteria isolated from the oral cavity of children adhered to a higher level to saliva-covered tooth surfaces than strains isolated from fermented meat products (Güngör *et al.*, 2013). Another issue is whether or not a mix of different strains works better than single strain concepts. Furthermore, the optimal daily dosage is yet to be determined and preliminary data from the global laboratories indicate that a clear dose–response relationship may occur. Maybe higher doses will be advocated in the future? To summarize, a better understanding of the molecular mechanisms of probiotics, the effect of probiotic mixtures versus single strains, the effect of formulation of probiotics and the fate of ingested probiotics should help to clarify the clinical value of probiotic supplements for oral health. Finally, the potential additive use of prebiotics and synbiotics, a combination of prebiotics and probiotics, needs to be clarified in the oral cavity.

Summary

In light of the new paradigm concerning prevention and management of biofilm-mediated oral diseases, the interest in probiotic bacteria is rapidly growing. While no information is available concerning prebiotics, to date, it is possible to conclude that daily intake of probiotic supplements may:

- influence and restore the oral biofilm balance and diversity;
- suppress salivary levels of mutans streptococci;
- prevent and arrest caries, especially in combination with fluoride;
- moderate mutans streptococci re-growth after full-mouth disinfection;
- reduce plaque amount and exert a beneficial effect on gingivitis and periodontal pockets, especially in combination with traditional periodontal therapy, i.e. scaling and root-planing;
- hamper oral mucositis and alleviate bad breath.

However, more high-quality research is needed to close gaps of knowledge concerning optimal species, dose regimens and long-term effects. It should therefore be stressed that probiotic therapy is not an alternative but a complement that always should be combined and added to evidence-based methods and good clinical practice.

Disclosure

Drs Rose Jørgensen and Keller have received PhD grants from Biogaia AB, Lund, Sweden.

References

Ahola, A.J., Yli-Knuuttila, H., Suomalainen, T., Poussa, T., Ahlström, A., Meurman, J.H., and Korpela, R. (2002). Short-term consumption of probiotic-containing cheese and its effect on dental caries risk factors. Arch. Oral Biol. *47*, 799–804.

Barfod, M.N., Christensen, L.B., Twetman, S., and Lexner, M.O. (2012). Caries prevalence in Danish pre-school children delivered vaginally and by caesarean section. Acta Odontol. Scand. *70*, 190–293.

Bosch, M., Nart, J., Audivert, S., Bonachera, M.A., Alemany, A.S., Fuentes, M.C., and Cuñé. J. (2012). Isolation and characterization of probiotic strains for improving oral health. Arch. Oral Biol. *57*, 539–549.

Burton, J.P., Chilcott, C.N., Moore, C.J., Speiser, G., and Tagg, J.R. (2006). A preliminary study of the effect of probiotic Streptococcus salivarius K12 on oral malodour parameters. J. Appl. Microbiol. *100*, 754–764.

Cagetti, M.G., Mastroberardino, S., Milia, E., Cocco, F., Lingström, P., and Campus, G. (2013). The use of probiotic strains in caries prevention: a systematic review. Nutrients 5, 2530–2550.

Çaglar, E., Sandalli, N., Twetman, S., Selvi, S., Ergeneli, S., and Kavaloglu, S. (2005). Effect of yoghurt with *Bifidobacterium* DN-173010 on salivary mutans streptococci and lactobacilli. Acta Odontol. Scand. *63*, 317–230.

Çaglar, E., Kavaloglu, S., Ergeneli, S., Sandalli, N., and Twetman, S. (2006). Salivary mutans streptococci and lactobacilli levels after ingestion of the probiotic bacterium *Lactobacillus* reuteri ATCC 55730 by straws and tablets. Acta Odontol. Scand. *64*, 314–318.

Çaglar, E., Kavaloglu, S.U., Kuscu, O.O., Sandalli, N., Lif Holgerson, P., and Twetman, S. (2007). Effect chewing gums containing xylitol and probiotic bacteria on salivary mutans streptococci and lactobacilli. Clin. Oral Investig. *11*, 425–429.

Çaglar, E., Kuscu, O.O., Cildir, S.K., Kuvvetli, S.S., and Sandalli, N. (2008). A probiotic lozenge administered medical device and its effect on salivary mutans streptococci and lactobacilli. Int. J. Paediatr. Dent. *18*, 35–39.

Caglar, E., Onder Kuscu, O., Selvi Kuvvetli, S., Kavaloglu Cildir, S., Sandalli, N., and Twetman, S. (2008). Consumption of probiotic ice-cream with Bifidobacterium Bb-12 and its effect on salivary mutans streptococci and lactobacilli. Acta Odontol. Scand. *66*, 154–158.

Campus, G., Cocco, F., Carta, G., Cagetti, M.G., Simark-Mattson, C., Strohminger, L., and Lingström, P. (2013). Effect of a daily dose of *Lactobacillus* brevis CD2 lozenges in high caries risk schoolchildren. Clin. Oral Invest. May 5, Epub ahead of print.

Cannon, J.P., Lee, T.A., Bolanos, J.T., and Danziger, L.H. (2005). Pathogenic relevance of *Lactobacillus*: a retrospective review of over 200 cases. Eur. J. Clin. Microbiol. Infect. Dis. *24*, 31–40.

Cannon, M., Trent, B., Vorachek, A., Kramer, S., and Esterly, R. (2013). Effectiveness of CRT at measuring the salivary level of bacteria in caries prone children with probiotic therapy. J. Clin. Pediatr. Dent. *38*, 55–60.

Chinnappa, A., Konde, H., Konde, S., Raj, S., and Beena, J.P. (2013). Probiotics for future caries control: A short-term clinical study. Indian J. Dent. Res. *24*, 547–9.

Chuang, L.C., Huang, C.S., Ou-Yang, L.W., and Lin, S.Y. (2011). Probiotic *Lactobacillus paracasei* effect on cariogenic bacterial flora. Clin. Oral Investig. *15*, 471–476.

Cildir, S.K., Germec, D., Sandalli, N., Ozdemir, F.I., Arun, T., Twetman, S., and Caglar, E. (2009). Reduction of salivary mutans streptococci in orthodontic patients during daily comsumption of yoghurt containing probiotic bacteria. Eur. J. Orthod. *31*, 407–411.

Cildir, S.K., Sandalli, N., Nazli, S., Alp, F., and Caglar, E. (2012). A novel delivery system of probiotic drop and its effect on dental caries risk factors in cleft lip/palate children. Cleft Palate Craniofac. J. *49*, 369–372.

Cummins, D. (2013). The development and validation of a new technology, based upon 1.5% arginine, an insoluble calcium compound and fluoride, for everyday use in the prevention and treatment of dental caries. J. Dent. *41* (Suppl. 2), S1–11.

Devine, D.A., and Marsh, P.D. (2009). Prospects for the development of probiotics and prebiotics for oral applications. J. Oral Microbiol. doi: 10.3402/jom.v1i0.1949.

Dewhirst, F.E., Chen, T., Izard, J., Paster, B.J., Tanner, A.C., Yu, W.H., Lakshmanan, A., and Wade, W.-G. (2010). The human oral microbiome. J. Bacteriol. *192*, 5002–5017.

Dodds, M.W., Johnson, D.A., and Yeh, C.K. (2005). Health benefits of saliva: a review. J. Dent. *33*, 223–233.

Ericson, D., Hamberg, K., Bratthall, G., Sinkiewicz-Enggren, G., and Ljunggren, L. (2013). Salivary IgA response to probiotic bacteria and mutans streptococci after the use of chewing gum containing *Lactobacillus reuteri*. Pathog. Dis. *68*, 82–87.

Glavina, D., Gorseta, K., Skrinjarić, I., Vranić, D.N., Mehulić, K., and Kozul, K. (2012). Effect of LGG yoghurt on Streptococcus mutans and *Lactobacillus* spp. salivary counts in children. Coll. Antropol. *36*, 129–132.

Güngör, Ö.E., Kırzıoğlu, Z., Dinçer, E., and Kıvanç, M. (2013). Who will win the race in childrens' oral cavities? Streptococcus mutans or beneficial lactic acid bacteria? Benef. Microbes. *4*, 237–245.

Gupta, G. (2011). Probiotics and periodontal health. J. Med. Life *4*, 387–394.

Hallström, H., Lindgren, S., Dahlén, G., Renvert, S., and Twetman, S. (2013). Effect of probiotic lozenges on inflammatory reactions and oral biofilm during experimental gingivitis. Acta Odontol. Scand. *71*, 828–833.

Harini, P.M., and Anegundi, R.T. (2010). Efficacy of a probiotic and chlorhexidine mouth rinses: a short-term clinical study. J. Indian Soc. Pedod. Prev. Dent. *28*, 179–182.

Hasslöf, P. (2013). Probiotic lactobacilli in the context of caries as biofilm-mediated disease. Academic dissertation, New series no: 125. Umeå University, Umeå, Sweden. Available at http://umu.diva-portal.org/

Hasslöf, P., West, C.E., Videhult, F.K., Brandelius, C., and Stecksén-Blicks, C. (2013). Early intervention with probiotic *Lactobacillus paracasei* F19 has no long-term effect on caries experience. Caries Res. *47*, 559–565.

Hatakka, K., Savilahti, E., Pönkä, A., Meurman, J.H., Poussa, T., Näse, L., Saxelin, M., and Korpela, R. (2001). Effect of long term consumption of probiotic milk on infections in children attending day care centres: double blind, randomized trial. BMJ *322*, 1327.

Hatakka, K., Ahola, A.J., Yli-Knuuttila, H., Richardson, M., Poussa, T., Meurman, J.H., and Korpela, R. (2007). Probiotics reduce the prevalence of oral candida in the elderly – a randomized controlled trial. J. Dent. Res. *86*, 125–130.

Hesselmar, B., Sjöberg, F., Saalman, R., Aberg, N., Adlerberth, I., and Wold, A.E. (2013). Pacifier cleaning practices and risk of allergy development. Pediatrics *131*, e1829–37.

Hillman, J.D., and Socransky, S.S. (1987). Replacement therapy of the prevention of dental disease. Adv. Dent. Res. *1*, 119–125.

Holgerson, P.L., Vestman, N.R., Claesson, R., Öhman, C., Domellöf, M., Tanner, A.C., Hernell, O., and Johansson, I. (2013). Oral microbial profile discriminates breast-fed from formula-fed infants. J. Pediatr. Gastroenterol. Nutr. *56*, 127–136.

Holz, C., Alexander, C., Balcke, C., Moré, M., Auinger, A., Bauer, M., Junker, L., Grünwald, J., Lang, C., and Pompejus, M. (2013). *Lactobacillus paracasei* DSMZ16671 Reduces Mutans Streptococci: A Short-Term Pilot Study. Probiotics Antimicrob. Proteins *5*, 259–263.

Iniesta, M., Herrera, D., Montero, E., Zurbriggen, M., Matos, A.R., Marín, M.J., Sánchez-Beltrán, M.C., Llama-Palacio, A., and Sanz, M. (2012). Probiotic effects of orally administered *Lactobacillus reuteri*-containing tablets on the subgingival and salivary microbiota in patients with gingivitis. A randomized clinical trial. J. Clin. Periodontol. *39*, 736–744.

Iwamoto, T., Suzuki, N., Tanabe, K., Takeshita, T., and Hirofuji, T. (2010). Effects of probiotic *Lactobacillus salivarius* WB21 on halitosis and oral health: an open-label pilot trial. Oral Surg. Oral Med. Oral Pathol. Oral Radiol. Endod. *110*, 201–208.

Jindal, G., Pandey, R.K., Agarwal, J., and Singh, M. (2011). A comparative evaluation of probiotics on salivary mutans streptococci counts in Indian children. Eur. Arch. Paediatr. Dent. *12*, 211–215.

Juneja, A., and Kakade, A. (2012). Evaluating the effect of probiotic containing milk on salivary mutans streptococci levels. J. Clin. Pediatr. Dent. *37*, 9–14.

Kang, M.S., Chung, J., Kim, S.M., Yang, K.H., and Oh, J.S. (2006). Effect of Weissella cibaria isolates on the formation of Streptococcus mutans biofilm. Caries Res. *40*, 418–425.

Kang, M.S., Kim, B.G., Chung, J., Lee, H.C., and Oh, J.S. (2006). Inhibitory effect of Weissella cibaria isolates on the production of volatile sulphur compounds. J. Clin. Periodontol. *33*, 226–232.

Karuppaiah, R.M., Shankar, S., Raj, S.K., Ramesh, K., Prakash, R., and Kruthika, M. (2013). Evaluation of the efficacy of probiotics in plaque reduction and gingival health maintenance among school children – A Randomized Control Trial. J. Int. Oral Health *5*, 33–37.

Keller, M.K., Bardow, A., Lykkea, J., Jensdottir, T., and Twetman, S. (2012). Effect of chewing gums containing *Lactobacillus reuteri* on oral malodour. Acta Odontol. Scand. *70*, 246–250.

Keller, M.K., and Twetman, S. (2012). Acid production in dental plaque after exposure to probiotic bacteria. BMC Oral Health. Oct 24;12:44.

Keller, M.K., Hasslöf, P., Dahlén, G., Stecksén-Blicks, C., and Twetman, S. (2012). Probiotic supplements (Lactobacilli reuteri) do not affect re-growth of oral microbiota after full mouth disinfection with chlorhexidine. Caries Res. *46*, 140–6.

Keller, M.K., Larsen, I.N., Karlsson, L., and Twetman, S. (*2014*). Effect of tablets containing probiotic bacteria (*Lactobacillus reuteri*) on early caries lesions in adolescents: a pilot study. Benef. Microbes. Jun *2*; *1–5*.

Kraivaphan, P., Amornchat, C., Triratana, T., Mateo, L.-R., Ellwood, R., Cummins, D., DeVizio, W., and Zhang, Y.P. (*2013*). Two-year caries clinical study of the efficacy of novel dentifrices containing *1.5*% arginine, an insoluble calcium compound and *1,450* ppm fluoride. Caries Res. *47*, 582–590.

Krasse, P., Carlsson, B., Dahl, C., Paulsson, A., Nilsson, A., and Sinkiewicz, G. (2006). Decreased gum bleeding and reduced gingivitis by the probiotic *Lactobacillus reuteri*. Swed. Dent. J. *30*, 55–60.

Laine, M.A., Tolvanen, M., Pienihäkkinen, K., Söderling, E., Niinikoski, H., Simell, O., and Karjalainen, S. (2014). The effect of dietary intervention on paraffin-stimulated saliva and dental health of children participating in a randomized controlled trial. Arch. Oral Biol. *59*, 217–225.

Laleman, I., Detailleur, V., Slot, D.E., Slomka, V., Quirynen, M., and Teughels, W. (2014). Probiotics reduce mutans streptococci counts in humans: a systematic review and meta-analysis. Clin. Oral Investig. *18*, 1539–1552.

Lang, C., Böttner, M., Holz, C., Veen, M., Ryser, M., Reindl, A., Pompejus, M., and Tanzer, J.M. (2010). Specific *Lactobacillus/Mutans Streptococcus* co-aggregation. J. Dent. Res. *89*, 175–179.

Lexner, M.O., Blomqvist, S., Dahlén, G., and Twetman, S. (2010). Microbiological profiles in saliva and supragingival plaque from caries-active adolescents before and after a short-term daily intake of milk supplemented with probiotic bacteria – a pilot study. Oral Health Prev. Dent. *8*, 383–8.

Li, D., Li, Q., Liu, C., Lin, M., Li, X., Xiao, X., Zhu, Z., Gong, Q., and Zhou, H. (2014). Efficacy and safety of probiotics in the treatment of Candida-associated stomatitis. Mycoses. *57*, 141–6.

Lif Holgerson, P., Harnevik, L., Hernell, O., Tanner, A.C., and Johansson, I. (2011). Mode of birth delivery affects oral microbiota in infants. J. Dent. Res. *90*, 1183–1188.

Loesche, W.J. (1976). Chemotherapy of dental plaque infections. Oral Sci. Rev. *9*, 65–107.

Marsh, P.D. (2004). Dental plaque as a microbial biofilm. Caries Res. *38*, 204–211.

Marsh, P.D. (2009). Dental plaque as a biofilm: the significance of pH in health and caries. Compend. Contin. Educ. Dent. *30*, 76–78.

Marttinen, A., Haukioija, A., Karjalainen, S., Satokari, R., Nylund, L., Holgerson, P., Öhman, C., Twetman, S., and Söderling, E. (2012). Short-term consumption of probiotic lactobacilli does not affect acid production in supragingival plaque. Clin. Oral Investig. *16*, 797–803.

Mayanagi, G., Kimura, M., Nakaya, S., Hirata, H., Sakamoto, M., Benno, Y., and Shimauchi, H. (2009). Probiotic effects of orally administered *Lactobacillus salivarius* WB21-containing tablets on periodontopathic bacteria: a double-blinded, placebo-controlled, randomized clinical trial. J. Clin. Periodontol. *36*, 506–513.

Montalto, M., Vastola, M., Marigo, L., Covino, M., Graziosetto, R., Curigliano, V., Santoro, L., Cuoco, L., Manna, R., and Gasbarrini, G. (2004). Probiotic treatment increases salivary counts of lactobacilli: a double-blind, randomized, controlled study. Digestion *69*, 53–56.

Näse, L., Hatakka, K., Savilahti, E., Saxelin, M., Pönkä, A., Poussa, T., Korpela, R., and Meurman, J.H. (2001). Effect of long-term consumption of a probiotic bacterium, *Lactobacillus rhamnosus* GG, in milk on dental caries and caries risk in children. Caries Res. *35*, 412–420.

Nelun Barfod, M., Magnusson, K., Lexner, M.O., Blomqvist, S., Dahlén, G., and Twetman, S. (2011). Oral microflora in infants delivered vaginally and by caesarean section. Int. J. Paediatr. Dent. *21*, 401–406.

Nikawa, H., Makihira, S., Fukushima, H., Nishimura, H., Ozaki, Y., Ishida, K., Darmawan, S., Hamada, T., Hara, K., Matsumoto, A., Takemoto, T., and Aimi, R. (2004). *Lactobacillus reuteri* in bovine milk fermented decreases the oral carriage of mutans streptococci. Int. J. Food Microbiol. *95*, 219–223.

Petersson, L.G., Magnusson, K., Hakestam, U., Baigi, A., and Twetman, S. (2011). Reversal of primary root caries lesions after daily intake of milk supplemented with fluoride and probiotic lactobacilli: a clinical and microbiological study on older adults. Acta Odontol. Scand. *69*, 321–327.

Reid, G., Younes, J.A., Van der Mei, H.C., Gloor, G.B., Knight, R., and Busscher, H.J. (2011). Microbiota restoration: natural and supplemented recovery of human microbial communities. Nat. Rev. Microbiol. *9*, 27–38.

Riccia, D.N., Bizzini, F., Perilli, M.G., Polimeni, A., Trinchieri, V., Amicosante, G., and Cifone, M.G. (2007). Anti-inflammatory effects of *Lactobacillus brevis* (CD2). on periodontal disease. Oral Dis. *13*, 376–385.

Romani Vestman, N., Hasslöf, P., Keller, M.K., Granström, E., Roos, S., Twetman, S., and Stecksén-Blicks, C. (2013). *Lactobacillus reuteri* influences regrowth of mutans streptococci after full-mouth disinfection: A double-blind, randomized controlled trial. Caries Res. *47*, 338–345.

Shah, M.P., Gujjari, S.K., and Chandrasekhar, V.S. (2013). Evaluation of the effect of probiotic (Inersan) alone, combination of probiotic with doxacycline and doxacycline alone on aggressive periodontitis – a clinical and microbiological study. J. Clin. Diagn. Res. 7, 595–600.

Sharma, A., Rath, G.K., Chaudhary, S.P., Thakar, A., Mohanti, B.K., and Bahadur, S. (2012). *Lactobacillus* brevis CD2 lozenges reduce radiation- and chemotherapy-induced mucositis in patients with head and neck cancer: a randomized double-blind placebo-controlled study. Eur. J. Cancer *48*, 875–881.

Shimauchi, H., Mayanagi, G., Nakaya, S., Minamibuchi, M., Ito, Y., Yamaki, K., and Hirata, H. (2008). Improvement of periodontal condition by probiotics with *Lactobacillus salivarius* WB21: a randomized, double-blind, placebo-controlled study. J. Clin. Periodontol. 35, 897–905.

Singh, R.P., Damle, S.G., and Chawla, A. (2011). Salivary mutans streptococci and lactobacilli modulations in young children on consumption of probiotic ice-cream containing *Bifidobacterium lactis* Bb_{12} and *Lactobacillus acidophilus* La5. Acta Odontol. Scand. *69*, 389–394.

Sinkiewicz, G., Cronholm, S., Ljunggren, L., Dahlén, G., and Bratthall, G. (2010). Influence of dietary supplementation with *Lactobacillus reuteri* on the oral flora of healthy subjects. Swed. Dent. J. *34*, 197–206.

Slawik, S., Staufenbiel, I., Schilke, R., Nicksch, S., Weinspach, K., Stiesch, M., and Eberhard, J. (2011). Probiotics affect the clinical inflammatory parameters of experimental gingivitis in humans. Eur. J. Clin. Nutr. *65*, 857–863.

Souza, M.L., Cury, J.A., Tenuta, L.M., Zhang, Y.P., Mateo, L.R., Cummins, D., and Ellwood, R.P. (2013). Comparing the efficacy of a dentifrice containing 1.5% arginine and 1450 ppm fluoride to a dentifrice containing 1450 ppm fluoride alone in the management of primary root caries. J. Dent. *41* (Suppl. 2), S35–41.

Staab, B., Eick, S., Knöfler, G., and Jentsch, H. (2009). The influence of a probiotic milk drink on the development of gingivitis: a pilot study. J. Clin. Periodontol. *36*, 850–856.

Stecksén-Blicks, C., Sjöström, I., and Twetman, S. (2009). Effect of long-term consumption of milk supplemented with probiotic bacteria and fluoride on dental caries and general health in preschool children. A cluster randomized study. Caries Res. *43*, 374–381.

Stensson, M., Koch, G., Coric, S., Abrahamsson, T., Jenmalm, M., Birkhed, D., and Wendt, L.K. (2013). Oral administration of *Lactobacillus reuteri* during the first year of life reduces caries prevalence in the primary dentition at 9 years of age. Caries Res. *48*, 111–117.

Strachan, D.P. (2000). Family size, infection and atopy: the first decade of the 'hygiene hypothesis'. Thorax *55* (Suppl. 1), S2–10.

Sudhir, R., Praveen, P., Anantharaj, A., and Venkataraghavan, K. (2012). Assessment of the effect of probiotic curd consumption on salivary pH and streptococcus mutans counts. Niger Med. J. *53*, 135–9.

Taipale, T., Pienihäkkinen, K., Salminen, S., Jokela, J., and Söderling, E. (2012). *Bifidobacterium animalis* subsp. *lactis* BB-12 administration in early childhood: a randomized clinical trial of effects on oral colonization by mutans streptococci and the probiotic. Caries Res. *46*, 69–77.

Taipale, T., Pienihäkkinen, K., Alanen, P., Jokela, J., and Söderling, E. (2013). Administration of *Bifidobacterium animalis* subsp. *lactis* BB-12 in early childhood: A post-trial effect on caries occurrence at four years of age. Caries Res. *47*, 364–372.

Takahashi, N., and Nyvad, B. (2011). The role of bacteria in the caries process: ecological perspectives. J. Dent. Res. *90*, 294–303.

Teughels, W., Durukan, A., Ozcelik, O., Pauwels, M., Quirynen, M., and Haytac, M.C. (2013). Clinical and microbiological effects of *Lactobacillus reuteri* probiotics in the treatment of chronic periodontitis: a randomized placebo-controlled study. J. Clin. Periodontol. *40*, 1025–1035.

Teughels, W., Newman, M.G., Coucke, W., Haffajee, A.D., Van Der Mei, H.C., Haake, S.K., Schepers, E., Cassiman, J.J., Van Eldere, J., van Steenberghe, D., and Quirynen, M. (2007). Guiding periodontal pocket recolonization: a proof of concept. J. Dent. Res. *86*, 1078–1082.

Toiviainen, A., Jalasvuori, H., Lahti, E., Gursoy, U., Salminen, S., Fontana, M., Flannagan, S., Eckert, G., Kokaras, A., Paster, B., and Söderling, E. (2015). Impact of orally administered lozenges with *Lactobacillus rhamnosus* GG and Bifidobacterium animalis subsp. lactis BB-12 on the number of salivary mutans streptococci, amount of plaque, gingival inflammation and the oral microbiome in healthy adults. Clin. Oral Investig. *19*, 77–83.

Tsubura, S., Mizunuma, H., Ishikawa, S., Oyake, I., Okabayashi, M., Katoh, K., Shibata, M., Lizuka, T., Toda, T., and Iizuka, T. (2009). The effect of Bacillus subtilis mouth rinsing in patients with periodontitis. Eur. J. Clin. Microbiol. Infect. Dis. *28*, 1353–1356.

Twetman, S., Derawi, B., Keller, M., Ekstrand, K., Yucel-Lindberg, T., and Stecksén-Blicks, C. (2009). Short-term effect of chewing gums containing probiotic *Lactobacillus reuteri* on the levels of inflammatory mediators in gingival crevicular fluid. Acta Odontol. Scand. *67*, 19–24.

Twetman, S., and Keller, M.K. (2012). Probiotics for caries prevention and control. Adv. Dent. Res. *24,* 98–102.

Twetman, S., and Stecksén-Blicks, C. (2008). Probiotics and oral health effects in children. Int. J. Paediatr. Dent. *18,* 3–10.

Wescombe, P.A., Hale, J.D., Heng, N.C., and Tagg, J.R. (2012). Developing oral probiotics from *Streptococcus salivarius*. Future Microbiol. *7,* 1355–1371.

Vicario, M., Santos, A., Violant, D., Nart, J., and Giner. L. (2013). Clinical changes in periodontal subjects with the probiotic *Lactobacillus reuteri* Prodentis: A preliminary randomized clinical trial. Acta Odontol. Scand. *71,* 813–819.

Vivekananda, M.R., Vandana, K.L., and Bhat, K.G. (2010). Effect of the probiotic *Lactobacilli reuteri* (Prodentis) in the management of periodontal disease: a preliminary randomized clinical trial. J. Oral Microbiol. Nov 2;2. doi: 10.3402/jom.v2i0.5344

de Vrese, M., and Schrezenmeir, J. (2008). Probiotics, prebiotics, and synbiotics. Adv. Biochem. Eng. Biotechnol. *111,* 1–66.

Wolff, M., Corby, P., Klaczany, G., Santarpia, P., Lavender, S., Gittins, E., Vandeven, M., Cummins. D., and Sullivan, R. (2013). *In vivo* effects of a new dentifrice containing 1.5% arginine and 1450 ppm fluoride on plaque metabolism. J. Clin. Dent. *24* Spec no A:A45–54.

World Health Organization. (2001). Report of joint FAO/WHO expert consultation on evaluation of health and nutritional properties of probiotics in food including power milk with live lactic acid bacteria. Cordoba, Argentina.

Yin, W., Hu, D.Y., Fan, X., Feng, Y., Zhang, Y.P., Cummins, D., Mateo, L.R., Pretty, I.A., and Ellwood, R.P. (2013). A clinical investigation using quantitative light-induced fluorescence (QLF) of the anticaries efficacy of a dentifrice containing 1.5% arginine and 1450 ppm fluoride as sodium monofluorophosphate. J. Clin. Dent. *24,* Spec no A:A15–22.

Yli-Knuuttila, H., Snäll, J., Kari, K., and Meurman, J.H. (2006). Colonization of *Lactobacillus rhamnosus* GG in the oral cavity. Oral Microbiol. Immunol. *21,* 129–131.

Zahradnik, R.T., Magnusson, I., Walker, C., McDonell, E., Hillman, C.H., and Hillman, J.D. (2009). Preliminary assessment of safety and effectiveness in humans of ProBiora3, a probiotic mouthwash. J. Appl. Microbiol. *107,* 682–690.

Cholesterol-lowering Effects of Probiotics and Prebiotics

29

Min-Tze Liong, Byong-H. Lee, Sy-Bing Choi, Lee-Ching Lew, Amy-Sie-Yik Lau and Eric Banan-Mwine Daliri

Abstract

Recently, the use of probiotics and prebiotics as a cholesterol lowering agent has become increasingly popular. This chapter will highlight some of the *in vitro* and *in vivo* evidence showing the potential of probiotics and prebiotics in improving serum lipid profile. Data revealing details at molecular levels have also been included in this chapter. The proposed mechanisms for cholesterol removal by probiotics include assimilation of cholesterol by growing cells, binding of cholesterol to cellular surface and incorporation into the cellular membrane, deconjugation of bile via bile salt hydrolase, coprecipitation of cholesterol with deconjugated bile and production of short-chain fatty acids from oligosaccharides. In this chapter, we have highlighted on a few more selected cholesterol lowering mechanisms that are feasible and supported by in-depth evidence. Although cholesterol lowering abilities of probiotics has been extensively reported; recently, controversies have risen attributed to the activities of deconjugated bile acids that repress the synthesis of bile acids from cholesterol. Using a molecular docking approach, we have demonstrated that deconjugated bile acids have higher binding affinity towards some orphan nuclear receptors namely the farsenoid X receptor (FXR), leading to a suppressed transcription of the enzyme cholesterol 7-alpha-hydroxylase (7AH), which is responsible for bile acid synthesis from cholesterol. Possible detrimental effects due to increased deconjugation of bile salts such as malabsorption of lipids, colon carcinogenesis, gallstones formation and altered gut microbial populations, which contribute to other varying gut diseases, are also included in this chapter. The effects of probiotics and prebiotics on other cholesterol-related disorders such as formation of abnormal erythrocytes are also discussed in this chapter. As described in the past studies, hypercholesterolaemia could induce alterations in the human erythrocyte plasma membrane. Administration of probiotics and prebiotics has improved erythrocyte membrane fluidity, decreased membrane rigidity and altered membrane lipid profiles. Probiotics and prebiotics is a new feasible approach to use natural interventions for cholesterol management.

Introduction

Cholesterol is one of the essential structural components crucial in establishing proper cellular membrane permeability and fluidity in our body. Two of the main lipoproteins that transport cholesterol are the low-density lipoprotein- (LDL) and high-density lipoprotein (HDL)-cholesterol. During hypercholesterolaemia, higher concentrations of LDL and lower concentrations of HDL is usually found in blood and the excess can accumulate on the walls of the arteries, and together with plaques, this leads to the narrowing of the arteries. There are many risks associated with hypercholesterolaemia, ranging from coronary artery disease to heart attack and stroke, causing morbidity and mortality. With the increasing development of various drugs to ameliorate this metabolic disease, pharmaceutical companies have pumped huge investments in the discovery of potent drugs. Amid these, natural approaches such as non-drug strategies have also increased. Dietary intervention such as monitoring the intake of saturated fat and replacing saturated fat with mono or polysaturated fatty acids have always been highlighted by healthcare practitioners to their patients. Recently, the use of probiotics as a cholesterol lowering agent has also become increasingly popular.

Hypercholesterolaemia is a metabolic derangement characterized by the presence of high levels of cholesterol in the blood (Ooi *et al.*, 2010a). Elevated blood cholesterol is known to be a major risk factor for coronary heart diseases (Kumar *et al.*, 2012a). It was predicted by World Health Organization (WHO, 2013) that cardiovascular diseases will remain as the leading cause of death by 2030, such that an estimation of 23.6 million people will die from the disease in the year 2030. Over 80% of cardiovascular disease-related deaths take place in low- and middle-income countries (WHO, 2013). According to a report, hypercholesterolaemia contributes to 45% of heart attacks in Western Europe and 35% of heart attacks in Central and Eastern Europe (Kumar *et al.*, 2012a). Those with hypercholesterolaemia have three times higher risk of heart attack compared with those with normal blood lipid profiles (Kumar *et al.*, 2012a). High blood cholesterol levels can be reduced by dietary management, behaviour modification, regular exercise or even drug therapy (Kumar *et al.*, 2012a; Ooi *et* al., 2010a). However, apart from the high cost of cholesterol lowering drugs, they may also present severe adverse effects (Kumar et al., 2012a). Probiotics and prebiotics, which could be a cheaper and safer alternative, have been studied to ascertain their ability to lower cholesterol.

The beneficial effects of probiotics have been extensively studied for the last few decades and many strains of probiotics microorganisms have been evaluated for their cholesterol lowering properties. However, those from the genera of *Lactobacillus*, *Bifidobacterium*, *Lactococcus*, and *Streptococcus* have dominated most research publications in such a health benefit. Scientific evidence on the cholesterol lowering effects initially surfaced in 1959, where Danielsson and Gustafsson observed that the cholesterol level in germ-free animals were lower than in normal conventional animals (Danielsson and Gustafsson, 1959). Later, Beher *et al.* (1964) and Wostmann *et al.* (1966) reported that the intestinal microbiota was responsible for the conversion of cholesterol to coprostanol and conversion of primary bile to secondary bile; leading to a decrease in serum cholesterol levels.

Mott *et al.* (1973) conducted an animal trial using cholesterol-fed piglets and further confirmed that serum cholesterol was lowered by intestinal bacteria. Meanwhile, Mann *et al.* (1974) observed that when African Masaai men consumed large amounts of *Lactobacillus* strain fermented milk for 21 days, serum cholesterol level decreased compared with their initial levels. In addition, Harrison *et al.* (1975) reported that serum cholesterol level decreased when the *Lactobacillus* count in the bowel microbiota of newborn was increased. Since then, there have been numerous studies conducted to evaluate the cholesterol lowering properties of probiotics.

Prebiotics, defined as 'indigestible fermented food substrates that selectively stimulate the growth, composition and activity of microbiota in gastrointestinal tract and thus improve hosts, health and well-being' (Roberfroid, 2007) has also acquired scientific recognition in cholesterol lowering effects in recent years. Prebiotics such as fructooligosaccharides (or oligofructose), inulin, lactulose, and galactooligosaccharides are fermented by gastrointestinal microbiota to further selectively stimulate the growth and/or activity of beneficial intestinal bacteria. In addition, prebiotics are also utilized by the intestinal microbial population to produce short-chain fatty acids which may lead to improvement of lipid profiles.

More recent studies do not only focus on the effects of probiotics and prebiotics on the serum total cholesterol levels, but also on the different types of lipoproteins and the triglycerides levels. In animal trials, the effects on the liver cholesterol and triglycerides content are also normally included. Expressions of genes responsible for cholesterol metabolism, catabolism and efflux have also been extensively studied in animals. This chapter will highlight some of these aspects.

In vitro and *in vivo* evidence

In vitro findings

The potential cholesterol-lowering effects of probiotic strains are usually screened using *in vitro* methods. Previous studies have reported that lactobacilli removed cholesterol *in vitro* by various mechanisms. In a study evaluating the ability of lactobacilli to remove cholesterol *in vitro* under the conditions that mimic the human gastrointestinal tract, Lye *et al.* (2010a) observed that strains of lactobacilli were able to remove cholesterol via assimilation of cholesterol during growth, binding of cholesterol to the surface of cells (Fig. 29.1), disruption of cholesterol micelle and deconjugation of bile salts. The authors also reported that the removal of cholesterol was strain dependent and bile dependent.

This was further corroborated by Pan *et al.* (2011) which reported that *Lactobacillus fermentum* SM-7 cells assimilated 61.5% and co-precipitated and absorbed 38.5% of the cholesterol in media containing 0.1 g/l cholesterol. Some bacteria including *Lactobacillus* species have been reported to have protease-sensitive receptors on the cell surface which may be responsible for tight contact with exogenous cholesterol/phosphatidylcholine vesicles. They then assimilate cholesterol into their own cell membrane to reduce membrane fluidity thereby influencing their membrane permeability (Efrati *et al.*, 1981; Kumar *et al.*, 2012a; Tani *et al.*, 1993). Other studies have shown that probiotics may produce exopolysaccharides which may bind cholesterol, or that

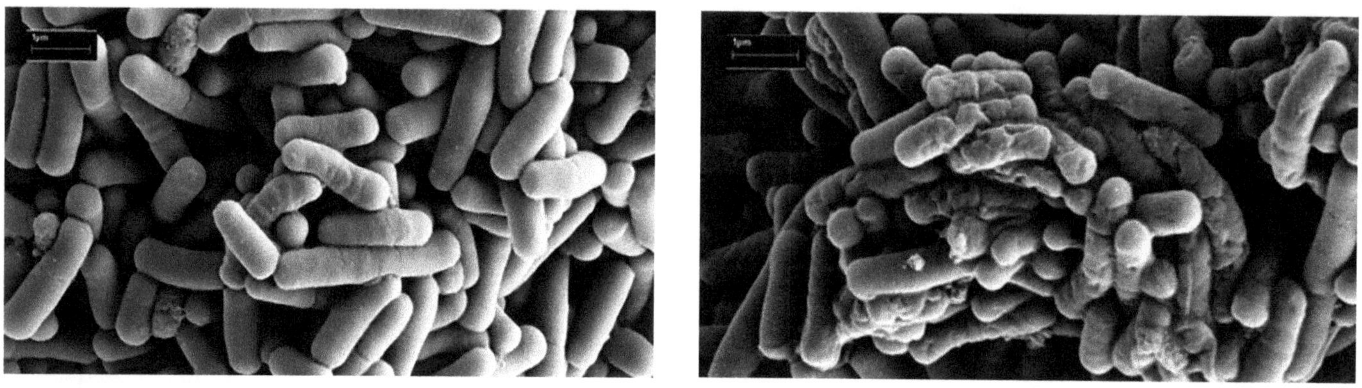

Figure 29.1 Scanning electron micrograph images of lactobacilli cells growing in the absence of cholesterol (left) and cells grown in the presence of cholesterol (right). In the latter case, a rough surface was observed, illustrating the attachment of cholesterol onto cellular surface. Magnification: 10k x.

the numerous amino acids and peptidoglycan in probiotic cell walls may account for the strong interactive forces with which both live and dead probiotics cells mop up cholesterol (Guo *et al.*, 2011; Kimoto-Nira *et al.*, 2007; Raghavan *et al.*, 2011).

In another *in vitro* study by Bordoni *et al.* (2013), it was reported that strains of bifidobacteria could also remove cholesterol by assimilating cholesterol during growth. It has also been reported that lactobacilli could reduce cholesterol *in vitro* via conversion of cholesterol to coprostanol and via incorporation of cholesterol into the cellular membrane during growth (Lye *et al.*, 2010b). Moreover, Ahire *et al.* (2012) reported that growing cells of *L. helveticus* CD6 possessed high intracellular cholesterol oxidase-like activity, thus leading to degradation of cholesterol *in vitro*.

In addition to cholesterol removal ability from cholesterol containing media, there have been increasing studies reporting on the ability of probiotics to interfere with intestinal cholesterol absorption. Niemann–Pick C1-like 1 protein (NPC1L1), a polytopic intestinal transmembrane protein containing a sterol-sensing domain, is essential in mediating cellular uptake of various sterols especially cholesterol (Yu, 2008). Huang and Zheng (2010) reported that probiotic *L. acidophilus* ATCC 4356 was able to reduce the expression of the NPC1L1 gene, leading to inhibition of cellular uptake of micellar cholesterol in intestinal Caco-2 cells. Similarly, it was reported that *L. plantarum* Lp27 isolated from Tibetan kefir grains inhibited cholesterol absorption through down-regulation of NPC1L1 expression in Caco-2 cells (Huang *et al.*, 2013b). Similar trends were also observed by Yoon and co-workers where *L. rhamnosus* BFE5264 isolated from Maasai fermented milk and *L. plantarum* NR74 isolated from Korean kimchi down-regulated the expression of NPC1L1 in Caco-2 cells (Yoon *et al.*, 2013). Another important protein that regulates cholesterol movement across cells is the heterodimeric transporter complex, ABCG5/ABCG8, the function of which is necessary for the majority of sterols secreted into bile (Kosters *et al.*, 2006). It has been reported that treatment of Caco-2 cells with *L. rhamnosus* BFE5264 and *L. plantarum* NR74 also resulted in up-regulation and elevated expression of ABCG5 and ABCG8, thus promoting cholesterol efflux in enterocytes, leading to increased secretion of cholesterol into bile (Yoon *et al.*, 2011).

Most prebiotics promote the growth of lactic acid-producing bacteria. Through this they have also been proposed to have beneficial effects on lipid metabolism. In addition, intestinal fermentation of prebiotics leads to the production of substantial amounts of short-chain-fatty-acids (SCFAs) such as acetate and propionate, which can be transported to the liver via the portal vein and enters the cholesterogenesis and lipogenesis pathways. It has been suggested that the SCFAs produced by the intestinal microbiota may regulate lipid metabolism in intestine, thus reducing plasma cholesterol levels. Alvaro *et al.* (2008) showed that incubation of SCFAs namely acetate, propionate and butyrate to human Caco-2 enterocytes down-regulated the expression of nine key genes involved in intestinal cholesterol biosynthesis according to cDNA microarray analysis and quantitative real-time PCR. Besides exerting effects via SCFAs, prebiotics with long chain non-digestible polysaccharides such as β-glucan have also been shown to down-regulate the intestinal cell expression of genes known to play critical roles in fatty acid and cholesterol synthesis and fatty acid transport (Drozdowski *et al.*, 2010).

In vivo findings

The use of animal models and humans to evaluate the effects of probiotics on serum cholesterol levels has been emphasized over the years. However, recent studies have also focused on the lipoprotein profile as well as triglycerides level and liver lipid profiles. New strains of probiotics have been evaluated in animal models for their potential in cholesterol lowering effects (Table 29.1), while human studies have shown promising evidence that well-established probiotics possess significant cholesterol reducing effects (Table 29.2). Many studies have used rats, mice, hamsters, guinea pigs and pigs as models due to their similarities with humans in terms of cholesterol and bile acid metabolism, plasma lipoprotein distribution, and regulation of hepatic cholesterol enzymes (Ooi and Liong, 2010). These animals also share an almost similar digestive anatomy and physiology, nutrient requirements, bioavailability and absorption, and metabolic processes with humans, making them useful experimental models for research applications (Patterson *et al.*, 2008). Hence, the positive cholesterol reducing effects shown in animal studies suggest a similar potential in humans. Human trial results that paralleled those obtained from animal studies have further attested to the transferability and reliability of results obtained in selected animal models.

Many studies have evaluated the effects of probiotics on serum lipid profiles, intestinal lipid uptake and even liver lipid deposition. A recent study has illustrated that the administration of *L. fermentum* ATCC 11976 to hamsters on a high-cholesterol diet, yielded lower fat deposition compared with the control that were not administered with lactobacilli (Fig. 29.2). However, there are increasing data revealing details at molecular levels by investigating on the gene expression profile of the host tissue when probiotic strains were administered. Zhong *et al.* (2012) reported that administration of *L. casei* Zhang to male Wistar rats ($n = 10$) for 67 days evokes a significant reduction of serum triglyceride (TG) and low density lipoprotein-cholesterol (LDL-C) by 31.3 and 30.8% ($P < 0.01$), respectively; they have also showed using microarray analysis that 22 genes involved in lipid metabolism were significantly up-regulated while another 8 genes involved in lipid metabolism were down-regulated. In another study evaluating the effect of *L. paracasei* (strain number not provided) in modulating cholesterol metabolism, Trasino *et al.* (2013) administered *L. paracasei* to Ossabaw pigs ($n = 5$) for 18 weeks. They found that two genes related to cholesterol uptake, i.e. low-density lipoprotein receptor (LDLR) and acyl-CoA:cholesterol acyltransferase (ACAT), were significantly down-regulated in alveolar macrophages of pigs fed with probiotics compared with control. Moreover, Hu *et al.* (2013) reported that administration of *L. plantarum* NS5 and *L. delbrueckii* subsp. *bulgaricus* NS12 to hypercholesterolaemic SD rats for 6 weeks resulted in significant down-regulation of ACAT and up-regulation of cholesterol 7a-hydroxylase (CYP7A1), leading to improved serum lipid profile.

Table 29.1 Cholesterol lowering effects of probiotics in animal models

Strains	Subjects	Groups diet and treatment dosage	Significant ($P<0.05$) effects compared to control	Reference
Lactobacillus plantarum Lp27	Male Sprague–Dawley rats ($n=10$)	Control: high-cholesterol diet		Huang *et al.* (2013b)
		Group 1: high-cholesterol diet + *l. Plantarum* lp27 (2×10^9 cfu daily for 4 weeks)	Serum-TC reduced 37%; serum-LDL reduced 68% Serum-TG reduced 54%	
L. brevis 119–2	Male Sprague–Dawley rats ($n=11$)	Control: high-cholesterol diet		Watanabe *et al.* (2013)
		Group 1: high-cholesterol diet + viable *l. Brevis* 119–2 (4.68×10^{14} cfu daily for 2 weeks)	Plasma-TC reduced 26%; plasma-LDL reduced 62%; liver-TC reduced 20%	
		Group 2: high-cholesterol diet + dead *l. Brevis* 119–2 (1.3g daily for 2 weeks)	Plasma-TC reduced 26%; plasma-LDL reduced 48%	
		Group 3: high-cholesterol diet + viable *l. Acidophilus* atcc43121 (9.23×10^{14} cfu daily for 2 weeks)	Plasma-HDL increased 50%; liver-TC reduced 26%; liver-TG reduced 34%	
Lactobacillus plantarum (BCRC 80061)	Male Syrian hamsters ($n=8$)	Control: high-cholesterol diet + adlay milk		Wang *et al.* (2010)
		Group 1: high-cholesterol diet + adlay milk fermented with *l. Plantarum* (100 mg/kg body weight daily for 8 weeks)	Serum-TC reduced 16%; serum-LDL reduced 24%; liver-TC reduced 18%	
Lactobacillus casei F0822	Male Wistar rats ($n=6$)	Control: cholesterol-enriched diet		Guo and Li, 2013)
		Group 1: cholesterol-enriched diet + ad libitum drinking water supplemented with f0822 (1×10^8 cfu/ml daily for 3 weeks)	Serum-TC reduced 26%; serum-LDL reduced 29%; liver-TC reduced 25%	
		Group 2: cholesterol-enriched diet + drinking water supplemented with f0822 (1×10^9 cfu/ml daily for 3 weeks)	Serum-TC reduced 39%; serum-LDL reduced 63%; liver-TC reduced 41%	
L. plantarum KCTC3928	C57bl/6 male mice ($n=7$)	Control: high-fat diet		Jeun *et al.* (2010)
		Group 1: high-fat diet + live *l. Plantarum* kctc3928 (10^9 cfu daily for 4 weeks)	Plasma-TC reduced 33%; plasma-LDL reduced 42%; plasma-HDL increased 35%; plasma-TG reduced 32%; liver-TC reduced 18%; liver-TG reduced 28%	
Lactobacillus rhamnosus GG	Male Wistar rats ($n=7$)	Control: high-cholesterol diet		Kumar *et al.* (2013)
		Group 1: high-cholesterol diet + milk fermented with probiotic lgg (10^8 cfu/ml) daily for 45 days	Serum-TC reduced 32%; serum-TG reduced 41%; serum-LDL reduced 51%	
Lactobacillus plantarum Lp09 and Lp45	Sprague–Dawley rats ($n=10$)	Control: high-cholesterol diet		Huang *et al.* (2013a)
		Group 1: high cholesterol diet + 2 ml (10^9 cfu/ml) of a *l. Plantarum* lp09 solution, daily for 4 weeks	Serum-TC reduced 30%; serum-TG reduced 24%; serum-LDL reduced 48%; liver-TC reduced 33%; liver-TG reduced 17%	
		Group 2: high cholesterol diet + 2 ml (10^9 cfu/ml) of a *l. Plantarum* lp45 solution, daily for 4 weeks	Serum-TC reduced 18%; serum-TG reduced 12%; serum-LDL reduced 22%; liver-TC reduced 34%; liver-TG reduced 19%	
L. fermentum SM-7	Male ICR mice ($n=10$)	Control: high-fat diet		Pan *et al.* (2011)
		Group 1: high fat diet + 10% skimmed milk with 2×10^{10} cfu/ml sm-7, daily for 28 days	Serum-TC reduced 14%; serum-TG reduced 5%; serum-LDL reduced 24%	

Although there have been many studies demonstrating the cholesterol lowering effects of probiotics, it is worth noting that the degree of the cholesterol lowering effects is strain specific. It has been reported that different strains need varying dosage to exhibit cholesterol lowering effects and higher dosage may not necessarily translate to better effects on cholesterol (Ooi and Liong, 2010). Probiotics have been delivered in the range of 10^8 to 10^{10} cfu/day in man and 10^8 to 10^{14} cfu/day in animals. Some probiotics have been shown to be efficacious at lower levels, while some require a substantially higher amount to exert cholesterol lowering effect. For instance, the administration of *L. fermentum* SM-7 to male ICR mice (2×10^{10} cfu daily, for 4 weeks) resulted in 13.6% reduction in serum total cholesterol and 24.1% reduction in LDL-C (Pan *et al.*, 2011). However, in another study, administration of *L. plantarum* KCTC3928 to male C57BL/6 mice at lower dosage (1×10^9 cfu daily, for 4 weeks) resulted in 33% reduction in serum total cholesterol and 42% reduction in LDL-C (Jeun *et al.*, 2010). Similarly in human trial, administration of *L. plantarum* 299v (2×10^{10} cfu daily, for 6 weeks) resulted in 2.5% reduction in serum total cholesterol (Naruszewicz *et al.*, 2002). However, administration of *B. longum* BL1 (3×10^8 cfu daily, for 4 weeks) resulted in 5.1% reduction in serum total cholesterol (Xiao *et al.*, 2003). Thus, clinically effective dosage of probiotics should only be established based on studies of the specific strains conducted in human trials.

While the cholesterol lowering effects of probiotics has been well-documented, prebiotics have also gained increasing attention in cholesterol studies, due to their role in promoting

Table 29.2 Cholesterol lowering effects of probiotics in human subjects

Strain	Experimental design	Dosage	Results	Reference
L. plantarum 299v	36 healthy subjects; randomized, placebo-controlled, double-blind, parallel,	2×10^{10} cfu daily for 6 weeks	TC reduced 2.5% LDL-C reduced 7.9%	Naruszewicz *et al.* (2002)
B. longum BL1	32 healthy subjects; randomized, single-blind, placebo-controlled and parallel study	3×10^8 cfu daily for 4 weeks	TC reduced 5.1%	Xiao *et al.* (2003)
L. acidophilus, *B. bifidum* and *L. rhamnosus GG*	31 mildly hypercholesterolaemic subjects; double-blind, placebo-controlled, cross-over study	1×10^8 cfu of each of *L. acidophilus*, *B. bifidum* and LGG daily for 5 weeks	TC reduced 4.7%	Larkin *et al.* (2009)
L. acidophilus La5 and *B. lactis* Bb_{12}	60 subjects with type 2 diabetes mellitus; randomized double-blind controlled trial	1.24×10^9 cfu La5 and 1.08×10^9 cfu Bb_{12} daily for 6 weeks	TC reduced 4.5% LDL-C reduced 7.4%	Ejtahed *et al.* (2011)
L. reuteri NCIMB 30242	127 hypercholesterolaemic subjects; randomized, double-blind, placebo-controlled, parallel-arm, multicentre study.	2.9×10^9 cfu; twice daily for 9 weeks	TC reduced 9.1% LDL-C reduced 11.6% non-HDL-C reduced 11.3% apoB-100 reduced 8.4%	Jones *et al.* (2012a)
Mixture of *L. plantarum* (CECT 7527, CECT 7528 and CECT 7529)	60 hypercholesterolaemic subjects; controlled, randomized, double-blind trial.	1.2×10^9 cfu daily for 12 weeks	TC reduced 17.4% LDL-C reduced 17.6% oxidized LDL-C reduced 15·6%	Fuentes *et al.* (2013)

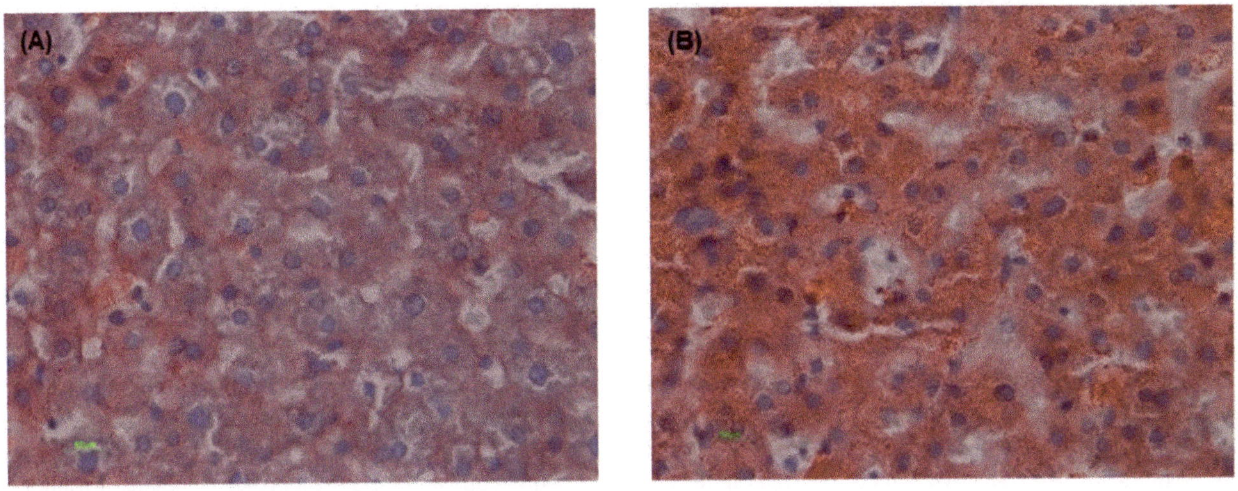

Figure 29.2 Representative photomicrographs of livers from hamsters after 12 weeks on a high-cholesterol diet (synthetic, purified hyperlipidaemic methionine-deficient and choline-devoid diet) and supplemented with (A) *L. fermentum* ATCC 11976 and (B) without treatment. *L. fermentum* ATCC 11976-supplemented animals (A) showed less fat infiltration in the hepatocytes whereas, control animals (B) were filled with 30% greater micro-vesicular fat deposits (Oil Red O staining, 400X). (Bhathena *et al.* (2013); open access article.)

the growth of probiotics. Brighenti *et al.* (1999) used a randomized, double-blind, placebo-controlled and parallel design trial involving twelve healthy male volunteers to study the effect of the prebiotic inulin on lipid profiles. In this 12-week trial, the authors found that the daily consumption of 50 g of a rice-based ready-to-eat cereal containing 18% inulin significantly reduced plasma total cholesterol and triacylglycerols by 7.9% (±5.4) and 21.2% (±7.8), respectively compared to the control. In addition, Causey *et al.* (2000) conducted a randomized, double-blind and crossover study involving twelve hypercholesterolaemic subjects to assess the effects of inulin from chicory root on blood cholesterol level. The 3-week study found that daily intake of 20 g of inulin significantly reduced 14% serum triglycerides when compared to placebo. Serum cholesterol levels were also reduced, however, with weaker correlation (p=0.14). The cholesterol lowering effects of other indigestible and fermentable compounds such as resistant starch and β-glucan have also been evaluated. In a study evaluating the cholesterol lowering effect of resistant starch in rats, Han *et al.* (2003) administered 50 g of resistant starch from tebou and adzuki daily to male F344/DuCrj rats for 4 weeks. Results from the study showed that resistant starch up-regulated the expression of hepatic scavenger receptor class B1 (SR-B1) mRNA and Cyp7A1 mRNA, thus leading to a significantly lower serum total cholesterol level (approximately 25% reduction)compared with control. Moreover, using a randomized, double-blind parallel group design, Queenan *et al.* (2007) investigated the effects of administration of oat β-glucan (6 g daily for 6 weeks) on seventy-five hypercholesterolaemic men and women. The authors reported that administration of oat β-glucan significantly reduced total cholesterol and LDL cholesterol level by 5% and 7% respectively, when compared to control group.

Mechanisms of actions

Although extensive studies have been carried out to prove the ability of probiotics to reduce cholesterol *in vivo* both in animal models as well as human subjects (Jones *et al.*, 2012b, 2013; Kumar *et al.*, 2011), most of the hypotheses raised to date are based on *in vitro* experiments, and few attempts have been made to evaluate the possible hypocholesterolaemic mechanisms based on *in vivo* trials. The proposed mechanisms for cholesterol removal by probiotics include assimilation of cholesterol by growing cells, binding of cholesterol to cellular surface and incorporation into the cellular membrane, deconjugation of bile via bile salt hydrolase, coprecipitation of cholesterol with deconjugated bile and production of short-chain fatty acids from oligosaccharides (Kumar *et al.*, 2012b). Some of these basic mechanisms have been covered in Section 2.0 above as we present data from *in vitro* and *in vivo* studies. Here, we will elaborate on a few selected mechanisms that are feasible and supported by in-depth evidence.

One major hypothesis has indicated that the cholesterol-lowering effect of probiotics may be linked to their bile salt hydrolase activity (Kimoto-Nira *et al.*, 2007; Lye *et al.*, 2010a). The major source of cholesterol in the human body includes biosynthesis by the liver and absorption by the intestines. It is known that liver hepatocytes produce bile salts which enhance dietary lipid absorption as well as cholesterol/fat transport across the intestinal epithelium. The de novo synthesis of bile salt involves the formation of primary bile acids from dietary cholesterol and conjugation with either glycine to form glycocholic acid or taurine to form taurocholic acid via an amide bond at the C-24 position (Fig. 29.3). More than 95% of bile salts are returned to the liver by enterohepatic circulation (McAuliffe *et al.*, 2005). Reabsorbed bile acids are taken up by hepatocytes, re-conjugated, and re-secreted into bile. During these processes, an important biotransformation that must take place before subsequent modifications is termed deconjugation catalysed by the bile salt hydrolase, BSH (Patel *et al.*, 2010).

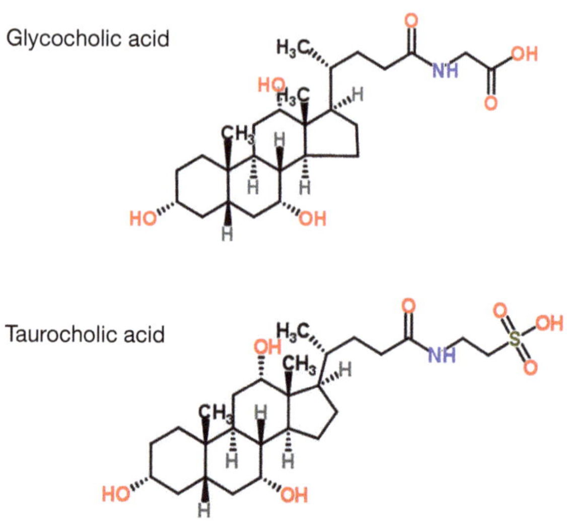

Figure 29.3 Glyco- and tauro-conjugated cholic acid.

Since most bile acids (or salts when still conjugated) are absorbed, there is no need for *de novo* synthesis from cholesterol. This could cause an accumulation of cholesterol in the body.

However, many *Bifidobacterium* and *Lactobacillus* species have been found to have bile salt hydrolase (BSH) genes, some strains having more than one BSH homolog, which are not identical (Chae *et al.*, 2013). The BSH genes have been cloned and characterized (Jiang *et al.*, 2010; Kim and Lee, 2005, 2008; Kim *et al.*, 2004). BLAST search analysis of the BSH DNA sequences showed that these enzymes are part of the N-terminal nucleophile hydrolase superfamily and therefore have an N-terminal catalytic nucleophile that can cleave amide bonds (Chae *et al.*, 2013). BSH has similar active sites of PVA (penicillin V amidase). Probiotics may therefore produce BSH which hydrolyses both glyco-conjugated and tauro-conjugated bile acids to liberate free primary bile acids. The deconjugated bile acids are less soluble and less efficiently reabsorbed from the intestinal lumen than conjugated bile salts (Lye *et al.*, 2010a). As increased levels of free bile acids are excreted in the faeces, the synthesis of new bile salts from available cholesterol obviously reduces cholesterol levels. Such probiotics containing active BSH therefore increase the rate of formation of bile salts from cholesterol, thereby reducing cholesterol associated problems. Cholesterol reduction with BSH active-LAB in host bile salt metabolism is comparable to that of drugs, bile salt sequestrants (e.g. cholestyramine and colestipol) that bind to bile acids and prevent their reabsorption into the enterohepatic circulation, thereby resulting in increased demand for cholesterol as a precursor of bile. About 4% of bile salts are lost in the faeces and about 10% of cholesterol is excreted in an unmetabolized form, thus de novo bile salt synthesis from cholesterol is used to maintain the pool size under steady state. Therefore, conversion of cholesterol to bile salts in the liver and their subsequent secretion and faecal excretion provides the major route (~ 90%) for elimination of excess cholesterol. Hence, probiotic bacteria targeting the metabolism of bile salts could provide beneficial effects to patients with CVD. Statins (or HMG-CoA reductase inhibitors) are also used to lower cholesterol levels by inhibiting the enzyme HMG-CoA reductase, which plays a central role in the production of cholesterol in the liver, which produces about 70% of total cholesterol in the body. HMG-CoA reductase inhibitors (statins) inhibit cholesterol absorption. Oral feeding of encapsulated BSH enzyme from *L. buchneri* ATCC 4005 alone in animal trials could accomplish about 50–58% reduction of serum cholesterol level in rats (Sridevi *et al.*, 2009), suggesting a potent role for BSH enzymes. Hypothetical cholesterol lowering mechanism by actions of BSH has been proposed in Fig. 29.4 (Jones *et al.*, 2013).

One other mechanism by which probiotics may reduce cholesterol levels is by the short-chain fatty acids (SCFAs) they produce from fermentation. SCFAs act as ligands that bind to activate peroxisome proliferator-activated receptors (PPARs). PPARγ controls the transcription and translation of angiopoietin-like protein 4 (ANGPTL4), a lipoprotein lipase (LPL) inhibitor, produced by brown and white fat as well as liver and intestine

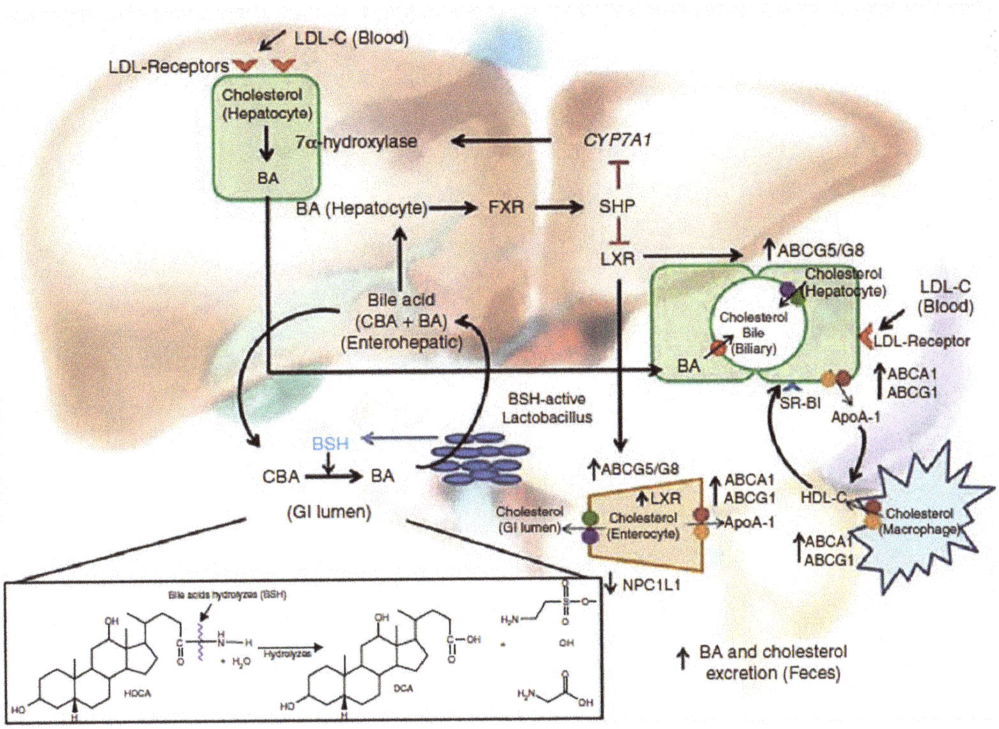

Figure 29.4 Hypothetical cholesterol lowering mechanism by actions of BSH (Jones *et al.*, 2013).

(Kersten, 2014). LPL promotes the release of fatty acids from circulating chylomicrons and very-low-density lipoproteins (VLDL), and results in their storage as triglycerides in the adipose tissue. Butyrate has been found to show the highest activation of PPARγ followed by propionate and then acetate (Alex *et al.*, 2013). SCFAs therefore suppress LPL activities via PPARγ activation and up-regulating ANGPTL4 levels thereby controlling fatty acid oxidation in muscle and adipocytes resulting in a reduction in fat storage (Alex *et al.*, 2013; Korecka *et al.*, 2013). In humans, mutations within the ANGPTL3 gene give rise to familial hypolipidaemia, characterized by low plasma levels of total cholesterol, LDL-C, HDL-C, and triglycerides, or combinations (Kersten, 2014)

According to Marcil *et al.* (2003), butyrate may also inhibit HMG-CoA reductase activity. HMG-CoA reductase catalyses the conversion of 3-hydroxyl-3-methyl-HMG-CoA to mevalonate (Fig. 29.5), the rate-limiting step in cholesterol synthesis (Parnell and Reimer, 2009).

Marcil *et al.* (2003) observed that incubating Caco-2 cells with butyrate caused a reduction in the levels of HMG-CoA reductase activity without any change in acyl-CoA:cholesterol acyltransferase (ACAT) activity (Figs. 29.5 and 29.6).

Butyrate may therefore reduce the levels or amounts of HMG-CoA reductase thereby reducing the conversion of 3-hydroxyl-3-methyl-HMG-CoA to mevalonate resulting in a reduction in cholesterol biosynthesis. The cholesterol lowering effects of SCFA may account for the ability of fermented milk and prebiotic fibres to reduce cholesterol levels as observed by Anderson and Gilliland (1999), Xiao *et al.* (2003) and Ooi *et al.* (2010a). However, further studies are required to establish the HMG-CoA reductase inhibition ability of butyrate.

Probiotics may also directly prevent intestinal absorption of cholesterol. Primarily, cholesterol is absorbed in the duodenum and proximal jejunum, with little absorption in the ileal segment of the intestine (Perova and Metel'skaia, 2008). The Niemann-Pick C1-Like 1 (NPC1L1) protein, which is expressed in the brush border membrane of enterocytes of the small intestine, plays a critical role in the absorption of intestinal cholesterol (Jia *et al.*, 2011). The liver X receptors (LXRs) in the liver mediate cholesterol metabolism by inducing the expression of SREBP-1c. SREBP-1c is a transcription factor involved in the control of expression of many genes including some genes involved in cholesterol biosynthesis. ATP-binding cassette subfamily G members 5 and 8 (ABCG 5 and ABCG 8) are major sterol transporters on the apical side of enterocytes (Gorenjak *et al.*, 2014). NPC1L1, ABCG 5 and ABCG 8 and lipoprotein lipase (LPL) genes are regulated by LXR. When LXR is activated, NPC1L1 is down-regulated resulting in decreased cholesterol uptake in the intestines (Duval *et al.*, 2006; Huang and Zheng, 2010).

It has been reported that probiotic strains such as *L. plantarum* PCS 26 may lower cholesterol levels by up-regulating LXR which causes an up-regulation of ABCG 5, ABCG 8 and LPL while down-regulating NPC1L1. This results in less absorption of cholesterol and promotes biliary cholesterol secretion (Gorenjak *et al.*, 2014; Yoon *et al.*, 2013; Yu *et al.*, 2002).

Controversies: two sides of a coin

The BSH-generating ability of probiotics microorganism has always been associated with a cholesterol-lowering capability in the battle against hypercholesterolaemia. In 2012, Cardioviva™ (containing *L. reuteri* NCIMB 30242) was launched, the first

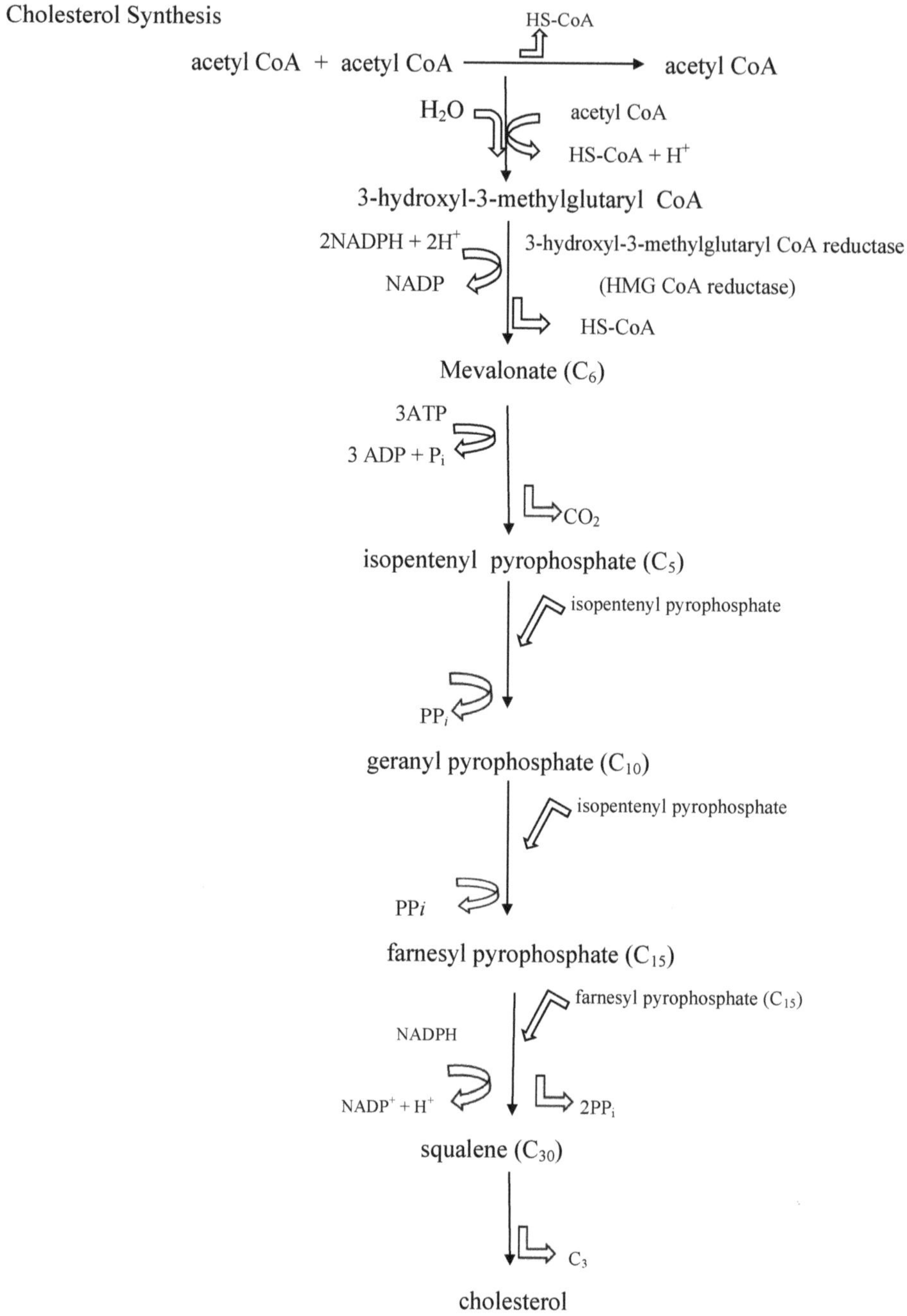

Figure 29.5 Cholesterol biosynthetic pathway. HMG-CoA reductase catalyses the rate limiting step in cholesterol biosynthesis (Marcil *et al.*, 2003).

and only probiotic on the market that has been clinically proven to naturally help reduce LDL cholesterol, a risk factor for heart disease, by 11.6% in hypercholesterolaemic adults. It works in two ways: by maintaining the amount of cholesterol the body produces and by maintaining the amount absorbed from food.

Two approaches had been adopted in clinical trials evaluating this product, i.e. using yoghurt incorporating *L. reuteri* NCIMB 30242 and capsules containing the strain. Results show that consumption of *L. reuteri* NCIMB 30242 containing yoghurts over a 6 week period results in significant reductions in total cholesterol (TC) (–4.8%), low-density lipoprotein cholesterol (LDL-C) (–8.9%), non-high density lipoprotein cholesterol (non-HDL-C) (–6.0%), and apolipoprotien B 100 (ApoB100) (–0.05 mmol/l). Serum concentrations of triglycerides and high-density lipoprotein cholesterol (HDL-C) were unchanged over the course of the study (Jones *et al.*, 2012a). Conversely, a total of 127 subjects completed a randomized, double-blind, placebo-controlled, parallel-arm, multicentre study conducted with the capsule (Jones *et al.*, 2012b). Subjects were randomized to consume *L. reuteri* NCIMB 30242 capsules or placebo capsules over a 9-week intervention period. *L. reuteri* NCIMB 30242 capsules reduced significantly LDL-C by 11.6%, total cholesterol by 9.1%, non-HDL-cholesterol (non-HDL-C) by 11.3% and apoB-100 by 8.4% relative to placebo. The

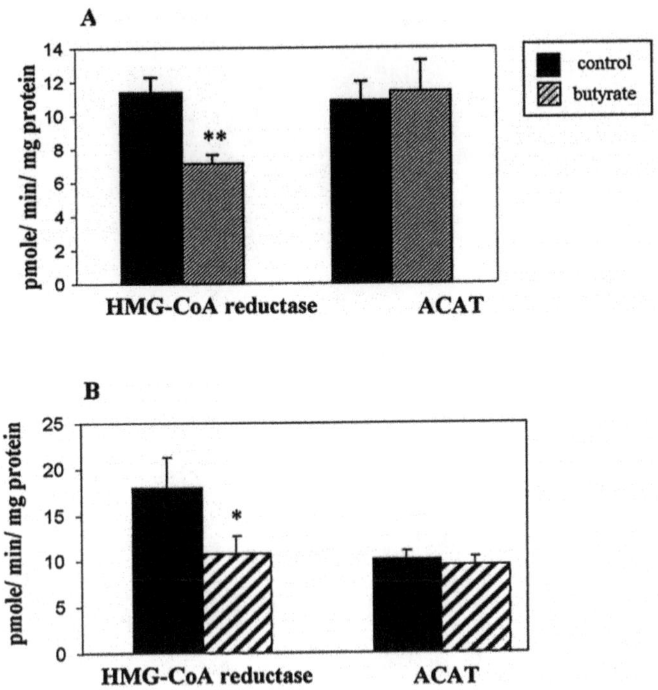

Figure 29.6 Activities of key cholesterol regulatory enzymes in Caco-2 cells incubated with (A) 10 mmol/L and (B) 20 mmol/L of butyrate for 20 h at 37°C. Cell homogenates were assayed for dl-3-hydroxy-3-methyl-glutaryl-CoA (HMG-CoA) reductase and acyl-CoA:cholesterol acyltransferase (ACAT). Values are means ± SEM for six independent experiments. *Different from control, $P < 0.02$; **different from control, $P < 0.0005$. Adapted from Marcil *et al.* (2003).

ratios of LDL-C/HDL-cholesterol (HDL-C) and apoB-100/apoA-1 were reduced by 13.4% and 9.0%, respectively, relative to placebo. Triglycerides and HDL-C were unchanged. High-sensitivity C-reactive protein and fibrinogen were reduced by 1.1 mg/l and 14.3% relative to placebo, respectively (Jones *et al.*, 2012b). The results of both clinical studies are an excellent indication of the cholesterol-lowering and cardiovascular health potential of this particular strain.

However, recent controversies have questioned the possibility of deconjugated bile binding to the FXR orphan nuclear receptor farsenoid X receptor (FXR), leading to altered bile acid metabolism. The FXR can be found in liver, intestine, kidney and adrenal cortex, and is a major bile sensor that protects the liver from bile acid toxicity by regulating the transcription of genes involved in bile acid homeostasis. Makishima and co-workers (1999) revealed that deconjugated bile salts, in particularly chenodeoxycholic acid and cholic acid, have a higher tendency of binding and activating FXR compared with other bile acids The DNA binding site of FXR-bile acid complex then attaches to the promoter region of Cy7a gene. The Cy7a gene is responsible for the transcription of 7-alpha-hydrolase (7AH) enzyme, which is known as the rate limiting enzyme in bile synthesis. The binding of FXR-bile acid complex to the promoter region of Cy7a gene thus leads to a suppressed transcription and subsequently reduced bile acid synthesis from cholesterol (Russell, 1999). Under such a circumstance, there is no such need to utilize cholesterol as a precursor to form new bile, and thus diminishing the cholesterol lowering effects.

In view of the molecular insight which can be achieved by computational approach, we utilized molecular docking to further substantiate the binding interaction of both deconjugated and conjugated bile acids towards FXR. Docking data indicated that deconjugated bile acids such as cholic acid, chenodeoxycholic acid, deoxycholic acid and lithocholic acids have higher propensities to bind with the FXR receptor compared with conjugated bile acids as in glycocholic acid, taurocholic acid, glycodeoxycholic acid and taurodeoxycholic acids, with the binding affinity ranging from −12.44 to −9.35 kcal/mol compared with conjugated bile acids which ranging from −10.59 to −7.88 kcal/mol (Fig. 29.7). The selected bile acids in this study were cholic acid, chenodeoxycholic acid (CDCA), deoxycholic acid and lithocholic acid (LCA). However, chenodeoxycholic acid and lithocholic acid demonstrated lowest free energy of binding (FEB) among other deconjugated bile acids, −12.11 and −12.44 kcal/mol, respectively, indicating better binding affinities (Fig. 29.8a and d). A positive correlation has been reported with a recent study showing that IC50 values of FXR receptor activation of both CDCA and LCA were lower compared to other bile acids such as glycodeoxycholic, taurodeoxycholic, glycocholic and taurocholic acids (Lew *et al.*, 2004).

Docking data also showed that the sterone core of the bile acids faced towards the hydrophobic core at the binding site of FXR whereby the carboxylic acid side chain protruded outward to the peripherals of the binding site (Fig. 29.7a-d). These orientations are beneficial and functional in hydrogen bonding interaction with the conserved binding residues such as His298, Ala295 and Ser336 (Fig. 29.9). A previous study by Cai and coworker (2007) had revealed that His298, Ala295 and Ser336 play a role towards the selective binding of FXR with CDCA and LCA across different organisms such as mammals, birds and reptiles.

Nonetheless, the binding affinity of conjugated bile acids was comparatively lower towards FXR receptor. The conjugated bile acid had adopted a different orientation. It was reverted compared with the binding orientation of deconjugated bile. The sterone ring of the conjugated bile acids (taurocholic, glycochenodeoxycholic and taurochenodeoxycholic acids) were extended out to the peripheral of the binding site and the carboxylic side chain was facing inward at the hydrophobic site (Fig. 29.7). This justifies the lower binding affinities for the conjugated bile acids on FXR compared with deconjugated bile acids which is attributed to a lack of catalytic activities.

Possible detrimental effects due to the deconjugation of bile have also been reported. Due to the low emulsifying capacity, high concentration of deconjugated bile has been associated with various detrimental health aspects, ranging from malabsorption of lipids and steatorrhoea to gall stone formation and disturbance of gut microbiota profiles (Table 29.3).

Thus, the deconjugation of bile by the actions of BSH, yielding free bile acids (and subsequent conversion into secondary bile), may seem to be a 'short term gain (possible cholesterol lowering properties), but long term loss' (alteration of various health parameters which contribute to other varying diseases).

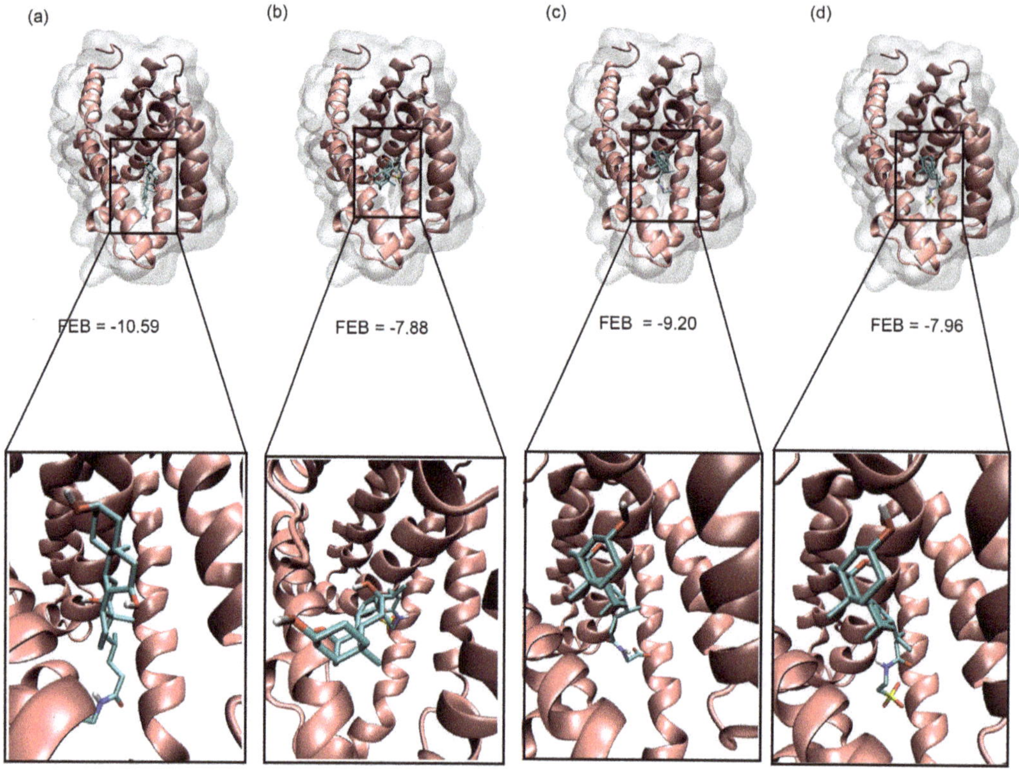

Figure 29.7 Binding of FXR receptor with conjugated bile acids: (a) glycocholic acid (b) taurocholic acid (c) glycochenodeoxycholic acid and (d) taurochenodeoxycholic acid. The sterone core of bile acids (b) (c) and (d) face the peripheral region of the binding pocket while the amino moiety and carboxylic side chain are embedded deep inside the hydrophobic pocket. Such orientations are unfavourable, possibly leading to a lower binding affinity compared with deconjugated bile acids.

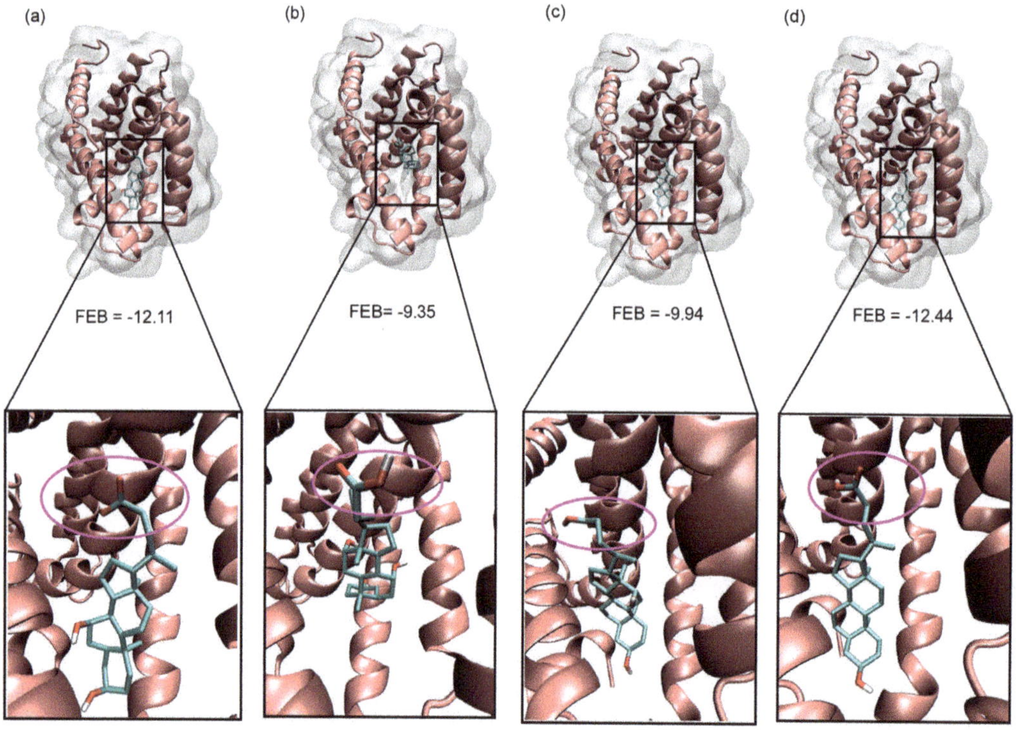

Figure 29.8 Binding of FXR receptor with deconjugated bile acids (a) chenodeoxycholic acid (CDCA) (b) cholic acid (CA) (c) deoxycholic acid (DCA) and (d) lithocholic acid (LCA). Similar binding orientation was observed within all docked conformation of deconjugated bile acids; sterone ring was embedded into hydrophobic pocket and carboxylic side chain was extended to peripheral region of the binding site. LCA bound deeper into the hydrophobic pocket compared with others and reflected better binding affinity.

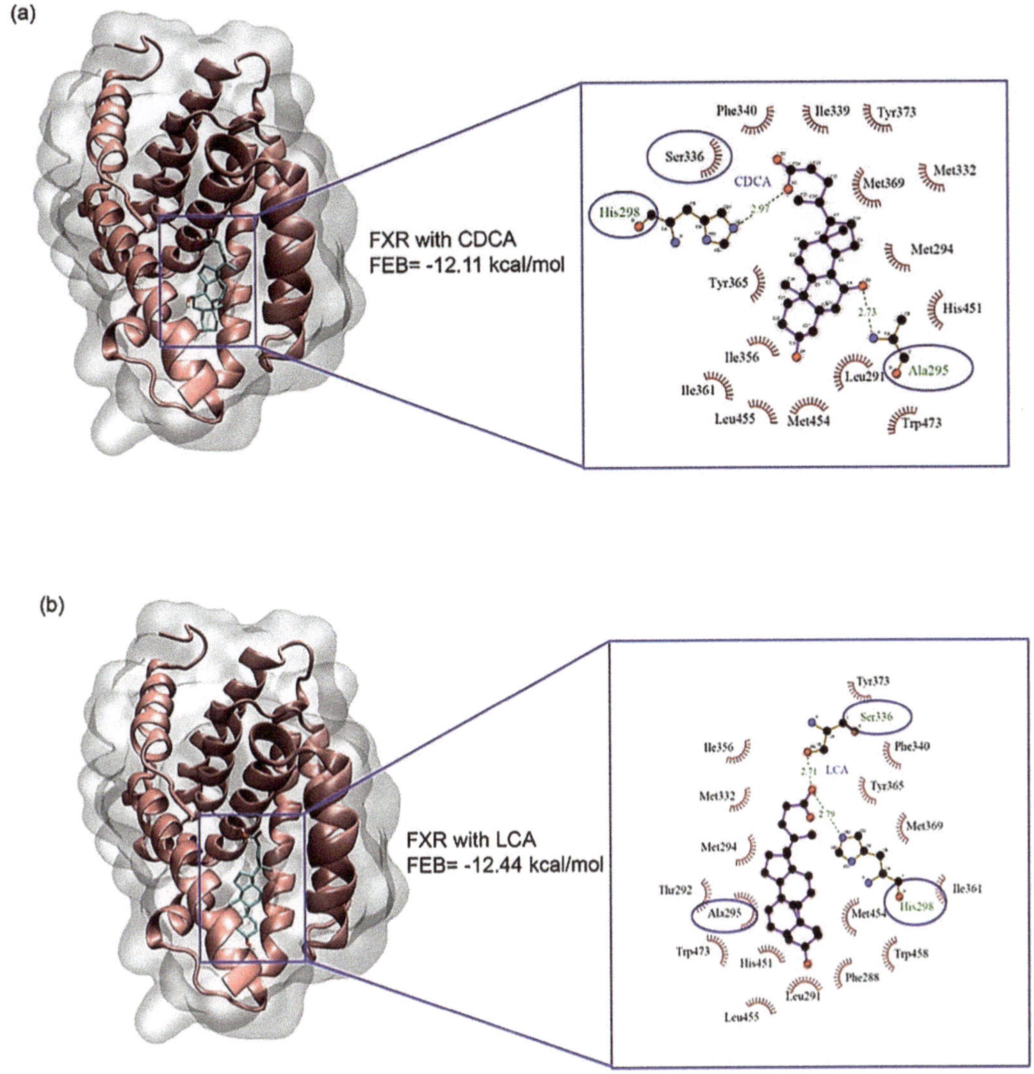

Figure 29.9 Docking of chenodeoxycholic acid (CDCA) and lithocholic acid (LCA) against FXR receptor. Conserved residues such as His296, Ser336 and Ala295 (blue circle) interact with both CDCA and LCA via hydrogen bond and hydrophobic interactions. These conserved residues may act as catalytic triad leading to enhanced binding affinity towards CDCA and LCA.

While probiotics have been widely recognized to play a vital role in cholesterol lowering properties, our results and review substantiate the need to look beyond BSH as a single mechanism for hypercholesterolaemia, and/or as a sole mean to justify a cholesterol-lowering property of a probiotic strain. Albeit all the controversies, other mechanisms may play a more essential role in explaining the functions and roles of probiotics in combating blood cholesterol. A need for a balanced gut microbial profile and physiological lipid/cholesterol cycle should be prioritized in the pursuit of a long-term cholesterol management.

Cholesterol-related disorders: alleviation by probiotics and prebiotics

Hypercholesterolaemia is normally associated with high levels of LDL, which is responsible for carrying free or esterified cholesterol from liver to other cells (Anandharaj *et al.*, 2014; Shepherd, 2001). With the high levels of LDL (4.41mmol/L) (Liong *et al.*, 2007), the amount of free cholesterol is also increased in the blood circulation. As described in the past studies, hypercholesterolaemia was found to have the ability to induce alterations in the human erythrocyte plasma membrane (Koter *et al.*, 2003; Narayanan, 2009).

Erythrocytes are particularly sensitive to changes in cholesterol concentrations in extracellular medium. As the free cholesterol in the plasma lipid readily diffuses into the erythrocyte membrane, accumulation of free cholesterol is observed in the erythrocyte membrane. It was proven that hypercholesterolaemic patients have concentrations of membrane cholesterol seven times higher than those without hypercholesterolaemia (Koter *et al.*, 2003). The accumulation of free cholesterol in the erythrocyte membrane will decrease the deformability of the erythrocytes and caused abnormal alterations in the erythrocyte membrane (Koter *et al.*, 2003). Also, hypercholesterolaemic conditions have been related with morphological abnormalities in erythrocytes which can subsequently increase the risk of impairing

Table 29.3 Detrimental health aspects as attributed to the actions of deconjugated bile

Detrimental effects	Reference
Deconjugated bile has low emulsifying capacity, leading to malabsorption of lipids, causing steatorrhoea in humans and growth depression in chickens	Kim and Lee (2005), Knarreborg *et al.* (2002)
Bile acid deconjugation by lactobacilli is one of the important factor in causing disturbed bile acid and lipid metabolism in patients with short bowel syndrome	Bongaerts *et al.* (2000)
BSH activity in *Listeria monocytogenes* is one of the virulence factors involved in intestinal and hepatic phases of listeriosis	Dussurget *et al.* (2002)
Colon carcinogenesis	
Deconjugation of taurine-conjugated bile acids by BSH releases hydrogen sulfide from the sulfonic acid moiety of taurine	Payne *et al.* (2008), Van Eldere *et al.* (1996)
Hydrogen sulfide is highly toxic; increase colonocyte turnover by inhibiting butyrate metabolism in the gut	
Faecal secondary bile acids are elevated in populations with high incidence rate of colon cancer	
Gallstones formation	
Accumulation of deconjugated bile, which in turns serves as substrates for other gut microbiota (such as *Clostridium*) and conversion into secondary bile	Farkkila and Miettinen (1990), Thomas *et al.* (2000), Berr *et al.* (1992), Venneman and van Erpecum (2010), Begley *et al.* (2005), Hussaini *et al.* (1995), Bernstein *et al.* (2005), Park *et al.* (2008)
Increased amounts of DCA in bile of cholesterol gallstone patients	
The more hydrophobic secondary bile acids promote cholesterol crystallization, producing cholesterol supersaturated vesicles	
Crystallized cholesterol together with calcium salts and bile pigments forms gallstones; obstructs biliary ducts	
Secondary bile increase oxidative stress via generating ROS, causing oxidative DNA damage; increasing risks of colon cancer	
10 days supplementation of BSH positive *Lactobacillus acidophilus* 43121 (7 log cfu daily) to 400 hypercholesterolaemia-induced male pigs, faecal DCA and LCA increased	
Alteration of gut microbial profile	
DCA exerts a strong antimicrobial activity; inducing membrane damages via a detergent effect	Floch (2002), Islam *et al.* (2011)
Compared to its primary counterpart (cholic acid), DCA has a 10-times higher bactericidal activity	
1 mmol/L of DCA into growth medium inhibited the growth of *Bacteroides fragilis*, lactobacilli and bifidobacteria *in vitro*	
In the gut, DCA increased Firmicutes and decreased Bacteroidetes. The increase in Firmicutes resulted in the expansion of Clostridia and Erysipelotrichi	

physical function which is referred to as erythrocyte's function in delivering oxygen and binding of oxygen molecules to iron in the cytoplasm thus causing microcirculation impairment of the erythrocytes (Koter *et al.*, 2003; Liong *et al.*, 2007; Narayanan, 2009; Ooi *et al.*, 2010b).

In normal erythrocytes, the major components of the erythrocyte plasma membrane are phospholipids and cholesterol (Narayanan, 2009; Ohvo-Rekila *et al.*, 2002; Yawata, 2004). Phospholipids are the most abundant membrane molecules with an equal portion of hydrophobic and hydrophilic portions. The membrane phospholipids are asymmetrically arranged into a lipid bilayer with a thickness of two molecules (Melzak *et al.*, 2012; Mohandas and Gallagher, 2008; Narayanan, 2009; Ooi *et al.*, 2010a; Yawata, 2004). Cholesterol, on the other hand, does not form a bilayer but dissolve readily into the phospholipids bilayer and intercalates between the phospholipids molecules in equal amount (Melzak *et al.*, 2012; Mohandas and Gallagher, 2008; Yawata, 2004). It was reviewed that the erythrocyte membrane cholesterol concentration is in equilibrium with the plasma cholesterol concentration. Therefore, the erythrocyte membrane cholesterol content directly reflects the cholesterol content in plasma and vice versa (Buchwald *et al.*, 2000).

Being the major lipids in human erythrocyte membranes, the interactions between phospholipids and cholesterol will exert a degree of immobility on the portion of the lipid molecules. Cholesterol acts to prevent crystallization of the fatty acid chains to maintain the bilayer matrix in an intermediate fluid state. As stated by Cress *et al.* (1982), the ratio of cholesterol: phospholipids are often used as convenient index of 'packing'. The normal molar ratio of cholesterol to phospholipids is 0.84–1.05 mol/mol to maintain the normal rheological properties of the erythrocyte (Liong *et al.*, 2007; Ooi *et al.*, 2010b; Yawata, 2004). It was reported that the cholesterol: phospholipids ratio is crucial for the fluid properties and membrane deformability of the erythrocytes membrane (Ooi *et al.*, 2010b; Yawata, 2004).

Erythrocyte membrane fluidity is an important biophysical property and is responsible for the mobility and rotational motion of membrane lipid molecules (Owen *et al.*, 1982). Enrichment or depletion of cholesterol content and the cholesterol: phospholipids ratio is correlated with the erythrocyte fluidity properties. Fluidity of erythrocyte is reduced with an increased of cholesterol: phospholipids ratio which lead to the stiffness of the lipid shell (Buchwald *et al.*, 2000; Koter *et al.*, 2003; Owen *et al.*, 1982; Parmahamsa *et al.*, 2004). The erythrocyte fluidity will affect the structure and properties of the erythrocytes and subsequently form shape irregularities. Studies have demonstrated that in hypercholesterolaemic condition with an increased ratio of cholesterol: phospholipids (1.19–1.30 mol/mol), a reduction in the erythrocyte membrane fluidity was observed (Liong *et al.*, 2007; Ooi *et al.*, 2010b; Owen *et al.*, 1982). In patients with liver disease, enrichment of cholesterol within the outer leaflet of the erythrocyte membrane caused the formation of irregular projections on the erythrocyte membrane. As a result, erythrocyte membrane fluidity was reduced, leading to the reduced in erythrocyte deformability, which caused the erythrocytes to be remodelled into abnormal morphology, such as acanthocytes (Fig. 29.10) as they passed through the spleen.

Membrane deformability is closely related to the fluidity of the erythrocytes. Erythrocytes deformability is an important rheological factor in controlling microcirculation in organs. Erythrocyte deformability refers to the ability of the erythrocytes to restore their normal shape without fragmentation or loss of integrity while entering and passing small nutritive capillaries (Berndt-Zipfel *et al.*, 2013; Kuwai and Hayashi, 2006). Deformation is needed in maintaining the normal circulation that delivers oxygen to the cells by allowing the erythrocytes to pass through the narrow capillaries in the microcirculation. The unique biconcave shape of the erythrocytes enables them to undergo deformation with a constant surface area (Storch and Schwarz, 2004; Yawata, 2004; Zhang *et al.*, 2009). However, studies verified that hypercholesterolaemia leads to decreased membrane deformability and increased in surface area. As a result, deteriorated erythrocyte deformability will shorten the survival of erythrocytes to less than 100 days compared with a normal erythrocyte survival of 110–120 days (Berndt-Zipfel *et al.*, 2013; Kuwai and Hayashi, 2006; Mohandas *et al.*, 1980).

Past studies have revealed that hypercholesterolaemic conditions caused deposition of cholesterol in the extracellular part of the erythrocyte membrane, leading to changes in the normal morphology of erythrocytes (Memon *et al.*, 2003). Iannuzzi (1997) reported that high concentration of LDL was also associated with erythrocyte shape abnormalities. Typically, a human erythrocyte has a biconcave shape with 8 μm in diameter and 2 μm in thickness (Storch and Schwarz, 2004; Yawata, 2004; Zhang *et al.*, 2009) (Fig. 29.11). Enrichment of cholesterol in the erythrocyte membrane caused the discocytes to loose their disc shape and changed in diameter and thickness. There are different types of abnormal erythrocytes due to the alteration of the erythrocyte plasma membrane, namely echinocytes, acanthocytes and spur cells (Aihara *et al.*, 2001; Foglia, 2010; Yawata, 2004). Echinocytes are seen as crenated cells which are spherical in shape with spicules which are uniformly distributed on the surface whereas acanthocytes are characterized by few, irregularly distributed spicules (Foglia, 2010; Weinstein, 1974). Spur cells, as reviewed by Cooper *et al.* (1969), having unusual thorny projections and are chemically dissimilar with acanthocytes. These irregularities of the erythrocytes were altered by cholesterol through mechanical deformation which causes transformation of erythrocyte from biconcave to other abnormal shapes.

Evidence showed that morphological defects of the erythrocytes were observed in hypercholesterolaemic rabbits, guinea pigs and dogs. A high proportion of echinocytes were present in hypercholesterolaemic rabbits whereas the cholesterol-fed guinea pigs and dogs showed the formation of spur cells – a sharp, horny and star-like shape erythrocyte (Iannuzzi *et al.*, 1997; Liong *et al.*, 2007; Ooi *et al.*, 2010b). In an animal trial, pigs on a high-fat diet

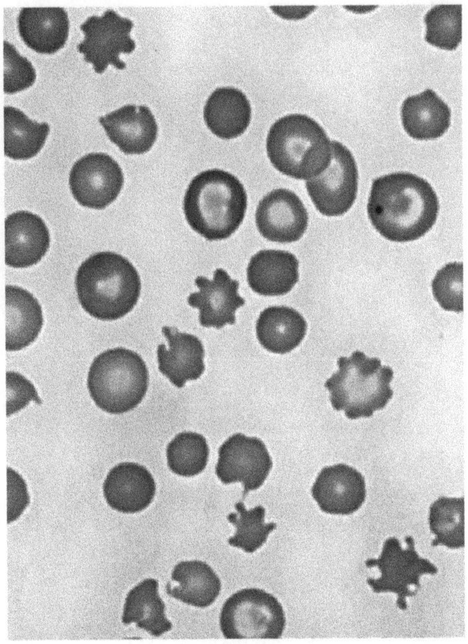

Figure 29.10 Acanthocytes. (Copyright from https://ahdc.vet.cornell.edu/clinpath/modules/rbcmorph/acanth.htm. Used with permission from Cornell University; original image located on 'eClinPath' website at Cornell University.)

had higher prevalence of spur cells compared with those on the low-fat diet.

The administration of probiotics or prebiotics have been demonstrated in lowering serum cholesterol levels in rats, pigs and human (Ooi *et al.*, 2010a). The combination of probiotics and prebiotics is known as synbiotics. Clinical evidences showed that with the administration of synbiotics, which consist of *L. acidophilus* ATCC 4962, fructooligosaccharide, inulin and mannitol, plasma LDL-cholesterol was lowered compared to those without supplementation (46.63–59.19%) (Liong *et al.*, 2007). Also, administration of synbiotic increased the membrane fluidity of erythrocytes, accompanied by a decrease in membrane rigidity by reducing cholesterol enrichment and alterations in erythrocyte membrane lipid profiles (Bengmark, 2008; Owen *et al.*, 1982). A human trial involving the supplementation of a synbiotic containing *L. acidophilus* CHO 220 and inulin also showed a reduced concentration of cholesterol in erythrocyte membrane by 21.02% over 12 weeks compared with the control group (Ooi *et al.*, 2010a). Under hypercholesterolaemic conditions, high concentration of unesterified cholesterol is present in the plasma, which is readily exchanged with the cholesterol within the outer leaflet of the erythrocyte membranes. With the reduction of the plasma cholesterol by the supplementation of synbiotic, the ratio of cholesterol: phospholipids in the erythrocyte membrane was decreased to a normal value. This was shown by a significant increase in concentration of phospholipids and a significant decrease in concentration of cholesterol in the erythrocyte membrane which contributed to a normal molar ratio of cholesterol: phospholipids at 0.89 after 12 weeks of supplementation of synbiotic (Ooi *et al.*, 2010a), which will increase the fluidity towards normal values (Owen *et al.*, 1982). The increased fluidity of the

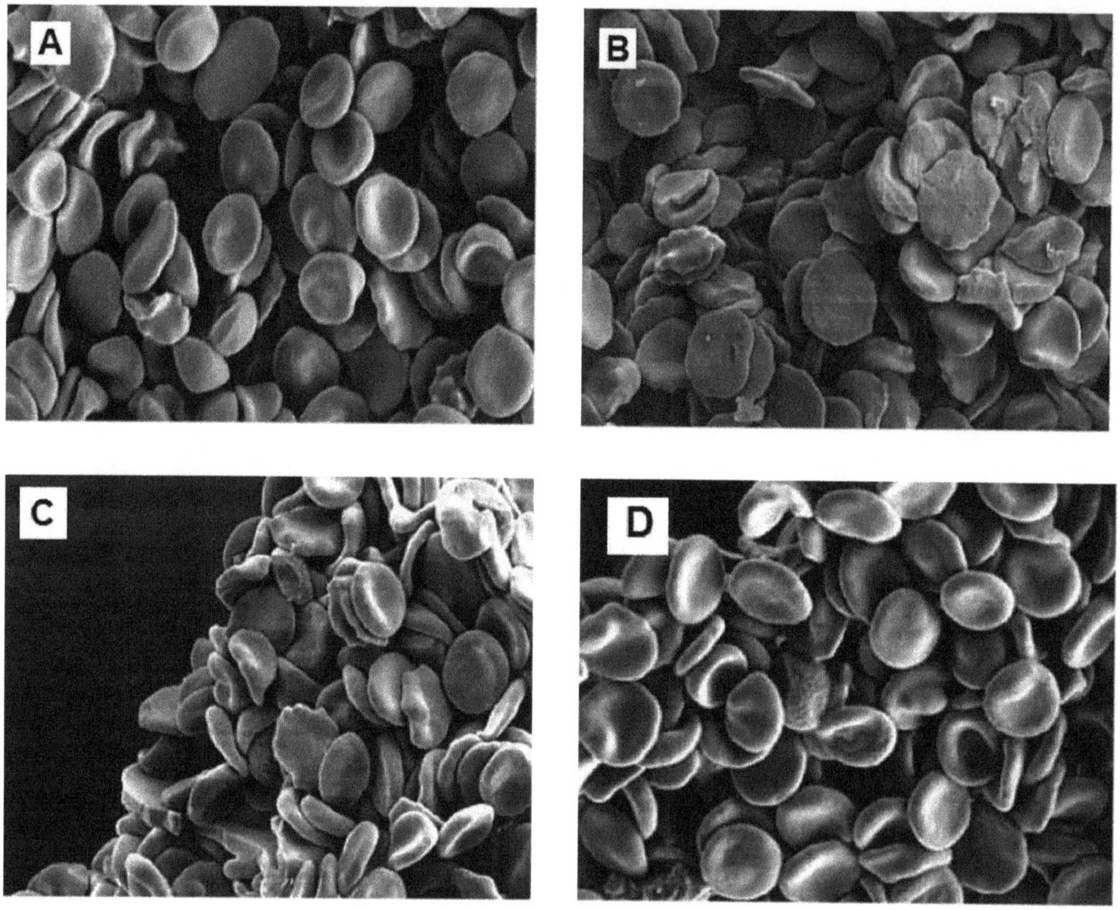

Figure 29.11 Erythrocytes from subjects that were healthy (A); hypercholesterolaemic (B); hypercholesterolaemic and administered with a placebo after 12-week (C); and hypercholesterolaemic and administered with a synbiotic containing Lactobacillus gasseri CHO-220 and inulin after 12-week (D). A reduced occurrence of spur cells were detected in the treatment group who was consuming the synbiotic. (Ooi *et al.*, 2010b; Copyright licence number from Elsevier: 3356420429784.)

erythrocyte membrane also directly increased the erythrocyte membrane deformability and further improved the morphology of the erythrocyte (Ooi *et al.*, 2010a).

Data from animal and human trials have illustrated that the combination of probiotics and prebiotics not only lowered blood cholesterol levels and improved lipid profiles, but also alleviated cholesterol-related disorders such as abnormalities in erythrocytes.

Future perspective for cholesterol-reducing probiotics

It is quite amazing that the small dose of probiotic bacteria administered daily into the body can have considerable impacts on the gut and the entire body at large. Indeed, a few strains of probiotics have shown positive results in lowering cholesterol levels in humans and may soon be accepted on the market. Though the hypocholesterolaemic mechanism of probiotics is not yet fully understood, it is an accepted fact that cholesterol and bile salt metabolism are closely related. Based on this, 'BSH hypothesis' has been proposed as a mechanism by which some probiotics with BSH genes to lower cholesterol.

Interestingly, the hypocholesterolaemic effects of *L. reuteri* NCIMB 30242, which showed high BSH activities in *in vitro* experiments, have been confirmed in animal studies as well as in humans (Jones *et al.*, 2012a; Jones *et al.*, 2012b). However, more research is still needed to validate the safety on secondary bile acids and other adverse effects for long term consumption.

The physiological impact of BSH activity on the host (mammalian) cells and on the enzyme (BSH) producing bacteria is worth exploring. Deconjugated secondary bile salts are thought to be a cause of gallstones (Thomas *et al.*, 2000). They are also reported to cause depression of growth in chickens due to the poor intake of lipid in the small intestine (Knarreborg *et al.*, 2002) and are suspected to cause colorectal cancer (Kumar *et al.*, 2012a).

Although many other hypotheses have been proposed about the cholesterol lowering ability of probiotics, further *in vivo* studies are required for their substantiation because laboratory conditions may not necessarily reflect human body conditions.

Over the years, proteomics has enhanced a better understanding of the molecular basis of probiotic functionality, providing a clear picture of cell membrane proteins involved in host epithelial binding (Siciliano and Mazzeo, 2012). This technology could therefore be applied to study the exact proteins that exhibit the proposed cholesterol binding and mobbing ability of probiotics. However, since different bacterial species or even strains may have different properties, an understanding of the species-specific mechanisms of action by which probiotics influence cholesterol biosynthesis and/or clearance is still open for investigation.

The identification of probiotic protein patterns closely linked to cholesterol lowering abilities could prompt the development of new and faster *in vitro* assays using biological tests and clinical trials. Genomics and transcriptomics may aid in further investigations on the conserved and variable regions of the *bsh* genes from various probiotic species. This could be useful for defining molecular biomarkers for the identification and selection of novel probiotic strains with predictable and improved functionality. For instance the specific probiotic metabolite or cell wall component that binds to LXR to regulate sterol regulatory element-binding protein 1c (SREBP-1c) expression (Gorenjak *et al.*, 2014; Yoon *et al.*, 2013; Yu *et al.*, 2002) is required for a better understanding of how probiotics reduce cholesterol at the molecular level.

It is obvious that there are differences in the metabolic activities of different cell lines (cancerous cells and normal cells), and thus the results obtained from some studies using these cell lines may differ as different mechanisms might have been used. Huang and Zheng (2010) and Gorenjak *et al.* (2014) showed contrasting effects. Huang and Zheng (2010) reported a significant down-regulation of NPC1L1 mRNA levels in CaCo2 cells in the presence of *Lactobacillus rhamnosus* GG, while Gorenjak reported an up-regulation of NPC1L1 mRNA levels in HIEC and HUIEC cell lines in the presence of *L. rhamnosus GG*. Probiotic studies should therefore be performed on systems that represent actual situations as much as possible so as to prevent disagreements in reports from similar experiments. Standards pertaining to what types of cells used for specific investigations will be found in future.

Investigations on how hypocholesterolaemic probiotics can help to maintain normal cholesterol levels in healthy individuals as well as their adverse effects is critical. Measures to counter such side effects, if any, will go a long way to protect consumers. Whether these probiotics have impact on cholesterol reduction alone or they may influence other unrelated molecules such as enzymes and proteins in cholesterol biosynthesis should also be considered.

As certain bacteria can influence the activities of others, future research should investigate if the activity of these cholesterol lowering probiotics may be antagonized by other bacteria already inhabiting the body especially in patients who may have multiple disease conditions. Manufactures will also have to indicate whether patients consuming some other probiotics to treat certain conditions may as well take their products and still experience the intended health effects.

Conclusions

The ability for probiotics and prebiotics to control cholesterol levels is fascinating and has attracted a lot of attention over the last decade. Many *in vivo* studies in animals as well as humans continue to confirm the cholesterol lowering ability of probiotics and prebiotic supplements. However, the mechanisms by which these control cholesterol levels in the body are unclear. Though there are many proposed hypotheses, most of these are yet to be confirmed *in vivo* and clinical trials. However, the understanding of the species-specific mechanisms of action by which probiotics influence cholesterol biosynthesis is still open for investigation. It is hoped that some of the postulates from *in vitro* studies may serve as guides to unveiling the actual mechanisms *in vivo* in the near future. The application of genomics, transcriptomics and proteomics may help reveal a deeper and clearer understanding of the molecular basis of probiotic and prebiotic functionality, providing e.g. a clear picture of proteins that are involved in cholesterol binding and mobbing. Probiotic and prebiotic research is still a young field and as the search for the cause and effect of their activities unveil, they may soon become better alternatives to cholesterol lowering drugs due to their safety and relatively low cost.

References

Ahire, J.J., Bhat, A.A., Thakare, J.M., Pawar, P.B., Zope, D.G., Jain, R.M., and Chaudhari, B.L. (2012). Cholesterol assimilation and biotransformation by *Lactobacillus* helveticus. Biotechnol. Lett. *34*, 103–107.

Aihara, K., Azuma, H., Ikeda, Y., Akaike, M., Abe, M., Sugihara, T., and Matsumoto, T. (2001). Successful combination therapy – flunarizine, pentoxifylline, and cholestyramine – for spur cell anemia. Int. J. Hematol. *73*, 351–355.

Alex, S., Lange, K., Amolo, T., Grinstead, J.S., Haakonsson, A.K., Szalowska, E., Koppen, A., Mudde, K., Haenen, D., Al-Lahham, S., *et al.* (2013). Short-chain fatty acids stimulate angiopoietin-like 4 synthesis in human colon adenocarcinoma cells by activating peroxisome proliferator-activated receptor gamma. Mol. Cell. Biol. *33*, 1303–1316.

Alvaro, A., Sola, R., Rosales, R., Ribalta, J., Anguera, A., Masana, L., and Vallve, J.C. (2008). Gene expression analysis of a human enterocyte cell line reveals downregulation of cholesterol biosynthesis in response to short-chain fatty acids. IUBMB Life *60*, 757–764.

Anandharaj, M., Sivasankari, B., and Parveen Rani, R. (2014). Effects of probiotics, prebiotics, and synbiotics on hypercholesterolemia: a review. Chinese Journal of Biology *2014*, 1–7.

Anderson, J.W., and Gilliland, S.E. (1999). Effect of fermented milk (yoghurt) containing *Lactobacillus* acidophilus L1 on serum cholesterol in hypercholesterolemic humans. J. Am. Coll. Nutr. *18*, 43–50.

Beher, W.T., Baker, G.D., Anthony, W.L., and Penney, D.G. (1964). Effect of hyocholic acid on cholesterol and bile acid metabolism in the mouse. Proc. Soc Exp. Biol. Med. *116*, 442–445.

Bengmark, S. (2008). Synbiotics in human medicine. In Therapeutic Microbiology: Probiotics and Related Strategies, J. Versalovic, and M. Wilson, eds. (Washington, ASM Press), pp. 307–321.

Berndt-Zipfel, C., Michelson, G., Dworak, M., Mitry, M., Loffler, A., Pfutzner, A., and Forst, T. (2013). Vildagliptin in addition to metformin improves retinal blood flow and erythrocyte deformability in patients with type 2 diabetes mellitus – results from an exploratory study. Cardiovasc Diabetol *12*, 59.

Bhathena, J., Martoni, C., Kulamarva, A., Tomaro-Duchesneau, C., Malhotra, M., Paul, A., Urbanska, A.M., and Prakash, S. (2013). Oral probiotic microcapsule formulation ameliorates non-alcoholic fatty liver disease in Bio F1B Golden Syrian hamsters. PLoS One *8*, 12.

Bordoni, A., Amaretti, A., Leonardi, A., Boschetti, E., Danesi, F., Matteuzzi, D., Roncaglia, L., Raimondi, S., and Rossi, M. (2013). Cholesterol-lowering probiotics: *in vitro* selection and *in vivo* testing of bifidobacteria. Appl. Microbiol. Biotechnol. *97*, 8273–8281.

Brighenti, F., Casiraghi, M.C., Canzi, E., and Ferrari, A. (1999). Effect of consumption of a ready-to-eat breakfast cereal containing inulin on the intestinal milieu and blood lipids in healthy male volunteers. Eur. J. Clin. Nutr. *53*, 726–733.

Buchwald, H., O'Dea, T.J., Menchaca, H.J., Michalek, V.N., and Rohde, T.D. (2000). Effect of plasma cholesterol on red blood cell oxygen transport. Clin. Exp. Pharmacol. Physiol. *27*, 951–955.

Cai, S.Y., Xiong, L., Wray, C.G., Ballatori, N., and Boyer, J.L. (2007). The farnesoid X receptor FXRalpha/NR1H4 acquired ligand specificity for bile salts late in vertebrate evolution. Am. J. Physiol. Regul. Integr. Comp. Physiol. *293*, R1400–1409.

Causey, J.L., Feirtag, J.M., Gallaher, D.D., Tungland, B.C., and Slavin, J.L. (2000). Effects of Dietary Inulin on Serum Lipids, Blood Glucose and the Gastrointestinal Environment in Hypercholesterolemic Men. Nutrition Research *20*, 191–201.

Chae, J.P., Valeriano, V.D., Kim, G.B., and Kang, D.K. (2013). Molecular cloning, characterization and comparison of bile salt hydrolases from *Lactobacillus* johnsonii PF01. J. Appl. Microbiol. *114*, 121–133.

Cooper, R.A. (1969). Anemia with spur cells: a red cell defect acquired in serum and modified in the circulation. J. Clin. Invest. *48*, 1820–1831.

Danielsson, H., and Gustafsson, B.E. (1959). On serum cholesterol levels and neutral fecal sterols in germfree rats. Bile acids and steroids 59. Arch. Biochem. Biophys. *83*, 482–485.

Drozdowski, L.A., Reimer, R.A., Temelli, F., Bell, R.C., Vasanthan, T., and Thomson, A.B. (2010). Beta-glucan extracts inhibit the *in vitro* intestinal uptake of long-chain fatty acids and cholesterol and down-regulate genes involved in lipogenesis and lipid transport in rats. J. Nutr. Biochem. *21*, 695–701.

Duval, C., Touche, V., Tailleux, A., Fruchart, J.C., Fievet, C., Clavey, V., Staels, B., and Lestavel, S. (2006). Niemann-Pick C1 like 1 gene expression is down-regulated by LXR activators in the intestine. Biochem. Biophys. Res. Commun *340*, 1259–1263.

Efrati, H., Rottem, S., and Razin, S. (1981). Lipid and protein membrane components associated with cholesterol uptake by Mycoplasmas. Biochim. Biophys. Acta *641*, 386–394.

Ejtahed, H.S., Mohtadi-Nia, J., Homayouni-Rad, A., Niafar, M., Asghari-Jafarabadi, M., Mofid, V., and Akbarian-Moghari, A. (2011). Effect of probiotic yoghurt containing *Lactobacillus acidophilus* and *Bifidobacterium lactis* on lipid profile in individuals with type 2 diabetes mellitus. J. Dairy Sci. *94*, 3288–3294.

Foglia, A. (2010). The acanthocyte-echinocyte differential: The example of chorea-acanthocytosis. Swiss Med. Wkly *16*, 13039.

Fuentes, M.C., Lajo, T., Carrion, J.M., and Cune, J. (2013). Cholesterol-lowering efficacy of *Lactobacillus plantarum* CECT 7527, 7528 and 7529 in hypercholesterolaemic adults. Br. J. Nutr. *109*, 1866–1872.

Gorenjak, M., Gradisnik, L., Trapecar, M., Pistello, M., Kozmus, C.P., Skorjanc, D., Skok, P., Langerholc, T., and Cencic, A. (2014). Improvement of lipid profile by probiotic/protective cultures: study in a non-carcinogenic small intestinal cell model. New Microbiol. 37, 51–64.

Guo, C.F., and Li, J.Y. (2013). Hypocholesterolaemic action of *Lactobacillus casei* F0822 in rats fed a cholesterol-enriched diet. Int. Dairy J. *32*, 144–149.

Guo, L.D., Yang, L.J., and Huo, G.C. (2011). Cholesterol removal by *Lactobacillus plantarum* isolated from homemade fermented cream in Inner Mongolia of China. Czech J. Food Sci. *29*, 219–225.

Han, K.H., Fukushima, M., Kato, T., Kojima, M., Ohba, K., Shimada, K., Sekikawa, M., and Nakano, M. (2003). Enzyme-resistant fractions of beans lowered serum cholesterol and increased sterol excretions and hepatic mRNA levels in rats. Lipids *38*, 919–924.

Harrison, V.C., and Peat, G. (1975). Serum cholesterol and bowel flora in the newborn. Am. J. Clin. Nutr. *28*, 1351–1355.

Hu, X., Wang, T., Li, W., Jin, F., and Wang, L. (2013). Effects of NS *Lactobacillus* strains on lipid metabolism of rats fed a high-cholesterol diet. Lipids Health Dis. *12*, 12–67.

Huang, Y., Wang, X., Wang, J., Wu, F., Sui, Y., Yang, L., and Wang, Z. (2013a). *Lactobacillus plantarum* strains as potential probiotic cultures with cholesterol-lowering activity. J. Dairy Sci. *96*, 2746–2753.

Huang, Y., Wu, F., Wang, X., Sui, Y., Yang, L., and Wang, J. (2013b). Characterization of *Lactobacillus plantarum* Lp27 isolated from Tibetan kefir grains: a potential probiotic bacterium with cholesterol-lowering effects. J. Dairy Sci. *96*, 2816–2825.

Huang, Y., and Zheng, Y. (2010). The probiotic *Lactobacillus acidophilus* reduces cholesterol absorption through the down-regulation of Niemann-Pick C1-like 1 in Caco-2 cells. Br. J. Nutr. *103*, 473–478.

Iannuzzi, A., Bianciardi, G., Faccenda, F., Gnasso, A., Scarpato, N., Di Marino, L., Iaccarino, G., Simoes, C., Sacchi, G., and Weber, E. (1997). Correction of erythrocyte shape abnormalities in familial hypercholesterolemia after LDL-apheresis: does it influence cerebral hemodynamics? Heart Vessels *12*, 234–240.

Jeun, J., Kim, S., Cho, S.Y., Jun, H.J., Park, H.J., Seo, J.G., Chung, M.J., and Lee, S.J. (2010). Hypocholesterolemic effects of *Lactobacillus plantarum* KCTC3928 by increased bile acid excretion in C57BL/6 mice. Nutrition *26*, 321–330.

Jia, L., Betters, J.L., and Yu, L. (2011). Niemann-pick C1-like 1 (NPC1L1). protein in intestinal and hepatic cholesterol transport. Annu. Rev. Physiol. 73, 239–259.

Jiang, J., Hang, X., Zhang, M., Liu, X., Li, D., and Yang, H. (2010). Diversity of bile salt hydrolase activities in different lactobacilli towards human bile salts. Ann. Microbiol. *60*, 81–88.

Jones, M.L., Martoni, C.J., Parent, M., and Prakash, S. (2012a). Cholesterol-lowering efficacy of a microencapsulated bile salt hydrolase-active *Lactobacillus reuteri* NCIMB 30242 yoghurt formulation in hypercholesterolaemic adults. Br. J. Nutr. *107*, 1505–1513.

Jones, M.L., Martoni, C.J., and Prakash, S. (2012b). Cholesterol lowering and inhibition of sterol absorption by *Lactobacillus reuteri* NCIMB 30242: a randomized controlled trial. Eur. J. Clin. Nutr. *66*, 1234–1241.

Jones, M.L., Tomaro-Duchesneau, C., Martoni, C.J., and Prakash, S. (2013). Cholesterol lowering with bile salt hydrolase-active probiotic bacteria, mechanism of action, clinical evidence, and future direction for heart health applications. Expert Opin. Biol. Ther. *13*, 631–642.

Kersten, S. 2014. Physiological regulation of lipoprotein lipase. Biochimica et Biophysica Acta *1841*, 919–933.

Kim, G.B., and Lee, B.H. (2005). Biochemical and molecular insights into bile salt hydrolase in the gastrointestinal microflora: a review. Asian Australasian J. Animal Sci. *18*, 1505–1512.

Kim, G.B., and Lee, B.H. (2008). Genetic analysis of a bile salt hydrolase in *Bifidobacterium animalis* subsp. *lactis* KL612. J. Appl. Microbiol. *105*, 778–790.

Kim, G.B., Miyamoto, C.M., Meighen, E.A., and Lee, B.H. (2004). Cloning and characterization of the bile salt hydrolase genes (bsh) from Bifidobacterium bifidum strains. Appl. Environ. Microbiol. *70*, 5603–5612.

Kimoto-Nira, H., Mizumachi, K., Nomura, M., Kobayashi, M., Fujita, Y., Okamoto, T., Suzuli, I., Tsuji, N.M., Kurisaki, J.I., and Ohmomo, S. (2007). *Lactobacillus* species as potential probiotic lactic acid bacteria. Jpn Agr. Res. Q. *41*, 181–189.

Knarreborg, A., Engberg, R.M., Jensen, S.K., and Jensen, B.B. (2002). Quantitative determination of bile salt hydrolase activity in bacteria isolated from the small intestine of chickens. Appl. Environ. Microbiol. *68*, 6425–6428.

Korecka, A., de Wouters, T., Cultrone, A., Lapaque, N., Pettersson, S., Dore, J., Blottiere, H.M., and Arulampalam, V. (2013). ANGPTL4 expression induced by butyrate and rosiglitazone in human intestinal epithelial cells utilizes independent pathways. Am. J. Physiol. Gastrointest. Liver Physiol. *304*, G1025–G1037.

Kosters, A., Kunne, C., Looije, N., Patel, S.B., Oude Elferink, R.P.J., and Groen, A.K. (2006). The mechanism of ABCG5/ABCG8 in biliary cholesterol secretion in mice. Journal of Lipid Research *47*, 1959–1966.

Koter, M., Franiak, I., Broncel, M., and Chojnowska-Jezierska, J. (2003). Effects of simvastatin and pravastatin on peroxidation of erythrocyte plasma membrane lipids in patients with type 2 hypercholesterolemia. Can. J. Physiol. Pharmacol. *81*, 485–492.

Kumar, D., Singh, G., Singh, M., Sharma, D., and Rathore, M.S. (2012a). Diet and functional foods in treatment and maintenance therapy of colon disorders. J. Appl. Pharmaceut. Sci. *2*, 177–187.

Kumar, M., Nagpal, R., Kumar, R., Hemalatha, R., Verma, V., Kumar, A., Chakraborty, C., Singh, B., Marotta, F., Jain, S., *et al.* (2012b). Cholesterol-lowering probiotics as potential biotherapeutics for metabolic diseases. Exp. Diabetes Res. *902917*, 3.

Kumar, M., Rakesh, S., Nagpal, R., Hemalatha, R., Ramakrishna, A., Sudarshan, V., Ramagoni, R., Shujauddin, M., Verma, V., Kumar, A., *et al.* (2013). Probiotic *Lactobacillus rhamnosus* GG and Aloe vera gel improve lipid profiles in hypercholesterolemic rats. Nutrition *29*, 574–579.

Kumar, R., Grover, S., and Batish, V.K. (2011). Hypocholesterolaemic effect of dietary inclusion of two putative probiotic bile salt hydrolase-producing

Lactobacillus plantarum strains in Sprague–Dawley rats. Br. J. Nutr. *105*, 561–573.

Kuwai, T., and Hayashi, J. (2006). Nitric oxide pathway activation and impaired red blood cell deformability with hypercholesterolemia. J. Atheroscler Thromb *13*, 286–294.

Larkin, T.A., Astheimer, L.B., and Price, W.E. (2009). Dietary combination of soy with a probiotic or prebiotic food significantly reduces total and LDL cholesterol in mildly hypercholesterolaemic subjects. Eur. J. Clin. Nutr. *63*, 238–245.

Lew, J.L., Zhao, A., Yu, J., Huang, L., De Pedro, N., Pelaez, F., Wright, S.D., and Cui, J. (2004). The farnesoid X receptor controls gene expression in a ligand- and promoter-selective fashion. J. Biol. Chem. *279*, 8856–8861.

Liong, M.T., Dunshea, F.R., and Shah, N.P. (2007). Effects of a synbiotic containing *Lactobacillus* acidophilus ATCC 4962 on plasma lipid profiles and morphology of erythrocytes in hypercholesterolaemic pigs on high- and low-fat diets. Br. J. Nutr. *98*, 736–744.

Lye, H.-S., Rahmat-Ali, G.R., and Liong, M.-T. (2010a). Mechanisms of cholesterol removal by lactobacilli under conditions that mimic the human gastrointestinal tract. International Dairy Journal *20*, 169–175.

Lye, H.S., Rusul, G., and Liong, M.T. (2010b). Removal of cholesterol by lactobacilli via incorporation and conversion to coprostanol. J. Dairy Sci. *93*, 1383–1392.

Mahrous, H. (2011). Probiotics bacteria from Egyptian infants cause cholesterol removal in media and survive in yoghurt. Food Nutr. Sci. *2*, 150–155.

Makishima, M., Okamoto, A.Y., Repa, J.J., Tu, H., Learned, R.M., Luk, A., Hull, M.V., Lustig, K.D., Mangelsdorf, D.J., and Shan, B. (1999). Identification of a nuclear receptor for bile acids. Science *284*, 1362–1365.

Mann, G.V. (1974). Studies of a surfactant and cholesteremia in the Maasai. Am. J. Clin. Nutr. *27*, 464–469.

Marcil, V., Delvin, E., Garofalo, C., and Levy, E. (2003). Butyrate impairs lipid transport by inhibiting microsomal triglyceride transfer protein in Caco-2 cells. J. Nutr. *133*, 2180–2183.

McAuliffe, O., Cano, R.J., and Klaenhammer, T.R. (2005). Genetic analysis of two bile salt hydrolase activities in *Lactobacillus* acidophilus NCFM. Appl. Environ. Microbiol. *71*, 4925–4929.

Melzak, K., Melzak, S., Gizeli, E., and Toca-Herrera, J. (2012). Cholesterol Organization in Phosphatidylcholine Liposomes: A Surface Plasmon Resonance Study. Materials *5*, 2306–2325.

Memon, L., Spasojevic-Kalimanovska, V., Jovic, P., Spasic, S., and Bogavac-Stanojevic, N. (2003). Determination of cholesterol in erythrocyte membrane. Jugoslovenska medicinska biohemija *22*, 213–219.

Mohandas, N., Clark, M.R., Jacobs, M.S., and Shohet, S.B. (1980). Analysis of factors regulating erythrocyte deformability. Journal of Clinical Investigation *66*, 563–573.

Mohandas, N., and Gallagher, P.G. (2008). Red cell membrane: past, present, and future. Blood *112*, 3939–3948.

Mott, G.E., Moore, R.W., Redmond, H.E., and Reiser, R. (1973). Lowering of serum cholesterol by intestinal bacteria in cholesterol-fed piglets. Lipids *8*, 428–431.

Narayanan, B. (2009). Influence of cholesterol on shape parameters of erythrocytes in hyperglycemic subjects. Turkish J. Hematol. *26*, 77–81.

Naruszewicz, M., Johansson, M.L., Zapolska-Downar, D., and Bukowska, H. (2002). Effect of *Lactobacillus plantarum* 299v on cardiovascular disease risk factors in smokers. Am. J. Clin. Nutr. *76*, 1249–1255.

Ohvo-Rekila, H., Ramstedt, B., Leppimaki, P., and Slotte, J.P. (2002). Cholesterol interactions with phospholipids in membranes. Prog Lipid Res. *41*, 66–97.

Ooi, L.G., Ahmad, R., Yuen, K.H., and Liong, M.T. (2010a). *Lactobacillus* acidophilus CHO-220 and inulin reduced plasma total cholesterol and low-density lipoprotein cholesterol via alteration of lipid transporters. Journal of Dairy Science 93, 5048–5058.

Ooi, L.G., Bhat, R., Rosma, A., Yuen, K.H., and Liong, M.T. (2010b). A synbiotic containing *Lactobacillus* gasseri [corrected] CHO-220 and inulin improves irregularity of red blood cells. J. Dairy Sci. 93, 4535–4544.

Ooi, L.G., and Liong, M.T. (2010). Cholesterol-lowering effects of probiotics and prebiotics: a review of *in vivo* and *in vitro* findings. Int. J. Mol. Sci. *11*, 2499–2522.

Owen, J.S., Bruckdorfer, K.R., Day, R.C., and McIntyre, N. (1982). Decreased erythrocyte membrane fluidity and altered lipid composition in human liver disease. J. Lipid Res. *23*, 124–132.

Pan, D.D., Zeng, X.Q., and Yan, Y.T. (2011). Characterization of *Lactobacillus fermentum* SM-7 isolated from koumiss, a potential probiotic bacterium with cholesterol-lowering effects. Journal of the science of food and agriculture *91*, 512–518.

Parmahamsa, M., Reddy, K.R., and Varadacharyulu, N. (2004). Changes in composition and properties of erythrocyte membrane in chronic alcoholics. Alcohol Alcohol *39*, 110–112.

Parnell, J.A., and Reimer, R.A. (2009). Effect of prebiotic fibre supplementation on hepatic gene expression and serum lipids: a dose–response study in JCR:LA-cp rats. British Journal of Nutrition *103*, 1577–1584.

Patel, A.K., Singhania, R.R., Pandey, A., and Chincholkar, S.B. (2010). Probiotic bile salt hydrolase: current developments and perspectives. Appl. Biochem. Biotechnol. *162*, 166–180.

Patterson, J.K., Lei, X.G., and Miller, D.D. (2008). The pig as an experimental model for elucidating the mechanisms governing dietary influence on mineral absorption. Exp. Biol. Med. *233*, 651–664.

Perova, N.V., and Metel'skaia, V.A. (2008). [Plant sterols and stanols as the dietary factors lowering hypercholesterolemia by inhibition of intestinal cholesterol absorption]. Kardiologiia *48*, 62–69.

Queenan, K., Stewart, M., Smith, K., Thomas, W., Fulcher, R.G., and Slavin, J. (2007). Concentrated oat beta-glucan, a fermentable fibre, lowers serum cholesterol in hypercholesterolemic adults in a randomized controlled trial. Nutr. J. *6*, 6.

Raghavan, C.M., Nanda, A., Yuvarage, R., Kumar, D.J.M., Murugan, A.S., and Raja, R.B. (2011). Assimilation of cholesterol by *Lactobacillus* species as probiotics. World Appl. Sci. J. *14*, 552–560.

Roberfroid, M. (2007). Prebiotics: the concept revisited. J. Nutr. *137*, 830S–837S.

Russell, D.W. (1999). Nuclear orphan receptors control cholesterol catabolism. Cell 97, 539–542.

Shepherd, J. (2001). The role of the exogenous pathway in hypercholesterolaemia. Eur. Heart J. Suppl. 3, E2–E5.

Siciliano, R.A., and Mazzeo, M.F. (2012). Molecular mechanisms of probiotic action: a proteomic perspective. Curr. Opin. Microbiol. *15*, 390–396.

Sridevi, N., Vishwe, P., and Prabhune, A. (2009). Hypocholesteremic effect of bile salt hydrolase from *Lactobacillus buchneri* ATCC 4005. Food Res. Int. *42*, 516–520.

Storch, A., and Schwarz, J. (2004). Diagnostic test for neuroacanthocytosis. In Neuroacanthocytosis Syndromes, A. Danek, ed. (Springer, Netherlands).

Tani, H., Sato, M., Tsuchiya, H., and Namikawa, I. (1993). Cholesterol incorporation intoBacillus megaterium without compositional modification of membrane lipids in response to changes of the membrane functions. Current Microbiology *26*, 253–256.

Trasino, S.E., Dawson, H.D., Urban, J.F., Jr., Wang, T.T., and Solano-Aguilar, G. (2013). Feeding probiotic *Lactobacillus paracasei* to Ossabaw pigs on a high-fat diet prevents cholesteryl-ester accumulation and LPS modulation of the Liver X receptor and inflammatory axis in alveolar macrophages. J. Nutr. Biochem. *24*, 1931–1939.

Wang, C.-Y., Wu, S.-C., Ng, C.-C., and Shyu, Y.-T. (2010). Effect of *Lactobacillus*-fermented adlay-based milk on lipid metabolism of hamsters fed cholesterol-enriched diet. Food Res. Int. *43*, 819–824.

Watanabe, S., Katsube, T., Hattori, H., Sato, H., Ishijima, T., Nakai, Y., Abe, K., and Sonomoto, K. (2013). Effect of *Lactobacillus* brevis 119-2 isolated from Tsuda kabu red turnips on cholesterol levels in cholesterol-administered rats. J. Biosci. Bioeng. *116*, 45–51.

Weinstein, R.S. (1974). The morphology of adult red cells. In The Red Blood Cell, D. Surgenor, ed. (New York, Academic Press), pp. 214–260.

WHO (2013). Cardiovascular diseases (CVDs).

Wostmann, B.S., Wiech, N.L., and Kung, E. (1966). Catabolism and elimination of cholesterol in germfree rats. J. Lipid Res. *7*, 77–82.

Xiao, J.Z., Kondo, S., Takahashi, N., Miyaji, K., Oshida, K., Hiramatsu, A., Iwatsuki, K., Kokubo, S., and Hosono, A. (2003). Effects of milk products fermented by *Bifidobacterium longum* on blood lipids in rats and healthy adult male volunteers. J. Dairy Sci. *86*, 2452–2461.

Yawata, Y. (2004). Composition of Normal Red Cell Membranes. In Cell Membrane (Wiley-VCH Verlag GmbH & Co. KGaA), pp. 27–46.

Yong, Z., and Heping, Z. (2013). The Effect of Probiotics on Lipid Metabolism. In Lipid Metabolism, R.V. Baez, ed. (InTech).

Yoon, H.-s., Ju, J.-h., Kim, H., Lee, J., Park, H.-j., Ji, Y., Shin, H.-k., Do, M.-S., Lee, J.-m. and Holzapfel, W. (2011). *Lactobacillus rhamnosus* BFE 5264 and *Lactobacillus plantarum* NR74 promote cholesterol excretion through the up-regulation of ABCG5/8 in Caco-2 Cells. Probiotics Antimicro Prot. 3, 194–203.

Yoon, H.S., Ju, J.H., Lee, J.E., Park, H.J., Lee, J.M., Shin, H.K., Holzapfel, W., Park, K.Y., and Do, M.S. (2013). The probiotic *Lactobacillus rhamnosus* BFE5264 and *Lactobacillus plantarum* NR74 promote cholesterol efflux and suppress inflammation in THP-1 cells. J. Sci. Food Agric. 93, 781–787.

Yu, L. (2008). The structure and function of Niemann-Pick C1-like 1 protein. Curr. Opin. Lipidol *19*, 263–269.

Yu, L., Li-Hawkins, J., Hammer, R.E., Berge, K.E., Horton, J.D., Cohen, J.C., and Hobbs, H.H. (2002). Overexpression of ABCG5 and ABCG8 promotes biliary cholesterol secretion and reduces fractional absorption of dietary cholesterol. J. Clin. Invest. *110*, 671–680.

Zhang, J., Johnson, P.C., and Popel, A.S. (2009). Effects of erythrocyte deformability and aggregation on the cell free layer and apparent viscosity of microscopic blood flows. Microvasc Res. 77, 265–272.

Zhong, Z., Zhang, W., Du, R., Meng, H., and Zhang, H. (2012). *Lactobacillus* casei Zhang stimulates lipid metabolism in hypercholesterolemic rats by affecting gene expression in the liver. Eur. J. Lipid Sci. Technol. *114*, 244–252.

Perspectives on Differences Between Human and Livestock Animal Research in Probiotics and Prebiotics

30

Tyler E. Askelson and Tri Duong

Abstract

Probiotics and prebiotics are used widely because of their reported benefits to digestive and immune health. While there is significant evidence to support their effectiveness in humans and livestock animals, interpretation of the results of this research is complicated by the wide differences in research performed in humans compared with livestock animals. This chapter will explore host-specific digestive physiology, experimental constraints, and probiotic and prebiotic functionality. The insight provided by an understanding of these critically important differences will provide a context in which results of host-specific studies and their broader implications to the science can be evaluated.

Introduction

Research in the application of probiotics and prebiotics in human and animal health has grown due to the increased interest in the gastrointestinal microbiota. The microbial community present in the gastrointestinal tract is thought to play an important role in host animal health and is a potentially important therapeutic target that can be manipulated in order to achieve positive health outcomes. Thus, the use of probiotics and prebiotics represents a powerful strategy for the manipulation of the microbial community in the gastrointestinal tract (Boyle *et al.*, 2006; Collins and Gibson, 1999; Preidis and Versalovic, 2009; Ziemer and Gibson, 1998).

Probiotics are live microorganisms, which when administered in adequate amounts confer a health benefit on the host (WHO/FAO, 2001), and prebiotics are selectively fermented ingredients that result in specific changes in the composition and/or activity of the gastrointestinal microbiota conferring benefit(s) upon host health (Gibson *et al.*, 2010). The benefits of probiotic and prebiotic application in humans and animals include inhibition of pathogens (Hoyos, 1999; Sgouras *et al.*, 2004), improved digestive function (Parvez *et al.*, 2006; Roberfroid, 2000), and modulation of immune responses (Kyriakis *et al.*, 1999; Rosenfeldt *et al.*, 2003; Sartor, 2004). In animal agriculture, probiotics and prebiotics are thought to be an important potential alternative to the use of antibiotic growth promoters (AGP) (Huyghebaert *et al.*, 2011; Yang *et al.*, 2009). When used in livestock, they have been demonstrated to promote growth performance at levels similar to AGPs (Mountzouris *et al.*, 2007; Shen *et al.*, 2009) and reduce gastrointestinal colonization by pathogens (Molbak *et al.*, 2007).

The benefits of probiotic and prebiotic use in humans and livestock animals have been well demonstrated (Table 30.1). However, the overall effectiveness of their application is thought to be mixed (Simmering and Blaut, 2001), and results of their use in livestock animals are perceived by some to be superior when compared to humans. A direct comparison of the overall effectiveness of probiotics and prebiotics between studies in humans and livestock animals, and between livestock animal species, is complicated by differences in the digestive physiology of the host animal, experimental constraints, and the desired experimental outcomes. Indeed, such a direct comparison may be inappropriate and irrelevant. An appreciation of host-specific differences and interpretation of research in this context will allow host-specific

Table 30.1 Examples of probiotic and prebiotic benefits

Type	Reference
Digestion	
Human	Anderson and Gilliland (1999), de Vrese *et al.* (2001), Mustapha *et al.* (1997)
Chicken	Askelson *et al.* (2014), Cavazzoni *et al.* (1998)
Cattle	Grand *et al.* (2013), Nocek and Kautz (2006), Nocek *et al.* (2003)
Pathogen Inhibition	
Human	Corr *et al.* (2007), Gueimonde *et al.* (2006)
Pig	Molbak *et al.* (2007)
Chicken	Neal-McKinney *et al.* (2012)
Immunomodulation	
Human	Kalliomaki *et al.* (2001), Rosenfeldt *et al.* (2003), Szajewska *et al.* (2001)
Pig	Kyriakis *et al.* (1999)
Chicken	Janardhana *et al.* (2009)
Cattle	Fleige *et al.* (2009), Sun *et al.* (2010)

studies to contribute to an improved understanding of probiotic and prebiotic functionality for application in both human and animal health.

Digestive physiology and microbiology

The autochthonous gastrointestinal microbiota of humans and animals is complex and dynamic with hundreds of microbial species coexisting in a web of interdependency and antagonism, both with each other and the host animal (Bradshaw *et al.*, 1994; Hooper *et al.*, 2002; Jami and Mizrahi, 2012). Host factors including age (Mackie *et al.*, 1999; Robert, 2012) and diet (Hemarajata and Versalovic, 2013; Robert, 2012) have been demonstrated to affect the composition of the gastrointestinal microbiota. Thus, introduction of allochthonous microbial species or specific dietary components to these already complex environments can cause shifts in the composition of the microbial community (Netherwood *et al.*, 1999; Tannock *et al.*, 2000) and potentially produce beneficial or negative effects on the host animal.

Ruminant and monogastric animals

The anatomic features and physiology of the digestive tract vary widely between humans, swine, cattle, and chickens (Fig. 30.1). As a result of these host-animal adaptations to dietary niches, not only is the gastrointestinal microbial community significantly different between species (Hopkins *et al.*, 2002; Jami and Mizrahi, 2012), but the digestive organs of primary research interest is species-specific as well (Table 30.2). While the predominant bacterial genera may be similar between host animal species, the microorganisms vary at the species and subspecies levels (Lu *et al.*, 2003; Suau *et al.*, 1999). Variation in the composition of the gastrointestinal microbial community between individuals of the same species is also significant (Robert, 2012). Thus, it is important to consider the interactions of probiotic cultures and prebiotics with the host-animal, the autochthonous microbiota, and the nutrition of the host animal.

The microbiota of cattle and other ruminant animals is considerably complex (Russell *et al.*, 1992) when compared with monogastric animals. The bovine rumen is populated by microorganisms which ferment indigestible plant material producing volatile fatty acids (VFA) and microbial crude protein for absorption by the host animal (Russell *et al.*, 1992). The predominant microorganisms found in the rumen include the Firmicutes *Butyrvibrio* and *Ruminococcus* and the Bacteroidete *Prevotella* (Jami and Mizrahi, 2012), while the predominant microorganisms of the lower intestines include strict anaerobic species from the genera *Bifidobacterium*, *Clostridium*, and *Bacteroides* (Dowd *et al.*, 2008). Additionally, differences in microbial composition of the rumen have been demonstrated between beef and dairy cattle. Beef cattle are populated by greater numbers of cellulitic bacteria including *Fibrobacter*, *Ruminococcus*, and *Succiniclasticum*, likely reflecting the need for greater fibre digestion of hay-fed beef cattle (Jami and Mizrahi, 2012; Robert, 2012). While *Prevotella* are the most abundant genus in the rumen of cattle, they are present in lower abundance in beef cattle than in dairy cows (Robert, 2012).

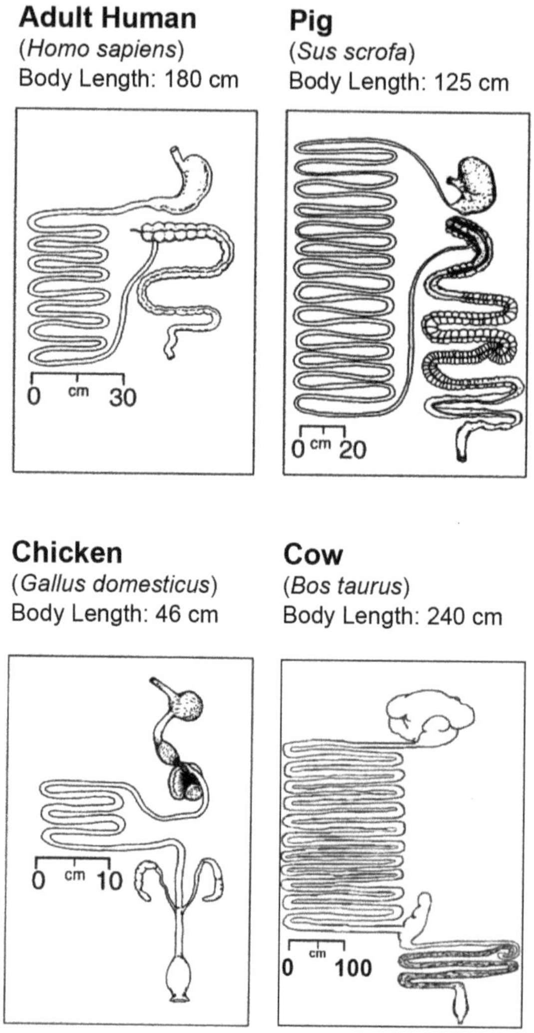

Figure 30.1 Digestive anatomy. Anatomic features of the human, pig, chicken, and cattle. [Human and chicken adapted from Comparative Physiology of the Vertebrate Digestive System with the permission of Cambridge University Press. Copyright (1995) Cambridge University Press. Pig adapted from the American Journal of Physiology (Argenzio and Southworth, 1975). Copyright 1976 American Physiological Society].

Differences between chickens, humans, and swine demonstrate interspecific variation in the gastrointestinal microbiota of monogastric animals. The chicken ileum is populated largely with *Lactobacillus*, followed distantly by *Clostridium* and *Enterococcus*, whereas *Clostridium* is more prevalent than *Lactobacillus* in the caecum (Lu *et al.*, 2003). In monogastric mammals, the human intestinal tract is dominated by *Bacteroides*, *Clostridium*, and *Bifidobacterium* (Hopkins *et al.*, 2002; Langendijk *et al.*, 1995; Suau *et al.*, 1999), while *Lactobacillus*, *Streptococcus*, and *Selenomonas* (Pryde *et al.*, 1999) are the predominant microbes in swine.

The administration of prebiotics has been demonstrated to stimulate growth of beneficial microorganisms in the gastrointestinal tract (Geier *et al.*, 2009; Gibson and Roberfroid, 1995; Haarman and Knol, 2006). However, the selectivity of specific prebiotic compounds and the host animal species should be considered. A study of prebiotic administration in poultry found *Lactobacillus salivarius* and *Lactobacillus acidophilus* to be the

Table 30.2 Features of relevant digestive organs*

Host	Organ	Cells per g or per ml	Organ contents	Reference
Human	Colon	3.2×10^{11}	220 g	Cummings *et al.* (1990), Van Houte and Gibbons (1966)
Pig	Colon	5.4×10^{10}	9 l	Butine and Leedle (1989), Dukes (1955)
	Caecum	2.8×10^{10}	1 l	
Chicken	Caecum	9.5×10^{10}	2 g	Barnes *et al.* (1972), Salanitro *et al.* (1974)
Cattle	Rumen	2.1×10^{10}	106 l	Habel (1975), Hungate (1966)

Adapted from Proc. Natl. Acad. Sci. U.S.A. (Whitman *et al.*, 1998) with permission of the publisher. Copyright (1998) National Academy of Sciences, U.S.A.

predominant lactobacilli in chickens administered fructooligosaccharide (FOS), while *Lactobacillus reuteri* was most prevalent in chickens administered mannanoligosaccharide (MOS) (Geier *et al.*, 2009). However, a human study found consumption of fructooligosaccharide resulted in an increase of *Bifidobacterium angulatum* (Langlands *et al.*, 2004). Additionally, host-specific differences in gastrointestinal transit time may limit the effectiveness of prebiotics. The effectiveness of prebiotics may be reduced in animals with shorter transit times due to the incomplete utilization of prebiotic substrates by gastrointestinal microorganisms (Child *et al.*, 2006).

Host specific experimental constraints

The limitations placed on research performed in livestock animals are different from those placed on research in humans. The effects of differences in these important experimental constraints likely contribute to perceptions of the effectiveness of probiotic and prebiotic application. Experimental conditions can be controlled more stringently in livestock animals than with human subjects. Thus, the confounding effects of uncontrolled or uncontrollable variables can be limited, reducing overall experimental variation, and increasing experimental power.

Experimental subjects

The nature of research subjects is an obvious factor that affects experimental design and results from probiotic and prebiotic research. Selective breeding of production livestock animals such as broiler chickens and Holstein cows has produced homogenous in-bred genetic lines with increased uniformity and production potential (Notter, 1999). As a result, homogeneity between livestock animal subjects is greater than in human subjects. An early study evaluating oligofructose administration in humans used eight subjects, seven men and one woman, ranging in age from 21 to 48 (Gibson *et al.*, 1995). A meta-analysis of sixteen studies investigating the effects of probiotic and prebiotic administration on lipid levels in humans (Pereira and Gibson, 2002) further demonstrates the wide variations in the gender composition, age range, and number of subjects that can be seen in human trials. A study of probiotic reduction of diarrhoea in weaning piglets used 256 piglets only days apart in age with an even gender distribution. Probiotic and prebiotic studies in chickens are able to use up to thousands of animals of the same age reared under identical experimental conditions (Timmerman *et al.*, 2006; Torres-Rodriguez *et al.*, 2007; Vicente *et al.*, 2007). Thus, experimental animals are typically nearly identical in age and have far greater size, weight, and genetic uniformity when compared to their human counterparts.

Protocol compliance

Strict adherence to experimental protocols is necessary for the control of variables and their influence on results. These variables are more easily maintained in livestock animals than with human subjects. Research animals are typically housed in facilities where environmental conditions including lighting, temperature, and access to food and water can be controlled. However, human volunteers are not typically subjected to similarly rigorous constraints. The host diet is a particularly important factor that must be considered in studies of probiotic and prebiotic effectiveness because of its effect on the gastrointestinal microbiota. Replicate groups of research animals are pen fed experimental rations formulated specifically for the study (Askelson *et al.*, 2014; Timmerman *et al.*, 2006), while the diet of human subjects is not nearly as easily controlled. Additionally, compliance with prescribed probiotic and prebiotic dose is more easily maintained for experimental animals than with human subjects. The removal of subjects found violating experimental protocols further reduces experimental power (Taylor *et al.*, 2007; Whorwell *et al.*, 2006).

Ethical considerations

Significant interest in the potential of probiotics and prebiotics to prevent and treat disease has developed. The differences in ethical constraints and their effects on experimental design and study results must be considered when evaluating probiotic and prebiotic effectiveness (Boyle *et al.*, 2006). When evaluating the potential of probiotics and prebiotics to prevent infection in livestock animals, direct challenge studies in which experimental subjects are administered infectious to lethal doses of pathogenic microorganisms can be performed (Asahara *et al.*, 2004; Pascual *et al.*, 1999). However, research in human subjects must rely on natural infection (Gionchetti *et al.*, 2003; Szajewska *et al.*, 2001) or direct challenge using attenuated pathogens (Fang *et al.*, 2000). Alternatively, challenge experiments may be performed using rodents (Asahara *et al.*, 2004; Chenoll *et al.*, 2011) as intermediate models prior to natural infection experiments in humans.

The health of human subjects and severity of the potential infection are additional factors for consideration in natural infection studies and an additional complication over research in livestock animals. Two particularly interesting studies serve as examples highlighting the varying effectiveness of studies performed in healthy compared with ill adults. A study using healthy human volunteers performed during the common cold season found consumption of a probiotic cocktail containing *Lactobacillus gasseri* PA 16/8, *Bifidobacterium longum* SP 07/3, and *Bifidobacterium bifidum* MF 20/5 reduced the duration and severity of cold symptoms (de Vrese *et al.*, 2005). A study evaluating the potential of probiotics to prevent infectious complications in patients with severe pancreatitis found no difference in the incidence of infectious complications (Besselink *et al.*, 2008). However, the number of deaths was almost three times greater in the probiotic treated group than in the placebo group. It has been reported that known complications of the experimental probiotic treatment were not disclosed to the study subjects (ScienceInsider, 2009).

Probiotic and prebiotic functionality in humans and livestock

Probiotics and prebiotics are used in order to derive a wide array of health benefits for the human or livestock animal host including increased resistance to intestinal pathogens, improved immune health, and improved digestive function. The specific functionalities which provide these benefits are similar regardless of host species. However, the outcomes desired from probiotic and prebiotic use and the corresponding experimental endpoints used to measure their effectiveness are not universal across host species. Additionally, while improved health is the primary motivation for the use of probiotics and prebiotics in humans, their use in livestock animals is motivated primarily by the economic need for improved livestock production and performance parameters.

Digestion and metabolism

Consumption of probiotics and prebiotics has long been known to exert positive effects on digestion and digestive function in humans and animals. These general benefits to digestion and their contribution to nutrition and general quality of life are perhaps the most widely understood. However, their application in digestion is continuing to expand, and novel applications of this functionality are being explored. Additionally, the effects of probiotics and prebiotics on host animal metabolism have only recently begun to be understood and represent an additional novel application area in human and animal health.

Probiotics have been demonstrated to improve digestibility of food, reducing negative effects of maldigestion while increasing the available nutrient content. Lactose intolerance is a common maldigestive disorder in humans. After the consumption of milk and dairy products, fermentation of undigested lactose by microorganisms in the large intestine causes discomfort to the host. The production of β-galactosidase by probiotic cultures in yoghurt has been demonstrated to improve lactose digestion and tolerance (de Vrese *et al.*, 2001; Martini *et al.*, 1991). Consumption of unfermented milk containing *B. longum* B6, *B. longum* ATCC 15708 (Jiang *et al.*, 1996), and *L. acidophilus* N1 (Mustapha *et al.*, 1997) has also been demonstrated to improve lactose digestion. Phytic acid is a phosphorus source in livestock animal feeds (Dalal, 1977; Reddy *et al.*, 1982) that is underutilized in non-ruminant livestock due to its poor digestibility (Jongbloed *et al.*, 1992; Nelson, 1967; Selle *et al.*, 2006). Additionally, phytic acid exerts anti-nutritive effects through strong chelation of divalent cations. Administration of phytate-degrading *Lactobacillus* was demonstrated to improve weight gain of broiler chicks fed a phosphorous-deficient diet to a level comparable to those fed a phosphorus-adequate diet (Fig. 30.2) (Askelson *et al.*, 2014). In addition to increasing bioavailability of phytate-phosphorus in monogastric livestock animals, phytate-degrading probiotic cultures may be useful in correcting malabsorption syndrome in human vegetarians (Famularo *et al.*, 2005).

Improved digestive function is an area that has seen significant increases in product advertising and attention in popular press. Administration of *Bifidobacterium animalis* DN173010 shortened intestinal transit time in women (Marteau *et al.*, 2002). Consumption of a probiotic beverage containing *Lactobacillus casei* Shirota (Koebnick *et al.*, 2003) improved gastrointestinal symptoms in patients with chronic constipation. Additionally, a study in children found treatment with *Lactobacillus casei rhamnosus* Lcr35 (Bekkali *et al.*, 2007; Bu *et al.*, 2007) to have similar efficacy to and causing less abdominal pain than a magnesium oxide laxative.

Results of several studies suggest probiotics and prebiotics can affect lipid metabolism and potentially reduce risk factors of coronary disease in humans. Probiotics have been reported to reduce serum cholesterol (Mann, 1974). While the mechanism is not clearly understood, it has been hypothesized that the probiotic microorganisms may metabolize cholesterol and bile salts (Pereira and Gibson, 2002); see also elsewhere in this book. In two separate studies, consumption of fermented milk containing *L. acidophilus* L1 reduced serum cholesterol (Anderson and Gilliland, 1999). Similarly, serum low density lipoprotein (LDL) cholesterol was reduced by 5.4% in male volunteers who consumed yoghurt containing two *L. acidophilus* strains (DN 112.053 and DN 112.096) and FOS (Schaafsma *et al.*, 1998).

While interest in using probiotics and prebiotics to alter lipid metabolism is not limited to humans, interest in livestock production is driven by production concerns. Excess fat deposition in broiler chickens is undesirable to producers because of reduced carcass yield and to consumers that prefer a leaner product. A probiotic mixture containing 12 *Lactobacillus* strains improved body weight gain and feed conversion and reduced abdominal fat deposition in broiler chickens (Kalavathy *et al.*, 2003). Similar benefits of probiotic administration have also been demonstrated in egg production. *Pediococcus acidilactici* MA18/5M supplementation in egg-laying hens reduced cholesterol content and increased concentrations of polyunsaturated fatty acids, including linoleic acid and linolenic acid, in egg yolks (Mikulski *et al.*, 2012). A probiotic product containing *L. acidophilus* NP51 and *Propionibacterium freudenreichii* NP24 increased milk fat percentage when given to dairy cattle (Boyd *et al.*, 2011). While similar lipid increasing effects were seen from *Bacillus subtilis* natto in a separate study (Sun *et al.*, 2013). Increasing, rather than decreasing milk fat percentage is desirable to dairy producers because of downstream processing needs.

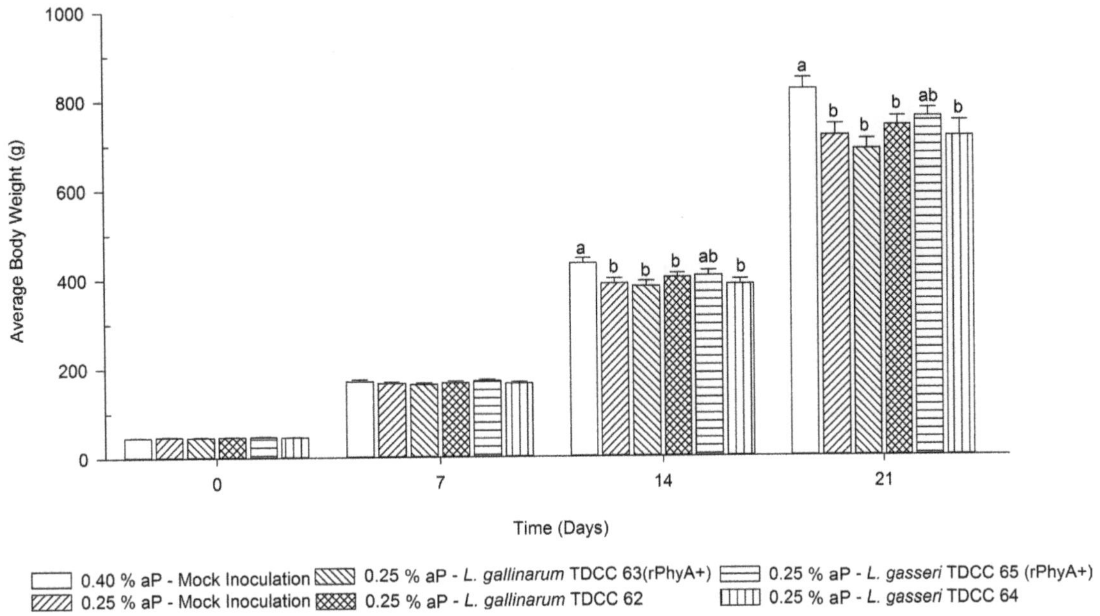

Figure 30.2 Phytate-degrading *Lactobacillus* improves growth of broiler chickens. Broiler chickens were fed a phosphorus adequate control diet (0.40% aP) and administered a mock inoculation, or fed a phosphorus-deficient diet (0.25% aP) and administered either a mock inoculation (MRD) or cultures of *L. gallinarum* TDCC 63 (rPhyA+), *L. gallinarum* TDCC 62 (Empty Vector), *L. gasseri* TDCC 65 (rPhyA+), or *L. gasseri* TDCC 64 (Empty Vector). Data shown are the mean body weight ± SEM of treatment groups. Different letters indicate means are significantly different ($P < 0.05$). [Adapted from Appl. Environ. Microbiol. (Askelson *et al.*, 2012)] with permission of the publisher. Copyright (2014) American Society for Microbiology)].

Pathogen inhibition

The ability of probiotics and prebiotics to reduce colonization of bacterial pathogens in the gastrointestinal tract is an important functionality useful for both human and animal health. Probiotic microbes are able to inhibit pathogens in the gastrointestinal tract through several mechanisms. Probiotic microorganisms competitively exclude pathogens from attachment to mucosal surfaces through competition for shared binding sites (Lu and Walker, 2001) and steric interference of protein adhesins located on the surface of pathogenic bacteria (Bernet *et al.*, 1994; Lee and Puong, 2002). In addition to preventing adhesion, an *in vitro* study demonstrated the varying ability of *L. acidophilus* TMC 0356 and *Lactobacillus rhamnosus* TMC 0503 to displace *Salmonella* Typhimurium, *Cronobacter sakazakii, Clostridium difficile*, and *Escherichia coli* which were already adherent to human epithelial cells (Gueimonde *et al.*, 2006). Inhibition of pathogen adherence is also seen in a pig intestinal mucosa model (Collado *et al.*, 2007). These studies demonstrate probiotic strain and host specific inhibition of pathogens, highlighting need for case-by-case selection of probiotic cultures to reduce adherence of specific pathogens.

The production of pathogen inhibiting compounds is a well understood probiotic mechanism (Oelschlaeger, 2010). Neal-McKinney *et al.* (Neal-McKinney *et al.*, 2012) demonstrated that the production of lactic acid by *Lactobacillus* cultures is an important mechanism for the reduction of *Campylobacter jejuni* in livestock animals (Fig. 30.3). Hydrogen peroxide production by *Lactobacillus* has also been shown to inhibit *Salmonella* (Pridmore *et al.*, 2008). *Campylobacter* (Duong and Konkel, 2009) and *Salmonella* (Guard-Petter, 2001) are common inhabitants of the gastrointestinal tract of poultry and important human foodborne pathogens. The use of probiotics and prebiotics to reduce the incidence of these organisms is motivated primarily by public health and food safety concerns rather than for the benefit of the animal. The production of bacteriocins by probiotics has the potential to prevent gastrointestinal infection in humans. A direct challenge study in mice demonstrated that bacteriocin production by *Lactobacillus salivarius* UCC118 reduced counts of *Listeria monocytogenes* by 80% in the liver and spleen of infected mice relative to a negative control (Fig. 30.4) (Corr *et al.*, 2007). Additionally, while *L. salivarius* UCC118 also protected mice from infection by *Salmonella* Typhimurium, the protection was not bacteriocin mediated. While bacteriocin production by probiotic cultures is hoped to be an important alternative to antibiotics in the treatment of bacterial infections, the effectiveness of this mechanism has not yet been evaluated in humans.

Immune modulation

Significant interest has grown in the immune modulating effects of probiotics and prebiotics. Probiotics are able to directly stimulate immune responses and regulate inflammation in a strain specific manner. Whereas prebiotic immunomodulation is thought to occur primarily in collaboration with commensal microorganisms (Janardhana *et al.*, 2009), there is evidence suggesting that carbohydrate polymers can directly interact with cells of the immune system (Wismar *et al.*, 2011). See also elsewhere in this book.

In addition to the direct action against pathogens described previously in this work, probiotics have been demonstrated to help protect the host from pathogens by enhancing host defences.

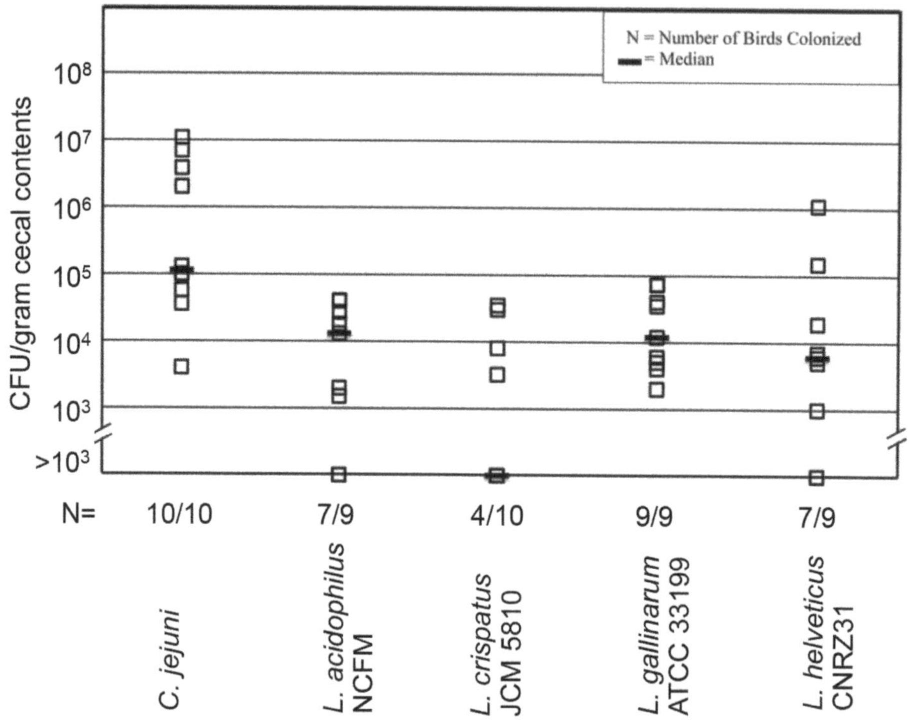

Figure 30.3 Probiotic *Lactobacillus* cultures reduce *Campylobacter jejuni* colonization of broiler chicks. Campylobacter was enumerated from the caecum of broiler chicks inoculated with probiotic *Lactobacillus* cultures or a mock inoculation and experimentally challenged with C. jejuni. [Adapted from PLoS One (Neal-McKinney *et al.*, 2012) under the terms of the Creative Commons Attribution License. Copyright (2012) Neal-McKinney *et al.*].

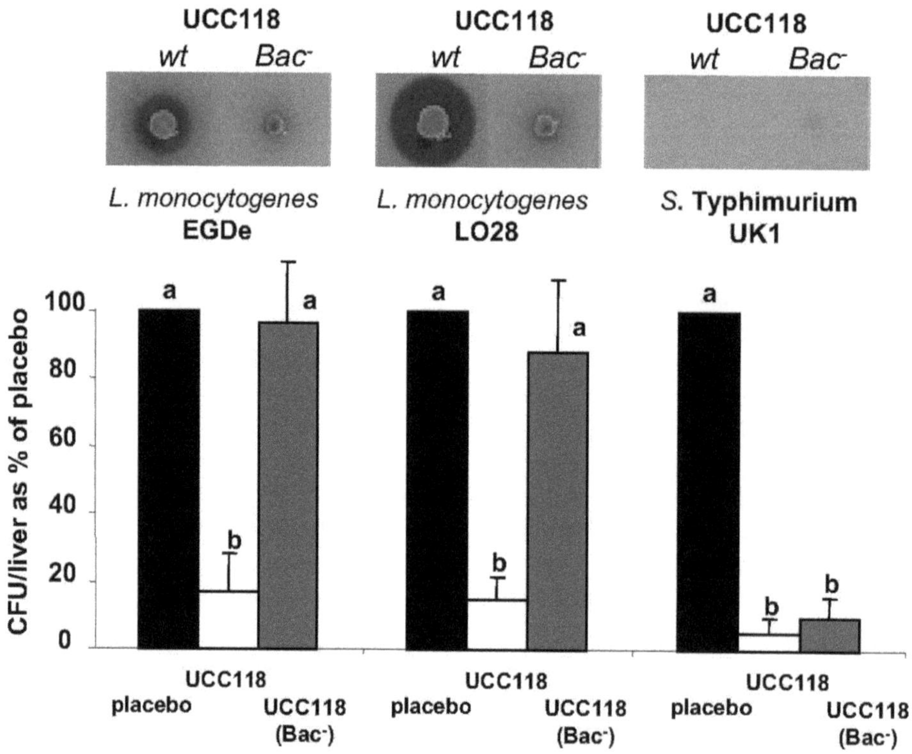

Figure 30.4 *Lactobacillus salivarius* UCC118 bacteriocin Abp118 mediates protection against *Listeria monocytogenes* infection. Upper, well diffusion assay. Wild type (wt) and Abp118-deficient (Bac⁻) *L. salivarius* UCC118 cultured in media seeded with *L. monocytogenes* and *Salmonella* Typhimurium. Lower, pathogens were enumerated from the liver of mice administered a placebo (filled bars), wild-type *L. salivarius* UCC118 (open bars), or Bacteriocin-deficient (Bac⁻) *L. salivarius* UCC118 (grey bars) and infected with *L. monocytogenes* EGDe, *L. monocytogenes* LO28, or *Salmonella* Typhimurium. Different letters indicate means differ significantly ($P < 0.001$; $n = 5$). [Reprinted from the Proc. Natl. Acad. Sci. U.S.A. (Corr *et al.*, 2007) with permission of the publisher. Copyright (2007) National Academy of Sciences, U.S.A.].

The probiotic cocktail VSL #3 improved epithelial barrier function and prevented *Salmonella* invasion of epithelial cells (Madsen *et al.*, 2001), while *L. casei* CHCC3139 stimulated the production of cytokines responsible for cell-mediated immunity (Christensen *et al.*, 2002; Hart *et al.*, 2004). A synbiotc combination of a commercial direct fed microbial containing *Enterococcus faecium* (strain unknown) and prebiotic lactulose increased mRNA expression of IgA Fc receptor when administered to calves (Fleige *et al.*, 2009). Human infants fed formula containing FOS or galactooligosaccharide (GOS) had greater faecal secretory IgA than infants fed control formula (Bakker-Zierikzee *et al.*, 2006).

Probiotic cultures including *Lactobacillus* species (Matsumoto *et al.*, 2005; Pelto *et al.*, 1998) and prebiotics including mannanoligosaccharide (Nelson *et al.*, 1991) have been demonstrated to reduce expression of inflammatory cytokines. Management of inflammatory bowel disorders (Vanderpool *et al.*, 2008) and atopic allergy are areas where the anti-inflammatory effects of probiotics and prebiotics show promise. Several studies demonstrate the potential of probiotic bacteria (Bibiloni *et al.*, 2005; Venturi *et al.*, 1999) and yeast (Guslandi *et al.*, 2003) to induce remission of active ulcerative colitis. Atopic diseases including atopic eczema, allergic asthma, and allergic rhinitis are allergic hypersensitivities. Immunomodulatory probiotics may be useful in preventing or reducing the severity of atopic disease (Kalliomaki *et al.*, 2001). One double blind placebo controlled study observed diminished clinical signs and symptoms of atopy in infants receiving *L. rhamnosus* GG and *Bifidobacterium lactis* BB-12 (Isolauri *et al.*, 2000). Oral bacteriotherapy with *L. rhamnosus* 19070–2 and *L. reuteri* DSM 122460 reduced proinflammatory eosinophil cationic protein in older children with atopic dermatitis (Rosenfeldt *et al.*, 2003). Whereas probiotics and prebiotics may exhibit anti-inflammatory properties in livestock animals, management of chronic intestinal inflammation and atopic disease is likely to be more applicable in humans than in livestock.

Future perspectives

Diarrhoeal disease

The morbidity and mortality in humans (Kyne *et al.*, 2002; Loo *et al.*, 2005) and losses to livestock production (Fairbrother *et al.*, 2005; Loneragan *et al.*, 2005) due to bacterial and viral enteric disease is significant, making the reduction of diarrheal disease in humans and livestock an important global health objective. The combined pathogen inhibiting and immune stimulating functionality of probiotics and prebiotics suggests they may be able to contribute to achieving this important goal. Nosocomial infectious diarrhoea is an important problem in children, prolonging hospital stays and increasing medical costs (Ponce *et al.*, 1995). Several studies have demonstrated the ability of probiotic *L. rhamnosus* GG to reduce the incidence of diarrhoea in children (Szajewska *et al.*, 2001; Szajewska and Mrukowicz, 2001). Additionally, the efficacy of several probiotics in preventing antibiotic-associated diarrhoea (McFarland, 2006) and Travellers' Diarrhoea (McFarland, 2007) has also been shown. Similar reduction of diarrhoea in swine and cattle have also been demonstrated (Abe *et al.*, 1995). Administration of probiotic *Bacillus* cultures reduced post-weaning diarrhoea syndrome related mortality (Kyriakis *et al.*, 1999) in piglets. Additional studies have shown probiotic-mediated reduction of diarrhoea and complications from pathogenic *E. coli* in piglets (Lee *et al.*, 2012; Schroeder *et al.*, 2006) similar to *L. rhamnosus* GG in infants.

Agricultural sustainability

Rapid growth in the use of renewable biofuels has led to a reallocation of arable land from food to fuel ethanol production (Donohue and Cunningham, 2009) and forced livestock producers to becoming increasingly reliant upon secondary feedstocks that are poorly digested with lower available nutrient content. The ability of probiotics to increase digestibility and nutrient utilization from poorly digested feed constituents through biocatalysis (Askelson *et al.*, 2014) in the gastrointestinal tract will become important as feed costs continue to increase. Additionally, the immune stimulating functionality of probiotics and prebiotics will continue to be important in disease management and potentially reduce losses in livestock production (Fuller, 1989; Patterson and Burkholder, 2003; Reid and Friendship, 2002). Thus, by increasing efficiency and overall productivity, probiotics and prebiotics have the potential to make important contributions to agricultural sustainability and global nutritional security.

AGP alternatives

AGPs have been widely used to increase weight gain (Moore *et al.*, 1946), improve feed efficiency (Emborg *et al.*, 2002; Gaskins *et al.*, 2002), and reduce mortality in livestock animal production (Callesen, 2003; Cromwell, 2002). Regulatory and consumer concerns over the development of antibiotic resistant microbes have led to a decline in the non-therapeutic use of antibiotics (Casewell *et al.*, 2003). However, the reduction of their use has led to reduced livestock productivity and increased disease in production animals. Supplementation of livestock feed with probiotics and prebiotics has been demonstrated to improve growth performance to levels similar to antibiotics, thus they represent an important alternative to the use of AGPs in livestock production (Huyghebaert *et al.*, 2011). However, while the use of probiotic cultures with growth promoting properties are desirable in livestock animal production, their use in humans is undesirable due to the growing worldwide obesity epidemic.

Vaccine and anti-infective delivery

Because of their long history of safe use, immune stimulating functionality, and importance in human and animal health, probiotic microorganisms have received considerable interest as potential vectors for the delivery of vaccines antigens to mucosal surfaces (Duong, 2010). An increasing number of studies have investigated the potential of live probiotic microorganisms as vaccines with several of examples in which they have elicited antigen specific immune responses (Table 30.3). Additionally, 'bioshield' strategies in which probiotic microorganisms confer passive immunity through expression of receptor proteins or antibody fragments are also being explored (Wells and Mercenier, 2008). The use of recombinant probiotic cultures for this novel

Table 30.3 Examples of vaccines delivered by probiotic microorganisms

Organism	Antigen	Host	Reference
Lactobacillus acidophilus NCFM	*B. anthracis* PA	Mouse	Mohamadzadeh *et al.* (2009)
	HIV-1 Gag	Mouse	Kajikawa *et al.* (2012)
Lactobacillus casei ATCC 393	CSFV epitope 290, Parvovirus VP2	Pig	Xu *et al.* (2011)
Lactobacillus plantarum NCIMB8826	Dust mite allergy Derp p 1	Mouse	Rigaux *et al.* (2009)
	C. tetani TTFC	Mouse	Grangette *et al.* (2001)
Lactococcus lactis E7	HPV E7	Mouse	Bermudez-Humaran *et al.* (2005)
Lactococcus lactis IL-12			
Lactococcus lactis NZ9000	Avian influenza HA	Chicken	Lei *et al.* (2014)
Pichia pastoris SMD1168	CSFV glycoprotein E2	Pig	Lin *et al.* (2009)
Pichia pastoris KM71H	*C. perfringens* α toxin	Chicken	Gil de los Santos *et al.* (2012)

functionality is expected to be widely applicable in both human and animal health and for improving food safety as well.

Concluding remarks

There is a growing body of work investigating the functionality of probiotics and prebiotics in human and livestock animal health. While there is strong evidence to support their efficacy, the complicating factors described in this work provide insight into the questions regarding their overall effectiveness. Host-specific, probiotic strain-specific, and application-specific differences further confound the already complex interactions which occur in the gastrointestinal environment. An understanding of these differences in research studies in humans and livestock animals is necessary for understanding the results of host-specific studies and their broader implications to the science. Additionally, while probiotics and prebiotics are sometimes viewed mistakenly as a universal solution to a wide array of health problems, review of the literature suggests that, similar to small-molecule therapeutics, specific probiotic cultures or prebiotic compounds are only beneficial when used for specific applications in specific host species. Because of the benefits they may provide in both human health and livestock animal production, novel applications for probiotics and prebiotics are being developed. Continued research as described in other chapters is needed to elucidate specific host, microbe, and environmental interactions important in the gastrointestinal tract. An improved mechanistic understanding of probiotic and prebiotic functionality in specific host-species contexts will lead to improved application of probiotics and prebiotics for the benefit of human and animal health and livestock animal production.

Acknowledgements

We would like to thank Dr Tryon Wickersham (Texas A&M University) and Dr Robin Anderson (USDA Agricultural Research Service) for insightful discussions during the preparation of this manuscript. We would also like to thank Dr Ana Paulo do Carmo and Dr Koen Venema, for the invitation to contribute a chapter to this work.

Preparation of this work was supported by Texas A&M AgriLife Research and USDA-NIFA Hatch project number TEX09405. Tyler E. Askelson was supported by the Dr David B. Mellor Endowed Scholarship and a graduate assistantship from the Texas A&M University Department of Poultry Science.

References

Abe, F., Ishibashi, N., and Shimamura, S. (1995). Effect of administration of bifidobacteria and lactic acid bacteria to newborn calves and piglets. J. Dairy Sci. *78*, 2838–2846.

Anderson, J.W., and Gilliland, S.E. (1999). Effect of fermented milk (yoghurt) containing *Lactobacillus acidophilus* L1 on serum cholesterol in hypercholesterolemic humans. J. Am. Coll. Nutr. *18*, 43–50.

Argenzio, R.A., and Southworth, M. (1975). Sites of organic acid production and absorption in gastrointestinal tract of the pig. Am. J. Physiol. *228*, 454–460.

Asahara, T., Shimizu, K., Nomoto, K., Hamabata, T., Ozawa, A., and Takeda, Y. (2004). Probiotic bifidobacteria protect mice from lethal infection with shiga toxin-producing *Escherichia coli* O157: H7. Infect. Immun. 72, 2240–2247.

Askelson, T.E., Campasino, A., Lee, J.T., and Duong, T. (2014). Evaluation of phytate-degrading *Lactobacillus* culture administration to broiler chickens. Appl. Environ. Microbiol. *80*, 943–950.

Bakker-Zierikzee, A.M., van Tol, E.A.F., Kroes, H., Alles, M.S., Kok, F.J., and Bindels, J.G. (2006). Faecal SIgA secretion in infants fed on pre- or probiotic infant formula. Pediatr. Allergy Immu *17*, 134–140.

Barnes, E.M., Mead, G.C., Barnum, D.A., and Harry, E.G. (1972). The intestinal flora of the chicken in the period 2 to 6 weeks of age, with particular reference to the anaerobic bacteria. Br. Poult. Sci. *13*, 311–326.

Bekkali, N.L., Bongers, M.E., Van den Berg, M.M., Liem, O., and Benninga, M.A. (2007). The role of a probiotics mixture in the treatment of childhood constipation: a pilot study. Nutr. J. *6*, 17.

Bermudez-Humaran, L.G., Cortes-Perez, N.G., Lefevre, F., Guimaraes, V., Rabot, S., Alcocer-Gonzalez, J.M., Gratadoux, J.J., Rodriguez-Padilla, C., Tamez-Guerra, R.S., Corthier, G., *et al.* (2005). A novel mucosal vaccine based on live Lactococci expressing E7 antigen and IL-12 induces systemic and mucosal immune responses and protects mice against human papillomavirus type 16-induced tumours. J. Immunol. *175*, 7297–7302.

Bernet, M.F., Brassart, D., Neeser, J.R., and Servin, A.L. (1994). *Lactobacillus acidophilus* La-1 binds to cultured human intestinal-cell lines and inhibits cell attachment and cell invasion by enterovirulent bacteria. Gut 35, 483–489.

Besselink, M.G.H., van Santvoort, H.C., Buskens, E., Boermeester, M.A., van Goor, H., Timmerman, H.M., Nieuwenhuijs, V.B., Bollen, T.L., van Ramshorst, B., Witteman, B.J.M., *et al.* (2008). Probiotic prophylaxis in predicted severe acute pancreatitis: a randomized, double-blind, placebo-controlled trial. Lancet *371*, 651–659.

Bibiloni, R., Fedorak, R.N., Tannock, G.W., Madsen, K.L., Gionchetti, P., Campieri, M., De Simone, C., and Sartor, R.B. (2005). VSL#3 probiotic-mixture induces remission in patients with active ulcerative colitis. Am. J. Gastroenterol. *100*, 1539–1546.

Boyd, J., West, J.W., and Bernard, J.K. (2011). Effects of the addition of direct-fed microbials and glycerol to the diet of lactating dairy cows on milk yield and apparent efficiency of yield. J. Dairy Sci. *94*, 4616–4622.

Boyle, R.J., Robins-Browne, R.M., and Tang, M.L.K. (2006). Probiotic use in clinical practice: what are the risks? Am. J. Clin. Nutr. *83*, 1256–1264.

Bradshaw, D.J., Homer, K.A., Marsh, P.D., and Beighton, D. (1994). Metabolic cooperation in oral microbial communities during growth on mucin. Microbiol-Uk *140*, 3407–3412.

Bu, L.N., Chang, M.H., Ni, Y.H., Chen, H.L., and Cheng, C.C. (2007). *Lactobacillus* casei rhamnosus Lcr35 in children with chronic constipation. Pediatr. Int. *49*, 485–490.

Butine, T.J., and Leedle, J.A.Z. (1989). Enumeration of selected anaerobic bacterial groups in caecal and colonic contents of growing-finishing pigs. Appl. Environ. Microb. *55*, 1112–1116.

Callesen, J. (2003). Effects of termination of AGP use on pig welfare and productivity. Beyond antimicrobial growth promoters in food animal production, 43.

Cavazzoni, V., Adami, A., and Castrovilli, C. (1998). Performance of broiler chickens supplemented with *Bacillus coagulans* as probiotic. Brit Poultry Sci. *39*, 526–529.

Chenoll, E., Casinos, B., Bataller, E., Astals, P., Echevarria, J., Iglesias, J.R., Balbarie, P., Ramon, D., and Genoves, S. (2011). Novel probiotic *Bifidobacterium bifidum* CECT 7366 strain active against the pathogenic bacterium *Helicobacter pylori*. Appl. Environ. Microbiol. *77*, 1335–1343.

Child, M.W., Kennedy, A., Walker, A.W., Bahrami, B., Macfarlane, S., and Macfarlane, G.T. (2006). Studies on the effect of system retention time on bacterial populations colonizing a three-stage continuous culture model of the human large gut using FISH techniques. Fems Microbiology Ecology *55*, 299–310.

Christensen, H.R., Frokiaer, H., and Pestka, J.J. (2002). Lactobacilli differentially modulate expression of cytokines and maturation surface markers in murine dendritic cells. J. Immunol. *168*, 171–178.

Collado, M.C., Grzeskowiak, L., and Salminen, S. (2007). Probiotic strains and their combination inhibit *in vitro* adhesion of pathogens to pig intestinal mucosa. Curr. Microbiol. *55*, 260–265.

Collins, M.D., and Gibson, G.R. (1999). Probiotics, prebiotics, and synbiotics: approaches for modulating the microbial ecology of the gut. Am. J. Clin. *69* Nutr. 1052S-1057S.

Corr, S.C., Li, Y., Riedel, C.U., O'Toole, P.W., Hill, C., and Gahan, C.G. (2007). Bacteriocin production as a mechanism for the antiinfective activity of *Lactobacillus salivarius* UCC118. Proc. Natl. Acad. Sci. U.S.A. *104*, 7617–7621.

Cromwell, G.L. (2002). Why and how antibiotics are used in swine production. Anim Biotechnol. *13*, 7–27.

Cummings, J., Banwell, J., Segal, I., Coleman, N., Englyst, H., and Macfarlane, G. (1990). The amount and composition of large bowel contents in man. Gastroenterology *98*, 2.

Dalal, R.C. (1977). Soil organic phosphorus. Adv. Agron *29*, 83–113.

de Vrese, M., Stegelmann, A., Richter, B., Fenselau, S., Laue, C., and Schrezenmeir, J. (2001). Probiotics – compensation for lactase insufficiency. Am. J. Clin. Nutr. *73*, 421S-429S.

de Vrese, M., Winkler, P., Rautenberg, P., Harder, T., Noah, C., Laue, C., Ott, S., Hampe, J., Schreiber, S., Heller, K., *et al.* (2005). Effect of *Lactobacillus gasseri* PA 16/8, *Bifidobacterium longum* SP 07/3, B-bifidum MF 20/5 on common cold episodes: A double blind, randomized, controlled trial. Clin. Nutr. *24*, 481–491.

Donohue, M., and Cunningham, D.L. (2009). Effects of grain and oilseed prices on the costs of US poultry production. J. Appl. Poultry Res. *18*, 325–337.

Dowd, S.E., Callaway, T.R., Wolcott, R.D., Sun, Y., McKeehan, T., Hagevoort, R.G., and Edrington, T.S. (2008). Evaluation of the bacterial diversity in the feces of cattle using 16S rDNA bacterial tag-encoded FLX amplicon pyrosequencing (bTEFAP). Bmc Microbiol. *8*.

Dukes, H.H. (1955). The physiology of domestic animals. The physiology of domestic animals.

Duong, T. (2010). Vaccines and Anti-Infective Agents: Delivery via Lactic Acid Bacteria. In Encyclopedia of Biotechnology in Agriculture and Food (Taylor & Francis), pp. 664–667.

Duong, T., and Konkel, M.E. (2009). Comparative studies of Campylobacter jejuni genomic diversity reveal the importance of core and dispensable genes in the biology of this enigmatic food-borne pathogen. Curr. Opin. Biotech. *20*, 158–165.

Emborg, H.-D., Ersboll, A., Heuer, O.E., and Wegener, H.C. (2002). Effects of termination of antimicrobial growth promoter use for broiler health and productivity. Beyond Antimicrobial Growth Promoters in Food Animal Production, 38.

Fairbrother, J.M., Nadeau, E., and Gyles, C.L. (2005). *Escherichia coli* in postweaning diarrhea in pigs: an update on bacterial types, pathogenesis, and prevention strategies. Anim Health Res. Rev. *6*, 17–39.

Famularo, G., De Simone, C., Pandey, V., Sahu, A.R., and Minisola, G. (2005). Probiotic lactobacilli: an innovative tool to correct the malabsorption syndrome of vegetarians? Med. Hypotheses *65*, 1132–1135.

Fang, H., Elina, T., Heikki, A., and Seppo, S. (2000). Modulation of humoral immune response through probiotic intake. Fems Immunol. Med. Mic *29*, 47–52.

Fleige, S., Preissinger, W., Meyer, H.H.D., and Pfaffl, M.W. (2009). The immunomodulatory effect of lactulose on *Enterococcus faecium* fed preruminant calves. J. Anim Sci. *87*, 1731–1738.

Fuller, R. (1989). Probiotics in Man and Animals. J. Appl. Bacteriol *66*, 365–378.

Gaskins, H.R., Collier, C.T., and Anderson, D.B. (2002). Antibiotics as growth promotants: Mode of action. Anim Biotechnol. *13*, 29–42.

Geier, M.S., Torok, V.A., Allison, G.E., Ophel-Keller, K., and Hughes, R.J. (2009). Indigestible carbohydrates alter the intestinal microbiota but do not influence the performance of broiler chickens. J. Appl. Microbiol. *106*, 1540–1548.

Gibson, G.R., Beatty, E.R., Wang, X., and Cummings, J.H. (1995). Selective stimulation of *Bifidobacteria* in the human colon by oligofructose and inulin. Gastroenterology *108*, 975–982.

Gibson, G.R., and Roberfroid, M.B. (1995). Dietary modulation of the human colonic microbiota – introducing the concept of prebiotics. J. Nutr. *125*, 1401–1412.

Gibson, G.R., Scott, K.P., Rastall, R.A., Tuohy, K.M., Hotchkiss, A., Dubert-Ferrandon, A., Gareau, M., Murphy, E.F., Saulnier, D., and Loh, G. (2010). Dietary prebiotics: current status and new definition. Food Science & Technology Bulletin: Functional Foods *7*, 1–19.

Gil de los Santos, J.R., Storch, O.B., Fernandes, C.G., and Gil-Turnes, C. (2012). Evaluation in broilers of the probiotic properties of Pichia pastoris and a recombinant P. pastoris containing the Clostridium perfringens alpha toxin gene. Vet. Microbiol. *156*, 448–451.

Gionchetti, P., Rizzello, F., Helwig, U., Venturi, A., Lammers, K.M., Brigidi, P., Vitali, B., Poggioli, G., Miglioli, M., and Campieri, M. (2003). Prophylaxis of pouchitis onset with probiotic therapy: A double-blind, placebo-controlled trial. Gastroenterology *124*, 1202–1209.

Grand, E., Respondek, F., Martineau, C., Detilleux, J., and Bertrand, G. (2013). Effects of short-chain fructooligosaccharides on growth performance of preruminant veal calves. J. Dairy Sci. *96*, 1094–1101.

Grangette, C., Muller-Alouf, H., Goudercourt, D., Geoffroy, M.C., Turneer, M., and Mercenier, A. (2001). Mucosal immune responses and protection against tetanus toxin after intranasal immunization with recombinant *Lactobacillus plantarum*. Infect. Immun. *69*, 1547–1553.

Guard-Petter, J. (2001). The chicken, the egg and Salmonella enteritidis. Environ. Microbiol. 3, 421–430.

Gueimonde, M., Jalonen, L., He, F., Hiramatsu, M., and Salminen, S. (2006). Adhesion and competitive inhibition and displacement of human enteropathogens by selected lactobacilli. Food Res. Int. *39*, 467–471.

Guslandi, M., Giollo, P., and Testoni, P.A. (2003). A pilot trial of Saccharomyces boulardii in ulcerative colitis. Eur. J. Gastroen Hepat *15*, 697–698.

Haarman, M., and Knol, J. (2006). Quantitative real-time PCR analysis of fecal *Lactobacillus* species in infants receiving a prebiotic infant formula. Appl. Environ. Microb. *72*, 2359–2365.

Habel, R. (1975). Ruminant digestive system. Sisson and grossman's the anatomy of the domestic animals *1*, 861–915.

Hart, A.L., Lammers, K., Brigidi, P., Vitali, B., Rizzello, F., Gionchetti, P., Campieri, M., Kamm, M.A., Knight, S.C., and Stagg, A.J. (2004). Modulation of human dendritic cell phenotype and function by probiotic bacteria. Gut 53, 1602–1609.

Hemarajata, P., and Versalovic, J. (2013). Effects of probiotics on gut microbiota: mechanisms of intestinal immunomodulation and neuromodulation. Therap Adv. Gastroenterol. 6, 39–51.

Hooper, L.V., Midtvedt, T., and Gordon, J.I. (2002). How host-microbial interactions shape the nutrient environment of the mammalian intestine. Annu. Rev. Nutr. 22, 283–307.

Hopkins, M.J., Sharp, R., and Macfarlane, G.T. (2002). Variation in human intestinal microbiota with age. Digest. Liver Dis. 34, S12-S18.

Hoyos, A.B. (1999). Reduced incidence of necrotizing enterocolitis associated with enteral administration of *Lactobacillus acidophilus* and *Bifidobacterium infantis* to neonates in an intensive care unit. Int. J. Infect. Dis. 3, 197–202.

Hungate, R.E. (1966). The rumen and its microbes. The rumen and its microbes.

Huyghebaert, G., Ducatelle, R., and Van Immerseel, F. (2011). An update on alternatives to antimicrobial growth promoters for broilers. Vet. J. *187*, 182–188.

Isolauri, E., Arvola, T., Sutas, Y., Moilanen, E., and Salminen, S. (2000). Probiotics in the management of atopic eczema. Clin. Exp. Allergy *30*, 1604–1610.

Jami, E., and Mizrahi, I. (2012). Composition and similarity of bovine rumen microbiota across individual animals. PLoS One 7.

Janardhana, V., Broadway, M.M., Bruce, M.P., Lowenthal, J.W., Geier, M.S., Hughes, R.J., and Bean, A.G.D. (2009). Prebiotics modulate immune responses in the gut-associated lymphoid tissue of chickens. J. Nutr. *139*, 1404–1409.

Jiang, T., Mustapha, A., and Savaiano, D.A. (1996). Improvement of lactose digestion in humans by ingestion of unfermented milk containing *Bifidobacterium longum*. J. Dairy Sci. 79, 750–757.

Jongbloed, A.W., Mroz, Z., and Kemme, P.A. (1992). The effect of supplementary *Aspergillus niger* phytase in diets for pigs on concentration and apparent digestibility of dry-matter, total phosphorus, and phytic acid in different sections of the alimentary-tract. J. Anim Sci. 70, 1159–1168.

Kajikawa, A., Zhang, L., Long, J.L., Nordone, S., Stoeker, L., LaVoy, A., Bumgardner, S., Klaenhammer, T., and Dean, G. (2012). Construction and Immunological Evaluation of Dual Cell Surface Display of HIV-1 Gag and Salmonella enterica Serovar Typhimurium FliC in *Lactobacillus* acidophilus for Vaccine Delivery. Clin. Vaccine Immunol. *19*, 1374–1381.

Kalavathy, R., Abdullah, N., Jalaludin, S., and Ho, Y.W. (2003). Effects of *Lactobacillus* cultures on growth performance, abdominal fat deposition, serum lipids and weight of organs of broiler chickens. Brit Poultry Sci. *44*, 139–144.

Kalliomaki, M., Salminen, S., Arvilommi, H., Kero, P., Koskinen, P., and Isolauri, E. (2001). Probiotics in primary prevention of atopic disease: a randomized placebo-controlled trial. Lancet 357, 1076–1079.

Koebnick, C., Wagner, I., Leitzmann, P., Stern, U., and Zunft, H.J.F. (2003). Probiotic beverage containing *Lactobacillus* casei Shirota improves gastrointestinal symptoms in patients with chronic constipation. Can. J. Gastroenterol. *17*, 655–659.

Kyne, L., Hamel, M.B., Polavaram, R., and Kelly, C.N.P. (2002). Health care costs and mortality associated with nosocomial diarrhea due to *Clostridium difficile*. Clin. Infect. Dis. 34, 346–353.

Kyriakis, S.C., Tsiloyiannis, V.K., Vlemmas, J., Sarris, K., Tsinas, A.C., Alexopoulos, C., and Jansegers, L. (1999). The effect of probiotic LSP 122 on the control of post-weaning diarrhoea syndrome of piglets. Res. Vet. Sci. *67*, 223–228.

Langendijk, P.S., Schut, F., Jansen, G.J., Raangs, G.C., Kamphuis, G.R., Wilkinson, M.H., and Welling, G.W. (1995). Quantitative fluorescence *in situ* hybridization of *Bifidobacterium* spp. with genus-specific 16S rRNA-targeted probes and its application in fecal samples. Appl. Environ. Microbiol. *61*, 3069–3075.

Langlands, S.J., Hopkins, M.J., Coleman, N., and Cummings, J.H. (2004). Prebiotic carbohydrates modify the mucosa associated microflora of the human large bowel. Gut 53, 1610–1616.

Lee, J.S., Awji, E.G., Lee, S.J., Tassew, D.D., Park, Y.B., Park, K.S., Kim, M.K., Kim, B., and Park, S.C. (2012). Effect of *Lactobacillus plantarum* CJLP243 on the growth performance and cytokine response of weaning pigs challenged with enterotoxigenic *Escherichia coli*. J. Anim Sci. *90*, 3709–3717.

Lee, Y.K., and Puong, K.Y. (2002). Competition for adhesion between probiotics and human gastrointestinal pathogens in the presence of carbohydrate. Brit J. Nutr. *88*, S101-S108.

Lei, H., Peng, X., Shu, H., and Zhao, D. (2014). Intranasal immunization with live recombinant *Lactococcus lactis* combined with heat-labile toxin B subunit protects chickens from highly pathogenic avian influenza H5N1 virus. J. Med. Virol.

Lin, G.J., Liu, T.Y., Tseng, Y.Y., Chen, Z.W., You, C.C., Hsuan, S.L., Chien, M.S., and Huang, C. (2009). Yeast-expressed classical swine fever virus glycoprotein E2 induces a protective immune response. Vet. Microbiol. *139*, 369–374.

Loneragan, G.H., Thomson, D.U., Montgomery, D.L., Mason, G.L., and Larson, R.L. (2005). Prevalence, outcome, and health consequences associated with persistent infection with bovine viral diarrhea virus in feedlot cattle. Javma-J. Am. Vet. Med. A *226*, 595–601.

Loo, V.G., Poirier, L., Miller, M.A., Oughton, M., Libman, M.D., Michaud, S., Bourgault, A.M., Nguyen, T., Frenette, C., Kelly, M., *et al.* (2005). A predominantly clonal multi-institutional outbreak of *Clostridium difficile*-associated diarrhea with high morbidity and mortality. New Engl J. Med. *353*, 2442–2449.

Lu, J.R., Idris, U., Harmon, B., Hofacre, C., Maurer, J.J., and Lee, M.D. (2003). Diversity and succession of the intestinal bacterial community of the maturing broiler chicken. Appl. Environ. Microb. *69*, 6816–6824.

Lu, L., and Walker, W.A. (2001). Pathologic and physiologic interactions of bacteria with the gastrointestinal epithelium. Am. J. Clin. Nutr. *73*, 1124s–1130s.

Mackie, R.I., Sghir, A., and Gaskins, H.R. (1999). Developmental microbial ecology of the neonatal gastrointestinal tract. Am. J. Clin. Nutr. *69*, 1035s–1045s.

Madsen, K., Cornish, A., Soper, P., McKaigney, C., Jijon, H., Yachimec, C., Doyle, J., Jewell, L., and De Simone, C. (2001). Probiotic bacteria enhance murine and human intestinal epithelial barrier function. Gastroenterology *121*, 580–591.

Mann, G.V. (1974). Studies of a surfactant and cholesteremia in the Maasai. Am. J. Clin. Nutr. *27*, 464–469.

Marteau, P., Cuillerier, E., Meance, S., Gerhardt, M.F., Myara, A., Bouvier, M., Bouley, C., Tondu, F., Bommelaer, G., and Grimaud, J.C. (2002). Bifidobacterium animalis strain DN-173 010 shortens the colonic transit time in healthy women: a double-blind, randomized, controlled study. Aliment. Pharmacol. Ther. *16*, 587–593.

Martini, M.C., Lerebours, E.C., Lin, W.J., Harlander, S.K., Berrada, N.M., Antoine, J.M., and Savaiano, D.A. (1991). Strains and species of lactic-acid bacteria in fermented milks (Yogurts) – effect on invivo lactose digestion. Am. J. Clin. Nutr. *54*, 1041–1046.

Matsumoto, S., Hara, T., Hori, T., Mitsuyama, K., Nagaoka, M., Tomiyasu, N., Suzuki, A., and Sata, M. (2005). Probiotic *Lactobacillus*-induced improvement in murine chronic inflammatory bowel disease is associated with the down-regulation of pro-inflammatory cytokines in lamina propria mononuclear cells. Clin. Exp. Immunol. *140*, 417–426.

McFarland, L.V. (2006). Meta-analysis of probiotics for the prevention of antibiotic associated diarrhea and the treatment of *Clostridium difficile* disease. Am. J. Gastroenterol. *101*, 812–822.

McFarland, L.V. (2007). Meta-analysis of probiotics for the prevention of traveler's diarrhea. Travel Med. Infect. Dis. *5*, 97–105.

Mikulski, D., Jankowski, J., Naczmanski, J., Mikulska, M., and Demey, V. (2012). Effects of dietary probiotic (Pediococcus acidilactici) supplementation on performance, nutrient digestibility, egg traits, egg yolk cholesterol, and fatty acid profile in laying hens. Poult. Sci. *91*, 2691–2700.

Mohamadzadeh, M., Duong, T., Sandwick, S.J., Hoover, T., and Klaenhammer, T.R. (2009). Dendritic cell targeting of *Bacillus anthracis* protective antigen expressed by *Lactobacillus acidophilus* protects mice from lethal challenge. Proc. Natl. Acad. Sci. U.S.A. *106*, 4331–4336.

Molbak, L., Thomsen, L.E., Jensen, T.K., Knudsen, K.E.B., and Boye, M. (2007). Increased amount of *Bifidobacterium thermacidophilum* and *Megasphaera elsdenii* in the colonic microbiota of pigs fed a swine dysentery preventive diet containing chicory roots and sweet lupine. J. Appl. Microbiol. *103*, 1853–1867.

Moore, P.R., Evenson, A., Luckey, T.D., Mccoy, E., Elvehjem, C.A., and Hart, E.B. (1946). Use of sulfasuxidine, streptothricin, and streptomycin in nutritional studies with the chick. J. Biol. Chem. *165*, 437–441.

Mountzouris, K.C., Tsirtsikos, P., Kalamara, E., Nitsch, S., Schatzmayr, G., and Fegeros, K. (2007). Evaluation of the efficacy of a probiotic containing *Lactobacillus, Bifidobacterium, Enterococcus,* and *Pediococcus* strains in promoting broiler performance and modulating cecal microflora composition and metabolic activities. Poultry Sci. *86*, 309–317.

Mustapha, A., Jiang, T.A., and Savaiano, D.A. (1997). Improvement of lactose digestion by humans following ingestion of unfermented acidophilus milk: Influence of bile sensitivity, lactose transport, and acid tolerance of *Lactobacillus acidophilus*. J. Dairy Sci. *80*, 1537–1545.

Neal-McKinney, J.M., Lu, X., Duong, T., Larson, C.L., Call, D.R., Shah, D.H., and Konkel, M.E. (2012). Production of organic acids by probiotic lactobacilli can be used to reduce pathogen load in poultry. PLoS One 7.

Nelson, R.D., Shibata, N., Podzorski, R.P., and Herron, M.J. (1991). Candida Mannan – chemistry, suppression of cell-mediated-immunity, and possible mechanisms of action. Clinical Microbiology Reviews *4*, 1–19.

Nelson, T.S. (1967). The utilization of phytate phosphorus by poultry – a review. Poultry Sci. *46*, 862–871.

Netherwood, T., Gilbert, H.J., Parker, D.S., and O'Donnell, A.G. (1999). Probiotics shown to change bacterial community structure in the avian gastrointestinal tract. Appl. Environ. Microb. *65*, 5134–5138.

Nocek, J.E., and Kautz, W.P. (2006). Direct-fed microbial supplementation on ruminal digestion, health, and performance of pre- and postpartum dairy cattle. J. Dairy Sci. *89*, 260–266.

Nocek, J.E., Kautz, W.P., Leedle, J.A., and Block, E. (2003). Direct-fed microbial supplementation on the performance of dairy cattle during the transition period. J. Dairy Sci. *86*, 331–335.

Notter, D.R. (1999). The importance of genetic populations diversity in livestock populations of the future. J. Anim Sci. *77*, 61–69.

Oelschlaeger, T.A. (2010). Mechanisms of probiotic actions – A review. Int. J. Med. Microbiol. *300*, 57–62.

Parvez, S., Malik, K.A., Kang, S.A., and Kim, H.Y. (2006). Probiotics and their fermented food products are beneficial for health. J. Appl. Microbiol. *100*, 1171–1185.

Pascual, M., Hugas, M., Badiola, J.I., Monfort, J.M., and Garriga, M. (1999). *Lactobacillus salivarius* CTC2197 prevents *Salmonella enteritidis* colonization in chickens. Appl. Environ. Microb. *65*, 4981–4986.

Patterson, J.A., and Burkholder, K.M. (2003). Application of prebiotics and probiotics in poultry production. Poultry Sci. *82*, 627–631.

Pelto, L., Isolauri, E., Lilius, E.M., Nuutila, J., and Salminen, S. (1998). Probiotic bacteria down-regulate the milk-induced inflammatory response in milk-hypersensitive subjects but have an immunostimulatory effect in healthy subjects. Clin. Exp. Allergy *28*, 1474–1479.

Pereira, D.I.A., and Gibson, G.R. (2002). Effects of consumption of probiotics and prebiotics on serum lipid levels in humans. Crit. Rev. Biochem. Mol. *37*, 259-+.

Ponce, M.F., Rial, M.J., Alarcon, N., Szefner, M., and Aguilar, M.D. (1995). Use of a prospectively measured incidence rate of nosocomial diarrhea in an infant/toddler ward as a meaningful quality assessment tool. Clin. Perform Qual Health Care 3, 128–131.

Preidis, G.A., and Versalovic, J. (2009). Targeting the human microbiome with antibiotics, probiotics, and prebiotics: gastroenterology enters the metagenomics era. Gastroenterology *136*, 2015–2031.

Pridmore, R.D., Pittet, A.C., Praplan, F., and Cavadini, C. (2008). Hydrogen peroxide production by *Lactobacillus johnsonii* NCC 533 and its role in anti-*Salmonella* activity. FEMS Microbiol. Lett. *283*, 210–215.

Pryde, S.E., Richardson, A.J., Stewart, C.S., and Flint, H.J. (1999). Molecular analysis of the microbial diversity present in the colonic wall, colonic lumen, and cecal lumen of a pig. Appl. Environ. Microbiol. *65*, 5372–5377.

Reddy, N.R., Sathe, S.K., and Salunkhe, D.K. (1982). Phytates in legumes and cereals. Adv. Food Res. *28*, 1–92.

Reid, G., and Friendship, R. (2002). Alternatives to antibiotic use: Probiotics for the gut. Anim Biotechnol. *13*, 97–112.

Rigaux, P., Daniel, C., Hisbergues, M., Muraille, E., Hols, P., Pot, B., Pestel, J., and Jacquet, A. (2009). Immunomodulatory properties of *Lactobacillus plantarum* and its use as a recombinant vaccine against mite allergy. Allergy *64*, 406–414.

Roberfroid, M.B. (2000). Prebiotics and probiotics: are they functional foods? Am. J. Clin. Nutr. *71*, 1682s–1687s.

Robert, W. (2012). The bacterial community composition of the bovine rumen detected using pyrosequencing of 16S rRNA genes. Metagenomics *2012*.

Rosenfeldt, V., Benfeldt, E., Nielsen, S.D., Michaelsen, K.F., Jeppesen, D.L., Valerius, N.H., and Paerregaard, A. (2003). Effect of probiotic *Lactobacillus* strains in children with atopic dermatitis. J. Allergy Clin. Immun. *111*, 389–395.

Russell, J.B., O'Connor, J.D., Fox, D.G., Van Soest, P.J., and Sniffen, C.J. (1992). A net carbohydrate and protein system for evaluating cattle diets: I. Ruminal fermentation. J. Anim Sci. *70*, 3551–3561.

Salanitro, J.P., Fairchilds, I.G., and Zgornicki, Y.D. (1974). Isolation, culture characteristics, and identification of anaerobic bacteria from the chicken caecum. Appl. Microbiol. *27*, 678–687.

Sartor, R.B. (2004). Therapeutic manipulation of the enteric microflora in inflammatory bowel diseases: Antibiotics, probiotics, and prebiotics. Gastroenterology *126*, 1620–1633.

Schaafsma, G., Meuling, W.J., van Dokkum, W., and Bouley, C. (1998). Effects of a milk product, fermented by *Lactobacillus acidophilus* and with fructo-oligosaccharides added, on blood lipids in male volunteers. Eur. J. Clin. Nutr. *52*, 436–440.

Schroeder, B., Duncker, S., Barth, S., Bauerfeind, R., Gruber, A.D., Deppenmeier, S., and Breves, G. (2006). Preventative effects of the probiotic *Escherichia coli* strain Nissle 1917 on acute secretory diarrhoea in a pig model of intestinal infection. Dig. Dis. Sci. *51*, 724–731.

ScienceInsider (2009). Report slams deadly dutch probiotic study (Science/AAAS).

Selle, P.H., Ravindran, V., Bryden, W.L., and Scott, T. (2006). Influence of dietary phytate and exogenous phytase on amino acid digestibility in poultry: a review. J. Poultry Sci. *43*, 89–103.

Sgouras, D., Maragkoudakis, P., Petraki, K., Martinez-Gonzalez, B., Eriotou, E., Michopoulos, S., Kalantzopoulos, G., Tsakalidou, E., and Mentis, A. (2004). *In vitro* and *in vivo* inhibition of *Helicobacter pylori* by *Lactobacillus casei* strain Shirota. Appl. Environ. Microbiol. *70*, 518–526.

Shen, Y.B., Piao, X.S., Kim, S.W., Wang, L., Liu, P., Yoon, I., and Zhen, Y.G. (2009). Effects of yeast culture supplementation on growth performance, intestinal health, and immune response of nursery pigs. J. Anim Sci. *87*, 2614–2624.

Simmering, R., and Blaut, M. (2001). Pro- and prebiotics – the tasty guardian angels? Appl. Microbiol. Biotechnol. *55*, 19–28.

Stevens, C.E., and Hume, I.D. (1995). Comparative physiology of the vertebrate digestive system, 2nd edn (Cambridge; New York: Cambridge University Press).

Suau, A., Bonnet, R., Sutren, M., Godon, J.J., Gibson, G.R., Collins, M.D., and Dore, J. (1999). Direct analysis of genes encoding 16S rRNA from complex communities reveals many novel molecular species within the human gut. Appl. Environ. Microbiol. *65*, 4799–4807.

Sun, P., Wang, J.Q., and Deng, L.F. (2013). Effects of Bacillus subtilis natto on milk production, rumen fermentation and ruminal microbiome of dairy cows. animal 7, 216–222.

Sun, P., Wang, J.Q., and Zhang, H.T. (2010). Effects of *Bacillus subtilis* natto on performance and immune function of preweaning calves. J. Dairy Sci. *93*, 5851–5855.

Szajewska, H., Kotowska, M., Mrukowicz, J.Z., Armanska, M., and Mikolajczyk, W. (2001). Efficacy of *Lactobacillus* GG in prevention of nosocomial diarrhoea in infants. J. Pediatr. *138*, 361–365.

Szajewska, H., and Mrukowicz, J.Z. (2001). Probiotics in the treatment and prevention of acute infectious diarrhoea in infants and children: A systematic review of published randomized, double-blind, placebo-controlled trials. J. Pediatr. Gastr Nutr. 33, S17-S25.

Tannock, G.W., Munro, K., Harmsen, H.J.M., Welling, G.W., Smart, J., and Gopal, P.K. (2000). Analysis of the faecal microflora of human subjects consuming a probiotic product containing *Lactobacillus rhamnosus* DR20. Appl. Environ. Microb. *66*, 2578–2588.

Taylor, A.L., Dunstan, J.A., and Prescott, S.L. (2007). Probiotic supplementation for the first 6 months of life fails to reduce the risk of atopic dermatitis and increases the risk of allergen sensitization in high-risk children: A randomized controlled trial. J. Allergy Clin. Immun. *119*, 184–191.

Timmerman, H.M., Veldman, A., van den Elsen, E., Rombouts, F.M., and Beynen, A.C. (2006). Mortality and growth performance of broilers given drinking water supplemented with chicken-specific probiotics. Poult. Sci. *85*, 1383–1388.

Torres-Rodriguez, A., Donoghue, A.M., Donoghue, D.J., Barton, J.T., Tellez, G., and Hargis, B.M. (2007). Performance and condemnation rate analysis of commercial turkey flocks treated with a *Lactobacillus* spp.-based probiotic. Poultry Sci. *86*, 444–446.

Van Houte, J., and Gibbons, R. (1966). Studies of the cultivable flora of normal human faeces. Antonie van Leeuwenhoek *32*, 212–222.

Vanderpool, C., Yan, F., and Polk, D.B. (2008). Mechanisms of probiotic action: implications for therapeutic applications in inflammatory bowel diseases. Inflamm. Bowel Dis. *14*, 1585–1596.

Venturi, A., Gionchetti, P., Rizzello, F., Johansson, R., Zucconi, E., Brigidi, P., Matteuzzi, D., and Campieri, M. (1999). Impact on the composition of the faecal flora by a new probiotic preparation: preliminary data on maintenance treatment of patients with ulcerative colitis. Aliment Pharm. Therap *13*, 1103–1108.

Vicente, J., Wolfenden, A., Torres-Rodriguez, A., Higgins, S., Tellez, G., and Hargis, B. (2007). Effect of a *Lactobacillus* species-based Probiotic and dietary lactose prebiotic on turkey poult performance with or without *Salmonella enteritidis* challenge. J. Appl. Poultry Res. *16*, 361–364.

Wells, J.M., and Mercenier, A. (2008). Mucosal delivery of therapeutic and prophylactic molecules using lactic acid bacteria. Nat. Rev. Microbiol. *6*, 349–362.

Whitman, W.B., Coleman, D.C., and Wiebe, W.J. (1998). Prokaryotes: The unseen majority. Proc. Natl. Acad. Sci. U.S.A. *95*, 6578–6583.

WHO/FAO (2001). WHO Expert consultation on evaluation of health and nutritional properties of probiotics in food including powder milk with live lactic acid bacteria. Córdoba, Argentina October, 1–4.

Whorwell, P.J., Altringer, L., Morel, J., Bond, Y., Charbonneau, D., O'Mahony, L., Kiely, B., Shanahan, F., and Quigley, E.M.M. (2006). Efficacy of an encapsulated probiotic *Bifidobacterium infantis* 35624 in women with irritable bowel syndrome. Am. J. Gastroenterol. *101*, 1581–1590.

Wismar, R., Brix, S., Laerke, H.N., and Frokiaer, H. (2011). Comparative analysis of a large panel of non-starch polysaccharides reveals structures with selective regulatory properties in dendritic cells. Mol. Nutr. Food Res. *55*, 443–454.

Wrong, O., Edmonds, C.J., and Chadwick, V.S. (1981). The large intestine: its role in mammalian nutrition and homeostasis (New York: Wiley).

Xu, Y.G., Cui, L.C., Tian, C.Y., Zhang, G.C., Huo, G.C., Tang, L.J., and Li, Y.J. (2011). Immunogenicity of Recombinant Classic Swine Fever Virus CD8(+). T Lymphocyte Epitope and Porcine Parvovirus VP2 Antigen Coexpressed by *Lactobacillus* casei in Swine via Oral Vaccination. Clin. Vaccine Immunol. *18*, 1979–1986.

Yang, Y., Iji, P.A., and Choct, M. (2009). Dietary modulation of gut microflora in broiler chickens: a review of the role of six kinds of alternatives to in-feed antibiotics. World Poultry Sci. J. *65*, 97–114.

Ziemer, C.J., and Gibson, G.R. (1998). An overview of probiotics, prebiotics and synbiotics in the functional food concept: Perspectives and future strategies. Int. Dairy J. *8*, 473–479.

The Use of Probiotics to Enhance Animal Performance

31

Juliana Teixeira de Magalhães, Luciene Lignani Bitencourt,
Marta Cristina Teixeira Leite, Ana Paula do Carmo and
Célia Alencar de Moraes

Abstract

Enteric infections by pathogenic bacteria and clinical expression of disease occur frequently in young animals. Multiple environmental changes usually lead to stressful conditions and trigger transitory inflammatory responses in the gut that can contribute to anatomical and functional intestinal disorders (Berge and Wierup, 2012; Heo *et al.*, 2013; Suda *et al.*, 2014). These diseases have often brought about significant economic losses in animal production. In order to solve this problem, antibiotics have been included in animal feeds either at subtherapeutic levels (acting as growth promoters-AGPs), or at therapeutic levels, to treat diseases. As growth promoters, they reduce competition for nutrients between the gastrointestinal (GI) microbiota and the host. The effects of such competition have often been at the cost of animal performance. Unfortunately, there has been considerable concern over the use of AGP's, because their long term and extensive use in animal production has resulted in selection for survival of resistant bacterial species or strains (Van der Fels-Klerx *et al.*, 2011; Thacker, 2013).

Moreover, consumer demand for organic and natural food continues to increase in the world because of an ongoing perception that organic or natural products are better than their conventional counterparts in terms of safety, taste, and increased health benefits. The general term organic foods is used to define foods that are produced without using chemical fertilizers, additives, and synthetic pesticides as well as not being processed with irradiation. Among organic foods, the overall organic meat market size is small compared with the conventional meat industries. However, according to the Organic Trade Association, the organic meat industry has grown $3.8 billion in 2014 (OTA, 2014). In general, with increases in organic meat products, new management approaches are needed to compensate for potential food safety concerns and animal health. Thus, the use of probiotics have been suggested as the most desirable alternative for livestock due to their beneficial effects. Probiotic bacterial species are included in the diet to promote health, protecting the intestine against pathogenic microorganisms and reducing inflammation (Ross *et al.*, 2010; Meng *et al.*, 2010, Cho *et al.*, 2011; Thacker, 2013).

This chapter will focus the use of probiotics as a feed additive discussing their beneficial effects in swine, poultry, aquaculture and ruminants.

Probiotics for swine

The period following weaning is characterized by a high incidence of intestinal disturbances with diarrhoea and depression of growth performance in piglets. Poor growth performance associated with weaning in pigs is a result of multi-factorial stressors including environmental, nutritional and psychological stressors. At weaning, pigs have to deal with the abrupt interruption in the established social interaction with sow and littermates, and the stress of adapting to a new environment, amongst which drastic changes in the diet: the piglet has to cope with the sudden withdrawal of sow milk and adapt to less digestible, plant-based dry diets containing complex protein and carbohydrate sources including various anti-nutritional factors. Hence, piglets have a sharp reduction in feed intake immediately after weaning. While approximately 50% of weaned pigs consume their first feed within 24h post-weaning, in approximately 10% weaning anorexia persists for up to 48h. Antibiotics and minerals, especially ZnO and $CuSO_4$, are often included in the diets for weaned pigs to control post-weaning diarrhoea (PWD) and optimize growth performance. PWD is usually associated with proliferation of one or more strains of enterotoxigenic *Escherichia coli* (ETEC) in the gastrointestinal tract (Heo *et al.*, 2013).

Antibiotics have played a major role in the growth and development of the swine industry for more than 50 years. Their efficiency in increasing growth rate, improving feed utilization and reducing mortality from clinical disease is well documented. However, consumers are becoming increasingly concerned about drug residues in meat products (Vondruskova *et al.*, 2010). In addition, it has been suggested that the continuous use of antibiotics may contribute to a reservoir of drug-resistant bacteria which may be capable of transferring their resistance to pathogenic bacteria in both animals and humans (Van der Fels-Klerx *et al.*, 2011). As a result, many countries have banned or are banning the inclusion of antibiotics in swine diets as a routine means of growth promotion. The European Union (EU) banned the use of antibiotics as growth promoters in swine and livestock production already on 1 January 2006.

In the past two decades, an intensive amount of research has been focused on the development of alternatives to antibiotics to maintain swine health and performance like organic acids, plant extracts and nutraceuticals such as copper and zinc, antimicrobial peptides, clay minerals, egg yolk antibodies, essential oils and recombinant enzymes (Thacker, 2013; Heo *et al.*, 2013; Berge and Wierup, 2012). Unfortunately, the vast majority of these compounds produce inconsistent results and rarely equal antibiotics in their effectiveness. The most widely researched alternatives include probiotics and prebiotics (Jacela *et al.*, 2010; Simon, 2010; Ross *et al.*, 2010; Heo *et al.*, 2013; Bednorz *et al.*, 2013).

Pre-requisites for the healthy and efficient growth of young pigs are the rapid maturation of the gut mucosa and the mucosa-associated lymphoid tissue, and the formation of a local stable and complex bacterial community. In neonatal pigs, suckling and the maternal environment shape the gut microbiota. Later, when weaning stress causes a transient drop in favourable bacteria, the oral supply of microbes in the form of probiotics could contribute to re-establish the microbiota balance. The possible main strategies necessary to obtain the settling of beneficial microbiota and stabilize it in the gut of young pigs are the manipulation of the gut microbiota in newborns, the re-equilibration of the microbiota after weaning and the development of a health-promoting effect in stressed animals. More knowledge is required on the mechanisms underlying the interplay between the commensal microbiota and the gut mucosa in healthy and stressed pigs (Bosi and Trevisi, 2010).

Many strains of bacteria have been used as probiotics in swine. The most commonly used species being acid lactic bacteria (LAB), including bifidobacteria. Commercial species of probiotics are usually isolated from the intestinal microbiota of the intended consumer (human, chicken or pig). Species other LAB which are currently being used in probiotic preparations include *Bacillus* species, yeasts (*Saccharomyces cerevisiae*) and moulds (*Aspergillus oryzae*). *Lactobacillus* and *Bifidobacterium* species have been used most extensively in humans whereas species of *Bacillus*, *Enterococcus* and *Saccharomyces* yeast have been the most common organisms used in livestock. However, there has been a recent increase in research on feeding *Lactobacillus* to livestock (Cho *et al.*, 2011).

There is still a need to clarify the effectiveness of probiotics in pigs and the underlying mechanisms through which they function. Several studies related to probiotics were performed on livestock and beneficial effects including growth enhancement, nutrient digestibility, stimulation of the immune system and disease prevention were observed (Cho *et al.*, 2011). The use of immunobiotics strains as supplemental additives for piglets feedings, could be used as a strategy to maintain and improve intestinal homeostasis, which is important for the development of the pig and for health and performance throughout the productive life of the animal (Suda *et al.*, 2014). Although all the mechanisms by which probiotics confer performance and enteric health benefits to piglets are not completely understood, a few have been proposed. They reduce the incidence of PWD by decreasing the number of enterobacteria. For example, LAB produced lactic acid, which lowers the pH of the intestinal content and helps to inhibit the development of invasive pathogens such as *Salmonella* spp. or strains of *E. coli* (Simon, 2010; Ross *et al.*, 2010). Moreover, some probiotic strains can reduce adherence of mucosa-adherent *E. coli* (Bednorz *et al.*, 2013). Also probiotic administration to animals could act against parasites. Bednorz *et al.*, have shown that intestinal tissue of animals of control group exhibited a greater number of eosinophils and a widespread cellular infiltration compared with samples of probiotic supplemented feed group. The presence of this cellular infiltration could involve parasite infection. An increase in eosinophils is usually related with parasite infection and it is also know that such infections cause growth retardation characterized by reduced weight gain or even weight loss in animals (Ross *et al.*, 2010).

The use of probiotics to improve meat quality has been questioned, and the results in pigs have been inconsistent. Some authors have reported advantages of probiotics on meat quality (Alexopoulos *et al.*, 2004; Meng *et al.*, 2010), whereas others have observed negative results (Quadros *et al.*, 2001). These discrepancies may be due to variations in the animals and the probiotics used in these studies. Energy intake is highly associated with feed intake, which may indicate that the intake of probiotics may be affected by the energy density of the diet. A relationship has been shown between dietary energy and probiotic intake. Meat quality is improved with addition of probiotics particularly when pigs were fed high-energy and high-nutrient-density diets (Meng *et al.*, 2010).

An important finding is that the mode of action for probiotics is unique, i.e. it is species or even strain specific. Most probably the studied probiotics act directly and/or via modifications of the intestinal microbiota on the immune system (intraepithelial lymphocyte population) (Simon, 2010).

Probiotics for poultry

The poultry industry has become an important economic activity in many countries and poultry meat has been associated frequently and consistently with the transmission of enteric pathogens, including *Salmonella* and *Campylobacter* (Food and Agriculture Organization of the United Nations and the World Health Organization (FAO/WHO 2002)). The protective microbiota which establishes itself in the gut is very stable, but it can be influenced by some dietary and environmental factors. The three most important are excessive hygiene, antibiotic therapy and stress.

Eggs have been implicated as vehicles in numerous outbreaks of salmonellosis; in particular, eggs are a major vehicle of transmission of strains of *Salmonella* serovar Enteritidis. Commercial poultry hatcheries are ideal environments for the contamination and dissemination of enteropathogens. Ideally, the fertile eggs received from farms should be enteropathogen free; however, this is not the commercial reality in many countries. The dust generated from contaminated eggs within a hatcher/incubator, a critical control point, can spread enteropathogens to other areas of the hatchery depending upon airflow. Also prolonged use of litter, with multiple batches of birds, is considered another, if not greater problem, in the spread of pathogens. In order to minimize transmission of enteropathogens, drinking water should be of potable quality. But there is a potential for biofilm formation in drinking systems, so regular cleaning and sanitation is critical. Poultry feed is obviously critical for growth of poultry, yet it is frequently cited as a major vehicle of transmission of *Salmonella*. Serovars found in feed mills can often be found in birds during rearing and/or at slaughter (Cox and Pavic, 2009).

Many strategies are aimed at good poultry meat production and to control of enteric pathogens throughout the whole production chain, focusing on *Salmonella* and *Campylobacter*. For example, organic acids as salts of propionic, formic, acetic and butyric acid are used most widely as feed additives to reduce the risk of survival of salmonellae in feed. Additionally, acid treatment protects feed from recontamination and horizontal spread of *Salmonella*. The use of vaccination in poultry against pathogens of public health concern is a more recent development and has been limited largely to serovars of prevalence (e.g. Enteritidis and Typhimurium). Antibiotics were used widely in the industry as growth promoters and as prophylactics to minimize the risk of colonization by enteric pathogens. However, the use of antibiotics for the control of pathogens of public health significance no longer is routine in many country of the world due the presence of multi-drug resistant commensal bacteria (*Escherichia* spp., *Enterococcus* spp.). Even therapeutic use is very carefully considered, given the concerns over the selection or generation of antibiotic-resistant strains and evidence that some antibiotics may facilitate colonization and increase shedding and prolong carriage of *Salmonella*. Bacteriophage therapy was shown to be effective in reducing carriage of salmonellae in live birds also. Phages and their lysins can serve as an alternative to chemical agents in processing and reducing pathogen populations after spraying onto the surface of the carcass. It will lead to a reduction or perhaps even elimination of the use of antibiotics and chemical treatments (Cox and Pavic, 2009; Diarra and Malouin, 2014).

Prebiotics like β-glucans or fructooligosaccharides (FOS) are used to enhance the natural intestinal microbiota, e.g. by promoting the growth of probiotic species, such as the endogenous LAB and bifidobacteria, thereby reducing the likelihood and persistence of colonization by enteric pathogens. Increased numbers of *Lactobacillus* and *Bifidobacterium* species correlate with reduced *Salmonella* prevalence. There has been considerable interest in the use of a combination (synbiotics) of beneficial microbes (probiotic) and fermentable carbohydrates (prebiotic) to improve gut health. The prebiotic would improve the viability of the probiotic during transit in the gastrointestinal tract and result in more probiotic fermentation to produce short-chain fatty acids (SCFA). The term 'fermbiotics' has been suggested as appropriate to describe this combination (Niba *et al.*, 2009).

Probiotics are being considered the better alternative for poultry and already some farmers are using them in preference to antibiotics and other methods. In the wild, the chicken would receive a complete gut microbiota from its mother's faeces and would consequently be protected against infection. However, commercially reared chickens are hatched in incubators which are clean and do not usually contain organisms commonly found in the chicken gut. There is an effect of shell microbiological contamination which may influence gut microbiota characteristics. Moreover, also HCl gastric secretion, which starts at 18 days of incubation, has a deep impact on microbiota selection. Therefore, an immediate use of probiotics supplementation at birth is more important and useful in avian species than in other animals (Kabir, 2009). In broiler nutrition, probiotic species belonging to *Lactobacillus*, *Streptococcus*, *Bacillus*, *Bifidobacterium*, *Enterococcus*, *Aspergillus*, *Candida*, and *Saccharomyces* have a beneficial effect on broiler performance, live weight gain, modulation of intestinal microbiota and pathogen inhibition, intestinal histological changes, immunomodulation, certain haematobiochemical parameters, improving sensory characteristics of dressed broiler meat and promoting microbiological meat quality of broilers (Kabir, 2009).

Although competitive exclusion (CE) fits the definition of probiotics it is one form of treatment with probiotics. The CE culture is usually administered as a single dose to the neonatal animal, such as the day-old chick or a newborn piglet. It can also be used to restore the digestive tract microbiota after antimicrobial treatment. It is usually composed of a mixture of non-pathogenic bacteria typically found in the gastro-intestinal tract of the animal. By using this model, a number of probiotics have been shown to reduce colonization and shedding of *Salmonella* and *Campylobacter*. Competitive exclusion is a very effective measure to protect newly hatched chicks, turkey poults, quails and pheasants and possibly other game birds, too, against *Salmonella* and other enteropathogens (Kabir, 2009).

In poultry, CE was initially developed and used to increase poultry production, where newly hatched chicks can be protected from subsequent *Salmonella* infections by accelerating the establishment of a complex, protective microbiota. Chicks hatched in sterile incubators are not colonized in a normal manner by intestinal microbiota compared with chicks hatched naturally under the hen in a nest containing faecal microbes. It is essential that the CE is administered before exposure to *Salmonella*. The *Salmonella* preventing effect is also shown in field situations. There are several commercial CE cultures on the market also with efficacy against *Campylobacter*, *Clostridium perfringens* and both *Salmonella* and *Campylobacter* (Cox and Pavic, 2009; Berge and Wierup, 2012).

Many studies have demonstrated efficacy in laboratory and field trials and there is a long list of probiotic products licensed in the EU as 'gut microbiota stabilizers' for different animals and production categories (European Commission, 2011). Probiotics are not destroyed by pelleting of food, and are shelf-stable and cost effective. This is especially true for spore-forming probiotics. So, a large variety of *Bacillus* spores have been investigated with regard to their stability and heat tolerance. LAB cultures have been extensively researched for their ability to reduce *Salmonella* infection in poultry. They are used commercially for *Salmonella* control in poultry and turkey production in many countries now (Chen *et al.*, 2012) for broiler growth performance and blood parameters (Alkhalf *et al.*, 2010), for potential protective effects on broiler chickens to *Eimeria tenella* that causes caecal coccidiosis (Giannenas *et al.*, 2012), to degrade aflatoxin (AFB1) and improve animal production in broilers (Zuo *et al.*, 2013). Also the yeast *Saccharomyces boulardii* has an ability to reduce populations of *Salmonella* and *Campylobacter* in broiler chickens subjected to transport stress, and it was able to reduce the frequency of *Salmonella* colonization to lower than the pre-stress (before transport) levels (Berge and Wierup, 2012).

Probiotics for aquaculture

Aquaculture industry has experimented remarkable developments as a result of intensification of large-scale production and globalization of seafood trade. The increasing demand in fish and shellfish culture has led to several changes in aquaculture practices requiring their cultivation at high densities and intensive production with addition of commercial diets, antibiotics, growth promoters and several other additives (Wang *et al.*, 2008; Newaj-Fyzul *et al.*, 2014). As a consequence of this intensive production, farmed aquatic animals have been exposed to stressful conditions that have been related to diseases and deterioration of environmental conditions. Due to this, infection diseases are a significant problem in aquaculture and many approaches have been used to prevent and/or mitigate pathogen effects in those animals (Bondad-Reantaso *et al.*, 2005; Wang *et al.*, 2008).

It has been proven that the microbiota of digestive tract plays an important role in the resistance to infectious diseases. Thus, probiotics have become widely used also in aquaculture for disease control and improvement of fish and shellfish growth, since the use of antimicrobial agents as a preventative measure has been questioned due to the evolution of resistance in pathogenic bacteria. This strategy offers several advantages to overcome the limitations and side effects of antimicrobial drugs and also leads to high production by enhancing growth and disease resistance. (Nomoto, 2005; Ringo *et al.*, 2007; Wang *et al.*, 2008; Nayak, 2010; Mehrim, 2009; Mishra *et al.*, 2001; Sahu *et al.*, 2008).

Nowadays, several preparations of probiotics, either as monospecies or multispecies supplements, are commercially available for aquaculture practices and have been introduced in fish farming as feed additives or incorporated in pond water (Wang *et al.*, 2005; Wang *et al.*, 2008; Najak, 2010; Decamp and Moriarty, 2006).

The range of probiotics evaluated for use in aquaculture is considerably wider than in terrestrial agriculture and numerous microbes have been identified as probiotics for aquaculture practices including *Lactobacillus, Lactococcus, Leuconostoc, Enterococcus, Carnobacterium, Shewanella, Bacillus, Aeromonas, Vibrio, Enterobacter, Pseudomonas, Clostridium and Saccharomyces* species (Nayak, 2010).

These microorganisms differ markedly in their mode of action with the most common effects proposed as being the improvement in feed conversion efficiency and live weight gain, protection against pathogens by competitive exclusion for adhesion sites, production of organic acids, hydrogen peroxide and several other compounds such as bacteriocins, lysozyme, siderophores and also modulation of physiological and immunological responses in fish (Gibson, 1999; Khattab *et al.*, 2004; Vine *et al.*, 2004; Balcazar *et al.*, 2006a; El-Dakar *et al.*, 2007; Al-Dohail *et al.*, 2009; Saenz de Rodrigues *et al.*, 2009; Nayak, 2010)., wuch as Among the numerous beneficial effects of probiotics, modulation of immune system is one of the most commonly reported benefits and many probiotics have been recognized as immunomodulators in finfish and shellfish. Immunological studies have been performed in several fish using different probiotics in order to evaluate their potency to stimulate immunity under *in vitro* and *in vivo* conditions and results indicate that probiotics either individually or in combination can enhance both systemic as well as local immunity in fish (Irianto and Austin, 2002; Ortuno *et al.*, 2002; Panigrahi *et al.*, 2004; Panigrahi *et al.*, 2005; Salinas *et al.*, 2005; Balcazar *et al.*, 2006b; Diaz-Rosales *et al.*, 2006; Kim and Austin, 2006a,b; Salinas *et al.*, 2006; Balcazar *et al.*, 2007; Panigrahi *et al.*, 2007; Aly *et al.*, 2008; Choi and Yoon, 2008; Kumar *et al.*, 2006; Pan *et al.*, 2008; Salinas *et al.*, 2008; Picchietti *et al.*, 2009; Nayak, 2010; Newaj-Fyzul *et al.*, 2014).

It has been demonstrated that probiotics often stimulate innate and cellular immunity by interaction with immune cells such as monocytes and macrophages leading to enhanced phagocytic activity. Besides that, interaction with polymorphonuclear leucocytes and natural killer (NK) cells has also been observed, enhancing innate immune responses. (Aly *et al.*, 2008; Kumar *et al.*, 2006). Some probiotics may induce an increase in the numbers of erythrocytes, leucocytes, lymphocytes, monocytes, and macrophages, as well as neutrophil adherence and migration (Taoka *et al.*, 2006; Aly *et al.*, 2008; Kumar *et al.*, 2006; Abd El Rhman *et al.*, 2009; Sharifuzzaman and Austin, 2010; Newaj-Fyzul *et al.*, 2014). Other effects observed with probiotics administration are the increase in haematocrit, plasma bactericidal and complement activity, respiratory burst activity, besides some enzymes indispensable to fight against pathogens, such as lysozyme, superoxide dismutase, anti-protease and peroxidase (Panigrahi *et al.*, 2004; Salinas *et al.*, 2006; Taoka *et al.*, 2006; Newaj-Fyzul *et al.*, 2007; Aly *et al.*, 2008; Pieters *et al.*, 2008; Sharifuzzaman and Austin, 2010; Sun *et al.*, 2010; Zhou *et al.*, 2010).

In addition to nutritional and health benefits, some probiotics can also be used as water additives playing an important role in decomposition of organic matter, reduction of nitrogen and phosphorous level and control of ammonia, nitrite and hydrogen sulfide. Improved water quality has been associated to probiotics, *e.g.* the use of *Bacillus sp.* to ameliorate bad water quality, and increase survival, growth rates and health of *Penaeus monodon*, besides reducing pathogenic vibrio load (Dalmin *et al.*, 2001; Wang *et al.*, 2008; Nayak, 2010).

The dominant probiotics used in aquaculture belong to Gram-positive bacteria, especially LAB, *Bacillus* and bifidobacteria groups. There has been considerable success with the use of spore-forming species of *Bacillus*, both as feed and water additives, improving health and water quality for finfish and shellfish (Nimrat *et al.*, 2012; Newaj-Fyzul *et al.*, 2014)

In aquaculture, *B. subtilis* and *B. licheniformis* are the most commonly used as probiotics (Moriarty, 1998; Moriarty, 2003; Nayak, 2010). In rainbow trout, immunity enhancement and significant protection against *Yersinia ruckeri* infection has been reported after administration of feed supplemented with a commercial probiotics containing spores of *B. subtilis* and *B. licheniformis* (Raida *et al.*, 2003). Furthermore, administration of this combination of probiotics in trout also resulted in significant improvement of feed conversation rate, growth rate and protein efficiency ratio (Bagheri *et al.*, 2008; Merrifield *et al.*, 2010). Similarly, *B. subtilis* spores when introduced into water eliminated *Vibrio* species from the larvae of snook (Kennedy *et al.*, 1998). Other species of *Bacillus* have also been reported as probiotic in aquaculture, including *B. cereus; B. circulans, B. toyoi, B. megaterium* and *B. clausii* (Gatesoupe, 1994; Irianto and Austin, 2002; Bandyopadhyay and Das Mohapatra, 2009; Li and Mai, 2009). *Bacillus circulans* PB7, was shown to have an important immunostimulation effect in *Catla Catla*, leading to higher survival against *Aeromonas hydrofila* infection (Bandyopadhyay and Das Mohapatra, 2009). Spores of *Bacillus toyoi* and other *Bacillus* species increased the growth of *Scophthalmus maximus* and common snook when used as feed additive (Gatesoupe, 1994; Irianto and Austin, 2002). Feed supplementation with *B. megaterium* also beneficially influenced the immune response, intestinal microbial populations and disease resistance in shrimp (Li and Mai, 2009).

Probiotics containing spore-forming bacteria display some advantages: due to the physical and biological characteristics of the spore they can resist adverse environmental conditions. Preparations containing spores are extremely resistant, heat-stable and have a prolonged shelf life. Moreover, spores can survive transit through the stomach with its hostile (acid) environment, properties that cannot be assured using probiotics in their vegetative form (Huang *et al.*, 2008; Nayak, 2010).

Despite such advantages of spore forming bacteria, several non-spore forming bacteria are used as probiotic in aquaculture as well. These are administrated as vegetative cells and exhibit very promising immunomodulatory results. Among the LAB, *Lactobacillus delbrueckii, Lactobacillus plantarum, Lactobacillus acidophilus, Lactobacillus rhamnosus* and *Lactococcus lactis* have been widely used in aquaculture. *L. rhamnosus* reduced mortality of rainbow trout challenged with *Aeromonas salmonicida* and also combated *Edwardsiella tardia* infection in tilapia, *Oreochromis niloticus* (Nikoskelainen *et al.*, 2001; Pirarat *et al.*, 2006). It has been demonstrated that *L. rhamnosus, L. acidophilus* and *L. lactis* can significantly stimulate phagocytic activity in several fish species (Brunt and Austin, 2005; Brunt *et al.*, 2007; Panigrahi *et al.*, 2004; Pieters *et al.*, 2008). Administration of some LAB probiotics, such as *Carnobacterium maltaromaticum, C. divergens, L. lactis, L. sakei, L. mesenteroides* and *L. rhamnosus* has been related to elevation of serum as well as skin mucosa levels of lysozyme in fish (Kim and Austin, 2006b; Song *et al.*, 2006; Taoka *et al.*, 2006; Balcazar *et al.*, 2007). Application of a mixture of *L. plantarum, L. salivarus* and *L. rhamnosus* to water has also resulted in improved survival of fish larvae (Talpur *et al.*, 2012).

Most research has focused on application of probiotics as single strains; however, combination of multiple species, such as spore forming and non-spore-forming bacteria are also found to increase immunity in fish (Salinas *et al.*, 2005; Salinas *et al.*, 2008; Taoka *et al.*, 2006). Mixtures of *L. plantarum* and *B. subtilis* stimulated higher haematocrit and serum bactericidal activity in Nile tilapia compared to the use of single strains. In addition, supplementation of Nile tilapia with a mixture of *L. acidophilus* and *B. subtilis* resulted in improved growth and increased neutrophil adherence and lysozyme activity, as well as resistance against *Pseudomonas fluorescens* and *Streptococcus iniae* infections (Aly *et al.*, 2008; Cao *et al.*, 2012).

Several studies have confirmed the beneficial effects of both forms of probiotics, either as monospecies or multispecies of microorganisms (Salinas *et al.*, 2005, 2008; Aly *et al.*, 2008; Gram *et al.*, 1999; Balcazar *et al.*, 2006b, 2007; Kumar *et al.*, 2006; Diaz-Rosales *et al.*, 2009). However, it has been demonstrated that multispecies or multistrain probiotics are more effective and greater effects on systemic innate immunity has been observed by using multispecies probiotics in fish (Aly *et al.*, 2008; Balcazar *et al.*, 2006b, Salinas *et al.*, 2006).

Although most probiotic strains used in aquaculture belong to Gram-positive bacteria, a diverse range of Gram-negative bacteria including some strains of *Aeromonas, Vibrio, Pseudomonas* and Enterobacteria species are also applied as probiotics (Gibson *et al.*, 1998; Gram *et al.*, 1999; Chythanya *et al.*, 2002; Irianto and Austin, 2002; Brunt and Austin, 2005; Panigrahi *et al.*, 2007; Kesarcodi-Watson *et al.*, 2008; Pieters *et al.*, 2008; Gobeli *et al.*, 2009; Hai and Buller, 2009).

Thus, a wide range of microorganisms has been identified as (potential) probiotics for aquaculture, with benefits ranging from a better feed conversion, enzymatic contribution to digestion, inhibition of pathogenic microorganisms, providing growth promoting factors, increased immune response and improved water quality (Nimrat *et al.*, 2012; Mohapatra *et al.*, 2013). Therefore, probiotic administration is an important strategy in the modern aquaculture exerting an important contribution to productivity, nutrition of cultured animals, disease control and environmental quality.

Probiotics for ruminants

The rumen is an 'open' fermentation chamber, inhabited by microorganisms that anaerobically digest complex components of feedstuffs and generate fermentation products, mainly acids, and microbial cell mass for utilization by the host. The absorption of fermentation products, mainly volatile fatty acids, from the rumen into the blood, and outflow of microbial cells to the abomasum and small intestine provide the animal with energy and protein, respectively. Interaction between the rumen microbes and the ruminant are the most perfect and complex example of symbiotic relationship between living beings (Russel, 2002).

However, feeding practices necessary to meet high nutritional requirement of ruminants with high performance are a challenge to rumen microbial population. Due to increasing safety concerns about the risk of releasing antibiotic resistance in the environment and the persistence of chemical residues in animal products, manipulation of ruminal fermentation by supplementation of more 'natural' products such as probiotics can be a route capable of increasing digestive efficiency of these animals, improving herd health and productivity (Chaucheyras-Durand and Durand, 2010).

The principal microorganisms used as probiotics for ruminants, as for other animals, are bacteria and fungi. Most bacterial probiotics have maximum efficacy in pre-ruminant calves, whereas fungal products have been shown greater benefits in adult dairy or beef cattle. According to Seo *et al.* (2010), bacterial probiotics species may be classified as lactic acid producing (*Lactobacillus*, *Bifidobacterium*, *Streptococus* and *Enterococcus*), lactic acid utilizing (*Megasphaera* and *Propionibacterium*) and other microorganisms (*Prevotella* and *Bacillus*). Among fungal probiotics, yeasts are commonly used, and commercial products contain species of *Saccharomyces* and *Aspergillus*.

Several mechanisms have been proposed to explain the effects of probiotics and it is likely that the positive results reported in the different animal studies are due to a combination of some, if not all, of these (Chaucheyras-Durand and Durand, 2010). The metabolic activities of the probiotic strains and survivability throughout the gut appear to be of great importance for an optimal efficacy (Chaucheyras-Durand *et al.*, 2008).

High levels of concentrates in feedstuffs are rapidly fermented in the rumen and give rise to a rapid accumulation of lactate, SCFA and other volatile fatty acids, which contribute to decrease ruminal pH. When ruminal pH is below 6.0, the activity of cellulolytic bacteria is seriously decreased and the number of protozoa declines. Among the microbial changes associated with low ruminal pH is an increase in the number of bacteria that are low pH-tolerant such as *Streptococcus bovis* (lactate producer) and *Megasphaera elsdenii* (lactate user). As long as lactate utilizing bacteria can metabolize the lactate that is produced the situation remains under control, although the animals are in sub-acute acidosis (rumen pH between 5.6 and 5.2) and experience fluctuations in dry matter intake resulting in decreased milk production. As pH continues to drop, the lactobacilli population will take over, entailing increased lactate concentration. This elevated lactate concentration contributes to decrease rumen pH below 5.2, and to the disappearance of lactate-utilizing bacteria. This serious condition known as acute or metabolic acidosis can lead to animal death if left untreated (Owens, 1998).

Lactobacillus and *Enterococcus* used as probiotics might prevent ruminal acidosis in dairy cows (Nocek *et al.*, 2002) by facilitating the growth of ruminal microorganisms adapted to the presence of lactic acid in the rumen (Yoon and Stern, 1995) and by stimulating and sustaining an active population of lactic acid utilizers, which in turn would result in elevated ruminal pH (Nocek *et al.*, 2003, 2002). Lactic acid utilizing bacteria have also been used as probiotics to decrease concentrations of lactate and maintain ruminal pH. *Megasphaera elsdenii* utilize lactate and prevent drastic pH drops caused by accumulation of lactate in the rumen when fed a highly fermentable diet (Kung and Hession, 1995) and supplementation of *M. elsdenii* was proposed as a means of preventing acute acidosis in transition animals. Propionibacteria ferment lactate to propionate. Since propionate is the major precursor for gluconeogenesis in ruminants (Reynolds *et al.*, 2003), increments of propionate production in the rumen result in increases of hepatic glucose production (Stein *et al.*, 2006), providing more substrates for lactose synthesis, improving energetic efficiency and reducing ketosis (Weiss *et al.*, 2008).

It is generally believed that fungal cultures mediate their effects on ruminal microbes, especially the fibre digesting and lactate utilizing bacteria. It is theorized that *Saccharomyces* and *Aspergillus* provide micronutrients that act as growth factors, such as organic acids or vitamins, for these bacterial population, thereby increasing bacterial numbers recovered from the rumen (Nisbet and Martin, 1991).

There is also an increasing amount of evidence that live yeast stabilizes ruminal pH and decreases the risk of acidosis (Chaucheyras-Durand *et al.*, 2008). *In vitro* studies have reported that live yeasts could influence the balance of lactate-metabolizing bacteria, by limiting lactate production by *Streptococcus bovis* and favouring lactate uptake by *M. elsdenii* or *Selenomonas ruminantium* (Chaucheyras *et al.*, 1996). Brossard *et al.* (2006) reported that a strain of *Saccharomyces cerevisae* could prevent pH decrease by stimulating certain populations of ciliate protozoa, which rapidly engulf starch and thereby effectively compete with amylolytic, lactate-producing bacteria. Rose (1987) suggested that yeasts remove oxygen in the rumen. Yeast cells in the rumen use available oxygen on the surface of freshly ingested feed to maintain metabolic activity. Jouany *et al.* (1999) observed a significant decrease in redox potential, up to −20 mV, in the rumen with yeast supplementation. This change creates better conditions for the growth of strict anaerobic cellulolytic bacteria, stimulating their attachment to forage particles (Roger *et al.*, 1990), which can improve fibre digestibility (Wallace and Newbold, 2007)

Reducing the environmental impact of livestock, for example by mitigating methane excretion by ruminants, which is estimated to represent 3–5% of global warming power, is also an increasing concern and probiotics may represent an interesting ecological tool to achieve this goal (Martin *et al.*, 2006). Methanogens and acetogenic bacteria compete for hydrogen. Chaucheyras *et al.* (1995) reported that hydrogen utilization and acetate production by a ruminal acetogen bacterial isolate was improved *in vitro* by the addition of a yeast strain, even in the presence of methanogens. In addition, increased propionate by *Propionibacteria* supplementation may reduce hydrogen available for methane production in the rumen (Stein *et al.*, 2006).

In young pre-ruminants, bacterial probiotics such as LAB generally target the small intestine, as the rumen is not yet developed. The proposed mechanisms of probiotics in the post-rumen are similar to those observed in monogastric animals, such as: production of antibacterial compounds (bacteriocins and acids), competition with pathogens for colonization of mucosa and/or for nutrients, production and/or stimulation of enzymes, stimulation of immune response by the host, metabolism and detoxification of desirable compounds and maintenance of a lower pH thereby inhibiting growth of pathogens (Seo *et al.*, 2010).

Chaucheyras-Durand and Durand (2010) suggest that the most significant effects of probiotics have been reported when they have been included in the diet during particularly stressful periods for the animal: at the beginning of the lactation period, at weaning, and after a dietary shift from high forage to high readily fermentable carbohydrates.

In general, the importance of bacterial probiotics fed to young and/or stressed calves has been to establish and maintain 'normal' intestinal microorganisms rather than as a production stimulant (Krehbiel *et al.*, 2003). Some researchers reported that supplementing calves with *Lactobacillus* and *Streptococcus* decreased the incidence of diarrhoea (Beckman *et al.*, 1977; Maeng *et al.*, 1987; Fox, 1988).

The weaning period during which solid feed is introduced is also challenging for the gut. Chaucheyras-Durand and Fonty (2001) reported that the addition of probiotics to the diet of lambs increased the rate at which different bacterial species became established.

Research on the effect of probiotics supplementation on milk yield and composition has been very limited. In dairy ruminants, probiotics have been shown to improve performance, the most consistent effects being an increase in dry matter intake and milk production (Sniffen *et al.*, 2004; Jouany, 2006) In general, an increase in the order of 0.75 to 2 kg milk/day was reported, although the effect on feed intake and milk composition has been more variable (Chiquette, 2009). According to Kung *et al.* (1997), a greater response is generally observed at the beginning of lactation when animals are fed a high-concentrate diet. Also, in beef cattle, growth parameters (average daily, final weight, intake, and feed to gain ratio) have been reported to be improved by daily probiotics supplementation (Lesmester *et al.*, 2004).

Although responses to probiotics can be positive in many cases, feeding practices, composition of animal diets, farm management, dosage, timing and strains of probiotics are parameters which can influence the microbial balance in gastrointestinal tract and consequently affect animal performance (Newbold *et al.*, 1995). These do require additional studies to optimize use of probiotics.

References

Abd El-Rhman, A.M., Khattab, Y.A.E., and Shalaby, A.M.E. (2009). *Micrococcus luteus* and *Pseudomonas* species as probiotics for promoting the growth performance and health of Nile tilapia, *Oreochromis niloticus*. Fish Shellfish Immunol. *27*, 175–180.

Aguirre-Guzmán, G., Lara-Flores, M., Sánchez-Martínez, J.G., Campa-Córdova, A.I., Luna- González, A. (2012). The use of probiotics in aquatic organisms: a review. Afr. J. Microbiol. Res. *6*, 4845–4857.

Al-Dohail, M.A., Hashim, R., and Aliyu-Paiko, M. (2009). Effects of the probiotic, *Lactobacillus acidophilus*, on the growth performance, haematology parameters and immunoglobulin concentration in African Catfish (*Clarias gariepinus*, Burchell 1822). fingerling. Aquac Res. *40*,1642–1652.

Alexopoulos, C., Georgoulakis, I.E., Tzivara, A., Kyriakis, C.S., Govaris, A., and Kyriakis., S.C. (2004). Field evaluation of the effect of a probiotic-containing *Bacillus licheniformis* and *Bacillus subtilis* spores on the health status, performance, and carcass quality of grower and finisher pigs. J. Vet. Med. A Physiol. Pathol. Clin. Med. *51*, 306–312.

Alkhalf, A., Alhaj,M., and Al-homidan, I. (2010). Influence of probiotic supplementation on blood parameters and growth performance in broiler chickens. Saudi Journal of Biological Sciences *17*, 219–225.

Aly, S.M., Ahmed, Y.A.G., Ghareeb, A.A.A., and Mohamed, M.F. (2008). Studies on *Bacillus subtilis* and *Lactobacillus acidophilus*, as potential probiotics, on the immune response and resistance of Tilapia nilotica (*Oreochromis niloticus*) to challenge infections. Fish Shellfish Immunol. *25*, 128–136.

Austin, B., Struckey, L.F., Robertson, P.A.W., Effendi, I., and Grifith, D.R.W. (1995). A probiotic strain of *Vibrio alginolyticus* effective in reducing disease caused by *Aeromonas salmonicida, Vibrio anguillarum* and *Vibrio ordalii*. J. Fish Dis. *18*, 93–96.

Bagheri, T., Hedayati, S.A., Yavari, V., Alizade, M., and Farzanfar, A. (2008). Growth, survival and gut microbial load of rainbow trout (*Onchorhynchus mykiss*) fry given diet supplemented with probiotic during the two months of first feeding. Turk. J. Fish. Aquat. Sci. *8*, 43–48.

Bairagi, A., Ghosh, K.S., Sen, S.K., and Ray, A.K. (2002). Enzyme producing bacterial flora isolated from fish digestive tracts. Aquac. Int. *10*, 109–121.

Balcazar, J.L., de Blas, I., Ruiz-Zarzuela, I., Cunningham, D., Vendrell, D., and Muzquiz, J.L. (2006a). The role of probiotics in aquaculture. Vet. Microbiol. *114*, 173–186.

Balcazar, J.L., Vendrell, D., de Blas, I., Ruiz-Zarzuela, I., Girones, O., and Muzquiz, J.L. (2006b). Immune modulation by probiotic strains: quantification of phagocytosis of *Aeromonas salmonicida* by leucocytes isolated from gut of rainbow trout (*Oncorhynchus mykiss*) using a radiolabelling assay. Comp. Immunol. Microbiol. Infect. Dis. *29*(5–6), 335–343.

Balcazar, J.L., de Blas, I., Ruiz-Zarzuela, I., Vendrell, D., Calvo, A.C., Marquez, I., Gironés, O., and Muzquiz, J.L. (2007). Changes in intestinal microbiota and humoral immune response following probiotic administration in brown trout (Salmotrutta). Br. J. Nutr. *97*, 522–527.

Bandyopadhyay, P., Das Mohapatra, P.K. (2009). Effect of a probiotic bacterium *Bacillus circulans* PB7 in the formulated diets: on growth, nutritional quality and immunity of *Catla catla* (Ham.). Fish Physiol. Biochem. 35, 467–478.

Bechman, T.J., Chambers, J.V., and Cunningham, M.D. (1977). Influence of *Lactobacillus acidophilus* on performance of young dairy calves. J. Dairy Sci. *60(1)*, 74.

Bednorz, C., Guenther, S., Oelgeschläger, K., Kinnemann, B., Pieper, R., Hartmann, S., Tedin, K., Semmler, T., Neumann, K., Schierack, P., Bethe, A., and Wieler, L.H. (2013). Feeding the Probiotic *Enterococcus faecium* Strain NCIMB 10415 to Piglets Specifically Reduces the Number of *Escherichia coli* Pathotypes That Adhere to the Gut Mucosa. Appl. Environ. Microbiol. 79 (24), 7896–7904.

Berge, A.C., and Wierup, M. (2012). Nutritional strategies to combat *Salmonella* in mono-gastric food animal production. Animal *6(4)*, 557–564.

Bogut, I., Milakovic, Z., Brkic, S., Novoselic, D., Bukvic, Z., 2000. Effects of *Enterococcus faecium* on the growth rate and content of intestinal microflora in sheat fish (*Silurus glanis*). Vet. Med. *45*, 107–109.

Bondad-Reantaso, M., Subasinghe, R.P., Arthur, J.R., Ogawa, K., Chinabut, S., Adlard, R., Tan, Z., and Shariff, M. (2005). Disease and health management in Asian aquaculture. Vet. Parasitology *132*, 249–272.

Bosi, P and Trevisi, P. (2010). New topics and limits related to the use of beneficial microbes in pig feeding. Beneficial Microbes. *1(4)*, 447–454.

Brossard, L., Chaucheyras-Durand, F., Michalet-Doreau, B., and Martin, C. (2006). Dose effect of live yeast on rumen microbial communities and fermentations during butyric latent acidosis in sheep: a new type of interaction. Anim. Sci. *82*, 829–836.

Brunt J., and Austin, B. (2005). Use of a probiotic to control lactococcosis and streptococcosis in rainbow trout, *Oncorhynchus mykiss* (Walbaum). J. Fish. Dis. *28*, 693–701.

Brunt, J., Newaj-Fyzul, A., and Austin, B. (2007). The development of probiotics for the control of multiple bacterial diseases of rainbow trout, *Oncorhynchus mykiss* (Walbaum). J. Fish Dis. *30*, 573–579.

Cao, H.P., He, S., Wang, H.C., Hou, S.L., Lu, L.Q., and Yang, X.L. (2012). Bdellovibrios, potential biocontrol bacteria against pathogenic *Aeromonas hydrophila*. Vet. Microbiol. *154*, 413–418.

Chang, C.I., and Liu, W.Y. (2002). An evaluation of two probiotic bacterial strains, *Enterococcus faecium* SF68 and *Bacillus toyoi*, for reducing edwardsiellosis in cultured European eel, Anguilla anguilla, L. J. Fish Dis. *25*, 311–315.

Chaucheyras, F., Fonty, G., Bertin, G., Salmon, J.M., and Gouet, P. (1996). Effects of a strain of *Saccharomyces cerevisae* (Levucell SC), a microbial additive for ruminants, on lactate metabolism *in vitro*. Can. J. Microbiol. *42*, 927–933.

Chaucheyras, F., Fonty, G., Bertin, G., and Gouet, P. (1995). *In vitro* H_2 utilization by a ruminalacetogenic bacterium cultivated alone or in association with an archae methanogen is stimulated by a probiotic strain of *Saccharomyces cerevisiae*. Appl. Environ. Microbiol. *61*, 3466–3467.

Chaucheyras-Durand, F., and Durand, H. (2010). Probiotics in animal nutrition and health. Beneficial Microbes *1*, 3–9.

Chaucheyras-Durand, F., Walker, N.D., and Bach, A. (2008). Effects of active dry yeast on the rumen microbial ecosystem: past, present and future. Anim. Feed Sci. Tecnh. *145*, 5–26.

Chaucheyras-Durand, F., and Fonty, G. (2001). Establishment of cellulolytic bacteria and development of fermentative activities in the rumen of gnotobiotically-reared lambs receiving the microbial additive *Saccharomyces cerevisiae* CNCM I-1077.Reprod. Nut. Dev. *41*, 57–68.

Chen,C.Y., Tsen, H.Y., Lin, C.L., Yu, B., and Chen, C.S. (2012). Oral administration of a combination of select lactic acid bacteria strains to reduce the *Salmonella* invasion and inflammation of broiler chicks. Poult. Sci. *91(9)*, 2139–2147.

Chiquette, J. (2009). The role of probiotics in promoting dairy production. WCDS Advances in Dairy Tecnhology *21*, 143–157.

Cho, J.H., Zhao, P.Y., and Kim, I.H. (2011). Probiotics as a dietary additive for pigs: a review. Journal of Animal and Veterinary Advances *10*(16),2127–2134.

Choi S.H., and Yoon, T.J. (2008). Non-specific immune response of rainbow trout (*Oncorhynchus mykiss*) by dietary heat-inactivated potential probiotics. Immune Netw. *8(3)*, 67–74.

Chythanya, R., Karunasagar, I., and Karunasagar, I. (2002). Inhibition of shrimp pathogenic vibrios by a marine *Pseudomonas* I-2 strain. Aquaculture *208*, 1–10.

Cox, J.M., and Pavic, A. (2010). Advances in enteropathogen control in poultry production. Journal of Applied Microbiology *108*, 745–755.

Dalmin, G., Kathiresan, K., Purushothaman, A., 2001. Effect of probiotics on bacterial population and health status of shrimp in culture pond ecosystem. Indian J. Exp. Biol. *39*, 939–942.

Decamp, O., and Moriarty, D.J.W. (2006). Probiotics as alternative to antimicrobials: limitations and potential. World Aquac. *37(4)*, 60–62.

Diarra, M.S., and Malouin, F. (2014). Antibiotics in Canadian poultry productions and anticipated alternatives. Frontiers in Microbiology *5*, article 282.

Diaz-Rosales, P., Salinas, I., Rodriguez, A., Cuesta, A., Chabrillon, M., Balebona, M.C., Morinigo, M.A., Esteban, M.A., and Meseguer, J. (2006). Gilthead seabream (*Sparus aurata* L.). innate immune response after dietary administration of heat-inactivated potential probiotics. Fish Shellfish Immunol. *20*, 482–492.

Diaz-Rosales, P., Arijo, S., Chabrillon, M., Alarcon, F.J., Tapia-Paniagua, S.T., Martinez- Manzanares E., Balebona, M.C., and Morinigo, M.A. (2009). Effects of two closely related probiotics on respiratory burst activity of Senegalese sole (*Soleasenegalensis*, Kaup) phagocytes, and protection against *Photobacteriumdamselae* subsp. *piscicida*. Aquaculture *293*, 16–21.

El-Dakar, A.Y., Shalaby, S.M., and Saoud, I.P. (2007). Assessing the use of dietary probiotic/prebiotic as an enhancer of spinefoot rabbitfish *Siganus rivulatus* survival and growth. Aquac. Nutr. *13*,407–412.

European Commission 2011. European Union Register of Feed Additives Pursuant to Regulation (EC) No. 1831/2003, Appendixes 3 and 4. Retrieved 31 July 2014, from http://ec.europa.eu/food/food/animalnutrition/feedadditives/comm_register_feed_additives_1831-03.pdf

FAO/WHO 2001. Health and nutritional properties of probiotics in food including powder milk with live lactic acid bacteria. Joint FAO/WHO Expert Consultation. Retrieved 31 July 2014, from http://www.who.int/foodsafety/publications/fs_management/probiotics/en/

Fox, S.M. (1988). Probiotics intestinal inoculants for production animals. Vet. Med. *83*, 806–830.

Gatesoupe F.J. (1994). Lactic acid bacteria increase the resistance of turbot larvae, *Scophthalmus maximus*, against pathogenic Vibrio. Aquat Living Resour. *7*, 277–282.

Giannenas, I., Papadopoulos, E., Tsalie, E., Triantafillou, E., Henikl, S., Teichmann, K., and Tontis, D. (2012). Assessment of dietary supplementation with probiotics on performance, intestinal morphology and microflora of chickens infected with *Eimeria tenella*. Veterinary Parasitology *188*, 31– 40.

Gibson L.F. (1999). Bacteriocin activity and probiotic activity of *Aeromonas media*. J. Appl. Microbiol. Symp. Suppl. *85*, 243S-248S.

Gibson, L.F., Woodworth, J., and George, A.M. (1998). Probiotic activity of *Aeromonas media* on the Pacific oyster, *Crassostrea gigas*, when challenged with *Vibrio tubiashii*. Aquaculture *169*, 111–120.

Gobeli, S., Goldschmidt-Clermont, E., Frey, J., and Burr, S.E. (2009). *Pseudomonas chlororaphis* strain JF3835 reduces mortality of juvenile perch, *Perca fluviatilis* L., caused by *Aeromonas sobria*. J. Fish Dis. *32*, 597–602.

Gram, L., Melchiorsen, J., Spangaard, B., Huber, I., and Nielsen, T.F. (1999). Inhibition of *Vibrio anguillarum* by *Pseudomonas fluorescens* AH2, a possible probiotic treatment of fish. Appl. Environ. Microbiol. *65*, 969–973.

Hai, N.V., Buller, N., and Fotedar, R. (2009). Effects of probiotics (*Pseudomonas synxantha* and *Pseudomonas aeruginosa*) on the growth, survival and immune parameters of juvenile western king prawns (Penaeuslatisulcatus Kishinouye, 1896). Aquac. Res. *40*, 590–602.

Heo, J.M., Opapeju, F.O., Pluske, J.R., Kim, J.C., Hampson, D.J., and Nyachoti, C.M. (2013). Gastrointestinal health and function in weaned pigs: a review of feeding strategies to control post-weaning diarrhoea without using in-feed antimicrobial compounds. Journal of Animal Physiology and Animal Nutrition. *97*, 207- 237.

Hong, H.A., Duc, L.H., and Cutting, S.M. (2005). The use of bacterial spore formers as probiotics. FEMS Microbiol. Rev. *29*, 813–835.

Huang, J.M., La Ragione, R.M., Nunez, A., and Cutting, S.M. (2008). Immunostimulatory activity of *Bacillus* spores. FEMS Immunol. Med. Microbiol. *53*,195–203.

Irianto, A., and Austin, B. (2002). Use of probiotics to control furunculosis in rainbow trout, *Oncorhynchus mykiss* (Walbaum). J. Fish Dis. *25*, 333–342.

Jacela, J.Y., DeRouchey, J.M., Tokach, M.D., Goodband, R.D., Nelssen, J.L., Renter, D.G., and Dritz, S.S. (2010). Feed additives for swine: fact sheets-prebiotics and probiotics, and phytogenics. J. Swine Health Prod. *18*, 132–136.

Jouany, J.P. (2006). Optimizing rumen functions in the close-up transition period and early lactation to dry matter intake and energy balance in cows. Anim. Reprod. Sci. *96*, 250–264.

Jouany, J.P., Mathieu, F., Senaud, J., Bohatier, J., Bertin, G., and Mercier, M. (1999). Influence of protozoa and fungal additives on ruminal pH and redox potential. S. Afr. J. Anim. Sci. *29*, 65–66.

Kabir, S.M.L. (2009). The Role of Probiotics in the Poultry Industry. Int. J. Mol. Sci., *10*, 3531–3546. doi:10.3390/ijms10083531

Kennedy, S.B., Tucker, J.W., Neidic, C.L., Vermeer, G.K., Cooper, V.R., Jarrell, J.L., and Sennett, D.G. (1998). Bacterial management strategies for stock enhancement of warm water marine fish: a case study with common snook (*Centropomus undecimalis*). Bull. Mar. Sci. *62*, 573–588.

Kesarcodi-Watson, A., Kaspar, H., Lategan, M.J., and Gibson, L. (2008). Probiotics in aquaculture: the need, principles and mechanisms of action and screening processes. Aquaculture. *274*,1–14.

Khattab, Y.A., Shalaby, A.M.E., Sharaf, S.M., El-Marakby, H.I., and Rizkalla, E.H. (2004). The physiological changes and growth performance of the Nile tilapia *Oerochromis niloticus* after feeding with Biogen® as growth promoter. Egypt J. Aquat. Biol. & Fish. *8*, 145–158.

Kim, D.H., and Austin, B. (2006a). Cytokine expression in leucocytes and gut cells of rainbow trout, *Oncorhynchus mykiss* Walbaum, induced by probiotics. Vet. Immunol. Immunopathol. *114*, 297–304.

Kim, D.H., and Austin, B. (2006b). Innate immune responses in rainbow trout (Oncorhynchus *mykiss*, Walbaum) induced by probiotics. Fish Shellfish Immunol. *21*, 513–524.

Krehbiel, C.R., Rust, S.R., Zhang, G., and Gilliland, S.E. (2003). Bacterial direct-fed microbials in ruminants diets: performance response and mode of action. J. Anim. Sci. *81*, E120-E123.

Kumar R., Mukherjee, S.C., Ranjan, R., Nayak, S.K.(2008). Enhanced innate immune parameters in *Labeo rohita* (Ham.). following oral administration of *Bacillus subtilis*. Fish Shellfish Immunol. *24*,168–172.

Kumar, R., Mukherjee, S.C., Prasad, K.P., and Pal, A.K. (2006). Evaluation of *Bacillus subtilis* as a probiotic to Indian major carp *Labeo rohita* (Ham.). Aquac. Res. *37*, 1215–1221.

Kung, E. Jr., Kreck, E.M., Tung, R.S., Hession, A.O., Sheperd, A.C., Cohen, M.A., Swain, H.E., and Leedle, J.A.Z. (1997). Effects of a live yeast culture and enzymes on *in vitro* ruminal fermentation and milk production of dairy cows. J. Dairy Sci. *80*, 2045–2051.

Kung, L., Jr., Hession, A.O. (1995). Preventing *in vitro* lactic acid accumulation in ruminal fermentations by inoculation with *Megasphaera elsdenii*. J. Anim. Sci. 73, 250–256.

Lesmeister, K.E., Heinrichs, A.J., and Gabler, M.T. (2004). Effects of supplemental yeast (*Saccharomyces cerevisiae*) culture on rumen development, growth characteristics, and blood parameters in neonatal dairy calves. J. Dairy Sci. *87*, 1832–1839.

Li, J., Tan, B., and Mai, K. (2009). Dietary probiotic *Bacillus* OJ., and isomaltooligosaccharides influence the intestine microbial populations, immune responses and resistance to white spot syndrome virus in shrimp (*Litopenaeus vannamei*). Aquaculture *291*, 35–40.

Maeng, W.J., Kim, C.W., and Shin, H.T. (1987). Effect of a lactic acid bacteria concentrate (*Streptococcus faecium*Cernelle 68). on growth rate and scouring prevention in dairy calves. J. Dairy Sci. *9*, 204–210.

Martin, C., Morgavi, D., Doreau, M., and Jouany, J.P. (2006). Comment réduire la production de methane chez les ruminants? Fourrages. *187*, 282–300.

Mehrim, A.I. (2009). Effect of dietary supplementation of Biogen (Commercial probiotic) on mono-sex Nile tilapia Oreochromis niloticus under different stocking densities. J. Fisher Aquat. Sci. *4*, 261–273.

Meng, Q.W., Yan, L., Ao, X., Zhou, T.X., Wang, J.P., Lee,J.H., Kim,I.H. (2010). Influence of probiotics in different energy and nutrient density diets on growth performance, nutrient digestibility, meat quality, and blood characteristics in growing-finishing pigs. Journal of Animal Science *88*, 3320–3326.

Merrifield, D.L., Bradley, G., Baker, R.T.M., and Davies, S.J. (2010). Probiotic applications for rainbow trout (*Oncorhynchus mykiss* Walbaum). II. Effects on growth performance, feed utilization, intestinal microbiota and related health criteria postantibiotic treatment. Aquac. Nutr. *16*, 496–503.

Mishra, S., Mohanty, S., Pattnaik, P., and Ayyappan, S. (2001). Probiotics: possible application in aquaculture. Fish Chimes. *21*, 31–37.

Moriarty D.J.W. (2003). Probiotics in aquaculture. Microbiol. Aust. *24*, 15–17.

Moriarty, D.J.W. (1998). Control of luminous *Vibrio* species in penaeid aquaculture ponds. Aquaculture. *164*, 351–358.

Mohapatra, S., Chakraborty, T., Kumar, V., Deboeck, G., and Mohanta, K.N. (2013). Aquaculture and stress management: a review of probiotic intervention. J. Anim. Physiol. Anim. Nutr. *79*, 405–430.

Nayak, S.K. (2010). Probiotics and immunity: a fish perspectives. Fish Shellfish Immunol. *29*, 2–14.

Newaj-Fyzul, A., Adesiyun, A.A., Mutani, A., Ramsubhag, A., Brunt, J., and Austin, B. (2007). *Bacillus subtilis* AB1 controls *Aeromonas* infection in rainbow trout (*Oncorhynchus mykiss*, Walbaum). J. Appl. Microbiol. *103*, 1699–1706.

Newaj-Fyzul, A., Al-Harbi, A.H., and Austin, B. (2014). Review: Developments in the use of probiotics for disease control in aquaculture. Aquaculture. *431*, 1–11.

Newbold, C.J., Wallace, R.J., Chen, X.B., and McIntosh, F.M. (1995). Different strains of *Saccharomyces cerevisiae* differ in their effects on ruminal bacteria *in vitro* and in sheep. J. Anim. Sci. 73, 1811–1818.

Niba, A.T., Beal, J.D., Kudi, A.C., and Brooks, P.H. (2009). Bacterial fermentation in the gastrointestinal tract of non-ruminants: Influence of fermented feeds and fermentable carbohydrates. Trop. Anim. Health. Prod. *41*,1393–1407.

Nikoskelainen, S., Ouwehand, A., Salminen, S., and Bylund, G. (2001). Protection of rainbow trout (*Oncorhynchus mykiss*) from furunculosis by *Lactobacillus rhamnosus*. Aquaculture *198*, 229–236.

Nimrat, S., Suksawat, S., Boonthai, T., and Vuthiphandchai, V. (2012). Potential *Bacillus* probiotics enhance bacterial numbers, water quality and growth during early development of white shrimp (*Litopenaeus vannamei*). Vet. Microbiol. *159*, 443–450.

Nisbet, D.J., and Martin, S.A. (1991). Effect of a *Saccharomyces cerevisiae* culture on lactate utilization by the ruminal bacterium *Selenomonasruminantium*. J. Anim. Sci. *69*, 4628–4633.

Nocek, J.E., Kautz, W.P., Leedle, J.A.Z., and Allman, J.G. (2002). Ruminal supplementation of direct-fed microbials on diurnal pH variation and *in situ* digestion in dairy cattle. J. Dairy Sci. *85*, 429–433.

Nocek, J.E., Kautz, W.P., Leedle, J.A.Z., and Block, E. (2003). Direct-fed microbial supplementation on ruminal digestion, health, and performance of pre- and postpartum dairy cattle. J. Dairy Sci. *89*, 260–266.

Nomoto, K. (2005). Prevention of infections by probiotics. J. Biosci. Bioeng. *100*, 583–592.

Organic Trade Association (OTA). (2014). U.S. organic industry valued at nearly $29 billion in 2010. Accessed Sept. 2014. http://www.organicnewsroom.com/2014/05/american_appetite_for_organic.html.

Ortuno, J., Cuesta, A., Rodriguez, A., Esteban, M.A., and Meseguer, J. (2002). Oral administration of yeast *Saccharomyces cerevisiae* enhances the cellular innate immune response of gilthead seabream (Sparusaurata L.). Vet. Immunol. Immunopathol. *85*,41–50.

Owens, F.N., Secrist, D.S., Hill, W.J., and Gill, D.R. (1998). Acidosis in cattle: a review. J. Anim. Sci. *76*, 275–286.

Pan, X., Wu, T., Song, Z., Tang, H., and Zhao, Z. (2008). Immune responses and enhanced disease resistance in Chinese drum, *Miichthys miiuy* (Basilewsky), after oral administration of live or dead cells of *Clostridium butyricum* CB2. J. Fish Dis. *31*,679–686.

Panigrahi, A., Kiron, V., Kobayashi, T., Puangkaew, J., Satoh, S., and Sugita, H. (2004). Immune responses in rainbow trout *Oncorhynchus mykiss* induced by a potential probiotic bacteria *Lactobacillus rhamnosus* JCM 1136. Vet. Immunol. Immunopathol. *102*, 379–388.

Panigrahi, A., Kiron, V., Puangkaew, J., Kobayashi, T., Satoh, S., and Sugita, H. (2005). The viability of probiotic bacteria as a factor influencing the immune response in rainbow trout *Oncorhynchus mykiss*. Aquaculture. *243*, 241–254.

Panigrahi, A., Kiron, V., Satoh, S., Hirono, I., Kobayashi, T., Sugita, H., Puangkaew, J., and Aoki, T. (2007). Immune modulation and expression of cytokine genes in rainbow trout *Oncorhynchus mykiss* upon probiotic feeding. Dev. Comp. Immunol. *31*, 372–382.

Picchietti, S., Fausto, A.M., Randelli, E., Carnevali, O., Taddei, A.R., Buonocore, F., Scapigliati, G., and Abelli, L. (2009). Early treatment with *Lactobacillus delbrueckii* strain induces an increase in intestinal T-cells and granulocytes and modulates immune- related genes of larval Dicentrarchuslabrax (L.). Fish Shellfish Immunol. *26*, 368–376.

Pieters, N., Brunt, J., Austin, B., and Lyndon, A.R. (2008). Efficacy of in-feed probiotics against *Aeromonas bestiarum* and *Ichthyophthirius multifiliis* skin infections in rainbow trout (Oncorhynchus mykiss, Walbaum). J. Appl. Microbiol. *105*, 723–732.

Pirarat, N., Kobayashi, T., Katagiri, T., Maita, M., and Endo, M. (2006). Protective effects and mechanisms of a probiotic bacterium *Lactobacillus rhamnosus* against experimental *Edwardsiella tarda* infection in tilapia (Oreochromis niloticus). Vet. Immunol. Immunopathol. *113*, 339–347.

Quadros, A.R.B., Kiefer, C., Ribeiro, N.L.C., and Zink, L.A. (2001). Características qualitativas da carne de suínos alimentados com rações contendo ou não probióticos. In Reunião Anual da Sociedade Brasileira de Zootecnia, Piracicaba. Anais, Piracicaba, Brazil, 794–795.

Raida, M.K., Larsen, J.L., Nielsen, M.E., and Buchmann, K. (2003). Enhanced resistance of rainbow trout, *Oncorhynchus mykiss* (Walbaum), against *Yersinia ruckeri* challenge following oral administration of *Bacillus subtilis* and *B. licheniformis*. J. Fish Dis. *26*, 495–498.

Reynolds, C.K., Aikman, P.C., Lupoli, B., Humphries, D.J., and Beever, D.E. (2003). Splanchnic metabolism of dairy cows during the transition from late gestation through early lactation. J. Dairy Sci. *86*, 1201–1217.

Ringo, E., Myklebustd, R., Mayhewe, T.M., and Olsen, R.E. (2007). Bacterial translocation and pathogenesis in the digestive tract of larvae and fry. Aquaculture. *268*, 251–264.

Roger, V., Fonty, G., Komisarczuk-Bony, S., and Gouet, P. (1990). Effects of physicochemical factors on the adhesion to cellulose Avicel of the ruminal bacteria *Ruminococcus flavefaciens* and *Fibrobacter succinogenes* subsp. *succinogenes*. Appl. Environ. Microbiol. *56*, 3081–3087.

Rose, A.H. (1987). Responses to the chemical environment. In: The yeasts (Ed. Rose, A.H., Harrison, J.S.). Vol 2, Academic Press, London, pp. 5–40.

Ross, G.R., Gusils, C., Oliszewski, R., de Holgado, S.C., and González, S.N. (2010). Effects of probiotic administration in swine. Journal of Bioscience and Bioengineering *109(6)*, 545–549.

Russel, J. B. Rumen microbiology and its role in ruminant nutrition. 1 ed. Ithaca, New York, 2002. 119 p.

Saenz de Rodriguez, M.A., Diaz-Rosales, P., Chabrillon, M., Smidt, H., Arijo, S., Leon- Rubio, J.M., Alarcón, F.J., Balebona, M.C., Morinigo, M.A., Cara, J.B., and Moyano, F.J. (2009). Effect of dietary administration of probiotics on growth and intestine functionally of juvenile Senegalese sole (*Solea senegalensis*, Kaup 1858). Aquac. Nutr. *15*, 177–185.

Sahu, M.K., Swarnakumar, N.S., Sivakumar, K., Thangaradjou, T., and Kannan, L. (2008). Probiotics in aquaculture: importance and future perspectives. Ind. J. Microbiol. *48*, 299–308.

Salinas, I., Cuesta, A., Esteban, M.A., and Meseguer, J. (2005). Dietary administration of *Lactobacillus delbrueckii* and *Bacillus subtilis*, single or combined, on gilthead seabream cellular innate immune responses. Fish Shellfish Immunol. *19*, 67–77.

Salinas, I., Diaz-Rosales, P., Cuesta, A., Meseguer, J., Chabrillon, M., Morinigo, M.A., and Esteban, M.A. (2006). Effect of heat-inactivated fish and non-fish derived probiotics on the innate immune parameters of a teleost fish (*Sparus aurata* L.). Vet. Immunol. Immunopathol. *111*, 279–286.

Salinas, I., Abelli, L., Bertoni, F., Picchietti, S., Roque, A., Furones, D., Cuesta, A., Meseguer, J., and Esteban, M.A. (2008). Monospecies and multispecies probiotic formulations produce different systemic and local immunostimulatory effects in the gilthead seabream (*Sparus aurata* L.). Fish Shellfish Immunol. 25, 114–123.

Seo, J.K., Kim, S.W., Kim, M.H., Upadhaya, S.D., Kam, D.K., and Ha, J.K. (2010). Direct-fed microbials for ruminant animals. Asian-Aust. J. Anim. Sci. *23*, 1657–1667.

Sharifuzzaman, S.M., and Austin, B. (2010). *Kocuria* SM1 controls vibriosis in rainbow trout (*Oncorhynchus mykiss*, Walbaum). J. Appl. Microbiol. *108*, 2162–2170.

Simon, O. (2010). An interdisciplinary study on the mode of action of probiotics in pigs. Journal of Animal and Feed Sciences *19*, 230–243.

Sniffen, C.J., Chaucheyras-Durand, F., Ondarza, M.B., and Donaldson, G. (2004). Predicting the impact of a live yeast strain (Levucell SC CNCM I-1077). on rumen kinetics and ration formulation. In: Proceedings of South West Nutrition Conference, Phoenix, Arizona, USA.

Song, Z., Wu, T., Cai, L., Zhang, L., and Zheng, X. (2006). Effects of dietary supplementation with *Clostridium butyricumon* the growth performance and humoral immune response in Miichthysmiiuy. J. Zhejiang Univ. Sci. B. *7*, 596–602.

Stein, D.R., Allen, D.T., Perry, E.B., Bruner, J.C., Gates, K.W., Rehberger, T.G., Mertz, K., Jones, D., and Spicer, L.J. (2006). Effects of feeding propionibacteria to dairy cows on milk yield, milk components, and reproduction. J. Dairy Sci. *89*, 111–125.

Suda, Y., Villena, J., Takahashi, Y., Hosoya, S., Tomosada, Y.,Tsukida, K., Shimazu, Y., Aso, H., Tohno, M., Ishida, M., Makino, S., Ikegami, S., and Kitazawa, H. (2014). Immunobiotic *Lactobacillus jensenii* as immune-health promoting factor to improve growth performance and productivity in post-weaning pigs. BMC Immunology *15*, 24.

Sun, Y.Z., Yang, H.L., Ma, R.L., and Lin, W.Y. (2010). Probiotic applications of two dominant gut *Bacillus* strains with antagonistic activity improved the growth performance and immune responses of grouper *Epinephelus coioides*. Fish Shellfish Immunol. *29*, 803–809.

Talpur, A.D., Memon, A.J., Khan, M.I., Ikhwanuddin, M., Danish, M.M., Daniel Abol-Munafi, A.B. (2012). Inhibition of pathogens by lactic acid bacteria and application as water additive multi isolates in early stages larviculture of *P. pelagicus* (Linnaeus, 1758). J. Anim. Plant Sci. *22*, 54–64.

Taoka, Y., Maeda, H., Jo, J.Y., Kim, S.M., Park, S.I., Yoshikawa, T., and Sakata, T. (2006). Use of live and dead probiotic cells in tilapia *Oreochromis niloticus*. Fish. Sci. 72, 755–766.

Thacker, P.A. (2013). Alternatives to antibiotics as growth promoters for use in swine production: a review. Journal of Animal Science and Biotechnology. *4*,1–12.

Van der Fels-Klerx H.J., Puister-Jansen L.F., Van Asselt E.D., and Burgers, S.L. (2011). Farm factors associated with the use of antibiotics in pig production. J. Anim. Sci. *89*,1922–1929.

Vine, N.G., Leukes, W.D., Kaiser, H., Daya, S., Baxter, J., and Hecht, T. (2004). Competition for attachment of aquaculture candidate probiotic and pathogenic bacteria on fish intestinal mucus. J. Fish Dis. *27*, 319–326.

Vondruskova, H., Slamova, R., Trckova, M., Zraly, Z., and Pavli, I. (2010). Alternatives to antibiotic growth promotors in prevention of diarrhoea in weaned piglets: a review. Vet. Med. *55*,199–224.

Wallace, R.J., Newbold, C.J. (2007). Microbial feed additives for ruminants. Rowett Research Institute, Bucksburn, Aberdeen, UK. Available online:http://www.old-herborn-university.de/literature/books/OHUni_book_8_article_9.pdf.

Wang, Y.B., Xu, Z.R., and Xia, M.S. (2005). The effectiveness of commercial probiotics in Northern White Shrimp (*Penaeus vannamei*) ponds. Fish. Sci. *71*, 1034–1039.

Wang, Y.B., Li, J.R., and Lin, J. (2008). Probiotics in aquaculture: Chalenges and outlook. Aquaculture. *281*,1–4.

Weiss, W.P., Wyatt, D.J., and McKelvey, T.R. (2008). Effect of feeding propionibacteria on milk production by early lactation dairy cows. J. Dairy Sci. *91*, 646–652.

Yoon, I.K., and Stern, M.D. (1995). Influence of direct-fed microbials on ruminal microbial fermentation and performance of ruminants: A review. Asian-Aust. J. Anim. Sci. *8*, 533–555.

Zhou, X.X., Tian, Z.Q., Wang, Y.B., and Li, W.F. (2010). Effect of treatment with probiotics as water additives on tilapia (*Oreochromis niloticus*) growth performance and immune response. Fish Physiol. Biochem. *36*, 501–509.

Zuo, R., Chang,J., Yin, Q., Wang,P., Yang,Y., Wang,X., Wang,G., and Zheng, Q. (2013). Effect of the combined probiotics with aflatoxin B1-degrading enzyme on aflatoxin detoxification, broiler production performance and hepatic enzyme gene expression. Food and Chemical Toxicology *59*, 470–475.

Pharmaceutical Aspects of Probiotics and Prebiotics

32

Indu Pal Kaur, Parneet Kaur Deol, Simarjot Kaur Sandhu and Praveen Rishi

Abstract

In recent times, increasing evidence of benefits of probiotics, for health restoration coupled with consumer's inclination towards safe, natural, and cost effective substitutes for drugs, has generated immense interest in exploring their therapeutic potential. Administering probiotics as such not only compromises their viability but also fails to guarantee their successful establishment in gut. In recent times emphasis has been laid on pharmaceutically developing probiotics, which will not only maintain their viability during storage but also assist in their successful delivery to, and establishment at, the site of action in a viable form. Further to above, supplementation of probiotics with prebiotics is also gaining predominance in the recent past. Numerous reports indicate that co-encapsulating prebiotics with probiotics, improves viability of the latter, apart from the independent growth promoting effect of the former on the inherent healthy gut microbiota of the host.

The current chapter aims to analyse probiotics as pharmaceuticals, their encapsulation techniques, existing pharmaceutical preparations, their regulatory status in some countries, and future research needs. The current status of prebiotics in health care is also highlighted.

Introduction

Increasing concerns like cost of treatment and adverse drug reactions with current therapies and consumer's inclination towards natural substitutes has intensified research to develop safe and cost effective alternative medical therapies (Foligne *et al.*, 2013; Reid *et al.*, 2003). Since the advent of mankind, some 10,000 years ago, the human genome has hardly changed (0.05%) however, the drastic dietary and lifestyle changes in human life have brought about an onslaught of chronic diseases like cancers, diabetes, depression, cardiovascular and neurodegenerative diseases (Eaton, 2006). Research findings have established a strong link between the diet and general well being (Khan and Ansari, 2007) and have clearly demonstrated the significance of nutrition in preventing the occurrence of various clinical pathologies (Broekaert and Walker, 2006). However, it is neither practically possible nor a preference of mankind to make high fibre, raw or fresh insecticide-free food a part of the routine diet, since tinned and fast foods have become a way of life for the modern man. Hence, a more practical solution to the issue is to promote these beneficial foods consumed by the earlier man as medicine. The concept was proposed by Hippocrates way back in 431 BC, when he stated that 'Let food be thy medicine'. The same idea forms the basis for a new class of agents proposed as 'nutraceuticals'. The term, first coined in 1989, was defined as a dietary substance that delivers concentrated form of a bioactive substance in quantities that exceed what can be obtained from food (Penner *et al.*, 2005).

There has been a growing appreciation of the role played by commensal microbes in human and animal health; be it through mediation of intestinal development and innate immunity, or digestion of food and protection of host against diseases (Reid, 2008; Sansonetti, 2006). However, a very fine balance exists between the beneficial and pathogenic microorganisms in the human gut. Whenever this homeostasis of healthy versus harmful microorganisms gets disrupted, the microbial equilibrium is skewed towards disease onset (Mombelli and Gismondo, 2000; Sarkar, 2013). The existing clinical attitude favours use of antibiotics for killing and eradication of pathogenic microorganisms, however, the biological relationship between multicellular organisms and the microbial world is now considered to be one of symbiosis rather than enmity. Hence, the belief that, preservation of the normal microbiota or its suitable modulation, rather than its deletion, can provide a greater clinical benefit, has led to the emergence of probiotics as a successful therapeutic option (Sreeja and Prajapati, 2013). Furthermore, rapid progress in the development and application of molecular techniques for assessment of the human gut microbiome has helped in establishing the true nature, diversity and metabolic potential of the gut microbiota, now also referred to as the 'hidden organ' (Quigley, 2011).

Probiotic bacteria are currently classified as nutraceuticals and account for billions in sales annually (Cook *et al.*, 2012). A recent global probiotic market analysis estimated a 7% annual growth, boosted mainly by a rising request from the Asian and European consumer, with a 48 billion dollars forecast in the next 5 years (Report, 2012).

In order, to produce health benefits, the probiotic has to be present in a sufficient number and in a viable form at the time of consumption. Further to this, it has to withstand stresses encountered in the gastrointestinal tract until it is established in the colon. Reports indicate that products containing free probiotic cells shows poor survival in the gut (De Vos *et al.*, 2010). Providing these cells in a physically protected form, to resist adverse environmental conditions, is therefore an approach currently receiving considerable interest (Kailasapathy, 2009). Encapsulation of bacterial cells by various means tends to provide them with a resistant enclosure protecting against the adverse environment of the gastrointestinal tract (GIT). Entrapment of live probiotic strains within a carrier material, not only protects them (Liserre *et al.*, 2007; Shima *et al.*, 2009; Thantsha *et al.*, 2009), but also extends their viability upon storage and helps present them in a conveniently consumable form (Lian *et al.*, 2003; Oliveira *et al.*, 2007a,b). The intelligent selection of encapsulating material (i) can avoid release of viable cells in stomach and upper GIT where conditions for growth and establishment are not favourable; (ii) can ensure slow release of viable cells over extended periods such that time required for adhesion to the mucosa is available; (iii) prevents fast wash out of probiotic otherwise observed with free probiotics (owing to bioadhesive property of encapsulated material); and (iv) may also co-incorporate suitable prebiotic to ensure viability during storage and easy establishment in GIT mucosa upon administration.

The present chapter emphasizes various pharmaceutical approaches which are reported to not only improve probiotic viability on storage, but also facilitate its successful delivery and establishment in the gut (especially the large intestine) where it exhibits its beneficial effects. We have also included information on pharmaceutical prebiotic preparations used alone or in combination with probiotics.

Probiotics

Definition and health benefits

The Ayurvedic document, Charaka Samhita (1000 BC), includes a term 'Jataragini'-fire in the stomach, which probably referred to the beneficial microbiota of the gastrointestinal tract. It is mentioned in these ancient records that this fire of the stomach is the sustaining force of all living beings. Another term, 'Takra,' i.e. fermented milk, is often referred to as 'Amrita' or elixir. It may thus be said that knowledge about the importance of natural and healthy gut microbiota existed even during the early times. However, it is now scientifically established that some *Lactobacillus* strains stabilize the intestinal microbiota simultaneously making it healthy while destroying or limiting various pathogenic strains. More than 100 years ago, in 1907, Elie Metchnikoff was the first to propose a scientific rationale for the role of lactobacilli in maintaining health and longevity (Caramia, 2004). The term probiotic was however coined for the first time in the year 1965, when Lilly and Stillwell (1965) used it to describe any substance or organism that contributes to intestinal microbial balance; however, Fuller (1989) in 1989 elaborately emphasized their role in health.

According to FDA working definition, probiotics are classified as 'live biotherapeutics' and are defined as 'live microorganisms, including bacteria and yeast, that are intended to have a therapeutic effect in humans and are used in disease prevention or treatment, regardless of the route of administration'. The latter may extend from oral to intra-vaginal, and topical (Hoffman, 2008; Ross, 2006).

The term 'probiotic' includes not only bacteria belonging to the genera *Lactobacillus* and *Bifidobacterium*, but it also extends to other lactic acid producing bacteria (leuconostocs, pediococci, lactococci, enterococci, and streptococci), non lactic acid producing bacteria (*Bacillus, Clostridium, Propionibacterium*, Gram-negative *Escherichia coli*), and yeast (*Saccharomyces*) (Burgain *et al.*, 2011). Although the use of spores of some *Bacillus* or *Clostridium* spp. may appear controversial from a safety point of view, however, their advantages like heat stability, resistance to low pH of stomach and to other deleterious conditions like bile secretions in the intestinal environment, long term storage without refrigeration, and no need for encapsulation, has increased research interest in them compared with the more vulnerable vegetative cells (Bader *et al.*, 2012; Foligne *et al.*, 2012). While a substantial number of microbial species have been reported to exhibit potential probiotic properties (Holzapfel *et al.*, 2001), only the most documented and robust strains make it to the market (Foligne *et al.*, 2013). Further to this, significant differences are observed not only in between the species but also among strains belonging to the same species attributing mainly to diversity in their phenotypes (Azais-Braesco *et al.*, 2010).

Probiotics in general are believed to provide positive health benefit via three major mechanisms, which include (i) production of antimicrobial compounds such as organic acids, free fatty acids, hydrogen peroxide, and bacteriocins (Balcázar *et al.*, 2007; Gollop *et al.*, 2005) and conditions (acidic pH due to release of SCFA) (O'Hara and Shanahan, 2007; Paton *et al.*, 2006) unfavourable for pathogen growth; (ii) competing with pathogens for adhesion site (Balcázar *et al.*, 2007; O'Hara and Shanahan, 2007) and nutrients (Meier and Steuerwald, 2005); and (iii) stimulation of host immune system (Bell, 2007; O'Hara and Shanahan, 2007; Rook and Brunet, 2005). The potential areas of their application widely ranges from gastrointestinal diseases like colon and gastric cancers, diarrhoea, enterocolitis, *H. pylori* infections, IBS, IBD, Crohn's disease, sucrose isomaltase deficiency, and lactase deficiency to non gastrointestinal disease applications like topical disorders, neurological disorders, hypertension, pain, allergy, obesity, kidney stones, cadmium toxicity, diabetic foot ulcers, infections, and immunomodulation (Fig. 32.1). The list is inexhaustible! However, it may be noted that not all these claims have been validated and the task lies with the researchers involved in these areas to confirm whether these are just pretentious opinions or have some truth in them. It will be important to scientifically explain and experimentally prove these statements.

Ideally a probiotic strain needs to meet the following criteria before it can be categorized as prebiotic:

- It should have proven beneficial effects in the host.
- Its ingestion must not present any risk to the host.

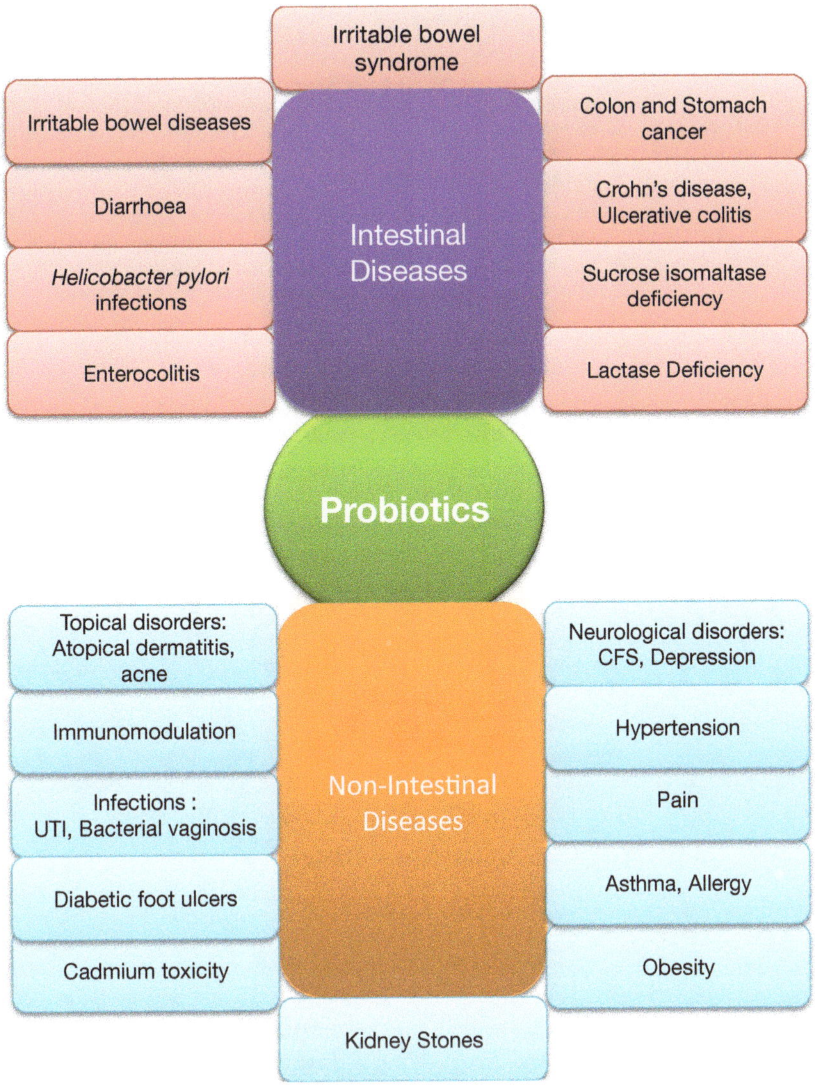

Figure 32.1 Diseases targeted by probiotics. CFS, Chronic Fatigue Syndrome; UTI, Urinary Tract Infection.

- It should be able to withstand and survive physiological stresses met by it during storage and upon ingestion. The latter include acidic pH of the stomach and presence of biliary salts.
- It should reach in sufficient number at the site of action, i.e. the large intestine or colon, in a viable form.
- It must maintain its characteristics and remain viable and stable during the manufacturing process and storage.
- It must be able to proliferate and colonize in the digestive tract so as to continue manifesting beneficial effects even after discontinuation of therapy (Kaur *et al.*, 2009).

The probiotic concept, though with us for decades (supported majorly by fad and folklore in the past) is now gaining support, from the scientific community not only in the form of experimental studies in animal models, but also, as, well designed clinical studies for various diseases (Quigley, 2011). The growing market interest in promoting health in a natural way has further intensified research and investments in this area. This has placed an overwhelmingly large number of new products comprising probiotics on the shelves of the supermarkets and pharmacies. Some of the well known probiotic preparations which are present in the market (as pharmaceutical/nutraceutical) and supported by human clinical trials for their efficacy in management of various clinical conditions are shown in Table 32.1. Food products containing probiotics, howsoever are not discussed here.

Pharmaceutical aspects

Probiotics have caught the fancy of the scientific community majorly in the past decade. This is witnessed by the appearance of hundreds and thousands of research and review papers on the internet at a single click of any relevant keyword in the area. Apart from scientists, many big food and pharma giants including Chr. Hansen, Abbott Nutrition, Pfizer, Nestle, and Merck are investing a lot of their resources (money, time, and manpower) in this area of nutraceuticals (O'Riordan *et al.*, 2001; www.isapp.net). As far as existing marketed probiotic products are concerned, most of them are available in the form of powders, capsules, tablets, drops, chewing gum, lozenges, straws, stick packs, and bottle caps generally containing combinations of two or more probiotic strains with or without vitamins or prebiotics.

Table 32.1 Examples of clinically proven probiotic preparations (Sreeja and Prajapati, 2013)

Probiotic preparation	Marketed by	Composition	Clinical indication	Dosage form
VSL#3 and VSL#3-DS	Sigma-Tau Pharmaceuticals, Inc., Gaithersburg, MD	*B. longum*, *B. infantis*, *B. breve* *L. acidophilus*, *L. casei*, *L. bulgaricus*, *L. plantarum* and *S. salivarius* subsp. *thermophilus*	Ulcerative colitis (UC), ileal pouch and irritable bowel syndrome	Powder
Probio-Tec	Chr. Hansen, Denmark	*B. animalis* Bb-12, *L. acidophilus* NCFB 1748, *L. paracasei* F19, *L. acidophilus* LA-5, *L. casei* 431, *L. rhamnosus* GG, *S. thermophilus* TH-4, *L. fermentum* PCC	Diarrhoea, in combating antibiotic associated side effects, urogenital disorder, skin disorders, constipation, antibiotic-associated diarrhoea in children, infantile diarrhoea, colic/irritability in infants	Powder, capsules, chewable tablets, stick packs
HOWARU, Dophilus, Bifido, Rhamnosus	Dupont, USA; Halal and Kosher Certification, ISO 9001, HACCP	*B. lactis* HN019, *B. lactis* Bi-07, *L. acidophilus* NCFM, *L. rhamnosus* HN001	Protect (for cold and flu), restore (for gut microbiota restoration) and balance (for improving digestion and immunity)	Powder
BioGaia	BioGaia, Sweden	*L. reuteri* Protectis	Improved gut and immune health in adults and infants and for oral health.	Probiotic straw, chewable tablets, drops, lozenges and gum
BLIS K12	BLIS Technologies Ltd., New Zealand	*S. salivarius* K-12	Oral health	Powder
ProbioKid	Institute Rosell-Lallemand, Canada	*L. helveticus* R0052, *B. bifidum* R0071, *B. infantis* R0033+FOS	Common infections in children during winter	Powder
Lalflor	Institut Rosell-Lallemand	*S. cerevisiae boulardii* CNCM I-1079	Diarrhoea including travellers' diarrhoea and diarrhoea occurring during tube feeding	Capsule
Protecflor	Pasteur Institute, France	*S. cerevisiae* variant boulardii CNCM I-1079, *L. rhamnosus* R0011, *L. helveticus* R0052, *B. longum* R175	Common infections	Powder

Presently commercialized probiotic products, be they the oil suspensions for infants, lozenges and chewing gums for oral health, or single dose sachets to be taken with food and beverages, are being designed to be more consumer friendly (easy to handle and use) (Czinn and Blanchard, 2009; Sreeja and Prajapati, 2013).

However the major concern that still needs to be addressed in reference to these preparations is the capacity of these strains to retain their viability during storage and transit through the gastrointestinal tract, so as to reach the desired site of establishment. The beneficial effects of probiotic microorganisms appear when they arrive in sufficient numbers, and in a viable form, in the gastrointestinal tract (Chavarri *et al.*, 2012; Gilliland, 1989). However, many reports in literature indicate poor survival of free probiotic cells both upon incorporation into a dosage form and in the harsh gut environment (Anal and Singh, 2007; Chavarri *et al.*, 2012).

It is generally considered that a daily dose between 10^6 and 10^9 colony forming units (cfu) is required for the elicitation of beneficial effects of probiotics (Ouwehand and Salminen, 1998; Rokka and Rantamäki, 2010). However, the use of multiple strains in a single preparation (considered better than using a single strain, by some workers (Meier and Steuerwald, 2005; Sarkar, 2013), may be in order to match the natural human microbiota), has made it difficult to correctly calculate the dose (Sreeja and Prajapati, 2013). Further to this, due to reports on high loss of viability on storage, manufacturers have started adding high overages to the preparation (Sreeja and Prajapati, 2013). This not only adds to the cost of the product but also results in administration of a variable dose of probiotics (Bora *et al.*, 2009).

Another concern is the transient nature of the administered probiotics. To produce reasonable impact, probiotics need to be retained in the GIT for longer durations, to allow sufficient time for adherence/effects. Clinical trials with probiotic strains have shown that the most of the probiotics usually disappear from the gastrointestinal tract within a couple of days after the ingestion is discontinued. To achieve the desired long-term effects, repeated administration is required (Saarela *et al.*, 2000), which not only adds to the cost of therapy but also leads to decrease in its acceptability as an alternative therapy both by the patients and the physicians, alike.

Approaches used to enhance viability

Considering the enormous potential of probiotics in a wide range of diseases and the need to maintain their viability for illustration of health promoting effects, several approaches have been tried to enhance the viability of probiotics. These include:

1 selection of resistant strains or spores over vegetative cells which elicit a better capacity to resist physical, mechanical and physiological stress (Foligne *et al.*, 2012);
2 exposing the existing probiotic strains to mild but controlled stress conditions for bestowing on them the capacity to adapt to stress (stress adaptation) (Mills *et al.*, 2011);

3 incorporation of micronutrients like prebiotics into probiotic formulations;
4 use of various encapsulation techniques using suitable polymers to entrap probiotics and protect them against adverse storage and physiological conditions.

Need of pharmaceutical couture: encapsulation of probiotics

The presence of diverse conditions (variation in pH, presence of enzymes and mechanical stress of gut motility and peristalsis) in the human digestive tract requires a systematically designed probiotic release system. Encapsulation is a process of physicochemical or mechanical restriction of a substance in a particulate material resulting in small particles containing an active agent in the core material surrounded by a coating or shell (Chavarri *et al.*, 2012). A variety of polymers, carbohydrates, fats and waxes can be used for encapsulating the actives depending on the active to be incorporated and/or protected. Encapsulation of probiotic bacteria for imparting protection against severe environmental factors has been explored extensively (Anal and Singh, 2007; Kailasapathy, 2006; Krasaekoopt *et al.*, 2003). The goal of encapsulation is not only to create a safe and favourable micro-environment for bacteria in which they can survive during processing and storage but to release them at the appropriate site (e.g. large intestine or colon) (Chavarri *et al.*, 2012), and in sufficient numbers.

The general aims of encapsulation are:

1 to protect the probiotic bacteria on storage and on passage through the stomach, since free probiotic is reported to have low survival at highly acidic pH (1.2–4) of stomach (Anal and Singh, 2007);
2 to release the probiotics at their target in the gut (Rokka and Rantamäki, 2010);
3 to extend the stay of probiotic in gastrointestinal tract so as to provide time for its establishment in the gut wall. This can also be achieved through mucoadhesion of the encapsulated probiotic to the mucosal membranes lining the epithelial tissue (Sogias *et al.*, 2008). Although, exact mechanism of mucoadhesion is not fully understood, but it is believed that mucoadhesive polymers interact with mucus through weak interactions (electrostatic, hydrogen bonding and hydrophobic) (Khutoryanskiy, 2011), diffusion effects, and friction-like mechanical effects (Cook *et al.*, 2012).

Generally the bacteria is entrapped in microparticles with a size ranging from 1 to 100 μm; thus, the process is called microencapsulation (Rokka and Rantamäki, 2010). Although a few reports claim the development of probiotic nanoparticles, however, these nanoencapsulating techniques invariably result in killing the probiotic population through chemical or enzymatic procedures (Feher and Lejto, 2012). The current microencapsulated probiotics in the market are typically between tens of microns to a few millimetres in size (Cook *et al.*, 2012; Ding and Shah, 2009b), with reports on survival improvement with increasing size (Muthukumarasamy *et al.*, 2006; Rokka and Rantamäki, 2010).

However, the success of encapsulation will depend on rate of survival of probiotic bacteria during the processing. The process has to be gentle enough to avoid any damage to the entrapped cells. In addition, the encapsulating material used must be non-cytotoxic and non-antimicrobial, so as to ensure safety of host as well as the bacteria (Chavarri *et al.*, 2012; Cook *et al.*, 2012). Furthermore, it has to be mechanically strong to protect against any adverse environmental storage conditions thereby ensuring that the entrapped cells are maintained alive. However, it must be semi permeable, so as to release the probiotic cells at a specific site of the human gut (Chavarri *et al.*, 2012).

Encapsulation techniques

Encapsulation techniques are broadly classified into four categories, namely extrusion, emulsification, drying, and coacervation techniques. All the techniques do suffice the purpose of improving the survival of probiotic on storage and delivery, but the selection of the best one depends upon the size and production scale desired, and maximum shear that probiotic cells can withstand (Chavarri *et al.*, 2012).

Extrusion techniques

Extrusion, the oldest and most common (King, 1995) approach, involves simple steps of dispersing the probiotic in a hydrocolloid/polymer solution, introducing this cell suspension into continuously stirred hardening/cross-linking solution at a fixed rate, and hardening of the introduced hydrocolloid solution in the hardening solution (Fig. 32.2). Factors which monitor the size of the developed microspheres includes diameter of the orifice, viscosity and flow rate of the polymeric solution, distance from the orifice to the cross-linking solution, concentration and temperature of the polymer solution, and speed of stirring (Brun-Graeppi *et al.*, 2011). The hydrocolloid cell suspension can be introduced to the hardening solution either dropwise (using needle) or by jet breakage (using sophisticated equipment which work on vibration technology, namely Encapsulator BIOTEC or jet cutter technology) (Lord Rayland, 1878; Pruesse and Vorlop, 2004). Regardless of selected techniques, the cell suspension drops need to solidify for successful entrapment. Solidification is achieved by a variety of mechanisms, namely formation of ionotropic and thermal gels, precipitation, polymerization, freezing, coextrusion, interfacial coacervation, or interfacial polymerization. The resulting hydrocolloid beads are highly porous and can be coated using oppositely charged polymers (e.g. coating negatively charged alginate beads with positively charged chitosan), transamination or spraying the polymer solution in fluidized bed coater in order to assure a better retention of encapsulated probiotic bacteria (Chavarri *et al.*, 2012).

Extrusion does not involve use of deleterious solvents and is gentle in operation. Therefore high viability of probiotic cells is maintained due to minimal losses during processing (Krasaekoopt *et al.*, 2003). Further, it can be performed both under aerobic and anaerobic conditions. However, extrusion processes fail to produce microspheres with size <500μm and requires use of low to moderate viscosity polymer solutions (Reis *et al.*, 2006). Rapid cross-linking and hardening at the surfaces of the

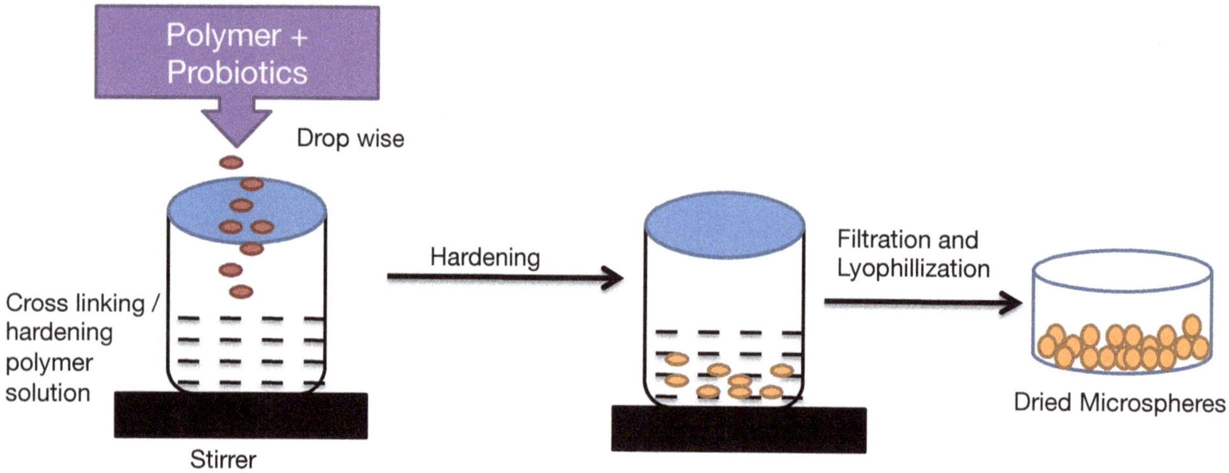

Figure 32.2 Steps involved in encapsulating probiotics using extrusion technique.

microspheres delays the movement of cross-linking ions into the inner core, resulting in less stable microspheres (Liu *et al.*, 2002). Furthermore, scaling up of the process is generally difficult due to the slow production of microspheres (Burgain *et al.*, 2011). Some recent literature reports on microencapsulation of probiotic strains by extrusion are enlisted in Table 32.2. The most commonly used polymer for this technique is alginate, probably because of its non-toxic nature, biocompatibility, cost-effectiveness and mild gelling conditions. However, alginate tends to form porous microparticles, which is responsible for lower probiotic entrapment due to their leakage. Hence, they are generally coated with another polymer like chitosan to reduce pore size.

Emulsification method

An emulsion is defined as dispersion of two immiscible liquids stabilized by adding one or more emulsifier(s). When the core phase is aqueous, it is termed as a water-in-oil (w/o) emulsion however, for a hydrophobic core, oil-in-water emulsions (o/w) results. The emulsification technique has been widely used for encapsulation of various microbial cells (Adhikari *et al.*, 2000; Ding and Shah, 2009a; ÖZer *et al.*, 2008). Aqueous suspension of cells and/or polymer (dispersed phase) dispersed in an oil/organic phase (continuous phase), are homogenized to form a w/o emulsion with the aid of suitable emulsifiers. The dispersed phase is then congealed by cooling or addition of a cross-linking agent to the emulsion (Fig. 32.3). The microspheres thus produced are subsequently harvested by filtration or centrifugation. The method is basically a chemical technique which involves microencapsulation by interfacial polymerization. Factors which majorly affect the microsphere size include speed of stirring, rate of addition of the cross-linking solution, and concentration of the emulsifier (Shah and Ravula, 2000). Table 32.3 lists some recent studies where emulsification is employed to encapsulate probiotic strains.

The emulsion technique is easy to scale up, and gives small sized microspheres with a high survival rate of the bacteria (Chen and Chen, 2007b). However, the main drawback lies in large variation in size and shape of microspheres produced by these techniques. Production of the desired size can be achieved by varying the speed of agitation and the water/oil ratio (Kailasapathy, 2009). Further to this, being a batch process, this technique is not highly acceptable by the industry (Chavarri *et al.*, 2012).

Drying

Spray drying

Spray drying is a commonly used method of encapsulation in the food industry (Picot and Lacroix, 2004). It involves rapid evaporation of water through atomization of a probiotic emulsion or suspension and carrier material in a dry gas (Fig. 32.4). Factors like product feed rate, air flow, feed temperature, inlet air temperature, and outlet air temperature need optimization in order to produce discrete well-formed microspheres (O'Riordan *et al.*, 2001; Vega and Roos, 2006). For example, if the inlet temperature is low, there is reduction in rate of water evaporation leading to formation of aggregates and cracks in the polymeric membrane (Brun-Graeppi *et al.*, 2011), while at high inlet air temperatures viability of probiotic cells gets adversely affected due to dehydration of the cells as well as inactivation of essential enzymes that maintain cellular balance (Ananta *et al.*, 2005; Doherty *et al.*, 2010; John *et al.*, 2011). Similarly, the feed temperature adjustment is crucial as it can modify the viscosity of the polymer solution which in turn compromises the spraying efficiency (Brun-Graeppi *et al.*, 2011).

The advantages of spray drying include rapidity, high reproducibility and relatively low cost of the procedure. However, on large scale production, high installation and operational costs pose a major challenge to the process (John *et al.*, 2011). The range of polymers that can be used for encapsulation is also limited (Brun-Graeppi *et al.*, 2011; Gouin, 2004).

Freeze drying/lyophilization

Freeze drying is a dehydration process which works by freezing the product and then reducing the surrounding pressure to allow the frozen water to sublimate directly from the solid phase to the gas phase. A suspension/emulsion of probiotics is frozen in the

Table 32.2 Recent examples of microencapsulation by extrusion

Strain	Material used	Biofunctionality	Reference
L. gasseri and *B. bifidum* (strain numbers not reported)	Alginate beads coated with chitosan	Increased acid and bile stability	Chávarri *et al.* (2010)
L. rhamnosus GG and *L. acidophilus* NCFM	Alginate beads coated with chitosan	Increased viability and acid stability	Sohail *et al.* (2010)
Streptococcus phocae PI80	Alginate beads coated with chitosan	Increased acid stability	Kanmani *et al.* (2011b)
L. acidophilus LA14 and *B. lactis* BI07	Alginate beads with xanthan gum; alginate beads with cellulose acetate phthalate	Increased acid stability	Albertini *et al.* (2010)
L. casei LBC81 LYO	Sodium alginate, amidated low-methoxyl pectin	Pectin incorporation (Alginate: Pectin:: 1:4 and 1:6) improves acid and bile stability	Castilla *et al.* (2011)
E. faecium MC13	Sodium alginate coated with chitosan	Increased viability and resistance to simulated gastric (pH 2.0) and intestinal fluids (pH 7.5)	Kanmani *et al.* (2011a)
L. plantarum NCIMB 8826	Sodium alginate coated with chitosan	Increased viability	Nualkaekula *et al.* (2012)
L. casei ATCC 393	Alginate, chitosan and carboxymethyl cellulose	Increased acid and bile stability	Lia *et al.* (2011)
L. casei 01	Sodium alginate and starch	Reduced porosity and hygroscopicity, increased viability	Chan *et al.* (2011)
L. plantarum CRL 1815 and *L. rhamnosus* ATCC 53103	Xanthan:gellan gum (1%:0.75%); Jamilan:gellan gum (1%:1%)	Xanthan:gellan gum (1%:0.75%) improves acid stability	Pranteda *et al.* (2012)
B. longum 15708	Alginate and O-palmitoylated alginate	Increased viability	Aminea *et al.* (2014)
Lactobacillus rhamnosus GG ATCC 53103	Whey protien	Increased acid stability and peptic-resistance	Doherty *et al.* (2011)
L. acidophilus La5	Sodium alginate and resistant starch	Increased survival rate of *L. acidophilus* La5 in Iranian white brined cheese after 6 months of storage.	Mirzaeia *et al.* (2012)
L. casei 01, *L. paracasei* LAFTI L26, *L. acidophilus* Ki and *B. animalis* Bb-12	Sodium alginate	Highest survival rates through exposure to lethal conditions of temperature and pH	Borges *et al.* (2011)

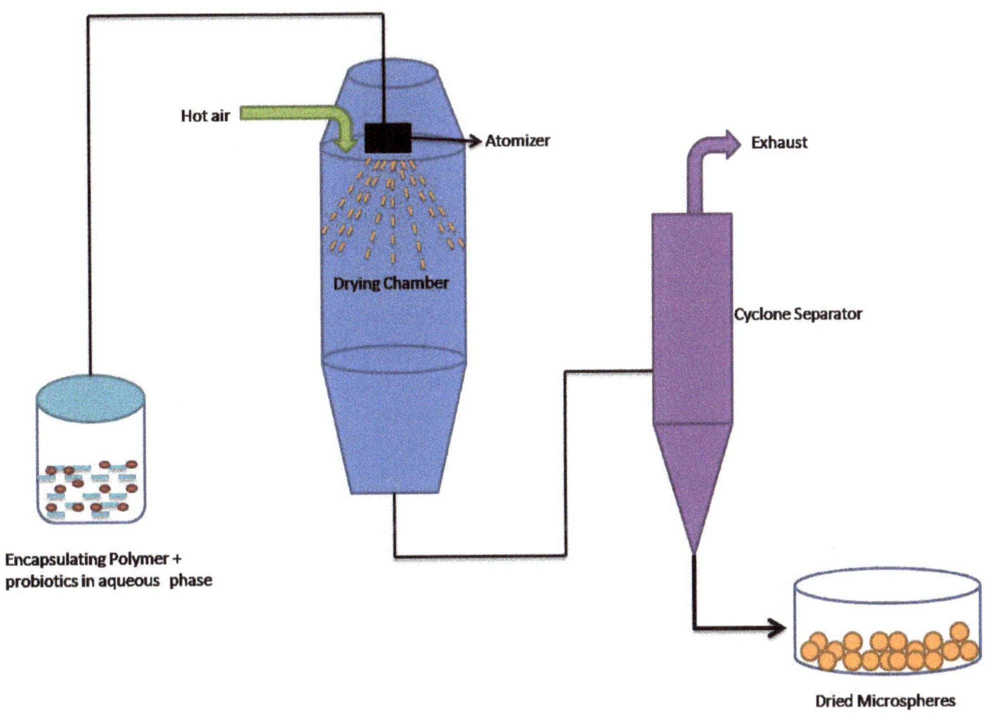

Figure 32.3 Steps involved in encapsulating probiotics using emulsification technique.

Table 32.3 Recent examples of microencapsulation of probiotics by emulsification

Strain	Material used	Biofunctionality	Reference
L. paracasei F19 and *B. lactis* Bb_{12}	Milk proteins	Increased viability and acid stability	Heidebach *et al.* (2009a)
L. casei 431	Sodium caseinate, gellan gum and glucono-δ-lactone	Increased viability and acid stability	Nag *et al.* (2011)
L. helveticus and *L. delbrueckii* spp. *lactis* (strain numbers not reported)	Sodium alginate	Viability of microorganisms decreased as stirring rate was increased	Flores *et al.* (2014)
S. cerevisiae Y235	Alginate–chitosan	Increased survival rate under harmful stress	Song *et al.* (2013)
L. acidophilus CGMCC1.2686	Alginate	Improved acid and bile stability	Cai *et al.* (2014)
L. rhamnosus LC705	Sweet whey	Increased acid and bile stability	Gonzáleza *et al.* (2009)

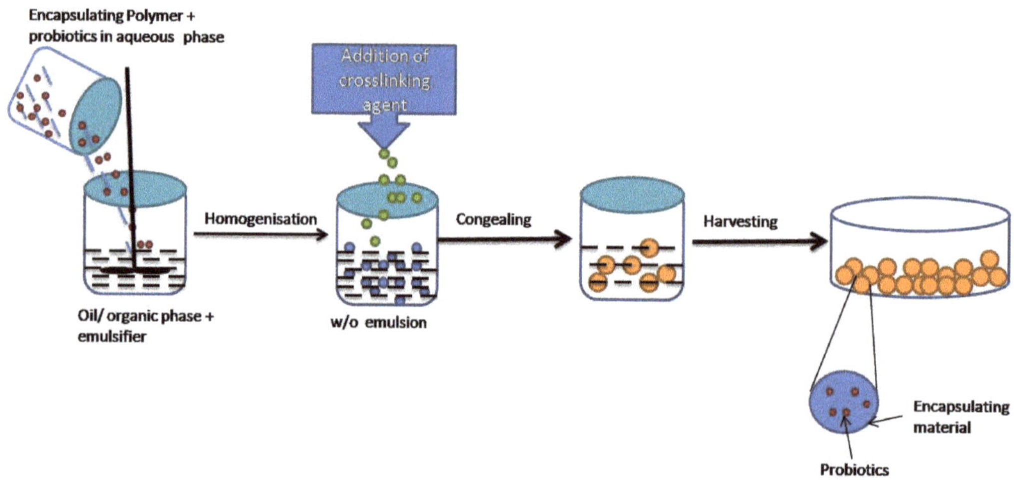

Figure 32.4 Steps involved in encapsulating probiotics using spray drying.

presence of carrier material at low temperatures. This will lead to sublimation of the water under vacuum. Cryoprotectants can be used during freeze drying in order to improve probiotic viability. A major advantage of this method is that water phase transition and oxidation are avoided (Capela *et al.*, 2006). However, freeze-drying is a very expensive technology (Chávez and Ledeboer, 2007) and is used as a second step of encapsulation process i.e. to dry probiotics previously encapsulated by other techniques, as emulsion (Heidebach *et al.*, 2010). In this way it is possible to improve the stability not only in the stomach but also during transit through the gastrointestinal tract, thus optimizing the beneficial effects of probiotic consumption.

Spray freeze drying

Spray freeze drying combines processing steps that are common to freeze-drying and spray drying. It involves atomisation of probiotic suspensions into a cold vapour phase of cryogenic liquid (liquid nitrogen) resulting in a dispersion of frozen droplets that are further freeze dried (De Vos *et al.*, 2010; Kailasapathy, 2009; Semyonov *et al.*, 2010; Wang *et al.*, 2006). The obtained microspheres have uniform size with large specific surface area, but the technique has a long processing time and involves high cost (30–50 times more expensive than spray-drying) (Zuidam and Shimoni, 2009).

Some recent studies using drying method for encapsulating probiotic are listed in Table 32.4.

Coacervation

Microencapsulation using the coacervation technique involves phase separation of one or more incompatible polymers from the initial coating polymer solution by altering the pH, temperature or composition of the solution (John *et al.*, 2011; Oliveira *et al.*, 2007a,b; Park and Chang, 2000). This leads to the separation of incompatible polymer and deposition of dense coacervate phase surrounding the core material resulting in formation of microspheres (Gouin, 2004; John *et al.*, 2011; Nihant *et al.*, 1995; Oliveira *et al.*, 2007a,b). To further strengthen the microspheres sometimes chemical or enzymatic cross-linking agents can be used. Crucial factors which can influence size distribution are volume of the dispersed phase, addition rate of the incompatible polymer to the coating polymer solution, stirring rate of the dispersion and core material to be encapsulated (Nihant *et al.*, 1995; Nihant *et al.*, 1994).

Coacervation technique for encapsulating probiotics is relatively simple and cost effective, it avoids use of high temperature or organic solvents, and allows higher loads of probiotic cell. However, being a batch process, its scale up and industrial applicability is limited. Furthermore, for extended shelf life an additional

Table 32.4 Examples of microencapsulation using drying methods

Strain	Method	Material used	Biofunctionality	Reference
L. paracasei LMG P-21380	Spray freeze drying	Maltodextrin and trehalose	Increased probiotic viability	Semyonov *et al.* (2010)
L. acidophilus La-5	Spray drying	Reconstituted sweet whey or skimmed milk	Increased acid and bile stability	Maciel *et al.* (2014)
L. acidophilus NCIMB 701748	Spray drying	Sodium alginate, chitosan and hydroxypropyl methylcellulose (HPMC)	Increased stability and viability	Yonekura *et al.* (2014)
B. animalis Bb-12	Spray drying	Reconstituted skimmed milk (RSM)/inulin and RSM/ oligofructose-enriched inulin	Increased stability of the ricotta cream samples	Freirea *et al.* (2013)
B. animalis Bb-12	Spray drying	Whey protein	Increased viability	Cislaghi *et al.* (2012)
B. animalis Bb-12	Spray drying	Gum arabic, gelatin and pectin	Increased viability	Behzadi *et al.* (2013)
L. casei Nad, *L. plantarum* com, *L. paracasei* A13, *L. plantarum* 8329 and *L. acidophilus* A9	Spray drying	Skimmed milk	Increased resistance to simulated gastric conditions	Paéz *et al.* (2012)
B. animalis Bb-12	Spray drying	Reconstituted skimmed milk (RSM) with the prebiotics inulin, oligofructose, and oligofructose-enriched inulin	Addition of inulin and oligofructose improves probiotic entrapment; oligofructose enriched inulin improves viability on storage	Fritzen-Freire *et al.* (2012)
L. rhamnosus GG	Spray drying	Whey protein isolate (WPI)–maltodextrin, WPI–maltodextrin–glucose, WPI–inulin, and WPI–inulin–glucose mixtures	Addition of glucose improves viability on storage	Ying *et al.* (2012)
L. reuteri DSM 17938 c	Spray drying	Sodium alginate and chitosan	Improved acid tolerance, allowing the cell population to remain constant after 3 h in simulated gastric conditions and increased heat tolerance	Malmo *et al.* (2013)
L. reuteri DSM 20016)	Spray drying	Whey protein	Increased stability during storage and adequate survival in simulated gastric juices and a distinct release in intestinal juices.	Jantzen *et al.* (2013)
L. casei 01	Spray drying	Alginate, fructooligosaccharide and chitosan	Increased resistance to simulated gastric conditions	Ivanovska *et al.* (2012)

drying process needs to be included on the final product which may be harmful to the cells.

Encapsulation materials

The biomaterials used for probiotic encapsulation include natural and synthetic polymers. Ideally the encapsulation materials should be cheap, biocompatible, biodegradable, non-toxic, and non-irritant. Various polymers used in microencapsulation along with the advantages and issues related to their use have been summarized in the Table 32.5.

Issues with the encapsulation approach

Although numerous studies on encapsulation of probiotics for better viability and survival have been reported using a range of polymers, the approach suffers from various issues, as discussed below.

Cost of microencapsulation

It is estimated that the cost of encapsulated probiotic bacteria is about two or three times that of free probiotics (Brownlie, 2007). The increase in cost can be attributed to added steps in the process and requirement of time and financial resources for product development. Encapsulation of probiotics using natural biopolymers is often difficult to scale up, and involves high processing costs (Anal and Singh, 2007; Doleyres and Lacroix, 2005). Above all, the encapsulation process suffers from losses in terms of low probiotic entrapment, as one log reduction in probiotic count on encapsulation is generally recorded in literature.

Reduction in cost can be achieved by employing easier manufacturing of products, lowering wastage of bacterial material, and improving the overall health impact of the product. Apart from this, it has to be kept in mind that these encapsulated probiotics are specialized products with applications different from those achieved with free probiotics.

Survival of bacteria during encapsulation and drying

Thermal and osmotic resistance of lactic acid bacteria is species dependent (Lian *et al.*, 2002; Picot and Lacroix, 2003; Rokka and Rantamäki, 2010). Survival rates in the spray-drying and freeze drying processes fall in range of 70–85%. Though such survival rates are acceptable, the prolonged storage stability of these products is often low.

Table 32.5 Commonly used encapsulating polymers

Polymer	Source	Chemical structure	Gelation mechanism	Advantages	Disadvantages	Reference
Alginate	Brown algae	Linear polysaccharide consisting of 1→4 linked β-(d)-glucuronic (G) and α-(l)-mannuronic (M) acids	Gelling upon contact with divalent metals (e.g. calcium, cadmium or zinc), forming what has been described as an 'egg box structure' between four G residues	Mild gelling conditions; lack of toxicity; GRAS status; biocompatible; low cost	Variable composition affecting its gelation; difficult scaling up; porous microparticulate systems	Gouin (2004), Rinaudo (2008), Stokke *et al.* (2000), Wahab *et al.* (1997)
Chitosan	Crustacean shells, insect cuticles and the membranes of fungi	Linear cationic polysaccharide containing both glucosamine and N-acetyl glucosamine residues	Ionotropic gelation in the presence of anions	Lack of toxicity; GRAS status; biocompatible; low cost	Variable composition; inhibitory effect on LAB; mostly used as a coating material for microspheres at low concentration; acid labile	Chávarria *et al.* (2010), Groboillot *et al.* (1993)
Milk proteins (such as casein, whey protein)	Milk	Composed of lactose and soluble proteins	Acid-induced gelation for caseins and heat induced gelation for whey proteins	Biocompatible; lack of toxicity; excellent gelation properties		Hébrard *et al.* (2006), Heidebach *et al.* (2009a,b), De Castro-Cislaghi *et al.* (2012), Doherty *et al.* (2011), Livney (2010), Millqvist-Fureby *et al.* (2001)
κ-Carrageenan	Red algae	Linear structure composed of d-galactose units alternately linked by α(1–3) and β(1–4) linkages	Thermal/ionotropic; gelation in the presence of potassium ions	Biocompatible; low cost	Brittle gels; does not withstand stresses; variable composition; used potassium ion has inhibitory effect on some bacteria	Audet *et al.* (1988), Chen and Chen (2007b)
Cellulose acetate phthalate	Chemical synthesis	Cellulose derivative polymer	pH dependent	Enteric drug delivery; physiologically inert	Cannot form gels by ionotropic gelation method; only capsules have been developed by emulsification method	Anal and Singh (2007), Fávaro-Trindade and Grosso (2002)
Gums						
Gellan	*Pseudomonas elodea*	Linear polysaccharide, consists of repeating units of glucose, glucuronic acid, and rhamnose	Thermal/ionotropic; gelation in the presence of cations	Able to withstand the high temperature of the autoclaving process without significant loss of gel strength; acid-resistant	High gel setting temperature (80 –90°C for 1 h) which results in high injury to probiotic cells.	Moslemy *et al.* (2002), Muthukumarasamy *et al.* (2006)
Xanthan	*Xanthomonas campestris*	Pentasaccharide units formed by two glucose units, two mannose units and one glucoronic acid unit.	Ionotropic gelation in the presence of divalent cations, commonly calcium ions	Excellent in protecting probiotic bacteria from acid environment of gastric juice at pH 2 and 3% taurocholic acid; highly resistant to enzymatic degradation	Variable composition	Ding and Shah (2009a)
Gelatin	Collagen	Consists of glycine, proline and 4-hydroxyproline residues	Thermal/cross-linking using formaldehyde or glutaraldehyde/ physical cross-linking using high pressure, irradiation	Makes a thermoreversible gel; amphoteric nature gives the ability of having synergistic effects with anionic polysaccharides such as gelan gum		Anal and Singh (2007), King (1995), Krasaekoopt *et al.* (2003)
Agar	Red Algae	Agarbiose backbone with alternating 1, 3-linked 3-d-galactopyranose and 1,4-linked 3, 6-anhydro-a-l-galactopyranose	Thermal gelation	Resistance to degradation by most microorganisms; lack of toxicity; biocompatible	Low mechanical strength; high cost	Olguín (2012), Rosevear (1984)
Starch	Maize, potato, barley and oat	Polysaccharide consisting of amylose and amylopectin	Thermal gelation	Resistant starch (RS) is not degraded by the pancreatic amylase enters the intestine in the indigestible form. This specification apart from giving the microbeads good enteric delivery characteristic (release of bacterial cells in the large intestine), also gives them prebiotic functionality as they can be used by the probiotic bacteria in the intestine	Decomposes under acidic conditions and presence of pancreatic enzymes in the GIT	Phillips *et al.* (1995), Sajilata *et al.* (2006)
Polyvinyl alcohol (PVA)	Chemical synthesis		Cross-linking using boric acid	Good chemical resistance, good thermal stability, good physical properties, excellent biocompatibility, and low cost	Microorganisms enclosed in the PVA matrix can be drastically damaged by boric acid during the bead preparation process	DeMerlis and Schoneker (2003)

Removal of water in drying techniques may damage cell membranes and associated proteins, because water is important in stabilizing biological molecules (Ananta *et al.*, 2005; Lian *et al.*, 2002; Rokka and Rantamäki, 2010). The presence of deoxidant and desiccant (e.g. sugars) in the dried probiotic products has been found to improve cell survival (Hsiao *et al.*, 2004; Lian *et al.*, 2002). Sugars are known to protect dehydrated biomaterials by acting as water substitutes. They can replace water molecules around proteins and polar residues of membrane phospholipids and form hydrogen bonds with the proteins when water is removed thereby preventing cell damage to a great extent (Ananta *et al.*, 2005). Further, in the freeze drying process, cryoprotectants are usually added to maintain probiotic viability. Sultana *et al.* (Sultana *et al.*, 2000) reported that glycerol improved probiotic survival during freezing by almost 100-fold.

The microencapsulation process utilizing natural polymers like alginates also suffers from viability losses on processing. Alginates form porous beads on ionotropic gelation with calcium chloride solution. Various approaches have been tried to minimize the leakage of probiotics from such a porous network. Coating alginate beads with positively charged polymers like chitosan tends to improve the probiotic entrapment. Apart from this, addition of Hi-Maize starch to alginate has also been found to result in a higher number of live bacteria in the microcapsules (Sultana *et al.*, 2000). Increasing the alginate concentration and capsule size also increases the survival of probiotics during heat treatment (Mandal *et al.*, 2006).

Erroneous method of studying survival rate

Due to the fact that probiotic cells need to be alive when they are consumed, it becomes important to have an error proof and reproducible method for studying the encapsulation efficiency of probiotics. Commonly used plate count method, although easy to perform, does not always give true picture about viable bacteria (Rokka and Rantamäki, 2010). A non-cultivable population of metabolically active probiotic cells, capable of producing health promoting effect, fails to get representation in this method (Lahtinen *et al.*, 2008).

The use of commercial live/dead kits (LIVE/DEAD® *BacLight*™) and techniques like flow cytometry can be used to get more relevant information about the metabolic status of encapsulated bacteria (Lahtinen *et al.*, 2006). These tests are based on fluorescent staining of nucleic acids and can distinguish live bacteria with intact cytoplasmic membrane from dead bacteria with a compromised membrane (Rokka and Rantamäki, 2010). Furthermore, few modifications are required in sample preparation of these encapsulated probiotics in order to correctly estimate the viable count using conventional total plate count method (Champagne *et al.*, 2011). Generally, in formulating an encapsulated system dried cells are used. These dried cells have passed through a number of stressful processes during their production which include pumping, centrifugation, freezing, and desiccation. As a result some cells are stressed (viable but non cultivable). They need to be rehydrated with at least 20–50% solids for 15–30 minutes at 30–37°C before going for dilutions and sample preparation. Furthermore, it is important to properly homogenize the first dilution for some time (need optimization depending on the strain) so as to break the chains of probiotic bacteria (Champagne *et al.*, 2011).

Status of probiotics: regulatory issues

Being a preventative and curative therapy, the status of probiotics is still not clear in the health industry. Clinical practitioners use probiotic pharma products mostly as supplements (Sreeja and Prajapati, 2013). Adding to the vow is the lack of well defined globally accepted regulations regarding probiotics. This has led to uncontrolled entry of unscrupulous probiotic products in the market.

The regulatory framework for probiotics differs from country to country. Probiotic foods and food supplements are covered by the Food Products Directive and Regulation (regulation 178/2002/EC; directive 2000/13/EU) among the European countries which considers them as foods/functional foods/novel foods/natural remedy. Some of the probiotic products which were marketed as herbal products not obtaining a drug registration as per the directive of Herbal Medicinal Products Directive (2004/24/EC) are considered under food supplements category. Prior to their use in the European Union, it is essential that all probiotic food and supplement health claims are scientifically evaluated by the European Food Safety Authority (EFSA). Presently, EFSA experts do not accept that adequate data are available to confirm health claims on the beneficial effects of any probiotic strains (Gibson *et al.*, 2011; Mercenier *et al.*, 2002; Saxelin, 2008). The panel on Dietetic Products, Nutrition, and Allergies of the EFSA has rejected all but one of over 300 health claims on the benefits of probiotic bacteria. As a consequence of this not even a single probiotic product, food or food supplement can mention the health benefits of the incorporated strain (Foligne *et al.*, 2013).

EFSA suggests that a multidisciplinary team focusing on study design, biomarkers and clinical endpoints of research should be constituted to design, evaluate, and execute future studies (Foligne *et al.*, 2013).

In Japan, FOSHU (Food for Specified Health Use) regulation was established in 1993. A FOSHU label can only be used on the product after its scientific dossier which fully substantiates the claim(s) is prepared and approved by the Japanese Ministry of Health and Welfare (Mercenier *et al.*, 2002). A major difference between the European and Japanese regulations is that in Japan an active, approved ingredient, shown to be present in the product, allows a producer to put FOSHU label where as in Europe a clinical trial with the final product, and not with the constituent ingredient is a prerequisite.

In the USA, probiotics are considered as dietary supplements and require FDA approval for use as biologicals. The US FDA is the key federal regulatory agency which has an impact on both the clinical research and the development of probiotics as biological drugs. Probiotic products may be regulated by several FDA centres depending on which regulatory category it is being used for. These centres include Center for Food Safety and Applied Nutrition, which is responsible for foods, dietary supplements, and 'foods for special dietary uses;' Center for Veterinary Medicine, which is responsible for animal feed products and animal drugs; and Center for Biologics Evaluation and Research (CBER), which regulates human biologics (Heyman and Menard, 2002).

As such, there are no guidelines that exist in India for use of probiotics as drugs. But for probiotics in food, the Indian Council of Medical Research (ICMR) along with the Department of Biotechnology (DBT) formulated their guidelines in 2011.

In Canada, probiotics are regulated by Natural Health Products Directorate (NHPD) which became a law in 2004. It covers therapeutic claims, risk reduction and structure/function claims. It allows for qualified claims based on weight of evidence provided. Strain specific evidence regarding identity, safety and efficacy needs to be attested and can be asked for or questioned by NHPD at any time.

Probiotics are thus majorly being used as food or food ingredients. These guidelines do not include probiotics to be delivered as pharmaceuticals or genetically modified microorganisms (GMOs) (ICMR-DBT, 2011). As far as Food Safety Standards Act is concerned, a special third category – 'Foods for Special Dietary Uses/Functional Foods/Nutraceuticals/Health Supplements' – is also being mentioned. But this act does not differentiate functional food, nutraceutical, dietary supplements; instead it classifies all these products as foods for special dietary application. This allows for nutraceuticals, functional foods and food supplements with therapeutic claims to be marketed without declaration of nutraceutical content, clinical trial results and health studies. Although selling of such products is being done via health food stores or internet, the strain of the microorganism used for the preparation of corresponding product is, generally, not mentioned on the product label. Many over the counter products available in health food stores are generally neither reliable nor efficacious, and many make false claims. The availability as food, food supplement or drug, is amplifying the challenges for the consumer or the health care worker in choosing the right product, for the right application and the right person. Given the above mentioned weaknesses, there is no doubt that a tight regulatory framework is a prerequisite for such claims, permitting the removal of products that have health claims but lack scientific proof.

Prebiotics

Definition

A prebiotic is a selectively fermented ingredient that allows specific changes, in the composition and/or activity of the gastrointestinal microbiota that confers benefits upon host well-being and health (Roberfroid, 2007).

The criteria for designating a food agent as a prebiotic are (Fooks and Gibson, 2002; Roberfroid, 2007; Wang, 2009) that:

- It can persist throughout the digestive tract and reach the colon. The proposed prebiotic needs to be resistant to gastric acidity, hydrolysis by mammalian enzymes, and to gastrointestinal absorption. Resistance does not necessarily mean that the prebiotic is completely indigestible, but it should guarantee that a significant amount of the compound is available in the intestine (especially the large bowel) to serve as a fermentation substrate (Gibson *et al.*, 2004).
- It is fermented by intestinal microbiota (Gmeiner *et al.*, 2000).
- It selectively stimulates the growth and/or activity of those intestinal bacteria that contribute to health and well-being. This requirement is however most difficult to fulfil (Fooks *et al.*, 1999).
- It can withstand processing treatments which include low pH and temperature variation (Bohm *et al.*, 2006).

Although it is difficult to satisfy all the criteria, some commonly used prebiotics are listed in Table 32.6. It needs to be highlighted here that the growth enhancing potential of each prebiotic is highly specific and one cannot extrapolate the results obtained from one prebiotic to another (Al-Sheraji *et al.*, 2013; Wang, 2009).

Present status: health benefits of prebiotics

There are numerous reports on the health benefits of prebiotics which include reduction in cancer risk, increase in mineral absorption, role in lipid regulation, acute gastroenteritis, immunomodulation, and obesity (Roberfroid *et al.*, 2010) (Fig. 32.5). Prebiotics do not act themselves, however, it is the change induced by these agents in composition of gut microbiota which is responsible for their effects (Wang, 2009). Prebiotics exert their effect by selectively feeding/stimulating one or a limited number

Table 32.6 Commonly used prebiotics for humans

Types of prebiotic	Reference
Carbohydrate type	
Inulin-type fructans including Fructooligosaccharides	Franck (2002), Roberfroid (2005), Sangeetha *et al.* (2005)
Isomaltulose	Lina *et al.* (2002)
Xylo oligosaccharides	Vazquez *et al.* (2000)
Galacto oligosaccharides	Alander *et al.* (2001)
Cyclodextrins	Singh *et al.* (2002)
Raffinose oligosaccharide	Johansen *et al.* (1996)
Soybean oligosaccharide	Mussatto and Mancilha (2007)
Lactulose	Villamiel *et al.* (2002)
Lactosucrose	Kawase *et al.* (2001)
Isomaltulose	Lina *et al.* (2002)
Palatinose	Lina *et al.* (2002)
Malto oligosaccharide	Kaneko *et al.* (1994)
Isomalto oligosaccharide	Kaneko *et al.* (1994)
Arabinoxylo oligosaccharide	Eeckhaut *et al.* (2008)
Enzyme resistant dextrins	Barczynska *et al.* (2012)
Gluco oligosaccharide	Chung and Day (2002)
Non-carbohydrate type	
Green Tea	Lopez de Lacey *et al.* (2014)
Ginger	Kuma *et al.* (2001)
Quercetin	Chávarri *et al.* (2010)

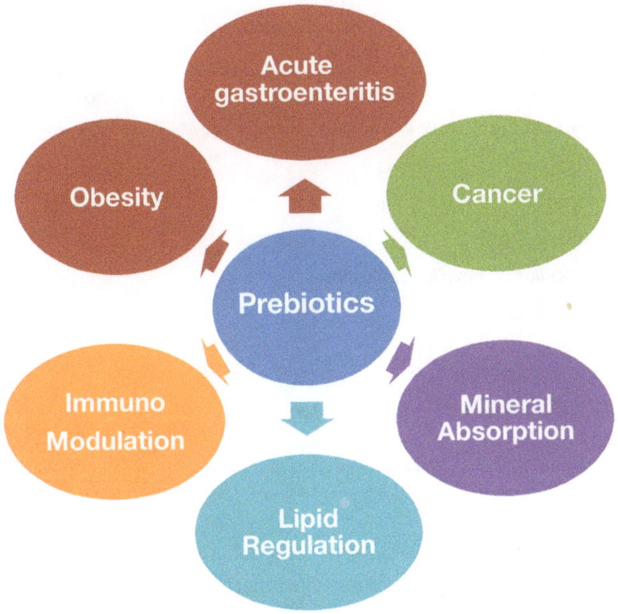

Figure 32.5 Health benefits of prebiotics.

of microorganisms. The most important bacterial genera targeted for selective stimulation are the indigenous bifidobacteria and lactobacilli (Teitalbaum and Walker, 2002).

The growth promoting effect of prebiotics has been extensively explored by administering them as such in a dosage form. Table 32.7 lists only a few examples of the never-ending list of available prebiotic products. However, a recent trend is to coencapsulate them with one or more probiotics to achieve a synbiotic preparation. By doing so, encapsulated prebiotic not only promote the growth of existing gut microflora but also act as growth supporter for entrapped probiotic on storage (Bello *et al.*, 2001; Patel *et al.*, 2014). The concept of synbiotic has been extensively explored for carbohydrate type prebiotics like inulin and fructooligosaccharides (FOS) (Kelly, 2008; Slavin, 2013). However, these prebiotics apart from encouraging the growth of probiotic bacteria may equally favour non-probiotic pathogenic bacteria residing in the gut (Bello *et al.*, 2001) and the latter trait is a concern. In order to selectively stimulate probiotic growth, exploring newer non-carbohydrate selective prebiotic sources need to be explored. Recently some polyphenols have shown promising results (Kuma *et al.*, 2001; Lopez de Lacey *et al.*, 2014).

Probiotics vs prebiotics: what to prefer

In the recent years, with the emergence of the concept describing the role of gut microbiota for healthy life, there is an ever increasing debate on the preference between probiotics and prebiotics. Despite sufficient literature on both concepts there is always a discussion on which of these plays a more important role in promotion of public health. There are a few claims that favour the use of prebiotics which are illustrated below:

- A fetus is basically born germ free, and it is only after birth that the human intestinal microbiota colonizes. Its composition basically depends on human milk, weaning, infants environment, maternal microbiota, and diet. Prebiotic present in the infants diet plays a major role in defining the gut microbiota of the newborn (Wang *et al.*, 2006).
- The colonic microbiota of each individual is highly specific and unique and depends majorly on the individual's nutritional pattern. Therefore, instead of introducing unfamiliar strains to the gut, it is better to amplify one's endogenous probiotics by providing prebiotics (Gibson *et al.*, 2004).
- Probiotics are more susceptible to environmental and manufacturing stresses compared with prebiotics.
- Prebiotics bypass the whole gut and are selectively fermented by the colonic microbiota, hence they are self targeted. Furthermore, they are an economically more viable option (Rad *et al.*, 2013).
- Prebiotics are better accepted by the public, who still have hesitation related with consumption of live microorganisms (Sreeja and Prajapati, 2013).

Table 32.7 Marketed prebiotics

Food product	Company	Prebiotics	Dosage form
Nutren Fibre®	Nestle	Inulin	Powder
Miracle Fibre	The Vitmin Shoppe, North Bergen, NJ	Inulin	Powder
Kal NutraFlora FOS	Nutraceutical Corp, Park City, UT	Short-chain fructooligosaccharides	Powder
Solaray NutraFlora FOS	Nutraceutical Corp	Short-chain fructooligosaccharides	Capsules
Fiber Choice	GlaxoSmithKline, Philadelphia,PA	Inulin	Chewable tablet/drops
Smart Fiber Stixx	Intelligent Nutrition LLC, New Orleans, LA	Oligofructose and inulin	Powder
Balance, Prebiotic Fibre	pHion	Inulin fibre; psyllium seed husk (*Plantago ovata*)	Powder
Organic, Inulin	Now Foods	Organic inulin (FOS)	Powder
Frutafit® HD	Sensus	Inulin FOS	Powder
Frutalose® L85	Sensus	Oligofructose	Syrup
Vivinal GOS	Friesland	Galactooligosaccharides	Syrup

Table 32.8 Examples of marketed formulations with inulin as a prebiotic

Food product	Probiotic (strain numbers usually not reported)	Dosage form
Bioplus	*B. bifidum* and *L. acidophilus*	Capsules
Actiflora	*B. bifidum*, *L. acidophilus*, *L. bulgaricus* and *S. thermophillus*	Capsules
Advanced Enzyme Optima, with Prebiotics and Probiotics	*Bacillus coagulans*	Capsules
Ark Naturals, Gentle Digest, Includes Prebiotics and Probiotics	*Bacillus coagulans*	Capsules
Yum-V's, Probiotic+Prebiotic Fibre, White Chocolate Flavour	*L. acidophilus* and *B. lactis*	Candy
Lane Labs, Flora 3, Probiotic, Prebiotic	*Saccharomyces boulardii*, *Bacillus coagulans*	Veggie caps
Dulce Nutritionals, Probiotic+Prebiotic Fibre, White Chocolate Flavour	*L. acidophilus* and *B. lactis*	Chewables
Optiflora Prebiotic and Probiotic	*B. longum* and *L. acidophilus*	Powder
Zenith Nutrition Probiotic blend	*L. acidophilus*, *L. rhamnosus*, *B. bifidum*, *B. longum*	Capsules
Inlife pre and probiotcs	*L. acidophilus*, *L. rhamnosus*, *B. bifidum*, *B. longum*	Capsules

Apart from this, there are reports claiming inability of exogenously administered probiotic bacteria to function with efficiency as desired in the absence of prebiotics (Sreeja and Prajapati, 2013). Thus it is advisable to at least combine a suitable selective prebiotic along with a probiotic to get the maximum out of the combined therapy. Some marketed synbiotic formulations are listed in Table 32.8.

Future perspectives: what needs to be done

It has been well accepted now that by modulating the intestinal microbiota of the host through administration of probiotic microorganisms and/or prebiotics, health benefits are conferred. This provides a good means of prevention and remediation against a range of disorders. Yet, each probiotic and prebiotic related health benefit is strain-specific or oligosaccharide-specific, and effects obtained with one cannot be extrapolated to others. Popularity of these products is quiet evident from the fact that numerous probiotics and prebiotics have reached the clinical or pharmaceutical field through a variety of 'over the counter' (OTC) products (Sreeja and Prajapati, 2013). In spite of all these advances, many of the marketed probiotic and/or probiotic products are spurious. They have not undergone the rigorous procedures required for development of modern therapeutic drugs. The areas which still need more research to elevate the status of these agents as successful pharmaceuticals are summarized below:

- The probiotic therapy is based on intestinal microbiota, therefore a deeper understanding of gut microbiota is required with respect to interrelationships of various species existing therein, the changes in the gut microbial composition at different age groups particularly in infants and the elderly, and impact of nutritional status on the gut microbiota (Sreeja and Prajapati, 2013).
- Regulatory authorities as well as the probiotic and prebiotic manufacturers should have a deeper understanding of the science. Regulatory bodies should come up with well defined procedures for conducting studies related to these formulations so that the manufacturers are guided to fulfil the required scientific and thus legal requirements. Regulations have a huge impact on the development and marketing of these products. Strict regulations can wipe out faulty products from the market and genuine products can gain more acceptance by consumers (Boyle *et al.*, 2006)
- Active involvement of medical practitioners along with probiotic manufacturers is required for their acceptance as alternative therapy. Most physicians still lack faith in the use of probiotics and prebiotics as drugs.
- Most of the probiotic products available on the market contain more than one strain. Hence, how a probiotic strain acts when used alone or in combination with other strains needs to be evaluated. Also, their dose, mechanisms of action, appropriate administrative regimens should be understood especially when the formulation is to be used for treatment of a disease condition (Sreeja and Prajapati, 2013).
- Efficacy of these formulations may change when used along with other drugs for a particular disease.

References

Adhikari, K., Mustapha, A., Grün, I.U., and Fernando, L. (2000). Viability of microencapsulated bifidobacteria in set yoghurt during refrigerated storage. J. Dairy Sci. *83*, 1946–1951.

Al-Sheraji, S.H., Ismail, A., Manap, M.Y., Mustafa, S., Yusof, R.M., and Hassan, F.A. (2013). Prebiotics as functional foods: A review. J. Funct. Foods *5*, 1542–1553.

Alander, M., Matto, J., Kneifel, W., Johansson, M., Kogler, B., and Crittenden, R. (2001). Effect of galacto-oligosaccharide supplementation on human faecal microflora and on survival and persistence of Bifidobacterium lactis Bb-12 in the gastrointestinal tract. Int. Dairy J. *11*, 817–825.

Albertini, B., Vitali, B., Passerini, N., Cruciani, F., Di Sabatino, M., Rodriguez, L., and Brigidi, P. (2010). Development of microparticulate systems for intestinal delivery of *Lactobacillus acidophilus* and *Bifidobacterium lactis*. Eur. J. Pharm. Sci. *40*, 359–366.

Aminea, K.M., Champagneb, C.P., Salmieria, S., Brittenb, M., St-Gelaisb, D., Fustierb, P., and Lacroixa, M. (2014). Effect of palmitoylated alginate microencapsulation on viability of *Bifidobacterium longum* during freeze-drying. LWT-Food Sci. Technol. *52*, 111–117.

Anal, A.K., and Singh, H. (2007). Recent advances in microencapsulation of probiotics for industrial applications and targeted delivery. Trends Food Sci. Technol. *18*, 240–251.

Ananta, E., Volkert, M., and Knorr, D. (2005). Cellular injuries and storage stability of spray-dried *Lactobacillus rhamnosus* GG. Int. Dairy J. *15*, 399–409.

Audet, P., Paquin, C., and Lacroix, C. (1988). Immobilized growing lactic acid bacteria with k-carrageenan-locust bean gum gel. Appl. Microbiol. Biotechnol. *29*, 11–18.

Azais-Braesco, V., Bresson, J.L., Guarner, F., and Corthier, G. (2010). Not all lactic acid bacteria are probiotics, ...but some are. Br. J. Nutr. *103*, 1079–1081.

Bader, J., Albin, A., and Stahl, U. (2012). Spore-forming bacteria and their utilization as probiotics. Benef. Microbes 3, 67–75.

Balcázar, J.L., Vendrell, D., de Blas, I., Ruiz-Zarzuela, I., Gironés, O., and Múzquiz, J.L. (2007). *In vitro* competitive adhesion and production of antagonistic compounds by lactic acid bacteria against fish pathogens. Vet. Microbiol. *122*, 373–380.

Barczynska, R., Slizewska, K., Jochym, K., Kapusniak, J., and Libudzisz, Z. (2012). The tartaric acid-modified enzymeresistant dextrin from potato starch as potential prebiotic. J. Funct. Foods *4*, 954–962.

Behzadi, S.S., Wu, S., Toegel, S., Hofrichter, M., Altenburger, I., Unger, F.M., Wirth, M., and Viernstein, H. (2013). Impact of heat treatment and spray drying on cellular properties and culturability of *Bifidobacterium bifidum* BB-12. Food Res. Int. *54*, 93–101.

Bell, S.G. (2007). Immunomodulation, part V: probiotics. Neonatal Network *26*, 57–60.

Bello, F.D., Walter, J., Hertel, C., and Hammes, W.P. (2001). *In vitro* study of prebiotic properties of levan-type exopolysaccharides from Lactobacilli and non-digestible carbohydrates using denaturing gradient gel electrophoresis. Syst Appl. Microbiol. *24*, 232–237.

Bohm, A., Kleessen, B., and Henle, T. (2006). Effect of dry heated inulin on selected intestinal bacteria. Eur. Food Res. Technol. *222*, 737–740.

Bora, P.S., Puri, V., and Bansal, A.K. (2009). Physicochemical properties and excipient compatibility studies of probiotic Bacillus coagulans spores. Sci. Pharm. *77*, 625–637.

Borges, S., Barbosa, J., Camilo, R., Carvalheira, A., Silva, J., Sousa, S., Gomes, A.M., Pintado, M.M., Silva, J.P., Costa, a., *et al.* (2011). Effects of encapsulation on the viability of probiotic strains exposed to lethal conditions. Int. J. Food Sci. Technol. *47*, 416–421.

Boyle, R.J., Robbins Browne, R.M., and Tang, M.L.K. (2006). Probiotics used in clinical practice. what are the risks? Am. J. Clin. Nutr. *83*, 1256–1264.

Broekaert, I.J., and Walker, W.A. (2006). Probiotics and chronic disease. J. Clin. Gastroenterol. *40*, 270–274.

Brownlie, K. (2007). Encapsulation and controlled release technologies in food systems. In, J.M. Lakkis, ed. (Ames, Blackwell Publishing), pp. 213–233.

Brun-Graeppi, A.K.A.S., Richard, C., Bessodes, M., Scherman, D., and Merten, O.W. (2011). Cell microcarriers and microcapsules of stimuli-responsive polymers. J. Control. Release *149*, 209–224.

Burgain, J., Gaiani, C., Linder, M., and Scher, J. (2011). Encapsulation of probiotic living cells: from laboratory scale to industrial applications. J. Food Eng *104*, 467–483.

Cai, S., Zhao, M., Fang, Z., Nishinari, K., Phillips, G.O., and Jiang, F. (2014). Microencapsulation of *Lactobacillus* acidophilus CGMCC1.2686 via emulsification/internal gelation of alginate using Ca-EDTA and CaCO3 as calcium sources. Food Hydrocolloids *39*, 295–300.

Capela, P., Hay, T.K.C., and Shah, N.P. (2006). Effect of cryoprotectants, prebiotics and microencapsulation on survival of probiotic organisms in yoghurt and freeze-dried yoghurt. Food Res. Int. *39*, 203–211.

Caramia, G. (2004). Probiotics: from Metchnikoff to the current preventive and therapeutic possibilities. Pediatr. Med. Chir *26*, 19–33.

Castilla, S.O., Calleros, L.C., García-Galindo, H.S., Alvarez-Ramírez, J., and Vermon Carter, E.J. (2011). Textural properties of alginate–pectin beads and survivability of entrapped Lb. casei in simulated gastrointestinal conditions and in yoghurt. Food Res. Int. *43*, 111–117.

Champagne, C.P., Ross, R.P., Saarela, M., Hansen, K.F., and Charalampopoulos, D. (2011). Recommendations for the viability assessment of probiotics as concentrated cultures and in food matrices. Int. J. Food Microbiol. *149*, 185–193.

Chan, E.S., Wonga, S.L., Leea, P.P., Leeb, J.S., Tic, T.B., Zhangd, Z., Poncelete, D., Ravindraa, P., Phana, S.H., and Yima, H.Z. (2011). Effects of starch filler on the physical properties of lyophilized calcium–alginate beads and the viability of encapsulated cells. Carbohydr. Polym. *83*, 225–232.

Chávarri, M., Marañón, I., Ares, R., Ibáñez, R.C., Marzo, F., and Villarán, M.C. (2010). Microencapsulation of a probiotic and prebiotic in alginate-chitosan capsules improves survival in simulated gastro-intestinal conditions. Int. J. Food Microbiol. *142* 185–189.

Chavarri, M., Maranon, I., and Villaran, M., C. (2012). Encapsulation technology to protect probiotic bacteria.

Chávarria, M., Marañóna, I., Aresa, R., Ibáñezb, F.C., Marzob, F., and del Carmen Villarána, M. (2010). Microencapsulation of a probiotic and prebiotic in alginate-chitosan capsules improves survival in simulated gastro-intestinal conditions. Int. J. Food Microbiol. *142*, 185–189.

Chávez, B.E., and Ledeboer, A.M. (2007). Drying of probiotics: optimization of formulation and process to enhance storage survival. Drying Technol. *25*, 1193–1201.

Chen, M.J., and Chen, K.N. (2007b). Applications of probiotic encapsulation in dairy products. In Encapsulation and Controlled Release Technologies in Food Systems, J.M. Lakkis, ed. (USA, Wiley-Blackwell), pp. 83–107.

Chung, C.H., and Day, D.F. (2002). Glucooligosaccharides from Leuconostoc mesenteroides B-742 (ATCC 13146): A potential prebiotic. J. Ind Microbiol. Biotechnol. *29*, 196–199.

Cislaghi, F.P.D.C., Silvab, C.D.R.E., Freireb, C.B.F.F., Lorenzb, J.G., and Annab, E.S.S. (2012). Bifidobacterium Bb-12 microencapsulated by spray drying with whey: Survival under simulated gastrointestinal conditions, tolerance to NaCl, and viability during storage. J. Food Eng *113*, 186–193.

Cook, M.T., Tzortzis, G., Charalampopoulos, D., and Khutoryanskiy, V. (2012). Microencapsulation of probiotics for gastrointestinal delivery. J. Controlled Release *162*, 56–67.

Czinn, S.J., and Blanchard, S.S. (2009). Probiotics in foods and supplement. In Nutrition and health: probiotics in paediatric medicine, S. Michail, and P.M. Sherman, eds. (NJ, Totowa, Humana Press), pp. 299–306.

De Castro-Cislaghi, F.P., Silva, C.D.R.E., Fritzen-Freire, C.B., Lorenz, J.G., and Sant'Anna, E.S. (2012). Bifidobacterium Bb-12 microencapsulated by spray drying with whey: survival under simulated gastrointestinal conditions, tolerance to NaCl, and viability during storage. J. Food Eng *113*, 186–193.

De Vos, P., Faas, M.M., Spasojevic, M., and Sikkema, J. (2010). Encapsulation for preservation of functionality and targeted delivery of bioactive food components. Int. Dairy J. *20*, 292–302.

DeMerlis, C.C., and Schoneker, D.R. (2003). Review of the oral toxicity of polyvinyl alcohol (PVA). Food Chem. Toxicol. *41*, 319–326.

Ding, W.K., and Shah, N.P. (2009a). Effect of various encapsulating materials on the stability of probiotic bacteria. J. Food Sci. *74*, M100-M107.

Ding, W.K., and Shah, N.P. (2009b). An improved method of microencapsulation of probiotic bacteria for their stability in acidic and bile conditions during storage. J. Food Sci. *74*, M53-M61.

Doherty, S.B., Gee, V.L., Ross, R.P., Stanton, C., Fitzgerald, G.F., and Brodkorb, A. (2010). Efficacy of whey protein gel networks as potential viability-enhancing scaffolds for cell immobilization of *Lactobacillus rhamnosus* GG. J. Microbiol. Methods *80*, 231–241.

Doherty, S.B., Gee, V.L., Ross, R.P., Stanton, C., Fitzgerald, G.F., and Brodkorb, A. (2011). Development and characterization of whey protein micro-beads as potential matrices for probiotic protection. Food Hydrocolloids *25*, 1604–1617.

Doleyres, Y., and Lacroix, C. (2005). Technologies with free and immobilized cells for probiotic bifidobacteria production and protection. Int. Dairy J. *15*, 973–988.

Eaton, S.B. (2006). The ancestral human diet: what was it and should it be a paradigm for contemporary nutrition? Proc. Nutr. Soc *65*, 1–6.

Eeckhaut, V., Van Immerseel, F., Dewulf, J., Pasmans, F., Haesebrouck, F., Ducatelle, R., Courtin, C.M., Delcour, J.A., and Broekaert, W.F. (2008). Arabinoxylooligosaccharides from wheat bran inhibit Salmonella colonization in broiler chickens. Poult. Sci. *87*, 2329–2334.

Fávaro-Trindade, C.S., and Grosso, C.R.F. (2002). Microencapsulation of *L. acidophilus* (La-05). and B. lactis (Bb-12). and evaluation of their survival at the pH values of the stomach and in bile. J. Microencapsulation *19*, 485–494.

Feher, J., and Lejto, T. (2012). Methods of preparing probiotic nanoparticles. Patent no. US 12/939, 395.

Flores, W.R., Ramírez, E.G.R., and Montoya, J.A.S. (2014). Microencapsulation of *Lactobacillus* helveticus and *Lactobacillus* delbrueckii using alginate and gellan gum. Carbohydr. Polym. *98*, 1011–1017.

Foligne, B., Daniel, C., and Pot, B. (2013). Probiotics from research to market: the possibilities, risks and challenges. Curr. Opin. Microbiol. *16*, 284–292.

Foligne, B., Peys, E., Vandenkerckhove, J., Van Hemel, J., Dewulf, J., Breton, J., and Pot, B. (2012). Spores from two distinct colony types of the strain Bacillus subtilis PB6 substantiate anti-inflammatory probiotic effects in mice. Clin. Nutr. *31*, 987–994.

Fooks, L.J., Fuller, R., and Gibson, G.R. (1999). Prebiotics, probiotics and human gut microbiology. Int. Dairy J. *9*, 53–61.

Fooks, L.J., and Gibson, G.R. (2002). Probiotics as modulators of gut flora. Br. J. Nutr. *88*, S39-S49.

Franck, A. (2002). Technological functionality of inulin and oligofructose. Br. J. Nutr. *87*, S287-S291.

Freirea, C.B.F., Prudêncioa, E.S., Ambonia, R.D.M., C., Pintoa, S.S., Murakamia, A.N.N., and Murakamib, F.S. (2013). Effect of the application of Bifidobacterium BB-12 microencapsulated by spray drying with prebiotics on the properties of ricotta cream. Food Res. Int. *45*, 306–312.

Fritzen-Freire, C.B., Prudêncio, E.S., Amboni, R.D.M.C., Pinto, S.S., Negrão-MurakamiA.N., and Murakami, F.S. (2012). Microencapsulation of bifidobacteria by spray drying in the presence of prebiotics. Food Res. Int. *45*, 306–312.

Fuller, R. (1989). Probiotics in man and animals. J. Appl. Microbiol. *66*, 365–378.

Gibson, G.R., Brummer, R.J., Isolauri, E., Lochs, H., Morelli, L., Ockhuizen, T., Rowland, I.R., Schrezenmeir, J., Stanton, C., and Verbeke, K. (2011). The design of probiotic studies to substantiate health claims. Gut Microbes *2*, 299–305.

Gibson, G.R., Probert, H.M., Van Loo, J., Rastall, R.A., and Roberfroid, M.B. (2004). Dietary modulation of the human colonic microbiota: Updating the concept of prebiotics. Nutr. Res. Rev. *17*, 259–275.

Gilliland, S.E. (1989). A review of potential benefits to consumers. J. Dairy Sci. 72, 2483–2494.

Global Industry Analysis Report (2012). www.esomar.org/uploads/industry/reports/global-market-research-2012/ESOMAR-GMR-2012_Preview.pdf

Gmeiner, M., Kneifel, W., Kulbe, K.D., Wouters, R., De Boever, P., and Nollet, L. (2000). Influence of a synbiotic mixture consisting of *Lactobacillus* acidophilus 74-2 and a fructooligosaccharide preparation on the microbial ecology sustained in a simulation of the human intestinal microbial ecosystem (SHIME reactor). Appl. Microbiol. Biotechnol. *53*, 219–223.

Gollop, N., Zakin, V., and Weinberg, Z.G. (2005). Antibacterial activity of lactic acid bacteria included in inoculants for silage and in silages treated with these inoculants. J. Appl. Microbiol. *98*, 662–666.

Gonzáleza, D.J.P., Montielb, R.G., Callerosc, C.L., Islasd, R.P., and Cartera, E.J.V. (2009). Encapsulation of *Lactobacillus rhamnosus* in double emulsions formulated with sweet whey as emulsifier and survival in simulated gastrointestinal conditions. Food Res. Int. *42*, 292–297.

Gouin, S. (2004). Microencapsulation: industrial appraisal of existing technologies and trends. Trends Food Sci. Technol. *15*, 330–347.

Groboillot, A.F., Champagne, C.P., Darling, G.F., and Poncelet, D. (1993). Membrane formation by interfacial cross-linking of chitosan for microencapsulation of *Lactococcus lactis*. Biotechnol. Bioeng. *42*, 1157–1163.

Hébrard, G., Blanquet, S., Beyssac, E., Remondetto, G., Subirade, M., and Alric, M. (2006). Use of whey protein beads as a new carrier system for recombinant yeasts in human digestive tract. J. Biotechnol. *127*, 151–160.

Heidebach, T., Först, P., and Kulozik, U. (2009a). Microencapsulation of probiotic cells by means of rennet-gelation of milk proteins. Food Hydrocolloids *23*, 1670–1677.

Heidebach, T., Först, P., and Kulozik, U. (2009b). Transglutaminase-induced caseinate gelation for the microencapsulation of probiotic cells. Int. Dairy J. *19*, 77–84.

Heidebach, T., Först, P., and Kulozik, U. (2010). Influence of casein-based microencapsulation on freeze-drying and storage of probiotic cells. J. Food Eng *98*, 309–316.

Heyman, M., and Menard, S. (2002). Probiotic microorganisms: How they affect intestinal pathophysiology. Cell. Mol. Life Sci. *59*, 1151–1165.

Hoffman, F.A. (2008). Development of probiotics as biologic drugs. Clin. Infect. Dis. *46*, S125-S127.

Holzapfel, W.H., Haberer, P., Geisen, R., Bjorkroth, J., and Schillinger, U. (2001). Taxonomy and important features of probiotic microorganisms in food and nutrition. Am. J. Clin. Nutr. *73*, 365S-373S.

Hsiao, H., Lian, W., and Chou, C. (2004). Survival of bifidobacteria after spray drying. J. Agric. Food Chem. *84*, 134–139.

ICMR-DBT (2011). Guidelines for evaluation of probiotics in food., I.C.o.M. Research, ed. (New Delhi, Director General).

Ivanovska, T.P., Petruševska-Tozi, L., Kostoska, M.D., Geškovski, N., Grozdanov, A., Stain, C., Trajče Stafilov, T., and Kristina Mladenovska, K. (2012). Microencapsulation of *Lactobacillus* casei in chitosan-Ca-alginate microparticles using spray-drying method. Maced. J. Chem. Chem. En. *31*, 115–123.

Jantzen, M., Göpel, A., and Beermann, C. (2013). Direct spray drying and microencapsulation of probiotic *Lactobacillus reuteri* from slurry fermentation with whey. J. Appl. Microbiol. *115*, 1029–1036.

Johansen, H.N., Glitso, V., and Knudsen, K.E.B. (1996). Influence of extraction solvent and temperature on the quantitative determination of oligosaccharides from plant materials by high-performance liquid chromatography. J. Agric. Food Chem. *44*, 1470–1474.

John, R.P., Tyagi, R.D., Brar, S.K., Surampalli, R.Y., and Pervost, D. (2011). Bio-encapsulation of microbial cells for targeted agricultural delivery. Crit. Rev. Biotechnol. *31*, 211–226.

Kailasapathy, K. (2006). Survival of free and encapsulated probiotic bacteria and their effect on the sensory properties of yoghurt. LWT-Food SciTechnol. *39*, 1221–1227.

Kailasapathy, K. (2009). Encapsulation technologies for functional foods and nutraceutical product development. CAB Reviews: Perspectives in Agriculture, Veterinary Science, Nutrition and Natural Resources *4*, 6.

Kaneko, T., Kohmoto, T., Kikuchi, H., Shiota, M., Lino, H., and Mitsuoka, T. (1994). Effects of isomaltooligosaccharides with different degrees of polymerization on human faecal bifidobacteria. Biosci. Biotechnol. Biochem. *58*, 2288–2290.

Kanmani, P., Kumar, R.S., Yuvaraj, N., Paari, K.A., Pattukumar, V., and Arul, V. (2011a). Effect of cryopreservation and microencapsulation of lactic acid bacterium Enterococcus faecium MC13 for long-term storage. Biochem. Eng J. *58–59*, 140–147.

Kanmani, P., Kumar, R.S., Yuvaraj, N., Paari, K.A., Pattukumar, V., and Arul, V. (2011b). Cryopreservation and microencapsulation of a Probiotic in Alginate-chitosan capsules improves survival in simulated gastrointestinal conditions. Biotechnol. Bioprocess Eng *16*, 1106–1114.

Kaur, I.P., Kuhad, A., Garg, A., and Chopra, K. (2009). Probiotics: Delineation of Prophylactic and Therapeutic Benefits. J. Med. Food *12* 219–235.

Kawase, M., Pilgrim, A., Araki, T., and Hashimoto, K. (2001). Lactosucrose production using a simulated moving bed reactor. Chem. Eng. Sci. *56*, 453–458.

Kelly, G. (2008). Inulin-Type Prebiotics – A Review: Part 1. Alternative Medicine Review *13*, 315–329.

Khan, S.H., and Ansari, F.A. (2007). Probiotics – the friendly bacteria with market potential in global market. Pak J. Pharm. Sci. *20*, 76–82.

Khutoryanskiy, V.V. (2011). Advances in mucoadhesion and mucoadhesive polymers. Macromol. Biosci. *11*, 748–764.

King, A.H. (1995). Encapsulation of Food ingredients: a review of available technology, focusing on hydrocolloids. In Encapsulation and controlled release of food ingredients, S.J. Risch, and G.A. Reineccius, eds. (Washington DC, American Chemical Society), pp. 213–220.

Krasaekoopt, W., Bhandari, B., and Deeth, H. (2003). Evaluation of encapsulation techniques of probiotics for yoghurt. Int. Dairy J. *13*, 3–13.

Kuma, Y., Akahoshi, R., Kudo, T., Kawami, K., Shibata, M., and Hashimoto, S. (2001). Fermented milk drinks and foods and process for producing the same.

Lahtinen, S.J., Ahokoski, H., Reinikainen, J.P., Gueimonde, M., Nurmi, J., Ouwehand, A.C., and Salminen, S.J. (2008). Degradation of 16S rRNA and attributes of viability of viable but nonculturable probiotic bacteria. J. Am. Diet. Assoc. *46*, 693–698.

Lahtinen, S.J., Gueimonde, M., Ouwehand, A.C., Reinikainen, J.P., and Salminen, S.J. (2006). Comparison of four methods to enumerate probiotic bifidobacteria in a fermented food product. Food Microbiol. *23*, 571–577.

Lia, X.Y., Chenb, G.X., Suna, W.Z., Park, H.J., and Chac, S.D. (2011). Preparation of alginate/chitosan/carboxymethyl chitosan complex microcapsules and application in *Lactobacillus* casei ATCC 393. Carbohydr. Polym. *83*, 1479–1485.

Lian, W., Hsiao, H., and Chou, C. (2003). Viability of microencapsulated bifidobacteria in simulated gastric juice and bile solution. Int. J. Food Micrbiol. *86*, 293–301.

Lian, W.C., Hsiao, H.C., and Chou, C.C. (2002). Survival of bifidobacteria after spray-drying. Int. J. FOOD Microbiol. *74*, 79–86.

Lilly, D.M., and Stillwell, R.H. (1965). Probiotics: growth promoting factors produced by microorganisms. Science *47*, 747–748.

Lina, B.A.R., Jonker, D., and Kozianowsky, G. (2002). Isomaltulose (Palatinose): A review of biological and toxicological studies. Food Chem. Toxicol. *40*, 1375–1381.

Liserre, A.M., Re, M.I., and Franco, B.D.G.M. (2007). Microencapsulation of *Bifidobacterium animalis* subsp. *lactis* in modified alginate-chitosan beads and evaluation of survival in simulated gastrointestinal conditions. Food Biotechnol. *21*, 1–16.

Liu, X.D., Yu, W.Y., Zhang, Y., Xue, W.M., Yu, W.T., Xiong, Y., Ma, X.J., Chen, Y., and Yuan, Q. (2002). Characterization of structure and diffusion behaviour of Ca-alginate beads prepared with external or internal calcium sources. J. Microencapsulation *19*, 775–782.

Livney, Y.D. (2010). Milk proteins as vehicles for bioactives. Curr. Opin. Colloid Interface Sci. *15*, 73–83.

Lopez de Lacey, A.M., Perez-Santin, E., Lopez-Caballero, M.E., and Montero, P. (2014). Survival and metabolic activity of probiotic bacteria in green tea. LWT – Food Sci. Technol. *55*, 314–322.

Lord Rayland, F.R.S. (1878). On the stability of jets. P Lond Math Soc *10*, 4–13.

Maciel, G.M., Chaves, K.S., Grosso, C.R.F., and Gigante, M.L. (2014). Microencapsulation of *Lactobacillus acidophilus* La-5 by spray-drying using sweet whey and skim milk as encapsulating materials. J. Dairy Sci. doiorg/103168/jds2013-7463.

Malmo, C., La Storia, A., and Mauriello, G. (2013). Microencapsulation of *Lactobacillus reuteri* DSM 17938 Cells Coated in Alginate Beads with Chitosan by Spray Drying to Use as a Probiotic Cell in a Chocolate Soufflé. Food Bioprocess Tech. *6*, 795–805.

Mandal, S., Puniya, A.K., and Singh, K. (2006). Effect of alginate concentrations on survival of microencapsulated *Lactobacillus* casei NCDC-298. Int. Dairy J. *16*, 1190–1195.

Meier, R., and Steuerwald, M. (2005). Place of probiotics. Curr. Opin. Crit. Care *11*, 318–325.

Mercenier, A., Pavan, S., and Pot, B. (2002). Probiotics as biotherapeutic agents: present knowledge and future prospects. Curr. Pharm. Des *8*, 99–110.

Millqvist-Fureby, A., Elofsson, U., and Bergenståhl, B. (2001). Surface composition of spray-dried milk protein-stabilized emulsions in relation to pre-heat treatment of proteins. Colloids Surf B Biointerfaces *21*, 47–58.

Mills, S., Stanton, C., Fitzgerald, G.F., and Ross, R.P. (2011). Enhancing the stress responses of probiotics for a lifestyle from gut to product and back again. Microb. Cell. Fact. *10*, S19.

Mirzaeia, H., Pourjafarb, H., and Homayouni, A. (2012). Effect of calcium alginate and resistant starch microencapsulation on the survival rate of *Lactobacillus* acidophilus La5 and sensory properties in Iranian white brined cheese. Food Chem. *132*, 1966–1970.

Mombelli, B., and Gismondo, M.R. (2000). The use of probiotics in medical practice. Int. J. Antimicrob. Agents *16*, 531–536.

Moslemy, P., Neufeld, R.J., and Guiot, S.R. (2002). Biodegradation of gasoline by gellan gum-encapsulated bacterial cells. Biotechnol. Bioeng. *80*, 174–185.

Mussatto, S.I., and Mancilha, I.M. (2007). Non-digestible oligosaccharides: A review. Carbohydr. Polym. *68*, 587–597.

Muthukumarasamy, P., Allan, W.P., and Holley, R.A. (2006). Stability of *Lactobacillus reuteri* in different types of microcapsules. J. Food Sci. *71*, M20-M24.

Nag, A., Han, S.K., and Singh, H. (2011). Microencapsulation of probiotic bacteria using pH-induced gelation of sodium caseinate and gellan gum. Int. Dairy J. *21*, 247–253.

Nihant, N., Grandfils, C., Jérôme, R., and Teyssié, P. (1995). Microencapsulation by coacervation of poly(lactide-co-glycolide). IV. Effect of the processing parameters on coacervation and encapsulation. J. Controlled Release 35, 117–125.

Nihant, N., Stassen, S., Grandfils, C., Jérome, R., Teyssié, P., and Goffinet, G. (1994). Microencapsulation by coacervation of poly(lactide-co-glycolide). III. Characterization of the final microspheres. Polym. Int. *34*, 289–299.

Nualkaekula, S., Lentona, D., Cooka, M.T., Khutoryanskiyb, V.V., and Charalampopoulosa, D. (2012). Chitosan coated alginate beads for the survival of microencapsulated *Lactobacillus plantarum* in pomegranate juice. Carbohydr. Polym. *19*, 1281–1287.

O'Hara, A.M., and Shanahan, F. (2007). Mechanisms of action of probiotics in intestinal diseases. The Scientific World J. *7*, 31–46.

O'Riordan, K., Andrews, D., Buckle, K., and Conway, P. (2001). Evaluation of microencapsulation of a Bifidobacterium strain with starch as an approach to prolonging viability during storage. J. Appl. Microbiol. *91*, 1059–1066.

Olguín, E.J. (2012). Dual purpose microalgae-bacteria-based systems that treat waste water and produce biodiesel and chemical products within a biorefinery. Biotechnol. Adv. *30*, 1031–1046.

Oliveira, A.C., Moretti, T.S., Boschini, C., Baliero, J.C.C., Freitas, L.A.P., and Freitas, O. (2007a). Microencapsulation of B. lactis (BI 01). and *L. acidophilus* (LAC 4). by complex coacervation followed by spouted-bed drying. Drying Technol. 25, 1687–1693.

Oliveira, A.C., Moretti, T.S., Boschini, C., Baliero, J.C.C., Freitas, O., and Favaro-Trindade, C.S. (2007b). Stability of microencapsulated *B. lactis* (BI 01). and *L. acidophilus* (LAC 4). by complex coacervation followed by spray drying. J. Microencapsulation *24*, 685–693.

Ouwehand, A.C., and Salminen, S.J. (1998). The health effects of viable and non-viable cultured milk. Int. Dairy J. *8*, 749–758.

ÖZer, B., Uzun, Y.S., and Kirmaci, H.A. (2008). Effect of microencapsulation on viability of *Lactobacillus acidophilus* LA-5 and *Bifidobacterium bifidum* BB-12 during kasar cheese ripening. Int. J. Dairy Technol. *61*, 237–244.

Paéz, R., Lavaria, L., Vinderolab, G., Auderoa, G., Cuatrina, A., Zaritzkyc, N., and Reinheimerb, J. (2012). Effect of heat treatment and spray drying on lactobacilli viability and resistance to simulated gastrointestinal digestion. Food Res. Int. *48*, 748–754.

Park, J.K., and Chang, H.N. (2000). Microencapsulation of microbial cells. Biotechnol. Adv. *18*, 303–319.

Patel, P.J., Singh, S.K., Panaich, S., and Cardozo, L. (2014). The aging gut and the role of prebiotics, probiotics, and synbiotics: A review. Journal of Clinical Gerontology & Geriatrics 1–4.

Paton, A.W., Morona, R., and Paton, J.C.S. (2006). Designer probiotics for prevention of enteric infection. Nat. Rev. Microbiol. *4*, 193–200.

Penner, R., Fedorak, R.N., and Madsen, K.L. (2005). Probiotics and nutraceuticals: Non-medicinal treatments of gastrointestinal diseases. Curr. Opin. Pharmacol. *5*, 596–603.

Phillips, J., Muir, J.G., Birkett, A., Lu, Z.X., Jones, G.P., O'Dea, K., and Young, G.P. (1995). Effect of resistant starch on fecal bulk and fermentation-dependent events in humans. Am. J. Clin. Nutr. *62*, 121–130.

Picot, A., and Lacroix, C. (2003). Production of Multiphase Water-Insoluble Microcapsules for Cell Microencapsulation Using an Emulsification/Spray-drying Technology. J. Food Sci. *68*, 2693–2700.

Picot, A., and Lacroix, C. (2004). Encapsulation of Bifidobacteria in whey protein-based microcapsules and survival in stimulated gastrointestinal conditions and in yoghurt. Int. Dairy J. *14*, 505–515.

Pranteda, M.L.J., Poncelet, D., Macías, N.E.M., Arcos, A., Aguilera, M., Sánchez, M.M., and Cormenzana, R.A. (2012). Stability of lactobacilli encapsulated in various microbial polymers. J. Biosci. Bioeng. *113*, 179–184.

Pruesse, U., and Vorlop, K.D. (2004). The jetcutter technology. In Fundamentals of Cell Immobilization Biotechnology, V. Nedovic, and R. Willaert, eds. (Dordrecht, Nederlands: Kluwer Academic Publishers), pp. 295–309.

Quigley, E.M.M. (2011). Gut microbiota and the role of probiotics in therapy. Curr. Opin. Pharmacol. *11*, 593–603.

Rad, A.H., Torab, R., Mortazavian, A.N., Mehravany, E.V., and Mehravany, L.V. (2013). Can. probiotics prevent or improve cold and influenza? Nutrition, 1–2.

Reid, G. (2008). Probiotics and prebiotics – Progress and challenges. Int. Dairy J. *18*, 969–975.

Reid, G., Jass, J., Sebulsky, M.T., and McCormick, J.K. (2003). Potential uses of probiotics in clinical practice. Clin. Microbiol. Rev. *16*, 658–672.

Reis, C.P., Neufeld, R.J., Vilela, S., Ribeiro, A.J., and Veiga, F. (2006). Review and current status of emulsion/dispersion technology using an internal gelation process for the design of alginate particles. J. Microencapsulation *23*, 245–257.

Rinaudo, M. (2008). Main properties and current applications of some polysaccharides as biomaterials. Polym. Int. *57*, 397–430.

Roberfroid, M. (2007). Prebiotics: The Concept Revisited. J. Nutr. *137*, 830S-837S.

Roberfroid, M., Gibson, G.R., Hoyles, L., McCartney, A.L., Rastall, R., and Rowland, I. (2010). Prebiotic effect: Metabolic and health benefits. Br. J. Nutr. *104*, 51–63.

Roberfroid, M.B. (2005). Inulin-type fructans as functional food ingredients. Boca Raton: CRC Press.

Rokka, S., and Rantamäki, P. (2010). Protecting probiotic bacteria by microencapsulation: Challenges for industrial applications. Eur. Food Res. Technol. *231*, 1–12.

Rook, G.A., and Brunet, L.R. (2005). Microbes, immunoregulation, and the gut. Gut *54*, 317–320.

Rosevear, A. (1984). Immobilized biocatalysts – a critical review. J. Chem. Tech. Biot. *34*, 127–150.

Ross, J. (2006). Regulatory considerations. In american society for microbiology colloquium report probiotic microbes: the scientific basis. In American Academy of Microbiology (Washington), pp. 8.

Saarela, M., Mogensen, G., Fonden, R., Matto, J., and Mattila-Sandholm, T. (2000). Probiotic bacteria: Safety, functional and technological properties. J. Biotechnol. *84*, 197–215.

Sajilata, M.G., Singhal, R.S., and Kulkarni, P.R. (2006). Resistant starch – a review. Comprehensive Reviews. Food Sci. Food Safety *5*, 1–17.

Sangeetha, P.T., Ramesh, M.N., and Prapulla, S.G. (2005). Recent trends in the microbial production, analysis and application of fructooligosaccharides. Trends Food Sci. Tech. *16*, 442–457.

Sansonetti, P.J. (2006). The innate signalling of dangers and the dangers of innate signalling. Nature Immunol. *7*, 1237–1242.

Sarkar, S. (2013). Potential of probiotics as pharmaceutical agent: a review. BFJ *115*, 1658–1687.

Saxelin, M. (2008). Probiotic formulations and applications, the current probiotics market, and changes in the marketplace: a European perspective. Clin. Infect. Dis. *46*, S76-S79.

Semyonov, D., Ramon, O., Kaplun, Z., Brener, L.L., Gurevich, N., and Shimoni, E. (2010). Microencapsulation of *Lactobacillus paracasei* by spray freeze drying. Food Res. Int. *43*, 193–202.

Shah, N.P., and Ravula, R.R. (2000). Microencapsulation of probiotic bacteria and their survival in frozen fermented dairy desserts. Aust. J. Dairy Technol. *55*, 139–144.

Shima, M., Matsuo, T., Yamashita, M., and Adachi, S. (2009). Protection of *Lactobacillus* acidophilus from bile salts in a model intestinal juice by incorporation into the inner-water phase of a W/O/W emulsion. Food Hydrocolloids *23*, 281–285.

Singh, M., Sharma, R., and Banerjee, V.C. (2002). Biotechnological applications of cyclodextrins. Biotech. Adv. *20*, 341–359.

Slavin, J. (2013). Fibre and prebiotics: Mechanisms and health benefits. Nutrients *5*, 1417–1435.

Sogias, I.A., Wiliams, A.C., and Khutoryanskiy, V.V. (2008). Why is chitosan mucoadhesive? Biomacromolecules *9*, 1837–1842.

Sohail, A., Turner, M.S., Coombes, A., Bostrom, T., and Bhandari, B. (2010). Survivability of probiotics encapsulated in alginate gel microbeads using a novel impinging aerosols method. Int. J. Food Microbiol. *145*, 162–168.

Song, H., Yua, W., Gaoa, M., Liu, L., and Maa, X. (2013). Microencapsulated probiotics using emulsification technique coupled with internal or external gelation process. Carbohydr. Polym. *96*, 181–189.

Sreeja, V., and Prajapati, J.B. (2013). Probiotic formulations: application and status as pharmaceuticals – a review. Probiotics & Antimicro. Prot. *5*, 81–91.

Stokke, B.T., Draget, K.I., Smidsrod, O., Yuguchi, Y., Urakawa, H., and Kajiwara, K. (2000). Small-angle X-ray scattering and rheological characterization of alginate gels. 1. Ca-alginate gels. Macromolecules *33*, 1853–1863.

Sultana, K., Godward, G., Reynolds, N., Arumugaswamy, R., Peiris, P., and Kailasapathy, K. (2000). Encapsulation of probiotic bacteria with alginate-starch and evaluation of survival in simulated gastrointestinal conditions in yoghurt. Int. J. Food Micrbiol. *62*, 47–55.

Teitalbaum, J.E., and Walker, W.A. (2002). Nutritional impact of pre- and probiotic as protective gastrointestinal organisms. Annu. Rev. Nutr. *22*, 107–138.

Thantsha, M.S., Cloete, T.E., Moolman, F.S., and Labuschagne, P.W. (2009). Supercritical carbon dioxide interpolymer complexes improve survival of *B. longum* Bb-46 in simulated gastrointestinal fluids. Int. J. Food Microbiol. *129*, 88–92.

Vazquez, M.J., Alonso, J.L., Dominguez, H., and Parajo, J.C. (2000). Xylooligosaccharides: Manufacture and applications. Trends Food Sci. Technol. *11*, 387–393.

Vega, C., and Roos, Y.H. (2006). Invited review: spray-dried dairy and dairy-like emulsions – compositional considerations. J. Dairy Res. *89*, 383–401.

Villamiel, M., Corzo, N., Foda, M.I., Montes, F., and Olano, A. (2002). Lactulose formation catalysed by alkaline-substituted sepiolites in milk permeate. Food Chem. *76*, 7–11.

Wahab, S.M.A., Ahmed, M.A., Radwan, F.A., Hassan, R.M., and ElRefae, A.M. (1997). Relative permittivity and electrical conductivity of some divalent metal alginate complexes. Mater. Lett. *30*, 183–188.

Wang, Y. (2009). Prebiotics: Present and future in food science and technology. Food Res. Int. *42*, 8–12.

Wang, Z.L., Finlay, W.H., Peppler, M.S., and Sweeney, L.G. (2006). Powder formation by atmospheric spray-freeze-drying. Powder Technol. *170*, 45–52.

Ying, D.Y., Sunb, J., Sanguansria, L., Weerakkodya, R., and Augustina, M.A. (2012). Enhanced survival of spray-dried microencapsulated *Lactobacillus rhamnosus* GG in the presence of glucose. J. Food Eng. *10*, 597–602.

Yonekura, L., Sun, H., Soukoulis, C., and Fisk, I. (2014). Microencapsulation of *Lactobacillus* acidophilus NCIMB 701748 in matrices containing soluble fibre by spray drying: Technological characterization, storage stability and survival after *in vitro* digestion Journal of Functional Foods *6*, 205–214.

Zuidam, N.J., and Shimoni, E. (2009). Overview of microencapsulates for use in food products or processes and methods to take them. In Encapsulation Technologies for Active Food Ingredients and Food Processing, N.J. Zuidam and Nedovic, V., eds. (New York, Springer-Verlag), pp. 3–29.

Part V

Future Perspectives

Future Possibilities for Pro- and Prebiotics: Is the Sky the Limit?

33

Koen Venema and Ana Paula do Carmo

Abstract

Over the past two decades significant progress has been made in both the probiotics and the prebiotics area. Whereas earlier the primary diseases for probiotics were inflammatory bowel disease and allergy, this has now expanded to multiple other diseases and disorders too. And whereas earlier a prebiotic effect was considered to be synonymous with an increase in bifidobacteria, the research field now encompasses much more, and for instance looks at butyrate production, or levels of *Faecalibacterium prausnitzii* or *Akkermansia muciniphila*. One can therefore wonder: Are pro- and prebiotics effective for everything? This chapter hypothesizes on aspects that are beginning to be addressed, or have not even been considered yet, but might become relevant in the future as well.

Introduction

The idea underlying the hypothesis that certain strains of probiotics and certain prebiotic compounds (not necessarily carbohydrates) are beneficial in a number of diseases and disorders, even in some that have not yet been studied in any great detail, stems from the fact that both pro- and prebiotics may influence the gut microbiota. And although probiotics perhaps may not always have a large influence on the composition of the microbiota, they do seem to influence metabolic activity of the microbial consortium (McNulty *et al.*, 2011). One of the ways pro- and prebiotics may affect systemic processes is through their interaction with the immune system (either directly, or indirectly through changing the gut microbiota). This explains why orally ingested bugs or dietary compounds (and drugs for that matter) can have effects distally from the gut, since the immune system in the gut associated lymphoid tissue (GALT) is connected with the immune system in other parts of the body through blood and lymph. Hence effects on such diverse disorders as infectious diseases of the respiratory and urogenital tract, behaviour and cognition, autism, and, for example, breast, bladder or prostate cancer have been shown, besides the more commonly studied problems such as allergy, inflammatory bowel disease, irritable bowel syndrome, obesity and diabetes, etc.

It should be emphasized that the effects are strain-dependent for probiotics, and also compound-specific for prebiotics. This has clearly been shown for probiotics: not all immune modulating probiotics are efficacious for the same disorders. This seems counter-intuitive with the previous paragraph though. If two strains have an immunomodulatory activity, and both are anti-inflammatory, why would they not both be beneficial, e.g. for allergy? It goes to show that we know too little of the exact mode of action of probiotics to be able to properly select the most efficacious strain for a specific disease or disorder. For prebiotics this is also the case. Very few studies have been done, certainly clinically, that compare different prebiotics. The additional complexity that prebiotics also work through the gut microbiota, which is individual specific, makes it more difficult to decipher beforehand which prebiotic to give to which patients or population groups, despite the fact that in terms of microbial activity there is great redundancy between the microbiota from different individuals. If even simply the length of a molecule (degree of polymerization; DP) matters how the gut microbiota handles the substrate, as has been shown for instance for inulin with different DP (van Nuenen *et al.*, 2003), one can appreciate that prebiotics with different monosaccharide composition or glycosidic linkages will be handled completely different by the microbiota, even within one and the same individual. Until we can predict the structure–function relationship, preferably based on preclinical data, we will have to rely on trial-and-error. And then we have not even considered pitfalls like dose, etc. (See Chapter 1 of this book).

And even then there is the host-side. We are all unique. Single-nucleotide polymorphisms (SNPs) in our genetic material may affect the efficacy of pro- and prebiotics, and this may be why there are non-responders in clinical trials, just as there are with drug therapy. It is clear that there is still much to be done in the field.

Nevertheless, the wealth of new developments described in this book shows that it is a very promising field. Numerous additional disorders and diseases beyond the classical ones (if we consider research areas that were studied only two decades ago as 'classical') are being studied. Several more are likely to be investigated too. And this is because pro- and prebiotics, themselves or through the gut bacterial organisms and their metabolites, interact with interrelated immunological, metabolic and neuroendocrine pathways.

Already incorporated in the individual chapters of this book are future outlooks. Below we hypothesize on future aspects that are beginning to be addressed, or have not even been considered yet, but might become relevant in the future as well, and that have not been covered (in detail) in the individual chapters. It should be interpreted as light reading, and examples are not (at) all scientifically substantiated (yet), but the reader will see that for every example, a possible link with the gut microbiota is given.

Relatively unexplored areas for pro- and prebiotics

Are you a lark or an owl? Are you at your best in the morning or in the evening? What is your chronotype? And is your microbiota influencing these circadian rhythms? It would not surprise us. It is intriguing that the gut microbiota may also affect our biological clock and processes regulated by that internal clock. The gastrointestinal tract displays biological rhythms in e.g. gastric acid secretion, epithelial cell proliferation, gastrointestinal motility, and appetite regulation (Hoogerwerf, 2006). Strikingly, the development of gastrointestinal complications after administration of aspirin depends on the time of administration. Changes in our biological rhythms have been associated with gastrointestinal symptoms.

The hormone in the body that appears to be important in helping regulate the internal body clock's cycle of sleep and wakefulness is melatonin. Melatonin is produced by the pineal gland in the brain. Melatonin levels vary over 24 hour cycles and are controlled by our circadian rhythm. Normally its production is reduced by being in bright light. Levels increase again at night. Melatonin is present in some plants that we eat as foods, but is produced in the body from serotonin, by conversion of serotonin to N-acetyl serotonin, which is the direct precursor of melatonin. Serotonin itself is produced from the amino acid tryptophan. Serotonin mediates several important physiological functions including neurotransmission and gastrointestinal motility, and is involved in pain-perception (O'Mahony *et al.*, 2014b; Vanhoutvin *et al.*, 2009). Bacteria can produce melatonin and serotonin, in addition to other molecules that have neuromodulatory or immunomodulatory activity such as GABA, histamine and acetylcholine (Forsythe and Kunze, 2013; Iyer *et al.*, 2004). Although we are unaware of studies that have actually shown production of these bioactives by the gut microbiota, it is clear that the gut microbiota may also influence host production of these neuromodulators. Therefore, it is not unlikely that the gut microbiota, and hence pro- and prebiotics as modulators of the microbiota composition and activity, play a role in sleeping behaviour, pain perception and other biological processes that are under the regulation of melatonin, serotonin and other microbially produced neuromodulators (O'Mahony *et al.*, 2014a).

Since the finding of melatonin production in the gastrointestinal tract by enterochromaffin cells of several vertebrate species (Bubenik, 2002), a growing number of studies have been involved in the relationship between this hormone and food intake processes. In zebra fish it regulates food intake (Piccinetti *et al.*, 2010). Molecularly it increases leptin and reduces ghrelin expression, molecules playing a key role in central appetite regulation. In this way, melatonin, and thus the gut microbiota producing or influencing the production of this hormone, may also contribute to obesity.

Indole-3-propionic acid, a melatonin-related molecule, has been studied for its effect as a potential disease-modifying therapy for Alzheimer's disease (Bendheim *et al.*, 2002). Although the evidence is scarce, this may point to a role of the gut microbiota in Alzheimer's.

Anecdotal stories states that a diet high in maize leads to more aggression and cannibalism. This was concluded from the fact that ancient cultures in South and Middle America, such as the Inca's, were cannibals. From epidemiological studies it is clear that countries above the median in maize consumption have significantly higher homicide rates than countries below the median in maize consumption. Maize is relatively low in the essential amino acid tryptophan, the precursor for serotonin and melatonin. Research from a variety of sources suggests the hypothesis that populations consuming maize-based diets may have high homicide rates due to a reduction in brain tryptophan and/or serotonin (Ernandes *et al.*, 1996; Mawson and Jacobs, 1978). There is a convincing body of evidence that serotonin inhibits aggressive behaviour (Mawson and Jacobs, 1978). Maize, low in tryptophan, may thus influence all biological processes eluded to above. Alternatively, maize (or rather the resistant starch fraction in it) may directly affect the activity of the microbiota and in this way influence biological behavioural processes.

Also symptoms of depression and anxiety in anorexia nervosa have been linked with plasma tryptophan and serotonin metabolism (Gauthier *et al.*, 2014). Since these can potentially be modulated by the gut microbiota, which in turn can be controlled by probiotics and prebiotics, it is worth studying this in more detail.

There is ample evidence in several rodent models that the gut microbiota influences anxiety, behaviour and cognition. A nice example is how faecal transfer from the more aggressive (less anxious) NIH Swiss mice into the more anxious BALB/c mice lead the mice to become more aggressive (Bercik *et al.*, 2011). Since the gut microbiota can be influenced by dietary components (including pro- and prebiotics), it seems evident that these anxiety, behaviour and cognition can be modulated through the gut microbiota (in addition to direct effects of food component on, for example, cognition).

An important feature of psychiatric conditions is the different prevalence rates reported in males and females. For instance, whereas autism occurs more frequently in males (4:1, male: female) depression and anxiety are more prevalent in females (Borre *et al.*, 2014). A role of the microbiota in autistic spectrum disorders has also been reported e.g. (MacFabe *et al.*, 2011). Could it be that a difference in microbiota between men and women is responsible for the observed differences in prevalence? To date, a limited number of studies have focused on the impact of gender differences of the gut microbiota, and further studies are needed to explore sex-related factors underlying the differentiating outcomes in brain neurochemistry between men and women as a function of microbiota–host interactions.

Oxytocin is a hormone and neurotransmitter that plays an important role in the neuroanatomy of intimacy. It is often referred to as the 'trust hormone' or 'bonding hormone'. However, oxytocin is increasingly being seen as a brain chemical that does a lot more than just bring couples closer together, especially its role in social recognition and anxiety. It is released in large amounts after hugging and kissing (Montiel-Castro *et al.*, 2013), and at lower doses when shaking hands, or supposedly even during gazing. Given its ability to break-down social barriers, induce feelings of optimism, increase self-esteem, and build trust, oxytocin is increasingly being seen as something that can help people overcome their social inhibitions and fears. Studies are showing that it may be effective in treating debilitating shyness, or to help people with social anxieties and mood disorders. Interestingly, its most recently discovered but not as well characterized roles include interactions with body energy balance and the immune system (Poutahidis *et al.*, 2013). Even more interesting is the recent report that oxytocin accelerates wound-healing in mice, and that the probiotic *Lactobacillus reuteri* ATCC-PTA- 6475 induces increased oxytocin expression (Poutahidis *et al.*, 2013).

Using a simple coin-toss prediction task in which participants could dishonestly report their performance levels to benefit their group's outcome, it was shown that oxytocin increases group-serving dishonesty (Shalvi and De Dreu, 2014). Oxytocin shifts the participant's focus from self to group interests. These findings highlight the role of oxytocin in trust and cooperation, and in shaping dishonesty. This provides valuable insight into when and why collaboration turns into corruption. So, next time you sit around a negotiation table make sure you serve a functional beverage with the proper probiotic to switch of the dishonesty in your competitors.

SNPs in the oxytocin receptor also play a role in trust behaviour (Krueger *et al.*, 2012), so could differences in expression of the receptor, influenced by the gut microbiota also play a role in this social aspect? It is not unlikely. In mice it has been shown that the microbiota influence behaviour and defines aggressiveness, as eluded to above (Bercik *et al.*, 2011). Both in mice and man a gene has been found that may be responsible for aggression and anger control (Richter *et al.*, 2011). So, if the gut microbiota can influence these processes, why not trust, social bonding, social inhibition and fears, etc. The example of *L. reuteri* increasing oxytocin perhaps lays the foundation of more research in this area.

There are also sex differences in the neural and behavioural response during human social interaction (Rilling *et al.*, 2014) For instance, oxytocin (and vasopressin) increase brain activity in men within areas playing a key role in reward, social bonding, arousal and memory. In women on the other hand oxytocin had no effect or in some cases actually decreased brain activity in these regions was observed. Rilling *et al.* (2014) suggested that the results showed opposing effects enabling accurate social responses for women, while men respond with increased egocentricity and less adaptive regulation. This has crucial implications for explaining gender differences in social skills such as empathy and prosociality. (Tomova *et al.*, 2014) Perhaps the microbiota could (or already does) play a role here. Can personalized probiotic or prebiotics affect this? This is certainly worth to investigate.

Another concept, based on the already discussed believe that the gut microbiota can affect functions in the body distally from the gut through affecting, for example, the immune system, is the concept of 'beauty from within'. It relies on the intake of pro- and/or prebiotics for improved skin (and hair) health. There have been a number of studies that have shown that pro- and prebiotics affect skin health. For instance, *Lactobacillus paracasei* NCC 2461 has been shown to decrease skin sensitivity and strengthen the barrier function, thereby decreasing transepidermal water loss (Gueniche *et al.*, 2014). Moreover, the same, orally ingested strain was shown to modulate the skin immune system leading to the preservation of skin homeostasis (Benyacoub *et al.*, 2014). A fermented milk containing *Bifidobacterium breve* strain Yakult and galacto-oligosaccharides was shown to improve skin health, by preventing both skin dryness and disruption of keratinization in healthy adult women. The mechanism was shown to be a reduced production of phenols such a *p*-cresol by the gut microbiota (Miyazaki *et al.*, 2014). Similarly, in mice, the intake of probiotics improves the glow of health of the fur of the mice (Erdman and Poutahidis, 2014). Imagine probiotics being able to cure or prevent baldness! In this respect one might also think of topical treatment with probiotics. Currently the scientific proof for this is clearly lacking, but opportunistic companies already sell soaps and shampoos containing what they call 'probiotics'.

It has been long known that the gut microbiota plays a role in colon cancer promotion. Besides the health beneficial metabolites that the microbiota produce, they can also produce compounds that are considered to be toxic at high doses, such as H_2S, *p*-cresol, phenol, ammonia, biogenic amines, etc. On the other hand, the gut microbiota also plays a role in immunomodulation and modulate regulatory T-lymphocyte phenotypes, maintaining systemic immune balance. This would determining the fate of preneoplastic lesions towards regression rather than further outgrowth. The paradox of these two opposing activities of the gut microbiota and regarding cancer prevention and perhaps (immuno)therapy has nicely been discussed recently (Poutahidis *et al.*, 2014a). Perhaps it will be possible to accomplish a tumour-suppressive environment in the future through modulation of the gut microbiota, e.g. using pro- and/or prebiotics.

It has long been known that stress affects the gut. Increased stress leads to disruption of the gut barrier function and there are indications that this may be one of the onset mechanisms for irritable bowel syndrome. One form of stress is extreme exercise. Recently it was shown that elite athletes, specifically Irish male professional rugby players, showed increased levels of plasma creatine kinase (a marker of extreme exercise), and inflammatory and metabolic markers, as expected (Clarke *et al.*, 2014). However, they also showed an unexpected higher diversity of the gut microbiota. Since extremes of exercise are often accompanied by extremes of diet (e.g. excess protein intake for muscle build up), the authors concluded that the relationship between extreme exercise, diet and gut microbiota is complex. Nevertheless, in preclinical screening studies, a number of probiotics have been shown to increase barrier function of a disrupted gut epithelial layer. If this could be accomplished in (extreme) athletes it would reduce the 'side-effects' of extreme activity and would improve their performance.

The microbiota has also been implicated in longevity (Biagi *et al.*, 2013; O'Toole, 2012). The original concept goes back to Elie Metchnikoff, who believed that Bulgarians lived to an old age due to the consumption of yoghurt with the beneficial microbes *L. bulgaricus* and *S. thermophilus*. Nowadays we see that the gut microbiota of elderly is different both in composition and activity. With respect to composition it has been observed that centenarians had a marked reduction in *F. prausnitzii* and relatives (Biagi *et al.*, 2010). In an effort to study changes in gut microbiota composition in elderly living in the community, versus those in day-hospital, or rehabilitation or long-term residential care, Claesson *et al.*, studied the microbiota of 178 elderly. They showed that the microbiota of these groups could be separated using statistical analyses, and correlated with measures of frailty, co-morbidity, nutritional status and inflammatory markers (Claesson *et al.*, 2011; Claesson *et al.*, 2012). With respect to the microbiota activity, the genes encoding for one of the enzymes responsible for production of the healthy metabolite butyrate (butyryl-CoA:acetate CoA-transferase) were assessed by Hippe *et al.* (2011) in individuals of different age. Faecal samples of young healthy omnivores, young vegetarians and elderly omnivores were evaluated. The elderly had significantly fewer copies of the butyryl-CoA:acetate CoA-transferase gene than young omnivores, while vegetarians showed the highest number of copies. The authors concluded that the gastrointestinal microbiota of the elderly is characterized by a decreased butyrate production capacity. These results suggest that the butyryl-CoA:acetate CoA-transferase gene is a valuable marker for gastrointestinal microbiota function. These are just a few recent examples of the research taking place in this area.

At the other end of the age spectrum, the effect of the gut microbiota in children is intensively studied. Relatively new are the studies on microbiota and epigenetics, and how this influences health and disease later in life. The concept allows dietary interventions when the fetus is still in the womb. That the gut microbiota does influence epigenetic expression of genes (Remely *et al.*, 2014c) should perhaps not be that surprising any more after all of the examples above. Epigenetic regulation by short-chain fatty acids (SCFA), particularly butyrate, on immune mediators and the SCFA-receptor FFAR3 was recently shown in patients with type 2 diabetes and/or obesity (Remely *et al.*, 2014a; Remely *et al.*, 2014b).

An interesting concept that gut microbiota drives eating behaviour has been put forward recently (Alcock *et al.*, 2014). The hypothesis is based on the idea that microbes in the gastrointestinal tract are under selective pressure to manipulate host eating behaviour to increase their fitness, sometimes at the expense of host fitness. Although at present perhaps difficult to believe (or perhaps we just refuse to acknowledge that we are no longer in control of our own body, but that everything we do is driven by the gut microbiota), the idea is not that far-fetched as we have seen above that the microbiota or certain probiotic strains do affect other aspects of our behaviour, such as anxiety, trust, aggressiveness, etc. So why not eating behaviour? Remember the connection between maize, gut microbiota and cannibalism? Cannibalism is a form of eating behaviour, right??

Some random thoughts: in the fruit fly (*Drosophila melanogaster*) *Lactobacillus plantarum* has been shown to reduce the time of progression from larvae to flies (Storelli *et al.*, 2011). Translated to humans this would mean that puberty is reduced. Would not that be something for parents with rebel children?? Part of the attraction between certain individuals is through compounds called sex pheromones. They are secreted and volatile, and surround the body of the host producing it. 'Love or, at least, sexual attraction is literally in the air' is how *Scientific American* phrased it (Hadhazy, 2012). Pheromones are produced and excreted through the skin and sweat glands, amongst others. Now, the skin also contains a complex microbiota, especially in areas that are also moist, such as the armpit. And we all know how smelly armpits can be! It is therefore highly conceivable that the skin microbiota can affect pheromone production, or may in fact metabolize pheromones that are produced by the host, degrading or modifying them, thereby affecting our choice in a lifelong partner. Alternatively, when affecting skin microbiota in existing couples one might interfere in happy marriages and introduce an increase in divorce, or more adultery... That commercial of a deodorant that attracts zillions of women is not far besides the truth! It just should contain the right pheromone. An intriguing study has shown that the same strain that improves wound healing, *L. reuteri* ATCC-PTA- 6475, also increases youthful serum testosterone levels and testicular size in ageing mice (Poutahidis *et al.*, 2014b). Since oxytocin has been shown to considerably improve sexual performance, the induction of this hormone by *L. reuteri* ATCC-PTA- 6475 might be the underlying mechanism. Perhaps the email spam about Viagra will be replaced by spam about probiotics...

It is clear from the above that the gut microbiota may be more in control of our body than we might wish for. With 10^{14} cells, they even outnumber our own cells by a factor of 10. However, we can modulate the (amount of) control they have over us, by modulating the composition of the microbiota through feeding e.g. probiotics and prebiotics.

References

Alcock, J., Maley, C.C., and Aktipis, C.A. (2014). Is eating behavior manipulated by the gastrointestinal microbiota? Evolutionary pressures and potential mechanisms. Bioessays *36*, 1–10.

Bendheim, P.E., Poeggeler, B., Neria, E., Ziv, V., Pappolla, M.A., and Chain, D.G. (2002). Development of indole-3-propionic acid (OXIGON) for Alzheimer's disease. J. Mol. Neurosci. *19*, 213–217.

Benyacoub, J., Bosco, N., Blanchard, C., Demont, A., Philippe, D., Castiel-Higounenc, I., and Gueniche, A. (2014). Immune modulation property of *Lactobacillus paracasei* NCC2461 (ST11). strain and impact on skin defences. Benef. Microbes *5*, 129–136.

Bercik, P., Denou, E., Collins, J., Jackson, W., Lu, J., Jury, J., Deng, Y., Blennerhassett, P., Macri, J., McCoy, K.D., *et al.* (2011). The intestinal microbiota affect central levels of brain-derived neurotropic factor and behavior in mice. Gastroenterology *141*, 599–609, 609 e591–593.

Biagi, E., Candela, M., Turroni, S., Garagnani, P., Franceschi, C., and Brigidi, P. (2013). Ageing and gut microbes: perspectives for health maintenance and longevity. Pharmacol. Res. *69*, 11–20.

Biagi, E., Nylund, L., Candela, M., Ostan, R., Bucci, L., Pini, E., Nikkila, J., Monti, D., Satokari, R., Franceschi, C., *et al.* (2010). Through ageing, and beyond: gut microbiota and inflammatory status in seniors and centenarians. PLoS One 5, e10667.

Borre, Y.E., Moloney, R.D., Clarke, G., Dinan, T.G., and Cryan, J.F. (2014). The Impact of Microbiota on Brain and Behavior: Mechanisms & Therapeutic Potential. In Microbial Endocrinology: The Microbiota-Gut-Brain Axis in Health and Disease, M. Lyte, and J.F. Cryan, eds. (New York, Springer), pp. 373–403.

Bubenik, G.A. (2002). Gastrointestinal melatonin: localization, function, and clinical relevance. Dig. Dis. Sci. 47, 2336–2348.

Claesson, M.J., Cusack, S., O'Sullivan, O., Greene-Diniz, R., de Weerd, H., Flannery, E., Marchesi, J.R., Falush, D., Dinan, T., Fitzgerald, G., *et al.* (2011). Composition, variability, and temporal stability of the intestinal microbiota of the elderly. Proc. Natl. Acad. Sci. U.S.A. *108* (Suppl. 1), 4586–4591.

Claesson, M.J., Jeffery, I.B., Conde, S., Power, S.E., O'Connor, E.M., Cusack, S., Harris, H.M., Coakley, M., Lakshminarayanan, B., O'Sullivan, O., *et al.* (2012). Gut microbiota composition correlates with diet and health in the elderly. Nature *488*, 178–184.

Clarke, S.F., Murphy, E.F., O'Sullivan, O., Lucey, A.J., Humphreys, M., Hogan, A., Hayes, P., O'Reilly, M., Jeffery, I.B., Wood-Martin, R., *et al.* (2014). Exercise and associated dietary extremes impact on gut microbial diversity. Gut *63*, 1913–1920.

Erdman, S.E., and Poutahidis, T. (2014). Probiotic 'glow of health': it's more than skin deep. Benef. Microbes 5, 109–119.

Ernandes, M., La Guardia, M., and Giammanco, S. (1996). Maize based diets and possible neurobehavioural after-effects among some populations in the world. Human Evol. *11*, 67–77.

Forsythe, P., and Kunze, W.A. (2013). Voices from within: gut microbes and the CNS. Cell. Mol. Life Sci. *70*, 55–69.

Gauthier, C., Hassler, C., Mattar, L., Launay, J.M., Callebert, J., Steiger, H., Melchior, J.C., Falissard, B., Berthoz, S., Mourier-Soleillant, V., *et al.* (2014). Symptoms of depression and anxiety in anorexia nervosa: links with plasma tryptophan and serotonin metabolism. Psychoneuroendocrinology 39, 170–178.

Gueniche, A., Philippe, D., Bastien, P., Reuteler, G., Blum, S., Castiel-Higounenc, I., Breton, L., and Benyacoub, J. (2014). Randomized double-blind placebo-controlled study of the effect of *Lactobacillus paracasei* NCC 2461 on skin reactivity. Benef. Microbes 5, 137–145.

Hadhazy, A. (2012). Do Pheromones Play a Role in Our Sex Lives? Scientific American, Feb 13, 2012.

Hippe, B., Zwielehner, J., Liszt, K., Lassl, C., Unger, F., and Haslberger, A.G. (2011). Quantification of butyryl CoA:acetate CoA-transferase genes reveals different butyrate production capacity in individuals according to diet and age. FEMS Microbiol. Lett. *316*, 130–135.

Hoogerwerf, W.A. (2006). Biologic clocks and the gut. Curr. Gastroenterol. Rep. *8*, 353–359.

Iyer, L.M., Aravind, L., Coon, S.L., Klein, D.C., and Koonin, E.V. (2004). Evolution of cell–cell signaling in animals: did late horizontal gene transfer from bacteria have a role? Trends Genet. *20*, 292–299.

Krueger, F., Parasuraman, R., Iyengar, V., Thornburg, M., Weel, J., Lin, M., Clarke, E., McCabe, K., and Lipsky, R.H. (2012). Oxytocin receptor genetic variation promotes human trust behavior. Front. Hum. Neurosci. *6*, 4.

MacFabe, D.F., Cain, N.E., Boon, F., Ossenkopp, K.P., and Cain, D.P. (2011). Effects of the enteric bacterial metabolic product propionic acid on object-directed behavior, social behavior, cognition, and neuroinflammation in adolescent rats: Relevance to autism spectrum disorder. Behav. Brain Res. *217*, 47–54.

Mawson, A.R., and Jacobs, K.W. (1978). Corn consumption, tryptophan, and cross-national homicide rates. J. Orthomolec. Psychiatry 7.

McNulty, N.P., Yatsunenko, T., Hsiao, A., Faith, J.J., Muegge, B.D., Goodman, A.L., Henrissat, B., Oozeer, R., Cools-Portier, S., Gobert, G., *et al.* (2011). The impact of a consortium of fermented milk strains on the gut microbiome of gnotobiotic mice and monozygotic twins. Sci. Transl Med. 3, 106ra106.

Miyazaki, K., Masuoka, N., Kano, M., and Iizuka, R. (2014). Bifidobacterium fermented milk and galacto-oligosaccharides lead to improved skin health by decreasing phenols production by gut microbiota. Benef. Microbes 5, 121–128.

Montiel-Castro, A.J., Gonzalez-Cervantes, R.M., Bravo-Ruiseco, G., and Pacheco-Lopez, G. (2013). The microbiota–gut–brain axis: neurobehavioral correlates, health and sociality. Front. Integr. Neurosci. 7, 70.

O'Mahony, S.M., Clarke, G., Borre, Y.E., Dinan, T.G., and Cryan, J.F. (2014a). Serotonin, tryptophan metabolism and the brain–gut-microbiome axis. Behav. Brain Res. 277, 32–48.

O'Mahony, S.M., Felice, V.D., Nally, K., Savignac, H.M., Claesson, M.J., Scully, P., Woznicki, J., Hyland, N.P., Shanahan, F., Quigley, E.M., *et al.* (2014b). Disturbance of the gut microbiota in early-life selectively affects visceral pain in adulthood without impacting cognitive or anxiety-related behaviours in male rats. Neuroscience 277C, 885–901.

O'Toole, P.W. (2012). Changes in the intestinal microbiota from adulthood through to old age. Clin. Microbiol. Infect. *18* (Suppl. 4), 44–46.

Piccinetti, C.C., Migliarini, B., Olivotto, I., Coletti, G., Amici, A., and Carnevali, O. (2010). Appetite regulation: the central role of melatonin in Danio rerio. Horm. Behav. 58, 780–785.

Poutahidis, T., Kearney, S.M., Levkovich, T., Qi, P., Varian, B.J., Lakritz, J.R., Ibrahim, Y.M., Chatzigiagkos, A., Alm, E.J., and Erdman, S.E. (2013). Microbial symbionts accelerate wound healing via the neuropeptide hormone oxytocin. PLoS One *8*, e78898.

Poutahidis, T., Kleinewietfeld, M., and Erdman, S.E. (2014a). Gut microbiota and the paradox of cancer immunotherapy. Front. Immunol. 5.

Poutahidis, T., Springer, A., Levkovich, T., Qi, P., Varian, B.J., Lakritz, J.R., Ibrahim, Y.M., Chatzigiagkos, A., Alm, E.J., and Erdman, S.E. (2014b). Probiotic microbes sustain youthful serum testosterone levels and testicular size in ageing mice. PLoS One 9, e84877.

Remely, M., Aumueller, E., Jahn, D., Hippe, B., Brath, H., and Haslberger, A.G. (2014a). Microbiota and epigenetic regulation of inflammatory mediators in type 2 diabetes and obesity. Benef. Microbes 5, 33–43.

Remely, M., Aumueller, E., Merold, C., Dworzak, S., Hippe, B., Zanner, J., Pointner, A., Brath, H., and Haslberger, A.G. (2014b). Effects of short-chain fatty acid producing bacteria on epigenetic regulation of FFAR3 in type 2 diabetes and obesity. Gene 537, 85–92.

Remely, M., Lovrecic, L., de la Garza, A.L., Migliore, L., Peterlin, B., Milagro, F.I., Martinez, A.J., and Haslberger, A.G. (2014c). Therapeutic perspectives of epigenetically active nutrients. Br. J. Pharmacol.

Richter, S., Gorny, X., Marco-Pallares, J., Kramer, U.M., Machts, J., Barman, A., Bernstein, H.G., Schule, R., Schols, L., Rodriguez-Fornells, A., *et al.* (2011). A Potential Role for a Genetic Variation of AKAP5 in Human Aggression and Anger Control. Front. Hum. Neurosci. 5, 175.

Rilling, J.K., Demarco, A.C., Hackett, P.D., Chen, X., Gautam, P., Stair, S., Haroon, E., Thompson, R., Ditzen, B., Patel, R., *et al.* (2014). Sex differences in the neural and behavioural response to intranasal oxytocin and vasopressin during human social interaction. Psychoneuroendocrinology 39, 237–248.

Shalvi, S., and De Dreu, C.K. (2014). Oxytocin promotes group-serving dishonesty. Proc. Natl. Acad. Sci. U.S.A. *111*, 5503–5507.

Storelli, G., Defaye, A., Erkosar, B., Hols, P., Royet, J., and Leulier, F. (2011). *Lactobacillus plantarum* promotes Drosophila systemic growth by modulating hormonal signals through TOR-dependent nutrient sensing. Cell Metab. *14*, 403–414.

Tomova, L., von Dawans, B., Heinrichs, M., Silani, G., and Lamm, C. (2014). Is stress affecting our ability to tune into others? Evidence for gender differences in the effects of stress on self-other distinction. Psychoneuroendocrinology 43, 95–104.

van Nuenen, M.H.M.C., Meyer, P.D., and Venema, K. (2003). The effect of various inulins and *Clostridium difficile* on the metabolic activity of the human colonic microbiota *in vitro*. Microb. Ecol. Health Dis. *15*, 137–144.

Vanhoutvin, S.A., Troost, F.J., Kilkens, T.O., Lindsey, P.J., Hamer, H.M., Jonkers, D.M., Venema, K., and Brummer, R.J. (2009). The effects of butyrate enemas on visceral perception in healthy volunteers. Neurogastroenterol. Motil. *21*, 952-e976.

Appendix I: Web Resources

Chapter 1

- FSTA®:
  - an intelligence service constantly monitoring developments in the areas of food science, food technology and nutrition
  - http://www.foodsciencecentral.com/
- IFIC (International Food Information Council Foundation):
  - dedicated to the mission of effectively communicating science-based information on health, nutrition and food safety for the public good
  - http://www.foodinsight.org/
- FAO/WHO views on probiotics:
  - http://www.who.int/foodsafety/publications/fs_management/en/probiotics.pdf
  - http://www.who.int/foodsafety/fs_management/en/probiotic_guidelines.pdf?ua=1
- International Life Sciences Institute (ILSI) task-force on probiotics:
  - http://www.ilsi.org/Europe/Pages/TF_Probiotics.aspx
- ILSI task-force on prebiotics:
  - http://www.ilsi.org/Europe/Pages/TF_Prebiotics.aspx
- ISAPP (International Scientific Association for Probiotics and Prebiotics):
  - http://www.isapp.net/
- American Gut:
  - World's largest open-source science project to understand the microbial diversity of the Human Gut.
  - Learn which microbes live in your gut, skin and mouth
  - http://humanfoodproject.com/americangut/

Chapter 2

- KEGG pathway:
  - an on-line database for wiring diagrams of molecular interactions, reactions, and relations
  - http://www.genome.jp/kegg/pathway.html
- The Human Metabolome Database (HMDB):
  - a freely available electronic database containing detailed information about small molecule metabolites found in the human body
  - http://www.hmdb.ca/
- The International Association for Probiotics and Prebiotics (ISAPP):
  - an international non-profit collaboration of scientists dedicated to advance scientific excellence in probiotics and prebiotics
  - http://www.isapp.net/Probiotics-and-Prebiotics/The-Science

Chapter 4

Bacterial genome sequences

The following sites contain genome sequence information of *Bifidobacterium*; some searching may be required:

- http://www.ebi.ac.uk/genomes/bacteria.html
- http://www.ncbi.nlm.nih.gov/genomes/MICROBES/microbial_taxtree.html
- http://www.genomesonline.org/organisms?Organism.Domain=BACTERIAL

Societies promoting the science behind *Bifidobacterium*:

- International Scientific Association for Probiotics and Prebiotics: http://www.isapp.net/
- International Probiotics Association: http://www.internationalprobiotics.org/about
- Global Alliance for Probiotics: http://www.gap-probiotics.org/
- International Life Sciences Institute: http://www.ilsi.org/Europe/Pages/TF_Probiotics.aspx

Sources of general information regarding probiotics:

- American Nutrition Association: http://americannutritionassociation.org/newsletter/science-probiotics
- American Gastroenterological Association: http://www.gastro.org/patient-center/diet-medications/probiotics

Chapter 5

- Accurate information about the current status of a name, synonyms, and other useful information concerning prokaryotes:
  - http://www.bacterio.net/propionibacterium.html
- General information from MicrobeWiki, the student-edited microbiology resource:
  - https://microbewiki.kenyon.edu/index.php/Propionibacterium
- Web page of the International Scientific Association for Probiotics and Prebiotics showing general information related to scientific advances in probiotics and prebiotics:
  - http://www.isapp.net/

Chapter 6

Functional foods

- http://www.isabellashealthybakery.com/functional.html
- http://www.sporegen.com
- http://www.sabinsa.com/newsroom/press-releases/pr20121002.html)

Probiotics manufacturers

- http://www.ganedenBC30.com

EFSA Journal

- www.efsa.europa.eu/efsajournal
- http://www.efsa.europa.eu/it/efsajournal/pub/773.htm

Chapter 7

- http://onlinelibrary.wiley.com/doi/10.1111/j.1365–2036.2007.03442.x/pdf
- http://www.dairyscience.info/index.php/probiotics/232-yeast-probiotics.html
- http://www.europeanreview.org/wp/wp-content/uploads/1000.pdf

Chapter 8

Animals

- EFSA: Additives and Products or Substances used in Animal Feed (FEEDAP)
- http://www.efsa.europa.eu/en/panels/feedap.htm
- American Association of Feed Control Officials: Ingredient Definitions Committee
- http://www.aafco.org/Directory/CommitteePages/IngredientDefinitions.aspx

Humans

- EFSA: Opinion on the Qualified Presumption of Safety on selected microorganisms
  - http://www.efsa.europa.eu/en/efsajournal/doc/587.pdf
- EFSA: Substantiation of health claims related to non-characterized bacteria and yeasts
  - http://www.efsa.europa.eu/en/efsajournal/doc/1470.pdf
- National Center for Complementery and Alternative Medicine: Oral Probiotics
  - http://nccam.nih.gov/health/probiotics/introduction.htm
- US Food and Drug Administration: Regulatory perspectives on pre- and probiotics
  - http://www.iom.edu/~/media/Files/Activity%20Files/PublicHealth/MicrobialThreats/Vaillaincourt17Mar05ProbioticsRegulation.pdf
- Agriculture and Agri-Food Canada: Function claims and probiotic claims for food
  - http://www4.agr.gc.ca/resources/prod/doc/prog/ai-ae/pdf/probiotics_probiotiques_eng.pdf
- FAO/WHO: Probiotic definition
  - ftp://ftp.fao.org/docrep/fao/009/a0512e/a0512e00.pdf

Trade Associations

- International Probiotics Association
  - http://www.internationalprobiotics.org/about
- International Scientific Association for Probiotics and Prebiotics.
  - http://www.isapp.net/

Chapter 9

General information about *E. coli* infections and outbreaks can be found at the following websites:

- World Health Organization:
  - http://www.who.int/topics/escherichia_coli_infections/en/
- Centers for Disease Control and Prevention (USA government):
  - http://www.cdc.gov/ecoli/
- European Centre for Disease Prevention and Control:
  - http://ecdc.europa.eu/en/healthtopics/escherichia_coli/pages/index.aspx
- About *E. coli*:
  - http://www.about-ecoli.com/

Genome databases and tools for colibiome analysis can be found at the following websites:

- Pathosystems Ressource integration Center (PATRIC):
  - http://patricbrc.org/portal/portal/patric/Taxon?cType=taxon&cId=561
- Virulence Factors of pathogenic Bacteria (VFDB-comparative pathogenomics):
  - http://www.mgc.ac.cn/cgi-bin/VFs/compvfs.cgi?Genus=Escherichia
- *E. coli* Database portal:
  - http://www.uni-giessen.de/ecoli/IECA/index.php
- -Colibri and Genolist World-Wide Web server (Institut Pasteur):
  - http://genolist.pasteur.fr/Colibri/
  - http://genodb.pasteur.fr/cgi-bin/WebObjects/GenoList.woa/1/wa/goToTaxoRank?level=Escherichia

- *E. coli* genome project (University of Wisconsin):
  - http://www.genome.wisc.edu/

Web resources related to MLST analysis according to the three main shemes are provided by the following sites:

- A multilocus sequence typing database system for pathogenic *E. coli* (EcMLST):
  - http://shigatox.net/new/tools/ecmlst.html
- Institut Pasteur's MLST website:
  - www.pasteur.fr/mlst
- Achtman scheme:
  - http://mlst.warwick.ac.uk/mlst/dbs/Ecoli

General information, newsletters about gut microbiota and probiotics and impact on health:

- The gut Microbiota for Health:
  - http://www.gutmicrobiotaforhealth.com/
  - http://gutmicrobiotaforhealth.com/c/probiotics
- Metahit, the human gut genes catalogue:
  - http://www.metahit.eu/index.php?id=360
- Metagenopolis:
  - http://www.mgps.eu/index.php?id=homepage
- ISAP (International Scientific Association for Probiotics and Prebiotics):
  - http://www.isapp.net
- Probiotics: Applications in Gastrointestinal Health and Disease:
  - http://www.usprobiotics.org

Information related to probiotic preparations containing *E. coli* strains can be found at the following websites:

- Mutaflor®:
  - http://www.mutaflor.de/cms/(German site)
  - http://mutaflor.ca/(Canadian site)
  - http://mutaflor.com.au/(Australian site)
- Symbioflor®2:
  - http://symbiopharm.de/en/products/symbioflor-2.html
- Colinfant newborn:
  - http://dyntec.cz.s3.amazonaws.com/232_eng_colinfantnb.pdf

Database related to clinical studies of human participants conducted around the world are available at the following website:

- http://clinicaltrials.gov/ct2/home

Chapter 11

GMO

- http://www.hse.gov.uk/biosafety/gmo/information.htm
- http://www.efsa.europa.eu/en/panels/gmo.htm

Company

- http://www.actogenix.com/
- http://www.vitherapharma.com/index.html

Clinical trial

- https://clinicaltrials.gov/
- International Scientific Association for Probiotics and Prebiotics:
  - http://www.isapp.net/

Chapter 12

- Metahit homepage:
  - http://www.metahit.eu/
- American Human Microbiome Project home page:
  - http://commonfund.nih.gov/hmp/index
- Vaginal microbiome consortium:
  - http://vmc.vcu.edu/about
- Mediateque INRA:
  - http://mediatheque.inra.fr/media/detail/246161/private
- Micalis homepage:
  - http://www.micalis.fr/micalis_eng/
- CDC page for IBD:
  - http://www.cdc.gov/ibd/
- Chrons and colitis foundation of America:
  - http://www.ccfa.org/
- What is a health benefit? Researchers issue probiotic guidance for EFSA applications. By Nathan Grey+, 11-Sep-2013:
  - http://www.nutraingredients.com/Research/What-is-a-health-benefit-Researchers-issue-probiotic-guidance-for-EFSA-applications
- Analytical Profile Index Biomerieux homepage:
  - http://www.biomerieux-usa.com/servlet/srt/bio/usa/dynPage?doc=USA_PRD_LST_G_PRD_USA_5

Chapter 14

Science Sites

- Centro de Referencia para Lactobacilos:
  - http://www.cerela.org.ar/
  - The Dairy Institute of Tucuman belongs to National Scientific and Technical Research Council -Argentina.
- Center for Gastrointestinal Biology and Disease:
  - http://www.med.unc.edu/cgibd
- Instituto de Productos Lácteos de Asturias:
  - http://www.ipla.csic.es/.
  - The Dairy Institute of Asturias belongs to Science and Food Technology Higher Council for Scientific Research.
- Laboratory of Microbiology Wageningen:
  - http://www.wageningenur.nl/en/Expertise-Services/Chair-groups/Agrotechnology-and-Food-Sciences/Laboratory-of-Microbiology.htm
- Microbiome Core Facility:
  - http://www.med.unc.edu/microbiome
  - metagenomic technologies to examine microbial communities
- Southeast Dairy Foods Research Center:
  - http://fbns.ncsu.edu/sdfrc/sdfrc.html.

  - The SDFRC has been in operation since 1988. North Carolina State University is the lead institution, joined by Mississippi State University. It is one of six National Centers funded and supported by Dairy Management Inc. (DMI), through the Dairy Research Institute (DRI).

Company sites

- American Type Culture Collection (ATCC):
  - http://www.atcc.org/
- Chr. Hansen:
  - http://www.chr-hansen.com/research-development/customization.html
  - development of bacterial cultures for the dairy industry.
- BioGaia is a Swedish company:
  - http://www.biogaia.com/
- BIOPOLIS, S.L | Tailor-made Biotechnology:
  - http://www.biopolis-biotech.com/en/agri-food-biotechnology/
- BioSidus Probiótico intestinal (Bioflora™):
  - http://www.biosidus.com.ar/biopharmaceuticals.php
- Food for Health Ireland:
  - http://www.fhi.ie/home/
- Lactina-ltd:
  - http://www.lactina-ltd.com/eng/products-zakvaski.php
- Lactosan GmbH and Co. KG manufacturers of products based on living lactic acid bacteria:
  - http://www.lactosan.at/company.en.34.htm
- Lalleman:
  - http://www.lallemand.com/
- Nestle Institute Health Sciences Lausanne:
  - http://www.nestleinstitutehealthsciences.com/
- TI Food Nutrition:
  - http://www.tifn.nl/partnershipforinnovation

Chapter 15

- KEGG:
  - Kyoto Encyclopedia of Genes and Genomes
  - http://www.genome.jp/kegg/
- CAZy:
  - Carbohydrate-Active enZYmes Database
  - http://www.cazy.org/
- HMP:
  - Human Microbiome Project
  - http://www.hmpdacc.org/
- MetaHIT:
  - Metagenomics of the Human Intestinal Tract
  - http://www.metahit.eu/
- RegPrecise:
  - Collection of Manually Curated Inferences of Regulons in Prokaryotic Genomes
  - http://regprecise.lbl.gov/RegPrecise/
- PharmacoMicrobiomics:
  - The Drug-Microbiome Portal
  - http://pharmacomicrobiomics.com/
- mPUMA:
  - microbial Profiling Using Metagenomic Assembly
  - http://mpuma.sourceforge.net
- BioCyc:
  - a collection of 2988 Pathway/Genome Databases
  - http://biocyc.org/
- MetaCyc:
  - a database of nonredundant, experimentally elucidated metabolic pathways
  - http://www.metacyc.org/
- Pathway Tools:
  - a comprehensive symbolic systems biology software system that supports several use cases in bioinformatics and systems biology
  - http://brg.ai.sri.com/ptools/
- American Gut:
  - World's largest open-source science project to understand the microbial diversity of the Human Gut. Learn which microbes live in your gut, skin and mouth
  - http://humanfoodproject.com/americangut/
- Rob Knight Lab, University of Colorado – Software:
  - https://knightlab.colorado.edu/wordpress/?page_id=28
- HMDB:
  - Human Metabolome Database
  - http://www.hmdb.ca/
- G6G:
  - Directory of Omics and Intelligent Software! This software directory lists data mining and additional software from the fields of biotechnology and artificial intelligence
  - http://g6g-softwaredirectory.com/index.php

Chapter 16

- http://qiime.org/:
  - Quantitative Insights into Microbial Ecology or QIIME is an open source software package used to compare and analyse microbial communities, usually using 16S rRNA gene or metagenomics NGS data generated from different platforms
- http://www.microbial-ecology.net/probebase/credits.asp:
  - the Department of Microbial Ecology at the University of Vienna runs probeBase, a database housing oligonucleotide probes usually targeting 16S rRNA and used for fluorescent *in situ* hybridization
- http://www.arb-silva.de/SILVA rRNA database:
  - an on-line resource riboromal RNA (both large and small unit rRNA) sequence data
- http://rdp.cme.msu.edu/RDP:
  - **an on-line resource riboromal RNA (both large and small unit rRNA) sequence data**
- http://greengenes.lbl.gov/cgi-bin/nph-index.cgi:
  - a web application providing access to the greengenes 16S rRNA gene sequence alignment for browsing, blasting, probing, and downloading
- http://www.hmpdacc.org/:
  - stores all the protocols, computational methods, and the results produced by the Human Microbiome Project.

Chapter 17

Table I.1 Resources available for metagenomic analysis

Type	Resource	Features	Website	Reference
Reference sequence DB	Greengenes	Quality-controlled microbial DNA data and tools	http://greengenes.lbl.gov/cgi-bin/nph-index.cgi	DeSantis *et al.* (2006)
	SILVA	Quality checked and regularly updated datasets	http://www.arb-silva.de/	Pruesse *et al.* (2007), Quast *et al.* (2013)
	RDP	Online data analysis, rRNA derived phylogenetic trees, aligned and annotated rRNA sequences	http://rdp.cme.msu.edu/ https://github.com/rdpstaff	Cole *et al.* (2009)
Annotation and analysis tools	MG-RAST (MetaGenomics Rapid Annotaton using Subsystem)	Annotation, taxonomic and functional assignment, pathway reconstruction single platform (depositing, locating, analysing, visualizing, data sharing)	http://metagenomics.nmpdr.org	Glass *et al.* (2010)
	IMG-M (Integrated Microbial Genomics/ Metagenomics)		http://img.jgi.doe.gov	Markowitz *et al.* (2012), Markowitz *et al.* (2014)
	MetaRep (METAgenomics REPorts)	High-performance comparative metagenomics	http://www.jcvi.org/metarep/; http://www.jcvi.org/hmp-metarep/	Goll *et al.* (2010), Goll *et al.* (2012)
	CAMERA (Community cyberinfrastructure from Advanced Micobial Ecology Research and Analysis)	Data analysis pipeline, single platform	http://camera.calit2.net	Sun *et al.* (2011b)
	MEGAN	Metagenome analysis software	http://www-ab.informatik.uni-tuebingen.de/software/megan	Huson *et al.* (2007), Huson and Weber (2013), Mitra *et al.* (2011)
	DIYA	Bacterial annotation pipeline	https://github.com/bioteam/DIYA	Stewart *et al.* (2009)
	RATT	Transfer annotation from reference genome onto assembly	http://ratt.sourceforge.net/ (now, part of PAGIT: http://www.sanger.ac.uk/resources/software/pagit/)	Otto *et al.* (2011)
	VMGAP	Virus-specific annotation pipeline		Lorenzi *et al.* (2011)
	CloVR	Virtual machine, automated sequence analysis from desktop, cloud computing.	http://clovr.org	Angiuoli *et al.* (2011), White *et al.* (2013)
	Galaxy	Web-based data analysis pipeline	https://usegalaxy.org/	Blankenberg *et al.* (2010)
	HUMAnN	Direct read alignment against protein DB, without prior assembly	http://huttenhower.sph.harvard.edu/humann	Abubucker *et al.* (2012)
	Ergatis	Genome annotation and comparative analysis	http://ergatis.sourceforge.net/index.html	Hemmerich *et al.* (2010), Orvis *et al.* (2010b)
	MetaPathways	Direct read alignment against protein DB, without prior assembly	http://hallam.microbiology.ubc.ca/MetaPathways/	Konwar *et al.* (2013)
Assembly tools	RaY	De novo metagenome assembly and profiling	http://denovoassembler.sourceforge.net/	Boisvert *et al.* (2012)
	Velvet	De novo assembly, short read alignment, Kmer DeBruijin graph traversal-based method	http://www.ebi.ac.uk/~zerbino/velvet/	Namiki *et al.* (2012), Powell and Seemann (2013), Zerbino (2010), Zerbino *et al.* (2009)
	SOAPdenovo	Short read assembly, Kmer DeBruijin graph traversal-based method	http://soap.genomics.org.cn/soapdenovo.html	Luo *et al.* (2012)
	Newbler	De novo assembler, designed for 454 platform	http://454.com/products/analysis-software/index.asp	Miller *et al.* (2010)
	ABySS	Kmer DeBruijin graph traversal-based method	http://www.bcgsc.ca/platform/bioinfo/software/abyss	Simpson *et al.* (2009)
	ALLPATHs	Kmer DeBruijin graph traversal-based method	http://www.broadinstitute.org/software/allpaths-lg/blog/	Butler *et al.* (2008), Maccallum *et al.* (2009)
	Genovo	Generative probabilistic model of read generation from environmental samples	http://cs.stanford.edu/group/genovo/	Laserson *et al.* (2011)

Table I.1 Continued

Type	Resource	Features	Website	Reference
	CLCbio	Kmer DeBruijin graph traversal-based method	http://clcbio.com	
	Meta-IDBA	Kmer DeBruijin graph traversal-based method	http://i.cs.hku.hk/~alse/hkubrg/projects/metaidba/	Peng *et al.* (2011)
	MetaVelvet	Kmer DeBruijin graph traversal-based method	http://metavelvet.dna.bio.keio.ac.jp/	Namiki *et al.* (2012)
	SMASHCommunity	Assembly and gene calling, data exploration, visualization	http://www.bork.embl.de/software/smash/	Arumugam *et al.* (2010)
Mapping/ Alignment tools	BWA	Mapping low-divergent sequences, based on Burrows-Wheeler aligners	http://bio-bwa.sourceforge.net/	Li and Durbin (2009)
	Bowtie	Memory-efficient short read aligner, based on Burrows-Wheeler index	http://bowtie-bio.sourceforge.net/index.shtml	Langmead *et al.* (2009)
	Novoalign	Single-ended and paired-end reads, Illumina GA platform, Needleman-Wunsch algorithm	http://www.novocraft.com/main/index.php	
	SOAP	Alignment of short oligonucleotides	http://soap.genomics.org.cn/soap1/	Li *et al.* (2008), Li *et al.* (2009)
	MrFAST	Mapping NGS reads, prediction of copy-number variation of duplicated segments and genes	http://mrfast.sourceforge.net/	Alkan *et al.* (2009)
	CloudBurst	Parallel read-mapping algorithm, SNP discovery, genotyping	http://sourceforge.net/projects/cloudburst-bio/	Nguyen *et al.* (2011), Schatz (2009)
	BFAST	Mapping of short reads to reference genomes	http://sourceforge.net/projects/bfast/	Homer *et al.* (2009)
	MUMer	Rapid genome alignment	http://mummer.sourceforge.net/	Kurtz *et al.* (2004)
	BLAST (Basic Local Alignment Search Tool)	Local similarity between sequences	http://blast.ncbi.nlm.nih.gov/Blast.cgi	
	MAQ (Mapping and Assembly with Quality)	Based on Burrows-Wheeler aligners	http://maq.sourceforge.net/	Kumps *et al.* (2010)
Reference DB for functional metagenomics	KEGG (Kyoto Encyclopaedia of Genes and Genomes)	Integrated database (16 main DB), broadly categorized systems, genomic, chemical, health information	http://www.genome.jp/kegg/	
	SEED	Comparative genomics environment, curation of genomic data, by annotator across many genes	http://www.theseed.org	
	CAZy (Carbohydrate Active enZyme database)	Genomic, structural, and biochemical information on CAZymes	http://www.cazy.org/	
	PFAM	Collection of protein families using multiple sequence alignments and hidden Markov models	http://pfam.xfam.org/	
	Phylofacts	Structural phylogenomic encyclopaedia for protein functional and structural classification	http://phylogenomics.berkeley.edu/phylofacts/	
	Uniprot	Protein sequence and functional information	http://www.uniprot.org/	
	Mascot	Search engine that uses mass spectrometry data to identify proteins from primary sequences	http://www.matrixscience.com/home.html	
16S based analysis tools	MOTHUR	16S-focused, quality-control, taxonomic profiling, data exploration, statistical analysis, visualization	http://www.mothur.org/wiki/Main_Page	Schloss *et al.* (2009)
	QIIME (Quantitative Insights Into Microbial Ecology)		http://qiime.org/	Caporaso *et al.* (2010), Kuczynski *et al.* (2011), Navas-Molina *et al.* (2013)
	VAMPS (Visualization and Analysis of Microbial Populations)		http://vamps.mbl.edu/	Huse *et al.* (2014)

Table I.1 Continued

Type	Resource	Features	Website	Reference
Data exploration, statistical analysis and visualization tools	bioBakery	Metagenome-focused, taxonomic and functional profiling, data exploration, statistical analysis, visualization	http://huttenhower.sph.havard.edu/biobakery	
	MetaPhlAn	Clade-specific marker genes, assign reads	http://huttenhower.sph.harvard.edu/metaphlan	Segata *et al.* (2012)
	PhymmBL	Classification method using interpolated Markov models	http://www.cbcb.umd.edu/software/phymm/	Brady and Salzberg (2009, 2011)
	STAMP (Statistical Analysis of Metagenomic Profiles)	Data exploration, quality control, taxonomic profiling, data exploration, statistical analysis, visualization	http://kiwi.cs.dal.ca/Software/STAMP	Parks and Beiko (2010)
	LEfSe		http://huttenhower.sph.harvard.edu/galaxy/	Segata *et al.* (2011)
	MaAsLin		http://huttenhower.sph.harvard.edu/galaxy/	
	MetaStats		http://metastats.cbcb.umd.edu/	Foster (2003)
	PICRUSt	Determination of virtual metagenome by using 16S data set	http://huttenhower.sph.harvard.edu/galaxy/	
	microPITA	Targeted selection of samples	http://huttenhower.sph.harvard.edu/galaxy/	Tickle *et al.* (2013)

Chapter 18

- Prebiotics:
  - http://www.prebiotin.com/science/
- Necrotizing enterocolitis:
  - http://www.nlm.nih.gov/medlineplus/ency/article/001148.htm
- Diabetes:
  - http://www.diabetes.co.uk/news/2013/Feb/prebiotic-supplement-linked-to-reduced-risk-of-type 2-diabetes-96394547.html
  - http://lowcarbdiets.about.com/od/nutrition/a/oligosaccharide.htm
- Obesity:
  - http://www.optibacprobiotics.co.uk/blog/2013/01/could-prebiotics-help-to-tackle-obesity
- Cosmetic:
  - http://www.henkel.com/press/2005–12028-prebiotic-cosmetics-restore-natural-bacterial-balance-5868.htm
- Allergy:
  - http://www.medscape.com/viewarticle/804455
  - http://www.cochrane.org/podcasts/issue-1-3-january-march-2013/prebiotics-infants-prevention-allergy
- Vaginal health:
  - http://vhessentials.com/products-prebiotic-suppositories.php
- Aquaculture:
  - http://www.thefishsite.com/articles/1280/probiotics-in-aquaculture-do-they-work

Chapter 19

- Prebiotics:
  - http://www.isapp.net/
  - http://www.prebiotic.ca/prebiotic_fibre.html
  - http://www.fda.gov/default.htm
- Inulin:
  - http://www.prebiotic.ca/inulin.html
- Fructooligosaccharides (FOS):
  - http://www.vrg.org/blog/2012/10/22/oligofructose-and-fructooligosaccharides-fos-derived-mostly-from-chicory-root-or-cane-sugar/
  - http://bodyecology.com/articles/understanding_fos.php
- Lactulose:
  - http://suppversity.blogspot.com.br/2014/01/supplement-review-lactulose-isomerized.html
- Galactooligosaccharides (GOS):
  - http://www.fortitechpremixes.com/research/galacto-oligosaccharides-increase-calcium-absorption-in-young-girls/
  - http://www.healthsnap.ca/blog/what-does-that-mean-galactooligosaccharides.html
- Resistant starch (RS):
  - http://www.precisionnutrition.com/all-about-resistant-starch
  - http://digestivehealthinstitute.org/2013/05/10/resistant-starch-friend-or-foe/
- Soybean oligosaccharides:
  - http://www.whfoods.com/genpage.php?tname=foodspice&dbid=110
- Xylooligosaccharides (XOS):
  - http://www.healthyfellow.com/1199/a-new-prebiotic-ingredient/
- Isomaltooligosaccharides (IMO):
  - http://blog.questnutrition.com/what-is-isomalto-oligosaccharide/

Chapter 20

- http://www.cazy.org/Genomes.html is a database providing genomic and enzyme information on the carbohydrate active enzymes of all organisms
- http://www.phenol-explorer.eu/Phenol-Explorer is a comprehensive database on polyphenol content in foods

Chapter 21

- HealthGrain:
  - https://www.healthgrain.org/
- Fibebiotics:
  - http://www.fibebiotics.eu/
- Dietary Guidelines:
  - www.DietaryGuidelines.gov
- Selected products:
  - BioBran (Arabinoxylan): http://daiwa-pharm.com/english/product/biobran.html
  - Oat bran (beta-glucan): http://www.med.nyu.edu/content?ChunkIID=104429\

Chapter 22

Fibres

- Fibres definition:
  - http://www.codexalimentarius.net/download/report/710/al32_26e.pdf;
  - http://eurlex.europa.eu/LexUriServ/LexUriServ.do?uri=OJ:L:2008:285:0009:0012:en:pdf

Polyphenols

- Comprehensive database on polyphenol content in foods, including comprehensive data on polyphenol metabolism:
  - http://www.phenol-explorer.eu/

Human milk oligosaccharides

- David Mill's lab with the description of the human milk project:
  - http://fgp.ucdavis.edu/prog/fg/proj/mo
- Bod's lab dedicated to research on human milk oligosaccharides:
  - http://www.bodelab.com/bode-lab/Welcome.html
- Company sites (HMO suppliers):
  - http://www.inbiose.com/products_hmo.html
  - http://www.glycom.com/product_samples.asp
  - http://www.elicityl-oligotech.com/client/document/elicityl---oligotech---human---mammalian-milk-oligosaccharides-hmos_24.pdf
  - http://www.carbosynth.com/carbosynth/website.nsf/(w-productdisplay)/CA9C3737BB8CEA8480256A6B00359924
  - http://www.jennewein-biotech.de/en_food.php

Polyols

- European Association of Polyol producers website monitoring the regulatory development surrounding polyol and providing scientific based information:
  http://www.polyols-eu.com/index.php

Peptides as novel prebiotics

- Examples of patents mentioned peptides as potential prebiotics?:
  - https://www.google.com/patents/
  - EP1355541A1?cl=en&dq=glycomacropeptide+prebiotic&hl=en&sa=X&ei=kZbHU86BDYLOygPvroJQ&ved=0CDMQ6AEwAw
  - https://www.google.com/patents/EP1228707A1?cl=en&dq=glycomacropeptide+prebiotic&hl=en&sa=X&ei=kZbHU86BDYLOygPvroJQ&ved=0CCwQ6AEwAg

Chapter 23

- FSTA®:
  - an intelligence service constantly monitoring developments in the areas of food science, food technology and nutrition
  - http://www.foodsciencecentral.com/
- IFIC:
  - International Food Information Council Foundation
  - dedicated to the mission of effectively communicating science-based information on health, nutrition and food safety for the public good
  - http://www.foodinsight.org/
- FAO/WHO views on probiotics:
  - http://www.who.int/foodsafety/publications/fs_management/en/probiotics.pdf
  - http://www.who.int/foodsafety/fs_management/en/probiotic_guidelines.pdf?ua=1
- International Life Sciences Institute (ILSI) task-force on probiotics:
  - http://www.ilsi.org/Europe/Pages/TF_Probiotics.aspx
- ILSI task-force on prebiotics:
  - http://www.ilsi.org/Europe/Pages/TF_Prebiotics.aspx
- ISAPP:
  - International Scientific Association for Probiotics and Prebiotics
  - http://www.isapp.net/
- American Gut:
  - World's largest open-source science project to understand the microbial diversity of the Human Gut. Learn which microbes live in your gut, skin & mouth
  - http://humanfoodproject.com/americangut/

Chapter 24

- Gut microbiota expert information exchange
  - http://www.gutmicrobiotaforhealth.com/
- Newly published obesity-related research
  - http://www.obesityandenergetics.org/

- Blogs written by prominent obesity prevention and management researchers:
  - Dr. Sharma's Obesity Notes (http://www.drsharma.ca/)
  - http://www.weightymatters.ca/
  - http://www.davidkatzmd.com/
- Video: Friends with Benefits: The Human Microbiome
  - http://www.youtube.com/watch?v=GAF-AblJpfM

Chapter 25

The MetaHIT (Metagenomics of the Human Intestinal Tract) is an EU-financed project aiming at establishing associations between the genes of the human intestinal microbiota and our health and disease: http://www.metahit.eu

The International Human Microbiome Standards (IHMS) project is an EU-financed project which seeks to coordinate development of standard operating procedures (SOPs) and protocols that will optimize data comparisons in the human microbiome field: http://www.microbiome-standards.org

The European Network for Gastrointestinal Health Research (ENGIHR) is a European Science Foundation Research Networking Programme (RNP) which promotes interactions between researchers interested in gut health research in Europe: http://www.engihr.eu

This consumer-oriented webpage provides a wide range of information about gut and health, including the gut–brain -axis and pre- and probiotics: http://www.gutandbowel.com

Chapter 26

Probiotics and their paediatric use

- Research on fundamental and applied aspects of probiotics and prebiotics:
  - http://www.isapp.net/
- Children intestinal disease:
  - http://www.webmd.com/children/news/20101128/children-may-benefit-from-probiotics-prebiotics
- Bifidobacteria:
  - http://naturaldatabase.therapeuticresearch.com/nd/PrintVersion.aspx?id=891&AspxAutoDetectCookieSupport=1
- Guidelines for the Evaluation of Probiotics in Food:
  - ftp://ftp.fao.org/es/esn/food/wgreport2.pdf
- Regulation for probiotics:
  - http://ylfa.org/images/file/Workable%20solution%20for%20probiotics-2012.pdf

Prebiotics and their paediatric use

- Research on fundamental and applied aspects of probiotics and prebiotics:
  - http://www.isapp.net/
- Children intestinal disease:
  - http://www.webmd.com/children/news/20101128/children-may-benefit-from-probiotics-prebiotics
- Allergies:
  - http://www.sciencedaily.com/releases/2013/03/130328075716.htm

Chapter 27

Useful web resources related to the management of irritable bowel syndrome:

- http://onlinelibrary.wiley.com/advanced/search/results
- http://www.springer.com/food+science/book/978-0-387-79057-2
- http://www.ncbi.nlm.nih.gov/pubmed/?term=probiotic+and+prebiotic
- http://scholar.google.co.in/scholar?q=probiotic+and+prebiotic+in+IBS
- http://www.worldgastroenterology.org/assets/export/userfiles/Probiotics_FINAL_20111128.pdf
- http://www.theromefoundation.org/rome_III_gastro/
- http://www.aafp.org/afp/2009/0615/p1108.html
- www.sciencedirect.com

Chapter 28

Science sites

- The human oral microbiome database:
  - http://www.homd.org/

Company sites

- Biogaia AB:
  - http://www.biogaia.com
- Chr Hansen:
  - http://www.chr-hansen.com
- Yakult probiotic milk drink:
  - http://www.yakult.co.jp/english

Chapter 29

Molecular docking

- Learn more about Molecular docking:
  - http://www.biomolecular-modeling.com/Software_Docking.html
- Free docking software:
  - http://autodock.scripps.edu/
- Complete docking suite:
  - http://www.schrodinger.com/Glide
- Guidelines for evaluation and safety of probiotics in food
  - http://www.who.int/foodsafety/publications/fs_management/probiotics2/en/
- More statistical and global brief on cardiovascular diseases:
  - http://www.who.int/cardiovascular_diseases/en/

Chapter 30

Livestock feed supplements

- BENEO:
  - http://www.beneo.com
- BiOWiSH Technologies:
  - http://www.biowishtechnologies.com
- Kemin Industries:
  - http://www.kemin.com
- Novus International:
  - http://www.novusint.com

Microbial culture repositories

- American Type Culture Collection:
  - http://www.atcc.org
- Japan Collection of Microorganisms:
  - http://jcm.brc.riken.jp

Animal research resources

- Federation of Animal Science Societies:
  - http://www.fass.org
- Guide for the Care and Use of Agricultural Animals in Research and Teaching:
  - http://www.fass.org/page.asp?pageID=216
- European Federation of Animal Science:
  - http://www.eaap.org
- Association for Assessment and Accreditation of Laboratory Animal Care International:
  - http://www.aaalac.org
- Guide for the Care and Use of Laboratory Animals:
  - http://www.aaalac.org/resources/theguide.cfm

Chapter 32

The International Scientific Association for Probiotics and Prebiotics (ISAPP):

- http://www.isapp.net/

Index

R

S

T

V

W

X

www.ingramcontent.com/pod-product-compliance
Lightning Source LLC
LaVergne TN
LVHW070830220926
848486LV00030B/266